RNA/DNA and Cancer

Joseph G. Sinkovics

RNA/DNA and Cancer

Springer

Joseph G. Sinkovics
Retired Professor, M.D. Anderson Hospital
Comprehensive Cancer Center
The University of Texas
Houston, TX
USA

Retired External Professor and Honorary Member
H.L. Moffitt Comprehensive Cancer Center
The University of South Florida
Tampa, FL
USA

External Professor, Department of Molecular
Medicine
The University of South Florida Morsani
College of Medicine
Tampa, FL
USA

Retired Medical Director; Senior Scientific
Medical Advisor,
The Cancer Institute
St. Joseph's Hospital
Tampa, FL
USA

ISBN 978-3-319-22278-3 ISBN 978-3-319-22279-0 (eBook)
DOI 10.1007/978-3-319-22279-0

Library of Congress Control Number: 2015946582

Springer Cham Heidelberg New York Dordrecht London

Cover illustration: "Background: Ernst Haeckel's Die Kunstformen der Natur (public domain); foreground: dividing lung cancer cells (ibbis.pbworks.com; scinerds.tumblr.com; pharmafile.com). Composition by Ildikó Tóth-Sinkovics graphic-artist)"

Printed on acid-free paper

Springer International Publishing AG Switzerland is part of Springer Science+Business Media (www.springer.com)

Acknowledgments

As for Ref. [37], the author owes gratitude to Iván Hollós for electron microscopy (preparation of the material and photographing the images of the author's choices, expressed in Ref. 36a); to Ferenc Fornosi for providing the membrane-filters and assisting in their use; to Department Head Elek Farkas for his support; and to Vincent Groupé for reviewing the entire material before its submission for publication in Nature (Ref. [37]).

The author dedicates his discussions in the text on, and references to, oncolytic viruses, in this article, to the memory of his medical school friend László K. Csatáry, who devoted the last decades of his professional life for the treatment of patients with "incurable" cancers ("unconventionally", as the need dictated) with inoculations of a British isolate of an attenuated Newcastle Disease Virus, imported to Hungary, which he renamed MTH ("More than Hope", without revealing its real origin). This virus was professionally purified and concentrated for him in preparation for clinical use in Budapest, Hungary (it was not licensed in the USA) (Ref. [71]).

The author expresses his gratitude to the long list of his colleagues and friends, who worked with him and supported him both in laboratory research and in clinical practice of infectious diseases and medical oncology/hematology, as listed in the postscript of the volume "Cytotoxic Immune Lymphocytes…" (Ref. [147]). The author is very much thankful to publisher János Schenk (Schenk Buchverlag), for his special attention to the editing and illustration of that volume, and for releasing some of those illustrations for this volume. All illustrations published by the author in the International Journal of Oncology were released free by the Editor-in-Chief Prof. Demetrios Spandidos.

To Prof. Dr. Márta Melles, the Director and Physician in-Chief of the National Epidemiological Center, Budapest, Hungary, for her permission and support; to staff virologist Orsolya Nagy, and to the head of the Division of Virology, Prof. Mária Takács of the same institution, for their editing the manuscript, reformatting and rearranging the tables prepared by the author in manuscript format, for the arrangement in order of the illustrations, according to the specifications of the publisher, and for the digital transfer of the entire material to the publisher, the author expresses his ever-lasting deep gratitude. The author is especially grateful to Dr. Orsolya Nagy and Dr. Mária Takács for their expert management of the figures and tables, and for their extensive negotiations with copyright owners in order to obtain their permission for the publication of the figures (and for the replacement of those figures, the first choice of which was not possible to publish). The author is sincerely apologetic to those authors and copyright owners whose e-mail-donated figures he could not publish consequentially to the complexity of the copyright law and Springer's interpretation of it. He and his helpers exerted an honest attempt at credentialing and financially satisfying the copyright owners; the author accepts the responsibility with sincere regrets for any inadvertent errors.

For the support of the publication of this volume, the author is very much obliged and grateful to Melania Ruiz and Ilse Kooijman Springer Editors. The Author especially formulates his gratitude to the Editorial Team of M. Venkatesan and Geethajayalaxmi Govindarajan for their good willed and expert investiture in the works of the final editing of this most complex manuscript. Ms. Lorrain Lutton, the current President of St. Joseph's Hospital of Tampa, FL provides her full support to the Hospital's Cancer Institute. In this institution of most distinguished founders and of highly laudable performance for over 30 years, this Author received library-, computer- and office-facilities allowing him to write and complete his MS. The Author expresses his gratitude to Ms. Lorraine Lutton.

The author has benefitted from the comments of and discussions on this topic with his audience at the special session arranged and held at the Hungarian Scientific Academy on July 27 2011. His lecture and discussion was entitled "Oncogenic Inflammatory Processes and the Awakening of the Primordial DNA". The session was chaired by Prof. Dr. György Berencsi; invited discussant was Prof. Dr. István Ember, who, and whose successor, Prof. Dr. István Kiss, published the author's presentation in the periodical Hungarian Epidemiology (Ref. [2224] last issue in 2014;11:55–69). The author was invited to pre-publish a brief article on the subject matter of this volume by Profs. Dr. Dóra Szabó and Dr. Ildikó R. Dunay in the European Journal of Microbiology and Immunology 2015;5:25–43.

Contents

Author's Credentials

In the late 1940s, the author served as a research fellow at the Institute of Pathophysiology, and as an assistant and adjunct professor at the Institute of Bacteriology/Immunology, Pázmány Péter Medical University (now Institute of Medical Microbiology and Immunology, Semmelweis Medical University, Budapest); and in the mid-1950s, as a senior staff virologist at the National Institute of Public Health, and as a board-certified clinical pathologist (the successor institute now is the National Epidemiological Center, Budapest, Hungary). In 1957, in the USA, he was a Rockefeller fellow at the Waksman Institute of Rutgers The State University, New Brunswick, NJ. Thereafter, he served as an intern and resident physician at Cook County Hospital and as a research fellow at the Hektoen Institute of Cook County School of Medicine in Chicago, IL (1958; 1960-1); and at the University of Texas M.D. Anderson Hospital, Houston TX (1959; 1962–1979). The author is USA specialty board-certified in medical laboratory virology, public health, internal medicine, infectious and tropical diseases, and in medical oncology (including malignant diseases of the bone marrow and lymphatic organs); he is board-eligible in hematology. He obtained licensure by examinations to practice medicine in the States of Florida, Indiana, Texas and Washington. After fulfilling services and the practice of academic and clinical medicine for six decades, he is now retired professor from the staff of the services of medical oncology-hematology at the Department of Medicine, The University of Texas M.D. Anderson Hospital, Houston, TX (1962–1980, on staff); and on the external professorial staff of the Departments of Medicine and Microbiology and Immunology (now Molecular Medicine), of the University of South Florida Morsani College of Medicine, Tampa, FL (1983–current). He is Honorary Member of the H.L. Moffitt Comprehensive Cancer Center, Tampa FL. He is retired Medical Director of the Community Cancer Center/Cancer Institute of St. Joseph's Hospital, Tampa FL, where he serves now in the position of Senior Scientific Medical Advisor.

The author designed the work, performed the research, and wrote the entire manuscript. The author reports no dependence on any financial rewards or payments; and there are no conflicts of interest.

Introduction

Background

The Mother Empress (RNA) and the Son Emperor (DNA) of all molecules. It is probably correct to envision the vast preponderance of ancient unicellular life forms as spheroplast-like entities forming communities and replicating by cell divisions in their mature age. Thus, the ancient cells were exempted of senescence and natural death. While their essential genes were propagated by vertical transmissions, horizontal transfers of useful genes were sought after, competed for, and avidly accepted. The precellular appearance of RNA, the ribozymes of the RNA World, predated the generation of DNA molecules. Inasmuch as single and organized nucleotides coexisted with polypeptides before the formation of cells, they remained in unison within the first cell membranes. Their story continues here in a narrative.

The deoxyribonucleic acid DNA consists of two polynucleotide chains wound around each other in the form of a double helix. It is the Watson-Crick left-handed Z-DNA double helix formed by dinucleotide repeats in sequences of alternating purine-pyrimidine CG/AT repeats. Its two helices are held together by hydrogen bonds formed between the alternating desoxyribose purine and pyrimidine bases. Each separate helix serves as a template for enzymes to resynthesize and re-unite the entire molecule. The master molecule revealed itself for the first time in the X ray diffraction photographs of Rosalind Franklin (Figure 1). The deep and narrow major groove, and the wide and shallow minor groove of the A-form DNA is accessible to various DNA-binding proteins. DNA-binding proteins provide the highest level of communication between cytoplasm and nucleus. Its versatile structure, organization in chromosomes, templated replication, unlimited mutability, repairs of its breaks, and its immense digitalized memory are described in volumes filling up entire libraries. The DNA molecule could condense all this enormous amount of information in a small capsule containing a roll of its strands.

J.G. Sinkovics, *RNA/DNA and Cancer*, DOI 10.1007/978-3-319-22279-0_1

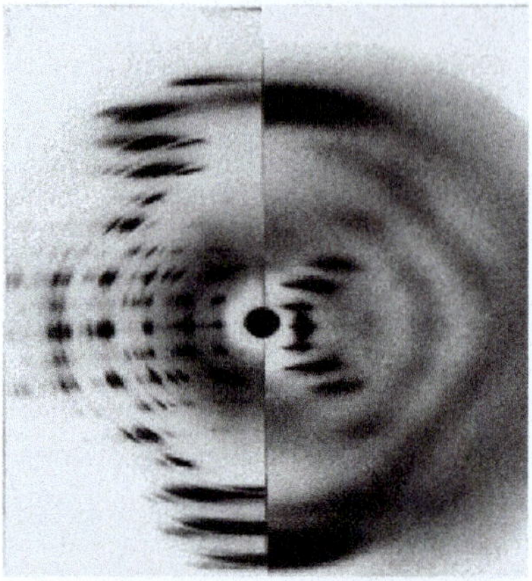

Figure 1 X-Ray Diffraction of DNA by Rosalind Franklin. The RNA/DNA complex. (A- and B-DNA X-ray diffraction patterns). http://en.wikipedia.org/wiki/Molecular_models_of_DNA#mediaviewer/File:ABDNAxrgpj.jpg by Bci21 http://commons.wikimedia.org/wiki/User:Bci21 is licensed under CC BY-SA 3.0 http://creativecommons.org/licenses/by-sa/3.0/). Reference Appendix 2, Explanations to the Figures

The primordial cellular genomes consisted of competing and cooperating RNA/DNA complexes. These ancient faculties of cooperation and competition remain preserved up to the present time. While DNA commands the genome, RNA remains the ruler of the epigenome. However, the RNAs of the spliceosomes begin their actions in the nucleus. MicroRNAs buzz all over the epigenome and in the cytoplasm.

Horizontal transfer of T-DNA (T-strand, Ti plasmid) by the Agrobacterium (the "tumefaciens") into plant cells causes the growth of large tumors recognized as "crown gall formations". The Ti plasmids incorporating the T-DNA and the *vir*-genes reproduce the natural phenomenon, as in Figure 12.1 and legend in Frederic Bushman's "Lateral DNA Transfer" (Cold Spring Harbor Laboratory Press, pp 448, 2002) [1]. The Agrobacterium oncogene transforms yeast cells. Same as in animal cells, the acetylated promoter keeps the oncogene silent; deacetylases activate the oncogene. Methylated genes are silenced; demethylation reactivates the genes. Plants are not exempted from carcinogenesis and its epigenetic control. Just now, at the concluding phase of this entire manuscript, in PLoS One, August, 2013, the emergence of the noncoding ncRNAs, small sRNAs, circular circRNAs, and antisense asRNAs of the tumefacient Agrobacterium is revealed. The inter-kingdom transfer of a bacterial oncogene into a plant cell is regulated by ncRNAs of the original host. In eukaryotes of multicellular vertebrate hosts, proto-oncogenes appear to rest or rise under the control of ncRNAs of the host (reviewed in the last issue of Int J Mol Sci, Sept, 2013). Accordingly, a postscript Table XXXV is being constructed and is attached in the text to the end of this volume.

The fundamental processes that transformed inanimate molecules into the state of the first animated biomolecules were the pairing of the purine and pyrimidine bases, adenine with uracil in the RNA, or with thymine in the DNA molecule, and guanine with cytosine in both RNA and DNA (the Chagraff rule); and the formation of a three-nucleotides codon for the encoding of an amino acid (aa). The four bases act as natural coding substances (the substitutes of C. R. Woese are cited). The sequence of the aa will then be arranged by transcription on tRNA templates into a linear strand of a polypeptid. "Organized DNA did not exist in the form of genes at the beginning of life", but even when it was formed, "the epigenetic information (was) not written in DNA and possibly (was) not coded in any symbolic fashion" and that DNA-guided reactions "gave way to life that would last for eternity". So presents, with quotations from Manfred Eigen's and R. Winkler-Oswatitsch's "Ludus vitalis", a superb alloy of philosophy and factual knowledge, the retired professor, Friedrich H. Schmidt (In: "Biological genesis; the first step from dead matter to life" Dovepress Journal: Research & Report in Biology, 2013;4:1–9).

The side by side hypercycling of the two biomolecules, Eigen's RNAs and Ghadiri's polypeptides, was not life yet, even though other ingredients, lipids and sugars, were present (generated anew, but not hypercycling). At one point, pre-cellular ribozyme-armed ribosomes (and pre-nuclear spliceosomes) must have had arisen. Heinrich Matthaei and Warren Nirenberg could work with extracellular *E. coli* ribosomes, rendering them to be functional *in vitro*. Ribozyme-armed cytoplasmic elements functioned long before the formation of nuclei. The "spark of life" of Eigen and Schmidt (*vide supra*) struck within a fragment of a second in a precellular ribosome, when some tRNAs captured some amino acids and started lining them up in chains. This occurred just once in the Universe and then only on Earth (*vide supra*)? However, once such an error-laden system was established, viroids could have been generated and from their repeated fusions, a precellular virus world and gene pool could have emerged (as envisioned by Eugene Koonin *et al*, see in the text). As protocells and cellular life emerged, the precellular viruses have become the parasites of the cells. Some of these viruses might have induced cell-to-cell fusions and even contributed to the formation of cell nuclei (see in the text).

This author's first encounter above the basics, his true acquaintance with, and admiration for, the DNA molecule, were provided and promoted by Peter J. Russell of Reed College, Portland, Oregon, in his fabulously written and illustrated "Fundamentals of Genetics" 1994 (HarperCollins College Publishers, New York); by William H. and Daphne C. Elliot of the University of Adelaide, South Australia, in their far above basics volume "Biochemistry and Molecular Biology" (Oxford University Press, Oxford, New York, Melbourne) reprinted with corrections, 1997; and by Seyhan N. Eğe's superbly basics "Organic Chemistry Structure and Reactivity" 3rd edition 1994 (D.C. Heath & Co, Lexington, Toronto). The monograph "RNA Life's Indispensable Molecule" by the Rockefeller University professor James Darnell appeared recently in 2011 (Cold Spring Harbor Laboratory Press, Cold Spring Harbor). This text ascends from the class room of college students to the most advanced laboratories of top PhDs. Its last chapter "RNA and the beginning of life" is the light switched on in a dark room (one may say "in a dark universe"). The "Catalog of Human Cancer Genes" published in 1999 by the University of Oklahoma professor of genetics, John J. Mulvihill, with the foreword of Victor A. Kusick (Johns Hopkins University Press, Baltimore, London) is the firm foundation of our subject matter, that never will be outdated, no matter how many new oncogenes will continuously be discovered. No matter from how many new angles the inexhaustible and inventorially rich biology of the cell is looked at, each approach reveals something new and unique. Robert J. Earl and Robert A. Wallace of the University of Florida, Gainesville, published the "Biology the Realm of Life" 3rd edition, 1996 (HarperCollins College Publishers); and Gerald and Teresa Audesirk of the University of Colorado, Denver, produced "Biology Life on Earth" 2nd edition 1989 (Macmillan Publishing Company, New York). Educating this author further, other splendently written and illustrated volumes were those of Gilbert F. Scott's "Developmental Biology" 5th edition 1997 (Sinauer Associates Publishers, Sunderland); and the volume entitled "Biology Concepts and Connections" by Neil A. Campbell, Lawrence G. Mitchell and Jane B. Reese (Benjamin/Cummings Publishing Company, Redwood City Menlo Park, 1994).

This author was impressed by the intimate relationship of the two progenitor molecules of life, the RNA/DNA complex. In some extant genomes, long non-coding ncRNA strands harbor within their stretch, a short one-exon sequence, for example, the human stem cell *sox*2 proto-oncogene, whose product is the SOX2 proto-oncoprotein (P.P. Amaril cited in the text). Was the complimentary cDNA originally engendered inside the primordial RNA, and they preserved the relics of their original relationship, or was the DNA inserted into the RNA from an outside source (for example, as the retrotransposon of the ancient Mauriceville plasmid) (Figure 2). The reverse transcriptase of the Mauriceville retroplasmid initiates cDNA synthesis *de novo* (without primer) at the 3′ end of the tRNA molecule (cited in the text).

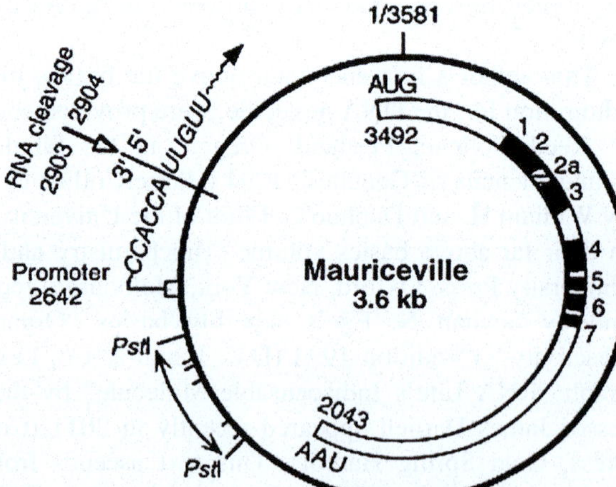

Figure 2 The Ancient Mitochondrial Mauriceville Plasmid in the Neurospora crassa. [2335] Reference Appendix 2, Explanations to the Figures

If the first DNA molecules originated within RNA strands, late in the precellular era, or thereafter within the first spheroplast-like proto-cells (described in the text), their relationship was and remains that of the "mother and son". Accordingly, this article impersonates them as "the Empress", the mother RNA, and "the Emperor", the son DNA, RNA → cDNA. A long noncoding RNA strand is the mother and within it the one exon Sox gene is the son. The name Sox derived from the Y chromosome-embedded *sox* gene (SRY, sex determining region, or box, in the Y chromosome) (Figure 3). The ancestral stem cells' *sox* gene families appeared in the protochordates, or even before. In the sea urchin and in the amphioxus, *sox* gene product proteins mediate germ-layer specification. In human malignant tumors, Sox2/Oc4/Nanog proteins abound (see embryonal carcinoma of the testis, and many others). When it pops up in the literature, that the Kaposi sarcoma virologists named the herpesviral-related shutoff exonuclease also as a SOX protein, quite different from the *sox*/SOX stem cell gene and its gene-product protein, this author's only recourse is to appeal for the readers' leniency toward the overburdened geneticists and virologists. The archaeal (*Sulfolobus acidocaldarius; Acidithiobacillus thiooxidans*) and the prokaryotic (*Aquifex aeolicus*) so-called *sox* genes and SOX gene product proteins should also be discounted from this context; those are apocytochrome B and cytochrome C oxidase subunits, and are thiosulfate sulfur-oxidizing, polysulphide reductase enzymes. However, the real primordial RNA/DNA complex remains the impersonator of the imperial majesties for all molecules, which exist and function under their reign. The DNA-binding Sox proteins de-differentiate cells into their stem cell stages. The Empress and the Emperor dance together, but whenever the opportunity arises, the Empress asserts her ancient leading role over her son, the Emperor, by issuing some ancient small

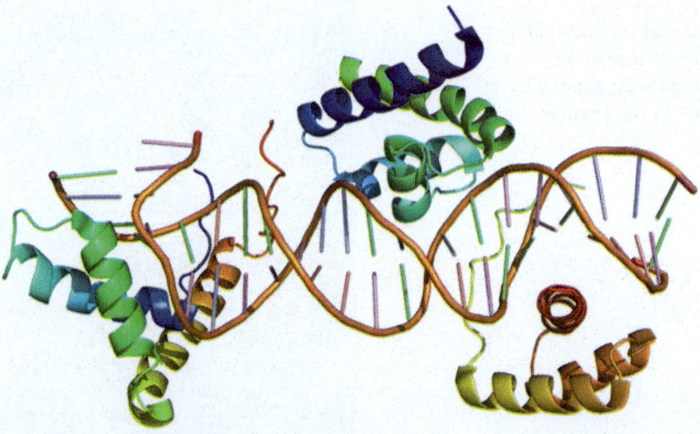

Figure 3 The sox Gene-product High Mobility Group Proteins in Stem Cells and in Tumor Cells. Reference Appendix 2, Explanations to the Figures (Structure of the SOX14 protein. http://en.wikipedia.org/wiki/File:Protein_SOX14_PDB_1gt0.png by Emw http://commons.wikimedia.org/wiki/User:Emw is licensed under CC BY-SA 3.0 http://creativecommons.org/licenses/by-sa/3.0/)

hairpin shRNAs, just as a warning: "No"! And the ribosomes will not translate the attacked mRNA into a protein.

Since the invasion of the first proto-cells by pre-existing virus-like molecules or viruses (the small viroids and large viruses formed from their repeated fusions, a presumption) up to the huge viruses that might have pre-existed and metabolized in the precellular era, multiple viral and host cell DNAs face each other in the infected cells. Different viruses interfere with each other as a rule; some viruses do away with interference and rather collaborate in order to keep the host alive for their protracted residence within. Elementary ancient virus-cell interactions replayed? (Timidly expressed in J Sinkovics: Die Grundlagen der Virusforschung, Verlag der Ungarischen Akademie der Wissenschaften, Budapest, 1955-6). There was then not even the inclination of circular herpesviral DNA genomes resting in the cellular epigenome, and periodically releasing microRNAs to neutralize host cell DNA-derived mRNAs, that would have been spliced and translated in the ribosomes into some antiviral proteins. An up-to-date text is now offered to elaborate on the battle, or reconciliation, of the viral and host DNAs and their missiles, the microRNAs.

Stem cells, haploid germ cells, somatic cells, transformed cells. The ancient unicellular life forms survived under the most adverse conditions in a chemically hostile, overheated (boiling water; pouring lava; volcanic fumes), chemically imbalanced as to excessive acidity or alkalinity, and heavily radioactive environment. The γ-rays-irradiated but radiation-resistant *Deinococcus radiodurans* (Figure 4) suffers genomic damage; it is its chaperone proteins in its proteome that perform the rescue of the genome (Mediterranean Institute for Life Sciences, Split, Croatia).

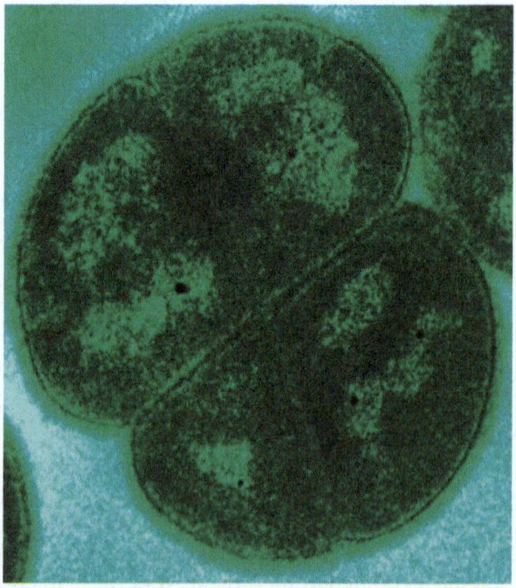

Figure 4 *Deinococcus radiodurans* and Its Survival in High Energy Radiation Fields. Reference Appendix 2, Explanations to the Figures

The heat shock proteins of the archaea chaperoned the protein molecules that were essential for life. The unicellular kinetoplastid (kinetoplastids are its mitochondria) protozoan, the trypanosoma (*T. brucei*) mobilizes heat shock proteins (Hsp) when stressed, as a cell survival pathway. No wonder, the most advanced eukaryotes undergoing malignant transformation in their multicellular hosts, enlisted these ancient proteins for their service: Hsp now protect and chaperon the oncoproteins! Geldanamycins deprive the trypanosoma of its heat shock proteins, and as the parasites die, the infected mice survive (as shown by K.J. Meyer and T.A. Shapiro of Johns Hopkins University Hospital, Baltimore, USA, in the J Infect Dis Aug 1, 2013 issue). Potent inhibitors of the human Hsp90 exist (the patented AUY922 molecules), that kill neuroendocrine cancer cells. PARP cleavage (for polyadenosine diphosphate ribose polymerase, see Table II in the text), and the suppression of ErbB, and the growth hormone-like insulin-like growth factor receptor (IGF-R) induce apoptotic tumor cell death. Similarities of the cell survival pathways between the trypanosoma (*T. brucei; T. cruzi*) driven by sugar to IGF-R, or activated Hsp, and the cancer cell offer themselves for a comparison (as Zitzmann K *et al* of Ludwig-Maximilian University of Munich published it in the Dec 2013 issue of the Internat J Oncology).

The most versatile units of the primordial RNA/DNA complex elongated, amplified, G-quadruplexed, and innovated themselves (when a duplicated extra gene mutated and became a new gene, as Susumo Ohno first proposed); further, diverged, tandem duplicated, aligned and merged, sealed end-to-end, broke and fused, or point-mutated, promptly and liberally in order to encode any new cell organelles, organs and their "cell survival pathways". The descendants of the first unicellular eukaryotes (example: the Trypanosoma) still repair their telomeres after each cell division, activate "cell survival pathways" (examples: the PI3K pathway in Giardia; the metamorphosis bordering trans-speciation in Dictyostelium, or Naegleria). The Oxytricha genome forms G4 quadruplex DNA complexes and carries the ancient piggyBacs, may be even ancestral piggyBats (see in the little brown bats in the text).

The mobile RNA elements invaded the genomes of the ancient uni- and multicellular hosts from the Volvox upward, through the choanoflagellates, Trichomonas, Dictyostelium, Saccharomyces and Candida, Tetrahymena, and Trypanosoma. Multicellular early life forms from the amphioxus (*Branchiostoma floridae*), *Suberites domuncula* Demospongiae (*vide infra*), to the heavily involved Schistosoma followed suit. In distress, the retrotransposons' provirus DNAs encoded (and still encode) the "mutator phenotype" for survival. In the human genome, in Lynch syndrome, or in chronic infections, the transposon/retrotransposon-proviral DNAs encode the "mutator phenotype", first reversibly by epigenomic interventions, then constitutively (irreversibly). The clinics diagnose "cancer." Accordingly, the text extensively reviews all forms of mutator phenotypes and retroviral oncogenesis.

In multicellular hosts, the stem cell compartments preserved most of the faculties of the ancient RNA/DNA complex. The healthy stem cells' RNA/DNA complex serves its multicellular host in its ontogenesis (in fertized egg cells, in the larvae, pupae and nymphs, or in the embryo). The pluripotent and asymmetrically dividing stem cells generate organs: first, anlage dissolution by apoptosis, followed by organ regeneration. Examples are the morphogenesis in the cocoons of insects; the tadpole's regression of gills and formation of lungs; the coelacanth's relatives' (Figure 5) sarcopterygian fins becoming legs crawling on dry land (as imagined, the most powerful hind fins becoming the hind legs of the amphibians, the jumping frogs); and the developmental stages in the ontogenesis of the mammalian kidney. Some basic morphogenic transformations are bordering trans-speciation. Examples are the life cycles of Dd, DiDi, Dicty, the Dictyostelium (Figure 6), or Naegleria, or Theileria (see in the text). Hieronymus Bosch saw the slime molds as evil spirits in the "Garden of Earthly Delights"; the Texas cowboys suspiciously diverted the run of their herds and horses to bypass widely the huge blebs of fluctuating yellow sponges feeding on cow-manure, the slime mold colonies, off Houston, TX; and the young PhD students at Rice University, Houston, love to tender to the subject matter of their curricula, Dicty, the "social amoeba".

The artists' alleged minor distortions do not abolish Ernst Haeckel's "ontogenesis recapitulating phylogenesis" of the clade. The stem cells' RNA/DNA complex remained fully capable of encoding practically any physiological need in the onto- and phylogenesis of its multicellular hosts. The RNA/DNA complex initiated the host's natural selection under changing environmental conditions. Examples are the flying reptiles (the Archaeopteryx); flying theropod dinosaurs (the feathered birds, Aves); flying mammals (the bats, Vespertiliones). Further achievements are, **1**, the evolution of the maternal organ, the uterus, and the fetal organ, the placenta (from the egg-laying platypus up to mammals). The return of land mammals (ambulocetus → balaena) to the sea (example: the ancestral line of the pre-hippopotamus becoming whales: from a plant-eater, a meat-eater). For **2**, the evolutionarily achieved ultimate task accomplished is the encoding of the human cerebral cortex. The RNA/DNA complex proves itself capable of resolving the most complicated tasks of biological engineering. Would not the master progenitor molecules carry inherently the basic elements for the sustenance of all-resistant, independent, immortal cells, too? If some cell communities survive several thousand years (Atacarma desert shrubs in Chile; the baobab tree in Kruger National Park, South Africa; the Posidonia Oceanica sea grass in the

Figure 5 The Coelacanth Is Alive. (Latimeria Chalumnae – Coelacanth. http://commons.wikimedia.org/wiki/File:Latimeria_Chalumnae_-_Coelacanth_-_NHMW.jpg by Alberto Fernandez Fernandez http://commons.wikimedia.org/wiki/User:Afernand74 is licensed under CC BY-SA 3.0 http://creativecommons.org/licenses/by-sa/3.0/) Reference Appendix 2, Explanations to the Figures

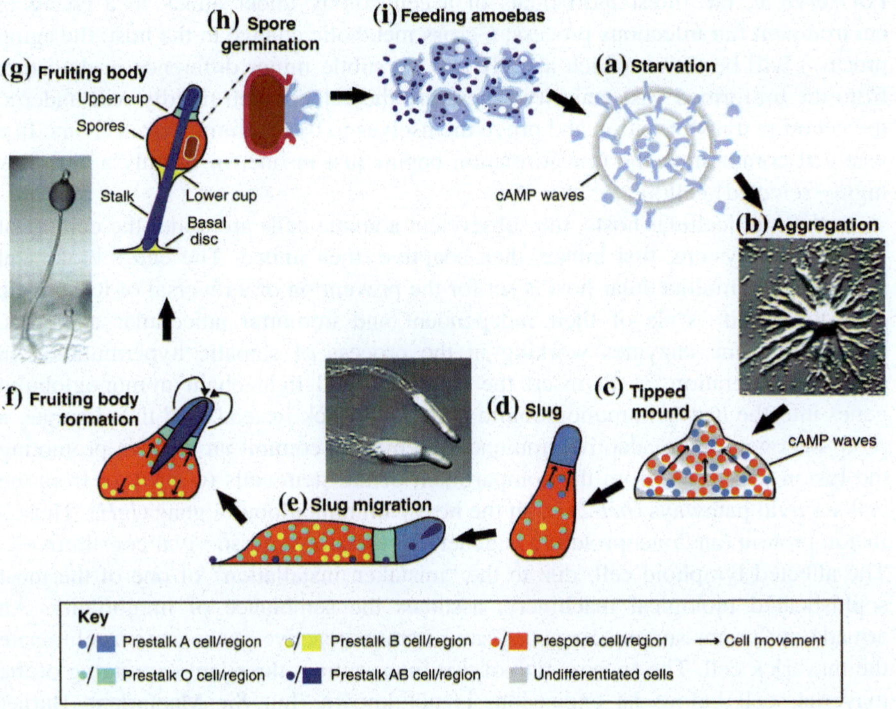

Figure 6 The Amoebozoan Dictyostelium discoideum and Its Cell Cycle. [2336] Reference Appendix 2, Explanations to the Figures

Mediterranean Sea; the cyanobacterial stromatolits of Western Australia) that does not necessarily mean that their individual cells were immortalized (see in the text); or are they living in a biological entity free of senescence, but different from "malignant transformation"? (Rachel Sussman: The Oldest Living Things in the World", University of Chicago Press, 2014).

It is not so that the RNA/DNA complex possesses a foresight, as to what environmental changes to expect; for these, it has no premonition to prepare for in advance. Evolution is not supposed to have any foresight. However, one of the multiplicities of its products is able to adapt to the changed environment, and select out its progeny, whereas, the other competitors perish. The elements of these new faculties appear to have been present in a small subpopulation, even before the full folding out of the new environmental changes. The close to omnipotent RNA/DNA complex refrains from, or fails to, catching up with the new environmental changes in retroactively re-adjusting the inadequate faculties of the left-behind hosts. The inexhaustible mutability, the driving force of the Darwinian selection process, rendered the host in advance to be best fitted to survive under any new circumstances. The faculties enabling the host to adapt happened to be "in the waiting."

For survival, the fittest individuals of a cell colony under attack in a changing environment (an infectious process; a gross metabolic change in the host; the aging process) will be those, which already carry a subtle minor difference in deviation from the uniformed vast majority. These are the cells, which rapidly will undergo the genomic transformation and prove themselves to be the immortal survivors. In a vast cell community, the transformation begins just in one (or in only a very few highly selected) cell(s).

In the multicellular hosts, the subservient somatic cells are under the control of the immune system, first innate, then adaptive, then united. The entire biological system of the multicellular host is set for the prevention of a reversal of its somatic cells to the life style of their independent and immortal unicellular ancestors. Ironically, some enzymes working in the process of somatic hypermutation in antibody generation, as to insert the newly created light chain immunoglobulin genes into the heavy immunoglobulin gene, miss-nick instead, and thus liberate, a proto-oncogene. The adaptive immune system may commit an error in promoting the fusion of a gene, from the compartment of the stem cells (c-*myc*), or from the cell survival pathways (*bcl*-2), with the heavy immunoglobulin gene (*IgH*). Thus, a fusion protein (an "oncoprotein") is generated dictating cell survival constitutively. The affected lymphoid cell, due to the "mistaken installation" of one of the most sophisticated biological machinery, assumes the semblance of immortality. An armada of "tumor suppressor genes" and immunoreactive genes arise to eliminate the maverick cell. The success rate of this intervention, the silent expunging of the maverick cell before its expansion, is not known (but Sir Macfarlene Burnet guessed it just right, in: "The concept of immunological surveillance" (Prog Exp Tumor Res 1970) (see in the text). In contrast, the success rate of the expanding transformed cells is well known, as it is measured as the incidence of malignant tumor generation in the life time of various multicellular hosts (prominently including *Homo*). Once the driving mutations (PI3K/Akt/mTOR; Wnt, see in the text) are set, the genome is triggered to initiate hundreds (or more than a thousand) somatic mono(uni)genic gain-of-function mutations. These are for any unforeseen environmental constellation into which just one in a thousand of the mutated cell will have to fit, in order to be able to faultlessly metabolize in it. Permission to re-publish pending, a text-figure will show how so-called defensive enzymes, the APOBECs, can also induce such mutations.

When the RNA/DNA complex can generate "by mistake" cells transformed into the life style of their ancient unicellular ancestors, it reveals its inherent aboriginal potency to generate such individual cells "intentionally". It is a display of a faculty for the preservation of cellular life in mortal multicellular hosts under stress (a chronic infection), or under any other threat of extinction. The clinics diagnose the inherent natural process as the fatal malady "cancer" (Siddhartha Mukherjee: "The Emperor of All Maladies" 2010, Amazon.com). In the multiple forms of gain-of-function re-arrangements, there are no exact criteria for the distinction between an accidental genetic deviation from norm, and an encoded inherent process initiated for the same purpose. An example of the former is the "balanced chromosomal rearrangement", due to misaligned chromosomes in the spindle (that may result in aneuploidy). An example of the latter is the strategy of the oncogenome with sequential gains of cellular vitality: elimination of auto-apoptotic processes, serial gain-of-function point-mutations and gene fusions meticulously avoiding nicks within protein-encoding sequences, and generating new proteins (the "oncoproteins") exempt of ubiquitination. Are here the characteristics of the cell known at the clinics as "the cancer cell" displayed?

The stem cell ascending in the crypts of Lieberkühn in the colonic mucosa is the star. This stem cell divides asymmetrically, thus creating the differentiating epithelial cells of the colonic mucosa, while it also renews itself. The ultimate result of differentiation is at the end of the trajectory. It is the epithelial cell of the intestinal lining, a subservient somatic cell. In the colon, this cell is bathing in liquid feces, that is populated by the natural microbiome, which is engaged in a balancing act with its phages, and with its host's tolerance. The epithelial lining is shed and regenerated practically daily. The clockwork of the genomics disallows any de-differentiation (for the time being). However, a clone of bacteria becomes invasive (an enterotoxigenic Bacteroides), and the environment suddenly changes. The host mobilizes a chronic and protracted immune reaction. Some of the $CD133^+$ stem cells in the bottom of the crypt sense a signal, conveyed by accident, or by the inherent rules. The stem cell refuses to ascend, and to asymmetrically divide; no differentiating somatic daughter cells will be released. This stem cell divides symmetrically producing two stem cells in each division. This stem cell puts all its ancient genome to full alert. By deacetylating their promoters, it liberates the cell survival pathways encoded by the proto-oncogenes. The activated genome dictates full symmetric mitoses (no asymmetric mitoses). Some genes try to encode inhibitors of the cell cycle; the promoters of these genes will be immediately silenced at their CpG residues by hypermethylation (the so-called "tumor suppressor genes" are inactivated, among them first APC). The guardian of the genome, the wild type p53 gene, rises to induce apoptosis. Its protein product will be neutralized by the MDM protein, and sent to ubiquitination. The rapidly dividing stem cells demand a blood supply by generating VEGF-R. The host obliges and provides a network of blood vessels. In response, the stem cells activate their β-catenin/Wnt/Hedgehog pathway (Figure 7). Accordingly, the suppressor gene dickkopf is silenced and β-catenin is released from the cytoplasm for transfer to the nucleus. The ancestral wnt/hedgehog proto-oncogenes exist in the universal eukaryotic genomes from the choanoflagellates (Monosiga), and the cnidarians (Nematostella) on, with a side way to the drosophilas' *notch* gene. The evolutionarily vertically transferred genes reach up to Homo. The cell survival pathway of the sea anemone (Nematostella), the genes regenerating the cut-off head of the hydra, become oncogenes in the human stem cell.

By now, the transformed cells form a large conglomerate in the colonic wall, from where they will disseminate (Figure 8). In addition to the stem cells, some of the somatic cells are also induced to de-differentiate into stem cells. At the bottom of the crypt, a set of stem cells continues to divide symmetrically, in a set sequence. Their *snail/slug* proto-oncogenes change their epithelioid cytoskeleton into a slick mesenchymal-sarcomatoid structure. Fibroblast-like cells (some of them of bone

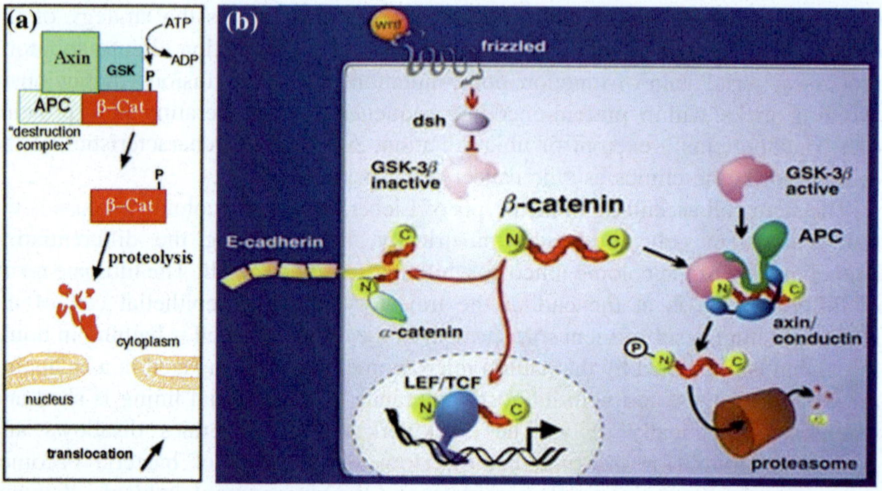

Figure 7 **a** The Wnt Pathway for Body Image in Ontogenesis, and for Malignant Transformation in Oncogenesis. (Wnt doesn't bind to the receptor. Axin, GSK and APC form a "destruction complex," and β-Cat is destroyed). http://en.wikipedia.org/wiki/Wnt_signaling_pathway#mediaviewer/File:Axindestructioncomplex.png by JWSchmidt http://en.wikipedia.org/wiki/User:JWSchmidt is licensed under CC BY-SA 3.0 http://creativecommons.org/licenses/by-sa/3.0/). **b** The Wnt Pathway for Body Image in Ontogenesis, and for Malignant Transformation in Oncogenesis. With kind permission © Walter Birchmeier, Max Delbrück Center. All Rights Reserved 2015. [2337] Reference Appendix 2, Explanations to the Figures

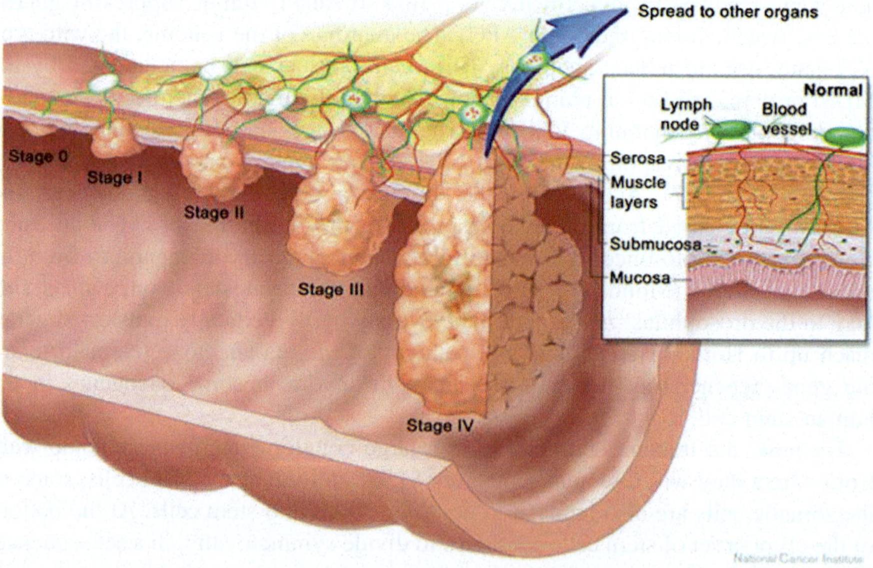

Figure 8 Staging in Colon Cancer. © Terese Winslow. All Rights Reserved 2015

marrow-derivation) induce further the epithelial-to-mesenchymal (ETM) transformation. The elongated cells sneak through lymph- and blood vessel walls, travel, and occupy lymph nodes and parenchymal organs, with preference to the liver. There, subverted macrophages, Kupffer cells, mast cells and vascular endothelial cells welcome and feed them; the T lymphocytes and NK cells recognize them as their own "self", and refrain from attacking them. If immune T cells reacted to them, chemokines (stromal cell-derived factor, SDF) calls in regulatory T (Treg) cells to put an end to that reaction. The invaders exhibit great resistance to chemical and radiation exposures (like their primordial ancestors on Earth did). These cells could exist long in the state of autophagy, and regain an accelerated malignant behavior years later (see in the text). The recurrent tumor lost its brake, the microRNA-451; it produces ATP-binding cassette ABC protein pumps (as their ancestral unicellular eukaryotes, the dinoflagellate algae, do, *vide infra*) to immediately exude ("pump out") chemotherapy drugs. Thus, a full resistance to FOLFOX/FOLFIRI (5-fluorouracyl; oxaliplatin; irinotecan) is gained. These cells eventually kill the host. Newly discovered antibiotics (salinomycin) and anti-CD133 monoclonal antibodies may eliminate them (Table I). To switch off billions of years of differentiation in a few cells living in the police state of a well-organized cell community is a challenge only primordial forces conserved in the RNA/DNA complex can measure up to.

The differentiated somatic cell of the integument (a keratinocyte) obeys proper biochemical orders (in the form of proto-oncoproteins with or without chemo-, cyto-, and lymphokines) to reverse its trajectory and become a stem cell, and then to re-differentiate into a nerve cell (Shinya Yamanaka's Nobel Prize, 2012). Improperly stimulated, it may become a basal cell carcinoma, or a squamous cell carcinoma. The ancient RNA/DNA complex will de-differentiate the mature somatic cell to its ancestral stage of existence. A fatal event mistakenly installed? A process of re-juvenation inherently installed? Is this a blind duty of the genomic retrotransposons for the maintenance of the living matter in whatever formation or shape (see in the text)?

In the highly chemo- and radiotherapy-resistant malignant peripheral nerve sheath tumors (MPNST), the tumor suppressor genes, the apoptosis-inducer p53; the PI3K/Akt oncogene-suppressor PTEN; the ras oncogene-suppressor NF-1 are lost (phosphatidyl inositol kinase 3; phosphatase tensin deleted on chromosome ten; AK mouse thymic lymphoma oncogene; rat sarcoma oncogene; neurofibromatosis), but the major promoter oncoproteins, the EGF-R and the IGF1-R (receptors, epidermal and insulin-like growth factors) are amplified without a gain-of-function mutation. Szokol B, Gyulavári P, Baska F et al of the Semmelweis University, Budapest, and the Max Planck Institute, München, just announced the discovery of new EGF-R small molecular inhibitors. Will they work in patients with lung cancer, where gefitinib and erlotinib have failed? The cascades of the ancient cell survival pathways of PI3K/AKT/MAPK (mitogen-activated phosphatase kinase), and those of EGF-R, c-MET (hepatocyte growth factor receptor) continue. Is it an ancient, and neither expunged, nor domesticated retroviral LTR of unknown relationship to microRNAs

Table I Oncogenic Stem Cells

In the crypts of Lieberkühn asymmetrically dividing stem cells release a somatic daughter cell and renew themselves. The somatic cell may de-differentiate into an asymmetrically dividing stem cell with mutated APC (in the GSK-3β/Axin complex) allowing cytoplasmic β-catenin to enter the nucleus to activate proto-oncogenes *tcf/lef* and notch. This is the Wnt ligand → Frizzled → Disheveled oncogenic pathway. The malignant cells (colorectal adenocarcinoma) respond to the FOLFIRI protocol: remissions are induced. Another population of stem cells in the bottom of the crypt begins to divide symmetrically (producing two stem cells) and rest long between divisions. These stem cells lose miR-451 control (miR-451 is downregulated), overproduce CD133, and ATP-binding cassette (ABC) transporter proteins, invite fibroblasts of bone marrow derivation, perform the conversion ETM, metastasize late; exhibit full resistance to FOLFIRI and XRT. May be forced to re-differentiate into asymmetrically dividing stem cells by salinomycin, BMP4 (bone morphogenetic protein) and anti-IL4 mcab.[1]

Similar transformed stem cell invasions occur in breast cancers induced by paclitaxel and avascular ischemic hypoxia due to bevacizumab.[2] Notch protooncogene emerges as a major luminal breast cancer stem cell up-regulator.[3] NFκB drives breast cancer stem cells.[4]

ATP-binding ABC proteins protected unicellular eukaryotes (dinoflagellate algae) from ingestion of toxic substances; same for vertebrate multicellular (fish) and mammalian hosts. Malignantly transformed cells revert to overproduction of ABC proteins. ABC proteins neutralize oncogene-antagonist curcumin. Verapamil, cyclosporin, erlotinib, marine products (agosterol, bryostatin, ecteinascidin, sipholane, triterpenoids) reverse the effects of ABC proteins in restoring cell susceptibility to cytotoxic drugs. Bromoditerpene parguerenes I/II of red algal derivation force the intracellular retention of doxorubicin, paclitaxel, vincristine better, than verapamil, thus promote cell death by toxins (chemotherapy in clinical oncology).[5]

1 Puglisi MA et al World J Gastroenterol 2013;19:2997–3006. ***2*** *Conley SJ et al PNAS 2012;109:2784–9; Polyak K & Vogt PK PNAS 2012;109:2715–7.* ***3*** *Lafkas D et al J Cell Biol 2013;205:47–56; Suman S et al Br J Cancer 2013;109:2587–96; Peng GL et al J Huazhong Univ Sci Technol Med Sci 2014;34:195–200.* ***4*** *Zhang W & Grivennikov SI Breast Cancer Res 2013;15(5):316; Yamamoto M et al Nat Commun 2013;4:2299.* ***5*** *Abraham I et al Mar Drugs 2012;10:2312–21. Hawley TS et al Am J Hematol 2013;88:265–72. Huang XC et al Biochem Pharmacol 2013;85:1257–68. Lainey E et al Cell Cycle 2012;11:4079–92. Liu S et al PLoS One 2013;8(5):e63895. Sakulterdkiat T et al Anticancer Res 2012;32:5337–42. Wang DZ et al PLoS One 2013;8(5):e3659. Yamada A et al Breast Cancer Res Treat 2013;137:773–82. Zhang DM et al Mol Pharm 2012;9:3147–59.*

(miR), that is responsible for these gene amplifications? Oncogene-activator host cell genomic LTRs (long terminal repeats) are discussed in the text.

The stem cell practices self-preservation by dividing asymmetrically in almost unlimited numbers of divisions, with the release of a somatic daughter cell on each occasion. Since stem cells are singularly distributed in the parenchyma, they must build up that daughter cell within their nucleus and cytoplasm from nutrients received by diffusion; unless, there are special stem cell compartments (in the Lieberkühn crypts in the gut), where stem cells may receive blood supply. In contrast, the differentiated somatic cells of the multicellular hosts are allowed only a limited number of cell divisions. The somatic cells of the multicellular host are instructed to leave their chromosome ends uncapped after each mitosis. This is in contrast to many extant free-living, or parasitic unicellular life forms, which diligently recap their chromosomal ends after each of their divisions (*Trypanosoma brucei* in culture). Theileria-infected host lymphocytes are forced to do so (see in the text). The ancestral unicellular eukaryotes probably practiced telomere re-capping after each cell divisions. The regular enlisted-for-service somatic cells of the multicellular hosts are allowed not more than approximately 30 cell divisions in their life span (the Hayflick rule). Thus, they undergo senescence and eventually die. They may be replaced by new somatic cells, as newly generated daughter cells of the ever-present resident stem cells provide them. In contrast, the somatic cells' gain-of-function point-mutations may reverse their course toward immortality. These cells now repair their chromosomal ends with an overactive telomerase, and activate in a constitutive manner their cell survival pathways. The notorious PI3K pathway is activated upon the elimination of its arch-enemy suppressor, the PTEN gene (see in the text). Loss-of-function mutations, or deletions, abolish the tumor suppressor genes, or by ubiquitination, their gene product proteins, respectively. These faculties were installed in multicellular hosts in order to prevent constitutively activated cell survival pathways, the rebellious stem or somatic cells may revert to. The clinical textbooks read: "loss of the tumor suppressor genes occurred, and the malignantly transformed cells replicate constitutively and unopposed".

In full activity, the somatic cells' glycolytic pathways are those of oxygenation through the mitochondria with high energy (ATP) yields. However, upon "malignant transformation", one form or the other of the ancient Warburg type pre-mitochondrial glycolysis is re-activated. The Warburg's switch is recalled from the past to engineer the return to one or to another form of the ancient Embden-Meyerhof-Warburg type glycolysis (Figure 9). This form of glycolysis

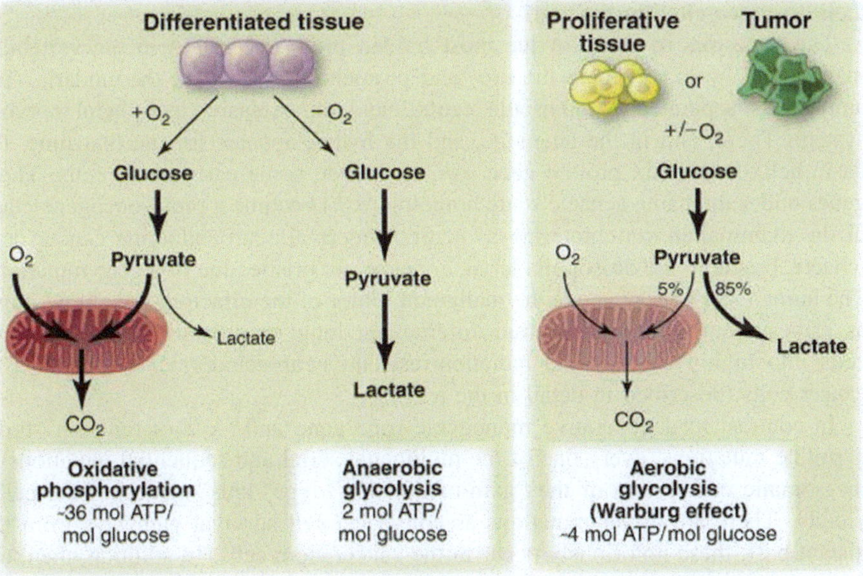

Figure 9 Warburg's Aerobic Not-mitochondrial Glycolysis. [2338] Reference Appendix 2, Explanations to the Figures

served unicellular life forms in the ancient anaerobic (or even aerobic) environments prior to the acquisition of mitochondria. Nevertheless, it is an oxidative form of glycolysis, but with a very low energy (ATP) yield. The transformed cells practice it even in the present aerobic environment by-passing their mitochondria. However, in many ways, the transformed cell will eventually switch to a form of glycolysis of high energy yield. Examples of such maneuvers are known in the glioblastoma-, or in the mantle lymphoma-cells (documented in the text).

The coding capacities of the cell-transforming RNA/DNA complexes follow a serial and consequential pathway in eliminating apoptosis-inducing genes first (the "tumor suppressor genes"), and activating "cell survival pathways" by expropriating all physiological growth factors (even erythropoietins) and their receptors in autocrine or paracrine circuitries in multicellular hosts. The oncogenic RNA/DNA complex of transformed single cells exhibits most of the faculties of the ancient primordial RNA/DNA complex of the stem cells. Cells of the marine hydrozoon cnidaria, *Hydractinia echinata,* are able to bud off new individuals, or to undergo meiosis resulting in either ovum- or sperm-forming cells, thus preserving either stem cell, or germ cell faculties. However, when the individual cells form colonies, some cells differentiate into epithelial cells of the "gastrointestinal or vascular network" shared by the entire colony. Retained in the interstitium are the stem cells. In these cells, the polynem (Pln) gene (equivalent to the Oct4 human stem cell gene) may suddenly dictate a reversal to the pheno-genotype of the primordial individual cells. The differentiated cells of the gastrointestinal and vascular network suddenly begin to de-differentiate. De-differentiated Hydractinia and Medusozoa cells practice reverse ontogenesis. They may re-differentiate into mature organs. They may uncontrollably proliferate, transform into large tumors and actually kill their host (see cited in the text).

The ancestors of some of the most hidden proto-oncogenes in the vertebrate mammalian hosts served in the uro- and protochordates and in the cnidaria. The amphioxus separated a minuscule central and an extensive peripheral nervous system. There, and in the anemone, and the hydra, appears for the first time, the basic helix-loop-helix protein gene *ash*, to encode some early nerve cells. These genes under the name achaete scute homolog (*ash*) become a proto-oncogene clade in the mammalian vertebrate hosts' neuroectodermal/neuroendocrine cancer cells (chaete, bristle in the drosophila larva; a-chaete, no bristle, due to a gene mutation). The human *ash* genes encode the malignant tumor of the olfactory ganglion known as esthesioneuroblastoma, or transdifferentiate lung or prostate adenocarcinoma cells into highly chemo- and radiation-resistant neuroectodermal-neuroendocrine cancer cells (described in detail in the text).

In contrast to degenerative monogenic (one gene and "loss-of-function" type) germline mutations, every one of the multigenic serial and sequential mutations in the somatic cell is one of the "gain-of-function" type, thus increasing the cell's vitality. There are driver mutations as conserved cell survival pathways from the distant past; these notoriously occur in the cancer stem cells. In addition, there are numerous somatic cell one-gene mutations, all of them the gain-of-function type. The textbooks say that these are fundamentally random events. May the question be asked: could not the DNA initiate such gain-of-function mutations by the thousands in its progeny made to fit into just one of millions of not foreseen environmental changes. These changes are to come, and in which just one in thousands of the mutated cells will be able to fit into and survive in it?

Some of the stressed unicellular life forms (the protists, the ciliate, *Oxytricha trifallax*, etc.), as well as the "oncogenes" in multicellular host cells (c-*myc*), convert their linear DNA into a G-quadruplex (propeller) formation. What occurs in a human oncogene (c-*myc*), occurs normally, or under 'stress', in the genome of a ciliate. In the "oncogenome", distant genes travel within the genome from chromosome to chromosome, to form unisons by fusion with another gene. The fused genes encode the oncoprotein for an irreversible cell survival pathway. So do the genes of the potato-pathogen oomycete Phytophthora for increased virulence and pathogenicity (see in the text). Did similar proteins drive the ancient unicellular life forms through blasts of UV and ionizing radiations (the DNA-repair enzymes of *Deinococcus radiodurans*), or in water close to its boiling temperature (the heat resistant polymerases of the *Sulfolobus acidocaldarius/solfataricus*, or *Thermus aquaticus*, Taq), and/or in potentially lethal chemical reactions? A chemical reaction of this type is the high concentrations of sulfuric acid and $pH < 3$ in pyrite ores, the environment in which the archaea, *Ferroplasma acidarmanus,* thrives (see in the text). The transformed extant cells (referred to as "cancer cells") possess extraordinary abilities, not only to pump out from the cytoplasm harmful chemicals, but also to neutralize them within the cytoplasm. Cisplatinum forms intrastrand cross-links (adducts) in the DNA molecule. Before that, the cancer cell neutralizes the cisplatinum molecule (diammine-dichloro-platinum) in the cytoplasm by conjugating it with metallothionein, or glutathione; after that, by enzymatically excision-repairing the platinum-DNA adducts. The defense mechanisms of unicellular "radiodurans" life forms and transformed single cells of multicellular hosts toward chemical- and radiation-induced damage are quite comparable.

In the oncogenes of the multicellular hosts, the elementary basic structure of the genes encoding cell survival pathways and accelerated cell divisions in the extant descendants of the primordial unicellular ancestors can be recognized. In a form of human leukemia cells, the descendant of the yeasts' cell-division accelerator *bub* gene (budding uninhibited by benzimidazole) and that of the trypanosoma's *nup* gene (nucleoporin) form a fusion oncoprotein. Bub protein overexpression in itself is an inducer of aneuploidy and Aurora B kinase hyperactivation (see text). Through these measures, the ancient unicellular life forms were endowed with the faculties of close to non-surmountable physico-chemical resistance, and the faculties of cell divisions in young mature age, thus escaping by prevention senescence and natural death. Some 'malignantly transformed' individual cells of the multicellular host are able to revert to the metabolism of their distant ancestors.

Some malignantly transformed cells in the multicellular hosts behave like those unicellular life forms, which have become the parasites of the multicellular hosts. These unicellular life forms (Plasmodia, Theileria, Trypanosoma) alter their cell surface antigenicity, induce tolerance by masquerading as "self", and subvert mesenchymal cells of their host for their active support in producing the appropriate chemo-, lympho-, and cytokines for the growth factor receptors of the invaders. The CD47 ligand of leiomyosarcoma cells in acting on the signal regulatory protein receptor of M1 naturally defensive host macrophages, convert them into tumor-friendly M2 Mϕs. At the Stanford University Medical Center in California, antibodies neutralizing CD47, secure that M1 macrophages remain naturally defensive and will engulf and digest leiomyosarcoma cells (PNAS April 24, 2012). The transformed cells (cancer cells) of the multicellular hosts (the human host) repel the host's immune defenses (especially those mediated by immune lymphocytes and NK cells) exactly the same way as unicellular parasites overcome their hosts. Epithelial cancer cells may undergo alterations bordering trans-speciation by assuming the phenotype of vascular endothelial cells ("vasculogenic mimicry"), or that of elongated mesenchymal cells (fibroblasts, monocytes) for accelerated locomotion toward forming metastases ("epithelial-to-mesenchymal transition"), as illustrated widely in copyrighted figures in the literature.

The malignantly transformed cell and the syncytiotrophoblasts learn from each other. The fetal trophoblast of the placenta utilizes ancient defensive reactions similar to those of unicellular eukaryotic parasites and/or tumor cells, in order to evade its rejection; more than that, it is frequently invasive in provoking full tolerance by the mother. The placenta utilizes defensive reactions similar to those of tumors and parasites, in order to evade its rejection: thus, it was referred to as a "pseudomalignant organ", and its features as such were tabulated. Here is that Table reproduced. At The University of Texas M.D. Anderson Hospital's 56th Annual Symposium in 2003, held on the topic "Cancer Immunity: Challenges for the Next Decade", J.C. Horvath, H. David Kay and this author presented the two extreme means by which ancestral NK cells protected the Botryllus colonies from fusion with incompatible colonies, or how placental trophoblasts deal with maternal invariable iNKT cells for the protection of the fetus against rejection. It is the opposite of what secured the individuality of the Botryllus colony; it is a reconciliation in the interest of the fetus. The fetal trophoblasts and tumor cells learned each others' defensive (and aggressive) strategies. In many malignant tumors, especially teratomas and lymphomas, or in autoimmune diseases, latent endogenous retroviruses become reactivated (as documented in the monograph of Joseph Sinkovics "Cytolytic Immune Lymphocytes..," Schenk Buchverlag, Passau & Budapest, 2008). Many of these endogenous retroviruses are fusogenic (promote cell fusions), a faculty probably preserved from primordial ancestors. Fusogenic retroviruses were instrumental for the evolution of the placenta. In the superb Springer Verlag textbook of the editor György Berencsi III, and his co-author, Mária Takács, under the title "Maternal Fetal Transmission of Human Viruses and Their Influence on Tumorigenesis", these concepts were presented a-new in 2012 in great details. What was not re-cited in that volume from the precursory literature is being summarized here (with the generous permission of the editor, Demetrios Spandidos). Indeed, endogenous retroviruses and the fetal trophoblasts changed the course of the evolving adaptive immunology, by imitating the tumor cell, and then, *vice versa*. The inserted RNA-to-DNA proviruses become from pseudogenes to functional genes (F Bushman) [1].

Table (a modified reproduction*) "**Placenta as a "Pseudo-Malignant Organ"**

Placenta features expressed/acquired	Trophoblast	Malignant Cell
Chronionic gonadotropin	++++	+++
TGFα↑EGF-R	++	+++
TGFβ antagonizing Immune T cell expansion	++	++
Placenta GF	++++	++
GM-CSF	++	++
VEGF/VEGF-R	+++	+++
bFGF/FGF-R	+++	+++
MMP (matrix lysis for invasiveness)	++	++
Cox (subversive prostaglandins)	++	++
PP-A-R (growth factors for deciduas/placenta/tumors)	++	++
Mϕ MIP 1 (attracts inhibitory NKT cells)	++	++
Th-1 induction (INFγ TNFα[a] IL-2)	↓	↓
Th-2 induction (IL-4 IL-10)	↑	↑
B cells antibodies, cytolytic	↓	↓
ADCC	↓	↓
enhancing	↑	↑
MHC-expression	↓	↓
Immune T cells	Evasive	Maneuvers
Treg cells (antagonizing immune T cells)	↑	↑
NK/NKT cells	↓	↓
FasL ('counterattack' on Fas $^+$ immune T cells)	↑	↑
IDO (depleting tryptophan)	↑	↑
Vascular mimicry (replacing spiral arteries)	↑	↑
IGF (activated by endogenous retrovirus in the placenta)	↑	↑
Proto-oncogenes (wnt, fms, etc. activated)	Reversibly ↑	constitutively ↑
Endogenous retroviruses (fusion protein syncytin)	++++	++

a, added ADCC, antibody-directed cellular cytotoxicity COX, cyclooxygenase FGF, fibroblast growth factor IDO, indol-amino-2,3-dioxygenase IGF, insulin-like growth factors MIP, monocyte inflammatory protein PP-A-R, peroxisome proliferator-activated receptor
* Sinkovics JG & Horvath JC Internat J Oncology 2005;27:5–47

The human cerebral cortex. The consummate bioengineer RNA/DNA complex finally encoded the human cerebral cortex, endowed with trillions of nerve cells (Figure 10). The human brain has evolved to the competence of encompassing a limited but rapidly expanding comprehension of the universe and life within it. There are some 5000 to 200,000 synapses per one mature neuron. The entire human brain is the site for expression of the widest scale of emotions, from generous altruism to devastating anger and revenge; to pleasant amusements, artistry, logical thinking, imagination (often beyond logics), but nevertheless with keeping the faith in the transcendental, planning in advance, premonitions, and driving curiosity.

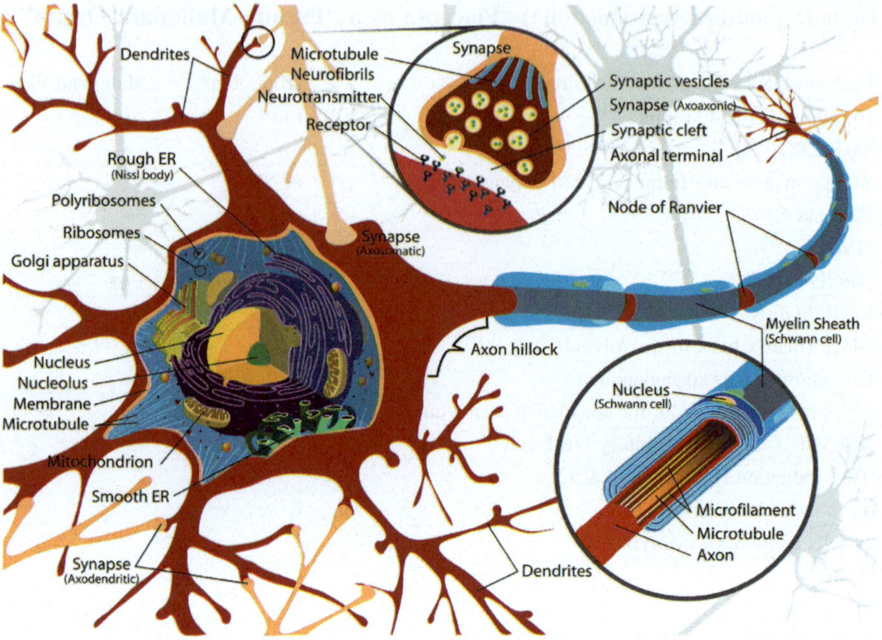

Figure 10 Neurons, Axons and Synapses in its surroundings

The "driving curiosity," is what fuels the passion for scientific research (István Hargittai, 2012, Amazon.com). The human cerebral cortex has come to understand the Universe/Multiverse and the living matter within.

The cassettes of memory for solid preservation of events, words (words and grammar of foreign languages), thoughts, past happenings, and large volumes of learned material, reside in an organ packed with highly specialized neurons, the "hippocampus library" (of György Buzsáki, New York University, NY). Remember Henry Molaison? We do remember him as we read his story again (Corkin, Suzanne: "Permanent Present Tense", Basic, 2013), as he was unable to remember anything. Henry was the man, whose severe epilepsy was cured by removing surgically part of his brain, including the hippocampus. By losing that organ, he had lost his entire storage of memory. Unfortunately, the aging process decimates the neurons, but some of them somehow preserve our oldest memories.

The pioneer of molecular immunology, Susumu Tonegawa, switched from the discovery of somatic hypermutation for antibody production in the plasma cell, to the study of the thinking and memory-preserving cells of the brain. His name is frequently printed first or last in the articles published in Science, and in the best journals of neurophysiology. Some hippocampal engram-bearing cells may be falsely induced to preserve a 'fear memory' of a non-existent threat or insult (foot shocks). For that, the c-*fos* proto-oncogene has to be activated by a synthetic ligand (Science July 26, 2013). The c-*fos*/Fos proto-oncogene/protein stands for the

Finkel-Biskis-Jinkins' FBJ murine osteosarcoma retrovirus, which snatched its proto-oncogene from the genome of its original host cell (c-*fos* → v-*fos*). False fear-feeling memory cells revealed themselves by expressing the light-induced marker channel rhodopsin-2 (implanted into them by a ChR2 gene-carrier adeno-associated virus). Superb experimentations reveal the wide physiological territory proto-oncogenes cover; is it a lapse, or is it an inherent faculty, that these proto-oncogenes can immortalize some of their host cells?

Individual cells of the brain cell community touch each other in synapses and communicate with solubilized chemicals through their receptors and with circuitries of insulated electric currents. The magnificent system may fall victim of faulty blood supply, plaques of misfolded protein molecules (Carleton Gajdusek's prions), mis-wired electrical circuitry, imbalanced ligands overacting, or not acting at all, truncated receptors incessantly signaling, senescence, prionic, viral, bacterial, fungal, parasitic infections, autophagy, apoptosis, oncogenesis and death. Its protector, the microglia, may fail to defend the brain, or it may directly attack it in the events of various forms of autoimmune encephalopathies. However, the end-differentiated neurons may reactivate their ancient stem cell genes, which turn these cells into tumors of the highest degree of de-differentiation and malignancy (a glioblastoma multiforme). Inside the cell nucleus, newly activated genomic changes may constitutively arise to dictate an irreversible overproduction of the progeny of those involved cells. In the most advanced system biological evolution ever created, hidden errors of installation remain? Or is it an inherent faculty of the universal eukaryotic genome to preserve the genes of the ancient cell survival pathways from the unicellular ancestors up through the evolutionary ladder to the end product, the vertebrate mammalian hosts, Homo, on the top. One of the most malignant tumors, the glioblastoma multiforme, originates in the human brain, in the most advanced system, as a consequence of a misdirected biological evolution? Are the cell survival genes of the unicellular ancestors still present in the human genome, at its highest level of evolution?

The ancient cell survival pathways are the ancestors of the proto-oncogenes. Some of the protozoal genes of the choanoflagellate (Monosiga), the p21-activated pak/Pak pathway, transfect and transform mammalian cells *in vitro* (see in the text). The siliceous marine sponge *Suberites domuncula* (Demospongiae) arising in the Cambrian sea about 500 mya, and becoming the ancestor of multicellular organisms, lives with ribosomal proteins closer in similarities to vertebrate mammalian proteins, than to those of invertebrates. View the springerimages for the evolution and molecular phylogeny, mitochondria and innate immunity, which is able to reject grafts, of Demospongiae in en.wikipedia.org/wiki/Demosponge). As shown by Helena Ćetković and her associates at the Rudjer (Ruđer) Bošković Institute in Zagreb, Croatia, among these proteins, acquired for the good of its own metabolism, the sponge operates ras-like and src-like genes (proto-oncogenes, see in the text) and has a 700 amino acid Tec family protein that is in 55 % similarity with the human Bruton tyrosine kinase (BTK). The Tec family kinases (see in the text) signal from the T-cell, B-cell lymphocyte receptors for inducing cell maturation, or the inhibition thereof, a maturation arrest; in it, constitutively expressed mutated kinases keep cells (lymphocytes) immature and replicating. The human oncoprotein BTK is a major inducer of human B-cell lymphomas, from the moderately malignant chronic lymphocytic leukemia (CLL) to the highly malignant mantle cell lymphoma MCL) (see text). The small molecular enzyme-inhibitor ibrutinib blocks the enzyme and induces spectacular remissions in both CLL and MCL. Would ibrutinib kill that innocent sponge, too, by blocking one of its physiological "cell survival pathways"?

The genomics of the drosophila leukemia and brain tumor are comparable to that of some human tumor*s* (as documented in the text). The ancient cell survival genes had been preserved throughout evolution and retained their potency to be reactivated, either by accidental errors (amplifications; mutations), or by an inherent faculty securing the continuity of the living matter in the form of "immortalized" cells.

Any attempt to eradicate by physico-chemical means the emerging and advancing unicellular clones of the transformed cells is met with a resistance equal to that, what the first primordial cells presumably must have expressed, in order to survive quite violent physico-chemical encounters on the ancient Earth (and they/we did survive). The asymmetrically dividing stem cell in the bottom of the Lieberkühn crypt expresses the ABC proteins and pumps out all toxic substances that might have entered its cytoplasm. The dinoflagellates' toxins kill the manatees in the Florida bays, but are harmless for the algal cells. The ancient unicellular eukaryotes were armed with the pumps to expunge toxins, and their ancestors preserved the faculty. The tumor cell descending to the level of its unicellular ancestors uses the ABC proteins to pump out the chemotherapeutical molecules. The tumor cell assumes the appearance and the mode of locomotion of an amoeba; it is a genetically encoded transformation of its microtubular cytoskeleton that enables this cell to return to the life style of its distant ancestors, as written with a remarkable foresight of discoveries to follow in molecular biology by Stuart R. Hameroff in his "Ultimate Computing" (Elsevier Science Publishers, 1987; electronic edition, pp 1–229, 2003).

A medical oncology textbook of 2062 pages in two volumes printed in 1986 has condensed over 20,000 references. This single-authored volume disappointedly documented the extraordinary resistance malignantly transformed human cells (the cancer cells) exhibited toward chemotherapy (Joseph G. Sinkovics' Medical Oncology, An Advanced Course, 2nd ed volumes I and II, Marcel Dekker, NY, 1986). Even after complete clinical remissions, relapses of leukemias and solid tumors were common events. However, the chiefs of services and teachers of academic medical oncology, prominent among them: Lodovico Balducci, H.L. Moffitt Cancer Center, Tampa, FL; Ephraim Casper, Memorial Sloan-Kettering Cancer Center, New York City; Fred Conrad, Carl Plager, C.C. Shullenberger, M.D. Anderson Hospital, Houston, TX; Karl Tornyos, Tulane University, Charity Hospital, Veterans' Administration Hospital, New Orleans, LA; Sándor Eckhardt, Országos Onkológiai Intézet, Budapest, Hungary, all expressed support and appreciation for the mode of factual reporting that characterized best that volume.

The transforming RNA/DNA complex is evolutionarily resistant to radiation and chemotherapeutical attacks. Are there any newer products of the human cerebral cortex, that will gain control of the transforming RNA/DNA complex? The text provides a cautious affirmative answer: yes, lenalidomide degrades the Ikaros IKZF (Ikaros zink finger) and Aiolos and ZPBP (zona pellucida-binding protein) oncogenes in myeloma cells (Science Jan 17 2014). Very well explained in Lessard C. J. et al in context with SLE (systemic lupus erythematosus) in Am J Human Genetics 2012;90; 648–660. The monoclonal antibody (mcab) ipilimumab creates a situation in which the masquerade of self of the tumor cells is pierced through by the attack of immune T and native NK cells of the host. Pembrolizumab surpasses ipilimumab in that (*vide infra*). Communicating with the transformed cell on its biological language, a proposal made long ago [2070], may be responded to by now, and favorably so for the multicellular host.

Research

The "cell survival pathways" of unicellular eukaryotes were subjected to a point by point comparison at the biomolecular level with the driving forces of the so called malignantly transformed single cells of multicellular hosts (vertebrate mammalians, prominently including Homo). The compared subjects were the descendants of the earliest uni- and multicellular life forms (archaea, eubacteria, algae and choanoflagellates; all orders of unicellular and early multicellular eukaryotes: amoebae, giardia, trichomonads; kinetoplastids; tunicate ascidians; yeasts; fungi; cnidaria, the protochordates, the amphioxus; the sea urchin; nematodes; insect larvae, pupae and adults; the vertebrates: cartilaginous (chondrichthyes) and bony fish, tadpoles and amphibians; aves, and mammalians). For example, the PI3K pathway drives the Giardia; it is the driver of practically every human tumor cell. In the sea anemone, NFκB/STAT (nuclear factor kappa B; signal transducer activator of transcription) is a cell survival pathway. In human malignant lymphomas, it is a prominent oncogene/oncoprotein (more in the Appendix; in Int J Oncology Oct 2015;47:1211–29). The achaete scute genes encoded the ancient nervous system in the amphioxus; their human homolog Ash is encoding neuroectodermal transition of adenocarcinoma cells; and it is the primary oncogene in esthesioneuroblastoma.

Gain of function mutations and gene fusions. A subject of investigation was approached repeatedly from different angles of inquiry, in order to extract most its contributions. This approach may appear redundant. However, the Subject Index lists the page numbers, where these inquiries considering one particular subject of interest are dispersed at different sites in the text. For example, the Florida amphioxus, or the choanoflagellates received an extensive scrutiny spelled out at different points in the Table of Contents. The comparative studies included events of metabolism, in particular glycolysis; frequency of cell divisions and preservation of telomere lengths; gene fusion and fused gene product proteins ("oncoproteins"); proliferating cell nuclear antigens; intronization of the oncogenome (losses or gains of introns); epigenomic enzymatic activities, including arginine-methylators, events with semblance to immortality; apoptosis resistance; and resistance to physico-chemical damage. Considering the type of mutations, for the one nucleotide-, one amino acid-changing point mutations, the emphasis was put on the "gain-of-function" type mutations. G-quadruplex DNA formations and DNA propellers (see in the text), high propensity mutability, loss-of-function mutations (involving the loss of the so-called tumor suppressor genes) leading to anti-apoptosis, *versus* gain-of-function mutations (involving the oncogenes) resulting in accelerated cell cycles over the anti-apoptotic events are analyzed. Events of aneuploidy and trans-speciation; and fusion oncogene-oncoprotein generations are comparatively presented. Events of trans-speciation, the parasitic unicellular microorganisms (Naegleria; Theileria), and the host cells infected by them, both undergo, are comparatively presented. Accidental DNA damage with success or failure of DNA repair during somatic hypermutation in the adaptive immune system is recognized. Endogenous retroviral virion reassembly and lymphomagenesis during the event of, and consequential to, somatic hypermutation are correlated. Preservation of ancient "pro-survival" genes in stem cells is recognized throughout. The contribution of the ancient proliferating cell nuclear antigen, including Ki-67, and the arginine methyltransferases to the regressive trans-speciation of the cell is documented. The immense power retained by the RNA (the interfering short hairpin and PiWi microRNAs) from the era when the RNA predated, and acted superior to, the DNA, is repeatedly pointed out. The DNA-binding proteins are guided to their targets by miRNAs/ncRNAs (reviewed in postscript Table XXXV).

Results

The mode of reporting. The chapters "The Introduction" at the beginning, and the "Making Cancer History" at the end, are written in a narrative style, yet adhering strictly to the facts. The text of the volume reports the works of the authors cited by names in numerical order in the List of References. The text bears the burden of numerous cross references; these are for the sake of viewing the subject matter in question from different angles and in different contexts. These cross references appear well organized in the Subject Index. Inserted are the author's (in third person, JGS) *Comments;* these are not sequence-numbered; the Comments are under the sequence numbers of the subject matters to which they are reflecting.

The "oncogenome". Single cells of multicellular hosts transformed by the so called "oncogenome" imitate the life style and the biological characteristics of primordial unicellular life forms (based on the genomics of their descendants, the extant unicellular life forms, especially eukaryotes). These extant cells possess and exhibit faculties conserved from the era of the ancient single cells (the world of the unicellular and early multicellular eukaryotes), that are re-played in evolving multicellular hosts (larvae, tadpoles, and embryos). Permanent cultures of transformed extant cells practice the locomotion of the amoeba, metabolize sugars in the

Warburg way, exercise “cell cannibalism” by phagocytosis, and display the spread of cytoplasmic and genomic material (in exosomes), for horizontal uptake by other cells, events that must have characterized the primordial cell colonies. Asymmetrically dividing stem cells preserve the faculties of unicellular life forms within their multicellular hosts. On occasions, the ancient “cell survival pathways” might be constitutively expressed in extant healthy unicellular life forms. In addition to their vertically transmitted genes to progeny, many of these faculties were horizontally transferred and avidly accepted in inter-species and even in inter-taxon settings. Sequential, coordinated and constitutively expressed “loss-of-function” events concern the so-called tumor suppressor genes; in contrast, sequential and engineered “gain-of-function” mutations characterize the oncogenome. The actions of the oncogenome can be interpreted as not necessarily entirely chaotic and disorganized. The fusion oncoproteins are formed with avoidance of cuts (nicks) within the genes of the functional protein domains, and the fused oncoproteins are built to avoid ubiquitination (the H. Hegyi et al article is cited in the text).

“Oncogenesis”, as recognized at our clinics to be a fatal illness for the multicellular host, may be viewed as the arousal of the transformed and trans-speciated single cells in a seemingly disorganized multicellular community. It may be the expression of an inherent faculty of the original primordial RNA/DNA complex, for the maintenance of immortalized cellular life in any form, and under any circumstances. The primordial RNA/DNA complex was the ancestor of all cell survival pathways, preserved and expressed in the extant “oncogenomes”. The origin of constitutively expressed cell survival pathways may be traced back as evolving gradually in several evolutionarily recognizable periods. Most of the so called tumor suppressor genes were acquisitions of the eukaryotes and the multicellular hosts. These genes function as opponents to the reversal of the subservient somatic cells into their ancestors’ independent and immortal unicellular life style. Malignant transformation of cells is frequently endogenously initiated. In these cases, the multicellular host extends support to the RNA/DNA complex. In-growth of blood vessels and provisions of nutrients and molecular mediators renders the host-tumor relationship similar to the mother’s relationship with the placenta of her fetus. Malignant transformation extrinsically initiated (viral, bacterial, parasitic oncogenesis) elicits vigorous defenses of the host. Vaccines work against it much better, than against the endogenously initiated malignant transformations. The genomes of the pathogens evolved the measures, that insinuate “selfness” to the host for acceptance. Some of the endogenously initiated events of malignant transformation appear as if the involved cells returned to the independent immortal life style of their ancient ancestors. The RNA/DNA complex preserved an inherent faculty within to extricate certain individual cells from a host, which is expected to succumb to life-threatening inflictions, for the rescue and restoration of the original geno-phenotype their individualized independent and immortalized primordial ancestors possessed. However, errors not yet eliminated in its continuing evolution and still retained in the adaptive immune system were not overlooked. The misguided nicks by the master enzyme activation-induced cytidine deaminase

(AID) during the process of somatic hypermutation may unite the the heavy chain immunoglobulin gene, not with a newly constructed light chain gene, but with a liberated proto-oncogene (c-*myc*; *bcl2*).

The "mutator phenotype" is reversibly encoded for survival by transposon-, or retrotransposon-derived DNA proviruses in the descendants of the ancient uni-, and multicellular hosts. The phenomenon is compared with the mutator phenotype encoded in cancer cells (Lynch syndrome; inflammatory carcinogenesis). The conclusion is documented; retroviral oncogenesis (malignant transformation of cells by inserted retrotransposon-derived DNAs) is widely spread, but remained concealed for a long time. The text attempts at providing a detailed and reasonable explanation for this phenomenon. Among the arguments listed, most recent data for the induction of pediatric acute myelogenous leukemia by hidden endogenous retroviral genomic reactivation are discussed. So is Hodgkin' s disease. The "DNA only, cut and paste transposons", like the recently discovered piggyBats (see in the text) emerge as suspect messengers for the activation of oncogenes. The immortal HeLa cells respond to them.

The consummate bioengineer. The RNA/DNA complex remains the consummate bioengineer of life in organized healthy, and in "malignantly transformed" cell communities. The number of cell survival pathways and their combinations are innumerable and amphisbaenic; the RNA/DNA complex utilizes and switches on readily any one or all of these pathways. Neither one of these processes is exempt of errors, inherent or acquired. The text distinguishes those forms of oncogenesis, that are **1**, initiated by external agents and life forms (oncogenic viruses; bacteria; protozoa) and are forcefully opposed by the host; and those that are **2**, due to acquired or innate errors remaining in the system and not yet corrected by on-going evolution (lymphomagenesis by the enzymes AID/RAG; mutated DNA-repair enzymes; radiation exposure). These oncogenic events are viewed as different in mechanisms and consequences from those that are **3**, gain-of-function point mutations of persistent proviral inserts of ancient retroviruses (for example, the V600E BRAF mutation, see in the text); and those **4**, that are due to the inherent mobilization of the ancient RNA/DNA complex for the rescue of some selected individual cells from a multicellular community for their immortalized unicellular independent existence after the death of their host. In the Appendix 1, a parallel is drawn as to the extraordinary LINE1 activity both in unicellular and early multicellular eukaryotic cells, and in the so-called malignantly transformed human cells. It appears as if the host welcomed these latter events, and extended all its available services and support for their full accomplishment.

However, the human cerebral cortex is encoded by the same RNA/DNA complex, that encodes the oncoproteins. These elements seek the disruption of an organized but doomed multicellular community for the rescue of immortalized unicellular individuals. The RNA/DNA complex mobilizes all its resources for the full control of its progenitor and inventor, serving now as its vehicle, the stem cell, both in the processes of onto- and oncogenesis.

It is pointed out throughout this text that the science and practice of medical oncology in a century long stepwise process has reached the point, where it can force (or biologically "convince", or coerce) a malignantly transformed cell to kill itself, or to reverse, or switch off, its course of self-transformation, without major harm to its host. The host may be enabled to reject its transformed cell population by induced immunobiological means, the same which are naturally available for the hosts exposed to alien invaders, especially unicellular protozoa.

However, the invaders possess and operate a counter-genome. Some ancient maladies were extinguished by "herd immunity." The Pithecanthropine hominids might have subdued the sarcomagenic retroviruses, which remained mature and active in the gibbon ape, koala and woolly monkey, and way below (down to the fish, as veterinary pathologists list them in Veterinary Pathology 2013;50;390–403). The human genome carries their proviruses. Smallpox was not eradicated by the natural means of the invaded hosts; it was external interventions that erased it (similarly to measles or poliomyelitis, but not HIV-1/AIDS). The influenza A virus just produced its new variant, the H7N9 strain. From H1N1 through H5N1 to H7N9, the flu viral genome shows that viruses naturally acquire increased virulence (pathogenicity) by means of gain-of-function genomic mutations and/or gene reassortments (*vide infra*). What the ancient viral genomes naturally do, is being proposed to do for them in the laboratory. In order to render the potential victim population protected in advance in expectation of the natural rise of such new mutant viral substrains, vaccinate the potential victims with laboratory-produced mutants, before the mutant viruses appear in nature. A proposal that is much debated in recent issues of Nature (Aug 8, 2013) and Science (Aug 9, Oct 18, and Nov 22, 2013). Will there ever be a vaccine to prevent cancer?

The viruses provide the proof that the ancient genomes constructed themselves so, as to carry the inherent intrinsic faculty of generating gain-of-function mutations. Already at the beginnings, the primordial genomes (RNA viral genomes) gained virulence not only by random gain-of-function mutations in tree-like out-branching within their basic vertical clonal evolution. These events were applied to multicellular organisms before the discovery of viruses, archaea and eubacteria in Darwin's "On the Origin of the Species", 1859; or in Haeckel's Tree of Life, 1879. Indeed, it has been so by the dramatic innovations of the recalcitrant living matter, in its reticulate (horizontal and recombinatorial) evolutionary events. It is evident even in viral triple reassortments (as reviewed from Columbia University, New York; Institute of Advanced Studies, Princeton, NJ; and Stanford University, CA, and published in PNAS November 12, 2913). The H7N9 genome is the result of not a triple, but a dual reassortment. Genomic segments uniting in H7N9 for H7 are from a Chinese (ZJ12), and for N9, not from a Korean (KO14), but from another Chinese (Jiangsu, JS5) locally pre-existing flu A strains (reviewed in Biology Direct 2013 Oct 26). Bacterial genomes mutate even in an unchanging environment (Lenski's *E. coli*, in Science, November 15, 2013).

The genomes of eukaryotic cells follow suit. This ancient faculty of mutability in the RNA/DNA complex, when it is practiced excessively, it prominently characterizes the entity recognized in the scientific literature and at our clinics as the "cancer cell." There, in addition to one, or a few sequentially induced, complex driver mutations (the oncogenic pathways), the enormous cell population of a cancer carries large numbers of somatically gain-of-function point-mutated genes. So, that just one cell out of many thousands, would fit metabolically into one particular environment, that is undergoing whatever unforeseen new changes. Thus, the primordial RNA/DNA genome has evolved with the inherent faculty to last "wherever" and "forever". In their space vehicle, the transformed cell, or even out of it, the driver rocket engineers, the RNA/DNA complex, stay alive for the duration of the physicochemical Universe, whether individually preserved in the ice core of a meteorite, or on the temperature of the liquid nitrogen in the outer space; or organized in cell colonies in an Earth-like planet, and there with or without the human cerebral cortex co-existing.

However, for the human race and its cerebral cortex it is essential to learn what the obstacles were that the Martian RNA/DNA complex has had failed to overcome.

Part I
The Primordial RNA/DNA Complex Evolving

The Primordial Genes Are Conserved as Stem Cell Genes and Proto-Oncogenes

The Ancient Earth

Amino acids and nucleotides coexisted before cells. Amino acids and nucleotides were formed readily from small organic molecules [2–4a, b]. Hypercyclic networks of polypeptides (L-enantiomers dominating), and nucleic acids (RNA, catalyzing ribozymes, and ribonucleoproteins (RNP) coexisted in the ancient Earth some four milliard (billion) years ago in a pre-cellular era [5–10]. Lipids and sugars could be synthesized without hypercycling. The primordial chemical reactions resulted in the formation of virus-like agents (viroids) different from extant cell-dependent viruses. Extant viroids are relics from the precellular era [11–14a, b]. Ribozyme-armed precellular ribosome-like units might have generated the ancestral viroids. Fused viroids might have formed ancestral large viruses. The first pre- or protocellular entities consisted of vesicles with double lipid membranes. These vesicles incorporated the interactive large molecules (Figures 11 and 12), including complex nucleic acids and "self-replicating and supercycling polypeptides" (Figure 13). The vesicles grew in size and complexity and were able to divide (chemotons of Gánti; coacervates/коацерват of Oparin/Опарин); and replicators of Szathmáry [15–17]. The primordial genomes and proto-ribosomes might have consisted of RNA complexes (mRNA, rRNA, tRNA, ribozymes, and RNP ribonucleoproteins), replicating with genetic variations due to copying errors. It is not known how and when DNA appeared first and if more than once, and by possible different mechanisms of origin: synthesis, like the original RNA, or conversion from RNA to DNA. The precellular origin of viruses (virus-like agents) is very likely, since ribozymes/RNPs and polypeptides existed and interacted before cells were formed. The double-walled vesicular entities were invaded by the pre-existing virus-like agents. The virus-like agents experienced losses of their independent genes, preserved their "virus hallmark genes" of Koonin (genes not shared with host cells), and acquisitioned new genes deriving from their host cells (*vide infra*). Their replication has become dependent on the metabolism of their host cells, which the

J.G. Sinkovics, *RNA/DNA and Cancer*, DOI 10.1007/978-3-319-22279-0_2

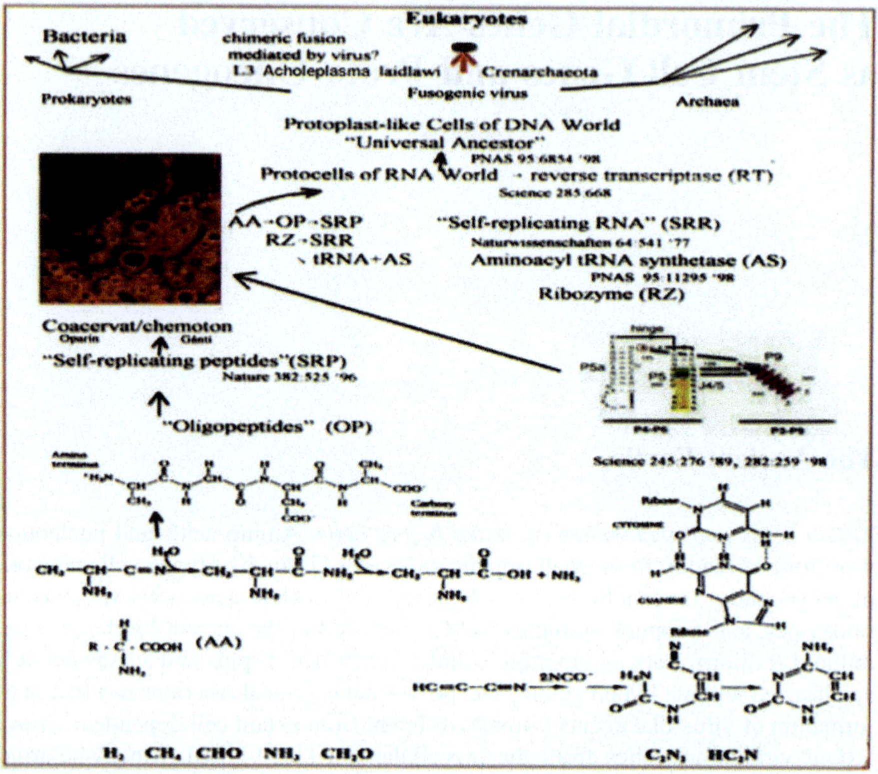

Figure 11 Biomolecules on the Ancient Earth. Reference Appendix 2, Explanations to the Figures

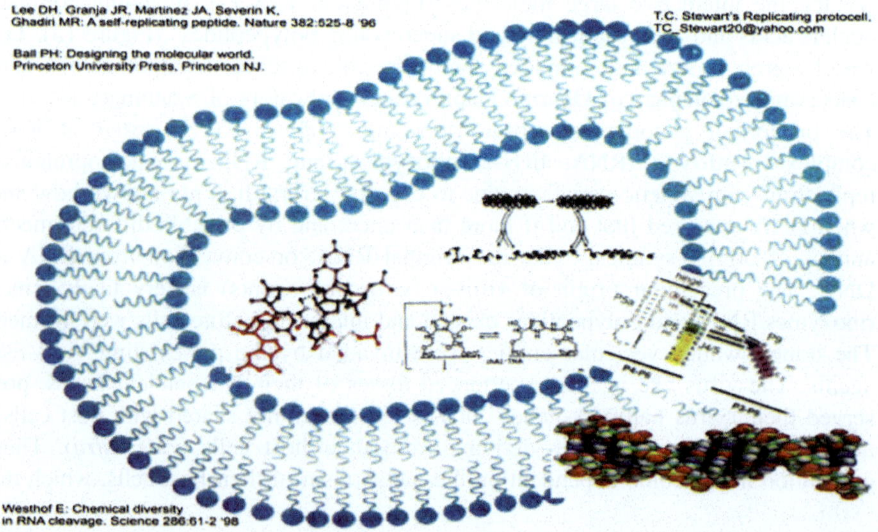

Figure 12 Chemotons, Koacervats, Protocells. Life in the Protocell. Reference Appendix 2, Explanations to the Figures

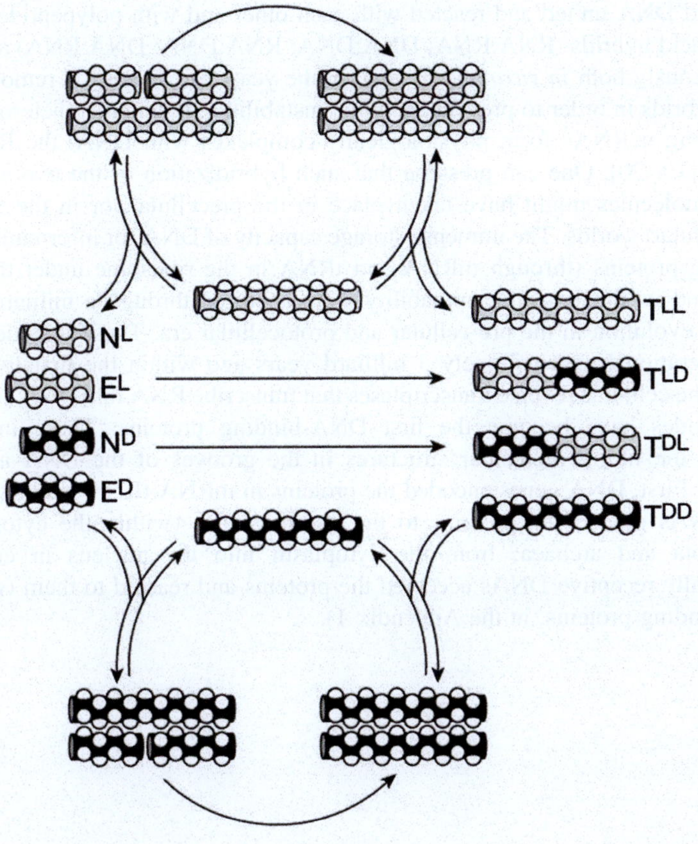

Figure 13 Autocatalytic Chiroselective Reactions of Hypercycling Peptides. [8] Reference Appendix 2, Explanations to the Figures

viral agents expropriated and ruled. The successful viral genomes could readily overtake and redirect the host cell genomes. Other viruses inserted their genomes into the chromosomes of their host cells, where these proviral units were retained and remained conserved (*vide infra*).

The primordial virus world and gene pool included the RNA/DNA complex. The precellular virus-like agents possessed either or both RNA and DNA strands forming the precellular "Virus World" and "Primordial Gene Pool" of Koonin, Senkevich and Dolja [18], dominated by ribozymes, and ribonucleoproteins (RNP). The fact that extant cells possess long non-coding RNA strands with RNA → cDNA structures being embedded in them, the human *sox*2 gene at 3q26.3–27 with an overlapping intron (P.P. Amaral *et al*, *vide infra*), suggests that RNA and DNA might have already co-existed in similar constellations in the pre-cellular and proto-cellular environment. The DNA either evolved from RNA, or these two master molecules co-evolved side-by-side, but RNA predating DNA. In the precellular era,

RNA and DNA united and reacted with each other and with polypeptides. Indeed, nucleic acid hybrids (RNA:RNA; DNA:DNA; RNA:DNA; DNA:RNA) are formed spontaneously both *in vitro* and *in vivo*. In the yeast cell, RNase H removes RNA: DNA hybrids in order to prevent "genome instability". In contrast, heterochromatic non-coding ncRNAs form physiologically complexes with DNA: the RNA:DNA hybrids [19, 20]. One can presume that such hybridization of the two co-existing master molecules might have taken place in the precellular, or in the immediate protocellular, worlds. The immense storage capacity of DNA for information and for encoding proteins (through mRNA and tRNA in the ribosome under the genetic code), and its inexhaustible mutability were acquired through a milliard (billion) years of evolution in the pre-cellular and protocellular era. These faculties became fully operational approximately 3 milliard years ago within the first living cells. Within the cell, it is reverse transcriptases that transcribe RNA into DNA [28]. Large polypeptides have become the first DNA-binding proteins. These interactions required the fitting molecular structures in the grooves of the DNA and in the proteins. First, DNA genes encoded the proteins in mRNA-tRNA interactions; then microRNAs guided the proteins to the chromosomes (within the cytoplasms in prokaryota and archaea; from the cytoplasm into the nucleus in eukaryota). Structurally receptive DNAs accepted the proteins and reacted to them (*vide infra*, DNA-binding proteins, in the Appendix 1).

The First Cells Ascend

Primordial Spheroplasts

The "Last Universal Common Ancestor(s)" (LUCA; the "C" may stand also for the first "Cellular") existed before the division of the first cellular elements into the three domains of life on Earth: archaea (C. Woese), prokaryotes (eubacteria), and eukaryotes. It is not known if archaea preceded prokaryota, or *vice versa*. In contrast, a form of regressive evolution from large spheroplastic eukaryote-like cells into microbes has been proposed and widely discussed [21–25]. The first cellular entities lived in colonies consisting of proto/spheroplastic individuals. Competition for nutrients and for genomic enrichment useful for metabolism in complacency with demands of a hostile environment (heat stable proteins/enzymes; gain-of-function mutations with constitutively installed "cell survival pathways"; efficient DNA repair) must have been relentless. These entities transferred their genes vertically, but also through horizontal (lateral) routes [26, 27]. The large DNA viruses (mimivirus, and its clade) and amoebae preserved an ancient relationship. Their genomes are separated, but they continue to exchange their genes (*vide infra*).

Metabolically active viral agents very likely predated the appearance of protocells. The viral agents invading these cellular entities lost the majority of their genes within these protocells and first living cells, but acquired new genes from the genomes of their new hosts. In the extant viral entities, remained recognizable the ancient "viral hallmark genes" of Koonin *et al* [18], which were not shared with cellular genomes. The first viruses (phages) contributed mightily through horizontal gene transfer events widely practiced below the Darwinian threshold [27]. Viral genes donated to the protocells might have contributed to the formation of genomes condensed into chromosomes. Both RNA and DNA remain embedded in chromatins and rounded around histones in archaea, prokaryota, and later in the eukaryotes. Some large DNA viruses might have been instrumental in the formation of nuclei in eukaryotes. Poxviruses encode α-polymerases; archaea operate with δ-polymerases, which start DNA synthesis on RNA primers; eukaryotic nuclei

J.G. Sinkovics, *RNA/DNA and Cancer*, DOI 10.1007/978-3-319-22279-0_3

possess similar α- and δ-polymerases. Poxviruses, ancestral or extant, replicate in the cytoplasm in "virus factories". Ancient poxviruses could have had readily infected a-nucleated archaea-like cells and formed cell nuclei, with separation of nuclei and soma (ribosomes) to follow [28, 29a, b]. The first eukaryote(s) might have been the fusion product(s) of a crenarchaeota spheroplasts and a mycoplasma-like prokaryota (eubacterium). Endosymbiotic proteobacteria engulfed later serve as mitochondria in animal cells, and cyanobacteria as plastids in plant cells (*vide infra*).

Viral Genes Exchanged

Basic virology. The newly acquired viral genes encode the synthesis of structural proteins of the viral particles for the package of the viral genome in them. Viral microRNAs switch off defensive reactions of the host cells. The first defensive reactions were the CRISPR and APOBEC pathways (clustered regularly interspersed short palindromic repeats; apolipoprotein B-editing enzymes catalytic). Later, in multicellular hosts, mRNA-neutralizing microRNAs, interferons, antibodies, NK cells, dendritic cells and immune T cells rise for antiviral defense. It was recognized early, that large granular lymphocytes (later: NK cells) attack virally infected cells (and autologous, or allogeneic cancer cells), and small compact lymphocytes (later: the immune T cells) actually develop to render the specifically directed attacks [30a, b]. It took long time for the scientific community to accept that the large granular lymphocytes (the NK cells) evolved to attack their targets without specific preimmunization (*vide infra*).

The viral genome (either RNA or DNA), which replicated in the host cells (in the cytoplasm and/or in the nucleus, or in both) will be encapsulated (packaged) in the viral particle. Other proteins guide the egress of the viral particles from the host cell. The viral envelopes will express molecular configurations, which specifically attach them to the membrane of potential new host cells. Chemokines will facilitate the entry of virus particles into the cytoplasm. The liberated viral genomes either replicate in the cytoplasm, or enter the nucleus either on their own, or with the help of host cell-derived enzymes and proteins. The viral genomes readily mutate to change the protein configuration of the viral capsid epitopes (prominently so in the influenza viruses and in human immunodeficiency virus HIV-1). Thus, viral particles escape host immune reactions, both lymphocyte- or antibody-mediated. If certain elements of the viral capsids remain unchanged (not mutable, neither by amino acid composition, nor in their three-dimensional structure) host immune reactions directed specifically to those unchanged epitopes could possibly eliminate the virus. The search is on in advanced virus laboratories for such permanent antigenic epitopes in HIV-1 envelopes. Broadly neutralizing monoclonal antibodies (mcab) could locate the permanent antigenic epitopes [31].

Advanced virology. RNA viruses and retrotransposons predated the formation of DNA viruses in the pre-cellular virus world of Koonin *et al* [18]. Precellular ribosome-like entities might have been operated by ribozymes and RNPs. Cellular ribosomes could remain operational extracellularly (Heinrich Matthaei and Marshall Nirenberg, *vide infra*). In the cellular world, there are two enzymes that transcribe RNA to DNA: telomerase and reverse transcriptase (RT). The gene of the enzyme RT is a component of the retroviral genome [32]. Template-encoded proteins of this complexity are not expected to have been formed in the pre-cellular era, unless precellular RNP-armed entities resembling ribosomes existed.

In its host cell nucleus, the retroviral RNA genome rests retrotranscribed by RT into a DNA provirus. As a pseudogene, it may remain inactive in the descendants of that species. However, in the processes of placentation, or somatic hypermutation, dispersed segments of proviruses may find one another and fuse into functional genomes endowed with the *env* gene. Thus, budding endogenous retroviruses may arise (*vide infra*). There is a long list of pseudogenes, which have become useful functional genes as reviewed by Bushman [1]. Proviral LTRs (long terminal repeats) are preserved in the genomes for millennia. Reactivated LTRs (immature and mature retroviruses) induce autoimmune reactions, or actually activate proto-oncogenes. Reactivated ancient LTRs emerge as inducers of the "mutator cellular geno-phenotypes". This may occur in invertebrate hosts in distress, and results in the switch-on/switch off activation of "cell survival pathways", or in the constitutive activation of these pathways in cells of large multicellular hosts. These cells eventually undergo the so-called malignant transformation. A large number of these malignant transformations begins with amplification of non-mutated genes, whose promoters were de-silenced by enzymatic de-acetylation, or de-methylation. An example is the highly malignant sarcoma of Schwann cells, the malignant peripheral nerve sheath tumor (MPNST), as referred to in the Introduction.

An extremely complex machinery is acting upon the retroviral LTR and its promoter. The process involves viral transactivating proteins (Tax for HTLV-I; Tat for HIV-1) and cellular factors (the nuclear factor kappa B, NFκB; the specific protein, Sp1; cyclic adenosine monophosphate, cAMP-response element, CRE; CRE-binding protein, CREB; the Tat-activated RNA structure, TAR) and kinase phosphorylations, epigenomic acetylations, methylations, deacetylations, demethylations, always with the appropriate enzymes. RNA polymerases carried out the essential first steps. Protein synthesis begins with the release of mRNAs from the cell nucleus into the cytoplasm. These mRNAs pass through intranuclear spliceosomes, and travel to the cytoplasmic ribosomes (*vide infra*), there to encode proteins under the guidance of tRNAs of the genetic code. The cells' ribosomes synthesize the viral structural proteins [33–35]. However, some mRNAs may be intercepted and destroyed by microRNAs, small interfering iRNAs and hairpin RNAs (mi, si, hp) (*vide infra*).

The First Eukaryotes Were Fused Cells

A malleable spheroplast. In an early observation, malleable, and thus potentially filterable, spheroplasts of aged *E. coli* cultures (or large cytoplasmic units thereof pouring out through torn cell membranes) [36a, b, c] (Figure 14) coalesced before regeneration into vegetative cells, with or without the support of *Sarcina lutea*, the Russian "метод кормилок" (metod kormilok). In this original process unfolding, viral particles (phages) were not specifically looked for (a mistake). However, electron-dense cytoplasmic ovoid particles were visualized and photographed (Figure 14). The small ovoid bodies originally were believed to be bacterial chromosomal fragments in the cytoplasm. In retrospect, the ovoid electron-dense particles might have been viral elements contributing to the fusion of the proto-spheroplasts [36a, b, c, 37–39]. The observation of the fusion of large spheroplastic fragments (proto-spheroplasts) in dark field-, or electron-microscopy, suggested to this author that ancient cells (proto-spheroplasts) might have gained vitality by fusion, and that fusogenic viruses might have mediated these events. This author proposed that the ancestor of the *Acholeplasma laidlawii* L3 virus (MLV phage) induced the fusion of crenarchaeota and prokaryota proto/spheroplasts, and thus created the first eukaryote [39]. It is not known if the fusogenic virus of *A. laidlawii* would work in a combination of ancient prokaryotic (*A. laidlawii; Aquifex aeolicus, A. pyrophylus; Thermotogales, T. maritima*) and crenarchaeotic proto-spheroplasts.

The Icelandic *A. pyrophylus*, a strict chemolithoautotroph, grows in hot water (67–95 °C) using electrons from hydrogen and sulfur for the reduction of oxygen and nitrates. By its 16S rRNA, it is the most primitive microorganism known, competing for the lowest rank with the *Thermotogales* (reviewed in Klenk *et al*) [40]. These microorganisms would be candidates for phage-mediated fusion experiments with spheroplasts imitating of ancient prokaryotes and crenarchaeotes. Thus, the laboratory recreation of the generation of the first eukaryotes may be possible.

Crenarchaea: divide like eukaryotes. The Mollicute, the *A. laidlawii* cell, divides by the prokaryotic FtsZ mechanism [41]. The euryarchaeotic (*Thermococcales and Thermoplasmales*) and prokaryotic FtsZs are filamentous tubulins with cytokinetic Z rings-forming sheets, and cytoskeletal septosomes with affinity to guanine triphosphatase, and exhibiting nucleotide binding sites. In contrast, the crenarchaeota (*Sulfolobales*, i.e. *Sulfolobus acidocaldarius*) initiate their cell divisions with the three gene operons, *cdv* (cell division), and mobilize over twenty genes to accomplish the process. The genes and gene product proteins cdvB and cdvC of crenarchaeota (with the exception of *Thermoproteales*) are related to the eukaryotic proteins active in cell divisions. These are the vesicle budding process, the ESCRT-III sorting complex (endosomal sorting complex required for transport) and CHMPIII (charged multivesicular body protein [42, 43]. *) The mitochondria and the chloroplasts (plastids) preserve prokaryote-like division-mechanisms (*vide infra*).

*) *Extraordinary additional material (beyond this author's capacity): Zillig, Wolfram Prangishvilli, David et al Max Planck-Institut Martinsried 82152 Germany Viruses, plasmids and other genetic elements of thermophilic and hyperthermophilic*

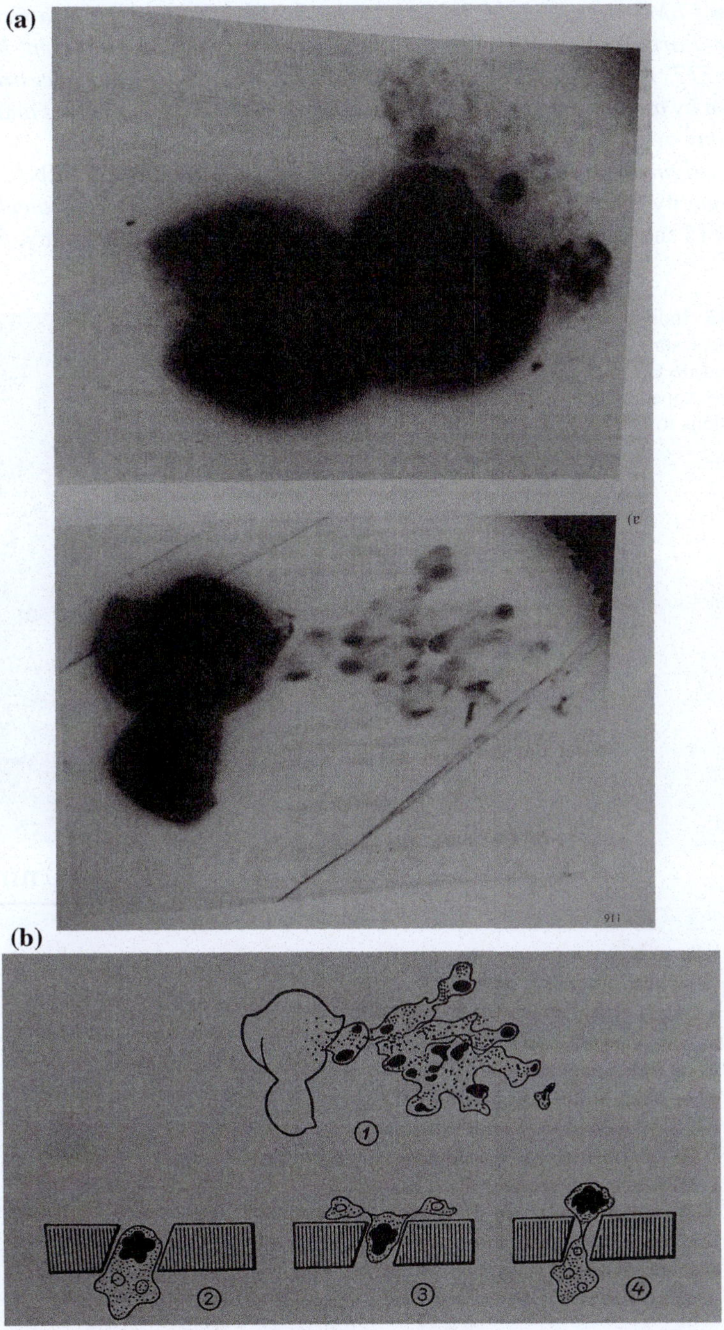

Figure 14 Cell Wall-free Prokaryotic Protoplasm Retains Its Vital Signs. Reference Appendix 2, Explanations to the Figures

Archaea FEMS Microbiology Reviews 1998;18:225–236. Forterre, Patrick Institut Pasteur Paris France The universal tree of life: an update. Front Microbiol 2015;6:717. Csűrös, Miklós University Montréal Québec Canada Ancestral reconstruction by asymmetric Wagner parsimony over continuous characters and squared parsimony over distributions (in print).

The origin of Mollicutes. A preliminary report based on 16S rRNA, catalogs evolutionarily connected the Mollicute *Mycoplasmatales* (including *Acholeplasma* sp) (Figures 15 and 16) with *Bacillaceae*, thus suggestive of a clostridial ancestry. According

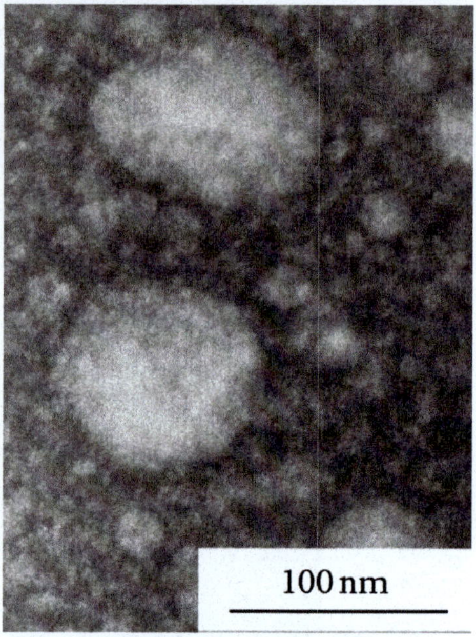

Figure 15 Individual Cells of Acholeplasma laidlawii. Its Complex Life Cycle [2339] Reference Appendix 2, Explanations to the Figures

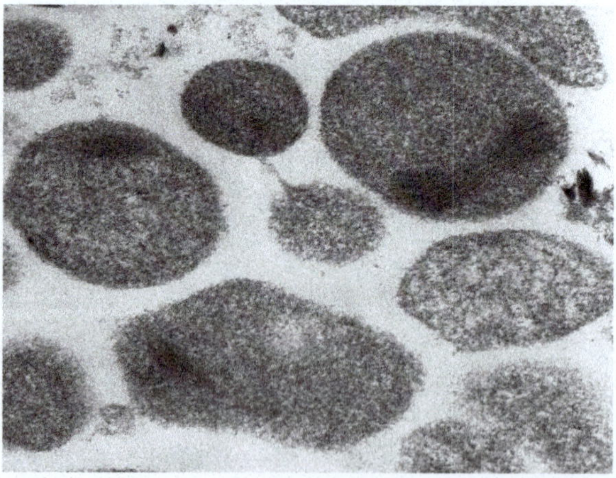

Figure 16 Transmission Electron Microscopy of Acholeplasma laidlawii Cells. With Kind permission © Dr Ponomareva Anastasiya Anatolevna. All Rights Reserved 2015. Reference Appendix 2, Explanations to the Figures

to this evaluation, the *Mycoplasmataceae* are not representative derivatives of a primitive prebacterial ancestor. Rather, they have undergone a "degenerative evolution" from clostridia (*C. ramosum, C. innocuum*) [44]. However, the 16S rRNAs cover only about 0.3 % of the average bacterial genome. The well conserved and larger (encoding 1200–1400 amino acids) subunits of DNA-directed RNA polymerases (RNAP) emerge as more reliable tools for constructing a phylogenic tree [40]. Accordingly, phylogenies recapitulated by 16S rRNA and RNAP can be incongruent, favoring for validity RNAP. In tests of higher sensitivity [40], *T. maritima* ranked lower (was deeper branching), than *A. pyrophylus.* Further, the RNAP-based phylogenic data suggest the position of *Mycoplasmataceae* to be even below *T. maritima.* "Bacteria invented cell wall biosynthesis after the mycoplasmas had branched off from the common bacterial line of evolution", as stated by Klenk *et al*) [40].

However, when prokaryotic evolution started with *Mycoplasmataceae*, not all spots on the ancient Earth were hyperthermic. Life on Earth might have started with mesophilic microorganisms. The hyperthermophilic microorganisms (*Thermotogales, Aquificales;* and hyperthermophilic archaea) populated other equatorial nests. The ancestral reverse gyrase, that the hyperthermophils evolved and modified [45, 46] led to the "secondary thermoadaptation" of Korolev [47]. At the other end of the spectrum, there exist the cryophilic crenarchaeotes in the Antarctica (reviewed in Klenk *et al*) [40]. These microorganisms may be found in Lake Vostok (Озєро Восток) in the Antarctica.

Phytoplasmas are considered to be insect-vectored plant-pathogenic descendants of acholeplasmas. Sequence-variable mosaics (SVM) in their genomes are considered to be remnants of ancient and recurrent phage-attacks. Cryptic remnants of mobile elements were identified as derivatives of phage proviruses. These phage proviruses originally inserted in the acholeplasma genome were the initiators of the phytoplasmic clade‘s divergence from the acholeplasma ancestral line. Extant *A. laidlawii* strains do not harbor these SVMs [48]. The gyrase A gene product proteins of several acholeplasma strains (*A. laidlawii; A. axanthum, A. granularum*) are antigenically related [49]. Maniloff's writings referred to *Mycoplasmataceae* as microorganisms, which "arose by degenerative evolution as a branch of the gram positive eubacteria" (clostridia) [50]. This view was contested (convincingly for this author) by Kenk and Zillig [40].

Achole: no cholesterol. The entire genome of *A. laidlawii* within a single 1,496,992-bp circular chromosome was sequenced. The genome contains four riboswitches, and in its T boxes aminoacyl-tRNA genes, including the glutamyl-tRNA aminotransferase. The Entner-Doudoroff carbohydrate metabolism of *A. laidlawii* is unique in that, it possesses the complete pentoso-phosphate pathway. It utilizes glucose, fructose and galactose for energy (ATP) yield. The pigmented *A. laidlawii* cell synthesizes carotinoids (lycopen precursors). The cell membrane consists of glycolipids and lipoglycans instead of cholesterol (while most other mycoplasmas have cholesterol-rich cell membranes). The terminology "achole" reflects to this fact, nevertheless the correct nomenclature "Acholeplasma" fell victim to frequent misspellings (overlooked by editors). The acholeplasma genome and proteome are rich in nucleotide metabolism and enzyme constituents, but this microorganism is unable to synthesize nucleotides *de novo*. No plasmids were found and there is no mention of phage genomes inserted in the host's genome [51]. It remains unknown how *A. laidlawii* cells protect themselves against viral infections (CRISPR, *vide infra*).

Acholeplasma viruses. This host-virus relationship probably represents one of the earliest scenarios played out by phages and archaea/eubacteria. As to virus-(phage-) mediated fusion of primordial prokaryotic and archaeal proto-spheroplasts, polyethylene glycol induces fusion of intergeneric mycoplasma cells (including those of *A. laidlawii*, as one of the partners) [52]. *A. laidlawii* viruses were isolated first by Gourlay *et al* [56–58]. Some viral particles upon attaching to mycoplasma cell membranes induced fusions. Fusogenic viruses (usually enveloped viruses) abound in nature [38]. The first fusogenic virus might have been a mycoplasma virus (phage). The extant mycoplasma *Acholeplasma laidlawii* (Figures 15 and 16) harbors a fusogenic virus ("phage") that is able to unite cell wall-free, cell membrane-coated mycoplasma cells, observed first as "giant cell formation" [53–58] (Figures 17 and 18). The *A. laidlawii* L3 (MLV or MVL3) dsDNA short-tailed phage acted as the fusogenic virus. Cell fusions were brought about by simultaneous attachment of viral particles to two adherent cells. The large fused cells assumed "tumor cell-like" appearance and retained the enveloped viral (phage) particles, as shown by Brunner et al [53] (Figure 18). The L3 phage is referred to as a T7-like virion with linear dsDNA [55]. The *A. laidlawii* phages' L1 and L2 integration of their genomes into the host cell's genome is very complex [59–61]. *A. laidlawii* cells first infected with the L3 phage still produced, upon a second

Figure 17 Acholeplasma laidlawii Phages: Lytic, Temperate and Fusogenic. [2340] Reference Appendix 2, Explanations to the Figures

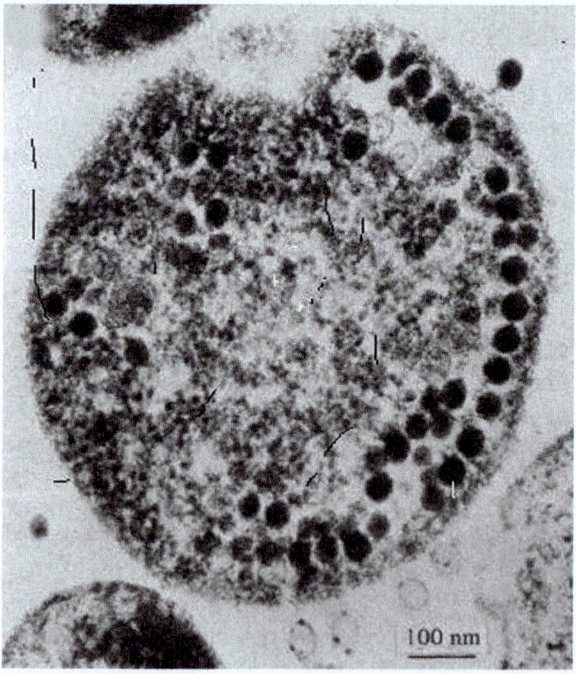

Figure 18 **a** A Fusogenic Virus (Phage) of Acholeplasma laidlawii. Possible Evolutionary Consequences. [2341ab] Reference Appendix 2, Explanations to the Figures. **b, c** A Fusogenic Virus (Phage) of Acholeplasma laidlawii. Possible Evolutionary Consequences. [53] Reference Appendix 2, Explanations to the Figures

(a)

(b)

(c)

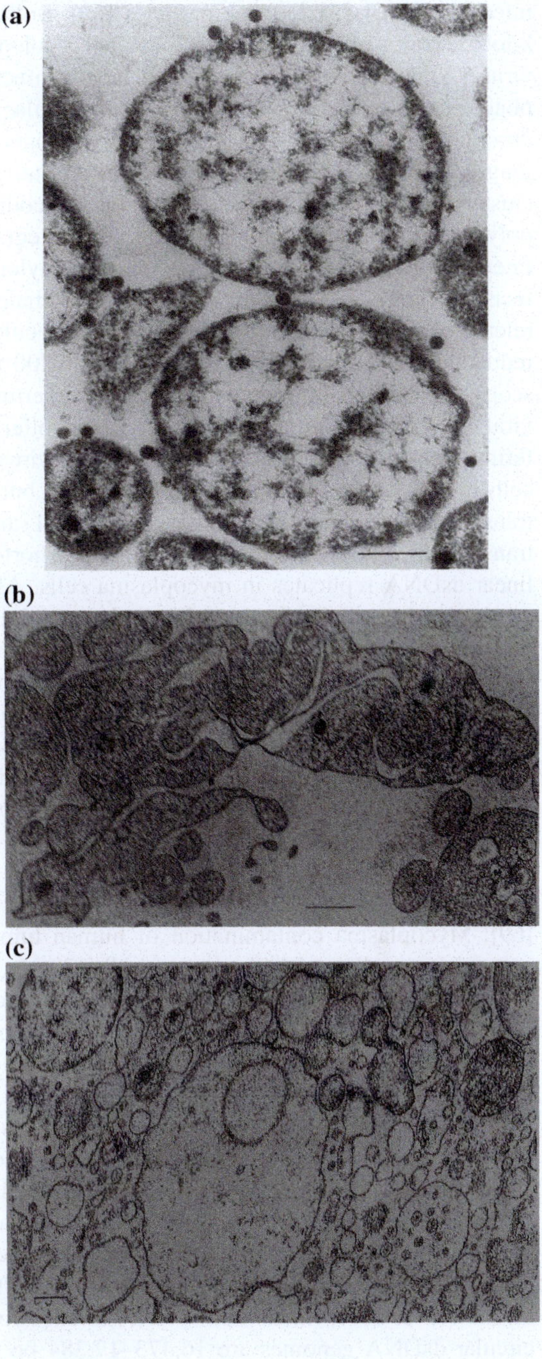

infection with L1, L1 phage particles [62]. Both the L1 and L2 phages insert 3.1 kilobase pairs of their genomes into the acholeplasma genome. These are "mini-viruses" that interfered with the full length infectious virus particles [60]. The L2 phage is a spherical enveloped virus with circular dsDNA. The infectious process is productive, not cytocidal and potentially lysogenic. The L2 genome is 11,965-bp long with 15 ORFs; its ORF 13 acting as the start codon. Gene expression occurs in clusters (the clusters are separated by noncoding intergenic regions) releasing polycistronic mRNAs for translation. The integrase is encoded from ORF 5 [61]. Phenol-extracted L2 viral DNA with polyethylene glycol added effectively transfects *A. laidlawii* cells, that results in immediate viral particle synthesis with the release of infectious viral particles [62, 63]. Pecularities of the L3 genome (terminal redundancy: length in restriction map is 36,200 *versus* 39,400 by electron microscopy, a 3200 bp difference); and circular permutations due to packaging of the viral DNA from a concatemeric precursor, offer further explanations for the differing measurements [64]. The genomes of viruses L1 and L2 infect *A. laidlawii* cells by transfection with polyethylene glycol, but the L3 viral genome could not be transfected by this route; however, both L1 and L3 DNA were infectious by transfection with electroporation [65]. The short-tailed 60 nm MV-L3 virus with linear dsDNA replicates in mycoplasma cells. After it shuts down host cell DNA synthesis, the cell synthesizes the viral structural proteins, releases viral particles, but does not die in a lytic process. This is the virus, which by attaching to cell surfaces induced acholeplasma cell fusions, but once inside the cell it replicated and could kill the cells ("non-lytic cytocidal infection") [54, 66, 67]. However, virus-producer cells could survive and grow [68]. The same laboratory from where mycoplasma (acholeplasma) cell fusions and replicating virus with eventual (slow) cell death were reported, did not explain the mechanism in this difference of viral behavior [54, 67, 68].

Mycoplasma cells may fuse with the cells of eukaryotic hosts: *M. fermentans* fused $CD4^+$ Molt, and $CD4^-$ 12E lymphocytes *in vitro* without visible cytotoxicity [69]. Mycoplasma contamination of human tumor cell cultures is frequent with mycoplasma cells attached to tumor cells [70a, b, 71]. However, *M. fermentans*, or *M. penetrans* in embryonic mouse cell cultures activated the *src*-related *vav* proto-oncogene, and induced malignant transformation (*src*, Rous sarcoma; *vav*, sixth letter of the Hebrew alphabet standing for linkage or connection) [72]; recently reviewed in [73].

Crenarchaea fuselloviruses. The dsDNA fuselloviruses of crenarchaeota are among the spindle-shaped viruses (SSV, "a tailed spindle", *fusé,* French; *fusus*, Latin) and are not known to fuse their host cells. Crenarchaeal viruses (phages) form a unique group of their own, sharing genes and plasmids, and proteins among each other, but not (or to a very limited extent) with phages and plasmids of other genera of prokaryotes or archaea. The viral sticky tail fibers attach the viral particles on the cell surface; membrane vesicles arrange four viral particles in rosettes. The circular dsDNA genomes are 16,473–17,384 bp with 31–35 ORFs; 18 ORFs are shared, the others show remarkable diversity. The viral genomes could integrate

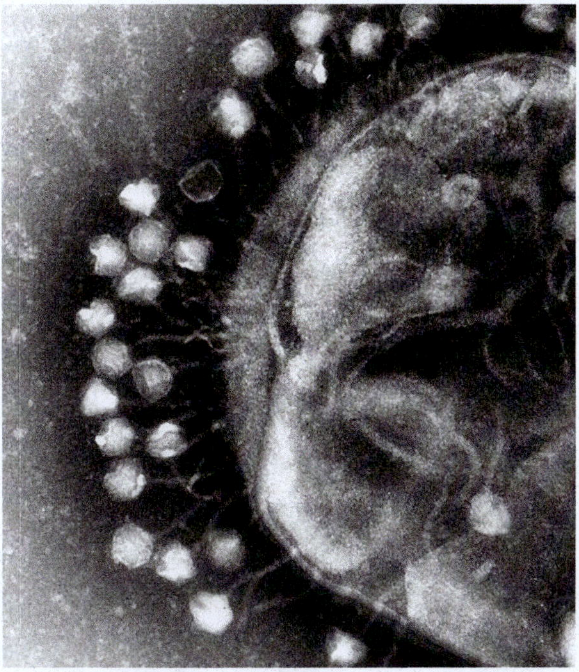

Figure 19 Bacteriophages Attach in Order to Penetrate Bacterial Cell. Basic Host Cell – Virus Relationship Established. (Phage. http://en.wikipedia.org/wiki/File:Phage.jpg by Dr Graham Beards http://en.wikipedia.org/wiki/User:Graham_Beards is licensed under CC BY-SA 3.0 http://creativecommons.org/licenses/by-sa/3.0/) Reference Appendix 2, Explanations to the Figures

into the tRNA host cell genes (attA, archaea recipient attachment site; for shared phage attachment site, attP) [74–76].

In response to invading phages (Figure 19), archaeal and prokaryotic cells incorporate viral and plasmid DNA sequences into the spacers within their CRISPR systems (clustered regularly interspaced short palindromic repeats) (Figure 20). The CRISPR-associated genes (CAS) transcribe these alien genomic sequences into guide RNAs. Within the CRISPR-associated complex for antiviral defense (CASCADE), the specific RNA strands cleave and disable the cognate foreign DNA sequences (crRNA to target DNA) before they could initiate coding for viral structural proteins. The CRISPR system preserves memory of alien viral or plasmid DNA invaders, a genetic record of all prior encounters with foreign transgressors, for prompt mobilization of the armada of cognate guide RNAs (CRISPR-derived crRNAs) (Figure 20). The *Sulfolobus solfataricus* CASCADE consists of a helical complex and a crescent-shaped structure for the modified RNA recognition motif [77–81].

The recreation of the first eukaryota. Viruses so far could not be fitted in the Tree of Life, except for the recent appearance from the depth of may be four billions (milliard) years of the clade of mimiviruses/megaviruses. For cellular life, the bi-lipids-surrounded vesicles were needed. The first eukaryotes might have been fused cells of pre-existing spheroplasts of crenarchaeons and prokaryotes. A fusogenic phage might have engendered this unison. LUCA was the single survivor of these experiments of Nature on Earth? The large fused cells engaged

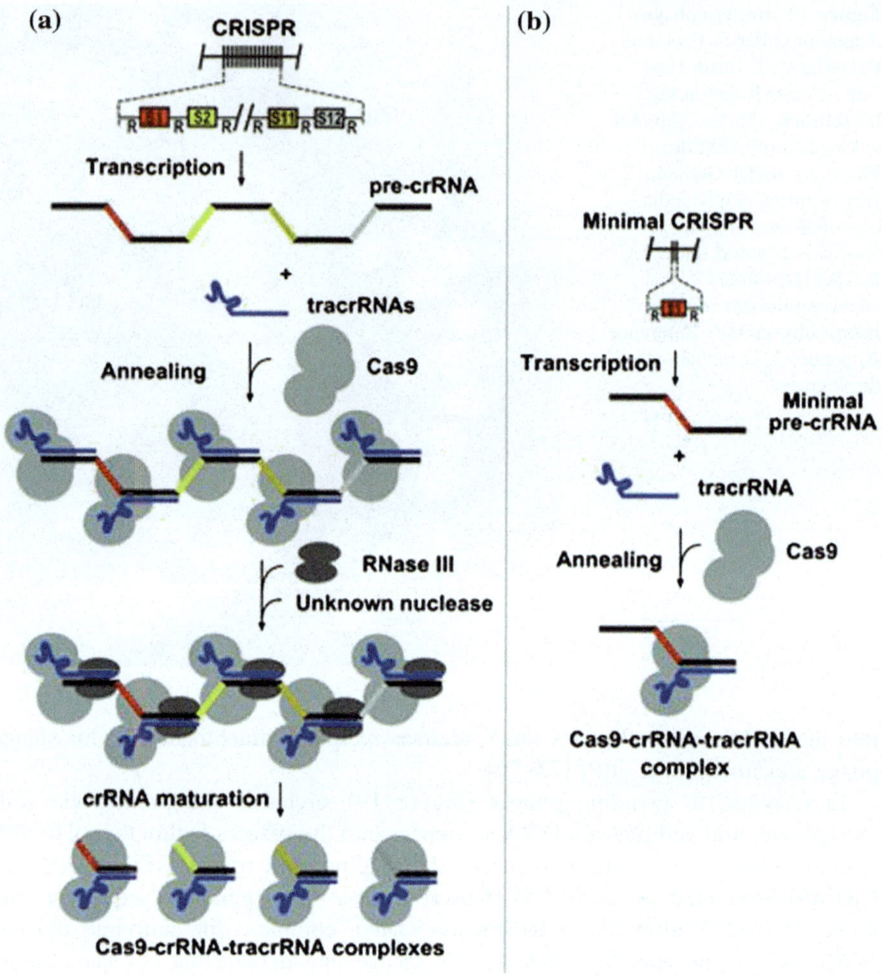

Figure 20 The First Antiviral (Antiphage) Defense of Bacteria: CRISPR. [2342] Reference, Appendix 2, Explanations to the Figures

proteobacteria and cyanobacteria for cytoplasmic endosymbiosis, which included the transfer of prominent genes from the endosymbionts into the nucleus of the host. Among the evolving spheroplasts, in addition to vertical gene transfers, horizontal (lateral) exchange of genes was widely practiced [26, 27].

The fuselloviruses of crenarchaeota are much less likely to be fusogenic for crenarchaeotic and prokaryotic spheroplasts, than the L3 *A. laidlawii* phage would be. The cell division mechanisms of crenarchaeota (*vide supra*), especially those of the thaumarchaeon ("*miraculous*") *Nitrosopumilus maritimus*, and eukaryotes are phylogenetically related (by the shared operon *cvd*; ESCRT; FtsZ) [42, 43], yet even these "ingredients" could not have created eukaryotic cells by their presumed fusions (so believes Forterre) [82]. The possible re-creation of the first eukaryote-like cells by fusing pairs of crenarchaeota (the thermoacidophil *Sulfolobus acidocaldarius*) and the ancient *A. laidlawii* cells, or other prokaryotic spheroplasts (those of the hyperthermophilic, hydrogen-oxidizing obligate chemolithoautotrophs) with viral mediation: the *Acholeplasma laidlawii* phage MV-L3 (*vide supra*) is here vividly envisioned.

There is no evidence whatsoever for spheroplasts of archaea and prokaryota naturally fusing in the present environment, so as to create new lines of eukaryota; however, their symbiotic relationships abound. After the "Darwinian threshold" established itself, the individual genomes are forcefully protected from horizontal gene transfers. However, the extant survivors of the ancient worlds may actually like each other, when brought together again in some fine media in the presence of fusogenic virus(es). This bold experimentation should be tried in a laboratory able to withstand bellicose and ironic criticism [39, 83]. For that laboratory a detailed account of *A. laidlawii* and its fusogenic phage L3 is provided here. Once the pair

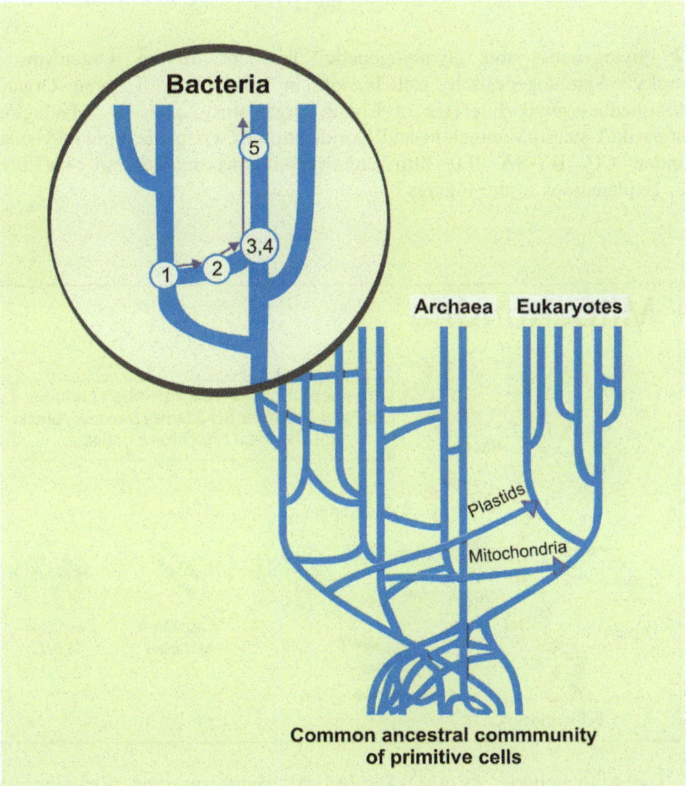

Figure 21 Network of Horizontally Transferred Genes In Prokaryota and Archaea. With kind permission © Barth F. Smets, Ph.D. All Rights reserved. 2015. Reference Appendix 2, Explanations to the Figures

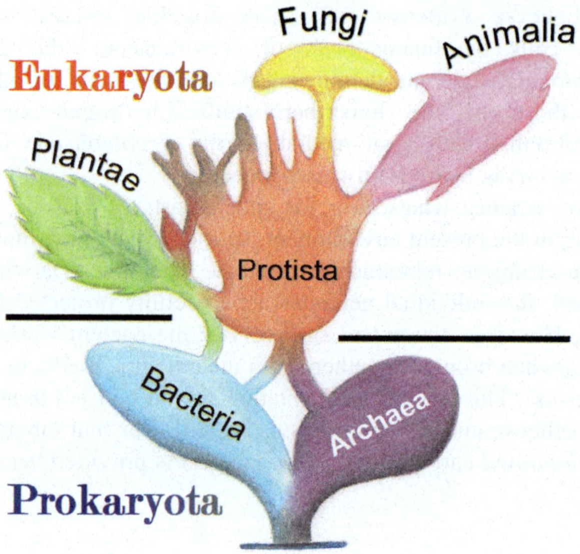

Figure 22 Phylogenetic and Symbiogenetic Tree of Living Organisms. Konstantin Mereschkowsky's Symbiogenesis by Cell Fusions in 1910. (Tree of Living Organisms. http://commons.wikimedia.org/wiki/File:Tree_of_Living_Organisms_2.png by Maulucioni http://es.wikipedia.org/wiki/Usuario:Maulucioni and Doridí http://es.wikipedia.org/wiki/Usuario:Doridí is licensed under CC BY-SA 3.0 http://creativecommons.org/licenses/by-sa/3.0/) Reference Appendix 2, Explanations to the Figures

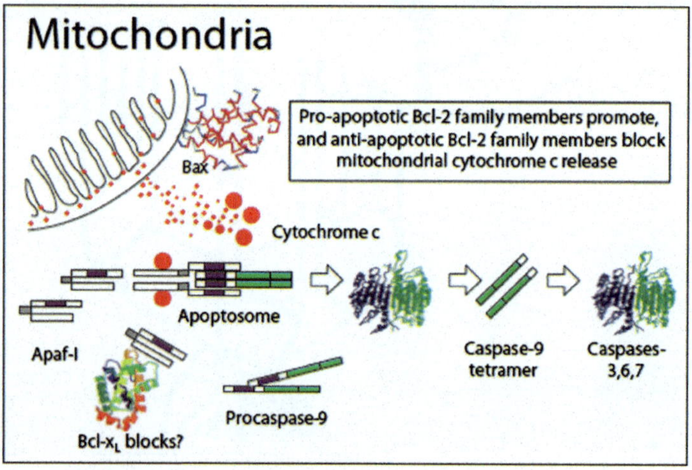

Figure 23 The Mitochondria. © AbD Serotec. All rights reserved. Reference Appendix 2, Explanations to the Figures

of prokaryotic and crenarchaeotic proto-spheroplasts are fused, MV-L3 replication could and possibly will be suppressed by a combined CRISPR/CASCADE system (Figure 20) shared by the archaea and the prokaryota [84–87].

Widely practiced horizontal gene transfers failed to "uproot the Tree of Life" (Figures 21 and 22). If a first pre-cellular domain of life existed in the form of viruses extending from viroids synthesized in pre-cellular ribozyme-armed ribosomes, and fused into very large, but inert viruses, that melted into the now existing three domains: prokaryota, archaea, and eukaryota, as their dependent parasite. It has been accepted that mitochondria and chloroplasts derived from endosymbiotic bacteria (Figure 23). The exact sequence of these events remains a speculative matter.

Eukaryotic Single Cells and the First Cell Communities Then and Now

The Organelles

Mitochondria and chloroplasts. The presumption is, that the first eukaryotic cells derived from the virally-mediated fusion of two equal partners, a crenarchaeal and a prokaryotic sphero(proto)plast, one of each. The viral agent might have been the ancestor of the Acholeplasma phage (Figures 15–18). These new large cells managed to separate their genome (condensed in chromosomes) from the ribosome. Each of the partners have already incorporated and operated a cytoplasmic ribosome. Presumably, ribozyme-engineered ribosomal units existed and functioned in the precellular Virus World. Two genomes and their protein derivatives, alien to each other, worked out a physiologically compatible relationship within the cellular membranes. Viral agents pre-existing in the outside environment, added alien new genomes encoding more incompatible proteins (if compatible self and incompatible non-self were recognized as such at that level of communal life characterized by quite promiscuous exchanges of genetic materials) [27]. Under these circumstances, only a very small number of the fused cells actually survived (if these events, virally mediated spheroplast fusions, could be reproducible in the laboratory). The survivors selected out the combinations of the best genes from an oversupply. The large cells thus established, now in a position of superiority in their environment, engulfed some proteobacteria and/or cyanobacteria (or both). Further conflicts of alien genes and proteins had to be accommodated. There emerged functionally most powerful cells with cytoplasmic mitochondria or chloroplasts (or both). Many mitochondrial or cyanobacterial genes have taken up positions in the host cell nucleus, as soon as it was formed. Eukaryotes packed their numerous chromosomes into nuclei surrounded by nuclear membranes. Membranes were already formed around cytoplasmic compartments as shown by Blobel [88a, b]. The chromatids of the chromosomes were anchored to some protein linings of the inner nuclear membranes. The long DNA strands were rolled over histones. The long DNA strands were interrupted by segments of RNA (the introns). In some constellations,

J.G. Sinkovics, *RNA/DNA and Cancer*, DOI 10.1007/978-3-319-22279-0_4

it appeared as if protein-coding genes were embedded in long noncoding RNA strands (referring again to the *sox*2 gene in the human chromosome at 3q26.3). The DNA genes released their mRNAs. The microRNAs deriving from the introns are able to specifically reacting with untranslated (UT) segments of the mRNAs. Powerful enzymes carried acetyl, methyl and phosphate groups for attachment/de-attachment to DNA and histone strands in the epigenome. Gene silencing and gene expressions depended on these reactions. The powerful nucleated large cells compartmentalized their cytoplasmic organelles. Mitochondria; chloroplasts; hydrogenosomes, peroxisomes (peroxisome proliferator-activated receptors PPAR) (Table II) (*vide infra*), Golgi apparatus, endoplasmic reticulum and ubiquitosomes could be separated with appropriate routes in between them for protein traffic [88a, b]. As to the origin of mitochondria and chloroplasts, Lynn Margulis' 50 pages manuscript was rejected 15-times before it was accepted as a valid assessment of actual evolutionary events. After her death (year 2012), her views still stand as the platform for all further developments [89a, b]. Now, this author proposes that the phagocytosis of proteobacteria and/or cyanobacteria, and thus the acquisition of mitochondria by the first eukaryotes, the protozoa, the first animal cells; and the acquisition of chloroplasts, by algae, the first plant cells of the seas and dry land, occurred after the first eukaryotes were already formed by the virally mediated fusions of two equal partners, one archaeal and one prokaryotic spheroplast.

The double-membraned mitochondria derive energy from catabolic oxidation of carbohydrates by acetyl-CoA (in the tricarboxylic acid Krebs cycle), lipids and proteins. A proton gradient process (protons passing through the ATP synthase) in the cristae results in ATP generation. Proton and electron leakages, sending electrons to oxygen, generate superoxide and oxidative stress. The redox protein, cytochrome C, receives electrons from the coenzyme Q, oxidized by cytochrome C-reductase. The ancient mitochondrial mtDNA (subject to unrepaired loss-of-function type mutations) produces its own tRNA; it replicates and divides. Mitochondria produce fused units mediated for the outer membranes by the mitofusin (Mfn) enzymes. The second mitofusin attaches (tethering) mitochondria to the endoplasmic reticulum (ER), a process essential for fission. Loss-of-function mutation of the mitochondrial gene for Mfn2 leads to impaired Ca^{++} metabolism. The inner membrane fusion is mediated by the protein OPA1 (optic atrophy). This 129 kDa dynamin-related mitochondrial protein is encoded by a nuclear gene (subject to autosomal loss-of-function type mutation). Dynamin-like proteins are microtubule-associated and hydrolyse guanine triphosphates. Mitochondria divide by fission mediated by dynamin-related proteins (mitochondrial fission factors, Fis and Drp), subjects to loss-of-function mutations and deletions with severe morbid consequences. If one fission product is defective, it is removed by the process of mitophagy (lysing in autophagosomes). Retention of defective mitochondria may lead to the autosomal recessive form of Parkinson's disease. Mitochondria initiate the intrinsic pathway of apoptosis by releasing cytochrome C through

Table II Distinguishing PPAR from PARP

Peroxisome proliferator-activated receptors (PPARαδγ) respond to natural ligand fatty acids, prostacyclins/prostaglandins. PPARδ is active in inflammatory processes including inflammatory carcinogenesis [73]. Suppressed by gene product protein of APC (adenomatous polyposis coli tumor suppressor gene, 5q21). APC-knock-out results in inflammatory carcinogenesis, especially within colonic polyps. Intranuclear ß-catenin translocation with activation of inflammatory genes and proto-oncogenes occur. T cell factor (TCL4)/ß-catenin complex targets c-myc and PPARδ, due to TCL-4 binding site on PPARδ promoter. Up-regulated PPARδ (high mRNA level) in colon cancer. In endometrial cancer, tumor tissue protein expression of PPARαß is high, that of PPARγ is low. PPARßδ could induce cell differentiation, thus PPAR agonists may be of therapeutic value. Rosiglitazone PPARγ agonist.[1,2] PGE2 signaling cascade targeting PPARγ inhibited growth of cholangiocarcinoma.[3] Obligatory for placentation; trophoblast cell-fusion into syncytio-trophoblasts is mediated by human endogenous retroviral envelope syncytin and PPARγ/RXRα (retinoid X receptor) signaling as regulated by MAP-kinase.[4] Epigenetic contact between PPARγ and bone morphogenetic protein (BMP) in mesenchymal stem cells regulate adipogenesis and osteoblastogenesis: in promoter of Runx2, H3 is acetylated and Lys4 is trimethylated.[5] PPARγ2 differentiates AR^+ benign human prostate epithelial cells (BHPEC); deficiency of PPARγ2 and PTEN induces in BHPEC either urothelial, or keratinizing squamous cell transdifferentiation.[6] γ-Tocotrienol and PPARγ antagonist are antiproliferative in breast cancer.[7]
Polyadenosine diphosphate ribose polymerase (PARP) initiates repair of ssDNA damage in BRCA mutated adenocarcinoma cells, but fails to restore tumor suppressor faculties of the gene. Inhibitors of PARP (olaparib; cordycepin) elicit the phenomenon of "synthetic lethality" (Table X) consisting of cell death due to high sensitivity of homologous recombination-defective cells to DNA-damaging agents. Lethal damage to tumor cell occurs upon double DNA deficiencies. Otherwise, tolerated inhibition of WEE kinase ("small", Scottish) in a $p53^{-/-}$ tumor cell, is lethal. Inhibition of a DNA repair enzyme in a tumor cell already suffering from a gene mutation (BRCA1/2) delivers a lethal insult.[8] PARP inhibitors leaving the ssDNA damage unrepaired promote synthetic lethality (Table X) in tumor cells upon the second insult. Cells in replicative senescence, with short telomeres, and with intact p53, exhibit resistance to malignant transformation. The telomere protein tankyrase 1 poly(ADP-ribose) polymerase regulates DNA damage responses at telomeres; suppresses telomere sister chromatid exchange. Localizes from cytoplasm to mitotic centrosomes for mitotic spindle pole formation. Immediate early viral proteins of HSV (herpesvirus) expropriate cellular tankyrase 1 for intranuclear viral replication. Tankyrase exerts inhibitory effects by directly binding EBV EBNA1.Telomere-associated PARP tankyrase inhibitor XAV939 reduces viral replication to a minimum.[8–12] Inhibitor olaparib is effective in ovarian, pancreatic and prostate carcinomas.[13] Tumors (adenocarcinoma pancreas, prostate; melanoma) with germ-line BRCA1/2 mutations respond to treatment with PARP inhibitors.[14]
SLX4 DNA repair protein; SYPCP synaptonemal complex protein. Mono-allelic and bi-allelic DNA repair gene mutations (BRCA1/2; RAD51C/D etc) induce chromosomal instability. PARP inhibitors induce dsDNA breaks. SYCP3 mutations did not cause aneupoidy. Sycp3-like X-linked SLX4 gene contributes rare cases of BRCA1/2 mutation-negative familial breast cancers. Will the SLX4-mutant chromosome resist olaparib?[15] SLX1-4 are Holiday junction resolvases/endonucleases, that repair endogenous DNA lesions. They may initiate DNA recombinations in stalled and collapsing replication forks.[16]
1 *Knapp P et al PPAR Res 2012;471524.* ***2*** *Peters JM et al Nat Rev Cancer 2012;12:181-9; Vyas D et al Lipids 2013 PMID:24781.* ***3*** *Lu D et al J Biol Chem 2013 MS M113.453886;* ***4*** *Ruebner M et al J Mol Med 2010;88:1143-56.* ***5*** *Takada I et al PPAR Res 2012:607141.* ***6*** *Strand DW et al Am J Pathol 2013;182:449-59.* ***7*** *Malaviya A & Sylvester PW PPAR Res 2014:439146.* ***8*** *Guo G-s et al Int J Oral Sci 2011;3:176-9.* ***9*** *Chang S Aging 2010;2:639-642.* ***10*** *Dregalla RC et al Aging 2010;2:691-708.* ***11*** *Li Z et al J Virol 2012;86:492-503.* ***12*** *Deng Z et al J Virol 2005;79:4640-50.* ***13*** *Kaufman B et al #11024 ASCO 2013.* ***14*** *O'Sullivan CC et al Front Oncol 2014;4:42.* ***15*** *Shah S et al PLoS 2013;8(6):e66961; Roos A et al Arch Gynecol Obstet 2013;288:1153-8.* ***16*** *Fekairi S et al Cell 2009;138:78-89.*

permeabilized outer membrane. The anti-bcl-2 molecule Bax permeabilizes the outer mitochondrial membrane. In the apoptosome, cytochrome C, ATP, Apaf and CARD (apoptotic protease activating factor; caspase activation and recruitment domain) enlists procaspase-9 → caspase-3, the executioner of apoptotic cell death. Mitochondrial loss-of-function gene mutations bypass Mendelian inheritance, as these entities are transmitted to progenies through the maternal line. Mitochondria replicate by fission. The process of mitophagy removes defective mitochondria. The GTPase mitofusins induce the fusion of mitochondrial outer membranes; the inner membranes are fused by the dynamin-like enzyme, OPA1. The intrinsic apoptotic pathway is initiated by the release of the mitochondrial enzyme cytochrome-C. Of the Bcl-2 anti-apoptotic family members, excel the pro-apoptotic enzymes Bax and Bak (Figure 23) [90a, b] (Table XXXII, *vide infra*).

The photosynthetic cyanobacterial endosymbiont yielded the chloroplast-derivative plastids in the cytoplasm of plant cells. Plastids preserved the bacterial mode of divisions by fission initiated by the tubulin structural-homolog, FtsZ (Z ring) protein (*vide supra*) [42, 43]. The FtsZ genes and the *min*B/D/E operons (minute chromosome segment) apparently reside in the plant cell's nucleus and not in the plastid anymore. The host cell's division and the plastid's division are mediated by two related but separate FtsZ systems. A chloroplast-targeting transit peptide delivers the plastid-dividing proteins into the plastid. The pre-dividing "dumbbell" plastids elongate and form a constricting isthmus with the plastid dividing ring structure. After the break at the isthmus, the two daughter plastids separate and seal the tear on their envelopes (Figure 24).

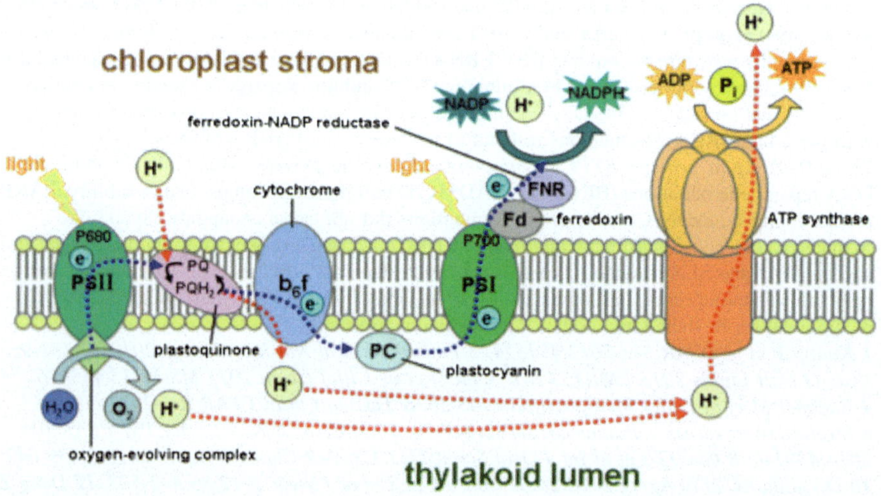

Figure 24 The Chloroplast. Reference Appendix 2, Explanations to the Figures

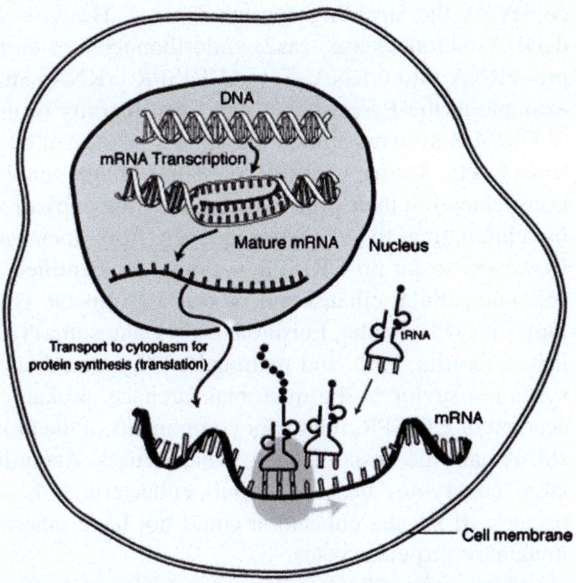

Figure 25 The Ribosomes. Reference Appendix 2, Explanations to the Figures. (MRNA-interaction. http://commons.wikimedia.org/wiki/File:MRNA-interaction.png by Sverdrup http://en.wikipedia.org/wiki/User:Sverdrup is licensed under CC BY-SA 3.0 http://creativecommons.org/licenses/by-sa/3.0/)

The ribosome. Of cytoplasmic organelles, the operational mechanism of the ribosome and the gene code reflects best to the preservation and transit of RNA/DNA cooperation from the pre-cellular to proto-cellular and cellular periods. Extant established ribosomes can function extracellularly [91, 92], thus the formation of operational proto-ribosomes with ribozymes and tRNAs might be dated back to the end of the precellular and the beginning of the proto-cellular era. This was the era, which induced the highly intellectual retrospective brain work for the generation of imaginary and experimentally featured molecular models (depicted in fascinating cartoons) of the pre-cellular proto-ribosomes. Those ribozymes and the tRNAs cooperating for the alignment of amino acids in pre-arranged orders, as events from the non-coded processes to the era of coded and templated protein synthesis, have arisen [93–95]. Small nuclear and nucleolar sn/snoRNAs, and small nuclear ribonucleic proteins generate the intranuclear spliceosomes. Messenger mRNA processing is accomplished in the cytoplasmic ribosome (Figure 25).

CRISPR

The first antiviral defense. While the ribosomes might have had precellular ribozyme-based ancestors in the primordial virus world, for the assemblage of the CRISPR/CAS/CASCADE system (clustered regularly interspaced short palindromic repeats; CRISPR-associated complex for antiviral defense), advanced cells (archaea, prokaryota) were essential. Once established, the valuable CRISPR operons travelling horizontally were avidly accepted. The CRISPR spacers are copies of the invading foreign DNAs. The *cas* genes encode enzymes (cas1, dsDNA endonuclease; cas2, endoribonuclease; cas6, endoribonuclease) cleaving pre-crRNA into crRNA. Some CRISPR crRNAs attack alien RNA sequences (for example in the *Pyrococcus*). The vast majority of archaeota possess a multiplicity of CRISPR systems amounting up to 1 % of their entire genome; prokaryota follow suit closely. Under CRISPR pressure, phage genomes may undergo point mutations, changing their proto-spacer gene (the copy of which is a CRISPR spacer), or by eliminating the copied sequence from their genome [79, 86] (Figure 20). However, so far no CRISPR system was identified in algae, dinoflagellates, plant cells, unicellular ciliates and protozoa (Amoeba, Giardia, Trichomonas) and kinetoplastids (Plasmodia, Leishmania, Trypanosoma) (*vide infra*), or in the vertebrates: fishes, reptilia, birds and mammals. Why would such a valuable antiviral defense system, a savior of the microbial (archaea, prokaryota) world, be abandoned? On account of CRISPR, the major components of the biomass in the seas or in dry land, still remain the hosts of phages and viruses. According to some views, primordial large eukaryotes descended into eubacteria [24, 25], thus eukaryotes predated bacteria. If so, the eubacteria could not have inherited their CRISPRs from those imaginary pro-eukaryotes.

Which primordial faculties are to be preserved? For the comparison of the malignantly transformed stem-, or somatic cells of a multicellular host with the ancient unicellular eukaryotic cells, the presumption has to be accepted that the descendants of the ancient unicellular eukaryotic life forms, the extant free-living, or hosts-parasitizing ciliates, or kinetoplasts preserved most of their original intracellular antiviral defense systems, and the "cell survival pathways", which were their resistance factors to viral, or physico-chemical insults. Transforming their life style from free-living existence to parasitism certainly resulted in the loss of many genes, and the acquisition of new genes from the genome of their hosts, or from successful fellow-parasites, if not vertically, then laterally transferred. The presumption should be, that extant unicellular life forms conserved and still practice a great deal of their ancestral life style, and thus offer a retrospective view at their extinct ancestors. Some of these primordial cell survival pathways may be observed in those cells of the multicellular hosts, which undergo the so called "malignant transformation". The initiators of these cells in the multicellular hosts are the asymmetrically dividing ancient stem cells. With this presumption, let the comparisons begin.

MicroRNAs

The Argonautes. There is no information on the CRISPR/CAS system (Figure 20) in the unicellular eukaryotes. In response to viral infection, even evidence for interferon production is missing in unicellular, or multicellular invertebrate hosts. The unicellular eukaryotes' first line antiviral defense is gained by not expressing cell surface receptors for viral entry. The eukaryotic interfering RNA (RNAi) and

PIWI systems appear as an analogy of the archaeal/prokaryotic CRISPR/CASS system (*vide supra*). These are basic defenses of the genome against intruder transposons, phages, plasmids and viruses. The eukaryotic RNAi system operates the Argonaute/PIWI protein families (Ago/PIWI, *vide infra*) (Figure 26), which are analogous to the archaeal/prokaryotic Argonaute homologs (pAgos). These are nuclease enzymes able to slice their target RNAs. The eukaryotic posttranscriptional gene silencing is under the control of small interfering siRNAs, microRNAs (miR), and the RNA interference (RNAi) and PIWI-associated RNAs (piRNAs). The archaeal/prokaryotic ancestor of the eukaryotic miR system consists of small antisense RNAs, which were present in *Archaeoglobus fulgidus* and *Sulfolobus solfataricus.* At the bottom of the prokaryotic lineage are *Aquifex aeolicus* and *Thermus thermophilus*; both possess pAgo proteins able to cleave RNA in DNA/RNA duplexes. Thus, the nucleases of pAgo system appear as the earliest defensive elements for securing the purity of the individual genomes [79, 86]. However, the system was shared between co-existing cells by lateral (horizontal) gene transfers. This faculty of the system must have allowed retrotransposons, phages, plasmids and viruses to imitate it, and thus find a way to insert themselves into alien genomes, up to the human genome, which is packed with innumerable ancient retroposon insertions. The processes of gene silencing begin with the actions of the intranuclear Drosha, and continue by the cytoplasmic Dicer enzymes in the RISC machine (RNA-induced silencing complex). Argonautes practiced gene silencing in enlisting microRNAs (miR) during the primordia of cellular life (Figures 26 and 27).

There appears in the untranslated region (UTR) in the genome of the ciliate *Paramecium tetraurelia* the motif of GUACAUUA, which is conserved in the metazoan ribosomal protein genes as a putative microRNA binding site. It is located in the genome as to accept mRNA-binding PUF-proteins (drosophila Pumillo protein and Fem-3 feminizing factor). The genome of *P. tetraurelia* is considered to

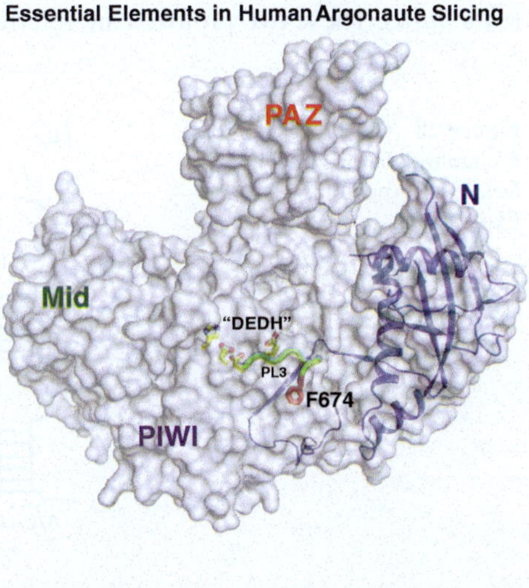

Figure 26 The Argonautes' RISC (RNA-induced silencing complex) [2343] Reference Appendix 2, Explanations to the Figures

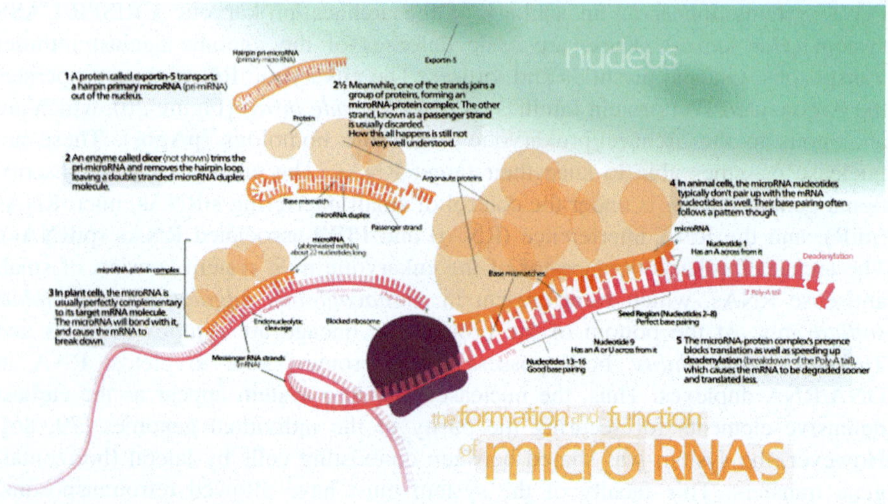

Figure 27 The MicroRNAs' Drosha and Dicer. (MiRNA. http://commons.wikimedia.org/wiki/File:MiRNA.svg by Kelvinsong http://commons.wikimedia.org/wiki/User:Kelvinsong is licensed under CC BY 3.0 http://creativecommons.org/licenses/by/3.0/deed.en) Reference Appendix 2, Explanations to the Figures

be the result of three-fold duplication. A similar sequence in the 3′UTR motif, yeast's UGUAUAUUA, destabilizes its host's mRNA [96]. Paramecia (*P. tetraurelia*) purge their genome of inserted alien elements with great accuracy, taking use of ancient but fully domesticated transposases as piggyback elements (later more).

The basic elements of mRNA and microRNA (miRNS, siRNS, RNAi) interactions must have appeared first in the ancestors of the deep-branching pseudouridinylating protozoon, Giardia (Figure 28). The system remains fully conserved in the extant parasitic descendants, the ancient protozoa, Giardia and Trichomonas (*vide infra*). These two protozoa are referred to as amitochondriate, because they do

Figure 28 The Process of Pseudouridylation. [2344] Reference Appendix 2, Explanations to the Figures

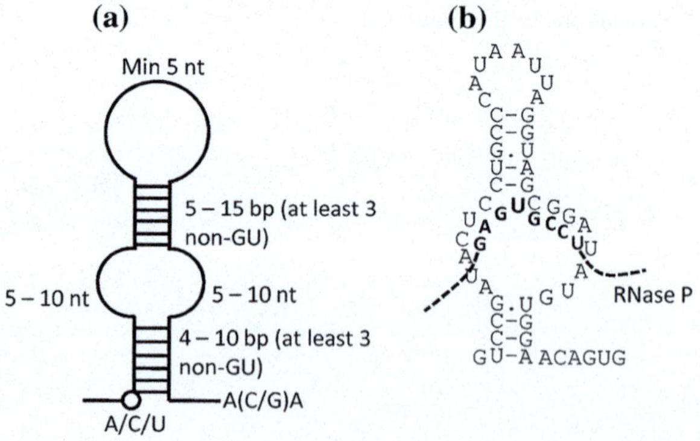

not have mitochondria. Very likely, formerly possessed mitochondria were lost after the transfer of mitochondrial genes into the cellular nuclei. Instead mitochondria, giardia has mitosome, and trichomonas has hydrogenosome. In these unicellular eukaryotes, microRNAs (small inhibitory siRNA, micro miRNA, piwiRNA, or piRNA) are processed by the enzymes Dicer and Drosha for incorporation into the RNA-induced silencing complex (RISC). RISC engages the argonaute (AGO)/Piwi elements. The abbreviations are overwhelming: drosophila P-element-induced wimpy testicle (piwi); P element, Rubin-Spradling P strain P cytotype drosophila, P transposon/transposase; PPD, PAZ/PiWi domain; PAZ (PiWi/Argonaute/Zwille domain, *vide infra*), standing for enzymes for cleaving the targeted mRNAs [97]; SZ Drosophila (Szeged, Hungary).

Giardia replicates a dsRNA virus (giardiavirus, *Totiviridae*), which is not cytopathic to the trophozoites. The Giardia microRNA repertoire consists of ancient plant and animal sequences, some of which are originating either in the nucleus, or in the nucleolus (miR-2 and miR-5 are snoRNAs). There is no identified miR that would protect the Giardia cell against giardiavirus (in contrast to Trypanosoma, Tetrahymena or Caenorhabditis cells, which can render themselves dsRNA virus-free). In Giardia cells, the Dicer and Argonaute systems are installed, but the Drosha is missing [98, 99]. It is not clear, if these unicellular ciliate eukaryotes possess also the RNAi system; however, Giardia is a strong candidate for RNAi possession [100]. In examining the Giardia genome, wild gene losses and host gene gains due to the parasitic life style should be considered.

The unicellular algae are the ancestors of multicellular plants in the sea, or in dry land. Among them, the biflagellate *Chlamydomonas reinhardtii* with light-sensitive eyespots within their chloroplasts, and with hydrogenosome, possess miR regulatory faculties [101–103]. The chlamydial derivatives, the Volvocales, bear their non-reproductive cells outside the colony (the swimmers), and protect their stem cells and gonidia within the gelatinous body of the colony.

In all multicellular hosts, the microRNAs have been recognized to be the key posttranscriptional regulators of gene expressions. The *gypsy/copia* retrotransposon-, or retrovirus-carrier flying insects [104–106], the nodavirus-carrier caenorhabditis nematodes [107, 108], or the fishes, amphibians, reptiles, birds and mammals all operate a very extensive microRNA apparatus. Very frequently, the microRNA and its 3′UTR (untranslated) binding site on the targeted mRNA molecule are evolutionarily conserved. An extensive microRNA network is operational within the well-known innate immune system. For example: TLRs and IL-6 and microRNAs promote proliferation and even transformation of B lineage lymphocytes, even memory cells, by Epstein-Barr virus (EBV) [109]. The adaptive antiviral immune system: NK cells in antibody-dependent cell-mediated cytotoxic (ADCC) reactions, immune T cells, virus neutralizing antibodies follow suit. MicroRNAs connect host cells with their infecting viruses, both in their latent and cytolytic relationships [110, 111].

The p53 protein induces miR-145. The ancient physiological posttranscriptional control of gene expression served the unicellular, and the non-vertebrate multicellular, hosts in environmental distress for survival. Stem cells of multicellular hosts may expropriate the activated system for the survival of individual cells in the distressed host. The microRNA-145 controls the physiological order of cell metabolism. The p53 protein binding the miR-145 promoter DNA, induces miR-145 production. Then miR-145 can neutralize the mRNAs of c-*myc*, *irs*-1 (insulin receptor signaling), and those of the stem cell genes (*sox*2, *oct*4, krüppel-like factor 4, *klf*/KLF). The miR-145 promoter gene harbors a binding site for the CCAAT enhancer-binding protein-ß (C-EBP). The C-EBPß-2 may bind also the p53 promoter. This interaction disables p53 so, that the cell cycles go through their S phase without interruption. However, the binding of C-EBPß-2 by the p53 promoter annuls miR-145 activation by p53. The antioxidant resveratrol inhibits the phosphorylation of C-EBPß, thus the inactive C-EBPß cannot neutralize p53 [112]. The cells of multicellular communities with re-activated proto-oncogenes (the "cell survival pathways" of ancient cells) achieve constitutive status of their proto-oncogenes, and thus become transformed (malignant tumor) cells (*vide infra*).

Viral genomic microRNAs against host cell genomic micro/messenger RNAs. Viral genomes also generate siRNAs and miRNAs and induce the host cells' genomes to release microRNAs of their own. Is it possible that eukaryotic cells, both of animalia and plantae, acquired the elements of the system from viruses, at the primordial time when pre-existing viral genomes invaded the protocells (*vide supra*)? A brief summary of miRNAs released during EBV and KSHV infections (Epstein-Barr human herpesvirus, HHV-4; Kaposi sarcoma-associated virus, HHV-8) to ensure their latent existence in the host cell's epigenome, was written in 2011 (Figures 29 and 30), but has become immediately outdated [73]. EBV-miRs can target EBV's own late membrane protein (LMP) mRNAs. LMP induces the

Figure 29 Kaposi Sarcoma's Herpesviruses. Reference Appendix 2, Explanations to the Figures

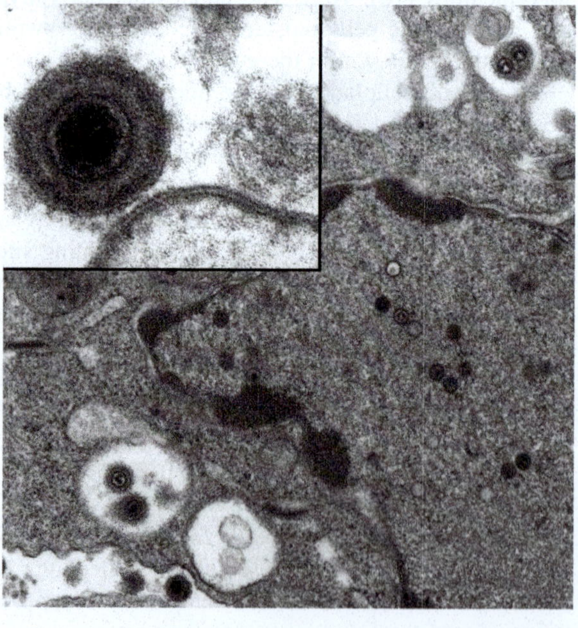

(a)

(b)

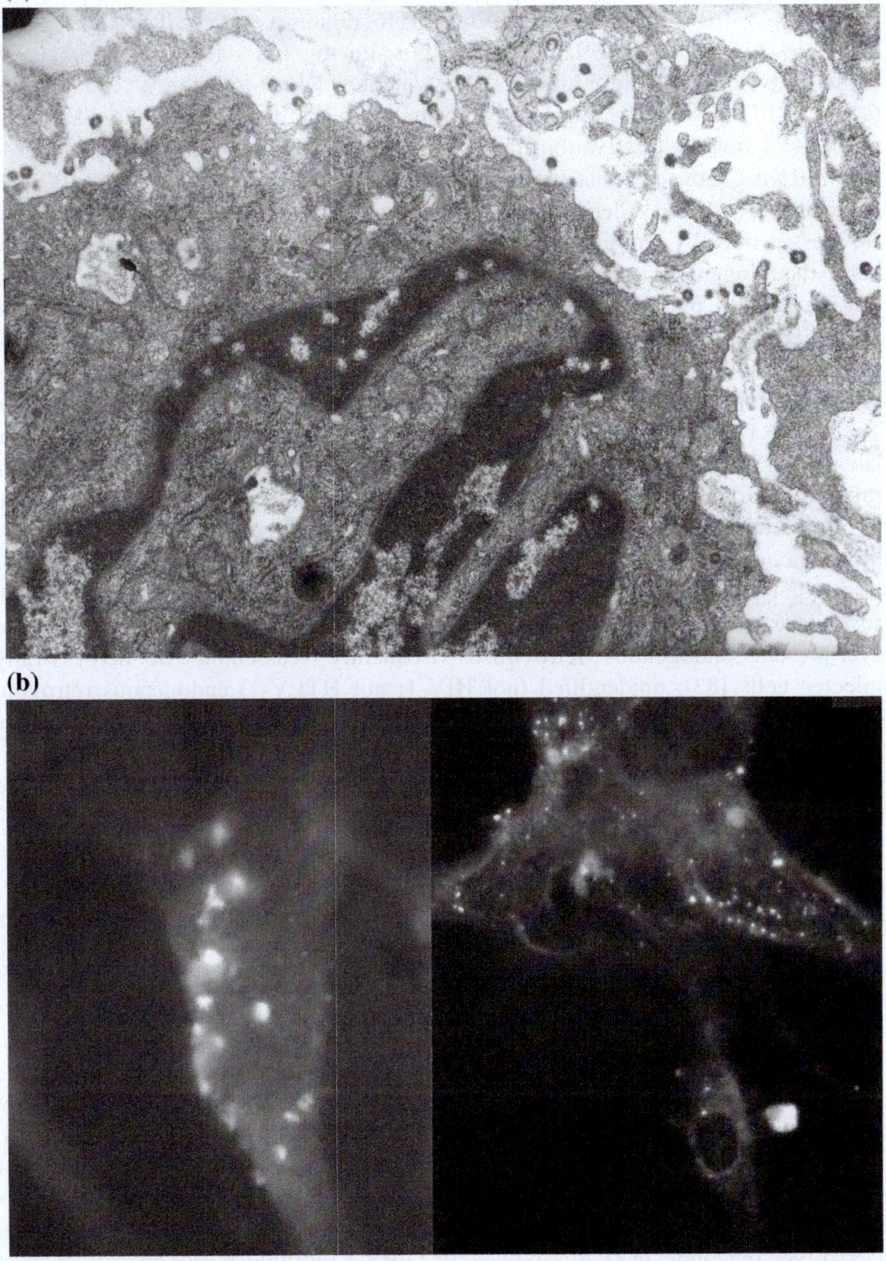

Figure 30 Kaposi Sarcoma's Endogenous Retroviruses. Reference Appendix 2, Explanations to the Figures

host cell-derived miR-29, the two-faced (oncogenic and/or oncosuppressor) micro-RNA. EBV's LMP1 mRNA's 3′UTR region is targeted by the c-*myc* oncogene-derived miR-17/92. LMP's expression by miRs-100b/25 is enhanced, while miR-17/20/106 are inhibited. LMP1 induces NFκB to act as a "cell survival factor" [113]. KSHV/HHV-8-derived 25 mature miRs, by targeting the 3′UTRs of mRNAs of host cell-derivation, degrade and thus inhibit the translation of these host mRNAs. Consider that human host cells release and operate over 1400 recognized miRs. KSHV-derived miRs down-regulate the host cell-derived tumor suppressor miRNA families miR-7b and miR-220/221. The tumor suppressor and anti-neoangiogenesis factor, the host cell thrombospondins, are over ten-fold downregulated by miRs of viral genome derivation. The genome of the KSHV/HHV-8 releases miR-K12-11, which is an ortholog of the host cell miR-155, a highly upregulated miR in B cell lymphomas. PEL cells (primary effusion lymphoma) contain high levels of KSHV-miR-K12-11 (alias miR-155). The miR-K12-11 attenuates type I IFN- and TGFß-production, and maintains viral latency [113, 114]. Discovered very recently, these extensive microRNA networks ruled the most ancient host cell/virus relationships. The existence and advancement of the transformed cell (the tumor cell) depend on the rulings of the proto-oncogene-derived microRNAs. The ancient heat shock proteins, protectors of primordial enzymes working in boiling water, are now recruited for chaperon services to be rendered to oncoproteins and to viral proteins (Figure 29). The ever-present endogenous retroviruses frequently reassemble in herpesvirally infected cells [83]; unidentified (not HIV-1; not HTLV-1) endogenous retroviral particles bud from KS cells (Figure 30). The reticuloendotheliosis viruses were found to have retroviral ancestry (see in the Appendix 1).

Amoebae: "melting pots of evolution" (Raoult). The situation in amoebae is quite extraordinary. Developing from single celled to multicellular and eventually spore-forming entity, the *Dictyostelium discoideum* (Figure 6) possesses some 12,500 genes in its 34 Mb genome. Its DIRS-1 (*D* intermediate repeat sequences) and Skipper retrotransposons release an abundance of short [18-26 nt] RNAs. Many of these short RNA sequences display antisense complementarity to mRNAs. Here, the distinction between microRNAs (miR) and small interfering RNAs (siRNA) is clear. The former derive from imperfectly base-paired hairpin transcripts; the latter are dsRNAs synthesized by RNA-dependent RNA polymerases (RdRp) using a ssRNA segment as template. Both miRs and siRNAs exert their effect within the Argonaute/Piwi complex. The *D. discoideum* Dicers are unique in that they do not contain the N-terminal helicase domain, that all other eukaryotic Dicers (fungi, plants, animals) possess. But the missing domain was found as the result of domain-swapping, within the *D. discoideum* enzyme RdRp. The helicase domain is missing also in the Giardia, Tetrahymena; and Trypanosoma Dicers function in RNAi-type reactions. In *D. discoideum*, its DIRS-1 retrotransposon releases most of the microRNAs (much more, than what is released by its Skipper retrotransposon). Small RNAs of Skipper retrotransposon-derivation were released in cells devoid of the second Dicer protein, and the enzyme RdRp; these microRNAs showed complementarity to non-retrotransposon-derived mRNAs (for an as yet unresolved relationship). Thus, several miR biogenesis pathways are in parallel operation in these amoebic cells for DNA gene expression control [115]. Farther down, bacteriophages engaged in tyrosine recombinase (YC), or DNA methyltransferase activities, were already DIRS-like elements. The LTR retrotransposons contain protein-coding internal region in two open reading frames (ORF), *gag* and *pol*. The ORF *pol* usually encodes the reverse transcriptase (RT) to copy the RNA into DNA, and an integrase to insert the new DNA segment into the genome of the cell. DIRS form the fifth group of LTR retrotransposons. The non-LTR retrotransposons encode also esterases, pass cell membranes, and thus facilitate horizontal gene transfers [116, 117].

Amoebae phagocytose bacteria for nourishment. The chlamydia *Parachlamydia acantamoebae,* and the bacterium *Legionella drancourtii,* resist intracytoplasmic digestion within the amoeba. The chlamydia enlarges its genome 3-fold within the amoeba due to unilateral gene gains received from the bacterium [118]. In the case of the nucleocytoplasmic mimivirus (mimicking microbes), horizontal gene transfers occur between the mimivirus and the amoeba, and with the bacteria also residing and replicating in the same amoebal cytoplasm, and between the mimivirus and its phage. In contrast to the chlamydia, the mimivirus loses a great deal of its gene compartment, to the extent that it changes from a sympatric (without loss of identity from interbreeding) to an allopatric resident. Due to the loss of viral fibers, the mimivirus of reduced size is referred to as the "bald" virus. The bald mimivirus cannot support the replication of its satellite virophage [119]. The giant nucleocytoplasmic Marseille virus collected genes from its host, the amoeba, and from digestion-resistant bacteria, co-residents of the same amoebal cytoplasm, and also from horizontal, but obscure, routes from phycodnaviruses (the *Paramecium bursaria chlorella* virus) [120]. The newly discovered *Megavirus chilensis* elevated itself to the rank of the largest virus so far ever isolated [121]. Its genome is loaded with host cell genes; it either developed endogenously from its host, or it has exchanged its own genuine viral genes with those of its host, once it has taken up residence within the cytoplasm of the large amoeba. The intraamoebal bacterium *Legionella drancourtii* benefited by acquiring sterol-reductase gene from its host, the amoeba [122]. The amoeba benefited from extensive transfer (of 18 genes) from the Legionella. There was no detectable gene exchange between the bacteria (Legionella and chlamydia) [118]. Pathologists of the last century in observing cancer cells in locomotion often referred to them as "amoebas". Indeed, amoebas use the same cytoskeletal transforming proteins, different forms of high mobility group proteins, cancer cells resort to in practicing local invasiveness (Figure 31) (more in the Appendix 1).

Comment. How the mimiviruses and the amoebae re-play the ancient scenarios, is not well understood. It is possible, that the first unicellular eukaryotes encountered those protoviral agents, which were generated from fused viroids, that replicated in precellular ribozyme-powered ribosomes. These agents transferred their residence into the first cells and there have become intracellular parasites. It is also possible that a cellular domain existed prior to the appearance of the current three domains of life (archaea, prokaryota, eukaryota) and fell victim of extinction. The

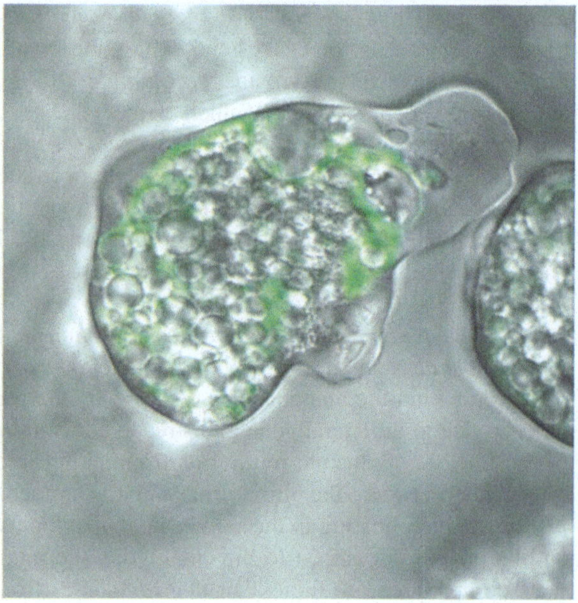

Figure 31 The Exemplary Amoeba (A. histolytica) Is Imitated by Cancer Cells. [2345] Reference Appendix 2, Explanations to the Figures

new proteins of mimiviruses and pandoraviruses alien to the current three domains of life raises the possibility that they derived from a pre-existing, but by now extinct, domain of life. The relationship between amoeba, bacteria and large viruses is so unique, that it will not be possible to analyze it further at this time in the context of oncogenesis. It is not known if the intra-amoebal parasites replicate independently within their host, and if their progeny bursts out; or if their divisions are obligatorily synchronized with that of their host's genome. It is not known what the interactions of the microRNAs of the host and the large viral genomes are. The amoebae do not possess an antiviral CRISPR system, but operate an argonaute system [123], whereas the chlamydia and the bacteria might be well endowed with CRISPR. In this context, if the large viruses harbor a virophage, how about the other intracellular residents of amoebae, phages for the chlamydia and the legionella? The amoeba-parasitizing chlamydia reveal significant phylogenetic diversity [124]. So much so, that they might have had the luxury of supporting a temperate phage of their own. In the amoeba, the megavirus lost, and the amoeba gained genes to the extent that the viral particle diminished itself into a phage-less obligatory intracellular parasite: an extant virus. As to the chlamydia, in them the evolutionary direction has taken the opposite course. The pre-chlamydia as a simple genome (may be a "plasmid") entered the amoeba, where it has enlarged its genome. If the gain of genes in the chlamydial genome continues, an independent free-living bacterium may exit from the amoeba (whereas, gene loss may reduce the Chlamydia to eventually become a large DNA virus). Once explored, these liaisons will reveal some of the deepest past evolutionary events.

AGO/PIWI/HIWI/YB

A major contribution from Tetrahymena. The free-living ciliate protozoan, *Tetrahymena thermophila* releases small RNAs to unify with the PAZ/PIWI domain (PPD) proteins (Piwi/Argonaute/Zwille), thus forming effector ribonucleoproteins (RNP). The RNPs exert catalytic and slicer/cleavage activities on their targets, thus achieving gene silencing by histone and DNA modifications, and by inducing mRNA decay. Piwi RNPs suppress transposon expression. *T. thermophila* ORFs generate 12 PPD proteins of the Piwi clade. Upon asexual fission, the cells preserve the silent diploid germline micronucleus, and the large polyploid somatic macronucleus (MIC; MAC). Upon conjugation, MAC differentiation requires the generation of 27-30 nt sRNAs, termed scan RNAs (scnRNA). The tetrahymena Dicer enzyme Dcl1 produces the heterochromatin-formatting and DNA-eliminating 28-29 nt sRNAs during the sexual cycle, while Dcl2 produces the 23-24 nt sRNAs for gene regulation during asexual growth (rapid replication by fission with telomere conservation in rich media) [125a, b, 126] (Figures 32 and 33). Of the Tetrahymena PIWI family (TWI) proteins, TWI12 is essential for cell divisions. The end position of the TWI12-bound tRNA 3′ fragment (a RNP) precisely matches RNAs detectable in embryonic stem cells. This mechanism is operational in rapidly dividing cancer cells, among them human hepatocarcinoma, or prostate carcinoma cells. In these cancer cells, the tRNA 3′ fragments, the 5′ end positions, the 3′CCA residues and the 5′ monophosphate terminus, are identical with the same arrangements as in the Tetrahymena TWI12/tRNA complex. These articles of the highest level, describe expressed sequence tags (EST), PAZ/PIWI domain-associated sRNAs, piRNAs, noncoding ncRNAs, micro-inhibitory miRNAs, transfer tRNAs, ribosomal rRNAs, small nucleolar and small nuclear snoRNAs and snRNAs. DNA-elimination in conjugating cells, and class 23-24 nt sRNAs replicating constitutively; the 23-24 nt

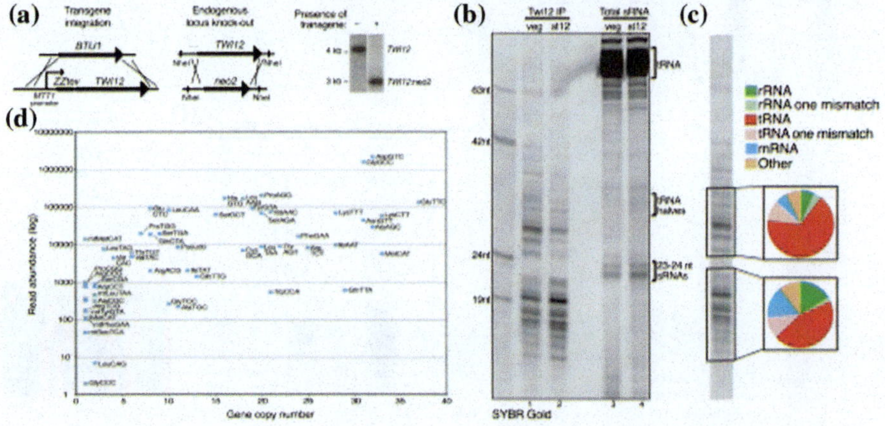

Figure 32 Tetrahymena Molecular Biology I. [125b] Reference Appendix 2, Explanations to the Figures

sRNAs derive from hairpin transgenes for the transcriptional silencing of cognate mRNAs; pseudogenes release sRNAs. G-rich strands of telomeric repeat sRNAs arise at cell divisions. TWI1/11 genes induce conjugation; TWI2/8/12 genes are robustly expressed in dividing cells without elimination of their mRNAs (shown in Northern blots by Couvillion *et al*) [125a, b] in her highly educational Figures 32 and 33. For constitutively replicating Tetrahymena histone genes, see Yu & Gorovsky [126]. For the tRNA CCA residues, see Wilusz *et al* [127]. For the hidden layers of human small RNAs in tumor cells (hepatocellular carcinoma), see Kawaji *et al* [128], introducing the bidirectional promoters in human hepatocellular carcinoma cells. The small RNAs are in antisense (opposite) orientation to the LTR and LINE inserts, the long terminal repeat and long interspersed repeat element. Intron-derived RNAs are 21 nt and 23 nt for the sense and antisense orientation. In the bidirectional promoters, two promoters lie on opposite strands. The overlapping transcripts are on different strands. Sense-antisense hybridized dsRNAs may be formed. It is pointed out, that in human cancer cell growth, the human PIWI (HIWI) proteins are upregulated to act. The PIWI (drosophila P-element-induced wimpy testis) equivalent human HIWI dictates an accelerated growth rate (*vide infra),* and as this author proposes: with prompt telomere repairs (maintaining telomere lengths, thus escaping senescence). The PIWI equivalent TWI12/tRNA assured the close to immortal survival of the tetrahymena clade. Is the human Hiwi oncogenesis a form of descent for the transformed cell of a multicellular host (the human host) to the life style of the primordial

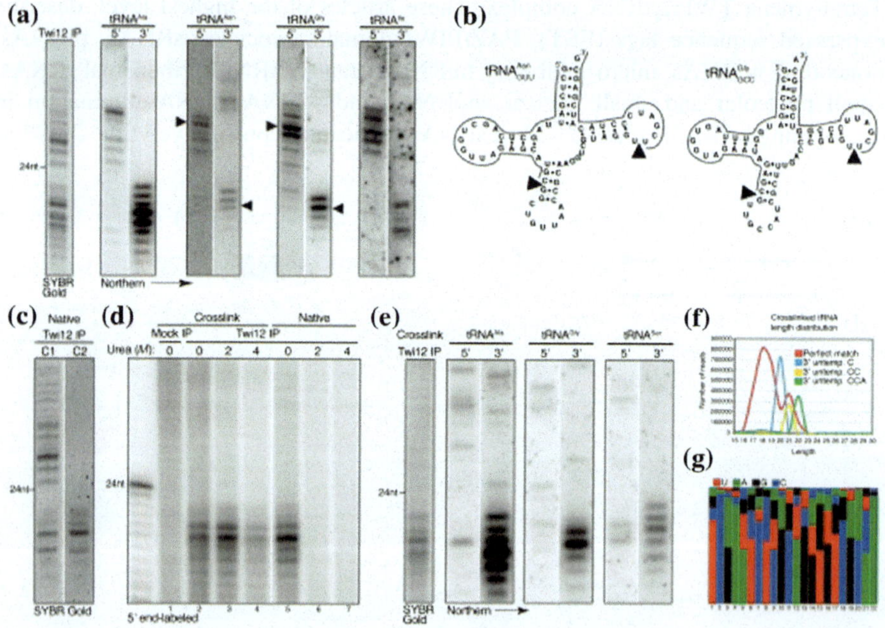

Figure 33 Tetrahymena Molecular Biology II. [125b] Reference Appendix 2, Explanations to the Figures

independent and close to immortal unicellular life forms (the tetrahymena)? "Physiological oncoproteins" (TWI12/tRNA) were essential for the cell divisions, replication without senescence and natural death, and for the rapid rate of reproduction in the archaic unicellular hosts, that the extant Tetrahymena preserved and conserved?

Comment. The AGO Dicer: to play games with dice; take chance with death. Drosha: class II primary priRNA-processing ribonuclease III, and the dsRNA-binding protein, DiGeorge syndrome chromosomal region 8, DGCR, cleave stem-loops of miRNA.

The drosophila's triangle between YP, Piwi and Hh. While the AGO proteins are ubiquitous and widely distributed in every cell (up to the differentiated somatic cells in the multicellular hosts), the *piwi* gene product proteins (PIWI) maintain stemness and self-renewal of germ cells and stem cells (for example, HIWI in $CD34^+$ human hematopoietic stem cells). The insect genome (drosophila) vulnerable to malignant transformation fed by constitutively replicating stem cells is the basic science lesson for medical oncologists.

This author has heard for the first time in 1970 John J. Trentin of Baylor at Houston to lecture on the "stemness" of hematopoietic cells regenerating in the bone marrow [129]. Now, the PIWI proteins are associated with the novel class of small RNAs, the 24-32 nt piRNAs. Tens of thousands of piRNA species are produced in the mammalian hosts (and exist in abundance in the drosophila, the DiWi, or in the zebrafish, the ZiWi, or in the mouse, the MiWi, and the human, the HiWI). The piRNAs derive from ssRNA precursors released from the intergenic sequences of DNA strands, are active in gametogenesis, as the stemness of germline cells depend on them in meiosis. Tissue stem cells depend on piRNAs in their asymmetric divisions maintaining self-reproduction for "stemness", while releasing a daughter cell for differentiation into the subservient somatic cell compartment. The system silences and regulates ribosomal mRNA translation (in silencing mRNAs by microRNAi) [130]. However, the PIWI (and YB) proteins are the first oncoproteins, or the intranuclear inducers of oncogenes, when the stem cell chooses to revert to the life style of its ancient ancestors. The YB proteins (Y box-binding protein), discovered in the drosophila [131, 132], regulate the proliferation of both germline and somatic stem cells and activate the *piwi* and *Hh* (hedgehog) genes. (The *Hh* gene was named by the students after the appearance of the fruit fly larvae, which were covered with denticles pointing forward and backward as encoded by the mutant gene; or named after the speedy 17 years old mischievous "sonic hedgehog" mascot escaping from the planet Moebius and appearing on Earth for more mischief, in the most popular Sega video of 1993). The drosophila YB proteins form an organelle in the cytoplasm of the ovarian and testicular somatic cells, positioned next to the mitochondria. The YB proteins maintain the germline stem cells and the somatic stem cells. The YB protein-regulated PIWI proteins form complexes with the Argonaute proteins and process single-stranded small RNA precursors; during this process PIWI proteins bind piRNAs, the PIWI-interacting RNAs [133, 134]. The difficult to find by histochemical technology, the YB proteins, are upregulated in malignantly transformed cells, including human cancer cells [135].

The human PiWi is HiWi. The equivalent of the self-renewing PIWI/AGO gene family is located on human chromosome 8; the human PiWi/Ago family consists of 1-4 subfamilies. The *piwi*2 gene is active in embryonic life, but it is silenced in all somatic cells of the adult host. The PIWI2 protein first re-appears in pre-cancerous stem cells. It is often bound to the piwi-interacting short RNA sequences (piRNAs), also called repeat-associated small interfering RNAs (rasiRNAs). Some adult cells with reactivated PIWI form "embryonic bodies" and eventually die apoptotic death. When a germline-to-precancerous stem cell (GSC → pCSC) constitutively expresses the PIWI2 proteins (the Piwi-like2 oncoprotein), the cell nuclei will also express the NFκB protein (whose oncogene was originally picked up by a reticuloendothelial virus, Rel: c-*onc* → v-*onc*). NFκB activates the cell survival pathway STAT-3, and the anti-apoptotic protein Bcl-2. The PiWi mRNA could be demolished by RNA interference (RNAi). For these assays, siRNAs were prepared by Integrated DNA Technologies, Coralville, IA [136]. The intranuclear co-expression of the NFκB and PIWI2 proteins is tantamount to oncogenesis.

The human HIWI proteins 1-4 work in the Argonaute system (Figure 26), as PPD/PAZ domain proteins. A wide variety of human cancers overexpress the PIWI2 (HIWI2) proteins. These proteins are overexpressed in adeno- and squamous cell carcinomas, in sarcomas (in gastrointestinal stromal tumors, GIST), and in testicular germ cell tumors. Molecular downstream targets of Piwi2 include the anti-apoptosis gene for Bcl-$_{XL}$ protein, the STAT (signal transducers and activators of transcription), and the Akt (Ak mouse thymus lymphoma retroviral oncogene) cell survival pathways. These entities can neutralize cell cycle inhibitors (cyclin D1) [137]. STAT3 is oncogenic in particular, because it is operational in a constitutive fashion, when activated through its tyrosine kinase residue at TYR705. The Src oncoprotein phosphorylates STAT. PIWI /HIWI binds STAT at its PAZ domain. The phosphorylated STAT3 protein translocates into the nucleus and by binding the promoter of p53 it suppresses it, thus annulling all p53 activity. The malignantly transformed cell will resist apoptosis induction [138]. Esophageal squamous cell carcinomas express PIW (HIWI) both in the cytoplasm and nucleus in histologically high grade tumor cells [139]. Hiwi expression in soft tissue sarcoma cells correlated with high mortality [140]. Overexpression of Hiwi in colon cancer cells was an adverse prognostic factor [141]. Intranuclear transfer of cytoplasmic proteins (drosophila wingless Wnt/ß-catenin), and the Akt pathway can be inhibited by sulforaphane and perfosine, respectively. Other small molecular inhibitors of the sHH (sonic hedgehog) pathway, the multidrug resistance pump, and the pro-tumor stromal interaction, are being tested (cyclopamine, defequifar fumarate, repertaxin) [142]. Anti-Hiwi monoclonal antibody is available in China for histochemical diagnosis.

The HiWi antigens provoke lymphocytes. Piwi Hiwi are cancer/testis antigens and as such they could be used in cancer vaccines for melanoma or sarcoma or adenocarcinoma (breast cancer; prostate cancer) [143–145]. Vaccinated patients develop immune lymphocytes reactive with NY-ESO-1 (New York esophageal) antigens, synergize with ipilimumab (probably through the cytolytic NKT cells working in the ADCC reactions), and induce tumor regressions [145, 146]. Human synovial sarcomas express fusion oncoproteins, the β-catenin oncoprotein, and NY-ESO antigens (Figure 34). Clinically effective cytotoxicity could be achieved against NY-ESO-1 antigen-expressing melanoma and synovial sarcoma tumors by adoptive immunotherapy with immune T lymphocytes expressing a transduced TCR directed against this antigen (Tables III/I and III/II). In the author's laboratory at M.D. Anderson Hospital, already in the late 1960s and early 1970s, patients' lymphocytes frequently attacked synovial sarcoma cells, but the target antigens were not identified (Sinkovics JG: "Cytolytic Immune Lymphocytes…" Schenk Buchverlag, 2008) [147]. Presumably, it was a NY-ESO-like fusion oncoprotein (Figure 34) that the autologous lymphocytes attacked.

This author is highly rewarded just by being able to see this achievement, inasmuch as it was in his laboratory at the M.D. Anderson Hospital in Houston, TX (*vide infra*), where both naturally immune, or viral oncolysate vaccine-induced small compact lymphocytes (later: immune T cells), or large granular lymphocytes

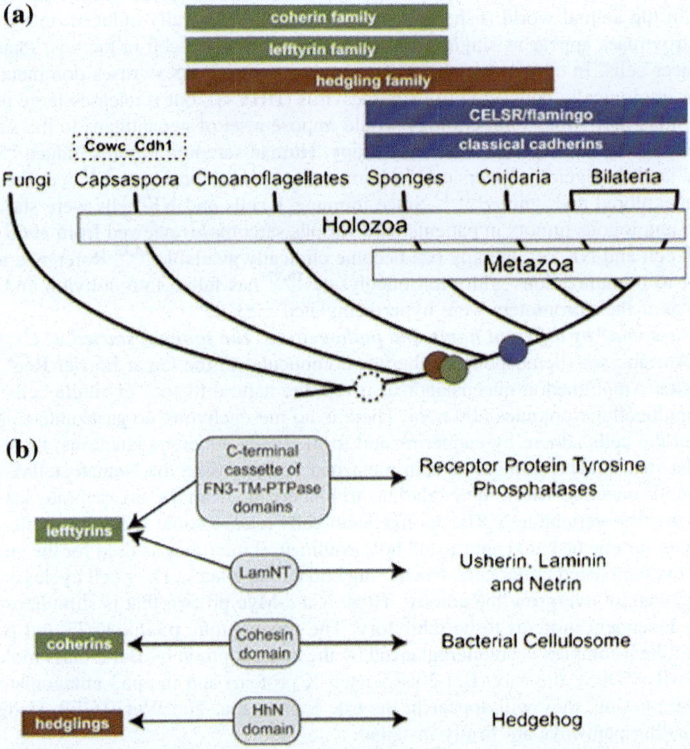

Figure 34 Synovial Sarcoma Fusion Oncoprotein. Cadherin/β-catenin. NY-ESO Antigens. [2346] Reference Appendix 2, Explanations to the Figures

Table III/I Sarcoma Immunobiology & Immunotherapy I

Adult patients with metastatic sarcomas seldom survive tumor-free over 5 years (even when doxorubicin/ifosfamide were given as the lead regimen).[1] Combination chemotherapy and viral oncolysate immunotherapy with or without BCG additively improved remission rate and duration (in not prospectively randomized patients).[2] NY-ESO (and other tumor-specific) antigens are, or can be, induced to be expressed in human sarcoma cells. Patients mobilize immune T cells and NK cells against autologous tumors.[3] Sarcoma cell viral oncolysate vaccines induced autologous host immune reactions. Autologous host lymphocyte (immune T cells; LAK cells) induce tumor lysis and remissions.[4] Sarcomas are chemotherapy-resistant, immuno-bio-therapy-susceptible tumors. Modern Melanoma-Sarcoma Services[5] are unaware of it that the immunotherapy-sensitivity of these tumors was revealed some 40 years ago at the Melanoma-Sarcoma Service of M.D. Anderson Hospital, Houston TX (last chapter in this volume).

NY-ESO antigen expression in synovial sarcoma (SS) cells.[6] Anti-CTLA4 ipilimumab renders synovial sarcoma cell to be immunological targets.[7] Patients were treated with adoptive therapy using autologous lymphocytes expressing genetically-engineered TCR to react with NY-ESO antigen. Patients with metastatic sarcoma experienced durable remissions (2/11 CR >1 y duration); 4/6 patients with metastatic SS entered partial remissions.[8]

Abundance of new ideas dominated by DC-vaccines and genetically modified immune lymphocytes. What are the best target antigens? Are human sarcomas virally induced? While sarcomas in the animal world (fish; aves, mammals) are retrovirally induced, in the human tumors retroviruses appear as single particles budding from one cell in the vast majority of the non-producer cells. In contrast, in the marmosets oncogenic herpesviruses dominate. Kaposi sarcoma is etiologically connected to its herpesvirus (HHV-8), but it releases large progenies of an endogenous retrovirus. Viral etiology would impose a set of neoantigens to the sarcoma cell, serving as a definitive target for immunotherapy. Human sarcoma filtrates induce "focus formation" and "antigenic conversion" in human embryonic fibroblasts.[9–11] Yet the matter remains unexplored and ignored.[12,13] Since immune T cells and NK cells were shown to be reactive to autologous tumors in patients with various sarcomas (reviewed from 1968 to 2008)[11], immune T cell and NK cell therapy has become clinically available.[14,15] Reference to oncolytic viruses, or to immunizations with viral oncolysates[16,17] has fallen into oblivion and remains silenced ('as if their promoters were hypermethylated me3').

"Lymphosarcoma" by different name and pathogenesis: the sponges started it! Extant sponges from the Adriatic sea (demosponge Suberites domuncula) to the Great Barrier Reef (demosponge Amphimedon queenslandica) reveal the natural history of single cells forming adherent multicellular colonies 600 mya. There is no mesenchyme, no gastrointestinal or neural system, yet the cells adhere by cadherins and form zonula adherens junctions; there is intercellular matrix. The sponges remain metazoan Parazoa (like the Nematostella), and do not become true Eumetazoa (from the cnidarians to bilaterians, from the invertebrate sea urchin and insects to the true vertebrates). The sponge stem cells release some special somatic cells differentiated for one task: to capture and hold commensal microbes as food for the entire colony. These are the flagellated collar cells resembling choanoflagellates. Their cell cycles are regulated so that they cannot overgrow the colony. There is a c-Myc protein that is stimulatory and there are cyclin-dependent proteins to be inhibitory. The proapoptotic p53/p63/p73 and Bak (Bcl-2 antagonist killer) complex is counterbalanced by the anti-apoptotic by Bcl-2 class molecules. The FasL-to-FasR (CD95), the Bax (Bcl-2-associated X protein) and the p53 antagonist MDM proteins are missing: they will appear in the true Eumetazoa. The Wnt, TGFβ, Hedgehog and Notch signaling pathways are firmly installed.

The sponges innate immune system is based on phagocytes and on Toll-like and interleukin receptors; as to alien cells approaching, the sponges immune system is alloreactive: distinguishes self from nonself.[18,19] In addition, sponges (S. domuncula) operate a tyrosine kinase that shows 55% similarity and 38% identity with the human Bruton kinase (BTK), a maturation factor of B lymphocytes.[20] Loss-of-function point mutation of the human Bruton kinase gene causes the

(continued)

Table III/I (continued)

X-linked agammaglobulinemia. Gain-of-function mutations of the BTK gene is the major causative factor of B-cell malignant lymphomas (including mantle cell lymphoma) and CLL. The recently constructed small molecular inhibitor of the human BTK protein is ibrutinib; by oral administration ibrutinib induces durable complete remissions.[21,22] Further gain-of-function point mutations of the BTK gene result in resistance of the disease to ibrutinib.[21–23]

1 *Yap BS & Sinkovics JG 15th ASCO New Orleans LA 20:352 #C250 1979. 3rd Internat Cancer Congress Seattle Washington 146 #2375 1982. Sinkovics JG 6th Chicago Cancer Symposium. Tumor Progression, Elsevier N Holland 315-331 1980.* ***2*** *Sinkovics JG & Plager C Chicago Symposium Immunotherapy Solid Tumors. Franklin Institute Press 211-219, 1977. Sinkovics JG & Horvath JC Internat J Oncol 2006;29:765-777. Sinkovics JG. Exp Rev Anticancer Ther 2007;7:31-56.* ***3*** *Sinkovics JG Cytolytic Immune Lymphocytes. Schenk Buchverlag Passau & Budapest Pp 391 2008.* ***4*** *Berghuis D et al Clin Sarcoma Res 2012;2(1):8. Finkelstein SE et al Immunotherapy 2012;4:283-90. Lai JP et al Mod Pathol 2012;25:854-8. Linehan DC et al Semin Surg Oncol 1999;17:72-7. Maki RG Curr Opin Oncol 2006;18:363-8. McClay EF & Slovin SE Semin Oncol 1989;16:328-32. Pedrazzoli P et al J Cancer 2011;2:350-6. Pollack SM et al PLoS One 2012; 7(2):e32165. Ratnavelu K et al Oncol Lett 2013;5:1457-60.* ***5*** *D'Angelo SP et al Sarcoma 2014;391967.* ***6*** *Lai JP et al Mod Pathol 2012;25:854-8.* ***7*** *Maki RG et al Sarcoma 2013:168145.* ***8*** *Robbins PF et al J Clin Oncol 2011;29:917-24.* ***9*** *Sinkovics JG Expert Rev Anticancer Ther 2007;7:31-56; 2007;7:183-210.* ***10*** *Györkey F et al Cancer 1971;27:1446-54.* ***11*** *Sinkovics JG Cytolytic Immune Lymphocytes... Schenk Buchverlag, Passau/Budapest, 2008.* ***12*** *Francescutti V & Skitzki JJ Surg Oncol Clin N Am 2012;21:341-55.* ***13*** *Goldberg JM Curr Opin Oncol 2013;25:390-7.* ***14*** *Pollack SM et al Sarcoma 2011:438940.* ***15*** *Ratnavelu K et al Oncol Lett 2013;5:1457-60.* ***16*** *Sinkovics JG & Horvath JC Int J Oncol 2006;29:765-777.* ***17*** *Sinkovics JG & Horvath JC Arch Immunol Ther Experiment 2008;56S:1-59.* ***18*** *Srivastava N et al Nature 201;466:720-6.* ***19*** *Yuen B et al Mol Biol Evol 2014;31:106-20.* ***20*** *Cetkovic H et al Genomics 2004;83:743-5.* ***21*** *Ma J et al Br J Haematol 2014; doi:10.1111/bjh.12974.* ***22*** *McDermott J & Jimeno A Drugs Today 2014;50:281-300.* ***23*** *Wodarz D et al Blood 2014 2014;123:4132-5.*

Table III/II Sarcoma Biology & Immunotherapy II

Interdigitating dendritic cell sarcoma: DCs are not exceptions, could undergo sarcomatous transformation.[1]

The *highly immunogenic NY-ESO gene* occasionally expressed, especially in synovial sarcoma, frequently is silenced, but could be reactivated by epigenomic de-methylation using 5-aza-2-deoxycitabine (azacytidine). Immune T lymphocytes attack the re-expressed NY-ESO protein and kill sarcoma cells.[2,3]

Ewing's sarcoma cells conceal NKG2D-ligand expression, thus escaping NK cell attack. Histone deacetylase inhibitors (SAHA, suberoyl hydroxamic acid, Vorinostat) enhance expression of NKG2D ligand and render Ewing's sarcoma cells vulnerable to NK cell attack. The balance of miR suppression (for let-7) and overexpression (for miR-17-92; miR-145) is under the control of the EWS/FLI-1 fusion oncoprotein (Friend leukemia insertion site).[4]

Co-evolution of p53 and its antagonist MDM. The gene(s) of p53 are over one billion years old. The ancient gene encoded the fused p63/p73 protein in the algae choanoflagellates. It is conserved and operational in the extant multicellular sea anemones (Nematostella). The original fused p63 and p73 genes duplicated in cartilaginous fish to generate p53. Bony fish operate the three genes separately (p53, p63, p73). The human "guardian of the genome" resides at 17p13. The genes p63 at 3q27-29, and the p73 at 1p36.3 acquired large new introns. The separated exons encode more diverse haplotypes. Both p53 and MDM ancestors may have co-existed in the invertebrate ancestor placozoans (Trichoplax adhaerens) 780 mya. The mdm/MDM genes could not be identified in choanoflagellates and sponges (also absent in archaea and prokaryota)

(continued)

Table III/II (continued)

In contrast to tumor suppressor gene p53, the MDM genes are oncogenes; the MDM protein adheres to the p53 (p63; p73) proteins marking them for ubiquitination (as MDM2 is a ubiquitin ligase). This is an early process in sarcomagenesis (Table VII). A single MDM gene/protein exists in invertebrates (sea squirt; amphioxus). Jawed vertebrates (first the sharks) operate MDM2 and MDM4, but the ancient gene duplicated before the appearance of gnathostomata (chondrichtyes). The human MDM2 gene is located at 12q14.3-q15, and the MDM4 gene at 1q32.[5,6] MDM2 amplification and PI3K exon 20 point mutation H1047R (histidine; arginine) drive rhabdomyosarcoma.[7]
The *MDM antagonist nutlin*. The differentiation-inducer protein nutlin (in acute myelogenous leukemia) is an MDM2 antagonist in MyLsrc (Table VII). It sensitizes sarcoma cells to monoclonal antibody death receptor DR5 agonist drozitumab.[8–13]
In MyLsrc, amplified CDK4/CyclinD complexes phosphorylate the Rb protein, which dissociates from E2F; intranuclear E2F activates mitotic cyclins and the cell cycle advances non-stop. MDM2 and CDK4 co-amplify. CDK4 antagonist flavopiridol and MDM2 antagonist nutlin reverse the pathway and stop the cell cycle.[14]
PPARγ antagonists rise in MyLsrc. c-Jun N-terminal kinase and ASK1 (apoptosis signal-regulating kinase) inhibit PPARγ (Table II), a differentiation-inducer of adipoblasts. Histologically and by molecular biology, PPAR-agonists (tro- and rosiglitazone) induce differentiation of liposarcoma cells, and reduce Ki-67 expression, but in clinical trials they were ineffective. However, there may be additive (or synergistic) interaction between PPARγ agonists and ET-743[15–17] (Table II).
Ecteinascidin. The tetrahydroisoquinolide alkaloid (ET-743, Trabectedin, Yondelis) of marine tunicate *Ecteinascidia turbinata* fits into the minor groove of the DNA molecule, inducing DNA strand breaks. With PARP-inhibition tumor cells may suffer synthetic lethality (Table X). Clinically effective remission-induced in MyLsrc[18,19] (Table VII). Extensive tumor necrosis in a case of pleomorphic sarcoma.[20]
PI3K and mTOR inhibitors. The antibiotic wortmannin-derivative PX-866; the mTOR pathway inhibitor ridaforolimus prolong relapse-free survival in pre-treated metastatic disease[21-23]
Insulin-like growth factor and neoangiogenesis. Monoclonal antibodies can target both ligand (MEDI-573) and receptor (MAB391; CP01-B02)[24]. Cituxutumumab and mTOR inhibition (temsirolimus) combined[25]. In embryonal rhabdomyosarcoma, high mobility group protein A2 (HMGA2) attaches to IGF-R and induced cascade N-RAS→MAP/ERK.[26]
1 Zhang J et al Int J Urology IJU-00535-2013. 2 Lai JP et al Mod Pathol 2012;25:854-8. 3 Pollack SM et al PLoS One 2012;7(2):e32165. 4 Berghuis D et al Clin Sarcoma Res 2012;2 (1):8; De Vito C et al PLoS One 2011;6(8):e23592. 5 Belyi VA et al Cold Spring Harbor Perspectives Biology 2010:P2(6):a110198. 6 Momand J et al Gene 2011;486:23-30. 7 Kikuchi K et al Sarcoma 2013:520858. 8 Ambrosini G et al Oncogene 2007;26:3473-81. 9 Müller CR et al Int J Cancer 2007;121:199-205. 10. Singer S et al Cancer Res 2007;67:6626-36. 11 Pishas KI et al Oncol Rep 2013;30:471-7. 12 Korotchkina LG et al Cell Cycle 2009;8:3777-8. 13 Secchiero P et al Neoplasia 2007;9:853-61. 14 Conyers R et al Sarcoma 2011;ID 483154. 15 Charytonowicz E et al J Clin Invest 2012;122:886-98. 16 Debrok G et al Br J Cancer 2003;89:1409-12. 17. Demetri GD et al Proc Natl Acad Sci USA 1999;96:3951-6. 18. Hoiczyk M et al Int J Oncol 2013;43:23-8. 19 Samuels BL et al Ann Oncol 2013;24:1703-9. 20 Schuler MK Case Rep Oncol Med 2013;320797. 21 Hong DS et al Clin Cancer Res 2012;18:4173-82. 22 Demetri GD et al J Clin Oncol 2013;31:2485-92. 23 Keedy VL Onco Targets Ther 2012;5:153-60. 24 Bid HK Clin Cancer Res 2013;19:2984-94. 25 Schwartz GK et al Lancet Oncol 2013;14:371-82. 26 Li Z et al Cancer Res 2013;73:3041-50.

(later: NK cells) were observed and photographed for the first time in the late 1960s as they attacked and lysed, or killed by "nuclear clumping" (later: apoptosis), autologous and allogeneic melanoma or sarcoma cells of patients treated for metastatic disease (Figures 35 and 36 [147]. At that time, the target antigens attacked by autologous (and allogeneic) lymphocytes were unknown. This author serving as healthy control lymphocyte donor, discovered in the late 1960s that his large granular lymphoid cells indiscriminately attacked and lysed allogeneic human cancer cells. In patients with cancer, both small compact (immune T cells) and large

(a) (b)

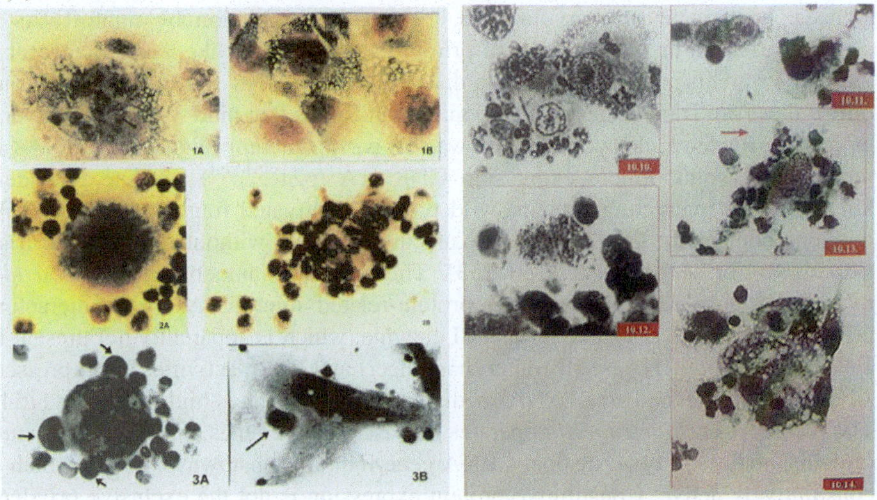

Figure 35 The Earliest Documentation of Cytotoxicity to Human Cancer Cells Mediated by Two Different Types of Lymphocytes. Reference Appendix 2, Explanations to the Figures

(a) (b)

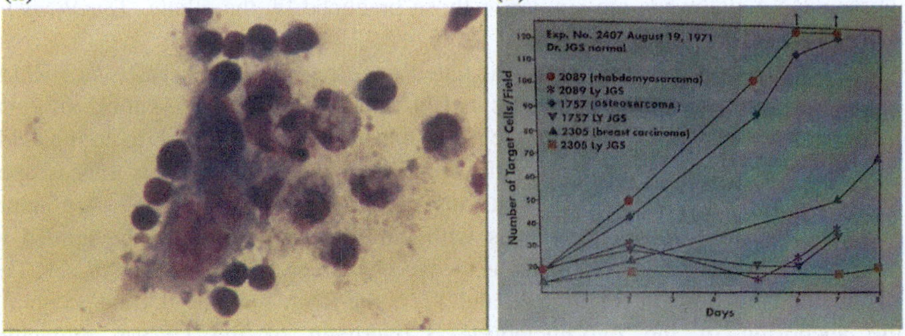

Figure 36 Large Granular Lymphoid Cells (Later: Natural Killer Cells) Kill Human Cancer Cells without Pre-Immunization. Reference Appendix 2, Explanations to the Figures

granular (NK cells) attacked cancer cells. The healthy donors possessed only the large granular cells (later: NK cells). Anticancer cell immune T cells were absent in healthy individuals, but the NK cells were active (Figures 35 and 36). A NIH/NCI project site visitor (who missed observing this obvious distinction), declared that this author (JGS) was creating "laboratory artifacts", because lymphocytes do not attack without pre-immunization. It took several years to prove that natural killer cells existed [147]. A factual narrative is included in the Appendix 1.

The ancient RNA is alive and ticking. Post-transcriptional modification of RNA (rRNA, tRNA) is a common event in thermophil archaea and prokaryota, however, may be by different mechanisms. The nine nucleosides common in tRNAs of the three domains of life indicate that this event has evolved before the separation of archaea, bacteria and eukaryota [148, 149]. Gene regulation by microRNAs is clearly present in the descendants of primordial eukaryotes, like the semianaerobic giardia. From its nucleolus, the giardia cell releases two snoRNAS: miR2 and miR5; these are Dicer-digested functional microRNAs that bind 3′-UTRs of several targeted mRNAs. Cell cycle control was an early function of the microRNAs in modulating cyclin protein expression CDKN1A ($p21^{cip1/waf}$) (cycline-dependent kinase; CDK- interacting protein; wild type p53-activated fragment). Some PiWi domains contain the DDH motif (asp.asp.hist) with or without the catalytic triad (asp.his.ser, or ser.glut.hist) [150–153]. The anaerobic amoeba (*Entamoeba histolytica*) (Figure 31) releases Piwi protein-related microRNAs (piwi-interacting piRNAs) [154]. Dicer is an RNAase III enzyme, which is apparently not present in *E. histolytica*, but RNAase III (other than Dicer) and RdRp activities are expressed. Abundant Piwi-related small RNA-mediated gene silencing is highly evident [154]. The paramecium *P. tetraurelia* operates at least 15 Piwi genes encoding Argonaute proteins, and at least two distinct siRNA gene-inhibitory pathways, one of which is constitutively active. Constitutive protein expression is not the exclusive privilege of malignantly transformed (cancer) cells: the paramecium is endowed with this faculty within its physiological metabolisms of its "cell survival pathways".

Dicer-processed siRNAS carry out gene silencing by cleavage of endogenous mRNAs. The second siRNA-mediated gene silencing process is active during sexual reproduction in the germline genome mediated by meiosis-specific Dicers. These siRNA pathways are similar to those practiced by the ciliate, *Tetrahymena thermophila* (*vide supra*) [155].

Of the 3 modes of small RNA-mediated gene regulation (microRNA, siRNA, piwiRNA), *Trypanosoma brucei* uses its RNA inhibitory system to knock out those mRNAs that are traveling through the ribosomes for the translation into amino acids to form the surface glycoproteins of the kinetoplast. The trypanosome, thus escapes the host's immune attack directed at those antigens [156]. *T. brucei*'s Argonaute protein (TbAGO1) forms complexes with the siRNAs in order to be guided to the targeted mRNAs. The dsRNA was cleaved by Dicer to become the siRNA. The Argonaute PAZ domain (Figures 26 and 27) is the RNA-binding module accepting the anchoring of the two-nucleotide 3′-overhang of the siRNA molecule.

The Argonaute C terminus is its Piwi domain. The Argonaute's N terminus, the RGG (arg.glyc.glyc) domain, is where arginine methyltransferase (*vide infra*) can deposit methyl (CH_3) groups. The N terminus accepts binding with polyribosomes [157]. The Argonaute/Piwi complex is not essential for the life of the host. This system is extensively simplified in *Trypanosoma cruzi*. There may be an alternative system in operation. It is possible that the small RNAs are released from mature tRNA and assemble in cytoplamic granules [158]. *T. brucei* lacks the enzyme RNA-dependent RNA-polymerase (RdRp), therefore it cannot amplify the transient reaction. Another kinetoplastid, the *Leishmania braziliensis*, performs this reaction (downregulation of gene expression) with a full set of RNAi genes. In contrast, Old World leishmania species show only the remnants of a non-functional system. However, Plasmodia generate dsRNA sequences for downregulating of gene expressions. The apicomplexan *Toxoplasma gondii* possesses genomics for Dicer, Ago and RdRp [159]. In permanent cultures, individuals of the *Trypanosoma brucei gambiense* colony display full length chromosomes due to constant telomerase activity, but may be on physiological and thus reversible, not proven constitutive, that is irreversible, terms,

Comment. There appear evidences that in the so called malignant transformation of somatic or stem cells in multicellular hosts, reversion takes place to an ancient life style of the cell. Trypanosoma can conceal its surface antigens, as transformed cells in a multicellular host also refrain from antigenic expression by withholding their histocompatibility complexes. In a nutrient rich environment, Trypanosoma cells replicate and maintain their telomere lengths after each cell divisions (especially when replicating in liquid media).

A major characteristic of the microRNAs is that new families of them are continuously generated in the metazoan genomes. Once integrated into the gene regulatory network, the mature sequences remain unchanged (seldom undergo mutations) and as such, they remain conserved. Thus microRNA networks can reliably be traced backward alongside the phylogenetic tree and its outbranchings. In mammalian taxa, microRNAs that originated before the divergence of vertebrates into the lines of jawless fish (lamprey and hagfish) and gnathostomata (the sharks), are preserved. Mammalian taxa preserve microRNAs that were generated in teleost fishes. Piscine taxa guard microRNAs originating in the ascidian urochordates (the Botryllus) and cephalochordates (the amphioxus). Chordata are characterized by operating miRNA families miR-216 and miR-217 [160]. Tetrapods (frogs, mammals) carry eight restricted miR families. Amniota (aves, mammalia) share two miR families. Eutherian mammalian taxa (mouse, human) possess two restricted miR families. It will be most instructive to compare some organ-specific miRs, that are available for the zebrafish's kidney, liver and pancreas, thymus, melanocytes, and brain [160], between the brain and the pharynx of Pan (chimpanzee) and Homo.

Donors of Argonaute (Archaea and Prokaryota), MicroRNAs (Caenorhabditis), and PiWi (Drosophila)

The Argonaute-like proteins might have existed in the precellular world as partners to primordial RNA molecules for their unison. AGO/RNA units were firmly established within the first archaea and prokaryota (Figure 26). There, they have become highly specialized, the AGOs in offering modules for the anchorage of small RNA sequences cognate toward untranslated regions of mRNAs. The process is as ancient as life itself. The archaeal-bacterial AGOs were and remain closely related. Among the bacteria, the *Aquifex aeolicus*, and among the archaea, *Pyrococcus furiosus,* possess unique (by crystal structure) PAZ domains (Figure 26). The Argonaute of *A. aeolicus* and the PiWi of *Archaeoglobus fulgidus* bind ssDNA rather than ssRNA. Eukaryotic AGOs underwent extensive diversifications due to increased numbers of AGO genes and their horizontal transfers. PAZ and PiWi domains and DDH motifs (*vide supra*) were subsequent additions, without exact evidence for their first appearance. All eukaryotic cells possess the entire system, but some components of the system could be lost without any major detrimental consequences.

There is enough amino acid motif-sharing between archaeal and eukaryotic AGOs to suggest phylogenetic relationship between these two domains of life [153, 161]. The common ancestor of eukaryotes must have operated at least one Argonaute protein (polypeptide), one PIWI-like protein, one RNAase III-like endonuclease (Dicer) enzyme and one RdRp. The *Giardia intestinalis* (Figure 28) has one RNAase III dicer and PAZ domain, PAZ-like proteins, Argonaute-Piwi and Dicer-like RNAase III. All six supergroups of eukaryotes inherited and/or horizontally acquired this machinery of ribonucleoproteins. These are (not in evolutionary order): 1. Amoebozoa (amoeba; slime molds); 2. Opisthokonta (fungi and animals); 3. Excavata (diplomonads, euglenozoa, rhizaria, foraminifera, cercozoa); 4. Archaeaplastida/Plantae (red and green algae, plants); 5. Chromalveolata (dinoflagellates, apicomplexan parasites); and 6. Stramenopiles (brown algae, diatoms, zoosporic fungi). For tabulated display, see Medina [162]. Some specific eukaryotic lineages experienced losses complete or partial from the AGO-Piwi-Dicer-RdRp machinery: *Saccharomyces cerevisiae* (Opisthokonta), *Trypanosoma cruzi, Leishmania major* (Excavata), *Plasmodium falciparum* (Chromalveolate).

A presumption exists that the RNAi mechanisms may be lost and they may be regained throughout eukaryotic evolution. The RNAi-negative *T. cruzi* and *L. major* possess PiWi-like polypeptides; *T. brucei* practices RNAi, but without utilizing a Piwi protein. The *Tetrahymena thermophila* utilizes three divergent Dicer sequences. *T. cruzi* and *Chlamydomonas reinhardtii* live without the RdRp enzyme. In rooting the phylogenetic tree of eukaryotes, Amoebozoa and Opisthokonta were unikonts (ancestrally monociliates), whereas Archaeaplastida, Chromalveolata, Rhizaria, and Excavata were bikonts (ancestrally biciliates). Thus, the root was the common eukaryotic ancestor, whose line diverged into unikonts and bikonts. Fungi, green algae and plants encode exclusively Argonaute proteins. *T. thermophila, Paramecium tetraurelia*, and Amoebozoa encode Piwi-like and HMG proteins (Figure 31). Opistokonta (animals) cells encode both Argonaute and Piwi proteins. Most animals encode one type of Dicer, but insect genomes encode two Dicers. The rapid evolution of insect Dicers is attributed to pressure from viral (both plant and animal) pathogens. Thus, there is an ancestral Dicer derived from the common ancestor, and a newly acquired Dicer. The other RNAi-involved RNAase III enzyme, Drosha, is present in the genome of animalia, but it is absent in other taxa, thus it is a new acquisition. Since, Drosha is distantly related to eubacterial RNASe III enzymes, it might have been horizontally acquired by some ancestral animalia cells. The eukaryotic RdRp domain reflects backwards toward a monophyletic origin (awaiting confirmation from Giardia); it is different from the viral RdRp. However, the Argonaute-Piwi-Dicer elements appeared in the archaea/prokaryota cells way before the rise of eukaryotic cells [163].

Choanoflagellates #1. The transmembrane teneurin proteins are important in tissue pattern formation. Their extracellular domain floats in the intercellular space, and their intracellular domain enters the nucleus. Teneurins contain epidermal growth factor (EGF) repeats, cystein-rich domains, large regions identical to the prokaryotic YD repeats (tyrosine/aspartate), and a core-associated domain of the retrotransposon-encoded "hot spot" proteins. Teneurins were discovered in the drosophila, in the caenorhabditis nematode, in the amphioxus, *B. floridae,* and in the sea urchin, *S. purpuratus.* Teneurins appeared first in the genome of choanoflagellates (phylum Choanozoa), but are absent in sponges (phylum Porifera), and in cnidarians (Coelenterata). The original teneurin protein (see it tabulated later) was composed by the fusion of two proteins, whose genes, one from a prokaryote, and the other one, the one carrying the cystein rich enzyme code, from an alga (blue-green algae, class Cyanophycea), or a diatom (phylum/class Bacillaryophyta; Heterokontophyta, Chrysophyta). The original genes were horizontally transferred into the ancestral choanoflagellate. It is presumed, that the teneurin intercellular proteins in the choanoflagellates contributed to the evolution of the first metazoans. In chordates, teneurins contribute to the pattern formation of the brain [164, 165]. Much later, a table (Table XVIII) reviewes teneurins in context with choanoflagellates #6.

Choanoflagellates #2 and associates. The cadherins, meaningful in cell-to-cell contact, appeared first in choanoflagellates [166]. A Wnt/ß-catenin (Wg, gene *wingless* drosophila ortholog to mouse mammary tumor provirus insertion site, int, integrated; ortholog to human at chromosome 12, oncogene *int*: united as *wnt*; cadherin associated protein) [167–169].

Ancient stem cells at work. Signaling was fully operational over 550 million years ago in the phylum Cnidaria. Extant representative descendants are the eumetazoan *Hydra* polyps, the colonial marine hydroids, *Hydractinia*, and the starlet sea anemone, *Nematostella vectensis* (Anthozoa, within Cnidaria) (Figure 37).

In the hydra (Figure 37) epithelial cells of the gastric region continuously divide creating a tissue growth permanently migrating toward the oral region, where these

Figure 37 The Ancient Stem Cells Regenerate the Mutilated Cnidaria Hydra. Reference Appendix 2, Explanations to the Figures

cells either differentiate into tentacles, or evaginate into buds to form daughter polyps. The process is mediated by Wnt/ß-catenin signaling (Figure 7), preventing the phosphorylation and degradation of cytosolic ß-catenin by GSK3 (glycogen synthase kinase), and thus facilitating its translocation into the nucleus. The Wnt ligand attaches to its low density lipoprotein receptor, that is releasing the receptor-related protein (LRP6). LRP6 degrades the scaffold protein Axin and blocks GSK3. GSK3 fails to phosphorylate ß-catenin, which escapes degradation and enters the nucleus [170]. The regenerative physiological pathway of the hydra appears to be the predecessor of the wnt-mediated oncogenesis in mammalian cells (Figure 7).

Other long hidden inhibitors of the process are the Dickkopf family proteins. The Dickkopfs were hidden in the Spemann organizer in the Xenopus and newts in the 1930s (Figure 38), but were discovered later as co-actor inducers of head development, and the natural antagonists of Wnt signaling [171]. After removal of the hydra's head, a Wnt/ß-catenin signaling cascade regenerates it (with Notch interacting).

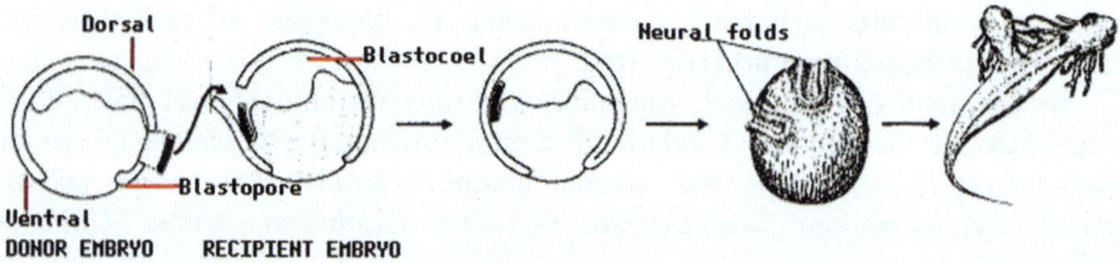

Figure 38 The Skill of Student Hilde Mangold and the Vision of Professor Hans Spemann United. [2347] Reference Appendix 2, Explanations to the Figures

The continuously proliferating epithelial cells are driven by the JNK protein. This is a proto-oncogene kinase in all multicellular hosts, known as *jun*-kinase. The c-*jun* proto-oncogene was incorporated into the genome of the avian sarcoma virus 17 (ASV-17): *c-jun* → *v-jun*. The Japanese word for 17 is jun. In Xenopus embryos, JNK inhibits the intranuclear transfer of ß-catenin [172]. This proto-oncogene is inhibited by the small molecule anthrapyrazolone SP600126. SP600126 inhibited phosphorylation of the JUN protein and suppressed bud formation in hydra polyps. Glycogen synthase kinase-3 (GSK) is blocked by alsterpaullone (A), a 2-cyanoethyl cycline-dependent kinase inhibitor. Colchicine-, or thapsigargin-induced apoptosis is prevented by GSK-3 inhibitors (like A) [173]. A is an activator of the *tcf/lef proto*-genes; the TCF/LEF proteins (T cell growth factor; lymphoid enhancer factor) form complexes with ß-catenin [174]. GSK-blockade comes with the intranuclear translocation of ß-catenin and results in the ectopic formation of tentacles in the hydra's head. Intranuclear ß-catenin induces tissue evaginations. The anthrapyrazolone, SP600126 added to A inhibited ectopic tentacle formations. The inhibitor of ß-catenin, the actinomycete-product ZTM000990, added to A, inhibited both ectopic tentacle and head formations [169]. In the hydra, inhibition of GSK-3 (meaning stabilization of ß-catenin) with lithium or with diacylglycerol, induces ectopic tentacle and head formations. Wnt ligand low-density lipoprotein receptors (Lrp5,6) also bind Dkk1, thus preventing Wnt signaling. The expression of genes *hydkk1/2/4* is essential for the regeneration of hydra's head. Activated Wnt/ß-catenin down-regulates *hydkk1/2/4* transcripts. The Wnt/Dickkopf antagonism is re-played in the vertebrate genomes [175], but Dickkopf proteins could not be found in insect and nematode genomes. However, the Wnt/ß-catenin signaling and ß-catenin /TCF/LEF protein complex formations are active processes in *C. elegans* [176]. In the drosophila, Wnt/ß-catenin signaling formats the dorsoventral axis and the wing and leg discs. In Nematostella, gastrulation and germ layer formations are directed by Wnt/ß-catenin signaling (reviewed by Broun *et al*) [177]. Many aspects of the physiological Wnt/Dickkopf relationship are replayed in the oncogenesis of mammalian cells.

Beta-catenin becomes a human proto-oncogene. The physiological Wnt pathway of the choanoflagellates (Monosiga), sea anemones (Nematostella), sea squirts (Ciona) becomes a most prominent proto-oncogene in the human genome (Figure 7). Translated into human oncogenesis, the Wnt/ß-catenin signaling is very prominent in colonic adenocarcinoma. In the colonic epithelial cells, cytosolic ß-catenin is phosphorylated. The adenomatous polyposis coli (APC) gene product protein forms complexes with GSK-3 and with casein kinase. The phosphorylated ß-catenin is degraded in proteasomes. The non-phosphorylated ß-catenin translocates into the nucleus, where it activates TCF/LEF, cyclin D1, matrix metalloproteinase 7, c-*myc* and survivin genes. The mutated *apc* gene encodes a truncated APC protein that fails to retain ß-catenin in the cytoplasm (thus the non-phosphorylated ß-catenin readily transfers into the nucleus) [178]. In a recent discussion, it was pointed out that in human colon cancers the Dickkopf family appears to be corrupted. The Dick-1 protein is not encoded due to the epigenetic silencing of its translation. The Dick-2,3,4 proteins appear as if they rather promoted, than antagonized tumor growth) [27]. For the Wnt pathway, see Table IV and Figure 7.

Table IV WNT Signaling; LiCl versus GSK-3; WNT Virology

The dicysteine motif HIV-1 Tat (transactivator of transcription) is inhibitory to Wnt/β-catenin signaling in astrocytes. Tat physically interacts with TCF (T cell factor). HIV Tat neutralizes the neuroprotective (anti-HIV) efficacy of Wnt/β-catenin signaling in astrocytes, thus allowing the neuro-pathogenicity of HIV-1.[1] Wnt/β-catenin signaling is a driving force for human extravillous cytotrophoblasts. Human betaherpesvirus CMV infection of the placenta is inhibitory to Wnt/β-catenin signaling; in the infected cell, β-catenin remains sequestered in the cytoplasm.[2] It is glucose synthase kinase GSK-3β, that mediates phosphorylation of cytoplasmic β-catenin for its ubiquitination. LiCl by inhibiting GSK-3β protects β-catenin, and thus activates Wnt signaling. ***Gammaherpesviruses EBV (HHV-4) and KSHV (HHV-8)*** are activators of Wnt signaling in the infected cells. EBV latent membrane protein 2A (LMP) activates PI3K/Akt, which is essential for the translocation of cytoplasmic β-catenin into the nucleus. LMP2A-activated Akt targets GSK-3β (the inhibitor of Wnt signaling) for inactivation. Intranuclear co-expression of LMP2A and β-catenin activate tcf/TCF, which in turn activates c-*myc* and cyclin D1. The infected cell responds with accelerated mitoses.[3] Latency-associated antigen of KSHV, among other functions, interacts with cellular GSK-3αβ, negative regulator of β-catenin in the cytoplasm. Instead of degradation, LANA promotes intranuclear transfer of GSK-3β. Intranuclear activation of the *tcf/lef* genes by β-catenin follows. Thus, Wnt signaling is activated in full.[4]

1 *Henderson LJ et al J Neurosci 2012;32:16306-13.* ***2*** *Angelova M et al PLoS Pathog 2012;8 (10):e 1002959.* ***3*** *Morrison JA et al J Virol 77:12276-84.* ***4*** *Fujimuro M & Hayward SFD J Virol 2003;77:8019-30.*

The breast cancer cell population is a mixture of tumorigenic (breast tumor-initiating cells, BTIC), and not tumor-initiating cancer cells. Biologically, the BTIC are a population of self-renewing transformed stem cells. This cell population is notoriously chemo-radiotherapy-resistant. This cell population undergoes epithelial-to-mesenchymal transition, gains propensity for locomotion and spreads by locoregional invasion and lympho-hematogenous routes to form distant metastases. The Wnt/ß-catenin signaling pathway is a driving force of this tumor cell population. Naturally, GSK3 and the DKK1 protein act as the antagonists, and Wnt3a ligand as an agonist for this tumor cell population. Inhibition of GSK with the small molecule BIO accelerated tumor cell growth (in mammospheres and tumorspheres). The investigational inhibitor PKF-118-310 of Wnt/ß-catenin signaling decreased the Ki-67, Axin2 and cyclin D1 expressions and irreversibly blocked *in vitro* sphere formation of the BTIC. BTIC pretreated with PKF-118-310 (curcumin-derivative) failed to grow in xenografts. It was not apoptosis induction that caused the failures of the transplants to take. BTIC so treated expressed the luminal lineage marker, but not the myoepithelial lineage marker. Upon further transplants, these cells failed to initiate tumor growth. Of the natural PKF compounds, curcumin (diferuloylmethane), only PKF-118-310 has been synthesized [179]. This compound acts by abrogating the binding of ß-catenin to Tcf4, or to GST-Tcf4 complexes, and inhibits the expression Wnt-activated genes for c-Myc or CyclinD1 [180].

Androgen-independent prostate cancer cell lines exhibit high levels of Wnt/ß-catenin signaling. The non-steroidal anti-inflammatory drug, sulindac sulfide, celecoxib, the cyclooxygenase-2 inhibitor, the nitric oxide-donor aspirin-derivative,

NO-ASA, and the curcumin-derivative, PKF-118-310 (*vide supra*) inhibited Wnt/ß-catenin signaling (the intranuclear binding of ß-catenin to activate the *tcf/lef* genes) in androgen-independent prostate cancer cell lines. When Wnt is not signaling, cytosolic ß-catenin is phosphorylated and is sent to ubiquitination, and thus, to its degradation in the proteasome. For this effect, the adenomatous polyposis coli (APC) tumor suppressor gene, the protein Axin, the GSK3ß, and the casein kinase-1 are necessary. Mutated and inactivated APC defeats this process. The oncogenic effector of this process is the Wnt ligand's binding to its receptors frizzled (Fz) and to the low density lipoprotein receptor-related protein (LRP). In this process, ß-catenin remains un-phosphorylated and instead of its ubiquitination, it enters the nucleus, where it activates cell cycle inhibitory and anti-apoptotic genes. It is presumed ß-catenin to be a co-activator of the androgen receptor (AR). However, curcumin inhibited both AR and phosphorylated ß-catenin expression in prostate cancer cells lines [181]. Not APC mutation, but the Wnt/ß-catenin pathway is activated in some prostatic cancer cells (especially in those that have metastasized). The above listed compounds to various degrees of efficacy inhibited the Wnt/ß-catenin pathway [182]. It is not well documented as yet, whether this is due to the cessation of cell divisions (but without apoptosis-induction), or due to induction of cell differentiation. Further, the oncogene/oncoprotein Sox2 is highly active in some prostate cancers [183]. The Wnt signaling can act through stem cells and proto-oncogene Sox2. In the ctenophores (Mnemiopsis leidyi), the Wnt pathway is physiologically 'defective', in that, β-catenin is readily transferred from cytoplasm to the nucleus to activate its target genes of cell proliferation, due to an incomplete 'β-catenin-destroyer complex'; and the antagonist of the system, the dickkopfs are non-existent. Oncogenic mutations in the human genome disable of the 'β-catenin-destroyer complex' and delete the dickkopfs. Thus adenocarcinomas are generated (Figure 7). Further, the ctenophore genome is well-endowed of sox/Sox pathways. In the human genome, sox genes maintain stemness, and induce malignant transformation (more in the Appendix 1).

In the retina of the tadpole, Wnt and Sox2 co-act to inhibit neural differentiation by inducing Notch, and thus sustain the proliferation of undifferentiated progenitor cells. The cHLH transcription factor gene and gene product protein, Xath5, switches off Sox2 (basic helix-loop-helix; xenopus atonal homolog gene), thus allowing the terminal differentiation of retinal cells [184–186]. Despite of their diversity, both in insect, amphibian, and mammalian retina, the target gene of Pax6 is *ath5*; the mammalian *ath5* is the ortholog of the drosophila *atonal* photoreceptor gene. Could an Ath5 protein switch off the oncogene *sox*2, or neutralize the Sox2 protein, by forming a complex with it? No such information exists at this time.

The elegant nematode in the spotlight. The gene expression-inhibitory faculty of small dsRNA sequences (RNAi) was discovered in 1998 [188], and an extensive class of microRNAs was identified in the nematode *Caenorhabditis elegans* (Figure 39) in 2001 [189]. siRNAs from exogenous sources guide endonucleases to silence targeted mRNAs deriving from endogenous sources, and *vice versa.* The dsRNAs are processed by the same Dicer-1 endoribonuclease into 21 nt siRNAs. The endogenous and exogenous systems use their own different Argonaute

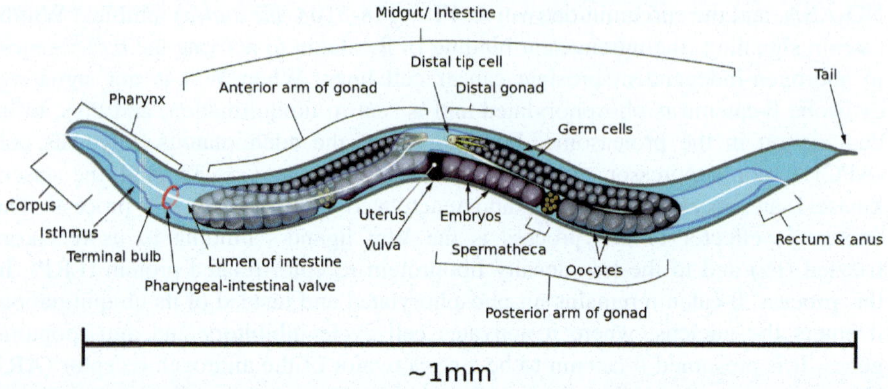

Figure 39 Activated Oncogenes in the Caenorhabditis Embryo (the Dauer State). Unopposed Replication of Undifferentiated Epithelial Stem Cells Kill the Adult Nematode. (Caenorhabditis elegans hermaphrodite adult). http://en.wikipedia.org/wiki/File:Caenorhabditis_elegans_hermaphrodite_adult-en.svg by KDS444 http://commons.wikimedia.org/wiki/User:KDS444 is licensed under CC BY-SA 3.0 http://creativecommons.org/licenses/by-sa/3.0/) Reference Appendix 2, Explanations to the Figures

proteins. Further, microRNAs of the miR-35-41 cluster expressed in germline and embryonic cells regulate the RNAi arm of the system. The retinoblastoma gene homolog of *C. elegans* is the *lin-35RB* gene (cell lineage abnormal). Low miR-35-41 expression results in the suppression of *lin-35Rb*, and *vice versa*, miR-35-41 by suppressing the exogenous RNA pathway, allows the rise of LIN-35RB protein levels. In hosts with disabling *lin-35* mutation, the exo-RNAi activity, including the Argonaute protein, is high [190]. The let-7 microRNA family of *C. elegans* has its mammalian counterparts. In eukaryotic (human) tumor cells, the let-7 miRNA (for "lethal") acts as if it were a suppressor of the RAS oncoprotein [191], thus elimination of let-7 may be lethal to its host.

The *C. elegans* genes *lin-4* and *let-7* (lineage; lethal) express short noncoding ncRNAs. The lin-4 miRNA represses *lin-14*, which regulates larval stage L1 cell divisions in embryonic life. Anything that promotes cell divisions in the embryo may pop up as an oncogene in the adult; anything that suppressed that entity may function in the adult as a tumor suppressor. What is physiological in embryonic life may be lethal when replayed in adult life. The genes *lin-14* and *lin-28* encode nuclear and cytoplasmic proteins, respectively, for direction of L1 pattern and conserved RNA-binding motifs, respectively. Gene *lin-4* targets these two genes directly through their 3′ UTR of their mRNA. If *lin-4* fails to act, L1 cell division pattern continues without differentiation into L2, and the larva overloaded with replicating L1 cells (simulating "oncogenesis") dies. The *let-7* genes regulate *hbl-1* (drosophila-hunchback homolog transcription factor in *C. elegans*), an important regulator of the L2 pattern. The failure of *let-7* to suppress *hbl-1* after the completion of the L2 phase, results in the continued proliferation of L2 cells within the L3 phase. Gene *lin-46* through miR-48, miR-84 and miR-241 could correct the matter and restore the switch from L2 into L3 patterns. LIN-14 and HBL-1 protein levels must be balanced for the orderly advancement of L1-to-L2-to-L3 patterns. There is a connection between *daf* (homolog of drosophila nuclear hormone receptor, DNHR; decay accelerating factor, CD55), *let-7* and *hbl.* DAF-12 induces the release of miR-84 and miR-241, which regulate let-7 expression, which is a regulator of *hbl-1* expression (through the 3′UTR of its mRNA). While DAF-12 up-regulates the *let-7* family, these down-regulate *hbl-1*, thus securing an orderly transit from L2 to L3 and to L4. In the adult transition L4, the seam cells fuse to form the cuticle and the lateral ridges (alae) of the mature worm. Genes conserved from worm to mammal are *lin-28* high in fetal life and low in adulthood, and *let-7*, low in the embryo and rising in the adult. Thus *lin-28* maintains stemness, and *let-7* is a cell differentiation inducer. Successful tumor cells eliminate *let-7* microRNAs. *Drosophilalet-7* differentiates the wing, but when mutated, the fly remains a wingless infant with juvenile cells replicating without maturation in the wing bud (wing imaginal disc). *C. eleganslet -7* competes with lin-28. Lin-28 targets pre-let-7, preventing its cleavage by Dicer, but let-7 family members miR-48, miR-84, and miR-241 cannot be regulated by LIN-28 [192].

Metazoan gametogenesis is regulated by small RNAs. Distinct PiWi-clade AGOs (WAGOs, W for worms, *C. elegans*) carry out the temperature-dependent process. Higher up on the scale, the male platypus has intraabdominal testicles, whereas male mammalians carry extraabdominal testicles in the scrotum. Spermatogenesis remains temperature-sensitive as well within the nematodes. Incorporated in the *C. elegans'* endogenous small RNA pathway, there exists a subset competent in targeting spermatogenesis-related mRNAs; these 26 nt small RNAs are specifically expressed in the male germline [193]. Thus, in *C. elegans* larvae, stem cells or germ cells may proliferate uncontrollably, or cease to divide and die apoptotic death, as regulated by sets of microRNAs. As long as the proteins directing cell proliferation are not constitutively activated, and their production may be decreased, or stopped, they fall short of being "oncoproteins" encoded by oncogenes. However, if mutated genes encode these proteins constitutively, the physiological process carried out in the interest of the multicellular host, switches to the exclusive interest of the single cell population driven to enlarge against any opposition, or even with the support, of the subverted host (Figure 39).

The lesson of Lin. "Stemness" of eukaryotic cells is maintained openly or latently by the single, or coordinated, actions of gene-product proteins LIN28, Oct4, SOX2, and NANOG (cell line abnormal; octamer-binding; SRY-related high mobility group transcription factor; SRY, sex/gender region in Y chromosome; eternally youthful Celtic tribe). The human *sox*2 stem cell pluripotency proto-oncogene is embedded into an intron within a long non-protein coding multi-exon RNA (ncRNA) mapping to chromosome 3q26.3-q27. This unison is referred to as SOX2OT, since an intron is overlapping the *sox*2 gene (Sox overlapping transcript, OT). SOX2OT is transcribed by RNA polymerase II from a highly conserved site (highly conserved element, HCE). Therefore, Sox2ot is a mRNA-like ncRNA. In fetal life, loss-of-function mutation of Sox2, results in the absence of lens and retina development (microphthalmia, anophthalmia). After fetal life, the promoters of this *sox* gene are blocked by triple methylations of lysine (K) 3, or 27, on histone 3 (H3 K4me3; H3K27me3). Sox2OT may be expressed in tissues in which Sox2 is silenced. Trimethylated lysins in the histones of the Sox2 promoters are different for active and silenced states: H3K4me3 and H3K27me4, respectively. Highly conserved noncoding sequence block of "transposons-free regions" are associated with the promoters. Orthologs of Sox2, Sox2OT and its novel isoform Sox2dot (distal overlapping transcript) are widely distributed in vertebrates [194a, b]. The SOX2 protein is overexpresssed in lung and sinonasal squamous cell carcinomas. Some of these tumors run a highly malignant clinical course. In other tumors overexpresssed SOX2 oncoprotein overlaps with a somewhat favorable clinical course (Figures 40 and 41).

Oct4, NANOG, and SOX2 stem cell genes may withhold protein encoding, but remain conserved in stem cells, thus existing fully suppressed; more so in somatic

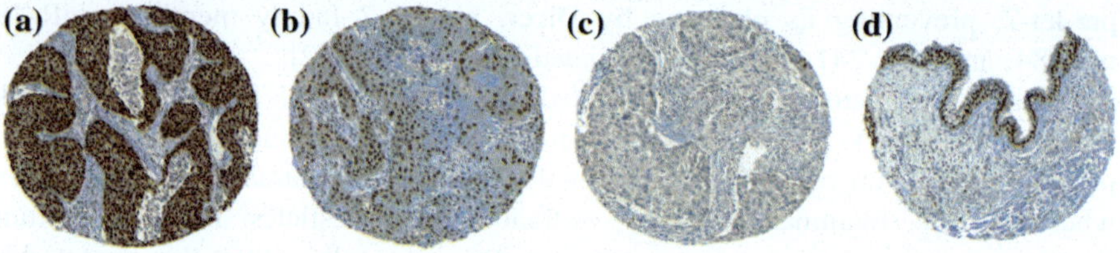

Figure 40 Human Sox Oncoprotein I. [2348] Reference Appendix 2, Explanations to the Figures

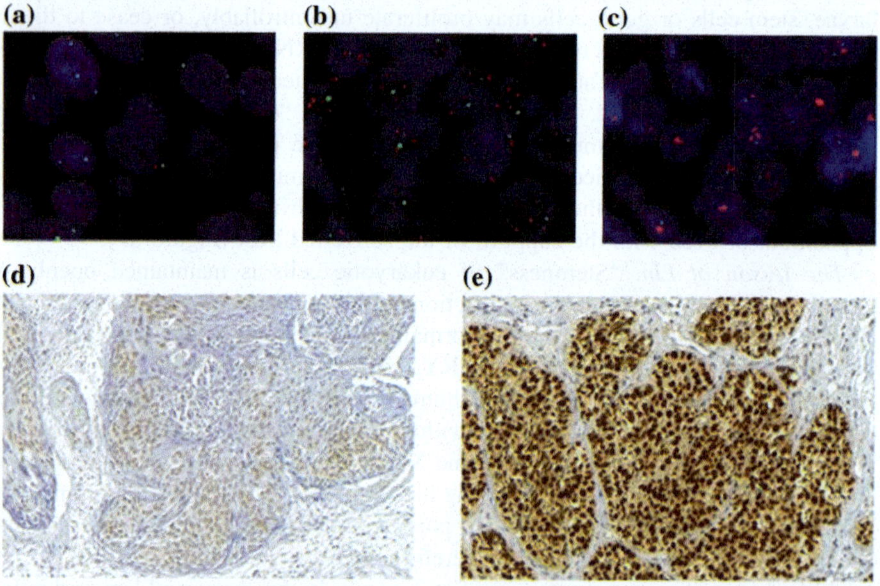

Figure 41 Human Sox Oncoprotein II. [2349] Reference Appendix 2, Explanations to the Figures

cells. Activating these genes in differentiated somatic cells would result in the re-expression of the faculties of pluripotential stemness (referred to as "de-differentiation"). Inducing pluripotency (induced pluripotent stem cells, iPS cells) by reprogramming differentiated cells through retroviral transduction of stem cell genes (Oct4/Klf4), or applying their protein products, a technology by now widely practiced in the laboratory. Only genuine stem cells can both self-renew and initiate somatic cells for differentiation, gene silencing and specified service. However, these cells preserve their stem cell faculties, which can be reactivated with the administration of appropriate growth factors. Constitutive re-expression *in vivo* of ancient "cell survival pathways", or generation of new cell survival pathways mediated by newly generated fused proteins (oncoproteins) encoded by fused genes (oncogenes) transforms the stem cell into an immortalized entity (recognized at the clinics as the "cancer cell").

Non-coding ncRNAs either lack ORF, or are functional as ribosomal, transfer, or small nucleolar r/t/snoRNAs. The small interfering, and small temporal si/stRNAs emerged in the archaea/prokaryota world and populated the vast background of the universal genomics of the entire living world. Functional ncRNAs, like mRNAs, are polyadenylated, spliced and transcribed by RNA polymerase II; are homologous with expressed sequence tags (EST) and display CpG islands. Complimentary cDNAs containing sense-antisense pairing, arise by internal oligo-dT priming. The chimeric cDNA libraries constructed from all the microRNAs ("cDNA library generation from ribonucleoprotein particles") suggests that microRNAs may transform into cDNAs *in vivo*. At the beginning (Introduction) and at the end, the postscript Table XXXV serves as an introduction to a rapidly expanding field of circular RNAs, newly explored (*vide infra*).

LIN28 expression is regulated (inversely correlated with) by microRNAs let-7, miR-9, miR-30 and miR-125. LIN-28 expressor tumor cells suppress these microRNAs. LIN28 can bind to the miRNA let-7 and induce its decay. Of over 1000 miRNAs released from the human genome, some 900 have been identified. miRNAs binding to the LIN28 mRNA 3′-UTR are 12 miR-125, 3 miR-9, and 6 miR-30 family members deriving from distinct chromosomal loci, or clustered together in the same locus. These miR-RNAs are not at all expressed in undifferentiated embryonic stem cells. The expression of these mi-Rs results in a dramatic drop of LIN-28 levels and the cell becomes a differentiated somatic cell. Otherwise LIN28 is able to initiate and maintain (continuously supply new) tumor cells [195a]. Tumor suppressor genes *let-7* and miR-125 are either intact, but epigenetically silenced, or mutated and thus disabled, or neutralized by destroying their gene product protein through the process of ubiquitination. The early discoveries concerning microRNAs and oncogenesis were summarized as presented at the June 10-15 2009 Keystone, Colorado, USA conference (summarized by SE Godshalk *et al* in Epigenetics Feb 16, 2010 [195b].

Nematode viruses. It was not easy to recognize viral pathogens of *C. elegans* (Figure 39). The laboratory strains of *C. elegans* probably were virus-free (if not, some conclusions of basic research may have to be revised). The positive RNA strand Nodaviruses infect the nematodes. The viral genome encodes its own RdRp.

PCR yields overlapping amplicons of two contigs: RNA1 (of 3,628 nt, encoding the enzyme), and RNA 2, (2,653 nt, encoding the capsid proteins). Two distinct strains, the Orsay virus and the Santeuil virus, induce natural and experimental infections. Naturally infected strains of *C. elegans* were derived from a rotting apple and from a snail on rotting grapes. Viral genomes are not stably integrated into host cell genomes, thus there is no vertical transmission. Nematode-infecting nodaviruses express the unique ORF σ (sigma). There appears to be no memory cells for anti-viral immunity: there are takes of re-infections of worms that were rendered virus-free. Antiviral immune reactions are carried out by the RNAi systems. RNAi defective worms remain virus carriers. Viral infections may kill the worms. The most involved site of the infection is the intestinal tract [196]. Three Dicer-related helicases initiate RNAi/siRNA-specific antiviral reactions (by destroying viral mRNAs) in *C. elegans*. In virally infected mammalian cells, RIG-like helicase (retinoic acid-inducible gene) triggers interferon (IFN, *vide infra*) production. The Dicer-related helicase (*drh*) genes are closely related to the RIG-like helicase receptors, inducers of innate antiviral immune reactions, but in *C. elegans* cells they do not induce IFN production (no IFN genes exist there as yet). *C. elegans* artificially deprived of its RNAi defenses accepted infection with replicons of the flock house insect viruses, which integrated themselves into the host's chromosomes. However, the defensive *drh-1* and RNAi system was "negatively regulated" by the wild type *drh-2* genes. Both *drh1* and *drh2* were dispensable for the biogenesis or mRNAs, siRNAs and piRNAs. For example, piRNA levels accumulated to wild type levels in *drh-1/drh-2* mutants. Thirty-five RNAi factors participate in the antiviral defense of this nematode [197].

Health care of the aging Caenorhabditis. Physiological longevity determinants operate in *C. elegans*. The host cell factor-1 (HCF) acts as a suppressor of DAF-16 (dauer formation, constitutive; drosophila decay accelerating factor) (Figure 39), as HCF neutralizes DAF upon encounter within the nucleus. DAF16/FOXO (forkhead box) transcription factors extend life span by improving stress resistance. HCF-1 and *sir* gene product protein SIR (silent information regulator), a histone deacetylator, adhere to each other and compete in order to down- or up-regulate DAF-16/FOXO. Insulin-like growth factor-1 (IGF) mediates the phosphorylation of FOXO, thus retaining it in the cytoplasm. In the absence of IGF-1 action, the dephosphorylated FOXO translocates into the nucleus. The physical association of HCF-1 with DAF-16/FOXO and SIR-2.1 is conserved and acts in the mammalian genome. There, SIR becomes SIRT (sirtuin, silent mating type information regulation 2 homolog), and another factors referred to as protein 14-3-3 and FOXO co-activator PGC-1α (peroxisome proliferator-activated receptor gamma, PPARγ co-activator (Table II) complicate the basic interactions. SIRT regulates FOXO by deacetylation (of its histone), and thus activates it; it is not proven, but suspected that SIRT regulates HCF as well by deacetylation [198]. In general, deacetylation of histones aimed at the promoters of proto-oncogenes activates them, and reacetylation deactivates them. As to the PPARs peroxisome proliferators, PPARδ may act as an inflammatory carcinogen (in colon carcinoma), whereas PPARγ may be a promoter of cell differentiation; reviewed in [73] (Table II) [199, 200]. The very well referenced *daf*-2 gene (decay accelerating factor, drosophila) is conserved up to vertebrate mammals (mouse; *Homo*). In caenorhabditis it is connected with sugar metabolism and encodes insulin-like receptors. Loss-of-function mutation of this gene retards the aging process and prolongs the life span of the elegant worm double-fold. Further prolongation of the life span is added if the *pept*-1 gene (intestinal di- and tripeptide transporter) is also silenced. The active DAF-9/DAF-12/DAF-16 prime pathway increases the expression of the *daf2/pept*-1 genes; in the opposite, loss of DAF-9/DAF-12 silences *daf2/pept*-1 genes. The signal for longevity is a rising glutathione level [201]. The human DAF (cluster of differentiation CD59) is expressed as an anchored glycosyl-phosphatidylinositol (GPI) molecule. The questions are if tumor cells eliminate DAF for the prolongation of their life span, and if GPI is related to PI3K, the PTEN-suppressed cell survival pathway?

The Van Gogh (VANG) caenorhabditis transmembrane protein shows sequence similarities with the drosophila (and mammalian) protein VANG. These proteins mediate correct intercalation of primordial cells during embryogenesis. The caenorhabditis homolog of the mammalian insulin receptor gene is *daf*-2. DAF-2 phosphorylates kinases that can negatively regulate DAF-16FOXO. However, dephosphorylated DAF-16 translocates into the nucleus. Its target genes activated, heat shock transcription factor HSF-1 among them, ameliorate reactions to stress; detoxify proteins, and increase life span. The *C. elegans* VANG-1 protein as a "gerontogene", increases stress resistance, slows metabolism, reduces lipofuscin storage, and thus extends life and reproductive span [202].

Drosophila's longevity assurance gene. In the central nervous system of drosophila larvae, neuroblasts express the neural precursor gene *asense* (*ase*). These stem cells divide asymmetrically to self-renew, and to produce the small daughter cells, the ganglion mother cells (GMC). The terminally dividing GMC progeny is either two neuron or glial cells. Tumor suppressor genes *brat* (brain tumor) and *numb* (negative regulator of Notch) and *prospero* control the physiological pathway. Numb induces endocytosis of Notch receptors, thus deprives the cell of Notch signaling. The reputable proto-oncogene product protein Notch would de-differentiate the cell (*vide infra*). Prospero translocating from cytoplasm to nucleus suppresses cell cycling genes and activates cell differentiation inducer genes. Brat applies brake to cell divisions. Knock-out of *brat, numb* and *prospero* results in an increasing population of undifferentiated neuroblasts driven by unopposed Notch proteins. Further, suppression of *ase* in these cells fixes the undifferentiated stage of these cells. To this effect, the activation of the transcription factor tamtrack (Ttk) may contribute. However, it is Notch, which encodes a transmembrane receptor, the cytoplasmic domain of which is cleaved off to migrate into the nucleus. This pathway is well preserved in the drosophila, in the caenorhabditis, and in the mammalian (including human) hosts. In the nucleus, the intracellular Notch domain binds CSL (consisting of CCAAT-binding factor; Su (H), suppressor of hairless; and Lag-1, longevity assurance gene). The equivalents of drosophila CSL are Lag-1 in *C. elegans* and CBF-1 (C-promoter binding factor) in *Homo sapiens*. CSL forms a tertiary complex with ICN (ictacalcin) and Mam (mastermind). This complex activates Notch target genes *hairy* and *enhancer of split* (HES). The Su(H)-binding drosophila gene product protein Hairless is an antagonist of Notch signaling. It works in unison with CtBP (C-terminal binding protein) and *groucho* (Gro). Proteins Hairless (to escape inactivation) and Notch (to keep it active) compete for Su(H) (the suppressor of hairless) [203, 204]. If Notch prevails, rapidly replicating neuroblasts will be "immortalized" (see above: the "longevity assurance gene"). If Notch is constitutively activated, the process will be irreversible (Figure 42).

Comment. While the Argonaute-Piwi-Dicer system defended the primordial genomes from intruders, it was pierced and penetrated by a multitude of horizontally intruding genes and operons. It has become the regulator of meiosis (provider of germ cells) and generator of stem cells. It has installed itself firmly into animal cells as a powerful anti-apoptotic protector of cellular life by suppressing pro-apoptotic p53, and inducing STAT3/BCL-$_{XL}$ anti-apoptotic pathways (Lee et al/Lu et al) [136, 137]. When the system reverts to its ancient faculties in the multicellular host,

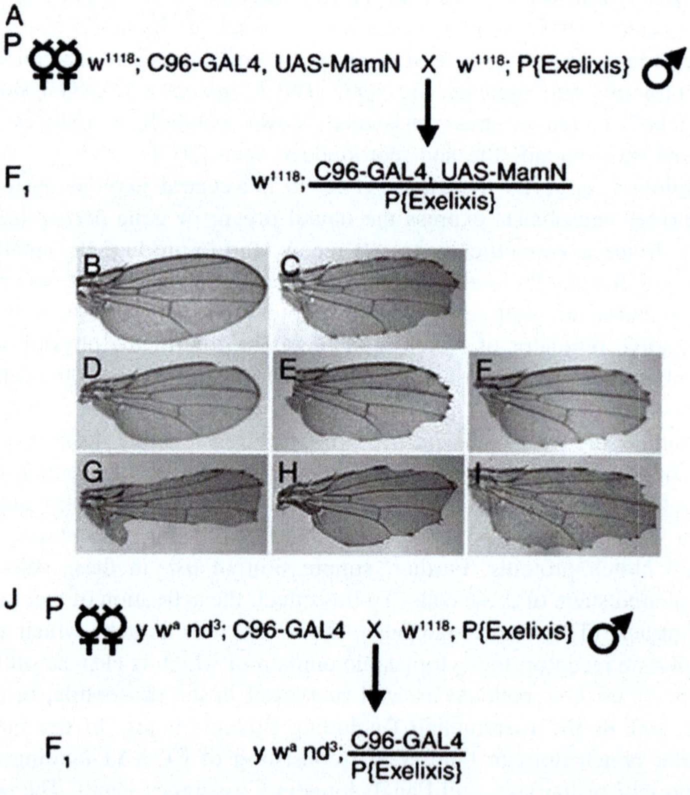

Figure 42 A Mutated Gene in the Drosophila's Wing Anlage Becomes a Mammalian Vertebrate Oncogene: Notch. [2350] Reference Appendix 2, Explanations to the Figures

the afflicted cells will proliferate, while escaping senescence and natural death, like their primordial ancestors did. At the present time, genomic studies are inadequate for the documentation of any sequence-congruity between oncogenic microRNAs of multicellular hosts (those microRNAs that eliminate the mRNAs of the tumor suppressor genes) and microRNAs operational in unicellular eukaryotic hosts. Such a congruence, if documented, would argue for the regression of the oncogenome to a much lower evolutionary level, i.e. to the level on which the primordial RNA/DNA complex operated in the independent and immortal ancient unicellular life forms. This faculty of the DNA might have been conserved in the stem cells and is re-expressed in the malignant transformation of the stem cells, or the somatic cells, of the multicellular host contributing to apoptosis-resistance and metastasis formation [205–207a, b]. Would malignant transformation of those cellular entities (NK cells; macrophages, dendritic cells, *vide infra*), which preserved most of their faculties from the era of innate immunity be different from that of those cellular entities (immune T cells; antibody producing immune B cells), which have arisen in the era of adaptive immunity?

The class III histone deacetylases SIRs/sirtuins (silent mating type information regulator in the yeast) emerge as antagonists of senescence. Sirtuins deacetylate histone H4K16 and some non-histone proteins (NFκB; PPARγ, acetyl coenzyme A, and α-tubulins). The vertebrate mammalian cells operate 7 sirtuins. Sir1 is a regulator of glucose homeostasis. Glucose homeostasis has a major promotional effect on the aging process and oncogenesis. Sir1 is a suppressor of IGF-R and its ligand (*vide infra*). Its anti-inflammatory effect is due to its down-regulation of NFκB. Over-expression of IGF-R and its ligand, and NFκB are both known to be inducers or promoters of carcinogenesis. The mitosis regulator, Sir2 acts in dependence of nicotinamide adenine dinucleotide. Sirtuins 3, 4, 5 are of ancestral mitochondrial derivation; Sir4 is a mono-ADP-ribosyltransferase. Sir6 is an auto-ADP-ribosyltransferase and modulator of the telomeric chromatin. Sir7 is an activator of RNA polymerase I (Pol I) and a protector against genotoxic stress. Sirtuins are protectors of the multicellular cell community against the emergence of constitutively replicating individual cells (oncogenesis), or are promoters of the survival of individual cells under adverse conditions (see Subject Index).

DCAMKL (Double-Cortin and Calmodulin Kinase-like)

Nanoparticle encapsulated siRNA. The tunicate urochordate, *Ciona intestinalis* (the sea squirt), is well endowed with Dicer, Drosha, Argonaute and two Piwi homologue proteins and over 80 miRs [208]. Operon gene clusters are composed of up to six genes sharing a common promoter. The promoter issues a polycistronic mRNA, that trans-splices into several monocistronic mRNAs. The *C. intestinalis* organizes its 2909 genes into 1310 operons [209]. The interaction of ciona miR-124 with Notch is essential for the development of its nervous system. Notch silences miR124 in the epidermal midline cells, whereas in the peripheral nervous system midline epidermal cells' miR-124 silences Notch. Notch and miR-124 antagonize each other: Notch silences miR-124 in epidermal midline cells and miR124 silences Notch in epidermal peripheral nerve cells [210]. In multicellular hosts, Notch-mediated oncogenesis involves microRNAs. In the human host, gastrointestinal tumor (adenocarcinomas) suppressor microRNAs are let-7a, miR-200 and miR144 targeting oncogene-derived mRNAS: c-Myc, K-Ras (Kirsten's rat sarcoma), ZEB2 (zinc finger E-box transcription factor), and Notch-1. The DCAMKL-1 (doublecortin, CaM: calmodulin, kinase-like) intestinal stem cell (ISC) marker is upregulated in many tumors (pancreas; colon), but its mRNA is attacked and annihilated by nanoparticle encapsulated siRNA-DCAMKL (with or without the gamma secretase inhibitor DAPT, difluorophenacetyl-alanyl-phenylglycine t-butyl ester). DCAMKL-1 knockdown and DAPT-induced Notch inhibition, and blocked tumor (colon adenocarcinoma xenografts) progression. These treatments resulted in significant rises of both miR let-7a and miR-144 levels. DCAMKL-1 negatively regulates (causes a decrease of) miR-144 levels, and DCAMKL knockout by siDCAMKL results in a rise of miR-144 and a drop of DCAMKL expression.

Notch. (Figure 42: drosophila wing with "notches").The Notch mRNA at base pair 189 of its 3′UTR has a binding site for miR-144. The epithelial-to-mesenchymal transition of colon carcinoma cells is preceded by a drop of miR-200a, miR-200c, miR-141 and mi-429 levels. However, DCAMKL-1 knockout (without DAPT treatment) results in a rise of miR-200a and in the cancelation of ETM transition. Oncoproteins Snail and Slug actively participate in the process of ETM (reviewed in [26, 27]). Treatment with siDCAMKL-1, DAPT, or their combination, did downregulate Snail and Slug expression [211]. In pancreatic adenocarcinoma induction, K-ras mutation (codon 12 encodes aspartic acid instead glycine: Kras$^{(G12D)}$) and p16 Ink4a/p14 arf (inhibitor cdk4a; alternating reading frame) deletion occur. Tumor cells, especially those that undergo ETM, up-regulate the Notch receptors and their ligands (jagged, delta-like); cytoplasmic NFκB translocates into the tumor cell nuclei. NFκB target genes are c-myc, mmp-9/cox -2, survivin, bcl-2, and cyclin D1. The cell cycle is uninhibited and is refractory to apoptosis. The TaqMan miRNA assay detected that with activated Notch the microRNA-200 family (miR-200abc) receded. With elevated miR-200b, the Notch pathway declined. Notch mRNA-cognate siRNA reversed the malignant course of murine model pancreatic carcinoma [212a, b].

Malignant glioma cells conceal the expression of genes neurabin-II and doublecortin (DCX). Lentiviral vector-mediated transfection of malignant glioma cells with neurabin II and DCX genes induced microtubule catastrophe and endomitosis due to stem cell differentiation in the glioma cells. In these cells, c-*jun* NH2-terminal kinase (JNK1) was phosphorylated and procaspase 3 was induced. Further treatment with simvastatin activated procaspase 3 to cause apoptosis of glioma cells [213].

GILT/IFNs

Interferon gamma predated IFN-alfa. In the amphioxus (*Branchiostoma floridae; B. tsingtauense/belcheri; B. japonicum*) there is a great deal of GILT activity (gamma-interferon-inducible thiol reductase).The amphioxus and the sea urchin possess and express GILT. Later in evolution, GILT will be constitutively expressed in antigen-processing cells (dendritic cells); it will be induced by interferon-gamma (*vide infra*), and regulated by STAT1 (signal transducer and activator of transcription). The STATs will have the reputation of being "cell survival pathways" constitutively expressed in cancer cells. The amphioxus uses the system in its daily life to activate its innate immune system, which is devoid of all interferons. For example, the mammalian IFNγ is a very weak activator of the amphioxus GILT, even though the amphioxus *gilt* gene's seven exons and six introns are very similar to those of the vertebrate *gilt* gene [214] (Table V). In vertebrate genomes, the GILT promoter is at 150 and 510 bp upstream of exon 1 of the *gilt* gene. The *gilt* promoter gene contains the IFNγ activated sequence elements [215]. Malignant melanoma cells switch off Gilt-mediated antigen procession in the class II MHC molecules (to render the malignant cell unrecognizable to immune T cells). Interferon-gamma can to a certain extent restore GILT-mediated antigen processing. Even STAT1, the negative GILT regulator ("STAT1 negatively regulates constitutive IFNγ-inducible GILT expression") [215], when reconstituted, it supported INFγ-inducible *gilt* gene expression in melanoma cells [216]. Devoid of interferons, the amphioxus expresses those genes that later will evolve to encode interferon-inducible proteins. The gamma interferon-inducible lysosomal thiol (GILT) reductase catalyzes thiol-bond reduction in those later cells that process

Table V AmphiGILT

Gamma-Interferon-inducible (IFNγ) lysosomal thiol-reductase (GILT) appears in the amphioxus (*Branchiostoma belcheri/floridae*) and in the purple sea urchin (*Strongylocentrotus purpuratus*). AmphiGILT gene has 7 exons and 6 introns similarly to the vertebrate GILT. Mammalian IFNγ does not induce expression of AmphiGILT.[1]

PSMB (proteasome subunit β-type) proteins appear first in the archaea (Haloferax; Pyrococcus) (University of Florida, Gainesville, FL)[2,3], continue to be present in amphioxus and agnatha[4], and become later in gnathostomata IFNγ-inducible PSMB proteins (example: zebrafish, *Danio rerio*).[5]

In the adaptive immune reaction, from gnathostomata upwards, PSMB immunoproteasomes produce the antigenic peptides that are fitted in the endoplasmic reticulum into the grooves of the peptide-binding domain MHC molecules. MHC molecules evolved by gene duplications, recombinations of pre-existing DNA segments and gene (exon) shuffling. From the agnatha downwards, there are no PSMB immunoproteasomes and MHC class molecules.[6,7] NK cell receptors co-evolved with MHC ligands.[8]

1 Liu N et al Mol Immunol 2007;44:2631-7. 2 Gil MA et al Microbiology 2007;153:3009-22. 3 Zhou G et al J Bacteriol 2008;190:8096-105. 4 Takezaki N et al Gene 2002; 282:179-87. 5 Murray BW et al J Immunol 1999;163:2657-66. 6 Klein J & Nikolaidis N Proc Natl Acad Sci USA 2005;102:169-74. 7 Zhao M et al BMC Evol Biol 2013;13:113. 8 Parham P & Moffett A Nat Rev Immunol 2013;13:33-44

antigens. STAT1 often expressed in cancer cells, regulates the expression of GILT reductase. Later, in antigen-processing cells (dendritic cells), the expression of GILT reductase becomes constitutive. STAT1 acting upon the GILT promoter can activate this enzyme even the absence of IFNγ. The sea urchin possesses the GILT system without MHC class II antigen expression. However, in the xenopus (*X. laevis*), GILT is an IFNγ-inducible antigen-processing molecule [217]. The IFNγ-inducible proteasome genes in vertebrates take their origin in the IFN-free amphioxus with an intermediate grade in the agnatha (Takezaki et al cited in Table V). The myeloid differentiation factor MyD88-dependent pathway of TLR activation also originated in the amphioxus' TIR adaptor (Toll/IL-1R), the TICAM (TIR-containing adaptor molecule-1). It activates NFκB in a MyD88-independent manner [218]. The amphioxus expanded its microRNA repertoire, as compared to protostomes and the tunicate ascidians, exemplified by the solitary urochordate, *Ciona* [219, 220].

In the amphioxus, poly(I:C) and some viruses induce the expression of those genes (guanine triphosphatase-, GTPase-induced genes, especially Mx, the myxovirus resistance gene, and the guanylate-binding protein, GBP gene), while ciona, sea urchin, caenorhabditis and drosophila do not express these genes anymore. From fish (zebrafish) through amphibians (frog), birds and mammals, the IFN-inducible Mx genes are well expressed. The marsupial opossum lost this gene from its genome (yet it survives through some other antiviral measures). Some GBP genes are induced by LPS, IL-1ß, or TNFα. Immunity-related GTPase (IRG) proteins exist in vertebrates, but also in amphioxus. The very large GTPase-inducible proteins (VLIG) appeared only in vertebrates (could not be induced in amphioxus, ciona, caenorhabditis and drosophila). These are the genes that later in evolution will be expressed by cells that are able to produce IFNs as well. However, the amphioxus is devoid of IFNs and IFN-R genes [221]. The amphioxus possesses the Toll/IL-1 receptor (TIR) adaptor, the precursor of the vertebrate TICAM 1,2 (TIR-containing adaptor molecule-1,2). The amphioxus' TIR adaptor activates NFκB independently from MyD88 [218]. Involved recently in studies of genomics, *Platynereis dumerilii* so far revealed a highly disciplined genome. Its immune system practices innate faculties and its stem cell so far refrained from uncontrolled proliferative activities. What are its genes "in waiting" for its phylogenetic descendants? (Figure 43).

Type I IFNαß. In the multicellular vertebrate hosts, the differentiated somatic cells' main antiviral defense is by type I interferon (IFNαß) production. It was not possible to demonstrate IFNαß production beneath the vertebrates. Some cartilagenous fish (the elephant shark) and certainly bony teleost fish (zebrafish, pufferfish) are able to produce their own type I IFNs and their receptors. The first IFNs appear to have originated in ancestral sharks from cytokine IL-10 genes consisting of 5 exons and 4 introns, and eventually binding to three different types of receptors (IFN-αR1/2 for type I; IFN-γR1/2 for type II; and IFN-λR and IL10R2 shared for type III IFNs) [219–221].

Fish type I IFN genes contain both exons and introns (as the mammalian IFNλ). Fish IFNγ shows high similarity in its exon-intron structure to its mammalian counterpart [222]. In the fish, the IFN-induced Mx (myxovirus resistant protein)

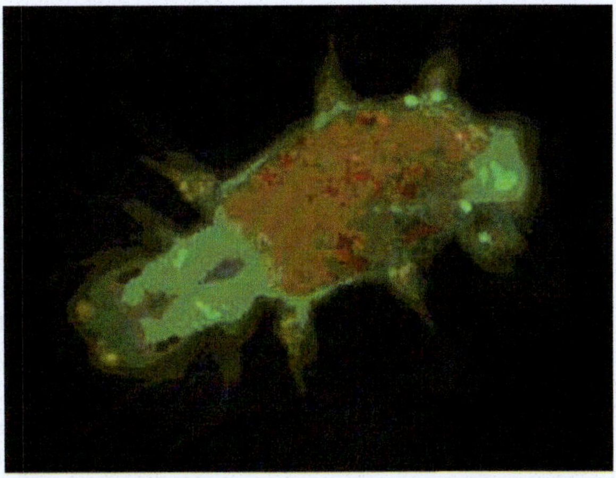

Figure 43 No Uncontrollably Replicating Stem Cells Emerged (so far) in the Marine Annelid. With kind permission © Nicole Rebscher. All Rights Reserved 2015. Reference Appendix 2, Explanations to the Figures

gene encodes the cellular antiviral state, in which the replication of the dsRNA hemorrhagic septicemia rhabdovirus is inhibited. The zebrafish IFNs have 34–39 % congruity with avian and mammalian IFNs, except for the number and location of the cystein residues: two in fish, four in avian-mammalian IFNs. The introductory sentence concerning zebrafish interferons reads: "Interferons (IFN) play a major role in orchestrating the innate immune response toward viruses in vertebrates" [223]. However, IFNs (Type αβ) were acquired in the adaptive immune system by vertebrates, and are not known to have existed below (but IFNγ appeared first). IFNs may act in connecting the adaptive immune system with its predecessor, the innate immune system. However, type I IFNs act through the STAT (signal transducers and activators of transcription) pathway, IRFs (IFN regulatory factor), and ISGF (IFN-stimulated gene factors). The zebrafish type I IFNs signal through receptors CRFB1-5 (cytokine receptor family B). The three-dimensional structures of zebrafish IFNs indicate that they were predecessors of the mammalian type I IFNαß (not the mammalian type III IFNλ of IL-10/IL-22-derivation) [223]. The amphibian (frog) type I IFN genes preserved their original (fish) structure of five exons, four introns. From the reptilian type I IFN genes upwards (avian, monotreme, marsupial, eutherian/mammalian) type I IFNs derive from intron-less genes. The Xenopus genome encodes type I and type III IFNs. The xenopus type III IFN shows 20–29.8 % amino acid (aa) identity with human and avian type III IFNs, respectively. The xenopus IFNλ shows 14–20 % amino acid identity with its own type I IFNs. The xenopus IFNλ attaches to receptors IFN-λR1 and IL10-R2. The loci of IFN-λRs and the IL-22R2 gene remained side-by-side in the xenopus, aves and human genomes. In both the xenopus and in the human genome the IL-10R2 is constitutively expressed. IFNs type I and type III both deriving from the original introns-containing genes co-exist in the amphibian xenopus. In both the xenopus and mammals, the IFNλ genes are activated by poly(I:C) treatment, or by viruses (but INFαß genes and IFNλ genes may respond to different viruses) [224, 225].

The divergences of IFNß occurred at the level of monotremes (the platypus, the egg-laying mammal), probably 130 million years ago (*vide infra*).

Intron loss. The retroposition event leading to the loss of introns from the mammalian IFN gene occurred before the emergence (divergence) of the bird line, probably 250 million years ago, because avian IFN genes are also intron-less. In the platypus, the type I IFN isoforms are considered to represent the alpha/beta prototypes. However, at the level of the platypus, there are NK cells, MHC antigens, T cells with receptors (TCR) undergoing the RAG1,2- and RSS-initiated V(D)J recombinations (*vide infra*). Of the granule proteases (neutrophil cathepsins, T cell granzymes, mast cell chymases), the platypus has them, but expresses only one chymase, while the opossum has two. The first chymase gene evolved over 200 million years ago [226]. From yeast to mammals, SIN1 (stress-activated protein kinase SAPK-interacting protein) arises in a very ancient pathway, predating IFN production. However, prokaryota, protozoa and plantae express no SIN; fungi and metazoa express it; in vertebrates SIN1 represent a highly conserved protein with 79–99 % aa sequence homology. In insects, the *sin* gene is intronless, thus it might have originated by retroposon insertion, such as reverse transcription of a mRNA. In vertebrates SIN1 possesses two sequences for interaction with type I IFN receptors, thus linking the SAPK and the IFN pathways [227]. Thereafter evolved the IFN subclasses: IFNδ, -ε, -κ, -τ, ω) as placentation in eutherian mammals replaced egg-laying (the trophoblast IFNτ in ruminants, *vide infra*).

Type I IFNs activate JAK/STAT: a dangerous liaison. Physiologically, type I IFNs activate the JAK/STAT3 complex; STAT3 is phosphorylated and translocates into the nucleus, but the *socs* (suppressor of cytokines) genes may, or may not respond. Histone deacetylases promote STAT3 upregulation. The histone deacetylase inhibitor, trichostatin A (TSA) suppresses JAK2/STAT3 signaling by re-acetylating the *socs* promoter histones. Upon treatment with TSA, colorectal adenocarcinoma cells undergo G1 arrest and apoptosis, due to down-regulation of *bcl-2*, survivin and *ink4a* p16 genes [228, 229]. When the ancient cooperation of stem cell genes switches to their constitutive (irreversible, persistent) expression, small molecular inhibitors can suppress the expression of the transformed gene-product proteins (recognized as oncoproteins). In the structure of the human IFNα, the four cysteins form two disulfide bonds to stabilize the molecule. The major producer of human IFNα is the plasmacytoid dendritic cell. Amphibians (the frog) are well endowed with type I IFNs and their receptors. Birds are good producers of type I IFNs, and their receptors [222, 224].

In sheep, in her pregnant uterus, IFN-tau (τ) of trophoblast-derivation induces temporarily a proto-oncogene cascade consisting of JAK/STAT, EGFR-, SRC-, RAS-, RAF-kinases, and ERK, taking place in the endometrial luminal epithelial cells (janus activated kinase; signal transducer and activator transcription; epidermal growth factor receptor; Rous sarcoma oncogene; rat sarcoma; rat fibrosarcoma; extracellular regulated kinase) [230]. This is a physiological event; the proto-oncogenes are not constitutively activated and are not suffering any mutations. The proto-oncogenes are activated in a prostaglandin-producer, differentiated somatic cell (not in a stem cell). However, in human breast and prostate adenocarcinomas, the tumor initiator cell population derives from luminal epithelial stem cells [231–233]. Type I IFNs are known to temporarily activate JAK/STAT in other systems (*vide infra*). Chinese cardiologists improve the healing of myocardial infarcts by activating temporarily JAK/STAT (sometimes with resuvastatin, or curcumin) [234, 235]. The physiological deactivator of the system is by the SOCS proteins (suppressor of cytokine signaling). In malignant tumors the *socs* gene promoters are often silenced by multiple methylations (for example in squamous cell carcinoma of the H&N, with or without HPV-association [236]). In another case of inflammatory carcinogenesis, gastric adenocarcinoma, *socs-1* genes were silenced by hypermethylation; in contrast, phosphorylated JAK/STAT3/MAPK, and IL-6 were overactive. Infection of cancer cells in xenografts with adenoviral vector of *socs-1* gene (AdSOCS-1) suppressed JAK/STAT3/MAPK and exerted antiproliferative effect [237]. In the pregnant ovine uterus, cessation of IFNτ production switches off the proto-oncogene cascade, while prostaglandin release is regulated, and luteolysis is accomplished [230].

Physiologically activated oncoproteins were inhibited: JAK/STAT by the tryphostin AG-490, RAS by the farnesyltransferase inhibitor manumycin from *Streptomyces parvulus*, RAF by RAF kinase inhibitor-1 (from EMD BioSci, San Diego, CA); SRC kinase by PP2 (pyrazolo-pyrimidine-derivative) [230]. ERK or MEK (MAPK and ERK) inhibition is usually achieved by upstream inactivation of BRAF, but it can be accomplished by direct attack on their mRNA by shRNA [238]. In colon adenocarcinoma cells driven by JAK/STAT, RAS and SRC oncoproteins, JAK was deactivated by AG-490 and STAT3 was depleted by siRNA. Cancer cells were arrested in G1 and underwent apoptosis. However, proteasome inhibitor MG132 antagonized AG-490 and protected JAK/STAT activity [239]. The *in vitro* growth of a pancreatic adenocarcinoma cell line driven by constitutively activated STAT3 and VEGF, MMP-2 (vascular endothelial growth factor; matrix metalloproteinase) and IL-6 was retarded by AG-490 [240]. The ancient tyrosine-phosphorylated and intranuclearly translocated STAT functions in the slime mold (*D. discoideum*) as a repressor of differentiation of stalk cells [241]. For JAK inhibitor ruxolitinib in myelodysplasia, for BRAF inhibitor vemurafenib in melanoma (*vide infra*).

JAK mutation in the drosophila. The human genome encodes four *jak* genes and a seven members of the STAT protein family. In contrast, flies (drosophila) have one *jak* gene (*hopscotch*) and operate one STAT protein (STAT92E). The physiological JAK/STAT signaling pathway is activated by cytokines (IFN type I) [242]. Janus-associated kinase mutation occurs in the drosophila with activation of STAT. The mutation involves nucleotide 617, that encodes in the gene product protein instead of valine, phenylalanine. In the haemocytes the mutated and phosphorylated JAK protein enters the nucleus and activates the genomics of the STAT pathway. The *stat* gene product proteins dimerize and activate their inflammatory target genes dictating serial rapid cell divisions [243]. The drosophila JAK homolog is the Hopscotch tyrosine kinase protein existing in a single copy, but associated with the drosophila homolog of STAT. The JAK/STAT cascade induces massive tissue overgrowth in the drosophila. The Wg protein overexpression results in the formation of multiple duplications of the wing discs. The JNK pathway induces apoptosis of the disc cells. The transforming growth factor-ß (TGFß) equivalent decapentaplegic (DPP) causes wing overgrowth and the formation of ectopic extra winglets. Both the DPP and the c-Myc are probable JAK/STAT target genes; their mRNAs are overexpressed [244]. Drosophila SOCS gene-product proteins (suppressor of cytokine signaling) are operational [245, 246], but it was not examined if the *socs* gene was silenced and if so, by what mechanism [239]. Further, unpaired (Upd), the d-JAK Hopscotch ligand's role, in JAK/STAT activation, if any, remains obscure. Indeed, in the drosophila ovary, up-activated *d-jak*/JAK induces proliferation and invasion of the formerly stationary epithelial cells [247, 248].

The human jak/JAK. The human JAK/STAT responds to IL-6 type cytokines; oncostatin M specifically induces IL-6 production in fibroblast-like synoviocytes. Overproduction of inflammatory cytokines (IL-1, TNFα, IL-6-related cytokine oncostatin M (OSM)), aggravate the clinical course of rheumatoid arthritis. The JAK inhibitor tasocitinib (Axon) suppressed the overproduction of IL-6 with clinical benefit to patients suffering from rheumatoid arthritis [249]. IFNβ induces various cytokines, and pro- and anti-inflammatory gene product proteins. In the microglia, NFκB is induced, together with is repressor, the chemokine RANTES (regulated upon activation, normal T cell-expressed and secreted). The JAK/STAT pathway is activated, which induces the release of SOCS proteins (suppression of cytokine signaling). In the pathogenesis of multiple sclerosis, an overproduction of TH-17 T lymphocytes is of a dominant role. The SOCS proteins are inhibitory to Th17 lymphocyte activities. Thus, the end result of IFNβ production or administration is alleviation of the autoimmune reactions in multiple sclerosois [250–253].

The Philadelphia chromosome-negative myeloproliferative neoplasms now include polycythemia vera and essential thrombocythemia. The JAK1/2 non-receptor tyrosine kinase genes undergo gain-of-function constitutive somatic mutations in the human bone marrow. In primary myelodysplasia/myelofibrosis (MF), the $JAK2^{V617F}$ mutation is the most common (in nucleotide 617 valine is replaced by phenylalanine). In the minority of patients with myelodysplasia-MF, the thrombopoietin receptor gene (TPO-R) receives gain-of-function mutation: MPL^{W515L} (ligand (L) to megakaryocyte promoter in TPO-R, in nucleotide 515 tryptophan is replaced by leucine). Both JAK mutations activate STAT, or MAPK, and fibrinogenic and angiogenic factors. In some cases of MF, JAK1/2 function in overactivity without demonstrable mutations. Roxulitinib (Jakafi, Incyte Pharmaceutical, Novartis) small molecular inhibitor (a derivative of cyclopentyl-pyrrolol-dipyrimidin-pyrazole-propan-nitrile) of the mutated JAK1/2 oncoproteins reduces extramedullary hematopoiesis in the spleen. With reduction in numbers (but not complete eradication) of JAK2 V617F mutated cells, blood counts may approach normalization, but bone marrow fibrosis may persist. Thrombocytopenia and/or anemia require supportive measures, dose reductions or treatment discontinuation. Patients remaining on therapy gain symptomatic relief and prolongation of life. On rare occasions, patients on therapy experienced bone marrow transformation into acute myelogenous leukemia [254–257].

The trump card of NK cells: IFNγ. The NK cells of innate derivation produce the type II IFN, IFNγ. GammaIFN production is usually associated with the appearance of T lymphocytes. However, NK cells lack *rag*1/2-induced somatic gene hypermutations, *i.e.* antibody production, or TCR rearrangement. However, the iNKT (invariant) subclass cells express a form of TCR. The subset of CD1d-restricted NKT cells with invariant Vαß TCR, produce IL-4, IL-10, IL-13, IFNγ, and TNFα. Type I human iNKT cells are with Vα24-JαQ TCR. The iNKT cells respond with cytokine release to stimulation with α-GalCer (galactosylceramide [258]. *In vivo*, αGalCer was toxic, but ß-glucosylceramide given intraperitoneally was tolerated (in mice). This treatment increased the cytotoxicity of iNKT cells to liver tumors [259]. Histone deacetylase inhibitors (valproic acid) down-regulated NK cell response to the oncolytic herpesvirus used to treat glioblastoma (in mice). The STAT pathway was inhibited and IFNγ levels dropped, just to recur in a robust rebound 72 hours later [260]. This intervention as described is not advantageous, but indicates strong epigenomic events in the virally infected tumor-bearing host. Patients receiving allogeneic HLA-identical sibling hematopoietic cell grafts may still develop graft-*versus*-host disease, as reviewed in [261a, b], even as high as grades II-IV, depending on the dosage of donor iNKT cells in the graft. The higher was the $CD4^{-}$ iNKT cell dosage, the lower was the incidence and grade of the G*v*H disease [262a]. The lack of signaling lymphocyte-activation molecule-associated protein (SLAM; SAP) in T and in invariant NKT cells causes the X-linked lymphoproliferative agammaglobulinemia; α-galactosylceramide-activated NKT cells may act as noncognate, or cognate helpers (Figure 1 in the article) to Ig-producer B cells. Physiologically, SAP expression is an obligatory helper to antigen-specific B cells [262b].

The trump card of pDCs: IFNα. Plasmacytoid dendritic cells (pDC) are the major producers of type I IFNs. Endogenous overproduction of type I IFNs (as well as type I IFN therapy) may induce a variety of "autoimmune" diseases of systemic (lupus erythematosus, SLE), or organ-specific (rheumatoid arthritis, RA; multiple sclerosis, MS) clinical manifestations. In the systemic form (SLE), pDC-induced Treg cells must also fail. In the organ-specific form, the role of overproduced TNFα surpasses that of type I IFNα. In this case, myeloid DCs activate immune T cells and macrophages for the overproduction of TNFα. In SLE, an active JAK/STAT pathway induces IFNα-regulated, -related and-signatured genes [263]. In Japanese patients with methotrexate-resistant RA, JAK pathway activated cytokines (IFNγ, IL-17, IL-6, IL-8), were elevated. The JAK pathway inhibitor tofacitinib (CP-690,550) in a double-blind clinical trial restored methotrexate sensitivity with clinical and laboratory improvement [264].

Some cytokines. The IL-6 class cytokine oncostatin M (OSM), IL-1 and TNFα of monocyte-macrophage derivation are activators of the *jak* gene and thus the JAK/STAT and/or JAK/MAPK pathways in fibroblast-like synoviocytes of RA. These cells overproduce IL-6 and MMPs, harbor an activated JAK gene, and allow intranuclear translocation of the phosphorylated STAT protein dimers. The specific inhibitor of OSM, CP690,550 may reverse the entire process. The p38 mitogen-activated protein kinase (MAPK) with the member of its group, the c-Jun

N-terminal kinase (JNK, *vide supra*) is another OSM-activated cell survival growth factor complex, also active in RA The MAPK inhibitor SB203580 almost completely suppress IL-6 production [249]. In multiple sclerosis, IFNß is a JAK/STAT activator, yet it is therapeutically effective in inducing objective responses; these are lost in patients who produce IFNß-neutralizing antibodies [250, 265]. In the microglia, IFNß activates JAK and phosphorylation of STAT at tyrosine 701 and serine 727 (Y701; S727). The S727 phosphorylation induces the maximal transcriptional activation of STAT. IFNß phosphorylates also ERK. Elevated OSM levels in MS lead to JAK/STAT, NFκB, MAPK, TNFα and inducible oxide synthase (iNOS) activation in the microglia. IL-12/IL-27 antagonize the OSM-induced reactions [251, 266]. In $CD8^+$ T lymphocytes, IFNß decreased JAK1, and increased MxA (influenza A myxovirus resistance) protein expression [267]. In Sjögren's syndrome, phosphorylated p38/MAPK and NFκB were highly expressed in the epithelial cells next to lymphocytic infiltrates, and activated Fas-mediated apoptosis of the epithelial cells occurred. Inhibition of p38MAPK by SB203580 and NFκB by caffeic acid phenylethyl ester abolished Fas-mediated apoptosis of epithelial cells [268]. It is most remarkable, that in LPS-induced sepsis, activation of p38-MAPK, JNK and NFκB are major contributors to the coagulopathy. Specific inhibitors (SB203580 for p38/MAPK; BAY117082 for NFκB; SP600125 for JNK) reversed LSP-induced coagulopathy [269]. Ligands activating TLR7/9 induce pDCs to produce IFNαβ through the TAK1-IKKβ signaling pathway (TGFβ-activated kinase; phosphorylation of inhibitory IκBα kinase, IKKα/β). Viral ssRNA stimulates TLR7; unmethylated viral DNA segments stimulate TLR9 (TLR3 responds to dsRNA; TLR4 responds to LPS). IFNβ-stimulated JAK/STAT pathway was essential for the production of IFNα. The TLR7/9-stimulated production of type I IFNs by pDCs can be abrogated by IKKβ inhibitors (listed in the article under "Compounds") and some chemical formulas are provided. For example, a TAK1 inhibitor is an oxozeaenol-derivative. These compounds may become therapeuticals for SLE and other autoimmune diseases [270].

The ancient JAK/STAT/MAPK complex serves as a "cell survival pathway", when temporarily activated; the constitutively activated pathway transforms the cell to immortality, with the eventual demise of its host. However, small molecular inhibitors already exist (*vide supra*), that may reverse the oncogenic process: the constitutively activated oncogene or its oncoprotein (including the overproduction of type I IFNs) may be de-activated (*vide infra*).

Elements of Natural (Innate) Immunity Encounter Viruses in Tumor Cells

Oncolytic viruses in IFN-deficient tumor cells. In the Type I IFN-deficient tumor cells' innate immunity fights oncolytic viruses, but vaccination with viral oncolysates induced strong anti-tumor cell immune reactions. Hilary Koprowski presented a lecture at the young M.D. Anderson Hospital in 1957 on viral oncolysate therapy of mouse ascites carcinomas (Bunyamwera virus against Ehrlich and Krebs ascites carcinomas) and proved that mice rejecting a virally infected tumor developed immunity against challenge with the same, but not virus-infected, autologous tumor cells [271]. Jean Lindenmann, the co-discoverer of interferon (Isaacs & Lindenmann) [272], launched a major study on the viral therapy of cancer in mice. Viral oncolysates of Ehrlich ascites carcinoma cells were prepared with an attenuated influenza A virus strain. Tumor cell deaths with tumor rejections occurred in viral oncolysate-treated mice. Viral oncolysates are probably containing fused viral and tumor antigens, which induced the production of tumor cell-lytic antibodies. In the monograph, photographs were shown of tumor cells surrounded by round cells (lymphocytes), but cell-mediated immunity was not recognized. Mice after rejecting their tumors were challenged with live not-virus-infected cells of the original tumor: there were no takes. Postoncolytic anti-tumor immunity was acquired and documented [273].

Independently from each other, but at the same time, and later co-operating, Cassel and Sinkovics followed suit in applying viral oncolysate therapy to patients with cancer. The National Cancer Institute USA of the "Conquest of Cancer" era (under the directorship of Frank "Dick" Rauscher) denied funding for "viral oncolysate therapy of metastatic human cancers (melanoma and sarcoma)". Neither W. Cassel at Emory University, Atlanta Georgia for Newcastle disease viral (NDV) oncolysates for melanoma, nor J. Sinkovics at the University of Texas M.D. Anderson Hospital, Houston, Texas, for PR8 influenza A viral oncolysates for melanoma and sarcoma, received NCI funding. However, the latter application was approved with this note: "approved without funding due to low priority." At both locations, private donations and funds supported the work. Both work stations were able to report tumor regressions, durable complete remissions in patients with metastatic melanoma at Emory University [274], and durable partial remissions and long stabilizations of disease in patients with metastatic sarcoma and melanoma at M.D. Anderson Hospital [71, 275–277]. In this latter clinical trial, antibody- and lymphocyte-mediated immune reactions, as tested *in vitro* to the patients' tumor cells, were documented [147].

Comment. In that era, cure of cancers with combination chemotherapy was the only acceptable successful result, and stabilization of disease was considered a failure due to its relapses, as a rule (*vide infra*). Since then, converting cancers into a chronic ailment to live with has been accepted as a positive result of treatments.

Tumor cell-killer lymphoid cells naturally exists. It was discovered at the M.D. Anderson Hospital laboratory "Section of Human Tumor Virology and Immunology" in the late 1960s (*vide infra*), that healthy people circulated large granular lymphocytes (later recognized to be natural killer cells), which killed non-specifically allogeneic tumor cells, whereas patients, in addition to these cells, mobilized, albeit often in depleted numbers, also small compact lymphocytes (later recognized to be immune T lymphocytes), which killed specifically tumor cells. Some of the photographs showing large granular lymphocytes (later: NK cells) attacking and killing human tumor cells either by cytoplasmic lysis, or by "nuclear clumping" (later: apoptosis) in the late 1960s and early 1970s are of historical value

[30, 147, 278, 279] (Figures 35 and 36). It was observed that the small, compact lymphocytes often disappeared in chemotherapy-treated patients (within a generalized leuko-lymphopenia). In contrast, patients immunized with viral oncolysates appeared to have mobilized higher numbers of the large granular lymphoid cells against their tumors (as visualized on the stained slides of the chambers in which the cultured tumor cells were exposed to the patients' lymphocytes [71, 147] (Figure 36).

Comment. The NCI project site visitors could not accept the existence of lymphoid cells, which were able to kill tumor cell without pre-immunization. The existence of such lymphoid cells would have contradicted one particular project site visitor's printed reports claiming strict acquired specificity for anti-cancer cell reactions of human lymphocytes. Investigators of that particular Seattle laboratory, in their colony inhibitory assay, overlooked antitumor cell cytotoxicity by large granular lymphoid cells of healthy donors. The project side visitors declared, that the chamber-slide assay (developed and in use at M.D. Anderson's Section for its value in offering clearly visualized results) [147] produced *in vitro* "artifacts". Therefore, Sinkovics's project was transferred to the NCI for 'clarification' of these "artifacts". There the misguided views of the project site visitors were overruled. It was the NCI intramural laboratory of Herberman, Djeu, and Kay (H.D. Kay, a former associate of Sinkovics), where the existence of these large granular lymphocytes, now referred to as "NK cells", which could kill tumor cells without pre-immunization, was confirmed [30a, 71, 277–279]. Marshall Lichtman in his chapter in Peter Wiernick's "Neoplastic Diseases of the Blood" gives credit to Sinkovics for being the first investigator, who recognized in the late 1960s NK cells (as "large granular lymphoid cells") in his own blood and photographed them in the act of killing allogeneic human cancer cells [147, 280]. The Herberman laboratory at the NCI characterized murine and human NK cells [280]. It is medical history, that academically trained immunologists upon their first encounter with NK cells refused to accept, that "immune reactions without pre-immunization" could possibly exist, and referred to the first photographs of tumor cell-attacking human NK cells as "in vitro artifacts" (Figure 36).

Viral oncolysates. Postoncolytic anticancer immunity was first reported by Koprowski et al [271]. Lindenmann and associates proved that viral oncolysates immunize mice against Ehrlich ascites carcinoma cells [273]. Independently, but contemporaneously, Cassel and Sinkovics prepared viral oncolysates for immunization of patients with metastatic melanoma and sarcoma. These two clinical trials conducted under the control of academic institutions (Emory University, Atlanta, GA; MD Anderson Hospital, University of Texas College of Medicine, Houston, TX) recorded documented proven remission inductions (partial remissions in Houston; partial and occasional complete remissions in Atlanta) in these clinical entities. However, these were not the so-called prospectively randomized clinical trials. The NIH/NCI highly enriched by President Nixon's funding attracted a new group of mutually self-supporting project site visitors, who declined grant support for both clinical trials [274, 275] (Figure 44). The official reasoning of the NIH/NCI was that tumor-specific immunotherapy is possible only against identified tumor antigens (and those were not yet discovered in the 1960s).

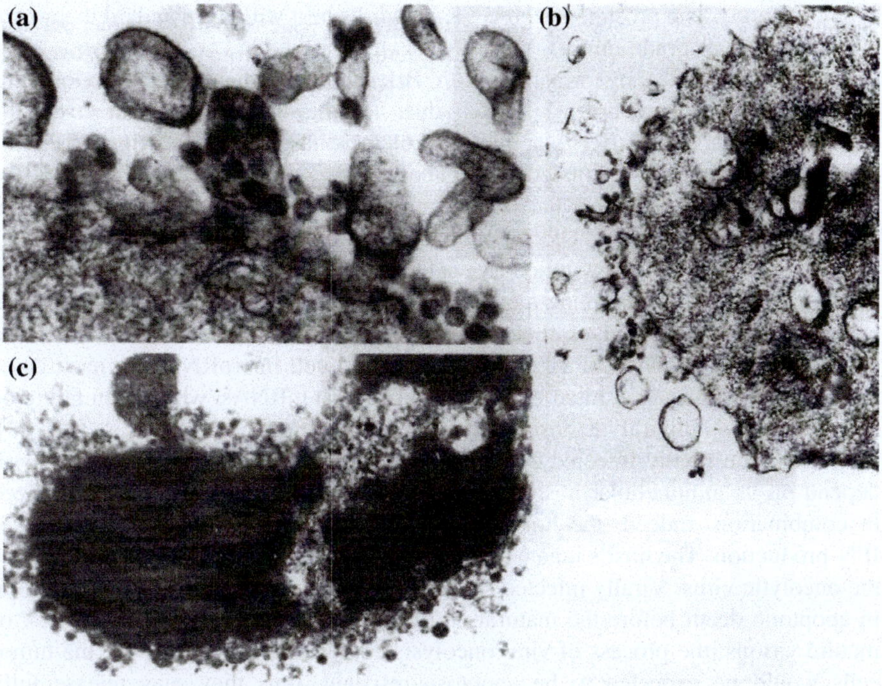

Figure 44 Viral Oncolysate Cancer Vaccines Immunized the Hosts against Fused Viral Envelope and Host Cell Membrane Proteins. Reference Appendix 2, Explanations to the Figures

Later, investigators at Baylor Medical School and M.D. Anderson Hospital determined that viral oncolysates induce antibody production against various tumor (and viral) antigens in human patients with melanoma and ovarian carcinoma cases [281]. Indeed, the immunodominant targeted antigen must be a protein essential for the tumor cell's survival, in order for the immune reaction to kill the cancer cell by knocking out that particular antigen. Reversal of the immunodominant hierarchy, and cross priming of the immune reactions, induced not only by infecting viruses [282a, b], but especially by viral oncolysates should not be left further uninvestigated. If the anti-tumor immunization induces the immunodominant reaction against an insignificant tumor antigen, then the tumor escapes harmless. However, the highly sophisticated genetically engineered, attenuated, or fusogenic oncolytic viruses now in use leave no room currently for further clinical trials with viral oncolysates. Even highly pathogenic viruses (adeno-, herpes-, measles- polio, and pox-viruses) could be attenuated for the purpose viral oncolysis [71, 283]. Of conditionally replicating adenoviruses (CRAd), the E1A-AR fusion gene-carrier CRAd is activated by androgens or non-steroidal anti-androgens. The virus activated by bicalutamide sensitizes tumor cells to intensity-modulated radiotherapy [284]. In most cases, oncolytic viral therapy for metastatic human tumors requires additional measures for success (usually, attenuated dosages of specific forms of chemotherapy). The prostate-specific CRAd works best with the standard therapy of the primary high grade tumor.

Upon malignant transformation with *ras* mutation, the tumor cell loses its capacity to produce type I IFNs, due to the dephosphorylation of the dsRNA-activated PKR (RNA-activated protein kinase) (reviewed in 275). This phenomenon certainly facilitated viral oncolysis.

Instead of type I IFNs, what innate immune faculties can the virally infected transformed cell mobilize? Did its oncogenic genome undergo a regression of its immune faculties on the evolutionary scale downwards to the pre-IFN level? As if the transformed cell's immune faculties were operating at the level of the ancient invertebrate hosts devoid of IFN genomics? A widely available line of defense would be the destruction of viral mRNAs by host cell microRNAs. In most cases the opposite occurs: viral microRNAs neutralize host mRNAs, which then fail to be translated into antiviral defensive proteins [285]; further reviewed in [71, 73]. Otherwise, the virally infected dedifferentiated somatic or stem cell's defense will depend on its multicellular host's innate and adaptive immune faculties mobilized in combination. Indeed, the host's not *ras*-transformed cells remain capable of IFN-production. The host's innate and adaptive immune faculties remain reactive to the oncolytic virus. Virally infected somatic cells may surrender to their destruction in apoptotic death before the maturation of the new viral progeny; thus in lack of mature virions, the process of viral oncolysis comes to a halt. However, the tumor cells would be expected to be apoptosis-resistant, thus they may release fully mature viral progenies before they die. Therapeutically administered oncolytic viruses certainly take use of their own evasive mechanisms and may or may not overcome the cell. The failure of oncolytic viral therapy, if it fails, is due to the host's effective innate and adaptive immune reactions, that eventually neutralize the oncolytic virus; reviewed in [71]. The virally attacked tumor cell can release "danger signals" for the mobilization of antiviral immune reactions of its host. Fusogenic oncolytic attenuated herpesviruses spread from one cell to the other without extracellular exposure to virus-neutralizing host antibodies [286–288]. Viral therapy of human cancers remains an investigational modality (except in Hungary, where an attenuated British strain of Newcastle disease virus was licensed for this purpose, and in China, where genetically engineered adenoviruses are licensed for human therapeutical use) [71].

Comments. In *éclat* contrast to the pre-conceived view of mutual antagonism between chemo- and immunotherapy, in the last decade clinically documented proof is being continuously provided for the additive, even synergistic, efficacy of combined, selected and properly synchronized chemo-immunotherapy. These treatment protocols are especially valid, if the immunotherapy modality is not a form of a non-specific immunostimulation, but a tumor-specific vaccine [276, 277].

Viral oncolysis was fitted into this volume written for the comparison of the ancient RNA/DNA complex signaling in unicellular life forms with that of the oncogenome-working cancer cells. The RNA/DNA complex of the cancer cell appears to have transformed those cells to the likes of unicellular life forms. Those life forms could not as yet make IFNs: the transformed (cancer) cells are also

IFN-defective. Within the cancer cell taken from its host, innate immune antiviral mechanisms would dominate (unless the transformed cell is a lymphocyte/plasma cell). However, within the multicellular host, interferons, antibodies and T lymphocytes will be mobilized against the oncolytic virus. To the host's eventual demise, these immune reactions may eradicate the virus, but fail to reject, rather they act to rescue, the tumor cell. At the M.D. Anderson Hospital's oncolytic virus conference of 1957, Albert Sabin rose to say, that "for success, various different oncolytic viruses will have to be used in succession" (cited in ref 71). Further, the replicating oncolytic viral genomes directly challenge the transforming oncogenic cellular RNA/DNA complex. The viral genome usually takes over the entire metabolism of the host cell for the construction of a large new viral progeny, with consequential cell death. For the toleration of the viral genome, the cell is subdued by viral genomic segments linearly inserted in the host cell's genome, or by the circularized viral genome placed in the cell's epigenome as a resident entity. The aim of successful viral therapy is oncolysis first, and active immunization of the host against fused antigenic viral and tumor cell proteins second, thus overruling tolerance to self. The persistence of a harmless virus would be acceptable, if it biologically (genomically) interfered with an oncogenic virus. In the tumor cell, the oncolytic viral genome is not against the tamed genome of a somatic cell, it is against the oncogenome of a constitutively transformed stem cell.

Selected Abbreviations:

ADCC = antibody-dependent cell-mediated cytotoxicity
CASCADE = CRISPR-associated complex for antiviral defense
CRISPR = clustered regulated interspaced palindromic repeats
PAZ/PIWI = P-element-induced wimpy testicle, drosophila; argonaute;
Zwille binocular; pinhead gene (PNH/ZLL-ZLL encoded. Member of Wuschel-related homeobox protein family. Sequences from green algae to angiosperms; Arabidopsis thaliana central shoots meristem.
PPAR = peroxisome proliferator-activated receptor (Table II)
SIR = silent information regulator deacetylase (activator) of p53 promoter
SZ = Szeged, 6726 Hungary for Drosophila Stock Center, Molecular Biology, University of Szeged

A Highly Effective Immune System Predating Adaptive Immunity

Agnatha

Stepwise diversification. The coeloma cavities of the deuterostome sea urchins (*Strongylocentrotus purpuratus*) mobilize monocyte-like cells, which respond to bacterial endotoxins. The diversity of the response of sea urchin coelomocytes to lipopolysaccharides was unexpectedly high: 81 full length cDNAs contained 67 unique nucleotide sequences (the 185/333 family) encoding 64 different proteins. Complement C3 was expressed, too. The amino acid sequence variability dominated by a RGD (arginine, glycine, aspartic acid) motif was a reflection to single nucleotide polymorphism, and deletion and insertions. The activated genes consisted of two exons and one intron [289, 290a, b] (Figure 45).

The early vertebrates, the lampreys, metamorphose from larvae to adults; the adults parasitize large fishes in sucking tissue fluids from their subcutis. The adult lamprey has a spinal cord, gills, kidneys, a circulatory system, eyes and rows of teeth in a jawless round mouth (*Cyclostomata*). The lampreys circulate blood with cellular elements such as red cells, myeloid cells and lymphoid cells. The blood cells are generated in the typhlosole in the larvae, which is an intestinal pouch considered to be a hematopoietic spleen. In the adult, the hematopoietic cells are formed in the dorsal fat body. The first two genes of the six-genes Ikaros family appear in the lamprey; this gene family will be essential for lymphocyte ontogenesis in mammalian hosts within the adaptive immune system [291, 292]. The agnatha (lamprey; hag fish) evolved a superb immunological defense system that secured their survival since the Cambrian explosion, after they diverged from their common ancestor. These are the combinatorial assemblages of the leucine rich repeat (LRR) cassettes within the variable lymphocyte receptors (VLR). The lamprey lymphocytes express the VLRs. The VLR gene assembly has been sequenced. The cDNA library generates the diverse leucine-rich repeats (LRR) consisting of an invariant signal peptide and highly variable modules, the LRRs. The crescent-shaped protein plates encoded by the VLR are not immunoglobulins, yet

J.G. Sinkovics, *RNA/DNA and Cancer*, DOI 10.1007/978-3-319-22279-0_5

Figure 45 The Coelomocytes of the Sea Urchin Carried in the Waiting the Primordial Genes of Macrophages and Dendritic Cells. [2351] Reference Appendix 2, Explanations to the Figures

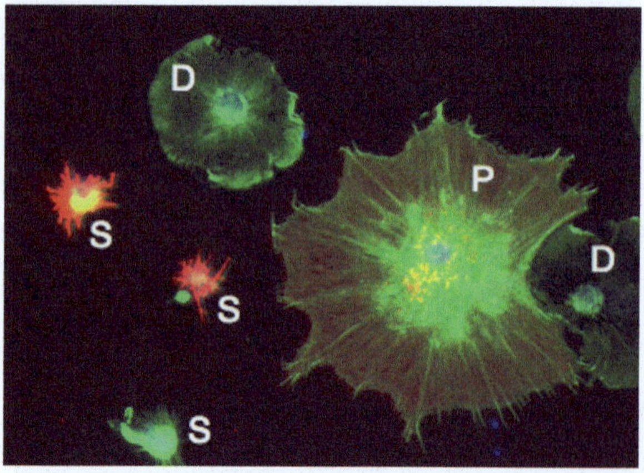

they specifically bind antigens. Upon antigenic stimulation, the lamprey lymphocytes undergo "blastic transformation." One subgroup of these lymphocytes produces no LRR, but appear to release chemokines and their receptors (CCR9, CXCR4, CXCR2, the IL-8 receptor) and lymphokines (IL-8, Il-17, and macrophage migration inhibitory factor) (reviewed with superb illustrations in color [293].

Missing are the ancestors of NK cells (they must be there, but in hiding: not yet discovered as such in the agnatha). It was in the colonies of the ascidian urochordate, the Botryllus (*B. schlosseri*), where the ancestors of the NK cells appeared first [294] (Figure 46). The iridovirus-infected and tumor cell-responsive lymphoid cell populations of the hagfish and the Xenopus were repeatedly studied, and the presence of "NK cells" is mentioned, but their pictures are not published, and their morphology has not been described [295–297]. At the level of amphibians (Xenopus), the innate and the adaptive immune systems are already working in unison. FV3 iridovirus-infected tadpoles with MHC class I deficiency succumb, but adult frog populations are well immunized against that virus displaying both antibody and cell-mediated adaptive immunity [298, 299].

Agnathan microRNAs. When the lamprey-gnathostomata lines diverged, the agnatha immune system was left behind. It was somewhat due to gene duplications, but more than that, due to the development of a much expanded microRNA network, that this divergence was induced. Deuterostomata released their microRNAs from the well conserved introns of their protein-coding genes; one such ancient cluster is that of the miR-1/miR-133 cluster [300, 301]. This newly acquired microRNA network remained preserved in the teleost and bony fishes, reptilians, frogs, birds, and mammals (referred to as eutherian taxa, mouse and human). Of 56 chordate-specific miR families of teleosts and mammals, 41 were present in the sea lamprey. The founders of the miR families miR-216-217, miR-126, and miR-155 established themselves in the agnatha (lamprey) [302]. The microRNA innovations resulting in the vast expansion of the metazoan microRNA repertoire appear to have occurred in three episodes: first in the protostomes and deuterostomes, second, in the first vertebrates (including the hagfish and the lamprey); and third, in the

Figure 46 The Botryllus' Large Granular Cells in the Haemolymph Express Lectins and Dissolve Parenchymal Cells of an Incompatible Colony Approaching for Uninvited Fusion. [294] Reference Appendix 2, Explanations to the Figures

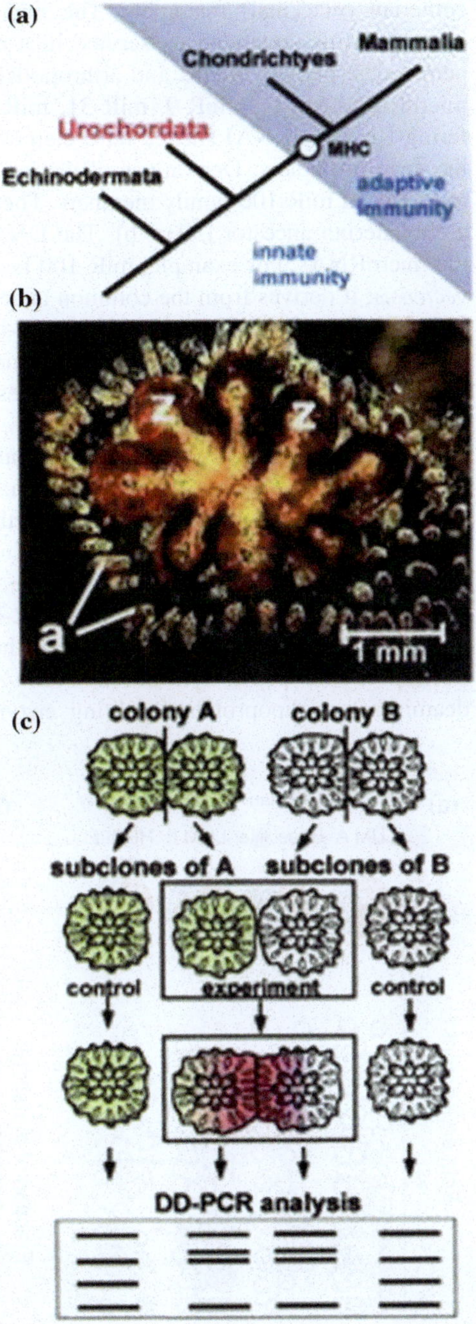

eutherian (placental) mammals. The microRNA cluster let-7 discovered in the caenorhabditis, extended its service in a range of hosts (echinoderms, mollusks, nematodes, platyhelminths and arthropods). The sea urchin operates 6 families of microRNAs (miR-1, miR-9, miR-31, miR-124, miR-125, miR-184). Small interfering RNAs (siRNA) silence retro-elements by DNA and histone methylation in the gene promoters. Derivatives of the let-7 cluster are the Hox cluster-associated miR-10 and miR-100 family members. They are the results of a gene duplication in a eubilaterian ancestor [303a, b]. The DNA genome remains under the control of the microRNAs. For example, miR-100 is operational in the cnidaria *Nematostella vectensis*; it derives from the common ancestor of plants and animalia living in the sea over 600 mya, and as an exemplary ur-bilaterian microRNA, it is very active in the human genome [304], exercising most versatile roles in the malignantly transformed cells, the prominent one of these is the inhibition of polo-like kinase in glioblastoma multiforme and other tumors.

An adaptive immune system without immune globulins. The agnatha possess an antigen-specific adaptive immune system, which culminated its course in these hosts and appear to have not evolved further. The agnathan immune system was replaced by a different immune system, which evolved independently further in the gnathostomata sharks (descendants of the extinct Placoderms), bony fishes (teleosts), reptiles, birds and mammals. VLR-like genes in the zebrafish and in the pufferfish create a "grey zone" between the two immune systems [301]. Another overlap is the possession of two AID-APOBEC (activation-induced cytidine deaminase; apolipoprotein B-editing enzyme catalytic) genes in the lamprey to

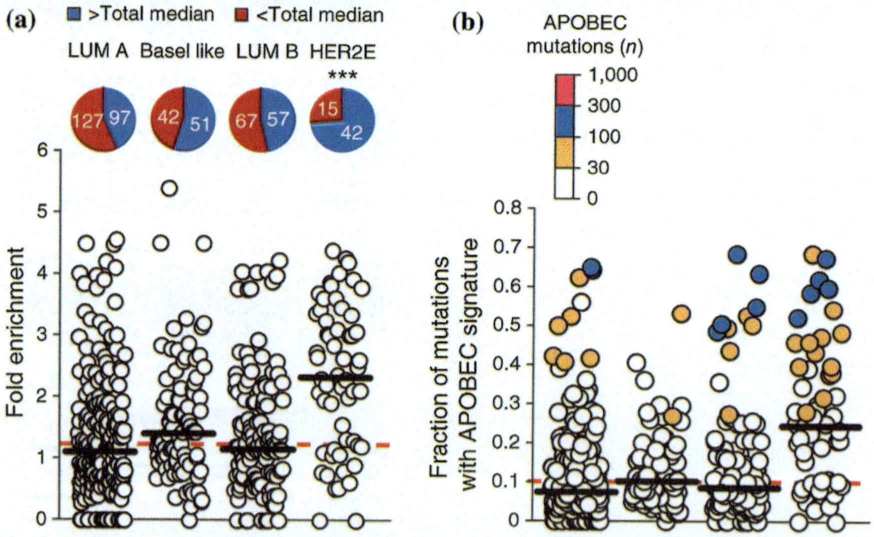

Figure 47 The Ancient Enzymes AID & APOBEC May Place Nicks to Unintended Genes, thus Promoting Gene Fusions and Oncoprotein Generation. [2352] Reference Appendix 2, Explanations to the Figures

encode proteins (enzymes) for conversion-type re-arrangements of the VLR loci [305–307]. The mammalian adaptive immune system has become the most complex defensive mechanism, the individual ancestors of which appeared separately dispersed from one another first in the amphioxus (the V genes) and the sea urchin (the RAG genes and the complement). The new system overlies in unison the ancient innate immune system (of phagocytes, NK cells, macrophages, monocytes and dendritic cells, Toll-like receptors, chemo- lympho- and cytokines and bactericidal substances). However, this superb system is potentially oncogenic, due to off-target mutagenic lymphomagenesis induced by the AID enzyme, or even by the APOBECs (Figure 47).

There is no evidence for immune function-related oncogenesis in agnatha. Hagfish living in the close-to-shoreline mud develop liver cancers and insulinoma/hamartomata tumors (lamprey), due to chemical pollution (which do not occur in agnatha living far from the shore line) [308, 309]. Would it be that the attack of an infectious agent on the host and the immune defense it induced invokes the ancient DNA to transform host cells into independent and immortal individual cells (recognized as lymphomagenesis at the clinics)? Or, is the process a consequence of a faulty evolutionary installment of the adaptive immune system and its initiator, the ultimate mutator enzyme, AID (reviewed in Mechtcheriakova et al) [310a, b].

Elements of Adaptive Immunity Self-assemble

Gnathostomata

Putting together the Team! The mode of installment of the unique genomic sequences (VDJ); recombination activating genes (RAG 1/2); recombination signal sequences (RSS); major histocompatibility complex (MHC), which started to work in unison for the first time in the lymph nodes along the intestinal tract of the ancient sharks is not known. These elements now encode the entire adaptive immune system in vertebrate hosts, which have bone marrow, spleen and thymus. The elements of the entire system were acquired by the ancestral jawed gnathostomata cartilaginous fishes, the primordial sharks, some 500 hundred million years ago. It is presumed that the adaptive immune system was acquired in part through vertical inheritance, and in part through horizontal gene transfers. The horizontal gene transfers were postulated to be those of ancient retrotransposons: Transib transposons encoding *rag*1 to join a pre-existing *rag*2 gene) [311] (Figure 45). Herpesviral infections by the ancestor of the EBV from the amphioxus and sea urchin to the chondrichtyes gnathostomata, and from there upward on the evolutionary ladder, as carriers of the V(D)J/RAG1-2 sequences was also proposed (*vide supra*). The single elements of the adaptive immune system united first in the Placoderms and sharks; reviewed in [27, 147]. The idea of ancient herpesviral (ancestral EBV) horizontal gene transfer was elaborated on by Dreyfus *et al* [312]. Neither Dreyfus, nor Sinkovics realized at the time of their publication, that H. H. Niller *et al* reported first in 2004 at an international conference in Balatonfüred, Hungary, that the EBV genome harbors a sequence identical (co-linear) with a part of the V(D)J genome (*vide infra*).

Vertical transfers of lymphocyte genes pre-existing in the agnatha (hagfish, and especially lamprey) are possible, but due to lack of intermediate hosts (the Placoderms are extinct) not proven. The evolutionary trajectory the genes encoding the growth factors for lymphogenesis (zink finger lymphocyte differentiation factor, Ikaros; B cell lymphoma, Bcl11a; ubiquitin conjugated enzyme; transcription factor E2A; B cell homeodomain-containing enzyme, PaxB; B cell-induced maturation

J.G. Sinkovics, *RNA/DNA and Cancer*, DOI 10.1007/978-3-319-22279-0_6

protein, Blimp-1) is not known well, as yet. Further, there are overlapping expressions of clusters of differentiation molecules, especially CD4 and CD45, between agnatha and gnathostomata lymphocytes. Further, chemokine ligands and receptors in lymphocyte traffic (CXCl12, CXCR4), interleukins (IL-8, IL-17), and TLRs (TLR2abc, TLR7, TLR10) overlap. Elements of the VJ sequences are in the genome of lamprey lymphocytes, but without encoding immunoglobulins. In the larval typhlosole, Thy1$^+$ (CD90$^+$) lymphocytes, and the forerunner of the thymic gene Foxn1 exist (but there is no thymus in the agnatha). However, the Foxn4 lamprey gene is co-expressed with one of the three Notch ligand delta4 genes; the Notch AB proteins are well expressed in lamprey lymphocytes. The agnatha VLR system is absent in the urochordates (*Ciona*), amphioxus (*Branchiostoma*) and arthropods (*Drosophila*) [305]. Thus, there are good reasons to consider vertical transmissions for at least some elements of the adaptive immune system, and probable horizontal (lateral) transfer for the others. As to the V(D)J system, an RSS-flanked DNA element might have been embedded in an ancestral V-type Ig domain gene as a substrate of the RAG1,2 enzymes [313a, b] (Figure 45). The somatic DNA rearrangement of the V(D)J gene segments is catalyzed by the enzymes encoded by the *rag*1,2 genes, beginning in gnathostomata and continuing up to mammalian taxa. While the sea urchin possesses the *rag*1,2 genes and the encoded six-bladed propeller proteins, caenorhabditis, drosophila and anopheles genomes are entirely devoid of these genes. The sea urchin *rag*1,2 genes are close homologs of the gnathostomata *rag*1,2 genes (except for the location of some introns). The coding regions of the sea urchin *rag*1,2 genes derive from multiple exons and these are co-expressed. The sea urchin RAG1,2 proteins function in a capacity (not catalyzing V(D)J segments) different from that of the gnathostomata RAG1,2 enzymes, which is to bind RSSs. The sea urchin RAG1,2 enzymes must have cognate target sites on the DNA genome, but these so far remain unidentified [313a, b] (Figure 45).

The enzyme AID targets non-Ig genes and induces lymphoma. Evolutionarily, APOBEC1 and AID are operational in yeast cells, and are involved in RNA/DNA editing and in lipid metabolism (apolipoprotein B-editing enzyme catalytic; activation-induced cytidine deaminase: cytidine → uridine, or cytosine → uracil resulting in uracil: guanine mismatches). Later, appear APOBEC3 involved in innate immune regulations, and APOBEC2 in muscle development [306, 307]. APOBEC3G protects the human host against leukemia- or sarcomagenic retrovirus insertion (Moloney virus), but failed to protect it against the insertion of HTLV-I, II, and the AIDS virus 1,2) (human T cell lymphotrophic leukemia virus; human acquired immunodeficiency); reviewed [261b] (Figure 47).

The possible events of lymphomagenesis are triggered into their initial positions in the physiologically activated AID/V(D)J system with consequential U:G mismatches (uracil: DNA glycosylase). Physiologically, the system endeavors to encode the antigen-specific configurations of the IgG light chains for antibody and TCR diversification (SHM, somatic hypermutation) and class switch recombination (CSR). However, off target actions of AID may induce lymphomagenesis. The five-axon AID gene maps to 12p13. Physiologically AID mediates the SHM in the immunoglobulin variable (V) region and the immunoglobulin CSR in mature B cells. The enzyme AID deaminates the ssDNA strand at WRC (cytidine) motifs, or WGCW "hot spots" (W = A/T; R = A/G). Within the chromatin, histones (H) 3/4 become activated by acetylations, or H lysins (K) are silenced by hypermethylations (for example: H3/K4/3me). For the transfer of AID from cytoplasm to nucleus to act upon its target genes, IL-4/anti-CD40 stimulation is necessary [314–316].

In addition to its RSS-directed physiological heptamer/nonamer-mediated cleavage of the V(D)J recombination in the antigen-specific antibody generation, in the somatic hypermutation (SHM), the RAG enzymes inadvertently may nick at two sites in the major breakthrough region (Mbr) of the *bcl*-2 gene residing at 18q21. The occurrence of dsDNA breaks both in the bcl-2, and in the IgH gene at 14q32 results by translocation in the fusion gene t(14;18)(q32;q21) [317, 318]. If the IgH fusion partner is c-*myc* released by dsDNA break from its locus at 8q24, the fusion oncogene will be t(8;14)(q24;q32). The RAG collaborator enzyme AID, initiator of SHM and CSR, by targeting DNA strands off-target can bring about these translocations [318–320]. In the case of the c-*myc* translocation, G-quadruplex DNA is formed (*vide infra*). The bcl2/IgH and c-*myc*/IgH translocations are characteristic for Burkitt's lymphoma. Burkitt's tumors often harbor EBV; the epigenically located latent circular viral genome releases microRNAs; reviewed in [73]. During its long evolutionary trajectory, EBV might have acquired from a host cell a *bcl*-2 gene, which is now expressed as the viral gene BHRF1 (Bam restriction endonuclease Bacillus amyloliquefaciens fragment H rightward open reading frame 1) [321]. The c-*myc*/IgH translocations can be induced by the activated AID [319], but ds-broken c-*myc* can fuse with several other ds-broken human genomic sites [322]. The MYC-IgH-BCL2 rearrangements are the most frequent: t(8;14)(q24;q32) in Burkitt's lymphoma, and t(14;18)(q32;q21) in follicular lymphoma, but a complex der(8)t(8;14;18) with both *bcl*-2 and c-*myc* translocated to a single IgH allele also may occur [323]. These fusion genes are complicons and when amplified in a genome with biallelic inactivation of p53, tumor cell growth takes a most aggressive course [324]. Myeloid and plasmacytoid DCs (especially DC-specific C-type lectin receptor-positive, intercellular adhesion molecule, ICAM-grabbing DC-SIGN cells) activate AID in plasma cells with resultant AID-induced dsDNA breaks. This process is inhibited by the blockade of the receptor activator of NFκB/receptor ligand (RANKL). AID mRNA-targeting siRNA blocked AID activity [325]. This cytokine triad consists of the receptor activator of the NFκB ligand (RANKL), its receptor (RANK), and the endogenous decoy receptor osteoprotegerin (OPG). RANKL is a member of the TNF family. Cells are expressing RANKL attached to RANK. In the bone, osteoblasts express RANKL and osteoclasts are $RANK^+$. The soluble pseudoreceptor OPG imitates RANK and occupies RANKL on the osteoblasts. The system rests (osteoclasts are not activated). Numerous tumors metatatic to bones express the RANK/RANKL (receptor activator of nuclear factor-κB ligand) pathway. Multiple myeloma plasma cells expressed no RANK/RANKL pathway; RANKL was expressed by vimentin-positive bone marrow stromal cells [326, 327]. The fully human monoclonal antibody (mcab) denosumab (Prolia) directed against RANKL becomes a treatment modality for postmenopausal osteoporosis, and bone metastases of malignant tumors (*vide infra*).

AID-induced dsDNA breaks and translocations. Beside the AID-induced physiological gene mutations resulting in antibody diversity, mistargeted dsDNA breaks bring about potentially oncogenic mutations due to gene translocations creating fusion oncogenes-oncoproteins. The first type of gene mutations (U:G mismatches) are either accepted, or if repaired, receive an error-prone attempt at it by low fidelity DNA polymerases. Mismatch repair of AID-induced U:G mispairs is essential for CSR and SHM for immunoglobulin diversity. In contrast, the second type of mutations are suppressed by error-free repairs [328]. The RAG1/RAG2 enzymes may induce mistargeted dsDNA breaks at V(D)J or TCR loci. Especially loss of the the "dispensible" RAG2 C-terminus impairs joining substrates that persists during the cell cycle. All (100 %) RAG2$^{c/c}$ p53$^{-/-}$ mice die rapidly (mean 12 weeks) with immature T cell thymic lymphoma, whereas RAG2$^{-/-}$ mice suffer impaired (partial, delayed) lymphopoiesis without lymphoma development and no death in 1 year. However, some 65 % of p53$^{-/-}$ mice die with thymic lymphoma in (mean) 23 weeks. These results suggest that RAG2 exerts a genome guardian role [329]. The RAG2 carboxy terminus is essential for maintaining genomic stability. The MutSαβ proteins bind to DNA mismatch sites. Mouse strains with mutated or deleted *mut*Sα genes are available for research; these are theMsh2$^{-/-}$ and Msh6$^{-/-}$ (microsatellite heterozygous) strains. These mice lost the MSH proteins (MutSα consists of MSH2 and MSH6 heterodimers) and with it the protection against T cell lymphomagenesis in Msh2-mutated mice, whereas Msh6-mutated mice develop B cell lymphomas. The repair of AID-generated U:G is done error-free, whereas the mistargeted A:T breaks receive an error-prone attempt at their repair [330].

Lymphomagenesis emanating from the c-*myc*/IgH translocation [331] is naturally suppressed by p53, p19 (p14, p19 ARF, alternating reading frame), and by the ataxia telangiectasia mutated (ATM) proteins, which are not inhibitory to the physiological Ig/IgH translocation; and by gene product protein PUMA (p53 upregulated modulator of apoptosis). BCL2 and BAF (B cell activating factor) promote the translocation; BAF and protein kinase Cδ inhibit the physiological Ig/IgH translocation [332]. Special phosphorylations of the AID protein (serine, S38; threonine, T140) and mutation from serine or from threonine to alanine, are inhibitory to CSR or to SHM [333].

The 3'UTR of the AID mRNA is a direct target for miR-181b, resulting in the downregulation of AID translation. In lymphoma cells (Burkitt's, mantle cell, diffuse B cell) high expression of miR-181b prevented the accumulation of high levels of AID protein, whereas endotoxin (LPS) and IL-4 stimulated AID production. Activated B cells express low miR-181b levels [334]. Further, post-transcriptional modulation of the AID gene expression is exerted by its repressor miR-155. The host cell's genomic site for the insertion of avian leukosis virus induced lymphoma gene is BIC (B cell integration cluster). The non-coding *bic* RNA is the precursor of miR-155. In the human genome, the *bic*-related miR-155 gene releases the miR-155 precursor. The *aicda* gene (activation-induced cytidine deaminase) encodes AID; the AID mRNA contains a single miR-155 binding site in its 3'UTR. By inhibiting

AID-induced c-myc translocation, miR-155 acts as a tumor suppressor microRNA [335, 336]. The promoter gene of miR-124-1 in chromosome 8p is frequently methylated (silenced) in human malignant lymphoma cells, including those of NK/T lymphoma cells. The miR-124-2 is a direct inhibitor of the CDK6 cyclin-dependent kinase mRNA [337]. Moderate and pronounced anti-AID mRNA activity is exerted by miR-155 and miR-93, respectively [338].

Patients treated with imatinib mesylate, the suppressor of the p210 $^{BCR/ABL}$ oncoprotein in chronic myelogenous leukemia (CML), sustain an excellent clinical response. This oncoprotein is antigenic in the hosts of origin and elicit weak antibody- and T lymphocyte-mediated immune reactions. It was considered that a ABL/BCR protein vaccine (free of nucleic acids) be developed for the immunization of the therapeutic bone marrow donor. Thus, the donor marrow would not only replace the recipient's bone marrow, but would deliver a set of immune memory cells addressed to this oncoprotein (261a, b). At that time it was not understood that the ABL/BCR oncoprotein induced an AID-mediated immune reaction. However, the highly effective imatinib therapy replaced bone marrow transplantation for the treatment of the disease. Patients on imatinib therapy experience the development of hypoglobulinemia. Imatinib mesylate was shown to be a proven inhibitor of the enzyme AID at two levels. Imatinib inhibited IL-4- and LPS-induced CSR (class switch recombination), reduced the sizes of splenic germinal centers, and suppressed the key of AID activation, transcription factor E2A and gene PAX5 (imunoglobulin enhancer binding; paired box gene) [339]. The fact that imatinib is also an inhibitor of PDGF and c-*kit* (CD117) was validated in the treatment of some sarcomas (gastrointestinal stromal tumor; dermatofibrosarcoma protuberans); reviewed in [340a, 341a]. In suppressing the "genomic mutator" AID, imatinib mesylate may emerge as a protector against "inflammation-induced carcinogenesis", its effects extending from protection against lymphomagenesis to the suppression of carcinogenesis in general, inasmuch as AID-induced mutations are suspect initiators of solid tumors (adeno- and squamous cell carcinoma [339].

Sarcomas, the "immortalized mesenchyme" (Tables VI and VII) are retrovirally induced in fish, birds, mice, felines, bovines, and some simians. However, no sarcomagenic retroviruses could be isolated from the human tumors; Kaposi sarcoma yielded the HHV-8 and an endogenous retrovirus [340a, b] (Figures 29 and 30). It remains unexplained what components of the filtrates induce the focus formation, and the antigenic conversion in the human sarcomas [147; 340a, b, 341a, b]. Especially, what is the origin of the enzymes that nick in between nucleotides to excise and to insert genomic segments in order to generate the numerous fusion oncogenomes/oncoproteins.

Reed-Sternberg cells engaged in a "ménage à quatre". Physiologically, light chain immunoglobulin genes fuse with the heavy chain IgH locus at 14q32: kappa (κ) from 2p11.2, and lambda (λ) from 22q11.2. In malignant transformation of B lymphocytes, the fusion partners of IgH are characteristic of the tumor type: c-*myc* in Burkitt's lymphoma: t(8;14)(q24.1;q32.3) [322, 342]; *bcl*-2 in follicular lymphoma: t(14;18)(q32.3;q21.3); and CCND1 *cylin* D, or PRAD (proline-rich attachment domain mapped to 11q13), in mantle cell lymphoma: [11;14](q13.2;q323).

Table VI Sarcoma Genomics

Embryonic formation of neural crest's dorsolateral mesodermal zone in the frog is classic example of physiological epithelial-to-mesenchymal (ETM) transition. It Is encoded by genes *snail*1, *slug* (*snail*2), *twist*1 and *dermo*1 (*twist*2) and TGFβ; *snail*2 controls bone morphogenetic protein/wingless (BMP/Wnt) induction; early c-*myc* and late *sox*9 expression. Antagonist proteins Dickkopf and Sizzled (to Wnt), Chordin and Noggin (to BMP) appear.[1] These are the genomics acting without inhibitors in the ETM of a human adenocarcinoma[2] and in the initiation of phylloides tumors of the breast.[3] Human neural crest cells express markers p75, Sox9/10, CD44, HNK1; differentiate into Schwann cells, which are able to myelinate neurons.[4]
Sarcomas: "Constitutively Immortalized Mesenchyme". Human chondrosarcoma cell differentiation induced by lymphocytes occurred *in vitro*.[5] MicroRNAs act: miR-199 & miR-121a upregulate chondrogenesis. The blockade of miR-221 reversed chondro-inhibitory actions of the JNK inhibitor (c-Jun N-terminal kinase). MDM2-targets chondrogenic Slug for degradation. Slug was liberated by miR-221-mediated suppression of MDM2.[6] Chondrosarcoma cell differentiation mediated by miRs?
Benign adipocytes accumulate. Bone morphogenetic protein acting upon focal adhesion kinase (BMP; FAK): phosphorylation at site Y397 (tyrosine), cascades into adipocyte genesis by C/EBPα/PPARγ (CCAAT-enhancer-binding protein; peroxisome proliferator activated receptor, Table II) acting in mesenchymal stem cells. shRNA can inhibit FAK mRNA.[7]
In synovial sarcoma (SS) transcriptional repression by Snail, or inactivating mutation of E-cadherin results in loss of glandular epithelial differentiation. Monophasic SS with E-cadherin missense mutations carried the SYT-SSX1 fusion oncoprotein. Biphasic SS expresses membranous E-cadherin and the SYT-SSX1 fusion protein.[8] Differentiation of osteoblasts and adipocytes regulated by cross-signals of PPARγ (Table II), runt-related transcription factor Runx2, myeloid elf-1-like factor MEF (retrovirus ETS E 26 transcription) and distal-less homeobox 5 (Dlx5). MEF suppresses Runx-induced osteoblastogenesis; activates PPARγ for adipocytogenesis. Dlx5 upregulates Runx2 for osteoblastogenesis and downregulates PPARγ for suppressing adipocytogenesis.[9]
Sarcomas follow the HMG protein pathway. Embryonal rhabdomyosarcoma utilizes the ancient high mobility group protein AT hook2 (Tables XI/I and XI/II) to activate insulin-like growth factor mRNA-binding protein, whose target is N-ras/NRAS; the chain reaction HMGA2 → IGFBP2 → NRAS culminates in the activation of the MAP/ERK pathway. Osteosarcomas gain chemotherapy resistance and survival in the state of autophagy under the effects of HMGB1.[10,11]
1 Shi J et al Development 2011;138:3135–45. 2 Micalizzi DS et al J Mammary Gland Biol Neoplasia 2010;15:117–34. 3 Do SI et al Tumour Biol 2013; 34:445–53. 4 Liu Q et al Stem Cells Transl Med 2012;1:266–78. 5 Sinkovics JG POR 2004;10:174–87. 6 Kim D et al J Biol Chem 2010;285:26900–7. 7 Lee JS et al Biochem Biophys Res Commun 2013;435:696–701. 8 Saito T et al Oncogene 2004; 23:8629–38. 9 Baek K & Baek JH Adipocyte 2013;2:50–4. 10 Li Z et al Cancer Res 2013;73:3041–50. 11 Huang J et al Autophagy 2012;8:27507; Cancer Res 2012;72:230–8.

In contrast, in RS cells of Hodgkin's disease, the IgH locus attracts many different fusion partners: *rel*/NFκB from 2p16 (avian reticuloendotheliosis virus c-*nfkb* → v-*nfkb*), *bcl*6 from 3q27, c-*myc* from 8q24.3 [324]. In Hodgkin's Reed-Sternberg (RS) cells (Figure 48), commonly either *bcl*2 from 18q21.3, or *bcl*3 from 19q13.2 may translocate to the IgH locus [343]. Once induced as a re-programmed germinal center B cell, the RS cell gains gene function upregulations of Notch1, TRAF2 (TNF-R-associated factor), NFκB/Rel, JAK/STAT, and PI3K/AKT, *versus* losses of gene functions of TNFAIP3 (TNFα-induced protein),

Table VII Genomics of Myxoid/Round Cell Liposarcoma (MyLSrc)

Benign lipomas display borderline "malignant" genotypes involving 12q13-q15 and 3q27-q28, the locus of LPP (lipoma preferred partner) gene. LPP encodes proline-rich protein with 3 LIM domains at its carboxy terminal region (Table XVI). Lipomas harbor LPP/HMG protein hybrids (Tables XI/I and XI/II) in mature differentiated adipocytes (lipoma cells), which persist often fused with LPP. Commonly HMGA2 exons 1–3 and LPP exons 9–11 are involved. HMGA2 is expressed in undifferentiated mesenchymal cells; it may persist in terminally differenciated adipocytes (lipoma cells). In t(12,13), HMGIC (isoform C) fuses with LHFP (lipoma HMG fusion partner). These fusion proteins do not induce adipocyte de-differentiation.[1–3]
MyLSrc occupies midposition **2** among liposarcomas (LSrc): 1 well-differentiated/de-differentiated; **2** myxoid/round cell; **3** pleomorphic. Initial genomics of LSrc generation are amplification of MDM for neutralizing the p53 protein for ubiquitination; and amplification of CDK4 for neutralizing pRb (retinoblastoma) by phosphorylation (Table XIX). The p53 deprived cell escapes apoptotic death; the phosphorylated Rb protein fails to inhibit the cell cycle. The chromosomes harboring the affected genes assume the "ring and giant rod" appearance. Truncated (deleted 3' end) HMG proteins move in and open up receptors for miR-Let-7. The tetraspanin gene TSPAN31 co-amplifies with MDM (in many different types of sarcomas). The ancient TSPAN genes date back to the urochordates (the amphioxus harbors a cluster of 23 TSPAN genes) (Table XXIV). They have become co-oncogenes in human sarcomata. Further, c-Jun terminal kinase is amplified with ASK1 (apoptosis signaling regulator kinase). The co-amplification of c-Jun and ASK1 results in the inactivation of PPARγ (Table II), which is a promoter of embryonic adipoblast differentiation. Against the background of amplified genes, MyLsrc carries the translocation t(12;16)(q13;p11) resulting in the FUS/CHOP*(GADD) oncoprotein (fused in sarcoma; CCAAT enhancer-binding protein homologous protein; growth arrest DNA damage). FUS and EWS are RNA-binding proteins. CHOP* inhibits and neutralizes the adipoblast-differentiating CCAAT/enhancer binding protein (C/EBP) and maintains adipoblasts in continuous cycles of mitoses. The suppressed PPARγ loses its control; immature adipocytes with fully active PPARγ do not de-differentiate into adipoblasts, or liposarcomas. Highly virulent malignant growth is driven by PI3K. Its antagonist PTEN (Table VIII) may suffer a disabling point mutation. In contrast, the PI3K gene harbors a gain-of-function mutation in its H1047R (histidine; arginine) domain. The genomics of pleomorphic liposarcoma may overlap with that of MyLsrc: total deletion of Rb gene at 13q14.2-q14.3 (Table XIX) and amplification of mitotic arrest deficient MAD2 gene. Further suppressor genes often eliminated are p53 at 17p13 and NF1at 17q11.2. The genomics, especially those which are miRs encoded from gene 14q32.2, are different in primary sarcomas and in their established cell lines (Table XXIII).[4–6]

1 Ashar HR et al Cancer Genet Cytogenet 2003;143:160–8. ***2*** *Kubo T et al Anticancer Res 2009;29:2357–60.* ***3*** *Petit MM et al Genomics 1996;36:118–29.* ***4*** *Conyers R et al Sarcoma 2011;ID483154.* ***5*** *Renner M et al Genes Chromosomes Cancer 2012;51:982–96.* ***6*** *Singer S et al Cancer Res 2007;67:6626–36.*

* Oncogene CHOP not to be confused with chemotherapy regimen CHOP (for oncogene CHOP see Suzuki K *et al* Oncol Lett 2012;3:293–296).

the NK-cell maturation factor, TXNIP (thioredoxin interacting protein), and the NFκB activation antagonist, PPARGCA (peroxisome proliferator-activated receptor-gamma coactivator) [344]. This most elaborate article does not mention AID, as a potential activator of mis-targeted DNA breaks, possibly involving these oncogenes. However, in these gains and losses, the gain-of-function mutations keep RS cells alive and proliferating (however, seldom seen in mitoses in the tissues).

The German school proved, that Reed-Sternberg (RS) cells are of B lymphocyte origin and are generated from germinal center cells without cell fusions [345, 346].

Large multinucleated RS cells in a cultured cell line were generated by endomitosis [347]. However, the possibility for at least some RS cells to be cell fusion product "natural hybridomas" continues to exist [348a, b, 349]. The transformed B lymphocyte pre-RS cell may fuse with a lymphocyte partner. A valid argument for this event is that the double nuclei of the RS cell may be different from one another in their p53 expression pattern (one positive, one negative for point mutation): a "discordant pattern", most intriguing and "cannot be explained". However, the possibility that the nuclei with normal p53 originated from non-RS contamination was considered [350, 351]. Why cell fusion with a non-mutated lymphocyte was not considered? Indeed, the cell fusion possibility for RS cell generation as published in 1991 [348] was not recognized in the later literature. Further, the cell fusion possibility was rejected, based on the lack of non-mutated (resting) genomic elements in the nuclei of RS [346]. However, resting nuclear genomic elements could be expected only, if the mononuclear RS cell actively phagocytosed a resting cell, or just fused with it. In contrast, if an immunologically reactive cell (a lymphocyte in blastic transformation reacting to EBV, latent retroviral, or another RS cell-expressed antigen) actively attached and or entered the RS cell, the united genome with doubled chromosome numbers would not contain resting elements in "germline configuration" [352], yet it would be a fused genome. At least some of the multinucleated giant cells expressing both EBV, and antibody-production against it in the same cell (the unpublished Dreesman experiment is cited) [348] could possibly be "natural hybridomas", especially if a fusogenic endogenous retrovirus interacted [147, 352–354]. In the classical RS cells there is no immunoglobulin production due to epigenetic silencing of the IgH gene: lysine K9 on histone 3 is methylated (and reversibly so: 5-aza-2′-deoxycytidine demethylated the histone). The role of AID has not been determined [355]. However, lymphocyte-to-lymphocyte fusion between cells (malignant and non-malignant) in Hodgkin's lesions can result in giant polykarion formation [356]. In Dreesman's large multinucleated lymphoid cells derived from an EBV^+ Burkitt's lymphoma (BL) lesions, cited in Sinkovics [147, 348] (*vide supra*), an EBV^+ lymphoma cell might have fused with a normal B cell, which was producing the antibody against EBV within those cells. Some large multinucleated lymphoma (RS) cells may be engaged in a "*ménage à quatre*" relationship: two cells (the malignant cell and a reactive B cell), and two viruses (EBV and a fusogenic endogenous retrovirus) uniting in one RS cell. The matter remains open: not all RS cells are equal, some are more promiscuous than the others. Indeed, it has been recently recognized by the German school, that multinucleated RS cells are generated by fusions of single-nucleated RS cells (Figure 48).

Immunological lymphomagenesis. The off-target ssDNA/dsDNA breaks caused by AID are potentially oncogenic. Indeed, these are lymphomagenic at the IgH site. There, the human IgH locus consists of 90 to 123 V_H, 27 D_H and 6 J_H gene segments, offering a wide variety of targets. In most sporadic BLs, c-*myc* fuses with IgH μS (*mu* switch) region: t(8;14)(q24;q32), while the breakpoint in endemic BLs is at the J_H region [342]. The *bcl-2* gene (and its associated anti-apoptotic partners) could be released for translocation by the RAS nuclease. Further, uni- or biallelic

(a)

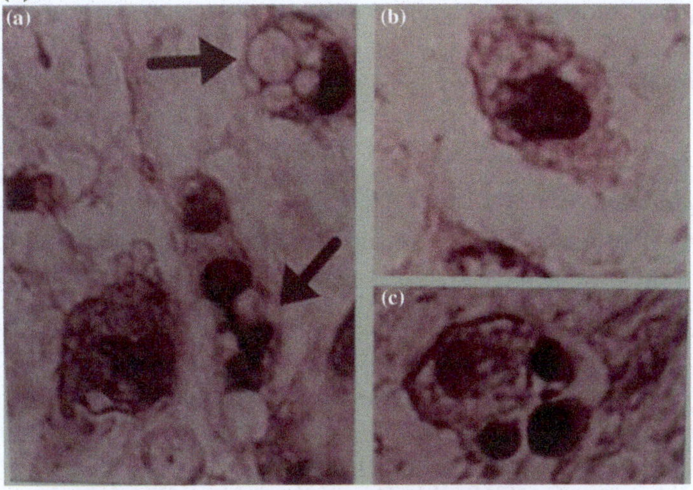

(b)

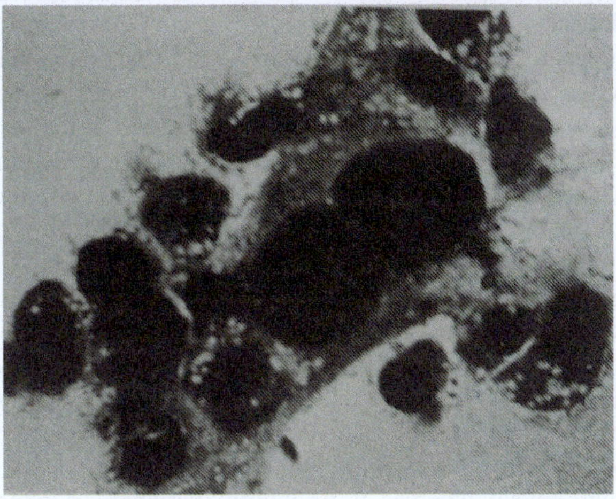

Figure 48 The Debated Origin of Reed-Sternberg Cells. Virological Studies for a Retroviral Associate to EBV Are Needed. Reference Appendix 2, Explanations to the Figures

mutations of TP53 occur (but p53 loss in both alleles occurred only in in DLBCL, diffuse large B cell lymphoma). The extraordinary plasticity of lymphocytes is expressed by the de-differentiation of terminally differentiated cells (somatic cells) by transcription factors CEBPαβ (CCAAT-enhancer-binding protein) into pluripotential stem cells expressing Oct3/4, SOX2, MYC, NANOG, LIN28, and KLF4 (Krüppel-like factor), when PAX5 (paired box gene) could be suppressed [344] (*vide supra*).

While in lymphoblastic malignant tumors PAX5 is frequently knocked out, gain-of-function mutations of CEBPαβ (CCAAT enhancer-binding protein) are difficult to document [357]. However, this most elaborate report does not name AID as the initiator of the SHM/CRS mutations. Further, clonal *bcl*-2/IgH rearrangements may persist in healthy individuals without lymphoma induction. This finding poses great difficulty in the determination, if a patient treated with follicular lymphoma is in complete (molecular) remission or not [358]. When NFκB induces aberrant (constitutive) AID expression in gastrointestinal epithelial cells, even if it is not a clearly accomplished malignant transformation, its high possibility exists [359]. Hepatitis C viral, or *Helicobacter pylori* infections trigger a chain reaction of TNFα → NFκB → AID expression in the gastrointestinal tract. Especially susceptible for AID overactivity are the patients with chronic inflammatory bowel disease or Barrett's esophagus [360, 361].

In the synovium of patients with rheumatoid arthritis, AID hyperactivity generates autoantibody production in ectopic lymphoid tissue [362]. The plasma cell-rich nasal polyps are "inflammasomes." Their immune environment is the Th2 type and immune globulins produced are of polyclonal IgE type. The 25 genes "AID signature" study in nasal polyps revealed a marked interpatient variability of AID mRNA expression [310a, b]. This occurs when AID expression is not constitutive and fluctuates. The AID variant with truncated C-terminus, due to the lack of exon 4 in the gene (AID-Δex4), is hyperactive in SHM (somatic hypermutation), but totally inactive in CSR (class switch recombination). Increased amounts of AID and IgE proteins and IgE-producer B cells appeared in the nasal polyps (by immunohistochemistry and immunofluorescence). AID overexpression occurred also in the tonsils. Thus, patients with nasal polyps also had chronic rhinosinusitis. AID expression in the thymus was considered normal and physiological. Of the genomics of inflamed nasal tissues, the B cell maturation factor PAX5 was overexpressed, and accordingly Notch1 was suppressed. Stromal cells in the nasal polyps overexpressed the IgE receptor (including the low affinity CD23 cluster). Unexpectedly, early growth factor family members underwent reduced expression. Expectedly, IL-13 levels were high. It remains to be determined if AID overexpression was the initiator of the nasal polyps and if so what factors kept the AID gene overactive. The physiological, or pathological nasopharyngeal flora of viruses and bacteria will reveal whether it is entirely variable, or rather uniform and characteristic of the condition [310a, b]. In the stomach, it is *Helicobacter*, in the colon it is enterotoxigenic *Bacteroides fragilis*, which initiate the inflammatory condition; reviewed in [27]. What is it in the case of nasal polyps?

Comment. Are the off-target ssDNA/dsDNA breaks by AID due to a misconstrued inherent error in the adaptive immune system, what all the installed preventive or repair mechanisms may fail to correct? Or, is the RNA/DNA complex responding to "danger signals" emanating from B cells engaged in a frantic antibody production (SHM) in order to overcome a threat by an invading pathogen? Was the primordial RNA/DNA complex inherently endowed by its structure and evolutionary position to create immortalized cells, for the preservation of the living matter in cellular form? Are the pluripotent stem cells of the multicellular hosts blind to the ultimate demise of the host, in trying to re-create the ancient scenario?

The RNA/DNA Complex at Work

The Unicellular Eukaryotes Preserve Their Primordial Resistence

Survival at the edges. The ancient unicellular life forms driven by the primordial DNA underwent mitoses in their mature age, thus escaping senescence and natural death. It is not possible to determine if the telomerases of ancient unicellular life forms acted constitutively in maintaining chromosome lengths (*vide infra*). The primordial DNA resisted physico-chemical damage due to heat, electricity, irradiation, and the extremes, such as hypersalinity, disorders the acid/base balance, and chemical reactions with metallic atoms (that are normally functional in some enzymes). The ancient cells (those of archaea, the first prokaryotes (eubacteria), the first unicellular eukaryotes) existed and metabolized in anaerobic niches in atmospheres of methane, ammonia, sulfur and CO_2 at temperatures close to the boiling point of water, at pH as low as 0, or in high salinity/alkalinity, while surviving in volcanic plumes at the periphery of pouring lava. The membrane-coated, iron atom-oxidizing *Ferroplasma acidarmanus* lives and metabolizes in a ph 0–<3 environment [363a, b]. Its archaeal enzyme, methionine-γ-lyase carries out the chemical reaction for the production of the volatile organic sulfuric compound methanethiol essential for survival and growth [364a, b]. The anaerobe *Syntrophus aciditrophicus* converts benzoate to acetate, bicarbonate to formulate and generates hydrogen from protons (of which its syntrophic archaea generates its own metabolism) [365, 366]. The enzymatic machinery of the *Deinococcus radiodurans* incessantly repairs radiation-induced DNA strand breaks [367, 368] (Figure 4). The thermoresistant DNA polymerase of *Sulfolobus acidocaldarius* or *Thermus aquaticus* (Taq) enabled Kary Mullins to carry out the first polymerase chain reaction [369].

J.G. Sinkovics, *RNA/DNA and Cancer*, DOI 10.1007/978-3-319-22279-0_7

Balancing Acts

G4 quadruplex in telomeres and in oncogenes. Barbara McClintock is cited for credit in the first description of capping eukaryotic chromosomes (in plant cells) for protection from ds-DNA breakage fusion catastrophes [370, 371].

Since the DNA molecule can form (single strands can fold into) G4 quadruplex structures *in vitro*, it is very likely that such DNA structures existed already in the pre-cellular world. In the laboratory, 4-guanine tracts spontaneously form four-stacked structures (G-quarters) held together by hydrogen bonding. In between the 4 G-quartets there are three ionized ammonium molecules (NH_4^-) [372–374]. G4 quadruplex DNA is an ancient formation extending from bacteria (archaea and prokaryota) [375] through yeast (*Saccharomyces*) [376–379], and protists (the ciliate, *Oxytricha trifallax*); so is the ciliate protozoan *Stylonychia lemnae* (Figure 49). Ciliates (paramecia, tetrahymena, oxytricha) command 3 genomes (or *vice versa*, the genomes command them): macro- and micronuclear, and mitochondrial. These genomes operate their own telomeres, including the mitochondrial genome. The *O. fallax* and *O. trifallax* genomes carry high copy numbers of ancestral transposable elements, some of which encode precisely functioning transposases [380–382]. The genomes preserve their own telomeres with G4 quadruplex DNA. The *O. trifallax* cells repair their chromosome ends with restoration of telomeres after each cell divisions (the telomeres are DNA-protein complexes; the protein is the reverse transcriptase, TERT). Reverse transcriptase templated by an RNA component polymerizes the short G-rich sequences into tandem repeats. Ciliates generate telomeres *de novo* in the somatic macronucleus developing from a mitotic copy of the germline micronucleus. This occurs after conjugation (mating), when each individual of the pair eliminates the old parental macronucleus (by apoptosis) and replaces it with a newly generated macronucleus. The precursor anlage of the new macronucleus is preserved as the new, but silenced, diploid germline micronucleus. The new individual divides by binary fission. After the cell division, the chromosome ends are capped with *de novo* generated telomeres. Telomerase adds telomeres to the protruding 3′OH ends of the chromosomes [370, 383, 384]. The anaerobe ciliate, *Nyctotherus* possesses also mitochondrium-derived hydrogenosomes with a genome of its own [385, 386]. The tetrahymena (*T. thermophila*) TERT gene releases extra number of mRNAs during conjugation; however, after the old macronucleus was eliminated by apoptosis, and the new macronucleus established itself, TERT production descends to its standard level [387].

DNA propellers. The a-mitochondriate *Giardia* is one of the most ancestral eukaryote. It has two telomeric G4 sequences: a propeller-type parallel-stranded one with three G-tetrads, and a basket type anti-parallel-stranded one with two G-tetrads. Components of the G-quadruplexes can be readily exchanged by a "cut and paste" principle; the chromosomal ends are kept capped [388]. The giardia genome is packed with retrotransposons (both LTR-, and non-LTR-types). The non-LTR-types are LINEs (long interspersed nuclear elements) with a single ORF

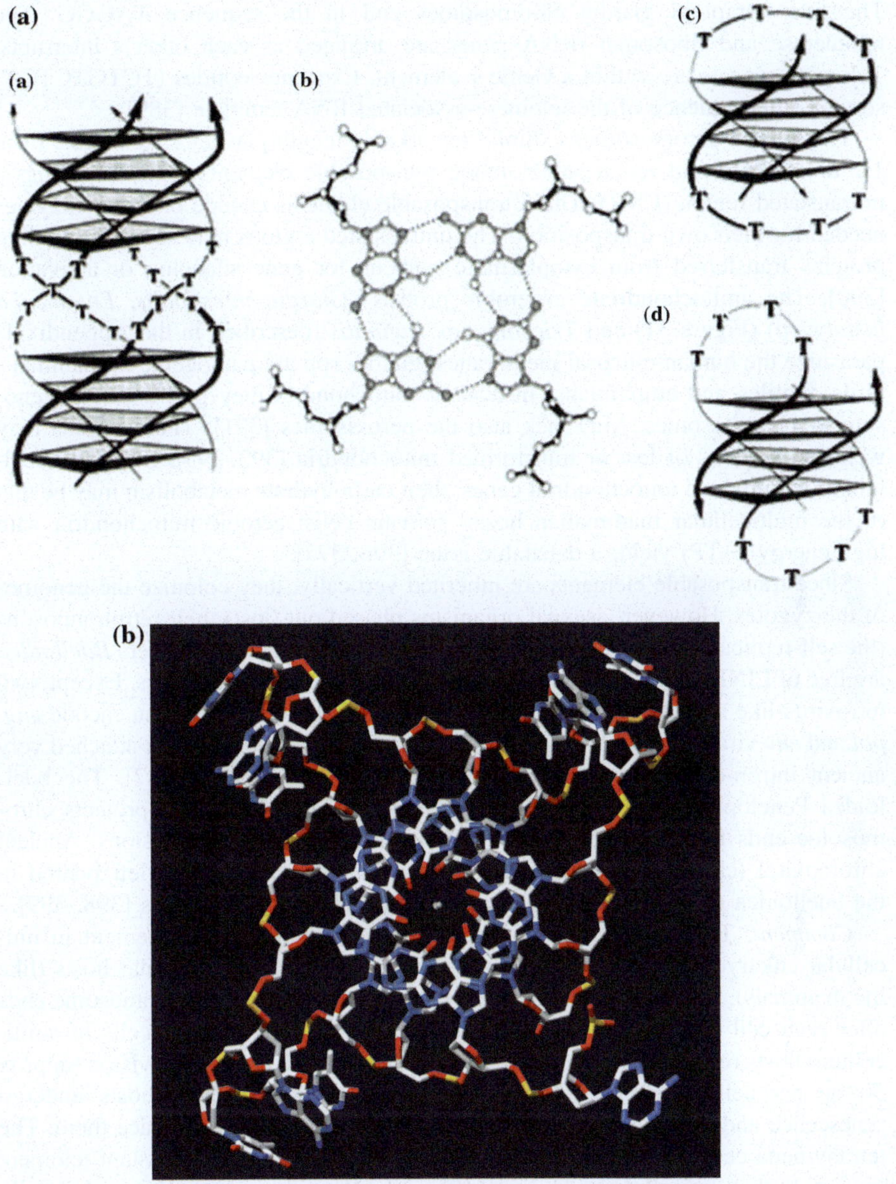

Figure 49 Newly Formed and Mutated-within DNA G4 Quadruplexes of Unicellular Eukaryotes and the Oncogenes of Advanced Multicellular Hosts (including Homo) [2353] Reference Appendix 2, Explanations to the Figures

and a C-terminal endonuclease. These retrotransposons are copied by reverse transcriptases. The genomic inscripts of non-LTR retrotransposons are inherited vertically, and thus become Genies (Giardia early non-LTR insertion elements).

The five tetraploid giardia chromosomes end in the sequence TAGGG. The telomerase and ribosomal rRNA genes are arranged as each other's interrupts. Telomere repeats are within a Genie 1 element. Extra nucleotides (TCTGTCTGT) serve as the sequence of the telomere-associated RNA template [389].

The giardia chromosomes exhibit "remarkable length plasticity" (elongate from 1.1 to 1.6 Mb) and retain subtelomeric transposable elements with very long 3′ untranslated region (UTR). These transposable elements encode only the enzymes needed for their own transposition. The untranslated regions may serve for binding proteins transferred from cytoplasm to nucleus for gene silencing or activation [390]. The amitochondriate anaerobic protists (*Giardia intestinalis, Entamoeba histolytica*) (Figure 31) and *Trichomonas vaginalis* (described in the Appendix 1) parasitize the human mucosal membranes; giardia spp are pathogenic in mammals, birds, reptiles and amphibians. Instead of mitochondria they possess hydrogenosomes and mitosomes. May lack also the peroxisomes [391]. These organs may well be remnants of lost or transformed mitochondria [392–395]. If these unicellular hosts retained mitochondrial genes, their carbohydrate metabolism may be that of the multicellular mammalian hosts' somatic cells: aerobic mitochondria with high energy (ATP) yield: a debatable issue [396, 397].

Since transposable elements are inherited vertically, they colonize the genomes of the zygotes. However, asexual organisms phased out (lost) their retrotranposons (the self-replicating "selfish genes"): the genomes of the asexual rotifers *Bdelloidea* are free of LINE-like and gypsy-like retrotransposons, due to deletions. Except, two retrovirus-like sequences persist in their telomeres. These elements can encode *gag*, *pol* and *env* viral genes [398]. Further, the rotifers harbor the telomere-attached very ancient intron-containing Penelope retroelements; reviewed in [147]. The bdelloidea Penelope retroelement is the telomerase-related Athena. It protects chromosome ends from shortening due to nucleotide losses at cell divisions. Ancient chromoviral (chromodomain-containing retrotransposon) insertions left behind in the bdelloidea genome are the Juno and Vesta LTR retrotransposons [398, 399].

Comment. There is an *éclat* difference between telomere replacement in unicellular eukaryotes (like the ciliates) and somatic cells of multicellular hosts (like the mammals). The ciliates immediately cap with telomeres their chromosome ends after each cell division, whereas the chromosome ends of somatic cells in multicellular hosts remain uncapped, and steadily shorten with each cell division (approx 50 bp per cell divisions). These somatic cells of multicellular hosts undergo senescence and die, at which time stem cells may (or may not) replace them. The tetrahymena cells are rejuvenated and escape senescence due to constant telomere repairs. In that, unicellular eukaryotes (example: the ciliates) resemble malignantly transformed somatic cells of the multicellular hosts, except that telomerase action is constitutive (irreversible) in the latter, whereas it can accelerate (at conjugation) and decelerate (after conjugation) in the former.

Those were the times in the Cambrian sea (or before), when horizontally acquired and inserted genes (viral and otherwise) were expropriated by unicellular organisms and preserved as part of their own genome. Self might have accepted non-self at those earliest stages of cellular evolution. There was no senescence consequential to chromosome shortening, and no natural death on account of cell divisions taking place in mature age. Some ancient unicellular eukaryotes, as judged from their extant descendants, displayed genomic and metabolic features that characterize the malignantly transformed individual cells of multicellular organisms.

The infamous PI3K "cell survival pathway". Giardia uses it in its normal life style (Figure 28). In the amitochondriate protists (giardin in giardia, the PI3K pathway), is physiological and reversible. Among other functions, it regulates the trophozoites' encystation and excystation. These processes are bordering the phenomena of "trans-speciation." The proliferative trophozoite internalizes its flagella, fragments its adhesive disk, and produces encystation vesicles, which fuse with the plasma membrane, forming the cell wall. The process is reversed at excystation. Calmodulin, calcium flow, protein kinases A, and B, cyclic AMP and MAPK regulate these processes. The encystation process is inhibited by the PI3K inhibitor LY294002, but not by another PI3K inhibitor, wortmannin, that allows the formation of highly pleomorphic cysts [400, 401]. In the giardia trophozoites, the genes *gipi*3k1 and *gipi*3k2 encode the PI3K enzyme. The slime mold, *Dictyostelium discoideum* (Figure 6) possesses a similar class III *gipi*3K1 gene. Extracellular release of *G. lamblia* PI3K inhibits the release of IL-12 by host's DCs [402]. The descendants of the free-living ancestral giardia must have added this faculty to their pre-existing PI3K system upon becoming a parasite.

Save and preserve your PTENs! The presence of an ancestral giardial PTEN gene is just hinted at [401]. The PTEN gene is active in *C. elegans* and is conserved in the mammalian lineages regulating organogenesis and cell traffic; it may undergo loss-of-function, and probably gain-of function mutations [403]. PTEN is active, not only by inhibiting PI3K; it may suppress the Wnt-1 oncogenic pathway (at least in *pten* transgenic mice). This latter effect occurs by inhibiting intranuclear localization of ß-catenin, and suppressing IGF-1 receptor expression [404].

The human PTEN, encoded from 10q23.3, acts as a protein/lipid phosphatase targeting PIP3 (phosphatidylinositol 4,5-biphosphate), the active product of PI3K. Loss-of-function mutation of PTEN results in uncontrolled PI3K/Akt "cell survival pathways", malignant if constitutively expressed. PTEN gene expression is frequently lost in aggressive adenocarcinomas. However, in prostate cancer cell lines, constitutively active Notch bound by its ligand Delta, preserved or restored PTEN function [405]. Intact PTEN in prostate cancer predicts better prognosis. Failure of PTEN to control PI3K results in activation of STAT3. The two systems are distinct and do not share components, but active PI3K phosphorylates STAT at tyrosine. Phosphorylated STATs dimerize, translocate to the nucleus, bind DNA in a sequence-specific way, and thus cause transcriptional activation. The PI3K inhibitor GDC-0941 blocks PI3K and Akt phosphorylation, especially phosphorylation of Y705 that is essential for PI3K activity. The Tec family kinases (TFK) are also controlled by PI3K, and as such are able to phosphorylate STAT. The TFKs are selectively inhibited by LFM-A13, thus LFM-A13 also inhibits STAT phosphorylation. The leflunomide metabolite analogue, LFM-A13 (α-cyano-β-hydroxy-β-methyl-N-(2,5-dibromo- phenyl propenamid) is also a Polo-like kinase inhibitor.

JAK and Src kinase inhibitors AG490 (tyrphostin, 2-cyano-3,4-dihydroxyphenyl)-N-(phenylmethyl)-2-propenamid) and Src-1, respectively, failed to inhibit STAT phosphorylation, so failed rapamycin, too. The Tec kinases emerge as collaborators in the maintenance of the STAT3 and PI3K pathways. Of the six TFKs, it is Bmx (bone marrow kinase on X chromosome) that is phosphorylated, and thus being the mediator between PI3K and STAT. However, the Tec kinase inhibitor LFM-A13 inadvertently phosphorylated oncoprotein Akt. Activated PI3K/Akt work together. It is questionable, if inactivated PI3K and activated Akt would be oncogenic? [406, 407] (Table VIII; Figure 50).

Table VIII Phosphatase and Tensin Homolog on Chromosome Ten (PTEN)

Some unicellular pathogenic eukaryotes Cryptosporidia; Theileria; Trypanosoma activate host cell PI3K molecules to promote their replication. Others (Dictyostelium, Giardia) possess their own PI3K genes/gene-product proteins. Giardia operates two PI3K genes (K-1, K-2) and one PI4K gene. Giardia PIKs inhibit DCs of the host's gut epithelium to react, and release IL-12; instead; induce the production of tolerogenic IL-10 (and the host will carry "asymptomatic" giardiasis). PI3K inhibitors act: wortmannin of fungal derivation inhibiting replication of Giardia and encystation of Entamoeba; flavonoid quercetin-based synthetic LY294002 inhibiting replication of Giardia by inducing the formation of pleomorphic multinucleated protozoal cells with disorganized cytoplasmic vesicles. The forerunner of natural inhibitors of PIKs, TEP1, tensin-like phosphatase, shows up in yeast (*S. cerevisiae*), dictating either meiosis, or sporulation. The drosophila DPTEN controls the PI3K homolog, Dp110, that converts PIP_2 into PIP_3. The cells of vertebrate hosts are armed with the PTEN/MMAC* inhibitors of the PI3K/Akt** pathway, fully operational in human cells. PTENs convert PIP_3 into PIP_2.[1–6]
Host cells undergoing malignant transformation knock out PTEN/MMAC at 10.q23.3 (frequently together with p53 and/or Rb). Germinal center diffuse large B cell lymphoma is driven by the PI3K/AKT pathway. MiR-214 posttranscriptionally eliminates PTEN in patients with gastric adenocarcinoma; miR-214 knockdown restores PTEN and inhibits tumor growth. PTEN-antagonist miR-205 raises the malignant phenotype of lung cancer cells (Table XXIII). In PTEN-deficient prostate cancer cells, chemokine CXCL12 (stromal-derived factor SDF-1) and its receptor CXCR4 reach high levels; these chemokines notoriously attract Treg cells. Silymarin, or difluorinated curcumin restore PTEN expression in squamous or adenocarcinoma cells.[7–12]
PTEN against PI3K/AKT. Germline mutation of PTEN results in Cowden syndrome. Somatic mutations of PTEN (with or without knock-out of p53) allow malignantly transformed cells to operate PI3K/AKT oncoproteins (Table XXIX). The intact PTEN is haplo-insufficient: a single copy of the gene is ineffective. The PTEN protein of 403 aa expresses in its C terminal one PDZ*** and two PEST (proline, glutamate, serine, threonine) domains. In contrast, malignantly transformed cells express the NEDD4-1 ubiquitin ligase proto-oncoprotein****. The PI3K protein co-acts with AKt (protein kinase B, cellular homolog of AKT8 thymic murine retroviral oncogene). AKT inactivates the tumor suppressor proteins BAD (BCl-2/BCL-XL-associated death promoter); FOXO3 (forkhead box O3, inhibitor of cycline-dependent kinases), inhibitors $p27^{kip}$ and $p139^{Rb2}$, GSK3 (glycogen synthase kinase, a c-*myc* proto-oncogene inhibitor), TSC2 (tuberous sclerosis complex), FasL (factor apoptosis stimulatory ligand), and Bim (bcl-2-interacting mediator of cell death). AKT activates oncogenes MDM2 (murine double minute ubiquitin ligase for destruction of p53 protein) and IKB kinase (for the release of anti-apoptotic NFκB). The uninhibited cyclin-dependent kinases Cdk1/2 dictate G_1/S-phase transitions non-stop. The disabled Rb protein $p130^{RB2}$ fails to downregulate E family transcription factor E2F, another mediator of the G_1/S transition. The PI3K/AKT axis activates

(continued)

Table VIII (continued)

mTOR (mammalian target of rapamycin), that is physiologically inhibited by the intact bi-allelic PTEN. Failure to inhibit mTOR results in the activation of the Ras/MAPK pathway. Rapamycin could kill mTOR signaling stem cells (cancer stem cells) with non-transformed stem cells spared. Further, adenoviral vectors could possibly replace the gene PTEN (or the p53 gene). The MDM protein antagonist nutlin could protect the p53 protein from ubiquitination (anti MDM action).[13] ***Intranuclear PTEN sumoylated*** (small 100 aa 12 kD ubiquitin-like modifier protein covalently attached to the classical 403 aa PTEN protein) is excluded from the nucleus by the ATM (ataxia telangiectasia mutated) protein. PTEN-transfected human embryonic kidney HEK239 cells, U87MG glioblastoma cells, and HCT116 colorectal carcinoma cells were used (referenced below). ATM phosphorylates both the PTEN and p53 proteins. If the PTEN phosphorylation site is mutated from threonine to alanine (T398A), PTEN is not phosphorylated by ATM. Mutated SUMO-PTEN/T398A was radioresistant and was not excluded from (remained in) the nucleus. DNA damage response was not mobilized (failure to recruit RAD51 to damaged sites) in PTEN-deficient cells. Only sumoylated PTEN functioned in the process of DNA damage response (DDR). PI3K induced cell survival signaling in PTEN-null ($PTEN^{-/-}$) cells. "Genotoxicity" induced by cisplatin, or doxorubicin, remained unrepaired. PTEN-deficient U87MG cells were highly sensitive to combined treatment with ionizing radiation and small molecular inhibitor of PI3K (BKM120). The $PTEN^{-/-}$ HCT116 cells were exquisitely susceptible to treatment with cisplatin and BKM120. HCT116 xenografts were resistant, HCT116 $PTEN^{-/-}$ xenografts were highly susceptible to combined treatment with BKM120 (patented) and one dose of cisplatin. Synthetic sensitivity of $PTEN^{-/-}$ tumor cells are targets to genotoxic and anti-PI3K therapy combined.[14] The PTEN protein exists in a 576 aa long variant. The $PTEN^{-/-}$ U87MG glioblastoma cells were transfected with PTEN-Long cDNA. PTEN-Long exits from the tumor cell into its microenvironment, from where it travels in the blood serum. PTEN has a cell-penetrating element similar to HIV's TAT (human immunodeficiency virus' transactivator of transcription) and thus enters cells. PTEN-Long entered tumor cells (U87MG; MDA-MB-468), inhibited PI3K signaling and induced caspase 3-mediated cell death. Intraperitoneally injected PTEN-Long induced regression of xenografts of these tumor cells.[15] More on PTEN. More on sumoylation (small ubiquitin-like modification) of inhibitor of growth (ING) proteins.[16]

1 *Bittencourt-Silvestre J et al Arch Microbiol 2010;192:259–65.* ***2*** *Cox SS et al BMC Microbiol 2006;6:45.* ***3*** *Kamda J & Singer SM Infect Immun 2009;77:685–93.* ***4*** *Goberdhan DCI et al Genes Dev 1999;13:3244–58.* ***5*** *Hernandez Y et al J Eukaryot Microbiol 2007;54:29–32.* ***6*** *Heymond J et al PNAS 2000;97:12672–7.* ***7*** *Conley-Lacomb MK et al Mol Cancer 2013;12 (1):85.* ***8*** *Pfeifer M et al Proc Natl Acad Sci USA 2013;110:12420–5.* ***9*** *Cai J & Li M et al Cancer Res 2013;73:8402–15.* ***10*** *Roy S et al PLoS One 2013;8(7):e68543.* ***11*** *Su CH et al J Med Food 2013;16:778–84.* ***12*** *Yang TS et al Cancer Cell Int 2013;13(1):68.* ***13*** *Liu W et al Anticancer Res 2008;28:3613–3620.* ***14*** *Bassi C et al Science 2013;341:395–99; Meng Y et al Nan Fang Yi Ke Da Xue Xue Bao 2013;33:956–61; Zhang Q et al J Pharmacol Sci 2013;122:223–31; Qian J et al Mol Med Rep 2013.* ***15*** *Hopkins BD Science 2013;341:399–402.* ***16*** *Salpathy S et al Carcinogenesis 2014 Jun 5 pii:bgu126.*

*Phosphatase tensin on chromosome ten; mutated in multiple advanced cancers
**Phosphatidyl inositol kinase; cellular homolog AKT 8 mouse thymic retroviral oncogene
***PDZ: postsynaptic density protein; drosophila disc large tumor suppressor; ZO: zonula occludens
****NEDD: neuronal precursor cell-expressed developmentally downregulated

Choanoflagellates #3. The first Tec kinase (tyrosine kinase expressed in hepatocellular carcinoma) appeared in the unicellular choanoflagellate, *Monosiga brevicollis*, whose sequenced genome is ancestral to that of the first metazoans [408].

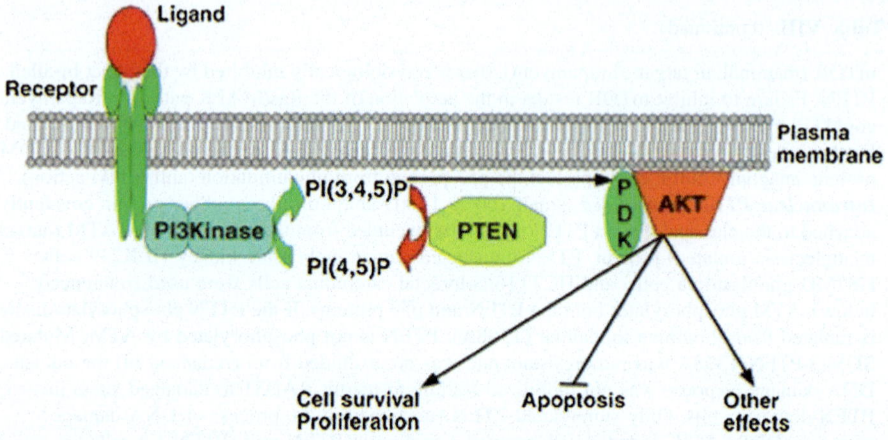

Figure 50 The Tumor Suppressor Gene PTEN and Its Ancient Relationship to PI3K/Akt [2354]

The choanoflagellates feed on oceanic bacteria, among them the algoriphagus Bacteroidetes, whose genome has also been sequenced [409]. Thereafter one *tec* gene was possessed in sponges, the hagfish, frogs and insects. The frog and the insect tec genes are unique. Fishes duplicated/triplicated their *tec*s. The platypus has four *tec* genes. Birds and mammals operate five tec family kinase (TFK) genes. These are, the *bmx* (bone marrow non-receptor tyrosine kinase gene on chromosome X); and the *btk* (Bruton X-linked immunodeficiency tyrosine kinase gene). An ancestral *btk* gene functional in a sponge from the Adriatic sea (see Introduction). Gene *btk* is essential in the human genome for T cell receptor signaling; it is vastly overexpressed in some lymphomas, including B-cell CLL, and therefore is being targeted for treatment with ibrutinib [410] (*vide infra*). Further, the *ltk* (IL-2 inducible T cell kinase); and the rlk/txk (resting lymphocyte kinase/T and X cell expressed kinase) form the tec family [411, 412].

Choanoflagellates #4. It was in the choanoflagellate and capsaspora genomes, where the c-*myc* and *max* genes appeared to encode the Myc and Max proteins for their heterodimerization (Young et al) [413]. The choanoflagellate *Monosiga brevicollis* operates fused carbamoyl synthase and transferase genes for the sake of gaining vitality. The hoglet gene in *Monosiga ovata* is the ancestor of the hedgehog receptors patched and smoothened (*vide infra*).

Comment. The trajectory of the proto-oncogenes of the multicellular hosts covered the distance close to a milliard (billion) years. In the stem cells of the multicellular hosts, these genes still may display their ancient features in dictating early cell divisions and instant telomere repairs, thus exempting their vehicles (the cells) from senescence and natural death.

Trypanosoma telomere lengths. A DNA base ß-D-glucosyl-hydroxymethyluracil ("J") is abundantly present in both strands of the GGTTA telomere repeats of the kinetoplastids leishmania (*L. donovani*) and trypanosoma (*T. cruzi*), and in other kinetoplastids. Kinetoplastids evolved over 500 million years ago as mitochondria-containing unicellular eukaryotes, which keep their mitochondrial DNA at the base of their flagellum. J maintains "the blocked restriction sites in silent telomeric variant surface glycoprotein (VSG) genes", thus there is a strict correlation between J and silenced VSG expression sites. Further, J suppresses recombination between repetitive sequences, including intragenomic retroviral proviral inserts (J acts as a "global repressor") [414]. The trypanosoma (*T. brucei*) surface antigenic expressions depend on the telomeric structural changes. The telomere G-overhang serves as the substrate for the telomerase in its *de novo* DNA synthesis, and the determination of the terminal nucleotide [415]. *T. brucei* evades host immune reactions by sequential expression of different surface antigens, since its genome consisting of eleven diploid megabase chromosomes (also some intermediate and many minichromosomes) encodes several hundreds of VSG. The actively transcribed VSG gene is located in subtelomeric units of the genome; these are the expression sites. Gene conversions and in situ switches regulate antigenic expressions. *T. brucei*'s hexameric telomeric TTAGGG repeats protect the chromosome ends from ds breaks. Cultured *T. brucei* telomeres grow at a steady rate of 6–12 bp/divisions (alternating with occasional truncations). The chromosome lengths of the extensively propagated *T. brucei* Lister 427 strain (in HMI-9 medium containing 1.1% methylcellulose, 15 % fetal calf serum and 5 % heat-inactivated human serum) grew rapidly: telomere gains were 6–12 bp/divisions [416a, b]. *T. brucei* and *T. gambiense* stabilize the length of their telomeres, even in absence of telomerase activity, thus they escape senescence and apoptotic cell death, and still maintain genomic integrity [417a, b]. The telomerases are active in extant kinetoplastids (preserved so from their ancient ancestors), while resting in the somatic cells of multicellular hosts.

The secret of the thousand years old trees and fungus colonies . As mentioned above, G4 DNAs occur in cells other than those of protists and "animalia". In *E. coli*, G4 DNA motifs have the predominance in promoter genes for transcription and signal transduction. G4 DNA motifs (G quadruplex aptamers, the small single-stranded nucleic acids folding into three-dimensional structures) operate in archaea, mycoplasma, clostridia, and mycobacteria [418, 419]. G4 DNA motifs are operational in plants; the chlorophyll synthetase G4 gene of *Arabidopsis thaliana* derives from a rhodobacter, and might have been acquired through horizontal transfer.

Plant cells (arabidopsis; oryza, etc.) contain chloroplasts and also mitochondria. Glycolate oxidation takes place both in the peroxisomes and mitochondria; J-proteins (*vide supra*) are protectors from heat-induced oxidative damage) [420–423].

The sequencing of the genome of the conifer (the Norway spruce) reveals unusually large numbers of repeated sequences in its 35–40,000 genes genome. There are large numbers of DNA sequences of no known function (P. Ingvarsson' newspaper interviews). The 1000 years old trees (*Pinus longaeva*) maintain telomere length in their somatic cells, in contrast to short lived (100–200 years life span) trees [424]. In the thousands of years old tree how long the individual cells live?

The same question arises concerning the thousand years old fungus colonies (*Armillaria ostoyae/bulbosa*, etc.): the 38 acres of fungus colony in Michigan, and the other one in Boca Raton, FL. These colonies consist of rapidly dividing cells, replacing naturally dead cells, or the individual cells acquired some phenomenal longevity? Would *in vitro* cultures of these fungal cells explain what is occurring *in vivo*? Underground symbiotic fungal hyphal network connects the roots of trees forest-wide, providing a secret molecularly digitalized communication between the individual trees [425]. Are sirtuins involved in the thousand years life spans of certain organized cell colonies (see in the Introduction)?

The oncogenes like to be G4 DNAs. The human telomeric DNA is in a G-quadruplex structure. Gellert et al; Ambrus et al; and Phan et al are among the major early contributors to the molecular biology of these structures [426–428]. In mammalian (including human) cells, the G4 DNA configuration is common in proto-oncogenes encoding oncoproteins BCL2, HIF1α, c-Kit, K-Ras, c-Myb, c-Myc, PDGF, Ret, and VEGF(B cell lymphoma; heat-induced factor; cellular tyrosine kinase; Kirsten rat sarcoma; myeloblastosis; myelocytosis; platelet-derived growth factor; re-arranged during transformation, thyroid cancer; vascular endothelial growth factor) and absent or rare in tumor suppressor genes (retinoblastoma, Rb) [429, 430]. Hypomethylations at the sites of somatic copy-number alterations (breakpoints) coincide with the appearance of G4 quadruplex formations [431].

Macrocyclic torands (telomestatin and derivatives) as natural products of *Streptomyces anulatus* emerge as agents targeting the G-quadruplexes; such a hexazole-designated oxazole (HXDV) is inhibitory to human leukemia lymphoblasts [432]. Telomestatin derivatives of macrocyclic structures (with 6 oxazole rings) and porphyrin-based ligands stabilize telomerase-driven cell divisions [433–435], but so far have not reached the clinics.

Selected abbreviations

SIR = silencing information regulator, yeast; SIRT (sirtuin) = silent mating type information regulator, yeast; mono-ribosyl transferase, nicotine-amide adenosine dinucleotide-dependent class III histone de-acetylase, desuccinylase; deacetylator of p53 protein. Suramine = SIRT inhibitor. For review on SIRTs, click Nerunkar PV & Nerunkar VR. For SIRT inhibitor inducing apoptosis in non-small cell lung cancer, click Hoffmann G Breitenbücher F Schuler M Ehrenhoffer-Murray AE. For SIRT inhibition inducing apoptosis in T cell lymphoma, click Nihal M Ahmad N Wood GS. For oncogenesis in mitochondrial SIRT3$^{-/-}$ mice, click Kim HS Patel K Muldoon-Jacobs K et al. For sirtuin inhibitor tenovin-6, click Ueno T, Endo S, Saito R et al. For resveratrol acting on SIRT1, click Yang Q, Wang B, Zang W et al. For tumor suppressor network p53/miR-34a/SIRT1, click Audrito V Vaisitti T Rossid et al.

YY Proteins

The G4 Quadruplex Promoters of YY Protein Genes

The Chinese symbol yin-yang (Figure 51). Telomerases extend DNA lengths after cell divisions; constitutively working telomerases contribute to the cell's escape from senescence, and thus to the "immortality" of cancer cells. Therefore, hTERT is considered to be a proto-oncogene. The DNA telomere ends are G4 quadruplex structures (*see above*). Telomerase expression is regulated by the YY1 proteins. The promoters of the YY1 (yin-yang) protein harbor genomic G4 DNA quadruplexes. These interact with the enzyme resolvase (G4R1), which unwinds G4 DNA (or RNA), and liberates YY1 release [436].

Viruses HIV-1 and EBV obey YY. The YY1 proteins established essential host cell/virus relationships for the regulation of the transmembrane entry and the epi- and intragenomic positioning of viral genomes. YY1 by downregulating the expression of the chemokine co-receptor CCR5 inhibits the transmembrane entry of HIV-1 into the cytoplasm of CD4 T lymphocytes [437, 438]. The two cellular proteins, YY1 and LSF (late simian virus SV40 factor), bind the LTR of HIV-1. This combination of YY1 and LSF induces histone deacetylases to act upon the transcription site of the LTR of the HIV-1 genome. The viral genome thus silenced, fails to initiate transcription. Overexpressed cellular YY1 comes with histone deacetylation (HDAC, histone deacetylase) and inhibition of the viral transactivator Tat pathway. These actions of YY1 are antagonized by the HDAC inhibitor trichostatin, bringing out HIV-1 of its quiescence, and thus restoring its susceptibility to antiviral drugs [439]. The MAPK (mitogen-activated protein kinases) phosphorylate the LSF, thus enabling it to bind the retrolentiviral LTR. The Wnt-induced (Wg, wingless, drosophila; Wg mouse homolog, int, integrated), and MAPK-inducer inflammatory protein p38 (and its agonist anisomycin) seal LSF attachment as a tetramer to the viral LTR. Thus HIV-1 remains in a non-replicative state in CD4 memory T cells. The p38 antagonist and MAPK and NFκB inhibitor (nuclear factor kappa B lymphoma, c-*rel* → v-*rel* turkey reticuloendothelial virus

J.G. Sinkovics, *RNA/DNA and Cancer*, DOI 10.1007/978-3-319-22279-0_8

Figure 51 The Ancient Chinese Yin-Yang. Reference Appendix 2, Explanations to the Figures

oncoprotein), SB239063 (hydroxycyclohexyl-fluorophenyl-methoxypyrimidine, Calbiochem) removed LSF attachment from the viral genome's LTR, promoted re-acetylation of histone H4 of the involved LTR nucleosome, and inhibited YY1's blockade on the viral genomic promoter. The aviremic carriers released HIV-1 particles for antiviral inhibitory drugs to act upon them [440]. The insertion of the leukemogenic retroviral genomes (Moloney virus; HIV-1; avian sarcoma virus) to form viral cDNA proviruses in the host cell genome (*vide supra; vide infra*) is carried out by retroviral integrases bound to the YY1 transcription factor. YY1 remains associated with the integrated viral cDNA. In host cells with YY1 silenced by small inhibitory siRNA, the extent of retroviral genomic integration is reduced [441]. In herpesviral infections (HHV-4 Epstein-Barr virus, EBV; HHV-8, Kaposi sarcoma associated herpesvirus, KSHV), YY1 regulates the transition from latent to lytic infections. EBV genes BZLF1 and BRLF1 induce the lytic phase in B lymphocytes, or epithelial cells, respectively (B, restriction enzyme Bam, Bacillus amyloliquefaciens; protein Z Zebra; left frame; early protein R). YY1 protein suppresses (negatively regulates) the promoters of these two genes, thus maintains latency. HHV-8 ORF50 gene product protein activates the viral lytic cycle. YY1 protein by binding to the promoter gene of ORF50, maintains viral latency [442, 443]. In the case of EBV in Burkitt's lymphoma cells, histone deacetylase inhibitors (HDAI) trichostatin (TST) and suberoylanilide hydroxamic acid (SAHA) and the DNA methyltransferase inhibitor azacytidine converted viral genomic dormancy to the lytic cycle. However valproic acid known to be a HDAI (in hyperacetylating H3), not only failed in inducing the viral lytic cycle, but it antagonized both TST and SAHA. This effect was the result of the activation by valproic acid of genes MEF, ZEB (myeloid Elf1-like factor; ELF, early flowering in arabidopsis; zinc finger E-box binding transcription factor) and YY1 [444]. In the case of HHV-8, COX2 and its metabolite PGE2 (cyclooxygenase; prostaglandin) create the inflammatory background for sarcoma induction. The eicosonoid PG receptors (EP) regulate the system. PGE2 and EP receptor agonists (AH6809 or

GW627368X, patented) induced the LANA (latency associated nuclear antigen) promoter and YY1 and maintained viral latency. In contrast, EP receptor antagonists downregulated both COX2 gene and LANA expression. EP receptor-induced LANA promoter activity was down-regulated by the inhibition of Ca^{2+} (hypocalcemia), PI3K (phosphatidyl inositol kinase) and NFκB. Whereas, down-regulated COX2 activity was reversed by activators other than the EP receptor. In KS cells the PGE2 and its EP receptors work both in autocrine and paracrine fashion. This system is essential for KS cells for maintaining viral latency [445]. COX2 and EP receptor antagonists could disrupt viral latency, and thus expose the tumor cells to the potent antiviral immune reactions of the host. YY1 levels are significantly elevated in HPV-induced squamous carcinoma cells. Arsenic compounds (arsenic trioxide, As_2O_3, ATO) suppress latent membrane protein expression by EBV [446], and induce p53 activation in human papilloma virus-induced squamous carcinoma cells, which overexpress the YY1 protein; in this latter target, apoptosis-induction by p53 resulted [447].

YY and papillomavirus. The papillomaviral oncogene E6 targets p53 for degradation and c-Myc for activation (E7 targets Rb for degradation and cJun for activation). Inhibitory p53 RNAi activates the EGFR promoter in human keratinocytes. YY1 suppresses p53 and also activates the EGFR. Sp1 (specificity protein) is a co-activator of YY1. Wild type p53 suppresses the Sp1 and YY1 proteins. Abolished p53 (by HPV E6) allows the rise of EGFR in keratinocytes (including cells transformed to squamous cell carcinoma) [448]. However, production of serpinB2, PAI-2 (serine protease inhibitor, plasminogen activator inhibitor type 2) restores Rb (retinoblastoma protein) in HPV-18-infected HeLa cells. The upstream regulatory region of the viral genome (URR) regulates E6/E7 expression. Rb protein is a repressor of URR. The PAI-2 binds the C pocket of the Rb protein. Rb silences the viral URR resulting in a drop of E7 levels. Further, Rb recruits the protein LIP (liver-enriched transcriptional inhibitory protein, related to uncleaved C/EBPß protein) to the URR binding site for C/EBPßYY1 (cytosine, cytosine, adenosine, adenosine, thymidine, CCAAT, enhancer-binding site at URR 7709-7719) resulting in transcriptional repression [449].

YY and adenovirus. The adenoviral oncogene E1A binds p300-YY1-HDAC3 complex in the c-myc promoter, thus the Myc-suppressor p300 is so disabled. Later HDAC3 dissociates from the complex allowing increased acetylation of c-myc [450]. When adeno-associated virus (AAV) integrated into chromosome 19, it switches in the presence of adenoviral E1a into its replicative phase. This is achieved by the binding of proteins Rep (viral self- replication initiator), YY1 and MLTF (major late transcription factor) to the p5 promoter at positions −80 for MLTF, −60 for YY1, and at −20 for the Rep binding element. In the absence of adenoviral E1a these three factors suppress p5 transcription. YY1 interacts with adenoviral E1a to switch from suppression to activation of p5 transcription (at response initiator site at +1). AAV excises its provirus in order to express its gene product viral structural proteins. The majority of viral protein transcription switches from p5 to p19. The Rep complex, while repressing the p5 promoter, instead of repressing also the p19 promoter, activates the p19 transcription [451].

The guardian of the genome encounters YY. From the short arm of human chromosome 17, the p53 gene encodes its 53-kD phosphoprotein (p53), which binds DNA and acts as a transcription factor. The healthy phosphoprotein from WTp53 inhibits cell divisions, acts quickly, and is generated in close to undetectable small amounts, whereas the mutated gene produces large amounts and easily detectable (by histochemistry), but ineffective phosphoprotein. The short-lived healthy phosphoprotein can be inactivated by binding with the MDM protein (the human equivalent of the double-minute chromosome product of mutated murine fibroblasts), and as such is sent to ubiquitination. In the cell, YY1 acting as a proto-oncogene gene product down-regulates p53 and promotes the ubiquitination of its protein product bound to MDM. YY1 prevents the acetylation of the lysins in the histones of the p53 gene by its coactivator p300. The de-acetylated p53 protein yields to MDM binding and ubiquitination, that the acetylated p53 protein resists. The p300 coactivator is a potent histone acetyltransferase, however subject to deactivation by YY1. YY1-inhibitory siRNA reverses the process: the co-activator p300 acetylates the p53 histones and stabilizes the p53 protein [452].

The Graffi mouse leukemia virus does not appear to have snatched a cellular oncogene to insert it into its genome. Instead, the virus inserts its genomic provirus into the promoter region of the YY1 gene located in mouse chromosome 12. In these mice YY1 levels rise and G-CSF levels drop (granulocyte colony stimulating factor), and neutrophil granulocytes remain undifferentiated. Since in some human acute myelogenous leukemias (AML) YY1 levels are high, the human YY1 gene (at 14q32) should be investigated for a human leukemogenic proviral DNA insertion [453]. For viral gene insertions into mouse and human chromosomes: *vide infra*.

MicroRNAs regulate YY. In the ribosomes YY1 protein activates the HRPL gene (hypersensitive response and pathogenicity, discovered in virulent bacteria) [454, 455]. Both YY1 and YY2 repress ribosomal gene transcriptions targeting primarily cell proliferation signals (which rise upon YY1 and YY2 knock-out by short hairpin RNAs). Ultraviolet light responsive genes are targeted by YY2, but not by YY1. Apparently DNA repairs by YY1 and YY2 were at different sites. Both YY1 and YY2 are regulators of mitochondrial genes and suppressors of the *myc* gene mRNA (thus Myc protein expression decreased 44 % by shYY1 and 38 % by shYY2 [456]. Of natural YY1 inhibitors, microRNA 34 (miRNA-34) emerges as the primary protector of p53 by decreasing YY1 expression [457]. By unknown mechanism, microRNA-125a-5p protected wild type p53 mRNA and its p53 protein expression, so much so, that p53-induced apoptosis was initiated in lung cancer cells. This effect disappeared when miR-125a-5 was blocked [458]. Chromosomal deletions 13q (miR-15a/miR-16-1 cluster), 17p (WTp53), and 11d (miR-34b/miR-34 cc cluster) worsen the prognosis of chronic lymphocytic leukemia (CLL). The derangement of the miR/p53 feedback circuitry resulted in the indolent replication and apoptosis-resistant accumulation of the CLL cells [459]. The 30 kD protein prohibition protects p53 from YY1 suppression by suppressing the YY1 promoter [460].

For long the kinase phosphorylating YY1 remained unidentified. The YY1 protein is a substrate of the Polo serine/threonine kinases. YY1 phosphorylation is cell cycle-regulated, peaks at G2/M and occurs at threonin 30 in the activation domain [461].

Elusive Alleles

The Living Fossil Remains a Successful Survivor

The "living fossil" successfully introduces gene imprinting and viviparity. The platypus' (*Ornithorhynchus anatinus*) genome has been sequenced [462], as referred to the multi-authored report in Nature. The chimera platypus, preserved parts of its reptilian genome, to which it added first avian and then some mammalian genes. The male platypus has testicles within the lower abdomen, but without descent into a non-existent scrotum. The female platypus possesses ten X chromosomes (5 X; and 5 Y chromosomes for the male platypus). She has a small uterine-like cavity for the fetus and for its small temporary placenta. However, eventually the foetus is encased in a parchment-shelled egg, and is delivered as such. Thus, the female platypus lays eggs encoded by avian-like genomics, but lactates, not from external mammary glands, but from subcutaneous glands oozing milk though the abdominal skin [463, 464]. The platypus genome provides retrospective insights into the evolution of gene imprinting. This epigenetic phenomenon is the parental-origin-specific mono-allelic expression (not imprinted) *versus* repression (imprinted) genes. Imprinting consists of gene methylation. The process is programmed during gametogenesis, and it is vertically heritable. In the mammalian line, it is evident in the marsupials (opossum and wallaby), and extensive in the eutherian mammals (including *Homo*). The monotreme platypus lies between egg-laying birds and placental viviparous eutherian mammals. The platypus sex chromosomes preserve the status that existed prior to the establishment of the viviparous eutherian mammalian XY system, in which gene imprinting is firmly installed. However, in 6 different autosomal genes of the platypus, gene imprinting is already present [465]. Insulin-like growth factor 2 (IGF) and mannose 6-phosphate/IGF2 receptor are not imprinted in birds and in the platypus, but are imprinted in marsupials and eutherians. The M6P/IGF2R is imprinted in the marsupial, opossum, but lacks the methylation of the CpG island in its intron 2. Higher up in lineages of rodents and artiodactyls (sheep, cow, pig) the M6P/IGF2RS is

J.G. Sinkovics, *RNA/DNA and Cancer*, DOI 10.1007/978-3-319-22279-0_9

imprinted. Scandentia (tree shrew), Dermoptera (flying lemur) and Primates (from ringtail lemur to *Homo*), M6P/IGF2R imprinting is absent. The higher mammalian M6P/IGF2R locus is involved in cell growth restriction and is probably pro-apoptotic in T cell-mediated immunity. Its non-imprinted bi-allelic expression is therefore advantageous to its primate hosts. In mice, one allele is imprinted, therefore just one (not Knudson's two hits, *vide infra*) mutation is needed to knock out the entire system [466, 467]. The ancestors of the eutherian imprinted gene clusters appeared in the common genome before the divergence of the bird line. This is most evident in the case of the IGF2 and the SDHD (succinate dehydrogenase subunit D) genes. The IGF2 gene is imprinted in the marsupials and eutherian mammals, but not in the monotremes (platypus). Most of the imprinted genes with allele-specific DNA methylation of their CpG islands are present in the germ line (germ line differentially methylated regions, gDMRs). Transposon-mediated increase in gene imprinting occurred in therian mammals, but PEG10 appeared as a new acquisition. The genes imprinted in therians, but not imprinted (are biallelically expressed) in the platypus are IGF2, IGF2R and UBE (ubiquitin protease E3A). In contrast to its abundance in primate genomes, the platypus genome contains very few if any LTRs [468].

The paternally expressed gene PEG10, mapped in the human chromosome at 7q21, is a retrotransposons-inserted gene adapted through mammalian evolution. The PEG10-RF1/Rf2 (ribosomal frameshift) proteins are inhibitors of apoptosis and inducers of cell mitoses. PEG's origin dates back to the yeast Ty3/gypsy retrotransposons. It contains the conserved "slippery" nucleotide sequence GGGAAACTC. PEG10 is highly expressed during embryonic life and in the placenta. PEG10$^{-/-}$ mouse embryos die (due to placental deficiencies). In adult individuals highly expressed PEG10 remains detectable in the testis and brain. Highly upregulated and constitutively expressed PEG10 proteins appear in hepatocellular carcinoma and in Wilms' tumor. Next to PEG10 gene in the human chromosome at 7q is head-to-head the paternally expressed sarcoglycan ε gene (SGCE). In mammals (humans) both the PEG10 and SGCE gene clusters are imprinted, but the androgen receptor (androgen hormones) and the c-MYC gene product proteins can promote PEG10 expression [469]. Thus PEG10 acts as a proto-oncogene.

Placental syncytiotrophoblasts. The division of Africa and South America provided for the separate further evolution of Afrotheria (examples: elephants, elephant shrews) and Xenarthra (examples: armadillos, sloths, anteaters). The Eurasian (and North American) taxons include the Laurasiatheria (examples: bats, carnivores). An ancestral fusogenic syncytin-encoding retroviral insertion was found in the carnivores of the Laurasiatheria clade (diverged from the Euarchontoglires: rodents, primates), preserved in dogs and cats. This event predated or paralleled the appearance of the murine, rabbit, simian and primate retroviral envelope genes (Figure 52), which generated the placental syncytiotrophoblasts; in primates and Homo reviewed [26, 27]. This fused cell network forms the maternal-fetal interface. In the guinea pig (Cavimorpha), the placental retroviral genome expressed intact Env protein, but not Gag and Pol proteins; the Env protein is not fusogenic, but promotes the invasiveness of the trophoblasts [470, 471].

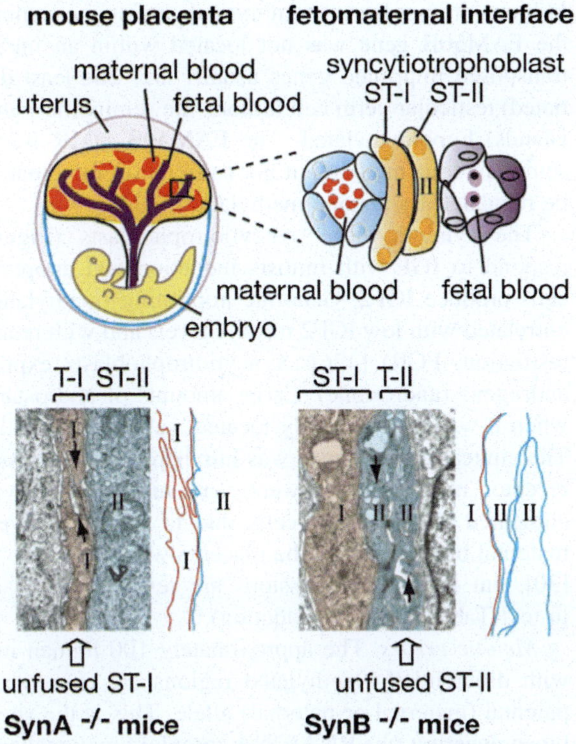

Figure 52 Endogenous Mammalian Retroviral Gene-Product Proteins Syncytins in the Syncytiotrophoblasts of the Placenta. [2355] Reference Appendix 2, Explanations to the Figures

Located in the Trapp9 (trafficking protein particle complex) introns 16 (mouse chromosome15), the Peg13 (paternally expressed) single exon gene undergoes germ line-derived maternal methylation in the CpG island in its promoter region. There are more CpG islands that are methylated only in the eutherians (placental mammals) and not in the marsupials. In the eutherian genomes novel CpG islands emerge in the introns, ready for the acquisition of gDMR [472]. The eutherian genome acquired these sequences within the immunobiological conflict between the maternal host and the fetus and its placenta: the viviparity-driven conflict [473]. The stable differentially methylated regions (DMRs) in the human genome are present within, or adjacent to, the imprinted gene, most often in the gametes, either of the female (mother) or male (father). The methylation is maintained in the somatic tissues throughout the lifetime the individual. At the promoter region of the human locus 6p25 of the FAM50B gene (sequence similarity 50, member family B, X-derived retrogene) a unique DMR at 5′ end is present. The gene FAM50B is non-existent in birds and the prototherian platypus, but present in the metatherian opossum (and above: the placental super-order, Afrotheria). However, in the promoter region of the mouse ortholog FAM50B, there is no DMR. Therefore, the human DMR of FAM50B is the consequence of a species-specific retrotransposition, which occurred after the sidewise divergence of the rodent lineages from the upward mammalian trunk. The imprinted gene is expressed in monoallelic fashion.

In human tumor tissues (embryonal carcinoma, seminoma, mixed germ cell tumors) the FAM50B gene was not located within an intron of a host gene, like retrotransposed imprinted genes usually are. The least differentiated (most undifferentiated) testicular germ cell tumors, the seminomas, are overall (globally and in CpG islands) hypomethylated. The FAM50B DMR is completely hypomethylated in seminomatous tumors, but not so in non-seminomatous germ cell tumors, that may be regulated by histone methylation [474].

The syncytin-fused syncytiotrophoblasts (Figure 52) express IGF-Rs and respond to IGF with mitosis induction and apoptosis-resistance [475]. Placental cells produce IGF2. Genomic imprinting (methylation) at the loci of IGF2 genes correlated with low IGF2 mRNA levels and with retarded fetal growth (fetal growth restriction, FGR) [476]. Syncytiotrophoblasts express cytoplasmic receptors for androgens (androstane). Large amounts of androstane receptor protein, especially when it was intranuclearly located, suggested constitutive expression of the gene. The clinical consequence was intrahepatic cholestasis. In these patients androstane receptor mRNA levels were very high [477]. Syncytiotrophoblasts have been compared to malignant cells due to their invasiveness in the decidua and the maternal blood stream. The placenta was referred to as a "pseudomalignant organ" [30], but its gene expressions are reversibly (not constitutively) expressed (see in-text Table in the Introduction).

Meiotic errors. The approximately 100 human imprinted protein-coding genes with differentially methylated regions are expressed exclusively from a specific parental (maternal or paternal) allele. This is the site where the ultimate master of bioengineering (the RNA/DNA complex) may make a grave mistake (in addition to some other loss-of-function degenerative mutations). A meiotic error in the ovary results in teratoma formation encoding the three embryonic germ cell layers exclusively from maternal chromosomes. A hydatidiform mole without a fetus may carry exclusively paternal genomes. The biparental origin is essential for the mammalian conceptus, including the properly imprinted and monoallelically expressed genes. Uniparental paternal disomy may be confined to one chromosomal segment (15q11;11p15) with survival, but with severe consequences [474].

Poecilia fishes. Attempts at viviparity dates back to poecilia fishes, and squamate reptiles (lizards and snakes) and amphibians. The immunobiology of parent-offspring conflict extends from mother and foetus to intrauterine intersibling competition, and to the co-expression *versus* imprinting of one of the bi-allelic maternal and paternal genes in the concepti [478–480]. Pregnant viviparous fishes carry large metabolic burdens due to multiple paternity of large brood (clutch) sizes [481]. Compared to mammalian species, gene imprinting during germ line development has not been well established in fishes and in birds. At this level, DNA (cytosine-5-) methyltransferase 1 (DNMT) maintains methylation, and DNMT3A/3B catalyze *de novo* methylations. In the chicken, the female germ cell-specific enzyme, DNMT3B is suppressed by microRNAs miR-29b (reducing global hypermethylation). The miR-222 inhibits cyclin-dependent kinase and PTEN, thus promoting cell proliferation by unopposed cell survival kinase PI3K. In chicken, miR-15, miR-29b, miR-222 and miR-383 down-regulate DNMT3b [482].

The expression of the IGF2 in placental fish (poeciliids) is in a balanced bi-allelic mode: there is no evidence for a parent-of-origin effect in the transcriptional regulation of the ligand IGF2. Therefore, IGF2 may not be involved in the development of the fish placenta, and the parent-offspring intragenomic conflict may not exist. Nevertheless, the IGF2 gene is highly expressed in the placenta. In the fish placenta, there are no trophoblasts. In contrast, in the mammalian placenta, IGF2 (in autocrine or paracrine circuitries) feeds the trophoblasts in their invasion of the maternal deciduas (*vide supra*). Leptin-R expression in the syncytiotrophoblasts added another circuitry to the feto-maternal relationship [483].

The Long Trajectory of the IGF-like Receptor and Its Ligands

Carbohydrates Provide Quick Energy

Ancient eukaryotes taste sugar. The amitochondriate primordial unicellular life form, the giardia (*vide supra*), possesses insulin-like receptor and its ligand, insulin-like peptides. Was the system installed into the earliest eukaryotes, or did a descendant of the free-living ancient giardia acquired it through horizontal gene transfers from its parasitized hosts? Giardia can be grown in L-cystein-rich media containing serum (or serum Cohn fraction). These sera contain insulin-like growth factors I and II (IGF). Not insulin, but IGF-II attached to its receptor and thus promoted the growth of the trophozoites. Recombinant human EGF or FGF failed to stimulate the growth. By serology, the giardia receptor was antigenically different from the mammalian IGF-R receptor. The giardia receptor signaled to the intrinsic phosphotyrosine kinase (PI3K?), when inducing mitoses. In response to sugar intake, the *Tetrahymena pyriformis* excreted an insulin-like peptide and synthesized its receptor; the ligand-to-receptor reaction was independent from ATP [485, 486]. *Ciona intestinalis* produces three insulin-like molecules encoded from three genes; these peptides are distant orthologs of the vertebrate insulin family members. Two putative insulin receptor sites were identified: one, the tyrosine kinase type, and the other, a new GPCR-type (G protein-coupled receptor), which probably is the ancestor of neurotransmitter and neurohormone receptors [487].

The tiny cerebral anlage and the notochord of the amphioxus attracted the attention of Russian, German, and British biologists in St.Peterburg, Russia, Königsberg, Germany, and London, Great Britain, in the mid- and late-1880s. The amphioxus uses its stem cells not for oncogenesis, but for tissue regeneration. Regeneration of amputated tissues (tail) begins with de-differentiation of existing tissues. Regeneration of fin, neural tube, notochord and muscle occurred in an orderly fashion. In older subjects (2-y old), the regenerative capacity diminished and if it occurred, it went on in a retarded pace [488]. The amphioxus yields cDNAs for an insulin-like peptide and for its receptor. Mammalian insulin induced the

J.G. Sinkovics, *RNA/DNA and Cancer*, DOI 10.1007/978-3-319-22279-0_10

phosphorylation of the amphioxus receptor. The amphioxus IGF-R is considered to be the ancestor of all vertebrate mammalian insulin receptors (and IGF-Rs) [489, 490]. Growth hormones activate insulin-like growth factor (the ligand) production. It is under suspicion that this combination may be proto-oncogenic.

Hyperglycemic drosophila, hypoglycemic caenorhabditis. In insects, the drosophila insulin-like receptor protein captures a family of insulin-like peptides (InR; ILPs). The InR is a most active "cell survival factor": it signals to activate the proto-oncogenes PI3K/Akt (Ak mouse thymic retrovirus oncogene). Phosphoinositide-dependent kinase-1 (PDK) is a cell protector anti-apoptotic factor. Akt is phosphorylated by phosphoinositide-dependent kinase-1 (PDK) and TOR (target of rapamycin) and vice versa. Further, Akt phosphorylates FOXO (forkhead box transcription factor). In the background, the expression of the *int* gene is regulated by ERK/MAPK, an activator of the *ets* proto-oncogene (the E26 replication-defective avian acute leukemia retrovirus captured the *ets* gene: c-*ets* → v-*ets*; it encodes cell transformation-specific factors: E**ts**, E-**t**wenty-**s**ix). MAPK/ERK activates Ets by phosphorylating it at threonine residue T151. The drosophila gene *pointed* turned out to be the ortholog of the *ets* gene. Through this chain reaction, Pointed (Ets) drives InR expression. InR activity clears circulating glucose. In flies with one copy of the InR gene missing, circulating glucose levels in the hemolymph rise high [491]. The insulin-driven extraordinary metabolic activity subsides with de-activation of the involved metabolic pathways; none of the proto-oncogenes are constitutively activated.

The genes *sna* (snail), *slu* (slug) and *twi* (twist) encode the drosophila mesoderm [492]. This very same 3-gene association carries major role in the epithelial-to-mesenchymal transition of human adenocarcinoma cells as they dowregulate epithelial (E-cadherin, cytokeratins), and upregulate mesenchymal (N-cadherin, fibronectin, vimentin) cellular genes (*vide infra*). Further, Slug is an anti-apoptotic protector of mammalian cells exposed to γ-irradiation. It is the mitochondria-induced apoptosis against which Slug acted as an inhibitor. While it has been recognized as a "cell survival agent" and an oncoprotein-collaborator, whose activities are antagonized by wild p53, unexpectedly, in the human prostate cancer cell lines (PC-3 and DU-145), Slug decreased cyclinD1 expression and halted cancer cell proliferation at G1. When anti-sense shRNA inhibited Slug synthesis, upregulated cyclin D1 expression promoted renewed cancer cell proliferation [493]. Of the inhibitors of the Wnt oncogenic pathway (frizzled-related protein, FRP; dickkopf proteins, Dkk; Wnt-inhibitory factor, WIF), WIF1 gene promoter is commonly silenced by hypermethylation of its promoter in prostate cancer. Functional WIF encodes epithelial cell markers and thus induces mesenchymal-to-epithelial transformation [494]. In Snail/Slug-induced epithelial-to-mesenchymal transformation, the chemokines CXCL12 (the ligand) and CXCR4 (the receptor) act as promoters. Androgens promote chemokine CXCR4 expression. Slug and these chemokines interact. Antisense shRNA-knockdown of Slug decreased the expression of these chemokines. CXCL12 blockade suppressed Slug-induced MMP9 production and thus tumor cell invasion (due to preservation of the intercellular matrix) [495]. Krüppel-like factor degradation liberates Slug; TGFß stimulated KLF degradation, thus it indirectly promotes Slug.

Slug initiated the ETM transformation of prostate cancer cells. Both KLF4 and FOXA1 (an epithelial cell determinant) remain Slug inhibitors [496a]. In PTEN- and p53-depleted prostate cancer cells, TGFß induces ETM, however, miR-1 and miR-200 were inhibitory to the transformation. In response, Slug emerged as the repressor of miR-1 and miR-200, thus accomplishing mesenchymal cell gene activations to complete the process of ETM transition [496b].

In caenorhabditis, TGFß regulates cell proliferation, differentiation, apoptosis, and sugar metabolism. Smad proteins (signaling mothers against decapentaplegic, *dpp*; decaplentaplegic: TGFß) interact with intranuclear DNA. Stagnant larvae, referred to as dauer formation, DAF, exit from their arrested stage upon TGFß stimulation: their *daf* (dauer defective) genes encode TGFß receptors. The receptor-associated Smad Daf-8 (the TGFß's signaling component) unites with nuclear hormone receptor-69 (which is the caenorhabditis' ortholog of the mammalian hepatocyte nuclear factor-4α) to increase insulin secretion. The hypoglycemic nematode becomes "lethargic" (reversible with glucose). Insulin-hypersecreting individual worms experience shortened life span [497a, b]. The PTEN lipid phosphatase homolog is DAF-18; it already antagonizes PI3K (AGE, in caenorhabditis) conserved so, up to mammalian cells. AKT phosphorylates DAF-16, the FOXO homolog (for intranuclear DNA activation). DAF 16 activity is modulated by PDP-1 (pyruvate dehydrogenase phosphatase). PDP is a negative regulator of TGFß and insulin gene expressions. DAF-2 is the insulin/IGF-1 receptor kinase, which initiates an AGE (PI3K) cascade, thus overruling PTEN. DAF7/TGFß binds receptors DAF1/DAF4, which activate by phosphorylation the receptor-associated R-SMAD [498a, b] (Figure 39).

Drosophila develops leukemia with genomics and dysregulation of hematopoietic differentiation comparable to that of human leukemia, as illustrated [499]. Loss-of-function mutations of the germ-line specific tumor suppressor gene *gld* (germ line development defective) loses its protective effect in caenorhabditis, and the elegant nematodes succumb to lethal germ line tumors. However, both in the drosophila and in the caenorhabditis, loss-of-function mutations of gene equivalents of *daf* (dauer defective; decay accelerating factor) result in prolongation of tumor-free life. The DAF-16/FOXO transcription factor (and other DAF proteins, DAF-2, etc.) downregulate the IGF-R and the insulin pathway, and positively target p53 and so-called tumor suppressor gene pathways. Some of the genes suppressed or upregulated by the DAFs in the drosophila and in the caenorhabditis are orthologs of human oncogenes or tumor suppressor genes [500] (Figure 39).

Another generation of proto-oncogenes arises. In caenorhabditis, the amino terminus of TGF fuses with the carboxy terminus of the neurotrophic tyrosine receptor kinase (NTRK); these proteins dimerise, autophosphorylate and activate the Ras-Raf-MEK-ERK pathway. Physiologically, this is a survival pathway. However, the chromosomally translocated TGF-NTRK fusion oncoprotein may be constitutively expressed, and thus transforms cells [501]. The NTRK proto-oncogene will be cited again in human cases of infantile fibrosarcoma (*vide infra*). Apoptosis regulation and mitochondrial fission or fusion are interrelated. Bcl-$_{XL}$ could cause either fission or fusion of mitochondria, depending on its concentration (overexpression resulting in fusion). Its opponent, Bax (Bcl-2-associated X protein), induces mitochondrial fission (apoptosis). However, the caenorhabditis genes *ced* (cell death) and *mfn* (mitofusin) have not as yet correlated their activities. The bcl-2 family influences also egg-laying (*egl*) [502].

The ancestor (ortholog) of the mammalian (human) anaplastic lymphoma kinase (Alk) fulfills physiological functions in the drosophila and it is expressed in the CNS (dAlk). In the embryonic gut, *dpp* (decapentaplegic/TGF) and dAlk interact. The Alk receptor's ligand, jelly belly, is secreted as encoded by gene *jeb* [503, 504]. Alk interacts with Ltk (leukocyte tyrosine kinase) in drosophila, caenorhabditis and in mammals [505]. The neurofibromatosis gene (*nf1*) is a downregulator of Ras signaling; drosophila carries an ancestral ortholog of the human *nf1* gene/protein (dNf). Mutated human *nf1* gene deficiency is the major contributor to neurofibromatosis; consequential unregulated Ras signaling is the direct cause of neurofibromatosis. The dAlk antagonizes organismal growth and olfactorial learning in the drosophila, while it acts as an activator of the Nf1 protein-regulated Ras/ERK signals; or is it the NF1 protein that co-localizes and inhibits the dAlk protein. The Nf1 signals promote organismal growth and olfactorial learning [506]. Descendants of genes of essential physiological functions in the drosophila (*alk, slug*) are activated to induce oncogenesis in mammalian cells distressed by chronic inflammation. In the chromosomal translocation t(2:5)(p23;q35). The genes nucleophosmin and anaplastic lymphoma kinase (alk) encode the NPM/ALK oncoprotein. This oncoprotein activates PI3K, STAT, JAK and NIPA (nuclear interacting partner of ALK). When the substrate of ALK, PSF (multifunctional pyrimidine tract binding protein-associated splicing factor) is recruited (transferred from nucleus to cytoplasm) and phosphorylated at its tyrosine residue Tyr293, PSF inhibits cell proliferation and induces apoptosis of NPM/ALK transformed cells [507]. Prostate cancer cells' invasiveness and EMT are driven by SNAI/SLUG oncoproteins encoding from locus 20q13.2 (*snai* has become *snail*) [508]. Knock-out of SNAI/SLUG by siRNA in these cells resulted in halt of invasiveness, and endomitoses without cell divisions, due to cell cycle arrest in G_0/G_1 [509].

The "newly" acquired mammalian hominoid gene TBC1D3 resides in chromosome 17q12, without known ortholog outside the primate lineage, with the AS160 (Akt substrate)/TBC1D3 domain (not tuberculosis). It is TRE-2/USP6-Bub2-Cdc16, originally functioned as a GTPase-activating protein (transformants receiving DNA from Ewing's sarcoma cells; ubiquitin specific protease; budding uninhibited by benzimidazole; cyclin-dependent kinase). The gene is considered to be a proto-oncogene. The gene complex, encoding Rab-GTA-P (Ras-associated binding; guanine triphosphatase activating protein) is engaged as insulin-stimulated (IS) glcucose transporter (GLUT). The insulin-responsive GLUT4 transports hexose to myocytes and adipocytes. The Tre-2-BUB2-CDC16 domain is an AKT2 (mouse retroviral thymic oncogene) substrate. Insulin-activated AKT2 phosphorylates TBC/AS160 at its GTPase-activating protein (GAP) and at several other sites. Signaling sequence steps are accomplished with the activation of insulin receptor substrate (IRS) and PI3K. Consequential to this cascade is the activation of epidermal growth factor (EGF $\rightarrow$ R) and extracellular signal-regulated kinase

(ERK) activation. TBC1D3 protects EGFR from degradation by ubiquitination. Through Raf1 (rat fibrosarcoma oncogene) activation, TBC1D3 indirectly activates *ras* (rat sarcoma oncogene, either Kirsten or Harvey). In turn, for its activation, TBC1D3 protects the insulin receptor substrate-1 (IRS) from ubiquitination by preventing serine phosphorylations of the molecule. TBC1D3 (*vide supra*: Tre-2; Bub2;Cdc16) actively interacts with the regulatory subunit of protein phosphatase 2A (PP2A), thus securing that IRS-1 serines remain de-phosphorylated [510, 511]. Cell survival proto-oncoproteins are in motion and become constitutively expressed in PTEN/p53-defective cells. It appears now that the prostate cancer gene 17, and gene *tre*-2 also reside in chromosome 17q12, and thus the progesterone receptor complex PRC/tre-2 are components of the TBC1D3 complex. These oncoproteins transformed the test cell line NIH3T3 [512]. Thereafter the *tre* genes became practically undetectable in lower mammalians, but reappeared in primates. Tre2 was found hiding as a fused gene in TBC1D3. TBC1D3 underwent segmental duplication and now exists in multiple copies in the human genome [513]. *)

*) *An attempt at simplification and explanation: TBC1D3 is a hominoid nucleocytoplasmic oncoprotein; its chromosomal locus at 17q12 is frequently amplified in various human cancers (adenocarcinomas). It is an activator of GTPases and the ras/Ras pathway; enhances EGF production and EGFR expression; can be eliminated by ubiquitination. Claimed to be a distant orthologue of the flagellar proteome of the alga Chlamydomonas reinhardtii. Some distantly ancestral genes for TBC1D3 were encoding flagellar proteins in the trypanosoma (Rab family GTPases, Ras-associated binding, rat sarcoma). The over 500 my trajectory of a human oncoprotein is a complicated but most exciting issue. The interested reader is advised to seek out original references for the details as published by Bhogaraju S & Lorentzen E: Crystal structure of Chlamydomonas reinhardtii flagellar RabGAP TBC-domain at 1.8 Å resolution. Proteins 2014;82:2282–2287. Compare with protozoal flagella: Michely C. Diniz et al from Brazil (Curr Protein Pept Sci 2012;13:524–546); Johanna L. Höög et al from the United Kingdom (eLife 2014; e01479); Michéle Lefebre et al from France (PLoS One 2013;8(9):e76380). Compare with archaeal and bacterial flagella (Ng SY et al J Mol Microbiol Biotechnol 2006;11:157–191; Muskotál A et al J Mol Biol 2010;403:607–615; Vonderviszt F et al Methods Mol Biol 2012;824:131–143). Compare chlamydomonas flagellae with cilia (Ishikawa T Cilia 2015;4:3).*

Uncertain relation to TBC1D3, but related to crenarchaeaceae and eukaryotic cell division mechanisms (*vide supra*), the endosomal sorting complex required for transport of endocytic cargo (ESCRT) are present in the trypanosoma [514, 515]. This system is involved in handling ubiquitylated intracellular cargo; it had been very well conserved within opisthokonta (fungi and animals), and the trypanosoma's multivesicular body may be an early manifestation of it [516].

The ancient Rab family GTPases. In the case of the dinoflagellates (zooxanthellae) living in endosymbiosis with cnidaria hosts, the active exclusion of Rab11 by the zooxantellae from the cnidaria phagosomes secures the maintenance of the established relationship [517]. In the yeast, Rab family GTPases (rat brain Ras-related GTPase), and yeast transport proteins (Ytp) control vesicular trafficking, exocytosis and cell polarity (the actin cytoskeleton). The GTPase activating proteins (GAP, Bub2/Bfa1) guard the spindle checkpoint pathway by allowing or preventing mitotic exit, depending if the spindle is oriented or mis-oriented. At the completion of mitosis, mitotic cyclins are disposed (degraded). Active Bub2 and Bfa1 negatively regulate the mitotic exit (MEN, mitotic exit network), if the spindle is mispositioned (misaligned) [518]. At the end of mitosis, the mitotic exit network (MEN) inactivates the cyclin-dependent kinases. The Bub2/Bfa1 GTPase-activating proteins (GAP) keep MEN inactive until the spindle is properly formed (aligned correctly along the mother cell's bud axis). Passage of the spindle pole body (SPB) through the bud neck (the constriction between mother and daughter cells) is the signal for the de-activation of Bub2/Bfa1 (brefeldin A), and the activation of MEN in the mother cell. A major component of MEN is the Ras-associated protein TEM1 [519]. The Bub2/Bfa1 GAP activity negatively controls TEM and thus the mitotic exit; however, Bub2/Bfa1 activity-deprived yeast cells may delay mitotic exit by a GTP/GDP switch of Tem1 (termination of mitotic exit). The SPB-located TEM1 is activated by Lte (lipid translocating exporter; GTPase exchange factor) as the SPB enters the daughter cell from the mother cell. The Lte-induced switch of GDP-to-GTP activates MEN [518]. The polo kinase Cdc5 by inhibiting Bfa through phosphorylation also activates MEN. In anaphase and telophase, the Cdc14 phosphatase is released from the nucleolus to reverse the phosphorylation of the Cdc substrates, thus inactivating them (until the commencement of the next cell cycle). However, when Bub2 complexes with Myc, MEN is inhibited (and the cell cycle remains open) [520]. Further, phosphatase 2A (PP2A) against expectation, did not dephosphorylate Bfa1; rather, it acted in concert with Bub2/Bfa1 in the negative control of MEN [521].

In the sea urchin embryo, cells moving from the mesoderm to the ectoderm secrete SNARE proteins (soluble NSF attachment protein receptor, N-ethylmaleimide-sensitive factor), syntaxin, vesicle-associated membrane protein (VAMP) and Rab3 in order to break cell adhesions (intercellular matrix) for their locomotion [522]. SNARE, syntaxins, VAMP and SNAP (soluble NSF-attachment protein), physiological directors of cellular traffic, are predecessors of the metalloproteinases (MMPs), that cancer cells secrete for the clearance of intercellular passageways for their locomotion (invasiveness).

In the caenorhabditis, similarly to a conserved mechanism in mammalian cells, Rab10 and EHBP-1 (Eps homology domain binding protein; Eps15, EGFR epidermal growth factor receptor pathway substrate) are the master regulators of intracellular protein traffic and recycling. Active GTP-bound Rab10 binds to the C-terminus of EHBP-1. In cells with mutated fusion protein, the protein traffic is misdirected (mis-sorted). In EHBP-defective cells (adipocytes), the insulin-regulated endosome failed to transport Glut (glucose transporter) to the plasma membrane [523].

In the human tumor, prostate adenocarcinoma, the PRC17 protein (polycomb group repressive complex) is encoded from the 17q12 amplicon [524]. The PRC (prostate cancer) protein contains a GTPase-activating protein GAP domain related to the yeast cell cycle regulators Bub2 and the Rab/Ytp-specific GTPase activator

Tre (*vide supra*). The PRC17 cDNA encodes the 549 aa protein sharing 84 % aa identity with Tre oncogenes (ubiquitin-specific protease, first cloned from transfected DNA of Ewing's sarcoma, *tre*). The Tre17 oncogene encodes GTPase Cdc42 [525].

In mammalian (human) cells, the GTP-activated Rabs (RabGAPs) are related to AS160/TBC1D1 and TBC1D4 (for TTBC1D3, *vide supra*). AS160 is the substrate of Akt proto-oncogene. GLUT4 (glucose transporter) is released from the cytoplasmic GSV (GLUT storage vesicle) to ascend to the plasma membrane from where GUT transports glucose. RabGAPs regulate this translocation of GLUT. The system is activated by insulin (insulin-like growth factor, IGF). All animal species operate TBC1D1; only vertebrates have AS160 [526]. The initiator of vertebrate DNA methylation, the amphioxus (whose genome encodes DNA methyltransferases and methyl CpG-binding domains and an IGF-R) [489, 490, 527], has anything to do with an ancient AS160 system?

Endogenous retroviral genes encounter Rab proteins and the mTOR pathway. A genomic segment of a human endogenous retrovirus-1 (HRES-1) has been mapped to chromosome 1q42. It is very likely a genomic sequence related to the *tat* gene of the human T cell lymphotropic virus (either type I or II). This DNA sequence encodes a 28-kD and a 24-kD proteins. Both sequences and proteins attract autoantibodies. The 24-kD protein is a GTPase, referred to as HRES-1/Rab4, which by endosomal recycling decreases the expression of cluster of differentiation CD4 and CD3/TCRζ (zeta) in the membrane of T lymphocytes. These proteins are overexpressed in lupus erythematosus (SLE) and induce antibody ("autoantibody") production [528].

Györkey and Sinkovics have shown (photographed) mature budding retroviral particles in tissues and cells (kidney glomeruli) in SLE, and very early proposed that the autoimmune reactions in these patients were directed to a latent human endogenous retrovirus (Figure 53) [147]. These authors also observed cytoplasmic filamentous structures in lymphocytes and in parenchymal cells of patients with SLE (and other autoimmune diseases) and connected these structures with the budding retroviral particles. The cytoplasmic tubuloreticular structures were proposed to be incomplete endogenous retroviruses. However, not so. The cytoplasmic filamentous structures proved to be nucleic acid-free products of the endoplasmic reticulum in reaction to IFNα overproduction, and were also present in pathological entities other than SLE (in Kaposi's sarcoma). From this point on, literature citations considering the retroviral etiology of SLE (and other autoimmune diseases) omitted referrals to both the tubuloreticular cytoplasmic structures and the maturing and budding retroviral particles. However, the early depiction of budding retroviral particles, now viewed independently from the cytoplasmic tubuloreticular structures, should have remained in the literature citations, inasmuch as latent retrovirus (es) are most probable etiologic factors in SLE (Figure 53). If the HRES-1 segment acquires an *env* sequence, it becomes a mature retroviral virion ready to be released by budding [147, 529–532, 533]. A connection between HRES-1, and Rab proteins (human endogenous retroviral sequence; Ras-related protein in brain) encoded in

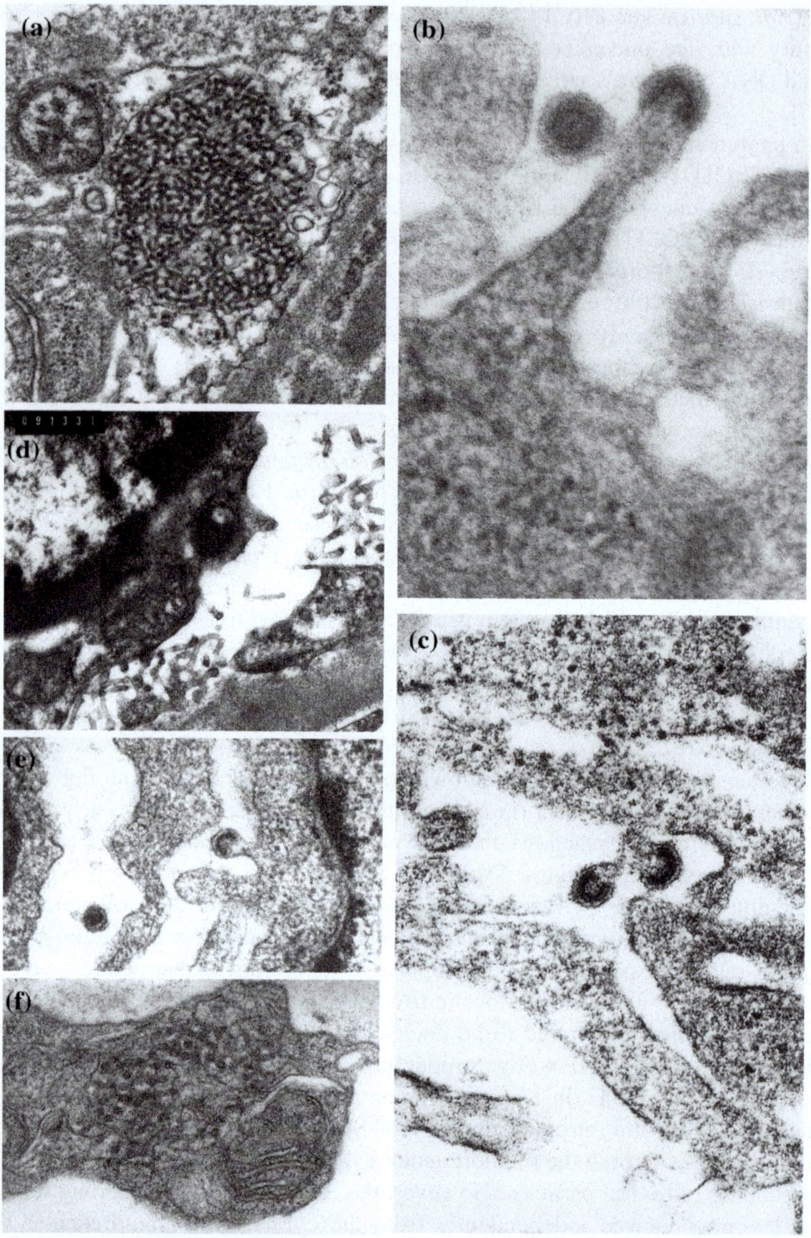

Figure 53 Decades Long Search for Human Endogenous Retroviruses. Reference Appendix 2, Explanations to the Figures

excess, and genomic mTOR activation in SLE is claimed to be the most likely etiological connections in SLE (and some other autoimmune diseases) [534a, b].

Comment. The IGFs and their receptors from primordial cell survival pathways advanced to be proto-oncogenes/oncogenes in the mammalian hosts. In the ancestors of the Giardia, the IGF-R signaled and activated its substrate the "cell survival pathway" PI3K. The activation of IGF-R receptor by IGFs is a most powerful cell survival pathway in single-celled life forms. It has been evolutionarily conserved. An example for *Homo* are the TBC1D1 pathways (*vide supra*). In the pregnant mammalian host, IGF is a driver of the placenta and fetal growth. The ancient IGF-to-R pathway is expropriated by malignantly transformed cells in practically all tumor categories (adeno- and squamous cell carcinomas, myelodysplasia advancing into myelogenous leukemia, malignant lymphoma, melanoma, and soft tissue and bone sarcomas), for the increase of the transformed cells' mitotic frequency and mass [535–569]. Signaling from the IGF-R can be targeted for its inhibition by siRNAs, mcab, mab (monoclonal antibodies, cixutumumab; figitumumab), bispecific antibodies, decorin, dopamine, lycopene, metformin, and small molecular kinase-inhibitors [570–582]. In contrast, high IGF-R expressing human cardiac stem cells exhibit superior therapeutic potential for healing myocardial infarcts [583].

Cancer cells revert to the ancient Warburg type carbohydrate metabolism, but need more energy than what that yields. Extant, free-living unicellular organisms preserved to a great extent the faculties of the overall resistance their predecessors practiced for defense against physico-chemical damage emanating from a hostile environment (*vide supra*). In the primordial anaerobic environment, these life forms practiced a Warburg type carbohydrate metabolism (Figure 9), and maintained their semblance to immortality due to constantly active telomerases, and thus intact chromosome lengths.

A search of the literature for the presence of hypoxia inducible factor (HIF) in the descendants of the ancient anaerobic unicellular eukaryotes (giardia; amoeba; trichomonas) ended with negative results. It appears as if mollusks (oysters), corals (coral-algal symbionts), and the tunicate sea squirt Ciona in the Cambrian sea, and the nematodes (Caenorhabditis) in dry land operated the first HIF molecules. In the cancer cell (clear cell carcinoma of the kidney), the tumor suppressor gene product protein, von Hippel-Lindau factor (vHL), loses its control over HIF-2α, which activates proto-oncogenes (for review, *vide supra*). The cancer cell generates energy from sugars in a Warburg type metabolic pathway, and lives in the state of immortality, due to constitutively activated telomerases. Thus, cancer cells are displaying clear signs of their regression to the life style of their distant ancestors. That was the life style driven by a primordial form of the RNA/DNA complex in the ancient anaerobic environment. The extant healthy cells through mitochondrial oxidative phosphorylation gain 36 mol ATP per mol glucose. In contrast, the cancer cell in its cytoplasmic glucose → pyruvate → lactate pathway, bypassing the mitochondria, generates only 2–4 mol ATP per mol glucose. The Warburg-type sugar metabolism consists of lactate production from glucose, what the cancer cells practice even in the presence of oxygen. The energy (ATP) yield is so low that the cancer cell must metabolize large amounts of glucose for its highly increased energy needs. The cancer cell overexpresses insulin receptors (*vide supra*) and captures insulin to be utilized as a growth factor in exocrine and autocrine circuitries.

In contrast to somewhat differentiated glioblastoma cells, using exclusively the Warburg glycolytic pathway, the glioblastoma stem cells (GSCs) were able to switch to some sort of an oxidative glycolytic pathway, giving a higher energy (ATP) yield. These cells used both, or either one, of the two isoforms of pyruvate kinases (PK-M1/2) and were able to switch from the low energy yield glycolytic, to the high yield, oxidative phosphorylation pathway. Due to their high mitochondrial reserve capacity, GSCs were radioresistant [584]. In Burkitt's lymphoma (BL) cells, the c-*myc* proto-oncogene (stem cell gene; myelocytic leukemia oncogene/oncoprotein) is often translocated to the IgH site. Under glucose deprivation and in hypoxia, BL cells switch to the glucose-independent tricarboxylic acid (Krebs) cycle by the activation of the glutaminase pathway. Glutaminolysis releases energy (other than ATP) as it yields glutamate, oxaloacetate, ketoglutarate and isocitrate. The glutaminolysis pathway of cancer cells can be inhibited by BPTES (bis-2-phenylacetamido-thiadiazol-ethyl sulfate [585]. However, glutaminase-silenced cancer cells enlist pyruvate carboxylase (PC) in order to derive energy from pyruvate. Inhibiting the mRNA of PC protein translation by lentivirus-vectored shRNA severely limited cell growth [586]. The Chinese herbal naphthoquinones, shikonin and its analogs, exert most powerful inhibitory activity on pyruvate kinase-M2 of tumor cell derivation [587]. These compounds arrested the cell cycle of melanoma cells, and induced reactive oxygen species in hepatocellular carcinoma cells, resulting in apoptosis and cell death *in vitro* [588, 589].

The precious carbohydrate metabolism genes were sought after in the archaic cell communities. Indeed, extant archaea, prokaryota (or eubacteria), and numerous unicellular eukaryotic life forms, utilize one of the Warburg type (Embden-Meyerhof-Parnas, Embden-Meyerhof-Warburg, Entner-Doudoroff) carbohydrate metabolic pathways (preferred even in the presence of oxygen). In this pathway, however, for better energy yield, mechanisms exist for a switch to the glutamate, glutathione, and glutaminolytic cycle in some extant unicellular life forms [590–593], just as in cancer cells (*vide supra*). Halophilic archaea (*Haloferax volcanii, Halobacterium salinarum*) contain in their cytoplasm glutamylcysteine, γGC (instead of glutathione), and the *gsh*A gene to encode glutamate-cysteine ligase for the synthesis of γGC. In contrast, *Pyrococcus furiosus* instead of glutathione, uses coenzyme CoA-SH (sulfhydryl) for free thiol [591]. The haloarchaeon *H. volcanii* derives its energy from fructose metabolism, but the genes for its fructose catabolism (phosphotransferases) are shared with bacteria (prokaryota) [594]. It is difficult to decide which one of them was first to evolve the system and how was the precious genomic segment acquired by the recipient.

The glycogen phosphorylase genes and the gene product enzymes of these a-mitochondriate eukaryotes resemble most that of the slime mold (*D. discoideum*) and appear to be related to fungal and plant phosphorylases [395, 595]. Glucokinases and glycosephosphate isomerases of the diplomonad giardia, and the parabasalid trichomonad (*T. vaginalis*) differ from the hexokinase family of enzymes widely spread among eukaryotes; they rather are related to cyanobacterial and chloroplast enzymes of this class. The ancestor of these two entities acquired the genes for these enzymes through horizontal routes [391, 396]. Several other giardia genes of carbohydrate metabolism (phosphomannosemutase, etc.) are not genuine eukaryotic genes either, but derivatives of horizontal gene transfer from bacterial donors (Mitra et al) [397].

In the anaerobic pathway, pyrophosphate may replace ATP as the phosphate donor [596]. Anaerobic carbohydrate metabolism is a primitive pre-mitochondrial trait in some ancient unicellular eukaryotes. *E. histolytica* lost its phosphofructose kinase and pyruvate kinase, and uses instead pyrophosphate-phosphofructokinase and pyruvate-phosphate-dikinase (PPDK), thus yielding 5 ATP molecules instead of 2. Both giardia and trichomonas possess the PPDK pathway, but the giardia enzymes appear to be different from those of the entamoeba and trichomonads. The PPDK pathway may co-exits with the ATP-dependent pathway; to this author this would indicate that previously possessed mitochondrial enzymes function now in the nuclei. However, the genes for these enzymes could have been acquired by horizontal transfer from bacteria [597]. Nevertheless, the carbohydrate metabolism of the three anaerobic protists (giardia, trichomonas, entamoeba) is one of the Warburg type [391], and thus different from the oxidative carbohydrate metabolism of the healthy somatic cells of multicellular hosts; rather, resemble the carbohydrate metabolism of the cancer cells. If the oncogenome in the cancer cell is switched to an archaic mode of glucose metabolism (the Embden-Meyerhof-Warburg variants) bypassing the mitochondria, then the oncogenic DNA is driving the cancer cell at the level of the primordial unicellular microorganisms. At that level telomeres were repaired after each cell division and those cells escaped senescence and natural death, but additional sources of energy had to be provided (*vide supra*).

The Embden-Meyerhof-Parnas-Warburg, or the Doudoroff-Entner, and the glutamate-glutathion pathways of glucose metabolism are widely in use by archaea (*Haloferax volcanii, Sulfolobus solfataricus*) and prokaryotes (*vide supra*). However, if mitochondria of the eukaryota are of alpha-proteobacterial origin, the sugar metabolism of bacteria may be similar to that of the mitochondriate eukaryotes, uni- or multicellular. Are amitochondriate unicellular life forms (examples Amoeba, Giardia, Trichomonas) restricted to the cytoplasmic pathways of glucose metabolism? However, if hydrogenosomes contain mitochondrial genes and/or, if mitochondrial genes were transferred to the nuclei before the resolution of the mitochondria, then some amitochondriate unicellular life forms may still metabolize carbohydrates the mitochondrial way.

What is the pathway of carbohydrate metabolism of the mitochondriate unicellular kinetoplasts (*Leishmania, Plasmodia, Trypanosoma*)? When *Ehrlichia, Naegleria,* or *Theileria* attack multicellular hosts, what form of carbohydrate metabolism they use? This matter appears to have been neglected for molecular studies. However, it is known that the trophozoites of the pathogenic *Naegleria fowleri*, the causative agent of the fatal amoebic meningoencephalitis, (Figure 54) express cell surface glycoproteins (αD-mannose, αDglucose and αL-fucose

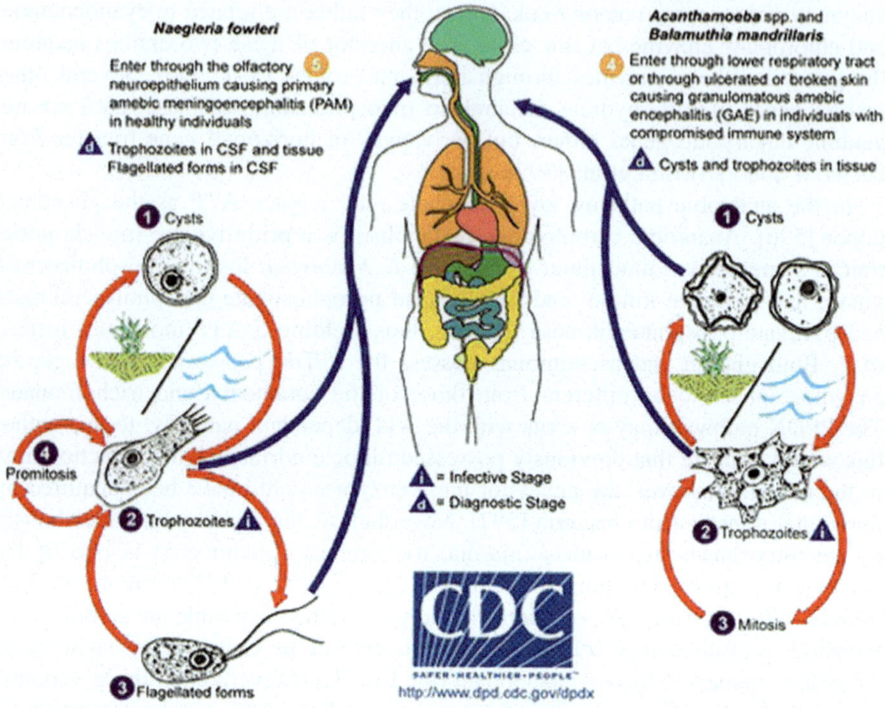

Figure 54 The Trans-speciating Transformations of the Naegleria. Reference Appendix 2, Explanations to the Figures

residues) for attachment to host cells. The nonpathogenic *N. gruberi* does not have these glycoconjugates in its armamentarium [598]. Theileria-parasites (*T. parva; T. annulata*) induce constitutive malignant transformation of infected bovine lymphocytes (and macrophages); the parasite-infected cells proliferate and spread in their host. The parasites activate first in the infected host cells the *jun* proto-oncogene (Japanese seventeen, after avian leukemia reticuloendotheliosis retrovirus 17) and consequentially raise anti-apoptotic Mcl-1 (member myeloid cell leukemia) and c-IAP inhibitor of apoptosis protein levels. The c-Jun NH2-terminal kinase (JNK) is a MAPK family member; transcription factor activator protein (AP-1) activates it by the phosphorylation of its serine residues Ser^{53} and Ser^{73}. The induction of AP-1 occurs via the PI3K pathway. The entire AP-1 (activator protein) transcription factor family is thus activated, and forms multiple dimers. The AP-1 family consists of c-Jun, c-Fos (Finkel mouse osteosarcoma retrovirus) and the ATF2 (activating transcription factor) family members. Further, the parasite activates the IkB signalosome; this results in the liberation of NFκB. The antiapoptotic Bcl2 family members compete with the pro-apoptotic Bim proteins (Bcl2-interacting mediator) [599]. Further, theileria schizonts-induced faculties of lymphomagenesis are those of the c-*myc* proto-oncogene, and the GM-CSF and

TNFα autocrine loops (with PI3K and NFκB activation) [600, 601]. In macrophages (Mϕ), theileria schizonts induce de-differentiation with loss of the immune faculties usually expressed by macrophages (superoxide and inflammatory cytokine production). Of the numerous gene product proteins that regulate differentiation of Mϕ, RUNX (runt-related transcription factor), the MAFs (musculoaponeurotic fibrosarcoma) proto-oncogenes and the colony stimulating factor-1 receptor (CSF1R) are suppressed most by the schizonts [602].

Innate and adaptive immune faculties could protect hosts against theileria infection, but infected lymphocytes outwit host defenses, exactly as malignantly transformed cells generally do [603, 604]. Even when host immune T cells kill the theileria-infected lymphocyte, the schizonts exit intact. The treatment of theileria infections is either preventive (vaccination against the parasite, not against the transformed lymphocytes) [605], or aimed at the kinetoplastids of ancient microbial acquisition. The sequenced theileria genomes allow to view a metabolic map of the apicoplast. Remarkably, special or commonly used antibacterial antibiotics by inhibiting the apicoplast can kill the parasite [606]. However, the most effective drug is the hydroxynaphthoquinone chemotherapeutic agent, the theilericidal buparvaquone (Buparvon) aimed at the sporozoites (not at the transformed cells). Upon the eradication of the parasites by chemotherapy, the infected cells die apoptotic death. In the transformed cells, upon treatment with buparvaquone, Bcl-2 is down-, but BAX and Apaf are up-regulated, and the p53 protein is allowed from its cytoplasmic sequestrated state to translocate to the nucleus, and there to exercise its pro-apoptotic activity [607].

All schizonts and the sporozoites (including those of malaria plasmodia) require a great deal of energy (ATP) in order to reside in the alien cells, and while undergo their own transformation, render the host cells "parasite-friendly." The carbohydrate flux leads to oxidative stress; re-routing the carbohydrate flux from glycolysis to the pentose phosphate pathway (PPP) by down-regulating the glycolytic cytoplasmic nicotinamide dinucleotide phosphate (altering the NADPH/NADP$^+$ ratio), and glyceraldehyde-3-phosphate dehydrogenase (GAPDH), and triose-phosphate isomerase (TPI). These processes are physiological responses of eukaryotic cells against reactive oxygen species (ROS). The glucose analog 2-deoxy-D-glucose (2DG) is a glycolytic inhibitor. Rapid decrease of cytoplasmic glutathione would cause oxidative stress, were it not so, that the cell mobilizes some of these defensive reactions to counteract ROS [608]. In contrast, naphthoquinones are inducers of oxidative stress (ROS), in part by the disruption of mitochondrial membranes. Multimeric naphthoquinones link together in different positions (binaphthoquinones) and thus are able to exert most variable biological activities. These extend from anti-retroviral actions to cytotoxicity to leukemic lymphoblasts. In this latter action, the mechanism is the induction of oxidative stress (generating ROS), or endoplasmic reticulum stress. The toxic effect of binaphthoquinones is mediated by the the disequilibrium of NADPH/NADP$^+$ that they induce. In mammalian cells, this action takes place not in the mitochondrium, but in the cytoplasm. Depolarization of mitochondrial membranes accelerates ROS production [609]. Naphthoquinones may suppress carbohydrate metabolism, similarly to shikonins, and thus act as anti-parasite (anti-kinetoplastids), and anti-cancer chemotherapeutic agents [610].

Comment. Ancient free-living unicellular life forms probably could not alter their environment and had to adapt themselves to "life on the edges." However, the unicellular eukaryotic parasites have means to render their multicellular hosts tolerant, or even "friendly" (supportive) toward their co-existence. In that, these parasites and cancer cells act alike.

Selected Abbreviations

Alk/ALK Anaplastic lymphoma kinase. Click, Eyre TA Eur J Haematol 2014; PR 26. doi:10.1111/EJH.12300. Alk gene rearrangement in non-small cell lung cancer detected by Vysis FISH (fluorescent in situ hybridization). Mutated alk/ros ALK/ROS (reactive oxygen species) respond to crizotinib. Co-mutated ret/RET (rearranged during transfection, also in medullary thyroid carcinoma); axl/AXL/αξλ/ΑΧΛ (Greek for uncontrolled, anexelekto, tyrosine kinase oncogene receptor on chromosome 19q13.1q13.2 for ligand Gas (growth arrest specific); platelet-derived growth factor-receptor-α, PDGF-R; neurotrophic tyrosine receptor kinase, NTRK, render tumors crizotinib-resistant. Click, Ou S-HI et al Front Oncol 2014 Apr 1;4:58.

hTERT human telomerase reverse transcriptase. Are tumor responsive immune memory T lymphocytes exempt of telomere shortening? Click, Qian Y et al Cell Immunol 2014;289:63–9. In Werner's progeria, rapid telomere shortenings and telomere attrition occur in mesenchymal stem cells due to decelerated synthesis of lagging strand telomeres; neural stem/progenitor cells are exempted of this lesion. Malignantly transformed fibroblasts or mesenchymal stem cells are reprogrammed to pluripotency with elongated telomere lengths restored. Click, Cheung HH et al Stem Cell Reports 2014;2:534–46.

Sarcoglycan gene sarcolemmal αγ-sarcoglycan protein deficiencies (loss-or-function point mutations familial, vertical; somatic, acquired). For mice with dystrophin and sarcoglycan deficiencies $Sgca^{-/-}$ developing embryonal rhabdomyosarcoma with excess of murine double minute MDM-2 protein and missense-mutated p53, click Fernandez K et al Am J Pathol 2010;176:416–434.

Primordial Fused Genes Become Oncogenes and Encode Oncoproteins

Primordial Fused Genes Roll Evolution Forward

Extant unicellular eukaryotes still do it. Predating the appearance of eukaryotes, archaeal and prokaryotic (bacterial) fused genes (or operons; or domains) were exchanged by horizontal transfer. These events were analyzed with great accuracy and show that the horizontal transfer of fused genes could transgress phylogenic barriers [611]. Fused genes were transferred from archaea to bacteria and vice versa. The distinction of self from non-self was not as yet imperative. The acetyl coenzyme CoA αβ subunits were extensively exchanged between archaea (*Aeropyrum pernix*) and bacteria (*Escherichia coli*; *Deinococcus radiodurans*) (Figure 4). The fused two domains of the purine biosynthetic enzyme phospho-ribosyl-formyl-glycin-amide were transferred from proteobacteria (the mitochondria) into the nucleus of the eukaryotes. An accelerated evolution of the eukaryotes followed. Thirty-one events of horizontal cross-kingdom fused gene transfers occurred in 51 examined scenarios. An extensive list on the evolutionary history of trans-kingdom gene transfers and fusions is given [611]. So far no evidence is provided for the acquisition of the virulence (toxin-encoding) genes of *Vibrio cholerae*, except for their transfer by phages originally from an unknown source (a dinoflagellate toxin gene)? The circumstances of the insertion of the *cag*A oncogene (cytotoxic antigen gene) into the ancestral *Helicobacter pylori*, and for the acquisition of a potentially oncogenic enterotoxin gene by the colonic *Bacteroides fragilis* remain unknown [27]. The C-terminal half of CagA protein carries the repeated motif EPIYA (glutamic acid-proline-isoleucine-tyronine-alanine). The EPIYA motif exists in two (or three) versions: Western, East Asian (and AmerInd). These motives undergo recombinations and fusions. The ancestral Helicobacter acquired the *cag*A gene by horizontal transfer from unknown source (s) [612]. If it were inserted into its genome by a phage, the extant bacteria have gotten rid of that phage, while retaining the *cag*A gene. In contrast, the *Vibrio cholerae* retained the phage as the carrier of the toxin gene [27].

J.G. Sinkovics, *RNA/DNA and Cancer*, DOI 10.1007/978-3-319-22279-0_11

While the first mitochondriate eukaryotes swam with one cilium (unikonta, opisthokonta), the bikonts grew another anterior cilium and acquired a myosin domain; further, innovated their locomotion also with pseudopodia. In this last common ancestor, two essential genes must have been fused: those for dihydrofolate reductase and thymidylate synthase [613]. However, there emerges the acquisition of the chloroplast, which induced the first major bifurcation of the ancient eukaryotic lines, thus dividing the lines of the photosynthetic and non-photosynthetic eukaryotes. Chloroplast-recipient cells acquired the shikimate pathway, consisting of the catalysis of erythrose-4-phosphate and phospho-enol-pyruvate to chorismate. Through photosynthesis, the chloroplasts provide NADH and ATP as well. The shikimate pathway has also been preserved, especially in fungi and plants; however, some apicomplexa (*Toxoplasma gondii*) operate a full or, a partial (*Plasmodium falciparum*) shikimate pathway. The shikimate pathway was exchanged by horizontal gene transfers (HGT), or acquired by endosymbiotic gene transfers (EGT, from chloroplast to nucleus). It is now possessed by *Phytophthora oomycetes* (*vide infra*), green and red algae (*Chlamydomonas* and *Cyanidoschyzon*), basidiomycete fungi (*Ustilago*), *Cryptococcus neoformans*, the zygomycete *Rhizopus*, the diatom *Thalassiosira*, and the ciliate *Tetrahymena thermophila*. The single-standing shikimate genes disappeared from the metazoan genomes. However, most of the preserved shikimate genes exist in fused formats, including the *arom* gene [614].

Recipients of genomic sequences of a new (post-mitochondrial) endosymbiont, the red algae, formed the taxon of Chromalveolata. The lineages bifurcated to photosynthetic and non-photosynthetic directions [615]. Plantae retained the shikimate pathway. The glyoxylate cycle genes, which are variants of the tricarboxylic (Krebs) cycle (malate synthase, MS; isocitrate lysase. ICL), were acquired by the first eukaryotes from bacteria/prokaryota (not from archaea) through HGT. These genes remained active synthesizers of succinate gluconeogenesis in arthropodes, echinoderms, nematodes (retaining the genes in fused forms), and vertebrates up to the platypus and opossum, but are missing in mammals [616]. Reptiles may operate the MS/ICL cycle, but birds do not have it. (The theropod dinosaurs lost it)? Despite their juxtaposition and their frequent horizontal transfers, the two genes (MS/ICL) ceased to function in the avian and mammalian genomes. The MS gene remains in existence as a pseudogene in placental mammals. Nematodes retained the fused genes, birds and mammals lost the single-standing genes.

The antioxidant NADPH (nicotinamide adenine dinucleotide phosphate) re-converts glutathione from its oxidized formula, glutathione disulfide, by electron donation to glutathione reductase, which splits the disulfited molecule into two glutathione molecules. Glucose-6-phosphate dehydrogenase (G6PD) is within the evolutionarily preserved pentose phosphate pathway, a producer of NADPH. A derivative by duplication and sequence divergence of the G6PD is H6PD (H, hexose), also known as glucose dehydrogenase (GDH), also a producer of NADPH. This gene originated from the duplication of the G6PD gene in the common ancestor of choanoflagellates and metazoans, after the line of this ancestor diverged from the fungi. In the choanoflagellates, the H6PD gene is not yet fused (*vide infra*).

In metazoans, in addition to its H6PD sequence, this gene has a 6-phosphogluconolactonase (6PGL) sequence, what it acquired by gene fusion. It is present in spongi, the first line diverging from choanoflagellates. There are two forms of this fused gene. In metazoans, the sequence is added to the C-terminus of the gene product protein. In Plasmodia, the same sequence is fused with the N terminus of the protein. The protein chemists (enzymologists) conclude that these two gene fusions occurred independently from one another. A third (probably the first) independent fusion of these genes had taken place in the species Giardia and Trichomonas, receiving these genes either separately, or already fused from prokaryotes, by HGT. The kinetoplastid parasites Leishmania and Trypanosoma express these genes separately (not fused) [617]. The *Plasmodium falciparum* hexose transporter (pfht/PfHT) is constitutively expressed, even in the mosquito, and it is essential for the parasite's intra-erythrocytic life cycle in its mammalian host. It is amenable to fusion protein formation. It is refractory to blasticidin chemotherapy [618]. Fusion oncogenes/oncoproteins in cancer cells are also notoriously chemotherapy-resistant.

Aspartic proteases expressed by *P. falciparum* (plasmodium export element, PEXEL) are related to those of the oomycetes (*vide infra*). The aspartic protease homolog plasmepsins are in the endoplasmic reticulum (ER) of the Plasmodia cytoplasm. Translocon proteins transfer these molecules into host cells (erythrocytes). The stepwise evolutionary course of secretory proteins (Secs) from the trypanosoma through yeast, caenorhabditis, and drosophila up to mammals and Homo has been deciphered [619]. The human Sec62 and Sec63 act as a complex (probably fused), whereas the trypanosoma does not have a Sec62 orthologue, and its Sec63 is devoid of the negatively charged carboxyterminal region. The human Sec62 interact with the ribosomal tunnel exit by the acquisition of two basic oligopeptide motifs, that cells of invertebrate hosts do not have [619]. Since mammalian cells do not produce plasmepsin enzyme, its specific targeting holds high therapeutic promise [620, 621]. The hosts' genetic background influences pathogenicity of different Plasmodia species; thus, malaria plasmodia might have influenced human evolution [622, 623]. NK cells are able to lyse plasmodia-infested (merozoits, trophozoits, schizonts) erythrocytes [624].

Comment. Fused proto-oncogenes habitually encode oncoproteins in the large multicellular vertebrate mammalian hosts. The end result is constitutive expression of the fused genome and increased virulence of the cell harboring and operating the resultant oncoproteins. The clinics diagnose malignant transformation and recognize (meticulously identify) growing cancer cells. These cells are removed *en mass* by surgery, and are targeted by chemoradiotherapy; only recently are small molecular inhibitors of the fused genomes and oncoproteins, or such modalities of immunotherapy used for treatment.

The exact causes and mechanisms of genome fusions determine the host response. In Burkitt's lymphoma's hallmark translocation, there is an external pathogen (EBV). The host mobilizes macrophages to engulf and digest the antibody-coated lymphoma cells. Whereas the ETV6/NTRK3 gene fusions in the congenital mesoblastic nephroma and/or infantile fibrosarcoma appear to have had been endogenously generated in the host cells' genome (ets protooncogene family; E26 avian acute leukemia retrovirus: Ets = E twentysix; neurotrophic tyrosine receptor kinase). Those tumor cells grow unopposed. The host may even promote their growth; or may eliminate them as if they were a left over undifferentiated anlage, by inducing their physiological differentiation (Figure 55). While an externally induced malignant tumor (African Burkitt's lymphoma) is attacked immunologically by its host (in the "starry sky" pathology antibody-coated transformed lymphocytes are digested by macrophages), the internally induced sarcoma is tolerated (if it ever turns out to be virally induced, the causative agent may be an endogenous retrovirus).

Fused Phytophthora genes encode virulence factors. Multifunctional genes have arisen from gene fusion events in the plant pathogene oomycete genus, *Phytophthora*. Phytophthora are chromalveolates (Figure 56); other well-known alveolates are the parasites Plasmodia and Toxoplasma, and the free-living Tetrahymena and Paramecia. The Phytophthora genome attests to a very complicated evolutionary history with multiple events of domain fusions, and horizontally (from bacteria) and endosymbiotically received genes (some 855 oomycete genes derive from the ancestral endosymbiont red alga). Oomycete cells practice active gene fusion events. The *P. infestans* (the Irish potato famine pathogen) genome is very rich in transposable elements (the short interspersed nuclear elements, SINE). However, these are under the control of small interfering RNAs [625]. Among the Phytophthora effector proteins, RXLR (Arg-any aa-Leu-Arg) are functioning as virulence or avirulence determinants. One of them is expressed as a fusion protein with a G-protein coupled receptor appearing as the C-terminal protein. The G-protein coupled receptor family protein is related to that of *D. discoideum* (Figure 6). These are pepsin-like aspartic proteinases, some causing rapid death of the attacked host cells. Phytophthora, together with other chromalveolates (free-living Tetrahymena and Paramecia, and pathogenic Plasmodia and Toxoplasma) share a distant common ancestor, which was harboring endosymbiotic red algae. Within this clade, oomycetes and diatoms shared a common ancestor. Some of their fused genes originated in their common ancestor; others have arisen after their divergence. The bifunctional oomycete metabolic enzymes are closely related to those of fungi, plants and animals. The fused aspartic proteinase appears to be related to that of Plasmodia. Some prominent multi- or bifunctional enzymes shared between oomycete (*P. sojae*) and other eukaryotes are methyl malonyl coenzyme A mutase, aspartate kinase-homoserine dehydrogenase, glutamine amidotransferase glutamine cyclase, guanine triphosphate cyclohydrolase-dihydropteroate synthase, PI3K, plekstrin (a widely ranged protein domain of 120 aa: "platelet and leukocyte C kinase substrate"), bifunctional dihydrofolate reductase-thymidylate synthase, serine biosynthesis, sulfate assimilation/metabolism enzymes, and a lysine biosynthetic pathway. The fused enzymes form large macromolecular complexes with highly increased efficiency in the biosynthetic processes; extensive metabolic channeling occurs between the catalytic domains [626].

The cause of root and stem rot disease of soybean plants, *P. sojae*, possesses virulence/avirulence genes *avr3l*, which are encoded from a single genetic locus.

Figure 55 Fusion Oncoproteins in Two Different Malignant Entities Confront the Hosts. [2356] Reference Appendix 2, Explanations to the Figures

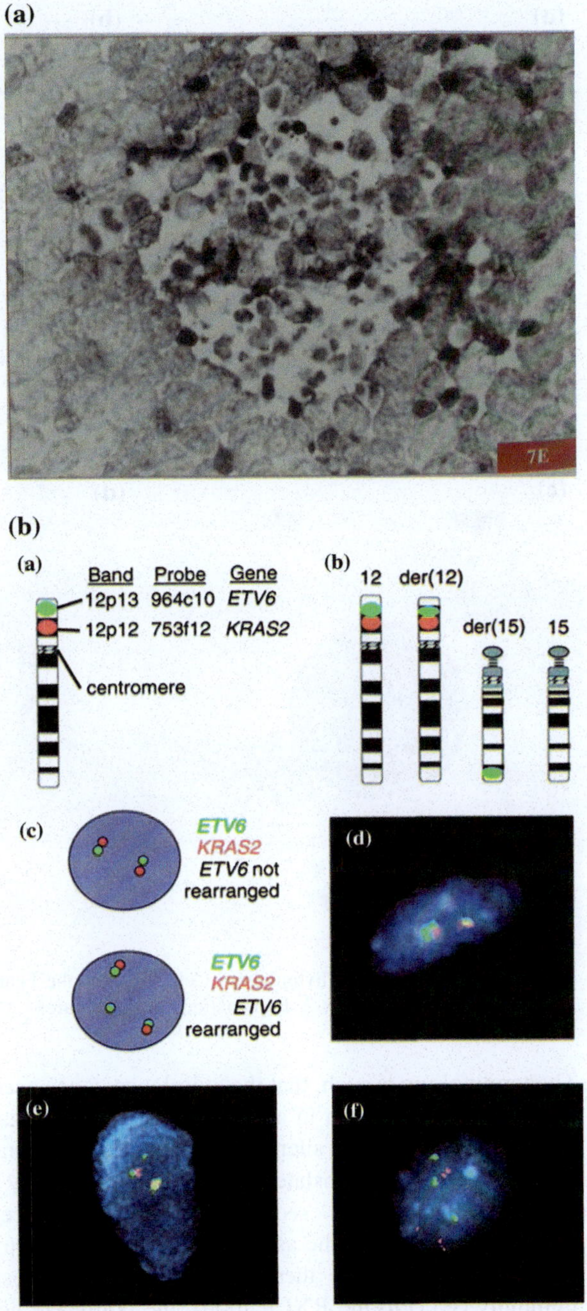

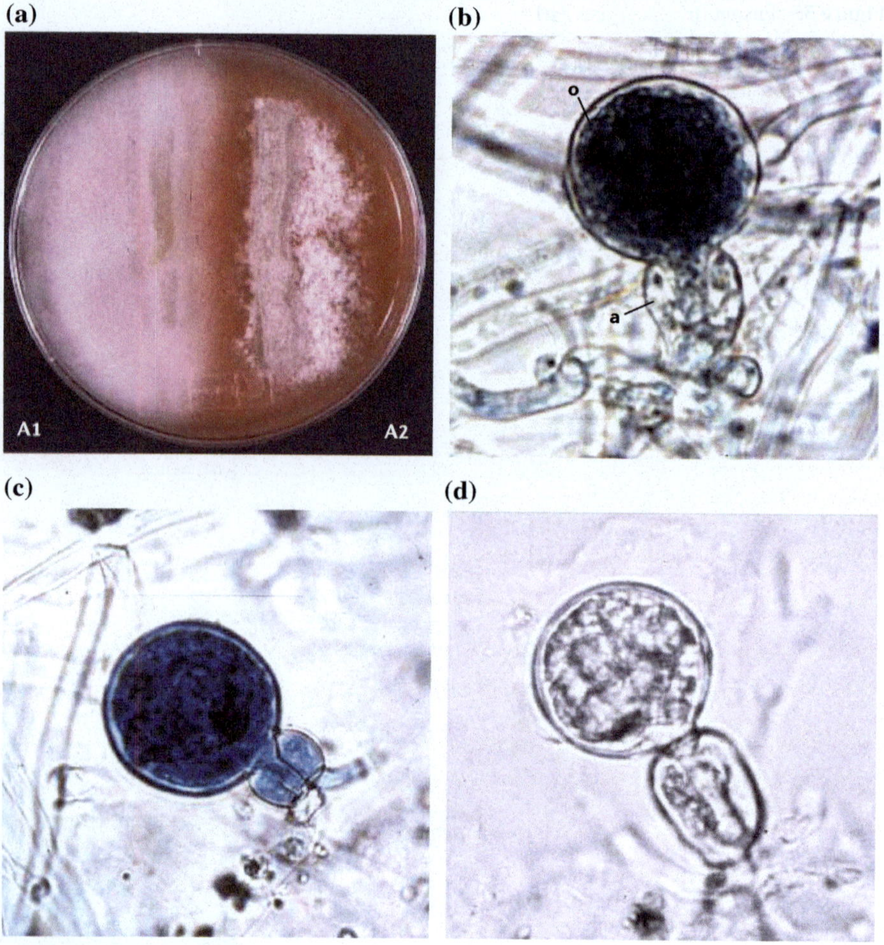

Figure 56 The Virulent Phytophthora Cell Assumes the Features of a Malignant Animal Cell. [2357] Reference Appendix 2, Explanations to the Figures

The terminology is such, that the avirulence genes encode effectors that promote the growth of the pathogen in a plant host lacking resistance. Small interfering RNAs can set the balance by silencing released *avr*3/5 transcripts. The pathogen's gene product proteins, if translated, may be confronted by the soybean resistance factor Rps. The consensus sequence of the oomycete protein is the RXLR (Arg-X-Leu-Arg) and the associated DEER (Asp-Glu-Glu-Arg) motifs. *Avr* homolog *Avh* genes encode these motifs. When the host develops effector-triggered immunity (ETI) to the RXLR motif, the oomycete gene is referred to as avirulent (*Avr*). Soybean resistance R-genes expressing ETI are referred to as *Rps* (resistance to *P. sojae*) [627]. Phytophthora pathogens encode the pepsin-like aspartic proteinases (AP) similarly to *P. falciparum*), but are devoid of the translocon complex.

These protein hydrolases are fusion proteins with a G-protein coupled receptor, as the C terminal partner (related to G-protein coupled receptors of Dikty) [628]. The necrosis-inducing Phytophthora proteins differed from those of bacteria and other fungi, indicating that these evolved within the Phytophthora clade [629]. A hypersensitive resistance response in the plants attacked by *P. infestans* is encoded be the R1 resistance gene and is directed at the Avr1 effector (avirulence; may stand for both virulence and avirulence) gene of the pathogene [630]. The phytophthora RXLR (arginine; leucine) effector protein AVRblb2 (alanine, valine, arginine; bulbous potato) suppresses in the host plant cells the production of papain-like cystein protease, that is able to digest phytophthora cells in the apoplast [631]. Plant cells by fusing the cell wall with the plasma membrane form a barrier against intracellular entry of a pathogen. Cell-to-cell adhesion is mediated by integrins, whose ligands in the extracellular matrix are the RGD (arginine-glycine-aspartate) motives. Vitronectin, the mammalian extracellular matrix ligand, preserved the RGD motif. Pathogens expressing the RGD ligand dissolve intercellular membranes, and separate cell wall to plasma membrane adhesions. Arabidopsis cell walls express lectin receptor kinase (LecRK-I), which connects the cell with the cell membrane and recognizes invasive *Phytophthora infestans* cell surface motif, IPI-O (intracellular proteinase inhibitor). The *P. infestans'* IPI-O, an RXLR-dEER effector (dE, dextro-glutamate, glutamic acid), disrupts the cell wall-plasma membrane adhesion, for its intracellular invasion. The Arabidopsis receptor LecRK-I.9 possesses two peptides antagonistic to phytophthora RXLR-dEER IPI-O's RGD motif, thus providing defense against invasion of the parasite (*P. brassicae*, in the case of Arabidopsis, which is naturally resistant *P. infestans*, but susceptible to *P. brassicae*). The gain-of-susceptibility cell phenotype supposed to provides the first line of defense [632]. The wild species *Solanum bulbocastanum* developed an effective RB (blight resistance, not retinoblastoma!) gene conferring resistance to most strains of *P. infestans* expressing IPI-O1/2. The RB gene encodes coiled coil, nucleotide-binding, leucine rich repeat proteins (CC-NB-LRR). These proteins express Toll-IL-1R (TIR) motifs. The activated RB proteins dimerise or polymerise. New *P. infestans* strains are emerging, which express new IPI-O4. This new effector protein not only overcomes the RB resistance gene, but is able to annul the recognition of IPI-O1 by RB. In the presence of IPI-O4, IPI-O1 fails to elicit RB-mediated resistance, the host plant's hypersensitivity reaction. The RB-mediated resistance, if induced by IPI-O1, could result in the induction of apoptotic death in the cells of the pathogen. The L129P mutation (L, leucine; P, proline) in the IPI-O1 molecule disables it (causes the molecule's failure to induce the hypersensitivity reaction in the "resistance" RB molecule). The L129P mutation in the IPI-O4 molecule alone, or as a double mutation with G135S (G, glycine; S, serin) deprive the molecule of its immunoresistance (render the molecule inductive to the plant's RB receptor to release the hypersensitivity reaction). Recently, the wild potato *S. stoloniferum* emerged with an RB that recognizes and responds to IPI-O4 [633]. Devastating losses of potato, tomato, cucumber, soy, etc. could be prevented with inhibitors of the Phytophthora proteases, however the fused enzymes (kinases) are usually resistant to chemotherapy.

Comment. In the oomycete, the protein products of the fused genes are chemotherapy-resistant, similarly to fused oncogene-product oncoproteins of malignantly transformed stem- or somatic cells of the higher multicellular hosts. Figure 56 and Table XXX (*vide infra*) refer to the similarities of *Phytophthora infestans* in action to malignantly transformed higher eukaryotes.

Exchange of some fused genes. The proteasome of *Trypanosoma brucei* was studied for fused gene product proteins (or for their fissions) in comparison for the same event in some other eukaryotic hosts. The *T. brucei* genes for hydroxymethyl-CoAlyase (leucine metabolism and ketogenesis) and a conserved protein (ubiquitylated protein sorting in the lysosome) were found fused in *Phytophthora infestans*. The *T. brucei* genes for glutamine hydrolyzing carbamoyl phosphate synthase and aspartate carbamoyltransferase were found fused in the choanoflagellate *Monosiga brevicollis*. The *T. brucei* genes "huntington" (misnomer: has no sequence homology with, or conserved domains related to, the mammalian Huntington gene; it has ankyrin scaffold-anchoring protein repeats for protein-protein interactions), and the GTPase activating proteins were found fused in the *Entamoeba histolytica*. The region of alignment between the two proteins extends for 323 aa and is related to the TBC GTPase activator domain (see above, at Ref [510, 511]). No TBC proteins (ref [510, 511], not tuberculosis!) could be found in archaea and bacteria. Proteins related to that of the *T. brucei*, DNA topoisomerase IB large subunit and DNA topoisomerase IB small subunit were found fused in the *Chlamydomonas reinhardtii*. Neither large nor small subunit topomerase IB were found in creanarchaea, but the large subunit is functional in bacteria. In *T. brucei*, electron transferring proteins and succinate dehydrogenase coexist, as in *Schizosaccharomyces pombe*; most bacteria operate this unit in fusion. Fission (split) of the two proteins occurred in *T. brucei*, *T. cruzi* and the Leishmanias. Gene fissions permit independent function for the two subunits, and the potential unison with another new partner. Horizontally acquired genes are commonly accepted in fusion with a pre-existing gene [634]. The *de novo* pyrimidine biosynthesis is catalysed by three enzymes: carbamoyl-phosphate synthetase II, aspartate transcarbamoylase, dihydroorotase (CSPII, ATC, DHO). The three enymes operate in fusion in fungi, amoebozoa and animals; and stand-alone in some protists and plants. *T. cruzi* fused the genes for the three enzyme [635]. The expression-site-associated gene 8 product protein of *T. brucei* localizes in the nucleolus, considered to be the domain of RNA polymerase transcription [636, 637]. The Euglenozoa clade consists of euglenoids, diplonemids, and kinetoplastids (whose mitochondria are referred to as kinetoplasts). The pyrimidine (*pyr*) biosynthetic gene cluster of trypanosomatids includes four stand-alone (*pyr*1-4) and 2 fused (*pyr*5/6) genes, shared between trypanosomids and bodonids (*Parabodo caudatus*). In the diplonemid, *Diplonema papillatum*, *pyr*6 remained a stand-alone gene. Kinetoplastids acquired *pyr*4 and *pyr*6 as stand-alone genes by HGT and trypanosomids united (fused) *pyr*6 to a pre-existing *pyr*5 [638].

The *Entamoeba histolytica* encodes a cytidyltransferase fusion protein horizontally transferred (at least one half of its gene) from the prokaryota, *Aquifex aeolicus*. The *A. aeolicus* and the *Thermofilum pendens* have already shared the fused tRNA synthetase proteins, specifically, methionyl-tRNA synthetase fusion proteins. Refined technology rejects false identification of so called fusion genes/proteins; valid and confirmed reports were cited [639].

What the yeast carries out in three steps, *Tetrahymena thermophila* accomplishes in one step. In the methionine salvage pathway, the *T. thermophila* fusion gene *mtn*BD product protein replaces the yeast's 3 separate *mtn* genes: *mtn*BCD (methylthioadenosine). While Tetrahymena scores a gain of function, the other ciliate, *Paramecium tetraurelia* failed to fuse these genes and lacks the entire pathway. Further, the apicomplexans (Babesia, Cryptosporidia, Plasmodia, Theileria, Toxoplasma) also lack these single genes, and so far failed to exhibit signs of this pathway [640, 641].

The third member of the alveolate clade (apicomplexans, ciliates, dinoflagellates), the dinoflagellates, so far did not undergo genome sequencing. Otherwise, those who operated the pathways then with single genes, could have lost them, whereas the Tetrahymena upon fusing two of its crucial genes, not only preserved them, but also achieved gain of function. Fused genes secure evolutionarily persistence of a function and gain of function, as the stages advance. A constitutively expressed sugar metabolism fusion protein in a parasite (plasmodium), that has become resistant to chemotherapy, presents a scenario reminiscent of the cancer cell transformed by a fusion oncoprotein and becoming resistant to chemotherapy. Cancer cells seldom if ever directly kill normal host cells. Cancer cells use natural means of their hosts (chemo-, cyto-, lymphokines; enhancing antibodies; growth factors) to turn host cells (fibroblasts; monocytes) from antagonistic to supportive ("cancer friendly"). The unicellular parasites (oomycetes) differ by killing their hosts' antagonistic cells with toxins (aspartic proteinases, plasmepsins). The only host cells readily killed by tumor cells are the host lymphocytes (example: tumor cell expressing FasL kills $CD8^+$ $FasR^+$ T lymphocytes) [147].

As the unicellular parasites escape innate and adaptive host immune reactions, so does the cancer cell. The difficulties in designing an effective preventive malaria vaccine (now genetically engineered adenovirus expressing the crucial plasmodia antigens; conjugated merozoit surface proteins) [642, 643], apply to the experience with preventive cancer vaccines (peptide vaccines; viral oncolysates; genetically engineered virus-mediated vaccines [71, 275, 276, 644–646]. Although preventive vaccines can successfully protect the host against infectious agents, preventive or therapeutic cancer vaccines may prolong life at the best, but usually fail to deliver a curative punch (knock-out for the cancer). Nevertheless: continue to try [647, 648].

Comment. For gains of viability or virulence, the unicellular eukaryotes frequently use fused gene product proteins (enzymes). For gains of viability, malignantly transformed cells of higher multinucleated hosts frequently use fused gene product proteins (tyrosine kinase oncoproteins).

Human fusion oncogenes and oncoproteins follow the primordial guidelines. The classical fusion oncogene/oncoprotein is the Philadelphia chromosome in chronic myelogenous leukemia (CML). Translocation and fusion of the genes between 9q34 (c-*abl*), the cellular proto-oncogene (usurped by the Abelson mouse leukemia retrovirus: c-*abl* → v-*abl*), and 22q11 (*bcr*), the break-point cluster region, were fused t(9;22), and encoded the $p210^{BCR/ABL}$ oncoprotein. The human disease, including its blast crisis, could be duplicated in mice [649, 650]. The human oncoprotein p210 is recognized in its host, and as such, it elicits weak immune reactions. These were possibly augmented by hIFNα. Under these circumstances, it was proposed that the healthy bone marrow (the progenitor stem cell) donor be immunized with the nucleic acid-free p210 protein [261a, b; 651]. Thus, the bone marrow transplant would have also provided a set of immune lymphoid cells. However, highly successful targeted therapy with imatinib mesylate (and its derivatives) canceled the bone marrow transplants for imatinib-treated patients. There are no p210 oncoprotein-immunized donors either (no such immunotherapy was as yet implemented). For prevention of relapses in targeted therapy, this project may be reactivated.

The platelet-derived GF receptor genes A/B (or α/β) may undergo fusion with gene TEL/ETV6 (TEL, telomere; c-*erbB/egfr* growth factor receptor gene of avian erythroblastosis retrovirus translocation variant), or gene FIP1L1 (Fip1-like1; F, factor interacting with poly(A)-like polymerase). The human homolog of *S. cerevisiae* polyadenylation factor, a subunit of CPSF (cleavage and polyadenylation specificity factor) binds RNA by its arginine-rich C-terminus. (Adenosine monophosphates at the 3′ end of mRNAs; over 100 adenylate residues form the [polyA tail; compete with alternative splicing; allow multiple transcripts from one single gene).

The human homolog Fip1 of *S. cerevisiae* Fip1p polyadenylation factor at 4q12 lies upstream of the *Rnf*32 (ring finger) gene. A 800 kb long chromosomal segment is deleted to bring the genes FIP1 and PDGF-Rα to close approximation, thus allowing the fusion into the F/P chimera. The constitutively expressed fused hematopoietic oncoprotein product exhibits the N-terminus of Fip and the C terminus of PDGF-Rα [652]. One of the common hematopoietic malignancies is eosinophilic leukemia. Co-activated are IL-5 (the IL-5 receptor is an eosinophil marker), and the JAK2 (Janus tyrosine kinase) proto-oncogene. JAK2 could be suppressed by siRNA or tryphostin derivative AG490. Cells with suppressed JAK downregulated their IL-5 production and STAT3, PI3K/Akt, c-Myc, NFκB, and survivin activation. The EOL-1 human eosinophilic leukemia cell line carries the F/P fusion oncogene (Fip1-like 1 FIP1L1; PDGF-Rα): F/P [653]. The fused P/C (primary chronic eosinophilic leukemia cells, CEL) could be suppressed with imatinib mesylate [654, 655]; the T674I (threonine; isoleucine) cells are imatinib-resistant.

The t(5;12) translocation consists of the gene fusion ETV6/PDGFRB (etv, ets tranlocation variant; ets, encoding transformation-specific transcription factors; related to E26, E-twenty-six replication-defective avian acute leukemia retrovirus; receptor beta). This chimeric oncoprotein activates STAT1/3/5, NFκB, PI3K, but responds with cell proliferation arrested, to bortezomib (inhibitor of the ubiquitin-proteasome system) and a mutant IκB (inhibitor of NFκB tranaslocation from cytoplasm to nucleus). The fusion oncoprotein further interacts with genes, which encode HES6 (hypereosinophilic syndrome), c-MYC and FOXO3. The IL-5Rα is overexpressed in eosinophil leukocytes; its expression is further driven by the PDGFR fusion oncoproteins. Overproduced with mutual upregulation are eosinophil peroxidase (EPX), thromboxane and its receptor, thromboxane/prostanoid (TXA; TPβ), and prostaglandin F2α (PGF2, FP). TPβ and FPα formed a fusion oncoprotein, which activated the STATs1/3/5 by phosphorylation of their tyrosine residues. TPβ/FPα activated the AKT proto-oncogene (also known as protein kinase B); this cascade may result in the inactivation of FOXO. The phosphorylation of IKK (IκB kinase) by PKB(AKT) leads to the release of NFκB from cytoplasm to nucleus, where it binds one of its target gene CCL2 (C-C motif chemokine ligand, monocyte chemotactic protein-1). The PI3K/PKB/NFκB cascade is inhibited by the morpholino-phenyl-benzopyran, LY294002. The release of NFκB is prevented by the amino-ethyl-amino-dimethyl-quinoxaline IKK inhibitor, BMS-345541, which blocks NFκB phosphorylation. Imatinib suppressed the fusion oncogene at dosage levels lower than what suppression of the ABL/BRC oncoprotein production required [656]. Fusion oncoproteins (fused DNA products) respond to the small molecular inhibitors better and with less toxicity, than the fused DNA oncogenes responded to combination chemotherapy.

Promiscuity in gene fusions. The mixed lineage leukemia (MLL) oncogene at 11q23 is a promiscuous fusion partner [657, 658]. It is related to the drosophila trithorax histone methyltransferase. In the human genome, a leukemia-inducer *mll* gene fusion partner is the *fmr*2 gene ("fragile mental retardation"; fragile X syndrome). In the human FMR, the gene and protein suffered germ-line mutation at ile304asn (isoleucin *versus* asparagine) in the second KH domain (*vide infra*) of the FMR protein [659, 660]. In addition, the gene and/or its promoter undergo a massive CGG trinucleotide repeat expansion with abnormal silencing methylation of the repeats and nearby CpG islands [661–663]. The healthy FMR protein functions in the nucleocytoplasmic transport and transcript of select RNAs. The N-termini of FMR proteins act as methyl substrate-binding proteins engaged in chromatin structure regulation. The production of the FMR protein may be knocked out by transcriptional silencing. The very active FMR protein harbors three RNA-binding domains, two KH-type and one RGG (arginine, glycine, glycine) box. The RNA-binding domains bind also G-quadruplexes (*vide supra*). The G-quadruplexes serve as high affinity ligands to FMRP [664, 665]. The K homology RNA-binding proteins (RBP) were first identified in the human heterogeneous nuclear riboproteins (hnRPN). In the evolutionary scale, RBPs are active in archaea, bacteria and protozoa (Trypanosoma) [666–669]. RBPs bind to specific sequences in the UTRs or ORFs of the targeted RNA molecules. While phosphorylation *versus* de-phosphorylation can determine RBPs' activity, RBPs could up- or downregulate the transcripts of tumor suppressor, or proto-oncogene mRNAs, respectively. Most prominent RBPs are Sam68 (Rous sarcoma, SCR-associated in mitosis), controlling Bcl-xL expression; present in all eukaryotes, except fungi; the cap-binding eIF4E (eukaryotic initiation factor); the La autoantigen (patient's initial), acts as an IRES-transacting factor (internal ribosome entry site).

The human cancer antigene R (HuR) translocates in tumor cells from the nucleus into the cytoplasm to stabilize cyclins in uninterrupted cell divisions; to force the overexpression of ERBB-2 in prostate cancer cells (c-*erb* → v-*erb* avian erythroblastic leukemia retrovirus oncogene; EGF-R) [670]. K usually used for lysine, but some investigators use K to signify "cleoproteins" (i.e. nucleic acid binding proteins). The KH module is an evolutionarily conserved motif of hydrophobic amino acids (the hydrophobic residue Gly-X-X-Gly). The GXXG motif may be diverged (interrupted, altered). The diverged GXXG motifs remain functional. The *Saccharomyces cerevisiae* Sc160p protein contains both conserved and diverged GXXG KH motifs. These proteins are evolutionarily conserved and function in higher eukaryotes from *Neurospora crassa* to human cells as vigilins [671]. The drosophila KH domain RNA-binding protein gene *dfmr* is equivalent to the human "fragile mental retardation" gene, but in the fly it functions in directing rapid pre-gastrulation cell divisions, neurogenesis, especially in the cellular neurotubular structure formation (actin-myosin; cytoskeleton). The very active DFMR protein's targets are the drosophila genes *ago* (Argonaute), *orb* (oocyte oo RNA-binding protein) *chic* (profilin *chickadee*), and others [672]. In the human host, the *mll*/fmr2 fusion oncogene and oncoprotein MLL/AF4 (proto-oncogenic segment of asymmetric *fmr*2 gene from 4q31) induces pro-B cell acute leukemia in newborn infants. The microRNA-142-3p aligns to the 3′UTR of the *mll*/*af*4 mRNA, and thus exerts anti-leukemia effect [658; 673, 674].

Potential fusion partners, other than leukemia-inducer oncogenes, of the *mll* gene include the breast cancer oncogene LASP of Lim and SH3 (Lim, caenorhabditis genes abnormal cell lineage, Lin, and mechanoreceptor promoter Mec; SH3, Rous sarcoma oncogene src homology 3 protein). The acute myelogenous leukemia-inducer MLL/LASP fusion oncoprotein is the product of the translocation t (11;17)(q23;q21) [675, 676]. The united LIM/SH3 protein LASP-1 (for LIM kinase, Table XVI) translocates from cytoplasm to the nucleus in breast cancer cells and dictates rapid cell divisions (high Ki-67, *vide infra*) and adverse prognosis [677].

Mesenchymal tumors from the benign lipoma to the most malignant liposarcomas (Table VII) utilize fusion oncogenes/oncoproteins (reviewed in [340a, 341]). In myxoid and round cell liposarcomas, the translocation t(12;16)(q13;p11) unites FUS-DDIT3 oncogenes/oncoproteins (fusion in sarcoma; DNA damage inducible transcript, formerly GADD, growth arrest and DNA damage). The cascade rolls by the CCAAT box (cytidine-cytidine-adenosine-adenosine-thymidine) enhancer-binding protein (C/EBP, CHOP, C/EBP-homologous oncoprotein), IL-8 promoter, and NFκB [678–680]. Trabectedine induces maturation of lipoblasts by restoring the terminal adipogenesis pathway; this may occur due to detachment of the FUS-CHOP oncoprotein from its targeted promoters, and/or by activation of C/EBP [681]. The translocation t(X;18)(p11;q11) in synovial sarcoma (SS) generates the fusion oncogene-oncoproteins SS18-SSXq1. In the cascade, if E-cadherin is activated, the sarcoma cell will be shaped in epithelioid form. In the background, the gene *snail* acted as an E-cadherin promoter repressor antagonist. If the gene *slug*, a weak suppressor of the E-cadherin promoter repressor, has the upperhand, the repressor prevails, and the sarcoma cells appear in a monophasic spindle cell shape. The cadherin-binding protein, β-catenin, connects the cytoplasmic tail of E-cadherin with the actin cytoskeleton. Upon the depletion of cytoplasmic β-catenin (as it translocates to the nucleus), the sarcoma cell undergoes transvestism into the spindle cell phenotypes. SS cells express therapeutic targets (platelet-derived growth factor, PDGF;

HER2/neu; IGF and its R) [340a, b, c, 682, 683]. From chromosome 18 (18q11-12) massive gene insertion into chromosome 8 increased the expression of the anti-apoptotic bcl2 gene 433-fold over [683]. In alveolar rhabdomyosarcoma, the PAX3 t(2;13), or a PAX7 t(1;13) gene fuses with FOXO1 (paired box; forkhead box protein 01; forkhead transcription factor receptor, FKHR). In these sarcoma cells, p53 and Rb tumor suppressor genes are eliminated. To this effect, a full length Japan "hypothetical protein FLJ10404, herein named IRIZIO", or "full length cDNA encoding part of the hypothetical protein FLJ10404, which we dubbed IRIZIO" was necessary [684], but "IRIZIO" remains a mistery, while complimentary DNA is a dsDNA product generated by RT from a mRNA. In N-Myc protein expressing alveolar rhabdomyosarcoma cells, ephrin-to-E-receptor, and scatter/hepatocyte growth factor-to-Met receptor are overexpressed [340]. RNA interference and "antigene peptid nucleic acid oligonucleotide" therapy blocked N-*myc* gene expression [685]. The proteasome inhibitor, bortezomib, and the Hsp90 inhibitor 17-DMAG (17-dimethylamino-17-geldanamycin-derivative alvespimycin, *Nocardia mediterranei*-derivative) induce autophagy in rhabdomyosarcoma cells in vitro. The lysosomal inhibitor chloroquine converts sarcoma cellular autophagy into apoptosis [686].

The EWS (Ewing's sarcoma) oncogene (22q12) is shared by many different tissue types of sarcomas, in addition to Ewing's sarcoma: extraosseous chondrosarcoma, desmoplastic small round-cell soft tissue sarcoma, myxoid liposarcoma, and clear cell sarcoma [340a]. For human neural stem cell mobilization, EWS may fuse with the FLI1 gene segment (Friend leukemia virus F-MuV proviral insertion site) (Figure 57), and this unison may induce the encoding of polycomb proteins from the BMi1 locus (B lymphoma Moloney leukemia virus M-MuLV proviral insertion site), accompanied by epigenetic silencing of the p16 protein (INK4a/Cdk4/6) [687, 688]. In desmoplastic small cell tumor, the exons 1–7 of the EWS gene fuses with the exons 8–10 of the Wilms' tumor (WT) gene. Here the WT gene surrendered its tumor suppressor faculties to the EWS oncogene [689, 690]. The NH2 terminus of the fusion product EWS/FLI1 engages the COOH terminus of the BARD1 gene (BRCA1-associated ring domain protein), a tumor suppressor gene, engaged in DNA repair, and/or in apoptosis induction in DNA-damaged cells. The EWS-FLI1 protein may further rescue p53- and INK4A-deprived cells (cancer cells) from apoptotic death (good for the transformed cell). The BRCA1/BARD1 heterodimer protein could mark for ubiquitination the EWS-FLI1 complex [691]. BRCA1 and EWS-FLI1 compete for unison with BARD1. The outcome of this competition may be the rescue of a DNA-damaged cell (cancer cell) from apoptotic death, or an attempt at its DNA repair, either error-prone or error-free. The EWS gene may fuse with the stem cell gene Oct-4 (octamer-binding protein in the POU transcription factor family, consisting of Pit1, pituitary-specific protein, octamer-binding Oct-4, and Uncoupled from the cell cycle protein, caenorhabditis). The fused gene is the result of a translocation: t(6;22)(p21;q12). The fused oncogene encodes a DNA-binding protein, which induces tumors in epithelial cells (hidradenoma; mucoepidermoid carcinoma). The physiological role of stem cell gene Oct4 protein is upregulated in testicular germ cell tumors (seminoma;

(a) (b)

(c)

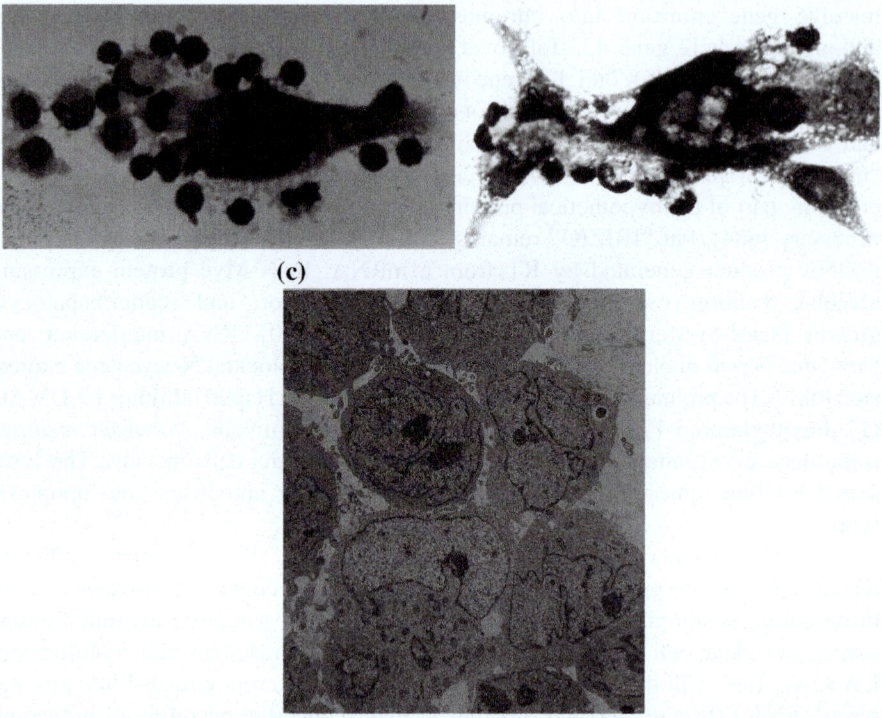

Figure 57 Further Examples of in vitro Cytotoxicity of Human Lymphocytes to Cultured Human Malignant Cells. Ewing Sarcoma Cells in Culture. Reference Appendix 2, Explanations to the Figures

embryonal carcinoma) claimed to be the result of a transition from mesenchymal to embryonic stage of the cell [692].

The so called bromodomain genes regulate the cell cycle and cell divisions; BRD4 overexpression inhibits G1-S phase transition. The double bromodomain-containing gene by translocation fuses with the nuclear protein in testis NUT gene: t (15;19)(q13;p13.1). In the fused gene, the translocation breakpoint in chromosome 19 transsects the longer BRD4 gene, but preserves the full transcript of the shorter BRD4 gene. Physiologically, the BRD4 transcripts are ubiquitous, but the NUT transcript is supposed to be confined to the testicle; thus, no NUT (nuclear protein testis) transcription should take place in the female gender. However, the fusion oncoprotein-induced highly malignant epithelial cell tumors were not confined to the male gender. If the long segment of the BRD4 gene is inhibitory, whereas, the sort segment is of promotional effect to cell division, then in the fused oncoprotein the former effect is lost. "NUT midline carcinoma" in female patient: 54 year old woman with left nasal cavity mass occupying the left orbit and directly invading the brain. By histopathology, the tumor is an undifferentiated carcinoma with keratinizing squamous differentiation. Fluorescent *in situ* hybridization (FISH) documented BRD4-NUT fusion oncogene; immunohistochemistry shows NUT protein overproduction. The tumor continued to advance despite chemo-radiotherapy [693, 694].

Comment. Is the generation of fusion genes and proteins a survival pathway, or is it the consequence of some mistaken nicks by enzymes AID or RAG on the wrong genes? Once the faculty of somatic hypermutation was based on the generation of new fused genes, the process proved itself to be not necessarily pathological, but also useful and physiological. Somewhat detailed descriptions are given to document the gain-of-function power of the fused DNA molecules in the human malignantly transformed cells. Ancient unicellular life forms gained "cell survival pathways" from fused genes. In the complex world of multicellular hosts, individual cells revert to the life style of the unicellular life forms exhibiting resistance not only to external physicochemical (chemo-radiological), but also to internal biological attacks. Increased vitality and survival for these cells is gained by fused genes-driven biochemistry.

It happens to good old men. The prostate cancer fusion oncoprotein TMPRSS2/ERG encoded from 21q22 consists of the transmembrane protease serine 2 and the *ets*-related genes (ERG). The fusion partner of ERG, TMPRSS2 is a matrix regulator. Ets stands for the E26 avian erythroblastosis retrovirally acquired cellular-to-viral oncogene c-*ets* → v-*ets* (*vide supra*). ERG directly targets the CRISP3 gene at 6p12.3 for up-regulation (cysteine-rich secretory protein). Another target of ERG is the Wnt signaling pathway, which is responsible for the initiation of the EMT of the tumors cells. Remarkably, refined curcumin-derivatives (3.3′-diindolylmethane) block this transition. The encoded gene product CRISP3 is an extracellular matrix protein with proteolytic role in matrix remodeling. In clinical tumor specimens, high CRISP3c levels correlated with locally invasive cancers [695–698]. The de-regulated tumor cell genome further overexpresses the genes TDRD1, and BMP (Tudor domain containing/mouse tudor repeat; bone morphogenetic protein). The germ line specification/determination (male *versus* female) tudor domain genes of the drosophila (*oskar, tudor, vasa*) and the mouse (*mouse tudor repeat 1,* Mtr-1) show evolutionarily conserved motifs. The stem cell gene *oskar* itself is the derivative of an ancient fusion process, which followed a horizontal gene transfer. In insect larvae (Holometabola/endopteryota, flies) some 300 million years ago a bacterial symbiont deposited a genomic segment encoding acetyltransferases, carboxylester hydrolases, glutamate synthases and lipases, that fused with a pre-existing tudor-like gene. The new gene, *oskar*, preserved its N-terminus and expressed the new sequence as its C-terminus. In the maternal oocytes the new gene encoded the germ plasm in the pole cells. Through meiosis, exclusively unigender haploid germ cells are generated [699]. It was Oscar Hertwig, who in 1876 described meiois in sea urchin oocytes. Loss-of-function mutation of the mouse *tudor* gene leads to a-spermiogenesis and sterility. Tudor-related gene domains (*Tdrd*) direct the male germ line development, as male germ cells derive from the pluripotent stem cells. In spermatogenesis, Tdrd7 directs ribonucleoprotein remodeling of the chromatoid bodies. Tdrd5 silences intruding retrotransposons and assembles chromatoid bodies in the process of spermiogenesis [700–702]. It has not been precisely explained how the overexpressed TDRD1 protein promotes carcinogenesis. Tudor-related protein is required for spermatogenesis.

In the TMPRSS2-ERG mutated prostate cancer cells (*vide supra*), the TGFβ, the Wnt pathways, and the process of EMT are activated, while DNA repair genes, among them BRCA2, suffer loss-of-function mutations [695, 696]. The bone morphogenetic protein (BMP) family members participate in PTEN regulation, and in IL-1β and IL-6 induction in macrophages. In the bone, BMP7 is secreted by stromal cells. The BMPR2 (receptor 2) induces in the prostate cancer's stem-like cells expression of the cell cycle inhibitor p21, and the metastasis suppressor N-*myc* downstream-regulated gene (NDRG1). In response, these stem cells assume the state of a reversible senescence [703]. Tumor cells constitutively expressing the BMP antagonist noggin suppress osteoblasts, thus their metastases appear as osteolytic. Prostate cancer cells harbor a silenced noggin pathway, thus uninhibited osteoblasts are allowed to attempt at some structural "blastic" bone repairs [704]. TMPRSS2-ERG-mutated prostate cancer cells may linger on in a state of a reversible senescence, or undergo EMT. There is cooperation between the mesenchymally transformed and the remaining epithelial tumor cells in that, the former supports the growth and metastases of the latter by providing the necessary growth factors [705]. Further, prostate cancer cells in the state of senescence may transdifferentiate into a PSA-nonproducer neuroectodermal cancer. This latter outcome may be consequential to androgen depletion therapy inducing the downregulation of the S-phase kinase-associated protein 2 [706].

The ETS (erythroblast transformation-specific; E-twenty-six) fusion transcript SLC45A3-ELK4 (solute carrier family 45, member 3; ETS-domain protein, ETS-related proto-oncogene, ETS-like kinase) occurs in benign, pre-cancerous, and cancerous prostatic tissue, but high levels of its mRNAs are produced only in the tumorous gland, and especially in metastatic tumors. The oncoprotein appears in the urine, when malignant transformation of the prostate occurred (but only in a distinct subset of prostate cancers). Exons of the two genes fuse after a 50 kb segment of chromosome 1p32.1 between them is deleted, and the *elk*-gene is pre-amplified consequentially to androgen effect. The chimeric mRNA is formed not by trans-splicing, but by cis-splicing referred to as "adjacent genes read through." The genes participating in this event do not have to be mutated or translocated [707, 708].

A very treacherous tumor. Clear cell sarcoma may be referred to as melanoma of the soft tissues. It grows painlessly next to tendons and aponeuroses. While it may contain melanin pigment and express antigens characteristic of neuroectodermal cells and melanomas, it is free of the *braf*/BRAF gene mutations characteristic of true melanomas (*vide infra*). Instead, clear cell sarcomas harbor the fusion oncoprotein EWSRT1/ATF1: t(12;22)(q13;q12) (Ewing's sarcoma and activating transcription factor). Usually codon 325 of the EWSR1 gene and codon 65 of the AFT1 gene are fused. ATF with the cyclic AMP-responsive element-binding protein and modulator (CREM) act as a transcription factor for gene expression. However, the fusion oncoprotein acts independently from cAMP, and incessantly, because its ATF component is truncated. The oncoprotein directly targets and constitutively activates several cAMP-inducible promoters. These are the genes which direct cell divisions and signal transductions. The elongated tumor cells have clear cytoplasm, large nuclei, and prominent nucleoli. The tumor cells express A100, melan-A, microphthalmia-associated transcription factor, and melanoma matrix protein Pmel17 (silver mouse homolog-like melanosomal protein, SILV); cytokeratins are negative. Genes normally activated (CREM, and protein kinase regulatory subunit type1 tissue-specific extinguisher, PRKAR1A) to switch off cAMP-induced signals are overexpressed in a "futile attempt" to counter the constitutively active EWS-ATF1 oncoprotein. In these tumor cells a significant (but unexplained) rise of the epidermal growth factor receptor (ERB-R) occurs. Tumor cells do spread by lymphangitic and hematogenous routes to lymph nodes and to the lungs. Sarcoma treatment protocols (doxorubicin, actinomycin, ifosfamide) are ineffective. Usually, melanoma-directed chemotherapy protocols are administered (nitrosourea carmustin, cisplatin, dacarbazine and tamoxifen). These regimens are also ineffective. However, EGFR overexpression offers itself as a therapeutic target [709].

The t(2;22)(q34;q12) EWSR1 and CREB1 (cAMP-responsive element-binding protein) fusion oncoprotein is a rare cause of clear cell sarcoma. Of 33 cases, 2 were of the EWSR1-CREB karyotype (31 EWSR1-ATF1), including the FLI (Friend leukemia insertion) site (Figure 57). This figure shows some of the earliest photographs (1968) of "large granular lymphocyte" (later NK cells) attacking human sarcoma cells [147, 278]. Tumor cell cytology displays some variability in presenting occasional rhabdoid and pleomorphic cells, and alveolar structures due to loss of cellular cohesion. Necrotic, but not apoptotic, tumor cells were present, and Bcl2 overexpression was of common occurrence. Epithelial membrane antigen or c-erbB2 overexpression occurred in 10-12/32 tumors [710]. Exon fusions may be variable: EWSR1 exon 8/ATF1 exon 4 (type 1 transcript), or EWSR1 exon 7/ATF exon 5 (type 2 transcript), or EWSR1 exon 10/ATF1 exon 5. Types 1 and 2 transcripts could be carried simultaneously [711]. The BRAF mutation characteristic of melanoma cells was not found in clear cell sarcoma.

The gain-of-function point mutation V600E (valine; glutamic acid) of the human *raf*B (BRAF) proto-oncogene is a driver of melanoma and colon cancer cells and hairy leukemia cells. The transformed cells do not transpeciate: they retain their original morphology (Figure 58). *)

*) *All these tumors respond to treatment with vemurafenib. Only hairy leukemia cells over-express the leukocyte transcript 1 and actin β genes, inducers of the cell membrane protrusions. In the IFNα-treatable but not curable hairy cell leukemia, bone marrow stem cells support leukemic hairy cells, more effectively against vemurafenib, than dabrafenib. The V600E BRAF mutation induces the MEK/ERK pathway, responsive to trametinib. Thereafter, NFκB still may induce relapses, but leukemic hairy cells respond further to monoclonal antibodies (rituximab, moxetumomab), or pseudomonas exotoxin immunotoxin. Hairy cell leukemia: best responding target to combined chemo-immunotherapy (Pettirossi V et al Blood 2015;125:1297-16; Robak T et al Expert Opin Investig Drugs 2015 robaktad@csk.umed.lodz.pl; Sári E et al Pathol Oncol Res 2014;20:973-89; Weldon JE Bioconjug Chem 2015;26:1120-8).*

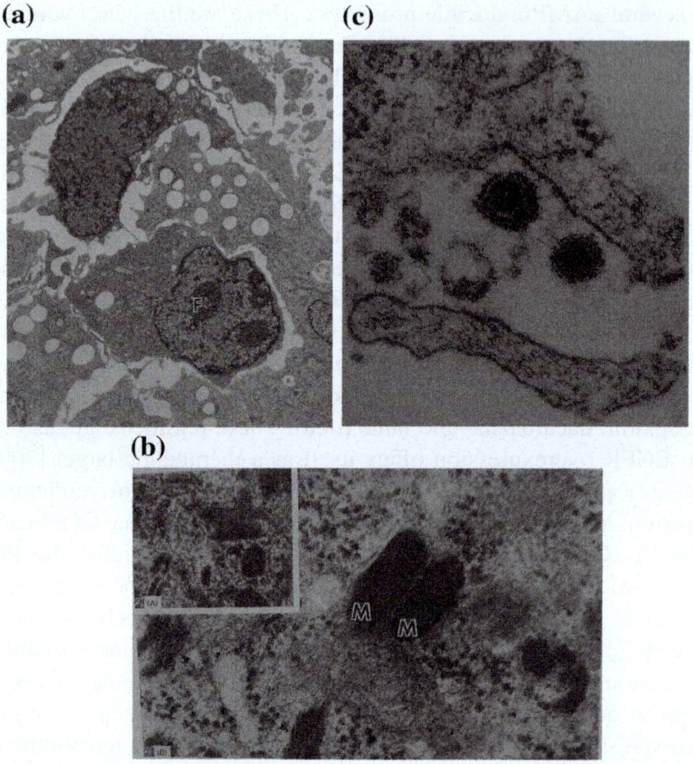

Figure 58 BRAF Oncoproteins Appear in Entirely Different Human Tumor Cells. Reference Appendix 2, Explanations to the Figures

The ATF/CREB oncogenes in the Ewing sarcoma may have their predecessor in *Neurospora crassa* (in Appendix 1). A phosphatase 2a (PP2A) gene undergoes a loss-of-function point mutation in *N. crassa* and fails to de-activate the MAPK pathway, which promotes excessive hyphal growth, resistant to its chemical inhibitors (*vide infra*). PP2A is a prominent element in the hominoid proto-oncogene TBC1D3 (*vide supra*). The operator of a most versatile genome, the sea squirt (*Ciona intestinalis*) in its transpeciation utilizes the ancestors of the human proto-oncogenes/oncoproteins ETS, WNT, TCF/LEF (intranuclear β-catenin-activated human oncogenes), Hh (hedgehog), and JAK/STAT, and the MAPK cascade for its "cell survival pathways" (in Appendix 1).

When the FasL induces mitosis. A travesty of justice. In checking human melanoma cell lines for FasL expression and for its use to kill $FasR^+$ (fragment apoptosis stimulator receptor) host T lymphocytes, Joseph Horvath and this author found a human melanoma cell line established at St. Joseph's Hospital, Tampa, FL, which expressed FasR, but upon binding FasL (purified), instead of the expected apoptosis, mitoses were induced. Since chromosomes 1 and 10 suffer breaks in melanoma cells, this author has come upon the most probable explanation. These broken chromosomes may misalign and fuse. The human FasR (CD95/APO-1) maps at 10q23. The locus of the human granulocyte colony stimulating factor receptor (G-CSF-R) is at 1p32-34 [712–715]. The high likelihood is that the broken extracellular and transmembrane FasR domain fused with the broken cytoplasmic G-CSF-R domain. Thus, FasR activation is transmitted into the nucleus by the G-CSF-R cytoplasmic tail as a signal for mitosis [716, 717] (Figure 59). This circuitry might function as an autocrine loop, if the $FasR^+$ melanoma cell releases FasL, or as an exocrine loop, if $FasL^+$ lymphocytes approach the $FasR^+$ melanoma cell [718]. The event of fusion of FasL and G-CSF-R could be reproduced in vitro, and the fused receptors could be expressed in vivo in transfected cells [719]. Such an external $FasR^+$ and intracellular G-CSF-R^+ chimera induced mitoses when stimulated with FasL. The external G-CSF-R^+ and intracellular $FasR^+$ chimera failed to induce apoptosis upon stimulation with the ligand of G-CSF [719]. The $FasR^+$ surface and G-CSF-R^+ cytoplasmic melanoma cells were driven into a mitotic response by binding FasL [716–718]. $FasR^+$ glioblastoma cells could also be stimulated to undergo either apoptosis or mitosis upon binding FasL [720]. These glioma cell lines expressed both FasR and FasL. The mitotic pathway was driven by MEK-ERK signaling (MAP and ERK, extracellular signal-regulated kinase). A caspase inhibitor (Z-Asp-CH_2-DCB, dihydrocytochalasin B) blocked both apoptosis and/or cell divisions, and inhibited phosphorylation of ERK by FasR. The MEK inhibitor PD98059 inhibited cell cycle progression, but not apoptosis. Under certain circumstances, FasL could act as an autocrine growth factor [720]. Thus, the FasR pathway running through the death domain of FADD to activate caspases, instead cell death, could induce a cell proliferation pathway even without generating fusion oncoproteins.

Gene amplifications, duplications and point mutations promote the selection of the fittest. Increased copy numbers (gene amplifications), gene duplications, and certain point mutations are the driving forces of evolution by creating "functional advantages". Highly educational reviews in original: "Darwinian evolution in the light of genomics" [721]; "rearrangement hotspots" within segmented duplications in the human genome [722]. G4 qudruplex DNA formations and gene fusions (*vide supra*) ascertain the long term establishment of new genes, as solidifying repeat orientations within G4 quadruplexes decreased the chances for loss due to mutagenicity [723].

A vision of cancer cells, as the "immortalized" lymphoblasts growing continuously in suspension cultures in the 1960s at Roswell Park, NY, and at MD Anderson, TX, as if individual cells extricated themselves from cell communities to resume the life style of the first highly resistant unicellular ancestral life forms, was presented in the 1970s: "…the cancer cell recaptured the immortality of unicellular organisms that existed before the social order of cell communities was established" [724]. An early attempt at recognizing the ancestors of oncogenes (*fes*, *myb*, *ras*, *wnt*) in unicellular eukaryotes, fruit flies and frogs, as cell survival pathways, was rendered in the monograph (Schenk Buchverlag pp. 40–42, Ref. [147]).

The short-lived caenorhabditis retains by high Wnt/β-catenin signaling (Wnt ligands to 'frizzled' receptors) an excess supplement of its asymmetrically dividing stem cells in adult life. Low Wnt/β-catenin signaling promotes the loss of stemness

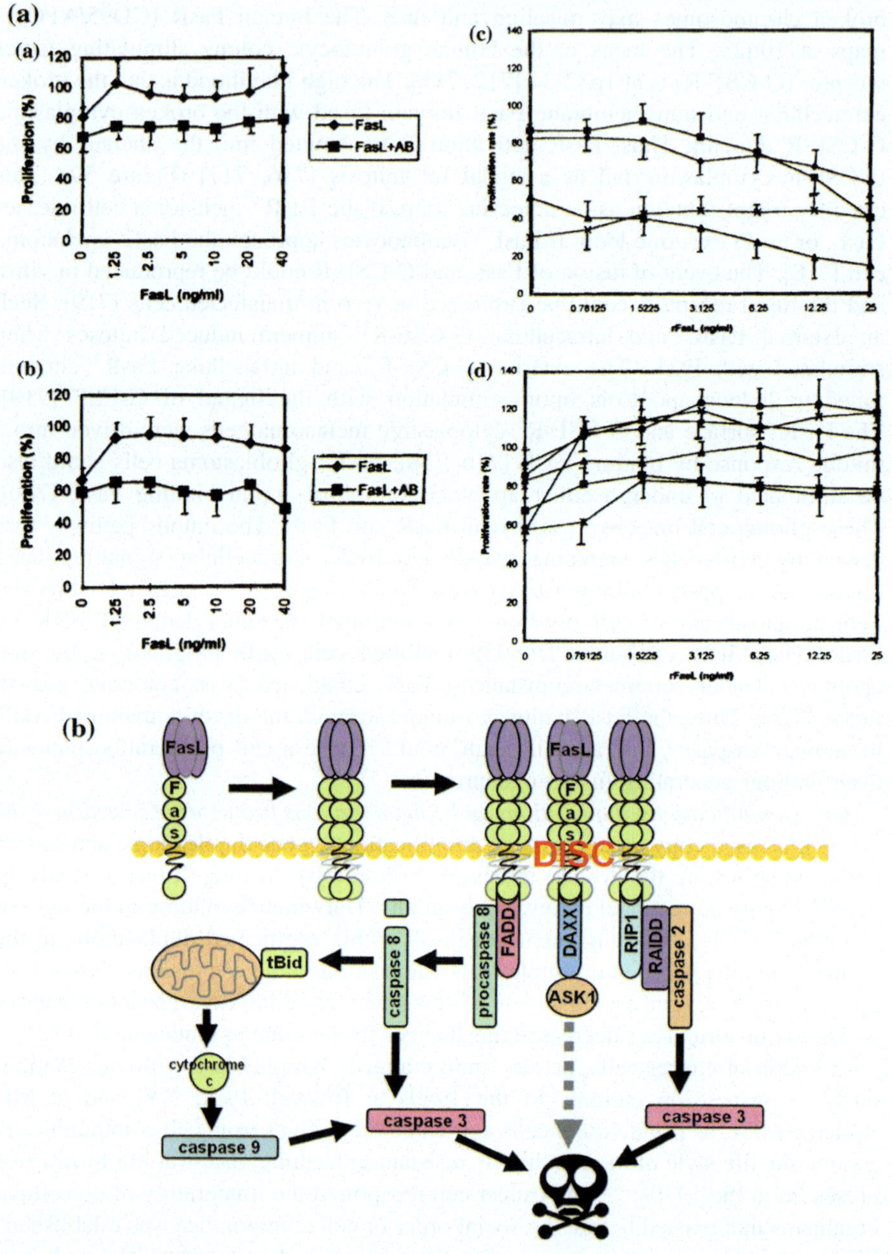

Figure 59 The Ultimate Paradox: the Cytotoxic FasL Is Used as a Mitogen in Human Melanoma Cells. Reference Appendix 2, Explanations to the Figures

(both daughter cells differentiate) [725]. In the drosophila, an excess stem cell supplement in adult life can lead to brain tumors, or overwhelming leukocytosis ("drosophila leukemia" with RAS^{V12}—overexpressing hemoblasts), in which processes, the ancestral oncoproteins first introduce themselves (Notch, Wnt/Hedgehog) as constitutively activated agents of cell survival [726, 727] (Figure 42). The drosophila gene ras85D, equivalent of the human *ras* family genes (H-, K-, Nras), undergoes the R68Q (arginine to glutamine) 'gain-of-function' mutation resulting in altered GTPase activity. The activated MAPK pathway suppresses by phosphorylation the proapoptotic Hid protein (head involution defective), which is the antagonist of the anti-apoptotic Diap (drosophila inhibitor apoptosis protein) [728]. Thus, physiologically apoptosis-prone insect cells are given a chance to survive 'immortalized'.

In contrast to point mutations of controversial value (either gain or loss) in evolutionary adaptation and selection, gene amplifications and duplications lead to evolutionary advancement. Rapidly expanding (and thus refining) their odorant (Or) and gustatory (Gr) receptor gene supplement, the pea aphids (*Acyrthosiphon pisum*) gain a highly specialized host-parasite relationship [729].

What are the flowering plants *Caryophyllaceae* (Silene, royal catchfly, narrow-leafed campoin) up to? In the last million years, in comparison to other related genera, they enormously enlarged their mitochondrial genomes. The large mitochondria are packed with recombined and mutated genes. Nuclear genes reversed the flow; not only originally mitochondrial genes now return from the nucleus to the mitochondria, but genuine nuclear genes as recombination partners follow them [730]. Are viruses (retrotransposons) or plasmids involved? What will be the evolutionary gain (or loss)? The tomato plant gained rapid enlargement of its fruit size. The progenitor wild tomato (*Lycopersicon pimpinellifolium; L. pennellii)* bore fruits of 1 cm diameter and of a few grams weight. The individual fruits of the domesticated plant (*L. esculentum)* have become large, fleshy and juicy. In the domesticated plant, the *fw2.2* gene's 663 nt ORFX from tomato chromosome 2 encodes a 163 aa polypeptide of approx 22kD. This ORFX is distantly related to other plant genes. The predicted ORFX sequence checked in the Protein Data Base resembled most that of the human chromosome 6q21, the locus of the Harvey *ras* proto-oncogene encoding H-ras protein [731]. Is the domesticated tomato a "plant sarcoma"? The FW2.2 polypeptide is a unique plant protein, which controls fruit size by activating cell cycle kinases. Tyrosine kinases (TK) operate through events of phosphorylation (transfer of phosphates from ATP to tyrosine residues of their targeted proteins).

The superfamily of metazoan tyrosine kinases (TK) has risen by modular recombinations (including gene duplications and shuffling) in pre-existing domains. Their increased activity begins in choanoflagellates and in the branchiostoma, amphioxus. Physiological life functions in the cnidaria (Nematostella), in the echinoderm sea urchin (Strongylocentrotus), in the ascidian sea squirt (Ciona), are based on rising families of TK. The amphioxus (*Branchiostoma floridae*) encodes 118 TK proteins, of which 29 becomes operational later in vertebrates. These molecules are the ancestors of constitutively expressed oncoprotein tyrosine kinases of advanced metazoans [732].

The novel oncogene to be, the tyrosine kinase nok/NOK, pops up in the amphioxus. The receptor-TK FGFR/PDGFR-related amphioxus NOK family already assumes the structure of an ectodomain (incomplete) for ligand capture, and cytoplasmic domain for signal cascade induction. The amphioxus harbors 22 NOK-related genes expressing full (not truncated) extracellular domains for various physiological functions. NOK-like TKs are operational in the cnidaria, echinodermata and ascidians [732]. The mouse and human homolog tyrosine kinase RTK NOK (novel oncogene tyrosine kinase) were also cloned. The 11+ exons human NOK maps to 12p12-21.8-27.2. Upstream lies its promoter with a CpG island. The NOK gene encodes a TK with a transmembrane protein of 23 aa at its NH2 terminus; its cytoplasmic domain consists of 270 aa at its COOH terminus. Its tyrosine, serine and threonine residues serve as potential phosphorylation sites. NOK shares 20–30 % aa identity with FGFR/PDGFR families. The human NOK is highly expressed in the prostate. A cDNA3-EPOR/NOK plasmid (erythropoietin-receptor) containing full length NOK with influenza virus hemagglutinin epitope tag, formed the pCDNA3-NOK. Transfected NOK gene malignantly transformed NIH3T3 cells, and induced highly malignant tumors in nude mice. From the inoculation site, tumor cells rapidly and widely metastasized in various organs including the spleen. The ERK1/2, RAS/MPK and the PI3K pathways were activated in the tumor cells. The MAPK inhibitor PD98059 and/or the PI3K inhibitor LY294002 inhibited tumor cell colony formation [733].

From the unicellular eukaryotes and fungi on (Chlamydomonas, Entamoeba, Phytophthora), presumably downward in their earliest ancestors, and heavily documented upward in the metazoan hosts (from the choanoflagellate, Monosiga to metazoans) tyrosine kinases operate the most complex and essential signaling pathways [734]. The tyrosine kinase families underwent great expansions in the caenorhabditis and drosophila, but into different directions. Thus, their orthologous relationships are not even evident. However, AXL (anexelekto, Greek for out-of-control, uncontrolled, *vide infra*) shows up in caenorhabditis. Platelet-derived growth factor receptor and hepatocyte growth factor receptor (PDGFR; HGFR) are shared between insects and caenorhabditis, and caenorhabditis and chordates (except, the HGFR extracellular domains are different in caenorhabditis and chordates). In vertebrate genera, tyrosine kinase gene duplications of different directions occurred between cyclostomata (lamprey) and gnathostomata (sharks), and then again between jawed fishes and tetrapods [732, 733].

Selected abbreviations

NOK, novel oncogene tyrosine kinase. A physiologically useful gene of the amphioxus becomes a human proto-oncogene/oncogene in itself, or with association of gene STYK (serine-threonine-tyrosine kinase) in targeting GSK-3β and AKT (glycogen synthase kinase; AK mouse retroviral thymic lymphoma; protein kinase B). For NOK or NOK/STYK in adenocarcinomas (breast; lung, prostate cancers), click Amachika T et al Lung Cancer 2007;56:337-40; Chung S et al Cancer Sci 2009;100:2100–14; Moriai R et al Anticancer Res 26:4969–73. For NOK/STYK in acute leukemias, click Kondoh T et al Exp Hematol 2009;37:824–30.

Wg wingless, drosophila. Wg-1 murine homolog is int-1 (integrated) with four coding exons in chromosome 15; Wnt, Wg and int (Wnt) gene families. Insertional mutagenesis: the endogenous mouse mammary tumor provirus is integrated within introns of the int genes; cDNA derives from the mRNA of gene int-1; int-2 is a fibroblast growth factor gene. Wg and int united in abbreviation Wnt proto-oncogenes; regulate normal tissue anlagen in fly and mouse embryo ontogenesis; normally highly conserved; expressed in the embryo, but not in the adult hosts. Similar physiological functions work in vertebrate mammalian embryos. Genes int inserted by MMTV (mouse mammary tumor virus) suffer truncations. Constitutively expressed in adult tissues, Wnt genes activate the E-cadherin to β-catenin pathway. The Wnt pathway is under microRNA control. In human cancers, miR-145 is a suppressor of the Wnt/β-catenin oncogenic pathway. (For drosophila Wg homolog mammalian int, click Cousin JP Bishop SA Martinez-Arias A; Nusse R & Varmus H; Swanstrom R Parker RC Varmus HE Bishop JM. For Wnt/β-catenin in cancer click Ochoa-Hernández AB Juárez-Vázquez CI Rosales-Reynoso MA Barros-Núnez P. For genistein downregulating onco-miR-1260b and inhibiting Wnt in renal cancer click Hirata H Ueno K Nakajima K et al. For tumor suppressor miR-145 targeting catenin in colon cancer, click Yamada N Noguchi S Mori T et al).

TBC1D3. Tre-2/USP6/Bub2/Cdc16 oncogene. Tre, transformants receiving DNA from Ewing sarcoma cells; tre oncogenes encode alternatively spliced proteins constitutively expressed in Ewing sarcoma cells. Ubiquitin-specific peptidase 6 is a tre-2 oncogene-encoded protein expressed in metastasizing aneurysmal bone cysts; expressed early in evolution (*Haloarcula marismortui*). Bub is a yeast gene encoding budding uninhibited by benzimidazole. Acts as an oncogene in malignant B and T cell lymphomas, including human lymphotrophic retroviral T cell malignancies. BUB oncoprotein targets (suppresses) protein MAD (mothers against decapentaplegic, TGFβ, drosophila), a participant of the MAD/MAX/MYC complex; MAD/MAX may displace MYC with loss of its transactivating function, thus canceling its oncogenic function. Cell division cycle-16 protein localizes in the centrosome to keep the cell cycle open in transformed cells. Click for TBC (Tre-2/Bub2/Cdc16) human oncogene/oncoprotein Gabernet-Castillo C O'Reilly AJ Dacks JB Field MC. Click for lung cancer oncogenes other than TBC1D3: Ostoros Gy Bajcsay A Baliko Z et al Magyar Onkologia 2012;56:114–132. For Tre-2/Bub2/Cdc16 oncogene: a riddle wrapped in a mystery inside an enigma, click Oliviera AM & Chou MM Front BioSci 2012;4:321–34.

Human cysteine-rich secretory protein (CRISP) cited on p 171 is a connective tissue growth factor inhibitor of adipocyte differentiation expressed under androgen control and as such acts as a prostate cancer biomarker (Udby L et al J Leukocyte Biology 2002;77;462–469). Important: Distinguish it from microbial antiviral CRISP (Figure 20) (Liang P et al Protein Cell 2015;6:363–372; A New Era in Molecular Biology Reiss A Hornblower B Robb B New England BioLabs, Inc http://neb.com/tools-and-resources/feature-articles/crispr-cas9-and-targeted-genom-editing).

Performance in the Laboratory

Exemplary Laboratory Technology

Obtaining cellular RNA and DNA. Transfer of genomic DNA from human cancer or leukemic cells into NIH 3T3 cells (established from mouse embryo fibroblasts) resulted in the transformation of 3T3 cells into focus-forming cells. Such focus-forming 3T3 cells first yielded the *ras* family (Harvey-, Kirsten-, N-ras) oncogenes. Co-transfection of such inoculated 3T3 cells into nude mice increased the sensitivity of the assay.

Frozen human tumor tissue was either homogenized in mortar and pestle in liquid nitrogen, or minced with razor blades, homogenized in phosphate-buffered saline in a homogenizer (Dounce homogenizer), and lysed at 37 °C in the lysis buffer of 0.5 % sodiumdodecyl sulfate $NaDodSO_4$/20 mM EDTA/40 mM Tris-hydrochloride pH 7.6/0.1 M NaCl/200 µg proteinase K (Merck)/mL. The cell lysates were extracted x2 with buffer-equilibrated phenol and x2 with chloroform/isoamyl alcohol. In the solution, ethyl alcohol precipitated the heavy weight DNA wound onto a glass rod. The ethanol-washed DNA was dissolved in 1mMTrisHCl pH 7.6/1 mM EDTA (ethylenediaminetetraacetic acid). Samples of 75 µg DNA was dissolved in 2.5 mL transfection buffer of 0.7 mM Na_2HPO_4-$7H_2O$/21 mM Hepes (hydroxyethyl-piperazine-ethane-sulfonic acid)/0.145 M NaCl pH 7. DNA Ca-phosphate precipitate was formed after adding (and gently shaking for 30′) 125 µL 2.5 M $CaCl_2$.

The 6×10^5 inocula of NIH 3T3 cells were seeded 24 h earlier in a 100-mm culture dish and fed with 10 mL Dulbecco/Eagle's medium. Into each culture dish 1.25 mL DNA-precipitate is to be added and incubated at 37 °C for 15–18 h. Thereafter the Dulbecco/Eagle's medium is changed with 5 % fetal calf serum added. Culture fluids are changed x2/wk. Foci of cultured cells are looked for in 2–3 weeks. Once focus-forming cells are isolated and secured, they are tested by Southern blots with the human Alu repetitive sequence probe BLUR-84 for hybridization. Repeated cycles of transfection remove those human DNA sequences

J.G. Sinkovics, *RNA/DNA and Cancer*, DOI 10.1007/978-3-319-22279-0_12

that are not that of the transforming gene. Specific probes for known human oncogenes recognize or rule out identity with them. Specific human oncogene probes commercially available are those of the three *ras* probes, and those that were picked up by retroviral genomes: *Blym, v-abl, v-erbA, v-erbB, v-fgr, v-fms, v-fos, v-fps, v-mos, v-myb, v-raf, v-rel, v-ros, v-sis, v-src, v-yes*. Thus a new oncogene can be recognized, sequenced, and its mRNA identified and properly named [735, 736].

The transforming gene *axl* (anexelekto, ανεξέλεγκτο, uncontrolled) was isolated in NIH 3T3 cells from the genome of *ras* mutation-negative human myelogenous leukemia cells deriving from patients in blast crisis. The Axl receptor's ligand is Gas (growth arrest specific, *vide infra*) [737–739]. First, tumor cell lines carrying transfected human oncogenic DNA were established in nude mice. Genomic DNA was digested by *M*bo (*Moraxella bovis*) digestive enzyme for cosmid cloning. Plasmid vectors carrying up to 57 kb DNA packaged into bacteriophage heads by Gigapack Stratagene were prepared. Bacteriophage-infected *E. coli* cosmid library was screened by BLUR8 for probing for repetitive Alu sequences. Screening yielded two Alu-positive clones: 1-1 and 4-2. A 2.8-kb EcoRI-BamHI (digestive enzymes) fragment in cosmid clone 1-1 was free of Alu and mouse sequences and contained the transcribed exons from the putative human gene; it was referred to as pcc-1. Both strands were sequenced by the dideoxy termination method (Sequenase). The *axl* transcribed in buffer (10 mM Tris pH 8.3, 50 mM KCl, 1.5 mM MgCl, and 0.001 % gelatin) with 200 U Superscript Moloney murine leukemia virus reverse transcriptase in 635 μM deoxynucleoside triphosphate (dNTP). 20U RNase (Promega), and 10 M dithiothreitol. This reaction required incubation for 10 min at room temperature, 45 min at 42 °C and 5 min at 95 °C. The cDNA product/microliter could be PCR amplified with 1.25 U Taq polymerase (Promega) in 50 μL 10 mM Tris pH 8.3, 50 mM KCl, 1.5 mM $MgCl_2$, 200 μM dNTP, and 0.001 % gelatin. The 30 cycles are set to be for 5 min at 94 °C, 1 min at 55 °C, 1.5 min at 72 °C, 1 min at 94 °C, 1 min at 55 °C, 1.5 min at 78 °C; 1 min at 94 °C, and 10 min at 50 °C for 28 cycles. The reaction mixture is to be electrophoresed on a 12 % polyacrylamide gel and stained with ethidium bromide. Primer sequences are selected at will [738].

Chromosomal localization of the *axl* gene is done by fluorescence *in situ* hybridization (FISH). The biotin-labeled *axl* probe is from a cosmid vector containing a 41-kb genomic fragment. The slides with human metaphase cells for chromosomal DNA to be denatured are immersed into a solution of 70 % formamide-x4 in saline sodium citrate buffer (SSC) at pH 7.0 (0.15 M NaCl; 0.015 M sodium citrate) at 70 °C for 2 min. For the slides, the hybridization mixture kept at 75 °C consists of 50 % formamide, x1 SSC, 10 % dextran sulfate pH 7.0, and 150 mL unlabeled sonicated total human genomic DNA; add 50–100 ng labeled probe. The mixture is incubated at 37 °C for 10 min. The slides are washed in 50 % formamide and SSC x3, 5 min each at 40 °C and in SSC x3 at 40 °C. The detection reagent consists of x4 SSC-0.1 % Triton, and 1 % bovine serum albumin for 3 washes 3 min each at 40 °C. The slides are incubated with fluorescein isothiocyanate-conjugated avidin at 37 °C for 30 min. Metaphase cells are counterstained with 4,6-diamidino-2-phenylindole dihydrochloride 200 ng/mL in 2x SSC for 5 min at room temperature. The slides are mounted in 20 mM Tris-HCl pH 8.0, 90 % glycerol containing 2.3 % DABCO (1,4-diazabicyclo(2,2,2)octane) antifade reagent (Sigma). Epifluorescence optics-equipped microscope is to be used.

For transfection and tumorigenicity assay, NIH 3T3 cells are grown in 100-mm dish. The 3 M sodium acetate-precipitated *axl* DNA is derived from the plasmid. After certain preparations (O'Bryan et al) [738], the DNA precipitate was retrieved from HEPES (N-2 hydroxyethyl piperazine-N-ethane sulfonic acid) and it was exposed to the 3T3 cells for 8 h at 37 °C. The cells were treated with 15 % glycerol and washed with phosphate-buffered saline. Cells were re-seeded in 100-mm dish containing 600 μg/mL G418 (Glaxo418, aminoglycoside polypeptide synthesis-blocker antibiotic). The G418-resistant 3T3 cells grown to confluence yield focus-forming colonies in 2 to 3 weeks. These cells in dosages of 1 x 10^5 to 5 x 10^5 grow into tumors in the flanks of nude mice [738, 739].

An attempt at clarifying the nomenclature. The K vitamin- and S protein-related Gas6 protein encoded by the growth arrest-specific gene 6 is the ligand of the Axl (anexelekto) receptor tyrosine kinase (RTK) family member (*vide supra*); Gas6 attaches to the C terminal of the Axl receptor. The S protein (Seattle, Washington) is a natural anti-thrombin anti-coagulant, since its deficiency results in deep venous blood clots. The Axl family members (TAM, Tyro, Axl, MER) were repeatedly re-discovered and called Ufo (unknown function protein), Ark (adhesion-related kinse), Tyro7; MER (macrophage expressed), Tyro12, Nyk (new tyrosine kinase), Flk (fetal liver kinase); Ryk re-named Eyk (retroviral tyrosine kinase; East Lansing tyrosine kinase) (*vide infra*). The murine homolog of Axl are Ark and Tyro7; the chicken homolog is Eyk (*vide infra*). The growth arrest-specific Gas6 ligand also activates receptors Rse/Sky (receptor sectatoris; S protein's receptor kinase tyrosine). Gas6-binding induces mitosis, which is blocked by wortmannin (a *Penicillium funiculosum* product). Wortmannin being a known inhibitor of PI3K/Akt, implies that Gas6 binding leads to PI3K/Akt activation. The human mammalian epithelial cell line C57MG continues to grow after reaching confluence upon Gas6 stimulation. The cascade leading to cell proliferation is initiated by the Gas6-bound Axl and Rse receptors. Gas6-induced stimulation of these receptors (Axl being the leader) overcomes density-induced growth inhibition. Testing molecular inhibitors LY294002 and wortmannin for PI3K, PD98059 for MEK, SB203580 for p38 MAPK, and rapamycin for mTOR/S6K (RSK, 40S 90kDa ribosomal ser/Thr S kinase), blockade of PI3K, S6K and MEK abolished the stimulatory effect of Gas6. Gas6 is able to activate Akt (protein kinase B) by phosphorylating it at serine 473 and threonine 308 residues. Rapamycin and inhibition of MEK fail, but wortmannin counteracts Akt activation by Gas6. The often constitutively activated GSK3 (glucose synthase kinase) is the downstream target of Akt. Paradoxically, growth factors insulin, EGF, HGF/SF (epidermal, hepatocyte, scatter factor) inactivate GSK3. So does Akt by phosphorylating serine residues 9 and 21 of GSK3; Gas6-activated Akt in turn inactivated GSK3 even in C57MG cells, where GSK3 works constitutively in doublets as αß isoforms. Wortmannin completely blocked Gas6-induced Akt-mediated phosphorylation of GSK3. Gas6-treated cells activated the Wnt/β-catenin pathway. Physiologically, GSK3 (within the APC/GSK3/Axin complex) marks β-catenin for ubiquitination by phosphorylating its serine residues. Escaping ubiquitination, in Gas6-treated cells, β-catenin activates the pro-oncogenic TCF/Lef pathway (T cell-factor; lymphoid enhancer factor). The β-catenin-induced proto-oncogenes TCF/Lef target for activation c-myc and cyclin D1. In contrast, the Gas5-related PDGF did not activate the TCF/Lef pathway. Cells with β-catenin excess resist detachment-induced (anoikis) programmed cell death (apoptosis) [739].

Gas6-driven Axl-expression was proven to be the driving force of human glioblastoma cell lines (SF126) subcutaneously and/or intracerebrally xenotransplanted into nude mice (mouse colony was not identified in the Supporting Materials & Methods; presumably University of Heidelberg colony). Subcutaneously xenotransplanted tumor cells rapidly invaded all tissues, whereas intracerebrally injected tumor cells grew locally. Tumor cells transfected with a dominant-negative mutant Axl receptor (truncated by loss of intracellular signaling domain) ceased mitoses and invasiveness. Mice xenotransplanted with such tumor cells gained significant prolongation of life [740].

In prostate cancer, expression of the the Gas6/Axl axis directs the tumor cells into the physiological "stem cell niches" of the bone marrow for a prolonged period of residence [741]. Osteoblasts/osteoclasts and endothelial cells in these niches express annexin II cell surface molecules made up from two 11-kDa (p11) and two 36-kDa (p36) subunits. The human gene anxa II encodes the annexin II proteins. Annexin II proteins assume tetramer formations, and as their substrate, attract growth factor tyrosine kinases and phospholipids. The gene organization of annexin A13 in protists, plants and vertebrates reveals that it is the founder of the annexin family [742]. The bone marrow stem cell niches sustain stem cells in the state of dormancy. Resting hematopoietic stem cells express the Gas6/Axl axis. In certain metastatic cancer cells (derivatives of transformed stem cells) annexin II induces their "homing" to the bone marrow. In particular, prostate cancer cells with the expression of annexin 2 receptors are apt to migrate to the bone marrow. In the endosteal stem cell niche, in these prostate cancer cells the expression of Gas6/Axl/Mer/Sky axis is activated. However, in this system, instead of proliferation (compare to breast epithelial cells, C57MG), the Gas6/Axl interaction induces dormancy. Healthy osteoblasts express annexin 2, which is an inducer of Axl expression in prostate cancer cells (as determined by flow cytometry). Osteoblasts (and osteosarcoma cells) are the major providers of Gas6 release for paracrine circuitries (as detected by ELISA assays). Gas6 induces ERK (extracellular signal regulated) phosphorylation in prostate cancer cells in 5 min (as determined by Western blotting). In matrigel assay, Gas6 exposure reduced the proliferation and invasiveness of Axl^+ prostate cancer cells. Annexin V staining detected a reduction in apoptotic prostate cancer cells in Gas5-exposed samples. Treatment of prostate cancer cells with annexin 2 and Gas6 extended protection from taxotere-induced cell death. In flow cytometry assays, Gas6 kept the prostate cell population in G_0 and reduced the number of tumor cells in $G_1 \rightarrow S/M/G_2$ stages of the cell cycle. In a murine model of human prostate cancer, the occurrence of osteogenic metastases favored the spine and the hindlimb. In comparison, the bone marrow of the hind limb expressed lower levels of GAS6, than the forelimb. In vertebral body transplants (vossicles) from Gas6$^{-/-}$ mutant mice, prostate cancer cells grew better, than in vossicles of Gas6$^{+/+}$ animals. In contrast, in subcutaneous sites of Gas6$^{-/-}$ immunocompetent mice, tumor inocula grew better than in Gas6$^{+/+}$ wild-type animals [741, 743].

Paget's "seed and soil" hypothesis of the year 1889 appears to be valid with the basic consideration, that the soil can accept or reject the seed. However, both the seed and the soil are subject to changes in their composition, which may significantly alter their reaction to each other. Gas6 may be an inducer, or an inhibitor of tumor cell mitoses. The role of immunological factors (tolerance *versus* rejection) may be the decisive underlying force. A Th1 immunological factor (IFNγ; TNFα) may be expropriated by a tumor cell as its essential growth factor. Osteoblasts, fibroblasts and vascular endothelial cells of the bone marrow constitutively produce the chemokine, stromal cell-derived factor-1 (SDF-1, CXCL 12); and its receptor, CXCR4, is frequently expressed in tumor cells [744, 745], as well as in tumor-infiltrating lymphocytes (be immune T, or T_{REG}, or B-CLL cells) [717].

The complex human Gas6 protein consists of an N-terminal-carboxyglutamic acid, four EGF-like, and a C-terminal sex hormone-binding globulin-like domains. Mass spectrometry and ELISA assay identifies and quantifies Gas6 in the human plasma. Human platelets were devoid of Gas6 supplement [746]. In the relationship of tumors and macrophages (Mϕ), Mϕ are instructed to excrete Gas6. Gas6 reacts with its receptors (Axl, MER, Tyro3) expressed by the tumor cells, and stimulates tumor cell proliferation [746, 747]. In Axl$^{-/-}$ mice, myelin oligodendrocyte glycoprotein in Freund's adjuvant-induced autoimmune encephalitis of increased intensity, and myelin debris accumulated due to the failure of their removal; activation of microglia and Mϕ was very much reduced. Due to the lack of Axl activation, the inflammatory reactions are not ameliorated [748].

The vascular smooth muscle cells in non-complicated carotid arterial plaques secrete increased levels of Gas6. In the lesions, TGFβ levels rise and TNFα levels drop. In complicated plaques, Gas6 and TGFβ levels drop and TNFα levels rise. Thus Gas6 and TGFβ exert anti-inflammatory effect [749]. In LPS-induced endotoxic shock, mice without functional MER signaling suffer high mortality (88 %), against high recovery rate in wild type mice with signaling MER pathway (7 % mortality). Anti-TNFα immune serum rescued mice without MER signaling by reducing their mortality rate. LPS-dependent activation of intranuclear NFκB was responsible of the high TNFα levels. The signalig MER pathway blocked this cascade [750]. In active multiple sclerosis, soluble MER and Axl receptors circulate in high levels. Acting as decoys, these soluble receptors bind Gas6, thus reducing its level. Cell surface Axl and MER receptors remain inactive. Thus, disease activity rises unmitigated [748, 751].

Assays show that shed and solubilized TAM (Tyro3, Axl, MER) receptors circulate in high levels in the blood of patients with active systemic lupus erythematosus (SLE). These high titers of solubilized receptors correlated directly with disease activity, especially with that of lupus nephritis of low C1q and high anti-DNA antibody levels. In contrast, solubilized sMER levels did not correlate with disease activity in rheumatoid arthritis [752]. In acute lymphoblastic leukemia, steady Gas6 production results in a Gas6-rich microenvironment in the bone marrow for the constant sustenance of leukemic cells. The leukemic cells form N-glycosylated (partially glycosylated) Mer, instead of the fully glycosylated receptor, which translocates into the nucleus (carried by importin/karyopherin), where it binds to chromatin. The fully glycosylated MER ascends to be expressed in the cell membrane. Brefeldin inhibits the partial, tunicamycin the complete glycosylation of MER. The MER glycoform-loaded leukemic cells gain chemoresistance. MER notoriously activates the "cell survival pathways" MAPK/ERK and/or PI3K/Akt. Investigationally available shRNA may inhibit the MER mRNA; such murine hosts may inhibit the growth of leukemia cell xenografts [753, 754].

The "macrophage expressed" Mer is considered to be the mammalian ortholog of the retroviral v-*eyk* proto-oncogene. The extracellular Mer receptor consists of immunoglobulin and fibronectin-like domains. Physiologically Mer directs phagocytosis and platelet aggregation. Pathologically, cells of the lymphopoietic system expropriated Mer to serve as their oncoprotein. Potential Mer inhibitors are staurosporine, SU9516 and BIM-9 [755].

The novel retroviral oncogene oncoprotein v-*eyk* was named after East (e) Lansing tyrosine (y) kinase (k). The former nomination of this oncogene, v-ryk, was discarded in favor of v-*eyk*. The donor chicken RPL30 retroviral isolate (ribosomal protein L30 family) derived from L.B. Crittenden of the Avian Disease and Oncology Laboratory, East Lansing, Michigan. The oncoviral genome and oncoprotein were isolated in Hidesaburo Hanafusa's laboratory at the Rockefeller University, New York City, NY. In this case, the viral RNA-to-DNA retroscribed *env* gene was inserted in the host cell's genome next to a tyrosine kinase locus, so that the two genes fused. The host genomic sequence was 1.39-kb long, inserted into the viral envelope glycoprotein gp37. In the fusion product, all viral sequences were preserved. In cells transfected with v-*eyk* oncovirus, first a 150 kDa precursor oncoprotein was synthesized, which was cleaved into the mature 69-kDa fusion oncoprotein expressed in the cell membrane. The cells underwent malignant transformation and grew as tumors in chicken. The c-*eyk* → v-*eyk* oncogene/oncoprotein exhibited high (67 %) aa identities with *axl/ufo/ark* (*vide supra*) oncogenes/oncoproteins. Two other chicken retroviruses (RPL25 and RPL28) deriving from L.B. Crittenden yielded the cellular erbA/B proto-oncogenes. In most viral-cellular fusion oncogene-oncoproteins retroviral genomic sequences were lost. In consequence, the transformed cells seldom if ever could replicate mature infectious retroviral particles. However, fusion with full retroviral genomes (of "helper retroviruses") could rescue the oncogenic retroviruses to full maturation. Peyton Rous' original Rous sarcoma virus and its high titer-derivative (Bryan' strain) might have been replication-defective (the oncogene replicated, but without a retroviral envelope). Upon later laboratory passages in the presence of a natural helper virus (Schmidt-Ruppin and Prague strains) could the viral genome encode the entire Rous viral envelope. The helper viruses (Rous-associated virus RAV-1/2) were separately isolated [756–758].

It was in 1976, when Stehelin, Varmus, Bishop, and Vogt reported that retroviral oncogenes derive from the host cells' genome [759]. The reverse transcriptase of Baltimore and Temin retrotranscribed the retroviral RNA genome into a viral-proviral DNA, which was inserted into the host cell's genome [32]. This author (JGS) in the mid-1960s could restore the full leukemogenic potency of attenuated retroviral inocula containing large numbers of non-leukemia-inducer retrovirus particles by co-inoculating extremely high dilutions of (containing just a very few) leukemogenic retrovirus particles. For explanation, it was offered that the retroviral genomes separately encode the viral structural proteins, and the leukemogenic potency, and that the gene(s) encoding the leukemogenic potency may be lost (in replicating, non-leukemogenic, but immunizing retrovirus particles). Further, it was proposed, that the leukemogenic potency may be regained from just a few co-infecting leukemogenic virus particles, which are replicating the leukemogenic part of their genome in excess [760–762]. This author (JGS) attempted to isolate human sarcomagenic retroviruses in the early 1970s by inoculating enveloped murine leukemia (Moloney or Rauscher) viruses into human sarcoma cell lines in vitro. When occasional cells released small colonies of enveloped retroviruses [147], the work was stopped (due to institutional regulations), and allowed to continue only in biohazard-safe laboratories (which were promised, but not installed). So far, no sarcomagenic mature retroviruses could be isolated from human sarcomas, even though focus formation and antigenic conversion could repeatedly be induced by cell-free materials [147, 340a, b, c, 341a, b].

Selected abbreviations

Gas6. The vitamin K-dependent cytoplasmic Gas6 growth arrest-specific protein 6, a ligand, and its receptor axl/AXL.

AXL, anexelekto uncontrolled, Greek, receptor tyrosine kinase member of the TAM (Tyro3-Axl-Mer) family. The extracellular domain is Ig-like; farther down fibronectin. Prominently co-expressed in acute myeloid leukemia, osteosarcoma, glioblastoma multiforme. In GMF, both in migrating tumor cells, in tumor-feeding vascular endothelial cells, and in reactive astrocytes next to the tumor, axl/AXL is overexpressed. (For anexelekto in GMF, click Hutterer M Knyazev P Abate A et al Departments Neurology Neurosurgery Pathology, Innsbruck Medical University, Innsbruck, Austria. For Gas6/Axl in renal cell carcinoma, click Gustafsson AM Martuszewska D Johansson M et al Department Laboratory Medicine Lund University, Sweden).

Inbuilt Errors Not Yet Corrected

Evolution Is Not Error-Free

The invasion of the tyrosine kinases can go out of control. The human genome operates 10 families of 32 nonreceptor, and 19 families of 58 receptor-connected TKs. Those TK families, which work in the ontogenesis of complex organs (ALK, anaplastic lymphoma kinase; EGF, epidermal growth factor; MET, hepatocyte growth factor receptor; RET, rearranged during transformation) are absent in the Nematostella. In contrast, the amphioxus possesses BRK and SRC (breast cancer; Rous sarcoma) non-receptor TK ancestors. Physiologically, BRK is involved in the fetal development of the pancreas and small intestine; SRC is a regulator of cell adhesions, motility and divisions. The metazoan ancestor possessed a BRK/SFK (Src family kinase) gene, which subsequently duplicated. The SFK gene was amplified in the sea urchin; the SRC gene split off to enter and follow the vertebrate lineage. The common ancestor of the receptor-TK (FGF-R, PDGF-R, VEGF-R) TK domains appear in the amphioxus and sea urchin (VEGF-R more dominant than the PDGF-R). The split between FGF-R and PDGF-R/VEGF-R occurred in the cnidarians before the emergence of the bilaterians. The PDGF-R/VEGF-R genes split by tanden duplication and acquired long stretches of hydrophobic amino acids in the vertebrates [732]. (Some hydrophobic aa: alanine, leucine/isoleucine, glycine, methionine, proline, tyrosine, valine).

The ancestral to endothelial tyrosine kinase proteins, the TIE-like proteins, TEK (tunica internal endothelial cell kinase), fulfilled functions other than blood vessel generation (angio-vasculogenesis) in the amphioxus [732]. These endothel-specific tyrosine kinase receptors first appeared in the amphioxus before any trace of vascularization, but are absent in the cnidarians (*N. vectensis*) [732]. The vertebrates enlisted these genes for encoding receptor signaling in true hemo- and lymphopoietic vascular endothelial cells [763, 764]. Angiosarcomas, the malignant tumors of the vascular endothelial cells (vascular and lymphangitic, including

J.G. Sinkovics, *RNA/DNA and Cancer*, DOI 10.1007/978-3-319-22279-0_13

Kaposi's sarcoma), overexpress TIE1/2 proteins (and angiopoietins, as ligands). TIE2 serves as the receptor of angiopoietins (in autocrine circuitries in hemangiosarcomas); the ligand of TIE1 remains unidentified. Angiopoietins act in complementary and coordinated fashion with VEGFs. Kaposi sarcoma cells bathe in a sea of growth factors (angiopoietins, IL-6, bFGF, GM-CSF, PDGF, TGFβ, and hepatocyte GF), and express receptors for all of these cytokines. In AIDS-related KS, the HIV-1 Tat protein (transactivator of transcription) induces the expression of VEGF receptors [765]. In addition to be a "breeding ground of herpesviridae" [83], where HHV-8 stands out as an etiologic agent, KS cells activate and release by budding enveloped particles of an endogenous retrovirus; the patients circulate antibodies reacting with the budding envelopes of these retroviral particles [147, 766, 767] (Figures 29 and 30).

Human recombinant HurIFNα emerged as weak, but not ineffective, neoangiogenesis inhibitor [768–770]. Some vascular endothelial proliferations (including hemangiosarcoma of the chest wall within radiation field of resected breast cancer) regressed upon HurIFNα therapy (author's female patient CS experiencing relapse-free remission over 10 years; unpublished). However, low response rate and high rate of tumor resistance characterized HurIFNα therapy of true hemangiosarcomas. The recommended standard therapy consists of paclitaxel and ionizing irradiation. Neither this regimen, nor the targeted small molecular VEGF/R inhibitors (sorafenib, sunitinib) secure high and durable response rates. The *tek* gene encodes TIE2 (tunica endothelial cell kinase; tunica internal endothelial cell kinase). The newly synthesized TIE2 inhibitor (a methoxy-naphthyl-methylsulfinylphenyl-piridyl-imidazole) alone or in combination with sunitinib induced significant partial remissions in xenografted VEGF-R2- and TIE2-expressor human angiosarcoma cell lines. Sunitinib alone induced tumor cell necrosis, and delayed growth progression, but failed to sustain durable remissions of these cell lines. The TIE2 inhibitor, or the combined regimen induced angiosarcoma cell death by apoptosis and major tumor volume reduction. However, a rapidly replicating subline of tumor cells emerged driven by the Ras/Raf/MAPK pathway [771], providing one of many examples of an oncogenic pathway switch from an inactivated to a newly activated oncogene cascade in a transformed cell.

Intron-less predecessors of the *ret/ptc,* Ret gene/protein (rearranged during transformation; papillary thyroid carcinoma) appear as pseudogenes acquired through inserted RNA sequences (ancient viral or host mRNA; retrotransposons) in the amphioxus. The amphioxus *ret* pseudogene contains TK (tyrosine kinase) and cadherin domains. The human RET receptor TK retains the extracellular cadherin domain. In the cytoplasm, RET forms chimeric oncoproteins (RET/PTC, rearranged during transformation, papillary carcinoma thyroid). In embryonic life, RET directs neuroectodermal development, including cell differentiation and migration. Its loss-of-function mutation results in Hirschsprung's disease (congenital aganglionic megacolon). Gain-of-function point mutations of the *ret* gene manifest in multiple endocrine neoplasias, or in medullary carcinoma of the thyroid. RET/PTC1 is the result of chromosomal inversion: inv (10)(q11.2q21.2). In RET/PTC-transformed cells the PI3K, the Ras/ERK and the β-catenin pathways become activated [772]. *)

*) *Further oncogenic associations of ret/RET include proto-oncogenes c-myc/MYC and c-kit/KIT combined with the downregulation of cyclin-dependent kinase inhibitor CDKN2C (c-Jun-N terminus and juxtaposed cysteine pair). The malignant tumor thus induced is an angiosarcoma (Styring E et al Br J Cancer 2014;111:407–412).*

The servant E-cadherin turns coat and releases the proto-oncoprotein β-catenin. E-cadherin is connected to the actin cytoskeleton by β-catenin. Free β-catenin is activated in the Wnt/β-catenin signaling pathway. Physiologically, excess free cytoplasmic β-catenin is phosphorylated on its serine-threonine residues and thus removed by ubiquitination in the proteasomes. This process requires the intact cooperation of the axin multiprotein complex, the APC protein and the Ser/Thr kinase GSK3β (*vide supra*). The Wnt ligand intervening, can inhibit the Axin/APC/GSK3 complex. Excess cytosolic β-catenin translocates into the nucleus, where it activates the the genes of Tcf/Lef (T cell factor; lymphoid enhancer factor). The DNA-binding gene product proteins TCF/LEF activate the genes c-Myc and cyclin D1. If constitutively activated, these proteins dictate incessant cell divisions. (TCF may be identical with, or related to, hepatocyte nuclear factor 4α, HNF). The specialized RT, telomerase enzymatic subunit (TERT) serves as direct target of β-catenin. β-Catenin up-regulates TERT mRNA expression. β-Catenin deficiency (suppression by siRNA) reduces telomerase activity and TERT expression resulting in shortening telomeres. β-Catenin activates histone methyltransferases resulting in tri-methylation of lysine (K) 4 of histone 3: H3K4me3 in the *tert* promoter. β-Catenin maintained telomere length in villus stem cells (in villus crypt structures), in embryonal teratocarcinoma cells, and colon cancer cells with APC loss-of-function mutation (resulting in excess cytoplasmic and nuclear ß-catenin). The partners of β-catenin in its positive interaction with *Tert* are Klf4 (stem cell master regulator, Krüppel-like factor) and TCF1. *Tert* gene transcription is initiated with the histone 3 lysine 4 trimethylation (H3K4me3) at the promoter. Gain-of-function mutations of β-catenin converts this stem cell factor into an oncogene [773] (Table IV).

The Wnt pathway is physiologically operational in the cnidarians (*Nematostella vectensis*) co-jointly with the TGFβ, FGF and Hedgehog (Hh) pathways. In the Hh pathway, the 'hedge' ligands activate the autocatalytic 'hog' domains. The choanoflagellate, *Monosiga ovata* possesses an ancestral Hh protein with C, and a threonin-rich, but without signaling, N, terminal. The *N. vectensis* operates signaling Hh proteins. The ligand-binding transmembrane Hh receptor 'patched' is expressed in the endoderm and signals to Hh-responsive genes. The desert-like Hh genes later underwent duplications to yield the indian and sonic Hh genes. In the drosophila larvae, the Hh gene acting as a morphogene directs segment polarity and wing disc, eye and primitive gut development [774]. The Wnt protein ligand binds its cell surface receptor 'frizzled.' Frizzled signals glycogen synthase kinase 3β. The GSK3β command inactivates the APC and Axin protein complex. Thus β-catenin escapes ubiquitination and translocates into the nucleus, where it forms a

DNA-binding complex with TCF. This dimer activates the genes of c-Myc, Myb, and NFκB. The activation of the *Tert* gene occurs through its promoter by c-Myc and β-catenin directly acting upon it [775]. Not listed other members of these pathways are FAK (focal adhesion kinase) and mTOR (mammalian target of rapamycin). In cells with APC loss-of-function mutation, the levels of both FAK and β-catenin are high. Key effectors of FAK signaling are PI3K/Akt/mTOR. In response to Wnt/β-catenin activation and c-Myc induction, FAK is a positive mediator of Akt/mTOR activation. FAK expression is high in regenerating, or malignantly transformed intestinal cells [776, 777]. Two FAK inhibitors (VS-4718; PF04554878), and PI3K/mTOR dual inhibitor VS-5584 are ready for clinical trials (PF, Pfizer; VS, Verastem) (Tables IV and IX).

Progeria (Werner syndrome): one of the inherent errors of bioengineering? The G4 quadruplex DNA is evolutionarily conserved from the unicellular life forms up to the human genome (*vide supra*). In mammalian (human) cells, G4 quadruplex DNA occurs in the telomeres, in the stem cell genes (which may transform into oncogenes: c-*myc*), in the promoter regions of rDNA repeats, and in unperturbed mitotic and meiotic cells. Unusually rich in G4 DNA formations are the

Table IX Cancer stem cells

Overlapping triangles. Nanog positively interacts with the promoter of focal adhesion kinase (FAK); FAK activates Nanog by phosphorylation. Their unison is suppressed by binding of WTp53 protein to the promoter of Nanog; in contrast, NFκB activates it. Suppressed p53 (by binding FAK-activated MDM) results in pluripotent stem cells with activated Oct-4 and Sox2; and/or in activated Hh/Gli pathway. In stem cells attached to extracellular matrix through integrin receptors (α-actinin, vinculin, talin, paxillin), FAK is autophosphorylated at Y397 (tyrosine).
Phosphorylated FAK activates Src, PI3K and Grb (growth factor receptor), from where activation signals cascade to *rac*, *raf*, *ras* resulting in MEK (MAPK/ERK) pathway. Migrating bone marrow stem cells offer chemokine stromal cell-derived factor, SDF-1, to further activate FAK[1]. In cells with an activated integrinβ1-FAK-PI3K/Akt/Wnt signaling pathway cytoplasmic telomerase is upregulated to inhibit apoptosis and maintain stemness. FAK could be suppressed by iRNA[2]. TGF-β bound to its receptor TGFβRII suppresses FAK activation; integrin signal-transducer FAK encodes squamous carcinoma cell de-differentiation in TGFβRII-null keratinocytes[3]. FAK emerges as the switch left on (disobeying off signals) in malignantly proliferating stem cells.
Inhibitors of FAK. Verastem Inc (founded by *Robert Weinberg*) produces small molecular inhibitors of the FAK/PI3K/Wnt/mTOR pathway: VS-6063 (administered with paclitaxel); VS-4718 (to follow) www.verastem.com.
Akt, Ak-mouse thymus retrovirus, protein kinase B; MDM, mouse double-minute chromosome (human equivalent); Gli, gliotactin, drosophila, glioma; Grb, growth factor receptor-bound 2; NFκB, nuclear factor B lymphocyte (Rel, reticuloendothelial virus); PI3K, phosphatydil inositol kinase 3; raf, rat fibrosarcoma; rac, ras-related C3 botulinum toxin substrate; ras, rat sarcoma (GTPases), Harvey or Kirsten proto-oncogenes; sdf, chemokine stromal-derived factor; Src, sarcoma (Rous); TGF, transforming growth factor; Wnt, wingless, drosophila; int, insertional mutagenesis gene, mouse.
1 *Golubovskaya VM Anticancer Agents Med Chem 2013;13:576–580.* ***2*** *Liu Z et al PLoS One 2013;8(1):e53576.* ***3*** *Schober M & Fuchs E Proc Natl Acad Sci USA 2011;108:10544–9.*

mitochondria. The G4 DNA quadruplex yields to unwinding (strand separation) by RecQ (recombination; hypo-rec, and hyper-rec in yeast) helicases. The mutated recQ gene is responsible for Werner syndrome; its malfunction induces the aging process. The conserved Q motif of the Rec helicase is the ATP binding site. The Q25A mutation (Q, glutamine; A, alanine) in helicase FANCJ deprived the molecule of its DNA-binding and ATPase activities and negated its ability to dimerize [778]. ATP hydrolysis yields the energy for the enzyme's translocation along DNA tracks [779]. Helicase gene mutations result in severe function-loss conditions (Fanconi's anemia J protein, FANCJ; Bloom syndrome, BLM; Werner's progeria, WRN). Helicases protect against replicative senescence (due to telomere repair); progeria is due to Gly574Arg and/or Met350Arg mutations in the appropriate helicase, or Lys125Asn/Lys135Glu (lysine; asparaginase; glutamic acid) mutations in the exonuclease domain. These are severe loss-of-function mutations inherited in autosomal recessive pattern.

In Werner syndrome, there is no growth spurt in adolescence; instead there is atrophy of the subcutis, graying and loss of hair, osteoporosis, atheromatosis, and senile cataracts. In Bloom syndrome, there is growth retardation, immunodeficiency and a multitude of malignant tumors, including acute leukemias. In contrast to the initial loss-of-function mutations, the FANC, BLM, and WRN helicase gene mutations lead to serial new gain-of-function mutations of the leukemogenic and solid tumor-inducing oncogenes (not yet precisely accounted for). Patients with the same syndrome have different tumors, or one patient may have two or three co-existing tumors of different histopathology. Patients with Fanconi or Bloom syndrome die of acute leukemias (myelogenous) and a broad range of solid tumors involving of all three germ lines. Patients with progeria eventually activate those fusion oncogenes/oncoproteins, which cause malignant tumors of the mesodermal germ line: soft tissue and osteogenic sarcomas and leukemias. Acral lentiginous melanoma, or follicular thyroid carcinoma, develop less frequently. Telomere-short cancer cells in Werner syndrome utilize the 'recombination-mediated alternative lengthening of telomeres' pathway. Nevertheless, cancer cells of Werner syndrome patients undergo senescence in culture (remain metabolically active, but refrain from mitoses). RecQ-defective cancer cells do not readily repair radiation- or chemotherapy-induced DNA breaks or damage (topoisomerase-inhibition by camptothecin or DNA cross-linking by mitomycin C or cis-platin), and therefore stop replicating. Sister chromatid exchanges (mitotic recombinations at mitosis or meiosis) do not occur in Werner syndrome cells, but anaphase bridges and thin threads connecting segregating chromosomes are formed. The Werner syndrome genome suffers gene deletions and translocations. The locus of the RecQL4 gene is on chromosome 8q, but instead of deletion, trisomy of this chromosome may be present. In Bloom and Werner syndromes, massive cell loss leads to short stature, and the growth hormone and IGF signaling pathways are suppressed [780, 781]. In Werner's progeria, the rapidly shortening telomeres in differentiated mesodermal/mesenchymal cells (fibroblasts) regain full lengths upon regaining de-differentiated pluripotency during their malignant transformation (*vide supra*).

Double agent enzymes. At normal cell divisions, newly synthesized histones would lack the modifications characterizing the pre-divisional status of the epigenome, were it not so, that chromatin modifying complexes copy from parental to newly synthesized nucleosomes the histone modifications. Translesional DNA synthesis (TLS) is practiced when low fidelity polymerases bypass sites of DNA damage. The deoxycytidyl transferase Rev1 polymerase bypasses sites of DNA damage (the gap) in order to resolve stalling and initiate TLS downstream of the gap. Rev1 (reverse transcription) recruits high fidelity TLS polymerases to the replication fork to re-initiate repair. The gap is filled thereafter by the Rad (radiation sensitive) recombinases. For the bypass, the sliding clamp, the proliferating cell nuclear antigen (PCNA) (*vide infra*), is to be ubiquitinated. Rad6/18 accomplishes this, and recruits high fidelity TLS polymerases/recombinases for the filling of the gap.

Rev1 is further involved in the replication of undamaged DNA in G4 quadruplex structures. Cells with mutated or deleted Rev1 polymerase fail to initiate DNA synthesis and stall until after the gap was filled. Further, the replication of the G4 quadruplex structures is interrupted. In consequence, the recycling of the parental histones will be uncoupled from the DNA synthesis and the newly synthesized histones will not receive their repressive modifications. Sets of helicases process G4 DNA quadruplex structures. The FANCJ, the Bloom and Werner RecQ helicases unwind G4 DNA quadruplexes into single strands to serve as templates for replication [782, 783]. In the case of Rev1 deficiency, mature B lymphocytes fail to express the transmembrane C-type lectin receptor; normally, these cells express CD72 constitutively. Lysine 4 in histone 3 is trimethylated, and lysines 9/14 in histone 3 are acetylated in the normal promoter (H3K4m3; H3K9/14ac). The transfer of these modifications does not occur in Rev1-deficient cells [783]. In cells with mutated or deleted Rev1 and RecQ enzymes, dsDNA breaks remain unrepaired due to stalling repair mechanisms, and the *sanctum sanctorum* of the epigenome will be severely violated. Haphazard alterations of the epigenome will set in.

In comparison to wild types, the counts (probe numbers) of gene dysregulations become very extensive in *fancj, blm* or *wrn* gene-mutated cells. These findings indicate the extreme importance of G4 DNA quadruplex helicases in the maintenance of epigenome stability. G4 quadruplex-induced uncoupling of DNA synthesis and histone recycling result in the loss of active chromatin marks. Cooperation between Rev1 polymerase and RecQ helicase enzymes is essential for the preservation of the pre-set balance [783].

The Y family polymerase, Rev1, is described as a contributor to tumor initiation and progression. The ubiquitinated form of the replication accessory protein, PCNA (monoubiquitinated proliferating cell nuclear antigen, *vide infra*), allows the access of Rev1 to the damaged DNA site. Rev1 and Rev7 (Polζ) interact with each other through noncatalytic subunits. The ubiquitin-binding motif (UBM) of Rev1 fits for attachment the Ub-PCNA site. In addition, Rev1 expresses the N-terminal of the BRCT domain (encoded in the C-terminal of the breast cancer tumor suppressor/susceptibility protein, BRCA) and the C-terminal polymerase interaction domain [784–786]. The activation of REV1 is further dependent on its physical binding with Hsp90, the chaperone of protein trafficking [786]. Hsp90's clients are the "cell survival proteins" especially when they act as constitutively expressed oncoproteins. Thus, Hsp90 is a target of cancer therapeuticals; 17-AAG (17-allylamino-17-demethoxy-geldanamycin) is a heat shock protein Hsp90/Rev1 antagonist. The FANCA (Fanconi) protein is another client of Hsp90. Hsp90 attracts Rev1 to DNA damage-induced stalled forks and postreplication gaps. In addition to Rev1, Hsp90 attracts also Polη to the Ub-PCNA site (ubiquitylation of lysine 164-PCNA). Inhibition of Hsp90 by17-AAG, or by its siRNA, reduced Rev1 accumulation at its targeted sites. Without active Hsp90, there was no Rev1-to-Ub-PCNA interaction. By promoting error-prone TLS, Hsp90 and Rev1 are major contributors to malignant transformation. The FANC-associated proteins (A-J) are also Hsp90-dependent. In the case of disabled Hsp90, the FANC proteins remain retained in the cytoplasm and eventually undergo proteasomal degradation [786].

Synthetic lethality. Chemical carcinogenesis by benzo(a)pyrene is mediated by Rev1. Rev1 gene-specific ribozyme (in plasmid complexed by polyethylene-imine and given by aerosol) depleted the production of Rev1 in mice. In the lungs of these mice, benz(a)pyrene lost over 90 % of its carcinogenic potency; in 27 % of the mice tumor-formation was completely abolished. Thus, error-prone translesional "repairs" of DNA breaks in the TLS (translesional DNA synthesis) pathway are highly carcinogenic [787]. Chemically induced DNA breaks mis-repaired in the TLS pathway result in cell populations resistant to the chemicals, which initially induced the DNA breaks. For the acquisition of this resistance, the activities of Rev1/Rev3/Rev7 error-prone repair polymerases in the TLS pathway are essential (Polζ = Rev3L/Rev7). Retroviral vectors expressing Rev3L-targeting shRNA depleted Rev3L production in Eμ-*myc*-B cell, and L5178Y/TK$^{-/-}$ lymphoma cells. In the absence of Rev3L, lymphoma cells recurrent after initial cyclophosphamide or ciplatin treatment, remained susceptible to these drugs. In contrast, lymphoma cells treated without Rev3L depletion, rapidly acquired resistance to these drugs. Chemoresistant recurrent lymphoma cells exhibited aggressive growth in mice. Thus Rev1/Rev3L suppression reduced chemically induced mutagenesis and acquired drug resistance in the targeted lymphoma cells. In control lymphoma cells, treatment with alkylators induced DNA mutations and elicited the overproduction of the Rev polymerases [788]. A treatment strategy emerges, in which the error-prone TLS DNA repair is inhibited, and another DNA break is induced in order to elicit the phenomenon of synthetic lethality (Table X).

Polyadenosine diphosphate ribose polymerase (PARP) (Table II) binds directly to PCNA for repair of ssDNA breaks. Inhibiting the repair would render the damaged DNA extremely vulnerable to a second unrepaired DNA damage, such as ds breaks (in BRCA1/2 gene; by radiation or chemotherapy). Thus PARP inhibitors (olaparib, 6(5H)-phenanthridinone), in combination with irradiation or chemotherapy, expose the DNA to "synthetic lethality" [789–791]. The ss- or dsDNA is vulnerable to breaks, but several mechanisms exist for the repair of ss- or dsDNA breaks: homologous recombination for dsDNA breaks (error-free), non-homologous

Table X Synthetic lethality

dsDNA repair. In order to replicate, the cancer cell must repair the very same dsDNA break(s) that initiated it. If an unrepaired ss/dsDNA break is followed by a second one, the redundant pathway, that replaced the function of the first dsDNA break, the cell becomes most susceptible to apoptotic death.[1] The most sensitive sensor of ssDNA breaks are the ataxia telangiectasia mutated (ATR) and Rad3-related protein kinases accumulating at the sites of stalled replication fork. dsDNA breaks are repaired by direct ligation of the broken free ends, the nonhomologous end-joinings, with possible deletions and translocations, or by the homologous recombinations (HR). The HR requires the synthesis on a template of the DNA strand to be replaced. Cancer cells enlist PARP1 for nonhomologous end-joining repairs. Polyadenosine diphosphate ribose polymerase 1 (PARP, Table II) acts in complex with DNA polymerase-β, DNA ligase III and XRCC1 (X ray repair cross-complementing gene/protein). Gene polymorphisms of XDCC1 occur (Arg399Gln; Arg194Trp; Arg280His) and may contribute to oncogenesis.[2–7] ***Inhibiting DNA break repair*** could kill a cancer cell. Failure of PARP1 to repair dsDNA break(s) leads to a mitotic catastrophe culminating in cell death. The PARP1 inhibitor olaparib (AZD2281) induces durable remissions in 30 % of BRCA1/2-carrier patients with breast and/or ovarian carcinoma. The PARP family member tankyrase (Table II) maintains telomeres. Its inhibition suppresses the WNT oncogenic pathway. In the germ-line Lynch syndrome (hereditary non-polyposis colorectal cancer) the genes involved are named according to their ancestral genes in the yeast, Saccharomyces. MSH2 and MLH1 are mutated tumor suppressor genes. The MSH2-deficient tumor cells are exquisitely susceptible *in vitro* to oxidative second DNA damage induced by methotrexate. P53-deficient cells show increased sensitivity to Polo-like kinase inhibitors. P53-proficient cells die if ATM (ataxia telangiectasia) is targeted. ATM-mutated CLL cells are susceptible to PARP inhibitors. Clinical trials needed. Caution: rapidly proliferating normal tissues (hematopoiesis, and mucosal linings) may suffer collateral damage.[8–10]
1 Dobzhansky T Genetics 1946;31:269–20. ***2*** *Kang S et al Asian Pac J Cancer Prev 2013;14:941–946.* ***3*** *Khlifi R et al J Genet 2012;91:375–384.* ***4*** *Li LP et al Asian Pac J Cancer Prev 2013;14:2033–6.* ***5*** *Niu Y et al Environ Toxicol Pharmacol 2013;36:311–9.* ***6*** *Sultana R et al PLoS One 2013;8(2):e57098.* ***7*** *Zhang L et al Exp Ther Med 2012;4:1057–6.* ***8*** *Curtin NJ & Szabo C Mol Aspects Med 2013;pii: S0098-2997–1.* ***9*** *Mladenov E et al Front Oncol 2013;3:113.* ***10*** *Shaheen M et al Blood 2011;117:6074–82.*

end-joining, nucleotide excision repair (for ssDNA breaks), base excision repair, or mismatch repair.

For a cell in the female breast tissue to turn malignant, both alleles of the BRCA1/2 (breast cancer) genes have to be mutated (or inactivated by promoter methylation). Error-prone attempts at repair (*vide supra*) add new mutations. According to the Dobzhansky principle (established in the drosophila in 1946), if two nonlethal mutations occur in combination and remain unrepaired, the cell dies: the principle of synthetic lethality (Table X). However, if the damaged DNA remains unrepaired, it may continue to suffer more mutations due to the action of error-prone repair mechanisms (*vide supra*), or becomes mismatch-repaired (broken fragments fused). Grossly malfunctioning protein molecules (in oncology terminology: oncoprotein kinases) will be encoded. In order to prevent this event, mutations that occurred in combination (two alleles for the same function) should be left unrepaired, so that the cell is allowed to die. If one allele is naturally inactive, the other allele should be induced to mutate, or become disabled/inactivated, so that the cell would die. Strong proof of principle exists for the killing of breast, ovarian and prostate cancer cells bearing homologous recombination (HR) DNA repair defects. Response benefits deriving from the addition of PARP inhibitors: approved: olaparib, iniparib, or investigational MK-4827; AG14361 to chemo-radiotherapy were achieved. When the tumor suppressor gene product protein, phosphatase and tensin homolog on chromosome 10 (PTEN) is disabled, PI3K/Akt become a constitutively expressed oncoprotein (Figure 50). These PARP inhibitor olaparib-resistant tumor cells may be still killed with the PARP-inhibitor AG14361b and PI3K-inhibitor LY294002 [792]. The lysophospholipid alkylphosphocholines in contrast to genotoxic chemo-radiotherapy, inhibit signal transduction within cellular membranes. Their representative, the perifosine-derivative, erufosine (ErPC3) induces mainly intrinsic mitochondria-mediated, and rarely extrinsic (FasL → FasR), apoptosis in various tumor cells, among them prostate cancer cells. ErPC3 exerts broad inhibitory activity by interrupting signal transduction in the PI3K/Akt, and MAPK/ERK cell survival pathways, when expressed constitutively in neoplastic cells [793].

PARP-1 (poly-ADP-ribose polymerase), the founder of the family, conserved the "PARP signature" in all vertebrates (Table II). All PARPs catalyze ADP-ribose unit polymerization. The specialty of the enzyme is the repair of ss- and dsDNA breaks. The PARG protein (poly-ADP-ribose glycohydrolase) is present in worms, flies, plants and animals. Animals with deleted PARG genes/proteins succumb to genotoxic agents and LPS-induced septic shock. The human BCRA1/2 tumor suppressor proteins carry out HR (homologous recombination) for the repair of ss- or dsDNA breaks. PARP-1 repairs these DNS breaks in BCRA1/2 deficient cells. However, BCRA1/2 deficient cells treated with PARP-1 inhibitors undergo severe further chromosomal aberrations. Similar DNA damage occurs in chemotherapy-treated, or X-irradiated cells. Histone H2AX (ataxia telangiectasia) and Rad51 proteins are mobilized for DNA repair. The phosphorylated H2AX, and γH2AX mark the dsDNA break sites. Molecules of the DNA repair enzyme Rad51 converge to these sites and these sites can be marked and numerically counted by immunofluorescence. The extent, or failure of DNA repair reflects to resistance, or susceptibility of tumor cells to the treatment given. PARP inhibitors significantly reduce the number of foci that would signify effective DNA repair. In the absence of functional BRCA1/2 repair, and/or in the presence of PARP inhibition, such tumor cells show collapsed DNA replication forks and die [794]. It will depend on its ATP/NAD (nicotinamide adenine dinucleotide) supply, whether the cell's death is due to apoptosis (no inflammatory response), or necrosis (with inflammatory response), inasmuch as apoptosis is energy (ATP) dependent. If excessive nuclear activation of PARP consumed ATP/NAD, the cell death will be due to necrosis [795]. Cells with depleted PARP, but especially $PARP^{-/-}$ $p53^{-/-}$ cells, suffer genomic instability.

The endurance of neoplastic DNA is shown by its recovery from mitotic catastrophe. The Burkitt lymphoma cell line Namalva suffered p53 mutation. The Namalva cells received 10 Gy X-ray irradiation. Large numbers of cells died in apoptosis. Surviving G1 arrested cells became endopolyploid, undergoing endoreduplication. Double-labeled γH2AX foci showed co-localisation of Rad51, and thus DNA repair by Rad51. The endopolyploid cells appeared in G2, expressed cyclin B1, and were exempted from apoptosis. The Annexin-negative apoptosis-free cells showed high Rad51 values; the Annexin-positive apoptotic cells were devoid of Rad51 accumulation. Tetra- and octaploid cells showed mitotic figures. Thus, "lethally" irradiated neoplastic cells retain the potency for recovery to a "potentially productive life cycle" [796].

Attempts at repairing the broken DNA; mismatch repair. PARP-5a (tankyrase-1/2) (Table II), the telomere-associated poly(ADPribose) polymerase, commits mismatch repair and non-homologous end joining repair errors. It interacts with TRF1 (telomere restriction fragment; telomeric repeat-binding factor), and thus indirectly with nucleostemin, the inhibitor of TRF1 dimerization and promoter of TRF1 degradation; otherwise, dimerized TRF1 binds telomere. Being a tankyrase1 inhibitor, its deficiency (FANCD2) thus increases polyADP-ribosylation of TRF. The telomere-capping protein shelterin contains (among others) the poly-ADP-ribose-polymerase tankyrase [797]. To another TRF1 domain, binds GNL3L (guanine nucleotide binding protein-like 3-like), which preserves TRF1 and prevents its ubiquitylation. Telomere-bound TRF1 blocks the access of telomerase to telomere ends to lengthen them [798]. Further, tankyrase acts upon the scaffold protein axin, which is an inhibitor of the proto-oncogene Wnt signaling cascade. Axin is naturally complexed with GSK-3β (glycogen synthase kinase) [170]. Axin is stabilized by Dab2 (disabled endocytic adaptor) by inhibiting its dephosphorylation by PP1 (protein phosphatase) [799]. Dab2 occupies the C-terminal domain of axin, thus blocking the the access of PP1 to the same site. If PP1 dephosphorylates axin, the destabilized axin liberates the Wnt cascade. The essence of Wnt signaling is the translocation of β-catenin from the cytoplasm into the nucleus. The multimeric complex holds β-catenin in the cytoplasm. This complex consists of the disheveled proteins, axin, adenomatous polyposis coli (APC) gene-product protein (encoded from locus 5q15), casein kinase-1, and GSK-3β. In the complex, the phosphorylated β-catenin is marked for degradation in the proteasome. However, the complex's holding down β-catenin in the cytoplasm, may be disrupted. Low density lipoprotein receptor-related protein 5/6 (LRP) and Frat (frequently rearranged in advanced T cell lymphoma; also named GBK for GSK-3-binding protein) unite; Frat dissociates LRP3/5 from axin. The LRP/Frat proteins subvert proteins "disheveled" to mediate their unison. They, and PARP5A (tankyrase-1), dislodge axin from GSK-3β, and liberate β-catenin; PP1 dephosphorylates β-catenin. If there were E-cadherin β-catenin unisons, those also will release β-catenin [800–804] (Table IV).

Selected Abbreviations

PCNA, proliferating cell nuclear antigen. Basic constituents of ancient genomes and proteomes. PCNA-associated factor (PAF15) physiologically recruits DNA replicative polymerases and switches to translesion synthesis polymerases to bypass replication blockade. Ectopically overexpressed in tumors in supporting oncogene replication (Xie C et al Int J Biochem Cell Biol 2014;50:127–31). Antisense long noncoding aslncRNAs stabilize PCNA mRNA by RNA hybridization; PCNA-AS1 promotes tumor growth by upregulating PCNA expression (Yuan SX et al Cancer Lett 2014. doi:10.1016/j.canlet.2014.03.029). Bone morphogenetic protein expression was inhibitory to hepatocellular carcinoma cell growth by arresting the cell cycle at G1 and inducing apoptosis; PCNA levels dropped, PI3K/Akt (phophatidyl inositol; AK mouse thymic retroviral lymphoma) activity was blocked (Zheng Y et al Tumour Biol 2014 Feb PMID 24573697).

Errors or Inherent Attributes for the Sustenance of Unicellular Life

Once Established, Life Is to Remain Present in This Universe

The ancient "cell survival pathways" of unicellular life forms become oncogenes in multicellular hosts. Proto-oncogenes are installed. The primordial physiological "cell survival pathways" PI3K; Wnt/catenin; Notch, as a "destabilizer" of catenin; NFκB; TGFβ, discovered as decapentaplegic, Dpp; Hedgehog, JAK/STAT were already operational in the cnidaria (a non-bilaterian). The *hoglet* gene of the choanoflagellate, *Monosiga ovata*, may be the ancestor of the Hh gene family. At the level of the drosophila, the Hh receptors 'patched' (*ptc*) and 'smoothened' (*smo*) are very similar to those that operate in vertebrates (however, the vertebrate *smo* gene duplicated) [774, 805]. Before the divergence into protostomes (the ancestors of molluscs, flatworms and arthropods), and into deuterostomes (the ancestors of chordata/vertebrates), these genomics/proteomics directed the phylogenesis and briefly revealed themselves in the ontogenesis of these extremely complex life forms (as deduced from the metabolism of their extant descendants) [806]. The temporarily active and reversible workings of the Wnt/β-catenin pathway in the tadpole developing into a frog [186, 807–810] commonly and readily convert themselves in advanced vertebrate mammalian hosts into constitutively installed oncogenes [811–815].

In the ontogenesis of the amphioxus, Wnt signaling establishes the anterior-posterior axis. Later, in vertebrate embryos, Wnt signaling patterns the dorso-ventral axis as well. The amphioxus Dkk3 protein is unique in that, it expresses a TGFβ-receptor domain (in addition to its serine-rich N terminal, and cysteine-rich C terminal domains). When the amphioxus Dkk protein was knocked out with anti-sense morpholino oligonucleotide, amphioxus larvae developed truncated heads. When a GSK3β inhibitor induced Wnt upregulation, the larvae developed similar truncated heads. Thus, inhibition of Wnt and overproduction by Dkk3 was essential to normal head development [816, 817]. The extracellular domains of the

J.G. Sinkovics, *RNA/DNA and Cancer*, DOI 10.1007/978-3-319-22279-0_14

Wnt receptors "frizzled" are activated by both LRP5/6 and β-catenins. Some of the dickkopf proteins (DKK1/2/4) bind LRPs (lipoprotein receptor-related protein), thus to inhibit Wnt pathway activation. Many MDA (M.D. Anderson) established breast cancer cell lines (*vide infra*) express both cytoplasmic and intranuclear β-catenin and DKK1; this combination in triple negative breast cancer was foretelling a bad prognosis. In contrast, DKK3 inhibited the proliferation and invasiveness of endometrial carcinoma cells [818–821].

The full-fledged Wnt (wingless/integrated) pathway consists of its frizzled (FZD) receptors and LRP5/6 co-receptors (lipoprotein receptor-related protein) for their cystein-rich glycoprotein ligands; cytoplasmic dishevelleds direct the impulse toward the APC/Axin/GSK/catenin complex [170]. Outside of this complex, β-catenin exists in a form adherent to the cytoplasmic tail of cell membrane cadherin. Upon loss-of-function mutation in APC, and tankyrase1/2 destabilizing Axin/GSK (*vide supra*), β-catenin released from the complex, translocates into the nucleus to act upon the TCF/LEF genes (T cell factor; lymphocyte enhancer factor). The c-*akt* gene expresses Tcf/Lef binding elements and undergoes activation. Joining PI3K, and constitutively expressed, the PI3K/Akt unison becomes an oncoprotein in colorectal adenocarcinoma [822]. Dickkopf-3 retained its ancient suppressive activity against overproduction of Wnt/β-catenin. In colon adenocarcinoma cells, Dkk3 induces apoptosis through the endogenous mitochondrial pathway [823], but in many colorectal adenocarcinoma cells, the *dkk*3 gene is loss-of-function mutated, or deleted (Table IV).

The ctenophores operate a physiologically deviant Wnt pathway without axin, thus promoting β-catenin's intranuclear transfer; further, the inhibitory dickkopf protein is also absent. These cells are physiologically functioning as if "malignantly transformed", like 'colon cancer' cells (see in the Appendix 1).

The ancient genes of AID/APOBEC relate to those of herpes- and retroviridae. Activation-induced cytidine deaminase (AID) (*vide supra*) causes the immunoglobulin gene diversification (V region-restricted Ig gene point mutations referred to as somatic hypermutation, SHM), and class switch recombinations (CSR, uniting the new light chain genes with the constant heavy chain genes). The creation of new antigen-specific light chain genes is absolutely essential for specific antibody generation, or antigen-specific T cell receptor conformation. During an accelerated antibody production and class switch recombination (CSR) in an infectious process, there is an increased risk for AID-induced lymphomagenesis; or even carcinogenesis. This occurs if aberrant AID expression would target a tumor suppressor gene and thus liberate a potential oncogene [310a, b]. For example, AID disabling PTEN on chromosome ten, would engender a constitutive activation of the ancient cell survival pathway, PI3K. Phosphatidyl inositols were (and are) the drivers of the giardia [824], the hepatitis C virus in the liver [825], and practically all human cancer cells [406, 407]. The nicking the wrong gene by "mistake" may liberate an oncogene, like *c-myc* (8q24), or *bcl*2 t(14;18)(q32;q21) in B lymphocytes of Burkitt's lymphoma (Figure 55), suffering reciprocal translocations with the IgH locus [826, 827].

The human host long ago (in evolutionary terms) has installed microRNAs miR-93 and miR-155 for protection against the mutagenic damage the misguided AID may inflict [338]. The germinal centers suffer genotoxic stress and ds DNA breaks during the diversification and assembly of new antigen-fitting TCRs and IgM/IgGs. Some of these cells die programmed cell death. In mice with AID$^{-/-}$ genotype, these cells are spared of apoptotic demise. It appears that AID-induced misplaced genomic nicks could impose constrains on the size of expanding germinal centers by eliminating cells with unwanted genomic damage [828]. In addition, the proto-oncogene *bcl*-6 is activated [829, 830]. The proto-oncoprotein BCL6 (B-cell lymphoma) blocks the tumor suppressor and B cell differentiation-inducer genes PRDM and BLIMP11 (positive regulatory domain I-binding factor-1; B lymphocyte-induced maturation protein) [831–833]. The direction of this cascade is toward lymphomagenesis. The complex issue of ds DNA break and defective CSR recognition sets in (detailed with numerous abbreviations of complex gene interaction in Sherman *et al*) [834]. The inactivation of the CREB co-activator gene CRTC2/TORC2 T (cyclic adenosine monophosphate response element binding regulated transcription coactivator; target of rapamycin) causes downregulation of the TCL1 oncogene (cAMP response element binding protein; CREB-regulated transcription coactivator/target of rapamycin; T cell lymphoma breakpoint protein). Adenosine monophosphate-activated protein kinase (AMPK) unites cytoplasmic chaperones with the CRTC2 protein. The mobilized DNA damage response (DDR) in recognition of dsDNA breaks (DSB) activates the tumor suppressor ATM signaling (ataxia telangiectasia mutated) [835]. In response, the tumor suppressor liver kinase B1 (LKB1) acts upon its substrate AMPK. The direction of this cascade is toward B lymphocyte differentiation [834]. Metformin (as expressed in N,N$(CH_3)_2$-imido-dicarbonimidic diamide-induced cancer) is a potential therapeutic stimulator of LKB1/AMP-activated protein kinase axis, as well an inhibitor of the mTORC1-controlled oncogenic protein translation pathway [836].

The proto-oncogene survival kinase Akt (*vide supra*) is an antagonist of LKB1 [837]. The origin of the PRDM zink finger gene family (positive regulatory domain modulator) can be traced back to protochordates (ciona, 2 genes; sea urchin, 7 genes; nematodes, 2 genes; and arthropods, 3 genes). No PRDM genes exist in fungi or plants. In vertebrates PRDM gene numbers reached, in fish (Fugu), up to 15, amphibians, birds and rodents up to 16, and primates up to 17. The highest numbers were reached by segmental gene duplications. The new gene loci underwent massive rearrangements, or even some exon deletions. Some tandem duplicants formed splicing variants. The human positive regulatory domain modulator, PRDM7, retained the entire sequence of the ancestral exon 3 (with intron gain). The ancestral 89 nt long sequence remained free of repetitive elements. Otherwise, PRMD7 underwent extensive structural rearrangements (a "newborn gene"). The alternative C-terminal region of PRDM7 is new and unique. Genes PRMD7 and 9 share some identical sequences. The mRNA generated from PRDM7 so far appears to be spared from miR-induced decay [838]. The tumor suppressor PRDM5 is an antagonist of the Wnt/β-catenin proto-oncogenic pathway (*vide supra*) [839]. The PR domain contains the retinoblastoma protein-interacting zinc-finger genes RIZ1,2. Gene RIZ1 at chromosome 1p36 with N-terminal domain is functioning as a tumor suppressor, while RIZ2 without an N-terminal domain is a proto-oncogene. This Yin-Yang gene uses alternative promoters and thus can encode either the RIZ1 or the RIZ2 protein. RIZ1 stops the cell cycle in G2/M phase and induces apoptosis. RIZ2 keeps the cell cycle open and promotes cell divisions. A histone methyltransferase within the PR domain can methylate lysine (K) 9 on histone 3 (H3K9me) and thus silence RIZ1. In advanced cancers (metastatic adenocarcinomas), the mRNA levels of RIZ1 are elevated. The RIZ1 protein activates insulin-like GF-binding protein 2 (IGFBP), glycoprotein SPARC (secreted protein acidic rich in cystein), and nuclear factor of activated T cells (NFATC). Both IGFBP and SPARC are notoriously elevated in cases of metastatic cancers. NFATC induces COX-2 production in breast cancer cells, thus promoting their invasiveness. Thus RIZ1 may lose its tumor suppressor faculties (inhibition of cell cycle and apoptosis induction), and transform into RIZ2 [840]. Follicular DCs (FDCs) induce in B-lymphocytes attached to them. The downregulation of micro-RNA miR-9, allows the upregulation of PRDM-1. The upregulated miR-30 is leading to the downregulation of BCL-6. These effects were achieved by the microRNAs interacting with the 3′UTR regions of the targeted mRNAs. Under physiological circumstances, FDCs educate antigen-exposed B lymphocytes by elevating their PRDM-1, and suppressing their BCL-6 expressions to clonally expand and/or persist as memory cells [833].

If enzyme AID is a potential activator of oncogenes, could APOBEC also induce oncogenesis? (Figure 47). The other cytidine deaminase, the founder of the APOBEC family (apolipoprotein B-editing enzyme catalytic), appeared on the scene together with the other faculties of innate immunity. The APOBECs specialize in antiviral immunity, targeting first non-LTR-retrotransposons and retroviruses. The APOBECs antiviral activities extended to encompass the territory of flavi- and paramyxoviruses, parvo- and hepadnaviruses, herpes- and papillomaviruses, and probably the TT (Torque teno) circoviruses. The encoded human APOBEC3 (A3) protein acts by deaminating cytidine into uridine in single-stranded DNA strands (Figure 47). Budding retroviruses incorporate into the viral particle A3 molecules. In newly infected cells, A3 molecules are released in the cytoplasm and aim to inhibit viral genomic replication. The Vif (virion infectivity factor) proteins of retrolentiviruses, or the Bet proteins of foamyviruses, or nucleocapsid protein of the human T cell lymphotrophic (HTCL) viruses prevent the encapsidation of the A3 protein [841].

The foamy spumavirus Bet proteins have a long history. African green monkeys (*Chlorocebus aethiops; Cercopithecus aethiops*) are carriers of foamy spumaviruses, but the best studied (in Heidelbeg, Germany) foamyvirus strain was isolated (in England) from tumor tissue of a human case of nasopharyngeal carcinoma [842].

In ***b***etween the standard ***env*** gene and the ***l***ong terminal repeat, two new genes characteristic to this virus were found and were termed *bel*1/2 [843, 844]. Between bel2 and the LTR, the canonical signal sequences characteristic of transcriptional regulation in retroviruses is located: the *tas* (transactivator spumavirus) sequence. The *bel* and the *tas* sequences together make the *bet* gene and its protein product Bet [845]. An open reading frame Orf3 protein of human HIV-2 shows sequence homology with the human spumaretrovirus Bel3 protein [846]. Foamy spumaviruses are closest to retroviruses [847]. The deep African ancestors of retroviruses, retro-lentiviruses and foamy spumaviruses residing in the same hosts bombarded the early hominoid genomes and must have exchanged a great number of their genes. They left behind a great number of inserted retrotransposons (may be pseudogenes) [848, 849]. The second article describes an early occurrence of non-LTR retrotransposons in the cephalochordate/urochordate (amphioxus; ciona) genomes [849]. This retro-transposon co-exists with an inserted herpesviral genome. This ancient herpesviral genome may be ancestral to Herpesviridae (*vide infra*).

Pseudogenes. Most pseudogenes derive from duplicated genes, which suffered mutations and/or deletions in the process and thus become non-functional (non-processed further). When reverse transcription (retrotransposition) created the cDNA from a RNA template, transcriptional errors disabled (rendered non-functional) the gene. Yet some "non-processed" peudogenes turned out to be processed and releasing DNA-templated either mRNAs or short or long non-coding RNAs. Some pseudogenes hide in the mitochondria. If the RNA sequence is relieved of its introns, it may function as a mRNA. A RT may transcribe the RNA into a dsDNA, ready for insertion just about anywhere in the genome. The inserted sequence may suffer deletions. But a processed pseudogene can eventually be transcribed. For the numerous pseudogenes of Trichomonas, see the Appendix 1.

A patient with melanoma yielded a $CD8^+$ T lymphocyte clone isolated from a lymph node carrying metastasis from the tumor. The non-cytolytic lymphocyte clone specifically released TNF and IL-2 upon contact with autologous tumor cells. The melanoma antigen responsible for the reaction was encoded by a pseudogene, that derived from a retrotranscribed mRNA. The antigenic aa sequence eliciting the reaction was MTQGQHFLQKV; V stands for valine, the stop codon. If F (phenylalanine) was followed by S (serine) replacing L (leucine) and V (valine) was followed by a Trp (tryptophan) stop codon, the mRNA was not translated into an antigen [850]. Estimates indicate that mammalian genomes contain 5000 (mouse) to 8000 (human) processed pseudogenes [848]. It is possible to calculate the likeli-hood of homology between pseudogenes and their functional paralogs. If during the long life time of a gene, thus during the time encompassing its evolution, it pro-duced many processed pseudogenes, the pseudogene sequences preserve the status of the gene at different time points of its evolution. Alignement of the pseudogene sequences with those of the modern gene will determine the dates of the non-functionalization event of the pseudogene. Nucleotide alignements in homologous position will refer to ancestral codons (Csuros & Miklos); the Hungarian spelling is Csűrös [851].

Comment. The ancient RNA World remained operational in the intracellular environment. The mRNAs, the transcribed (processed) pseudogenes, and micro- and long noncoding RNAs established a communicational network referred to as the "competing endogenous RNA" activity (ceRNA), as it re-plays its phylogenesis in its ontogenesis, in health and disease. The ceRNA system is actively participating in oncogenesis, both *pro* and *contra* [852, 853].

Selected abbreviations

AID. Activation-induced deaminase converts DNA cytidins into uridine. Transactivates antigen-driven Ig light chain gene point mutations in the event of somatic hypermutation for antibody diversification. Switches IgM to Ig expression. Off-target AID nicks release proto-oncogenes c-myc and bcl-2 for fusion with Ig heavy chain (IgH) gene, thus creating oncogenes. (Click for activation-induced deaminase biochemistry/immunology in Cogné M Limoges University Institut Universitaire de France, Limoges, France).

APOBEC family apolipoproteinB mRNA-editing enzyme catalytic. Inhibits LINE (long interspersed nucleotide elements) retrotransposition in vertebrates. (Click for RNA-editing enzymes in Lindič N *et al* Chair Biochemistry Faculty Chemistry Chemical Technology University Ljubljana, Ljubljana, Slovenia).

Viral Genomic Insertions in the Host Cell's Genome

The Retroviral Inheritance of the Human Race

"Retroviruses R Us": the retroviral inheritance of the human race. "Retroviruses R Us" was the title of the author's presentation given at the Kaposi Foundation's Award Conference in Budapest, Hungary (2010, *vide infra*). Ancient deposits of retroviral genomic repetitive elements (LTRs) occupy some 40 % of the human genome.

Foamy spumaviruses of distant (over 12–14 million years ago) African origin persist in gibbons (*Hylobates agilis*) and orangutans (*Pongo pygmaeus*) populating Borneo and Sumatra. The orangutan's foamy spumaviruses are closer to those of the African green monkeys, than those of the chimpanzees (*Pan troglodytes*) ('gnawing dungeon-dwellers'), Over 69 % of wild-caught Bornean orangutans circulate antibodies reactive with African green monkey foamy spumaviruses [854]. While APOBECs (or other immune faculties) protect chimpanzees against their own retroviruses (SIVcpz; foamy spumaviruses; leukemic retroviruses; and HHV8-related herpesviruses), the Australopithecines and their descendants (*Homo*) appear to have gotten rid of their inherent active retroviral flora, and might have lost their innate anti-viral immune faculties as well. In contrast to numerous retrovirally induced leukemias and sarcomas in aves and murines, no retroviral etiology of human sarcomas could so far be documented. Antibodies to human foamy spumaviruses are undetectable in Blood Bank blood samples [855]. Exposure from the outside to human lymphotrophic leukemia virus I/II, HIV-1/2, and foamy spumaviruses find a naturally defenseless host. On human chromosome 22 seven cytidine deaminase genes exist that could protect human lymphoid cells from *in vitro* infection by the Moloney murine leukemia virus. HIV-1 Vif could switch off and send to proteasomal degradation (ubiquitylation) the APOBEC3F/3G proteins. However, APOBEC3DE (A3DE) managed to undergo encapsidation and was capable of deaminating cytosins to uracils on viral plus strand DNA. A3DE inhibited the replication of HIV-1, and some simian SIV, but not the murine

J.G. Sinkovics, *RNA/DNA and Cancer*, DOI 10.1007/978-3-319-22279-0_15

leukemia virus. Further, APOBEC3DE is expressed more extensively, than the other APOBECs (Dang et al) [856]. Thus, a vertically inherited gene from the Australopithecine era might have been just very recently reactivated. "Herd immunity" to the herpesvirus HHV-8 persists, as latent viral carriership was not eliminated (as proven from the Mediterranean races on to the Amazonian and non-Amazonian Indian tribes); reviewed in Sinkovics' Monograph, and by Nascimento *et al* [147, 857].

The lamprey lymphoid cells encode diverse modular leucin-rich repeats that undergo rearrangements in response to pathogen exposure [858]. In the zebra finch, and in the lizard, the APOBEC enzymes were engaged in RNA editing. However, the lizard APOBEC deaminates bacterial DNA [859]. The human genome harbors eleven A3 ORFs in seven genes [841]. The retroviral effect of A3G is achieved by targeted deamination of cytosine in the viral cDNA. The mutated viral proteins may become nonfunctional. In the average, A3G-WT confers a 6-fold decrease in viral cDNA transcript numbers. A3G loop variants or catalytic mutants fail to reduce viral transcript numbers over 2-fold. Loop variants targeted different sequences within the viral genome. Thus, deamination-proficient A3Gs achieved significant reduction of retroviral replication. The A3G loop graft variants allowed (or promoted) the escape of mutant HIV particles. The viral biased A-rich nucleotide composition resulted from an A3G-mediated mutation. Codon changes result in aa exchanges in the viral antigens, as the viral envelope and the host's IgG repertoire co-evolve [860].

The HIV-1 viral infectivity factor (Vif) is the antagonist of the APOBEC (A3G/A3F) system. A functional Vif is essential ("absolutely required") for the productive viral infection. The 192 aa 23kDa Vif protein instructs the cellular E3 ubiquitin ligase complex (elongins, cullin and Rbx2 RING box protein) to liquidate A3G/3F in the proteasome; even A3G mutants succumb to Vif-mediated ubiquitination. The VIF molecules form dimers, trimers and tetramers. Vif's N terminus is RNA-binding aimed at the 5′ end of the HIV-1 genomic RNA (for protection against A3G/F viral RNA "editing"). Vif also binds to the cellular elongation factors and cullin (to activate them). The Vif aa proline-rich region PPLP activates elongation factor E3 ubiquitin ligase. The Vif aa ^{144}SLQYLA^{149} domain (beginning with serine and ending with alanine) fits the A3G/F configuration for interaction [861]. The host's innate and adaptive immunity so far failed to take control of the HIV-1-induced AIDS. Small molecular inhibitors of the retroviral enzymes (including HIV-1 proteinases and the Vif protein) converted the inexorably fatal disease AIDS into a practically symptom-free chronic carriership of non-replicating HIV-1 inserted DNA proviruses [862–864].

The xenotropic murine leukemia virus-related virus (gammaretrovirus, XMRV) was mistakenly identified as an infectious agent in the human prostate cancer, in myalgias, and chronic fatigue syndrome; however, it proved to be a murine passenger retrovirus, and as such, a laboratory contaminant of cultured human tissues [865a, b, 866]. It could not readily replicate in human tissues, because human A3G/A3F inhibit it. It is also susceptible to RT inhibitors zidovudine and tenofovir and integrase inhibitor raltegravir [865b]. The XMRV replication is well inhibited by the human APOBECs (in pigtailed macaques) [867]. APOBEC-1 has been referred to as the apo-B mRNA-editing enzyme catalytic polypeptide 1 [868]. Transgenic rabbit and mouse liver cells expressed APOBEC-1. The apo-B mRNA underwent extensive editing and the low density lipoprotein blood levels dropped at a low level. Livers increased in size due to hypercellularity and increased lipid storage. Histopathology of the liver showed dysplasia followed by hepatocellular carcinoma. The conclusion was proposed that the mRNA-editing APOBEC-1 enzyme was oncogenic in this artificially created system [868]. A wide range of naturally occurring human malignant tumors (different adenocarcinomas, hepatocellular-, neuroendocrine-, and medullary thyroid carcinomas, soft tissue sarcomas, and neuroblastoma) were examined for mRNA-editing by activated APOBEC-1. The results were negative. "Currently, there is no evidence that aberrant editing" (mRNA-editing) "mediated by APOBEC-1 contributes to the tumorigenesis of natural human carcinomas" [869]. However, more recent data indicate otherwise (Figure 47).

Chimpanzees gained good APOBEC-mediated protection against their AIDS- and Kaposi's sarcoma-inducer retrolenti-SIV/HIV and herpes-8 viruses. But the descendants of these retrolenti- and herpesviruses act in collusion in Homo. The mammalian leukemogenic-sarcomagenic retroviruses (abundant in aves and mice) could not be isolated from human sarcomas: the Australopithecines reduced them to pseudogenes? Endogenous human retroviruses pop out of lymphocytes in patients with lymphoma or SLE; or appear in teratoma, melanoma and adenocarcinoma cell [147] (*vide infra*). The human APOBECs failed to extend protection against HIV-1/2 in Homo. A retrolenti- and herpesviral collusion is evident in the human host. While the dominant leukemogenic-sarcomagenic avian and mammalian viruses are retroviruses, the Marek's herpesvirus in turkeys (and other birds), and *Herpesvirus saimiri* and related herpesviruses in New World (marmoset and related) monkeys are lymphomagenic. Some herpesviruses incorporated retroviral genomic segments (Marek's virus; Epstein-Barr virus, HHV-4). Kaposi's sarcoma herpesvirus (HHV-8) has a close relationship with HIV-1, and with an endogenous human retrovirus, which is activated and appears budding from Kaposi sarcoma cells [147] (*vide infra*).

Comment. Are there inherent errors in the installment of AID, APOBEC, RAG1/2 and Rev1 in the superb adaptive immune system and DNA repair machinery, that could not be corrected as yet by the still evolving vertebrate mammalian (human) host? Are reactivated ancient retroviral LTR inserts pathogenic (inducers of autoimmune and malignant diseases)? Detailed assessment follows later.

Insertional mutagenesis: If resident proto-oncogenes are the recipients. The RNA retro(lenti)viral genomes transcribed by their reverse transcriptase into DNA proviruses are readily inserted in the genome of their host cells [870]. Consequentially, insertional mutagenesis may result in malignant transformation of the host cell, if the new genomic insertion deregulates a silent proto-oncogene. The process of enhancer-mediated activation consists of the interaction between viral enhancer sequences and cellular promoters. The activated promoters will dictate the release of the mRNAs from the proto-oncogene for the synthesis of oncoproteins. Further, chimeric fusion transcripts of vector sequences and cellular mRNA can be formed. For this event to occur, the retroviral LTR has to integrate downstream of the cellular promoter in the transcriptional orientation same as that of the promoter. Instead of the silent cellular promoter, the 5′ or 3′ sequences of the viral LTR promoter will take command. In the case of self-inactivating (SIN) viral LTRs, the viral polyA signals induce the mRNAs' aberrant polyadenylation. Thus aberrant protein production may reach clonal dominance [870]. The retrovirally inserted IL-2 gene for the gene therapy of severe combined immunodeficiency (SCID) activated a T cell proto-oncogene, and thus induced clonal T cell proliferation terminating in acute T cell leukemia; discussed elsewhere [27]. Lentiviral gene therapy of β-thalassemia resulted in the activation of host gene HMGA2 (high mobility group A protein) by the fusion of the splice donor sequence with the third exon of this host gene. In the patient, a dominant myeloid cell clone developed; the genome of this clone harbored the copy of the integrated vector sequence. The use of lentiviral vectors with SIN, LTR avoids the aberrant splicing process [870].

It has recently been recognized that the high mobility HMGA2 (high mobility group A2) gene fuses with the EFCAB6 gene (EF-hand calcium binding domain; elongation factor thermo-instable helix-loop-helix protein family) creates the fusion oncogene/oncoprotein t(12;22)(q14.3;q13.2) [871]. The HMGA2 pathway is induced by Wnt/β-catenin; it maintains Ras protein in ETM transition; promotes re-caping of telomeres after each cell divisions; it is highly expressed in exocrine pancreatic carcinoma; active in angio- and liposarcoma; but its mRNA is suppressed by miR-let7. The formation of the fused oncoprotein results in a gain-of-function point mutation of the proto-oncogene: the JAK2V617F-positive (valium *versus* phenylalanine). BCR-ABL1-negative myeloproliferative neoplasms arise. The fusion oncoprotein abolished the let-7 (lethal) tumor suppressor miR. The let-7 miR normally, by aligning with the UTR region of the mRNAs for the Ras H, K, N, and/or c-Myc, and/or HMGA2 oncoproteins, inhibits the ribosomal translation of these mRNAs into the oncoproteins. Further, the complete transcripts of the EFCAB gene are the E and F alpha helices of parvalbumin/oncomodulin. The arrangement of the two perpendicular 10–12 residue α-helices reminded H. Kretsinger of his right hand: helix E index finger, helix F thumb, and the bent middle finger is the 12 aa Ca^{2+} binding loop. Ca^{2+} binding EF-hand proteins are in the background. Finally, the JAK2 is constitutively point-mutated. The myeloproliferative neoplasm thus generated is highly malignant [871].

The parvalbumin-like EF-proteins (EF-hands, E and F alpha-helices, helix-loop-helix proteins) are ancient installments. The original Ca^{2+}-binding EF-proteins appeared in the cilia and flagella of chlamydomonas, tetrahymena and trypanosoma (as calflaglins, calcimedines in the trypanosoma, together with the different annexins, calmodulin and protein kinase C isoenzymes). In the platyhelminth liver fluke (*Fasciola gigantica*), the EF-proteins of the tegument induce the parasitized host's IgE immunoglobulin response. In vertebrate mammals, the parvalbumins/oncomodulins (high mobility HMGA2 proteins) appear in the fetus and subside in adult life. In a thyroid carcinoma cell line, the HMGA2 gene was activated by a retroviral LTR. In the human genome, the HMGA2 gene resides at 12q13-15. In human malignant tumors (adenocarcinoma of the exocrine pancreas) increased amounts of the HMGA1 (encoded from a different gene) and HMGA2 protein were found [872a, b, 873]. The HMGA2 proteins, physiological at lower levels of evolution, are expropriated by murine and human cells to induce, or sustain, the so-called malignant transformation (Tables XI/I and XI/II).

Due to the abrogation of the disrupted EFCAB6 pathway, there is no, or there is only an altered DJ-1 expression pathway [871]. It is very difficult to obtain a concise view of the normal and pathological DJ-1 pathways. The DJ-1 homodimeric protein is relatively small: 189 aa monomers of 20 kD; homologous to some bacterial enzymes (ThiJ of thiamine biosynthesis, and Pfpl protease). Since the late 1990s, the DJ protein is referred to as the product of the PARK7 (parkinsonism) gene mapped at human chromosome 1p36.2-p36.3 (PARK6 is PINK1, the PTEN-induced putative kinase). Loss-of-function mutation of gene PARK7 leads to the early onset autosomal-recessive Parkinson's disease. In this condition, striatal dopamine is depleted and protein clumps of Lewy bodies appear in the cytoplasm of nerve cells. The point mutated DJ-1 protein harboring the L166P (leucine; proline) mistake becomes insoluble and forms clumps in the cytoplasm of nerve cells. The clumped protein resists removal by ubiquitination. However, the PARK7/DJ-1 gene (together with H-ras) transforms the NIH3T3 cells, thus acting as "a mitogen-dependent oncogene". The transformed NIH3T3 cells carry the L166P (leucine; proline) translocation. This is a gain-of-function mutation. However, the mutated protein induces the JNK/p38MAPK pathway (c-Jun terminal kinase; mitogen-activated protein kinase), which is in this case pro-apoptotic. That would qualify the L166P aa exchange as a loss-of-function mutation; even metformin can induce this pathway for apoptosis induction. DJ-1 protein was recognized as an inhibitor of tumor suppressor gene PTEN (phosphatase and tensin homologue); when PTEN is inactivated, the "cell survival pathway" PI3K/Akt kicks in as an oncogene (when constitutively activated). For the normal function of the "dopaminergic molecule", the DJ-1 protein-mediated (experimenting in C57BL6/J mice) removal of cytosolic reactive oxygen species (ROS) and maintenance of mitochondrial membrane integrity is claimed. Further, DJ-1 protein is expressed in the head of spermia. Further, DC-1 protein is an up-regulator of the androgen receptor (AR). Further, DJ-1 protein is expressed in normally functioning somatic and germ cells. Finally, DJ-1 protein serves as a scavenger for reactive oxidative species (ROS) by oxidation of cystein residues. DJ-1 maintains the integrity of the mitochondrial membranes. In its correlation with the anti-apoptotic survivin and cancer, DJ-1 co-expression was described as "associated with tumor stages and tumor cell differentiation" (up or down?), but not with lymph node metastasis induction, and "positively correlated to survivin gene expression" and "the expression of two genes are closely related to clinicopathological parameters" (whether up or down). In order to understand in terms of translational medicine to practice, the meaning of the JNK/p38 MAPK pathway (oncogenesis *versus* apoptosis), the M.D. Anderson Hospital article is recommended (especially Figure 8 in the article, explaining the involvement of the FasL → FasR connection) [874]. The

Table XI/I High mobility group (HMG) proteins of unicellular eukaryotic ancestors become oncoproteins in human cancer cells. I Introduction

Eukaryotic DNA-binding proteins of small MW move fast in Triton-acetic acid-urea, or sodium dodecyl sulfate polyacrylamide gel electrophoresis and to the nucleosomal chromosomal core histones. These are the architectural DNA-bender, nucleosome-binder enhancers of gene transcription. B family HMG proteins accelerate the V(D)J immunoglobulin recombination. Subfamily HMG A/B/N proteins exist as non-histone intranuclear molecules with acidic carboxy-terminal tail highly conserved in evolution. HMGB homologs are inflammatory inducers of TNFα and IL-6.[1] Ancient: present in extant unicellular/multicellular eukaryotes inherited from first eukaryotic ancestors. HMG proteins attach to receptor for advanced glycation end products (RAGE). RAGE and its ligand HMGPB1 are involved in autoimmune and oncogenic processes, in particular in myasthenia gravis and thymic epithelial tumors.[2]
HMG proteins of unicellular extant eukaryota. *Saccharomyces cerevisiae* in its chromatin remodeling complex directs HM group proteins to sites of dsDNA breaks for repair, and to broken chromosome end for protection from exonucleolytic degradation. HMG proteins serve the recovery of stalled replication forks and in nucleotide excision repair of UV light- or hydroxyurea-damaged DNA. Another HMG protein facilitates rDNA transcription analogously to mammalian HMG proteins.[3] *Tetrahymena thermophila* operates two distinct, but coordinated HMG proteins. One of the HMG proteins is active at early stage of conjugation and meiosis, when H1 is hyperphosphorylated; after conjugation these proteins are prominent in the differentiation process of the new macronucleus. Another HMG protein is expressed during starvation; its germ-line gene resides in the micronucleus. The cell cycle is under the control of the HMGB/C proteins. HMGB protein functions in the micronucleus to wrap DNA (shown in spectacular immunofluorescent figures). The DNA binding HMG1 box may be ancestral to that of vertebrate HMG14/17 proteins. The serine and threonine-containing tryptic peptides of a purified 11,626 MW HMG nonhistone chromosomal tetrahymena protein is similar to the vertebrate HMG1 protein.[4] *Oxytricha nova* yielded its first HMG protein; the encoding gene was interrupted by an intron[5] (thus its DNA-derived mRNA may be inhibited by an intron-derived microRNA). *Trypanosoma cruzi/brucei* operate kinetoplastid intranuclear HMG box domains not yet endowed with the C-terminal acidic tail of the mammalian HM group B proteins, yet able to function as an architectural protein in binding distorted cruciform DNA. The four trypanosoma HMG proteins are immunologically related to mammalian HMG proteins. The trypanosomal D protein (TDP) functions as a HMG protein facilitating encoding of variant surface glycoproteins through transcription by RNA polymerase I (Pol I). Another HMG protein practices rDNA transcription similarly to the mammalian upstream binding factor (UBF). The Istituto Biofisica Carlos Chagas Filho, Rio de Janeiro, Brazil, provides the genome map of 5013 *T. cruzi* genes for help in the search for effective medication for Chagas disease.[6] Entamoeba, plasmodia, schistosoma and toxoplasma cells all operate the HMGB1 subfamily proteins.[7]
Metazoan HMG proteins. Metazoan cells co-express the replication-dependent tightly controlled histone mRNAs and the HMG protein mRNAs; both are ready to pass through the spliceosomes/ribosomes, and/or are subjected to miR-degradation. The HBP histone hairpin-binding proteins are structures located in the UTR of the cell cycle histone-dependent mRNAs.[8] The cnidaria (*Nematostella vectensis*) uses HMG proteins for gastrulation and morphogenesis.[9] The amphioxus Branchiostoma, *B. belcheri; floridae; tsingtauense; mediterranean* release archetype cordate HMG B family (B1,2) proteins in their guts. In the gastrula, the protein expression is high; in the adult, it is confined to the intestinal tract. The AmphiHMGB cDNA and protein shows identities of 40−50 % with HMGB proteins of sea urchin, lamprey, trout, frog, chicken, mouse, and Homo. HMG box A, box B and tail aa sequences are highly conserved within the vertebrate clade. HMG protein is highly expressed in the amphioxus embryo in the cerebral vesicle, neural tube and pharynx. AmphiHMGB induces differentiation of neural ectoderm, mesoderm and endoderm.[10] In the sea urchin (*Strongylocentrotus purpuratus*) blastomeres, the intranuclear β-catenin interacts with the

(continued)

Table XI/I (continued)

Tcf/Lef genes to regulate mesenchyme, endoderm and ectoderm distribution. (This interaction in mammalian/human cells is oncogenic: the Wnt pathway). PPARγ ligands (Table II) repress β-catenin signaling. DNA-binding HMG proteins are the substrates for the M-phase-specific protein kinase; the phosphorylated HMG proteins condense chromatin at the G2 → M transition of cell cycle.[11] Cells of the parasitic filaria, *Wuchereria bancrofti* and *Brugia malayi,* harbor and tolerate intracellular symbiotic Wolbachia bacteria, while in the human host of the filaria, the Wolbachia proteins induce the release from host mϕ/monocytes TNFα, IL-6 and GM-CSF. The filarial HMGB1 subfamily protein has only the A box (the B box is missing as in yeast, plasmodia and drosophila), but it binds supercoiled DNA. The filarial A box protein shows 75 % sequence homology with the human A box protein. It is secreted in the human host and is responded to with an inflammatory reaction.[12]
HMG protein-induce erythroleukemia cell differentiation. Friend virus-induced mouse erythroleukemia cells differentiate under the effects of hexamethylene bisacetamide (HMBA) and HMG1 protein. Ca^{++} ionophore induced excess HMG1 protein and promoted leukemia cell differentiation. Further inducers of leukemia cell differentiation are HDAC inhibitors hyperacetylating N-terminal tails of core histones H3/4 (trichostatin), DMSO, phorbol myristate acetate (PMA), the activator of PKC.[13]
1 McCauley MJ et al Nucleic Acids Res 2013;41:167–81; Stros M et al Cell Mol Life Sci 2007;64:2590–606. Van Gent et al EMBO J 1997;16:2665-2670. 2 Moser B et al PLoS One 2014;9(4):e94118. 3 Ray S & Grove A. Biochemistry 2012;51:1825–35; Gadal O et al EMBO J 2002;21:5498–507; Kamau E et al J Biol Chem 2004;279:55234–40;Takaishi H et al J Gastroenterol 2012;47:969–77. 4 Hayashi T et al J Biochem 1989;105:577–81; Schulman IG et al Mol Cell Biol 1991;11:166–74; J Cell Biol 1987;104:1485–94; Wang T & Allis CD Mol Cell Biol 1993;13:163–73; Xu J et al Gene 2013;526:87–95 5 Callejas S & Guttiérrez JC Mol Biol Rep 2003;30:215–222. 6 Cribb P et al Int J Parasitol 2011;41:1149–56; Morales M et al J Cell Biochem 1992;50;279–64;Narayanan MS & Rudenko G Nucleic Acids Res 2013;41:2981–92; Porcel BM et al Genome Res 2000;10:11037; Urményi PP J Eukaryot Microbiol 199;46:542–4. 7 Abhyankar MM et al Eukaryot Cell 2008;7:1565–72; Gnanasekar et al Mol Biochem Parasitol 2006;145:137–46; Kumar K et al Parasitol Int 2008;57:150–7; Zhao YO et al PLoS Pathog 2009;5:e1000288. 8 Jaeger S et al Biochimie 2005;87:827–34; Landsman D et al J Biol Chem 1986;261:7479–84; Zhao X et al J Cell Sci 2004;117:6043–51. 9 Magie CR et al Dev Genes Evol 2005;215:618–30. 10 Huang X et al Int J Dev Biol 2005;49:49-52; Liu Z et al Comp Biochem Physiol B Biochem Mol Biol 2004;137:131–8. 11 Gebeshuber CA et al Cells Tissues Organs 2007;185:51–60; Huang L et al Dev Genes Evol 2000;210:73–81; Iwai S et al Int J Oncol 2010;37:1095–103; Lu D & Carson DA Eur J Pharmacol 2010;636:198–202; Mao CD & Byers SW Crit Rev Eukaryot Gene Expr 2011;21:207–36; Meijer L et al Eur J Biochem 1991;196:557–67; Yoshioka S et al Mol Cancer Res 2012;10:469–8. 12 Thirugnanam S et al Parasitol Res 2012;111:619–27. 13 Ishiguro K & Sartorelli AC Leuk Res 2008;32:89–96; Sparatore B et al Biochem J 1996;320:253–6.

phylogenetic trajectory of the DJ-1 superfamily proteins is extending from eubacteria up to transformed (cancerous) eukaryotic cells in multicellular hosts. DJ-1 and HSP90α proteins act as antagonists of PTEN, and thus liberators of the PI3K-p11/pAkt cell survival pathway. Example: urothelial carcinoma cells' invasiveness and metastastatic potential increases upon DJ-1/HSP90α overexpression in PTEN-depleted cancer cells [875]. DJ-1 emerges as a negative regulator of PTEN in many (both adeno- and squamous cell) cancers (Figure 50; Tables VIII and XII).

Table XI/II HMG proteins induce and/or promote human cancer growth

Growth factor and cell survival pathways of unicellular eukaryotes become human oncoproteins. *Chlamydomonas reinhardtii* utilizes the SET domain histone methyltransferases in its enhancer of zeste homolog (EZH): the Su(var)-related protein; the trithorax-like H3K4 histone methyltransferase. The *C. reinhardtii* transcriptional silencer *mut*11 gene (mutant defective) is homologous with the human *wdr*5 gene (tryptophan, aspartic acid repeats).[1] In sea urchin blastomeres, Wnt-ligands 1/8 induce Wnt-signaling; intranuclear β-catenin activates genes *tcf/lef*; TGFβ synergizes with transcription, and Wnt receptor antagonist dickkopf1 provides the negative feedbak. Gene *nodal* directs dorsal-ventral development of the sea urchin embryo. The *nodal* regulatory element is a highly conserved 350 bp region within the first intron of the gene; the intron contains SMAD-binding sites.[2] Zebrafish and mouse homologs of the sea urchin *nodal* gene (or its promoter) have Tcf/Lef protein binding sites. In the human genome, Activin/Nodal, bone morphogenetic protein, fibroblast growth factor and Wnt/β-catenin-activated *tcf/lef* genes work in unison; *tcf/lef* may emit mRNAs to encode oncoproteins.[3] In the era of innate immunity, extracellular high motility group (HMG) proteins activated TLR4 to mediate NFκB release as a pro-inflammatory defense reaction. This reaction had been conserved up to human epidermal tumors (basal cell/squamous cell carcinomas).[4]
Wnt10b/β-catenin upregulates HMG proteins and induces malignant transformation in the human genome. Embryonic switched-on/switched-off pathways constitutively reactivated in adult life are recognized as a form of oncogenesis. The embryonic morphogene of the TGFβ family, *nodal,* is active in fetal life and is silenced in the adult host; activin-like kinases (ALK) are the receptors for ligands TGFβ and nodal. miR-376c represses TGFβ and nodal. Nodal is re-activated in malignantly transformed cells including those of breast carcinoma (the triple-negative MDA-MB-231 cell line). The *nodal* mRNA can be demolished by shRNA with attenuation of the malignancy. Histone CH_3-transferase enhancer of zeste homologue 2 (EZH2, drosophila), active in fetal life, but mediates polycomb gene silencing in adult life by trimethylating histone 3 at lysine 27 (H3K27me3). This is the polycomb repressive complex (PRC) of EZH2 in action, due to its possession of the evolutionarily conserved SET domain (suppressor of variegation, Su(var); enhancer of zeste; trithorax histone CH_3-transferase, drosophila). The human EZH2 gene interacts with oncoprotein *vav*/VAV (the sixth letter of the Hebrew alphabet for 'link' or 'connection').[5] β-Catenin signaling upregulates HMG box A2 protein generation in high histological grade, chemotherapy-resistant triple negative breast cancers. In these tumor cells cullin (CUL4A) oncoproteins degrade tumor suppressor $p27^{kip}$ (kinase inhibitor p27). *Hmga2*/HMGA2 is a CCN-A2/B2 regulator (Cyr61, Ctgf, Nov family cell differentiation regulators: cystein rich protein; connective tissue growth factor; nephroblastoma overexpressed). CCN-A2/B2 is the downstream target of HMGA2. The CCN genes are cancer cell proliferation-related cyclins. The MDA-MB-231 TNBC cells assume stem cell faculties under the effect of HMGA1. HMGA1-treated somatic cells become pluripotential stem cells expressing the four Yamanaka factors (Oct4, Sox2, Klf4, c-MYc). Wnt/β-catenin inhibitor ICG-001 suppresses translation of HMGA2 mRNA; atorva-, fluva-, lova-, rosuva-, and simvastatins are HMG CoA reductase inhibitors.[6] Use of lipophilic statins did not change (neither increased nor decreased) the risk of any types of breast cancers (including HER2-negative TNBC). Statins downregulated NFκB, and slowed growth rates of breast cancer cells including the TN and MMTV antigen-positive MDA-MB-231 tumor cell line.

(continued)

Table XI/II (continued)

Simvastatin de-repressed PTEN in TN breast cancer cells MDA-MB-231; NFκB, AKT and BclXL were suppressed.[7]
Ancient HMG proteins: drivers of human cancers. The high grade "castration-resistant" $PTEN^{-/-}$ prostate cancer cells phosphorylate the EZH2 CH_3-transferase protein at serine 21 by the active PI3K/Akt pathway in order to switch it from a transcriptional repressor (H3K27me3) to a gene activator. Among others, genes EED and Suz12 (embryonic ectoderm development protein; suppressor of zest) are activated. Oncoproteins EED and SUZ12 now suppress gene DAB2IP, which was repressed by EZH2 (DAB2, disabled; DAB2-interacting protein, GTPase-activating). DAB2IP directly acting on the p85 subunit of PI3K, downregulates it; thus, EED and SUZ12 by repressing DAB2IP, liberate PI3K.[8] The HMGA2 protein is highly active in the fetus as a transcriptional regulator. It expresses three hook domains that attach it to the minor groove of targeted DNA molecules. In the mesenchyme, it closely interacts with growth factors EGF and PLDGF. However, its overexpression in prostate cancer cells is constitutive and independent from mesenchymal growth factors. Serves as ligand to RAGE receptors for 'advanced glycation end products'. These receptors downregulate PPARγ (Table II). Among other targets, invariant NKT cells express these receptors. In prostate cancer (PC) cells, HMGB2 proteins are implicated in ETS (E26) oncogene activation. PC cells with silenced HMGB1 lose their proliferative, invasive and metastatic potentials. Anti-HMGB1 monoclonal antibodies could neutralize it; and HMGB1 peptides could induce antibodies in vaccinated patients. The A-box peptide of the HMG protein molecule can occupy the RAGE receptor and prevent its activation by the full molecule. 18α-glycyrrhetinic acid of licorice derivation root is inhibitory to HMG proteins (but no clinical validity of any of these interventions has been documented).[9] The targets of miR-let-7 include EZH2 and HMGA1/2 proteins, but miR-let7 is downregulated in androgen deprivation-resistant patients with active prostate cancer (Table XXIII). The mRNA of oncoprotein SUZ 12 is inhibited by miR-19a; in the same report, miR-19a, miR-27a and miR-133b promote AR-driven androgen-dependent PCs.[10] Atorvastatin induced autophagy, downregulated Bcl2, and upregulated miR-182 in prostate cancer cells.[11] Would autophagic tumor cells consume their HMG proteins and recover in remission? Lung cancer metastases are inhibited indirectly by thyroid transcription factor 1 (TT-1), which raises miR-33a levels, repressors of HMGA3 mRNA. High HMGB1/3 levels come with adverse prognosis in gastric carcinoma. In hepatocellular carcinoma, HMGA2 dictates ETM transition (Table XXVIII/I).[12]
Down Syndrome. Trisomy at 21q22 contributes to leukemogenesis, most commonly B cell acute lymphoblastic leukemia. In the triplicated chromosome, trimethylation of histone 3 (H3lys(K)27 me3) is suppressed; the de-methylated gene of HMGN1 is liberated in progenitor B cells to overproduce the protein. HMGN1 turns off gene/protein PRC2 (polycomb repressive complex). In consequence, some 100 genes (not yet sorted out for the most responsible ones) customarily silenced by PRC2 become activated and encode cell cycle activation.[13] In *Arabidopsis thaliana*, a H3 histone lysine 27 methyltransferase suppresses the outgrowth of lateral roots. The lateral roots arise from asymmetrically dividing stem cells. The downegulating suppressor protein is a PRC2. In the same plant, retinoblastoma and polycomb gene-encoded proteins regulate cell differentiation.[14,15]
Viral therapy of cancer involves HMG proteins? The genetically engineered oncolytic measles virus replicates in human melanoma cells, while the host mobilizes first innate, then adaptive immune reactions against the virus and virally infected tumor cells. The appearance of HMGB1 protein and its release from virally infected (dying) tumor cells is observed. HMGB1 is released in inflammasomes. Lipopolysaccharides induce HMGB1 protein release, that is inhibited by rosiglitazone activated PPARγ. The advanced glycation end product receptors RAGE of

(continued)

Table XI/II (continued)

HMGB1 proteins induce VEGF-A production. Is HMGB1 neovascularizes tumors? It is not known if HMGB1 protein was antiviral, or proviral, in the measles virus-infected melanoma cells.[16]
1 *Strenkert D et al Nucleic Acids Res 2013;41:5273–89; Zhang C et al Plant J 2002;31:25–36.* ***2*** *Range R et al Development 2007;134:3649–64; PLoS Biol 2013;11(1):e1001467; Saudemont A et al PLoS Genet 2010;6(12):e1001259.* ***3*** *Lupo G et al 2013;3(4):120167; Mao CD & Byers SW Crit Rev Eukaryot Gene Expr 2011;21:207–36;* ***4*** *Weng H et al BMC Cancer 2013;13:311.* ***5*** *Polyak K & Weinberg RA Nat Rev 2009;9:265–273; Quail DF et al PLoS One 2012;7(11):e48237; Grzenda A et al Epigenetics Chromatin 2012;6:3. doi:10.1188/1756-8935-6-3; Fu G et al Hypertension 2013;61:864–72.* ***6*** *Goppelt-Struebe M et al Mol Pathol 2001;54:176–9; Győrffy B et al Breast Cancer Res Treat 2010;123:725–731; Wend P et al EMBO Mol Med 2013;5:264–79.* ***7*** *Campbell MJ et al Cancer Res 2006;66:8707–14; Ghosh-Choudhury N et al Cell Signal 2010;22:749–58; Litvinov SV & Golovkina TV Acta Virol 1989;33:137–42; Shah SN et al PLoS One 2013;8(5):e63419; Walker-Nasir E et al J Natl Cancer Inst 1982;69:371–80; Woditschka S et al Cancer Epidemiol Biomarkers Prev 2010;19:2479–87.* ***8*** *Binda O Epigenetics 2013;8:457–63; Qiao S et al PLoS One 2013;8(1):e53635; Xu K et al Science 2012;338:1465–9.* ***9*** *Chen YJ et al PLoS One 2013;8(6):e66611; Gnanasekar M et al Prostate Cancer 2013:157 193; Szmolarczyk R et al Arch Immunol Ther Exp 2012;60:391–9.* ***10*** *Peng X et al PLoS One 2013;8(8):e70442; Mo W et al PLoS One 2013;8(2):e56592; Schubert M et al PLoS One 2013;8(6):e65064.* ***11*** *He Z et al J Radiat Res 2012;53:225–33; Toepfer N et al Cancer Biol Ther 2011;12:691–9.* ***12*** *Rice SJ et al J Biol Chem 2013;288:16348–60; He W et al World J Surg Oncol 2013;11:161; Tang HR et al World Gastroenterol 2012;18:7319–26; Luo et al Oncol Lett 2013;5:1353–6.* ***13*** *Lane AA et al Nat Genet 2014 in print. doi:10.1038/ng.29.* ***14*** *Gu X et al Mol Plant 2014;7:977–88.* ***15*** *Kuwabara A Gruissem W J Exp Bot 2014;65:2667–76.* ***16*** *Huang B Gene Ther 2011;18:164–172; Hwang JS et al Mediators Inflamm 2012:352807; Lee JJ Mol Vis 2012;18:838–50; Lu B et al Nature 2012;488:670–4; Donnelly O et al Gene Ther 2013;20:7–15.*

The drosophila embryo expresses the ortholog of the mammalian EF-hand domain-containing proteins, referred to as swiprosins (Dm-Swip-1). In myotubule formations in the drosophila embryo, these membrane-expressed proteins mediate the physiological myoblast fusions [876]. In the mouse, swiprosin-1 travels in lipid rafts in the membranes of B lymphocytes. Human mast cells and cytotoxic lymphocytes express swiprosin (EF-hand coiled coil) proteins. Since swiprosin-1 induces apoptotic death in B cell receptor-stimulated immature B lymphocytes, it probably is a suppressor of lymphomagenesis. However, Swip-1-induced apoptosis is counteracted by LPS, Bcl_{XL}, and NFκB [877].

If the inserted genes render great new services. Hundreds of millions of years bombardments of the genomes with retrotransposons evoked the mechanisms of "domestication, exaptation and co-option". Frederic Bushman's [1] Table 10.3 printed in 2002 lists those genes, which were incorporated into the human genome and have become useful functional entities. The telomeres, syncytins, mammalian centromere associated proteins, and Rag1/2 are on the top of the list (Lander *et al* cited in Bushman) [1]. Genomic imprinting (*vide supra*) revealed the inequality of expression attached to the maternally and paternally transferred alleles. The paternally expressed PEG 10/11 and retrotransposon-like 1 (RTL1) Gag and Pol

Table XII DJ-1/ThiJ/Pfpl protein superfamily

Proteasomes. From ancestral proteasome-specific inhibitor clasto-lactacystin-β-lactone (clβL) of *Halofex volcanii* evolved to eukaryotic proteasomes for translation, transcription, metabolism, DNA repair, and cell cycle control. Multicatalytic protease proteins targeted for proteasome-mediated degradation. clβL-treated proteome targets are the DJ-1/ThiJ/Pfpl superfamily proteins (drosophila protein homologue of bacterial enzymes; thiamine biosynthetic protein; pore-forming protein-like amphiphilic bacterial protease). Crystal structures of human dimeric DJ-1 and *Escherichia coli* heat shock Hsp31 (*yedu* gene product) proteins share evolutionarily conserved domain: type I glutamine amidotransferase (GAT). On substrate luciferase, DJ-1 proved its chaperone activity. Both DJ-1 and Hsp31 possess hydrophobic patches for chaperone activity. In addition to be chaperones, DJ-1/ThiJ/Pfpl/Hsp31 are proteolytic suppressors of protein aggregation. Loss-of-function L166P (leucine; proline) mutated DJ-1 results in loss of its neuroprotective faculty and thus contributes to neuron-lethal protein-aggregation of Parkinson's disease.[1]

DJ-1 oncogene transforms NIH3T3 cells by itself and more efficiently with *ras/myc* combination. DJ-1 oncoprotein translocates from cytoplasm to nucleus. The human 189 aa 20 kD DJ-1 protein is encoded from 1p36.2-36.3. DJ-1 binds AR-binding region of PIAS (protein inhibitor of activated STAT); both proteins localize in the nuclei. PIAS inhibits AR; inhibitory PIAS annulled by DJ-1 binding. L130R (lysine; arginine) mutated DJ-1 fails to remove PIAS. DJ-1 by downregulating PTEN (phosphatase tensin analog deleted on chromosome 10 (Table VIII) in glottic squamous cell carcinoma (and in several other cancers) liberates onco-proteins PI3K/Akt and contributes to the development of lymph node metastases. An ancient protein of physiological cell metabolism has become an oncoprotein in multicellular hosts (vertebrate mammalians).[2]

1 Bandyopadhyay S & Cookson MR BMC Evol Biol 2004;4:6. Kirkland PA et al Microbiology 2007;153:2271–80. Lee SJ et al J Biol Chem 2003;278:44552–9. Trempe JF & Fon EA Front Neurol 2013;4:38. ***2*** *Nagakubo D et al Biochem Biophys Res Commun 1997;231:509–23. Taira T et al Gene 2001;263:285–292. Takahashi K et al J Biol Chem 2001;276:37556–63. Zhu XL et al J Exp Clin Cancer Res 2012;31:94.*

protein-encoding genes are so expressed in the eutherian and marsupial species (PEG10) and in the eutherian species (PEG11/RTL1). These genes were essential in the evolution of viviparous placental mammals [878]. The inserted retrotransposons commonly fall victim of the enzymes DNA methyltransferase: the hypermethylated genes are silenced. Darwinian forces of the environment are able to demethylate genes and/or alter on and off their histone acetylation status. Whether such a change may occur in a single individual of a species, who is able to vertically propagate the change, or simultaneously, or consequentially, in many members of the same species, is a matter of hypothetical scenarios.

The functional human endogenous retroviruses (HERV) exist in the human genome as transposable elements (TE). Classes I, II, and III of HERVs depend on their relatedness to gamma-, beta, or spumaretroviruses. The HERV-K could mature into enveloped particles primarily in teratocarcinomas. Its primer binding site matches a lysine (K) tRNA [879]. HERV-K encodes an immunosuppressive transmembrane envelope protein in cytotrophoblasts but not in syncytiotrophoblasts [880]. In prostate cancer cells, the *ets*-variant-1 (ETV1) fuses with HERV-K (*vide infra*). The RNA retroelements are reversely transcribed into DNA and inserted into the genome by self-encoded, or host-provided endonucleases. The HERV DNA copies may replicate with the genome, or through their RNA intermediates. It is estimated that >42 % (or according to some estimates up to two-thirds) of the human genome consists of retroelements [881, 882]. The LTR elements are inserted retroviral genomes with a defective *env* gene. The non-LTR elements are the long and short interspersed elements (LINE, SINE). The RNA polymerase type II transcribes LINE-1 (L1). L1 insertions in lung cancers coincided with genome hypomethylation of unknown significance: methylation signature distinguishes L1-permissive and L1-non-permissive tumors. L1 inserts into the MET oncogene (hepatocyte growth factor receptor) and the promoter is hypomethylated. In childhood undifferentiated germ cell tumors, the L1 ORF1 protein was overexpressed. The SINE elements are SVAs: SINE-R/VNTR/Alu (variable number tandem repeats; short DNA cut by *Arthrobacter luteus* restriction endonuclease, Alu). The RNA polymerase type III transcribes the Alus. Homologous Alu sequences may recombine, insert into, or next to, exons, undergo aberrant splicing from introns, cause the dry/atrophic macular degeneration when accumulate due to RNAase III Dicer enzyme deficiency. Indeed, inserted transposons, retrotransposons may induce genetic instability and contribute to loss-of-function mutations. DNA transposons from non-mammalian species may become genotherapeutic vectors. The Ivics & Izsvák works are reviewed in [27] (*vide infra*, Figure 62).

Inserted retroelements may become useful genes (reviewed in Bushman) [1]. It is theorized that the insertion of the V(D)J and Rag1 elements into the genome of the ancient sharks was accomplished by retrotransposons (*vide supra*). Other theories claim that an ancestral herpesvirus (a pre-EBV) spread these genes from the amphioxus and sea urchin to the ancestral sharks (*vide infra*); (reviewed in [27]. However, the expression of syncytins in the human trophoblats is encoded by endogenous retroviral genomes. This retroviral insertion might have promoted placentation in the evolutionary replacement of egg-laying [883]. The retroviral Env protein syncytin-1 and its receptor fuses placental cytotrophoblasts into the barrier of syncytiotrophoblasts. The complex process includes the gap junctional intercellular communication mediated by connexin and gap junction proteins (CX43/GJA1) and various receptors surrounding the syncytin receptor ERVW-1, SLC1A5, ASCT-2, RDR/ATB (endogenous retrovirus W-1; solute carrier; alanine, serine cysteine transporter; retrovirus D receptor; amino acid transporter B) [884].

The physiological placental trophoblast-fusing endogenous retroviral envelope protein syncytin is expropriated by breast cancer cells for the mediation of fusion between the cancer cells and endothelial cells; syncytin is expressed also in endometrial cancer cells [885, 886]. Human breast cancer cells expressed syncytin and both the breast cancer cells and the edothelial cells expressed the syncytin receptor. In endometrial carcinoma cells, steroid hormones increase TGFβ1/3 production and consequentially mRNA and syncytin protein level expression. Silencing the syncytin gene and/or neutralization of TGFβ1/3 by antibody canceled tumor cell fusions [886]. In somatic cells, HERVs are silent with well methylated promoters in their 5′/3′ LTRs. In various forms of testicular cancers, six HERV-W loci were overexpressed and their U3 promoters were hypomethylated [887].

The Syncytin-1 fusion protein (Figure 52) is encoded by a 56 kb segment from the human chromosomal locus 7q21. The ancient HERV-WE1 gene was inserted into the long arm of the 7th chromosome in Catarrhine monkeys sometime 25–40 mya. This gene is located between the PEX1 and ODAG genes (peroxisome biogenesis factor; ocular development-associated factor). In the human genome, the intergenic distance (sometime referred to as intronic region) is 30.9 kb. In the Gorilla/Chimpanzee/Homo subgroup the LTR elements MaLR (mammalian apparent LTR-retrotransposon) and HERV-WE1 (W for tryptophan tRNA) have been strictly conserved in 7q21.2 [888]. The proviral *env* gene encodes the 538 aa Env protein. The insertion of this gene in the ancestral genome occurred less than 40 mya, thus after the divergence of catarrhines and the New World monkeys (NWM), but before the divergence of Hominoids and Old World monkeys (OWM). The chimpanzee and human HERV-W proviruses are almost identical with similar inactivation mutations in the *gag* and *pol* elements. In the African OWM, the syncytin-1 gene suffered a great deal of inactivation, but the ERV3 and HERV-FRD genes (*vide infra*) remain conserved both in OWM and in hominoids. The NWM marmoset genome shows the integration site prepared for the future insertion in the African line [889].

The human endogenous retrovirus HERV-WE2 family envelope locus on chromosome Xq22.3 is active in peripheral blood mononuclear cells and encodes a 475 aa truncated Env protein. HERV-W uses tryptophan (W) tRNA as its primer. Monoclonal antibodies 6A2B2 and 13H5A5 detect the Env protein. Only the glycosylated Env protein is expressed on the cell surface; the unglycosylated Env protein remains in the cytoplasm [890]. The glycosylated Env protein is detectable by the mcab 6A2B2 on the surface of trophoblasts (physiological in pregnancy) and on nerve cells in patients with multiple sclerosis (*vide infra*). Of the four placental HERVs (HERV-K for lysine; erv3/HERV-R, for arginine; HERV-W, for tryptophan; HERV-FRD for phenylalanine, asparagine, aspartic acid), HERV-W and HERV-FRD encode envelope proteins, envW for syncytin 1 and envFRD for syncytin 2, with the highest fusogenic potency [891, 892]. In the HERV-FRD transcript, syncytin 2 is immunolocalized in the villous trophoblasts and cytotrophoblasts [893]. The ancient syncytin 2 appears to have been conserved for over 40 my. The glycoprotein syncytin 2 receptor is a carbohydrate transporter encoded from human chromosome 1p34.2 and is referred to as major facilitator superfamily domain 2 (MFSD2). The human choriocarcinoma cell line BeWo overexpresses MFSDS2, especially upon stimulation with the diterpene forskolin. However, the mouse and rat MFSD2 fail to induce syncytin 2-mediated cell fusion [894].

When two different viruses collaborate. Endogenous retroviruses have been the "usual suspects" inducers of autoimmune diseases. Latent herpesviruses readily induce the maturation of endogenous retroviruses; reviewed in [261b]. This interviral relationship was documented and discussed at the Department of Medical Microbiology & Immunology of the University of South Florida College of Medicine as a "criminal collusion" [895a, b, c]. During the events of somatic hypermutation, dispersed segments of endogenous proviral DNA sequences may encounter each other in order to fuse into replication capable genomes. Thus retro- and herpes-like viruses (EBV) frequently coexist in lesions of autoimmune diseases (especially in multiple sclerosis) and malignant lymphomas (primary effusion lymphomas), including Hodgkin's Reed-Sternberg cells, and in KSHV/HHV-8-induced Kaposi sarcoma cells both pre- and post-AIDS KS cells. Figure 30 shows abundantly replicating and budding retroviral particles (not HTLV-I; not HIV-1). In the clinical material and human tissue cultures of this author, the co-existence of herpes-like (EBV; KSHV) and endogenous retroviruses were frequently observed, documented and extensively reviewed [147]. The retroviral particles, even when appeared budding, failed to replicate in tissue cultures (characteristically, due to loss-of-function mutations, and inserted stop codons). In these events, the role of HERVs is equal, if not superior, to that of EBV; together ("criminal collusion") these two master viral agents of latency entrap their host's immune system in an endless, but futile, pursuit (*vide infra*).

The HERV-W genome (probably identical with or closely related to the multiple sclerosis associated retrovirus, MSRV) from chromosome Xq22.3 encodes syncytin 1 proteins in the nerve cells (astrocytes and microglia) in patients with multiple sclerosis (a condition in which EBV was repeatedly called upon as another latent inducer). The syncytin-like protein induces polyclonal replication of VβT cells. This protein is recognized in the patients' nerve cells by the syncytin 1-specific antibody 6A2B2 [896]. Human endogenous retrovirus type W envelope expression was studied with quantitative (for copy numbers) PCR for Env DNA/RNA in blood and brain mononuclear cells, and with three monoclonal antibodies for the protein. Patients with multiple sclerosis expressed the highest levels, often against negative controls, in all parameters. The association between MS disease and MSRV-type W element (probably syncytin) appears quite strong [897].

It is well known that acutely transforming retroviruses possess an "oncogene" of host cell origin, but are replication defective (very likely, the original Rous sarcoma virus first isolate was oncogenic, but replication defective). Whereas, retroviruses devoid of usurped cellular oncogenes replicate rapidly and may serve as envelope-donor vectors of the acutely transforming retroviruses. It was recognized quickly that the retroviral genes encoding viral structural proteins were different from the "leukemogenic genes" [757, 758, 898–900]. These viral genomes (the provirus) generate malignant transformation by insertional oncogenesis. The situation is highly oncogenic, when the inserted provirus is that of a leukemogenic/oncogenic retrovirus with the propensity to insert its provirus into, or next to, a cellular proto-oncogene. Huebner and Todaro in formulating their "oncogene theory" stated, that the activation of vertically transmitted endogenous type C RNA viruses by chemical carcinogens, radiation, and even by carcinogenic DNA viruses is the cause of most (all) cancers; and that cancer therefore is "a natural biological event" [901]. However, the prior observations that the retroviral genome contained the virus' own genes encoding its structural proteins and enzymes, and an additional gene that encodes the cell's malignant transformation; further, that this additional gene could be lost and regained from and by the viral particle [70a, b, 760–762] were not recognized in the literature. It was reported before, that the leukemogenic type C virus particle could replicate without that additional gene, and exert immunogenic effect in its host without inducing any malignant transformation [899, 900].

When the wnt proto-oncogenes cooperate. The wnt genes (wingless, drosophila; integrated int, mouse) appeared in invertebrate hosts (hydra; amphioxus; drosophila; caenorhabditis) to fulfill various physiological functions. In the amphioxus, head and body axis development and neuroectodermal signaling are directed by genes homologous to vertebrate Hh and wnt. In the embryonic body patterning of the drosophila, Wnt, EGF, FGF and BMP (bone morphogenetic protein, without any skeletal structures) already interact. In the caenorhabditis' synaptic motor circuitry in the nervous system and in segmental patterning of the trunk, the Wnt/β-catenin and the MAPK pathways interact [902–904]. In the injured tadpole (frog), the regeneration of the cord, muscle and tail depends on the Wnt/βcatenin/FGF interaction [905]. The mouse mammary tumor virus, reviewed in [83], inserts its proviral DNA into the *wnt*-1 (formerly *int*-1) gene (drosophila ortholog *wingless*). In the mouse embryo, the normal Wnt-1 pathway is active in the CNS (the formation of the neural plate and tube) and is silent in adult life, except for the testicles. There are some ten members of the Wnt gene family in the mouse, some resting silently in the mammary gland. A MMTV proviral DNA is inserted into the 5′ nontranslated region of the Wnt gene (*wnt*-1/3). This inserted DNA is in the same translational orientation as the original Wnt gene segment, thus it replaces the promoter functionally and activates the Wnt gene. The Wnt transcription for exon 1A starts from the right LTR of the provirus. Another DNA provirus is inserted in orientation opposite of the natural translational orientation of the Wnt segment. This proviral insert disrupts its transcription site (exon 1B) of the Wnt gene, but not that of the other first insert. The MMTV genes encode the viral structural proteins *gag*, *pol* and *env* [906]. Wnt/β-catenin-induced oncogenic pathways continue in the form of H-*ras*/K-*ras*, ERK, and PI3K/Akt activations. It is the co-activation of FGR/FGF-R genes that induce the target genes for the Wnt pathway. In Wnt-induced oncogenesis, the Wnt-antagonists, SFRP and DKK1 (secreted frizzled-related protein; dickkopf), are silenced [907]. Glycogen synthase kinase 3β (GSK) is another Wnt/β-catenin antagonist (Table IV).

Charlotte Friend, at Sloan-Kettering Cancer Center, New York City, in the late-1950s. The murine erythroleukemia virus discovered by Charlotte (F-MuLV) is devoid of expropriated cellular oncogenes. This erythroleukemia virus inserts its proviral DNA into the murine *ets*-related gene, locus Fli-1 (Friend leukemia insertion). The site was named after the E26 twenty-six replication-defective avian acute leukemia retrovirus encoding transformation-specific (Ets), transcription factors, which were usurped from the avian host cell of the original isolate; reviewed [70a]; and cited [340a, 341a]. In the cells induced by this virus, insertional inactivation of the p53 and/or p45 NFE2 (nuclear factor erythroid-derived) genes follows. The spleen focus-forming Friend erythroleukemia virus inserts its provirus into the Spi-1 locus. The Spi-1 protein (gene spitz, drosophila) blocks erythroblast differentiation. The miR-17-92 interacts with erythropoietin (Epo) signal transduction pathway. Red blood cell differentiation was blocked and erythroblast proliferation was enhanced. The Fli3 F-MuLV proviral insertion site is the murine homologue of the human C13orf25 gene. This gene from chromosome 13q31-32 encodes the polycistronic miR-17-92. RNA polymerase II transcribes this miR in a long hairpin structure. It is then subjected to processing by dsRNA-binding protein and RNase III. The enzymes Drosha and Dicer release the 21 nt mature miR. The RNA-induced silencing complex unites the miR with the mRNA expressing complementary bases to the miR. Thus, the miR incapacitates the mRNA (no translation to protein takes place). The c-Myc upregulated lymphoid cells produce increased amounts of miR-17-92. The inserted Friend's, or SL3-3 retroviral, proviruses activate the release of this miR; miR-17-92 is highly upregulated in lymphomas and in other malignant tumors [908, 909]. The NFE2 gene (nuclear factor, erythroid-derived) represses the leukemia-unrelated *gcm*1 gene (glial cell missing) in trophoblasts via regulating JunD binding, by acetylating the JunD promoter [910] (c-*onc* → v-*onc* avian sarcoma virus ASV-17; in Japanese 17 is ju-nana; "jun"). Malignant B lymphoma cells overexpress c-Myc and miR-17-92. The SL3-3 murine T cell lymphoma-leukemia retrovirus derived from AKR mice (Jacob Fürth's albino K strain mice with thymic lymphoma, akt/Akt, maintained at the Rockefeller Institute, AKR) [911, 912]. SL3-3 proviral insertion occurs next to the promoter of the NFAT gene (nuclear factor of activated T cell). The activated NFAT gene is a tumor suppressor. However, the inserted proviral sequence may disrupt the transcription of the gene resulting in a loss-of-function mutation of a tumor suppressor gene. Overproduction of the oncogenic miR-17-92 with loss of function of a tumor suppressor gene promotes lymphomagenesis [909, 913]. This author (JGS) spread the word in the 1960s that Charlotte's virus must be related to the human Di Guglielmo's erythroleukemia retrovirus; without any evidence for it, this view remains imaginary.

John Moloney, at the National Cancer Institute, Bethesda, Maryland, in the 1960s. The human homologue of the murine *bmi*-1 gene is mapped at chromosome 10p13. The gene was discovered in mice overexpressing the *bmi*-1 gene due to proviral insertion. The murine *bmi*-1 gene is the insertional site of the Moloney mouse leukemia-sarcoma retroviral (MMLV) provirus. The activated *bmi*-1 gene acts as a positive regulator of the c-*myc* gene, and negative regulator of the $p16^{INK4a}$ and $p14/p19^{ARF}$ gene (kinase 4a inhibitor; alternative editing frame), In various human tumors the *bmi*-1 gene is amplified (by Southern blot for genomic DNA digested with EcoRI and HindIII restriction enzymes). BMI-1 mRNA was measured by real-time quantitative reverse transcription PCR (total RNA transcribed into cDNA with MMLV-RT, probes and primers designed and labeled and TaqMan reagents used). Protein analysis was by Western blot (whole cell protein extracts from frozen tissue samples loaded on polyacrylamide gel; electroblotted to nitrocellulose membrane; blocked membranes incubated with mcab BMI-F6, conjugated to horseradish peroxidase; detected by chemoluminescence). All tests indicated a significantly amplified *bmi*-1 gene and protein expression in CLL and MCL (mantle cell lymphoma) [914]. The MCL translocation t(11;14)(q13;q32) includes the constitutive overexpression of cyclin D1; the cell cycle inhibitors are inactivated by homozygous deletion, while CDK4 and BMI-1 are overexpressed. Separately p53 and Rb1 (in blastoid MCL) are inactivated [915a, b]. Various adenocarcinoma cells (breast, ovary, pancreas) overexpressing BMI-1 proliferate and metastasize rapidly [916–918]. Antisense Bmi-1 RNA and siRNA retarded the growth of BMI-1 overexpressing tumor cells [919a, b, 920a, b]. Head&Neck (H&N) squamous cell carcinomas express highly ALDH11 (aldehyde dehydrogenase), Bmi1 and Snail proteins [919b]. Active ALDH1 stem cell gene is retained also in sarcoma and adenocarcinoma (colo-rectal, prostate) cells. This author (JGS) spread the word in the 1970s that John's virus was related to human lymphoma-leukemia-sarcoma retroviruses; this view persists unproven.

Of whatever etiology, the standard therapy of the most malignant mantle cell lymphoma has been until recently the R-CHOP chemotherapy regimen (rituximab, cyclophosphamide, doxorubicin, vincristine, prednisone) with or without consolidation with yttrium-90, and ibritumomab tiuxetan [921]. New targeted therapy (especially ibrutinib) may complement (or even replace) combination chemotherapy (*vide infra*). Histone methyltransferase and deacetylase hyperactivity, when counteracted with 3-deazaneplanocin and/or panobinostat for the inhibition of these enzymes, subsided with resultant apoptotic deaths of the tumor cells so induce [922a, b]. An extraordinary combination of translocated and activated oncogenes, silenced tumor suppressor genes; and epigenomic miR alterations could be subdued, or compensated for, in this entity (mantle cell lymphoma). The inactivated oncogenes are INK4-Cdk4 and c-Myc and/or Bax interactions. Myc-repressed miR-15a/16-1; inactivated Rb/p53; and translocated fusion oncogenes t(11;14)(q13; q3), AID-initiated translocations; STAT3; Sox 11; and induced Burkitt's-like transformation (blastoid variants). This is a rare example of control by contemplated chemo-immuno-radiotherapy over the highly proliferative transformation of immortalized lymphoid cells. In this case, the cell-transforming RNA/DNA complex could be switched off.

Adipose tissue-derived stromal cells are in use for tissue repair in regenerative medicine. Attempts at prolonging the replicative life span of these cells by introducing with lentiviral transduction technology human telomerase reverse transcriptase (hTERT), murine Bmi-1, SV40 viral large T antigen genes. These interventions resulted in the creation of an immortalized and senescence-free hTERT/Bmi-1 carrier cell line [923]. The medical profession is well aware of the potential dangers (sarcomatous growth) inherent in the clinical application of such cell lines.

Retroviruses spread in the laboratory. The controversial xenotropic murine leukemia virus-related retrovirus (XMRV) is considered to be a laboratory contaminant of human prostate cancer cell lines [865a, b]. However this virus infects human prostate cancer cells *in vitro*. XMRV-infected low grade human prostate cancer cells acquired accelerated growth rate due to downregulation of the cyclin-dependent kinase inhibitor $p27^{Kip1}$. The downregulation of the inhibitor was miR-221- and mi-222-mediated. Among the mechanisms of action, insertional mutagenesis is being considered [924].

Selected Abbreviations

PARP, mono/polyADP (adenosine diphosphate) ribose polymerase. Binds PCNA (proliferating cell nuclear antigen) for translesion DNA repair/synthesis. Click for adenosine diphosphate-ribose polymerase, Nicolae CM Aho ER Vlahos AM et al The Pennsylvania State University College of Medicine, Hershey, PA, USA).

PCNA, proliferating cell nuclear antigen. Hyperthermophilic archaeal virus (phage) utilized host cell DNA sliding clamp for its replication and DNA repair. Click for archaea PCNA, Gardner AF Bell SD Wehute MF et al New England Biol Labs, Ipswich MA 01938 USA. The C-terminal region of the xeroderma pigmentosum group G gene (XPG) carries a PCNA binding site. Endonuclease Rad2 homolog of XPG expresses binding site for PCNA. The complex exercises cell cycle arrest allowing accumulation of damaged DNA. Click for xeroderma antigen and PCNA Yu SL Kang MS Kim HY et al Inha Research Institute Medical Sciences, Inha University, Incheon 400–712, Republic Korea.

XMRV, xenotropic murine leukemia virus-related virus derives from the recombination of two murine leukemia viruses, which due to inadvertent spread in the laboratory contaminated human prostate cancer cell lines. Susceptible to restriction by APOBEC. In fresh human tonsillary lymphatic tissue, XMRV replicates yielding non-infectious viral progeny. Click for inadvertent retroviral contamination of cells, Curriu M Carrillo J Massanella M et al Irsi Caixa Institut Recerca Ciènces Salut Germans Trias i Pujol Hospital Germans Trias Universitat Autònoma Barcelona, Barcelona Spain.

Genomic Intrusions of the Three Most Dangerous Herpesviruses

Indirecte oncogenic herpesviruses EBV and KSHV. EBV. Some protein products of herpes viral genomic segments (EBV) encounter c-Myc proteins, form host cell and viral protein complexes, thus inducing immune reactions to host cell proteins via the phenomenon of molecular mimicry, so claim Niller et al [925, 926].

The gammaherpesvirus' EBV's latent membrane protein 2A (LMP) keeps infected B cells alive and activates the PI3K pathway; in the pleckstrin homology domain, PI3K phosphorylates the serine473/threonine308 Akt kinase pathway. In epithelial cells, this pathway inactivates GSK3β by phosphorylation. As a result, β-catenin escapes ubiquitination, accumulates in the cytoplasm and translocates into the nucleus, there to inactivate its target genes. The intranuclear chain reaction induced by β-catenin includes the activation of the T cell factor/lymphoid enhancer factor genes, whose gene product proteins induce cyclin D1 (CND1) and c-*myc* [927]. EBV contributes to the pathogenesis of Burkitt's lymphoma, Reed-Sternberg cells of Hodgkin's disease, diffuse B cell lymphoma, nasopharyngeal and mediastinal NKT cell lymphoma, AIDS-associated B cell lymphoma (brain tumor), hemophagocytic syndrome, nasopharyngeal squamous cell carcinoma, pediatric leiomyosarcoma. and gastric adenocarcinoma. It blocks apoptosis in many ways (genetically imitating cellular bcl2; blocking p53; expressing latent membrane protein 2A). The latency-associated nuclear antigen (LANA) of KSHV/HHV-8 inhibits the cytoplasmic anti-β-catenin activity of GSK3β, thus allowing β-catenin accumulation and intranuclear activity in lymphoma cells. Anti-apoptotic cell-proliferative effects prevail. Those were primary effusion lymphoma cells, in which EBV and KSHV co-exist (Table IV).

The widely spread human pathogenicity of EBV and its epigenetic control have been very well elaborated on by the team, Takács et al [928]. However, the importance of RBP-J in herpesviral (EBV, HHV-8) pathogenicity is increasingly recognized (recombination signal binding protein for immunoglobulin kappa J region, RBJ-κ). EBNA2 and EBNALP (leading protein) are expressed in the latency III state. EBNA2 up-regulates the *hes*1 gene (hairy and enhancer of split, in drosophila) and the c-*myc*. Association of RBP-Jκ with EBNALP (EBV nuclear leading protein) is essential for malignant transformation of B cells. EBNA2 activates RBP-Jκ. Nuclear receptor coreceptor (NCoR) and RBP-Jκ form a repressive complex aimed at EBNA2 and its activation of transcription. NCoR recruits histone deacetylase3 for further repression of transcription. The RBP-Jκ/NCoR repressive complex recruites histone deacetylases and RNA polymerase II, but EBV2 escapes the repression. EBNALP is an activator of EBV latency III LMP1 (latent membrane protein). EBNALP moves HDAC4 (histone deacetylase) into the cytoplasm: an act of derepression. At the end, EBNA2 stabilized NCoR-deficient RBPJ, and EBNALP dismissed the repressive NCoR-RBPJ complex and activated the *hes*1/*arglu*1 genes (hairy/enhancer of split, drosophila; arginine-glutamine rich 1). Further, EBNA2 recruits and stabilizes NCoR-free RBPJ. EBNALP dismisses the NCoR-RBPL repressive complexes from the gene promoters and enhancers (thus relieving them from under repression) [929]. The EBV immediate-early gene BZLF1 encodes the lytic cycle initiator trasnscription factor (*Bam* leftward reading frame, protein Zebra). The gene product protein downregulates CD74 lymphoblast cell surface antigen. The multiple function CD74 antigen serves as a receptor for Mϕ migration inhibitory factor, a chaperone for MHC-II antigen presentation to $CD4^+$ T cells, and as an upregulator of the anti-apoptotic Bcl2 proto-oncogene. Its downregulation by the BZLF1 protein allows the escape of EBV-infected B cells from recognition by $CD4^+$ T cells. The EBV genome compensates for the loss of the anti-apoptotic protection of the now downregulated Bcl2. Its BHRF1 gene encodes a vBcl-2 analog protein (*Bam*H rightward reading frame, Bcl2). Gene product protein BZLF1 (Zebra) inhibits the host cells' expression of the INFγ and TNFα receptors [930].

KSHV. The KSHV (HHV-8) lytic switch replication transcriptional activator Rta protein, encoded from viral ORF50, stimulates the cellular RBP-Jκ protein (recombination signal sequence binding protein), a Notch pathway component. RBJ-κ, product of the homologue of drosophila suppressor of hairless gene, Su(H), attaches to the promoter of the Rta protein (R transactivator). Thus, tertiary (ternary) Rta/RBP-Jκ/DNA complexes are formed. KSHV ORF57 encodes the *mta*/MTA gene product protein (mRNA transcript accumulation). Repetitive elements in the promoter of the *mta*/MTA gene are copied in the Rta protein (the long repetitive element is abbreviated as CANT). RBP-Jκ is not bound to the latent KSHV genome; it is its interaction with Rta (the ternary complex) that allows that binding. Rta binds those regions of the Mta promoter, which contain 7 copies of the 14 bp

CANT repeat sequences. RBP-Jκ enhances Rta's binding affinity. Rta transactivates the Mta promoter. The CANT repeats are depicted in the article and the last figure is a cartoon showing the interaction of the CANT repetitive elements with the Rta/RBP-Jκ complex [931].

The spectrum of KSHV infectivity spans from vascular/lympho-epithelial cells (Kaposi's sarcoma) to B and T lymphocytes; the latter two entities are the primary effusion lymphoma and the multicentric Castleman's disease. In these lesions, the virus may exist in latent and lytic life cycles. The viral Rta promoter gene has binding sites for the host cell RBP-Jκ protein. Deprivation of the viral particle at these binding sites, either RTA_{first} or RTA_{all}, renders it permanently latent. Without RBP-Jκ → Rta interaction, the RTA protein will not activate the lytic cycle. Virus particles RTA_{first} and RTA_{all} induce enhanced colony formation and proliferation of the infected cells. Latently infected cells express the LANA (encoded by viral gene ORF73), characteristic of the persistent viral episome. In contrast, the master switch for the lytic cycle, *rta*/RTA (replication and transcription activator) is encoded by ORF50. The RTA protein activates the viral lytic genes: ORF6 for single-stranded DNA binding, ORF21 for thymidine kinase, ORF57 for mRNA transcript accumulation (as *mta* gene, *vide supra*), ORF59 for polymerase processivity factor, ORF74 for G protein-coupled receptor (vGPCR), etc. [932a, b]. The very active viral genome encodes IL-6, miR-2, Mϕ inflammatory proteins (chemokines vCCL2/3), basic region leucine zipper (K-bZIP) protein [933], IFN regulatory factor (vIRF), kaposinA/B/C, etc. Replicating KS cells are driven by the PI3K/MAPK pathway. The KSHV-encoded viral FLICE-inhibitory protein (FLIP) induces transdifferentiation of B lymphocytes (FADD-like interleukin 1-converting enzyme; Fas-associated death domain; factor apoptosis stimulator). Ancestral to HHV-8, the rhesus monkey rhadinovirus already encodes the anti-apoptotic vFLIP. However, the transcriptional machinery of HHV-8 far exceeds that of the rhadinovirus [934, 935]. HHV-8-infected lymphocytes express Mϕ- and DC-antigens (among them myeloid suppressor cell antigens): Ki-67, PU1 (purine-rich sequence 5′-GAGGAA-3′); Mϕ transcription factor; B cell-specific transcription factor; *spi*: spleen focus forming Friend erythroleukemia virus proviral insertion site (Fli); and ETS member antigen (E26 avian erythroleukemia reticuloendotheliosis retrovirus oncogene) [936, 937]. Further expressed are vimentin, CD68, lysozyme, Mac2, CD11c, Gr1 (granulocyte, glutathione reductase), CD19. This transdifferentiation and antigen-expression pattern is similar to that of the natural human KS [938].

KSHV-infected vascular endothelial cells transdifferentiate in the process of the endothelial to mesenchymal transition, ETM. KSHV ORF57 (in synergism with ORF50) releases a mRNA transcript accumulation regulator for the expression of the viral lytic gene product proteins [939, 940]. In KSHV-infected endothelial cells, viral proteins LANA and Kaposin B upregulate the cellular gene PROX1 (a homologue of drosophila's Prospero gene, which is responsible, among other functions, to generate lymphoid cells). The Kaposin B protein stabilizes the PROX1 mRNA. The control of mRNAs for stabilization, or destabilization, depends on proteins attaching to the 3′ untranslated region of the mRNA (UTR). The 50-159 nt long AU rich element (ARE) forms the platform for the protein attachment. Most AREs are AUUUA pentamers. Class I AREs are on the mRNAs of c-myc, c-fos, cyclins A, B1, D1, and IFNγ. Class II AREs consists of overlapping AUUUA motifs on the mRNAS of TNFα, IL-1β, IL-2, IL-3, GM-CSF, COX-2, and VEGF. Class III AREs are devoid of the AUUUA motif on the mRNAs of c-jun, p53, hsp70. In the KSHV-infected endothelial cell, the PROX1 protein acting on the mRNAs encoding cell phenotypes, suppresses the BEC, and promotes the LEC direction of differentiation pathways (blood vascular endothelial cell; lymphatic endothelial cell).

The human RNA-binding protein (HuR) stabilizes the mRNAs encoding the LEC pathway. Kaposin B proteins stabilize these (including PROX1) and the GM-CSF- and IL-6-encoding mRNAs (both Class I ad II AREs). The reprogramming of endothelial-to-lymphoid cells may also involve the activation of the Akt proto-oncogene (murine retroviral thymic lymphoma; protein kinase B). A further interaction is the suppression of TGFβ by LANA; TGFβ being an inhibitor of PROX1, its suppression leads to the upregulation of PROX1. The remaining question is, if PROX1 acts as a proto-oncogene (as in colon cancer), or as a tumor suppressor gene (transforming a sarcomatous elongated cell into a lymphoid cell with probable reduced malignancy)? [941]. From its 12 miRNA genes, KSHV releases miRNAs, among them miR-K12-11, which is a homologue of the human miR-155. The differentiation of PEL cells is blocked; PEL cells express high levels of miR-K12-11. KSHV and EBV miRs target some 2000 host cell mRNAs, at same or different binding sites. Two miRs of herpesviral derivation (miR-155; miR-142-3p) mimic the same cellular miRs [113, 942]. These miRs target gene host cell gene transcripts, among them BACH-1, the antagonist of Blimf, master regulator of plasma cell maturation (*vide infra*).

HHV-6 attracted to the telomeres. The armada of herpesviruses co-evolved in hosts shared with retro-, retrolenti- and spumaviruses. Viral interference between these agents was replaced by several forms of close collaboration, in which the agents "agreed" to sustain the host alive as long as possible for their mutual benefit. The genus Lymphocryptovirus speciated following its hosts in Africa; reviewed in [147]. Retroviruses and retrotransposons and EBV are referred to repeatedly elsewhere in this text. EBV/HHV-4 and KSHV/HHV-8, when they retire in latency, they circularize their genomes, and position them in the epigenome. The epigenome is an active site in the cell nucleus. From the epigenome, the viral genomes maintain close communication with their host cells through the release of microRNAs; reviewed in [73]. However, EBV viral genomic DNA integrates into host chromosomes, both in lymphocytes, and in epithelial cells (in nasopharyngeal carcinoma). The linear EBV genome may integrate into host cell genes, which are involved in oncogenesis: gene MACF1, (microtubule actin crosslinking factor, Golgi localized plakin) in Namalva cells; gene BACH2 (B cell factor antibody class switch, basic region leucine zipper, heme receptor, antagonist of Blimp, B lymphocyte-induced maturation protein, master regulator of plasma cell differentiation), in Raji cells; c-REL and BCL-11 (reticuloendothelial retrovirus, nuclear factor kappaB; B cell lymphoma anti-apoptotic proto-oncogene) in NAB cells. In

EBV-infected cells, both episomal and integrated viral genomes may co-exist [943]. It is not understood well, if directed integrations occur and how. The viral genomes encode their own enzymes (integrases, transposases, endonucleases) and/or utilize host cell enzymes for their integration. Is it the structures of the integrating viral, and/or the recipient host cell genomes, or is it selectivity, if any (engineered homing endonuclease) particular enzymes are performing the act of integration that determine the target site for these genomic unisons? Can directed integrations be designed to replace random integration? [944]. While natural viral genomic integrations continue to follow their own established pathways, viral vectors for gene therapy may have to be specifically directed to the desired integration sites in intact and replication-ready condition. So far, the integrated viral provirus (HIV-1) could be selectively incapacitated [945], and the gene therapeutic integrations remain random.

The HHV-6 inserts its ancient herpesviral genome consisting of a 160 kb dsDNA into its host cells' genome; so does the avian Marek's lymphomagenic herpesvirus in turkeys; reviewed in [27, 946, 947]. HHV-6AB are co-factors in the etiology of lymphomas (including Hodgkin's disease) and multiple sclerosis. When HHV-6 co-infects T4 lymphocytes with HIV-1, it promotes the replication of the latter pathogen. The HHV-6 can replicate in B lymphocytes, endothelial cells, and astrocytes. In the replicative cycle, the viral entry site is the CD46 cell surface complex, a complement activation receptor. The intracellular virus starts its replication in the nucleus for the full lytic cycle, by expressing its immediate early, early, and late (IE. E, L) genes. The rolling cycle replication leads to the formation of two catenanes (knots) uniting as concatamers (inseparably interlocked macrocyclic circular DNA ring structures). The circularized viral genome with a head-to-tail junction is integrated episomally. The linear dsDNA HHV-6 genome harbors a sequence homologous with the human adeno-associated virus (AAV) type 2 *rep68/78* (replication) gene. This gene segment of HHV-6 was acquired in the past in a host cell co-infected by both viruses; reviewed in [27]. The gene product protein from viral ORF U94 has some 24 % sequence sharing with the rep68/78 gene. Without this gene product protein, the viral integration is hindered. The ORF U94 protein is a helicase and its N-terminus is similar to the DNA-binding endonuclease domains of the rep68/78 protein. Telomere lengthening may be contributed to, by the alternative mechanism, if the circular extrachromosomal telomere repeats are used as templates through a rolling circle replication [948, 949]. When the virus replicates its TTAGGG repeats, that are identical with those of the cellular telomere ends, it contributes to the maintenance of telomere lengths (Figure 60).

The HHV-6 genome was found first to be integrated in the long arm of human chromosome 22q13 in an EBV^- Burkitt's lymphoma cell line. Stimulating this cell line with phorbol acetate and Ca^{++} ionophore induced a full replicative cycle of the HHV-6 virus [950a, b]. The HHV-6 genome could be integrated into other human chromosomes, and as such, could be vertically transmitted to progeny. The HHV-6 genomes as a rule are integrated into one allele of the telomeric segments of the targeted chromosomes in every cell of the body. The telomeric repeats TTAGGG

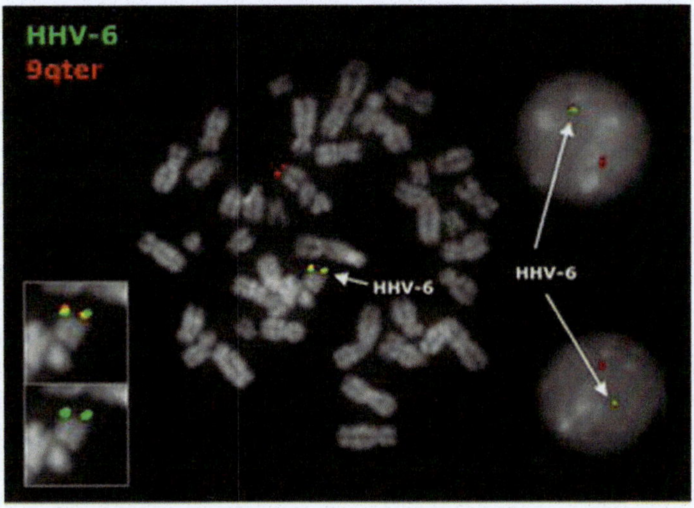

Figure 60 HHV6 travels from Isao Miyoshi to Peter Medveczky. [2358] Reference Appendix 2, Explanations to the Figures

are the most favored integration sites in chromosomes 17p13.3, 18q23, and 22q13.3. The viral genome at its both ends harbors these sequences identical to the telomeric repeats. Thus, homologous recombinations (HR) between the viral and the telomeric sequences secure the integration of the viral genome into the telomeres [948]. The U94 ORF of the HHV-6 genome encodes the AAV type 2 Rep 68/78 protein, which expresses DNA-binding, endonuclease and helicase-ATPase activities. In chromosomes with integrated HHV-6 genomic sequences, the cellular epigenomes were devoid of circularized viral genomes. Germ cells have either vertically transferred HHV-6 genome, or may be infected in utero in a mother with viremia. HHV-6 viral DNA is detectable in the blood plasma in 19^4 to 10^5 copies/ml. Hair follicle HHV-6 DNA settles the diagnosis of chromosomally integrated human herpesvirus (CIHHV), in distinction from viremic acute infection [943, 948] (Figure 60).

The integrated HHV-6 genome (CIHHV) may induce loss-of-function mutations, due to chromosomal segmental deletions at 17p13.3, 3q subtelomeric, and 8q23 del. Direct proto-oncogene activation has not been proven, but immune tolerance to fetally acquired HHV-6 may be induced; or the opposite, an autoimmune reaction may be generated [943]. Dendritic cells (DCs) infected with HHV-6B expressed TLR4, but reaction to endotoxin (LPS) was much decreased; so was NFκB activation [951]. HHV-6-infected monocytes failed to mobilize their usual immune reactions to intracellular cryptococci, which survived [952]. The antibody supplement of patiens with lytic HHV-6 infection contains virus neutralizing antibodies, but patients with latent HHV-6 infection (viral genomes inserted) are poor producers of virus-neutralizing antibodies. Host cells lytically infected with HHV-6B released at least four 60-to-65 nt pre-microRNAs, which mature into

18-to-19 nt microRNAs. In the AGO-containing RNA-induced silencing complex (RISC), the miRs align with the 3′ UTR region of targeted mRNAs and neutralize them. No convincing evidence was so far provided that NK cells or immune T cells attack those host cells that harbor inserted HHV-6 genomes. Upon viral reactivation (lytic infection) these patients remain quite defenseless [953a, b]. The HHV-6b-miR-R06-2 is an ortholog of the human miR-582-5p; it is of viral genomic and not of host cell genomic derivation. In the human host, miRNA-582-5p downregulates TGFβ. A new viral transcript from the minimal efficient origin of lytic replication (OriLyt) releases an abundance of RNA species of 18-to-19 nt [954].

The herpes simplex virus (HSV) 152-kb dsDNA genome encodes some 80 gene product proteins. The infected cell early viral proteins (ICP) act as ubiquitin ligases for the degradation of some host cell proteins: the promyelocytic leukemia protein (PML) that would suppress the replication of viral genes, and the interferon-inducer proteins. The viral genome replicates its DNA (and RNA), and eventually encodes viral structural proteins for the capsid assembly. Of several cellular proteins engaged in the viral replication is the proliferating nuclear antigen (*vide infra*).

Tankyrase, herpesviruses, telomeres, Wnt. PARP5a (tankyrase-1/2) (Table II) interacts with the telomere repeat binding factor (TRF1), for telomere maintenance, and with mitotic spindle pole formation in dividing cells. At the initiation of HSV infection, tankyrase 1 is phosphorylated by ERK and is transferred into the nucleus. In the nucleus, tankyrase binds the viral ICP (infected cell protein). Knocking out tankyrase protein translation by iRNA (RNA interference at the tankyrase mRNA level), or inhibition of tankyrase by its specific inhibitor XAV939, reduces viral replication to a minimum. Cytoplasmic tankyrase 2, and intranuclear tankyrase 1 are necessary to the full replicatory cycle of HSV. In the nucleus viral protein ICPo interacts with tankyrase 1, TGF-activated protein kinase 1, and Jun H proto-oncogene kinase. However, the ICP-attached tankyrase 1 is ultimately degraded [955]. The infected host cell escapes malignant transformation, due to its death consequential to the large load of the new viral progeny. Tankyrase by its ankyrin-domain attaches to the aminoterminal domain of EBV's EBNA1.

Tankyrase disrupts binding of telomere repeat factor 2 (TRF) to telomeric repeats, and separates sister telomeres in anaphase [956]. The telomeres in virally uninfected cells interact with tankyrases. The TTAGGG repeats are associated with the shelterin complexes of telomere repeat binding factors. The rapidly and frequently dividing cancer cells preserve their telomere lengths by alternative mechanisms, such as homologous recombination (ALT; HR). The HR consists of telomere sister chromatid exchange (T-SCE), that is increased in HR. Removal of tankyrase results in accelerated T-SCE; thus, tankyrase emerges as a T-SCE inhibitor, similarly to PARP1 (polyadenosine diphosphate-ribose polymerase) (Table II). Tankyrase *via* poly(ADP-ribosyl)ation removes TRF1 from its association with the telomere. In tankyrase-deficient cells (treated with PARP inhibitor XAV939, Novartis/Selleck), sister telomeres remain coherent, instead of separating. These cells are highly sensitive to irradiation-induced DNA damage with consequential fork stalling. A posttranslational degradation of DNA-dependent protein-kinase (DNA-PK) occurred (with no reduction of DNA-PK mRNA levels); DNA-PK is the non-homologous end-joining (NHEJ) protein. Uncapped leading strand telomeres in tankyrase-deficient cells would fuse their different chromatids. Further, dsDNA breaks in telomeres would also result in fusions [957].

Tankyrase stimulates axin degradation (sending the molecule to ubiquitination by ubiquitin E3 ligase); inhibition of tankyrase preserves and elevates axin levels. Tankyrase inhibitors antagonize the Wnt/β-catenin signaling pathway. If overaccumulated cytoplasmic β-catenin is allowed to translocate into the nucleus, there it activates LEF/TCF proto-oncogenes (*vide supra*). The LEF/TCF proteins activate oncogene c-*myc* and *cyclin* D. Axin is the major component of the β-catenin destruction complex; tankyrase emerges by targeting axin for destruction as the promoter of the Wnt/β-catenin oncogenic pathway, If so, could tankyrase inhibitors XAV939 ($C_{14}H_{11}F_3N_2OS$ Mw 312.31) act as a Wnt/β-catenin signaling inhibitor and protector of axin? Tankyrase inhibitors are referred to as IWRs (inhibitors of Wnt response). The tankyrase inhibitor XAV939 blocks the Wnt pathway in human breast cancer cells (basal cell line MDA-MB-231), and human colon cancer cell lines. The Wnt inhibitor kinase inhibitor 4 (WIKI) blocks Wnt/β-catenin signaling in numerous different human cancer cells (including colorectal adenocarcinoma cells) via first inhibiting tankyrase [958, 959]. Tankyrases "parsylate" by cleaving nicotinamid and ADP-ribose, which is then binds its substrates. The tankyrases' molecular pockets for the attachment of their inhibitors (IWRs) are compared in crystal structure to the other targeted molecular pockets of Abl and Kit kinases for imatinib, and BRAF's V600E mutation for vemurafenib [960] (*vide infra*).

Insertions of the Genomics of Adaptive Immunity

Ancient retrovirus and pre-EBV herpesvirus co-operated in donating the genetic elements of adaptive immunity? Major hypotheses have been formulated concerning the installation of the individual elements of the adaptive immune system at the level of the Placoderms (extinct) and the chondrichtyes gnathostomata sharks. The proto-elements of the adaptive immune system (VC, Rag1/2, MHC, AID) existed singly and engaged in functions other than immunological in the amphioxus, ascidians (Botryllus) and sea urchins [961]. These elements of the adaptive immune system worked for the first time in unison in the primordial sharks. The entire system with some modifications and duplications (or deletions) re-emerged in vertebrate bony fish, amphibians, reptiles, the descendants of the theropod dinosaurs, aves, and mammals. The major ancient constituents of the system are the lymphatic organs in the form of a chain of lymph nodes encircling the trachea and the intestinal tracts from the thymus downwards; the bone marrow and spleen, or the bursa of Fabricius in birds. The lymphatic organs harbor the thymic lymphocytes with TCRs capable of changing configurations by performing hypermutations, and producing chemokines, lymphokines, and cytokines; and the equivalents of bursal lymphocytes, which mature into plasma cells. Underlying are the minor and major histompatibilty complexes (MHC), and the intact and cooperative elements of the entire innate immune system, the TLRs, chemokines, cytokines, phagocytes (monocytes, leukocytes), dendritic cells, and NK cells, including the NK cells with invariant TCRs (NKT).

The primordial elements of the adaptive immune system are detectable singly and engaged in functions other than immunological in the amphioxus (immunoglobulin V/Rag1) [962], sea urchin (complement, rag1/2) [311, 313, 317, 963], and the ascidians (Botryllus) (Figure 46) with primordial NK cells, reported first by Khalturin et al [294], reviewed in [27, 30a], and presented in a poster at M.D. Anderson Hospital Symposium "Cancer Immunity: Challenges for the Next Decade" (Abstract #1-33 p 117, 2003) by H. David Kay, Joseph C. Horvath and Joseph G. Sinkovics (*vide infra*). The locus and the role of the fusibility/histompatibility locus, *fuhc* remains debatable [964, 965]. The unidentified elements of the MHC system induce "graft-rejection" in the botryllus (rejecting colony fusions mediated by special circulatory cells, probably ancestral to NK cells) [294], and in nemertean worms (genus Linnaeus), helminths, and snails [966–969a, b]. An ancient contributor to tolerance for the conversion of an invader into a symbiont is TGFβ. The cells of the sea anemone Cnidaria, *Aiptasia pallida,* are invaded by the dinoflagellate alga, the zooxanthella Symbiodinium. The alga fixes carbon and produces O_2. The cnidaria could reject the invader by the process of "bleaching." However, its cells release TGFβ, which switches off all defensive reactions. For example, when the cnidaria responds to LPS with nitric oxide production, TGFβ suppresses that reaction, too [970].

Some individuals of the cnidaria species, *Hydractinia symbiolongicarpus* form colonies in abandoned hermit crab shells. Individuals of two different colonies may unite by fusion, or activate a rejection reaction. The allorecognition genes *alr*1 and *alr*2 reside in the allorecognition complex (ARC) of the genome. These genes encode hypervariable transmembrane proteins expressing immunoglobulin V-like domains. The transmembrane Ig-like proteins are anchored by a cytoplasmic tail (for tyrosine-based intracellular signaling). The transmembrane protein encoded by the nine exons of the *alr*2 gene expresses three Ig-like variable domains. Intergenic sequences (introns) separate the exons. The ARC complex harbors copia-like retrotransposons and repetitive DNA tracts. Repetitive DNA tracts, *alr*2-related sequence duplications, sequence recombinations and pseudogenes contribute to haplotype diversity. Pseudogenes induce sequence variability in their related functional genes. Allelic diversity is known to be generated by the phenomena of gene conversion. Gene conversion remains the prominent mechanism of immunoglobulin diversity induction much later on, in the bursa of Fabricius in birds. These faculties of allodiversity discernible in the ARCs suggest that inter-allelic recombinations are transcription-linked somatic hypermutations induced by the B cell receptor-like transmembrane domains, again much later). The genetic diversity of Hydractinia colonies is high, because rejections are frequent, and fusions are rare (2–4 % of encounters) [971]. In this cnidaria, the pseudogenes contribute to *alr*2 diversification by sequence donations. The large numbers of *alr*2 alleles, the appearance of rare new alleles secure an extraordinarily wide genetic polymorphism and gene diversification of Hydractinia cells [972]. The colonies of this cnidaria consist of differentiated cells and undifferentiated (multipotent) stem cells. One differentiated cell type, the nematocyte, carries out the act of the rejection of approaching incompatible individuals. Another epithelial differentiated cell type serves in the gastrointestinal/vascular tract shared by the entire colony. The colonies maintain stem cells in the interstitial spaces: referred to as interstitial *i*-cells. The *i*-cells are round or slightly elongated with large nuclei and basophilic cytoplasms. The stem cell genes are referred to as members of the POU family (Pit, pituitary, identical with growth hormone factor 1, GHF1; Oct, octamer; Unc, nematode gene uncoordinated), but in reality in Hydractinia the POU family consists of the classical pluripotent stem cell genes: Oct4, Sox2, Nanog, c-Myc, Klf4. The *i*-cells express the polynem (*Pln*) stem cell gene (in addition to Nanos, Vasa, Piwi, and Myc). The Pln gene is a primordial Oct gene; its gene product protein reacts with immune sera directed against the human Oct4 gene product protein (POU5F1). When cnidaria cells were transfected with a Pln gene expression vector, the epithelial cells transformed into malignant tumor cells, morphologically resembling *i*-cells, which overgrew and killed the colony. Tumor cells expressed PLN (Oct4) proteins intranuclearly. Neoplastic cells upregulated their own endogenous *Pln* gene and suppressed the transgene; also, suppressed the differentiated epithelial cell markers (these, Hydractinia borrowed from the Hydra: cytosolic malate dehydrogenase and tubulin alpha chain are the same). The malignant phenotype was really inscribed into the cells' genotype: eggs derived from transgenic cells and fertilized with wild type sperm developed into larvae succumbing to malignant transformation. However, retinoic acid (RA) placed in the sea water induced tumor cell differentiation and rescued the colony. RA is a known antagonist of Oct4 oncogene/oncoprotein. The cells rescued by RA differentiated toward the nematocyte line reacting with the nematocyte antibody NCol-1, and re-expressed their epithelial cell markers TAC (*vide supra*). Interfering RNA (iRNA) treatment aimed at the Pln gene reversed the course of malignant transformation and directed it toward epithelial/nematocyte re-differentiation. The gene *polynem* product protein (POU protein PLN) received its name after many (poly) nematocytes [973]. In the colonies, *i*-cells usually reside in the stolons (septate hypha, tentacle, fastening the cnidocytes) and proliferate as driven by the Wnt pathway. In the context of stem cell proliferation, heat shock protein and its related gene Hsc1 activate the Wnt pathway, securing the mitotic cycle and inhibiting apoptosis. Apoptosis represented by DNA fragmentation kills metamorphosing cnidocytes before the development of stolon tentacles [974, 975]. These very complex interactions (a stem cell tumor in a cnidaria) in one sentence: when cnidaria stem cells are induced to differentiate into epithelial or vascular somatic cells, a primordial stem cell gene reverts them into immortal individuals with the propensity to replicate incessantly.

At the level of the Japanese eel, treatment with glycans activated the innate immune reaction of opsonization of phagocyosis by complement [976]. However, intraperitoneal immunization with inactivated Edwardsiella bacteria induced rising agglutinating antibody titers [977]. If this is confirmed (that the "antibodies" were truly immunoglobulins), than the eel's immune system is that of the vertebrate adaptive immunity (and not that of variable lymphocyte receptors with leucine-rich repeats, like in the cyclostomata lamprey (*vide supra*).

In opposition to vertical routes of inheritance, the sudden appearance of these elements (MHC, V(D)J, RAG1/2, RSS (recombination activating genes; recombination signal sequences) in a functional unity in the primordial shark, generated theories for horizontal (lateral) gene transfers. The vectors of the elementary constituents (the genes) of the entire adaptive immune system were first considered to be retroviruses/retrotransposons (Kapitonov & Jurka) [116, 311]; reviewed in [27], and this view persists from the 1990s to 2012 [311, 313a, b, 317, 878]. The antigen receptor proteins are not encoded by established germline genes. The specific antigen receptors are individually assembled from the V(D)J gene segments recombining in the process of the somatic hypermutation of Tonegawa and Hozumi [978]. The heptamer and nonamer sequence motifs of the recombination signal sequences (RSS) are separated by 12 or 23 bp spacers. The recombinations are facilitated by specific dsDNA breaks at the 12 or 23 spacers. The coding gene segments fuse into permanently retained coding joints and RSSs fuse into termporary signal joints. The Rag1/2 complex recognizes the 12 and 23 RSSs and performs a ssDNA nick between the RSS and the neighboring coding sequence leaving hairpin structure on the coding DNA and a blunt-cut RSS end. The RAG tranposases are retroviral integrases. Occasional inappropriate diversions of the transpositional pathway may result in DNA translocations with lymphomagenic consequences [979–982]. In the V(D)J recombination, the antigen receptor loci (IgH, IgK, IgL and the TC receptors B, G, D) are the legitimate major targets. Nonantigen receptor loci may be targeted by "mistake" resulting in intrachromosomal interstitial deletions and consequential segmental translocations. If proto-oncogenes (c-*myc; bcl-2*) are translocated, or tumor suppressor genes are deleted (PTEN), the involved cells may respond with "malignant transformation", most often with lymphomagenesis. In the RAG1/RAG2 complex, RAG1 is the V(D)J recombinase, and the accessory factor RAG2 restricts DNA cleavage to the G_0/G_1 phase of the cell cycle, and stabilizes the RAG1/RAG2/RSS complex. RAG2 in its C terminus expresses a highly conserved plant homeodomain (PHD). By this homeodomain, RAG2 binds to three-methylated lysins in the tails of nucleosomal histones (H3K4me3). RAG1 transposase genes operate in invertebrate hosts, other than the sea urchin: the Transib transposase family, of Kapitonov & Jurka (*vide supra*). Among invertebrate hosts, so far, only the sea urchin genome is known to harbor a *rag*2 sequence (other Transib *rag*1 sequences are *rag2*-free). The sea urchin RAG1 transposase protein shows aa similarities to the vertebrate RAG1 proteins. The sea urchin RAG2 protein has much less aa similarity to vertebrate RAG2 proteins (deriving from different ancestors), but both sea urchin and vertebrate RAG2 proteins express the PHD domain. However, the sea urchin RAG2 protein prefers to bind di-methylated lysines on the histone 3 tails (H3K4me2). The RAG1/RAG2 proteins in the sea urchin do not connect to a V(D)J) system, yet they fulfill not clearly understood defensive function in the coelomocytes (Figure 45) [963].

Antigen presentation and recognition is mediated by the MHC/HLA complex, whose genomic sequences are located on human chromosome 6.21.3. This chromosomal region retains duplicated copies of inserted ancient retroviral DNA proviruses representing at least 12 different HERV-K family members [983]. Monocytes/Mϕs and dendritic cells (DCs) are the major professional antigen-presenting cells (APCs). Family B7 proteins are essential costimulatory ligands for T cell activation in antigenic recognition; T cells express receptors for B7 proteins. These are basic elements of the adaptive immune system and are fully operational in teleost fish, fugu (tiger pufferfish, *Takifugu rubripes*) [984]. The MHC class II heterodimer glycoproteins present peptides to $CD4^+$ T cells. The class II α and β chains assemble in the endoplasmic reticulum and incorporate their chaperone, the invariant chain. This article depicts the organization of B7 genes and the phylogenetic analysis of the B7 family up the human B7 proteins. The invariant chain of the MHC class II heterodimer was functional in the ancestors of the sharks over 450 mya [985]. It appeared suddenly in cartilaginous fish, because no homologs or potential ancestors of the gene have so far been found in the lamprey or hagfish, or in the urochordate Ciona, or in the cephalochordate amphioxus (in Appendix 1). A precursor locus of the MCH class II transcription activator protein (CIITA) appeared also in the sharks. Lack of evidence for vertical transmission suggests acquisition of these genes through horizontal transfer, very likely by retrotransposons. A search for viruses circulating in the vast aquatic environment might reveal the suspect genomic sequence in a potential vector. The NK cell receptor NKp30 co-evolved with the B7 family protein B7H6. The NKp30 receptor is encoded from the MHC genomic complex (except not entirely so in the Xenopus). Most B7 family members harbor Ig superfamily V and C sequences. The B7 precursors originated from the proto-MHC. B7H7 is encoded from the gene HHLA2 [986].

It remains unexplained why the teleost Atlantic cod (*Gadus morhua*) failed to acquire MHC class II system with its invariant chain, and reactive $CD4^+$ T cells. Was it never acquired in the cod lineage, or was it lost to deletion, or is the cod operating an unusually effective innate immune system? Indeed, it appears that the cod expanded its entire innate and the MHC class I adaptive immune system. There is adequate response to endotoxin, and to bacterial (*Aeromonas salmonicida; Vibrio anguillarum*) and viral (pancreas necrosis virus) pathogens [987–989].

The discovery of H-H Niller and Associates. This author (JGS) was impressed by the reports of Dreyfus that EBV's transposable and signal sequences were similar to the mariner/TCL-like (total cell lysate) transposons, that EBV harbors V(D)J-and TCL-like sequences, and encodes a RAG-like recombinase [147, 990a, b, c]. How the EBV could have acquired retroviral-like sequences? The Marek viral genome incorporated an entire retroviral sequence; reviewed in [27]. This author promptly proposed (even before reading the papers of Niller *et al*, and Dreyfus) that a herpesviral ancestor through infectious passages transferred these elements from amphioxus and sea urchin to the primordial sharks [147]. He then proposed experiments in which a herpesvirus with *rag*1-negative genome grown in sea urchin cells could possibly have incorporated the *rag*1/2 gene complex into its genome [27].

The alpha/beta/gamma herpesviruses originated from a common ancestor well over 400 mya (since ancestral herpesviruses exist in the amphioxus and mollusks). By now herpesviruses cover a wide range of particle sizes from 120 kbp (human varicella/zoster, HVα) to 240 kbp (human cytomegalovirus, HVβ). Accordingly, herpesviral ORFs range from 70 to >170. The human oncogenic HHV-4 (EBV) and HHV-8 (KSHV) are gammaherpesviruses; the human roseolavirus HHV-6 is a betaherpesvirus (*vide supra*). The extensive McGeoch review (restricted to mammalian herpesviruses) indicated first the existence of a herpesvirally encoded immunoglobulin-like protein: the extreme left side of the HHV-8/KSHV genome harbors the gene for the K1 protein, which is an Ig-like highly variable receptor [991].

Niller lectured in 2003 at the 14th International Congress of the Hungarian Society of Microbiology on a 30 kb region of the EBV genome, which is co-linear with human immunoglobulin gene loci [992] (Figure 61). Neither this author, nor Dreyfus in his "paleo-immunology" article were cognizant of this most important announcement of Niller et al [27, 147, 312]. Niller and associates first showed that a 19 bp sequence of tRNA-like EBER1 promoter genome contains within its Ig lambda locus a twin sequence, which is a binding site for the c-Myc protein. It was further proposed, that these cellular and EBV sequences, including the EBER (the left part of the EBV genome) and the Ig variable region (containing variable, joining, switch and constant, VJSC, gene segments), derive from a common ancestor. Lymphocryptovirus herpesviral genomes (*H. papio, H. rhesus, H. saimiri*) harbor similar sequences [925, 926, 992]. It may be added, that the ancestors of New World monkeys left Africa some 30 mya; therefore these sequences were added to the genome of a common ancestor of Old and New World monkeys. It is not known if it were one solid genomic transfer, or sequentially acquired and later assembled genomic segments, that constructed the left part of the EBV genome. However, articles practically memorized by this author in the mid-2010s, failed to recognize the Niller *et al* report; there was no mention of its cognizance anywhere [993].

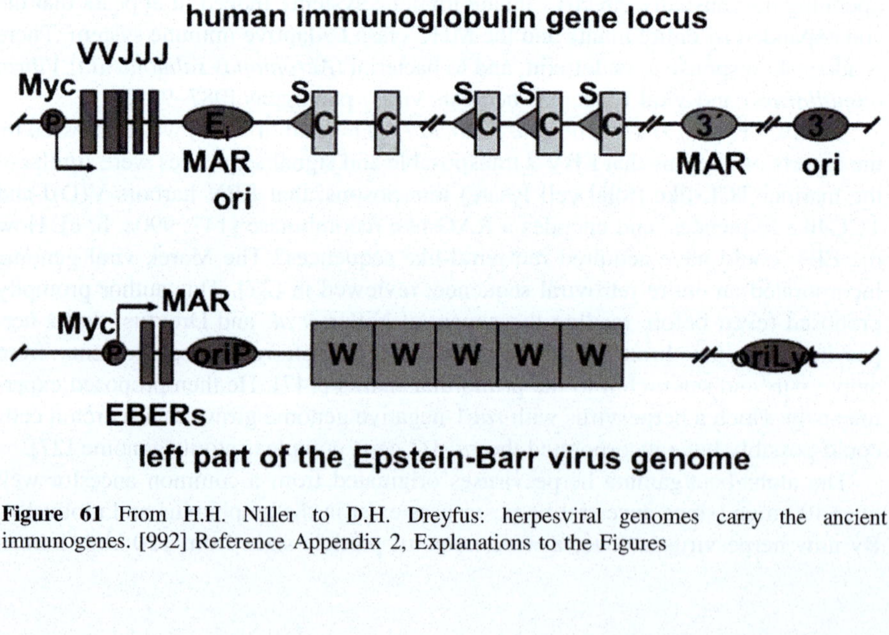

Figure 61 From H.H. Niller to D.H. Dreyfus: herpesviral genomes carry the ancient immunogenes. [992] Reference Appendix 2, Explanations to the Figures

In theory, the ancestral EBV picked up from an unidentified host the Ig segments and deposited them much later in a somewhat changed format into a host evolutionarily more advanced, than the original donor of the gene was. In the RNA World, tRNAs evolved into transposable elements and retroviruses; DNA replicons and DNA viruses followed [18]. In the first cells, soma-germline separation was accomplished. The gene exchanges between viral and cellular units resembled a ping-pong match; these interrelationships form the basis of the Ping-Pong Evolution Hypothesis of Niller et al [992, 994a, b]. EBV induces malignant transformation with association of c-Myc and by inhibiting apoptotic cell death (Burkitt's lymphoma; Hodgkin's Reed-Sternberg cells). EBV may immortalize B lymphocytes without malignant transformation; these lymphoproliferative conditions may recede under adequate T cell control (infectious mononucleosis; post-transplant lymphoproliferative disease). In a protracted course with loss of T cell control, the afflicted cells may transform into malignant entities [925, 926, 994a, b]. The microRNA146a exerts regulatory function in B lymphocytes. The dinucleotides CpG of its promoter are hypermethylated (silenced) in EBV^+ BL cells, whereas nasopharyngeal cell lines carry the miR-146 promoter CpG dinucleotides unmethylated. In EBV^- BL cell lines the miR-146 5′ regulatory sequences were unmethylated; the demethylated euchromatin histones H3/H4 remained acetylated, thus miR-146 is active (Szenthe et al) [944b].

Seven years after its publication, the Niller's et al priority report [992] (Figure 61) received full acknowledgement by Dreyfus et al [995]. EBV motifs were found for c-Myc binding [994a], and for NFκB and p53 binding [990a, b, c]. The carboxyl terminal of EBV is the ankyrin protein ZEBRA (BZLF-1, BamH1 Z encoded fragment left reading frame; Z Epstein-Barr replication activation). The ZEBRA protein is related to the drosophila's *cactus* gene product protein, also referred to as IκB-like (kappa B kinase) protein, the inhibitor of NFκB. NFκB is also referred to as c-Rel, after reticuloendothelial retrovirus capturing the cellular proto-oncogene (c-*onc* → v-*onc*). In general, ankyrin anchor proteins are of action at specific other sites (like spectrin-actin cytoskeletal proteins to cell membranes' ion channel sites). The ancient ankyrin regulatory proteins contain the *fos/jun* proto-oncogenes' DNA binding factor. The EBV ZEBRA carboxyl terminus forms the polymorphic structure ZANK (ZEBRA ANKyrin-like region; Bam Zta leftward; *bzlf*-1/BZLF-1 encoded lyic activation protein). The possession of ankyrin-like proteins by EBV enables the virus to bind cellular proteins NFκB and p53. As long as the EBV genome preserved VJSC sequences (*vide supra*), it also possesses TCR recombination sites. Dreyfus proposed that the original EBV ZEBRA/BZLF exons were acquired from two different sources: the ancient host cell's *fos/jun* transcription factor and the IκB immune regulatory protein, and that these elements were united by fusion. In the viral genome, ZEBRA encodes the lytic switch protein expressing in its amino terminal the Jun/Fos transcription factors; the carboxyl region of the ZEBRA protein binds NFκB. Binding NFκB inactivates the host cell's (the B lymphocyte's) innate and adaptive inflammatory defense reactions. Through CD21 receptors, EBV can infect T lymphocytes. Inactivation of NFκB in T cells triggers apoptosis; thus, T cell-mediated immune reactions are switched off. When the interaction of ZEBRA with p53 tends to bean inactivating (anti-apoptotic) state, cell immortalization takes place. The by now separately functioning NFκB and p53 genes and gene product proteins presumably derive from a common ancestral transcription factor. The "spandrel effect" (espandre, to spread out) is the result of the reaction of both proteins with ankyrin-like motifs. The ZEBRA ZRE DNA-binding site activates the viral lytic cycle (from epigenomic rest to full replication) in an immunologically suppressed NFκB-free host environment. Thus EBV genomic changes secure anti-apoptotic immortality to the B lymphocyte, due to the neutralization of the WTp53 [990a, b, c, 995].

Is it possible to recapitulate the trajectory of the single major elements of adaptive immunity on the phylogenic scale? These major elements would be the genes, which were eventually united in the process of somatic hypermutation: the V (D)J complex, and RAG1/2 and the recombination signal sequences (RSS). Which were the original hosts? Is there a vertical route of transmission conceivable? If horizontal (lateral) gene transfers united these elements, what were the gene vectors? In the mid-2000s "genes in the waiting" for recruitment in the adaptive immune system were recognized in the amphioxus and in the sea urchin; reviewed in [27, 147].

Paleogenomics (*viral and cellular*). The variable region-containing chitin-binding proteins (VCBP) of invertebrates contain two tandem immunoglobulin variable V-type domains. The V-type domains were characterized in the Florida amphioxus, the lancelet (*Branchiostoma floridae*). These structures are highly expressed in the guts. Features of the highly diversified immunoglobulin superfamily are already recognizable in these structures. VCBP genes 1, 2, 4 and 5 form a continuous chain. VCBP gene 3 occupies a separate locus [996a, b]. To this interpreter, these genomic structures were "unsettled" in the process of evolving. Extensive haplotype and copy number variations, indel polymorphisms, sequence polymorphism, and inverted repeats indicated strong genomic diversity. This genomic diversity was expressed in allelic variations, pseudogene formation, and in genomic polymorphism. These are the genomic segments that are already, or will soon be able to, encode immunoglobulins specifically fitting some antigens. Newly appearing alleles suggested hypervariability and an increasing repertoire for the encoded proteins in order to recognizing antigenic targets. A structure emulating a J region was identified. Genes VCBP 1 and 4 co-hybridized, but genes VCBP1, 2, 4 or 5 did not hybridize with gene VCBP 3. The VCBP gene cluster 2/5 appeared as paralogues 5a, 2b, 5b sharing similar exon/intron organization. The VCBP gene 5′s N-terminal V domain is VCBP2-type, but its C terminal V domain is its own VCBP5 type; it has a D3 chitin-binding domain (CBD). The VCBP2/5 gene cluster exhibited high degrees of haplotypic variations, unrelated coding regions, ORFs, large indels, inserted mobile elements, and 2 pseudogenes with disrupted exons. Some ORFs related to non-LTR LINE-like elements are encoded within three introns. The tightly linked VCBP 1 and 4 genes are inserted in opposite transcriptional orientation. VCBP gene 3 appears singly with one allele. Its V segment is encoded in two exons (in contrast to the other VCBP genes harboring V in three exons each). Its CBD domain is encoded by three exons, which are divergent from each other. Its cDNA possesses a premature stop codon, but also a pair of polyadenylation signals able to release mRNAs for encoding VCBP5 N-terminal (D1) and C-terminal (D2) V domains. The VCBP2/5 cluster expresses more inverted repeats (IR), than any other segment of the entire genome: this may indicate instability of this cluster. This cluster encodes a hybrid transcript of a ribonucleotide reductase. Introns of the 5a gene contain a 1.2 kb, a 3 kb and a 6 kb inserts. These ORFs encode RT/endonuclease gypsy retroelement sequences. A third haplotype of the VCBP cluster contains non-LTR LINE-like elements [997]. The amphioxus genome contains active Ngaro-like tyrosine recombinase retroposons, (DIRS, Dictyostelium intermediate repeat sequences) [197], next to the VCBP2/5 gene cluster and elsewhere. The retrotransposons encode Gag, RT, ribonuclease and tyrosine recombinase proteins (but no *env* is evident). The report concludes: "it appears as if the evolution of the VCBPs parallels that described for the vertebrate MHC region, as well as the NK cell KIR receptors" [996a, b].

Herpesviruses of vertebrate hosts (bony fish, amphibia, reptilia, aves, mammals) up to the highest ranks, all are well known. Much less characterized are the herpesviral DNA sequences embedded in the amphioxus genome; and the mollusks' and corals' herpesviruses. The gastropod abalone's herpesviral agent matures to full infectious virions. A dendrogram of the Herpesvirales order based on concatenated DNA polymerase and protein sequences from 34 herpesviruses, lists in its bottom section the amphioxus-associated herpesvirus, standing alone; the abalone and ostreid herpesviruses forming their own clade of Malacoherpesviridae; and the clade Alloherpesviridae consisting of the Ictalurid, Ranid 1 and 2, Cyprinid and Anguillid (eel) herpesviruses. The abalone herpesvirus is proposed to be the first recognized member of the genus Haliotivirus (ear-shell, auriform). Upstairs are the Herpesviridae subdivided into Alpha- (containing HHV1, 2, 3), Beta- (containing HHV5, 6, 7) and Gamma- (containing human HHV4, 8) herpesviruses. All known herpesviral genomes are deposited in the Genbank. The introns-free amphioxus herpesviral genome shows sequences overlapping with those of the abalone and oyster herpes viruses (AbHV-1; OsHV-1). The mollusc herpesviral genome is also intron-free. However, the Bilateria evolving in the Cambrian sea (*vide infra*) moved toward Protostomes (mollusks, flatworms), or toward Deuterostomes (protochordates, sea urchins) [998]. Herpesviruses following their hosts might have been still mixed by infections transgressing across species [999], as protochordates might have preyed upon mollusks, and primordial jawed fish (Placoderms; chondrichtyes) might have eaten sea urchins. A DNA-packaging (into preformed capsids) ATPase (hydrolase) of these (and cyprid) herpesviruses show distant relationship to that of bacteriophage T4. The DNA of ostreid herpesvirus1 of bivalve mollusks was characterized and amplified by PCR [1000], and thus it is available for further analysis (if it already contains sequences co-linear with those of mammalian variable sequences). The gastropod herpesvirus (AbHV-1) proved itself (by histochemical and electron microscopic analysis) to be a neurotropic virus. The first signs of neurons appear the hydra, but primordial neurogenic circuitry may exist in the mollusks, inasmuch as they express the Notch-δ, and coupled with it, the basic helix-loop-helix genes. The gene bHLH is ancestral to the atonal/neurogenin-related proneural gene family. An ancestral herpesvirus with the tendency to be neurotropic, would infect, and/or insert its genome into, the bHLH cells [998, 1001]. The Anguillid herpesviruses appear to occupy an intermediary position within the Alloherpesviridae between the cyprid and ictalurid herpesviruses. The AngHV1 virions are those of the classical herpesviruses: the core, the hollow capsomers of the icosahedral nucleocapsid encoded by the structural genes of the viral genome, the proteinaceous tegument and a host-derived envelope containing also virally encoded glycoproteins encasing the entire virion of about 200 nm. The ATPase subunit terminase is highly conserved and is related to that of bacteriophage T4. The completely sequenced herpesviral genomes are the ictalurid IcHV1, ranid RaHV1/2, cyprinid CyHV3, and anguillid AngHV1. This last genome gives the closest insight into the genome of an early Alloherpesviridae. The ds non-segmented DNA AngHV1 genome is 248531 bp in size and expresses 136 protein-coding ORFs. Its capsid morphogenesis 12 genes are conserved within Alloherpesviridae. Other genes are involved in DNA replication and DNA packaging. Five genes conserved in Alloherpesviridae (ORFs22, 52, 82, 98, 100) are of unknown function. Forty genes are shared with CyHV3 (but it is not entirely clear if all of these genes encode recognized proteins). Recognized shared genes are those expressing TNF and IL-10-related factors and other "immune response modulating factors" (not clearly stated if elements of adaptive immunity appear over the well established elements of innate immunity). There are no "somatically rearranged receptors", or "variable lymphocyte receptors of leucine-rich repeats" (like in the lamprey) mentioned. The AngHV1 and CyHV3 are representatives of an Alloherpesviridae subgroup positioned above Malacoherpesviridae. IcHV1 and RaHV1/2 form another Alloherpesviridae subgroup, a sister clade, "infecting other fish and frogs". CyHV3 and IcHV1 capsid protein-, envelope protein-, tegument protein-, and tegument-associated protein-encoding genes were well recognized. Immunoglobulin-like genes are not mentioned [1002a, b, c, d, 1003a, b, c, 1004]. Now in the AngHV1 particles five capsid, thirty envelope (eleven glycoproteins) and 32 tegument proteins are recognized. Some of the AngHV1 envelope and tegument proteins show low sequence homology with those of the CyHV3. The AngHV1 ORF95 encodes an envelope protein related to that of the salmon infectious anemia orthomyxovirus (ISAV). The AngHV1 ORF67 encodes an envelope protein of sequence homology with the spike protein of the white bream virus (order Nidovirales). These proteins serve attachments of viral and cell membranes. Some 28 host cell proteins were incorporated the AngHV1 virions (without genomic representation in the virions). None of the proteins listed by name appear to be immunoglobulins (van Beurden et al) [1003a, b, c]. However, the AngHV1 particles derived from eel kidney cell cultures. Native virions isolated from hosts capable of antibody production, would reveal adsorbed immunoglobulins. So far the basic genomic sequences encoding the elements of the adaptive immune system (rag1/2, RSS, VDJ) do not appear to be present in the ancient herpesviral genomes.

The classical experiments from Robert Good's Minneapolis laboratory [1005] included hagfish and lamprey (no immunoglobulin antibody production) and horned shark, teleost fish (antibody production), but no eel. In addition to glycans enhancing phagocytosis (innate immunity), agglutinin antibody production to inactivated bacterial antigens (adaptive immunity in the Japanese eel, *Anguilla japonica*) were reported.

The anguillid herpesviruses may be related to those of snakes. The herpesvirus of Lucké's frog kidney carcinoma (ranid herpesvirus 1, RaHV-1) was passaged, transmitted with filtrates of ascites fluid, and submitted to genomic studies. The 220 kbp genome yielded a 39,757- bp DNA insert containing genes belonging to the fish herpesvirus clade, rather than show relation with avian or mammalian herpesviruses [1002c, 1006]. The ancestor of some of the avian herpesviruses might have been carried once by theropod dinosaurs. Even closer to that lineage might be the herpesviral flora of crocodiles, alligators and turtles; reviewed in [27, 1007–1009]. The aquatic mammals probably harbor herpesviral flora deriving from their ancestors once populating estuaria and dry land. The herpesviral flora of the African hippopotamus (*H. amphibius*) deserves a closer comparison with that of the whales [1010–1012]. The majority of the members of a given herpesviral clade preferred the use of one of the optimal codons (CUG, CAC, CAG, AAC, GUG, GAC). Unfortunately, codon usage patterns of herpesviruses could not be tied to phylogenic viral species evolution, due some notable individual exceptions [1013]. The Aquatic Animal Health Program of the College of Veterinary Medicine at the University of Florida, Gainesville, FL, could not so far fit the herpesviruses of the monophyletic cetacean clade into an established mammalian gammaherpesviral genus [1014]. In the Orient (Japan), a monophyletic alphaherpesvirus clade showed up in the false killer whale (*Pseudorca crassidens*) and in the melon-headed whale (*Peponocephala electra*), in addition to cetacean gammaherpesviruses in the sperm whale (*Physeter macrocephalus*) [1015]. These extant herpesviral genomes should reflect to openly displayed, or concealed faculties of adaptive immunity (in reference to possible carriership of genomic sequences for MHC, RAG1/2, RSS, VDJ). However, the herpesviral pathogens so far sequenced at the Malacoherpesviridae and Alloherpesviridae dividing line, or below, so far did not reveal possessions of ancient adaptive immunity-related genomic sequences [1002a]. By re-examination of these genomics data, an oversight should be recognized, or ruled out. If Herpesviridae show the first appearance of DNA sequences related to those of adaptive immunity well above the Alloherpesviridae clade, then these viruses might be late recipients of these sequences, and not the ancient original donors of them. In that case, retroviral and related inserts will remain the suspect initiators of the genomics for adaptive immunity [311, 313a, b].

It is entirely possible for a large herpesvirus to incorporate retroviral genomic segments, as is in the Marek's oncogenic turkey herpesvirus; reviewed in [27], but it has not been determined at what time of the virus-host co-evolution did this event take place.

Additional chitinase CBD proteins (chitin-binding domains) encoded by the amphioxus genome are of antimicrobial substances. The other urochordate *Ciona intestinalis* possesses three VCBP3-like genes (*vide supra*) and releases some 40 CBDs. In the Cnidaria phylum (anemones, corals, jellyfish), *Nematostella vectensis* operates with 12 known CBDs. The genome of the Deuterostome, sea urchin (*Strongylocentrotus purpuratus*), encodes no CBD proteins [996a, b]. Prior to the full sequencing of the amphioxus genome, a sequence of a cDNA clone was found to be homologous with 45 % identity to the mammalian RAG1 gene activator. Seven transmembrane spans and two MtN domains (methylthioadenosine nucleosidase) characterized the amphioxus protein [1016]. The Rag2 protein (united with Rag1) was absent in the amphioxus, but showed up in the sea urchin [313a, b].

The Harbingers. Of the dormant, ancient transposons, Sleeping Beauty was the first to be reconstructed (Figure 62). The DNA superfamily of eukaryotic Harbinger autonomous transposons has been identified in protists, worms, insects and plants prior to Harbingers entering the vertebrates. Are these archaic DNA sequences retrovirally inserted proviruses? Harbinger transposons are responsible for the encoding of two protein superfamilies: specific transposases and DNA-binding proteins. The DNA-binding proteins carry the conserved SANT/myb/trihelix motif. The abbreviation SANT stands for the following elements: SWI3/SNF, switch, surcrose non-fermentable; ADA, adenosine deaminase activity; NcoR, nuclear receptor co-repressor; TFIIIB, transforming factor [311, 1017–1025]. The telomere-binding protein TRF2 carries the Myb/SANT domain as it reconstitutes telomeric nucleosomal array fibers (telomeric TTAGGG-repeat binding factor). The nucleosomal core consists of four pairs of basic histones in the formation of an octamer wrapped within superhelical dsDNA turns. The guanine-rich tandemly repeated 5′-TTAGGG-3′ telomeric sequences protect the chromosome ends from being recognized and processed as dsDNA breaks. Telomere length is maintained by trimethylated (3me) lysine (K) 9 of histone 3 (H3) and K20 of H4. The telomeric nucleoprotein complex is the shelterin. Dominant negative TRF2 expression would induce chromosome end-to-end fusions and reduced G-strand overhangs; this act would result in cellular senescence with apoptosis. The p53/ATM complex induces the apoptosis (ataxia telangiectasia mutated). Overexpressed (positive) TRF2 maintains telomeres in the cells, which do not undergo senescence. The TRF2 DNA-binding site Myb/SANT domain within the nucleosome interact with the ss 5′-d(TTAGGG)-3′oligonucleotide [1026]. The DNA-binding domain of the Chd1 chromatin-remodeling enzyme, a SNF protein, contains the SANT motif. The Chd1 and ISWI (imitating switch) enzymes generate regularly spaced arrays in the nucleosomes. The human ISWI homologue SNF2 acts in dimer formation on either side of the nucleosome, equalizing the lengths of linker DNAs. These most complicated processes entail histone methylation/de-methylation, acetylation/de-acetylation, phosphorylation/de-phosphorylation, ubiquitination/sumoylation (degradation of proteasomal substrate proteins by substrate recognition, ubiquitin-activating and conjugating enzymes; small ubiquitin-related modifiers), and ADP-ribosylation (histone adenosine diphosphate ribosylation, DNA repair, PARP, Table II).

The genomes including those of insect, fish, amphibians and vertebrates harbor long ago inserted SINE/LINE transposons (Figure 62). In the human nucleosome,

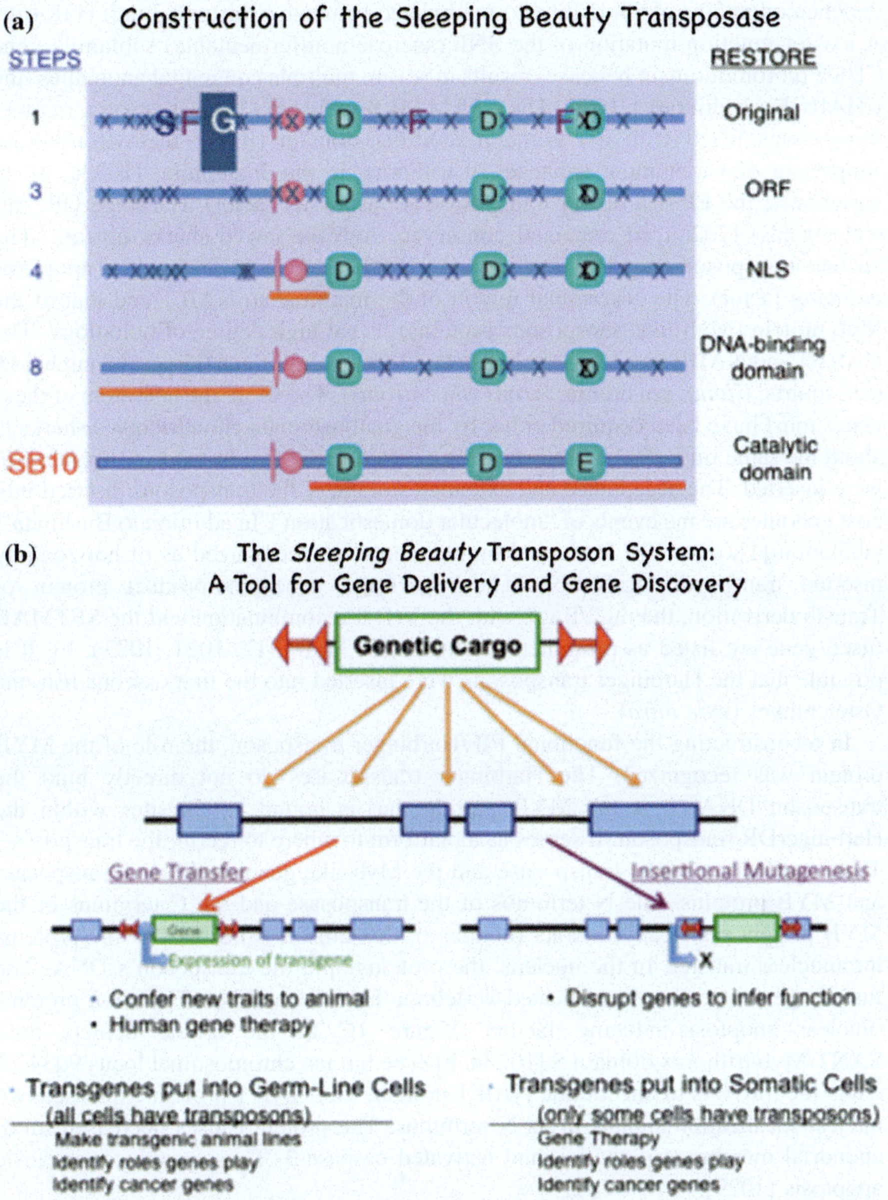

Figure 62 Zoltán Ivics and Zsuzsanna Izsvák Awaken the Sleeping Beauty. (SBTS3 http://commons.wikimedia.org/wiki/File:SBTS3.png by Perryhackett http://en.wikipedia.org/wiki/User:Perryhackett is licensed under CC BY 3.0 http://creativecommons.org/licenses/by/3.0/deed.en) Reference Appendix 2, Explanations to the Figures

the chemodomains of Chd1 bind to methylated lysine 4 of histone 3 tail: H3K4me. A loss-of-function mutation of the SNF (sucrose non-fermentable) subfamily gene CHD7 (chromodomain helicase) results in severe multiple congenital anomalies (the CHARGE syndrome) [1027]. The SETMAR protein is a derivative of a mariner transposons fused with a chromatin modifier domain (lysine methyltransferase suppressor of variegation, enhancer of trithorax, in the drosophila) [1023a, b]. In vertebrates, the PIF/Harbinger transposase (P instability factor), HarbingerDR, (direct repeats) [1028a, b] remained conserved from the jawed sharks upward. The ancient transposon-encoded proteins are HARB1 and NAIF1 (nuclear apoptosis inducing factor). The N-terminal region of the mammalian NAIF1 and that of the Myb protein (Myb-like transposons proteins) reveal high degree of homology. The HARB1 and NAIF1 proteins are not detectable in cyclostomata (lamprey) and below (sea squirts, *Ciona*; sea urchin, *Strongylocentrotus*). Therefore the ancestors of these genes must have been acquired either by the gnathostomata chondrichtyes sharks at about the same time (500 mya), when the genes for adaptive immunity (*rag1/2, vdj*) were inserted. The acceptance and extended storage of the transposons inserted into host genomes are the events of "molecular domestication". In addition to Bushman's tabulation [1], over 100 functional human genes are recognized as of horizontally inserted transposons-derivation. The *rag*1/Rag1 gene/gene-product protein of Transib derivation, the rag2/Rag2 with the V(D)J recombination and the SETMAR fused gene are listed as prominent examples [313a, b, 317, 1021, 1023a, b]. It is possible that the Harbinger transposons were inserted into the first osseous fish, the Osteichthyes (*vide infra*).

In reconstructing the functional PIF/Harbinger transposon, the role of the MYB protein was recognized. The Harbinger transposases do not directly bind the transposon DNA. It is the MYB protein that is bound to six sites within the HarbingerDR transposon. It serves as a platform to where to recruit the transposase. There are fusions of the transposase and the Myb-like genes and fused transposase and MYB proteins. The N terminus of the transposase and the C terminus of the MYB protein physically interact (bondage); this unison is needed for the proteins intranuclear transfer. In the nucleus, the proteins bind the transposon's DNA. The fusions generated the domesticated vertebrate HARB1 and NAIF1 fusion proteins (nuclear apoptosis-inducing factor). Figure 1C of the article depicts their SANT/Myb/trithorax domain S [1023a, b]. The human chromosomal locus 9q34.11 emits the mRNAs to encode the NAIF1 protein. The NAIF1 protein expresses two nuclear localization signals in its N terminus. The protein causes decreased mitochondrial membrane potential, and activated caspase-3. Thus, it induces intrinsic apoptosis [1029].

The *harbi*1 gene of Harbinger transposon-derivation (*LatiHarb*1) appeared first in the ancestral bony fish (Osteichthyes), and remains well conserved in the genome of coelacanths (Latimeria) (Figure 5). It is recognizable, that large families of SINEs in the coelacanth genome are the predecessors of functional genes later on in the vertebrate lineages. All originate from the canonical transposase referred to as the raw material myb-like gene, MYB-like DNA-binding protein [1030] (in the Appendix 1). MYB DNA-binding proteins are abundant in plants; the plant MYBs possess two motifs (R2, R3). The MYB and helix-loop-helix (HLH) protein transcription factors regulate anthocyanin genes that color flowers, fruits and vegetables [1031]. MYBs of animals have three (R1, R2, R3) DNA-binding domains.

The original *myb* gene was acquired by the avian myeloblastosis retrovirus from its host (c-*onc* → v-*onc*). This author (JGS) had the good opportunity to observe the abundance of avian myeloblastosis viral progenies in infected tissues in 1959 in the laboratory of Leon Dmochowski at M.D. Anderson Hospital, Houston, TX, and the extraordinary virulence of this virus in 1962 in the laboratory of Joseph W. Beard, Dorothy Beard and Ursula Heine at Dukes University (Durham, NC) [1032–1034a, b].

In mammalian embryonic life, the Myb proteins direct hematogenesis and spermatogenesis. In the human genome, the myeloblastosis (MYB) protein is encoded from locus 6q23. The point mutated or *mll*-fused (mixed lineage leukemia), or GATA1-fused (guanine-adenine-thymine-adenine) *myb* gene t(X;6)(p11; q23) is a leukemogenic oncogene [1035, 1036], often appearing in G4 DNA quadruplex formations. The ancient GATA1 gene is present in the X chromosome of the platypus. The G4-quadruplex ligand, telomestatin, is an inducer of apoptosis of glioblastoma, or mutated c-*myb*-induced stem cells, by targeting G4 DNA quadruplex formations within the malignant cells [1037]. The *T. vaginalis* genome encodes a very large number (>400) Myb proteins expressing DNA-binding domains and functioning as transcription factors within the nucleus. A < 40 kD sized protein may pass through the pores of the nuclear membrane. The Myb2 transcription factor, embedded in its DNA-binding domain (DBD, size aa 40–156), carries the organized structure nuclear localization domain (NLD, aa 48–143), that mediates Myb2's nuclear localization. The *T. vaginalis,* and the human Myb DBD share structural modules [1038] (in the Appendix 1).

Selected Abbreviations

RBP-J recombination signal binding protein for immunoglobulin kappa J region. Contributes to transactivation of Epstein-Barr virus genome to exert anti-apoptotic effect within host cell (Pegman PM et al J Virol 2006;80:8133–44). For Notch and RBP-J co-signaling for chemokine and receptor CXCR4 activation in order to induce bone marrow mesenchymal stem cells migration, click Wang YC et al Stem Cell Res 2013;11:721–35; Xie J et al Cell Immunol 2013;28:68–75. Notch/TGFβ/RBP-J (transforming growth factor) maintain differentiated prostatic basal stem cell lineage. Jagged ligand-induced Notch intracellular domain protein translocates into nucleus to bind RBP-J. Disrupting Notch induces basal stem cell proliferation (due to release of RBJP-J ?). In absence of Notch, TGFβ maintains basal cell quiescence. In order to malignantly transform, proliferating basal cells transdifferentiate into luminal cells. Luminal cells are naturally proliferative under the effect of Notch.

The Wnt pathway is suppressive to Notch. For deciphering who is the oncogene, and who is the suppressor, click Valdez et al Cell Stem Cell 2012;11:676–88. For failure of RBP-$J^{-/-}$ dendritic cells to respond to TLR T lymphocyte receptors, click Chen YR et al Mol Biol Rep 2013;40:1531–9.

Viral and Cellular Proteins Interact

Viral and Host Cell Genomes Become Intimate

Viral chromatin-interacting proteins. The genomes of dsDNA viruses encode host cell chromatin-interacting proteins [1039]. Examples are the HPV oncoproteins E6 and E7 (targeting cellular p53 and Rb, respectively), the SET-domain methyltransferases of large nucleocytoplasmic DNA viruses (mimivirus, Marseillevirus, Lausannevirus) [1040–1042], and the BEN domains of poxviruses (printed in Figure 1 of cited article: "Multiple sequence alignment of the BEN domain") [1043].

Papovaviruses and adenoviruses express proteins (E1A) with the LXCXE motif (leu-X-cys-X-glu), which fit into a pocket of the Rb protein, where they interact with the endogenous histone deacetylase-binding cleft. The herpesviral UL97 kinases phosphorylate the Rb protein, thus disrupting Rb interaction with HDAC and E2F, and driving the cell cycle through S-phase. Genes for the viral proteins with LXCXE module of dsDNA viruses appear to have been laterally transferred and shared. In polyomaviruses, the LXCXE motif is fused with the T antigen. Other dsDNA viruses express host DNA-binding proteins (adenoviruses, E2A; bacteriophages, helix-extension helix, HEH). The herpesvirus latency-associated nuclear antigen (LANA) is homologous and functionally the same as the HPV E2 protein. The polydnavirus BEN motif is a host cell chromatin modifier; for their own chromosomes, these viruses use proteins different from their host cell chromosome-modifying proteins. The polydnaviral histone methylases and demethylases are different from those of the host cell. The viral and cellular BEN domains BANP/SMAR1 (B cell translocation gene-associated nuclear protein; scaffold matrix attachment region) are chromatin proteins. Further viral chromatin proteins are the vaccinia virus poxvirus e5R, protein gene/protein NAC1, nucleus accumbens; BTB/POZ gene family domain, bric-à-brac (miscellaneous); broad complex tamtrack; and poxvirus zink fingers. The SMARs interact with histone deacetylases. The chordopoxviral (vacciniavirus early protein E5R) and the

J.G. Sinkovics, *RNA/DNA and Cancer*, DOI 10.1007/978-3-319-22279-0_16

polydnaviral SMAR1 proteins are different, but both work in the virosomes. The "new" molluscum contagiosum E5R gene/protein may be of host cell derivation. In contrast, the wasp genome may contain a virally inserted BEN domain gene. The papillomavirus E6 oncoprotein recruits an ubiquitin ligase for the degradation of the p53 protein. The RNA-recognition motif-like fold (RRM) is expressed in the E2 papillomavirus, and in the LANA/EBNA herpesvirus DNA-binding proteins. These proteins tether viral DNA to host chromosomes in mitosis. The herpesvirally infected cell protein 8 (ICP) expresses an OB-fold (oligonucleotide-oligosaccharide binding) DNA-binding domain. The ostreid class amphioxus herpesvirus (*vide supra*) was associated with its host cell's chromosomes by histones of host cell derivation. However, the tails of the virus DNA-serving host cell histones, diverged from those of the host cell configurations; reviewed in deSouza et al [1039].

The ABC ATPase SMC superfamily proteins (ATP-binding cassette; adenosine triphosphatase; structural maintenance chromosome) complexed with nuclease heterodimers Rad50/Mre11 (radiosensitive; meiotic recombinase). The *ruv*A/B/C genes and gene product proteins RUV-A/B/C (UVR, UV damage sensitive repair; Holiday junction migration and replication fork reversal, error-prone Holiday junction helicase-resolvase, RuvA sliding collar preventing unwinding of DNA) are preserved in all living cells from the last universal common ancestor upward. The genomic relationship of these proteins, frequently exchanged by lateral (horizontal) routes, extend from phages to the mimivirus [1044, 1045], through the archaea, *Pyrococcus furiosus* [1046] and eubacteria, *E. coli, B. subtilis, Deinococcus radiodurans* [1047–1053], to all eukaryotes upward.

The signal transduction ATPases (STAND) include gene response regulators characterized by the fusion of the terminal helical bundle to the NTPase (nucleoside triphosphatase) domain [1054]. The protein domains SET (Su(var) suppressor of variegation; enhancer of zeste homolog 2 polycomb gene (EZH2) are histone methyltransferases (targeting lysines, K) [1055, 1056]. The first SET dimer proteins were generated in eukaryotes and from there spread to bacteria by lateral transfer; reviewed by Aravind *et al* [1057]. The *Paramecium chlorella* phycodnavirus aimes its SET domain at H3K27me of its host [1058]. Study of the original articles is recommended.

Oncogenome teacher: acute promyelocytic leukemia. The herpesviral-infected cell's ring finger proteins (ICP) work as ubiquitin ligases and attack the PML (promyelocytic leukemia) protein ("PML nuclear bodies"). Even though the PML gene acts as a tumor suppressor, the fused PML-RARα gene is potentially oncogenic. Thus, could attenuated herpesviruses be oncolytic to PML cells? The PML-RARα (retinoic acid receptor, 15q21 & 17q21) fused oncogene/oncoprotein t (15;17) generates the fibrinolytic enzyme plasminogen → plasmin [1059]. In the fusion oncoprotein, the PML protein loses its tumor suppressive properties (inhibition of cell divisions, cessation of tumor cell migration and neoangiogenesis, and induction of replicative senescence, or apoptosis). Tumor cells frequently succeed in degrading the PML protein by ubiquitination [1060]. In contrast, IFNγ up-regulates PML expression and synergizes with all-trans-retinoic acid (ATTA) administration [1061]. Granulocyte colony stimulating factor (G-CSF) induces

PML cell differentiation (in synergism with arsenic trioxide and/or all-trans retinoic acid [1062]. Arsenic trioxide (A_2O_3, ATO) suppresses survivin, c-Myc, telomerase (telomeres remain short), and Pin1 transcription factor-induced NFκB (peptidyl-prolyl isomerase) [1063]. In a promyelocytic variant, the leukemia zinc finger proteins (PLZF) fuse with the RARα gene: t(11;17)(q23;q21) (*vide supra*). The PLZF protein has been evolutionarily highly conserved. It is related to the poxvirus zink finger domain (POZ), and the drosophila BTB complex domain (bric-à-brac tamtrack broad complex) (*vide supra*). The C-terminus of the protein contains the nine Krüppel-like zink C_2H_2 fingers for DNA binding. Krüppel-like proteins are encoded by stem cell genes; thus, PLZF may be an inhibitor of cell differentiation. Some promoters of the genes induced by IFNs (STAT/JAK) express PLZF binding sites. NK cells require PLZF expression for cytotoxic reactions (including granzyme generation) to MHC class I-deficient targets. PLZF-deficient iNKT cells fail in IL-4 and IFNγ production. The PLZF-RARα, and the BMi-RARα (B cell Moloney virus insertion site) oncoproteins may be resistant to all-transretinoic acid treatment.

Normal uterine cervical cells express high levels of the PLZF and CCS3 (cervical cancer suppressor, eukaryotic translation elongation factor 1α) proteins, which are absent in cancer cells. The CCS3 protein is an inhibitor of the PLZF-target genes, cyclin A2 and Rho (ras oncogene homolog). Androgen-independent prostate cancer cells (cell line DU145) abolished PLZF expression, and maintained PBX1 overexpression (pre-B-cell leukemia transcription factor). Forced re-expression of PLZF protein in these cells suppressed PBX1 expression, halted cell proliferation and induced cell death. However, PLZF protein failed to down-regulate the growth rate of androgene-independent prostate cancer cells (cell line LNCaP); reviewed in Suliman *et al* [1064].

Comment (paleovirology). When, in the pre-cellular Virus World (*vide supra*), the first a-nucleated protocells appeared, pre-existing viral agents invaded those cells. An extraordinary exchange of viral and cellular genetic elements (the "ping-pong match" in Niller's *et al* words with the genes being the ping pong balls) was initiated and concluded by the cells gaining nuclei and the viruses acquiring those genes that allow them to parasitize the cells (Figure 61). The original *v* and *rag*1 elements were deposited in the ancestors of the amphioxus by retroelements. The predominant view maintains that these elements with *rag*2 added later were further transferred either (or both) by vertical or horizontal transmissions to the ancestral cartilaginous fishes (Placoderms and gnathostomata Chondrichthyes).

A divergent view now emerges claiming herpesviral (ancestral EBV) transmission of the basic elements of adaptive immunity, but without being able to pinpoint the exact evolutionary stage of the virus and its host, when the herpesviral acquisition of the *v(d)j* co-linear, and *rag*1/2 sequences must have occurred. For that herpesviral acquisition of these elements, either herpesviral genomic insertion into, or retroviral-herpesviral co-infection of, the donor cells must be postulated. If Alloherpesviridae, but especially Malacoherpesviridae, already containted any of the *v(d)j* or *rag*1 elements, or retrotransposon-like ancestors thereof, one could possibly postulate that these elements were acquired through horizontal gene transfers at the level of the amphioxus, traveled with the virus throughout the evolutionary ladder until the Placoderms and cartilaginous fishes (gnathostomata Chondrichtyes), where these genomic sequences were inserted into the genomes of those new hosts. From there, these genomic sequences were transferred trough vertical routes, during which time they underwent duplications, deletions and additions, into the evolutionary descendents of the original hosts (amphibian, reptilian, dinosaur and aves, mammalians of dry lands and the oceans).

However, if the late-comer herpesviruses of bony fishes (zebrafish; cyprinids and above) exhibit for the first time the possession of these structures (DNA sequences encoding adaptive immune faculties), then the facts support the original retrotransposons-mediated transfer of these sequences [1021]. The herpesviruses were not the donors, but the late recipients of these genetic elements, acquiring them from the genomes of their vertebrate hosts, for whatever advantage these virions might have derived from this acquisition. In this case, it was retroviral elements, which implanted the ancestral genes into the protochordates and the adaptive immune faculties were forwarded from that level on through vertical (or horizontal) routes into the evolutionary descendants of those ancestors.

Somatic hypermutation generates new fused genes for the production of specific antibodies and TCRs. The hitherto dispersed gene segments now uniting may bring together retroviral proviral sequences, even an *env* gene. These events may coincide with misplaced nicks by AID (fusing the genes c-*myc,* or *bcl*-2, instead of new light chain gene, with the IgH gene). The result is the frequent appearance of budding endogenous retroviral particles in lymphoma cells, including the RS cells of Hodgkin's disease; reviewed in [147]. Would this event be "oncogenesis (lymphomagenesis) by mistake" in the system installed as adaptive immunity? Or is the host, while defending itself by antibody production against an infection, automatically signaling the RNA/DNA complex (the LTRs thereof) to transform the threatened mortal cells into an immortalized population?

Light Cast to Broken DNA, Its Repair, and Consequences

The DNA Sliding Clamp

PCNA is an ancient DNA-proliferator. Referred to as "DNA sliding clamp", the proliferating cell nuclear antigen (PCNA) is present in the a-nucleate Haloarchaeal (phylum *Euryarchaeota*) genomes [1065]. How it appeared there first is not clear at all. *Halobacteriales* (Haloarchaea) received and accepted laterally transferred genes even from eubacteria. These genes underwent duplications and lateral transfers within the 29 genera of the halobacterial clade. Another enzyme, the basic leucyl-tRNA synthetase is also an ancient acquisition of Euryarchaeota origin. Thus, at least one cluster of this enzyme was received from outside source and it did not evolve dependently of the halobacterial tRNA [1066]. Crenarchaeota possess another distinct PCNA. The chromatine-bond PCNA in the form of homotrimeric rings encircles DNA double helix, serving as platform for other recruited proteins, especially the auxiliary factor for DNA polymerase (DNA polymerase δ processivity factor). The replisomes of replicons form clusters known as replication factories, or foci. These foci contain the DNA polymerases, their anchor, the PCNA, replication factor C, replication protein A, DNA ligases, and cyclin A. PCNA acts in cell cycle control, and in the processes of DNA elongation, recombination, methylation and damage repair. Viral DNA replicating in host cell cytoplasm (vaccinia virus) or nucleus (polyoma virus) also forms large replication factories ("virus factories"). The PCNA is a homotrimer DNA-sliding clamp, that interacts with (it recruits) the DNA ligase, that also encircles the DNA nick (the DNA-binding domain; DNA replication foci) [1067–1071].

In unicellular eukaryotes (kinetoplastids), two distinct PCNA are operational. *Toxoplasma gondii* separates an exclusively intranuclear, and a both intranuclear and a cytoplasmic PCNA; in the S phase, both PCNAs serve inside the nucleus. In *Plasmodium falciparum*, one PCNA interacts with the original recognition complex (ORCs) of *S. cerevisiae* homologue, and co-localizes with it within the nucleus [1072]. The *Trypanosoma brucei* pre-RC protein is a Cdc45, which interacts with a

J.G. Sinkovics, *RNA/DNA and Cancer*, DOI 10.1007/978-3-319-22279-0_17

pre-BRCA1 protein (breast cancer sensitivity). Even closely related species (*T. brucei*; *T. cruzi*) distribute their PCNA granules within the nucleus in the S phase of the cell cycle differently: diffusely *versus* peripherally. The PCNA granules are degraded in the *T. brucei* nucleus in the G2 phase of the cell cycle [1073]. The conserved residues between the yeast and human PCNA (located in chromosome 13) are only 34 % [1074]. In *Leishmania major*, so far only one PCNA orthologue was identified, sharing close to 100 % identity with PCNAs of other leishmania species, but 75 % identity with trypanosoma PCNAs. In *L. donovani*, the actively proliferating promastigotes in their replication factories express PCNA in the G1 and S cycles of the pro- and metacycle [1075]. Leishmania exist as flagellate promastigotes in their insect hosts, where in the salivary glands they become metacyclic infectious forms. In the macrophages in the blood stream of the mammalian host, the Leishmania promastigotes transform into a-flagellate amastigotes, which reproduce a-sexually by binary fission. These forms express ORC1 and MCM proteins (origin replication complex; microchromosome maintenance, from yeast, *vide infra*). The intranuclear phosphorylated MCM4 protein interacts with PCNA, and they drive a fast G2/M cell cycle of the promastigotes. The role of MCM2-7 proteins in DNA replication and fork elongation is that of initiation and maintenance [1076]. Polymerase δ and PCNA associate through the PCNA interacting protein (PIP) [1077].

Potential oncoproteins other than PCNA are the proto-oncogene products Aurora kinase and Polo-like kinase. Both proteins serve as cell survival pathways in unicellular life forms, like in the Trypanosoma [1078, 1079]. In the drosophila, Aurora proteins are involved in the formation of the mitotic spindle, centrosome separation and in the G2 to M transition of the cell cycle [1080]. The Aurora and Polo kinases become full-fledged oncogenes/oncoproteins in multicellular vertebrate mammalian hosts. Examples: Polo-kinase oncogenesis in gastric or prostatic carcinomas [1081–1084]. Aurora-kinase oncogenesis is carried out by the phosphorylation and activation of the mTORC (mammalian target of rapamycin complex) and Polo-kinase pathways [1080]. The oncogenic Aurora-C kinase is point mutated at T191D (threonin to aspartic acid) and is constitutively activated (shown first in NIH-3T3 cells) [1085]. Aurora ABC are overexpressed in human tumors. Aurora gene activators are the c-Myc and the Bub1 proteins (human ortholog of the yeast *bub*1 gene (*vide infra*, budding uninhibited by benzimidazoles) [1086, 1087a, b].

The genome is protected by RNAase H1/2 against misincorporation of ribonucleotide triphosphates (rNTP) in place of deoxyribonucleoside triphosphates (dNTPs). Yeast cells deficient in RNAase H accumulate ubiquitylated PCNA and keep their postreplication repair pathway constitutively activated. RNAase H cleaves off the RNA moiety from RNA: DNA hybrids allowing the reconstruction of dsDNA molecules. When topoisomerase I cleaves off the RNA moiety, no dsDNAs are formed and the ssDNA suffers breaks. Yeast cells can survive genomic rNTP incorporation. Replication stress ensues upon the encounter of replicative DNA polymerase with rNMP (mono) in the template strand. Replicative polymerase cannot effectively replicate rNMP-containing templates. Replication forks stall at these sites. The genomes of cells in chronic postreplication repair pathway require replacement of error-prone translesional DNA synthesis with error-free template switch. Pol δ stimulated by PCNA bypasses without repair four consecutive rNMPs. Mono- and polyubiquitylated forms of PCNA abound in cells with unrepaired rNMP insertions. In replication-blocking lesions, PCNA is ubiquitylated; Rad18 (radiation sensitive error-prone ubiquitin ligase gene) encodes the ubiquitin ligase for conjugating PCNA. Even in RNAase-deficient cells, postreplication repair by polymerase ζ and template switch can effectively bypass rNMP in the DNA template. Poly ζ can insert a nucleotide opposite the lesion (adding dNTP opposite the lesion) [1088].

Dictyostelium (Figure 6) cells operate DNA repair enzymes ancestral to those of vertebrates. These are the non-homologous end-joining (NHEJ) kinases recruited by KU (Kansas University) proteins to sites of dsDNA breaks. During DNA replication, dictyostelium cells showed diffuse nuclear distribution of GFP-PCNA (green fluorescent). Dictyostelium cells express ATM/ETR kinases (ataxia telangiectasia mutated; ELAV-type RNA-binding protein; embryonic lethal abnormal visual protein, drosophila). These enzymes phosphorylate histone H2AX at sites of DNS breaks (used as a signal of DNA repair sites); but are inhibited by caffeine. Bleomycin-treated dictyostelium cells may recover dsDNA breaks, but caffeine and bleomycin-treated cells suffer mitotic catastrophe; GFP-PCNA recruited to sites of dsDNA breaks failed to conclude DNA repair [1089]. Changes (mutations; deletions) of PIP (phosphatydilinositol) resulted in flexible PCNA partner selection in fungi in the last 300 my and dictated a bidirectional divergence from the common ancestor [1090].

The phytoplanktons *Dunaliella* and *Isochrysis* doubled their PCNA protein abundance in darkness, when G2 and M phases dominate, than in light, which is the rest phase of their life cycle [1091]. The dinoflagellate *Pfiesteria piscicida* possesses PCNA protein of 258 aa residues with the untranslated 5′ and 3″ ends. The *P. piscicida* cells usually carry 41–48 copies of their PCNA gene, but the gene copies almost double in replicating cells [1092]. In the sea urchin embryo, green fluorescent protein-labeled GFP-PCNA accumulates in the nucleus during DNA synthesis and cell cycle progression to M phase. However, if the active MAP kinase ERK1 (mitogen-activated protein; extracellular receptor kinase) is inhibited by recombinant ERK phosphatase X(enopus)CL-100, or ERK inhibitor U9126, or DNA polymerase inhibitor aphidicolin, is administered, GFP-PCNA disappears from the zygote nucleus, incorporation of [3H]-thymidine into DNA ceases to occur, and chromatin decondensation does not take place. In contrast, the ERK activator, Ca^{2+} ionophore A23187 induces GFP-PCNA accumulation in the nucleus and [3H]-thymidine incorporation into the DNA [1093].

The central nervous system of the ascidian tadpole larvae consist of 370 cells with PCNA proteins in the nuclei. An antisense oligonucleotide can inhibit the PCNA mRNA. Inhibition of PCNA resulted in deformed head development with cessation of DNA synthesis and nuclear DNA fragmentation resembling programmed cell death [1094]. The drosophila or mosquito dacapo is a Cip/Kip family cell cycle inhibitory protein of 261aa residues (cycline-dependent kinase inhibitory protein; kinase inhibitory protein). The serine-threo/any aa/glutamate/aspartic acid motifs are the protein kinase phosphorylation sites and/or binding site for PCNA) [1095]. The drosophila operates two PCNAs: DmPCNA2 is localized in the nuclei as a homotrimeric complex and associates with Dm polymerases δ/ζ. DmPCNA2 is the sliding clamp for DNA repair [1096]. In caenorhabditis cells (also in yeast and mammals) homolog protein Rad17 (*hpr*-17 gene product protein, replication factor C, in caenorhabditis) loads the sliding clamp PCNA to sites of DNA damage (and to telomere ends, for telomere maintenance pathway, in caenorhabditis) [1097] (Figure 39).

The human PCNA was discovered as an autoantigen in patients with SLE [1098]. The gene for the human PCNA maps to 20p13; the related genes at 11p15.1 and Xp11.4 are probable pseudogenes. A total of four human PCNA pseudogenes could be found; one pseudogene is mapped to Xq13 [1099–1101]. Post-replication DNA repair involves mono-, or polyubiquitination of PCNA catalyzed by Rads 6-1. Monoubiquitination stimulates translesional DNA synthesis (TDS); polyubiquitination promotes template switching (*vide supra*). For template switching, among other factors, Rad18, the E3 ligase human helicase-like transcription factor (HLTF) is recruited.

In human breast cancer, the PCNA protein phosphorylated at tyrosine 211 (Y211) associates with the non-receptor tyrosine kinase proto-oncoprotein Abl (discovered in the Abelson murine leukemogenic retrovirus: c-*abl* → v-*abl*). The ABL protein's SH2 domain aa 160 to 163 with aa motif QXXI (glutamine-any aa-isoleucine) is the attachment site (and/or another attachment site is ARSA: alanine-arginine-serine-alanine) (He *et al* cited by Zhao *et al*) [1102]. The EGF-R phosphorylates into activity the PCNA protein. The phosphorylated PCNA protein exerts enhanced activity in DNA replication. In the BCR/ABL translocation, this fusion oncoprotein causes CML (*vide supra*). In human breast cancer cells, the Abl oncoprotein contributes to the transformed phenotype, anchorage-independent growth, inhibition of apoptosis and accelerated cell proliferation. The survival of patients with breast cancer worsens if they carry phosphorylated PCNA [1102]. Chromatin-bound PCNA is frequently phosphorylated at Y211 in advanced hormone-dependent and hormone-independent prostate cancer. Normal-appearing epithelial cells next to cancerous cell colonies already could contain Y211-phosphorylated PCNA molecules.

The cell-permeable synthetic peptide in which tyrosine (Y) is replaced by phenylalanine F (Y211F) inhibits phosphorylation of the native PCNA protein. In human prostate cancer cell line xenografts, treatment with the Y211F synthetic peptide resulted in the inhibition of DNA synthesis, S phase arrest, and reduced tumor growth rate. The non-phosphorylated native PCNA proteins are expected to escape inhibition by the synthetic peptide [1102]. In both breast and prostate cancer cells, the MEK (MAPK/ERK) pathway is highly and constitutively active; and its inhibition arrests tumor cell growth [1103, 1104]. In endometriosis, the chemokine ligand CCL2 of stromal cell origin increases the expression of PCNA together with that of the anti-apoptotic survivin and activates the MAPK/ERK/Akt pathway [1105]. High-grade malignant gliomas express much higher levels of PCNA proteins than low-grade gliomas [1106]. These tumors are promoted by the overactive

MAPK/ERK pathway [1107], but the connection, if any, betweeen PCNA and ERK remains to be investigated.

In the sea urchin embryo, the ERK pathway drives PCNA activity, and ERK inhibitors switch off PCNA-driven DNA replication [1108] (*vide supra*). In human cells, as well as in other eukaryotes, small ubiquitin-like modifiers (SUMO) sumoylate PCNA at multiple sites. PCNA-SUMO1 fusion prevents dsDNA breaks and recombinations due replication fork collapse at stalling DNA lesion sites [1109].

The mammalian cancer cell may take use of the ancient physiological cell survival pathway that was established in the Cambrian sea 500 million years ago (or even before). Indeed, ERK and PCNA interaction exists in human hepatocellular carcinoma cell lines. Cardiotonic steroids ouabain and cinobufagin inhibit the phosphorylation of ERK. A cascade of the reaction followed: downregulation of PCNA and c-Myc, S phase cell cycle arrest due to the inhibition of cyclin A and cyclin-dependent kinase 2, and upregulation of cyclin-dependent kinase inhibitor 1A (p21, Cip1) [1110]. The natural cytotoxicity receptor of primate NK cells is known as NKp44/ITIM (immunoreceptor tyrosine-based inhibitory motif). This NK cell receptor recognizes PCNA protein-expressor cells. Priority given to DNA replication, subverted the NK cell in that, the immunological synapse between PCNA and NKp44 signals "self" "negative" and "no reaction". The PCNA over-expressing cancer cell expropriated this reaction for its own benefit, and escapes NK cell-mediated cytotoxicity [1111].

The poly-adenosine diphosphate-ribose polymerase (PARP) (Table II) was essential for DNA repair in ancient fungi and remains in use in Aspergillus and Neurospora species. The PARP protein interacts with the entire protein metabolism (catalytic activities) of the cell in a stimulatory fashion. Haploid fungal cells with defective PARP die; diploid cells survive, but undergo increased mutational events. In one case, loss-of-function phosphatases fail to deactivate MAPK, resulting in hyphal hyperprolifreation (Figure 63). Human cancer cells use PARP to initiates

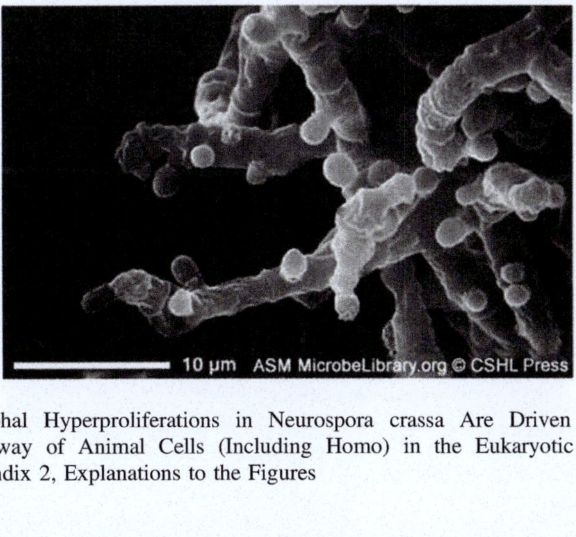

Figure 63 Hyphal Hyperproliferations in Neurospora crassa Are Driven by MAPK: an Oncogenic Pathway of Animal Cells (Including Homo) in the Eukaryotic Fungus. [2359] Reference Appendix 2, Explanations to the Figures

repair of their DNA breaks. PARP inhibitors render cells highly susceptible to additional DNA damaging agents. Synthetic lethality due to mitotic catastrophe sets in (Table X).

In some cells DNA replicates better in light than in dark. Cyanobacteria, Chlamydomyces and Volvocales first, and unicellular fungal eukaryotes (*Neurospora crassa*) thereafter rendered their life style light-dependent (Figure 63). The ancient circadian rhythm remains preserved in the vertebrate mammalian pineal gland. In the tumor, human pinealoma, descendents of Neurospora genes remain functional (see in Appendix 1).

Comment. A key actor of DNA replication (PCNA) appeared at the dawn of life in the archaea (before the cell nucleus developed, or was acquired), and then in the unicellular eukaryotes to sustain their DNA replication (error-prone, or error-free). Its pathway constitutively reappears in the "malignantly transformed cells" of the mammalian hosts, providing them with an irreversible stimulus to sustain their life, disorganized as it may be, but nevertheless, immortalized.

The light-sensitive cell survival pathways of some ancestral fungal eukaryota (Neurospora) remain conserved in vertebrate mammals and are found to be functional in the human pineal gland without or with its immortalized transformation (*vide infra*).

Undisciplined Introns

Debates at the Highest Level

Intron-poor and introns-rich genomes in the descendants of the earliest cellular life forms. Group I introns are ribozymes. Group II introns are self-splicing mobile ribozymes spanning in the proto-organelles from archaea/prokaryota/eubacteria to eukaryota. They may encode proteins for their own folding into secondary or tertiary structures. Spliceosomal nuclear introns presumably derive from group II introns of the mitochondrial/chloroplastic endosymbionts (*vide infra*). These introns are not present in a-nucleate archaeal or prokaryotic genomes, but are eukaryotic nuclear-genomically encoded pre-mRNA sequences (in the spliceosomes). In the gene, they reside between protein-encoding DNA segments (exons). The spliceosomes catalyze the removal of these introns from the mRNAs. Extranuclear (cytoplasmic) processing of group II intron-containing transcripts undergo transitional repression and nonsense-mediated decay (NMD). Group I introns, or nuclear spliceosomal introns are processed without NMD, or translational repression [1112]. Introns vary not only in numbers from genera to genera, but also in size from twenty nt (in the nucleomorphs) to kilobase sizes in some eukaryote metazoans (*Homo*). The common consensus is 5′ GT at one end ("donor" site), and 3′ AG at the other end ("acceptor" site). The branch point is somewhere within the intron's sequences.

The reason for the evolutionary diversities between intron sizes, positions, and numbers remains very much unexplained. Some genomes with the shortest individual introns are intron-rich (*Paramecium tetraurelia*). In some genomes with very low intron densities, one particular chromosome may exhibit an intron-rich segment (in chromosome 2 of the ultra-small green alga *Ostreococcus lucimarinus*). The dinoflagellate nuclei extracted most of the genes from their organelle plastids. A great deal of alternative and transplicing is going on (differential splice site pairing removes introns and joins exons to encode different proteins, thus generating protein diversity). The dinoflagellate *Amphidinium carterae* has many

J.G. Sinkovics, *RNA/DNA and Cancer*, DOI 10.1007/978-3-319-22279-0_18

intron-free genes and four intron-rich genes; they lack the canonical polyadenylation AAUA signals. The highly expressed genes displayed a mini polyA region and a spliced leader motif. They were in the tandem repeat category with short intergenic regions at both 5′ and 3′ ends. The less expressed, or silenced genes were intron-rich. The polyketide synthase gene harbored 18 introns; the *psbO* gene (photosynthetic light-harvesting plastid-associated oxygen-evolving enhancer) has 9 introns, but at many intron-exon boundaries the customary GT−AB splicing sites were lacking [1113]. The spliceosomal introns divide the coding exons, but are removed from the mRNAs in the spliceosomes (Table XIII).

The guardians of the most ancient canonical eukaryotic introns are the deep branching giardia, trichomonas and amoeba (Figure 31); for a newer version of this class, visit Adl et al [1114]. The parasitic giardia reduced its genome size containing four short introns. It was recognized early that the giardia ferredoxin gene harbors CT-AG introns. Giardia operates spliceosomes trans-splicing its four split-introns in the mRNAs issued from two of its genes: a heat shock protein gene and a flagellar motor dynein protein (sliding motor proteins) gene. Some giardia protein-coding genes are split and transcribed in separate fragments. The transcripts are assembled post-transcriptionally [1115]. The cartoon of Kamikawa et al [1116a, b] depicts the intron between exon 1 and exon 2, its pre-mRNA, its transit through the spliceosome, and its mature mRNA in the cis-splicing process. In the trans-splicing process, there is a left splintron piece, and a right splintron piece between exon 1 and exon 2. On the left side, 5′poly(A)3′, and on the right side, 3′ poly(A)5′ are positioned. The splintrons pass through the ribonucleoprotein complex spliceosomes to become from pre-mRNAs to mature mRNAs, with a non-canonical splice site at their 5′ end (CT), and a canonical splice site at their 3′ end (AG) [1115–1117]. The giardia may not have a Golgi apparatus, but the ribosomal protein L7 gene (Rp17a) is shared by a diplomonad (giardia) and a parabasalid (trichomonas). It is the highly conserved 12-nt 3′ splice-site motif that is shared. The descendants of the ancient trichomonad intron can be recognized in the same positions in the genomes of yeast and metazoa [391, 1118]. This intron is only 35 nt in length and its 5′ intron boundary sequence is CT [1119]. The noncoding RNAs are either small nuclear, or small nucleolar RNAs (snRNA; snoRNA). The snoRNAs modify specific nucleotides in rRNA and target pre-mRNAS for cleavage. The removal of introns from the pre-mRNA occurs in the RNP complex spliceosome: spliceosome-mediated intron recognition and excision. The functionally important sequences in the giardia spliceosomal snRNAs are conserved strictly; but different giardia strains exhibit mutational differences in their secondary spliceosomal structures. Figures 4 and 5 in the cited article [1120] show the divergent structures in the *G. lamblia* spliceosomes. As in all eukaryotes, the snRNAs contain the 5′ $3CH_3$-guanosine caps. The giardia U1 snRNA candidate SLII sequence CGCAUAC is closely resembles the U1 binding site sequence of *S. cerevisiae* U1 snRNA: CACAUAC. This sequence is also related to the U1 snRNA sequence UGCAUAU of *T. vaginalis*. The giardia U4 snRNA 5′ terminal is devoid of the Sm protein binding site; this site is also absent in the yeasts (*Saccharomyces cerevisiae*; *Candida albicans*). Seven heptameric ring Sm proteins form the core of

Table XIII The Introns of LECA. The Mauriceville Plasmid. Ancient LTRs Return as Oncogene Activators

The earliest ancestral eukaryotes presumably possessed intron-dominated genomes[1]; that scenario presumably applies to LECA: if possessing 5000 fusion-product genes and 3 introns per gene, the yield would be 15,000 introns. Thereafter a gradual intron loss has occurred and persists in the extant unicellular eukaryotes. Presuming, that the so-called cancer cell represents a regression to an ancestral unicellular eukaryotic life style, the oncogenome should contain substantially more introns (like at the level of the LECA), than that of the resting and/or only periodically dividing normal somatic cell genome. Such intron gain in the oncogenome would indicate a very deep regression event in the oncogenome. Lack of comparable intron counts (intron number, size and position measurements) between resting and transformed protein-coding genes renders this conclusion "pending". Superb computer work, the probabilistic Markov chain Monte Carlo model, but not the actual comparisons of intron counts between resting and activated genes have been provided (LECA is not available for actual intron assessment). The difference if any of intron counts between a somatic cell performing one or two special tasks, and the stem cell of the same multicellular host has not been revealed. The intron count of the same gene, be somatic or stem cell gene, at rest, or in the frantic oncogenic transformation remains unknown. Loss of a current relatively intron-rich genome of a cell in a multicellular eukaryotic host could indicate a descent to the relatively intron poor state of the extant unicellular eukaryotes (the ancestral line of Amoebozoa and Dictyostelia before and after its divergence); it would not be as deep as the one to the intron-rich LECA. LECA's genome was not resting comfortably; probably it was constitutively activated. If LECA was intron-rich,[1] then constitutively activated oncogenomes should also operate with an intron-rich genome?

Mobile group II introns of bacterial origin are autocatalytic RNA ribozymes; they "retrohome" by encoding a RT to retrotranscribe them into cDNA for insertion into the top strand DNA of the cellular genome, between two DNA exons. The host's replicative DNA polymerase performs the insertion. *Lactococcus lactis* intron RNA for its insertion into *E. coli*, but also into the drosophila genome, uses host non-homologous end-joining enzymes. Non-LTR retrotransposons in eukaryotic nuclear genomes insert themselves by this mechanism. LTR-containing retro-transposons (retroviruses) are reverse transcribed in the cytoplasm by RT with RNase II domain and DNA-dependent-DNA polymerase activity. Their pre-integration complex contains dsDNA.[2]

The Mauriceville retroplasmid (mRP) resides integrated in the mitochondrial DNA of the *Neurospora crassa* Metzenberg.[3,4] The 3.6 kb closed-circular DNA plasmid encodes its own RT. It recognizes 3′ tRNA structures; it synthesizes full length cDNA copies; circularized positive strand DNA is the final product. The RT is a non-LTR LINE retroelement; it is able to initiate DNA synthesis using 3′ tRNA for cDNA synthesis de novo. The first RT evolved from RNA-dependent RNA polymerase in the ancestor of all retroviruses.[5abc,6] The biology of *Neurospora crassa.*[4,7ab]

Transposable elements in the human genome are able to activate proto-oncogenes and initiate the mutator transformation of their host cells.[8]

1 *Koonin EV. J Hered 2009;100:618–623; Koonin EV et al Wiley Interdiscip Rev RNA 2013;4:93–105; Csűrös M et al PLoS Comput Biol 2011;7(9):e1002150.* ***2*** *Kulper MTR & Lambowitz AM Cell 1988;55:693704; Wang H, Lambowitz AM. Cell 1993;75:1071–1081.* ***3*** *Wu C et al Genetics 2009;181:1129–1145.* ***4*** *Fu C et al Eukaryot Cell 2011;10:1100–1109.* ***5a*** *Chang C-C et al Mol Cell Biol 1994;14: 6419–6432.* ***5b*** *Chiang L & Lambowitz AM Mol Cell Biol 1997;17:4526–4535.* ***5c*** *Chen BL & Lambowitz AM J Mol Biol 1997;271:311–332.* ***6*** *Yao J et al PLoS Genet 2013;9(4):e100346.* ***7a*** *Antal Z et al Plasmid 2002;47:148–152.* ***7b*** *Wu C et al Genetics 2009;181:1129–1145.* ***8*** *Chénais B. Biochim Biophys Acta 1835:28–35; Gene 2012;509:7–15.*

small nuclear ribonucleoprotein particles; named after patient Stephanie Smith's survival of motor neurons' Smith's survival (Wikipedia).

Metazoan eukaryotic cells express the Sm protein binding site in their U4 snRNAs. Many spliceosomal giardia introns are hybrids due to fusions at their branch points and 3′ splice sites) [1120]. The GTPase family *rab*1a gene (*ras*-associated guanoside triphosphate class B) of *T. vaginalis* has a 25-bp spliceosomal intron with splice sites at 5′ (GT), and 3′ (AG) ends, and the consensus sequence ACTAAC. The descendants of this ancient spliceosomal intron are operational in Amoebozoa, Opisthokonta, Excavata, Chromalveolata, and Plantae [1115, 1121].

In its relatively large genome, *T. vaginalis* harbors may be 65 short introns. The regulatory motifs at the 5′ and the 3′ boundaries are shared with the giardia introns. However, what is conserved in eukaryotic evolution, it is the spliceosomal catalytic interactive motifs. The trichomonad snRNAs lack a cap at their 5′ends (for the formation of this cap, the enzyme trimethylguanosine synthase would be needed). However, RNA polymerases II/III transcribe the uncapped sequences. The ancestors of the spliceosomal snRNAs probably were the self-splicing ribozyme introns. The RNA polymerase transcripts are capped and the cap is subsequently hypermethylated (by trimethylguanosine). The U6/U2 stable snRNA complex of *T. vaginalis* is the one phylogenically most conserved structure in other eukaryotes [1122].

T. vaginalis is listed as one of the etiologic agents of prostate cancer. It is pointed out that *T. vaginalis* possesses an exceptionally large and unusual genome with gene duplications, but poor in introns; expresses microRNAs, and has a hydrogenosome encoding the enzyme hydroxymethyl transferase. Hydrogenosomes lack cytochromes, perform no electron-transport-linked oxidative phosphorylation, or a tricarboxylic acid cycle; instead, convert pyruvate or malate under anaerobic conditions to hydrogen acetate and CO_2; but produce ATP [1123]. *T. vaginalis* received many of its genes through lateral (horizontal) gene transfers. Thus, it shares metabolic pathways (caffeine metabolism; N-glycan biosynthesis, glycosaminoglycan degradation, etc.) with a large number of bacteria (*B. anthracis, B. subtilis, H. pylori, K. pneumoniae, Pseudomonas, S. aureus, S. pyogens, V. cholerae*), fungi (*Aspergillus niger, Candida albicans*), protozoa (*Giardia lamblia, Entamoeba histolytica, Plasmodium vivax*) and their host, *Homo sapiens* [1124]. The TATA box (thymine-adenine thymine adenine) is located in the metazoan genomes 25–30 bp upstream of the transcription start site (TSS), from where it directs accurate transcription initiation. The DNA initiator element (Inr) located 5–15 bp upstream of the start codon ATG; it is present in 75 % of the protein coding genes of *T. vaginalis*. Analogous to the metazoan TATA box, are the M3/M5 elements in the genome of *T. vaginalis*. M5 contains the TSS and it is in the locus of the Inr. Protein-coding genes use either Inr or M5, and M3 is a synergist of Inr. The M3 consensus sequence resembles closely the classic eukaryotic Myb-recognizing element (MRE). The aa sequences of the Trichomonas myb/MYB proteins show close identity to their human counterparts; however they are evolutionarily possessed members of the ancient myb/MYB family (in the Appendix 1). In *G. lamblia*, in *Entamoeba histolytica*, the MRE regulates encystation genes [1125]. In human cells, Myb proteins (*vide supra*) recognize the MRE.

In the amoeba, *Dictyostelium discoideum* (Figure 6), the non-LTR retrotransposon TRE5-A (the original tRNA gene-associated repetitive element) targets the transcription factor (TF) complex. The full length TRE5-A element encodes two ORFs. The plus strand RNA TRE5-A serves as the template of its retrotransposition. The ORF1 protein binds TFIIIB (τ factor) subunit, and positions it at the 5′ end of the tRNA gene. The 3′ untranslated region of TRE5-A was essential for its retrotransposition. The process of target-primed reverse transcription uses the retrotransposon-encoded ORF2 enzyme to bind to the 3′ end of the retrotransposon RNA, while it nicks the DNA targeted for the integration. The emerging 3′ hydroxyl group will serve as the primer for the cDNA synthesis. A construct consisted of the blasticidin resistance gene, *bsr*, which is expressed in the minus strand of the TRE5-A element (*mbsr*, TRE5-A^{bsr}). An intron in an inverse transcriptional direction was placed in the *mbsr* gene, but it was in the plus-strand orientation in the TRE5-A retrotransposon. When the artificial intron was removed by splicing, the TRE5-A^{bsr} element could amplify and generate blasticidin-resistant clones. Without ORF1, the TRE5-A^{bsr} element failed to express any retrotransposon activities. The TRE5-A^{bsr} elements discovered new TFIIIC binding sites on the extrachromosomal DNA element-carriers of the ribosomal RNA genes [1126]. For more on Dd: *vide infra*. The group I intron of the free-living amoeba, Naegleria, resides in the ribosomal gene and consists of two ribozymes and an intron-homing endonuclease (the Nae.S516 twin-ribozyme intron). Ribozyme2 performs the excision, ligation, and intron RNA circularization. Ribozyme1 carries out the expression of the endonuclease by generating the 5′ end of the enzyme's mRNA [1127].

The freshwater-living ciliated protozoan, *Tetrahymena thermophila*, allowed insight into its in-group ribozyme compartment, including its *in vivo* machinery of RNA-folding by RNA-binding proteins and chaperones. A spectacular depiction of the secondary structure of the group I tetrahymena ribozyme is provided [1128, 1129]. It was in 1962 when Gellert *et al* first described the G-4 quartets formed by guanine-rich DNA molecules [426], and received credit for it in 2007 [1130]. The tetrahymena G-4 DNA structure formed *in vivo* and the protein binding to it was reported in 1994. It is a ssDNA region adjacent to the G4 structure that the protein binds [1131].

The characteristic 4-stacked G-tetrads of the telomeric d(TGGGGTTGGGGT) tetrahymena sequences differ from the human telomeric sequences only in A for G replacement: d(TAGGGTTAGGGT). The two forms are similar in reactions to temperature changes, in their unfolding rates and in their activation enthalpy (energy of activation) of 153 hJ/mol [1132]. A tetrahymena group I-related intron functioning in mammalian telomerase reverse transcriptase transcripts (TERT) (mouse and human) specifically spliced and replaced RNAs. Treatment of telomerase overproducing mouse and human cancer cells with the tetrahymena ribozyme group I introns resulted in non-functional trans-spliced transcripts. The tetrahymena ribozyme group I introns could induce the HSV thymidine kinase gene, thus activating ganciclovir-induced cancer cell death [1133]. The tetrahymena intron-rich genome is expected to contain about 20,000 genes. Mucocysts under the cell surface could discharge their contents and surround the individual cells with a thick layer of mucin. The cytoplasm is equipped for the digestion of bacteria and fungi serving as nutrients. Nutrient starvation induces conjugation: meiosis, gamete exchange and the formation of the zygotic nucleus. Constitutive formation and release of granule lattice (GRL) proteins may occur in large individuals growing into high density by rapid divisions in appropriate temperature and nutrient supply. Neither the genomic changes, nor the specific morphological conversions accompanying constitutive protein secretion were examined further. However, starvation-enhanced secretion in exosomes of cytosolic proteins was observed. The exosome/secretome proteins revealed no functional features [1134]. This author (JGS) is observing similarity between these tetrahymena cells and the transformed cancer cells. The cancer cells reverted to a unicellular life style, mobilized one or several genes for constitutive function, and release, believed for intercellular communication, exosomes/secretomes [1135].

Of kinetoplastids (class, Kinetoplastea, order Kinetoplastida/Bodonida, phylum Euglenozoa), the free-living trypanosome-like *Bodo saltans* operates an intron-poor trypanosomatid genome of 0.4 Mbp. Its evolutionary trajectory is well outlined from the Ichthyobodo (Prokinetoplastina) through Metakinetoplastina to Neobodo (*N. designis*) [1136, 1137]. Trans-splicing of some exons and cis-splicing of the poly(A) polymerase intron of *Trypanosoma brucei* was reported [1138a, b]. *T. brucei* exists in two life forms: the blood stream form in mammalian hosts, and the polycyclic form in the mid-gut of the tsetse fly (*Glossina* sp). The extent of the cellular remodeling performed between the two life forms border the DNA phenomenon of "trans-speciation". The genes examined represent 90 % of the trypanosoma genome operating with approx 20,000 mRNA molecules. Splicing and polyadenylation sites are recognized, but the introns received very limited attention. Introns are present in the poly(A) polymerase and in the DNA/RNA helicase genes [1139].

Yeasts (*S. cerevisiae*) conserve a great deal of G4 quadruplet DNA motifs. These motifs appeared in unperturbed mitotic and meiotic sites, and in the AT-rich mitochondrial mtDNA [378]. The *S. paradoxus* mt genome is 71,355 bp long with high AT content (86 %). It encodes cytochrome oxidase, ATPase, and ribosomal proteins. The mt genes are interrupted by nine group I and II introns, and two "unusual" GI introns [1140]. Intron-rich organisms evenly distribute their introns within coding sequences of their genomes. Intron-poor organisms unevenly increase their intron densities towards the 5′ ends of their genes. The intron-poor *S. cerevisiae* (less than one intron per gene) exhibits this latter pattern. In contrast, the Cryptococcus genome harbors seven introns per gene. Is there a difference in intron density between individual cryptococcci isolated from spinal fluid of patients with meningitis and those taken from pigeon droppings, or from culture vessels? Cryptococci delete their elongated introns from untranslated regions more frequently, than their short introns positioned close to protein-coding genes [1141]. Fungal genes encoding proteins related to the DNA repair enzymes in xerodema pigmentosum will be mentioned later (*vide infra*).

In filamentous fungi (*Aspergillus nidulans, Fusarium graminearum, Magnaporthe grisea, Neurospora crassa*) (Figure 63), a peculiar form of intron loss at sites other than the 5′ end of a gene was not due to homologous recombinations (HR) of poly-adenosine-primed reverse transcripts. An undefined selective pressure to preferentially conserve introns at the genomic ends 3′ and/or 5′ should be considered to be operational [1142]. Multiplication of introner-like elements (long spliceosomal introns with stable secondary structures) within fungal genes account for one form of intron gains. The introner elements undergo slow structural simplification and end up as regular spliceosomal intron [1143]. New sequences inserted during dsDNA breaks may assume the appearance (and function, if any) of new introns. Localized tandem duplications increase intron numbers. Transposable elements may convert into introner-like structures, as in the tunicate protochordate, *Oikopleura dioica* [1144]. These "cryptic elements" not only in algae, but also in fungi (eukaryotes) may become introns [1145].

Do introns mediate enhancement or repression of transcription? Transcription of the natural splicing-competent intron-containing gene ASC (apoptosis-associated speck-like protein with caspase recruitment domain, CARD) with and without (after the removal of) its intron showed intron-mediated enhancement of its transcription. Intron-dependent gene looping occurred (the juxtaposition of the promoter and terminator regions of the gene). Physical interaction of the promoter and the terminator regions occurred, or the promoter region interacted with the 5′ splice site and the terminator with the 3′ splice site [1146].

How far can we see backwards? The trichomonads possess a proline-rich protein, encoded by the *prp8* gene, which is a major spliceosomal component. Would the possession of this gene indicate that the ancestors of the deep branching diplomonads and parabasalids were endowed with spliceosomal introns, which they have lost (as their pre-existing mitochondria might have been lost as well)? If so, the diplomonads and parabasalia should be in the ranks of excavata (amoeboflagellates, kinetoplastids, euglenids, oxymonads, but this last class without the ventral groove). Some oxymonads might have retained the ancient spliceosomal introns that were lost from the diplomonads and parabasalia. Such spliceosomal introns were found in the oxymonad *Streblomastix strix*. *S. strix* populates the hindgut of a dampwood termite, as a symbiont. The *S. strix* genome so far yielded a number (twenty-one in 17 genes) of canonical spliceosomal introns (some single genes yielding 5 introns). Twenty introns showed the canonical GT-AG boundaries; one intron was exceptional with AC–AG splice sites. The gene for ribosomal protein 9 harbored the introns. Some of the algal genes travelled into and were built into the ancestral chromalveolate's nuclear genome flanked by AC-AG splice sites. The *S. strix* introns show homologues to spliceosomal introns of eukaryotic supergroups (plants) and in retrograde direction, very likely date back to the introns of the last common ancestor of all eukaryotes [1147].

The diatom *Thalassiosira pseudonana* emerges as a unique case in this context. Its highly intron-dense genome harbors introns that are not conserved in other divergent eukaryotic genera (only 8 % of its introns appear to be conserved elsewhere). It is therefore presumed that its current introns are recently acquired; that is, this diatom lost its introns that were vertically inherited from its diatom-apicomplexan common ancestor. Thereafter, it has gained a new set of introns. However, its current intron supplement does not match the position of introns in other species. This implies that intron-rich genomes may have arisen more than once in the course of evolution of this particular host [1148]. The original eukaryotic intron-rich genome might have developed from intron-donations, as the newly acquired mitochondrial genome was converted from that of a symbiont to that of an organelle. The mitochondrial genes transferred to the protonucleus of their new host carried and deposited there their group II type introns [1149].

Chimeric archaeabacterial and prokaryotic (eubacterial) ribosomes within the cytoplasm of a primordial cell synthesized from mRNAs deriving from the two types of genomes two different types of proteins (enzymes), very likely as fusion proteins. The cell gradually in successive stages separated by membrane-bound compartmentalization its chimeric 'nucleus' from its chimeric ribosomes [1150]. The reviewers readily accepting this new proposal of nucleus formation are those (Koonin EV; Forterre P), who proposed elsewhere that a pre-cellular Virus World existed and that newly formed protocells received and accepted the protoviruses as their first alien invaders. Similarly to the relationship of the amoeba with the mimivirus and marseillevirus, viral genes were donated to the cell nucleus, and cellular genes were incorporated into the viral genomes [13a, b, 14a, b, 18, 120]. It was also proposed, that some DNA viruses, practically in toto, have become the first cell nuclei. It is the ancestral α-polymerase of poxviral derivation that synthesized DNA on RNA primers. Adeno- and herpesviral δ-polymerases do not fit into the theory. Further, the cytoplasmic full replication cycle of orthopoxviruses might have allowed them to infect and live in a-nucleated proto-cells, like an ancestral archaea, already possessing δ-polymerases, but not α-polymerases [28, 29a, b]. The reviewers in supporting the new theory of cell nucleus formation as proposed by Gáspár Jékely [1150] refrained from mentioning any viral contributions to the process (neither did so Jékely). In this fundamental genomic and proteomic arrangements, the role of introns is not addressed.

Based on probabilistic models, it appears that the Last Eukaryotic Common Ancestor (LECA) was an intron-rich cell (Table XIII). LECA's immediate derivatives, the deep` branching unicellular eukaryotes, emerged with intron-poor nuclear genomes. Thereafter, the complex multicellular organisms gained introns and operate a much variable, but relatively intron-rich genome. The positions of ancient introns within the genes are evolutionarily conserved. The evolutionary history of complex multicellular organisms is characterized by steady slow events of intron losses interrupted by episodic intron gains. In some cases the intron gain was in excess of intron loss. For example, green algae lost, but land plants gained introns. LECA's presumed >5 introns per kb were diminished in its unicellular eukaryotic descendants, but have been regained first in the common ancestor of metazoa, and again in the genomes of mammalian cells. The intron density is related to the strength of the splice signals: highly efficient splice signals reduce intron density. The appearance of alternative splicing and increased transposon activity for active dsDNA break repairs resulted in the spurt of intron gains, as expressed in the computers of Csűrös *et al* [1151a, b]. Eukaryotic genomes in a hurry (genes whose transcripts require rapid adjustments for survival of environmental challenges) possess short introns. Genes (yeast, plant, mammalian) with rapid expression levels in response to stress appear to contain significantly lower introns densities than resting genomes (still questionable, *vide infra*). Spliceosomal introns can be regarded as brakes of regulatory responses. Half-lives for splicing reactions are 1 min for the first introns, and increasingly longer time (2–8 min) for subsequent introns. The increased transcript length comes with higher energetic cost.

The most ancient genomes might have appeared in the form of "RNA continuity". The emerging protein-coding exons interrupted the order of RNA and RNP sequences. Thus genomic introns are the "relics" of the RNA/RNP World. Accordingly, highly evolved and regulated cell cycle genes are introns-poor [1152a, b]. The so-called intron-rich LECA must not have been comfortably resting in a stress-free environment. How its introns are comparable with those of oncogenes in rest, or in constitutive activity? Introns containing nested (embedded) protein-coding genes are expected to be retained. The marine picoeukaryote alga, *Micromonas pusilla* has been experiencing an episode of massive introns gain; these introns contain newly inscribed repeat sequences (introners) that are absent in orthologous genes of close algal relatives; the introners are not transposable elements (in contrast to group II introns). This microorganism (*M. pusilla*), is infected with the chlorella strain dsDNA phycodnavirus, and RNA viruses (Mimoreovirus; and a dsRNA Prasinovirus, infecting also green algae) [1153–1156]. How viral infection of a host cell may increase its introns number?

Both U2 and U12 spliceosomes must have been present in the last eukaryotic common ancestor (LECA), U2 being the most ancient. The fact that prokaryotes are devoid of spliceosomal introns, renders the introns first theory unlikely. Intron acquisition has taken place in the first eukaryotes. However, the descendant unicellular free-living eukaryotes (amoeba, choanoflagellates, tetrahymena) are considered to be relatively intron poor. Gaines of introns by unexplained mechanisms occur in the *Daphnia pulex* off the shores of Oregon, or in the picoeukaryote *Micromonas pusilla*; reviewed in Rogozin *et al* [1157]. In comparison, would the parasitic unicellular life forms invading and subduing their hosts (Amoeba, Giardia, Naegleria, Theileria, Trichomonas; Plasmodia; Leishmania, Toxoplasma, Trypanosoma) activate an introns-rich, or an even more introns-poor genome? In contrast to unicellular eukaryotes, the individual cells of multicellular eukaryotes (both in animals and in plants) are considered to be intron rich [1157]. In events listed above, actual intron counts are not available.

The first nucleotides of the exon are separated accurately from the adjacent introns in the spliceosome. Most spliceosomal introns contain |GT at the donor splice site and AG| at the acceptor splice site (intron excision and exon splicing). Some special introns express different dinucleotides: |AT and AC|. These latter introns are spliced not in the major U2, but in the minor U12 spliceosomes. It is debated which one of these introns were the ancestral introns of the LECA. U2, U12, uridyl-rich small nuclear snRNA/RNPs associated with the branchpoint region of the introns. The U12 introns occupy the 5′ positions, whereas U2 introns favor the 3′ positions. Protein-coding exons might have emerged from the primordial introns. The extant eukaryotic genomes differ in a great extent as to their intron numbers (density), introns' sizes, positions, and distribution. The unicellular eukaryotic organisms usually are intron-poor, and the multicellular organisms are intron-rich (with many exceptions). Because of many exceptions, it remains very difficult to correlate intron density with the length of individual introns (density plotted against length). The intron-poor single-celled eukaryotes keep their introns in the 5′ position; the genomes of the intron-rich multicellular organisms distribute their introns evenly. The algal tiny symbionts, the nucleomorphs, and the intracellular parasite microsporidia live an intron-free life. Intron-poor genomes retain a few introns in those genes, which encode ribosomal proteins. The loss of these introns is with deleterious consequences (in yeast). *S. cerevisiae* lost and replaced most of its ancestral introns. Orthologous genes preserve the intron positions of their ancestors (with a few exceptions).

Parabasalid (*T. vaginalis*) and excavate (*Naegleria gruberi*) intron positions are remarkably conserved in exact positions in orthologous higher eukaryotic genes. Bursts of new intron insertions appear to accompany major evolutionary transitions. Nematodes (caenorhabditis) display recently inserted introns. One strain of *Daphnia pulex* (in Oregon) has recently acquired two new polymorph introns from unknown source, inserted into a GTPase encoding gene. In primates, the ring finger RNF113B retrogene encoding a ubiquitin ligase shows the appearance of a new intron (*vide infra*). This new intron was formed in the 3′ UTR from 59 nucleotides of a former exonic coding sequence. The relocation of intron/exon boundaries, the intron sliding, explains how orthologous introns may shift their positions. Major new intron genes occurred in the earliest metazoan. Analysis of 7000 intron positions in the highly conserved genes of eukaryotes revealed an evolutionary scenario, in which the common ancestor of modern eukaryotes was an intron-rich unicellular organism, whose immediate descendants suffered intron losses; vertebrates and flowering plants were compensated by the acquisition of new introns. The protein-coding genes of chromalveolates (the eukaryotic superphylum of T. Cavalier-Smith), including among others apicomplexa (cryptosporidia, dinoflagellates, ciliates, leishmania, plasmodia, trypanosoma, toxoplasma) exhibit low intron densities. The common ancestor of chromalveolates incorporated into its cytoplasm a red alga and vertically transmitted this plastid to all its descendants. Some of the algal genes travelled into and were built into the ancestral chromalveolate's nuclear genome. Especially, the ancestral alveolate lineages (paramecia, plasmodia, Tetrahymena, Theileria) are the most suspect candidates for harboring originally intron-rich genomes. However, the ancestral species of the chromalveolates are calculated to have had highly intron-rich genes. The inferred intron densities of the alveolar ancestors matched that of extant plant or animal cells (with a ≈ 19 % coefficient of variation due to uncertainty of estimates). Orthologues of some of these introns show up in the intron-rich genes of animals and plants. An evolutionary early intron gain was followed by a substantial intron loss (persisting in the extant descendants of the chromalveolates; in fungi and in green and red algae), which was compensated for with an intron gain in the the extant multicellular hosts (plants and animalia, which share some orthologous introns). These inferences of intron densities of ancestral unicellular eukaryotes were based on a flexible maximum-likelihood approach of calculations; reviewed in Csűrös *et al* and Rogozin *et al* [1151a, b, 1157].

Comment. This author (JGS) wonders, that in the age of high promiscuity, when horizontally transferred genes were sought after and promptly accepted, how much evolutionary pressure was exerted for the separation by membranes versus ready coalescence and cooperation with intruder gene-product proteins in the chimeric nuclei and ribosomes? During the process of compartmentalization of organelles, is there any change of the number and position of the introns in the involved genes? For compartmentalization of the cell, see Blobel [88a, b]. No distinction has been made between the intron content of somatic cells and stem cells of the multicellular organisms. Within the somatic cells, certain selected genes are highly operational, whereas other genes are silenced. In the somatic and stem cells, is there a change in introns count between resting proto-oncogenes and constitutively activated oncogenes? What is the intron count of a differentiated somatic cell, which under the effect of lympho- and cytokines reverts into a stem cell and activates Sox2, Oct4, Nanog, BMi1, c-Myc, and Kfl-genes? These are some of the retrolentivirally vectored stem cell genes in Sinya Yamanaka's work (Nobel Prize 2012; Yamanaka Discussion, Tuxillaplanet Video, Google).

Comment. If the oncogene-activated cells of the multicellular organisms assumed the life style of their ancient unicellular ancestors, as they resisted physico-chemical damage, or parasitized a host, the oncogenome of a transformed cell in a multicellular host would be expected to operate a genome alike those of the aggressive parasitic unicellular life forms living a stressful life.

Resting gene transforms into oncogenes: what happens to its introns? If one now considers intron loss or gain upon the transformation of a single eukaryotic cell's proto-oncogene into an oncogene to be characteristic to a reversal to an ancient life style, the level of descent still remains uncertain. If the depth of the regression encompasses the ancestral level antedating the divergence into evolutionary branches: the transformed cell would retain its intron supplement, or even increase it. If, however, the descent reaches and stops at the level of the extant kinetoplastids, a substantial intron loss should have appeared. The answer would come from accurate intron counts of resting proto-oncogenes in comparison with that of constitutively activated oncogenes (what is not available).

Twenty-one completely sequenced eukaryotic genomes showed preferential intron losses due to reverse transcriptions (RT-mRNA-mediated) and homologous recombinations (HR) at the 3′ genomic ends, where highly expressed genes reside; whereas more stable protein-coding genes at the 5′ end tended to retain their introns [1158]. The stable introns are those of group I, encoding ribozyme, and thus are restricted to vertical inheritance. In contrast, introns encoding from functional homing endonuclease (HE) genes are highly invasive. The chloroplast gene *rbcL* of green algae (large subunit RuBisCO ribulose-bisphosphate-carboxylase-oxygenase oxygen fixation enzyme) contains group IA1 and IA2 type introns. Introns IA1 and

IA2 are inserted at position 699 and 462, respectively. Intron IA1 derived from ribosomal RNA and encodes homing endonucleases [1159]. The homing nicking endonucleases carry HNH (hist-asp-hist) and GIY-YIG (gly-isoleu-tyr) amino acid motifs [1160, 1161]. The ORFs of group I introns encode DNA endonucleases. The phylogeny of group *rbcL* IA2 introns indicates vertical transmission (except for that of the *Gonium multicoccum*: there acquisition through "recent" horizontal transfer is considered). These group IA2 *rbcL* introns fall victim of degenerations and deletions [1162a, b]. Intron IA2 is a pseudobeadex intergenic spacer photosystem psbA intron. Good examples are the *Chlamydomonas reinhardtii* and *C. moewusii* psbA introns. These latter introns of 2363 and 1807 nt, respectively, encode proteins of 345 aa (photosystem II reaction center), and 312 aa, respectively. Depicted in the article are Intron 1 and Intron 2 with discussion how they may relate to mitochondrial introns [1163].

In the asexual amoeba, Naegleria (Figure 54), group I Nae.S516 intron harbors two ribozyme domains and an endonuclease-encoding sequence. Ribozyme1 manages the endonuclease expression; ribozyme 2 excises, circularizes and exon-ligazes the intron. The Nae5516 intron originated in the naeglerian ancestor and was vertically transmissible [1164]. The self-splicing introns are able to "parasitize" genomes. After excising themselves, they ligate the exons they leave behind. Their endonucleases make ds cuts in the DNA targeted for their insertion. Once inserted in a new location, they may delete their homing endonuclease (HE) sequence and become the persistent property of the new genome, being transferred vertically in the form of Mendelian inheritance [1165]. The ORFs of mobile group II introns are reverse transcriptase-related. Attempts at rooting, lead to the RT of LTR retroelements of mitochondrial- and chloroplast-derivation. Group II introns are readily transmissible through horizontal routes [1166].

Fungal genomes (aspergillus, botrytis, histoplasma, saccharomyces) carry inteins (internal protein sequences) inserted within host proteins and then excised post-translationally. Ligations of the flanking external protein (extein) sequences form the peptide bonds. In addition to means of protein splicing, inteins also operate DNA homing endonucleases. The full size inteins preserved their HEs; the mini-inteins lost the enzyme. The inteins appear to have originated in ancestral prokaryotic genomes and were transferred both vertically and horizontally following the divergences of evolutionary lineages. Of eukaryotes, the fungal genomes carry inteins. The spliceosomal PRP8 gene (pre-mRNA processing factor 8; self-catalyzing spliceosomal protein product) of the basidiomycete, *Cryptococcus neoformans*, encodes a set of inteins, referred to as VMA inteins (vacuolar membrane ATPase), related to the VMA inteins first discovered as restricted to yeasts (Saccharomyces), whereas the PRP8 inteins populate ascomycetes and basidiomycetes. The VMA inteins lost most of their HEs, the PRP8 inteins withheld their active enzymes. Against expectations, the VMA inteins are of more recent origin than the PRP8 inteins [1167, 1168].

The Rho family *ras*-related small GTPases (CDC42, Rho, Rac, *vide infra*) are fully active in yeasts as Dock proteins (dedicator of cytokinesis). Rac1a is a growth factor in yeasts (dimorphic fungi, the haploid amoeba, *D. discoideum*; and

Yarrowia lipolytica). In the latter microorganism, its *YlRAC*1 gene-product protein (genomics, Southern blot; proteomics, Northern blot) is essential for hyphal growth. In Dd, there are six chromosomes and the extrachromosomal element with rRNA genes. Fifteen contigs harbor the genes encoding Rho-related proteins; these genes are interrupted by 1–4 introns, some with alternative splicing, releasing more than one mRNA per gene (in Northern blot measured mRNAs). In the *rac* gene (*ras*-related C3 botulinum toxin substrate), the introns were in the 5′ half of the gene. These introns in this position were conserved from the ancestral *rac* gene. In yeasts, Rho proteins are involved in cell wall synthesis and budding [1169]. The Dd mitochondrial genome contains intronic genes (intronic ORFs), rRNAs and tRNAs and fused *cox* (cytochrome oxidase) genes with intronic ORFs for encoding HE homing type DNA endonuclease [1170]. An rRNA-encoding gene at its 3′ end carries a 547-bp self-splicing group I intron. A green alga is known to carry a similar (if not identical) intron at the same insertion site [1171].

In the nucleosomes of the chromatins, DNA strands are wrapped around the core histones. Histones are synthesized through their mRNAs for the package of the newly template-replicated DNA strands. In fungal evolution, the descendants of intron-rich ancestors are losing their introns. Both the filamentous ascomycetes and the yeasts (basidiomycetous cryptococci) retained just a few introns in their histone genes. The fastest evolving subjects (*S. cerevisiae* vs. *Yarrowia lipolytica*) eliminate most of their introns from their protein-coding genes [1172, 1173].

The most striking observation concerns some helicases encoded by the Dictyostelium genome. These helicases are related to those encoded in patients with xeroderma pigmentosum (Hebra and Kaposi 1874). Those mutated helicases fail to repair (by nucleotide excision) the UV light-damaged DNA. The Dd *rep*E gene (represent replication) is the ancestral homolog of the human excision repair cross-complementing repair deficiency UV-DDB/XPE gene. The Dd *rep*B/D genes are homologs of the human RAD25/XPB and RAD3/XPD genes (DNA damage-binding protein; radiation sensitivity DNA helicase). The *rep*B gene contains one intron, the *rep*D gene is intron-less. The original report depicts in two full pages the sequence homology of *rep*B and *rep*D proteins with the conserved functional domains underlined [1174].

Comment. Are intron counts is an expression of competition or cooperation between the ancient RNA and its derivative the DNA? Intron-rich genomes reflect to an ancient RNA/DNA complex? Intron-poor genome would indicate an evolutionary transformation into a DNA/RNA complex? The fastest evolving fungal genomes (*vide supra*) eliminate their introns? Would this mean that the resting proto-oncogene re-expresses ancient introns, when it transforms into an oncogene? If oncogenesis is a process of return to an ancient life style (the activation of the cell survival pathways), then the ancient unicellular eukaryotes in their daily struggle for survival should have been intron-rich (the intron-rich LECA in Table XIII).

The fate of the introns remains hidden in the process of virally induced leukemogenesis. New introns (intronization) may derive from retrogenes; retrogenes may derive from retropositioned processed mRNAs. Primate retrogenes RNF113B (flavoprotein-like ring-type zink finger protein; redox-driven ion pumps in bacteria;

Rnf-type nicotinic acid dehydrogenase) and DCAF12L2 (DDB, damage-specific DNA-binding protein; cullin-associated factor) yielded new introns from their splice variants [1175]. These genes exist in two transcript forms, one with spliced out, and the other one with retained introns. In the mammalian genome, RNF113A exists in an intron-less form. It is presumed that a copy of the RNF113 gene was retroposed into the intronic region of the NDUFA1 gene (NADH, nicotinamide adenine dinucleotide dehydrogenase; ubiquinone alpha subcomplex) residing on the X chromosome and encoding the MwFe protein (mitochondrial FeS protein, molecular weight 8072 Da). The small (5 kb) NDUFA1 gene with two introns (1.5 and 3 kb) encodes the 70 aa polypeptide/protein. The ubiquinone oxidase MWFE enters the mitochondria in support of the function of complex 1 [1176]. The MWFE proteins promote cell growth. Transgenes with constitutively active promoters encode large amounts of MWFE proteins in cells rapidly replicating *in vitro* (but not known if these cells would have grown into tumors *in vivo*). Dysfunctional (not understood well how) mitochondria protected cells from death by ionizing radiation. The p53 protein entering mitochondria (guided by nutrin) encounters the genes Bcl-2/Bcl-XL. This encounter is hindered in dysfunctional mitochondria. Dysfunctional mitochondria in tumor cells are protagonists for the survival of their transformed cells (Figures 23 and 64; Table XXXII). The Cul4-DDB1 E3 ubiquitin ligase regulates its DNA-methylation via histone 3, lysine K9 trimethylation: H3K9me3 [1177]. The primate-specific ring finger gene RNF113B was copied into an intron of the FARP1 gene [1175]. The FARP genes are listed as F for FERM (founded by band 4,1, ezrin, radixin, moesin), A for ARHGEF (Rho-specific guanine nucleotide exchange factor), R for RhoGEF (ras homolog guanine nucleotide exchange factor), and P for pleckstrin (platelet and leukocyte C kinase substrate).

FERM is a target of Janus kinase JAK. In the complex pathogenesis of the Caribbean adult T cell lymphoma/leukemia (ATLL) induced by the human T-lymphotrophic virus-1 (HTLV-1), the viral Tax oncoprotein constitutively activates the IL-2 → IL-2R pathway. Co-activated are the non-receptor Janus kinases JAK1/3; within JAK3, the N terminal of FERM is missense mutated: it is the pseudokinase mutation A572V (alanine; valine). The cascade continues in the constitutive activation of STAT5 and PI3K/Akt. Blockade of the cell cycle inhibitors and apoptosis inducers follows. Bcl2 does not appear translocated; it appears rather vastly amplified. The subverted host produces IL-9 (from monocytes) and IL-7 (from keratinocytes) for further stimulation of T cell proliferation. Stimulation of the erythropoietin receptor induces JAK2 activation; JAK2 may suffer Y119F point mutation (tyrosine; phenylalanine). However, in the Japanese (Oriental) variant of ATLL, may be (not proven), HTLV-1 genomic integration into the IL-9R gene induces the IL-2 IL-2R cascade. The status of JAK/FERM mutations is not documented. Treatment with type I IFNα and zidovudine prolongs life, but fails to cure the disease; targeted JAK inhibitors (tofacitinib) await clinical trials (Elliott *et al*; T.A. Waldmann, personal communication) [1178]. (Added at proofs: Pfizer marketed tofacitinib citrate under the trade name Xeljanz, a highly immunosuppressive drug, and advertised so, for the treatment of rheumatoid arthritis).

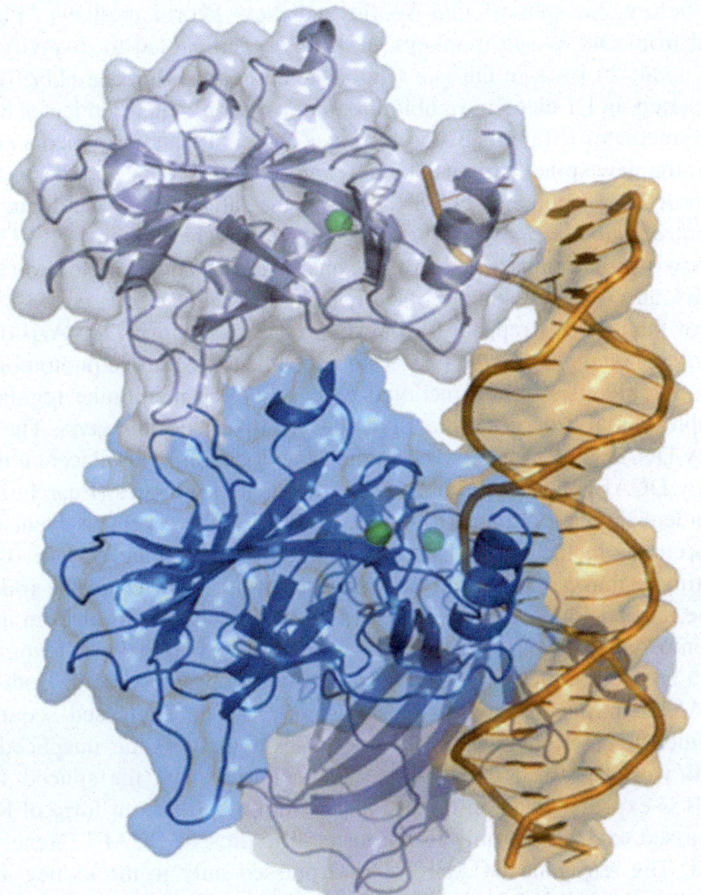

Figure 64 The Pro-apoptotic p53 Was Counterbalanced Early by the Anti-apoptotic MDM. [2360] Reference Appendix 2, Explanations to the Figures

Comment. Here, essential, physiologically installed elementary ancient genes (the FERMs, the JAKs) are exogenously enforced to malignantly transform and immortalize the T-lymphocytes of a highly evolved multicellular host, held defenseless in the process. It is not known if there were any changes in the numbers and positions of the introns from the resting to the activated state of the involved genes.

The fate of introns, cont'd. After its retroposition, the primate RNF113B gene gained its intron, but not by de novo insertion. The intron's donor site was a double point-mutated sequence: AG → GT. The acceptor site was created by another point mutation: GG → AG. The 105 bp intron contains 59 nucleotides from its previous sequence, and 46 nucleotides from the 3′ UTR untranslated region. The new intron is accepted by the U2 spliceosome. The new intron was formed in the primate genome before the split of Old World and New World monkeys. Platyrrhines separated from Old World monkeys ca 35 mya, and started to diversify in South America about 26 mya, in the late Oligocene and early Miocene [1179]. The new intron inserted an L1 element (a ubiquitous transposon which replicates through an RNA intermediate) within its 3′ end for the acceptor site and provided a new polyA signal for the new splice variant. The new intron exonized an Alu element at the 5′ end (a short DNA sequence characterized in the laboratory by endonuclease Alu (from *Arthrobacter luteus*) for regulatory function). The human RNF113B cDNA produces two variant transcripts: a two-exons transcript and a single exon transcript. The DDB1 and CUL4 associated factor, DCAF12 (*vide supra*) encodes a protein with more than 4 WD repeats (β-transducins terminating in Trp-Asp dipeptide). These proteins interact with the COP9 signalosome (constitutive photomorphogenic homolog, Arabidopsis; multifunctional eight-units sub-membrane regulatory protein complex). Only placental mammals have retrocopied this gene. The ancestral gene copy DCAF12L2 is conserved in the same location in all placental mammals. Gene copy DCAF12L1 is present only in rodents and primates (Euarchontoglires); it is a tandem duplication derivative of the DCAF12L2 gene. An event of intronization occurred in the DCAF12L2 gene altering its splicing pattern (the shared rodent-primate intron). A second intronization event occurred in the rodents only gene. The retroposition of genes RNF113 and DCAF12 occurred from autosomes to X chromosome. The parental RNF113 gene, and the parental multiexon DCAF12 gene both are detectable by sequence similarity on chromosome 9. Both RNF113 and DCAF12 retrogenes are on chromosome X. The expressed sequence tags (EST) indicate that RNF113A is expressed in all tissues; the unspliced form of RNF113B was expressed in all tissues, except testis, but the spliced variant of RNF113B is expressed also in the testis. The intron-spliced out form of RNF113B was expressed exclusively only in the testis. The present DCAF12 gene is widely expressed. The retrogene DCAF12L1 is expressed only in the kidney and testis. The retrogene DCAF12L2 is expressed in the eye and testis. The retrogenes favored their expression in the testis. Intron gains occur by exon intronization. Both genes, those escaping from the X chromosome, or retroposed from autosome to the X chromosome, favor expression in the testis. Retrogenes are not necessarily deprived of their introns: they may retain them [1175, 1180a, b].

Comment. This author immediately asks if there is a connection between the testicular retrogene-encoded proteins and the cancer testis antigens (CTA)? The prototypes are the NY-ESO-1 (esophagus) antigenic protein, and the MAGE (melanoma) antigens, following suit. The CTAs exert potent immunogenicity in the tumor-bearing hosts. The immunologic response consists of long lasting Th1-type CD4 T memory cell generation specifically reacting to tumor antigens. Utilizing a broad panel of CTAs as vaccines, would disallow the escape of those tumor cells, which block the expression of some of the targeted antigens, but continue to express the others [1181, 1182] (Tables III/I and III/II). (Observe the close to identical abbreviation CTLA4, for cytotoxic T lymphocyte-associated protein).

Cont'd *The fate of introns*. In the mammalian genomes long non-coding ncRNAs encompass many introns subjected to alternative splicing. In ancient eutherians (placental mammals), the X chromosome-inactivating Xist RNA (X-inactive specific transcript) derived from the LNX3 protein-coding gene, *Lnx* (ligand of numb protein-X), which encodes ubiquitin ligases. These ligases are targeted to their substrates by interactions with PDZ proteins (postsynaptic density, PSD-95/discs large/ZO-1 zona occludens, from drosophila). The human example is syntenin (syndecan-binding protein; transmembrane heparan sulfate proteoglycan). Syntenins express two PDZ domains. Syntenin is phosphorylated by proto-oncogene *src* kinases. PDZ proteins, syntenin and syndecan, regulate fibroblast GF-R, Sox4 oncogene proteosomal degradation, and tumor cell locomotion. Examples are: melanoma cell metastases in combination with melanoma differentiation associated gene-9-product protein, *mda*-9/MDA-9; and human breast cancer cells in athymic mice [1183a, b, c–1186]. However, the elimination of syntenin transcription by the HPV oncoproteins [1187] confirms tumor suppressor efficacy of some PDZ proteins. Some of the Xist introns were inherited from the *Lnx*3 gene, others were newly acquired. The splice sites of many mouse long noncoding lncRNAs |GT-AG| are conserved in the human genome; reviewed in Rogozin *et al* [1157].

The amphioxus re-emerges to display its introns. The ancient cephalochordate amphioxus clade was already diverging well over 100 mya (in the Appendix 1). Of the genomes of the three subdivisions, the European/Mediterranean *B. lanceolatum*, the Caribbean/Floridian *B. floridae*, and the Oriental/Chinese *B. belcheri*, (*Amphioxus tsingtauense*), the genome of the Mediterranean amphioxus carries the ancient inserted herpesviral genome [998] (*vide supra*). Obviously, the discovery of this herpesviral gene requires special attention. But the presence of this herpesviral gene is not noted in two detailed reports (one on the *B. floridae* pre-, and one on the *B. lanceolatum* (mediterraneum) postdating the Savin report) on the total genome sequencing of the amphioxus [1188]. Elements of the hedgehog (Hh) genomic complexes appear in the amphioxus. The ancestral genomes of (sonic, desert, Indian) Hh are already detectable in the amphioxus [1189]. The Hh genes direct the development of the CNS, notochord, tail bud and pharyngeal endoderm (for the gill slits). The derivatives of these genes can be recognized in the vertebrate Hh genome, except for the amphioxus' CNS: the driver enhancers of the anterior part and the *zona limitans intrathalamica* of the amphioxus brain disappeared (however, the initiators of the notochord remained). The amphioxus Hh genome underwent two rounds of whole genome duplications in the ancestor of the vertebrates (it must have occurred in the Cambrian sea). The first round of duplication generated the Indian Hh locus and the ancestors of the desert and sonic Hh loci. The second round of duplication established the desert and sonic Hh loci. Among the descendants of the amphioxus genes *Lmbr*1, *Rnf*32 and *Nhej*1 (limb region; ring finger; non-homologous end-joining) strong tendencies are recognized in directing various ontogenic tasks in the zebrafish and in the mouse embryo: notochord; epithelial cell sheet from the oral cavity to the hindgut; and limb development. Highly conserved non-coding regions are preserved in the introns. The driver enhancer of the pharyngeal lining expression is conserved in the introns of the *Rnf*32 gene from amphibians to mammals. The vertebrates conserved non-coding regions embedded into introns 1 and 2, as appear in the zebrafish. These sequences also found in the coelacanth are similar to those of the tetrapods. The enhancer for the Indian Hh endochondral bone formation (there are no bones in the amphioxus; how about the actin "cytoskeleton"?) is located in the intron of the *Nhej*1 gene. Of the two conserved introns, those of the amphioxus are twice the length of vertebrate Hh introns, even though the vertebrate introns preserved the enhancers embedded in them. Highly conserved noncoding regions are preserved in the introns of the tetrapod amniotes' Indian Hh *Nhej*1 gene. The *Nhej*1 gene may be phylogenetically unrelated to the *Lmbr*1 and *Rnf*32 genes. *Nhej*1 is very active in invertebrates (amphioxus, sea urchin, sea anemone and corals), but declined in vertebrate desert and sonic Hh genomes, while remained preserved in the Indian Hh genome); later in desert Hh, the *Rnf*32 gene declined. The phylogenic distance (evolutionary divergence of 535–550 million years) between the amphioxus' and the vertebrate mammalian Hh genomic loci proved to be much too long for the close preservation of its elements. The noncoding sequences are conserved in the last 100 million years among the 3 amphioxus species, but the cephalochordate noncoding sequences are absent in the vertebrate mammalian genomes. Even between the zebrafish and the mouse, the sequence motifs preserved in the introns are distant; however, these motifs are closer between reptilians and mammals [1189]. This article does not mention the herpesviral genome embedded in the amphioxus genome [998]. However, in the Supplementary Material of another article, there is a print-out of the ostreid herpesviral genome as it is integrated into the amphioxus genome [1039]. The genomic sequences from the EBV genome, that are similar to fragments of the V(D)J and Rag sequences [312, 992, 994a, b] could not be recognized in this print-out (and a special search for those sequences is underway). In the amphioxus genome, the Rel (NFκB) and one alternatively spliced Pax (paired box) have gained prominence [1190, 1191]. The Pax3-FKHR (paired box; forkhead homolog) in alveolar rhabdomyosarcoma functions as a fusion oncoprotein (*vide infra*). Before the amphioxus, the Limulus' ancestors operated an active NFκB pathway. The Nematostella united its NFκB/STAT cell survival pathways. In the human genome, the united NFκB/STAT pathway assumes the function of proto-oncogenes (in lymphomagenesis). In both subjects latent viral flora is present (the Appendix 1 refers to a manuscript describing the details of these unisons).

In the amphioxus, the cluster of eight *hairy* genes is the result of gene duplications and intron losses; duplicated genes also lose their *cis*-regulatory regions. The intracellular domain of Notch, as it was cleaved upon its ligand-binding, targets the *hairy* genes. The amphioxus *hairy* genes also express binding sites for the RBP-Jk protein (recombination signal-binding protein jk immunoglobulin kappa), even though the RBP-Jk protein has not as yet been identified, as such. These sites are located in the preserved non-coding cis-regulatory boxes. The amphioxus *hairy* genes (*HA* and *HE*) harbor coding exons interrupted by three preserved introns. The first two introns are located within the bHLH domain (basic helix-loop-helix), and the third intron separates the bHLH domain from the *Orange* domain. This very same intron-exon arrangement is present in the urochordate *Ciona intestinalis* (in the Appendix 1), and has been conserved through vertical inheritance to birds and mammals. The other amphioxus *hairy* genes lost their introns, especially remarkable is the loss of intron III (if it occurred at the divergence of urochordates and cephalochordates) [1192].

Vetulicolians, the Cambrian metazoans (*V. cuneata*, *Yuyuanozoon magnificissimi*, Yunan, China), presented with an arthropod-like body; but their pharyngeal openings (gill slits) at their complex lateral structures of their anterior section classifies them as the ancestors of bilaterian deuterostomes (amphioxus, tunicates, vertebrates), while a lateral divergence might have led to the echinoderms and hemichordates [1193]. Genomics and proteomics of the vetulicolians hold the secret, if the amphioxus herpesvirus [998] already entered them, or infected later their descendants.

Comment. One could speculate that the unicellular and the early multicellular life forms existed in a highly stressed life style, and replicated rapidly without rest, or senescence. In contrast, in the mature multicellular hosts, the overwhelming number of the somatic cell populations encode just one particular task, divide seldom if ever, and keep the rest of their genomes silenced (under the control of silenced promoters). Stem cells are in the minority, divide asymmetrically, but seldom, otherwise they rest. Haploid germ cells exert little activity until after become zygotes (the united sperm and the ovum). Thereafter, they dictate the ontogenesis of the individual. How their intron counts reflect to their life style?

Were the LUCA/LECA resting individuals (for LECA, Table XIII)? Rather, they were engaged in an incessant struggle for their survival. Are their overactive genomes and proteomics comparable with the rapidly replicating oncogenomes of the transformed cells of a multicellular host, descending to the life style of their unicellular ancestors? However, the exact comparative intron counts between resting, and or partially silenced (the somatic) cells, and the constitutively activated transformed cells are still missing (not available).

The hidden introns of human oncogenomics. Rac1 (*vide supra*) stimulates polymerization of actin filaments, and-Jun kinase. The latter action cascades into the mobilization of activating transcription factor (ATF), ETS-like transcription factor (kinase, ELK), and activator protein 1 (AP1). Rac1 induces the phosphorylation and proteolytic degradation of IκBα, thus liberating NFκB. Rac1 directly binds to phosphorylated STAT3/5A, thus promoting their translocation into the nucleus. The B cell lymphoma protein 6 (BCL-6) competes with STAT5 for the same DNA-binding sequence TTCNNNGAA. Rac1 activates p21-activated kinase (PAK1), which phosphorylates BCL-6. Thus, indirectly Rac1 activates these two competitors (activator STAT5, and repressor BCL-6), as depicted in Figure 5 of the original article [1194]. Rac1 may be constitutively activated and as such suppresses matrix metalloproteinase activity. Constitutively activated ERK yielded to suppression by Rac1 [1195a, b]. Smo is an inhibitor of Suppressor of Fused (SuFu) [1196, 1197]. The human tumor suppressor SuFu gene is mapped to 10q24–25. The SuFu protein contains the PEST motif (Pro-Glu-Ser-Thr). Due to alternative splicing, the SuFu protein exists in two isoforms of 433 aa (48 kDa) and 484 aa (54 kDa). Different malignant tumors (glioblastoma, prostate cancer, melanoma) often delete the SuFu gene. The SuFu proteins' primary targets for suppression are the Gli family genes 1–3. SuFu is the major negative regulator of the vertebrate Hh signaling pathway. Figure 1 of the original article is a depiction of predicted protein sequences of the human, mouse and drosophila SuFu genes aligned with phophorylation sites for protein kinase C and kasein kinase, and the PEST motifs marked (proline, P; glutamine, E; serine, S; threonine, T) [1196]. Indirectly, Smo induces reformatting of the cell's cytoskeleton; it inhibits SuFu and suppresses gene patched PTC1-induced apoptosis [1197] (*vide supra*). There is no intron count reported.

While the ancestral Hh genes appear in the amphioxus (*vide supra*), the first phylogenic appearance of the Smo-Ptc pathway has not been established; it is fully functional in the drosophila. The drosophila proto-oncogene kayak is homologous with the mammalian (human) proto-oncogene c-*fos* (c-*fos* → v-*fos*, Finkel-Biskis-Reilly murine osteogenic sarcoma retrovirus: c-*onc* → v-*onc*). One c-*fos* gene is nested in an intron (*fig*). Three promoters create four different transcripts of the c-*fos* gene; two promoters of the phosphatase-encoder *fig* induce two transcripts [1198a, b]. Polycomb gene/protein homologs are also operational in the drosophila and caenorhabditis. Of intron-rich genomes, the caenorhabditis genome is biased the most for the retention of its 5′ end introns [1199]. The polycomb protein homolog *bmi1*/BMI1 is active in the CNS (but not in the somites) of Xenopus embryos (tadpoles) [1200].

The human Hh genes (sonic Hh gene mapped to 7q36.2) exert high morphogenetic pattern in formatting stem cell-like activity in fetal life and maintain hair bulb activity up to well after middle age. In the adult host, interacting with the Wnt/β-catenin pathway (*vide supra*), with the BMi1, PCNA and GFAP pathway (B-cell Moloney murine leukemia virus insertion site, 10p13; proliferating cell nuclear antigen, locus human chromosome 20 with pseudogene 1 at Xq13, pseudogene 2 at 6p12; glial fibrillary acidic protein, 17q21), the Hh genes act as oncogenes. Seven transmembrane smoothened (Smo) proteins serve as Hh ligands and inducers of the patched (Ptch, PTC) receptors. Smo moves from intracellular membranes to the cell surface in order to encounter Ptch. When Smo is marked for ubiquitination by tagging it with the small protein ubiquitin, ubiquitin-specific protease 8 (USP8) interacts by removing ubiquitin from the Smo molecule; it acts as the major inhibitor of Smo ubiquitination. USP8 thus promotes cell surface accumulation of Smo. Overexpression of USP8 secures high Smo levels. This constitutes a prelude to oncogenesis, because Smo signal through its C terminal tail inhibits the tumor suppressor SuFu (*vide supra*). Free Ptch would inhibit Smo; Hh protein attached to Ptch releases Smo for Hh activation of Hh signaling. The phosphorylation of Smo by protein-coupled receptor kinase 2 (Grpk2) establishes Smo as a G protein-coupled receptor-like protein. At the cell surface, Hh induces other kinases (protein kinase A, casein kinase 1) to phosphorylate Smo. Thus, Smo remains the activator of the Hh proteins [1201]. Smo through its C-terminal tail activates the oncogene family Gli protein (glial, glioma, glioblastoma, gliotactin) transcription factors, and small GTPases RhoA and Rac1 ras homolog; (ras-associated C1 botulinum toxin substrate) [1202]. A comparative intron count of the ancestral, the highly evolved physiological, or the malignantly transformed Hh genomes is not available.

In a rare BCR/ABL1 t [9;22](q34;q11.2) transcript in human CML, the BCR intron 13 joined the ABL intron1a. Hydroxyurea not, but imatinib mesylate did induce the disappearance of the transcript [1203]. The 5′ region of the large first intron of the BCR gene contains two exons that are outside of the BCR breakpoint [1204]. The fibroblast growth factor receptor 1 (GFR) gene at 8p11 may form a translocated oncogene: t [8;22](p11;q11) and a FGFR1/BCR chimeric transcript. It may form several other chromosomal translocations with genes ZNF198 at 13q12 (zink finger protein); with gene CEP110 at 9q34 (congenital erythropoietic porphyria); and gene FOP at 6q27b (fibrodysplasia ossificans progressiva). In the translocation with BCR, in the fused oncogene, the breakpoint is within introns 4 and 8 of the BCR and FGFR1 genes, respectively, and results in introns-fusion [1205].

Secretive sarcoma genomics sharing Hh. Embryonal and fusion oncogene-negative alveolar rhabdomyosarcomas of children and young adults are driven by the Hh pathway [reviewed in 340a, 341a]. Another Hh-driven tumor is basal cell carcinoma ("carcinoma epitheliale adenoides") first described in Hungary in 1900 by Edmund (Ödön) von Krompecher [1870–1926], appearing as UV-light induced single cutaneous tumor, or occurring against the background of xeroderma pigmentosum of Ferdinand von Hebra and Moritz Kaposi, 1874. Currently recognized as the autosomal dominant hereditary basal cell nevus tumors (Robert Gorlin's syndrome) [1206]. The signaling pathway consists of Hh ligands' inactivating their transmembrane receptor, the tumor suppressor patches (PTC, Ptch, loss-of-function mutation); then, another transmembrane receptor liberated from under Ptch control (gain-of-function mutation), Smo (smoothened) signals the intracellular oncogene Gli, which signals various transcription factors (*vide supra; vide infra*). The small molecular inhibitor vismodegib (GDC-0449, Genentech) abolishes Smo signaling, and thus prevents Gli activation. In early clinical trials (including patients with medulloblastoma, *vide infra*) with "acceptable toxicity". Vismodegib is effective in the control of Hh pathway driven tumors prominently including basal cell carcinoma. Vismodegib interacts with cytochrome gene CYP2C8 and is considered to be teratogenic in pregnant women. Co-administration of vismodegib with other compounds interacting with CYPs, did not significantly alter vismodegib blood levels [1207a, b]. Vitamin D3 calcitriol binds its receptor, Vdr. Calcitriol (vitamin D-derivative by hydroxylation) inhibits the growth of basal cell carcinoma, an Hh-driven tumor (*vide supra*). Patched (Ptch) is secreting a calcitriol-related compound acting as a natural inhibitor of Hh signaling by keeping Smo (smoothened) suppressed. Loss-of-function mutation of Ptch leads to a deficiency of this substance and to the activation of Smo/Hh signaling. Silenced (unoccupied) vitamin D receptor leads to the development of de-differenciated basal cells. Gain-of-function mutations of Smo independently from Ptch may activate Hh/Gli signaling. Cyclopamine and GDC0449 are the specific Smo/Gli inhibitors. Vdr-signaling through its target genes Cyp24a1 (cytochrome hydroxyvitamin hydroxylase) and Tgm1 (transglutamase) differentiates basal cells [1208a, b].

Embryonal and fusion-negative alveolar rhabdomyosarcomas are driven by the Hh signaling pathway. Rhabdomyosarcoma cells highly express Vdr (vitamin D receptor). The VDR target genes are Cyp24α1 (cytochrome enzyme encoding gene family) and Gli; VDR upregulated Cyp24α1 and downregulated Gli. The expression of muscle cell differentiation markers p27, MyoD and MFR4 (muscle regulatory factor) is increased. Calcitriol inhibited the Hh pathway and activated the Vdr; Activated VDR induced differentiation of sarcoma cells. *In vitro*, rhabdomyosarcoma cells ceased replication with drop of the Ki-67 marker. Epigenomic interventions with demethylators (5-aza-2′-deoxycytidine) to re-activate tumor suppressor gene promoters, and re-acetylators (valproic acid) of oncogene promoters to re-suppress them, exerted antitumor effects in cases of rhabdomyosarcoma and medulloblastoma in Ptch mutant (platelet-derived growth factor receptor mutation) mice [1208a, b, c]. *)

*) *Here, it is worthwhile to view the rich Google collection of spectacular illustrations for acetylation-deacetylation of histone lysins for the regulation of gene expression; A. Csordas' pioneering report from 1990; and from the late 1990s the first descriptions of deacetylase recruitment by the RB protein for the repression of cell cyclin promoters as published in Science (Brehm A et al; Magnaghi-Jaulin L et al).*

Unfortunately, the intron density, introns' size and distribution in these genes at rest and after activation, is not readily available (Tables III/I, III/II, VI and VII).

The interrelationship of ancient unicellular prokaryotes to UV irradiation and calcitriol-like molecular metabolism is unknown. UV-light might have induced genomic activation (such as an elementary Hh-like cascade for accelerated replication), and calcitriol-like protective mechanisms for survival. In comparing the atmospheric conditions of the ancient Earth with the Martian situation in terms of the effects on unicellular organisms in their hydrated, metabolically active, frozen, or dried state, UV-irradiation induced mutations, DNA-repair, and resistance mechanisms [1209].

In the human cerebellum, granule neuron precursor cells are the potential medulloblastoma progenitors. In medulloblastoma, the stem cell renewal gene *bmi*1 is reactivated (B lymphoma Moloney mouse leukemia retrovirus insertion site). This gene, is an antagonist of the senescence- and apoptosis-related genes, $p16^{Ink4a}$ and $p19^{Arf}$ (inhibitor cyclin kinase; alternative reading frame) functioning as tumor suppressors. Naturally and experimentally, the GFAP is the driver of the sonic Hh ligands smoothened, the oncogenic Hh effector (*vide supra*). All SmoA1; $Bmi1^{+/+}$ and SmoA1;$Bmi^{+/-}$ mice developed fatal tumors; SmoA1;$Bmi^{-/-}$ mice, if they developed ectopic tumors, those were non-proliferative and were eventually eliminated. The ectopic tumor cells were GFAP-positive (glial fibrillary acidic protein), but nestin-negative and highly apoptotic. These cells expressed high levels of cDK-inhibitors and reduced cyclin D1 levels [1210]. The Gli protein binds the Bmi1 promoter, and Hh protein and Bmi1 mutually activate one another. Medulloblastoma-initiating cells express the marker CD133/Hh receptor able to receive ligands produced by CD133-negative cells [1211]. The report comes from Heidelberg, that TRAIL (TNF-related apoptosis-inducing ligand) and the chemical KAAD-cyclopamine (3-keto-N-amino-aminoethyl-aminoethylcaproyldihydrocin amoyl) in combination induce apoptosis of TRAIL-resistant medullocarcinoma cells, without affecting normal astrocytes. Cyclopamine is an antagonist of anti-apoptotic proteins Bcl-2 and c-FLIP (FLICE-like inhibitory protein; FADD-like IL-1β converting enzyme; Fas-associated death domain; factor apoptosis stimulator) [1212]. However, vismodegib (GDC-0449) is promising to be more efficient than cyclopamine (*vide supra*) [1207a, b, 1208a, b]. Glioma-initiating cells maintain stemness (self-renewal) and replenish the tumor cell population in glioblastoma. The inhibition of the CXCR4 chemokine pathway (stromal cell derived factor-1, SDF-1), and silencing the sonic Hh/Gli/Nanog network by miR-302-367 switched off the faculties of stemness, self-renewal and tumor cell replenishment [1213].

In synovial sarcoma (reviewed in 340a, 341a) (Figure 34), translocation of the SYT gene (synovial sarcoma translocated) from chromosome 18 to the SSX gene on X chromosome creates SYT-SSX2, a fusion oncogene/oncoprotein: t(X18)(p11; q11). The fusion oncoprotein interacts with the polycomb repressive complex and inactivates the *bmi*1 oncogene. The BMI1 oncoprotein normally activates the ubiquitination of histone H2A. In result, the monoubiquitination of H2A is highly attenuated. Consequentially, the activation/de-activation of polycomb target genes (Wnt, Notch, ephrin) are altered; even the derepressed genes are further up-regulated. Among these genes are jagged2, delta-like 1, nerve growth factor receptor, ephrin receptor, Bcl-2-like 1, ras family Rab40, MSX (muscle segment homeobox, Hox7), and PLDGF [1214]. It is not known if the switching on and off these faculties involved intron-gains in the involved genes.

In the human genome, the pro-apoptotic tumor suppressor gene WT p53 at 17p13.1 is driven by two promoters. Promoter 1 is in the first, not protein-coding exon; the more efficient promoter 2 is in the first intron of the gene. Figure 3 in the cited article shows the sequences required for the activities of p53 P1 and P2 [1215, 1216]. In the human genome proteome, the ret/RET proto-oncogene oncoprotein (re-arranged during transformation) is mapped at the pericentromeric region of chromosome 10 (in mitosis, kinetochores attach the spindle fibers to the centromeres, which link the sister chromatids). Single nucleotide polymorphism of intron 1 of the gene is associated with the onset of Hirschsprung disease (colonic a-ganglionosis) [1217, 1218].

It has been estimated that 35 % of human genes contain elongated introns within the 5′ UTR (untranslated region). While 3′ UTRs are longer than 5′ UTRs, they contain much lower numbers of introns, than the 5′ UTRs. The highly expressed genes contained the shortest 5′ UTR introns and tended to eliminate their introns entirely. Genes of regulatory roles retained an enriched 5′ UTR intron supplement (5′ UTR introns overrepresentation in regulatory genes). Genes of non-receptor protein tyrosine kinases retained 5′ UTR introns with a conserved DNA motif within [1219]. The introns produce distinct mature mRNAs through alternative splicing, thus could contribute to protein diversification. However, the 5′ UTR region is non-coding, and the splicing of 5′ UTR introns may have no effect on the aa sequences encoded into proteins. Almost 95 % of human genes are alternatively spliced.

Among the non-receptor tyrosine kinases (NRTK, lacking transmembrane domains) are the most highly overexpressed genes. Some of these genes are proto-oncogenes, such as the Rous sarcoma *src*/SRC gene/protein, or Janus kinase, JAK, or the tumor suppressor genes, cell cycle inhibitors, *wee*/WEE (cited in [73]). Others are genes engaged in cytoskeleton (actin) organization and transcription regulation. The NRTK 5′ UTRs appear to be evolutionarily conserved, and the human 5′ UTRs (listed in Table 1 of the cited article) may share a common regulatory mechanism acting via a shared motif. Such a representative DNA motif matched best the MAZ transcription factor. The myc-associated zink finger protein (MAZ) transcription factor regulates VEGF (and vice versa). Neoformed blood vessels in tumors (glioblastoma) overexpress MAZ and downregulate miR-125b. VEGF suppresses miR-125b production, which results in overexpression of MAZ; through transcriptional activation, MAZ induces VEGF production. Suppression of MAZ by shRNA renders vascular endothelial cells non-migratory and deprived of the faculty of tubule formation. The VEGF-R inhibitor vandetanib breaks the circuitry [1220].

The genes of the MHC protein complex, cytosolic ribosomes, hemoglobin complex, glutathione transferase, and transmembrane transporters were most enriched by the 5′-proximal coding region introns (5PCIs). The most 5′ proximal introns in their transcript are the 5 UIs; the most highly expressed genes were enriched by short (<1 kb) 5UIs (5′ UTR introns). Splicing-dependent enhancement in gene expression is regulated both by introns' position and size. The most highly expressed genes may be under selective pressure to operate with short introns. These align their enhancer elements closer (in the first introns) to the transcription site. The most proximal intron to the transcription start site contains most of the regulatory elements. Most nucleotide level regulatory elements are short (<15 nt) and could be contained within short (<1 kb) 5UI. The translational efficiency of upstream AUGs (initiation start codon adenine, uracil, guanine) of 8–12 nts for methionin tRNA is slow; highly expressed genes use downstream AUGs. Intronic AUGs are spliced out, thus blocking the message to reach the ribosome. However, this event alone cannot fully explain the overrepresentation of short 5UIs in the most highly expressed genes.

Intron-affected splicing-dependent translation efficiency is working in the mTORC (mammalian target of rapamycin complex) pathway, where exon junction complex follows splicing. In the nutrient-, stress-, and energy-sensing checkpoint kinase, mTORC, activated ribosomal protein S6K1 reacts with mRNA of SKAR (S6 kinase; Aly/REF-like target: *S. cerevisiae* mRNA export adaptor, Yra1/RNA export factor), also known as polymerase delta-interacting protein (POLDIP). The enhanced translation efficiency of the spliced over the nonspliced, mRNA takes place in the exon junction complex (EJC). The S6K1/mRNA SKAR is deposited at EJC during splicing. The SKAR-recruited activated S6K1 in the mRNP (ribonucleoprotein) serves as a conduit between mTORC signaling and the first pioneer round of translation and protein synthesis [1221–1224].

Carcinoma cells hide their introns. Aberrant alternative splicing was recognized in renal cell carcinoma, involving serine-arginine-rich (SR) proteins. The genes encoding the classical SR proteins are well identified. These proteins bind splicing enhancers located in exons (ESE), or in introns (ISE). The splicing silencers bind also heterogenous nuclear ribonucleoproteins (hnRNP). In the spliceosome, the ESE/ISE compete with the hnRNP. This competition may result in alternative splicing of mRNAs of tumor suppressor genes, or oncogenes. The mRNAs of oncogene Gli and tumor suppressor gene CEACAM (carcinoembryonic antigen-related cell adhesion molecule) in kidney cancer cells were subjected to alternative splicing (up- and downward regulation, respectively). Separate tumor deposits from the same primary tumor, due to alternative splicing, acquire different genomics. In the same tumor, the progenitor tumor cells may be preserved together with newly evolved daughter cells, which underwent alternative splicing. The kinase pathway, in which phosphorylation of the splicing factors takes place, is that of the oncoprotein PI3K/Akt. When splicing factors alternatively splice oncosuppressor mRNAs (loss-of-function effect), or oncoprotein mRNAs (gain-of-function effect), they themselves become oncogenes. The archetypal ASF/SF2 (alternative splicing factor/splicing factor) competes with hnRNP and can suppress CEACAM, or positively correlate with it. The splicing regulatory element of the CEACAM molecule is in its exon 7; one of these motifs is reactive with ASF/SF2. Alternative splicing of proto-oncogenes Gli-1, Rac1 (Rho-like GTPase, *vide supra*), and Ron by ASF/SF2 is not yet proven [1225, 1226]. The expression of the Ron oncogene (recepteur d'originenantais), both as its mRNA and oncoprotein is high and persistent in many cancers, and especially in lung adenocarcinoma cells [1227].

The complex prostatic adenocarcinoma cells (frequently inflammation-induced, and rarely vertically inherited), recently reviewed in Ref [73], are usually viewed first in the transrectal needle biopsy specimens. This intervention carries a high risk of septic infections, and was declared to be a "barbaric procedure" by Gettman, Mayo School of Medicine, Rochester, Minnesota (Bloomberg News online 2013). The more elaborate, but safer transperineal biopsy is recommended. The genomics/epigenomics of the primary tumor could immediately be determined. Blood tests would reveal the balance of the microRNAs. In advanced disease, the genomics of circulating tumor cells taken repeatedly from the blood and captured in highly specialized chips would unmask the changing genomics of the evolving tumor, as a guidance to targeted therapy. The tumor cells are very likely stem cell-derived, and miR-106b-25-sustained [1228]. Stem cell maintenance genes Oct3/4, Sox2, Nanog, Krüppel-factors, and Bmi1/Hh, the TMPRSS2 fusion oncoprotein, and the v-*ets* avian erythroblastosis virus E26 transformation-specific (Ets) family genes (Ets-related genes, ERG) (*vide supra*) provide the underlying layer of basic activity, beneath an overlying layer of hypoxia-inducible factors, PI3K/Akt, NFκB, Wnt/β-catenin, SDF-1/CXC/R4, and many other factors operating in a Rb-, p53- and PTEN-free (already eliminated) environment. The joint effect of eliminated PTEN and activated c-MYC spell out high risk for lethal outcome (online report by J. Xu, Wake Forest, Winston Salem, NC). Tumor cells with inactive PTEN, immediately activate the ancient PI3K/Akt "cell survival pathway" (*vide infra*).

BRCA-mutated prostate cancers await their turn. Increased lipooxygenase activity indirectly activates the Src classes of oncogenes. An array of highly individualized tumors is the result. Hh receptor PTC (patched) and glioma-associated oncogene homolog 1, GLI (subfamily Krüppel-like zink-finger protein) contribute to the stemness of $CD44^+$ tumorigenic cells. The $p63^+$ tumor cells sustain an overexpressed sonic Hh cascade. Epithelial-to-mesenchymal transition, or neuroectodermal transdifferentiation, may set in, as the androgen-deprivation (AD) hormone- and radiation-treated tumors become resistant to further AD. The peptidoglycan perlecan is invoking Hh signaling. Matrix metallo-proteinases (MMPs), induced by cyclooxygenase 2 (COX2) overproduction, are promoting tumor cell locomotion. The tumor cells switch from mitochondrial oxidative phosphorylation of mature somatic cells to Warburg-type glycolysis (practiced by free-living unicellular ancestors in the anaerobic environment of the ancient Earth, *vide supra*). Upregulated phosphoglycerate kinases perform enhanced glycolysis to provide energy for the continuous cell proliferation [1229]. In this complex and highly individualized process, intronic changes have not been as yet pinpointed. Primary therapy in 2011–2013 (if the elderly patient qualifies) is not yet one of the individualized targeted modality (even though the genes predicting lethal outcome are claimed to have been recognized). The standard therapy remains radical (robotic) surgery with the highest cure rate for primary tumors of Gleason <6 grade; non-curative hormonal testosterone-deprivation (at the risk of metabolic syndrome induced; stem cell accumulation with neuroectodermal transdifferentiation) (Table XXIX) and radiotherapy with external electron beams of precision-localization to the prostate, and interstitial brachytherapy. Radiotherapy carries the risk of radiation-induced late carcinogenesis (undifferentiated urothelial adenocarcinomas, neuroectodermal carcinomas, soft tissue and bone sarcomas). The incidence of radiotherapy-induced malignant tumors is as high as 1/250 at 5 years, and 1/70 at 10 years [1230–1232]. The incurable metastatic disease is treated for prolongation of life with sipuleucel immune dendritic cell therapy; docetaxel or cabazitaxel (Jevtana) chemotherapy; enzalutamide (Xtandi), or ARN-509 (Aragon), for androgen receptor blockade; abiraterone (Zytiga, Janssen) for anti-CYP17 gene blockade; Prostvac (Bavarian Nordic) active immunization vaccine therapy; and various forms of investigational oncolytic viral therapy. The intronic changes in hypertrophic benign prostatic cells in comparison with primary untreated tumor cells, or late radiation-induced tumors, are not known; a comparison of introns in normal cells at the margins of the specimen with that of the malignantly transformed cells could not be found (even though very large numbers of surgical specimens would be available for the study).

The phosphoinositide phosphatase of the tumor suppressor gene PTEN antagonizes PI3K/Akt. Otherwise, the phosphorylated phosphatidylinositol (from biphosphate to triphosphate, $PIP_2 \rightarrow PIP_3$) activates a number of proteins with pleckstrin homology domains (platelet-proteins of leukocyte kinase C substrate with lysine-serine-threonine-arginine KSTR string). Prominent among these proteins is the serine/threonine kinase phosphorylated Akt oncoprotein. In a cascade, Akt activates the entire ancient cell survival pathway for nutrient uptake, mitoses, and locomotion. The PTEN phosphatase (*vide supra*) dephosphorylates PIP3 → PIP2, and thus de-activates the cascade. Of several PTEN-targeting prominent microRNAs are mi-R 17, mi-R 19, miR-22, miR-25, and miR-302. In the cytoplasm ribonuclease III Dicer converts pre-mi-Rs into mature mi-Rs. Most mi-R genes are located in introns flanking protein-coding exons. The seed sequence of the mi-R's corresponding to the seed match of the targeted mRNA's 3′ UTR establishes their union. The mi-R-targeted mRNA will not be translated into polypeptide chains. PTEN-targeting mi-Rs prevent production of the PTEN phosphatase, and thus allow the $PIP_2 \rightarrow PIP_3$ reaction to take place, and so initiate the AKT cascade. This reversible reaction of healthy cells becomes oncogenic upon mutational deletion of the PTEN gene, or constitutive activation of the anti-PTEN mi-R cluster. In prostate cancer tissue, mi-R22 emerged as a proto-oncogene c-*myc* activator. Exogenous 3′ UTR-less PTEN did overcome mi-R22, proving that mi-R22 acted by PTEN de-activation. The mi-R106b-25 cluster emerged as another PTEN de-activator proto-oncogene. The miR-106b-25 cluster in combination with c-Myc or MCM7 transformed wild type mouse embryonic fibroblasts. The genes of this mi-R cluster reside in intron 13 of the DNA helicase MCM7 gene (minichromosome maintenance protein). Exogenous 3′ UTR-less PTEN could overcome mi-R106b-25. Even though the MCM gene contains oncogenic mi-Rs, in itself, alone, it was not oncogenic in the mouse prostate. Chromosome 7, with intronic residence of miR-25, miR-93 and miR-106b, expresses amplified genomes in patients with prostate cancer; chemokine CXC12 (stromal-derived factor, SDF-1) and its receptor CXCR4 are at high levels [1228, 1233].

There emerge the newly discovered fusion oncoproteins FGF-R1/TACC1 or 3 (transforming acidic coiled-coil protein) in glioblastoma multiforme. The TACC proteins were known for their involvement in the formation of the mitotic spindle. This structure ascertains that at cell division chromatin pairs are segregated so, that each daughter cell receives one of each chromosomes. In the human genome, the FGFR3 and TACC3 genes are located 48 kb apart on the short arm of chromosome 4p16; the locations of FGFR1 and TACC1, and FGFR2 and TACC2 are mapped in close association at 8p11 and 12q26, respectively [1234]. The TACC proteins (one in drosophila and caenorhabditis, each; three in the human genome) interact with clathrin (heavy and light chain protein sequences coating insides of cell compartments), ch-TOG (colonic-hepatic tumor overexpressed gene) product protein, and Aurora A kinases. Further, the TACC proteins interact with the Notch4/Int3/Ankyrin repeat cluster complex.

Natural products contain biologically active molecules: any intron changes? A long list of natural products is available for alleged preventive or even curative purposes; most evidence for this is empirical and derives from ancient to modern Chinese medical practices. For the maintenance of remission with occult tumor dormancy, these empirical medications are recommended by dietitians, nurses and paramedical practitioners, but not by clinicians of academic medicine. However, investigators and clinicians of academic medical institutions on occasions provide the list, the description, and the graphic presentation of these compounds [1229]. It is extremely interesting to read that several natural substances, curcumin being prominent among them, are claimed to switch off mammalian oncogenes. However, Reuthers Health Report online on April 26, 2013 claims that patients with prostate cancer on dietary supplements in addition to standard treatment gained no extra benefits (P. Posadzki, Korean Institute of Oriental Medicine in Daejeon). This particular tumor of extremely high incidence is stalling in the past era based on the beliefs that antihormones and radiotherapy kill all tumor cells, whatever their genomic background is. This tenet applies to the low grade Gleason <6 tumors with over 10 years possible survivals, and disregards the Gleason >8 tumors with their rapidly lethal further mutations. This tumor displays an extraordinary tenacity of the transforming RNA/DNA complex to prevail in the state of autophagy, and to re-emerge with several sequential new mutations (the neuroectodermal transition is one of them); or with gene amplifications, the most treacherous of these being the physiologically silenced, but now newly activated testosterone-production within the tumor cell itself for the physiologically active androgen receptor, in autologous circuitries (as pointed out in every oncology textbook). The tenacity of the "malignantly transformed RNA/DNA complex" in this tumor is such, that the metastatic castration-resistant disease remains incurable. Once transformed (regressed to its primordial state), the RNA/DNA complex resists in order to sustain the living cell. At the Tumor Immunology and Biology department of the Center of Cancer Research at the NCI/NIH, Bethesda MD, the new metastatic prostate cancer immunotherapy protocol combines CTLA4-blockade with ipilimumab and the Prostvac vaccines (*vide infra*). The natural anti-oncogenic agents (switching off oncogenes in tumor cells *in vitro*) gallo-catecholamines, curcumin, lycopene, genistein, quercetin, elagitannin polyphenols, resveratrol, silymarin, sulforaphanes, tocotrienols, etc. administered singularly may not work against an established tumor *in vivo*; have they been tested sequentially, or in combination, preferably in purified preparations in academic medical institutions?

Introns in ontogenesis. In embryonic life, ligands delta and jagged and receptors Notch direct part of the development of the nervous (including the eye), and hematopoietic (including NK and T cell differentiation) systems. The mouse *notch*4/*wnt1* genes (and also the fibroblast growth factor, FGF, genes) were targeted by the mouse mammary tumor virus (MMTV) for the insertion of its retrotranscribed provirus (*vide supra*). The provirus activated the intracellular domain (ICD) of the *notch*4 gene. The ICD of the *notch*4 gene is the *int*3 oncogene. Notch was known as an inhibitor of differentiation in the female mammary gland. Contradictory reports indicate that Int3 may suppress mammary gland development, or it may mediate branching morphogenesis of mammary epithelial cells. One of the intranuclear targets of ICD (Int3) is the protein-coding *rbpjk* gene (encoding recombination signal binding protein for immunoglobulin kappa J region-like). This transcription repressor gene is related to the drosophila gene 'suppressor of hairless', *su*/SU. Tacc3 inhibits the Int3/Rbpjk transcription of the Hes1 reporter gene (drosophila hairy and enhancer of split, *hes* gene homolog). These genes are members of the highly conserved IGKJRB family (immunoglobulin kappa J-region recombination signal binding protein). The 13 exons of the 67 kb functional IGKJRB gene are localized in the human genome at 3q25; and the locus of its two pseudogenes is at 9p13. In exons 2–11 the mouse and human IGKJRB genes are 98 % homologous; the homology of their exon 1 sequences is 75 % [1235a, b–1238]. However, it is not known how many introns these genes harbor, and how the intron numbers change when a silenced gene becomes activated. It is widely known that the mRNAs are deprived of their spliceosomal introns before entering the ribosome.

Early descriptions of the RBP gene (recombination signal binding protein) failed to mention any introns. The primary structure of the RBP-Jx (drosophila) gene is printed out with its termination codon, its DNA binding site and the aa identities marked; but the introns are not. In the mouse, the vertically inherited *hairy* genes (*hes*1) express RBP-Jk binding sites in their 5′ region. In the developing mouse brain, glutamatergic neurons express RBP-L exon 1 transcripts. RBP-J interacted with GST-ANK fusion protein (glutathione-S-transferase; ankyrin), but RBP-l did not. The RBP-L gene is depicted with its 12 exons together with the gene transcript (Northern blot for RBP-l mRNA; Southern blot with cDNA probe). Only the 494 bp half of the first introns is mentioned (as it was replaced with a nuclear localization signal peptide gene) [1239].

Of the RBP-JL mammalian gene product proteins, the homologs of the drosophila suppressor of hairless, *su*/SU gene, RBP-J tethers the intracellular region of the Notch receptor/Delta ligand complex for signal conduction. It fulfills this function in nematodes, insects, Xenopus and mammalia. In the DNA strands, RBPs recognize the (C/T)GTGGGAA consensus sequence for their attachment. In immortalized human B lymphocyte blasts, RBP-J tethers the EBV EBNA2 antigen. In the KSHV-infected cell, the latent-lytic switch transactivator, the ORF50 protein, activates the ORF46 promoter at nucleotide 69,425 of the viral genome. The ORF46 promoter gene expresses three binding sites for the RBP-Jk protein, thus conferring responsiveness to the ORF50 protein. A negative regulatory region in the ORF46 promoter confers resistance to the ORF50 protein. The lytic cycle inducer, sodium butyrate, relieves the negative regulator in the ORF46 promoter. The ORF46 gene encodes a uracil DNA glycosylase DNA repair enzyme [1240]. The KSHV LANA protein promotes viral genome replication in latently infected cells. There are interactions between LANA and cellular proteins: HMGA/B1 (high mobility group A/B), TRF1 (telomere repeat binding factor), XPA (xeroderma pigmentosum group A), PYGO2 (pygopus homolog), PPA/B2 (protein phosphatase 2A/B subunit), TIP60 (Tat-interactive protein), and RPA1/2 (replication proteins). However, "LANA expression in telomerase-immortalized endothelial cells resulted in telomere shortening" [1241]. How could LANA contribute to malignancy, if it sustains telomere shortening? The adenovirus capsid protein promoter has a RBP-J-binding sequence (TGGGAAAGAA); bound RBP-J suppresses the promoter. Unfortunately, the RBP-J interactions in the early generations of genetically engineered oncolytic adenoviruses did not receive any special attention; reviewed in [1242]. It was first described in association with Notch signaling, the CSL family of DNA signaling proteins: CBF1, Su(H), Lag1 (CCAAT-binding factor; suppressor of hairless, drosophila; longevity assurance gene, yeast, caenorhabditis). The CBF gene was equated with RBP-J [1243]. The human *hes*/HES gene (mammalian homolog of the hairy enhancer of split; *hes* gene in the drosophila), the Notch protein's target gene, HES-1 [1244], also binds RBP-J proteins. In ontogenesis, the Notch/RBP-Jk interaction is essential from placentation to individual parenchymal structure formation, such as chondrocyte regulation in enchondral bone formation [1245]. When a human chondrosarcoma cell line was observed to undergo differentiation *in vitro* [279], these genomic interactions were not available for studies, or for considerations.

The length of the EGFR intron 1 reflected to the clinical outcome of the cancer: cancers with shorter introns were more lethal and less responsive to erlotinib

[1246]. However, the polymorphism of EGFR intron 1 failed to correlate with EGFR expression and with the prognosis of the cancer [1247]. The pancreas transcription factor1 (PTF) maintains the differentiated state of the exocrine organ. It contains the RBP-L paralog [1248]. The stem cell maintenance gene and proto-oncogene/oncoprotein *sox*2/SOX2 is a target of the Notch/RBP-Jκ signaling. The neural stem cells generate new neurons throughout the life time of the host. Premature differentiation exhausts the stem cell pool. RBP-Jk signaling was essential for the development and maintenance of the hippocampal stem cell pool. Notch signaling enhances the activity of the *sox*2 promoter. Retrovirus-transduced RBP-J neurospheres, and NucleofactorII electroporationally plasmid-transfected *sox*2 luciferase reporter constructs were used in the assays designed to determine this activity. Into expression vectors, the cDNA of the Notch intracellular domain (NICD), or that of the dominant-negative RBP-Jk protein were cloned. The results were evaluated in part by Leica confocal microscopy. Immunoprecipitated chromatins were examined by Master Mix Agilent. Western blotting for proteins in nuclear extracts was performed in polyvinylid difluoride membranes with appropriate primary antibodies for tubulin, Notch1, and vimentin. It was established that Notch signaling remained active in stem cells. Stem cell maintenance subsided after RBP-Jk inactivation. RBP-Jκ activity was essential for stem cell maintenance in the hippocampus. The dominant-negative RBP-Jk expression construct inhibited NICD-induced activation of the *sox*2 gene promoter. Cells overexpressing NICD contained increased amounts of endogenous Sox2 protein. Sox2 restored neurosphere formation of RBP-Jκ-defective stem cells. In restoring stem cells, Sox2 used sonic Hh and FGF signaling; Sox2 utilized the proto-oncogenic Wnt/β-catenin and PI3K pathways [1249]. It should be noted, that these growth factors are not activated in a constitutive manner in the stem cells (author's comment).

Intron changes in reactive DCs and immune lymphocytes. RBP-J expression-deficient DCs failed to induce antitumor immunity in a system, in which normal DCs co-inoculated with tumor cells exerted antitumor immunity. Tumors containing RBP-J-deficient DCs failed to attract immune T and NK cell infiltrates [1250]. In these cases, tumor-draining lymph nodes remained small and non-reactive. In these tumors, the expression of chemokine receptor CCR7 was downregulated; this chemokine receptor attracts the migration of antigen-activated DCs. It is the Notch receptor responding to its ligands jagged1/2 and delta-like Dll1/3/4 that signals for cleavage of the NICD. Translocated to the nucleus, NICD activates the RBP-Jk gene, whose protein product activates the *hes*/HES system (*vide supra*). The RBP-Jκ-defective DCs failed to mature properly, as shown by their weak expression of MHC II [1250].

The proto-oncogene *rel* was incorporated by a reticuloendothelial retrovirus (c-*rel* → v-*rel*), the causative agent of lymphoma in turkeys. The gene product protein of this proto-oncogene is the nuclear factor kappa B (NFκB, B cell lymphoma). This gene product protein is activated in the cytoplasm, when its inhibitor IkB is eliminated by ubiquitination. The liberated NFκB protein translocates into the nucleus, where it activates several genes of pro-inflammatory and proto-oncogenic faculties. The LTR of the avian reticuloendothelial virus is naturally incorporated into the genome of Marek's avian (turkeys) oncogenic herpesvirus; reviewed in [27].

The RBP-JL proteins activate the intracellular domain of the Notch receptor. The NICD protein (Notch intracellular domain) entering the nucleus activates the *hes*/HEs system (*vide supra*). The RBP-Jκ protein and negative transcriptional regulator (Rep-κB) acting on the NFκB promoter appear to be closely related, or actually identical [1251]. The Rep-κB binding motif CATGGGAA (*vide supra*) is almost identical with the RBP-Jκ consensus site CGTGGGAA. Rep-κB binds the *hes*/Hes promoter, like RBP-Jκ does (hairy enhancer of split, drosophila). Specific antibody to RBP-Jκ reacts with Rep-κB. Both RBP-Jκ and Rep-κB suppress the NFκB promoter. However, when NFκB was constitutively activated in lymphoma cells, large amounts of Rep-κB/RBP-Jκ proteins in the nucleus failed to suppress the promoter. Truncated Notch (NICD) transfers to the nucleus and acts constitutively on the NFκB promoter: it may induce very high NFκB production [1251].

Acute allograft rejection in RBP-J deficient mice was very much accelerated, consequentially to the rise of T cells in all categories (Th1, Th2, Th17). A normal number of $CD4^{+}CD25^{+}Foxp3^{+}$ Treg cells was sustained, but these cells failed to suppress immune T cells [1252]. Gain-of-function mutation of the Notch receptor and its association with RBP-J transforms human T lymphocytes, and with association of EBV, human B lymphocytes into T cell lymphoma/leukemia, and B cell lymphoma. The mutated Notch protein binds RBP-J, ZNF143, ETS and RUNX genomic sites (zink finger protein; E26 avian leukemia virus oncogene; runt-related transcription factor). The ZNF143 is the human homolog of the Xenopus' transcriptional activator, Staf. Co-binding of RBP-J to Notch/ZNF143 sites frequently occurs. Notch may bind to promoters or enhancers of genes MYC, DTX, IGF1R, IL-7R, and the GIMAP cluster (human homolog of Xenopus gene deltex; GTPase of immunity associated protein) [1253]. The vertebrates-acquired GIMAP cluster consists of nucleotide-dependent dimerized GTPases. GIMAP3 regulates mtDNA segregation in leukocytes including T cells.

The latest on intron biology. Spliceosomal introns showed up in eukaryotes; neither spliceosomes, nor spliceosomal introns exist in bacteria. Inherited introns may be conserved through a very long evolutionary trajectory, or may be eliminated (lost). Intron losses occur by genomic deletions, or by reverse transcriptase-mediated DNA generation (reverse transcription). Spliceosomal introns are removed by excision from pre-mRNAs in the spliceosome. Eukaryotes are able to gain new spliceosomal introns. Intron gains occur by several, often overlapping mechanisms, listed as intron transposition, transposon insertion, tandem genomic duplication, intron transfer, intron gain during dsDNA breaks, insertion of group II introns, and intronisation due to unknown mechanisms [1254]. In the phenomenon of intron transposition, an excised spliceosomal intron reinserts itself (splices back into) its own (spliceosomal retrohoming), or into another (reverse transcription template switching) mRNA. These mRNAs with the transposed intron are usually reverse-transcribed into cDNAs. These cDNAs recombine with the genes in the original locus, during which process loss of other introns may or may not occur. The transposed introns retain their donor, acceptor and lariat branch point splicing sequences (BPS) located upstream at the 3′ splice site UACUAAC, or UCCUUAAC, or AAUUAAC with conserved adenosine residue and associated polypyrimidine tract, PPT. Transposon insertions occur into AGGT (protosplice site) sequences: donor site AG|GT, acceptor site AG|G [1254].

A new intron was created by segmental genomic duplication in the ancestor of jawed fish some 500 mya; this intron was absent in invertebrate ancestors (Florida lancelet, *B. floridae,* and sea urchin, *S. purpuratus*). In vertebrates, this intron is present in the ATP2A1 gene, but absent in the ATP2A2 gene. These genes encode sarco-endoplasmic reticulum calcium ATPases. The ATP2A genes remained intron-less outside the jawed vertebrates. The ancestral chordate genes underwent two rounds of duplication, and then lost one copy. The vertebrate tetrapodes possess three paralogues: ATP2A1, ATP2A2, and ATP2A3. All vertebrate ATP2A1 genes contain a new intron between the first band second nucleotides of the G310 codon at an AGGT motif. The ATP2A2 and ATP2A3 genes remained intron-less. In the human genome, the intron in the ATP2A1 gene splits a single ancestral exon into two exons. This intron exists conserved from teleost fish to tetrapods; thus it must be over 420 million years old. As reproduced in the laboratory, a mini-gene construct contained the duplication of exon 8 of the ATP2A2 gene. Intragenic tandem replication, the segmental duplication of the AGGT site, created the new intron so, that exon 8 remained exon 8, and exon 9 separated by the new U2-type intron, but without disrupting the protein-coding sequence of the gene. Autosomal recessive loss-of-function mutation of the human ATP2A1 gene results in the absence of the gene product protein SERCA1 (sarco-endoplasmic reticulum calcium). Consequentially, in Brody's disease, the exercised muscles twitch fast instead of relaxation [1255].

The DNA ds-breaks severe phosphodiester bonds in both strands of the DNA double helix. These lesions evoke the repair mechanisms homologous recombination and/or non-homologous end-joining. An undamaged template of a sister chromatid is required for the homologous recombination. Non-homologous end-joining is carried out without a template. This latter process may result in intron gain. *Daphnia pulex* utilizes this mechanism for its intron gains; the new introns are recruited from the mitochondria. Mitochondria represented by the α-proteobacteria, which were engulfed by, or fused with, an ancestral archaea (*vide supra*).

Mitochondria yielded the group II introns, the mobile ribozymes, transferred from the mitochondria into the ancient host cell genomes. In theory, these group II introns became the spliceosomal introns, as the united archaeal-proteobacterial fused cell formed its nucleus, and thus became a eukaryote (*vide supra*). Group II introns, as such, thus disappeared from the eukaryote genomes. In practice, the insertion of group II introns (from *Lactococcus lactis*) into the nucleus of a eukaryote (*S. cerevisiae*) abolished gene expression. The pre-mRNA harboring a group II intron splices in the cytoplasm, and then succumbs to nonsense-mediated mRNA decay (NMD), thus it is not translated. Insertion of group I intron in a Tetrahymena pre-mRNA, or another yeast's spliceosomal intron, escape NMD and are translated.

Comment. It appears as if eukaryotes protect their genome against the invasion by group II introns by expunging them. Nuclear membranes separated the soma (cytoplasm; ribosomes) from the genome. The nuclear genome harboring group II self-splicing ribozyme introns was under selective pressure to express itself through spliceosomal introns. The *L. lactis* group II intron commenced its functioning in the yeast (*S. cerevisae*), but it was subjected to nonsense-mediated mRNA decay, and the mRNA rendered intron-free in the spliceosome was not translated into protein. In contrast, a Tetrahymena group I intron functioned well in yeast [1112]. Some virulent lactococcus phages (*Lactococcus lacti* s Orf14 of phage bIL67) carry genes to encode ss nucleic acid-binding proteins (SSB) of archaeal derivation, replaying a unique ancient situation. Phages usually pick up genes from their own host bacterium; however, this Siphoviridae gene derived from an archaea. In general, bacteria received genes horizontally from archaea, but not vice versa (bacterial genes were not readily if at all transferred horizontally into archaea) [1256]. This leads to the problem, if ancient DNA metabolism of archaea and eukaryota were similar, but different from that of eubacteria/prokaryota. What was the derivation of the lactococcal group II introns? Actually, how did eukaryotic nuclear spliceosomal introns and cytoplasmic spliceosomes/AGO-ribosomes evolve: by transforming, or by replacing the group II introns? These remain unanswered questions.

Eukaryotes conserved the spliceosomal proteins from the yeast up to vertebrates and mammals. The highly conserved largest Prp8 protein (pre-mRNA processing) is in central position among the over 100 spliceosomal proteins. It expresses a reverse transcriptase RT-related domain. Very similar RT domains are encoded by prokaryotic retroelements, but prokaryotes do not possess spliceosomes; spliceosomes are eukaryotic acquisitions. The Prp8 RT domain lacks nucleotidyltransferase activity; instead, it expresses a fungal RT-related helical segment and the thumb domain of retroviral RTs. The fold of the C-terminal of the Prp8 protein resembles that of an H RNase. The retroelement evolving into the gene of the eukaryotic Prp8 protein might have been a mobile retrotransposons in the genome of a prokaryotic (eubacterial) ancestor. After gaining nucleic acid- (RNA-) binding domain(s) from another inactive retroelement(s) and so encoded RNA-binding proteins, the pre-Prp8 gene was transferred by the α-proteobacterium into the archaeal host for the unison that created the first eukaryote, where it performed in the cytoplasm its first transesterification reactions. It is much less likely that Prp8 evolved from RT gene of a retrovirus [1257].

Comment. In the primordial cells, and in the early multicellular eukaryotes, the RNA/DNA complex ruled the intron-rich primordial cells (LECA/LUCA of Csűrös et al). In multicellular eukaryotes, the single stem cells retain the ancient ruler, the RNA/DNA complex. So are the zygote-derived early embryonic cells. The dominant somatic cell mass in the advanced multicellular hosts is under the strict rule of a DNA/RNA complex. Once transformed, in these cells the RNA/DNA complex regains its primordial vitality manifested by the expression of introns and their derivative multiple RNAs. The presumably intron-rich transformed cell re-gained the immortality, independence and persistence endowed to the original unicellular eukaryotes, which survived the most severe conflicts in the environment of the ancient Earth. However, the accurate intron counts in support of this theory are not yet available.

Selected abbreviations

TERT, telomere reverse transcriptase. Click TERT promoter mutations in skin cancer by Pópulo H et al. TERT promoter mutations with BRAF V600E mutation thyroid cancer by Liu X et al. TERT promoter mutations cutaneous melanoma by Heidenreich B et al. Effect of four chemotherapeutic agents, bleomycin etoposide cisplatin cyclophosphamide, on DNA damage in spermatogonial cell line by Liu M et al.

DCAF, damage-specific DNA-binding protein cullin-associated factor. Click proliferating cell nuclear antigen PCNA-dependent regulation via ubiquitin ligase complex by Abbas T et al.

NDUFA gene, nicotinamide adenine dinucleotide dehydrogenase ubiquinone alpha subcomplex, encodes mitochondrial protein MwFe (ferrous sulfate). Click Heterozygous mutation in the X-chromosomal NDUFA gene in a young female patient by Mayr JA et al. A novel NDUFA1 mutation in neurodegenerative disease by Potluri P et al. Downregulation of NDUFA in basal cell carcinoma by Mamelak AJ et al.

Limulus polyphemus horseshoe crabs, carriers of Rel homology NFκB domains and their inhibitors IkappaB (IκB) Basith S et al PLoS One 2013;8(1):e54178; Wang XW et al Proc Natl Acad Sci USA 2006;103;4004–4009.

Elementary Epigenomics

The DNA Genome Issues a mRNA-Encoded Order; A microRNA Cancels It

The restless epigenome. Most histones form nucleosomes (octameric DNA compaction units). An array of transcription factors share DNA-binding sites in forming complexes with the chromatin modifying enzymes. These enzymes catalyze the give and take of low MW adducts (methyl, acetyl, phosphate groups). Some eukaryotic chromatins conserved motifs of archaeal and bacterial precursors of eukaryotic cells; even the phage of the mitochondrial 5α-proteobacterium might have contributed the gene for the HEH fold domain (helix-extended-region-helix) to its eukaryotic fusion partner. The HEH fold tethered the phage DNA to the inside of the viral capsid. In the eukaryotic nucleus, the HEH fold domain tethers chromosome ends to the inner lining membrane of the nucleus. The eukaryotic type I topoisomerase might have derived from ancient DNA-viral site-specific recombinases. The human TOP1 gene (topoisomerase DNA) is mapped at 20q11.2-q13.1. In somatic cells of vertebrate mammalian hosts (humans), TOP1 catalyzes the relaxation of supercoiled DNA during replication and transcription. In these hosts, the TOP1 gene may undergo increases of its gene copy numbers (amplification in colorectal adenocarcinoma), or gain of function point mutations. An example is the E418K (glutamic acid to lysine) mutation in Chinese nasopharyngeal carcinoma cells. Cancer cells with this mutation gained resistance to chemotherapy (camptothecin) [1258, 1259]. In cancer cells with defective DNA repair mechanisms (BRCA1/2 mutation), inhibitors of PARP (polyADP-ribose, Table II), with another form of DNA damage inducer (chemo-radiotherapy) cause synthetic lethality [1260] (Table X). The oncogenic DNA suffers unrepaired ds breaks, as monitored and quantitatively expressed by γH2AX (ataxia telangiectasia) staining [1261]. Here, the descendent of an ancient gene converted the somatic cell into an independent and immortal entity; its virulence could be broken only with a most elaborate attack of chemoradiation.

J.G. Sinkovics, *RNA/DNA and Cancer*, DOI 10.1007/978-3-319-22279-0_19

Double-stranded DNA viral genes encode chromatin-related proteins. In archaea, the bicaudi-, fusello-, lipothrix, and rudiviruses, in eubacteria, small caudate (tailed) bacteriophages (pico- and tectiviruses) and Lactococcus phage P4 (*vide supra*) encode chromatin-binding proteins. The adeno-, herpes-, and papovaviral LXCXE module (leu-X-cys-X-glu) fits into the pocket of the Rb gene; the papillomavirus E2/6/7 proteins express similar DNA-binding domains. The latency-associated nuclear antigen (LANA) of herpesviruses is homologous with the papillomaviral E2 protein. The BEN domain enzymes (*vide infra*) of polydnaviruses are chromatin-related. The large nucleocytoplasmic dsDNA viruses (NCLDV, the mimivirus clade) encode the chromatin remodeling SWI2/SNF2 ATPases (switch; sucrose non-fermentable). The ATPases SWI2/SNF2 unwind and loop-format the DNA. The histones and histone methylases, demethylases of NCLDVs are their own, different from their host cells' histones and their enzymes. The herpesvirally encoded early protein ICP8 (infected cell protein) binds ssDNA and recruits chromatin remodeling ATPases (SWI2/SNF2) to keep the viral genome in the euchromatin state: H3K9Ac, H3K14Ac, H3K4me2/me3 (acetylated and methylated lysine, K, of histone 3). It is presumed that LECA (the last eukaryotic common ancestor) possessed the SWI2/SNF2 ATPases that were acquired repeatedly horizontally from prokaryotes to function as DNA helicases. The Strawberry Notch and HARP-like ATPases (annealing helicases HepA-related proteins) to rewind unwound DNA, were eukaryotic acquisitions occurring much later in Figures 3 and 4 in the cited article [1263] (*vide infra*). Figure 4 depicts a chain of events for the presumed acquisition of the ubiquitin system by early eukaryotes from archaea (*Caldiarchaeum*), or bacteria. The core helical domain of the eukaryotic anti-apoptotic protein, BCL2, is related to bacterial toxins. Diverged derivatives of this protein (reviewed in ref [147], pp. 35–39) act on mitochondrial membranes, as anti-, or proapoptotic agents, either promoting, or inhibiting the release of cytochrome C. The genes of primordial caspases were transferred from bacteria to eukaryotes. Meta- and paracaspases of dictyostelid slime molds (Dd, *D. discoideum*) are of bacterial derivation. Members of the AID/APOBEC family (Figure 47) of cytosine deaminases of vertebrates disable retroviral genomes (*vide supra*). The primordial elements of the system probably appeared first in algae, and some activity of it is evident in cnidaria and nematodes. Some nucleic acid deaminases are present in some bacteria (the insect endosymbiont Wolbachia, plant-pathogen Pseudomonadacea, and in some filamentous fungi), but no evidence exists for a bacterial to eukaryote transfer. However, the sudden appearance of all the elements of adaptive immunity in the cartilaginous sharks strongly suggests horizontal transfers from various donors in a rapid (in evolutionary terms) succession (*vide supra*). The presumably intron-rich LECA, and its new nucleus, prevented the formation of chimeric ribosomes and further genomic conflicts between the host archaea and resident mitochondrion(ria). In possessing RdRP (RNA-dependent RNA-polymerase), LECA released siRNAs for gene silencing. Further decorated, LECA is to have had PARPs (from mono- to polyADP-ribosyl polymerase) (Tables II and XIII).

The eukaryotic HINT domain (Hh inteins) acquired from bacterial genes for polymorphic toxins, switched to the performance of autoproteolytic cleavage for the release of the signaling messengers. These are the ancestors of the Hh signaling pathway, which begins with the signal cascade for the activation of transcription factor Gli. The resting Gli factor is tethered in the cytoplasm to the SuFu protein (suppressor of fused). Gli released from SuFu constitutively activates the entire oncogenic Hh pathway (*vide supra*). SuFu itself pre-exists in bacteria as neutralizator of toxins. These activities constitute the "epigenetic code". In large nucleo-cytoplasmic viruses, SET-domain methyltransferases (Su(var), suppressor of variegation; enhancer of zeste gene, zeste-white Zw eye spot, trithorax, drosophila) and jumonji (cruciform) demethylases operate. The SET domain is often referred to as 'proto-oncogenes'. A set of eukaryotic enzymes catalyze $tRNA^{Phe}$ modifications. The SET domains are conserved from the viruses to all three domains of life. In plants (from the moss Physcomitrella to the flowering plant Arabidopsis), Su(var)3–9 homologues and related proteins (SUVH/SUVR) act in concert for H3K9 methylation, gaining function due to molecular divergence by mutations and gene duplications. While the genome of *Chlamydomonas reinhardtii* is intron-rich, the SUVH genes are almost completely intron-less. In poxviruses (molluscum contagiosum), and in polydnaviruses (shrimp white spot syndrome virus, WSSV), the BEN binding factor early enhancer protein transcription factor TFII family members (SMAR1; BTG3-associated nuclear protein, B-cell translocation gene; scaffold matrix-associated/attached region) react with histones. Further, E5R (Copenhagen vaccinia virus open reading frame), NAC1 (nitrogene assimilation control protein) carry out reactions with histones. During active DNA replication in the poxviral virosomes in the cytoplasm, the E5R protein actively organizes DNA structures. Like its predecessors, the human SMAR1 interacts with histone deacetylases HDAC3/4. The SMAR1 protein is listed as a tumor suppressor. The BEN proteins often fuse with the broad complex POZ domains (pox virus and zink finger) for a wide range of poorly defined functional contexts. In eubacteria (prokaryota) and in certain dinoflagellates, the DNA-packaging proteins are that of the HU/IHF (hyperthermic eubacterial; integration host factor), or DNABII family members. The long eukaryotic histone tails express positively charged residues in order to attach to the negatively charged DNA backbones. These histone tailes also serve as substrates for chromatin modifying enzymes SWI/SNF (switch; sucrose not fermented). Swi1 propagates in the budding yeast as a prion. The energy for these enzymatic reactions derives from ATP hydrolysis. DNA methylases encode major epigenetic information. These enzymes by covalently modifying histones (methylating and acetylating lysins and methylating both lysins and arginins in histones) inscribe "the histone code" [1019, 1039, 1043, 1262–1268].

The epigenome preserves acquired information: is this faculty heritable? It is the DNA the template of heredity claimed to be identical in all the cells of the multicellular hosts. Is it possible that DNA is a subject of environmentally induced changes? Instead of equicellular identity, somatic cell mosaicism (cells of different karyotypes in an individual) introduces changes in the DNA molecules in some somatic cell clones quite frequently, as if it were a physiological process. The epigenome (the site where the ancient RNAs began interacting) forcefully regulates gene expressivity. Are epigenetic changes heritable in the germ line?

The sanctity of the germ line is being pierced through. Paradigm shift in genetics is taking shape. The DNA can no longer be considered to be the sole agent of inheritance [1269]. The October 2012 issue of the Behavioral and Brain Science is devoted to this issue. The brain-derived neurotrophic factor (BDNF) is the mediator of neuronal plasticity. The BDNF gene (located in chromosome 11p15.5–p11.2) is under heavy epigenetic environmental regulation. Environmental experiences are encoded within the neurons [1270]. Women with history of longer oral contraceptive pill use reduce the number of their ovulatory cycles and the occurrence of maternal trisomy 21 ovarian mosaicism; the chance for conceiving a fetus with Down syndrome is reduced. Shorter overall mean length of oral contraceptive pill use allows higher numbers of ovulatory cycles and the chances for pregnancy with a trisomic fetus [1271].

The genome of female patients with Turner syndrome is with 45,X frequently with mosaicism. Their single X chromosome is usually maternally driven. The instable Y chromosome may be lost during meiosis. The 45,X cells transform into gonadoblastomas and other gonadal germ cell tumors. The Y and the X chromosomes carry a shox gene. With loss of the short arm of the X chromosome, due to deletion of SHOX (stature homeobox-containing) gene, the state of genetic haploinsufficiency sets in [1272]. In addition to mosaicism, the growth factor receptor-RAS signal transduction pathway of spermatogonial cells in aged males undergoes a process of dysregulation that expands clonally. Advanced paternal age carries the risk to the concepti for the transfer of "paternal age effect" disorders [1273].

The BTG3 gene (B-cell translocation) is an antiproliferative target of the p53 protein and as such it is referred to as a tumor suppressor gene. Within the cell cycle, it inhibits cyclin E (E2F1 inhibitor). Epigenetic increase of histone H3K27me3 demethylase JMJD3/KDM (jumonji/lysine K demethylase) downregulates the BTG3 protein, and induces $p16^{INK4a}$ and ERK-AP1 (extracellular signal-regulated kinase; activator protein factor) signaling, among others, in prostate cancer [1274]. The antiproliferative BTG/Tob (transducer of ErbB2) gene family protein products inhibit Src (Rous sarcoma) tyrosine kinase. BTG3 mRNA expression in prostate cancer cells is suppressed (determined by TaqMan quantitative real-time PCR). DNA-silencing methyl transferases recruit histone deacetylases. The result can be the double (DNA and histone) silencing of the promoter of a tumor suppressor gene. However, the gene is not suffering a loss-of-function mutation, or a deletion, thus its activation could be restored. The BTG3 mRNA was suppressed in prostate cancer cells, mainly through reversible hypermethylation of its promoter (determined by bisulfate-modified PCR). MethPrimer software selected primers in the CpG dinucleotide rich regions of the promoter and the selected amplicons were subcloned. Cytosines in methylated or unmethylated state were looked for. Treatment with the DNA methyltransferase inhibitor 5-aza-2′-deoxycytidine (DAC), or the isoflavone genistein, removed the CH3 adducts and restored BTG3 mRNA expression. Deacetylated histones were found; demethylator treatment with genistein or 5Aza-C (DAC) increased re-acetylation of histones H3/H4. Thus, the silenced BTG3 tumor suppressor gene could be reactivated *in vitro* (in prostate cancer cell lines) [1275].

The archaeal-bacterial CRISPR-Cas antiviral (anti-phage) defensive system employs DNAse under the guidance of RNAs to seek out, attack and cleave cognate intruder RNAs/DNAs (*vide supra*) [86] (Figure 20). The system works under Lamarckian principles. Random variations and fixations of their results in the workings of evolution appear to be a process of determinism ("occurrences in nature are causally determined by preceding events and natural laws; the combination of environmental and genetic events" in Webster's 9th New Collegiate Dictionary). Reverse transcription-mediated phage mutagenesis (diversity generating retroelements, DGR) for bacterial host tropism switching, and for the gene transfer agents (GTAs), which are the bacteriophages packing host cell DNA rather than phage genomic DNA. The phage genomic DNA is left behind integrated in the host cell's genome. These are the dedicated agents of HGT. HGT is widely viewed as a form Lamarckian inheritance [1276]. Indeed, parasites trans-speciate in order to be able to exist in different hosts (viruses and kinetoplasts traveling in insects and infecting vertebrates).

Comment. It is not known just exactly what rules apply to the introns either in free-living, or in parasitic unicellular eukaryotes, while swimming, or during encystations for resting, during meiosis, or when acting within the attacked host. Similar uncertainty applies to silenced or to actively replicating genomes, including the reactive immunological faculties, or the oncogenes in higher eukaryotic cells in multicellular hosts. The post-meiotic haploid germ cells, the resting stem cells, the somatic cells engaged in one particular function in the multicellular vertebrate hosts alter their intron numbers, apparently according to individualized and not generalized rules. May it be concluded, as expected, that silenced genes are intron-rich, and activated genes (oncogenes) are intron-poor? Some resting human oncogenes and their introns received attention [1219], but most of them not.

What are the intron numbers and positions within the constitutively active oncogenes? Is there any similarity as to the numbers and activities of the introns within the constitutively active oncogenes of the mammalian hosts (in the process of oncogenesis), and in the genomes of unicellular eukaryotes actively replicating in a nutrient rich environment, singularly (the Trypanosoma), or in communities (the Dictyostelium; the Botryllus)? Is there anything inherited from the pre-vertebrate multicellular hosts (from the amphioxus) upward?

The text detailed above names the genes that are involved in most complicated interactions. Sometimes the number of exons was given, but in this extremely active territory, offering itself as a very rich source of information, neither the number, size, position, nor the gene function-related increase, or decrease of the intron numbers (intron density) and positions are accounted for. Is a silenced genome intron-rich or intron-poor, or a hyperactive genome (an oncogenome) gains or loses introns? One may presume it, but so far without proof, that active tumor suppressor genes and oncogenes assume an intron-poor state, probably retaining only the first intron, or the mTORC spliceosome. If tumor cells of the most complex multicellular hosts undergo a retrograde descent on the evolutionary scale, do they reach the level of some primordial multicellular hosts (amphioxus), or go farther down to the level of the unicellular eukaryotes? Would this retrograde descent to the ancient RNA/DNA complex stop beyond the deep branching first single eukaryotes (the giardia), or reach the presumedly intron-rich genome of LECA? Was the LECA comfortably resting with an intron-rich genome in its hostile environment? In the complex multicellular hosts, the somatic cells performing unique special tasks, silence the rest of their genomes by increasing intron density? May the germ cells, and the stem cells occupy an intermediary position? The activated oncogenes duplicating at a fast rate, and encoding oncoproteins constitutively, may have to perform as intron-poor entities. However, the introns-harboring protein-coding DNA sequences, be oncogenic, or tumor-suppressive, or mi-Rs-regulated, must be regarded as actors in a much privileged position (and should be attracting the greatest possible attention).

Attempts were made to compare the number of genomic introns (spliceosomic introns) of activated oncogenes in malignantly transformed cells of metazoa (especially human oncogenes): **1**, with that of free-living unicellular entities (Amoeba; Naegleria; kinetoplastids, choanoflagellates), and **2**, with that of unicellular parasites (giardia, trichomonads, kinetoplastids, Theileria) (in the Appendix 1).

The number and position of introns in the resting human stem cell, the asymmetrically dividing stem cell, the malignantly transformed stem cell, and that of the differentiated or malignantly transformed somatic cells were to be compared. The presumption was that the genomic structure expressed in the number and location of the introns in malignantly transformed cells of metazoa (especially of human cancer cells) will be comparable to that of the unicellular entities, inasmuch as free living unicellular entities utilize "cell survival pathways", divide frequently and in mature age, and re-cap their telomeres, thus avoiding senescence and natural death. Further, the parasitic unicellular eukaryotes activate genes and mobilize microRNAs to counteract host defenses, and in that, they act similarly to cancer cells (or to the fetal trophoblast). Special interests were to be given to the G4 quadruplets occurring both in unicellular entities and in metazoan oncogenes (and in the healthy human genome): how do they handle their introns?

Great difficulties were encountered in the case of fused oncogenes in metazoa, because of the difficulty of recognizing for comparison such formations in the free living, or parasitic unicellular entities. However, natural cell (and gene) fusions occur during the fetal life of multicellular hosts (metazoa). The retrovirally mediated fusion of the syncytiotrophoblasts is the one most spectacular example. However, the neoplastic fused genes emerging postnatally encode oncoproteins of a new antigenic structure, which is not necessarily recognized by the host as self (because it was not expressed due to the absence of those fused gene during fetal life). The gene product proteins (oncoproteins) of the first recognized fusion oncogene in the Philadelphia chromosome (*bcr/abl* p210$^{BCR/ABL}$), the result of translocation and fusion of genes *c-abl* at 9q34 and *bcr* at 22q11) (reviewed in [651]), and the last recently discovered human fusion oncogene/oncoprotein, the (FGFR/TACC (transforming acidic coiled-coil protein-3) in glioblastoma multiforme [1234], serve as examples. Native and adaptive immune recognition of the oncoproteins enlist an armada of cyto-, chemo- and lymphokines and growth factors, fixed and circulating DCs, circulating blood cells (monocytes/Mϕ, granulocytes, NK/NKT cells, and CD4/CD8 T lymphocytes. It is not known just how the reactive genomes re-arrange their "from resting to active" exon/intron ratios.

Even the best reviews on the introns' long biological trajectory avoid any conclusions as to their changes, if any, concerning their numbers, length, or genomic location in the somatic cell, germ cell, stem cell, or in the great variety of brain cells, in rest, or in action. There is no methodically carried out comparison on intronic activity in the descendants of the ancient unicellular life forms in rest (encysted), or in activity (swimming, feeding; copulating); or the same in urochordates, ascidians, or in plants living in the seas, or flowering over dry land [1145, 1151a, b, 1157, 1165]. What happens to the introns of macrophages, dendritic cells, NK cells, T lymphocytes and B plasma cells when immunologically reacting? At the present time, it is not possible to conclude that intronic activities of the malignantly transformed cell of a multicellular host resemble that of an ancient unicellular life form, as if the cancer cell assumed the life style of its primordial unicellular ancestors, either living as independent immortal individuals, or as unicellular parasites of multicellular hosts.

Selected abbreviations

HINT Hedgehog inteins. For exteins and inteins in archaeabacterium, Pyrococcus abyssi, click Du Z Liu J Albrecht CD et al.

FGF/TACC fibroblast growth factor; transforming acidic coiled-coil protein. Click for TACC in yeast Schizosaccharomyces by Tang NH Takada H, Hsu KS Toda T. Click for TACC in glioblastoma genomics by Olar A & Alpade KD; and Singh D Chan JM Ceccarelli M et al. Click for TACC in EMT (epithelial-mesenchymal transition) by Ha GH Kim Jl Breuer EK. Click for oncogenic FGF-R (fibroblast growth factor receptor) in bladder cancer by Williams SV Hurst CD Knowles MA.

SMAR scaffold/matrix attachment region.

RNA/DNA: Bioengineers Supreme

Malignant Transformation of Some Selected Cells by Mistake?

Is it spectacular startling performance of great agility, or chromosomal chaos? Aneuploidy (abnormal number of chromosomes) was proposed to be the cause of cancer (Figure 65). However, the causes of aneuploidy remained unexplained. It has been recognized that the changing chromosomal numbers and genetic changes within, may encode a phenotype drastically different from what was pre-expressed by the pre-transformed cell. Some of the cells encoded by aneuploid chromosomal sets may appear as if derived from a species different from the pre-aneuploid host. Thus carcinogenesis may be viewed as a form of speciation (trans-speciation). Misdistribution of chromosomes into daughter cells at mitosis results in aneuploidy. Aneuploidy is connected with oncogenesis as a form of "speciation" as elaborated on by Duesberg *et al* [1277].

The mitotic kinase Bub1 missegregates chromosomes. In the yeast (*Saccharomyces,* budding uninhibited by benzimidazole), Bub proteins keep the cell cycle open (*vide supra*). Constitutive expression of bub1/Bub1 in eukaryotes activates proto-oncogenes c-*myc* and *aurora* B. Upregulation of c-MYC resulted in lymphomagenesis. Upregulation of AuroraB generated aneuploidy and chromosome missegregation with resultant oncogenesis at various sites (in mice) [1278a, b]. The AuroraA protein in the drosophila formats the mitotic spindle. In the human genome, AuroraA is located at 20q11-13, Aurora B at 17p13. This locus is frequently amplified in a number of human tumors (adenocarcinomas). In addition to regulatory cyclin-dependent kinases, the phosphorylated serine/threonine AuroraA kinase drives the cell cycle through G_2-M and amplifies centrosomes. Further, AuroraA activates another serine/threonine kinase, the mTOR complex 1/2. mTORC2 is an activator of Akt proto-oncogene. Akt reformats the cell's cytoskeleton. Once the cell is transformed, it survives the inhibition of AuroraA [1279]. While AuroraA/B are amplified, AuroraC is a constitutively activated mutant in human cancer cells [1280].

J.G. Sinkovics, *RNA/DNA and Cancer*, DOI 10.1007/978-3-319-22279-0_20

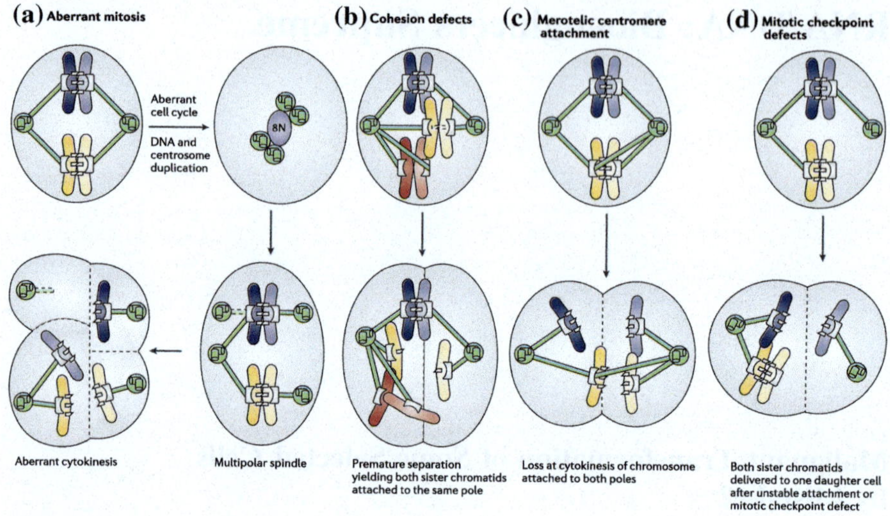

Figure 65 Cells in Aneuploidy may Die, or May Recover Immortal (as chemo-immunotherapy-Resistant Tumor Cells). [2361] Reference Appendix 2, Explanations to the Figures

Polo-like kinases (Plk) stand by to induce centrosome maturation and regulate bipolar spindle assembly; they are activated in G2 and degraded during the mitotic exit. It is the Aur-A and protein Bora (*aurora borealis*) that activate Plks by phosphorylation and complex formation (Plk1/Bora). in the complex, PLK phosphorylates Bora. In malignant tumors, the genes *plk*1 and *aurora*-A at 16p12.1 and 20q13, respectively, are much amplified, thus overproducing their protein products. The Plk1 and Aur-A proteins can be inhibited by small molecules BI-2536 and MLN-8054 (patented), respectively, and mRNA for Aur-A can be depleted by RNAi. However, the Plk1 inhibitors must not suppress Plk3, acting as a Plk1-antagonist tumor suppressor [1281].

The nucleoskeletal protein NUP1 organizes heterochromatin in the inner surface of the nuclear envelope in the trypanosoma. The chromosomal ends anchor in the lamina protein network. The laminin-analogue, NUP1 maintains the silenced state of genes located at the nuclear periphery. It acts upon the genes of telomere-proximal silencing variant surface glycoprotein (VSG). This region regulates the expression of cell surface antigens, which determine host immune reactions, or the lack thereof [1282–1286], and if the telomeres are re-capped or not capped (silenced) after each cell division.

How are the telomere ends treated in the trypanosoma cells rapidly (close to constitutively) replicating in liquid media? While the mammalian blood stream form of the parasite may revert to that one residing in the insect vector (simulating trans-speciation) and virulence (mouse pathogenicity) may decline upon prolonged *in vitro* cultivation, the speed of replication is retained: the telomeres are capped after each cell division [1287]. Unicellular life forms are able to sustain re-capped chromosome ends, and Nup proteins may interact in the capping process. However, the free-living kinetoplastid *Bodo saltans* is devoid of the Nup-1 protein. The dictyostelid (Dd) Nup proteins are known as interaptins. The nuclear porin Nup proteins might be instrumental to sustain an anti-scenescent state [1288, 1289]. In the inner membrane of the nuclear envelope, in the cells of multicellular organisms (except plants), the nuclear lamina form intermediate filaments, the lamins. Inside the nucleus, the lamins interact with the heterochromatins, histones and their modifiers, the chromatin-related proteins, and the enzymes that add or take adducts (methyl or acetyl groups). Transmembrane proteins link the lamins to cytoplasmic proteins (intermediate filament proteins, IF). These IF proteins are already functional in the Hydra, but are absent in plant cells. At the level of the Dictyostelium, lamin network for internal lining of the nuclear membrane is already present [1288, 1289].

In the human bone marrow, fused to the nuclear protein and phosphatase PP2A inhibitor, SET (*vide supra*) and Nup-214 act together as an oncoprotein in acute myeloid leukemia [1290], as another example of cellular hyperploidy.

Aneuploidy is consequential to mitotic errors; malignantly transformed cells frequently commit mitotic errors. In fetal life, the developing human brain suffers of chromosomal segregational defects during mitoses of glial cells and neurons. Aneuploid brain cells are eliminated by apoptosis. Areas not cleared by apoptosis, remain confined to sites of chromosomal mosaicism. The presence of functionally active aneuploid neurons "negatively affect" neuron-to-neuron and neuron-to-glial interactions [1291].

Recent data indicate that chromosomally abnormal (aneuploid) cells induce a T-, NK cell-, and Mϕ-mediated immune response. Unfolded proteins accumulating in the endoplasmic reticulum induce the ER stress, or unfolded protein response. Materially, calreticulin, in preventing egress of unfolded proteins, accumulates on the cell surface. In the absence of the inhibitory CD47, these cells may die apoptotic death. Otherwise, immune T cells expressing Fas ligand (and outnumbering Treg cells), and NKG2D ligand-responsive NK cells, attack hyperploidizated calreticulin-overexpressing cells. In syngeneic immunocompetent hosts (mice), these cells grow very slowly, or not at all, whereas in immunoincompetent hosts, they form multiple large tumors [1292, 1293].

Comment. Permission is asked for some liberal imagination. Some oceanic fish carry and shed large sarcomatous tumors. Sarcoma cells thus liberated could possibly survive individually in the nutrient-rich sea water? Some transformed cells of the clam Ma travel from host to host (in the Appendix 1). The Tasmanian devil with a huge aneuploid facial tumor, that is spread amongst its hosts by transferred cells, falls in a body of water and dies there. Could those tumor cells immortalized and replicating uncontrollably in their natural host, continue living in that body of water "rich in nutrients"? Would they live singly or could they eventually differentiate and form colonies?

A patient with a fungating incurable cancer, "immortalized" and growing uncontrollably in the form of aneuploid cells, leaves the human society, unties the boat and sails deep in the ocean, where falling in the water, dies. This malignant tumor, like all others, during its life time in its multicellular host, evolved into many subclones of malignant cells, as if different metabolic pathways "in waiting" were readied to fit into any new environmental situation, so, that the "fittest will survive". Phrased differently, in addition to any of its basic major genomic drive (BCR/Abl; Wnt/Hh; mTOR), the tumor cells undergo close to innumerable somatic mutations of metabolic types (*vide infra*), as if the RNA/DNA complex without a specific foresight would be preparing individual cells for survival in an ever changing microenvironment. Could therefore some of those cancer cells survive in the "nutrients rich" sea water, form colonies and eventually become organized as a society of cells? Or would they die and serve as nutrients for established species living in the sea? Would metagenomics of the oceans identify free-living tumor cells of the fishes (piscine sarcoma cells) or those of sea-living mammals? Or a billion year of living in multicellular hosts eventually expunged from the genomes those primordial sequences that made the amoeba, the dictyostelia, the tetrahymena, or the Naegleria "free-living"? The individual cells of a vertebrate multicellular community cannot regress on the evolutionary scale that far backwards. No evidence for such events of cancer cell survival exists. Even the HeLa cells maintained for decades in the laboratory (*vide infra*) under utmost care and balance of nutrients, are not differentiating into an organized society of cells; they rather phagocytose each other, practice autophagy, die of apoptosis, or just overgrow other cell cultures.

When natural death is approaching, in the police state of the multicellular organized cell community (a vertebrate mammalian host), the RNA/DNA complex presumably receives "signals" (from inserted transposons, *vide infra*) to rescue individual cells, in order to sustain cellular life in whatever form or shape. Due to this inherent faculty, the RNA/DNA complex immortalizes cells by re-activating those genes, whose ancestors sustained the first unicellular life forms in the hostile environment of the primordial Earth. The RNA/DNA complex creates thousands of gain-of-function point mutations in those cells in expectation of any (now unknown) environmental change in which that cell will have to establish its existence. It appears that these selfish individual cells could not repopulate their environment anymore, and thus eventually would become extinct. Unicellular entities, just exactly what cancer cells are, are not known to live freely on Earth anymore, neither singly nor in colonies (except in captivity in tissue cultures in academic laboratories). The first unicellular eukaryotes have had differenciated into millions of directions. Only their distant extant descendants are available for comparison. Is that a valid comparison? Many similarities to cancer cells the free living or parasitic unicellular entities display. Let us wait and see what remnants of extinct life shall we encounter on Mars, or elsewhere in the Universe.

Professor Preisz's Pettenkoferia. The famous Hungarian professor of bacteriology, Hugo Preisz, observed in the 1930s never seen before formations of most peculiar malleable large cells in his pure cultures of bacteria: "sonderbaren Formen wie ich sie vordem noch niemals gesehen hatte". These large cells might have appeared "spontaneously", but emerged as a rule in bacterial cultures grown on lithium-agar ("Lithium-chloride Agar gewachsener Bakterienkulturen"). The large cells ("Riesenformen") displayed a peculiar internal structure in the form of cytoplasmic particles and granules, and on occasions disintegrated into amorphous units and granules (Figure 66; Table XIV). He concluded that "protozoa" invaded his cultures and phagocytosed the bacteria. He accepted that these peculiar "protozoa were the Pettenkoferia" of Kuhn and Sternberg, named after the highly respected Professor, Max Josef Pettenkofer of Munich (München), Germany. Professor Preisz's well illustrated publication is preserved in the archives of old libraries (not cited anywhere anymore) in original form (Zentralblatt für Bakteriologie I. Abteilung 1937; 39:225) [1294]. The original publication was reviewed and some of its illustrations were reproduced in the author's monograph Die Grundlagen der Virusforschung [36a, b, c]. This author (JGS) viewed the Preisz illustrations as a scenario brought back from the depth of 3 billion years or more:

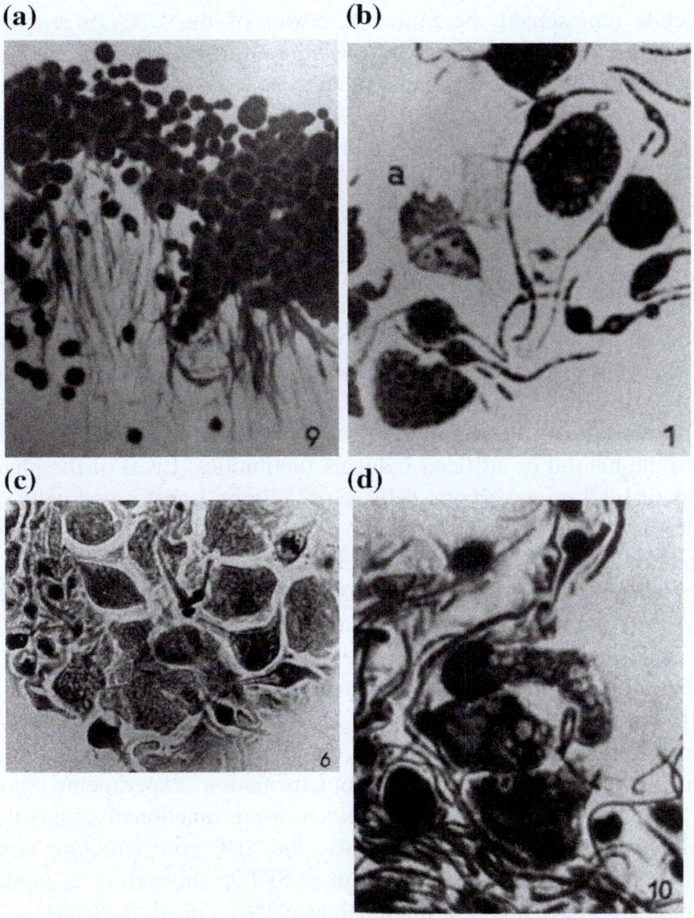

Figure 66 The Fate of Primordial Spheroplasts in Darwin's Pond Could Be Viewed Now in Bacterial Cultures Grown in LiCl Media. Reference [1294] Appendix 2, Explanations to the Figures

the first spheroplasts responding with a series of mutations in order to stay alive in an unstable and hostile (the LiCl) environment.

It took several decades to recognize by L. Dienes and E. Klieneberger-Nobel, that the large bodies were part of the life cycle of the bacteria themselves covering morphologically and biochemically an unusually wide span (and not the imaginary Pettenkoferia protozoa invading the bacterial cultures). Klieneberger-Nobel named these cells the "L-forms", after the Lister Institute; reviewed in [36a, b]. Joshua Lederberg and J. St.Clair referred to the large malleable forms as "spheroplasts"; cited in [73]; this author (JGS) called them "Pleuropneumonie-ähnliche Eigenheiten" in aged cultures of *E. coli* [36a, b, c]. Some small thus filterable (or fused) cytoplasmic units could regenerate into the original vegetative bacterial forms (especially, when *Sarcina lutea* co-existing yielded some "growth factors") (*vide supra*). The morphologically and biochemically widely diverse forms of bacterial cells represented the immense power of the DNA in encoding these deviant morphological and biological formations for the same cell, and eventually preserving and restoring the original structures as well. More recently, the stringent control of ribosomal RNA synthesis in *E. coli* spheroplasts was disrupted by LiCl treatment. The loss of the stringent control was due to the release of unidentified proteins from the cytoplasmic membrane [1295]. Unfortunately, no illustrations of the spheroplasts (if they resembled Preisz's Pettenkoferia) and no genomics studies on them were made available.

The large body of bacterial L-forms, proto- and spheroplasts offer themselves for the study of their proteomics and genomics. These studies may reveal some factual information on the life of ancestral spheroplast colonies. This author suspects that ancestral *ras* genes were involved in the extraordinary plasticity of these transformations (*vide infra*). The phenomena of complex spheroplast formations (in the extreme: formation of "Pettenkoferia") have not been investigated further, especially as to the natural or artificial inducers (antibiotics; LiCl) of the phenomenon. The effects of LiCl on eukaryotic cells, especially on transformed cells (malignant tumor cells) is summarized in Table XIV. Observe LiCl-inhibited β-catenin interacting enzyme GSK (glycogen synthase kinase).

The ancient proto-oncogene family ras. The signal recognition particle (SRP) is a cytoplasmic ribonucleoprotein; its domain is present in archaea as tRNA-like element, which has been conserved in eukaryotes. The 7S non-ribosomal RNAs from archaea closely resemble mammalian (human) SRP RNA (but the small domain III became deleted in bacteria). Domain IV of SRP RNA has been universally conserved. Domain I SRP RNA underwent major variations of its size throughout its evolution. Genetic complementation experiments showed that archaeal, bacterial, and eukaryotic SRP RNAs were functionally equivalent (could functionally replace one another), because the SRP core structure is conserved through evolution. A *C. elegans* and a human SRP9p showed 47 % aa identity and 66 % similarity. Especially the mammalian cDNA encoding Srp54 is well conserved: it shows 31 % aa identity and 60 % similarity with that of *E. coli*. The conserved core of SRP is a 54-kDa signal sequence-binding protein and its receptor is SRα. The SRP receptors SRαβ are heterodimer membrane GTPase proteins

Table XIV Lithium chloride (LiCl): Inducer of Pettenkoferia; Acting on Genomics of Bacteria, Unicellular Eukaryotes and Transformed Cells

Professor Preisz. Another morphological study on bacterial growth in lithium-containing media commensurate to that of Preisz's[1] could not be found. Growth inhibitory effect of such media on plant pathogenic Clavibacter spp was reported.[2] The major effect of LiCl is glycogen synthase kinase (GSK-3β) inhibition. Consequentially the effects of LPS are neutralized (that of gingival Porphyromonas) by GSK-β inhibition. GSK-3β inhibitors, including glutarimid antibiotics, were growth inhibitory to unicellular eukaryotes (Plasmodia, Leishmania, Trypanosoma).[3] Under the effect of LiCl, the ancient scenario of a spheroplastic colony undergoing non-lethal genomic mutations for survival opened up for Professor Preisz, allowing an insight into events as they were 3 billion years ago.[4] Later, in some eukaryotic cells, lithium becomes a protector in averting cell death by inducing autophagy.[5]

In cancer cells, LiCl blocked cyclin-dependent kinase gene E factor 2 (E2F) and thus inhibited prostatic cancer cell proliferation.[6] The growth of xenografted human prostatic adenocarcinoma cell lines was suppressed by inhibitors of GSK-3, prominent among them LiCl.[7] In prostatic adenocarcinoma cells, both hormone-dependent and independent, in combination with chemotherapy (doxorubicin, etoposide, vinblastine), LiCl was additively inhibitory and apoptosis-inducing, due to GSK-3β suppression.[8ab] In gastric adenocarcinoma cells, WNT-inhibitor paclitaxel, and GSK3β-inhibitor LiCl vie for and against multinucleated giant cell formation, expressed as a cell survival pathway under attack.[9] The growth of the human type I estrogen-stimulated endometrial Ishikawa adenocarcinoma cell line was inhibited by the combination of imatinib, LiCl and medroxyprogesterone. These tumor cells are PTEN-deficient (PI3K-active) (Table VIII), carry K-ras mutations and are driven by the anti-apoptotic, pro-angiogenic oncoprotein midkine.[10] LiCl with HDAC inhibitors (valproic acid; suberoyl bis-hydroxamic acid, SAHA) arrested growth due to apoptosis induction in medullary thyroid carcinoma cells.[11] In c-Myc transformed cells, LiCl blocked GSK3, a key component of PI3k/Akt(PKB) pathway. ERK levels reduced to non-transformed normal. The Ras effector pathway Raf/MEK/ERK (MAP) stopped in G1 (cell cycle stand-still).[12]

1 *Preisz H Studien über Pettenkoferien; Zbl Bakteriologie Parasitenkunde Infektionskrankheiten; XIII Tafeln 1937;139:225–270.* ***2*** *Smidt ML & Vidaver AK Appl Environ Microbiol 1986;52:591–3.* ***3*** *Die L et al Mol Immunol 2012;52:38–49.* ***4*** *García I et al Mol Divers 2011;15:561–7.* ***5*** *Motoi Y et al ACS Neurosci 2014 PMID 24738557.* ***6*** *Sun A et al Prostate 2007;67:976–88.* ***7*** *Zhu Q et al Prostate 2011;71:835–45.* ***8a*** *Azimian-Zavareh V et al Indian J Pharmacol 2012;44:714–21.* ***8b*** *Hossein G et al Avicenna J Med Biotechnol 2012; 4:75–87.* ***9*** *Kim SM et al Exp Cell Res 2005;308:18–28.* ***10*** *Bilir et al J Gynec Oncol 2011;22:225–32.* ***11*** *Adler JT et al J Surg Res 2010;159:640–4.* ***12*** *Al-Assar O & Crouch DH Mol Cancer 2005;4:13.*

anchored to the endoplasmic reticulum (ER). The mammalian receptors contain conserved bacterial elements. The conserved SRP core consists of the 54-kDa protein. Its carboxy-terminal is methionin rich (M domain) forming a complex with domain IV of SRP; domain IV consists of an RNA and homologs of the Srp68 protein. This, and the Srp54p signal recognition homologs express GTPase consensus motifs. All M domains express the aa motif RXXROA(R/K)GSGXSXX (D/E)V (X, any aa; O, hydrophobic). The Srp54 proteins are active GTPases. The Srp68 protein is a guanine nucleotide dissociation stimulator (guanine nucleotide release protein). There is sequence similarity between Srp54p, the GDS proteins (guanine nucleotide dissociation stimulator) and small Ras-related GTPases in the three branches of cellular evolution (archaea, bacteria, eukaryiota) [1296]. These proteins were present in the anchor of the Tree of Life.

The Ras protein evolutionary tree is traced back to pre-eukaryotic cells. The basic SRPR-B cell membrane signal recognition particle receptor B subunits were the initiators of the Ras superfamily [1297]. The *rap/ras* genes and gene product proteins in *Trichomonas vaginalis* suggested an ancestral gene and a divergence point for two separate lines [1298, 1299]. In plasmodia (*P. falciparum/vivax*), the rhoptry-associated Rap1 protein promotes the invasion of erythrocytes [1300]. Rhoptries are the pear-shaped membrane-bound vesicles located at the apical end of the merozoite. Coelomates and ascidians lost many *ras* family genes. Dictyostelium cells encode 14 Ras protein family members, and 6 PI3K "cell survival pathways". In rapidly replicating dictyostelium cells, the Ras proteins activate the PI3K pathways [1301]. RasG proteins are highly active in cells growing in suspension, or engaged in locomotion (chemotaxis). In insect-pathogenic *Beauveria bassiana* fungi, GTP-bound Ras1/2 proteins dictate hyphal growth. Mutant GDP-bound D126A (asparagine to alanine) Ras1 protein induced vigorous hyphal growth; the virulent rapidly growing hyphal cells were resistant to oxidative damage, fungicidal toxins and UV-irradiation. The GTP-bound G19V (glycine to valine) Ras suppressed conidial germination, hyphal growth and reduced UV tolerance. Most of the transcriptionally induced downstream genes involved in conidiation and multi-stress response are identified. Thus, Ras1 and Ras2 protein acted antagonistically in these fungal cells [1302]. In the low-branching unicellular eukaryote, Trichomonas (as preserved in its descendant *T. vaginalis*), the *ras* gene is intron-less. It encodes a protein of 181 aa length, expressing the characteristic G domain. It hydrolyses GTP to GDP in gene transcription. Its aa sequence indicates that it is not a human gene acquired recently by horizontal transfer; it shows cca 35 % aa sequence homology with *ras/rap* (rat sarcoma; receptor associated protein) genes of other species [1299]. Trypanosoma growing in liquid media in the laboratory must re-cap its telomeres after cell divisions, thus, escaping senescence, and by dividing in mature age, natural death, too. *T. brucei* expresses a GTP-binding Ras protein: *tbrlp* (GTA-binding Ras-like protein). The 227 aa protein contains six conserved subdomains designating it as an ancestral Ras/Rap family member predating the divergence of these genes [1298].

The phylogenetic analysis of the *ras* genes indicates that their existence predated the origin of eukaryotes, but that somehow they have become absent in plants (Arabidopsis), where Rho family members operate. In unicellular ancestors, the *arf* gene and the SRPR-B protein were the most likely founding members of the ras/Ras superfamily [1297]. In the drosophila, the Rap protein is referred to as Ras-like "another small GTPase". Notch signaling engages both Ras and Rap [1303]. In the caenorhabditis, the Ras-like GTPase Rap1 is identified as a regulator of integrin activation and cadherin-mediated cell contacts [1304]. In the mammalian hosts, Rap GTPases dephosphorylate cofilin, the actin-serving protein in the cytoskeleton of lymphocytes, as they spread out on the surface of DCs in order to contact antigens, which are presented to them [1305]. In synaptogenesis in the hippocampus, Rap protein regulates the actin-binding protein afidin by recruiting N-cadherin/β-catenin [1306]. In human kidney cancer, prostaglandin E2 and small GTPase Rap1 signaling promote cancer cell locomotion [1307]. When the divergence of the *ras* genes superfamily is listed (Ras for oncogenesis, absent in plants and alveolates; Rho for actin-signaling and cell cycle progression, highly expanded in plants; Arf/Sar for vesicle trafficking, Ran for nuclear transport, Rab for vesicular cargo trafficking), Rap is not included [1297]. Abbreviations: ras, Harvey or Kirsten rat sarcoma; Moloney mouse leukemia/sarcoma retrovirus in rats captured host cell gene: c-*ras* → v-*ras*; rap, ras (receptor)-associated protein; arf/sar, APD ribosylation factor/secretion-associated ras-related; rab, ras arabidopsis; Rho, ras homologue; ran, ras-related nuclear.

The hermaphrodite egg-laying mollusk sea slug, the "sea hare" *Aplysia californica/vaccaria* Cooper 1863 (Alacrity Marine Biological, Redondo Beach, CA) is studied in Saint-Petersburg, Russia, and in several laboratories (Marine & Atmospheric Science, Miami; F & WL McKnight Brain Institute, Gainesville; Whitney Laboratory Marine BioScience, St. Augustine) in Florida, USA. Aplysia circulates haemolymph with lectins; the lectins display unique juxtaposed N-terminal domains similar to those of the Ig superfamily and C-terminal fibrinogen [1308]. Aplysia also carries the *ras*-related gene, ARHI (aplysia ras homolog I), which is considered to be a tumor suppressor gene in the human genome. As such, it is suppressed (or deleted) in patients with papillary carcinoma of the thyroid [1309]. The human DiRas3 tumor suppressor gene-product protein (GTP-binding protein, imprinted gene, distinct subgroup of Ras family member 3; direct inhibitor of STAT3) is identical with ARHI. It is expressed or suppressed in several human cancers (hepatocellular, breast, gastric, ovarian, pancreatic, prostatic adenocarcinomas). When expressed, it induces autophagy and/or apoptosis in the cancer cells. It is an inhibitor of STAT3/FAK/Rho, or the mTOR/VEGF pathways, and sustains G1 cell cycle arrest. The Y397-phosphorylated FAK is the focal adhesion kinase, which is an activator of RhoA signaling. ARHI-bound STAT3 protein is immobilized in the cytoplasm, and thus is deprived of its DNA-activating activity within the nucleus; among the STAT3 target genes, those of E-cadherin and vimentin are prominent [1310, 1311a, b, c]. The DiRAS3 protein inhibits ancient pathways Ras/RAF/MEK/ERK (rapidly accelerating rat fibrosarcoma) and the Ras/MAPK/PI3K by forming multimeric complexes with H-Ras, as H-Ras binds to its effector protein C-RAF. The multimeric complex prevents B-RAF/C-RAF dimerization, thus inhibiting the activity of these oncoprotein kinases [1310]. ARHI/DiRAS3 also binds glutathione transferase importin proteins, thus inhibiting the transport of DNA-binding oncoproteins (STAT3) from cytoplasm to nucleus [1311a, b, c].

In prostate cancers, microRNA-221 and miR-222 attack the 3′UTR regions of the ARHI/DiRAS3 mRNA; in high grade prostate cancer cells these microRNAs are upregulated and ARHI/DiRAS3 mRNAs disappear, consequentially ARHI/DIRAS3 protein levels drop. The isoflavonoid genistein downregulates miR-221/222 production and restores ARHI/Di-RAS3 expression ([1312, 1313]; also reviewed early in Ref. [284]). Genistein deactivates (reduces its mRNA expression) the metastasis-associated gene 3 (MTA) by binding estrogen receptor beta (ERβ). The invasiveness of the malignant trophoblastic choriocarcinoma cells (cell line JAR in matrigel *in vitro*) was inhibited by genistein. Genistein-treated JAR cells expressed increased amounts of the cell adhesion molecule E-cadherin.

However, the mRNA of the Snail protein was also increased by genistein treatment [1314]. However, Snail is a potential inducer of mesenchymal transition of adenocarcinoma cells (*vide infra*).

The squamous cell carcinoma of the oral cavity arises due to viral (HPV, EBV), bacterial (*Porphyromonas gingivalis*) and chemical (tobacco smoke; ethanol) etiological factors; reviewed in [73]. The *ras* genes are prominent inducers of the PI3K/Akt, Mek/MAPK oncogenic pathways in head&neck (intraoral) squamous cell carcinomas, and the involvement of microRNAs is recognized, but not as well discerned, as in the adenocarcinomas [1315].

Unreliable enzymes of the adaptive immune system defend the host against infections? The genomic integrity of the first life forms was disrespected and commonly violated by horizontal gene transfers. However, useful genes were probably vigorously sought after and when acquired put to good use immediately. Parasitic elements emerged to take advantage for their own continuous existence in a well-established system. Not welcome, but nevertheless skillfully inserting themselves into some defenseless hosts, these elements induced the establishment of a genomic-defense system in these hosts. One particular form of the ancient genomes eventually were condensed into chromosomes, but not packaged as yet into nuclei. When nucleated and well-armed large cell appeared, the a-nucleate cells were simply engulfed in order to surrender their precious genes to their new masters. The full machinery of the innate immune system was in motion and sustained uni- and multicellular life on Earth for some 3 billion years [1316]. The confrontation between parasitic free DNA (retrotransposons), or RNA/DNA packaged into viral forms (the phages) and single cells (the archaea) continues well over 3 billion (milliard) years. The clustered regularly interspersed short palindromic repeat (CRISPR) (Figure 20), and its associated systems (Cas), with its adaptively operational protospacer adjacent motifs (PAMs), secured the survival of the single cells (Solfolobus, Pyrococcus, Haloferax etc. archaea, and all eubacteria/prokaryota) [1317–1319], but failed to eradicate the "pathogens". The competent periodontal pathogen, *Aggregatibacter actinomycetemcomitans* descends from CRISPR-Cas competent *Pasteurellaceae* ancestors. In non-competent *A. actinomycetemcomitans* strains, the CRISPR-Cas system is lost, or impaired. The genomes of these bacterial strains carry an increased population of parasitic DNA elements. The remnants of the CRISPR-Cas system are engaged in the regulation of the reduced sized genome. This self gene regulation is similar to the microRNA-regulated eukaryotic genomes [1320].

In jawless vertebrates (lamprey and hagfish), segregated genes for receptors with leucin-rich repeats (LRR) evolved for the recognition of foreign antigens. These genes somatically assemble to encode variable lymphocyte receptors (VLRs); these proteinaceous plates are entirely different from immunoglobulins, but variable enough to bind a great diversity of antigens. VLR-type receptors are expressed by "T-like lymphocytes" in thymus-like tissue in the gills. Plasma cell-like "B lymphocytes" secrete VLRB-type receptors in the hematopoietic organ. The agnathan genomes are devoid of the rag1/2 genes. An AID-APOBEC gene family (Figure 47) catalyzes the assembly of the VLR genes into antigen-responsive receptors. Thus, the activation-induced deaminases (AIDs) appear to have emerged before the installment of the adaptive immune system in the sharks. In the agnathans, AID is a demethylator of the CpG islands. Unmethylated CpG islands are numerous (85–90 %) in the amphioxus and tunicates; AID is demethylating agnathan CpD islands. In contrast, mammalian vertebrates keep the majority of their CpG islands methylated (silenced), thus their gene expressions are more rigorously regulated. It is unknown why agnathans expunge some of their chromosomes during their ontogenesis [1321]. May be a new competitor to their immune system is emerging, while their ancient innate and LRR/VLR mechanisms served them so well that installing a new system is not needed. May be the predecessors of the sharks, the Placoderms, became extinct, as they were unprepared to the chaos that the coexistence of the regulated old, innate, and the supremacy-demanding, but yet irregularly operating new adaptive systems together inflicted?

The entire adaptive immune system was acquired by the immediate ancestors of the sharks (the extinct Placoderms), either by Transib retrotransposons, or by a herpesviral ancestor of the EBV lineage, or both (*vide supra*). Constituents of the adaptive immune system (MHC, Rag1/2, RSS, V(D)J) united first in the ancestral sharks; the innate and adaptive immune systems were extensively reviewed [27, 147].

Shared ancestry and co-evolution with elements of the innate immune system are emphasized [1322]. It is being investigated if retrotransposons and/or a herpesviral agent ancestral to EBV installed horizontally the constituents of the adaptive immune system; or some of its elements were actually inherited vertically. Proof is needed if genomic segments of the adaptive immune system were acquired by a herpesviral ancestor of EBV at the level of the amphioxus and mollusks, or were incorporated into the EBV genome by its immediate predecessors infecting mammalian vertebrates in Africa (*vide supra*). In this latter case, the EBV is just a recipient of retroviral sequences derived from a host already in possession of these operons well established. Retroviral genomic segments in herpesviruses are accounted for; the Marek's herpesvirus possesses avian leukemogenic retroviral sequences; reviewed in [27].

Somatic hypermutations generated the light chains of the specific immunoglobulin antibodies, and re-formatted TCRs to fit the challenger antigens. Retained in the supreme mechanisms of the system, there is an inherent error-prone DNA repair, and a potential wrong-target DNA nicking error by AID resulting in lymphomagenesis (splendid reviews of Mechtcheriakova) [310a, b]. In the process of SHM or class switch recombination, the ability of AID to cause dsDNA breaks is not strictly enough restricted to one specific target. "Genome-wide off target" DNA nicks resulting in dsDNA breaks, point mutations and chromosomal segmental recombinations occur during fervent immunological defense expressed as specific immunoglobulin generation [1323]. The protective mechanisms installed by certain microRNA mobilizations [338] are inadequate. The pathogenesis of African Burkitt's lymphoma (BL) shows that the congregation of diverse factors brings about the background favorable for lymphomagenesis. There is epidemic malaria resulting in defensive B cell hyperplasia. Among the immunological responses to plasmodia, TLR9 directly activates AID. Memory B cells already harbor in their epigenome the circular EBV genome (not expressing LMP1). The defensive enzyme AID is working in the germinal centers in the processes of SHM and CSR. By mistargeted nicks, AID releases from under its control the c-*myc* gene. The c-*myc* gene is able to fuse with the *bcl*-2/6, or with the IgH gene, the IgH-myc translocation [1324]. The c-*myc* proto-oncogene juxtaposes itself from its locus at 8q23 to chromosome 14 next to IgH gene (or to chromosome 2 next to the κ, or to chromosome 22 next to the λ light chain genes). The DNA-binding c-Myc protein is constitutively produced from its new location. In the t(14:18) translocation, the bcl-2 anti-apoptotic proto-oncogene from chromosome 18 translocates to the IgH site on chromosome 14, a translocation most common in follicular lymphoma. In BL, and in other B cell lymphomas (diffuse large B-cell lymphoma, DLBCL), the EBI3 gene (EBV-induced gene 3) is activated. This gene encodes p40, a subunit of IL-12. The EBI3 gene is usually activated by EBV's LMP1, but it is expressed in DLBCL without EBV LMP1 expression. Overexpressed c-Myc protein suppresses EBI3 expression. The EBI3 expression in DLBCL is induced by NF-κB. It is very seldom expressed even in EBV^+ BL [1325] (Figure 55). The pro-inflammatory transcription factor c-Rel (NFκB) is essential for T cell differentiation, including Treg cell generation in the thymus. From the cytoplasm, active NFκB dimers are imported into the nucleus. The physiological role of NF-κB is to activate pro-inflammatory genes for cytokine production (IFNγ, IL-12) [1326]. The cited article provides a rare and most welcome list of its abbreviations (but totally omits the role of constitutively produced NF-κB in oncogenesis). The evolutionarily distant but biologically overlapping NFκB/STAT pathways in the Nematostella and in the human B cell lymphomas (including the Reed-Sternberg cell) is referred to in the Appendix; in Int J Oncology Oct 2015; 47:1211–1229.

The genomics of the pediatric T- and NK-cell lymphomas offer themselves for the study of their exon-intron re-arrangements (not yet analyzed or reported). The malignant T- and NK-cells express their innate and adaptive immune faculties (except for antibody production). Here the TCRβ/TCRαδ undergoing somatic hypermutation with high AID activity, induce gene tanslocations and fusions, with or without TCR translocations, but with extraordinary genomic re-shufflings [1327].

Inflammatory carcinogenesis is still reversible through switches in the epigenome. Chronic or specific infectious processes enter the epigenome, where they methylate, and thus silence the promoters of tumor suppressor genes, and de-acetylate, and thus activate the promoters of proto-oncogenes. A large network of microRNAs rules over mRNAs, either allowing, or disallowing translation. The term "epigenetic oncogenesis" is introduced. By demethylation, the tumor suppressor genes can be restored. By re-acetylation (inhibition of histone deacetylases) the proto-oncogenes can be de-activated. However, the chronic unrelenting infectious processes eventually delete the tumor suppressor genes (often by ubiquitination of their protein products), and point-mutate or fuse proto-oncogenes, thus facilitating their constitutive expression. Events of malignant transformations of the attacked, or of the defensively overproduced inflammatory cells irreversibly ensue.

It is cited how the cell cycle inhibitor Wee ("small" in Scottish) is downregulated by miR-155. Induction of the *vav* (6th letter in Hebrew alphabet) family proto-oncogene in mycoplasma-infected cell cultures is reported. Inflammatory squamous carcinogenesis in the oral cavity, or in the skin (xeroderma pigmentosum) is described. Inflammatory carcinogenesis in various tumor categories (glioblastoma, papillary thyroid carcinoma, Barrett's esophagus, breast, colonic, endometrial, gastric, pancreatic, prostatic adenocarcinomas) is individually reviewed. Melanoma arising in UV-light-induced dermatitis receives special attention. Lymphomas and sarcomas are suspected of viral etiology, and are treated accordingly (specific oncogenesis by some of the viral genes). So are reported bacterially-induced malignant transformations (*Chlamydia psittaci*-induced periorbital lymphomas; *Helicobacter pylori*-induced gastric MALT lymphomas or adenocarcinomas; enterotoxigenic *Bacteroides fragilis*-induced colonic adenocarcinomas). The genomic background of inflammatory carcinogenesis has been emphasized throughout; extensively reviewed [73].

The innate (CD11 myeloid cell-related genes, TLRs, ARG1 arginase, defensins) and adaptive (CTLA4 cytotoxic T lymphocyte-associated antigen, CD19/20 B lymphocyte genes, Treg cells) immune systems interact [1328, 1329]. The triggering receptor expressed on myeloid cells, the TREM gene family, emerges from its locus at 6p21.1 as a master regulator of adaptive immunity. The cluster contains MHC and NKp44 receptor genes, as well [1330]. The TREM1 protein is a leukocyte cell surface receptor belonging to the Ig superfamily. The TREM1 mRNAs and TREM protein expression rise in cases of pneumonia and sepsis [1331]. The anti-inflammatory TREM1 controls IL-1β, TNFα, GMCSF and several pro-, anti-inflammatory interleukin production. It suppresses Kupffer cells in the liver and inhibits inflammation-induced hepatocellular carcinogenesis [1332, 1333]. It is under study if loss-of-function mutation, or single nucleotide polymorphism of the TREMs allow inappropriate excessive inflammatory processes to continue and possibly contribute to autoimmune, malignant, or neurodegenerative diseases.

What was not available for the first unicellular life forms. In the fully subverted victim, conversion of the immune defenses of the multicellular host into the full support to the invaders (Table XV) was not available.

In a comparison of the arising of cancer cells in a multicellular host with that of the first single-celled eukaryotes appearing in a hostile physico-chemical environment, the first unicellular life forms had to generate their own "cell survival pathways" (like PI3K, PCNA, arginine methylatig/demethylating enzymes, etc.). In contrast, in the multicellular host, the malignantly transformed cell finds itself in a love-hate relationship with its host. It can induce healthy mesenchymal cells to provide its growth factors (VEGF, chemokines, cytokines, etc.) and receive shelter and full support (by imitating a fetus). However, it must withstand a hostile confrontation with the immune faculties of its host (by practicing the defensive maneuvers invented by the placenta). Of many examples, ovarian carcinoma receives growth factors from healthy adipocytes and fibroblasts [1334, 1335]. Ovarian cancer cells rapidly develop multidrug resistance toward vinblastine and taxanes. The let-7 gene product microRNA (its absence is lethal in the

Table XV Are Cancer Cells of Multicellular Organisms Comparable to Unicellular Eukaryotes with Activated Cell Survival Pathways?

Free living or parasitic unicellular eukaryotes with activated cell survival pathways (amoeba; giardia) use these faculties (PI3K) extensively, but seldom constitutively. Tumor cells in movement and phagocytosis ("cell cannibalism") activate genes derived from ancient unicellular eukaryotes (in text).[1] Pathogenic kinetoplastids and apicomplexans (plasmodia; trypanosoma) propagated in liquid media cap their telomeres after each cell division and grow in semblance of immortality (trypanosome clone Lister 427). In their host, trypanosoma cells constantly alter their variant surface glycoprotein antigenicity.[2] Oxytricha express G quadruplex DNAs, harbor piggyBacs, and activate mutator phenotypes. In that, tumor cells imitate ancient ancestors, or extant unicellular, eukaryotes (in text).
Theileria close to immortalize bovine host lymphocytes in order to sustain its life cycle within. Theileria-infected bovine lymphocytes express constitutively NFκB and c-Jun NH2-terminal kinase; neutralize pro-apoptotic p53 protein by MDM protein; grow in culture as immortalized lymphoblasts driven by an autocrine IL2-to-IL2R circuitry; and grow as lymphoma tumors in xenografts. All these events of "malignant transformation" are reversible after chemotherapy-killing of the parasite schizonts[3] (probably, because infested lymphoblasts were not shown to kill $FasR^+$ Immune T cells via FasL overproduction, or that they generate Treg cells). It is not the unicellular eukaryote, which emulates malignant transformation, but its target, the host lymphocyte. Unicellular eukaryotes, like tumor cells, are wide open for acceptance of horizontal gene transfers.[4] Naegleria readily undergo trans-speciation (in text).
Tumor growth enhancing host immune reactions benefit the tumor cell of the multicellular host, but were not available to the unicellular eukaryotic ancestors in distress. Tumors create Th2 environment and IgG4 dominance; generate chemokines for the invitation of Treg cells; and conceal target antigens. It is not only exemption from immune attack, but actual growth promotion (mitosis-induction) by host immune reactions that modern tumor cells had gained.[5–9]
Angioneogenesis The VEGF (ligands) and their receptors encoded in the tumor's hypoxic microenvironment provide nutrients and waste removal. VEGF-R autocrine circuitries actually activate proto-oncogenes (PI3K) in the tumor (melanoma) cells. Erythropoietin has been expropriated by tumor cells as a growth factor.[10] The cell membrane-bound or soluble heparin-binding EGF-like GF (HB-EGF) released in the tumor's microenvironment, or within the tumor cell in paracrine and autocrine circuitries does not initiate malignant transformation, but it promotes the growth of already transformed cells.[11]
Neoneurogenesis Prostate cancers invite the in-growth of parasympathetic-sympathetic nerve endings (synapses). The adrenergic sympathetic synapses are connected with β-adrenergic receptors of stromal smooth muscle cells outside the tumors. Xenografted prostate tumor with their adrenergic receptors deleted failed to grow. The parasympathetic synapses are formed within the tumors on muscarinic receptors of epithelial cells and release acetylcholine. In response, the epithelial cells generate TGFβ1, promoting their migration. Blockade of the muscarinic receptor rendered prostate cancer cells immobile (non-invasive; non-metastatic).[12] Androgen-deprivation therapy of PC combined with β-blockers (BB) reduced PC-related mortality. A muscarinic antagonist inhibited the cell cycle and stopped the growth of non-small cell lung cancer cells. BB administration prolonged survival of patients with breast cancer, and reduced relapse rates. Patients with triple negative breast cancer receiving chemotherapy benefited from addition of BB.[13] Overall, a UK data base failed to document any survival benefit of β-adrenoreceptor blockers (BB) in patients with various cancers; in cases of pancreatic and prostate cancers, shorter survivals occurred.[14] A BB example (of several): propranolol.

(continued)

Table XV (continued)

1 Sandhu R et al Cell Res 2013;23:537–51; Novel-Miner GA et al PLoS Pathog 2012;8(8): e1002900; Moraes Barros RR BMC Genomics 2012;13:220; Horn D & McCullochR Curr Opin Microbiol 2010;13:700–5; Hertz-Fowler C et al PLoS One 2008;3(10):e3527. 2 Dreesen O & Cross GA Exp Parasitol 2008;118:103–10. 3 Hayashida K et al Exp Mol Pathol 2013;94:228–38; Branco S et al J Vet Sci 2010;11:27–34; Ivanov V et al Mol Cell Biol 1989;8:4677–86. 4 Gilbert C & Cordaux R Genome Biol Evol 2013;5:822–32; Kishore SP et al BMC Evol Biol 2013;13:37; Parent KN et al Mol Bio 2013;4(2) pii:e00056–13. 5 Prehn RT Science 1972;176:1701;TheorBiol Med Model 2006;3:6; 2007;4:6. 6 Chiarella P et al Clin Med Oncol 2008;2:237–45. 7 Ellyard JI et al Tissue Antigens 2007;70:1–11. 8 Ladanyi A et al Cancer Immunol Immnunother 2011;60:1729–38. 9 Karagiannis P et al J Clin Invest 2013;123:1457–74. 10 Acs G Acs P et al Cancer Res 2001;61:3561–5; Chatterjee S et al J Clin Invest 2013;123:1732–40; Kumar SM et al Oncogene 2012;31:1649–60; Tóvári J et al Curr Mol Med 2008;8:481–91. 11 Ongusaha PP et al Cancer Res 2004;64:5283–90. 12 Magnon C et al Science 2013;341:143–4; Isaacs JT Science 2013;341:134 13 Grytli HH et al Prostate 2013;73:250–60; Hua N et al PLoS One 2012;7(12):e53170; Powe DG et al Oncotarget 2010;1:628–38; Melhem-Bertrandt A et al J Clin Oncol 2011;29:2645–52. 14 Shah SM et al Br J Clin Pharmacol 2011;72:157–61.

Commentary. Malignantly transformed cells in a multicellular host are intimately interrelated with antagonistic or supportive host factors that were not available to independently living unicellular eukaryotes with constitutively or reversibly operated cell survival pathways.

caenorhabditis) destabilizes the mRNAs of the MDR1 and IMP-1 proteins (multidrug resistance; inner membrane peptidase subunit 1) and maintains drug-susceptibility of the cancer cells. Downregulation of let-7 miR results in rising MDR1 and IMP-1 levels and consequential drug resistance (except for carboplatin, insensitive to the MDR protein) [1336–1338]. The let-7 miR family and miR-18 differentiate stem cells, promote mesoderm differentiation over that of endoderm and ectoderm, and suppress de-differentiated (tumor) cell growth. Let-7 and miR-18 downregulate the Acvr1b and Smad2 genes (activinA receptor type B; signaling mothers against decapentaplegic, identical with TGFβ, drosophila). Let-7 in targeting anti-apoptotic protein Bcl-xL is a pro-apoptotic factor. Further, the let-7 microRNA family members target and deplete mRNAs for oncoproteins c-MYC, RAS, and high mobility group antigen-2 (HMGA) [1336–1341]. Of microRNAs of oncogenic (miR-155, miR-17-5p, miR-21), or oncosuppressive (miR-15a, miR-16-1, miR-let-7) faculties, miR-200 was found to be the protector of the epithelial-type phenotype of cancer cells and an antagonist of EMT (epithelial-to-mesenchymal transition). This miR family sustains E-cadherin expression in epithelial cancer cells. When miR-200 was downregulated, E-cadherin levels dropped and vimentin expression increased. Consequentially, cancer cells assumed the mesenchymal phenotype. The EMT transformation is a form of de-differentiation manifesting in increased tumor cell mobility toward invasiveness and metastasis formation; reviewed in [26, 27]. The major oncoproteins involved in EMT are SNAI1/2 (Snail and Slug), Twist, SIP (SMAD-interacting protein), and ZEB1/2 (E-box-binding zink finger). The microRNA let-7 is a prominent protector of the epithelial phenotype. Let-7 targets the gene for HMGA2 (high mobility group A). Observe that the title of one article reads "Let-7 prevents early cancer progression by suppressing of the embryonic gene HMGA", while the text of another article by the same author reads "The expression of let-7 and concomitant up-regulation of its main target HMGA2 was found to be a characteristic of early neoplastic transformation in ovarian cancer". Further, it is miR-200 that is responsible for the retention of E-cadherin, thus antagonizing EMT [1337, 1338]. The mRNAs for ZEB1/2 are depleted by miR-200bc (miR-200b coming from chromosome 1, and miR-200c from chromosome 12). The ZEB proteins are suppressors of E-cadherin and promoters of vimentin [1337, 1338].

The oldest miR, emerging before bilateria evolved, is miR-100. The genome of the sea anemone, *Nematostella vectensis*, harbors it (in the Appendix 1). A restructured vertebrate-ancestor let-7 appeared in the urochordates. The amphioxus, *B. floridae,* already possesses a second set of let-7. In early vertebrates, the let-7 locus underwent a tandem duplication. Bilateria acquired let-7 and miR-125. The ancient let-7s working in deuterostomes, nematodes and platyhelminths were independently restructured. The originally discovered caenorhabditis miRs let-7 and lin-4 are closely related. Intron 9 of the protein-coding caenorhabditis gene F59G1.4 expresses lin-4 (opposed by an antisense transcript). It may be of great importance, that most of the deuterostoma miRs were transferred to gnathostomata. These include the let-7 sequences. The mode of transfer may be related to the adaptive immunity's transposons *rag*1/2 (*vide supra*). The vertebrate genomes underwent two rounds of duplications (including those of the cyclostomata lamprey). The genomes of teleost fish were subjected to the "fish-specific genome duplications". These events concern the zebrafish and pufferfish (*Danio rerio*; *Takifugu rubripes*), etc. The extant tetrapods possess 20 let-7 clusters. Probably the latest acquisition from Old World monkeys is on the human MIRLET7BHG non-protein coding locus (B host gene). One of its introns harbors miR-3619 expressed by stem cells [1342]. The phylogenetic curiosity is the deuterostome Xenoturbella *Hostenia miamia* (Diva Diniz Corrêa; the Marine Laboratory, Virginia Key, Miami, FL). Its genome so far yielded none of the established miRs [1343].

The receptor for hepatocyte growth factor (HGF) and scatter factor (SF) is c-Met. These ligands activate the c-Met receptor, but ligand-independent high expression of c-Met is known to occur. Physiologically, the HGF/SF ligands induce and regulate the development of the liver and the placenta [1344, 1345]. In the placenta, soluble c-Met (sMet) binds HGF and exhausts the ligand's supply for native c-Met binding on the trophoblast cell membrane [1346]. Dendritic cells migrating from skin to lymph nodes use c-Met-induced matrix metalloproteinase for their intercellular locomotion [1347]. The c-Met kinase induces the PI3K/Akt/mTOR proto-oncogene pathway; the constitutive expression of this pathway is highly oncogenic. For PI3K to run freely, PTEN knock-out is necessary. Overexpressed c-Met induced supernumerary centrosomes, multipolar spindle formations and aneuploidy (*vide supra*). PI3K inhibitor (LY294002) and/or Intact TP53 expression protected cells against aneuploidy [1348]. The serine-protease matripase (MT-SP1, epithin) appears in high titers in epithelial (squamous) cell cancers, which harbor overexpressed c-Met. In these cancer cells, zymogen autoactivation is increased, and transmembranic serine protease inhibitors are decreased. The signaling-incompetent single chain pro-HGF/SF molecules are converted into two chain signaling-competent HGF/SF ligands by MT-SP1 (matripase). These ligands may be released through auto- or paracrine circuitries. In H&N squamous cell carcinomas, stromal cells (fibroblasts, endothelial cells, lymphoid cells) release the single chain pro-HGF/SF molecules to be captured by monomeric c-Met receptors of the tumor cells. The tumor cell surface membrane excretes matripase molecules. The MT-SP1 molecules perform the conversion of the single chain pro-HGF/SF molecules into two chain ligands effective in activating c-Met [1349]. The intracellular microRNAs allow the translation of c-Met mRNAs into the c-Met kinase protein. MicroRNA-449 depleted c-Met mRNA, and inhibited the growth of c-Met-dependent hepatocellular carcinomas cells. Histone deacetylases (HDAC) repressed miR-449 production, and activated HGF signaling. In hepatocellular carcinoma cells, trichostatin (HDAC-inhibitor) increased miR-449 production and reduced c-Met expression. Extracellular signal-regulated kinases (ERKs) remained inactive (non-phosphorylated), tumor cell growth stagnated and apoptosis set in. Similar effects were elicited in miR-449-transfected tumor cells [1350].

Cancers other than H&N squamous cell carcinoma driven by c-Met are breast, colon, gastric, and ovarian cancers (adenocarcinomas). Returning to the gynecologic tumors, the HGF/SF-driven c-MET pathway induces and drives endometrial [1351], and ovarian (endometrioid, clear cell, mucinous and serous-papillary) carcinoma [1352]. The natural history of ovarian carcinomas is that of the silent onset, extensive loco-regional spread (invasion of pelvic organs and omentum), and ascites induction with tumor stem cell spheroids floating. By histological subtypes, there is a variation of oncosuppressor gene losses (PTEN, TP53, RB) and functional gains of the co-inducer oncogenes/oncoproteins (BRAF, KRAS, ERBB2, PI3K/Akt). At this point, extensive surgical interventions followed by chemotherapy (carboplatin and taxanes) interrupt disease progression. In short time, intraabdominal recurrence, and EMT of the individual tumor cells with loss of E-cadherin and rise of vimentin set in; MMP-9 cleaves cell surface E-cadherin, which is replaced by the fibronectin receptor ($\alpha_5\beta_1$-integrin) [1353]. Formation of distant hematogenous metastases characterizes the end phase of the disease. A most circumspect and advanced review is by Lengyel [1352]; it omits references to the defensive inflammatory cells, which co-invade the malignant ascites. Instead, a small molecular inhibitor of c-Met and VEGFR-2, foretinib (GSK1363089), is tested against human ovarian carcinoma xenografts in nude mice. Foretinib almost completely inhibited the local growth and distant metastases of human ovarian carcinoma xenografts [1354]. Accordingly, clinical trials have been initiated.

The anti-apoptotic BCL-xL (extra large) oncoprotein is a protector of malignantly transformed cells and atherosclerotic blood vessels. Oxidized low-density lipoproteins (ox-LDL) are inducers of apoptosis of endothelial cells, especially atherosclerotic blood vessels. The miR let-7 family (*vide supra*) upregulates ox-LDL-induced apoptosis of endothelial cells [1340], presumably those cells of the neovascular network, which feed tumor cell colonies.

Tumor cells eventually engage all established physiological growth factors for their own growth advantage. Paradoxically the "pro-apoptotic" death receptor FasR/CD95/APO-1 (factor apoptosis signaling) is so engaged. In an ovarian carcinoma cell line (HeyA8), a lentiviral vector delivering anti-CD95 shRNA, reduced cell surface expression of CD95 and established the state of apoptosis resistance. However, the CD95-deprived cancer cells ceased to enter the cell cycle for complete mitoses. In the CD95 apoptosis-resistant SKOV3 (Sloan-Kettering) ovarian carcinoma cell line, abolishing CD95 expression resulted in growth arrest, restoration of CD95 (FasR) expression, and resumption of cell cycling with full mitoses. It was possible to induce the same phenomena in human adenocarcinoma cell lines. In these cell lines, the Fas ligand, FasL/CD95L-neutralizing antibody NOK-1, or anti-FasL siRNA close to arrested (reduced) cell growth [1355]. The $CD95^+$ ovarian carcinoma cells acquire resistance to FasL-induced apoptosis, and consequentially overexpress FasR/CD95. The ovarian carcinoma-induced ascites fluid contains high levels of the ligand FasL. $CD95/FasR^+$ apoptosis-resistant tumor cells (or regenerating hepatocytes) driven by FasL mobilize the proto-oncogene Jun kinase-to-c-Jun pathways. The JNK small molecular inhibitor SP600125 stopped tumor cell growth. The CD95/FasR agonist antibody Jo2 activated the expression (phosphorylation) of JNK-c-Jun, and c-*fos* (Finkel osteosarcoma murine retrovirus-captured cellular proto-oncogene), a gene downstream of JNK. FasL-driven FasR can activate the proto-oncogene *scr* in glioblastoma cells [1356]. There is a difference how Fas/CD95 responds to soluble (s), or membrane-bound (m), FasL. The soluble sFasL delivers a non-apoptotic message, and induces FasR/CD95 to activate proto-oncogenes, including Fos. This unison may induce histiocytic sarcomas in the liver of mice. Whereas, mFasL conveys the standard pro-apoptotic message [1357]. The author's laboratory was the first to report tumor cell growth (melanoma) induced by a FasL-to-FasR/CD95 reaction [717, 718], followed by Shinohara, reporting FasL-driven growth of glioblastoma [720] (Figure 59).

Physiological growth factors for ovarian carcinoma. The ovarian carcinoma cells protect themselves against immune rejection. In peripheral blood mononuclear cells collected from healthy individuals, and from patients with ovarian carcinoma, IL-2, IL-12, and IFNγ production was quantitatively measured. Autologous tumor cell killing by peripheral blood mononuclear cells (PBMCs) of the patients with or without GCSF-treatment (*in vitro*) was compared. Healthy subjects produced higher levels of lymphokines, than did patients with cancer. GCSF-treatment abolished tumor cell killing faculties of the PBMCs [1358]. Ovarian carcinoma cell lines released soluble substances, that were inhibitory to IFNγ and IL-2R expression in $CD8^+$ T cells; instead IL-10 production was induced; lymphocytes were arrested in G_0/G_1; JAk3 and tyrosine phosphorylation of STAT5 were inhibited [1359]. In contrast, anti-estrogenes (tamoxifen/toremifen) and IFNα sensitized cytotoxic T cells and NK/LAK (lymphokine-activated killer) cells *in vitro* to react with anti-tumor cell-directed (ovarian carcinoma cells) cytotoxicity [1360]. Ovarian cancers generate tumor-protective TGFβ-dependent Treg cell populations. Treatment *in vitro* with IL-2 downregulated the expression of FoxP3, and induced the production of IFNγ, TNFα, and the expression of TH17 markers [1361]. The anti-idiotypic antibody 6B11 (patent ZL01130756.0) imitates the antigenicity of ovarian cancer antigen OC166-9. Dendritic cells stimulated with 6B11 antibody react and activate immune T lymphocytes. Immune T lymphocytes suppressed ovarian cancer cells' colony formation *in vitro* and inhibited the growth of human ovarian cancer cell lines in SCID mice [1362]. Peripheral blood lymphocytes of patients with ovarian cancers confined to the peritoneal cavity were stimulated *in vitro* with ovarian cancer antigen Mucin-1. The activated immune T cells were repeatedly administered intraperitoneally. Temporary responses were observed, except in one patient who achieved disease-free status. The anti-tumor immune response achieved by the first course of treatment was not further enhanced by repeated treatments [1363]. The c-Met receptor-directed monoclonal antibody CE-355621 suppressed to up to 98 % the growth of C-Met$^+$ human ovarian and gastric carcinoma, and glioblastoma xenografts in nude mice [1364]. Another anti-MET monoclonal antibody MetMab induced a complete remission in a patient with chemotherapy-refractory gastric carcinoma metastatic to the liver. This patient had MET gene mutations and very high serum HGF levels [1365].

The case history of RF 41 year old patient with serous ovarian carcinoma stage IIIC diagnosed in 1995 at St. Joseph's Hospital, Tampa, FL, was presented in 2004 at the 5th MD Anderson/Sloan-Kettering Hospitals Joint International Conference on Ovarian Cancer in Houston, TX [1366a]. After initial and second surgery and carboplatin/taxane chemotherapy, the patient relapsed with massive intraperitoneal disease, malignant cytology-positive ascites, and CA-125 rising to over 330 U/ml. Doxil (liposomal doxorubicin) chemotherapy induced partial response with cytology-positive ascites persisting. Lymphoid cells (CD4/8 T,-and NK-cells) were collected from her ascites and blood and were grown in IL-2-containing media in large plastic bags (preparations of Joseph C. Horvath). In an *in vitro* assay, her lymphocytes induced lysis of her native tumor cells. She received repeatedly intravenous infusions of her lymphoid cell preparations containing mixtures of CD4/8 T- and LAK- cells in approx 10^8–10^9/mL dosages at 2–3 weeks intervals with IL-2 (Proleukin, aldesleukin) given daily intravenously, then subcutaneously, but in rapidly reduced dosages. Dosage-reduction and actual discontinuation of IL-2 became mandatory (on request of the patient, her family, and psychiatry consultant) on account of severe suicidal psychotic reactions to it. An attempt was made to replace IL-2 with IFNγ (Actimmune). The patient experienced clinical complete remission with resolution of the ascites (gynecological examination by gynecologic oncologist surgeon Hector Arango on Feb 27 2003; pre- and post-treatment contrasted CT scans re-read and compared by diagnostic radiologists Gary Smith & John Rasmussen on Nov 22 2004 for the presentation at the conference). After the discontinuation of IL-2, the patient relapsed and died. The poster presentation was favorably commented on by gynecological oncologists James Taylor Wharton and Ralph Freedman of the professorial staff of MD Anderson Hospital. Conclusion: Immunological therapy (adoptive immune lymphocyte therapy with IL-2) of ovarian carcinoma disseminated, but confined to the abdominal cavity, will induce durable complete remissions with the potential of cure, barring adverse reactions to the therapy [30a, b].

In addition to the presentation of a patient with relapsed ovarian carcinoma entering remission on T cell NK combined infusion therapy [1366a], another patient provided proof for similar effects of combined T and NK cell therapy for massively relapsed ovarian carcinoma, however her treatment was combined with carboplatin and doxorubicin chemotherapy [1366b].

Since then, Immune T lymphocyte-therapy of metastatic ovarian carcinoma was elevated to potentially curative levels at the Ovarian Cancer Research Center, Perelman School of Medicine, University of Pennsylvania, Philadelphia, PA. Patient-derived (autologous) T cells by lentivirus gene transfer receive a chimeric antigen receptor (CAR) aimed at the tumor cells' folate receptor-α (FRα) antigen, highly overexpressed in ovarian adenocarcinoma cells. The attack by these genetically engineered lymphocytes on ovarian carcinoma cells is so violent that it borders "tumor lysis syndrome". After massive destruction of ovarian carcinoma cells, the CAR-bearing replicating autologous lymphocytes attack healthy cells expressing FRα at a low normal level. Before the first doses of CAR-bearing lymphocytes, the patient is depleted of her autologous lymphocyte population by apheresis in order to invite exponential expansion of the CAR-bearing immune T cells. The patients need normal lymphocyte re-infusions to lower the extraordinarily rapid expansion rate of the CAR-bearer immune lymphocytes [1367a, b].

Comment. Instead of presumed antagonism, certain combinations of chemo- and immunotherapy become additive and even synergistic in anti-tumor efficacy (*vide infra*).

Phenomenal Physiological Tasks Accomplished

The genes beneath the morphological metamorphoses accomplish any transformation in time. The unicellular eukaryote, Naegleria (*N. gruberi*) (Figure 54) under stress (changes of temperature and nutrient availability) undergoes a very rapid process of morphological and biochemical changes. The amoeboid form changes into a cyst, develops basal body (centriole) and emerges as a flagellated and swimming amoeboid cell with a network of cytoplasmic microtubular cytoskeleton [1368a, b]. The self-assembling gene product protein SAS-4 localizes in the centrosome and directs the duplication of that organ. A procentriole forms over each centriole in each cell division, so that two centrosomes are formed, one for each daughter cells [1369, 1370]. Practically all the genes encoding the transformation (centriole gene cluster; flagella gene cluster) have been identified. The orthologs of these genes farther down on the ladder (cyanobacterial *gnd* genes encoding 6-phospho-gluconate dehydrogenase, plant-like mitochondria) [1371, 1372] and higher up (Caenorhabditis, Drosophila; mosses, flowering plants, Arabidopsis) are recognized. Some of the ancient cyanobacterial genes have transferred long ago into the host cell nucleus, others remain in the acquired organelle. Even though 1.5 billion years separate the heterolobosean protist Naegleria from extant plants, the naeglerian mitochondria harbor RNA editing sites for pentatricopeptide repeat proteins (PPR) with DYW domains (aspartic acid-tyrosine-tryptophan) characteristic of plants. Highly variably, these proteins appear in mosses and remained the RNA editing entities in all higher plants. The Arabidopsis mitochondrial RNA editing factor 1 (MEF1) belongs to the DYW subfamily of PPR proteins. The Arabidopsis nuclear gene encodes DYW domain-carrier protein RARE1 (required for editing chloroplast accD, aminocyclopropane-1-carboxylate deaminase). The accD chloroplast transcript-encoding acetyl carboxyltransferase subunit of acetyl coA carboxylase is for fatty acid biosynthesis in the chloroplast [1373–1376].

Free-living amoebae (Naegleria, Hartmannella) select as their suitable niches hot springs and tolerate high temperature and 5 % salinity. In the Polish hot springs used for balneotherapy, *Legionella pneumophila* coexisted with Naegleria and Hartmanella [1377, 1378].

The archaeal and bacterial nanomotor self-assembling protein-structure rotates the flagellum filament like a propeller with extraordinary speed. Ras-like small GTPases (*vide supra*) form fusion proteins with the DnaJ domains and function as such. The flagellum's ultimate energy source is protons. LECA might have been endowed with a eukaryote-like flagellum(a), but whether it was acquired *in toto* through vertical inheritance of gene clusters and/or operons from archaeal-eubacterial ancestors, or if the flagellar genome was assembled from individually and horizontally transferred genetic elements in the first eukaryotes, is not known (but heavily debated). Euryarchaeota and Crenarchaeota cells might have exchanged horizontally flagellar genes. Flagellar genes were lost and regained (may be related to those of cilia). Flagellar genes may be silenced and re-activated, but the epigenomics of these phenomena have not been revealed [1379–1385]. The flagellar proteome of *Chlamydomonas reinhardtii* reappears in the human genome in the form of TBC1D3 protooncogene (*vide supra*).

The apicomplexan protozoa, Theileria (*T. annulate, T. parva*), are equipped with encoded faculties to overcome the immune defenses of their bovine hosts. Theileria schizonts convert their host cells (lymphocytes) into continuously replicating entities, which are exempt from apoptotic death in order to be supportive breeding grounds to the parasite. Once the parasite entered lymphocytes (or monocytes), its schizonts fuse into syncytia, which divide in the host cell cytoplasm and align with the spindles of the dividing cell, thus to be transferred to both daughter cells. The host cell's pivotal mitotic kinase, polo-like kinase-1 (Plk-1), supports the transfer of the parasites with the separating chromosomes to the daughter cells. This parasite practically "immortalizes" the infected lymphocytes. The infected lymphocytes grow in cultures indefinitely. Constitutively activated IL-2 and NFκB are produced in the cytoplasm of schizont-infected host lymphocytes. The production of MDM2 protein is increased and the WTp53 protein is neutralized. The schizont induces those host cell genomic faculties that are active in malignant lymphoma cells. Schizont-derived mRNAs encode subtelomere-encoded variable secreted proteins that enter host cell nuclei and localize in the nucleoli. It is not known if cytoplasmic ribosome biogenesis, or any of the nucleolar faculty is targeted by these proteins. The bovine host mobilizes immune T cells cytotoxic to the infected and transformed lymphocytes. However only that subclone of the immune $CD8^{+}T$ lymphocytes is effective, which is directed to immunodominant antigen(s), probably the Tp [214–224] epitope [1386–1390] (in the Appendix 1).

The gills transform to lungs. The book gills and the book lungs of the horseshoe crab (*Limulus polyphemus*) in comparison with those structures in the scorpion consist of page-like lamellae originating from a common ancestor, respectively, but differing for gas exchange in the different environments, aqueous and terrestrial. However, both circulatory systems flow haemolymph. While horseshoe crabs (Xiphosura) are aquatic (semi aquatic), the structure of their arachnid book lungs suggests terrestrial ancestry. The ancestral horseshoe crabs in the Lower Ordovician have already obtained their present day anatomy. The scorpions (Araneae) also preserved from the Silurian period their major body structures ("living fossiles") [1391]. Xiphosura and Arachnida diverged approx 470 mya. The limuluses remain on the shores for the night to mate in the moonlight (enviable creatures). Their book gills can manage gas exchange in, or out, of the water. The male limuluses' sperm cells contain the actin-binding catalytic proteins "αβ-scruins" [1392]. (This author presumes that the mating pairs tried to communicate in English, but they did not know how to spell). Silurian sea scorpions invading dry land developed spermatophore appendages (male genitalia, or "horn organs") for direct transfer of sperm into the body of the females [1393]. Spermatophores are absent in horseshoe crabs (sorry).

The oxygen-carrier molecules both in the limulus and in the scorpion (Euchelicerates) are the copper-protein hemocyanins [1394]. The hemolymph vascular system of the scorpion takes its anatomical origin in the form of two anterior aortae from a large pumping heart within a pericardium; arteries branch off from their aortae [1395, 1396].

The innate immune system supported well Xiphosura and Araneae during their long trajectory of life including the transfer from sea to land for Araneae and Arachnida. The granular phagocytic amoebocytes protect the Limulus by releasing a series of immune effector molecules (lectins, serpins, cystatins). The granular hemocytes recognize bacterial lipopolysaccharides that trigger the pathogen-associated molecular patterns. Protease zymogens form first a coagulum with the endotoxin and the excreted protease zymogens, terminating in a proteolytic cascade [1397, 1398, 1399]. Green algae of the Ulvaceae class infect, degenerate its developing organs, and eventually kill the Limulus [1400].

Fish breathe with internal gills. Upon the encounter of water and blood in the gills, a balance of O_2- and CO_2-sensing receptors is established. Terrestrial acclimation of Kryptolebias fish induces gill remodeling and a transitory stage of skin-breathing [1401, 1402]. Fish in the Amazon in warm slow moving and algae infested waters obtain more oxygen by vascularizing their lower lips, stomach, guts and swimbladder. Intraerythrocytic levels of ATP/GTP regulate oxygen uptake, transport and discharge [1403]. Larvae and tadpoles of amphibians gill-ventilate underwater, but by day 5 post-fertilization are able to carry out lung ventilation

[1404]. Pyrosequencing of the transition of larval gill to lung respiration in the tiger salamander yielded precisely identified transcripts: expressed in both tissues 721, in gills 1668, and in lungs 1210, total 3599 in the transcriptome [1405]. As the first amphibians and reptiles moved out of the sea, the ancient evaginated external gas exchanger gills were replaced by invaginated gas exchanger lungs. Bimodal (transitional) breathers at the water-air interface may temporarily (tadpoles), or permanently (newts, Urodela) possess both gills and lungs. In both systems, the respiratory surfaces are specially vascularized. In the pre-metamorphic tadpole, brain stem synapses regulate gill ventilation. As gills regress (get absorbed) and lungs expand, in the post-metamorphic frog, a switch of neuromodulators (serotonin and norepinephrine) in the brain stem takes place [1406]. The pleiotropic hormone, leptin, is essential in the conversion of the tadpole gills to lungs [1407]. In neuroepithelial cells of the frog the O_2 and CO_2 receptors regulate respiratory frequency [1404]. The bullfrog tadpoles' brainstems acutely respond to nicotine exposure. Paradoxically, the response to CO_2 exposure was blunted and the respiratory rate was rather attenuated than stimulated [1408]. The ancient interaction of the tyrosine kinase receptor EphB4 with its ligand EphrinB2 is reactivated in malignant tumors of practically any histopathology [1409, 1410]. How far downward on the evolutionary scale could the beginning of the Ephrin-to-EphB interaction be found? It might have been active in the vascularization of the gills in the ancient fish.

Insect respiration is tracheal, meaning there is no lung and no blood vessels are involved, but tracheal micro-branchings (tracheols) carry O_2 directly to tissues and cells; cellular mitochondria take up the oxygen [1411].

Non-avian theropod dinosaurs and early avian fossils found in China (*Confuciusornis sanctus*; *Zhongjianornis yangi*, *Yixianornis grabaui*) possessed anatomical features of their rib cage (uncinnate processes) preserved in extant birds. The theropod dinosaurs breathed with a flow-through respiratory system. The uncinnate processes and lung air sac structures diverged in those lines which transformed into aves [1412]. The avian lungs are the most complex and efficient gas exchange systems. The avian lung in flight takes up 2,5 times the maximal oxygen uptake of the mammalian lung in fast run; it may increase its volume 36 %; it tolerates cold air without distress; it extracts the maximum amount of oxygen from hypoxic (7 % O_2) air [1411, 1413a, b].

The evolutionary distance from gill to lung-breathing experienced its most critical transition in scorpions and amphibians. An imaginary anatomical description of the lungs in the first tetrapods is given [1414]. The mammalian lung is constructed to combine three fractal conduits: arterial, venous, and bronchoalvelolar. The fundamental principles and design of gas exchangers are inscribed in their stem cells. Need is the best inventor, and the stem cells' RNA/DNA complex responds and obliges [1411].

Enseal yourself into and fly out of the cocoon. The three stages of insect metamorphosis are pro-nymph, nymph and adult; within the cocoon: larva, pupa and adult [1415]. The cocoon shell of the silk worm *Bombyx mori* is soaked with antibacterial proteins [1416]. During metamorphosis, apoptosis and autophagy terminating in death of the cell, remove the obsolete ones. In the fat body of the mulberry silkworm (*B. mori*), the molting hormone 20-hydroxyecdysone (20E), product of the prothoracic gland, attaches to its receptor E20 → EcR-USP to induce apoptosis of these cells. The receptor component, ultraspiracle (USP) is pro-apoptotic. Its knockdown with antisense USP (anti-USP RNAi), results in cessation of the apoptotic cascade, due to unopposed action of IAP. Generally dispersed genes, IL-1 converting enzyme (ICE), apoptotic protease-activating factor (APAF), the caspase II, the neuronal precursor cell-expressed developmentally downregulated protein-2, the actin-related protein (ARP), and the inhibitor of apoptosis (IAP) interact in a precise sequence [1417a, b]. This ligand-to-receptor interaction induces the early apoptotic cascades, as above, and then the late genes rise for chitin biosynthesis.

What happens in the insect metamorphosis is replayed elsewhere in the host-tumor relationship: the pro-apoptotic genes (like USP in the worm) are silenced, and the anti-apoptotic genes (IAP) prevail to rescue the tumor cell. In the drosophila ovary (*D. melanogaster*), oogenesis begins with widely spread apoptosis. This apoptotic cascade specifically involves the ovary for the removal of mutated cells, even the nurse cell of these mutants disappear. Caspases dronc (drosophila nedd-2-like caspase, neuronal expressed down-regulated) and strica (serine/threonine-rich caspase) exhibit redundancy in action and work with caspases dcp-1 and drice (death caspase-1; drosophila ICE, interleukin-1β converting enzyme). These caspases can destroy an entire defective egg chamber, but rescue intact oocytes. Caspases may fail due to their own mutations, but upon disruption of strica, signaling through dronc is upregulated [1418]. The oncologist wishes that caspases of the patients with ovarian carcinoma would work this well against incipient tumor cell colonies.

In the musculature of the metamorphosing drosophila, multinucleated myoblasts dominate, both in the striated and smooth muscle bundles. The circular smooth muscle rings form syncytia. One of the genes involved in muscle multinucleation is *duf* (dumbfounded) [1419, 1420]. The Dumbfounded protein is fusogenic in the mouse for myoblasts. The mammalian homologue of this gene is TBX (T box transcription factor). The *rel* gene/Rel protein (NFκB), when recruited from the cytoplasm into the nucleus, targets the *tbx* gene. One pathological consequence of this reaction in human patients is cardiac hypertrophy and myocardial fibrosis. The intranuclear NFκB protein activates its muscular hypertrophy- and fibrosis-related genes, *mef*2a/Mef2a and *gata*4/GATA4 and TBX (mouse embryonic fibroblast; guanine, thymine adenine) [1421]. The TBX protein is a physiological activator of the *wnt*/Wnt and *fgf*/FGF pathways in the mouse for the initiation of mammary gland development [1422]. The TBX4/5 proteins activate the Wnt/FGF pathway in the chicken embryo and thus initiate the outgrowth of limbs (legs) [1423]. The Rel/NFκB protein in different setting is oncogenic. While c-Rel is not able to activate oncogenes c-*fos* and c-*jun*, the original v-Rel (the reticuloendothelial virus) readily induces proto-oncogenes including AP-1 and ATF2 (activating protein; activating transforming factor), induces Fos/Jun heterodimers, but the transformed cells may not express high Fra2 (*fos*-associated/related antigen). ATF2 acts through p38 MAPK phosphorylation, but may be inhibitory to the MEK/ERK pathway. v-Rel is an activator of H-Ras, while ATF2 may counteract by inhibiting the Ras-Raf-MEK-ERK pathway [1424]. The drosophila gene product protein Jeb (gene *jelly belly*) attaches to its receptor to induce a physiological Ras/MAPK cascade for the migration and differentiation of visceral muscle precursors. The human homologue of the Jeb receptor is ALK (anaplastic lymphoma kinase) (*vide supra*). In the drosophila, the Jeb receptor activates downstream the *duf* gene [1425]. Decapentaplegic (Dpp) is identical with TGF-β and is inhibited by Smad and Dad (signaling mothers againt dpp; daughters against dpp) [1426]. In the drosophila larval brain, the predecessors and homologues of the human gene adenosis polyposis coli (APC1/2) in the neuroblasts mediate asymmetric cell divisions. The cell cortex harbors the APC2 protein and the APC1 protein is localized in the centrosomes. While in the larval brain the APC genes are operational in asymmetrically dividing cells [1427], in the human colon the APC proteins are tumor suppressors. Of the retroelements populating the drosophila genome, the Penelope sequences encode a RT, an endonuclease and the GIY-YIG motif (glycine, isoleucine, tyrosine); reviewed in [147]. In the process of hybrid dysgenesis, several unrelated retroelements of the genome are mobilized. Penelope clone could be transferred by a full length construct into the genome of *D. melanogaster* [1428]. Penelope-related ancient retroposons (Neptune, Poseidon) are widely spread in deuterostomes (sea urchin, Strongylocentrotus), shrimp (Panaeus), piscine and amphibian genomes (pufferfish, Fugu and Tetraodon, frog, Xenopus), but vertebrate mammalian (human) genomes must have expunged them [1429] (for drosophila brain tumor and leukemia, see in the Appendix 1).

The polyembryonic wasp, *Copidosoma floridanum,* produces thousands of morulae from one fertilized egg. The morulae replicate like a package of stem cells; however the genomics of this most remarkable phenomenon is unknown. When the morula divides, one progeny will possess the primordial germ cells (PGC) and this progeny remains fertile. The other progeny, without PGC, will serve as soldiers in an infertile caste [1430, 1431].

The epigenomics of the metamorphosing lepidoptera great wax moth, *Galleria mellonella,* was studied by injecting histone acetyltransferases and their inhibitors (HAT, HDACI) into larvae before pupation. The effects on wound healing, pupation, and infections (LPS, Listeria, fungus Metarhizium) were analyzed by morphological and molecular studies. Insect genes encode molecules protecting against sepsis (the antimicrobial peptide gallimycin; the inhibitor of metalloproteinase, IMPI) and the genes for HAT and HDAC. The HDAC inhibitor SAHA (suberoylhydroxamic acid) increased survival after injury and hemolymph loss, whereas HAT inhibitors delayed pupation and accelerated larval death. HDAC inhibition resulted in high MMP levels (a pro-septic event), and also in the earlier and stronger activation of IMPI (an anti-septic event). However, in HDAC-inhibited larva, gallimycin production was attenuated. The pathogenes acted upon the epigenome so, that the induction of gallimycin was reduced. LPS-pretreated larvae gained some resistance against lethal doses of Listeria infection; Listeria infection in itself delayed the transformation of larvae into pupae (except for killed Listeria, which accelerated pupation). In that, Listeria infection imitated the effects of HAT inhibition. In conclusion, HDAC inhibition appears to have impaired the faculties of innate immunity [1432]. Lepidopteran insects (larvae to aduts) circulate in the hemolymph granular cells and plasmatocytes. Plasma proteins are mobilized in a serine protease cascade. The antimicrobial peptides and proteins coat invasive pathogens. The pathogens are either phagocytized or encapsulated in nodules, where they undergo melanization and digestion. The TLR pathways initiate the release of the antimicrobial peptides. The serpin family proteins conclude these reactions [1433].

The wasp *Ephidius ervi* parasitizes aphids by implanting yolkless eggs, that must hatch rapidly. The embryo is still covered with a rapidly cellularizing serosal membrane, until its instar larval stage. Until then, its blastocell may derive diffusible nutrients from its host. Thereafter, special large cells line up between the larva and the host. These cells have irregular nuclei, large ribosomes, many mitochondria, and accumulate small vesicles. The cell surface toward the larva is even and uninterrupted. The vesicles are exuded toward the larva. The cell surface facing the host is crowded with microvilli penetrating deeply into their host's tissues. The arrangement of this cellular monolayer is referred to as the "syncytial phase". Some of these cells attach to the host's trachea and oxygenate the larval environment. In this case, the yolk-less insect embryo develops a placenta-like structure to tap and transfer nutrients from the host to the parasite [1434]. The genomics of this maneuver has not as yet been clarified (if a fusogenic retrovirus is involved; no virus particles are visible in the published sections). This cellular structure suggests as if insects discovered the syncytiotrophoblast.

The microRNAs were discovered in the drosophila embryo [1435]. The embryos, larvae, pupae and adults of the large tobacco hornworm, *Manduca sexta,* yielded 176 microRNAs (and more to be identified) differentially expressed in all stages of its development. The Manduca miRs undergo the standard processing in the Argonautes, and silence targeted mRNAs in the RISCs. Some of these miRs (miR-965) are related to those of the other Bombycoidea, *Bombyx mori* [1436].

Comments. These relationships are given to envision the millions of years confrontations between the host and parasite DNAs behind the scenes, and the importance of the stem cells for the preservation for both participants.

"*When the shark bites...*" (Die Dreigröschenoper by Bertold Brecht). The shark has two rows of dozens of teeth and replaces frequently lost teeth in some 18 days. Gnathostomata grow may be thousands of teeth during their life time. The genomics of the sharks' tooth buds must consist of a pack of "proto-oncogenes". In order to identify these proto-oncogenes, the dentition of fish, reptiles and mammals is to be comparatively examined. The proto-oncogenes switch on and of (gain-of-function and loss-of-function activations alternating) in response to their start and stop codons. If the enamel and the calcification processes are of epithelial-line products, the dental pulp is of mesenchymal origin with stem cells retained in it. In the shark, the amelogenesis and the odontoblastic process were examined electron microscopically stage by stage [1437–1439]. In general, the Krüppel-like stem cell factor epiprofin is the leader of the Wnt/β-catenin/ameloblastoma ameloblastin dentition pathway [1440]. However, some strong stop codons prevent constitutive activation, or gain-of-function mutations, inasmuch as malignant jaw tumors in sharks do not appear.

The human malignant tumor of the odontogenic epithelium is the ameloblastoma, a most destructive, frequently recurring, and occasionally metastatic entity. The tumor cells always express the tooth enamel matrix genes, but in addition, the Wnt/sonic Hh pathway (Wnt10A at 2q35) is activated. This activity is attributed to high TNFα levels. Especially the Wnt region with the MMTV (mouse mammary tumor virus) insertion site is overexpressed; of the sonic Hh pathway, the Patched (PTCH) element was overexpressed. Other overexpressed gene-product proteins were FOS (Finkel mouse osteosarcoma), and the TNF-R; in contrast, TGFβ levels were low [1441]. The epigenomics of ameloblastoma are unknown. Mesenchymal stem cells could command epithelial cells of the gum to grow new teeth. The shark knows how.

Comment. And Bertold of East-Berlin knew how the shark bites and Mac the knife cuts: he was a Stalinist communist!

Phylogenic shifting of taxa. Phylogenetic analyses of DNA (nuclear milk casein and γ-fibrinogen genes, and mitochondrial cytochrome b) sequences suggest a common ancestor for hippopotamid artiodactyls (even toed ungulates) and cetaceans. Members of these taxa are hairless, lack sebaceous glands, nurse their newborns underwater, communicate even copulate underwater, and the males have no scrotum for their testes (inguinal in hippos and intraabdominal in cetaceans) [1442]. Consequentially to the marine environment due to an enigmatic transition from terrestrial to full aquatic habitat, the entire set of stress response proteins had to change and adapt. The innate TLR4 poised at the interface between host and an entirely new microbial environment has undergone a profound "adaptive evolution" [1443]. New SINE insertions appear in the genomes of Cetacea (dolphins, porpoises and whales). The common ancestor of the whales, the Mysticeti (filter-feeding whales) and Odontoceti (echolocating toothed whales) diverged in the late Eocene approximately 34 mya. The mitochondrial genomics contradicted this long-accepted taxonomic subdivision. The superfamily Delphinoidea (dolphins, porpoises, white whales) appears monophiletic. The Amazon river dolphin proves to be very much divergent from the marine dolphins (as the sperm whales differ from the balen whales) [1444]. The line that diverged to the Cetaceans and to the Hippopotamidae has not been clearly identified. The earliest Cetaceans appeared approximately 53 mya; whereas hippopotamids appeared approximately 16 mya, followed by the Pleistocene expansion of the hippopotamus [1445, 1446]. The common mammalian ancestor of cetaceans and hippopotamids probably lived in the shallow waters along shorelines. The line to become cetaceans probably diverged early (50 mya) and its members lived on "seafood" (aquatic carnivory). The quadritubercular dentition, auditory bulla for underwater hearing, and backward movement of the nasopharyngeal opening and trachea in the fossils are to be looked for; the key fossils are still to be discovered [1442]. The line left behind to become hippopotamids existed in the same environment, but became vegetarian. Its fossil record is incomplete. Still, it had to be a hippopotamid ancestor, which returned to the sea to become a cetacean. Gills could be replaced by lungs (*vide supra*), but lungs could not (so far) be replaced by gills: the cetaceans breath with lungs.

Archaeopteryx (*A. lithographica*), from the late Jurassic limestones represents the first attempt of the reptilian/dinosaur RNA/DNA complex at encoding wings to fly. The archaeopteryx feathers were colored black by melanosomes of the skin and their barbules' microstructure matches that of modern bird feathers [1447]. These feathered and volant dinosaurs traveled a long distance to become endothermic modern birds. The avian line diverged from theropod dinosaurs in the Jurassic period [1448]. The bony anatomy of avian skulls is dinosaur-like (paedomorphic, descendants resemble ancestral juvenals), except for the beak [1449]. The early avian radiation separated from the theropod dinosaur lineage over 150 my ago. Neoavian bird ancestors appeared in the late Cretaceous some 65 million years ago. It is attributed to retroposon insertions, which brought about the divergence of these lineages [1450]. The IFN-inducible transmembrane protein IFITM10 is the most highly conserved entity, shared in over 85 % aa identity between reptiles, birds and mammals [1451]. Melanin-containing organelles (melanosomes) in dinosaurs and in aves started the coloring of the iridescent plumage from the Microraptor in the early Cretaceous to that of extant birds [1452]. The total extinction of the avian dinosaurs/pterosaurs, and their gradual replacement with birds and mammals must have been the consequence of an overwhelming cataclysm. A major extinction of birds (with that of some reptiles and early mammals) at the Cretaceous-Paleogene (Tertiary) boundary did occur [1453], yet taxa of the avifauna survived and radiated. Aves is a highly speciose amniote clade counting over 10,000 species. The reproductive output (clutches *versus* litter size) of egg-laying dinosaurs far surpassed that of viviparous early mammals [1454], yet it was the birds and the mammals and not the dinosaurs that survived and radiated after the cataclysm.

Front limbs-powered flapping flight characterized pterosaurs, and continued in birds and bats. However each clade evolved anatomically different forelimbs. Some pterosaurs used feathered hind limbs as well. Re-shaping the humerus, ulna and radius forms feathered bird wings. Elongated fingers II-V connected with wide segments of featherless membranes form bat wings for a flying mammal. Genes encoding the elongated bat fingers II–V could be named. Prominent genes encoding this development are four homeobox genes, two SMAD genes, several Tbx (T box) and BMP (bone morphogenetic protein) pathway genes [1455], Sonar sensors and retina for sharp eye sight in dim light in bats evolved separately from the wings. The echolocation of flying insects enables bats to distinguish between insects (moth versus butterfly). Some moth wings evolved stealth installments (scales) to increase sound absorbance [1456]. Aerodynamic efficiency favors the birds' alternating flap and gliding flights over the bats' continuous flapping flight [1457, 1458a, b]. Physiology fulfills the demands imposed on it. The floating nectar-sucking hummingbirds (Trochilidae) perform up to 150 wing flaps per second and up to 1250 heartbeats per minute!

The origin of flight-less penguins is traced to the early Pliocene in the coastlines of South Africa (where the descendants of the ancestral birds still persist). In millenia, practically all other original taxa became extinct. The remnants (Sphenisciformes) might have been carried by the South Atlantic currents in the late Cenozoic to Antarctica [1459]. There the wings have become the most powerful fore-limbs for underwater swimming [1460]. There the formidable challenge of frozen glaciers and snow storms is counterbalanced by an abundance of fish with antifreeze blood chemistry (cryoprotective glycoproteins) proving plentiful of nourishment. The penguins' genome encoded the most remarkable heat retention structure located in the humerus of the fore-limbs. The structure is an arterial plexus with anastomoses and counter currents, CCHE (counter-current heat exchanger). One would readily assume that the genome responded to an environmental challenge. However, fossil records indicate that the CCHE was installed some 40 mya, during a "greenhouse Earth" interval [1461], but provided survival advantage later to those birds which had it. "Evolution has no foresight" (so teach the textbooks).

Comment. It must have been the adaptability of the genomes of the survivors, due to pre-inserted SINE and other retrotransposons, in the post-cataclysmic environment, that supported the newer generations, while the lack of these faculties doomed the prior inhabitants, which flourished in a different antecedent environment. The power of the RNA/DNA complex once installed in the primordial cells is to generate just about any biological change either in advance (but without precise foresight), or in response to demand, and remain just about inexhaustible.

Asymmetric mitoses of stem cells. Both in the larval or embryonic stages of healthy hosts, or in the cancer cell populations, pluripotent stem cells exist. The asymmetric mitosis of a stem cell results in the preservation of the mother stem cell: the self-renewal, and in the production of a differentiating daughter cell. The healthy adult cell populations preserve pluripotent stem cells characterized by asymmetric mitoses. The differentiating somatic cells format all the organs in the multicellular (metazoan) hosts. The healthy stem cells serve their host in tissue regeneration. In the self-renewing stem cell certain undifferentiated cell proteins (Nanog, Sox2, Oct4) are actively produced, whereas the encoding of cell differentiation proteins is silenced. In contrast, in the daughter cell differentiation prevails, and the proteins for the sustained undiffereniated state are silenced. Mitoses of the not self-renewing stem cells may result in two different daughter cells. Each daughter cell assumes a distinct pattern of differentiation. These forms of asymmetric stem cell mitoses are widely practiced in all multicellular organisms, and are very well documented in the caenorhabditis and in the drosophila. In the drosophila larva, Notch signaling maintains an overproduction of undifferentiated neuroblastic stem cells, whereas Aurora-A induces terminal differentiation of the neuroblastic stem cells. These cells divide in the not self-renewal mode, thus releasing a neuronal and a glial cell [1462, 1463]. In caenorhabditis larva, the lateral epithelial seam cells divide asymmetrically: the mother cell retains stem cell features, and the daughter cell terminally differentiates. In stem cells with up-regulated *wnt* gene, the mother cell produces two stem cells; whereas reduced Wnt signaling results in two daughter cells both undergoing terminal differentiation [1464]. In the human genome, both *notch* and *wnt* have become proto-oncogenes/oncogenes.

The asymmetric cell division in unicellular eukaryotes very likely was discovered first in 1969 in the frog-pathogenic *Trypanosoma montezumae* by R. Pérez-Reyes in Mexico [1465]. Indeed, the life cycle of *T. brucei* includes asymmetrical cell divisions regulating cell morphology, the location of organelles, and the appearance of the new flagellum. Also, trypanosoma and leishmania do not practice gene regulation at the transcriptional level; gene regulation for chromatin remodeling, RNA processing, DNA repair etc. is carried out posttranslationally (commonly by methylation on arginine residues of the targeted proteins). The transferred methyl groups derive from S-adenosyl-L-methionine molecules. The targeted substrate is an RNA-binding protein, or a histone. The target is the guanidine nitrogen of arginine. A unique faculty for this function is the family of four classes of protein arginine methyltransferases (PRMTs), functioning in RNA processing, transcription and signal transduction. This class of arginine methylases is absent in archaea, bacteria and in the giardia. *Saccharomyces pombe* is richly endowed with the enzymes protein arginine methyltransferase, PRMT1/5. The trypanosoma PRMTs show homology to the human PRMTs. The TbPRMT5 methylates the mitochondrial gene regulatory protein (RBP), and the omega mono- and dimethylarginine. The bloodstream form, and the procyclic insect-host form of the trypanosome, constitutively expresses its TbPRMT5 in the cytoplasm [1466a, b, 1467]. The PfPRMTs (*Plasmodium falciparum*) form monomethyl- and asymmetric dimethylarginins occupying either cytoplasm, or nucleus, or both. The H4Arg3me2 was essential for life. Small molecular blocking of PfPRMT1 inhibited growth of plasmodia cells. in the plasmodium genome, histone acetylation results in transcription activation. Trimethylation of lysine 9 on histone 3 (H3K9me3) silences, whereas H3K4me2/3 activates *var* gene transcription [1468]. The plasmodia *var* genes encode erythrocyte membrane proteins (PfEMP) in the invaded erythrocytes. Loss-of-function mutation of the *prmt*-1 gene in the caenorhabditis results in the silencing of longevity-related genes; the PRMT-1 protein functions as a cell longevity factor [1469]. The glycine- and arginin-rich GAR motifs in the telomeric shelterin complex are substrates for arginine methyltransferases. Loss of PRMT1 induces cell growth arrest; in healthy cells telomere doublets are formed. In cancer cells, PRMT1 knockdown results in telomere shortening [1470]. Thus, the classical telomere re-capping after each cell division, interpreted as sign of the cell's immortality, is canceled.

The human genome encodes at least 11 PRMTs (HsPRMT); type I PRMT1 and type II PRMT5 are in use in all eukaryotic cells. In the human genome, PRMT1 methylates arginine 3 in histone 4, and PMT5 methylates arginine 8 in H3 and arginine 3 in H4. The PRTM5-mediated symmetric dimethylation of histones results in transcription repression. Histone acetyltransferases and chromatin remodelling enzymes interact with PRMTs [1468]. PRMT1v2 promotes the survival and invasiveness of breast cancer cells [1471]. The PRMT6 promotes the cell cycle by repressing cyclin-dependent kinase inhibitors [1472]. In prostate cancer, the PRMT5 protein in cytoplasmic location is a tumor promoter. In resting healthy prostate cells, the PRMT5 protein assumes intranuclear location. In this position it is considered to be a tumor suppressor: it inhibits cell divisions in a

PRMT5-independent manner. In the prostate, PRMT5 forms complexes with a newly identified androgen receptor protein, p44. In developing prostate glands, p44 is of cytoplasmic location; in differentiated resting prostate cells, p44 assumes intranuclear position. Prostate cancer cells growing *in vitro*, or in xenografts were arrested in growth and proliferation, when p44 was forcefully translocated from cytoplasm to nucleus. The protein p44 apparently blocks the genes which promote cell cycle and causes cell cycle arrest in phase G_1/G_2. PRMT5 and p44 are co-expressed either in the cytoplasm or in the nucleus. The cytoplasminc co-expression of PRMT5 and p44 can be blocked by shRNA. The growth of human prostate cancer cell lines was arrested by this maneuver, same as the forced separate transfer of these two proteins from cytoplasm to nucleus. The enzyme PRMT5 catalyzes the transfer of CH_3-groups from S-adenosyl-methionine to the arginine (R) residues of histones. It is also highly expressed in lung cancer cells, with no activity in healthy lung tissue. Silencing the expression of PRMT5 in lung cancer cells resulted in the cessation of tumor cell growth and in the failure of the tumor to grow in xenografts [1473]. The RGG box (arginine-glycine-glycine) of the proto-oncoprotein EWS (Ewing's) is a substrate of the S-adenosyl-L-methionine binding domain-expressor PRMT8, for asymmetric dimethylation [1474]. Are proto-oncoproteins substrates of PRMTs?

Comment. The ancestral arginine methyltransferases serving the survival and life cycles of their unicellular hosts, and their descendants, but now constitutively expressed, promote and sustain the malignantly transformed cells in their multicellular hosts, in the state of their incessant mitoses, locomotion and invasiveness.

Volvox, the early host of stem cells. Volvox evolved from the unicellular alga chlamydomonas (*C. reinhardtii*). Volvocales exist in two different cell types: the small motile somatic cell, which eventually dies, and the large asexual reproductive gonidium. The two cell types derive from the gonidia-derived juvenile spheroids containing the somatic cells and the gonidia. The somatic cells and the gonidia are produced by asymmetric divisions. The acquisition of asymmetrical cell division is related to the gene *gls* (gonidia-less); the *gls*A gene encodes chaperone proteins, that direct the formation of the asymmetric spindle. In the gonidia *lag* genes prevent differentiation into the somatic lineage. In the somatic cells, *reg*A genes containing three intronic enhancers encode a somatic cell-specific nuclear protein, that prevents de-differentiation into gonidia [1475]. A colonial volvocalean possesses a group IA2 intron. This intron expresses an ORF, which encodes an endo-excinuclease [1476]. The vertically inherited intron may degenerate, or it may be lost, thus it might have served the ancestors, but the descendants may exist without it. Homologs of the gene product protein *gls*/Gls extend upward to eukaryotes from fungi to plants, and to vertebrates. The ancestor *C. reinhardtii* does not practice asymmetrical cell divisions. However, it contains a *gls* ortholog, the gene *gar*1 (gls-related). The Gar protein is 70 % identical to the Gls protein. Asymmetric dimethylation of Arg90 and Arg92 within the GAR motif is carried out [1477–1479a]. Thus, asymmetrical cell division is an ancestral event, which takes its quite exceptional origin in the volvocaleans (for plants), and kinetoplasts (for animalia).

The formate what eukaryotic stem cells of multicellular hosts are practicing derives from unicellular kinetoplasts.

Comment. The infectious process within a host is the battlefield, where the RNA/DNA complecti of the host and the parasite confront each other: DNA against DNA. The weaker system is allowed to lose and perish. The stronger system prevails—for a while. There is no final victory. If all pathogens were eliminated, what would challenge the genome to evolve further? Then, a standstill would be acceptable. However, any system not in use is eventually lost.

Without proto-oncogenes, there would be no stem cells, no germ cells, no ontogenesis for larvae and nymphs (no cocoons in which to transform); reptiles and aves would not lay eggs (there would be no eggs), tadpoles would not change from gills to lungs (there would be no tadpoles); and female mammalians could not carry their embryos in their uteri: bisexuality would have remained undiscovered and unexplored. Stable biological entities born by budding in the state of their original shape and form would populate a thermodynamically and physicochemically immobile universe. How could the living matter metabolize, in an immobile universe? It is unimaginable, that an immobile universe could possibly gender and sustain life (and death). Blasts of radiation must have induced point mutations. The living matter awakens to its immense and inexhaustible potentials. Its dynamism defies death. Review the relationship of entropy (the second law of thermodynamics) and life. Is the biomass of microbes built up in the hot springs with abundant nutrients available under the rules of a metastable equilibrium [1479b], the same that may rule the immortal growth of HeLa cells in culture, or the cancer cells in a subdued multicellular host? The growth curves and the mathematics are overwhelming, but the RNA/DNA complex evidently fitted the living matter into the thermodynamics of the physical universe.

It is most remarkable, when the RNA/DNA bioengineers encode oncogenes in an independent and immortal cell. There, control for organized cell communities may be lost, while elementary life is preserved. However, the supreme bioengineers resolve tasks of no less complexity in formatting the healthy living matter, fitted into constantly changing environments.

Selected abbreviation

PRMT, protein arginine methyltransferase (not postmastectomy radiotherapy). Click for p53-independent regulation of p21WAF1/Cip1 (wild-type p53-activated fragment/cycline-dependent kinase interacting protein) in scenescence by Phalke S Mzoughi S Bezzi M et al. Click for PRMT inhibitors by Vhuiyan M Thomas D Hossen F Frankel A. Click for PRMT in malignant gliomas by Han X Lin R Zhang W Nabors LB. Ckick for PRMT therapeutic targets in glioblastoma by Yan F Alinari L Lustberg ME et al.

Part II
The Oncogenome

Seamless Unity of the Nucleus and Cytoplasm: The Membranes Are Permeable

The Cancer Cell Nucleus

Divide like a stem cell. It is the centrioles of Theodor Boveri ("Ergebnisse über die Konstitution der chromatischen Substanz des Zellkerns", 1904) that are made up by microtubules of symmetrical configurations, and so engineer cell divisions. The centrosome is composed of two centrioles: the mature mother cell centriole and the immature daughter cell centriole. The centrioles give rise to the basal bodies (Mihály von Lenhossék: "Centrosom und Sphäre in den Spinalganglienzellen des Frosches", 1895). Centrioles may or may not participate in the spindle formation in eukaryotic cells. In some mitotic cells, spindles are formed without visible centrioles. In flagellate unicellular life forms, centrioles were essential cell organelles. In the cells of multicellular organisms (metazoans) centrioles are involved in the formation of bipolarity of the spindles. The centrioles are strong candidate engineers of the asymmetric cell divisions. In cancer cell lines propagated for decades in culture, centrioles persist [1480].

Common stem cell markers are the expression of CD44, CD133, CD117, CD34 and the aldehyde dehydrogenases (ALDH). Some of the clusters of differentiation (CDs) characteristically combine with chemokines (example: CD133 with CXCR4 in a pancreatic cancer cell population). Biologicaly, it is self-renewal, continuous propagation, asymmetric mitoses, and formation of tumor spheres that characterize stem cells (including transformed stem cells). CD117 is the tyrosine kinase receptor Kit of the c-kit → v-kit Hardy-Zuckerman feline sarcoma virus 4 oncogene homolog. Most natural stem cell populations are heterogeneous, and thus the expressions of stem cell markers by the individual cells of a given stem cell population is not uniform. Upon serial passages, stem cell populations may undergo clonal evolution [1481]. Based on three major cell types in the heterogeneous tumor cell populations (cancer stem cells; transit amplifying progenitor cells; terminally differentiated cells), a mathematical model for the outcome of such a cell population is proposed [1482].

J.G. Sinkovics, *RNA/DNA and Cancer*, DOI 10.1007/978-3-319-22279-0_21

In $CD44^+$, $CD133^+$ and $ALDH1^+$ (aldehyde dehydrogenase) human prostatic adenocarcinoma cells, the most prominent self-renewal gene product protein NANOG (*vide supra*) and its retrogene derivative NANOGP8 are often dominant in sustaining the stem cell status of the tumor cells. NANOG protein activates the insulin-like GF and ligand, CD44, ALDH1, and oncoprotein c-Myc. It reduces the expression of AR (rendering the tumor cells androgen-independent) and PSA (thus concealing the early phases of relapse). If the tissue culture media contain doxycycline, additional upregulations of IGF with chemokine CXCR4, CD133, c-Kit and that of the ATP-binding cassette transporters subfamily G (ABCG2) genes take place [1483]. This is very remarkable, because a previous report describes inhibitory activity of doxycycline to $Sox2^+$ human prostate cancer xenografts [1484]; however, these two reports appeared simultaneously, therefore without any cross-reference. The NANOG protein mRNA emanating from the human gene locus at 15q14 is susceptible to a knockout by shRNA. NANOG-deprived tumor cells lose their clonogenic and tumorigenic capabilities. The better differentiated PSA^+ tumor cells grow slower and are more androgen-depletion sensitive, than the PSA^- tumor cells. This latter tumor cell population is of low PSA-expressor, AR^-, $ALDH^+$ and $CD44^+$, and initiate robust tumor cell re-growth; express anti-stress genes, gain chemo- and hormone-therapy resistance; and divide asymmetrically producing one high PSA^+ daughter cell and a self-preserving low PSA stem cell [1485]. The oncogenic process in multicellular hosts appears to be a recall of the ancient stem cell-based life style.

Yamanaka *et al* induced human pluripotential stem cells from mature skin cells (keratinocytes). By cell fusion, or γ-retrovirally vectored genes encoding their protein products in a set sequence, such as Oct4, Sox2, Klf4, c-Myc reprogrammed mature fully differentiated cells into pluripotent naïve or primed stem cells. These genes active and essential in ontogenesis, are recognized as proto-oncogenes. Constitutively expressed, they become oncogenes. Generation of induced pluripotential human stem cells, and their maintenance on isogenic parental fibroblast: these cells continued to express undifferentiated pluripotent cell markers and could be induced to differentiate into any one of the three germ layers [1486] (in the Appendix 1).

The Cancer Cell Cytoplasm

Vasculogenic mimicry. In the mammalian pregnant uterus, the fetal trophoblasts not only invade the maternal decidua and enter the maternal blood circulation, but as a rule, line up in tubular structures with blood-filled lumina: a physiological case of pseudo-blood-vessel formation for additional blood supply to a target organ, the fetus and its placenta; reviewed with Table, in [83]. Choriocarcinoma cells follow suit in acquiring self-supply of blood circulation. In fact, practically all malignant tumors practice vasculogenic mimicry (VM). The tumor cells' DNA silences some of the oncogenes and activates the genes that physiologically encode vasculogenesis (endothelial cells) now to engage in (rather imitate) neo-angiogenesis. Signals from the cell surface molecules integrin αvß5 activate mitogen inducible genes (Mig). Mig-7 is activated in embryonic cytotrophoblasts. The Mig-7 protein co-localizes on the cell surface with factor VIII-associated antigen (the Willebrand factor), vascular endothelial (VE) cadherin, and laminin fragments. The laminin 5 γ2 chain-domainIII fragment is an activator of the EGF-R. In an internal circuit, EGF further activates Mig-7 expression. Mig-7$^+$ cytotrophoblasts assume endothelial cell-like phenotypes. Mig-7$^+$ colon carcinoma, or endometrial carcinoma cells form "leaky" blood vessel-like structures [1487]. However, the MIG-6 is a tumor suppressor in papillary thyroid carcinoma [1488]. Trophoblasts of gestational choriocarcinoma (originating from hydatidiform moles) formed blood vessel-like structures either by replacing endothelial cells in a preexisting blood vessel, or by forming blood vessels a-new, which anastomosed with endothelial cell-formed natural blood vessels. The article states "choriocarcinoma represents one of a few human tumor types that utilizes vasculogenic mimicry" [1489]. However, practically all human tumors properly examined show the capacity for VM, as one form of many other mechanisms activated for the acquisition of blood supply [1490]. VM stands out as an example of transformation of a tumor cell (be melanoma, Ewing's sarcoma, rhabdomyosarcoma, all adeno- or squamous cell carcinomas, hepatocellular carcinoma, glioblastoma, etc) in which the "species" characteristics of the original tissue origin are replaced (in great part, but not completely) by those of vascular endothelial cells. In addition to the molecular mediators activated in vasculogenic tumor cells as listed above (VE cadherin, ephrin receptor and ligand, EphA2), tissue factor pathway inhibitor 2/2 (TFPI) contributes to the endothelial transdifferentiation of melanoma cells. Further, tumor cells engaged in VM, frequently express excessive amounts of MMPs. Tumor cells in VM are driven by the PI3K pathway [1491]. Some tumor (melanoma) cells in VM are further promoted by the FAK-mediated signal transduction pathway (focal adhesion kinase) [1490].

Bevacizumab anti-neoangiogenesis therapy is aimed at sprouting new blood vessels. It could be effective in VM only, if from the vast repertoire of its genome, the tumor cell undergoing VM would have reactivated the silenced VEGF-R genes. Then, either by autocrine, or paracrine supply of the ligands, the tumor cell in VM would receive its major growth factor. Vascular tubule forming breast cancer cells ceased tubule formation, but continued to migrate upon treatment with bevacizumab; the combination of bevacizumab with MMP-9 inhibition stopped both VM and tumor cell migration. Phorbol myristate acetate induced VEGF-R2/3 expression and rendered the tumor cells bevacizumab-responsive [1492]. VM-forming melanoma cells were driven by VEGF-R1 and PKCα: their inhibition resulted in the cessation of VM formation (but inhibitors to VEGF-R2 and PKCα did not work) [1493]. Melanoma cells overexpressing VE cadherin (CD144), VEGF-R1, and TEK tyrosine kinase gene/receptor (or TIE2) at 9p21 (http://ghr.nlm.nih.gov/gene/TEK) with NGF-R CD271, immunoglobulin and EGF-like homology domains, angiopoietin receptor agonist Ang I were active in VM formation. VEGF-R1 was constitutively expressed. VEGF-R1 knockdown with shRNA inhibited VM formation and brought the cell cycle to halt [1494]. Uveal melanoma cells express nerve

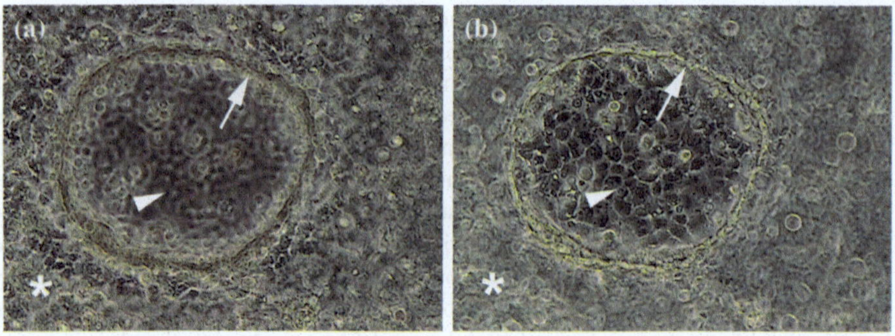

Figure 67 The "Malignantly Transformed Cell" Trans-speciates Itself as the Circumstances Call for: Vasculogenic Mimicry. [1495b] Reference Appendix 2, Explanations to the Figures

growth factor receptor (NGF-R, CD271) and become active in forming VM. These tumor formations are chemotherapy- and oncolytic herpes simplex virus-resistant [1495a, b] (Figure 67). The notorious liver metastases of this tumor express c-Met (for ligands hepatocyte growth factor and/or scatter factor), EGF-R and IGF-R1. MET knockdown failed, but EGF-R and IGF-1R inhibitors succeeded to induce tumor cell apoptosis [1496]. In hepatocellular carcinoma cell lines, inhibition of IGF-R1 with mcab AVE164 upregulated the epidermal growth factor receptor HER3 and AKT pathway. Co-targeting HER3 with gefitinib, or rapamycin, provided double blockade with complete arrest of tumor cell growth [1497]. In the 2013 ASCO meeting in Chicago, it was announced that bevacizumab and irinotecan prevented the re-growth of glioblastoma multiforme better than temozolomide. However, the overall survival of patients receiving bevacizumab was not prolonged (Gilbert et al for M.D. Anderson Hospital; awaiting publication).

Epithelial-to-mesenchymal transition, and vice versa. Transformed epithelial cancer cells (adeno- and squamous carcinoma cells) form large solid tumors, in which cell adhesion is still prominent. Due to the silencing of certain genes (tight junction proteins, occludins, cytokeratin, E-cadherin), and activation of hitherto silent other genes (vimentin, fibronectin, N-cadherin, myosin light chain kinase), the adherent cuboid cells transform into another direction in assuming an elongated configuration with re-arrangement of their cytoskeleton, loss of their apico-basal polarity and adhesiveness. These cells produce matrix metalloproteinases (MMP) and gain motility and invasiveness. The initiator of these changes frequently is TGFß. In this process, the first newly activated genes are *snail, slug, twist* and *zeb* (reviewed in [26]). Most of these genes were discovered in drosophila larvae, as they laid down mesoderm: *sna (snail*); *slut (slug*), and *twi (twist*). The human *sna (snail)* genes are mapped to the distal end of chromosome 20q; the human *slug* to 8q, and the human *twist* to 20p13. The human 3-exon and 29 kDa snail/SNAIL gene/protein contains three classic zink fingers and one atypical zink finger elements. Its 264 aa sequence is identical in 87 %, 59 %, 51 %, 51 %, and 31 % to mouse, chicken, zebrafish (or pufferfish), frog and drosophila SNAILs [1498]. The human genome harbors less characterized genes, which promote epithelial-to-mesenchymal transition. Cripto-1, member of the EGF family, and the TGFβ ligand Nodal, are active during fetal life in mesenchyme formation; Cripto-1 displays some stem cell characteristics. It downregulates E-cadherin, and upregulates mesenchymal markers vimentin, N-cadherin and Snail protein [1499].

In human breast cancers, the pituitary tumor-transforming gene-1 (PTTG), which was discovered in rats, encodes such an oncoprotein, termed secuin. This oncoprotein constitutively activates the PI3K/Akt pathway. The PTTG oncoprotein activates the *snail* gene (but not the *slug, twi* and *zeb* transcription factor genes), which induced EMT of the tumor cells. Targeting PTTG1 mRNA by shRNA reduced the tumor cells' ability to form colonies, or to grow in xenografts [1500]. In the ovarian carcinoma cell line #A2780, probably identical to a cell line established in the author's laboratory at M.D. Anderson Hospital [1501], and borrowed from there by gynecology fellow Philip DiSaia [1501a], the PTTG oncoprotein (encoded by an adenovirally delivered PTTG cDNA) upregulated TGFß, Slug, Snail, Twist and vimentin proteins and suppressed E-cadherin. These actions were reversed (gene deactivations occurred) by treatment of the cells with adenovirally delivered anti-PTTG siRNA [1502, 1503]. Human breast cancer cells exhibit extraordinary faculties for their transpeciation from sedentary cuboid to slender elongated and by lymph- or blood-flow traveling sarcomatoid cells, also using amoeboid locomotion.

In hepatocellular carcinoma cells, the Bcl2 and Twist proteins fuse into a complex that enters the nucleus and activates the genes, which encode EMT [1504]. Epigenetic deacetylation and hypermethylation of CpG islands in the E-cadherin gene commonly occur in cancer cells. Silencing E-cadherin allows cytoplasmic ß-catenin to translocate into the nucleus. The human E-cadherin gene expresses binding sites for the Snail oncoprotein. The hepatitis B virus X antigen downregulates the release of miR-373, the E-cadherin positive regulator. Consequentially in hepatocarcinoma cells E-cadherin is suppressed and the intranuclear ß-catenin contributes to the malignancy of hepatocellular carcinoma cells [1505a, b]. The originally chemically induced Met proto-oncogene is the receptor of ligands scatter factor and hepatocyte growth factor. However, the Fas-ligand traveling in lipid rafts could attach to Met and activate it by phosphorylation; in turn, Met signals through the activation of the STAT pathway [1380]. The cell surface-expressed FasL is known to induce external apoptotic death of $FasR^+$ immune T lymphocytes approaching in a multicellular host the FasL-expressor tumor cell; reviewed in [27, 147]. Hepatocellular carcinoma cells undergoing hepatocyte growth factor-induced EMT, over-express fibronectin, collagen-1, vimentin, cyclooxygenase-2, and Snail/Slug oncoproteins, while downregulate E-cadherin expression. The phosphorylated Akt pathway (without ERK) is the driving force of the increased mitotic activity [1506a, b, c]. The polycomb protein lysine methyltransferase zeste homolog (EZH-2) is a tumor promoter usually over-expressed in replicating cells, such as nasopharyngeal cancer cells. The oncoproteins Snail and EZH-2 form a linear complex, which additively suppresses E-cadherin through histone deacetylase actions on the E-cadherin promoter: HDAC molecules connect the EZH-2 and Snail proteins [1507]. A Twist-antisense plasmid exerted growth (cell mitoses) inhibitory effect in gastric cancer cells by inhibiting activator protein-1 and cyclin D1 mRNA [1508].

Synovial sarcoma cells are driven by the fusion oncoprotein SYT-SSX (synovial sarcoma translocated; synovial sarcoma X breakpoint) in the translocation t(X;18)(p11.2;q11.2); reviewed in [343, 344] (Tables III/I and VI; Figure 34). The tumor suppressor gene COM1 (candidate of metastasis) is a direct target of the SYT-SSX oncoprotein. COM1 expression is down-regulated in synovial sarcoma cells; synovial sarcoma cell transfected with COM1 cDNA succumb to apoptotic death [1509]. In biphasic synovial sarcoma cells, the elongated fibroblast-like tumor cells express the Slug and Snail oncoproteins, the activated Wnt pathway with ß-catenin accumulating from cytoplasm to nucleus, decreased E-cadherin, the surface adherens junctional protein, and an over-active PI3K/Akt pathway. Cells of the glandular (epithelial) compartment preserved E-cadherin and down-regulated Snail expression. The HDACI romidepsin reverses the tumor suppressor early growth response from negative (repressed by SS18-SSX through trimethylation: H3K27me3) to positive. Polycomb-mediated tumor suppressor gene repression can be reversed by HDACI with cessation of tumor growth resulting [1510]. Secernin proteins promote granule secretions (mast cell granules). The higher secernin-1 levels are in synovial sarcoma tissue (by gel electrophoresis), the longer extended the survival appears to be: 77 % 5 year survivals were recorded for patients with secernin-positive tumors *versus* 22 % survival for patients with secernin-negative tumors [1511].

Melanocytes migrating from the neural crest to the periphery, reversibly (not constitutively) upregulate Slug. The canonical Slug target, E-cadherin is not down-regulated in the migrating melanocytes. Slug is regulated by the microphthalmia-associated transcription factor (MITF). The MITF antagonists are gliotactins, the cholinesterase-like adhesion molecules, referred to as Gli. GLI2 is a Krüppel-like transcription factor targeted by the TGFß/SMAD pathway (signaling mothers against decapentaplegic, *dpp* in the drosophila). The MITF promoter expresses binding site for GLI2 (genetic linkage; gliotactin), for inhibitory interaction. Since TGFß induces GLI2, it is indirectly a suppressor of MITF. Melanoma MITF promotes malignant transformation, but its loss may not result in tumor cell death, inasmuch as mutated B-RAF and N-RAS oncoproteins sustain the malignant transformation. The pathways $PI3K^+/Akt^+/PTEN^-$ cooperated with E- or N-cadherin and Slug, Snail, Twist transcription factors. In $PTEN^+$ melanoma cells E-cadherin level dropped, Akt and N-cadherin levels rose. In $PTEN^+$, or wortmannin-treated cells (PI3K inhibition), E-cadherin levels rose and N-cadherin levels dropped. The PI3K pathway regulated the activity of the *snail* and *twist* genes [1512–1514]. The mRNAs to be translated to melanoma-promoting proteins are intercepted and deleted by miR-15/16, miR-96/182 and miR-141/200. Introduced into melanoma cell lines, miR-16, miR-96, miR-182 and miR-497 inhibited the expansion of these cell lines by interfering with melanoma cell proliferation [1515]. Vemurafenib inhibits BRAF-mutated (V600E) (valine; glutamic acid) melanoma cells. However, the stroma responds with the production of hepatocyte growth factor resulting in activation of its receptor c-MET. The MET-pathway of MAPK and PI3K/Akt restores melanoma cell proliferation. The c-MET pathway conveys resistance to EGF-R inhibitors, those may fail to control adenocarcinomas (breast, colon, ovary) and other tumors [1516–1518]. “Exquisitely selective” chemical inhibitors of epidermal-to-mesenchymal transition (EMT) may (or may not) provide long time relief [1519]. Natural products (phloroglucinol and sesquiterpenoid) are waiting to prove themselves in clinical trials [1520].

Pancreatic and prostate cancer cells undergo EMT by mechanisms seemingly different from the straight Slug/Snail/Twist pathway. There, one of the SLC (solute carriers) genes encodes ZIP14, a member of the LZT (LIV-1, leucine, isoleucine, valine subfamily, ZIP zink finger) protein transporters. These proteins differentiate mesenchymal soft tissue; their loss-of-function mutation causes spondylocheiro (SLC) dysplasia. Their gain-of-function mutation, or constitutive expression, induces mesenchymal de-differentiation of epithelial cells. LIV-1 directs the nuclear localization of the Snail protein. The LIV-1 mRNA could be attacked by antisense siRNA resulting in drop of LIV-1 protein levels with reduction of intranuclear Snail protein and increase in cytoplasmic E-cadherin. LIV-1 upregulates HB-EGF (heparin-binding epidermal growth factor) and MMPs. The MMPs cleave HP-EGF that phosphorylates its receptor and induces ERK signaling; constitutively activated EGFR drives EMT. These prostate cancer cells over-express also IGF-1 and TGFß [1521, 1522]. The epithelial-specific entities, and their homolog transcription factors (EHF) maintain differentiated epithelial cells, but when these factors are silenced, epithelial cells de-differentiate into mesenchymal-like stem cells; such events characterize some prostate cancer cell populations [1523]. The crypto gene product protein (better known as teratocarcinoma-derived growth factor-1) is a major contributor to EMT in breast, pancreatic and prostate adenocarcinoma cells, which assume the geno-phenotype of mesenchymal-like cancer stem cells [1499].

The nicotinamide adenine dinucleotide-dependent histone deacetylase, SIRT (sirtuin, silent information regulator; silent mating type information regulation homolog, Saccharomyces) is a deacetylator of histones of the androgen receptor gene. Deletion of *sirt*1 gene results in intraepithelial neoplasia (PIN); SIRT1 protein induces autophagy of prostate cancer cells, thus may be it is acting as a tumor suppressor [1524]. When SIRT1 activates the *zeb*1 gene (zink finger E-box binding) and SIRT1 and ZEB1 act together in prostate cancer cells, the epithelial cells undergo EMT consequentially to E-cadherin suppression and upregulation of mesenchymal markers. It is SIRT1, which deacetylates histone 3 of the E-cadherin promoter. Thus SIRT1 promotes the locomotion and metastasis-formation of mesenchymally transformed prostate cancer cells [1525]. In metastatic pancreatic adenocarcinoma cells, ZEB1, 2 were the inducers of the cancer cells' EMT [1526].

Neither PTEN loss, nor the RAS/MAPK pathway alone could result in EMT of prostate cancer cells; however, in combination the loss-of-function (for the tumor suppressor PTEN) and gain-of-function (for the oncogene) mutations resulted in EMT [1527]. Androgen-deprivation therapy suppresses prostate cancer cell growth for a while, with return of virulent malignancy later. The ZEB1 oncoprotein and the feedback loop of the androgen receptor induce the relapse of cancer cells in the form of EMT [1528], or PSA non-producer neuroectodermal cancers. In this latter case, androgen-deprived prostate cancer cells overexpress vimentin and undergo senescence; they do not replicate, but remain metabolically active; may have short telomeres, lost PTEN, but active Ras. These cells supply in paracrine circuitries neuroendocrine growth factors to androgen-deprived prostate cancer cells inducing their PSA^{-} neuroendocrine transformation [706, 1529]. The miR-1 and miR-200 levels drop in prostate cancer cells undergoing EMT; Slug expression is high in these cells. Forced re-expression of miR-1 and miR-200 inhibited EMT [1530]. Resveratrol was found to inhibit the expression of stem cell genes (Nanog, OCT4, Sox-2, c-Myc) in pancreatic cancer cells [1531]. Genistein acting either as a demethylator of hypermethylated tumor suppressor gene promoter CpG, or as a histone re-acetylator with deacetylase inhibitor Vorinostat (suberoylanilide hydroxamic acid, SAHA). In this case, it targeted deacetylated histone H3K9 of tumor suppressor gene promoter in prostate cancer cells, thus induced tumor cell death [1532]. Basically, an epigenetically efficient agent should methylate and/or reacetylate oncogene promoters, thus silencing them; and it should demethylate and deacetylate tumor suppressor promoters, thus activating them.

Breast cancer cells are induced to undergo EMT by NFκB and ZEB1/2 [1533]. Silencing by hypermethylation of histone H3 on lysine (K) 9 (H3K9me2) of the E-cadherin promoter by methyltransferase G9a, results in decreased E-cadherin expression. It is the Snail protein that activates the G9a gene; in turn, SNAI1 was activated by TGFß. The EMT-inducers ZEB1/2 and SNAI1/2 (Snail) are antagonized by miR-200 and miR-203, respectively, at the posttranscriptional level [1534, 1535]. It is not yet clear how the oncoproteins SNA and ZEB suppress the release of miR-200/miR-203. Breast cancer stem cells undergoing EMT up-regulate HIF-1α and miR21. An antagomir to miR-21 reversed these processes [1536a, b, c]. Tumor hypoxia (↑HIF1α) induced by antiangiogenic agents (sunitinib; bevacizumab) invited the accumulation of breast cancer stem cell progenitors [1537, 1538].

Comment. Oncogene/oncoprotein-targeted therapy with small molecular inhibitors, and oncoprotein-targeted therapy with monoclonal antibodies are rapidly advancing, Active tumor-specific immunization with genetically engineered breast cancer vaccines, adoptive genetically engineered immune lymphocyte (or NK/LAK cell) therapy, and genetically engineered oncolytic viral therapy are poised for a break-through. However, the defensive immune faculties of the multicellular host frequently switch from attacking to supporting the transformed cells. In a human breast cancer host relationship, "immune" T cells induced the EMT of the breast cancer cells. The mesenchymally transformed breast cancer cells expressed stem cell markers ($CD24^{-/lo}$; CD 44^{+}); by asymmetrically dividing, these stem cell retained stem cell faculties, and also yielded highly malignant tumor cells of epithelial cell phenotype. The Treg cells neutralize immune T cells (by chemokine-mediated, or FasL-to-FasR reactions). In contrast, the $CD8^{+}$ tumor-protective T cell clone induces the EMT by producing TGF-β and ILEI (IL-like EMT-inducer). It is the Family Member C (FAM3C) gene that encodes the interleukin-like EMT-inducer. EM-transformed cells resisted chemotherapy (mitoxantrone) and irradiation and produced an excess of DNA repair enzyme [1539]. Since exogenously added active Ras protein promoted EMT and stem cell formation [1540], the ILEI-producer T cells are suspect activators of the H/K *ras* proto-oncogenes. These T cells are treacherous traitors (ttT-cells) of their host; reviewed in [26]. Anergic (clonal or adaptive) T cells are tolerant toward emerging tumors. Autoimmune T cells are blocked by their CTLA4 or PD-1 antigens and PD-1-activator ligands. New mcabs neutralize these antigens and their ligands, and thus reactivate autoimmune cytotoxic T cells. Self-masquerading tumor cells (melanoma) are killed by these mcabs (mAbs).

The RNA/DNA-transformed cells regained the extraordinary vitality of ancient primordial unicellular organisms and readily mobilize all possible "cell survival pathways". On occasions, the cancer cell fuses with a mesenchymal cell (the tumor-supportive subverted fibroblast or macrophage). These cell fusions were recently discussed in a Springer publication of authors/editors Dittmar & Zänker [1541]. This author proposed that endogenous retroviruses that frequently appear budding from malignantly transformed lymphoma, melanoma and adenocarcinoma cells, mediate these cell fusions [26, 27].

Anti-apoptosis spelled out. The ancient pro-apoptotic guardian is the WT p63 gene predating p53 and p73. The triad forms the p53 superfamily. Originally the elementary triad might have been embedded in one single gene. The ancestral gene (s) appeared in the unicellular eukaryotes, while fungi were left out (Figure 64). The placozoa, Trichoplax, possess one p53 member, but the Nematostella already encodes three p53-like proteins (in the Appendix 1). Same applies to leeches (Helobdella), bivalves (Mya) and snails (Lottia). The invertebrate p53 family member proteins (cephalochordate amphioxus, Branchiostoma; urochordate sea squirt, Ciona, do not readily cluster with the proteins of the vertebrate p63 lineage. The triad (p53, p63, p73) underwent monophyletic evolution in the vertebrates. Among many contradictions, the Nematostella, Drosophila, and Caenorhabditis p53 family genes appear to be closer to the vertebrate p63-like family genes, than to the invertebrates' p53-like genes. Most of the invertebrate p53 superfamily proteins contain in their carboxyl terminus the SAM motif (sterile alpha motif), whereas the vertebrate p63 family proteins are devoid of it. Mollusks conserved the p53 line with its MDM inhibitors, and are experiencing the development of leukemia-like neoplasms (*vide infra*). The pro-apoptotic p53 equivalent in *Caenorhabditis elegans* is the CEP-1 protein, and in *Drosophila melanogaster*, the Dmp53 protein. The anti-apoptotic protein in caenorhabditis is the CED-9 (cell death defective) protein, and in drosophila is the DIAP (inhibitor of apoptosis protein). The Bcl-2 anti-apoptotic protein co-exists with its pro-apoptotic homologs (Bax, bcl-2-associated X protein; Bak, bcl-2-antagonist killer). Bax induces permeabilization of the mitochondrial membrane for the release of cytochrome C oxidase. Cytochrome C and Apaf-1 trigger the caspase cascade for nuclear DNA fragmentation. They are present in the amphioxus, sea squirt and sea urchin. The drosophila DIAP antagonist is Buffy; the activity of cytochrome C-induced caspases are uncertain. Rather, the Apaf-1 homolog ARK (APAF-related killer) activates the caspases, DCP-1 (death caspase protein), DRONC (drosophila initiator caspase; mammalian caspase 9 homolog; dro for drosophila, nc for Nedd2-like caspase), and the executioner caspase DRICE (drosophila ICE: interleukin-1β converting enzyme). DRONC activates DRICE. DRONC is inhibited by DIAP1, and is activated by Apaf-related killer, ARK. DIAP1 removes DRONC and DRICE by ubiquitylation; the cells so rescued are referred to as "undead" at Baylor College of Medicine and

M.D. Anderson Hospital, Houston, TX [1542, 1543]. The apoptosome-forming protein (Apaf, apoptotic protease-activating factor) in caenorhabditis is CED-4, and the caspases are CED-3 (cell death abnormal). The caenorhabditis cep-1/CEP is the ortholog of the human pro-apoptotic killer of transforming cells, WT p53. Negative regulators of CEP-1 are the mRNA-binding protein GLD-1 (germ line development defective), the worm DIAP (inhibitory apoptosis stimulating protein p53, ASPP), and the PRMT-1 (arginine methyltransferase) [1544–1549]. The *iaspp* gene has been conserved in higher vertebrate eukaryotes: its protein product iASPP actively neutralizes p53 in human lung cancer cells [1550].

The *Botryllus* colony (Figure 46) shares a vascular network for circulating hemolymph. One particular cellular component in the hemolymph expresses lectins related to those that NK cell express throughout their evolutionary scale extending up to Homo. These cells can kill the cells of another histoincompatible intruder botryllus colony [294]. The attacked cells die by nuclear clumping due to DNA fragmentation. The process of programmed cell death (apoptosis) is energy consuming. This type of cell death occurs in entirely autologous cells colonies, when useless cells or tissues are eliminated during ontogenesis (in the cocoons of insects; in the tadpoles absorbing their tails and gills; in the mammalian host, absorbing its pro- and mesonephros). Programmed cell death initiated in the mitochondria are intrinsic (*vide supra*), whereas cell death initiated by the Fas receptor Fas ligand unison (CD95 FasR; FasL) are extrinsic (fragment apoptosis stimulating); reviewed in [147] (Figure 23).

Some virally infected single cells self-activate the process of programmed cell death in order to die before the new viral progeny matures; thus, the entire cell colony escapes a viral infection ("cell altruism"). Many viral genomes (herpesviruses; poxviruses) contain genes, which encode anti-apoptotic proteins of ancient host cell origin, securing a living cell for the support of the full maturation of the new viral progeny [1551, 1552]. The Meq oncoprotein of Marek's disease herpesvirus directly interacts with the p53 protein and inhibits its apoptosis-inducer transcriptional activity [1553]. The genome of the Sloan-Kettering-isolate retroviruses carries the cellular oncogene/oncoprotein *ski*/SKI. The SKI protein bound to NAD-dependent histone H3K9 deacetylase SIRT1/Sirtuin (nicotinamide adenine dinucleotide; silent information regulator, silent mating type information regulation 2 homolog, *vide supra*) rendering the p53/SIRT1 interaction stable. The deacetylated p53 is silenced [1554] *).

*) *The oncosuppressor p53 is antagonized by MDM and Ski (Sloan-Kettering retroviral isolates c-onc ↔ v-onc); Ski downregulates the TGF-β, and upregulates the Wnt/β-catenin pathways; nutlin is a MDM antagonist. Ski recruits histone deacetylator SIRT1. Phosphorylated (the N terminal domain) and acetylated (the C terminal domain) p53 protein is posttranslationally active; the acetylated p53 protein gains DNA-binding ability. Deacetylation represses the transcriptional activity of the p53 protein. The oncoproteins SIRT1 and Ski collaborate in suppressing p53 [1554]. Ski is an example of an inserted and retained retroviral messenger, that initiates malignant transformation in a cell by switching off a tumor suppressor (see in Self Defense, the Cancer cell Reactivates Its Primordial Genome…).*

The early gene product proteins E6 and E7 of oncogenic HPVs acting with ubiquitin ligases and/or inhibiting keratinocyte involucrin production degrade the human pro-apoptotic Rb and p53 genes [1555]. Bacterially infected cells succumb to apoptotic death, but obligately intracellular Rickettsia and bacteria (Chlamydia) protect their host cells against apoptosis. Chlamydia-infected host cells (amoeba) activate the cell survival pathways PI3K/Akt and NF-κB. Customarily, PTEN is supposed to be knocked out, for PI3K/Akt to be activated (Figure 50). Mitochondrial membrane permeabilization by pro-apoptotic BAX and BAD (Bcl-2 antagonist of cell death) should be inhibited. Indeed, Akt is known to inhibit Bad by its phophorylation, and activates cIAP-1. Therefore, the chlamydia *Simkania negevensis* may secure the survival of its host cell by activating proto-oncogenes. However, when the Chlamydia grows in malignantly transformed cells (in HeLa cells in the laboratory), constitutively activated cell survival pathways are already provided for it [1556].

Cells in the process of "malignant transformation" activate anti-apoptotic genes, whose encoded proteins inhibit programmed cell death. Some of these genes have been exchanged between unrelated cells through horizontal routes; reviewed in [26, 27, 30]. Complex eukaryotic cells have sets of pro- and anti-apoptotic genes/gene-product proteins [147]. The NK cell compartments primarily generated for the elimination of invader alien cells, must have undergone a major paradigm shift in placental mammals. There the paternal antigens of the fetus are recognized, yet the fetus is not to be rejected [30a, 1557, 1558]. The placenta may succumb to viral, bacterial, and protozoal pathogens penetrating it; and some of the invaders are potentially oncogenic; reviewed in the Springer Verlag textbook of G. Berencsi, M. Takács *et al* [1559].

The pro-apoptotic WTp53 gene is in particular the target of the cell's anti-apoptotic activities. The ancestors of the MDM2/4 proto-oncoproteins (encoded by murine double minute extrachromosomal DNAs) existed as single genes in invertebrates (*C. intestinalis, B. floridae, M. trossulus*, etc. in the Appendix 1). After the gene duplication event approximately 440 mya (before the emergence of vertebrates), MDM2 and MDM4 appeared in vertebrates; the invertebrate and vertebrate MDM2/4 contain well conserved RING domains. The gene for the human MDM2 is located on chromosome 12q14.3-q15; the MDM4 gene is mapped at 1q13. These proteins are ubiquitin ligases that bind and neutralize the pro-apoptotic gene product proteins of the p53 gene family (p53, p63, p73) [1560] (Figure 64). The ancestral single p63/p73 gene emerged in the unicellular choanoflagellates and remained operational in the early multicellular sea anemones, conserving these genes up to cartilaginous fish (Chondrichtyes). The p63/p73 gene protected the germ line DNA (and still does in the sea anemone, *N. vectensis*) from "damage" (in the Appendix 1). Worded more accurately: the ancestral p53 superfamily gene product proteins originally ensured the fidelity and integrity of their hosts' germ cell genomes. Cartilaginous fish duplicated the ancient gene and possesses separate p53 and p63/p73 genes. Bony fish experienced another gene duplication event yielding separately the three p53, p63 and p73 genes. The p53, p63/p73 genes remain conserved in amphibians, reptilians, and mammalians (including *Homo*). In aves, the bursal-derived pentapeptide, BPP-II, activates the p53 pathway [1561]. The locus of the 19,198 nucleotide 11 exons human p53 gene is at 17p13. The 265,822 nucleotide 14 exons p63 gene is at 3q27-29. The 80,728 nucleotide 14 exons p73 gene is mapped at chromosome 1p36.2. Larger introns increase the sizes of the p63, p73 genes. The p53 protein moderates its pro-apoptotic activity in the pregnant uterus (the "sexually dimorphic roles"), thus fetal cells are protected. The LIF (leukemia inhibitory factor) and ARF (alternate reading frame) are upregulated. High LIF levels are essential for the implantation of the blastocyst. LIF levels are correlated with estrogen and progesterone dosages. The ARF protein dimerizes with MDM2 and inhibits it. The LIF gene was recently discovered to be a target of the p53 protein. The p53 protein may be subjected to single nucleotide polymorphism (SNP). Its codon 72 may encode either arginine (R72) or proline (P72). R72 induces higher endometrial LIF levels, than the proline allele. P72 may contribute to the failure of *in vitro* fertilization. The Caucasian and Asian races frequently (>70 %) express the R72 allele, whereas P72 allele expression in African females exceeds 60 %. The p53 sister gene product proteins p63 and p73 contribute to healthy egg formation in the ovary and to normal spindle checkpoints in the eggs. As to protection against transformed cells, the P72 allele is weaker in apoptosis induction than the R72 allele. In the *mdm2* gene, SNP may involve encoding T or G (threonine, glycine) at the intron nucleotide 309. The G residue potentiates MDM2 expression, thus attenuates p53 activity. Further, estrogens act upon G309 increasing its anti-p57 potency. Consequentially, the SNP309G allele and high estrogen levels are associated with increased cancer incidence [1562–1564]. *Caenorhabditis elegans* expresses a p53 homologue gene-product protein: *cep/*CEP, but the MDM2 family proteins are missing; some other proteins substitute for it. The situation is similar in the drosophila. Somatic cells rest in worms and fruit flies; it is only the germ cells that replicate. In these organisms, p53 still protects the germ line, as its main function was in the Nematostella. In the worm and in the fruit fly, p53 responds to inflammatory stimuli (nitric oxide), thus it may have had a role in innate immunity [1564].

Anti-p53 MDM. The human *mdm*2 and *mdm*4 genes each consist of eleven exons; the protein-coding exons are 3–10. The nucleotide units are most numerous in the frog *mdm*2/4 genes: 12,373 and 149,800, respectively) due to large numbers of LINEs inserted into their introns. In comparison, the mdm2 and mdm4 genes of zebrafish, mouse and human contain 9,501, 21,871, 37,259 and 25,578, 35,504, 41,738 nucleotide units, respectively. Consider the common ancestor of *Danio rerio* and *Homo* in the Paleozoic era. The human *mdm*2 and *mdm*4 genes encode proteins of similar aa content: 491 and 490 aa, respectively. The MDM2 protein RING domain at residues 438–479 with six cysteins, two histidines and two zink atoms attaches to the p53 protein for the purpose of marking the p53 protein for ubiquitination. The proteasome 268 carries out the degradation of the p53 protein. Thereafter the RING domain catalyzes the degradation of the MDM2/4 dimer [1560]. In pediatric acute lymphoblastic leukemia cells, the MDM2 protein is overexpressed and neutralizes the WTp53. Nutlin-3, the small molecular antagonist of MDM2 liberates p53 and enables it to induce apoptosis. In addition to p53-activated pro-apoptotic proteins, p53 induced repression of the anti-apoptotic protein survivin [1565] (Tables VI and VII; Figure 64).

Intronic mutations occur in the WTp53 gene in childhood acute lymphoblastic leukemia, without the incapacitation of the gene, The p53 gene product protein is neutralized by binding the MDM2 protein (*vide supra*). Gene amplification of the proto-oncogene at 12q13-14 results in the overproduction of the MDM2 protein. The C terminus of the MDM2 protein acts as a E2 ubiquitin ligase tagging the p53 protein for degradation in the proteasome. However, an entirely different mechanism can suppress the p53 gene and the production of its gene product protein. The phosphorylation-activated MDM2 protein can enter the nucleus. It is PI3K/Akt that phosphorylate for intranuclear localization the MDM2 protein. If the small molecular inhibitor nutlin-3 binds the p53-binding pocket of the MDM2 protein; the p53 apoptotic pathway remains uninhibited. All tumors (including adenocarcinomas), in which p53 is neutralized by MDM2, respond to nutlin-3. In responding tumor cells, the pro-apoptotic genes p21 and Bax become upregulated. A second function of p53 is the downregulation of the anti-apoptotic gene survivin. Survivin was downregulated in tumor cells treated with nutlin-3. ALL cells with upregulated MDM2 expression are resistant to doxorubicin. However, treatment with nutlin-3 restores the doxorubicin susceptibility of these ALL cells. Tumor cell with mutated or deleted p53 fail to respond to nutlin-3 [1565]. Nutlin-3 restored apoptotic tumor cell death both in EBV^+ and EBV^- Burkitt's lymphoma cells, except for the setting of EBV^+ latency III cells. In this case $p21^{WAF/CIP}$ overexpression conveyed the resistance (wild-type p53-activated fragment; cyclin-dependent kinase interacting protein) [1560]. Nutlin effectively restored p53-mediated apoptosis in rhabdomyosarcoma cells [1567]. Nutlin-3 is C30H30C12N4O4 (Cayman Chemical).

The *M. trossulus* mussel's p53 protein expresses MDM-binding sites. The ancestral protein for MDM2/4 exists in the mussel and binds its NH2 terminal to the C-terminus of the mussel's p53 protein. The *mdm* gene yields a 2852 nt cDNA with ORF of 1683 nt encoding a 561 aa protein of 62.6 kDa. The mussel's MDM mRNA was cloned. The highly conserved region match those in the vertebrate MDM2: the p53 binding N-terminal and the Zn-binding RING domain. The central acidic domain exhibits five phosphorylation sites. The MDM-binding sites on the p53 protein are also highly conserved. The MDM NH2 terminus forms the cleft for the binding of the p53 protein. The mussel's MDM binding site to p53 is 39 % identical with the vertebrate counterpart. While neoplastic hemocytes expressed elevated levels of p53 and MDM, Ras mRNA level was not elevated. The p19 H-Ras oncoprotein can obliterate the MDM2-p53 inhibitory interaction in the cells of vertebrate hosts. However, Ras protein expression was not elevated in the mussel's transformed hemocytes. Another inhibitor of the reaction, ARF, was absent in the mollusk's genome [1568]. In human acute lymphoblastic and myeloblastic leukemias, the mRNA levels for p53 and MDM2 both were elevated and BCL2 protein was overexpressed [1569]. It has been well established that the MDM-to-p53 reaction is quite ancient and that it has been well preserved from mollusks to vertebrates.

Relda M. Cailleau of M.D. Anderson, Houston TX, in the 1970s. In the article entitled “A naturally occurring cancer with molecular connectivity to human

(a)

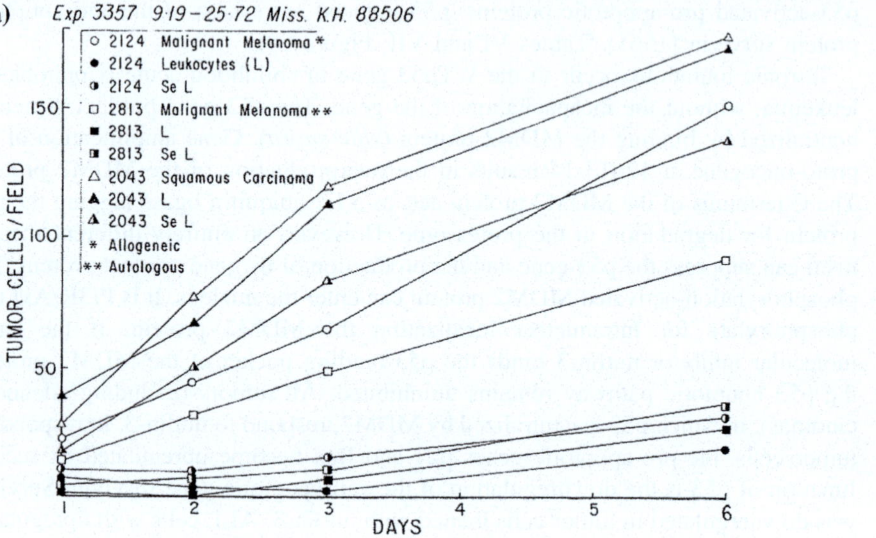
Exp. 3357 9·19–25·72 Miss. K.H. 88506
○ 2124 Malignant Melanoma *
● 2124 Leukocytes (L)
◑ 2124 Se L
□ 2813 Malignant Melanoma **
■ 2813 L
◨ 2813 Se L
△ 2043 Squamous Carcinoma
▲ 2043 L
◭ 2043 Se L
* Allogeneic
** Autologous
TUMOR CELLS/FIELD
150
100
50
1
2
3
4
5
6
DAYS

(b)

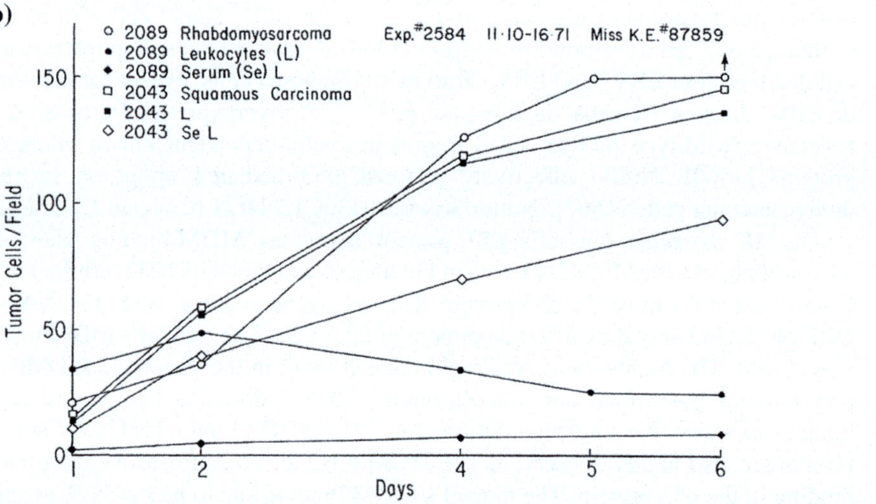
○ 2089 Rhabdomyosarcoma
● 2089 Leukocytes (L)
◆ 2089 Serum (Se) L
□ 2043 Squamous Carcinoma
■ 2043 L
◇ 2043 Se L
Exp.#2584 11·10-16·71 Miss K.E.#87859
Tumor Cells / Field
150
100
50
0
1
2
4
5
6
Days

◀ **Graph 1** **a** Graph showing inhibition of T cell cytotoxicity by pretreatment of target human tumor cell with antibodies. In *Experiment* #3357, female patient K.H. (MDAH#88506) with malignant melanoma yielded small compact lymphocytes (later: immune T cells), which inhibited the growth (were cytotoxic to) both allogeneic and autologos melanoma cells; pre-treatment of the target cells with the patient's serum in both cases exerted a weak, but consequent, inhibitory effect. The patient's lymphocytes did not inhibit (were not cytotoxic to) squamous carcinoma cells; pretreatment of these target cell with her serum induced cell-mediated inhibitory (cytotoxic) effect, mediated by large granular lymphocytes (later: NK cells). **b** Graph showing augmentation of cytotoxicity of large granular (NK) cells by pretreatment of target tumor cells with antibodies. In *Experiment* #2584, female patient K.E. (MDAH#87859) with rhabdomyosarcoma yielded a mixed lymphocyte preparation consisting of small compact and large granular lymphocytes. The lymphocytes inhibited the growth of (were cytotoxic to) the allogeneic rhabdomyosarcoma cell line #2089 [2174], and pretreatment of the target tumor cells with the patient's serum intensified this effect. The lymphocyte preparation failed to inhibit the growth of (was not cytotoxic to) squamos carcinoma cells (#2043) [2174], but pretreatment of the tumor cells with the patient's serum induced cytotoxicity mediated by large granular lymphocytes (later: NK cells). Graph is reproduced with permission from Sinkovics' "Cytolytic Immune Lymphocytes" Schenk Buchverlag, Passau & Budapest [147]

diseases" by Walker & Böttger [1570], Jessani *et al* receive credit for propagating the human breast cancer cell line MDA-MB-231 in xenografts and showing that the new stroma, the mouse mammary fat pad, profoundly changes the entire proteomics of the transplanted cancer cells, as compared to the *in vitro* grown original cells [1571]. While the credit for the documented results is fully deserved, this author misses quotation in the references of the Jessani article to Relda M Cailleau, who originated this cell line (together with several other widely used M.D. Anderson Hospital human breast cancer cell lines). Relda was working in the laboratory of this author (JGS), at the Section of Human Tumor Virology and Immunology, Department of Medicine, MD Anderson Hospital, Houston TX, in the 1970s, as a much welcomed guest investigator. The Section fully equipped for human cell culture work [1501] was privileged to assist her. Instead of listing Relda Cailleau's several original reports, one of her last report is cited here, which appeared just after the Section was closed in 1979 (due to lack of NIH/NCI funds). This report lists the cell line MDA-MB-231 referred to above [1572]. Staff medical oncologist on the Breast Cancer Service at M.D. Anderson, W.J. Reeves, provided this author (JGS) with fresh tumor tissue and blood samples from his patients. Accordingly, primary tumor cell cultures could be set up and the patients' immune reactions to autologous and allogeneic tumors could be tested *in vitro*. These tests confirmed that autologous reactions were mediated primarily by small compact lymphocytes (later: immune T cells), and allogeneic reactions were carried out by large granular lymphoid cells (later: NK cells). Antibodies (patients' sera) could block the cytotoxicity of the small compact T cells, whereas these sera could enhance the cytotoxicity of the large granular lymphoid cells [147, 1573a, b, c] (Graphs 1 and 2). These observation remained unexplained until after the discovery and reporting of the ADCC reaction (*vide infra*).

(a)

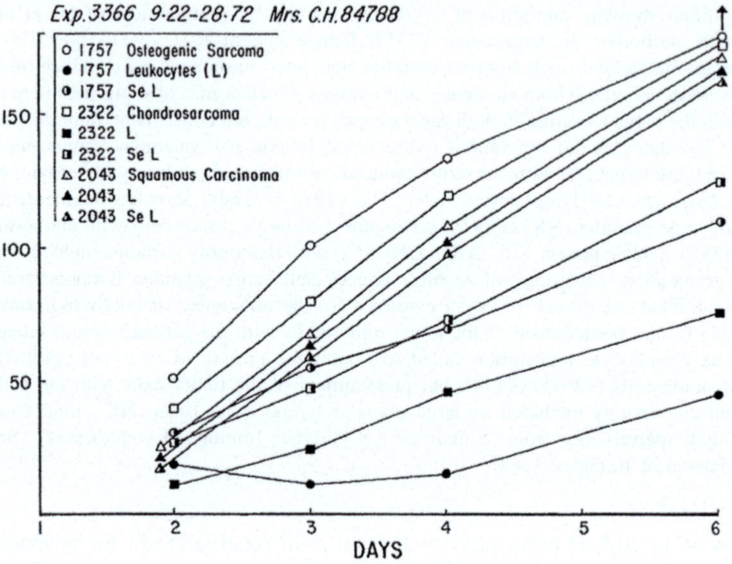
Exp.3366 9-22-28-72 Mrs. C.H.84788
1757 Osteogenic Sarcoma
1757 Leukocytes (L)
1757 Se L
2322 Chondrosarcoma
2322 L
2322 Se L
2043 Squamous Carcinoma
2043 L
2043 Se L
TUMOR CELLS/FIELD
150
100
50
1
2
3
4
5
6
DAYS

(b)

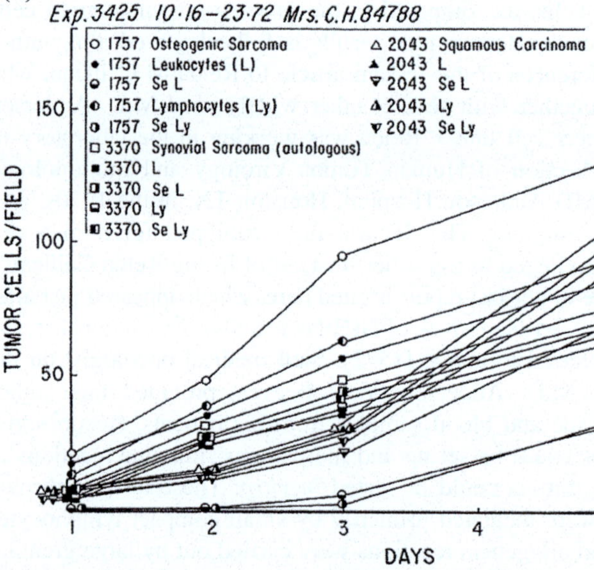
Exp.3425 10·16-23·72 Mrs. C.H.84788
1757 Osteogenic Sarcoma
1757 Leukocytes (L)
1757 Se L
1757 Lymphocytes (Ly)
1757 Se Ly
3370 Synovial Sarcoma (autologous)
3370 L
3370 Se L
3370 Ly
3370 Se Ly
2043 Squamous Carcinoma
2043 L
2043 Se L
2043 Ly
2043 Se Ly
TUMOR CELLS/FIELD
150
100
50
1
2
3
4
7
DAYS

◀ **Graph 2 a, b** Graphs of *Experiments* #3366 and #3425 show immune reactivity to allogeneic tumor (sarcoma) cells, but abolished reactivity toward autologous tumor (sarcoma) cells. The young female patient C.H. (MDAH#84788) with widely metastatic synovial sarcoma circulated lymphocytes cytotoxic to allogeneic osteo- and chondrosarcoma cells, but not to squamous carcinoma of the uterine cervix cells. Her serum blocked the cytotoxic effect of lymphocytes to sarcoma cells. After active immunization with irradiated allogeneic osteosarcoma cells, her lymphocytes continued to kill osteosarcoma cells, but her serum now increased the cytotoxicity of her lymphocytes against the osteosarcoma cells (by contemporary slang: converted from "blocking" to "unblocking"). Neither her lymphocytes, nor her serum and lymphocytes exerted any inhibitory (cytotoxic) effect on (were not cytotoxic to) her own autologous synovial sarcoma cells. The reaction against squamous carcinoma cells remained negative (*Experiment* #3425). In 1972, it was without explanation why the small compact lymphocytes of this patient were able to react to allogeneic sarcoma cells, but not to autologous sarcoma cells. The serum factors blocking cytotoxicity of small compact lymphocytes might have been specific antibodies covering (concealing) the epitopes that the T lymphocytes were to react to. The post-immunization serum factors might have been antibodies acting in an ADCC reaction through the Fc receptors of the patient's NK cells. However, if this experiment were done today, the patient's lymphocyte preparations would be tested for $CD4^+CD25^+FoxP3^+$ T_{REG} cells and it would probably be shown that T_{REG} cells were dominant in the patient's lymphocyte population, and that T_{REG} cells specifically inhibited her immune $CD8^+$ cytotoxic T cells, which would have been otherwise reactive to her autologous synovial sarcoma cells. Graph is reproduced with permission from Sinkovics' "Cytolytic Immune Lymphocytes" Schenk Buchverlag, Passau & Budapest [147]

A naturally occurring cancer: "oyster leukemia". The natural leukemia-like fatal disease of molluscan species is the neoplasm of the mussel *Mytilus trossulus*. The soft shell clam, *Mya arenaria* on the NW coast of the USA and Canada develops a fatal leukemia-like disease ("clam hemocyte cancer") originating in its hemolymph cells. In the "leukemic cells" the p53 gene-product protein is inactivated (sequestered in the cytoplasm) by mortalin, a mitochondrial Hsp70 protein. The report claims that "fly and worm *in vivo* models for adult human somatic cell cancers do not exist because their adult somatic cells do not divide" [1570, 1574]. However, this volume refers to "drosophila leukemia" and "drosophila brain cancer" (*vide supra*). The clam leukemia may be due to chemical, or viral inducers. The healthy clam hemocytes appear as granular, or a-granular motile round cells, some of which are phagocytic. The transformed ("cancerous") hemocytes are somewhat larger, grow (divide) *in vitro* and are serially passable; yet they are referred to as "not immortalized." The disease is transmissible by inoculation of the diseased hemocytes [1574]. "Viral etiology" is considered but without experimental documentation (transmission EM for viral particles; disease transmission with filtrates; genomics for inserted retro- or herpesviral genes; proteomics for retro-, or herpesvirally encoded proteins). For naturally cell-transplanted leukemia of Ma clams, see the Appendix 1.

Stanley J. Korsmeyer, 1950–2005 (Sidney Farber professor of pathology & medicine at Dana Farber Cancer Institute & Harvard). The fundamental contributions of Stanley to the works of the BCL-2 family of anti-apoptotic oncoproteins is widely appreciated (Editorial) [1575a, b]. The system is installed in the Nematostella, even though the sea anemone is not a triploblastic metazoan: it has only ectoderm and endoderm without a mesoderm in between (a diploblast). The Nematostella NFκB protein shows 50 % identity with the Rel homology domain (RHD, reticuloendothelial retrovirus) of the human NFκB RHD. The Nv-NFκB proteins are encoded from two alleles. While the hydra (*H. magnipapillata*) does not, the sponge Amphimedon (*A. queenslandica*) operates an ankyrin repeat domain-containing NFκB. However, the proto-oncogene-like IKK enzyme is expressed in the hydra. The drosophila genome encodes three, the human genome five NFκB proteins. The importance of the NFκB proteins is the wide range of DNA-binding sites the proteins target within the nucleus. Those are innate immunity genes (cytokines, MyD88, tumor necrosis receptor-associated factors, TRAFs), antioxidizing enzymes (superoxide dismutase, nitric oxide synthase), and the anti-apoptotic genes (BCL-2 family genes; TNFR-associated factor; IAP), and the IκB kinase(IKK) gene. The Nematostella IKK gene shows a great deal of sequence homology with the human IKK gene. The IκB protein anchors the NFκB protein within the cytoplasm. The IKK, encoded upon the activation of its gene by NFκB, phosphorylates the IκB protein, the negative regulator of NFκB, and condemns it to degradation by ubiquitination. The liberated NFκB protein is free to translocate into the nucleus. The Nematostella Bcl-3 fails to block the nuclear localization of Nv-NFκB. The NFκB pathway has been conserved in all bilaterians for some 600 my [1576]. For further discussions including the NFκB/STATcell survival pathway in the Nematostella, its human descendants' lymphoma-inducing potency, and their possible dependency on viruses, see the Appendix 1.

Apoptosis-prone transformed cells. The histone H2A variant H2AX (ataxia telangiectasia) is vulnerable to phosphorylation at serine 139 by PI3K at its locus on 11q23-3. The gene of H2AX may appear in several copy numbers (amplification). The regulator of H2AX expression is miR-24-2; this miR downregulates H2AX expression. Transformed cells (HeLa; breast cancer cell lines) expressing miR-24-2 become highly apoptosis-prone. The miR-24-2 targets the mRNAs of Fas-associated factor 1 (FAF), BCL-2, MDM2, and H2AX. Downregulation of H2AX introduces an impediment to DNA repair, and stimuli toward the apoptotic pathway. The mRNAs of the anti-p53 MDM2, and the anti-apoptotic BCL-2 proteins are not translated. Pro-apoptotic modulations appear to be through p53 activation and MDM2 downregulation. In this case, pro-apoptotic p53 proteins enter the mitochondria to initiate the endogenous form of apoptosis (in the Appendix 1). The cycline-dependent kinase inhibitor (CDKN1A) does not have a miR-24-2 binding site, yet its level drops; thus, the cell is not at standstill, but advances in the apoptotic pathway [1577].

Navitoclax. The specific BCL2/BCL-XL inhibitor navitoclax induces apoptosis of CLL cells, even in fludarabine-resistant, del(17p) and del(11q) adverse prognosis disease. The clinical response is long partial remissions with tolerable thrombocytopenia; no CRs were as yet observed [1578]. Mistaken nicks during somatic hypermutation by enzyme AID promote the fusion of genes *bcl*-2 and gene IgH at t (14:18). Resulting constitutive overproduction of anti-apoptotic BCL-2 protein follows. The pro-apoptotic proteins BAD, BAK and BAX contain three BH (BCL2 homology) domains. BAK and BAX release cytochrome C from the mitochondrium by boring pores in its membrane. Cytochrome C initiates the caspase cascade for DNA fragmentation. The reversal of these mechanisms reveals that BCL-2 originated in the ancient eukaryotes to protect the mitochondria from membrane damage and dissolution. The low branching Giardia and other "a-mitochondriates" (Trichomonas, Trimastix) might not have had this mechanism installed, and lost their mitochondria (leaving as its remnant a mitosome behind) [1579]. The navitoclax small molecules imitate the BH3 "death helix". Navitoclax molecules occupy unoccupied sites reserved for BCL-2/BCL-XL; and liberate the sequestered and trapped natural BH3 molecules. Since another role of BCL-2 is the maintenance of proper thrombocyte levels, its inhibition results in thrombocytopenia. Elevated levels of MCL-1 (myeloid cell factor) in leukemic lymphocytes mediate some resistance to navitoclax, counteracted by elevated BIM (BCL-2-interacting mediator of cell death) expression [1580].

Autophagy. Serum-starved cells may be just quiescent. Physiologically, cells remove useless proteins and even organelles: mitochondria undergoing excessive reactive oxygen generation (ROS) can be eliminated by mitophagy in the autophagosomes (Figure 68). Both autophagic and apoptotic cells have to overcome Bcl-2. The yeast autophagy-related factor, Atg6, beclin-1, acts against Bcl-2 in human cells, and BEC-1 protein acts against CED-9 (cell death defective) in caenorhabditis cells. The two processes are closely related [1581]. The development of autophagy, senescence, or apoptosis depends on the extent of p53-mediated reactions [1582]. Deprived of nourishment, or oxygen supply, cells resort to a form of auto-digestion. Adenosine triphosphate depletion may drive a cell into the state of autophagy [1583]. Inflammasomes are generated by bacterial flagellins, cell wall proteins, and lipopolysaccharides (endotoxins). Pre-autophagic cells refrain from the generation of inflammasomes. In septic shock, cardiac myocytes undergo autophagy [1584]. Just the activation of an oncogene may induce cellular stress reaction resulting in the conversion of an autophagic state [1585]. In autophagy, proteins without vital importance are consumed. Hypoxia inhibits mTOR and induces autophagy. Intact

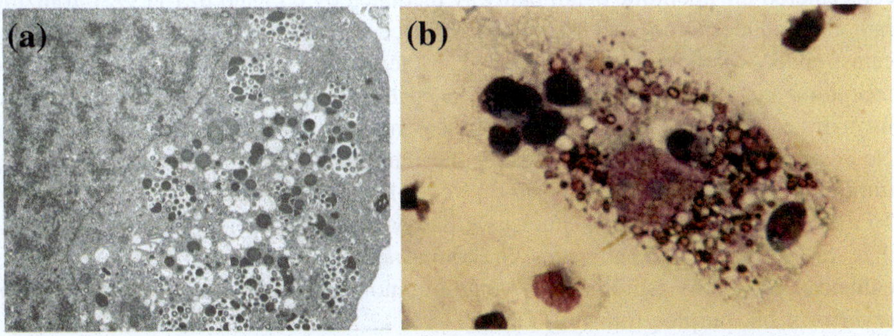

Figure 68 The Autophagic Tumor Cell (photographed as a 'subject of interest' before the discovery of the phenomenon 'autophagy') lies dormant, Dies, or Resuscitates Itself for Immortality: Another Chemo-immunotherapy-resistant Tumor Cell. Reference Appendix 2, Explanations to the Figures

mTOR inhibits the initiation of autophagy, but promotes geroconversion. Cells with shortened telomers undergo geroconversion. Senescent cells retain an intact mTOR pathway. Hypoxia suppresses gerogenic conversion (senescence) [1586]. The pro-apoptotic p53 or Rb stop short of apoptosis induction, and with p53/p21^{CIPWAF} (cyclin-dependent kinase interacting protein; wild type p53-activated fragment), or Rb/p16^{INK4a} (cyclin-dependent kinase inhibitor) induce senescence.

Androgen depletion induces senescence in prostate cancer cells. The senescent phenotype is characterized by the absence of telomerase activity, and the expression of β-galactosidase, the presence of phosphorylated H2AX, and heterochromatin protein 1β. The senescent prostate cancer cells did not divide, but continued to release IGF-binding protein 3 and cathepsin. The S-phase-associated protein 2 (Skp2) protected the cancer cells against senescene. Depletion of Skp2 with siRNA allowed the initiation of senescence, even in PTEN-active prostate cancer cells. Androgen-depleted prostate cancer cells expressed vimentin (in the process of EMT, or underwent neuroendocrine transdifferentiation [706, 1529]. Genetically engineered senescent, autophagic, CDK-inhibitor-overexpressing fibroblasts exerted increased tumor cell (MDA-MB-231 cell line, *vide supra*) growth-promoting properties. Such fibroblasts naturally arise in the human primary breast cancer bed [1587, 1588].

Prominent autophagy markers are acridin orange-stained vacuoles, autophagy proteins (Atg) and Unc-like kinases (uncoordinated, Caenorhabditis), the lipidated light chain3B, and beclin1 (Atg6) expressions. Beclin is a suppressor of "spontaneous" tumors in mice. Beclin proteins are downregulated in several human cancers (adenocarcinoma of breast, ovary, prostate). Autophagy-deficient cells accumulate damaged mitochondria (practice no mitophagy), and suffer from overproduced ROS. However, point-mutated *ras*-gene transforms autophagy competent cells, while autophagy-incompetent cells are able to resist [1589–1591]. Anti-apoptotic Bcl-XL is an antagonist of beclin1. Deviating the pathway from cell death and directing it toward autophagy is PI3K within the Atg14 complex [1582]. Quinacrine induces these markers in autophagic breast and colon cancer cells. These autophagic cancer cells entered the apoptotic death pathway mediated by p53 and p21 [1592]. Ionizing radiation induces tumor cell autophagy, senescence, apoptosis, necrosis, and mitotic catastrophe. Apoptosis may be induced in the exogenous (through cell membrane receptors) and endogenous (mitochondrial) pathways. Necroptosis is a programmed form of necrosis mediated by caspase 8 cleaving receptor interacting protein1 (RIP). Necrostatin-1 (Nec-1), the small molecular inhibitor of RIP1, can protect irradiated tumor cells from death [1593, 1594]. Nec-1 by inhibiting RIP1, antagonizes also NFκB (its translocation to the nucleus), reactive oxygen species (ROS) and MAPK, that are RIP1-activated substrates. Shikonin-induced apoptosis of leukemia cells is enhanced by Nec-1 (shikonin, a naphthoquinone derivative from *Lithospermum erythrorhizon*, purple gromwell) [1595].

The ataxia telangiectasia mutated gene is a promoter of the autophagic state. Irradiation induces miR-18a expression in colon carcinoma cells. These cells were increasingly prone to enter the state of autophagy. These autophagic colon cancer cells showed upregulated ETM, and downregulated mTORC genes/gene product proteins [1596]. Irradiation-induced senescence is an irreversible form of cell growth arrest, while the resting cells remain metabolically active. The senescent cell is (and remains) in G1 cell cycle arrested state. Instead of telomere restoration, the chromosomal ends undergo "end-to-end" fusions. Shuffling cell surface antigens in dying tumor cells may induce host immune reactions. Radiotherapy of cancer may induce immunogenic cancer cell death [1597]. There is potential recovery (with error-prone DNA repair) from the states of autophagy, senescence and even the mitotic catastrophe [796]. This author is interested in the evolution of necrostatins: if these proteins existed in, and protected the primordial eukaryotes from blasts of radiation bombarding the juvenile Earth.

Cellular viability gained by oncogenic fusion oncoproteins. These phenomena (already elaborated on in this treatise), lead back to the term "intrinsic structural disorders" of Hegyi *et al* [1598].

The fusion proteins gain DNA-binding and transcription regulatory functions, while lose Pfam domains (Pfms contain multiple alignments and hidden Markov models) [1598]. The Andrei Markov's model is a stochastic process skillfully aimed, but allowing an element of probability (based on random probabilities, random variables, X1, X2, X3 etc). The Pfam Genomic DNA can be directly searched against the Pfam library [1599, 1600]. In most fusion proteins with truncated domains (CDK6/MLL), the domain retains at least 40 % of its original structure ("a clever strategy for cellular survival"). Whereas, the IL2/TNF-R-SF17 (selle francais) fusion transcript lost too many aa from its IL-2 component, leaving a non-functional hydrophobic surface behind in the truncated domain. Fusion proteins with novel functions should have a close-to-normal ratio of hydrophobic residues. In the "disordered" regions, remote regulatory elements are able to perform autophosphorylations, Src signaling tyrosine kinases being the frequent targets.

On other occasions, antibodies directed at fused receptors could deliver caspases, which upon dimerization could induce apoptotic death in the targeted cell [1598]. The newly fused proteins did not exist in their fused form in fetal life, at the time when tolerance to self is established [1601], However, as single (pre-fused) proteins, they might have induced some degree of tolerance. In their fused form, their neoantigenic impact is quite moderate, but not entirely negative. Good evidence exists that the BCR/Abl oncoprotein is recognized by the patients' immune system, inasmuch, as it induces some cell- and antibody-mediated immune reactions; reviewed in [651]. In the BCR/ABL oncogene, nucleotides of the BCR tyrosine kinase phosphorylation motif fuse with ABL tyrosine kinase domain. In the fusion oncoprotein, Tyr177 becomes constitutively phophorylated: "intramolecular phosphorylation" [1598]. The adapter protein complex GRb2 and Shc (growth factor receptor-bound protein; Rous sarcoma src homology2 domain containing) attaches to the ABL Tyr kinase domain and connects the BCR/ABL fusion oncoprotein with p21Ras, PI3K, and proto-oncogene c-cbl (cysteine-S-conjugate β-lyase, ring finger ubiquitin ligase, active in *Dictyostelium discoideum*, Kaposi sarcoma herpesvirus receptor translocator in lipid rafts, AKT/ERK signal inducer, gain-of-function mutated in myelodysplasia/acute myelogenous leukemia) [1602–1605]. Thus, the

BCR/ABL oncoprotein induces an oncogenic cascade at multiple levels, some quite ancient.

In the TFG/ALK fused oncoprotein (tyrosine kinase fused gene, neurotrophin receptor TRK, *Trypanosoma cruzi* tyrosine kinase, Trypanosoma entry site into neural cells, anaplastic lymphoma kinase) intermolecular phosphorylation takes place [1606a, b]. The activated ALK expresses binding site for SH2 (Rous sarcoma homologue). The SH2 proteins are phosphotyrosine binding. The Src tyrosine kinases were active from the first unicellular eukaryotes upward on the evolutionary scale to vertebrates, aves and mammals. The TRK-A/B family proteins induce neovascularization, increasing both vascular and lymph vessel microdensity within tumors (oral squamous cell carcinoma). The NPM/Alk (nucleophosmin) fusion oncoproteins dimerize by their N-terminal, and their C terminal part. It is the Tyr tyrosine kinase domain, where two Tyr tyrosine kinases phosphorylate each other. The intermolecular phosphorylation remains permanent [1598, 1607].

In metastatic melanoma cells, the TFG gene (TFG-fused gene) is considered to be a putative tumor suppressor. In metastatic melanoma cells regularly deleted are the TP53, PTEN, PTPRD (protein tyrosine phosphatase receptor type D) and CDKN2A (cyclin-dependent kinase inhibitor) genes; while genes CCND1 (cyclinD), MITF (microphthalmia-associated transcription factor), MDM2 (murine double minute, human homologue) are amplified. Non-synonymous mutations are present in ACTA2 (actin assembly-inducing protein, *Listeria monocytogenes*), BMPR1A (bone morphogenetic protein receptor type 1A), and TFG genes. The TFG mutations are point-mutations: P380S and P380L (proline, serine, leucine). It is to be seen how a proto-oncogene with loss-of-function mutation becomes a tumor suppressor gene [1608]. Still, it is the constitutively activated BRAF gene product proteins that drive the MAPK pathway. The mRNAs of the oncoproteins exposed to miRs may undergo short term (by miR-200a), or long term (miR-141) inhibition. The miR-182 acts by suppressing MITF protein synthesis. The skin differentiation-inducer miR-203 suspends melanoma cell growth by inducing senescence. The locus of tumor-suppressor mIR-15/16 and miR-497 on chromosome 13 is frequently deleted in transforming cells. Proto-oncogene mRNA targets (cyclinD, AKT3, BCL2) of these miRs are liberated and act unopposed (BCL2 and MITF in synergism). Synthetic miRs mimics emerge as potential therapeutic agents [1609].

When the oncogene EWS (Ewing sarcoma) (Figure 57) fuses with activated transcription factor 1 (ATF1), and Friend leukemia virus insertion site gene (Fli1) in the protein product, there will be highly repetitive EWS activation domains (EAD). The transactivation-like function will be exerted by the N-terminal of EWS. This oncoprotein will deregulate the transcription of its target genes. The mixed-lineage leukemia (MLL) oncogene promiscuously fuses with several different gene partners. The CREB-binding protein (CBP) gene is one of the fusion partners (cyclic adenosine monophosphate-response element binding protein). The CBP is a histone acetyltransferase (HAT) with additional bromodomain. If an oncogene promoter is activated by de-acetylation, a HAT could silence it by re-acetylation. However, if aberrant DNA-binding (the EWS oncogene) instructs otherwise, the acetylation site will be mis-targeted and the oncogene remains active. Mis-targeting instructions derive from the fused DNA-binding domains of MLL, the three AT hook motifs and a DNA-methyltransferase (DNMT) homology region. In the MLL/ENL (transcription factor enolase) fusion oncoprotein in myelogenous leukemia cells, the AT hooks, the DNMT homology domain of MLL, and the C-terminal domain of ENL cooperate in emitting false transactivation signals [1598].

AT-rich chromosomal DNA attracts protein binding (AT-hooked proteins; proteins containing AT hook motifs). The AT hooks consist of short repeats of glycine, arginine, proline (GRGRP) and the central RGR inserts into the minor groove of the AT-rich DNA molecule. The strand-switch regions (SSRs) of the genome of *Leishmania amazonensis*, from where transcription initiates, are >50 % AT-rich (other parts of the genome are GC-rich). The EBV EBNA1 protein contains AT-binding sequence GRGRGRGRGRGGGRP. The yeast's replication origin recognition complex (ORC) attaches to AT-rich DNA. Mammalian cells contain the high mobility protein HMGA1a armed with three AT hooks. The AT hooks also bind G-quadruplex RNA. AT hook proteins also can displace histones from repressed chromatins (repressed chromatins are those with acetylated histones H3/4). AT hooks participate in gene expression, DNA duplication and repair [1610]. In the Arabidopsis, AT-hook protein AHL22 (AT-hook motif nuclear localized22) initiates flowering [1611].

Comment. The readiness of the genome to activate such mechanisms, but not necessarily as "disorders," suggests possibly otherwise. The genome may possess an inherent faculty to respond to stress and potential harm with a fundamental reorganization of its structure. The results prove that eukaryotic cells driven by fusion oncoproteins gain survival advantage bordering immortality, and remain practically indestructible. These cells resist physico-chemical damage. In their existence within a multicellular host, they evade and counteract even the most sophisticated immunological attacks. Peculiarly, in many instances, the host appears to welcome and supports these cells with growth factors. Indeed, these cells appear not to be the vulnerable products of some accidental internal turmoil. First, in the fusion process, cutting into expression domains of the genes is meticulously avoided. Some disorganized-appearing sequences may occur up- and downstream of the fused nucleotides (referred to when translated into proteins as intrinsically disordered or unstructured proteins, IDP/IUP), but the protein-coding domains are preserved. Second, the newly created and encoded proteins are structured so, that the host eventually accepts them as self, and thus exempts them of destruction by ubiquitination.

The Maya goddess AKNA. The Maya honored AKNA (moon goddess Awilix, Xbaquiyalo, in aztec), the lunar goddess of childbirth and fertility. The gene abbreviation acronym *akna* stands for AT-hook transcription factor. A research group receives credit for deleting the *akna* gene from newborn mice. The small frail mice died with severe inflammatory reactions consisting mainly of granulocytic infiltrates. These AKNA-less mice were like the newborn mice abandoned by their mother: motherless. AKNA genes are active in healthy lymphocytes, NK cells and DCs. AKNA proteins bind AT-rich regions of CD40 ligand promoters. The human chromosomal locus of AKNA at 9q32 is frequently submitted to loss-of-function mutation or is deleted in malignantly transformed cells. Genes repressed by AKNA are liberated in AKNA-knockout hosts. MMP-9 reached high level activity. Granulocyte count doubled [1612]. A single-nucleotide polymorphism at AKNA results in an a change in the encoded protein at the DNA-binding AT hook motif: R1119Q (arginine, glutamine). The consequence may be accelerated HPV activity and uterine cervical intraepithelial neoplasia [1613].

The high mobility group protein AT-hook 1 (HMGA1) regulates androgen dependence in prostate cancer cells. The protein rises to high level in androgen-independent prostate cancer cells. Androgen deprivation induced the rise of the HMGA1 protein levels. Overexpressed HMGA1 level promoted cancer cell growth under androgen-deprived condition, Knockout of HMGA1 protein production resulted in the cessation of cancer cell growth [1614].

Selected abbreviations

iASPP = PPP1R13l apoptosis-stimulating proteins of p53; phosphatase 1 regulatory subunit 13-like. Negatively regulates p53 (anti-apoptotic oncogene). Click for iASPP stimulating self-renewal of hematopoietic stem cells by Jia Y Peng L Rao Q et al. Click for miR-124a iASPP inhibitor by Lin L Gun X Liu S Wang X (in Chinese PMID 24463112). Click for downregulating miR-124 promotes tumor (glioblastoma) growth by Zhao WH Wu SQ Zhang YD. Click for miR-124 overexpression downregulating iASPP and inhibiting cancer cell (colorectal) growth by Liu L Zhao H Yao H et al.

MDM, murine double minute protein with human homolog, neutralizes p53 protein marking it for ubiquitylation (Figure 64). Click for MDM/p53 protein complex rheostat by Eischen CM & Lozano G. In FAK (focal adhesion kinase) protein complex, p53 protein is inactive. Click for R2 (Roslin) disrupting FAK/p53 protein complex and liberating p53 protein for action (to block tumor growth) by Golubovskaya VM Ho B Conroy J et al.

MDM = SLURP Mal de Meleda palmoplantar keratosis due to mutation in chromosome 8q24-qter in gene of secreted lymphocyte antigen-6 urokinase-type plasminogen-activator receptor-related protein, that is not to be mistaken for MDM as above. Click for SLURP by Eckl KM et al and/or Wajid M et al.

Endogenously Initiated Immortalization Is Welcome. Enforced Carcinogenesis by Alien Initiator Is Resisted

False Alarms May Sound

Cancers arising in healthy young hosts. If there exists an inherent aptitude in the RNA/DNA complex for the rescue of some individual cells from a host in demise, how could events of this sort be reasonably applied as an escape mechanism, or rescue mission, for the individual cells taken from a healthy juvenile host, who will die in the process? The existence of such signals may be presumed to exist in a doomed multicellular host, or in an infected individual, who will eventually succumb to "inflammatory carcinogenesis"; reviewed in [73]. There, the *wee* gene-product protein is knocked out by miR-155, and the cell cycle is rolling forward non-stop. Mycoplasma-infected cell cultures activate the *src*-related proto-oncogene *vav*. In Chlamydia-infected periorbital lymph nodes, or in Helicobacter-infected gastric mucosa, MALT lymphomas arise. MALT lymphomas arise due to the fusion of the IgG heavy chain locus (IgH) with the MALT gene: t (14;18)(q32;q21); references cited in [73]. These are mistaken V(D)J-mediated translocations of the CpG-type [1615]. Some of these entities may not culminate in an irreversibly advancing malignancy. Early processes of transformation are not due to irreversible gene mutations or translocations. Epigenetic silencing by DNA methylations of the promoters for the tumor suppressor genes, or activation by histone deacetylations of the promoters for proto-oncogenes are the introductory, but reversible, changes of cell transformations. These early reactions are reversible by epigenetic interventions (demethylations, re-acetylations). Some of these changes may regress upon the removal of the infectious pathogen; references cited in [73]. However, in a healthy newborn or juvenile host none of these recognizable events exist for the theoretical explanation, as to how malignant transformation of a select cell colony was initiated. Unrecognized chemo-radiation exposure, or horizontally or vertically, but silently, transferred oncogenic factors are possible. However, the alarming call for endogenous carcinogenesis may be erroneously activated.

J.G. Sinkovics, *RNA/DNA and Cancer*, DOI 10.1007/978-3-319-22279-0_22

In certain pediatric B-, T-, or NK-cell lymphomas (all subtypes of lymphoid cells being susceptible to EBV infection), neuroblastoma, anaplastic ependymoma, embryonal rhabdomyosarcoma, the healthy juvenile host is not likely to have generated danger signals forecasting its immediate demise to its RNA/DNA complecti. In an infectious process, during the frantic events of somatic hypermutation and IgG switch, the enzymes AID or RAG might perform mis-targeted nicks bringing the genes c-*myc* and/or *bcl*-2 together with the gene of IgH (on chromosome 14), thus creating fusion oncoproteins (African Burkitt's lymphoma; follicular lymphoma) (*vide supra*) [1616, 1617]. Notoriously, the widely spread EBV carries a *bcl*-2 gene homologue. The viral anti-apoptotic BHRF1 gene (restriction enzyme BamHI rightward fragment) immortalizes EBV-associated tumor cells [1618, 1619]. One must presume that the RNA/DNA complex on occasions is recruited into action to transform cells by bioengineering mistakes. In addition to the physiological somatic hypermutation and immunoglobulin switch, these bioengineering mistakes may occur in the processes of differentiation, and germ line lay-out and separation, during ontogenesis. Gene amplifications and reversible epigenetic changes occur, rather than gene point mutations, dsDNA breaks and gene fusions. Tissues transforming in these settings characteristically retain their susceptibility to physiological chemo-, lympho- and cytokines, microRNAs, and host immune reactions. The "cell survival pathways" (PI3K, MEK, STAT) even if they appear to be constitutively installed, prove to be reversible upon biological interventions.

Neuroblastomas of childhood (Figure 69) arise from sympathetic ganglia commonly in the adrenal medulla. On occasions, this tumor may differentiate into a benign ganglioneuroma. In contrast, it may de-differentiate into a widely metastatic and fatal tumor. Olfactory or esthesioneuroblastoma originating in the sphenoid sinus is another malignant variant, due to its location and invasive character. This tumor is highly malignant, if induced by the achaete scute homologue *ash* gene. Treated with surgery, and radiochemotherapy at the Sarcoma Service of M.D. Anderson Hospital in the early 1970s, this tumor now responds to the anti-neovasculogenic therapy (bevacizumab) and "gamma knife radiosurgery"

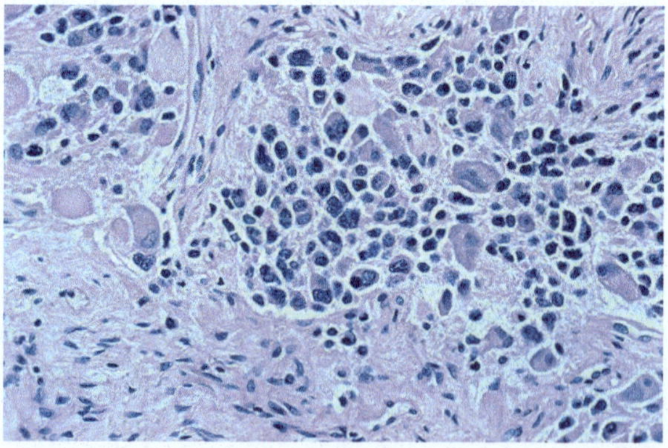

Figure 69 The Biology of the Neuroblastoma Cell Accommodates Malignant Transformation and Reversal to Full Differentiation. Reference Appendix 2, Explanations to the Figures

[1620]. N-Myc (on chromosome 2p23-24) [1621, 1622], and Bcl-2 (on chromosome 18q32) are amplified; so is the MDM2 gene (on chromosome 12q13). These latter genes are not mutated, not translocated or fused, but are overexpressed (amplified) in response to erroneously and temporarily installed epigenetic changes. However, neuroblastoma cells do not necessarily express receptors for TNFα, and thus escape another route to apoptotic cell death. The phosphorylated stathmin protein is upregulated to re-arrange microtubules for impending mitoses. However, after the mother- and daughter-cells are separated, the neuroblastoma stathmin is not de-phosphorylated by protein phosphatases for resting. Instead, it remains active in preparing the microtubules for another rounds of cell divisions, non-stop [1623]. Further, these tumors overexpress neurotrophin receptors (NT-R). The family of neurotrophins include nerve growth factor (NGF), brain-derived neurotrophic factor (BDNF), and NT3/4. The neurotrophins are the ligases of the NT-Rs, also referred to as the tropomyosin-related kinase receptors (Trk-R). Moreover, the host supports these tumors with a network of newly generated blood vessels. Neoangiogenesis in general is initiated and sustained by overproduced VEGF, FGF, PDGF, EGF, TGFα, and angiopoietin1/2. Naturally produced angiogenesis inhibitors (angiostatin, endostatin, activin-A, MMP-inhibitors) are usually downregulated in the vicinity of growing tumors, whereas MMP2/9 are usually overproduced. The N-Myc protein is an inhibitor of activin-A, and a downregulator of hyaluronan receptor CD44. The direct and the indirect drivers of the cell cycle (MAPK, PI3K, HGF/c-Met, MEK, or Akt/PKC) and neoangiogenesis, respond to small molecular inhibitors synthesized in the laboratory (AZD2171, SU1498, SU11657, SU6668, SB202190, LY294002, SU11274, PF2341066, PD98059, bisindolylmaleimide XI). Physiologically, there are small inhibitory RNAs or microRNAs (mi-Rs) mobilized for the neutralization of the mRNAs of the tumor growth factors. These mi-Rs can also be synthesized in the laboratory. Tumor cells usually silence these microRNAs.

The anti-thrombin Argatroban (listed in Physicians' Desk Reference as L-arginine-derivative for the treatment of heparin-induced thrombocytopenia in medical practice) suppresses thrombin-induced VEGF production in patients with neuroblastoma. The histone deacetylase inhibitors (HDACI), trichostatin and SAHA (suberoylanilide hydroxamic acid) keep the histones of tumor suppressor gene promoters acetylated, thus silenced. Silence is favored for the oncogenes, but it is disadvantageous for the suppressors (reviewed in context of inflammatory carcinogenesis) [73]. Trichostatin and IFNα in combination suppress neoangiogenesis and induce tumor regressions. VEGF-induced neoangiogenesis is responsive to the mcab bevacizumab. Active cyclin kinase inhibitors (CKI) keep the cell cycle under control. The expression of N-Myc and its heterodimerization with Max could be inhibited with IFNγ and retinoic acid treatment (*vide infra*).

However, the heterodimer formed by N-Myc and Max and/or the homodimers TrkA/NGF drive the "cell survival" PI3K/Akt (PKC) pathway. Expression of the TrkA receptor (tropomyosin-receptor kinase) occurred in tumors stagnating without vascular supply; and prone to differentiate; these tumors could not be readily xenografted (due to lack of viability). In contrast, tumors expressing the TrkB receptor induced neoangiogenesis, exhibited more aggressive behavior. TrkB-expressing tumors produced also hepatocyte growth factor (HGF) and its receptor Met. After binding HGF and Met, the cytoplasmic tail of the oncoprotein phosphorylates several kinases resulting in the activation of Src, STAT3, PI3K/Akt and Ras/MEK (MAPK/ERK) pathways. *)

*) *Inadvertently omitted reference discovered at final proof-reading. For review of Met-induced oncogene activations read Gusenbauer S et al Mol Cancer 2015;14 (1):54. doi:10.1186/s12043-015-0319-z.*

Some neuroblastoma cells overexpress the c-Kit receptor for its ligand, stem cell factor (SCF), and release SCF in an autocrine loop for self-stimulation of mitoses. Imatinib mesylate suppresses PDGF and tyrosine kinase c-Kit (CD117) expression. Somatostatin inhibited PDGF-induced phosphorylation of PDGFR and inhibited *ras* gene amplification. For local invasion, these tumor cells release MMP2/9. The synthetic MMP inhibitor, prinomastat, suppresses MMP production and tumor cell locomotion. However, MMP expression is promotional for neuroblastoma cell differentiation. The presence of MMP is necessary for the neurite formation of retinoic acid-treated neuroblastoma cells; neurite formation is the first sign of differentiation induction (*vide infra*) [1624].

The vitamin A-derivative, 13-cis-retinoic acid (CRA), or all trans-retinoic acid (ATRA), captured by its receptor (RARα, on 17q21.2) induces the differentiation of neuroblastoma cells. The mRNA and the protein of manganese superoxide dismutase (MnSOD) appear first in ATRA-treated cells. Due to the suppression of the microtubule-associated proteins, the lissencephaly proteins (mutated Lis1 gene) stop migration of neuroblastoma cells. Due to upregulation of the neurofilament protein-68, neuroblastoma cells emit filaments (neurites). Mobilized are PI3K, PKC-dependent α-secretases, and disintegrin metalloproteinases. Further, IFNγ and CRA/ATRA-treatment suppresses the transcription of the N-Myc protein; the N-Myc target gene of telomerase (hTERT) will not be activated after cell mitoses. If ATRA activates (phosphorylates) CREB (cAMP response element-binding protein), which in turn induces the transcription of caspase 8, the neuroblastoma cell will not rest and differentiate, but it will undergo apoptosis and die. If ATRA activates the anaphase-promoting complex, which will degrade the F-box protein SKp2 (S-phase kinase associated protein), the neuroblastoma cell will undergo the differentiation pathway. The retinamide (fenretinide), 4-HPR, N-(4-hydroxyphenyl, induces the pro-apoptotic genes (BAK and Bax) of the *bcl2* family; NFκB, GADD153 transcription factor (growth arrest DNA damage), 12-lipooxygenase, and reactive oxygen species (ROS), cannot inhibit, but actually lead to apoptosis. The combination of 4-HPR, INFγ and the the demethylator 5-aza-cytidine activate the promoter of caspase 8, again killing the neuroblastoma cell. Fenretinide (N-4-hydroxyphenyl retinamide, 4-HPR) inhibits VEGF and FGF generation. Neuroblastoma cells retain reactivity to biologicals (see monoclonal antibodies, mcab, later) [1624]. Post-transcriptionally regulatory micro-RNAs, and genome-wide DNA methylation alterations are involved in the process of neuroblastoma cell differentiation. Embedded in the HOX (homeobox) gene cluster, miR-10a expresses a RA-responsive element. ATRA-treatment rises miR-10a/10b levels and dramatically reduces N-Myc mRNA levels. Neuroblastoma cells differentiating toward ganglioneuroma, and ATRA-treated neuroblastoma cells, undergo genome-wide demethylations. Among the demethylated genes, stands out NOS1 on chromosome12. This gene catalizes nitric oxide generation. Nitric oxide generation induces neurite extensions, inhibits cell divisions, and is a criterion of neuroblastoma cell differentiation. Methyltransferase to DNA (DNAMT) generation is reduced in ATRA-treated neuroblastoma cells. The DNAMT1 mRNA is targeted for elimination by miR-152. Active N-MYC suppressed miR-152 production. ATRA dowregulates N-Myc. In response, miR-152 levels rise and consequentially DNAMT1 levels drop. Similarly, ATRA-upregulated miR-26a/b, and miR-125a/b target the DNAMT3 mRNA for degradation [1625]. In the neuroblastoma cell methylome, tumor suppressor genes silenced by hypermethylation, regain function due to demethylation. Moreover, RARα at its promoter complexes react with histone deacetylase 2 (HDAC2) and Krüppel-like factor 5 (Klf5). The promoter of the receptor is P21, a cycline kinase inhibitor (CKI). RARα agonist cyclin kinases (CK2α; Am80) can phosphorylate HDAC2; the phosphorylated HDAC2 detaches from RARα. Thus RARα regains its capacity to bind its ligand RA. This union results in the de-acetylation of Klf5, and in its dissociation from promoter P21. Among targets other than P21, Klf5 interacts with p53, and PPAR-lambda (Table II). Among targets other than RARα and Klf5, P21 interacts with p53, STAT1 and CREB [1626].

Laminas (LMN-AC0) serve as scaffolds for the anchoring of nuclear chromatin. Neuroblastoma cells with lamina genes deleted and lacking nuclear lamina, fail to differentiate in response to retinoic acid treatment. Lamina-defective neuroblastoma xenografts released cells with accelerated locomotion; lamina-free tumor cells formed large tumors rapidly, and exhibited resistance to chemotherapy (doxorubicin, etoposide, paclitaxel). Laminary structures in place were necessary for organ differentiation [1627].

The fundamental immunological statement, that "Antibody-dependent cytotoxicity (ADC) by normal non-T lymphoid cells bearing receptors for complement and for the Fc fragment of immune globulins has been demonstrated……as a natural defense reaction against virally or chemically induced mouse tumors…" with the appropriate literature cited [1628, 1629] was reported in the mid-1970s and was so acknowledged [724]. Prior to that, in the early 1970s, it was observed in the author's (JGS) laboratory, but remained unexplained [1630–1635] until after the above cited literature reference, that tumor cell-attacker lymphocytes could be inhibited or enhanced by patients' sera. Primary, or established human tumor cells were attacked by autologous or allogeneic lymphocytes *in vitro*, and that the autologous attacker lymphocytes usually derived from populations of small compact round cells with variable T cell receptors; sera were inhibitory to these lymphocytes. The allogeneic attacker lymphocytes derived from a population of large granular lymphoid cells without B, or only occasionally with a non-variable T, cell markers. While pretreatment of target tumor cells with patients' sera could inhibit the attack by the small compact lymphocytes, pretreatment of target tumor cells with patients' sera (or with healthy volunteers' sera) often significantly enhanced the cytotoxicity of the large granular lymphocytes (Graphs 1 and 2 *vide supra*) [276, 1573a, b, c, 1630–1637]. Large granular lymphocytes taken from the blood of healthy donors (first this author, JGS) with or without the pretreatment of tumor cells with sera, often attacked allogeneic tumor cells *in vitro*; reviewed in [30a, b, 147, 279]. It was seen immediately in the chamber slide preparation, that the small compact lymphocyte represented a tumor-specific clone of immune T cells mobilized only in patients with a particular tumor, but devoid of Fc immunoglobulin receptors, whereas the large granular lymphocytes were "natural killer cells" not requiring specific pre-immunization for attacking tumor cells. On account of their Fc receptors, sera (unidentified immunoglobulins, Ig), enhanced their cytotoxicity [147] (Figures 35 and 36 historically re-assessed in the Appendix 1).

Neuroblastoma (Nbl) cells are highly susceptible to the ADCC reaction. Di-sialo-ganglioside (GD2) cell surface antigens bind monoclonal antibody 3F8. Host NK cells attack these antibody-coated NBl in an antibody-dependent cell-mediated cytotoxic reaction (ADCC). If Nbl cells express their HLA class I antigens, the NK cell killer immunoglobulin-like receptors (KIRs) will refrain from the killer attack. The killer attack consists of the discharge of perforins/granzymes for cytoplasmic lysis, or the release of FasL to activate the FasR (CD95) for the induction of target cell's pro-apoptotic caspases; and/or the TNF-related apoptosis-inducing ligand (TRAIL). However, toward the mcab-coated Nbl cells, the host-nonreactive NK cells lose their inherent restrain: they will attack and destroy the targeted self Nbl cells. Moreover, the antiGD2 mcab induces also complement-mediated cytolysis. Reactive NK cells release IFNγ, which induces upregulation of HLA antigens. Also, RA-treated differentiating Nbl cells re-express their HLA molecules. At this point, NK cell disengage (refraining from attack on "self"). Without the mcab coat, these tumor cells will escape attacks from self-nonreactive ("unlicensed") NK cells. Allogeneic NK cells increase the efficacy of the 3F8 mcab [1638]. The anti-GD2 mcab also exist in an IL-2 fused preparation (Hu14.18-IL2). IL-2, GM-CSF and lenalidomide increase NK cell activity [1639, 1640]. Nbl cells resist TRAL (TNFα-related apoptosis-inducing ligand) due to silencing caspase-8 and TRAIL-R2. IFNγ-treated Nbl cells IL-2- and IL15-treated NK cells could attack and kill Nbl cells by releasing perforin. Lethally irradiated IL-21-producer K562 cells (blastic crisis-derived human myelogenous/ erythroleukemia cells, which differentiate into megakaryocytes upon phorbol-myristate acetate treatment) in the presence of IL-2 activated healthy donor-derived NK cells exert cytolytic attack on Nbl cells [1541]. The Penn State Children's Hospital in Hershey developed a neuroblastoma DC vaccine, that will be referred to later.

Anaplastic lymphoma of childhood with Notch is driven by amplified, not mutated, not translocated EPHB2-protein (ephrin receptor). In these cells, the INK4A/Arf (Cdkn2a) protein is deleted (signaling ephrin receptor; cyclin-dependent kinase inhibitor/alternative reading frame). The INK4A (Cdkn2a) genes are claimed to be tumor suppressors. Ependymoma cells maintain active telomerases (to cap telomeres) and amplify the EGFR (encoded from 7p11.2). Moreover, the PTEN gene-regulator miR-21 is overexpressed in these tumor cells.

The PTEN gene-product protein being the suppressor of the PI3K proto-oncogene, is deleted. Further, ependymoma tumors contain glia-like stem cell populations. These cells were capable of self-renewal in multipotent neurospheres. Overexpressed miRs in ependymoma cells are very likely epigenetic promoters of the tumor, whereas the deleted miRs would have been tumor suppressors. Actually, active and prominent miRs are miR-135a and miR-17-5p, whereas the downregulated miRs are miR-383 and miR-485-5p. The cluster of miRs emanating from chromosome 14q32.31 regulates positively the onset and the time of relapse after the removal (or radiotherapy) of the primary tumor (predicting adverse outcome). Especially, low level miR-203 (or is absence) is a predictor of relapse. This very same miR-203, with its gene, or promoter, methylated or demethylated, is involved in the regulation of the BCR/ABL oncoprotein expression in Ph^+ CML. It is very likely a stem cell differentiation-inducer miR, frequently suppressed in tumors. When miRs let 7d, miR-596 and miR-367 remained active, or recurred after treatment, they improved the patient's relapse-free prognosis; let7d could act as a negative regulator of proto-oncogenes *ras*, c- (or N-) *myc*, and the gene for HMGA2 (a high mobility protein). Gene amplifications of the TGF-β family are involved in the growth rate regulation of ependymoma cells. The mRNAs of these genes are under the control of miR-485-5p [1642, 1643].

This childhood tumor appears to carry no irreversibly mutated oncogenes, and no fusion oncoproteins; it is epigenetically regulated, and as such, it does not appear to be a refugee from a host otherwise doomed by unrelated life-threatening health conditions. The epigenetic errors could possibly be reversible and correctible (demethylation of silenced tumor suppressor gene promoters, re-acetylation to silence the de-acetylated and thus activated proto-oncogenes/promoters). Further, EPHA2, IL-13Rα2 and anti-apoptotic Survivin proteins are overexpressed, thus offering ependymoma chances to respond to for peptide vaccine therapy in order to maintain remissions induced by surgery (and radiotherapy) [1644].

The EPHB2 family genes (ephrins) encode proteins that promote tumor cell (pancreatic adenocarcinoma; glioma/glioblastoma) locomotion. Downstream of ephrin-B2 signaling lie STAT3 and MMP genes, and both are upregulated [1645, 1646]. In contrast, EPHB4 acted as tumor suppressor in colon carcinoma. Just one allelic loss of EphB4, significantly increased tumor growth in subjects with loss-of-function mutations of the Apc tumor suppressor gene. The Apc-mutated tumor cells activate Wnt signaling, β-catenin translocation into the nucleus with activation of genes for TCF/LEF, c-Myc, cyclin D and EPHB receptor expressions. The expression of EPHB4 either inhibited tumor growth, or if it was deleted, allowed tumor growth acceleration [1647].

Anaplastic childhood brain tumors (astrocytoma, ependymoma, medulloblastoma) harbor Notch proto-oncogene-encoded pathways. Physiologically, switched on and off Notch pathways drive the cytotrophoblasts of the fetal placenta by activating NFκB and MMP. The Notch proto-oncogenes act as receptors of their ligands Delta-like 4, Jagged (JAG) and Serrated. Delta-like 4-induced Notch activation is vasculogenic. The ligands could be mobilized by the yes/YES proto-oncogene/oncoprotein (Yamaguchi's c-onc → v-onc). The lunatic fringe sugar transferase (Lfng/LFNG) prevents the activation of the Notch proto-oncogene by its ligand Jagged. Loss-of-function point mutated Lfng liberates Jagged, thus contributing to the activation of the Notch receptors. The Notch receptor pathways activate TGFβ to encode the ETM transition of a pre-existing tumor (adenocarcinoma); another Notch protein target gene is the proto-oncogene c-myc/MYC. The JAG/NOTCH complex activates hepatocyte growth factor receptor MET/Caveolin in basal type triple-negative breast cancers (and very likely in all other tumors in which Notch receptors are activated). Further Notch target genes/proteins activated are Hes and Hey (Hairy enhancer of split; Hes-related YRPW motif, tyrosine, arginine, proline, tryptophan). The retroviral MMTV genome inserts itself into host cell genomic notch genes. The miR-124 is a Notch promoter. In medulloblastoma, Notch, sonic Hh and WNT are co-activated. In ependymoma, ERBB2/EGFRs are upregulated. Thus, ependymoma is a tumor treatable and potentially responsive to EGF-ligand and receptor targeting mcab and lapatinib. Notch activation requires γ-secretase; thus, γ-secretase inhibitors (such as DAPT) act as Notch inhibitors. In septic shock (in rats), Notch and PARP (Table II) are activated, and apoptotic death of neuronal cells ensues. DAPT averts neuronal cell death in septic shock. This extremely complex and newly explored issue is further explained in the Selected Abbreviations (*vide infra*).

Pediatric rhabdomyosarcoma (Figure 70). The two most common forms in which rhabdomyosarcoma (RMS) appears are the embryonal/botryoid (e), and the alveolar (a) subtypes. The malignant cells differentiate incompletely toward the skeletal muscle lineage and express actin, myosin, desmin, myoglobin, myogenin and the Z-band protein MyoD (myogenic differentiation antigen). The MyoD protein enters the nucleus to bind and so activate genes engaged in myoblast differentiation (RUNX, runt-related transcription factor; MEF, myocyte-specific enhancer factor; NFIC, nuclear factor I, C-type; JDP2, jun-dimerization protein; ju-nana Japanese for seventeen, after avian sarcomavirus ASV-17). However, due to defective contact between protein and its binding site on the gene MEF2C, or hypermethylated promoter of JDP2 (J domain containing protein; jun-dimerization protein), the genes are not necessarily activated [1648]. Virologists know the JDP protein as an inhibitor of EBV replication (it maintains viral latency) [1649].

The appearance of the embryonal eRMS is that of the embryonic skeletal muscle. This tumor afflicts children less than 10 years old (with a few exceptions in adolescents). It grows into tumor masses in the periorbital and genitourinary/retroperitoneal anatomical regions. It responds to treatment (if treated early); the 5 years tumor-free survival exceeds 80 %. The deficiency of the WTp53 pathway (as in Li-Fraumeni syndrome) is the first most common etiologic factor of eRMS [1650]. Other genetic deficiency syndromes with increased incidence of RMS are the Gardner syndrome with autosomal dominant APC mutation at 5121; Werner syndrome with recessive mutation at 8p12 (*vide supra*); nevoid basal cell carcinoma syndrome with autosomal dominant mutation at 9q22.3-q31 (the PTCH gene, patched); and the FAMM (familial atypical multiple mole melanoma) syndrome with autosomal dominant mutation of CDKN2A/p16^{INK4a} (cyclin-dependent kinase) at 9p21. Loss-of-function mutation, or neutralization of the p53 protein by

Figure 70 The Sarcoma (rhabdomyosarcoma) Cells Use the Sirtuin Longevity Proteins of the Yeast for the Prolongation of Their Life Span. Reference Appendix 2, Explanations to the Figures

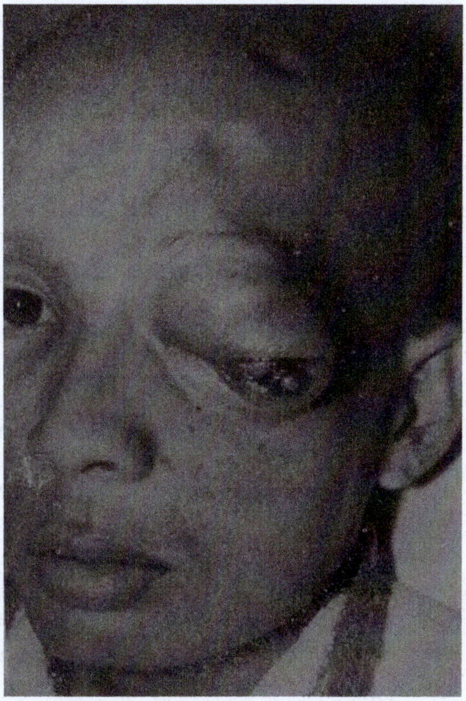

binding with MDM2 protein is one of the main mechanisms of p53 loss [1651, 1652] (in the Appendix 1). The p53 and MDM2 proteins are physically associated and co-localized in the nucleus. In this early observation, it was not recognized that the p53/MDM2 protein complex may undergo ubiquitination already in the cytoplasm. Allelic loss at 11p15.5 suggests the deletion of another tumor suppressor gene, the Beckwith-Wiedeman syndrome gene (in sarcoma cell line SCL22A18/BWR1A) [1653]. In the hierarchy of the cell subpopulations in eRMS tumors, one entity stands out by expressing CD133 (prominin), usually co-expressed with progenitor stem cell marker CD34. The $CD133^{+}$ cells grew in rhabdospheres. Rhabdosphere cells co-expressed other stem cell antigens (Oct4, Nanog, c-Myc, Pax3, Sox2). Some rhabdosphere cells differentiated into adipocytes, myocytes and neuronal cells; retinoic acid and/or DMSO (dimethylsulfoxide) induced the differentiation. Neuroblastoma cells (*vide supra*) were known to express CD133 and to downregulate it upon RA-induced differentiation. Ewing's- and osteosarcomas also harbor $CD133^{+}$ subpopulation of cells. Enriched $CD133^{+}$ cell population were highly tumorigenic and resisted chemotherapy (in xenografts); whereas, $CD133^{-}$ cells failed to induce tumors. In the human eRMS tumors, CD133 overexpression indicated worsening prognosis. The same "stem cell medium" in which eRMS cells were grown did not support the rise of tumorigenic subpopulation of aRMS cells [1654]. In the genome of eRMS cells, gene gains (in chromosomes 2, 8, 12, 13) appear suggesting gene amplifications. When Ras activation occurs, it is unlikely to be due to gain-of-function mutation; gene amplification could well be responsible for the *ras* gene signature in eRMS cells [1655]. In both embryonal and alveolar RMS tumor tissue, the levels of miR-1 (miR-1-1/1-2) and miR-133a (miR-133a-1/a-2) are practically depleted. (CD133 and miR-133 appear to be coincidental unrelated entities). Experimentally re-introduced into eRMS cell lines, these two microRNAs proved to be cytostatic to the tumor cells. The exposed cells stagnated in the state of a $G_{1\text{-}S}$ arrest. Tumor cells exposed to miR-1 expressed the muscle cell differentiation antigens myogenin and MyL1. The thymidine incorporation of tumor cells ceased upon exposure of these two microRNAs. The mechanism of this effect is the neutralization of those mRNAs which travel through the ribosome to be translated into oncoproteins. As studied in the RD (eRMS) cell line, some 545 gene-derived mRNAs presented binding sites for miR-1, and 404 such genes underwent reduced expression. Some 457 gene-derived mRNAs reacted with MiR-133, and 308 of these genes were downregulated [1656]. Of many miR-133a-repressed targets is the SMARCD1 (SWI/SNF-related matrix-associated actin-dependent regulator of chromatin subfamily D member 1; switch/sucrose non-fermented). Another miR-133a-repressed gene is the LASP1 (LIM and SH3 proteins; LIM: Lin11, Islet1, MEC-3; Lin: cell lineage, caenorhabditis; MEC: mechanosensation, mechanotransduction microtubules, caenorhabditis; SH, SRC Rous sarcoma homologue). For LIM kinase (Table XVI). A miR-repressed gene is TWF2 (twinfilin, actin-dependent, drosophila) [1657–1665]. The SMARCD1 is an ATPase regulator protein acting HDAC-like in epigenetic chromatin and HLH (helix-loop-helix) modifications, and in muscle cell differentiation. The HEY genes (hairy and enhancer, drosophila) encode the basic HLH transcriptional repressors [1666, 1667]. Of the large numbers of genes first upregulated in eRMS tumor cells, and then downregulated by miR-133 (Table 3 in the article) are ANXA2 (annexin), HNRNPU 9 (heterogenous nuclear riboprotein U), MYLK (myosin light chain kinase), PDGFA (platelet-derived growth factor A), TWF (twinfilin), SMARCA4 (*vide supra*), SOX9, STX6 (syntaxin), PCDC4 (programmed cell death neoplastic transformation inhibitor), TNFRSF10B (tumor necrosis factor receptor superfamily member 10B), YES1 (c-*yes* → v-*yes*, Yamaguchi sarcoma retroviral oncogene), CDCA8 (cell division cycle associated 8), PITPB (phosphatidylinositol transfer protein, β) LASP1 (LIM&SH3, *vide supra*) [1656].

This author (JGS) points out that the classical stem cell genes are not listed (Klf, Oct4, Sox2, Nanog, c-Myc); the Pax gene (*vide infra*) involved in myocyte differentiation, is also missing from the list. Therefore, the mRNAs of those genes must not have binding sites for miR-1 and miR-133. However, the SWI/SNF complex, LIM cofactors and SMARC are involved in carcinogenesis elsewhere (pancreas and breast cancers) [1668, 1669]. The chemokine receptor CXCR2 and LASP-1 co-immunoprecipitate and co-localize at the edges of migrating cells. It is the LIM domain (Lin/Isl/Mec) of LASP-1 (*vide supra*) that binds to the carboxy terminal of the chemokine receptors CXCR1/2, CXCR3/4. Without LIM domain

Table XVI Human cysteine-rich protein (CRP) LIM kinase

CRPs evolutionarily conserved (schistosoma; caenorhabditis; drosophila; arabidopsis).[1] Characterized by 52 aa double zink finger motifs. Named after caenorhabditis genes *lin* (cell lineage), Isl-1 (insulin-like, rat *isl* insulin gene enhancer binding protein), and *mec-3* (mechanosensory): LIM. These genes contain the LIM motif: (CXXC-$X_{17\text{-}19}$-HXXC)-XX-(CXXC-$X_{7\text{-}11}$-CX_8C).[1] LIM genes were exposed to "intron invasion" occurring subsequently to evolutionary divergences. Rhombotins (rhoms) are primary response genes (like c-*fos*, c-*myc*) promptly activated by serum in serum-deprived cell.[1] Related to RAG-1, VDJ, IL-2R . BMI-1. *Rhom*2 resides at human chromosome 11p13 and produces fusion oncoprotein t(11;14)(p15;q11) in acute T cell lymphoblastic leukemia. *Rhoms* encode 17.7 kDa proteins with two sets of RNA- and DNA-binding LIM motif.[2]

In cell locomotion: EGF-phosphorylated cofilin (actin-depolymerizing factor, ADF) is prominently expressed in mobile cancer cells. Cofilin promotes cancer cell migration and invasiveness.[3] The neurofibromatosis type 1 gene (NF1) from 17q11.2 encodes the 2818 aa protein neurofibromin, a suppressor of *ras* proto-oncogene. $NF1^{-/-}$ Schwann cells replicate to form multiple neurofibromata (Recklinghausen disease). In neurofibromin-depleted cells, cofilin is phosphorylated by the Ras-activated Rho-ROCK-LIMK2 pathway. ADF/cofilin is phosphorylated at serine 3 (Se3) by LIM kinases 1/2, which are phosphorylated by PAK (p21-activated kinase) and ROCK (Rho-associated coiled-coil kinase).[4] The hammerhead ribozyme could de-activate LIM kinase2 in human fibrosarcoma cells and thus induce colony formation.[5] MAPK and PI3K pathways are constitutively activated in neurofibromin-depleted cells. Overexpressed NF1-GRD (GTPase-activating protein-related domain) stops locomotion and promotes colony formation. Slingshots (SSH) are active in insulin-induced membrane protrusion by dephosphorylating cofilin. Cofilin SSH are stimulated by insulin and mediated by PI3K. Physiologically, neuronal cells utilize the ROCK-LIMK-cofilin pathway in switching it on and off for directing axon growth of neuronal cells. View locomotion of cofilin-driven HeLa cells microfilmed.[5–8] Amoeboid tumor cell locomotion in Appendix 1. LIM kinase1 and ROCK drive locomotion of human breast (MDA-MB-231) and prostate (PC3; LNCaP) cancer cells. ROCK activates LIMK1/2. Amplified PAK1-6 isoforms (p21-activated kinase) mobilize MAP kinase cascade (Ras/Raf/Mek/ERK); anti-apoptotic Bad/Bcl-2; and LIMK. PAK1 (at 11q13.5-q14) and PAK4 (at 19q13.2) collaborate in transforming cells. The PAK inhibitor PF-3758309 (patented) acts additively with other small molecular oncoprotein inhibitors. The ROCK inhibitor Y-27632 (patented) decelerates cancer cell locomotion. Pyr1 inhibition of LIMK stabilized microtubules in multidrug-resistant tumor cells, blocked cofilin-induced actin microfilament dynamics; inhibited cancer cell mobility.[9–13] Tetraspanins (Table XXIV) direct membrane protrusions of crawling cells.[14] Cofilin directs membrane protrusion at the leading edge and retraction of the tail for the amoeboid movement of individual cells from the age of the amoeba in Darwin's pond to the detached human cancer cell in the labyrinth of the body. The tumor cell expropriates the ancient physiological pathway.

1 *Wang X et al J Biol Chem 1992;267:9176–84.* ***2*** *Boehm T et al Proc Natl Acad Sci USA 1991;88:3927–3; Oncogene 1991;6:695–703.* ***3*** *Chung H et al BMC Cancer 2013;13:45.* ***4*** *Ozawa T et al J Biol Chem 2005;280:39524–39533.* ***5*** *Suyama E et al J Gene Med 2004;6:357–363.* ***6*** *Amano T et al Biochem J 2001;15:149–159.* ***7*** *Horita Y et al J Biol Chem 2008;283:6013–21.* ***8*** *Nishita M et al J Biol Chem 2004;279:7183–8.* ***9*** *Sumi T et al J Cell Biol 1999;147:1519–32.* ***10*** *Li R et al Clin Exp Metastasis 2013;30:483–95.* ***11*** *Prudent R et al Cancer Res 2012;72:4429–39.* ***12*** *Ye DZ & Field J Cell Logist 2012;2:105–16.* ***13*** *Yoshioka K et al Proc Natl Acad Sci USA 2003;100: 7247–52.* ***14*** *Bar R et al Biochem Biophys Res Commun 2011;415:619–626.*

attached, the chemokine receptors do not mediate chemotaxis. The C-terminal of LASP is the SH3 domain; the N-terminal LIM (Lin/Isl/Mec) domain of LASP contains two nebulin repeats [1670] (Table XVI).

The human nebulin repeats are remarkable, because the amphioxus' muscle sarcoma Z-line contains related actin- and connectin/titin-binding nebulins, and an SH3 domain [1671]. The Spongiae and the amphioxus genomes carry the pp60c-src kinase for signal transduction. The Ewing'sarcoma oncogene commonly fuses with the Fli proto-oncogene (Friend leukemia virus insertion; an *ets* proto-oncogene encoding transformation-specific transcription factors). However, the EWSR1 gene also fuses with the SMARCA5 gene, which encodes the chromatin remodeling protein hSNF2H (sucrose non-fermenting 2 homolog) [1672].

Epigenetic chromatin remodeling regulates gene expression. Rhabdoid tumors harbor a mutated SMARCA4/SNF5 gene [1673, 1674]. SNF5 is a tumor suppressor protein, that malignant tumors (rhabdoid tumors) often eliminate. After elimination of SNF5, BRG1 (Brahma-related gene) remains active, and aggressive SNF5-deficient tumor formation occurs. Upon the knock-down of BRG1, instead of tumor formation, cell cycle arrest at G0-G1 followed by apoptotic death sets in [1674]. The acetylated BRG1 protein complexes with p53 protein *in vivo*, leading it to polyubiquitination [1675]. The SNF2H chromatin remodeling enzyme induces latent HSV 1 early gene activation, and the initiation of intranuclear viral replication [1676].

Epigenetics of tumor immunology. Attempts at biological therapeutics. These events frequently are reversible. Protein arginine methyltransferases (PRMT) silence the RB gene (RB1 p105, RBL1 p107, RBL2 p130) by methylating histones H3R8 and H4R3 (R, arginine) in its promoter. Reducing PRMT protein production by exposing its mRNAs to microRNAs restores full tumor inhibitory RB gene activity [1677]. After the knock-out of p53, RMS cells inhibit apoptosis further by over-expressing survivin. The human survivin gene is very active during ontogenesis, expressing its five alternatively spliced transcripts. The gene is silent in healthy cells of the adult host, but becomes highly overexpressed in malignantly transformed cells. Mitochondrial survivin inhibits caspases 3/9. Small hairpin (sh) microRNAs aimed at the survivin mRNAs significantly reduce survivin expression in xeno-grafted RMS cells [1678]. The TNF-related apoptosis-inducing ligand (TRAIL) binds death receptors (DR4/5) and induces the Fas-associated death domain (FADD) activating caspases 8/10. Sarcoma cells may knock out p53, but continue to express DR5 (not DR4), thus retain their potential vulnerability to TNFα and its related ligands (TRAIL and FasL). Because of its fulminant, wide-spread (and localized: hepatocytes) actions, FasL-induced apoptosis is much too toxic for targeted clinical trials. Tumor cells are protected against apoptosis induction by the BCL-2, IAP and XIAP (X-associated inhibitor of apoptosis) proteins. Pro-apoptotic proteins BAK and BAX (bcl-2 antagonist killer; bcl-2-associated X protein) permeabilize the mitochondrial membranes for the release of cytochrome C, that forms the apopto-some in the cytoplasm. Apaf-1 (apoptotic protease-activating factor) contains procaspase-9. The activated caspase-9 cleaves into activity the DNA-aiming caspases-3/6/7/8. The mitochondrial protein, Diablo-associated SMAC (second mitochondria-derived activator of caspases) is also an inhibitor of IAP. The DR5 agonist mcab drozitumab enhances the activity of caspase 8. However, the gene for caspase 8 is frequently silenced in tumor cells (more in neuroblastoma, than in RMS cells) by hypermethylation. This could be reversed by treatment with the methyltransferase inhibitor, 5-dAzaC; 5-dAzaC restored the apoptosis-sensitivity of neuroblastoma cells (*vide supra*). Recombinant human TRAiL (rhTRAIL, dulanermin) despite its short half life, stabilizes (or promotes partial, even complete, regression of) some human tumors (relapsed follicular lymphoma in combination with rituximab). While antibodies to induce ADCC reactions (*vide supra*) have to bind the cells' FcγR; dulanermin acts without such restriction. Further, DR4/5 agonistic mcab are in clinical trials (lexatumumab, conatumumab, TRA-8/CS-1008). For their efficacy, binding to FcγR is essential. Conatumumab (AMG655) with IGF-R antibody (AMG479) and/or doxorubicin is in clinical trials for metastatic sarcoma [1679]. The Bcl-2-related anti-apoptotic protein Mcl-1 (myeloid cell leukemia sequence 1) is antagonized by the pro-apoptotic protein Noxa (noxious; noxa, harm). RMS cells express both, and the glycolytic inhibitor 2-deoxyglucose (2-DG) could downregulate Mcl-1, and upregulate Noxa expression in aRMS cells. The tumor cells of aRMS carry the fusion oncoprotein PAX3/FOXO1a (*vide infra*). In these cell the hexokinase inhibitor 2-DG dowregulated the fusion oncoprotein and activated BID (Bcl-2 homology domain 3-interacting death agonist), BAK and BAX (Bcl-2 antagonist killer; Bcl-2-associated X) proteins. The pro-apoptotic efficacy of BAK/BAX in itself was inadequate to oppose Bcl-2/Bcl-xl/Mcl-1. Apoptosis was induced when 2-DG upregulated protein Noxa; Noxa is the powerful antagonist of Mcl-1. The most profound 2-DG-induced cell metabolism changes included phosphorylation of elF2α (ephrin ligand family), and inactivation of the mTOR pathway [1680] (*vide infra*).

Most tumor cells (not all) switched back from mitochondrial oxidative phosphorylation (yield: 36 ATPs) to Warburg's aerobic glycolysis (yield 2 ATPs) (Figure 9). Mantle lymphoma cells can retain oxidative phosphorylation and switch back and forth from it to glycolysis. Mantle lymphoma cells and CLL cells are notoriously TRAIL-resistant. However, 2-DG could sensitize these cells to TRAIL. Under glucose-free conditions, these cells lost their sensitivity to TRAIL, reduced TRAIL receptor expression, and formed mal-functioning DISCs (death-inducing signaling complex). In glucose-deprived cells, the Akt pathway was hyper-activated. Akt inhibits the mTOR pathway by phosphorylation at $Ser^{939/989}$ and Thr^{462} of the TSC2 (tuberous sclerosis) protein, a negative regulator of mTORC (complex). At the same time, Akt phosphorylates PRAS40 (proline-rich Akt substrate), an activator of mTORC. Further intervenes AMPK (adeninmonophosphate-activated protein kinase) by phosphorylating Raptor, an inhibitor of mTORC. In glucose-free cells, Akt dominates and mTORC is downregulated. While aerobic glycolysis is blocked, switching to oxidative phosphorylation supports the maintenance of the Bcl-2 family (an anti-apoptotic event). Akt certainly is an ancient "cell survival pathway" (see in Giardia). Paradoxically and indirectly, 2-DG could protect against cell death via Akt phosphorylation. In general, ATP production in tumor cells is further depleted by 2-DG.

TRAIL-induced apoptosis is much enhanced in these cells. Mcl-1 levels were downregulated by 2-DG treatment (a pro-apoptotic event) [1681, 1682]. With Mcl-1 inhibition via 2-DG, Bcl-2 inhibitors in combination (ABT-737, Obatoclax, Navitoclax, *vide supra*) might irreversibly block the anti-apoptotic machinery of most tumor cells. Agent ABT-737 binds to the proteins Bcl-2, Bcl-xL and Bcl-w, but spares the Mcl-1 protein, a resistance that Noxa and bortezomid could overrule [1683]. A review elaborated on the constitutively active STAT3 pathway in RMS cells, which contributes to the maintenance of the Bcl-2 family proteins and to apoptosis-resistance. Accordingly, the small molecular STAT3 inhibitor XZH-5 (patented) suppressed also the Bcl-2 family proteins and survivin. The anti-apoptotic XIAP protein can be neutralized by the antisense oligonucleotide AEG35156 (patented). RMS cells so treated died apoptotic deaths. Another anti-apoptotic protein to be targeted is c-FLIP (c-FLICE-inhibitory protein; FADD-like IL-1β-converting enzyme; Fas-associated death domain; factor apoptosis stimulator) [1684, 1685].The isoflavonoid genistein induces suppression of mRNA production by core regulatory genes, PBK, BUb1, CDC20 (PDZ-binding kinase; postsynaptic density protein disc-large zonula occludens 1, zonulin, drosophila; budding uninhibited by benzimidazole, yeast; cell division cycle), and overexpression of GADD45A (growth arrest and DNA damage) and pro-apoptotic p53. Among the responsive tumors were the eRMS cells [1686].

In a preclinical adoptive lymphocyte therapy trial, RMS cells were sensitive. SCID-mice received human e/aRMS cell line xenografts (or human leukemia cells) and were treated with IL-15-expanded cytokine-induced killer cells (CIK). Engraftment with tumor cells occurred in 100 % of the animals. Equal numbers of inoculated CIK cells reduced RMS (or leukemia cell) tumor growth, even though the expansion of RMS (and especially leukemia) cells occurred much faster, than that of the CIK cells, and some of the RMS cells were FasL- and TRAIL-resistant. The CIK cells released the TRAIL ligand for the cytolytic attack. Attacker CIK cells expressed the NKG2D molecule and the RMS cells expressed its ligands, MHCA-related A, B, and UL16-binding proteins (unique long, ULBP). Target tumor cells were killed by the release of perforins/granzymes. The proportion of the precultured $CD3^{+}CD56^{+}$ NKT cells in the CIK cell mixture gradually increased up to 31 % by dat 21. One target tumor cell line was re-classified from RMS to Ewing's sarcoma due to its fusion oncoprotein EWS-FLI1. In the first clinical trial, donor-derived CIK cell induced remissions with grades I/II GvH disease in 36 % of the recipients [1687, 1688].

The ULBPs are ligands for the NKG2D receptor. ULBPs often expressed by tumor cells and on EBV-infected lymphocytes. Both NKG2D (natural killer group2, member D) and T cell receptors γб (TCRγб, gamma/lambda) recognize the ligand ULBP4 [1689]. The ORF 33 of Kaposi sarcoma HHV-8 and its simian ancestors, the baboon and rhesus rhadinoviruses, encode UL16-binding proteins appearing on the cell surface (referred to as "tegument proteins") [1690]. The genome of KSHV is kept repressed by trimethylations of the histones (H3K9me3; H3K27me3) associated with the latent viral genome [1691]. While H3K27me3 is repressive, H3K4me3 and the acetylated H3 is activating. Demethylases acting on

H3K27me3 induce the viral lytic genes. KSHV ORFs at the translational start sites (TSS) contain regulatory introns. The reversible heterochromatin on the viral lytic genes is maintained in a repressive fashion by the host cell's polycomb group proteins repressive complex (PRC). The small molecule 3-deazaneplanocin inhibited the PRC (inhibiting repression equals upregulating PRC), demethylated H3K27me3, and induced the lytic phase of KSHV. In a primary effusion lymphoma cell line coinfected with KSHV and EBV, only KSHV was, but EBV was not, reactivated. The PRC-suppressed MYT1 gene (myelin transcription factor 1) was reactivated in these cells (unexpectedly; the gene MYT1 became famous when it interacted in fibroblasts transforming them into neuronal cells) [1692]. Staurosporin by cleaving PARP (polyadenosine diphosphate ribose polymerase Table II) induces apoptosis in primary effusion lymphoma cells. This intervention did not affect H3K27me3 and did not reactivate KSHV. Without the reversal of H3K27me3, the KSHV, the genome cannot reactivate; H3K9me3 appears to maintain H3K27me3, and for KSHV reactivation both trimethylated histones drop simultaneously. For this effect histone demethylases are to be recruited into the gene promoters. Histone acetyltransferase and the chromatin remodeling complex SWI/SNF (*vide supra*) cooperate [1691]. The human cytomegalovirus (HCMV) encodes an immunoevasive UL16 ligand. The MICs are MHC class I chain-related molecules. HCMV UL16 complexes with the NKG2D ligand MIC-B. Ligand MIC-B is the activator of the NKG2D receptor. The central motif of the HCMV UL16 mimics the NKG2D protein, thus it is switching off any activation that would be directed to self. While HCMV-infected cells downregulate MHC class I molecules to prevent their presentation of HCMV peptide antigens to immune T cells, the "missing self" condition must not trigger NK cell attacks. Regularly, ULBP would trigger the NK receptor for NKG2D to induce target cell lysis. Some of the ULBPs are called RAETs (retinoic acid early transcripts), and UL16 is ULBP2 and UL142 is MIC-A. The HCMV gene UL112 releases a microRNA to neutralize the mRNA for MIC-B. UL16 expresses a NKG2D binding mode to attach to MIC-B. HCMV UL16 expresses five hydrophobic core residues that mimic the residues of the central region the receptor NKG2D for binding its ligands. However, the HCMV UL6 binds only some of the NKG2D ligands (MIC-B, ULBP1/2/6) and cannot bind other NKG2D ligands (MIC-A, ULBP3/4/5). Only those NKG2D ligands that carry a glutamine/glutamate at position 169, and a small hydrophobic side chain at position 162, bind UL16; all other ligands with arginine at position 169 fail to bind. Convergent evolution means that the HCMV and the human cells independently evolved two structurally different receptors (NKG1D and UL16), but these share an identical central ligand binding motif [1693]. If anything like this repeats itself between NKG2D and a human tumor cell, NKG2D may refrain from attacking that tumor cell.

Tumor vaccines or monoclonal antibodies should be directed against tumor antigens generated in somatic mutations, thus the antigens were not present during fetal life for tolerance induction. Rituximab kills leukemic/lymphoma B cells by ADCC reaction, but also depletes normal B lymphocytes. In contrast to the efficacy of rituximab, mcab directed at the acute lymphoblastic leukemia antigen CD19 were ineffective. The CD19- and CD3-directecd mcab blinatumomab activates T cells for attack against CD-19^{+} leukemia lymphoblasts, and thus it was able to induce remissions. T cells with chimeric antigen receptors (CARs) efficiently attack malignant B cells. The CD22^{+} blast cells of pediatric ALL outlived epratuzumab mcab therapy. Haploidentical NK cell infusions in addition to epratuzumab induced the intracytoplasmic concealment of CD22 antigens. The pseudomonas immunotoxin moxetumomab penetrated cells, attached to CD22 antigens and killed the targeted cells. Brentuximab vedotin carrying the antitubulin auristatin E molecule penetrates CD30^{+} Reed-Sternberg cells, arrests the cell cycle, and induces cell death. The NY-ESO antigen overexpressing synovial sarcoma cells (Figure 34) can be attacked by immune T cells with genetically engineered NY-ESO-specific receptor. However, the NY-ESO antigen expression is HLA-A2-restricted. The highly desirable cancer vaccine targets, the cancer-testis antigens, remain expressed throughout the entire adult life in testicular germ cells. MHC-restricted T cell receptors will recognize only the antigenic epitopes presented in the grooves of the MHC molecules that are restricted to that MHC allele [1694]. Sarcomas other than synovial sarcoma seldom if ever express the NY-ESO antigen [1695]. The synovial sarcoma fusion oncogene/oncoprotein (SS18/SWI/SNF) is activated by histone deacetylation. The HDAC inhibitor romidepsin inhibited the growth of synovial sarcoma cells. The synovial sarcoma fusion oncoprotein silences by hypermethylation the tumor suppressor early growth response 1 (EGR1) promoter DNA (H3K27me3), and attached polycomb group proteins to the promoter. Romidepsin reverted these epigenetic modifications [1996]. The Wilms tumor (WT1) peptide vaccine induced complete remission in a 6-year-old girl with metastatic alveolar rhabdomyosarcoma [1697a]. A similar result was achieved with a neuroblastoma vaccine co-administered with decitabine [1697b].

When the critical NY-ESO target antigen is not expressed because its gene promoter is silenced by methylation, demethylation of the critical CpG by DAC-treatment (5-aza-2′-deoxycytidine) may re-establish the synthesis of the antigenic protein. However, the antigenic peptide needs to be expressed by the appropriate MHC. The antigenic peptide needs to be presented to CD4^{+} T cells in an immunogenic manner; tolerogenic mode of presentation would induce an overwhelming production of Treg cells instead of immune CD8^{+} T cells. Genetic engineering by retroviral transduction of the polyclonal TCR αβ chains renders the CD4/CD8 T cells to be specifically NY-ESO-reactive. Of the genes expressed by this RNA to cDNA technology, the assessment of newly expressed genes, among them the highly important NY-ESO-1, MEGA/C, SSX1, TAG1, HLA-A, B2 M, and the reference gene GAPDH (New York esophageal carcinoma, melanoma antigen gene, synovial sarcoma gene, tumor antigen gene, human leukocyte antigen-A, β2-microglobulin, glyceraldehydes 3-phosphate dehydrogenase) needs to be proven by real time polymerase chain reaction (qPCR). The NY-ESO antigen was reacted in Western blots with the mouse monoclonal antibody, and the reaction was verified by reacting the mouse immunoglobulin with horseradish peroxidase-conjugated goat anti-mouse IgG. Polyclonal CD8^{+} T cells were generated to be specifically NY-ESO-1-reactive by repeatedly stimulating HLA-A*0201^{+} cells with autologous irradiated dendritic cells pulsed with NY-ESO-1 peptide. The reactive cells were separated by cloning. Specific lymphocyte-mediated cytotoxicity to targeted cancer cells was measured in chromium-release assay. DAC-induction induced cancer cells to express the NY-ESO antigen. The NY-ESO-specific immune T cells destroyed the NY-ESO-expressing colon cancer cells [1698]. Tumor cells usually silence by methylation of its promoter DNA the expression of the NY-ESO antigens MAGE-1. A 6 year old boy with widely spread neuroblastoma in relapse after heavy chemotherapy received autologous dendritic cells pulsed with MAGE-peptides with Toll-like receptor agonist imiquimod administered at the vaccination site; he was treated with decitabine (5-aza-2′-deoxycytidine, DAC). The patient developed $CD3^{+}CD8^{+}CD137^{+}$ T cells reactive with MAGE-A3 peptides. The patient cleared large solid tumor masses (a mediastinal tumor) and yielded bone marrow aspirates negative for tumor cells [1697b]. However, in other patients with metastatic neuroblastoma, a combination of DAC with chemotherapy (doxorubicin, cyclophosphamide) failed to induce durable remissions. Autologous neuroblastoma cell vaccines expressing interleukins induced IFNγ production, but no remissions. The attenuated varicella-zoster vaccine may attack and kill neuroblastoma cells, inasmuch as the virulent virus is cytotoxic in neuronal cells; however, it is immunizing, and virus neutralizing antibodies may curtail any effects of viral oncolysis (Figure 44; Table XVII and XXI) (*vide infra*).

T cell-depleted NK cells of healthy donors were stimulated with the repeatedly transduced CML cell line K562. Ewing's sarcoma cells were the most susceptible targets of these cytotoxic NK cells followed by RMS cells (see movie in the report). Tumor xenografts in mice were eliminated by activated NK cell infusions [1699]. Haploidentical stem cell transplants for pediatric tumors (neuroblastoma, Ewing's sarcoma, eRMS) deliver NKG2D-carrier NK cells. All target tumors expressed ULBP2. The immediate responses were complete remission in 3, partial remission in 1, and stabilization of disease in 2 patients. At 14 months, 3 patients remained alive in complete remission; 2 patients died with advancing disease, and 1 patient died due to treatment-related complications [1700].

Haploidentical hematopoietic stem cell transplants delivered to heavy conditioning regimen- pretreated patients with acute high risk leukemias. The patients selected out KIR and CD94/NKG2D expressing donor NK cells, which attacked and killed leukemic cells. The KIRs left HLA class I alleles unrecognized, and induced no GvHD [1701]. A second generation chimeric antigen receptor (CAR) bearing T lymphocyte used for adoptive therapy failed to eradicate RMS cells, but performed better against lymphoma and adenocarcinoma cells [1702]. Macrophages were shown to phagocytize and digest RMS cells. Cytoplasmic calcireticulin transferred to the cell surface render sRMS cells digestible by the Mϕ. Doxorubicin- (anthracyclin-) treated sarcoma cells transfer calcireticulin from the endoplasmic reticulum to the cell surface. Macrophages engulf these tumor cells.

Table XVII Epigenomics of cancer immunotherapy

Cancer-testis antigens ectopically expressed and spontaneously immunogenic in human tumors. Best examples NY-ESO (esophagus), MAGE (melanoma antigen), SSX (synovial sarcoma X chromosome breakpoint). Patients with SS developed lymphocytes cytolytic to their tumors (shown in late 1960s, text reference 147).
DNA methylation inhibitors increase response to treatment in myelodysplasia and acute and chronic myelogenous leukemia.[1] Demethylated acute myeloid leukemia cells express SSX.[2] Decitabine* and Vorinostat** increased CD95 FasR expression in colon cancer cells rendering them susceptible to $FasL^+$ T lymphocyte-induced cytolysis.[3] 5-aza-2′-deoxycytidine (DAC)* induced NY-ESO expression in colon adenocarcinoma cells rendering these cancer cells susceptible to NY-ESO-specific CD8 T cells.[4] Patients with prostate cancer express SSXs in metastatic (not in primary) tumors. SSX-specific IgG and T cells rise after DAC* treatment.[5] Patients with triple-negative breast cancer express NY-ESO; develop anti-NY-ESO IgG/CD8 T tumor-infiltrating lymphocytes.[6] Decitabine* increases NY-ESO expression in glioma cells rendering them susceptible to FasL-induced apoptosis.[7]
Valproic acid or Vorinostat** and/or decitabine* did not increase NK cell ligand expression. Pediatric B-lineage leukemia target cells treated with decitabine* showed increased sensitivity to NK cell cytolysis.[8]
Epigenetic interventions render neuroblastoma cells antigenic and immunotherapy sensitive. Decitabine* pretreatment increased response rate up to durable complete remission in DC MAGE-vaccinatecd patient with stage IV metastatic disease.[9,10] Neuroblastoma "archetype" tumor for immunotherapy. McAb 3F8 against disialoganglioside GD2 and barley-derived glucan-stimulated FcR^+ leukocyte- induced remissions (Sloan-Kettering protocol). *In vitro* activated autologous NK cells adoptively re-infused killed these tumor cells.[11abc]
Exceptionally malignant triple negative breast cancers (TNBC). Commonly originating in undifferentiated basal- , but occasionally, in luminal-cell types. A subtype expresses the Core Basal (EGFR and cytokeratin CK5/6) pathways.[12] Cituximab suppresses the EGFR/HER3-overexpressed tumor cells; it acts synergistically with dasatinib, IL-2 and IL-5, due to NK cell activation.[13–15] When mutated (not translocated) c-Myc and cyclin-dependent kinase CDK5 are co-expressed, CDK-inhibitor dinaciclib induced the second requirement for synthetic lethality[16] (Table X). Testosterone-driven androgen-receptor expression reduces the metastatic potential of TNBC cells.[17–20] Exception is the apocrine TNBC, which is dependent on androgen signaling. Accordingly, it responds to treatment with anti-androgen bicalutamide and anti-CYP17 abiraterone.[17] Protein serine/threonine phosphatase type 2 (PP2A) inactivates mTOR effectors (incuding PI3K/Akt), except when it suffers loss-of-function mutation, as in TNBCs.[21] Observe: in osmotic stress in Neurospora crassa cells, PP2A class phosphatases induce MAPK and excessive hyphal growth, whereas in TNBC cells PP2A appears to be tumor suppressive (unless loss-of-function point-mutated).[21] Observe: MAPK-driven MDA breast cancer cell xenografts were eradicated by oncolytic replication-competent herpesvirus NV1066.[22] Further antigens expressed in TNBC cells: sarcoma antigens vimentin and chondroitin sulfate proteoglycan;[23] PI3K/AKt/mTOR;[24] Wnt signaling; Notch; ephrin receptor A10 (see text); gene tyrosine kinase "uncontrolled" at human chromosome 19q13.1 (Greek, anexelekto) capturing ligand Gas6 (growth arrest specific);[25] hypoxia-inducible factor.[26] For patients with TNBC entering remission on chemotherapy, chemotherapy-resistant relapses occur as a rule.
Two vaccines expressing the ESO-NY antigen MAGE are expected to induce immunity protective against such relapses: the Astuprotimut-R vaccine of GlaxoSmithKline, and the genetically engineered attenuated Listeria vaccine.[27,28] The Listeria vaccine induced the immunosuppressive and STAT oncogene-activator IL-6; co-administration of the IL-6 inhibitor curcumin (diferuloylmethane) eliminated this detraction, induced IL-12 production, and reduced the numbers of tumor-induced tumor-protective myeloid-derived immunosuppressor cells (in mice; not yet in patients).[27] TNBCs are recognized by tumor-infiltrating lymphocytes, which improve their prognosis[29,30] Yet. the prognosis of TNBC remains unfavorable. It is the highly expressed miR-21, that is the suspected demolitor of tumor suppressor gene-derived mRNAs.[31]

(continued)

Table XVII (continued)

The expression stem cell factor focal adhesion kinase (FAK) is present in both tumor cells and vascular endothelial cells may persist at the latter location even in remission.[32] Attempts at inhibiting DNA repair by suppressing PARP (Table II) with olaparib or veliparib, and inhibiting neo-vasculogenesis with suniinib are in clinical trials.[33] If TNBC cells masquarede like "self" enities by expressing PD-1 ligands (see text), they will be attacked by PD-1 lymphocytes exempted from blockade by anti-ligand antibodies (pembrolizumab), as projected by Lajos Pusztai at Yale (Yale Cancer Center, Centerpoint, spring/summer 2014 pp 4–7).

1 Issa J-P J & Kantarjian HM Clin Cancer Res 2009;15:3938-46. ***2*** *Atanackovic D et al Am J Hematol 2011;86:918–22.* ***3*** *Yang D et al J Immunol 2012;188:4441–9.* ***4*** *Chou J et al J immunother 2012;35:131–41.* ***5*** *Smith HA et al Cancer Res 2011;71:6785–95.* ***6*** *Ademuyiwa FO et al PLoS One 2012;7(6):e38783.* ***7*** *Konkankit VV et al J Transl Med 2011;9:192.* ***8*** *Pfeiffer MM et al Front Oncol 2013;3:99.* ***9*** *Bao L et al Cancer Immunol Immunother 2011;60:1299–307.* ***10*** *Moreno L et al J Clin Oncol 2013;31:649–50.* ***11a*** *Krishnadas DK et al Pediatrics 2013;131:e336–41.* ***11b*** *Modak S et al Oncoimmunology 2013;2(3):e28402.* ***11c*** *Liu Y et al Clin Cancer Res 2013;19:2132–43.* ***12*** *Gasparini P et al 2014. doi:10.1371/journal/pone.0088525.*
13 *Baselga J et al J Clin Oncol 2013;31:2586–92.* ***14*** *Kim EM et al J Surg Res 2013;185:231–9.*
15 *Roberti MP et al Breast Cancer Res Treat 2012;136:659–71.* ***16*** *Horiuchi D et al J Exp Med 2012;209:675–96.* ***17*** *Fioretti FM et al J Mol Endocrinol 2014; 52:R257–65.*
18 *Mohammadizadeh F et al Adv Biomed Res 2014;3:132. doi:10.4103/2277-9175.133260.*
19 *McGhan LJ et al Ann Surg Oncol 2014;21:361–7.* ***20*** *McNamara KM et al Breast Cancer Res Treat 2014;145: 281–93.* ***21*** *Baldacchino S et al EPMA J 2014;5(1):3. doi:10.1186/1878-5085-5-3.* ***22*** *Gholami S et al Cancer Gene Ther 2014. doi:*10.1038/ctg.2014.28. ***23*** *Wang X et al J Natl Cancer Inst 201;102:1496–512.* ***24*** *Miller TW Breast Cancer Res 2013;14(1).301.*
25 *D'Alonso TM et al J Clin Pathol 2014; pii:jclinpath-2013-202161.* ***26*** *Wang T et al Proc Natl Acad Sci USA 2014;pii:201410041.* ***27*** *Cabezón T et al Mol Cell Proteomics 2013;12:381–294.*
28 *Singh M et al Cancer Med 2013;2:571–82.* ***29*** *Loi S Oncoimmunology 2013;2(7): e24720-014pii:mdu191.* ***30*** *Ali HR et al Ann Oncol 2014 pii:mdu191.* ***31*** *Dong et al Med Oncol 2014;31:57. doi:10.1007/s12032-014-0057-x.* ***32*** *Axelopoulo AN et al BMC Cancer 2014p;14: P237. doi:10.1186/1471-2407-14-237.* ***33*** *Pal SK et al Breast Cancer Res Treat 2011;125:627–36.*

*5-aza-2-deoxycytidine, 2-deoxy-azacytidine, decitabine DAC methyltransferase inhibitor, demethylator of CpG dinucleotide sites; releases silenced genes (tumor suppressor genes) from CH_3-effect.

**Vorinostat suberoylanilide hydroxamic acid SAHA histone deacetylase inhibitor; restores histone acetylation; acetylated proto-oncogenes are silent; deacetylated proto-oncogenes are active

The CD47 receptor is inhibitory to phagocytosis by Mϕ. RMS cell overexpress CD47. With signal regulatory protein α (SIRP), it protects healthy cells from Mϕ phagocytosis. GM-CSF-treated Mϕ increased their RMS cell phagocytic activity [1703].

In contrast to eRMS, aRMS tumors carry the translocated fused t(2;13)(q35;q14) or t(1;13)(p36;q14) oncogenes producing the fusion oncoproteins PAX3 or PAX7/FOXO1/FKHR (paired box homeotic gene; forkhead box O1) In addition, the amplified chromosome 12q13-15 contains the sonic Hh pathway, and the MDM2 oncogene. The MDM protein binds the p53 protein for ubiquitination. The amplified chromosome 2p24 contains the N-Myc gene. An intron of the amplified 13q31.32 segment contains the microRNA cluster miR-17-92. The list of

PAX3/FOXO1 target genes takes up two pages in small print. The receptor of hepatocyte growth factor/scatter factor (HGF/SF), the MET, the N-Myc, the IGF2, the FGF-R4 are among the prominent upregulated target genes [1704, 1705]. The miR-9 is the invasiveness and metastases promoting factor [1706]. The PAX3/FOXO1 oncoprotein stimulates the CPT1A gene (carnitine palmitoyltransferase); the CPT1A protein induces wide spread motility of aRMS cells [1707]. Cells of aRMS run on the PI3K/Akt/mTOR pathway and are notoriously chemo-radiotherapy resistant. The loss of p53 amounts to ATP loss as well. ATP loss affects sugar metabolism, ATP-deficient cells cannot tolerate sugar deprivation.

The glycolysis inhibitor 2-DG (*vide supra*) arrested the growth of aRMS cells. Cell lines of eRMS could outgrow 2-DG treatment. However, 2-DG-treated aRMS cells underwent a fusion process that differentiating healthy myotube cells normally practice. The fusion oncoprotein PAX3/FOXO1 is the inhibitor of myotube differentiation. Thus, 2-DG treatment downregulated the fusion oncoprotein. Treatment with 2DG-activated pro-apoptotic proteins BAK and BAX, which promote the release of cytochrome C from the mitochondria. However, upregulated Bcl-2, BCL-xL and MCL-1 resisted apoptosis induction. Agent 2-DG downregulated Mcl-1, induced cleavage of PARP (Table II), phosphorylated eukaryotic initiation factor Elf2α, inactivated mTOR (by dephosphorylation), and upregulated NOXA. NOXA is activated via phosphorylation by CDK5 (cyclin-dependent kinase). The NOXA protein specifically binds MCL1 and neutralizes it. Mannose in the medium by preventing ER stress, alleviated 2-DG-induced aRMS cell death, whereas ER stressor tunicamycin promoted 2-DG-induced aRMS cell death. Cell death with ER stress is mediated by 2-DG and the Bcl-2-interacting mediator (of death) BIM, but in aRMS cells mannose prevented 2-DG induced death without completely eliminating BIM [1680].

Comment. It appears that endogenously activated de-differentiation (trans-speciation; malignant transformation) is readily acceptable even in healthy young hosts. The host rather cooperates, than resists. In contrast, exogenously enforced malignant transformation (by a pathogen; viral oncogenesis) evokes immunological resistance, that is occasionally efficient (HPV uterine cervicitis), but more often overcome by the pathogen (EBV lymphoma; Kaposi sarcoma; Merkel cell carcinoma).

However, a biologically most sensitive tumor, devoid of fusion oncoproteins, and yielding readily to impulses to differentiate, might have arisen due to epigenetic errors in its genomic control. These processes (when clearly understood) are correctible. The irreversible determination of the RNA/DNA complex to drive the initial neuroblastoma, or eRMS cell, out of the cell community of its host for its own independent immortal life, is not evident. The youthful host is not dying due to any other reason, than the epigenetically mistargeted errors. These events leave the RA receptor undifferentiated (but not mutated) in neuroblastoma. Amplified proto-oncogenes (incuding *ras*), or the absence of the NF1 gene (neurofibromatosis) and its product protein, the Ras-inhibitory factor, and other reversibly silenced, but not eliminated, tumor suppressor genes, are the initial inducers of eRMS. The proto-oncogenes may be amplified, but are not irreversibly mutated (no gain-of-function mutations, no fused oncogenes/oncoproteins occurred). The epigenetic changes are reversible. The initial neuroblastoma or eRMS cells have not as yet extricated themselves out of their cell communities. The transformed cells preserve a great deal of their physiological-biological faculties and responsiveness to these stimuli; they still yield to the immunobiological control of their host. The initial drivers of the neuroblastoma cell are amplified proto-oncogenes (without gain-of-function mutations, or gene fusions) unopposed by reversibly silenced (but not 'loss-of-function mutated' or eliminated) tumor suppressor genes. These tumors originated in youthful basically healthy hosts. It is extremely unlikely that in the initiation of these tumors such a principle, as a signal that a host in threatening demise, chronically infected, devastated, or aged, would have sent to its RNA/DNA complex indicating that individual cells could or should be rescued in order to sustain cellular life. The transforming processes in these tumors at their initiation are different from what the constitutively and irreversibly activated 'gain-of function mutated' or fused oncogenes and fusion oncoproteins dictate in a cell community irreversibly deprived of its tumor suppressor genes. Those cells are already immortalized (the physiological apoptotic processes are irreversibly blocked), and resist physiological impulses to differentiate. Living in a multicellular community, the transformed cells already extricated themselves from that community, are individualized, and utilize sophisticated mechanisms to overcome host immune reactions. As the clue to their ultimate superiority, these cells subvert the multicellular community for gaining its full support, and in taking use of all physiological exocrine circuitries for the receipt of growth factors. Eventually they will produce their own growth factors in autocrine circuitries, as they could live as individual independent unicellular organisms outside the multicellular community. The initial transformation in neuroblastoma, or eRMS cells is different in its causation and course, from those transformations that multiply mutated genomes in despair undergo. These, and the mistargeted nicks of AID or RAG and the lymphomas so induced, are accidents in a most sophisticated system, evolutionarily not yet corrected.

Are there other causes of cell transformations, probably activated by signals from the multicellular host to its RNA/DNA complex in calling for a rescue mission? Is the sustentation of independent living cells after the disintegration of organized cell communities an inherent biological requirement? Could the immense amount of potential bioinformation (Ewan Birney, DNA Bioinformation), that the DNA molecule stores, be activated to exercise its preserved primordial sustentative faculties for the keeping of individual single cells alive under the most adverse circumstances?

It could be expected that danger signals from an endangered multicellular host to its RNA/DNA complexes would be numerous, continuous and penetrating. In contrast, against such expectations, it appears that only one signal penetrates barriers and is taken. In one particular organ, and in one site within that organ is the RNA/DNA complex able to respond. There, a single primary tumor arises. In contrast to endogenous oncogenesis, oncogenic viruses induce multifocal carcinogenesis (leukemic retroviruses; HPV; EBV; KSHV; polyoma viruses). Macfarlane Burnet proposed that multiple incipient primary tumor cells may be silently eliminated by surveillance mechanisms of the host [1708]. These surveillance mechanisms could be the cell's own apoptotic defense mechanisms, NK cells patrolling, or stromal cells (fibroblasts) practicing the "neighbor suppression" maneuver. Latent oncolytic viruses (reovirus, NDV) may exercise a silent oncolytic function. There may be a long "tumor cell dormancy" before one transformed cell succeeds in establishing itself as an advancing new cell population. These incidents would depend on the biochemical makeup of that niche, in which the candidate tumor cells reside. There is a strong natural resistance toward exogenous initiation of oncogenesis, whereas the endogenously initiated and alleged special signal-induced engagement of the RNA/DNA complex is accepted as more natural (as an "inherent faculty") and as such it is to be supported.

In addition to the biological danger signals emitted by the host in impending demise, that particular recipient cell may be co-exposed to physico-chemical factors at the same time, rendering it more receptive to those danger signals, than perfectly intact cells in its vicinity would be. Some signals may be inadvertent, or misread, and one of the targeted cells may be transformed "by a mistake" of the ultimate bioengineer RNA/DNA complex. Once a tumor has originated, its local invasion is immediately recognized, but its silent distant spread is often overlooked. However, metastatic tumor cells may rest in regional lymph nodes and in distant organs (the bone marrow being a common site for harboring silent metastases). The immediate fate of these tumor cells again is not that of prompt replications. Commonly there is another long period of dormancy in the state of very slow proliferation balanced by apoptotic depletion of tumor cells, or existence of tumor cells in the state of autophagy. The microenvironment may contain metastatic tumor cells for a decade. In contrast, molecular mediators released by the tumor may elicit the release from the microenvironment growth factors and vascular ingrowths for blood supply to the tumor cells. The tumor cell, which was overlooked by macrophages in its niche (tumor-friendly Mϕs may rise), must now withstand attacks by antibodies, NK cells and immune T-cells. Once the tumor reached this stage, it is usually immunoresistant, proliferates rapidly and has gained resistance to physicochemical attacks as well. By now some or all those genes that preserve stemness are constitutively activated. These are the stage IV lethal tumors [1709, 1710]. These are the tumor cells, which when put in culture will live in a disorganized community, will practice cannibalism, and will exist in immortality.

Selected Abbreviations

Notch. Anaplastic astrocytoma, ependymoma, medulloblastoma, Jag-activated Notch. Click Feletti A et al Brain Tumor Pathol 2014 March 19 in print; Fouladi M et al J Neurooncol 2013;114:173–9; Killela PJ et al Oncotarget 2013 Oct 16 in print; Mascaro Cordeiro B et al Childs Nerv Syst 2014 Apr 3 in print.

Cytotrophoblast driven by Notch. Click for Haider S et al Endocrinology 2014;155:263–74.

Delta-like4 to Notch ligand Jagged for glioblastoma vascularization, click Jubb AM et al Histopathology 2012;60:740–7. For Jag signal in neovascularization, click Kwon S-M et al Circulation 2008;118:157–165.

ETM, epithelial-to-mesenchymal transition.

Jag, Jagged ligand to Notch receptor. For reference click, Blacklow SC Structure 2013;21:2100–1. For JAG1/2 in medulloblastoma, click Fiaschetti G et al Acta Neuropathol Commun 2014;2(1):39. For Jagged-2/Notch-1/Hes-1 pathway, click Chen G et al PLoS One 2013;8(10):e76274.

DAPT. Gamma γ-secretase inhibitor DAPT N-[N-3,5-difluorophenacetyl)-1-alanyl]-S-phenylglycine t-butyl ester. In lung cancer, click for Hassan KA et al Clin Cancer Res 2013;19:1972–80. In sepsis, click for Huang M et al Chin Med J 2014;127:924–8.

Hes/HES hey/HEY, Hairy enhancer of split / Hes-related YRPW motif (tyrosine, arginine, proline, tryptophan). Click for epigenetic inactivation of Notch-Hes pathway in human acute lymphoblastic leukemia, Kuang SQ et al PLoS One 2013;8(4):e61807. For bone metabolism, click Zanotti S & Canalis E J Biol Chem 2013;288:21547–57. For insulin degrading enzyme, click Leal MC et al Biochim Biophys Acta 2012;1823:227–35.

Lunatic fringe sugar transferase, Lfng/LFNG. For reference click, Xu K et al Cancer Cell 2012;21:626–41.

miR-124 microRNA regulating Notch signaling pathway, click Liu XS PLoS One 2011;6(8):e23461.

MMP, matrix metalloproteinase.

sHh/Wnt/Notch, Sonic Hedgehog, Wingless/integrated /Notch pathway.

TGFβ, transforming growth factor beta; decapentaplegic, drosophila.

Yes/YES proto-oncogene/oncoprotein, Yamaguchi sarcoma retroviral oncogene homolog non-receptor tyrosine protein kinase, Src (Rous sarcoma) family kinases. For Yes/Jagged/Notch in hepatocellular carcinoma, click Tschaharganeh DF et al Gastroenterology 2013;144:1530–42.

The Return of the Primordial RNA/DNA Complex in the Form of Oncogenes

The ancestors of the proto-oncogenes. Many examples of these events are repeatedly referred to in the forerunning text. The PI3K cell survival pathway drives the giardia. The hoglet gene of the choanoflagellates is the ancestor the hedgehog pathway. The invertebrate chordate, amphioxus, already separated a minuscule dorsal central nervous system (CNS) from an extensive peripheral ventral nervous system consisting of epidermal sensory neurons. These cells are inserted in the epidermal ectoderm and are connected via long axons with the notochord and CNS. The axons contain the neurotransmitter γ-aminobutyric acid (GABA). The helix-loop-helix achaete-scute homolog (ASH) (Table XXIX, *vide infra*) is present and is connected with a delta ligand, the Notch receptor, and the bone morphogenetic protein, BMP (while there is no trace of osteoblasts, none whatsoever). Further search revealed that the ancient ASH exists also in the sea anemone (*Nematostella vectensis*), the hydra (*Hydra vulgaris*), and the sponge (*Amphimedon queenslandica)*, but without frank nerve cells [1711, 1712] (*vide infra*). The amphioxus operates a *wnt* gene and its antagonist the positive regulatory domain modulator. Among its dorsal/ventral patterning genes, amphioxus embryos express BMP, dickkopf and secreted frizzled-related/binding proteins (sFRP). The Wnt antagonists sFRBP/FRZB1 (secreted frizzled-related protein) are conserved up to vertebrate mammals. In the mammalian embryos overexpressed Frzb1 may abolish skeletal myogenesis [1713a, b]. This author asks if sFRZB/sFRP could possibly antagonize rhabdomyosarcoma cells?

The genomes of mesenchymal stem cells (MSC) may chose the TGFβ/BMP pathway in continuing the activation of the runt-related osteogenic master transcription factor Runx2. This pathway culminates in the upregulation of osteoblast collagen, alkaline phosphatase, and osteocalcin. The MSCs assume the geno-phenotype of the osteoblast. Were the adipocyte master transcription factor, PPARγ2 (Table II) activated first, the MSCs would have ended up as adipocytes [1714]. The element cerium, the flavone glycoside poncirin, estrogenes, linoleic acid and metfomin all influence MSC differentiation in favor of osteoblasts over adipocytes.

In the embryonal life of invertebrates and vertebrates, two prominent ligands activate through cell surface receptors the Smads (signaling mothers against decapentaplegic, drosophila). The ligands are Dpp (decapentaplegic, TGFß) and BMP (bone morphogenetic protein). The phosphorylated Smads translocate to the nucleus to regulate their target genes (highly specific protein-to-DNA attachments). The ligands Dpp/BMP react extracellularly with cystein-rich domains of their antagonist, the short gastrulation protein (Sog/chordin). The twisted gastrulation protein (Tsg) and the protein Tolloid form a complex with BMP/Sog/chordin, and Tolloid cleaves BMP4 off the complex readying it to attach to its cell surface receptor. The cystein-rich drosophila protein crossveinless-2 (C-2) is an activator of Dpp (TGFß), in opposition of Sog, it is a Dpp inhibitor [1715]. (Note: the drosophila does not have venous blood circulation). Many of these ancient genes (Smads) reappear in vertebrate mammalian hosts as proto-oncogenes, involved in the initiation and maintenance of the malignant transformation of their host cells [1716–1719].

Another internal error: the Okazaki fragments. The first traces of the proliferating cell nuclear antigen (PCNA) (*vide supra*) appear in the archaea and eubacteria before the formation of the nucleus in the eukaryotes. The PCNAs and the PRMTs (protein arginin methyltransferase), if not initiate, they certainly maintain the malignant transformation of the eukaryotic cells. Plasmodia (*P. falciparum*) utilize PCNA for encoding the biochemical changes necessary for their existence in the mosquito or in the mammalian erythrocyte [1720]. DNA damage results in stalled replication forks. The mono- and polyubiquitylated PCNA allows for direct nucleotide incorporation and copying from the undamaged sister strand. The

Spartan protein binds both mono- and polyubiquitylated PCNA and inhibits its de-ubiquitylation by the complex of ubiquitin-specific protease/USP-associated factor, thus opening access of translesional synthetic polymerases to the replication fork [1721]. The histone lysine methyltransferase SETD8 (SET domain containing: Su(var), S = suppressor of variegation, E = enhancer of zeste, T = trithorax, drosophila) methylates PCNA at lysine K248me (a non-histone protein). This lysine (K) methylation sets PCNA after the flap endonuclease FEN1. Normal FEN1 exerting DNA repair prevents carcinogenesis. Loss-of-function mutation, germ line variation, or inhibition of FEN1 removes this protection [1722]. Demethylation inhibited the maturation of the Okazaki fragments. Okazaki fragments are 100–200 nt short DNA strands newly synthesized in eukaryotes in the lagging strand DNA, proceeding in the direction opposite of the replication fork. In prokaryotes, Okazaki fragments are >1000 nt long. The Dna2 genes, in the yeast (yDna2) and in the human (hDna2) cells are closely related. Both enzymes are multifunctional (DNA-helicase, ATPase, exo/endonuclease). Okazaki fragment processing consists of the removal of RNA/DNA flaps at the terminus of O-fragment. Principally, this latter function is performed by the Rad27 gene-product protein FEN1 (first crystallized from *Archaeoglobus fulgidus*). Enzyme Dna2 renders long structured flaps to be acceptable substrates for FEN1. Replication protein A protects long flaps from cleavage by FEN1. The long flap is approached by its base, threaded and then cleaved. Preferred substrates for FEN1 are short (<30 nt) double flaps; the DNA2 enzyme performs ATP-independent strand exchanges and strand annealings. Thus, replication fork arrest is avoided [1723, 1724]. Demethylation of PCNA brought DNA replication close to a standstill. Cancer cells notoriously express methylated PCNA [1722].

In dsDNA replication, the leading strand is synthesized continuously by polymerase ε (Pol ε), and the lagging strand is formed by the hetero-tetramer RNA and DNA polymerase (Pol α) synthesizing the RNA primer and a short DNA strand. The lagging strand is extended by polymerase δ (Pol δ), allowing the Okazaki fragments to form. Proliferating cell nuclear antigen (PCNA) replaces POL α in the leading strand. The Pol α/primase complex puts 7–14 nt long RNA primers into the lagging strand. Pol δ begins DNA synthesis in the nascent lagging strand. The remaining RNA primers are the O-fragments that need to be removed by DNA2 and FEN1 and annealed into another lagging strand. The frequency of Okazaki fragment formation (millions of fragments per cell cycle) requires a rapid processing mechanism. Inefficient primer removal results in unligased nicks and dsDNA breaks. With efficient O-fragment processing, DNA replication proceeds error-free. If defective O-fragment processing is allowed to accumulate, the results are the cessation of DNA replication and cell death. Otherwise, surviving cells may enter the stage of "mutator phenotype". Indeed, loss-of-function mutation of FEN1, or faulty editing of Pol α, promotes malignant transformation of the cell [1725].

Ubiquitination of PCNA proteins from bulky template lesions allows for the recruitment of translesionally synthetic polymerases. The PCNA molecule undergoing removal by monoubiquitination is unable to cooperate with FEN1 and Pol δ, but it clears the platform for the entry to the site the translesional polymerases

[1726]. The germ line variation -69GG FEN1 genotype significantly correlated with lung cancers of coke oven workers and with various other cancers (hepatocellular, esophageal, gastric and colorectal) of the GI tract [1727].

Comment. This mode of carcinogenesis is akin to the misguided nicks of AID/RAG in the plasma cell during somatic hypermutation. Here, it is not the primordial RNA/DNA complex exercising their inherent function to rescue individual cells from a multicellular host with approaching demise. It is a genuine error in the installation of a system that is still evolving with protective measures already initiated.

Ki-67. Of the numerous mammalian cell nuclear antigens appearing in dividing nuclei, referred to as chromosome segregation proteins, Ki-67 (after the city of Kiel, Germany) received the highest practical usage: it has become the standard indicator, and quantitatively expressed so, for cells in divisions, as the proliferation-associated (not necessarily causative) antigen (pKi-67). The antigen Ki-67 is detected by monoclonal antibodies directed to the mammalian (human) Ki-67 protein. Early studies in Borstel Forschungsinstitut Germany, and at the Universities of Perugia and Bologna, Italy [1728-1730] find that these human Ki-67-directed mcab fail to detect the Ki-67 antigen in all mammalian species, not to mention pre-mammalian vertebrates, but highly reliably detect mammalian tumor cells in division. The IgG1κ mouse anti-human Ki-67 mcab reacts in histochemical staining with the human Ki-67 protein structured from its cDNA.

The Brazilian report showing Ki-67 antigen in dividing cells of the sporocysts and cercariae of *Schistosoma mansoni* within its vector, the snail, *Biomphalaria glabrata* (while resting schistosoma cells and all snail cells were negative) stands out as unique. This report lists hosts, whose dividing cell nuclei showed the presence of Ki-67 protein. In addition to all mammalian cells, the list includes plants (tomato, *Lycopersiconesculentum*; potato, *Solanum tuberosum,* grapes, *Vitis vinifera*), fish (zebrafish, *Danio rerio;* salmon, *Salmo salar;* trout, *Onchorhynchus mykiss*), and frogs (*Xenopus laevis/tropicalis*) [1731].

Is the Ki-67 gene a hidden proto-oncogene, that served the first multicellular eukaryote in their "cell survival pathways", and remained physiologically active in the trophoblasts of healthy mammalian embryos, or in those of hydatidiform moles [1732]? Was it present in unicellular eukaryotes? If plant cells have the Ki-67 genes, did already the algae (*Ecklonia cava*) acquire it? Were the ancestors of these widely spread genes transmitted vertically or horizontally? While all dividing healthy cells in embryonic life switch-on and off Ki-67 [1733], malignantly transformed cells constitutively express it. Clinically tumor-free human female breast biopsies showing positive stain for Ki-67 protein in >2 % of the cells indicated transformation into cancer in 14 % of the cases in 10 years [1734]. The relationship of the Ki-67 protein with Bcl-2 and VEGF in breast cancer tissue sections is positive [1735, 1736]. However mammalian genomic locus of the Ki-67 gene and its epigenomic control remain poorly studied. Kidney carcinoma cells transfected with genetically engineered adenovirus expressing anti-Ki-67 siRNA, ceased proliferating and died with apoptosis [1737]. Does this mean that Ki-67 acted as an oncoprotein in the renal cancer cells? *)

*) *The E2 family ubiquitin-conjugating enzyme (UbcH10) is evolutionarily conserved in the human genome. It functions in degrading those mitotic checkpoint inhibitory cyclins that regulate spindle assembly and mitotic exit at the conclusion of the cell cycle (probably similar to HPV16's 45E7/INK4A proteins' neutralization of RB and p53). UbcH10 is subject to gain-of-function point mutation, yet unknown as to the nucleotide involved, and the amino acid switch (but probably similar to BRAF V600E). This results in its constitutive overexpression. The persistent destruction of physiological cell cycle inhibitors allows the excessive accumulation of cell cycle initiator cyclins, prominent among them Ki-67. Irreversible cell cycles manifest in tumor formation with Ki-67 accumulating to high levels in the nuclei. Lentivirally vectored anti-UbcH10 smallhairpin iRNA aligning with the UbcH10 mRNA inhibits its translation to proteins in the ribosomes. Thus, consequentially to depleted UbcH10 and Ki-67, cell proliferations cease and tumors regress. The receding tumors also lose their former pro-migratory MMP production (Wang S-T et al Oncology Letters 2015;9:2171-6). The basic elementary cell nuclear antigen, Ki-67, used in the process of normal cell division, is quantitatively and constitutively dependent on the ubiquitin-conjugating enzyme UbcH10. The evolutionary antecedents of the E2 family ubiquitin enzymes, known as the ubiquitin-proteasome pathway, include the basic virulence pathways of a long list of low-branching unicellular eukaryotes, now functional as saprophytes or parasites (Muñoz C et al Biomed Res Int 2015:141526; Pereira RV et al Parasit Vectors 2015;8:329). A cell survival pathway of ancestral unicellular eukaryotes becomes an oncogene/oncoprotein in Homo.*

mtDNA. The posttranslational arginine protein methylation by R-CH3-transferases (PRMT) modifies proteins in the cytoplasm and remodels chromatin and DNA repair enzymes in the nucleus. A third class of PRMTs is evolutionarily conserved and remained active in the mitochondria, most prominently so in the Trypanosoma [1738]. This early branching kinetoplastid easily adapts from insect tissues to the mammalian bloodstream and readily modifies its ATPase complex accordingly (synthesizing ATP in the insect, and hydrolyzing ATP in the mammalian blood stream). In the mammalian host, mitochondria are engaged in ATP production (bioenergy supply). In esophageal squamous carcinoma cells, mtDNA copy numbers increase during the cells' epithelial to mesenchymal transition. Laryngeal squamous carcinoma cells produce D-loop mutations in their mtDNA. Mutated D-loop mtDNA copy numbers increase not only in the tumor cells, but also in the tumor-associated paracancerous normal cells [1739, 1740]. Single nucleotide polymorphism of the D-loop region in the mtDNA is a sign of lung cancer risk. Specifically, minor alleles of the nucleotides 151C/T, 200A/G, 524C/CA and 1627G/A indicated risk for squamous cell carcinoma [1741].

Comment. The mtDNA, supporting the trypanosoma in its normal life style, is enlisted by malignant tumors for source of extra energy.

Achaete-scute. These basic helix-loop-helix (bHLH) proteins appear in primordial nerve cells (chaeta, chaetē, long filament, or bristle; scute, scale or horny plate; dipteran bristles, lepidopteran wing scales, *Drosophila melanogaster*, *Bombyx mori*, mutant scale-less). These proteins bind the E box DNA sequences (with the consensus sequence CANNTG). The Florida amphioxus (*B. floridae*) operates the basics of a separate central and a peripheral nervous system (*vide supra*). The epidermal sensory neuronal (ESN) cells utilize the Delta ligand Notch receptor signaling. The Delta ligand gene is co-expressed with the achaete-scute homolog gene, *ash*. Both the development and positioning of the dorsal neural tube (notochord) and generation and placement of the progenitor cells of the ENS (epidermal sensory neurons) depend on BMP signaling. These cells express bHLH proteins and Ash. The amphioxus ESN cells are predecessors of the vertebrate placode-derived neurons [1742] (Table XXIX).

Even though the hydra (*H. vulgaris*) possesses only one gene for the bHLH domain of the achaete-scute family, the gene product protein, cnidarians' Ash, CnASH, is highly homologous for aa identity with the bHLH proteins of the drosophila and vertebrates. The conservation of the cnidarians Ash protein up to the drosophila means approx 600 my distance in evolutionary time. The CnASH protein readily dimerizes (forming heterodimers) with the drosophila achaete-scute binding partner Da (daughterless gene-product protein). The thin space between the basal metazoan hydra's two epithelial layers, the ectoderm and endoderm, is filled out by the basement meso-membrane. Cells of the body column (nematocytes) express CnASH, but not the cells of the head or the foot. As cells are continuously lost, these are replaced from stem cells [1743]. The remarkable regenerative capacity in the hydra is stimulated by the mitotic signal caspases that apoptotic cells emit [1744]. The meiotic synaptonemal complex structure (SC proteins, SYCP) of the hydra have been evolutionarily conserved up to vertebrate mammals: SC arose only once in evolution and formed monophyletic groups of orthologous proteins in all metazoans [1745].

In the hydroid, Hydractinia, the pluripotent stem cells are endowed with the classical stem cell genes, which are conserved up to the vertebrate mammalian hosts: Myc, Nanos, Oct4, Piwi. The upregulated Oct4-like transcription factor polynem (Pln) transforms hydractinia epithelial cells into tumors growing in the form of Frizzled-coupled Wnt receptor-expressing neoplasms [1746]. The cnidaria sea anemone (*Nematostella vectensis*) expresses eight neurogenic genes (four Ash genes, NvashA-D) in its embryonic (gastrula, larva and polyp) life and will have in its adult life both endodermal and ectodermal nests of nerve cells [1711]. The starlet sea anemone precedes the bilaterians in possessing a primordial p53 gene, and its sister, p63, which protect the progenies of the host from UV light-induced germ cell mutations via apoptosis induction in the afflicted germ cells, but not in the somatic cells [1747]. (If you are the lab technician: The experimental *N. vectensis* polyps are to be kept in artificial sea water and are to be fed daily with brine shrimp morsels (*Artemia salina*), freshly hatching from its eggs).

The annelid, *Platynereis dumerilii* (Figure 43) of the phylum Annelida (earthworms, ragworms, leeches) expresses achaete-scute and neurogenin genes, neurogenin being the dominant proneural gene for the formation of the trunk nervous system. The achaete-scute genes are involved in the development of the serotonergic neural cells, closer to those of the vertebrates, than to those of the drosophila

[1748]. The lepidoptera *Bombyx mori* expresses four ASH homologs in its larval tissues and in the wing/wing discs [1749].

Drosophila yielded in its large thoracic bristles (macrochaete) the first discovered achaete-scute genes and their HLH proteins. The zink finger proteins of the pannier gene antagonize the achaete-scute proteins. The drosophila pannier is homologous with the vertebrate transcription factor GATA-1 (guanine adenine thymine adenine) [1750]. The four drosophila achaete-scute genes form a complex (AS-C). The AS-C encodes the differentiating neuroectoderm in the gastrula, including the sensory mother cells. The drosophila Hox (homeobox) family and achaete-scute family proteins show high aa homology with the appropriate vertebrate proteins, since they have been conserved from the beginning time of metazoan evolution [1743].

Human primitive neuroectodermal tumors (PNETs) originate frequently in a bone (see in Introduction), show the Ewing's sarcoma gene translocation (22q12), express the achaete-scute gene-product proteins (mammalian achaete-scute-like homolog, ASH, ASCL, Mash), and have the bad reputation to be combination chemotherapy-resistant with frequent relapses.

Some poorly differentiated human sinonasal neoplasms (neuroblastoma, neuroectodermal tumors) express the ancient achaete-scute genes/gene-product proteins. This finding was interpreted as a "nonspecific marker for diagnosis" [1751, 1752]. Esthesioneuroblastoma of the nasal cavity and the paranasal sinuses express upregulated Mash1 [1753, 1754]. The Mash proteins may be the driving force of these tumors.

In the human lung, the achaete scute genes are essential for the migration and function of peripherally functioning neuroectodermal cells. Small cell carcinoma of the lung, or non-small cell carcinomas of the lung with neuroectodermal features express highly upregulated Mash proteins. Lung cancers of non-smokers display these features. Nonsmall cell or small cell lung cancers recur after remission induction; these tumors often undergo "neuroectodermal differentiation" and return in the form of this different histopathology. These tumors express highly upregulated HLH Mash proteins, the products of the achaete-scute complex homolog. The aldehyde dehydrogenase (ALDH) gene of lung cancer cells is the target for upregulation by the Mash proteins. Human lung cancer cells overexpressing $CD133^{hi}$, $ALDH^{hi}$ and $ASCL^{hi}$ are the most tumorigenic in nu/nu mouse xenografts. These tumor cells, after knockdown of ASCL1 with lentiviral vector expressing the specific shRNA, lost their capacity to induce tumors in xenografts [1755]. Some colorectal adenocarcinomas express significantly upregulated bHLH/ASCL1 proteins; Wnt signaling targeted the ASCL1 complex. Tumor cells with shRNA interference-downregulated ASCL1, lost all malignant characteristics: stopped proliferating and migrating, formed no colonies, and failed to grow in xenografts. All stem cell genes, CD133, Sox2, Oct4, BMi1, c-Myc and Lgr5 (leucine-rich repeat containing G protein-coupled receptor-5) became inactivated. In microRNA assays, miRNA-302b proved to be responsible for the induction of stemness characteristics [1756].

In inducing neuroectodermal cancer cells, the ASH1 protein repressed the Wnt-antagonist tumor suppressor dickkopf (DKK1/3) genes. DKK inhibits Wnt/β-catenin signaling by disrupting attachment of Wnt to its receptors, LRP (low density lipoprotein receptor-related protein) and Frizzled. The DKK genes undergo ASH1-mediated acetylation/deacetylation and inhibitory trimethylation of lysine 27 of histone 3 (H3K27me3) in the promoter region of DKK1 and E-cadherin; E-cadherin and integrinβ1 are natural inhibitors of the Wnt signaling pathway. The neuroendocrine cell marker synaptophysin rises in neuroendocrine transformation. DKK1 being a polycomb target gene, the polycomb repressive complex 2 (PRC2) is another inducer of the H3K27m3 intervention. Colorectal cancer cells silence the DIKK1 gene by DNA methylation. Lung cancer cells acetylate/deacetylate the DKK1 gene, and methylate the histones in their gene promoter. DKK1 is more often silenced in SCLC cells, than in NSCLC cells. The ASH1 protein expands the population of these histone-repressed cells. Lung cancer cells in the process of EMT, suppress E-cadherin by gene-product proteins SNAI (Snail) and bHLH protein TWIST [1757].

Mauriceville plasmid, the retroviral progenitor (Figure 2). The mitochondria of the filamentous ascomycete, *Neurospora crassa* (Figure 63) carry the circular DNA Mauriceville plasmid (Table XIII). the plasmid contains a reverse transcriptase (RT) sequence encoding a 81-kD protein. This gene resembles the bacterial retrons or group II introns. Another sequence within the plasmid resembles a group I intron. A mt RNA polymerase transcribes the plasmid; the 3.6-kb transcripts are reverse transcribed in RNP particles to yield a minus strand cDNA. The full length transcripts resemble tRNA sequences of some plant viruses (brome mosaic virus). The plasmid linear RNAs serve as mRNAs for encoding the RT protein. The plasmid RNA template could be degraded by a mitochondrial mtRNAase. The Maur-RT recognizes the 3′ end sequence of the tRNA and synthesizes de novo or primer-mediated cDNA. The RT may use template switching and thus produce hybrid cDNAs (incorporating mt tRNA copies) [1758a, b, c, d]. The mitochondria of the *Trichoderma harzanium* (the pathogen of the green mold disease in mushroom farms in Croatia and Hungary) yielded the plasmid pThr1, whose sequences were different from those of its mitochondrial genome, but its ORF for its RT was closely related to that of other retroplasmids (Mauriceville, Varkud, pFOXC2/3 of *Fusarium oxysporum*) [1759]. The linear minus strand cDNA of the pFOXC mitochondrial retroplasmids replicates autonomously. The pentameric terminal repeats of 5′-ATCTA replicated by RT resemble the ribonucleoprotein complex telomeres. The pFOXC plasmid uses base-paired or RNA primers (those are not needed for the Maur-RT/TERTs). Ancestrally, the Maur-RT is related to RNA-dependent-RNA polymerases, structurally related to RT. Materials and Methods are given in great details for mtRNA/mtDNA isolation; endogenous reverse transcription reaction and cDNA analysis; RT activity on templates; Southern hybridization; synthesis of RNA templates; and oligonucleotide purification from high percentage polyacrylamide gels [1760].

The telomeric chromosomal ends of plant cells (TTTAGGG) slightly differ from that of mammalian and filamentous fungal cells (TTAGGG). *Neurospora crassa* lacks the subtelomeric thermolabile hemolysin (TLH) gene of other filamentous fungi, and the short tandem repeats in its telomere-associated region, but without any detectable functional defects. *N. crassa* also encodes a number of cell fusion-inducing (hyphal anastomosis) genes, in order to unite the protoperitecium (immature female mating structure) with the long specialized hypha, trichogyne (Figure 63) (in the Appendix 1). The nuclear genome functions independently from the mitochondrial genome. The importance of the Mauriceville (and related) plasmids is that they may be the representatives of the retroviral progenitors [1758b].

Endogenous retroviruses are not inert. The selfish DNA, the "ultimate parasite", may undergo "domestication", whereby it gains an active LTR/ORF and encodes proteins for the improved/altered metabolism of its host. Bushman provided a tabulated list of such genes (Bushman) [1]. The original retroviral *gag* genes (group antigens) encoded matrix, capsid and nucleocapsid proteins. The *gag* genes of the Ty3/gypsy LTR retrotransposons distributed themselves in the genomes of fish, reptilians and insects, but the mammalian genomes expunged them. Other *gag* gene-derivatives persist in the mammalian genomes. The MART proteins (melanoma antigen recognized by T cells) are encoded from the X chromosomes and are expressed in the paternal alleles. Physiologically the MART proteins are involved in the formation of the placenta and the mechanism of apoptosis. The Pnma ancient *gag* gene-derived proteins (paraneoplastic Ma antigens) induce autoimmune reactions. The SCAN family proteins (*vide infra*) serve as transcription factors in hematopoiesis.

Suspect retroviral derivation is claimed for the ribonucleoprotein telomerase (a reverse transcriptase). There are DNA-dependent polymerases directing the pathway of DNA → RNA → protein, and RNA-dependent RNA- polymerases to replicate RNA viral genomes. The reverse transcriptases (RT) use RNA templates to polymerase DNA. The telomere DNA sequence covers the chromosomal ends; it is resynthesized after each cell division. The RT telomerase (TERT) provides the RNA template within it in the synthesis of DNA for the capping of the chromosomal ends.

The RTs have an ancient retrotransposon-retroviral origin (Belfort M et al [1761–1763]). The Ty1-copia retrotransposons appeared in the early eukaryotic genomes and were widely spread by horizontal transmissions. These routes of spread are best documented in angiosperm plants. Evolutionarily unrelated species, *Solanum melongena* and *Phytophthora infestans* (Figure 56; Table XXX) possess closely related such retrotransposons [1764]. The extensively laboratory-propagated Lister 427b clone of *Trypanosoma brucei* telomeres replicate indefinitely. It is the highly mutable subtelomeric chromosomal region that encodes the highly variable cell surface glycoproteins (VSG); an inherently unstable DNA sequence provides the rapid changes, that outrun the speed of the host's immune reactivity [416a, b–417a, b, 1765] (imitating a malignantly transformed higher eukaryocyte of a multicellular host: a "cancer cell").

The drosophila telomeres derive from two different non-LTR retrotransposons [1766]. The arabidopsis telomerase has undergone a recent (in evolutionary terms) duplication and exhibits extreme nucleotide sequence divergence in the template domains across *Brassicaceae* [1767]. The human telomeres in the somatic cells are not regularly re-capped after cell divisions, allowing senescence to set in (after the

Hayflick rule is fulfilled). The telomerase TERT protein is encoded through the mRNA-ribosome pattern, thus it may go through alternative splicing. In this process, TGFβ may degrade, while c-Myc may stabilize TERT, and posttranslational phosphorylations of the TERT protein regulate its activity. Other regulators of TERT are Hsp90, IGF-1, NFκB and protein kinase C (PKC). The c-Abl protein phosphorylates TERT on tyrosine residues and thus decreases its activity. Protein phosphatase2A (PP2A) is another TERT inhibitor ("negative controller"). The oncogene Akt (protein kinase B) is a TERT activator in the PI3K/Akt/mTOR pathway, but PP2A dephosphorylates Akt kinase on its serine473 and threonine308 residues, and inactivates it (and the PP2A inhibitor okadaic acid abolishes this effect). Signaling proteins, TNFα, and NFκB, are instrumental in the translocation of TERT from cytoplasm to the nucleus. The nucleolus protein nucleolin complexes with TERT and directs it to the nucleoplasm, where the TERT substrates are. The telomerase RNA is transferred into the TERT by the Cajal bodies, the omnipresent intranuclear membrane-less organelles. In mid-S phase, the Cajal bodies direct TERT to the telomeres. Figure 1 in the cited article is a concise depiction of the human TERT complex. The proteins shelterin exert control over the synthesis of telomeric DNA by TERT. Negative feedback regulation on telomere length is exercised, among others, by tankyrases 1/2 (*vide supra*) [1768].

The retroviral reverse transcriptases were present in the first living cells on Earth, the archaea and prokaryota (eubacteria). A retron gene (*ret*) is postulated for the origin of the complex consisting of RNA, multi-copy ssDNA and the enzyme reverse transcriptase. The complex was operational in archaea and then in eubacteria, both entities presumably acquiring it through vertical routes from their common ancestor [1769]. The genome sequences of the mesophilic thermotogales bacterium *Mesotoga prima* include mobile genetic elements, 19 copies of two different group II intron-carrying reverse transcriptases, and other related insertion sequences. Many horizontally acquired genes operate in the *M. prima* genome. These genes derive from Clostridia and from archaea. There is no mention of inserted phage genes [1770].

In contrast, hyperthermophilic archaeal Fuselloviridae (*vide supra*) insert their genomes into the genome of their host cells. The crenarchaeota Sulfolobus spindle-shaped viruses (SSV1/2) have circular dsDNA genomes. This virus (SSV1) maintains its genome integrated into its host's tRNA gene; the SSV2 gene is integrated into the the glycyl tRNA gene of the host cell's genome. These are not lytic, but temperate phages, whose genomes are activated by UV light exposure. That was the basic procedure for activating latent phages, as practiced by André Lwoff in the 1950s. At the Pasteur Institut, Paris, research fellow, Sándor Koch, introduced this author to André Lwoff, who explained the mechanisms of the process, as UV-sensitive suppressor genes release their hold on UV-resistant phage-replicatory genes, implying that the ancient phage replicatory genes were UV light-resistant, while more recently installed newer bacterial host genes, the suppressor genes, were UV-light-sensitive (S. Koch and A. Lwoff, personal communication, year 1963).

It is not the classical reverse transcriptase that integrates these viral genes (if DNA and not RNA genes). Virally encoded integrase functioning as an endonuclease and ligase carries out the insertion. At insertion, the host cell genome's tRNA *attA* sequence pairs with the viral integrase gene *attP* site [1771]. The ΨCh1 virus-isolate with an RNA/DNA genome integrates itself into the genome of its host *Natrialba magadii*. The circular ds DNA genome of the Euryarchaeota *Pyrococcus abyssi* virus (PAV) does not integrate itself into its host cell's genome [1772]. Another Euryarchaeota virus isolated from *Thermococcus prieurii* (TPV1) also harbors circular dsDNA genome apparently not integrated [1773]. Reverse transcriptase-mediated genomic insertions are the privilege of retrotransposons/retroviruses.

The intronless *gag* gene of some ancient mammalian retroviruses occupy the human X chromosome. These genes are the Mart1/2 or PEG10/11 (melanoma antigen recognized by T cells/lymphocytes; paternally expressed imprinted gene). Other *gag*-derived genes on the X chromosome are the *ma* genes (paraneoplastic Ma antigens, Pnma), which are involved in apoptosis regulation. The vertebrate-specific *gag*-derived three SCAN zink finger proteins are encoded from human chromosome 6p21.3. The SCAN genes span 1750 kb. The SCAN box encodes four proteins: **S**RE-ZBP, **C**Tfin-51, **A**W-1, Z**N**F-165 (serine, arginine, glutamic acid zink finger binding protein; cysteine, threonine; alanine, W for tryptophan), that specifically bind intranuclear DNA in those stem cells, which are engaged in hematopoiesis [1774–1777]. The human endogenous retroviral (HERV) elements are reverse transcriptase-transposed proviral DNA segments of the *gag* (group-associated antigen), *pol* (polymerase) and *env* (envelope) genes, or fragments thereof, inserted into the human genome usually several million years ago. HERV constitute about 4.8 % of the human genome. Long terminal repeats (LTR) containing *gag* and *pol* segments may regain the faculties of a vertically spreading retrovirus when they acquire an *env* gene segment. HERV particles most frequently recover in tumor tissues as "passengers", which may resume some pathological functions, such as cell fusions or induction of autoimmune host responses.

Several attempts (dominantly theoretical and literature-based) at re-enacting ancient events of retrovirology were made on the annual Herman Friedman & Steven Specter Clinical Virology Symposia of the Department of Medical Microbiology, The University of South Florida College of Medicine and the PanAmerican Society of Clinical Virology at Clearwater Beach, FL, and at the 92nd Annual Meeting of the American Association of Cancer Research in 2001 in New Orleans [895a, b, 1778a, b]. Eventually, during millennia of coexistence in the same hosts in Africa, a collusion for mutual advantages evolved between herpes and retro (lenti)viruses. These viruses co-exist without viral interference and induce mutually advantageous cyto- and lymphokines and microRNAs in their hosts [895ac]. Herpesviruses frequently activate latent retroviruses; reviewed in [261b].

Viral infections may induce an outpour of $CD4^+$ Treg cells, which outnumber $CD8^+$ immune T cells. Many years before the discovery of antiviral immune T cells and their antagonists, the Treg cells, and that the murine genome may express a mouse mammary tumor virus-related superantigens (sag), it was observed that lymphocytic choriomeningitis virus (LCMV) could cause "immune paralysis" in mice. These mice had high viral titers with pathological reactions in their lymph nodes. Despite no virus neutralizing antibody production, these mice experienced a prolonged survival [1779]. Now it is proven that certain attenuated LCMV strains (Armstrong clone 13) induced a forceful TCR V$\beta 5^+$ FoxP3^+ Treg cell reaction in the spleens of C57BL/6 mice. These CD4^+ T cells also overexpressed CD101/103, CD134, CD 137, ICOS (inducible T cell costimulator), and PD-1 (programmed death). These reactive Treg cells appeared to have derived from an expanding pre-existing pool and exerted no specific reactions to LCMV antigens, especially not to peptide GP66-77-MHC class II tetramer specific to an immunodominant LCMV epitope. Since LCMV does not express a *sag*, it was found that an endogenous murine *sag* induced the rise of this Treg cell population. The endogenous murine *sag* is encoded by the inserted mouse mammary tumor virus genome, Mtv8/9Sag. The fetal thymus failed to eliminate Sag-autoreactive Treg cells, which respond to some chronic infections. Mice genetically deficient in B cell production responded, but mice genetically deficient in DC cells failed to respond with Treg cell generation to infection with the clone 13 LCMV strain. The variables in the superantigen-induced Treg cell reactions are the state of the mRNAs derived from the provirus Mtv and the levels of expression of the MCV class II on the DC membranes. It is the presentation of the Mtv antigen by the DCs that induces the process, in response the FoxP3+ Treg cells express their costimulatory molecules [1780]. The endogenous human retroviral provirus HERV-18 induces a similar Treg cell response to EBV infection [1781].

Comment. This author asks, if arising tumors could activate dormant superantigen genes, and thus induce an outpour of Treg cells in the tumor-bearing hosts? Further, "proviral" DNA transcribed from retrotransposon-retroviral RNA may encode the mutator phenotype, both in uni- and multicellular hosts (*vide infra*).

PiwiRNA. The piwi/argonaute proteins (*vide supra*) are morphogenic determinants within sea urchin egg cells/oocytes. The outer layer (cortex) of unfertilized egg cells/oocytes is a storage compartment for mRNAs, (the seawi mRNA, for Seawi proteins, *vide infra*) ribosomal proteins, and granular assemblages of ribonucleoproteins. Studied in sea urchins, morphogenic determinants orient the embryo's animal-vegetal axis. At the animal pole, Sox and Ets proteins assemble to form ectoderm. At the vegetal pole, cells with intranuclear β-catenin (activating the genes of Tcf/Lef transcriptional regulators) will form the mesodermal structures including the coeloma. In between the two poles, endodermal cells are mixed with ecto- and mesodermal cells. These cells yield the microtubule MT-RNP complex consisting of two plasmids with 2605-bp and 1722-bp inserts and the 100-kD piwi/argonaute protein. Sea urchin ribosomes isolated by sucrose gradients, contained the 100-kD Seawi protein. The Seawi protein binds bep4 mRNA (bep4, butanol-extracted protein; bep1, blastula embryonic protein). The Bep4 protein, if produced, retains β-catenin in the cytoplasm and averts the intranuclear activation of the *tcf/lef* genes (T cell factor; lymphoid enhancer factor) [1782]. The only one *tcf/lef* genes of invertebrates (including *Hydra magnipapillata*), have become four distinct *tcf*1-4/*lef*1-4 genes in vertebrates, acting as proto-oncogenes within the Wnt/β-catenin pathway [1783], and are antagonized by the p53 family proteins [1784]. The TCF/LEF high mobility box proteins bind the Wnt-responsive elements, WRE; WWCAAAG (nucleotide symbol W stands for A or T) on their target genes. (In the classroom, it may pose some disturbance that amino acids are also symbolized with capital letters: A, alanine; C, cysteine; G, glycine; R, arginine; W, tryptophan).

NEMO-like kinase (NFκB essential modulator) activates the TCF/LEF proteins by phosphorylation. The homeodomain interacting protein kinase (HIPK2) activates the p53 protein by phosphorylation for suppression of the anti-apoptotic factor galectin [1785]. In recapitulation, the destruction complex consisting of APC protein, axin and GSK-3β protects the cell by phosphorylating, ubiquitylating and thus destroying β-catenin. In the Wnt pathway, the Wnt ligand induces a cascade leading to the sequestration of axin out of the destruction complex. The WNT ligand is captured by its receptors Frizzled and LRP5/8. LRP and Dishevelled sequestrate the axin protein; consequentially β-catenin accumulates and eventually transfers into the nucleus in order to activate genes Tcf/Lef, and thus begin transcriptions for incessant cell proliferation. GTPases Rac and Rho (the *ras* pathway) may be co-activated. Cells with overexpressed D vitamin receptors bind and immobilize β-catenin, thus preventing its transfer from cell membrane to cytoplasm and from there to nucleus.

Of the PIWI proteins of the platyhelminth, *Schmidtea mediterranea*, SMEDWI-3 shows 34 % identity and 54 % similarity with the human HIWI protein, while SMEDWI-1/2 are distinctly planarian proteins. The distinctly planarian SMEDWI-1/2 proteins (but not SMEDWI-1) process their host's transposable RNA and iRNA elements in the ping pong model of biogenesis [1786]. RNA polymerase II (or from Alu repeat elements, pol III) transcribe the primary microRNAs. The intranuclear Drosha/Pasha enzymes process these into precursor miRNAs. The exportin-moved pre-miRNAs are developed into ds miRNAs by the cytoplasmic enzymes Dicer/Loquacious. Within the RNA-induced silencing complex RISC, the argonaute proteins under the guidance of the mature microRNA either translate, or degrade, the targeted mRNA. In the stem cells, miRNAs can up- or downregulate stem cell genes. The miR-290-295 cluster maintains pluripotency by upregulating gene *lin*-28. In reciprocity, in mesenchymal stem cells, miR-335 expression upholds the canonical Wnt signaling pathway, but IFNγ suppresses it. The three basic elements of the piRNA/piwiRNA pathway (Argonaute, Aubergine, PIWI), are conserved from the platyhelminth *Schmidtea mediterranea* through drosophila and Mus to Homo. The expression of miRNA let-7a is low in mammalian embryonic stem cells, and rises with differentiation. Embryonic stem (ES) cells express pre-let7 miRNA, but differentiated cells express mature let7. The suppressor of the processing of let7 miRNA, lin-28, was upregulated in mammalian ES cells, but it was highly expressed in differentiated planarian cells, as if a reversal of function occurred during the co-evolution of lin-28 and miR-let7a (*vide supra*). The transposable element-reactive small RNAs (24–32 nt) in the cnidaria (Nematostella, Hydra) and drosophila are referred to as rasiRNAs (repeat associated small interfering). These RNAs escape Dicer-processing, and are processed exclusively by PIWI proteins, thus acquiring the designation piwi/piRNAs. PIWI proteins possess PAZ (PIWI, Argonaute, Zwille) and endonuclease domains. In the drosophila germ tissues, without PIWI proteins, there is no oogenesis or spermatogenesis. In the germ tissues operate sense piRNAs and antisense piRNAs. The argonaute protein (AGO3) binds and cleaves a sense piRNA and releases an antisense piRNA. In preference, AGO3 interacts with sense RNA strands, while AUG/PIWI pick the antisense strands. Partial complementarity of the 10 nts at the small RNAs 5′ ends play ping pong with the interacting enzymes. The antisense piRNA unites AUB and basepairs with an active transposable element, thus generating a secondary sense piRNA. The piRNAs map to and silence transposable elements (and non-transposon mRNAs). The mammalian (mouse) piRNAs are most active (binding and silencing mRNAs) in the meiotic process of spermatogenesis. There, MIWI 1/2 maintain the germline stem cells [1786].

The genomics/proteomics of the hyperthermophil annelid, *Alvinella pompejana,* withstands 50 °C under high pressure in the environment of hydrothermal vents [1787]. The degradation of the inserted alien transposons → proviral DNAs by piRNA/Piwi/Argonaute protein complexes, or their modifications (by arginine methylation), or the introduction of adequence-specific modification in chromatin domains are explosively unfolding [1788–1791]. These events, delicate as they are, must have withstood most adverse physico-chemical conditions. The primordial (probably pre-cellular) RNA/ribozyme, DNA, and Ghadiri polypeptide interactions must have withstood temperatures just below 100 °C. The RNA, "life's indispensable molecule" (James Darnell, see in Introduction), continues to exercise its ancient precellular priority and superiority over DNA, now within the cell, as well.

Choanoflagellates #5. The p21-activated serine/threonine kinases (PAK) appeared first in choanoflagellates: MoPak (*Monosiga ovata*), and continued their existence in fungi and multicellular sponges (*Ephydatia fluviatilis*). Mammalian cells transfected with MoPak constitutively express the oncoprotein and undergo malignant transformation. The Pak genes may install structural alterations (resembling loss-of-function mutation) and thus encode an autoinhibitory domain; the mutational loss of this domain results in enhanced cell-transforming activity of the MoPak gene. A relationship of Pak kinases with Rho (Ras homolog) family GTPases is conserved from uni- to multicellular hosts. Further, oncogenic Pak-induced cell transformation required association with Rac/Cdc42 (Ras-related C3 botulinum toxin substrate; cell division control) [1792a, b]. *Acanthamoeba castellani*, *D. discoideum* and *Entamoeba histolytica* all operate PAK genes and gene-product proteins endowed with p21-binding N terminal and catalytic C terminal domains, which are closely related. The DdPAK-A is regulated by the PI3K/Akt (protein kinase B) pathway. In multicellular mammalian eukaryotes this is an oncogenic pathway. DdPak-B encodes the myosin I heavy chain kinase (MIHCK) in dependence of the Rac/Cdc42 genes/gene-product proteins. Phosphorylated MIHCK promotes cell migration and phagocytosis. The DdPak-C is also regulated by PI3K; its activated Rac-GTPase-racB further activates GTPase-binding Cdc42/Rac domains, which establish cell polarization and pseudopodia formation. The yeast *S. cerevisiae* operates three Pak enzymes; they are activated by Cdc28/42. Enzyme Ste20 (sterile) activates MAPK for the mating pathway (in higher multicellular somatic cells the MAPK pathway drives transformed cells). The fission yeast *Schizosaccharomyces pombe* operates two Pak kinases: Pak1/2. Pak1 (also called Shk1) is regulated by the Src-related SH3 domain. Pak2 (also called Shk2) is regulated by the Ras1/Cdc32 complex (in higher eukaryotic cells, *src* and *ras* become proto-oncogenes). The drosophila genome encodes three PAk proteins, but only DPAK1 (related to Mbt, mushroom bodies tiny) has been well studied. It is insulin-like receptor-related and GTP-related, Rac-regulated. The drosophila photoreceptor Mbt is in the hippocampus region of its brain and can be constitutively activated [1793]. However, the lethal malignant brain tumor gene of the drosophila is also abbreviated D-l(3)mbt. It is considered to be a tumor suppressor gene; its loss-of-function mutation leads to malignant transformation of neuroblasts in the larval drosophila brain. The structure of this gene shows features that of a polycomb gene, therefore it has homologs in the vertebrate mammalian genomes. The human homolog of this gene is on chromosome 20q12; it is referred to as H-I(3)mbt. When the H-I(3)mbt gene-product protein is overexpressed, chromosomes fail to properly segregate during mitosis and the cell develops multinucleation due to endomitosis [1794]. In the caenorhabditis, the three Pak isoforms are working under Rac GTPases in regulating locomotion by motor neuron control. In the Xenopus, the four XPak1-4 genes were studied in detail. One major function of PAKs is to guide oocytes through meiosis. The complex process involves MAPK activated by gene c-*mos*; in mammalian cells c-*mos* is the proto-oncogene of the Moloney mouse leukemia/sarcoma retrovirus (c-*mos* → v-*mos*, *vide infra*). In the human genome, six *pak* genes were so far counted. Pak1-3 form group I and Pak4-6 form group II. Similarly to all other Paks, the human Pak protein isoforms contain the N-terminal regulatory, and the C-terminal conserved domains. Group I Paks contain GTPase-binding and autoinhibitory domains. Autophosphorylation of dimeric Group I Paks occurs only after the uncoupling of the inhibitory domain. Activation is GTPase-, Cdc42- and Rac-dependent. Group I Paks regulate cytoskeletal dynamics. Intranuclearly translocated Group I Paks may activate the forkhead trasnscription factor (FKHR), the estrogen receptor, and the gene *snail* (SNAI1). Group I Paks are anti-apoptotic activators (phosphorylators) of Bad, the MAPK pathway, NFκ-B, and the Jun N-terminal kinase. The monomeric Group II Pak4 are often found constitutively activated; Pak4 moves to the Golgi apparatus, transforms fibroblasts, and phosphorylates Bad to inhibit apoptosis. Pak5 is localized in brain cells, activates mitochondria and JNK. Pak6 is involved with hormone receptors (ER; AR) [1793].

The choanoflagellate, *M. brevicollis* possesses within its most elaborate receptor tyrosine kinase network Src/Abl kinases activating their substrate, the STAT "cell survival pathway". Combinations analogous with RasGEF-SH2 and RhoGAP-SH2 (*ras*-guanine nucleotide exchange factor; *ras*-homolog guanine triphosphatase activating domain) take place. At that level, for physiological functions, the Src kinase and its negative regulator, the C-terminal kinase Csk-mediated regulatory system, was already established. Monosiga cells mobilize PI3K and MAPK pathways (for the latter, very likely they possess Raf kinases). Further, choanoflagellates even though they are unicellular protists, they already express cadherins and lectins that will mediate intercellular interactions in multicellular organisms. Accordingly, Monosiga cells can form colonies. Fibroblasts from vertebrates transfected with the Scr ortholog from *M. ovata* undergo malignant transformation [1795, 1796]. The ancestors of the scaffolding proteins MAGUK (membrane-associated guanylate kinase) of cell communications in multicellular organisms (Metazoa) appeared first in the protist Monosiga. The MAGUK proteins contain, among other domains, the PDZ postsynaptic density protein drosophila disc large zonula occludens [1797]. Membrane guanylate kinases (MAGUK) may initiate malignant transformation of human epithelial cells, or suppress the process [1798]. The high risk HPV16/18 E6 oncoprotein expresses PDZ-binding domain. This unison is responsible for the Wnt signaling pathway so induced. The result is the intranuclear translocation of β-catenin [1799]. The c-Src tyrosine kinase expresses a ligand for the PDZ protein. When the C-terminal aa sequence Gly-Glu-Asn-Leu binds PDZ in human breast epithelial cells, the cells rest. When the ligand is mutated Leu/Ala, it induces locomotion and migration of the cells [1800, 1801]. The substrate of protein phosphotyrosine phosphatase H1 is the p38γ MAPK. The Ras effector MAPK acts independently from phosphorylation, but requires PDZ binding in order to induce colon cancer cell transformation. The PDZ-mediated p38 γPTPH1 (protein-tyrosine phosphatase) complex mediates Ras-dependent carcinogenesis [1802]. The human proteome contains over 250 proteins with PDZ domains (some with multiple PDZ-binding domains). Of these large number of cytoskeletal, cell polarity, and signal transduction pathways, some of these proteins mediate cell proliferation and act as oncoproteins [1803]. The physiological scaffold proteins of the choanoflagellates return in vertebrate mammalian cells as oncoproteins.

The *mos* proto-oncogene (*vide supra*) in the ascidian *Phallusia nigra* acts within the activated Mos/MEK/ERK cascade (mitogen-activated protein kinase; extracellular signal regulated) and with downregulated Cdk1 (cyclin-dependent kinase). This cascade creates the milieu favorable for Ca^{++} oscillation in meiosis of the fertilized eggs [1804]. The Mos protein kinase in urochordates activates MAPK and mediates the maturation of oocytes for meiosis; it regulates both cell cycle and asymmetric cell divisions. The pathway is active in cnidaria, echinodermata, and jellyfish (*Clytia hemisphaerica*). In contrast to somatic cells, in which the process of mitotic exit prohibits an immediate second cell division due to the destruction of cyclin B and the setting in the interphase, in the oocyte the meiosis I is followed immediately by meiosis II. In urochordate eggs Mos/MAPK cancel the interphase and generate supernumerary meioses. Inactivation of the Mos/MAPK pathway is necessary for the meiotic cycles to stop [1805a, b]. The cell division-driver mos/Mos act as an ancestral proto-oncogene in the urochordates.

In the mouse the Moloney retrovirus is fully mature, enveloped, and induces lymphocytic leukemia. Deprived of its envelope, the Moloney retrovirus induces soft tissue sarcomas, the most dominant of these tumors being the rhabdomyosarcoma [1806a, b, 1807a, b]. In tissue cultures, murine leukemogenic viruses lose their leukemogenic potency; is this due to loss of *env* gene expression? However, they remain immunogenic and mature into budding retrovirus particles; further, they may transform cultured cells, which grow out as sarcomas in mice [760, 1808a, b]. The *mos* gene in the human genome is mapped at 8q11, and low levels of the Mos protein is expressed in human cells [1809]. This author and his associates tried to induce human sarcomagenic retroviruses by treating human sarcoma cell cultures with the Moloney leukemogenic retrovirus (received from John D. Moloney as a gift), and with sera and tissue extracts of human leukemic specimens in expectation that an occult truncated human sarcomagenic retrovirus would be rescued by acquiring a Moloney virus *env* gene. These experiments failed to visualize large numbers and/or to generate replicating human sarcomagenic retroviruses. Human sarcoma cell cultures grew in foci with or without such treatment, but yielded no visible replicating retroviral isolates [147, 1810a, b].

Some therapeutic interventions may aggravate the situation. The niche of the cells in the process of malignant transformation is often subverted and induced to render supportive services to those cells. Should the niche be hypoxic, and further aggravated by reduced blood supply (if the patients received treatment with bevacizumab, or sunitinib), then the tumor stem cells will react by self-activation. The VEGF-neutralizing bevacizumab, or small molecular tyrosine kinase inhibitor sunitinib, instead of an intended therapeutic effect, could bring about an adverse situation. The tumor stem cells activated further by hypoxia, would advance by the PI3K/Akt/β-catenin (or another) pathway. Here is a transition from an indirect anti-tumor attack (reduction of vascular supply) to the tumor stem cell population with a gain of virulence resulting [1537a, b, 1538, 1811]. All potential hostile reactions (immune attacks by the host) would be averted and/or turned to the advantage of the transformed cells. The switch to a higher level of malignancy and the maintenance of such tumors by a continuous supply of tumor stem cells are triggered by hypoxia due to reduced blood supply. While ischemic tissue necrosis may be evident as somatic cells die, the tumor stem cells remain alive and active. The transformed cells would gain full independence by expressing their growth factors and autocrine receptors for their capture. The ultimate transformed cell is ready to invade its host and if taken out in culture, to begin its own independent life, immortal and indestructible. These rescued transformed cells do not appear to form new organized colonies. Instead, they exist as ancestral "amoebas" and practice "cell cannibalism." Their genomes continuously undergo further point mutations. Should their environment change, there would be cells with an altered metabolism "in waiting" and so fitted to survive under any changing circumstances.

The extracellular matrix glycoprotein tenascins (TNs) are abundant in embryonic tissues, but retreat to very moderate expression in mature organs, just to be re-expressed again heavily in tumor tissues; there, they are indicators of bad prognosis and increased aptitude for the transition ETM. Tenascin and fibronectin appear in the early chordate lineages: amphioxus and ciona. The pufferfish genome harbors five tenascin genes (some expressing fibrinogen domains). In the human genome, the physiological expression in fetal life, and the silencing in adult life of the tenascin genes is reversed in malignantly transformed cells. As unicellular eukaryotes evolved toward multicellular organisms, their amoeboid movement was inhibited. Human glioblastoma cells over-expressing tenascin, inhibit the amoeboid movement (emperipolesis) of T lymphocytes (Table XVIII).

If not by chemical structure, then by physiological functioning, teneurins are related to tenascins. Teneurins (TNe) are distantly related to corticotrophin-releasing factors and/or calcitonins. Teneurins appeared first in the pre-metazoan era in the choanoflgellates, while sponges and cnidarians do not have teneurin genes. The choanoflagellate teneurin gene is an extraordinary acquisition: "Choanoflagellate VI". The choanoflagellate (*Monosiga brevicollis*) gene encoding the first teneurins is the product of a fusion of a resident algal (or diatom) gene with a horizontally acquired prokaryotic gene. The cell surface protein thus acquired, firmly encased the bacteria to be the nutrients. In metazoans, teneurins became intercellular signaling molecules (in neuromuscular synapses). In contrast to tenascins, teneurins are not known to be over-expressed in malignantly transformed tissues. TNs-positive, TNs-negative and HER2/-positive breast cancers were studied for splicing signatures by RNA sequencing. Coding regions are influenced by expression of alternative pre-RNAs, the state of the promoters, mRNA and iRNA encounters for splicing and polyadenylation. Changing transcriptosomic dynamics result in protein-product isoforms. Cancerous tissues produce protein isoforms different from those generated in healthy normal tissues. In breast cancers, TN-C alters the behavior of stromal and periductal fibroblasts and myoepithelial cells. Breast cancer-associated genes that produce splice variants were identified. Hybrid isoform gene product proteins derived from breast cancer genes CDK4 (cyclin-depedent kinase), LARP (low affinity LA-related RNA-binding protein), ADD3 (adenosine deaminase), and PHLPP2 (tumor suppressor network leucine-rich Ser/Thr protein phosphatase, pleckstrin homology (PH) domain). PHLPP is an Akt phosphorylating and mTOR's rapamycin sensitivity-controlling gene-product protein. In the TNBC tumors, 560, in the HER2-positive tumors, 333 overlapping genes underwent exon inclusion events, including splicings, resulting in hybrid isoform production [1812] (Table XVIII).

Comment. An argument is presented for the wide-spread existence of a form of oncogenesis, whereby the primordial RNA/DNA complex is reactivated via signals from a host in severe distress. The primordial-like RNA/DNA intervention would be a rescue mission for individual cells of a multicellular host in distress/demise. The process recognized as "carcinogenesis" may be an inherent original faculty reactivated in the RNA/DNA complex. The involved cells resist death. The process is carried out in a meticulous sequence of events. First, a cancelation of cell suicidal (apoptotic) programs is endorsed, or is replaced by the long-term event of autophagy. Then a series of gain-of-function mutations are initiated. Further, the engendering of new genes by dsDNA breaks and gene fusions follows. The protein-coding genomic sequences would be spared from direct breaks. In the fused oncogenes, the newly formed fusion proteins would be constructed so that they will be exempt of ubiquitination. The drive for cellular immortality is in high gear, because the cells in the process of transformation are often welcomed and supported by their host.

Table XVIII Alternatively Spliced Tenascins (TNs) and Teneurins (TNe) in Cancer Cells

Extracellular matrix. The matrix glycoproteins appeared first in the amphioxus and ciona. In vertebrate mammalian embryos, tenascins (TNs, distinguish from same abbreviation for triple negatives) are abundantly expressed, but in adult life retreat to moderate restricted appearance. In malignantly transformed tissues TNs resume their abundant expression. In single cells, TNs expression inhibits amoeboid movement. TNs-coated glioblastoma cells inhibit amoeboid movement of T lymphocytes over the cell surface.[1,2]

Breast cancer cells exist in TNs^+/TNs^- and $HER2^+/HER2^-$ subclones. Extracellular matrix TNs communicates with cytoplasmic, epigenomic and intranuclear faculties of tumor cells. Illumina platform RNA sequencing revealed those breast cancer genes, which are involved most in TNs interactions. These are CDK4, LARP, ADD3 and PHLPP (see text). The same genes acted in normal breast tissue samples (NBS), and tumor samples, but novel hybrid isoform production in tumor tissues was in excess over NBS. The gene for X box-binding protein (XBP1) regulates endoplasmic reticulum stress response and anti-estrogen responsiveness; the isoform gene lacks exon2 and it occurs in TNs^- tumors. The isoforms of breast cancer anti-estrogen resistance gene (BCAR1) and high mobility group nucleosome-binding domain-containing gene (HMGN3) (Tables XI/I and XI/II) lack their exon 6. Nonsense-mediated mRNA decay (NMD) occurred in the case of XBP1 expression, while HMGN was exempt of it. TNs^- tumors produced CDK4 isoforms. The LARP1 gene produces several isoforms, which are not expressed at the protein level, except for one 1096 aa hybrid protein. Gene PHLPP produced a short novel hybrid isoform. The cancer-specific splice variants alter cell proliferation, adhesion *versus* invasion, angiogenesis induction, apoptosis-resistance and potential DNA damage repair (such as in BRCA1). Specifically alternatively spliced genes of breast cancer cells include well known FGF2, and Notch3. Overexpressed exons in TNs^+ and $HER2^+$ breast cancer cells far exceed that occurring in NBS. Some isoforms are specific for the tumor (TNs^+ or TNs^- or $HER2^+$). Some ORFs of the novel isoforms are degraded by NMD and stop codons. Only 5% of novel ORFs had properly located polyadenosyl signals, thus are functional; these encode novel proteins (in LARP1 and PHIPP2). Exon-switches between genes (XPB1, BCAR1) increased the tumor cells' invasiveness.[3]

Choanoflagellates #6. Transmembrane glycoproteins teneurins (TNe) were discovered in drosophila nerve cells and in Caenorhabditis. Single non-cordate TNe genes exist in ciona, amphioxus and sea urchin, but the gene and protein appeared first in choanoflagellates, as preserved in *Monosiga brevicollis*. The TNe gene does not exist in cnidaria (*N. vectensis*) or in fungi. All cordates (from fish up) possess TNe genes. Avian TNe derive from the dinosaurs and undergo alternative splicing, so do the human TNe genes. The human TNe-4 is encoded either by five, or by four exons, as its variant lacks the third exon. The choanoflagellate TNe gene is large (6,829-bp) and contains prokaryotic (the YD-repeats) and diatomal (*Phacodactylum tricornutum*) sequences; accordingly, it encodes a fusion protein. TNe proteins contain EGF repeats with cystein residues, prokaryotic YD-repeats (tyrosine, aspartic acid), retrotransposon hot spots, and proline-rich Src homology. Choanoflagellates collected some 1000 genes through lateral transfer; some of these genes ascended on the evolutionary ladder up to vertebrate mammals. TNe genes encode embryonic pattern formation (especially in the brain in vertebrates).[4] TNe is overexpressed in mesothelioma cells and is considered to contribute to chemotherapy resistance.[5] Latrophilins G-protein-coupled receptors bind teneurins for cell-to-cell adhesions (heterophilic binding); hemophilic TNe-to-TNe adhesions are unstable.[6]

1 Tucker RP and Chiquet-Ehrismann R Int J Biochem Cell Biol 2009;41:424–44. 2 Huang JY et al J Immunol 2010;185:1450–9. 3 Eswaran J et al 2013;3:1689. doi:10.1038/srep01689. 4 Tucker P J Exp Zool B Mol Dev Evol 2013;320:1–9; Tucker RP et al Mol Biol Evol 2012;29: 1019–1029. 5 Ziegler A et al Cancer Lett 20122;326:1–7. 6 Boucard AA et al J Biol Chem 2014;289:387–402.

If such an inherent mechanism exists, that in a large multicellular community a few selected stem cells are turned back to a stage of their primordial existence in order to survive the demise of their host, would individuals suffering heart attacks, strokes and other life-threatening emergencies (infections; septic shock) harbor incipient colonies of these "pre-cancerous cells"? In case of death, only a very meticulous postmortem examination could possibly reveal such an as yet hidden process. In case of the host's survival of the acute threat: is the incidence of malignant tumors increased, or are the initiated processes regressing (by Burnet's surveillance)?

However, the supreme machinery is not faultless. Faulty installments may induce a semblance of the cell rescue mission, but without a signal from a distressed host to initiate it. During the process of somatic hypermutation, the enzymes AID/RAG might nick oncogenes (c-*myc* or *bcl*-2) and unite these oncogenes by mistake with the IgH gene, instead of the newly created light chain immunoglobulin genes (*vide supra*). The fusion oncoproteins will drive the cell to replicate incessantly and to resist apoptotic death. If the cell suffered dsDNA damage (ds DNA breaks; over-accumulated Okazaki fragments not removed) and mobilized error-prone enzymes for repair, or faulty O-fragments removal, its unstable genome will mutate further and most of these mutations will be of the gain-of-function type, constitutively expressed. The cell remains aneupolid, and recovers from the state of autophagy, as a transformed, oncogenes-activated tumor cell. There are inflammatory processes, which silence the promoters of the so-called tumor suppressor genes by hypermethylation, and activate the promoters of cell survival pathways by histone deacetylation; reviewed in [73]. These genes are not yet mutated (the oncogenes), or deleted (the suppressor genes), thus the process at this stage would be reversible (by epigenetic methylation/demethylation and deacetylation/reacetylation) [73]. However reactive $CD8^+$ T cells enter the scene. These TGFβ-producer lymphocytes would be carriers of the activated interleukin-like epithelial-to-mesenchymal transition inducers (EMT; ILEI). The ILEI genome acts in cooperation with one of the *ras* proto-oncogenes. The enlisted ras proto-oncogene acquires a gain-of-function point mutation. Consequentially the epithelial cells in this inflammasome (breast or prostate parenchyma) emerge as irreversibly transformed cells [1539a]. These are the host lymphocytes this author proposed to be referred to as traitor/transforming (ttT) T cells [26, 27].

Whichever mechanism induces the return of the primordial RNA/DNA complex in the form of oncogenes: **1**, an inherent rescue mission of selected cells from a doomed multicellular host; **2**, or an error in the evolving adaptive immune system, not yet corrected (faulty installation of the enzymes, AID/RAG, that carry out the somatic hypermutation, or those enzymes that perform error-prone repairs of dsDNA breaks), or **3**, genomic attack by an "oncogenic virus" (HPV) and the response to it [1813a, b], the end results are the same. These interventions, internal or external, generate the rise of disorganized colonies of independent, immortalized and close to indestructible single cells. The rescue mission fails as the host dies with the "rescued cells" within. Should the rescued transformed cells arrive into an outside environment, in which they are able to metabolize and replicate, there they will prove their immortality and indestructibility.

The theory states that "malignant tumors" arise when the RNA/DNA complex restores the primordial metabolism of the ancient unicellular life forms in some selected cells of a well- organized multicellular community. The multicellular community is experiencing a pathological process (chronic infections prominently included) [73], threatening its existence. The RNA/DNA complex is exercising its inherent faculty for the selection of the fittest for survival. Under these circumstances, it is the cell expressing its "survival pathways" constitutively and alternatingly (by switching from one pathway to the other as opposing forces demand), which proves itself to be the fittest. Colonies of these newly arisen cells emerge, act as parasites of their host, the well-organized cell community, and under favorable conditions are able to survive the demise of their host. It is another question, if these highly individualized, independent and immortalized cells will ever be domesticated enough to form again an organized multicellular community in another environment. Examples will be provided for the role of genomically inserted LTR retrotransposons acting as the initiators of the process of oncogene activation (*vide infra*).

The malignant tumors, which may not fit into the new concept. Germ line mutations vertically transmitted appear to fall into a category best characterized as an error within the system. The BCRA I/II cancer suppressor gene mutations are the examples. Even though the results may be read as if the ancient RNA/DNA complex were driving the malignantly transformed cells, the initiation of the process cannot be viewed as "a rescue mission" for some individualized cells of a doomed host. This is a vertically transmitted loss-of-function-mutation in somatic cells (not an acquired damage to some somatic cells; not an act of stem cells, whereby stemness is constitutively expressed in response to some valid or erroneous messages received).

Alfred Knudson at *M.D. Anderson* in the 1960s explained for the first time the genetic difference between sporadic (one hit) and familiar (two hits) retinoblastoma. Losing one allele in a cell cycle-controlling tumor suppressor gene in one particular cell of the retina resulted in a malignant tumor of that particular cell line. The sporadic tumors therefore are curable with surgical removal. Tumors resulting from the loss of both alleles are multifocal, bilateral and not readily curable. Occasionally, a cell with one missing allele receives an externally acquired second hit. If treated with radiotherapy, these tumors recur or are replaced by newly generated mesenchymal tumors (osteosarcoma). Retinoblastomas arise consequentially to inbuilt errors of the RNA/DNA complex (gene conversion; nondisjunction; mitotic recombination). Therefore this tumor cannot be considered to be the manifestation of an act of rescue, even though the transformed cells have acquired immortality. In contrast, the Rb tumor suppressor gene product proteins arrest the cell cycle and prohibit the lengthening of telomeres, thus inhibiting it creates one of the "hallmarks of cancer" (Table XIX).

Viral oncoproteins generated by certain adeno-, papilloma-, and polyoma viruses by physical attachment of virally encoded proteins to the retinoblastoma gene product tumor suppressor protein mark that protein for degradation by ubiquitination. Those cells rendered defenseless due to the loss of the *rb* tumor suppressor gene or its protein RB, succumb to malignant transformation driven by an

Table XIX Retinoblastoma

Retinoblastoma tumor suppressor genes (Rb) are well established in dictyostelia amoeba, fish (zebrafish) and amphibians (frogs and newts).[1–3] In the drosophila, the chimno gene (chronologically inappropriate morphogenesis) activates Janus kinase JAK/STAT & E2F (cyclin-dependent E proto-oncogene factor 2).[4] In chicken, the *src* gene/oncoprotein induces proliferation of normally quiescent neuroretinal cells.[5] The mammalian (mouse) homolog of the chimno product is the 757 aa zink finger protein ZFP509.[4] Rb protein suppresses the cell cycle; enters the nucleolus; hyperphosphorylated Rb protein interacting with α-polymerase inhibits telomere elongation.[6]

The pediatric malignant tumor of the retina, retinoblastoma, retains a large stem cell population; its treatment with radiotherapy frequently induces secondary soft tissue or bone sarcomas. Retinal stem cell population is under Wnt (wingless, drosophila; *int*, mouse) proto-oncogene control. Progenitor cells' stemness maintained *versus* differentiation. Retinal stem cells express Nanog, Oct 3/4, ABCG2 (ATP-binding cassette genes) and Musashi (MS/1) stemness genes 1-3. siRNA knock-down of SFRP2 (Wnt inhibitor, secreted frizzled-related protein-2, decoy receptor) increases expression of stem cell marker genes. The Wnt signaling involves glycoprotein ligands-binding receptor frizzled and low-density lipoprotein receptor-related protein 5/6 (LRP). Ligand-bound receptors release cytoplasmic Disheveled. B-catenin is held phosphorylated by APC-Axin-GSK-3β in the cytoplasm; Disheveled dephosphorylates β-catenin. Mutated, blocked, or deleted APC/Axin/GSK-3β fail to retain β-catenin, which translocates into the nucleus. Wnt-activator LiCl (Table XIV) inhibits GSK-3β, downregulates SFRP2, and arrests the cell cycle. Hypermethylation of the SFRP2 promoter silences the gene and ceases its tumor suppressor activity (secreted frizzled-related protein). In the nucleus, β-catenin activates its targets *tcf/lef* (T cell factor/lymphoid enhancer factor). Protein products TCF/LEF further activate genes in control of the cell cycle (positive), apoptosis (negative), stemness (positive) and cell locomotion (positive). Prominent activated genes are c-*myc*, metalloproteinases MMP2/9, *notch*. Of stem cell genes, Nanog and Oct are expressed.[7–9]

1 *Gyda M et al PLoS Genet 2012;8(1):e1003106.* ***2*** *Destrée OH et al Dev Biol 1991; 153:1412–9.* ***3*** *Thitoff AR et al Anat Rec A Discover Mol Cell Evol 2003;271:185–8.* ***4*** *Flaherty MS et al Dev Cell 2010;18:556–68.* ***5*** *Pasteau S et al Oncogene 1997;15:17–28.* ***6*** *Takemura et al Biosci Biotech Biochem 2008;72:630–635.* ***7*** *Barbouti A et al Cancer Res 2003;63:1202–6.* ***8*** *Wuebben EL et al PLoS One 2012;7(4): e34824.* ***9*** *Silva AK et al Mol Vis 2010;16:36–45*

uncontrolled cell cycle. The 27 exon retinoblastoma gene *rb* mapped at 13q14.2 encodes its 110-kD 928 aa phosphoprotein product. Classically, the unphosphorylated protein associated with E2F allows the cell cycle to proceed from G1 to S phase in a highly controlled manner, by phosphorylations and dephosphorylations of the RB protein. The constitutively phosphorylated RB protein dissociates from E2F and E2F dictates uncontrolled mitoses. When the mutated or deleted *rb* gene fails to control E2F only in the retinal cells (deriving from a defective stem cell), retinoblastoma occurs. The permanently neutralized RB protein in any cell, fails to control the cell cycle. In the papillomavirus-infected uterine cervical squamous cell, the binding of viral oncoprotein E7 to the RB protein recruits cullin 2 ubiquitin ligase for the degradation of the RB protein; this, with other oncogenomic activations (apoptosis resistance, telomere recapping), results in the malignant transformation of the cell characterized by uncontrolled mitoses (illustrated descriptions in all oncology/virology textbooks; recent publications from the University of Debrecen, Hungary [1813a, b].

Hepatoblastoma. A newborn infant has hepatoblastoma. The lipid-modified hedgehog (Hh) ligands initiate the cascade leading to the development of hepatoblastoma. Receptors Patched and Smoothened (Ptch and Smo, from the drosophila) capture their ligands. Ptch is an inhibitor of Smo, except when Ptch is de-activated by its ligand. Receptor Smo is the proto-oncogene; when it is liberated from control under Ptch, it will initiate a cascade with the end result of the activated Gli oncoprotein (gliotactin). Gli is able to induce several histopathologically different tumors, other than hepatoblastoma (glioblastoma and medulloblastoma; basal cell carcinoma, intestinal stromal tumors). The nuclear expression of the Gli protein is characteristic for hepatoblastoma cells [1814]. The Wnt pathway is open and active in hepatoblastoma. The end result of this pathway is the cytoplasmic accumulation of β-catenin with (or without) aberrant expression of the β-catenin gene (CTNNB1). Some hepatoblastoma tumors retain fetal embryonic morphology (set in after the stage of fetal hematopoiesis in the liver passed). Some hepatoblasts differentiate either toward epithelial or mesenchymal directions. On occasions, the CTNNB1 gene is deleted. Exon 3 of the CTNNB1 gene exhibits a GSK3β-binding region (glycogen synthase 3). The Wnt pathway (Figure 7) allows the over-accumulation of β-catenin. The Wnt complex of APC/Axin/GSK3β marks β-catenin for degradation by ubiquitination. The malfunctioning Wnt complex (Axin disabled) fails to mark (no phosphorylation) β-catenin for degradation. The over-accumulated β-catenin either binds TCR/LEF proteins in the cytoplasm (T cell factor; lymphocyte enhancer factor), or binds TCR/LEF genes and other genes by special DNA-binding in the nucleus for activation (c-Myc, CyclinD1, Survivin, MMP and VEGF). In the HGF-to-Met pathway, the cytoplasmic tail of the ligand-bound Met receptor undergoes tyrosine autophosphorylation and encodes for hepatocyte regeneration, hepatoma cells' epithelial-to-mesenchymal transition, and vascular angiogenesis. Met binds β-catenin at the cell membrane and it releases β-catenin by phosphorylating it at tyrosine residues Y654 and Y670. Tyrosine-phosphorylated β-catenin is protected from proteosomal degradation [1815, 1816]. Alpha-fetoprotein (AFP) produced in the yolk sac and fetal liver is elevated in patients with hepatoblastoma [1817]. The epithelial type hepatoblastoma responds best pre- and postresection to chemotherapy with cisplatin, 5-FU and vincristine (or cisplatin and doxorubicin), or high dose cisplatin (with kidney protection). Chemotherapy-resistance exists, even against high dose cisplatin. In patients in remission, chemotherapy-induced late secondary malignancies may arise (due to mutations induced by chemical exposures in newborn age). Surgical resection (even of lung metastases) and liver transplantation are practiced and favored [1818–1820].

Gestational choriocarcinoma. Placental villous cyto- and syncytiotrophoblasts naturally fuse, as mediated by the endogenous retroviral Env proteins syncytins (*vide supra*) (Figure 52). The genomics of gestational choriocarcinomas are studied in established cell lines (BeWo, JAR, JEG). The Ras homolog Rho GTPase proteins work with guanine nucleotide exchange factors and GTPase-activating proteins (GEF; GAP). The GTPase-deficient RhoE/Rnd3 stands for the poorly understood resistance-nodulation-division. It is understood that RhoE contribute to the fusion of BeWo cells [1821]. The JAR and JEG choriocarcinoma cell lines are driven by amplified oncogene/oncoprotein Sox2 (sex determination Y- box). The specific physiological antagonist of the Sox protein is miR-145. The 3′-UTRs of the stem cell genes Oct4, Krüppel-like factor4 and Sox2 mRNAs align with miR-145 and get degraded (no Sox2 protein is translated). The growth in xenografts of the choriocarcinoma cell lines transfected with miR-145 was inhibited [1822]. In BeWo and JEG-3 cell lines, the Delta-like ligand 1 (DLL1) of receptor Notch works in an activated unison. Upon binding DLL1, the Notch receptor signals for the cleavage of Notch's cytoplasmic tail, that is now free to translocate to the nucleus for the activation of its target genes. The cell will produce and release matrix-lysing enzymes and thus gains invasiveness. The DLL1 mRNA is physiologically inhibited by miR-34. Inhibition of DLL1-activated Notch resulted in the reduction of MMP9 and urokinase-type plasminogen activator in the choriocarcinoma cells. As tumor cell locomotion and invasiveness are curtailed, the incidence of distant metastases also subsides. There is a positive cooperation between p53 and miR-34a activation; possibly, many incipient tumors are inhibited this way, and are never recognized (Burnet's immunosurveillance). In choriocarcinoma cells, not the p53 gene, but the apoptosis-stimulating proteins (ASPP) activated by the p53 protein are silenced by hypermethylation of their promoters. Forced expression of the ASPP reduces tumorigenicity of choriocarcinoma cells [1822, 1823]. The healthy cytotrophoblasts express the homeobox stem cell gene Nanog (for eternal youth of the mythological tribe Tir na nOg). Nanog protein is encoded from chromosome 12p; is highly expressed in embryonic stem cells. Krüppel-like factor regulates Nanog's withdrawal from the cells of the differentiating blastocyst. However, strong Nanog expression (high Nanog mRNA and protein levels) remains in the hydatidiform moles. Nanog protein expression is inhibited by short hairpin miRNA degrading the Nanog mRNA [1824, 1825]. Gestational choriocarcinoma cells kill the untreated host (the mother). However, in contrast to neoplastic stem cell, choriocarcinoma cells are highly sensitive to chemotherapy; metastatic choriocarcinoma is the metastatic cancer that is readily curable with chemotherapy (methotrexate; actinomycin D).

There are physiological proto-oncogene expressions in healthy trophoblasts: EGF (c-ERB1/HER1)+→EGF-R, HGF → MET, VEGF → VEGF-R, TGFαβ, IGF-2, placental GF, and platelet-derived GF. Expressed in exocrine and autocrine circuits. Further, increased integrin receptors, decreased cell adhesion molecule E-cadherin, increased MMP-9 and activated PI3K/Akt/mTOR and MAPK pathways are activated. Proto-oncogenes switched on and off are c-*abl*, c-*ets*, c-*fos*, c-*jun*, c-*myc*, c-*ras,* c-*sis:* "an analogous behavior of normal cytotrophoblasts and cancer cells". Of hormones, active are a growth hormone variant, prolactin, somatostatin, IGF-R, oxytocin, leptin, corticotrophins, and human choriogonadotrophin hCGαβ, and immune modulator TNFα. Of "tumor antigens", weak expressions of PSA and CEA are detectable. The Akt pathway blocks the pro-apoptotic protein, Bcl-2 antagonist, BAD. The trophoblasts refrain from the expression of MHCA classes I/II, while expressing HLA-G to deter NK cells. FasL expression induces external apoptotic death in $FasR^+$ immune T cells of the host (the mother) [1826].

Gestational choriocarcinoma cells over-express p21-activated kinase, PAK4. Human chorionic gonadotropin (hCG) acts as the upstream regulator, and PI3K as the promoter of PAK4 expression. PAKs are downstream effectors of Cdc42 (cell division cycle) and *ras* homolog Rho GTPase Rac1 (*vide supra*). The PAK oncoproteins promote anchorage-independent invasive and metastatic cell growth, modulate the cytoskeleton of traveling tumor cells, and protect them against apoptotic death. Gestational choriocarcinoma cell lines JEG3 and JAR encode through mRNA from chromosome 19q13.2-13.3 the amplified PAK 4 oncoprotein. The oncoproteins PAK1/2/4 promote the growth of ovarian cancer cells through the c-*src*/SRC and EGF overproduction pathways. In prostate cancer cells, HGF (hepatocyte growth factor) activates and PI3K promotes PAK4 expression. Expression of PAK4 protein was reduced in tumor cells treated with PI3K inhibitor LY294002 (patented) Knockdown of PAK4 protein expression by siRNA-to-mRNA interaction resulted in the cessation of choriocarcinoma cell migration and MMP production [1827, 1828].

The Haynal Imre University for Health Sciences (Budapest, Hungary) in collaboration with Brigham & Women's Hospital, Harvard Medical School (Boston, USA) recognized the DOC-2 tumor suppressor gene in the choriocarcinoma cell lines JAR, JEG and BeWo [1829–1831]. The ancestors of the human DOC-2 gene (deleted in ovarian cancer) are the drosophila and mouse disabled dDAB, mDAB genes and gene product phosphoproteins. These genes are alternatively spliced to produce two isoforms (p96, p97). Mouse macrophages activate mDab2 in response to mitoses induced by CSF-1. The human equivalent of mDAB is the Doc-2 gene product protein, that is well expressed in healthy ovarian cells and is deleted in ovarian carcinoma cells. The first box of the DOC-2 protein is the PID (phosphotyrosine interacting domain). The second box expresses the SH3 (Src homolog) binding motifs. The physiologically constructed transformed ("cancer") cell, the trophoblast, expresses the DOC-2 protein, and at the postpartum removal of the placenta it does not replicate any further (but may persist in a resting stage at distant sites, in the lungs, in the mother, for a while). Choriocarcinoma cells delete DOC-2. Malignantly transformed trophoblasts and ovarian carcinoma cells do not express DOC-2 [1830a, b–1831]. The choriocarcinoma suppressor gene product protein NECC (not expressed in choriocarcinoma clone 1) is encoded from chromosome 4q11-q12. In health, several organs (brain, lung, musculature, bladder and kidney, spleen) express NECC; in this context, it is important that healthy uterus and placenta express NECC. The NECC protein is, or is inductor of, the chorionic somatomammotropin (CSH) hormone 1. Choriocarcinoma cells transfected with the NECC gene undergo differentiation into syncytiotrophoblasts and lose their malignant phenotype and fail to grow *in vivo* in xenografts [1832]. The anterior pituitary-expressed growth hormone and several placental hormones (placental growth hormone, placental lactogene, human growth hormone, chorionic somatomammotropin hormone (hGH/CSH) interact in pregnancy [1833]. Pre-eclampsia and gestational diabetes is associated with hormonal imbalances

[1834], but hormonal promotional or inhibitory effects on tumor formation are not well understood.

There is a distinction between the germ cell tumors and the gestational choriocarcinoma. In the former, the genome is a pure host genome (female or male). In the latter, the genome is composed from the unification of the maternal and paternal haploid genomes. In a germ cell tumor arising in the kidney in a female patient, the presence of the paternal Y chromosomes proved that it was an extragonadal manifestation of a gestational choriocarcinoma [1835]. The host's immune reactions against the pure germ cell tumor of the gonads (all "self"), and against a gestational choriocarcinoma, in which the fetal cells carry both maternal and paternal chromosomes and express gene product proteins of both genders (not pure "self") must be different.

Comment. Could hepatoblastoma of a healthy newborn child, germ cell tumors of adolescents, or the gestational choriocarcinoma, be included into those theoretically conceived categories of carcinogenesis, that are manifestations of a rescue mission for individual cells, as it is carried out by the reactivated primordial RNA/DNA complex in a multicellular host, which is doomed to demise? The RNA/DNA complex would act as their primordial ancestors did in the unicellular eukaryotes maintaining them in a most hostile environment without senescence and natural death. In one fetus, a vital organ stops the process of differentiation in an immature stage, expresses "cell survival pathways" in a constitutive manner, and actually would overcome and kill its host with a hepatoblastoma. In another fetus, the physiological cellular barrier of fetal cell-derivation (the cyto- and syncytiotrophoblasts) preserves all faculties of pluripotent stem cells and uses them to overcome the mother (as the fetus turns into a "malignancy"). Were these embryos/fetuses under the threat of immediate death *in utero*, which activated some inherent primordial faculties preserved in their RNA/DNA complexes in order to mobilize a rescue act for the survival of the cells that will be immortalized into ancient unicellular eukaryotes? Could have the RNA/DNA complex received from its fetal host such an impulse for action? Rather, are there inherent errors in the system as installed, and evolutionarily not yet corrected, that bring about these events? Some temporarily activated cell survival pathways fail to switch off. Some genes, which normally initiate a rest period, become silenced (epigenetically so, by DNA methylation), and allow free flow of those genes, which fail to switch off, as they should, the cell survival pathways (epigenetically so, by de-acetylation to activate, and by the failure to re-acetylate to silence, their promoters)? In these cases, as in practically all cases of "malignant transformation", the immortalized cells are akin to their primordial ancestors, but have arisen by errors of the system, and not by an intended, organized, and evolutionarily installed rescue mission of the primordial RNA/DNA complex. Whether due to the exercise of an inherent faculty for the restoration of the ancient unicellular life forms, or due to some errors of installation, the primordial RNA/DNA complex is returning in these hosts in the disguise of "oncogenes". Intended, organized, and evolutionarily installed rescue mission of the primordial RNA/DNA complex returning in these healthy young hosts in the disguise of "oncogenes" is difficult to imagine or apply.

Selected Abbreviations

BMP, bone morphogenetic protein, acting as differentiation-inducer tumor suppressor. Click for BMP-induced osteosarcoma or rhabdomyosarcoma cell differentiation: Geng S et al Mol Cancer Ther 2014 Apr 10 PMID 24723453; Wolf S et al Int J Oncol 2014;44:1727–35. In metastatic prostate cancer cells, their contact with bone marrow stromal cells induced BMP production resulting in tumor cell proliferation in androgen-free media via protein kinase C/nuclear factor kappa B (NFκB (Lee GT et al Br J Cancer 2014;110:1634–44).

Fas, fragment apoptosis stimulator; FADD, Fas-associated death domain; FLICE, FADD-like Interleukin-1beta (IL-1β) converting enzyme. Click for human B-cell lymphoblasts producing constitutively Fas-ligand (FasL) to kill Fas-receptor positive ($FasR^+$) $CD4^+$ (cluster of differentiation) immune T cells (Klinker MW et al Front Immunol 2014 Apr 4. doi:10.3389/fimmu.2014.00144.

MMP matrix metalloproteinase, enzyme digestive to intercellular matrix; promoting cancer cell local invasion. Induced by 12-myristate 13-acetate and inhibited by naphthoquinone shikonin, Chinese herbal product (Jang SY et al Oncol Rep 2014 Apr 25. doi:10.3892/or.2014.3159).

PTPH, protein tyrosine phosphate, PRMT, protein arginine methyltransferase, spliceosome regulator and transcriptional cofactor, acting as a virulence factor in unicellular eukaryotes, and as a proto-oncogene in vertebrate mammalian (human) cells. In Plasmodium malariae cells, PRMT-methylation of arginines in histone H4 and H2 is essential for life and virulence, so much so that PMRT is selected to be a therapeutic target for small molecular inhibitors (Fan Q et al Biochem J 2009;421:107–18). Constitutively expressed PRMT is a promoter of blood stream forms of Trypanosoma brucei, by methylating mitochondrial protein regulatory genes as a high molecular weight, but non-essential, virulence factor (Pasternack DA et al Eukaryot Cell 2007;6:16665–81). Click for reactivation of retinoblastoma gene/protein and silencing of polycomb repressor complex in dying lymphoma cells with inhibited PRMT5 (Chung J et al J Biol Chem 2013;288:35534–47). Click for upregulated PRMT in metastatic melanoma cells (Nicholas C et al PLoS One 2013;8(9):e74710).

PHLPP pleckstrin homology PH-domain leucine-rich repeat protein phosphatase Bradley EW et al J Biol Chem 2013;288:9572–9582.

A Condensed Compendium

Looking Back Once More

A retrospective glance (in narrative style). A look at the first cellular life forms, especially at those which have not as yet reached the stage of being eukaryotes, reveals in the distance of three milliard (billion) years or more, entities dividing without senescence and natural death (archaea; prokaryota, eubacteria). The Ras-Rho-Rab clade appeared in archaea and prokaryota. It is the Roadblock domain in archaea where small GTPases first appeared and then reappear in the first eukaryotes; as discussed by Jékely [1150, 1836a, b]. Membrane-based compartmentalization in the first eukaryotes depended on the cooperation of Rabs and RanGTPases (Ras-related nuclear protein). Whether acquired by vertical inheritance (from fused archaeal-prokaryotic ancestor) or from the archaeal roadblock operon through horizontal transfer, ancestral DENN proteins (the longin domain) are functional in the Giardia and in the Trichomonas (differentially expressed in normal and neoplastic cells). An apparent agreement was reached between Jékely, the author, and the reviewers of his manuscript, on that, that the first eukaryote derived from a prokaryotic-archaeal cell fusion, and as if the chimera acquired later a proteobacterium for its mitochondrion. Those mitochondria still replicate on their own in the cytoplasm of eukaryotes. However, the duplicate ribosomes synthesized chimeric proteins. How sharply was self and non-self distinguished at that stage of cell promiscuity? The nucleus was formed within pre-existing cytoplasmic endomembranes. Blobel contributed to that phase of cell evolution [88a, b]. The nucleus acquired a great number of the mitochondrial genes and the type II introns. Were there any pre-existing large DNA viruses contributing to the formation of the nucleus? Villarreal debated that event [29a, b].

These faculties have been preserved in most free living extant unicellular eukaryotes (Amoebae, the Naegleria, or the kinetoplasts), but became lost in multicellular organisms, except for their gonads. The unicellular parasites (plasmodia, trypanosoma, theileria) possess an RNA/DNA complex genome that could

J.G. Sinkovics, *RNA/DNA and Cancer*, DOI 10.1007/978-3-319-22279-0_23

readily “trans-speciate” its host. Some eukaryotic genes, which remained conserved as stem cell genes, are being recognized as the inducers of the first generation of “cell survival pathways”, or proto-oncogenes of extant multicellular organism. These genes arose in the innumerable taxa and species of the first unicellular living entities (the arginine protein methylases, the proliferating nuclear proteins antigens, the enhancers of zeste; the ras family). Proteins encoded by first generation proto-oncogenes are the most established constituents of the cell, with well established, and the widest of connections to other components. The “cell survival pathways” (PI3K/Akt, Pak/MAPK, and many other ancient stem cell gene product proteins) are the examples. An immense armada of TLRs, chemokines and cytokines obeyed to, or antagonized with, these gene product proteins. Complex RNA-networks (siRNA → mRNA; piwiRNA → retrotransposons) regulated gene expression (were enabled to overrule the DNA-derived mRNAs). The RNA/DNA complexes preserved the imprints of prior gene/antigen intrusions (CRISPR) (Figure 20) and protected their hosts from alien invaders, including the “selfish” retrotransposons (with many failures, as the retrotransposons still prevail).

The phylogenetic distance (evolutionary divergence of 535–550 my) between the amphioxus’ and the vertebrate mammalian Hh genomic loci proved to be much too long for the close preservation of its elements. The noncoding sequences are conserved in the last 100 million years among the 3 amphioxus species, but the cephalochordate noncoding sequences are absent in the vertebrate mammalian genomes. Even between the zebrafish and the mouse, the sequence motifs preserved in the introns are distant; however, these motifs are closer between reptilians and mammals. The Rho family *ras*-related small GTPases (CDC42, Rho, Rac) are fully active in yeasts as Dock proteins (dedicator of cytokinesis). Rac1a is a growth factor in yeasts (dimorphic fungi *D. discoideum*; *Yarrowia lipolytica*). In the latter microorganism, its *YℓRAC*1 gene-product protein (genomics, Southern blot; proteomics, Northern blot) is essential for hyphal growth. In *C. albicans*, it is an increased mRNA level for CDC42 that precedes hyphal growth [1837]. Thus, the ancient intron RNAs appear to have cooperated with genomic DNAs and remained aboard during the vigorous genomic activities in the ontogenesis of the embryo.

The establishment of multicellularity imposed the strongest rules and regulations upon the cell community. The subservient diploid somatic cells accepted differentiation, service limited to one or two tasks and life ending in senescence and death. The haploid germ cells upon the unison of their two genders, differentiated into subservient and organized masses of diploid somatic cells. Some stem cells remained preserved as such, and by asymmetric mitoses preserved themselves and yielded one daughter cell for differentiation into a somatic cell compartments (the organs). Genome duplications and gene fusions generated new genes. Some of these genes were to prevent the “rebellion of the somatic cells”, that is, an effort of somatic cells to de-differentiate back to the stages of their ancient immortal ancestors had to be curtailed. If a population of cells can be generated by gain-of-function mutation and replicate out of order, the entire multicellular community would be disrupted. These reactions therefore would automatically trigger first within the mutated cell itself an energy-consuming process of self- destruction, the programmed cell death (apoptosis), practiced as external (originating at the cell surface) or internal (originating in the mitochondria). Should a stem cell genome continue to express its cell survival genes constitutively, that cell would out-populate all orderly cell communities. Should the stem cell genes be reactivated in a somatic cell, that cell would engage in a conduct most disorderly for an organized cell community. The stem cell is programmed to be long-lived, and thus be defended against apoptosis. Healthy host cell genomes could activate introns to release short RNA sequences (iRNA siRNA) to neutralize the mRNAs emanating from the constitutively replicating stem cell or somatic cell genes (oncogenes), thus no kinases (oncoproteins) to drive "cell survives pathways" would be synthesized. These genes of healthy cells are recognized as the "tumor suppressor genes". For example, PTEN rose to antagonize the constitutive "cell survival pathway" PI3K. In turn, the emergence of tumor suppressor genes elicited the generation of more newly formed (or altered pre-existing) genes encoding "trans-speciation backward". The gene product proteins of these genes first eliminate the tumor suppressor genes. These genes may be recognized as the second generation of oncogenes. The MDM gene product protein eliminating WT 53 protein, or the HPV E5 and E7 oncoproteins neutralizing p53 and Rb are the examples. The MDM/p53 protein complex is eliminated by ubiqitination (*vide supra*).

Freely differentiating clusters of fertilized and deposited eggs provided for the progenies of fishes, reptiles, dinosaurs, and insects. However, fertilized eggs of the species, which were to become mammalians, developed embryos individually encased in a capsule of parchment-like thin walls (membranes, in the platypus) within a new organ of their females, the uterus. By then, all vertebrate taxa and species possessed and practiced a unified innate and adaptive immune system. The fetal organ, the placenta, expresses all paternal antigens of the fetus. Its trophoblasts invade the maternal uterine decidua and even her blood stream. The maternal immune system was to be switched off, specifically and temporarily, in order to accept the placenta and the fetus. New mechanisms of tolerance-induction were installed and worked. They worked so well, that some stem cells, or even somatic cells, expropriated the survival tactics of the placenta, the "pseudomalignant organ" [30a] for their own de-differentiation into senescence-free immortal stages of cellular evolution. Thus, extant cells could recapitulate the life style their primordial ancestors originally established on Earth. Just before, or at the time of placentation, a new generation of Medawar's tolerance-inducer genes must have arisen to secure this process, by forcing the multicellular host to respond not only with passive tolerance, but actually with full support to the newly arising de-differentiated cell colonies. The tolerogen proteins encoded by this most recent generation of immune tolerance genes (retroviral syncytins) are expected to have a limited and not wide connections with the cell's ancient protein network (in contrast to the most ancient first generation oncogenes, like the *ras* family). These are descendants of the genes, which secure the success of a pregnancy, protect the placenta from immune rejection, and provide active support to the fetus. Vasculogenesis way predates placentation. Some unicellular ascidian tunicates (the Botryllus) evolved a communal vascular system. Even plants have arborating channels to circulate body fluids.

However, the new VEGFs and their receptors in constitutively working pathways represent a third generation oncogenes. The subverted host responds to an invader (the fetus in the uterus) by providing nutrients and by eliminating wastes by diffusion and neovascularization. Host defensive cells switch sides and instead of a defensive attack, render supportive service. In paracrine circuitries the host provides chemo-, lympho-, cytokines and full hormonal support for the support of the new cell colonies (the fetus). The RNA/DNA complex transforms individual members of the new colonies into slick mesenchymal type cells for locomotion and long distance travel within the host (fetal trophoblasts in the maternal blood and lymph circulation). Among others, the SLUG, SNAIL, and TWIST (onco)proteins endow the targeted cell with the mesenchymal phenotype. Many levels deeper in evolution, these once immobile, cuboid/epithelioid and well-differentiated somatic cells are now crawling and phagocytosing like an amoeba (*vide infra*). Aggressive new cells fuse with subservient supportive cells (reviewed in Dittmar & Zänker's Cell Fusions in Health and Disease, Springer Verlag, 2012) [1541]. Here, fusogenic retroviral envelope proteins render the service. Distant chromosomal genes are able to translocate and fuse to one another, in order to generate new aggressive genes. The fused genes, upon their excision and insertion avoid nicks within protein-coding domains. These maneuvers are carried out with meticulous care not to cut into protein-coding domains. The newly encoded proteins are structured so that they escape removal by ubiqitination. An oncogene (EWS, Ewing's sarcoma) may fuse with a tumor suppressor gene (WT, Wilms' tumor) as in small cell desmoplastic sarcoma. This act proves that the oncogene neutralized and overcome the tumor suppressor gene. Fusion oncogenes/oncoproteins abound in soft tissue and bone sarcomas [340a, 341a]. All customary innate and/or adaptive defense reactions of the host become futile: their target antigens have disappeared in the transformed cells.

Embedded doctrines contradicted: *"Without pre-immunization, there are no immune reactions. These are in vitro artifacts"*! Unbeknownst to the NIH/NCI project site visitors, the Botryllus mobilized hundreds of million years ago large granular cells circulating in its hemolymph, which rejected an unwanted intruder upon the first encounter with it: an "immune reaction" without pre-immunization. The first encounter of the NIH/NCI project site visitors with NK cells: "they cannot possibly exist; the Hellströms have not seen them; immune reactions are induced by pre-immunization; something in your chambers activates them; these are "*in vitro* artifacts" (Figures 35 and 36, also in the Appendix 1).

If in the process of cellular transformation, the expression of the MH complex is withheld, the NK cells will attack these "alien"-appearing cells. The NK cells should have proved to the eminent NCI project site visitors, that "yes, there is immunological attack without preimmunization" and "the events photographed by Joe in 1967–1969 were not 'artifacts', but immune T cells ("small compact lymphocytes") in patients only, and NK cells (then referred to as "large granular lymphocytes") both of patients and of the healthy donor (myself), attacking human tumor cells, autologous-allogeneic for the patients, allogeneic only for the healthy donor, as observed in the late 1960s [30a, 147, 279, 1630a, b–1637]. The very first NK cells appear to have protected the ascidian Botryllus' colonies from fusion with histoincompatible colonies. Their single fusibility/histocompatibility (Fu/HC) locus distinguished pairs of compatible from incompatible colonies. The incompatible colony is attacked and lysed by round hemolymph cells of the prospective recipient colony. In these colonies, individuals share a vascular network and a cloaca under a tunic. The BsCD94-1 (*B. schlosseri* cluster of differentiation) genes encode the histompatibility antigens; in addition, the cells express the lectin receptors CD94 and NKR-P1. Up to *Homo*, NK cells express CD94, NKR-P1 and NKG2. Thus, the Botryllus possessed and operated an ancient NK cell with a C-type lectin-binding domain (CTLD), which has been conserved up to the human NK cells [294, 1838] (Figure 46). Further, the killer KIR Ig-like human NK cell receptors express a piwi-like small RNA (piRNA). Further, the $CD56^{+}$ human NK cells operate the microRNA-181, which targets the mRNA of the NEMO-like kinase (Nlk, NFκB essential modulator), that is a Notch signal inhibitor. NEMO is also involved in the activation of the Wnt/ß-catenin signaling cascade in the zebrafish, and upward. There, Nlk phosphorylates and activates also the lymphoid enhancer factor-1 (Lef-1) [1839].

When the NK cells undergo malignant transformation, it is through the up-regulation of miR-21 and miR-155, and the phosphorylated AKt (AK mouse thymic lymphoma retroviral oncogene) cell survival pathway. Treatment with anti-miR-21 and anti-miR-155 oligonucleotides resulted in the restoration of tumor suppressor gene PTEN, PDCD4 and SHIP1 (phosphatase tensin homologue chromosome ten; programmed cell death; src homology inositol phosphatase). Lentiviral vectors of miR-21 and miR-155 sequences rendered resting NK cells $PTEN^{-}$, $PDCD^{-}$, and $SHIP^{-}$ with re-activated AKt pathway. The highly malignant NK cells in nasopharyngeal lymphoma, or aggressive NK cell leukemia, carry also the epigenomically located EBV genome [1840]. Silencing the miR-146a promoter gene by hypermethylation, or reactivation of the gene by demethylation, resulted in the rise or drop of NFκB levels, respectively. NKT lymphoma cells proliferated under the effect of NFκB. Under the effect of miR-146a, the cells ceased to produce the NFκB-activator TRAF-6 (tumor necrosis factor receptor associated factor), NFκB levels dropped and NKT cells become apoptosis-prone [1841]. The interaction of TGFβ and NFκB concerning apoptosis inductions has been clarified. Ligand-occupied receptor of TGFβ acts through TAK1 (TGFβ-activated kinase) inducing Smads, Rho-like GTPases, PI3K/Akt, p38/JNK MAP (Jun terminal kinase) pathways. In a major interaction, TGFβ promotes the association of TAK1 with the ubiquitin ligase TRAF6 (TNFR-associated factor). The TRAF/TNFR associated protein is TTRAP. TTRAP promotes TGFβ-induced apoptosis, thus favoring the elimination of tumor cells, whereas NFκB/PI3K/Akt act as an antagonist, favoring the persistence of tumor cells [1842].

The "criminal collusion" of retro(lenti)- and herpesviruses. As a rule, viruses interfere with one another. Not always: They may reach a cooperative agreement: a criminal collusion [895a, c], from the point of view of their hosts.

Millions of years of co-existence in the same evolving hosts, especially in Africa, resulted in the replacement of viral interference with viral collaboration between herpes- and retroviruses in the interest of sustaining the infected host organisms alive [147, 895a–c]. This collaboration culminated in the

HIV-1/EBV, and HIV-1/KSHV co-infected cells of body cavity effusion lymphomas [1843a, b, 1844], The cell proliferation pathway activated by KSHV is that of the PI3K/Akt/mTOR. KSHV proteins (K13) were inhibitory to FLICE (vFLIP effect) and mobilized the NFκB subunit RelB for the sustenance of c-Myc and the pro-BCL-2 anti-apoptotic MCl-1 [1845, 1846]. Actually, a long list of herpesvirally activated retro- and retro(lenti)viruses could be provided [261b]. In the coinfected hosts, some large herpesviruses appear to have acquired retroviral sequences actually inserted into their genomes. The Marek's disease herpesvirus incorporated genomic sequences of an avian reticuloendothelial virus (Rel; REV); reviewed in [27]. The human EBV possesses sequences reminiscent of VJ/RAG genomic sequences, which regulate the somatic hypermutations in the genomes of plasma cells during specific antibody production. The discovery of the presence of such sequences in the EBV genome was announced first by Niller and associates in 2003 at Balatonfüred, Hungary, and was published first in the Acta [992], as recited recently [1847] (Figure 61). This author learned of this monumental event much belatedly from the recently published Berencsi volume [1559, 1847], since the first Dreyfus publication was also unaware of this event [312]. An ancient herpesviral genome is envisioned to have been inserted into the genome of the amphioxus, and/or sea urchin. Herpesviruses are known to have travelled vertically and horizontally upwards in the evolutionary ladder (clams; fishes; reptiles; aves; mammals). Accordingly, a strong claim has been formulated for the early acquisition of the elementary genes of adaptive immunity (VDJ;RAG1/2) by an ancient herpesvirus (an ancestor of EBV) from their original hosts (amphioxus, clams; sea urchins), from where the ancestral EBV transmitted these genes into the first cartilaginous fishes, the sharks (gnathostomata; chondrichthyes) [147, 312, 992, 1847].

In opposition stand firm the claims that the ever-present retrotransposons implanted the first genes of the adaptive immune system into the primitive sharks [313a, b, 317]. The RAG transposases are recognized to be retroviral transposases of the Transib transposase family of Kapitonov and Jurka; Fugmann [311, 317]. If the genes of the adaptive immune system were acquired by retrotransposon insertions, the ancestral EBV was not the donor. The system was already installed and functional, when the ancestral EBV might have acquired later its VJ/RAG-like sequences from its mammalian hosts for its own use. Nevertheless, the relationship between the ancestral retro-, retro(lenti)-, and herpesviruses must have been, and still is, the most intimate. However, it might not have been a mutually advantageous coexistence. The avian and mammalian leukemia and sarcoma viruses are retroviruses (or reticuloendothelial retroviruses).

In opposition, leukemias of the New Word monkeys (marmosets, tamarins) are caused by herpesviruses. The reticuloendothelial virus sequence-carrier Marek's avian (turkeys) herpesvirus represents a unique unison of the two viral taxa; reviewed in [27]. The genome of the gammaherpesvirus of squirrel monkeys (*Saimiri sciureus*), *Herpesvirus saimiri*, operates from its epigenomic position by emitting small RNAs inhibitory to host cell mRNAs. The *H. saimiri* genomic tip-484 (tyrosine kinase interacting protein) constitutively activates host cell Lck (Src family member lymphocyte-specific p56 protein tyrosine kinase) and STAT to drive the replication of infected T cells [1848a–c]. The human lymphoma-leukemia tumors so far did not yield proven etiological viral agents, except for the only one human T cell lymphotrophic virus-I/II of Poiesz *et al* (HTLV) [1849a–d]. Of herpes viruses, EBV and KSHV stand out as etiologically contributory, but not necessarily the sole causative viral agents (this author considers them to be the causative agents). The rare budding retroviral particles seen in cells from human leukemia, lymphoma (including the Reed-Sternberg cell) belong to various classes of endogenous human retroviruses, which could not so far be established in culture. Long term lymphoblast suspension cultures originating from patients with hairy cell leukemia existed, and contained EBV-carrier B cell lymphoblasts [1850a–d]. However, not only *in vitro*, but also *in vivo*, patients with the initial hairy cell leukemia ended up with EBV-carrier large B cell lymphoma [1851a–c, 1852, 1853]. The silenced EBV genome in hairy cell leukemia becomes a producer of mature virions in large B cell lymphoma. An unclarified relationship exists between human herpes (EBV in particular) and endogenous retroviruses. The HTLV type II most elusive retrovirus exists in too few isolates to consider it a pathogen (and generates no antibodies in patients with hairy cell leukemia) [1854].

Have human APOBECs eliminated the lymphoma- and sarcomagenic retroviruses at the level of Australopithecines from the human race? The human APOBECs failed to protect *Homo sapiens* from HTLV-I/II and HIV-1/2. Hairy cell leukemia deserves special attention, since it carries EBV (Figure 47) [1850c], but its most likely causative point-mutation V600E is in its BRAF oncoprotein [1855a, b, 1856]. If a T cell variant of hairy cell leukemia exists, those cells may carry HTLV-II [1857]. This author is for a continuous pursuit for viral etiology of Burkitt's lymphoma, Reed-Sternberg cells, hairy cell leukemia, and Kaposi sarcoma cells: these cells may hide a special herpes- and retroviral unison.

Rebels rise in the Police State. The organized colony of cells in the multicellular hosts could not possibly tolerate the archaic metabolism of highly individualized, independent and basically immortal unicellular life forms exercising an inexhaustible mutability. Cell communities would be disassembled by these maverick cells. New genes had to be installed in order to induce programmed cell death upon the attempted reversal of a somatic cell to its ancient immortal life style. In the multicellular hosts, these genes are recognized as "tumor suppressor genes". The germ cells preserving some of the ancient genes become differentiating somatic cells upon their unison. The stem cells preserve a great number of the genes with which the ancient DNA operated. In an organized cell community, these genes are allowed to function in organ rejuvenation and regeneration. Epigenomics regulate by acetylations and deacetylations, methylations and demethylations of the histones (most often the lysins within the histones) for inactivation and reactivation of these genes (often acting on the gene promoters, rather than of on the target gene itself). When an ancient gene of cell survival was permanently deleted, the genome can regain equivalent substitutes of those genes by mismatch repairs of broken DNA strands, thus creating new fusion genes acting constitutively as "cell survival genes" In cell communities replicating readily to form their own independent colonies. These genes and gene product proteins are recognized as "oncogenes and

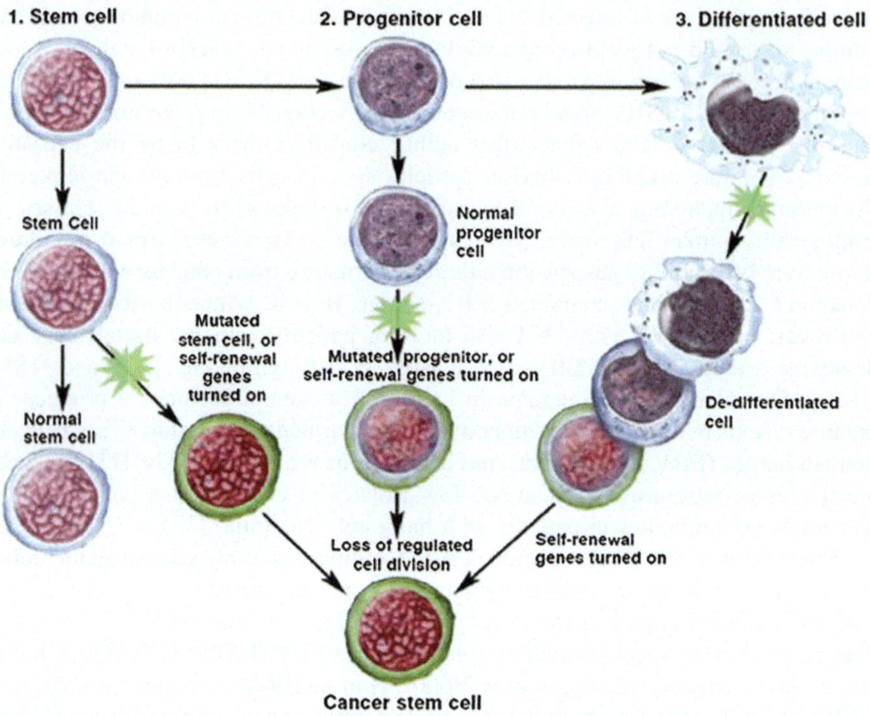

Figure 71 The Cancer Stem Cells Practice Reversal of Ontogenesis and Readily Trans-speciate Themselves as Unicellalar Eukaryotes (the Medusozoa) Do Within Their Life Span. © Terese Winslow. All Rights Reserved 2015. Reference Appendix 2, Explanations to the Figures

oncoproteins." In Lynch syndrome (colonic carcinogenesis), the fusion oncogenes are named after ancient cell survival genes of *Saccharomyces cerevisiae* (once dubbed "*S. sapiens*") [1858a, b].

It remains unresolved if malignant transformation ("re-immortalization") is the exclusive privilege of the preserved stem cells, or if germ cells, or somatic cells could undifferentiate ("de-differentiate") into a stem cell status (Figures 41 and 71).

The Computers Win the Game of Chess and Out-Perform Their Inventor, the Human Brain

The "Computer Biologists" are able to predict which oncogene will strike and how. How to counteract the invasion of the oncogenes (no sarcasm meant). External interventions with radiation and chemicals may decimate the new colonists (and harm the host as well), however resistant new mutants of the new colonists arise to re-occupy the host. The extraordinary resistance of the primordial ancestors to physico-chemical damage and their inherent inexhaustible mutagenicity are regained by the return of the primordial RNA/DNA as "oncogenes."

At this point, the supreme "computer biologists" opened up an immense repertoire (Francesca Ciccarelli and Matteo D'Antonio *et al*) [1859a, b]. Of the close to 1500 recognized human oncogenes, their strategy will have to be understood practically one by one. May be, first generation oncogenes act differently from second or third generation oncogenes. Thus, treatment could be designed through somewhat less complicated considerations, that is not against 1500 oncogenes individually, but against their grouping according to the generations of their historical arousal. Special attention is to be directed to the epigenetics of the individual oncogenes. The mRNAs of DNA oncogenes respond to their natural epigenetic microRNAs (or man-made antisense siRNA oligonucleotides). Antagomirs *versus* oncomirs may become the closest to natural defensive/therapeutic agents. This conflict is to be replicated at the clinics in favor of the antagomirs (Table XXIII) (*vide infra*).

Special attention is addressed to the ancient DNA methyltransferases (DNMT1/3) and their genes [1859a, b]. The approximate loci of these genes in the human genome are 4p16.3 and 4q35.1. In Chinese patients with acute myelogenous leukemia DNMT1 is amplified during the disease and its expression declines in complete remission. Zebularine is a specific inhibitor of DNMTs. In xenografted gastric carcinoma cells zebularine inhibited mitoses and activated mitochondrial apoptosis with suppressed Bcl-2, and elevated Bax and caspase-3 [1860, 1861]. The JAK/STAT3 and leukemia inhibitory factor (LIF) pathways interact with stem cell genes of Oct4 and Nanog. These stem cell genes are silenced in terminally differentiated fibroblasts (mouse), Activated Jak/Stat3 and LIF induce reprogramming of these cells into pluripotent stem cells. Reprogramming is preceded by the expressions of DNMT1/3, HDAC, lysine-specific histone demethylase-2, and CH3-CpG-binding protein-2. Inhibition of Jak/Stat and LIF abolishes the reprogramming pathway [1862].

Supreme "computer biology" [1859a, b] places PTEN also in a central position with many interconnections. The hindgut endoderm differentiates into cells phenotypical for rectal and bladder lining and prostate parenchyma under the effect of PPARγ2 (peroxisome proliferator activated receptor, Table II), and the prostate epithelium gains AR expression. In case of silenced PPARγ2, the prostate-directed cells transdifferentiate into urothelial cells. In case of silenced PPARγ2 and PTEN in the mesenchyme, the endodermal cells de-differentiate into keratinizing squamous metaplasia [1863]. Deletion of chromosomal locus 10q23,3 is tantamount to the loss of the PTEN gene. At its locus, the PTEN gene is flanked by genes BMPR1A (bone morphogenetic protein receptor) and FAS (fatty acid synthase). The breakpoint usually is within these two fragile genes [1864]. Deletion of PTEN liberates the PI3K/Akt pathway. The AKT cascade downregulates miR-199a-5p, and the FasL (factor apoptosis signaling; ligand), and upregulates Hif1α (hypoxia-inducible factor). FasL rises upon inhibition of Akt by wortmannin. In contrast, rising microRNA miR-21 results in dropping FasL levels [1865]. In PTEN-deficient prostate cancer cells, TGFß inhibits the expression of the metastasis suppressor N-myc downstream-regulated gene 1 (NDRG). NDRG1 prevents actin filament-polymerization, inhibits the locomotor pathway of Rho-associated coiled-coil protein kinase 1- (ROCK-) phosphorylated myosin light chain 2 protein (pMLC2). Consequential fiber assembly and contraction in iron-depleted cancer cells with NDRG1 expression renders tumor cells immobile [1866]. In contrast, silencing NDGR1 allowed the activation by phosphorylations of proto-oncogenes AKT, ERK1/2 and SMAD2L. The iron chelator di-2-pyridylketone 4,4-dimethyl-3-thiosemicarbazone (Dp44mT) *via* the activated NDRG1 inhibits the ROCK/pMLC2 pathway. It also inhibits the iron-containing enzyme, ribonucleotide reductase (RR). There is no actin filament-polymerization and fiber assembly; the tumor cells rest standstill. The agent Dp44mT could reactivate the silenced PTEN gene and reduce Ras/MAPK signaling that leads to the phosphorylation of proto-oncogenes ERK1/2 and SMAD2L (signaling mothers against decapentaplegic, TGFß, drosophila) [1867, 1868a, b].

The single exon Sox2 gene (sex determining region Y, SRY box 2) ranks very high among the toti-pluripotency regulator stem cell genes. An ancestral Sox RNA/cDNA sequence might have coexisted in the precellular world within a long RNA sequence and travelled through evolution highly conserved in that relationship. It remains embedded in a long noncoding intron RNA in the human genome (3q26–3q27) [194a, b]. Its expression is regulated in the control of its mRNA by mi-RNAs. In ontogenesis, it encodes neural cells from the ectoderm. Its loss-of-function mutations in the embryo result in micro- and anophthalmia. In adult life by amplification of its expression (not necessarily by gain-of-function mutations) induces adenocarcinomas (endometrium; prostate) and lung squamous cell carcinomas by upregulating markers p63 and keratin 6 [194a, b, 1869–1872]. The Sox2 mRNA, together with mRNAs of several other stem cell genes (Nanog, Oct4) are under the control of microRNAs [1873, 1874] (Figure 27).

Double agent IKBKE (inhibitor kappa light chain enhancer B cells). It was known that tumor suppressor gene PTEN controlled PI3K, and PI3K activated Akt by attaching to its pleckstrin-homology (PH) domain, thus allowing its phosphorylation. The IKBKE gene product protein (inhibitor of nuclear factor kappa B kinase subunit epsilon) is a PI3K-independent direct activator of proto-oncogene Akt. Activation of Akt occurs by phosphorylation of Akt-Thr308 and Ser473, but not through the PH domain, because PH inhibitors perifosine and triciribine do not inhibit IKBKE-mediated phosphorylation of Akt; Akt inhibitors MK-2206 and GSK690693 (patented) canceled IKBKE-mediated phosphorylation [1875]. Throughout its extremely long evolutionary activity working as a cell survival pathway in unicellular protozoa (giardia), the PI3K/Akt pathway gains phosphorylation at Ser 473 also by mTORC, but only when Akt is translocated to the plasma membrane. This translocation is accomplished by small GTPase Rac1 (*vide supra*). In inflammatory carcinogenesis; phorbols and LPS activate Akt, but the activated Akt is exempted from the inhibitory effect of the classical PI3K inhibitors LY294002 and wortmannin. Phorbols and LSD also activate IKBKE, endowing it with a role in inflammatory carcinogenesis. The IKBKE protein activates inflammatory cytokines (IFN-response factors), and STAT. Not directly, but through cyclin D, in a pleiotropic relationship, it may activate NFκB (*vide infra*). However, proto-oncogene Akt is a substrate of IKBKE. Its Akt-activatory effects make IKBKE an oncogene. In several human cancers (breast, prostate, non-small cell lung cancers), STAT3 and IKBKE activate each other vice versa. In breast cancers, IKBKE acts as an antagonist of tamoxifen. In lung cancer, nicotine and nicosamine ketones activate IKBKE. The IKBKE mRNA yields to regulatory miRs: it is targeted by miR-296 released from human chromosome 20q13.3, where it is imprinted by the paternal allele (and repressed by the maternal allele) [1876–1878a, b, 1879].

Originally, the IKBKE protein appeared in the early era of innate immunity, and has been engaged in TLR signaling regulation. There is an ancient relationship between TLRs and viral infections. Intrauterine CMV infections of the fetus and newborn babies cause deafness and mental retardation. The search for an effective CMV vaccine included the live attenuated Towne strain CMV, canary pox virus/CMV recombinants, or DNA plasmids, to express CMV antigens pp65 and glycoprotein gB [1880–1883]. The glycoprotein gB-MF59 vaccine induced in child-bearing young women anti-CMV immunity, efficient in preventing the maternal CMV infection [1884]. Single nucleotide polymorphisms in the TLR7 and IKBKE genes influence human immune reactions to CMV viral infections. Against such genetic background, the immune response to gB CMV vaccination was altered. In human subjects, CMV infection is recognized byTLR1/2/7/9 leading to signal protein-1 SP-1/NFκB, and type I IFN response. Plasmacytoid DCs produce most of the type I IFNs. Patients receiving liver transplant against the background of TLR2 SNP are highly susceptible to virulent CMV infection. To CMV gB vaccination, homozygous carriers of SNPs in TLR7 responded with the highest virus neutralizing antibody titers. In these subjects, CMV neutralizing antibodies prevented the SP-1/NFκB activation. The IKBKE I-kappa-B kinase attaches to the complex NFκB forms with its inhibitors. The IKBKE protein phosphorylates the inhibitor and the dissociated NFκB is liberated, while its inhibitor is degraded. The IKBKE protein phosphorylates IRF-3 as well (IFN-regulatory factor). In the maturing DC, IRFs and NFκB work in an transcriptional activation complex with a regulatory element, a promoter and an enhancer expressing multiple IRF- and NFκB-binding sites. As the DC reaches full maturation, it expresses CD83 [1884, 1885]. Vascular smooth muscle cells infected with CMV (with or without co-infection with chlamydia) become proatherosclerotic plaques [1886–1888]. In these smooth muscle cells, I-kappaB kinase, IKKepsilon, IRF, and chemokine genes for RANTES and CXCL10 (regulated upon activation normal T cell expressed and secreted; IFNγ-inducible protein ligand) are expressed [1889]. CMV overcomes the innate defense before the rise of the adaptive immune faculties. Out of these evolutionarily deep interrelations with basic innate immune faculties, emerges IKBKE in the human genome, as a proto-oncogene/oncogene.

The old Master Horologist may misdirect a nick? In the frenzy of somatic hypermutation, the enzymes AID or RAG may deliver DNA nicks at the wrong sites, thus bringing proto-oncogenes together for fusion by targeting the IgH locus. Not only the ever-present herpesviruses (EBV-HHV4, HHV-6, KSHV-HHV8) may go epigenetic, or insert themselves into the telomere, unisons of hitherto dispersed endogenous retroviral genes may also occur resulting in encoded budding virus particles [147]. If tumor suppressor genes suffer silencing, or inactivating mutations, germinal center B cells of naturally high proliferation rate yield to processes of malignant transformation (lymphomagenesis) [1889].

Physiologically, the enzyme, activation-induced cytidine deaminase (deoxycytidine deaminase, cytidine → uridine, AID) keeps the mammalian plasma cell's IgH heavy chain locus' switch region constitutively transcribed, as the heavy chain CH genes undergo constitutive germ line gene transcriptions. The 5-exons AID gene maps to chromosome 12p13 by FISH (fluorescence in situ hybridization) [1890]. The enzyme AID mediates the initiation of the somatic hypermutations in the immunoglobulin variable (V) regions and in the immunoglobulin class switches of mature B cells. DNA breaks, mismatched repairs, and somatic DNA recombinations occur. The enzyme AID deaminates ssDNA at WRC cytidine motifs, or WGCW "hot spots" (W = A/T, R = A/G, or could be purine). Even DNA/RNA hybrids are formed between the template DNA strand and the germ line gene transcript in order to stabilize the ssDNA substrate for the action of AID. The process of class switch recombination is not exempt of errors. Histone (H) modifications occur within the chromatin, while H4 lysines (K) 3, 4, 9 become silenced by trimethylation (H4K4me3/H3K4me3; H3K9me3/H3K27me3). Upon IL-4/anti-CD40 stimulation, H3 K9 3 m from silenced state becomes activated. Further, for the transfer of AID from cytoplasm to nucleus to act upon its target genes, IL-4/anti-CD40 stimulation is necessary [314–316].

All events of "oncogenesis" are in action in the physiologically stimulated V(D)J (variable, diversity, joining) system, in order to encode the antigen-specific configurations of the new IgG light chains for antibody diversification. However, off-target actions of AID may induce lymphomagenesis. AID-dependent *c-myc/IgH* translocation occurs in Burkitt's lymphoma cells. This translocation triggers p53 by the miR-142 promoter involved in the translocation (*c-myc/miR-142*) to induce cell death by apoptosis. The oncogene *bcl6* (encoding oncoprotein B cell lymphoma, BCL6) is another direct target of AID. The BCL6 protein is a nuclear transcriptional repressor and as such, an antagonist of p53. When expressed constitutively in germinal center cells, the resting proto-oncogene *bcl6* becomes an oncogene. Translocation t(3;14) of the *bcl6* gene carries adverse prognosis for the lymphoma it causes [1891a, b]. Uni- or multicellular hosts without an adaptive immune system do not show the presence of a recognizable *abl*6 gene; it is present in hosts of adaptive immune systems. However, the BCL6 system interacts with two elements of the innate immune system. It interacts with NKT cells, which form chemokine $CXCR5^+$ follicular NKT cells, acting as helper cells to IL-21-dependent antigen-specific B cells [1892]. Follicular dendritic cells secrete ß-catenin-independent Wnt5a, NFκB and BCL6 for the protection of germinal center $CD40^+$ B cells from apoptotic death [1893]. The natural antagonist of the BCL6 protein is the protein FBXO11 (F-box protein, orphan), that binds the BCL6 protein, thus directing it to ubiquitination [1894].

The 3′ UTR of the AID mRNA is a direct target for miR-181b, resulting in the down-regulation of AID translation. In lymphoma cells (Burkitt's, mantle cell, diffuse B cell) high expression of miR-181b prevented the accumulation of high levels of AID, whereas endotoxin (LPS) and IL-4 stimulated AID production. Activated B cells show low miR-181b levels (334). Further post-transcriptional modulation of AID gene expression is exerted by the repressor miR-155. The host cell genomic site for the insertion of avian leukosis virus-induced lymphoma is BIC; the non-coding BIC (B cell integration cluster) RNA is the precursor of miR-155. In the human genome, the *bic*-related MIRN155 gene releases the miR-155 precursor. The *aicda* gene (activation-induced cytidine deaminase) encodes AID; the AID mRNA contains a single mi-R155 binding site in its 3′ UTR. By inhibiting AID-induced c-*myc* translocation, miR-155 acts as a tumor suppressor microRNA [335, 336]. The promoter gene of miR-124-1 in chromosome 8p is frequently methylated (silenced) in human malignant lymphoma cells, including those of NK/T lymphoma cells. The miR-124-2 is a direct inhibitor of the CDK6 mRNA (cyclin-dependent kinase) [337].The AID/APOBEC (apolipoprotein B RNA editing) family co-evolved in vertebrates with adaptive immunity. The APOBECs are RNA cytidine deaminases; APOBEC3 is most active in placental mammals and co-exists with AID [1895]. APOBEC3 restricts retroviral insertion (but failed to protect the human genome against retrolentiviral HIV-1 insertion). APOBEC5 was discovered in the bacterium *Xanthomonas oryzae*. APOBEC1,2,5 are ancient innate family members (Lada et al) [306] (Figure 47).

Trans-speciated defensive cells turned out to become the alien invaders. Dendritic cell sarcomas (previously referred to as histiocytic sarcomas, or Langerhans tumors) appearing as follicular or interdigitating, and deriving often as the result of some "trans-speciation"from a pre-existing lymphoblastic tumor. The $CD52^+$ histiocytic sarcoma may regress upon alemtuzumab therapy [1896]. Tumor necrosis factor-α-inducible protein (TNFAIP2) is expressed in follicular and interdigitating dendritic cells, whether benign or malignant. This antigen (TNFAIP2) is overexpressed in histiocytic sarcoma cells, in Reed-Sternberg cells of Hodgkin's disease, and in large B cell thymic lymphomas [1897]. An argument for ancestral DNA genomes working in these tumors is the ability of the RNA/DNA complex to transdifferentiate tissues in elderly male patients from chronic lymphocytic leukemia, follicular lymphoma, or mantle cell lymphoma, into histiocytic sarcoma, interdigitating dendritic cell sarcoma, or Langerhans cell sarcoma, while clonal immunoglobulin gene rearrangements remain the same. Other gene expressions were negative for PAX5, weak for CEBPβ, and strong for Spi/PU.1 (C-enhancer-binding protein; avian erythroblastosis retrovirus E26 E-twenty-six oncogene *ets* family proteins Spi/PU spleen focus forming erythroleukemia virus insertion site; proto-oncogene PU.1 (purine-rich) encoding DNA-binding transcription factor regulating CD11b myeloid-specific promoter; recognizing for binding to purine-rich DNA sequence containing 5′-GAGGAA-3′ core box) [1898–1900a, b]. Latent EBV may or may not be present in the epigenome. The $CD21^+$, $CD68^+$, $S\text{-}100^+$ follicular dendritic cell sarcoma of the breast was free of EBV genomic sequences [1901]. The fact that the preceding lymphoid cell tumors and the subsequent dendritic cell tumors exhibited identical clonal immunoglobulin gene rearrangements (but different phenotypes) attests to the thaumaturgic capacity of the DNA molecule, already proven multifold, in the unicellular eukaryote Naegleria, in the metamorphosis of insects, amphibians (from the tadpole gills to the frog's lungs), flying reptiles (Archaeopteryx) and mammals (bats), and the return of dry land mammals to the sea (ancestral hippos to whales).

The malignant transformation in interdigitating dendritic cell sarcomas, and in macrophages (or Langerhans cells) may invoke the ancient DNA genome, however the rarity of these tumors so far prevented detailed studies on the genomics of these events. The follicular and interdigitating dendritic cells in lymph nodes are referred to as accessory cells. Accessory cell tumors morphologically imitated giant Reed-Sternberg-like cells positive for EBV genome, and EBV-generated microRNAs [1902], but were free of *bcl2* gene chromosomal translocations. Such translocations occur in African (or outside Africa) Burkitt's lymphomas from the locus of *bcl-2* at 18q21.3 as t(2;18)(p11;q21.3), or t(18;22)(q21.3;q11), and from the locus of *c-myc* at 8q24 as t(8;14)(q24;q32). These lymphoma cells usually (but not necessarily) are carriers of epigenetic EBV genomes. The EBV latent membrane protein 1 (LMP) activates in B lymphoblasts anti-apoptotic proteins Bcl-2 slowly (in 48–72 h), but permanently, and Mcl-1 (myeloid cell leukemia) factor immediately; the activated Mcl-1 is sustained by LMP1 [1903]. A review of the published morphology of the multinucleated giant accessory cell tumors strongly suggests to this author that these cells may be "natural hybridomas". Events of this sort have been predicted and eventually illustrated [147, 1904a, b].

Endogenous and Exogenous Oncogenesis

Viral oncogenesis exogenously enforced elicits immune defenses. If the oncogenes are exogenous and alien (oncogenic viruses), the initiated transformed cells grow in multifocal cancers (KS- and/or EBV-associated vascular and leiomyosarcomas; lymphomas; human papilloma virus-induced uterine cervical, or H&N squamous cell carcinomas; polyoma virus-induced carcinomas of the genitourinary tract, or Merkel cell tumors); or takeover of the entire lymphatic and hematopoietic system by the human T-cell lymphotrophic leukemia virus (HTLV). It is estimated that approximately 20 % of human malignant tumors are induced by oncogenic viruses. At the time when Ludwik Gross published several editions of his textbook "Oncogenic Viruses" (1961–1983) [1905], most of the oncologists believed that this figure was close to the 100 % range. However, the first droplets of doubts were instilled by H. Hanafusa, and H.E. Varmus and J.M. Bishop claiming that proto-oncogenes and their gene product proteins were present and functional in healthy cells [1906a–f, 1907a–c].

Polyomaviruses. The emergence of a few SV40 simian polyomavirus-induced human cancers occurred after the ingestion of the first batch of the attenuated live poliomyelitis virus vaccines; reviewed in [1908]. Some slowly replicating SV40 virus particles containing only one 72 bp enhancer remained much less detectable in the original poliomyelitis vaccines, than the rapidly replicating virions containing two 72-bp enhancers; thus some products declared to be SV40-free still might have been infected [1909]. Cancer-free carriers of the SV40 virus still live in human communities with poor hygienic conditions. In these communities this virus is transmissible through the fecal/urine-oral route [1910]. In mouse tissue cultures, SV40 induced aneuploidy and tumorigenicity even in telomerase-deficient cells, as these cells gained reproductive autonomy and immortality. The carcinogenic murine polyomavirus was discovered twice, first and originally in murine salivary gland carcinoma by Ludwik Gross [1911], and thereafter in tissue cultures established at the NCI [1912]. The polyomaviruses were known to cause human infections in the GU tract and in the brain (progressive multifocal leukoencephalopathy [1913, 1914].

Merkel cell carcinoma-associated polyomavirus (and before that, the HHV-8/KSHV) were discovered by Yuan Chang and her associates [1915]. The human Merkel cell carcinoma-associated polyomavirus strains were isolated by the digital transcriptome subtraction technology. Viral large T proteins (T1, T3) fused with a human tyrosine phosphatase receptor, and capsid proteins (VP1) were found in these tumor cells. The viral genome consisted of 5387-bp and in several cases the viral genome was integrated in the host cell genome. These tumor cells expanded clonally [1915]. The presence of integrated polyomavirus in Merkel carcinoma cells was immediately confirmed [1916–1918]. UV light exposure promoted the development of these tumors [1919]. The tumors induced lymphocytic infiltrates [1920–1922], and the extent of these infiltrates positively correlates with the prognosis. The polyomavirus large T antigen is a strong immunogen. Merkel cell carcinoma vaccines and anti-Merkel cell polyoma virus monoclonal antibodies are in use in investigational clinical trials. Merkel cell carcinomas, even metastatic tumors, were observed to regress "spontaneously"; reviewed in [1920].

In inflammatory carcinoma of the breast, the lymph channels are occluded by tumor cell emboli. The entire breast is imbibed in a massive lymphedema. However, the interstitium displays calor (hot) and color (red), without frank purulence. It does not yield specific bacterial or viral isolates. The luminal subtype A, and the HER2-enriched genomics are overexpressed (amplified), while TGFβ is attenuated. Presumably, natural TGFβ exerts protection against metastases; that faculty is lost in the inflammatory breast cancer cells. These cells expressed homozygous common missense variant single nucleotide polymorphism in the RNase L gene. Inflammatory breast cancer cell failed to show mouse mammary carcinoma-like retrovirus expression (*vide supra*) [1923, 1924]. In the absence of isolated pathogens, this clinical entity appears to be endogenously induced.

Comment. It was a response to physico-chemical inducers,that the ancient cells managed to survive by activating their cell survival pathways. Antiviral self-defense mechanisms were installed. In the end, the descendants of the single cells, and their viruses both remain in existence, forming the largest biomass on Earth. The primordial cell survival pathways remained conserved in the form of much more elaborate, and interrelated sequences In the stem and somatic cells of multicellular hosts. In addition to physico-chemical inducers, it is an armada of biological agents (inducers of acute to chronic infections and inflammations), that the multicellular hosts must endure and respond to. It is entirely compatible with the *modus operandi* of evolutionary biology, that some of these stimuli activate the successors of the ancient cell survival pathways in stem and somatic cells of the multicellular hosts, if persistent, constitutively. "Inflammatory oncogenesis" is either recognized to be exogenously induced, or it is evoked by endogenous "danger signals" in the chronically infected host to mobilize an inherent primordial "rescue mission" for the survival and immortalization of some transformed single cells. By natural selection, these cells are fitted most to survive under the changing circumstances.

In relatively immunocompetent patients, the clinical course of virally induced or virally-associated malignant tumors (EBV-related follicular lymphoma; KSHV-related Kaposi's sarcoma, HPV-induced squamous cell carcinoma; Merkel cell carcinoma) may run a protracted indolent clinical course with possible "spontaneous" remissions. In contrast to tumors induced by exogenous agents, endogenously induced tumors advance stealthily, insinuating of being "self". Endogenously induced tumor cells may be the results of a process of natural selection for survival, as individual cells are extricating themselves from the organized cell community, which is in the state of impending demise. This type of "endogenous oncogenesis" is often supported by its multicellular host organism. In contrast, oncogenesis imposed and enforced from the outside ("exogenous oncogenesis"), for instance, by an oncogenic virus, is opposed and often elicits an immunological defense reaction of the host.

Selected Abbreviations

DENN proteins, differentially expressed in normal and neoplastic cells. The N-terminal Rab GTPase (guanine-triphosphate; ras-like GTPase-binding protein) longin domain is of ancient prokaryotic derivation, that regulates vacuolar-vesicular transmembrane trafficking in eukaryotic cells and participates in autophagy (Jekely G Bioessays 2003;25:1129-38; Adv Exp Med Biol 2007;607:38-51; Biol Direct 2008;3,31.10.1186. Zhang D et al Front Genet 2012;3:283).

IKBKE, inhibitor nuclear factor kappa-B kinase subunit epsilon. IKBKE directly activates proto-oncogene Akt; Akt inhibits FOXO3a (forkhead box O) by phosphorylating its serines 32, 253, 315t; IKBKE degrades FOXO3a transcription factors in adenocarcinoma cells (breast and lung cancers). IKBKE interacts with NEMO and TRAF (*vide infra*), with interferon response factors (IRF), and with signal transduction and activation of transcription (STAT). FOXO target genes include reactive oxygen species (ROS), cyclin D, and FasL (factor apoptosis stimulation ligand to receptor Fas/CD95/ APO, apoptosis) (Guo JP PLoS One 2013;3(5):e63636).

NDRG, N-myc downregulated gene. Acts as tumor suppressor in colon adenocarcinoma and in glioblastoma, thus it is downregulated in actively growing tumor tissues; suppresses matrix metalloproteinase (MMP); downregulates Wnt/β-catenin (wingless; integrated) (Ding W et al OMICS 2012;16:263-7; Liu W et al EMBO Mol Med 2012;4:93-108; MA JJ et al Med Oncol 2012;29:3306-13; Melotte V et al J Natl Cancer Inst 2009;101:916-27).

NEMO kinase (regulatory subunit IKKγ), NFκB (nuclear factor kappa B lymphoma) essential modulator, inhibitor of kappa kinase gamma (IKKγ). Protein kinase C-interacting protein p62/sequestosome-1, is activated by interleukin IL-1β; in sequence, it activates of nuclear factor NFκB in TNFα-stimulated cells (Zotti T et al Mol Immunol 2014;58:27-31). NEMO promotes vFLIP (Fas-associated death domain-like interleukin-1-converting enzyme inhibitory protein) expression by Kaposi sarcoma associated herpesvirus (KSHV) (Tolani B et al J Virol 2014 March 26 PMID 24672029).

TRAF, tumor necrosis factor-associated factor; TTRAP, TRAF/TNF-R receptor associated protein. System was operational in ancestral mollusks (Yang J et al Dev Comp Immunol 2011;35:827-34). TRAF6 and p62 sequestosome suppress NEMO by ubiquitination; depletion of p62 by siRNA abrogates TRAF activity. NFκB activation depends on the ubiquitinational state of NEMO (Zotti T et al MoI Immunol 2014;58:27-31). TTRAP inhibits NFκB activation (Pype S et al J Biol Chem 2000;275:18586-93).

The Strategy of the Oncogenome

Some Selected Cells Are Subjected to the Transformation Process

Attempts at rescuing selected cells of a doomed host. Is it an inscribed ancient faculty inherent to the basic nature of Life? The "cell survival pathways" activated in unicellular eukaryotes in distress are the predecessors of PI3K, STAT, MAPK, Hh, Wnt, mTOR, etc. oncogenic pathways. The pluri(multi)cellular host defends itself against exogenous carcinogens, that act at different levels. Nuclear exposures to chemoradiation, or events of hyperoxidation (reactive oxygen species, ROS) induce dsDNA breaks and point mutations, that may be repaired (may remain unrepaired); oncogenic viruses take over the entire cellular metabolism, both in the nucleus and in the cytoplasm. The dsDNA damage is to be repaired; and the virus is to be neutralized by immunological means. In addition to exogenous carcinogenic forces, presumably, endogenous carcinogenesis exists as a rescue mission initiated by the conserved remnants of the primordial RNA/DNA complex. This inherent mechanism is envisioned as to be responsive to some danger signals emitted in a multicellular host undergoing the processes of a threatening demise. Reactivated endogenous retroviruses (*vide infra*) may represent such danger signals. In contrast to carcinogenesis induced by exogenous forces, this mechanism of oncogenesis appears to be accepted without resistance; frequently, it is even supported by the multicellular host. It appears as if the host actually promoted and actively cooperated with the endogenous form of oncogenesis. The newly arising endogenously generated transformed cell colonies receive blood supply and all the exocrine growth factors needed for their thriving; the immune reactions mobilized are more promoting than inhibitory to the welfare of the proliferating colonies of the newly arisen cells.

The oncogenomes of multicellular hosts and the cell survival pathways of the first unicellular life forms, as viewed retrospectively, appear to be very similar, with the probable difference being that the ancient cell survival genes in contrast to the

J.G. Sinkovics, *RNA/DNA and Cancer*, DOI 10.1007/978-3-319-22279-0_24

oncogenome, were not necessarily activated constitutively (inconvertibly). The ancient cell survival pathways (genes and gene product chemo-, lympho-, and cytokines) are identical, readily amplified, but are not necessarily constitutively activated, or deriving from point-mutated genes. In early phases of inflammatory carcinogenesis, the tumor suppressor genes are reversibly silenced (by epigenetic hypermethylation of their promoters), whereas, the proto-oncogenes are reversibly activated (by epigenetic de-acetylation of their promoters); reviewed in [73]. In a protracted process, whereby the somatic cells differentiated, or the stem cells provided differentiating somatic cells, cell survival pathways become silenced, and senescence and natural death set in as the inevitable conclusion of the life span in the somatic cell population. In order to dislodge these mechanisms off their set course, irreversible (constitutive) activation of the cell survival pathways must be installed. Thus bacteria or protozoa (Trypanosoma) in culture may imitate immortality by replicating before senescence [416a, b–417a, b]. The mechanisms of "immortality" existed in the primordial cells.

The organized multicellular communities are served by the large colonies of end-differentiated somatic cells. The somatic cells are allowed of a limited number of cell divisions as advancing to the unavoidable event of senescence: the telomeres are left uncapped after each cell division. As compared to the readiness of the stem cell to assume a primordial phenogenotype, such reversal of the somatic cell is much more hindered (by inbuilt cell-suicide mechanisms; by killer cells policing the cell community). However, the ancient genomic elements are conserved. They could be reactivated by the appropriate growth factors in the differentiated keratinocytes. These cells regressively transformed into stem cells, and then differentiated into nerve cells in the Shinya Yamanaka laboratory [1925] (in Appendix 1).

One form of endogenous "oncogenesis" may be considered to be an attempt of the ancient RNA/DNA complex to induce a reversal of some selected host cell to its ancient life style, however, in a constitutive biochemical setting. The process may be initiated in just one cell of a large multicellular host. Incipient colonies of such cells may even be eliminated with, or even without, some memory cell-preserved immunity left behind (Burnet's surveillance) [1708]. There is a switch, when the host accepts and begins to support the rising new colony. There appear to be two colonies of stem cells. One, is the de-differentiated somatic cell. It is not fully de-differentiated: the pathologist recognizes the germ line, from which it originated. It is rapidly dividing, producing large masses of tumors. It retains some chemo-radiotherapy susceptibility. It may be driven to retreat into a remission. Two, the second is the genuine stem cell, undifferentiated, not in need of blood supply, as it survives on diffusion in hypoxic avascular environments. It divides seldom; its divisions become symmetrical, yielding two stem cells (no new somatic cells are produced anymore). It is totally resistant to chemicals (expels them by an ancient pumping mechanism, like toxic sea-algae, dinoflagellae), and has the radioresistance of *Deinococcus radiodurans* (Figure 4). The mutated stem cell either promptly repairs its DNA (but not necessarily error-free), or unites trunked genes by fusing them into new genes (the fusion oncogenes). The process of oncogenesis is readily initiated in the genomes of stem cells, where the genes of stemness are retained in an active state; they will be expressed now constitutively without switch off. The promoter of focal adhesion kinase (FAK) initiates the action by activating Nanog through phosphorylation, and MDM for the neutralization (ubiquitination) of the p53 protein. Hormone-dependent breast and prostate adenocarcinoma cells in response to hormone deprivation or hormone receptor neutralization mobilize stem cells for transdifferentiation into another cell type (mesodermal/mesenchymal, vascular endothelial, or neuroectodermal) for hormone-independent growth (Table XX; Figures 67 and 71). Or, it may activate its androgen hormone-encoding genes for an autocrine loop. The state of the epigenome (histones and chromatins) and the microRNAs will be essential to maintain this state of existence.

In the endogenously initiated process of malignant transformation, the "transformed cell" activates events occurring serially and consequentially. With every new somatic mutation, the transformed cell gains further virulence (increased proliferation, locomotion and resistance). In contrast to the "loss-of-function" degenerative, vertically inherited one-gene mutations, oncogenesis consists of a series of somatic mutational gains of cell viability ("gain-of-function" mutations). Mutations generated in "inflammatory carcinogenesis" [73] are exogenous or endogenous? Oncogenic viruses, bacteria and protozoa induce the exogenous form of oncogenesis. But not directly oncogenic causes of inflammation (without an oncogenic pathogen) involving TGFβ-producer $CD8^+$ T-cells armed with ILEI (interleukin-like ETM transition inducer), if working with point-mutated *ras* genes (H-Ras^{V12}/K-Ras^{V12}), induce the stemness of the cells in the environment. The malignant transformation and the ETM-type locomotion of breast cancer cells may arise in this environment. The classical Snail and Twist proteins induce with (or without) *ras*-activation the ETM reaction, and the assumption of the stemness pheno-genotype [1539a, 1540a, b]; reviewed in [26, 73].

When the living cell communities found themselves in circumstances unsustainable for survival, they either adopt by natural selection the best fitted formation, or perish. The RNA/DNA complex acquired its faculties to sustain life throughout milliard years of evolution by practicing natural selection, even if it takes "transspeciation" (sea-living mammals; flying reptiles, or mammals). In a multicellular host, circumstances may arise (infections and inflammatory processes prominently included), that call for the natural selection of the fittest. Under those circumstances, the ultimate achievement of bioengineering, the so called cancer cell is the fittest to survive. Aneuploidy, the introduction to trans-speciation, sets in [1277, 1926].

The primordial "oncogenome", the RNA/DNA complex, performs what its native structure dictates. The major faculties expressed by the transformed cell that resemble those of the ancient life forms driven by the primordial RNA/DNA complex are as follows.

1. Elimination (silencing; deleting) of genes inhibitory to cell divisions, or to "cell survival pathways" (the PTEN to phosphatidylinositol kinase PI3K antagonism), referred to as the "tumor suppressor "genes: WTp53, Rb, PTEN, SuFu,

Table XX The Suppressed Stem Cells of Hormone-Sensitive Cancers Strike Back

Breast carcinoma (BC). Signals from ductal BC cells *in situ* eliminate the barrier of myoepithelial cells, and invite the migration of granulin-expressing stromal cells of bone marrow derivation for the receipt of some as yet unidentified growth factors. The carcinoma cells transit into the locally invasive stage. In the microenvironment, $CD8^+$ tumor-reactive immune T cells are gradually replaced by Treg cells. Some cancer cells activate NFκB through the RANK pathway (receptor activator of NFκB). The ligand for this receptor is provided by Treg cells.[1] Subclones of the cancer cell population undergo cytoskeletal and cell surface transformations allowing them to travel, settle and colonize distant sites (lungs; bones; liver; brain). Avascular hypoxia (induced by bevacizumab) increases stem cell population.[2] ILEI (IL-like ETM inducer), and TGFβ-expressing $CD8^+$ T cells induce the transition ETM from epithelial to mesenchymal geno-phenotype, thus creating invasive and metastasizing tumor cells.[3, 4] Hormone-receptor positive (ER^+) lymph node-negative tumors expressing the favorable HOXB13/IL17BR ratio (higher HOXB13, lower IL17BR) remain relapse-free upon protracted antihormonal therapy (letrozol following tamoxifen), whereas otherwise they would relapse. Normal mammary stem cells become tumorigenic consequentially to genomic and epigenetic changes (activated Sox2 and Nanog; amplified HER2; silenced ER/PR; activated stage-specific embryonic antigen SSEA, and integrin-α6 CD49/ITGA6; $CD44^{hi}/CD24^{lo}$. Cis-platin (CP) acts upon DNA, but the majority of CP molecules react with cytoplasmic protein structures: actin, tubulin, and zink-fingers, thus inducing changes in the cytoskeleton. CP negatively alters CD49 mRNA. Stage-specific embryonic antigen 4 (SSEA4) expression was reduced by CP, not by mRNA blockade, but by posttranslational action. CP induces differentiation of BC cells.[5,6]

Prostate carcinoma (PC). Differentiated basal epithelial cells may de-differentiate into pluripotential embryonic stem cells expressing Oct4, Sox2, c-Myc, Klf4 after treatment with retrolentiviral vectors of these genes. These cells of normal karyotype express also Nanog, SSEA-3, and tumor recognized/rejection/related Tra-1-81antigen genes. These cells could be induced to re-differentiate into AR- and PSA-secreting mature prostate cells. miR-let-7 promoted differentiation by suppressing EZH2 (enhancer of zeste polycomb protein, drosophila). miR-1 and miR-145 suppressed stem cell renewal (downregulated in prostate cancer).[7,8] Suppressing AR, standard androgen-deprivation therapy (ADT) of prostate cancer induces $CD133^+$ $integrin^+$ $CH5^+$ $CK8^-$ cancer stem/progenitor (S/P) cell invasion. Functional AR maintains cancer non-stem/progenitor (non-S/P) cell population. The highly invasive S/P cells exhibited very high tumorigenicity, far exceeding that of non-S/P cells. The AR gene was silenced in S/P tumor cells (no AR mRNAs released), whereas non-S/P tumor cells expressed AR. The less AR the S/P tumor cells expressed, the higher was their Ki-67 score. Forced AR expression suppressed the growth of S/P tumor cells *in vivo* in xenografts. In contrast, the higher the AR expression was in non-S/P cells, the higher was their growth rate (by Ki-67 stain and tumor sphere formation). ADT suppressed non-S/P tumor cells, but promoted S/P tumor cells. The S/P tumor cells are suppressed (actually killed) by the de-methylator 5-azathioprine (5-AZA) and γ-tocotrienol (γ-TT, tocopherol-derivative differing from vitamin E). α-Vit E failed to kill S/P cancer cells. γ-TT acts as an inhibitor of PI3k/Akt and Bcl2. The cancer non-S/P cells are suppressed by AR degradation enhancer ASC-Jq (patented, AndroScience Corp). The combination of the three agents effectively treated (killed) castrate-resistant human PC xenografts, or induced differentiation of cancer S/P cells into $CD133^-$ $integrin^-$ $CK5^-$ $CK5^+$ luminal epithelial cells).[9] Of the two components of vitamin E, not tocopherols, but tocotrienols (T3, 3TT) exert anti-cancer effects consisting of suppression of neoangiogenesis and transformation from epithelial to mesenchymal phenotype, and chemosensitization of stem cells.[10]

PC cell cells resting in autophagic state are induced to undergo apoptotic death by the PARP-mediated (Table II) mitochondrial release and intranuclear transfer of apoptosis-inducing factor (AIF) elicited by α-tocopherol in combination with vitamin K3 and ascorbate.[11] Bicalutamide (Casodex) or enzalutamide induce the late rise of prostate cancer stem cells (PCSC) driven by the TGFβ/Smad3/MMP9 pathway. PCSC may grow and metastasize without

(continued)

Table XX (continued)

rising PSA (but with rising N-cadherin) even during treatment with ADT or abiraterone (the "bone scan flare"). AR degradation enhancers are ASC-J9^{R} or crypto, anshinone.[12] Pamoate pyrvinium induces death of autophagic adenocarcinoma (prostatic) cells.[13] Comment: breast and prostate adenocarcinoma cells readily "trans-speciate"
1 *Place AE et al Breast Cancer Res 2011;13:227; Polyak K. J Clin Invest 2011;121:3786-8; Polyak K, Vogt PK. PNAS 2012;109:2715–7.* ***2*** *Conley SJ et al PNAS 2012;109:2784–9.* ***3*** *Santisteban M et al Cancer Res 2009;69:2887–95.* ***4*** *Ali HR et al Breast Cancer Res 2011;13: R118.* ***5*** *Ghebeh H et al BMC Cancer 2013;13:289.* ***6*** *Prabhakaran P et al Front Oncol 2013;3:134.* ***7*** *Chen X et al Cell Cycle 2013;12:579–86.* ***8*** *Zhao H et al PLoS One 2013;8(5): e64503.* ***9*** *Lee SO et al J Mol Cell Biol 2013;5:14–28; Yamashita S et al Neoplasia 2012;14:74-83.* ***10*** *Ling MT et al Carcinogenesis 2012;33:233–239.* ***11*** *Tomasetti M et al PLoS One 2012;7(12):e52263.* ***12*** *Lai KP et al Am J Pathol 2013;182:460–473; Lin T-H et al J Biol Chem 2013. doi:10.1074/jbc.M113.477216.* ***13*** *Deng et al Cell Death Disease 2013;4,e614; Venerando A et al Biochem J 2013;452:131–7. CD cluster of differentiation CH cadherin CK cytokeratin.*

Wee, WWOX, BRCA, NF, TSC, VHL, BHD, RASSF1A, LKB/STK, SET/PBRM, etc. (wild type p53 "guardian" of the genome; retinoblastoma; phosphatase and tensin homolog deleted in chromosome 10; suppressor of fused, at 10q24-24; small (in Scottish); double tryptophane oxydoreductase; breast cancer-associated; neurofibromatosis; tuberous sclerosis). The loss of suppressor gene von Hippel-Lindau (VHL) results in clear cell kidney carcinoma; Birt-Hogg-Dubs gene loss in chromophobe cell, chromophobe/oncocytic hybrid kidney carcinoma; the Ras-associated family member 1 blocks cyclin D1; deleted in several cancers, incl. renal cell carcinoma. The mouse lung cancer Kras-driven, develops when LKB/STK liver kinase B/serine-threonine-kinase is deleted; deletion or loss-of-function mutation results in GI tract polyposis/hamartomatosis in Peutz-Jeghers syndrome, and in phacomatoses (Cowden syndrome), and in many cancers incl. endometrial and renal cell carcinoma. The protein kinase encoded by LKB1/STK phosphorylates, and thus activates AMP-dependent kinase (AMPK), a major physiological inhibitor in the mTOR signaling pathway. AMPK inhibits mTOR by activating TSC (tuberous sclerosis tumor suppressor gene) and inactivating by phosphorylation mTOR scaffold protein Raptor. Lack of LKB/STK and AMPK activates hypoxia inducible factor HIF-1α, and mTOR complex signaling, consisting of constitutive activity of PI3K/Akt [1927, 1928]. Dual suppressors of mTOR and PI3K are in clinical trials (*vide infra*).

The SET domain proteins emerged in the yeast (Saccharomyces) and appear highly developed in the drosophila, as "suppressor of variegation, enhancer of zeste, trithorax" (SET). MLL, the promiscuous human gene product protein often fused, is a SET domain protein. In SHM (somatic hypermutation), the SETD3 gene by mistake may translocate into the igλ (lambda) locus and emerges in transformed B lineage lymphoma cells. The physiological function of SETD2 is that of a nucleosomal H3 lysine (K) 9-, or K36-specific methyltransferase (H3 K36me3).

SETD2 represses transcription, and as such, it may function as a tumor suppressor gene. The polybromo gene product SET domain protein (PBRM1) protects cells against mutational induction (loss of VHL gene) of kidney carcinoma, clear cell subtype [1929–1933]. *)

*) *Lysine methyltransferases harbor the SET domain. SETD2 emerges as a tumor suppressor in breast and kidney carcinoma, where its mRNA is demolished by miR-106b-5p (Xiang W et al Oncotarget 2015; 6:4066–4079). Histone methyltransferase SETD3 contributes to transfer of lambda light chain in p53-deficient B lymphomagenesis (Chen et al Haematologica 2013; 98:739–743).*

2. Major interventions of small hairpin and interfering (sh/i) short microRNAs are mobilized. These genomic elements act by silencing those mRNAs, which are to be translated into oncosuppressive and/or immunoreactive proteins. In the absence of these proteins, the cell is rendered defenseless toward the oncoproteins. In the opposite, events of mobilizing micro-RNAs to attack mRNAs, which are to be translated into oncoproteins, these miRNAs may preserve the integrity of the cell. The malignantly transformed cell switches off this second set of microRNAs. The mRNAs encoding oncoproteins are exempted from the attack of those microRNAS that could demolish them. The transformed cell under the protection of its favored microRNA, regains the faculties of the stem cell, prominent among them is the resistance to physicochemical (chemo-radiotherapy) insults. Thus gains oxaliplatin resistance the colon adenocarcinoma cell under the effect of microRNA-625-3p [1934].
3. Constitutive activation of "cell survival pathways" to induce frequent (incessant) cell divisions: STAT, MAPK, PI3K/Akt, mTOR, Wnt/Hh, etc. (signal transducer and activator of transcription; mitogen-activated protein kinase; phosphatidyl inositol kinase 3; Ak mouse thymic lymphoma oncogene; mammalian target of rapamycin; wingless/hedgehog).
4. Acquisition of growth factors as ligands for autocrine, or exo(para)crine circuitries. Sedentary solid tumors grow within stromal niches from where ligands are provided: epidermal growth factors (EGF), fibroblast growth factors (FGF), insulin-like growth factors (IG), keratinocyte growth factor (KGF); platelet-derived growth factors (PDGF), nerve growth factors (NGF), red cell growth factor (erythropoietins, EPO). All chemokines, lympho-cytokines, interferons, interleukins, NFκB-activated gene product proteins have been and are used by the leukemia, lymphoma, sarcoma or cancer cell in auto- or paracrine circuitries for its growth advantage. Examples (selected out from many): B-CLL cells grow *in vitro* under the effect of chemokines and IFNγ inside or outside fibroblasts [717]; adenocarcinoma cells of the breast and squamous carcinoma cells of the uterine cervix express EPO receptors and undergo mitoses upon capturing the ligand erythropoietin [1935ab].
5. Reversal to ancient Warburg type and/or glutaminolytic, or pyrophosphate metabolism. For high energy yield (ATP), some tumor cells revert to mitochondrial glycolysis. Burkitt's lymphoma cells, or glioblastoma cells readily acquire high energy-yield glycolytic pathways [584, 585]. Glucose-starved

mantle lymphoma cells, or glioblastoma cells are more vulnerable to chemoradiotherapy. Statins hinder glucose uptake by mantle lymphoma cells (may be: "take your chemotherapy without eating breakfast"…) [1936, 1937].

6. Generation of new genes by fusion of broken DNA strands, seeking unison even from distant sites within the genome. Even when an oncogene (Ewing sarcoma, Ews) fuses with a tumor suppressor gene (Wilms' tumor, WT) in the desmoplastic small round cell tumor (DSRCT) of Gerald and Rosai, the malignant features prevail: t(11;22)(p13;q12). The DNA-binding splice variant oncoprotein is deficient in a three-amino-acid residue (Lys-Thr-Ser). It induces the oncogene TALLA-1 (T-cell acute lymphoblastic leukemia-associated antigen). The TALLA-1 antigen is expressed in the DSRCT [1938]. In acute myeloid leukemia (AML) the fused oncogene/oncoprotein is at t(8;21)(q22;q22). The chromosomal segment ETO (eight-twenty-one) alleviates this condition and secures high (98 %) complete response rates to chemotherapy. However, gain-of-function point mutations of the c-kit proto-oncogene (D816V: valine for aspartic acid at 816) worsens the prognosis: extramedullary disease, lower response rates to treatment, and resistance to imatinib with higher relapse rates set in. This mutation has a male predilection. The c-KIT tyrosine kinase is the receptor for stem cell (also: mast cell) growth factor. The breakpoint region of chromosome 21 human DNA sequence is 67 % identical with that of the drosophila segmentation gene, runt [1939, 1940].
7. Transdifferentiation of geno-phenotypes. A highly recognized promoter of neuroendocrine transdifferentiation of adenocarcinoma cells is the achaete scute pathway [1711, 1712, 1755] (Table XXIX).

More recently recognized are the midkine and pleiotrophin pathways. The midkine gene is retinoic acid responsive and the protein is a heparin-binding growth- and differentiation factor. In birds it is known as the retinoic acid-induced heparin-binding protein (RIHBT). The avian and the mammalian/human midkine genes diverged about 250 mya. These ligands in the drosophila react with the receptor tyrosine phosphatase zeta (RTPζ). In the innate immune systems of caenorhabditis and drosophila appear the ancestors of the structurally different midkine and pleiotrophins (no information is available on their appearance in organisms positioned farther down on the evolutionary scale). In the drosophila larva, the *miple* genes (midline pleiotrophin family) regulate the layout of mesoderm. The ligand-to-receptor contact activates the MAPK pathway. If overactive, severe defects (scattered hypertrophy) of the mesoderm result. Accordingly, the drosophila gene product protein HOW (held out wing) switches off *miple*. At that level and later transfer into the united innate/adaptive immune system (Xenopus; Danio zebrafish), the heparin-binding growth factor midkine practices strong bactericidal activity (acting as an "innate antibiotic"). In higher vertebrates (avians, mammalians including humans), midkine reacts with several receptors, including receptor tyrosine kinase ALK (anaplastic lymphoma kinase). In the human lung, respiratory epithelial cells secrete midkine for antibacterial protection of the respiratory tract. In the human skin, keratinocytes constitutively produce midkine for antibacterial protection of the epidermis. Blood levels of human midkine are elevated in colitis, and rheumatoid arthritis. The midkine gene is expressed in differentiated bronchial cells for the anti-bacterial protection of the respiratory tract. In the skin, keratinocytes constitutively produce midkine for the antibacterial protection of the epidermis [1941–1944]. For the multiple biological activites of midkine in the mammalian host, its attachment to multiple cell surface receptors signaling different cytoplasmic/nuclear pathways, is the explanation.

In the mammalian (including human) host, midkine promotes B lymphocyte survival; it appears in leukoplakia and in the resulting squamous cell carcinoma; after its attachment to ALK, the cell activates NFκB and insulin receptor substrate-1 [1945–1949]. Midkine is overexpressed in several human tumors and its antisense knock-out frequently results in apoptotic death of the tumor cells (in gastric or hepatocellular carcinoma) [1950–1952].

Release of growth-promotional hormones by the subverted host to breast, endometrial, ovarian and prostatic adenocarcinomas commonly occur. Prostatic adenocarcinoma cells actually produce testosterone-like hormones for autologous growth via their AR (androgen receptors) in autocrine circuitry. Combined anti-androgen hormone- and radiotherapy induce senescence of prostate adenocarcinoma cells. The senescent cells remain alive without division, but provide growth factors to newly emerging non-PSA-producer neuroendocrine cancer cells [706, 1529]. Chromogranins appearing as products of these cells offer a diagnostic blood test [1953]. Prostate cancer cells transform into PSA-nonproducer antihormone-, radio-, and chemotherapy-resistant neuroectodermal/neuroendocrine cancers. The transdifferentiated neuroendocrine tumor cells overexpress midkine, Foxo2 (forkhead) and the human equivalent of Mash 1 (mouse achaete scute homolog), chromogranin, and synaptophysin. In human prostate cancer cells IL-1β, TNFα and NFκB together induced midkine production. Midkine protected prostate cancer cells against TNFα-induced apoptosis. When si RNA degraded the midkine mRNA, TNFα induced apoptosis in those prostate cancer cells [706, 1946, 1948, 1949].

The switch to a higher level of malignancy, and the maintenance of such tumors by a continuous supply of tumor stem cells are triggered by hypoxia due to reduced blood supply. The VEGF-neutralizing bevacizumab, or the small molecular tyrosine kinase inhibitors (sorafenib, sunitinib), instead of an intended therapeutic effect, could bring about this adverse situation. In a tactical error of the tumor, hypoxic tumor cells may release immunizing antigens. VEGF-suppression results in COX-2 overexpression in carcinomatous prostate glands [1954]. Here are examples for a transition from an indirect anti-tumor attack (suppression of the blood supply) to the selection of a tumor stem cell population with a gain of virulence [1537a, b, 1538, 1811, 1812] tolerant hosts and immunoresistant tumor cells. Unicellular entities, be tumor cells or parasitic kinetoplasts (plasmodia, leishmania), evade host immunity by similar mechanisms. In some unicellular pathogens (Naegleria; Theileria) their defense reactions are bordering "transspeciation" of the parasite, or of the targeted cell [73]. Some tumor cells induce host tolerance similar to tolerance-induction and defense of the placenta against ejection (see text Table in Introduction: "the placenta as a pseudomalignant organ") [30].

It is the ultimate travesty of justice: melanoma and glioblastoma cells (and probably other transformed cells as well) are able convert the FasL → FasR death domain into a mitosis-inducing pathway [716–718] (Figure 59). By FasL expression, certain tumor cells could kill Fas receptor-positive (CD95) immune T cells and by Fox3 expression could imitate Treg cells, thus escaping host immune reactions (reviewed in [27, 147]. Observe fabulous cartoon in Ref. [1955]. However, the extent of FasR and FasL in pancreatic carcinoma specimens did not significantly influenced the outcome of disease [1956]. Rescue of p53 by disrupting MDM-p53 protein interaction, liberated p53 to induce apoptosis in human laryngeal carcinoma cells by the FasL → FasR pathway [1957]. FasL-expressing lung cancer cells counterattacked host immune T cells; activated immune T cells were especially vulnerable to the counterattack. Elimination of FasL expression by siRNA abrogated counterattacks. Inhibition of the FasL → FasR ligation between tumor cells and immune T cells by an antibody appears to have liberated immune T cells rendering them able to attack tumor antigens. Thus enhanced anti-tumor cytotoxicity by immune T cells occur [1958]. FasL-expressing human melanoma cells killed CD95 $FasR^+$ immune T cells. FasL attached to CD95 $FasR^+$ human melanoma cells switched from an apoptotic to a mitotic pathway. The explanation is offered that broken chromosomes 1 and 10 fuse in these melanoma cells. Chromosome 10 expresses FasR at 10q23-26 and captures the FasL. Chromosome 1 expresses G-CSF at 1p32-34. In the fused protein product signaling runs from the FasL-bound FasR to G-CSF. The result is a cell mitosis [717, 1959] (Figure 59).

In their host, tumor cells and lymphocytes (immune T cell; NK cells) encounter each other, and both subjects of observation emit exosomes. The exosomes are engulfed by various host cells. Even in the case of an incipient and localized tumor, the entire host is made aware that a malignantly transformed cell colony is on board. It is under extensive investigation what biological effects exosomes may convey. With considerable oversimplification, it appears that exosomes of tumor cell-derivation are more immunosuppressive than immunostimulatory. Whereas exosomes derived from reactive DCs and immune T cells convey immunostimulatory effects. Tumor cells communicate to re-assert their presence by releasing microvesicles in the host's blood and lymph circulation [1960–1964].

8. *Tumor-nest vascularization*. Subversion of the stroma for the generation of exocrine growth factors (in multicellular hosts illustrated in some superb cartoons in the case of pancreatic carcinoma) [1955]. VEGF production by the stroma for the ingrowth of vascular endothelial cells, promoted tumor growth. Accordingly, VEGF inhibition by bevacizumab was expected to retard tumor growth. This result is observed in some cases (colon cancer), but in the case of breast carcinoma (and what subtype of breast carcinoma?), avascular hypoxia invite the invasion of tumor stem cells [1537a, b].

"Vasculogenic mimicry": Transformation of tumor cells into cells of vascular endothelial cell phenotypes. (i) "Pseudovasculogenesis" is occurring by fetal cyto- and syncytiotrophoblastic cells in the placenta, in particular when gestational choriocarcinoma is present; reviewed in [1489]. Vasculogenic embryonic cytotrophoblasts expressed the MIg-7 protein (migration inducing) [1487]. (ii) "Transspeciation" of cells: oncogenes transform malignant and stromal cells into vascular endothelial-like cells for blood supply to the rest of the tumor. Vasculogenic mimicry occurs in adeno- and squamous cell carcinomas, glioblastoma; melanoma; osteosarcoma [1490, 1495a, b] (Figure 67).

9. *Epithelial-to-mesenchymal transition* by the activities of gene-product proteins of Snail, Slug, Twist; extensively reviewed in [26, 27]. If the cell continues its process of tansspeciation, its next phenotype is the amoeboid formation (*vide infra*). Mesodermal (mesenchymal) malignant cells can transform into one another: example is the transition of hemangiopericytoma as it appears as a solitary fibrous tumor. Here, the early growth response domain of the transcriptional repressor NAB2 (nerve growth factor-binding protein) fuses with the activation domain of the transcriptional activator STAT6, thus forming the NAB2/STAT6 fusion oncoprotein (signal transducer and activator of transcription). The activator overrules the repressor. The pericyte becomes a solitary fibrous tumor cell [1965]. Physiologically, NAB2 responds positively to IL-2 induction and represses the formation of the pro-apoptotic TRAIL in expanding $CD8^{+}$ T cell clones (tumor necrosis factor-related apoptosis-inducing ligand) [1966]. Solitary fibrous tumors are often hypervascular and overexpress IGF-R and mTOR [1967]. Accordingly, clinical responses can be elicited to type I IFNαβ, bevacizumab, sunitinib; figitumumab; and everolimus [1968, 1969].
10. *Endogenous retroviral reactivation*. Numerous tumors (teratoma, lymphomas, melanoma, breast and ovarian adenocarcinomas, Kaposi sarcoma) (Figures 30 and 53) reactivate and release endogenous retroviruses for the promotion of epithelial-to-mesenchymal transition by cell fusions and other growth advantages) [26, 147]. Cell fusion in disease (cancer) and health are extensively reviewed in the volumes I/II written and edited by Dittmar and Zänker and co-authors [1541]. Anciently inserted and genomic LTRs retain the faculties for self-reactivation and induction of proto-oncogenes. If some cancers are the result of a rescue mission of individual cells from an endangered multicellular host, reactivated LTRs may be the initiator messengers for this process (*vide infra*).
11. *Release of circulating tumor cells, tumor cell endosomes, DNA oncogenes and oncoproteins* for distant metastasis formation in a process, whereby the established primary tumor switches from neoangiogenesis induction for itself to neoangiogenesis inhibition for its distant metastases, thus suppressing its own metastases: the angiostatins and endostatins of O'Reilly and Folkman [1970–1971a–d]. The presentations of J. Folkman at M.D. Anderson Hospital Symposia remain preserved in writing [1970a–d] and in the memory of his large audiences, this author among them. Showers of previously metastatic, but silenced tumors (dormant lung metastases) inducing self-neoangiogenesis after the standard surgical removal of the primary tumor, which suppressed neoangiogenesis to its own metastases, show up in crops. Loss of endostatin (a collagen XVIII fragment) and angiostatin (a plasminogen fragment) production

due to the removal of the primary tumor allows the hitherto silenced metastases to induce their own hypervascularity. Example: rapidly "awakening" formerly dormant lung metastases after the removal of the primary osteogenic sarcoma. Case presentations of patients rapidly relapsing with hematogenous metastases in the lungs after surgical removals of large primary tumors [1972]. Gene therapy has already been attempted by viral delivery of the angiostatin, or endostatin genes into the blood supply of the stroma or the tumor metastases [1973, 1974]. But what worked in the xenograft-carrier murine subcutis, so far failed in the human host.

Highly individualized versus haphazard somatic mutations in the evolution of individual human cancers. Malignant tumors exhibit extraordinary individual heterogeneity. Each tumor appears to consist of multiple and individually mutated cell colonies. During its life time, the given tumor continues to alter its genomics through new gain-of-function somatic point mutations in single nucleotides, thus generating numerous differing cell subcolonies. Beneath and in addition to the original genomic patterns (oncogenic pathways) that grossly characterize the primary tumors, large numbers of randomly dispersed somatic mutations including deletions and insertions arise in a continuous pattern. To this author, it appears as if the RNA/DNA complex were preparing cells in advance to metabolically fit into any one of the innumerable probable environmental conditions, in which those cells may find themselves.

A patient's renal cell carcinoma was grossly characterized by the loss of the VHL tumor suppressor gene at 3p, and the activation of the PI3K/Akt/mTOR pathway. PI3K converts PIP2 by phosphorylation to PIP3 (phoshatidylinositol-4,5,-triphosphate). Active PTEN is the antagonist of this reaction. PIP3 recruits Akt to the inner plasma membrane, where mTOR phosphorylates it at its serine residue 473; Akt then counter-phosphorylates the mTOR complex, which activates cyclin D1 and c-MYC to drive an uninhibited cell cycle. Everolimus may stop the cycle, or the cycle may continue resistant to everolimus. Exon-capture multiregion sequencing of pretreatment and posttreatment primary and metastatic tumor samples revealed an extensive array of additional somatic mutations characterizing this tumor individually. The number of additional mutations exceeded 100 nonsynonymous point mutations including 32 deletions. The so-called "private mutations" involved specific regions and were consequential to the clonal evolution of the tumor from its primary to its metastatic stages. The mutations were of rather branching, than of linear character. Ultra-deep sequencing revealed the number of new mutations occurring in the metastases, thus creating different tumor subclones. The new mutations reflected rather to the evolution of the tumor, than to mutagenesis by everolimus. Most primary tumor cells were diploid, but in the metastases the number of tetraploid tumor cells increased. Tumor cells with heterogeneity of allelic imbalance and aneuploidy differed from one another ("no tumor regions shared identical allelic-imbalance profiles") [1975]. Apart from the standard genomic characterization (loss of VHL, activation of mTOR), especially genes SETD2 and KDM6A/5C were mutated (H3K3 methyltransferase; X-chromosome-encoded histone H3K4 demethylase).

The genes of VHL, SETD2 and PRDM1 reside in the short arm of chromosome 3. Loss of the SET family domain histone methyltransferase PRDM1 is a predictor of renal cell carcinoma progression, thus PRDMs may either promote or inhibit tumor growth. The polybromo PBRM1 gene is lost, together with the VHL gene, in advancing kidney carcinomas [1930–1933].

Targeted genes suffer distinct different somatic mutations in different tumor cells. The loss of the SETD2 histone lysine methyltransferase (*vide supra*) renders the genome susceptible to proto-oncogene activation. According to a Chinese study, the SETD histone lysine methyltransferases induce cell differentiation, thus, are tumor suppressors (Dai JP in Acta Academiae Medicinae Sinicae. doi:10.3881/j.issn.100-503X.2009.06.007). In contrast to normal cells, the H3K36 in the tumor cells is demethylated. The missense mutation of L2431P (leucine to proline) of mTOR kinase results in the constitutive activation of the gene, which is aligned with structurally related PI3K gene (human chromosomal locus at 3p). The mutated, and thus disabled PTEN fails to control PI3K/Akt. Several other tumors exhibit similar highly individual somatic mutations evolving during the life-time of the tumors [1975]. The Akt gene on chromosome 14 is exposed to several single nucleotide polymorphisms (SNP); it is subject to activation by HSP90 [1976]. The mTOR complex pathway and HIF signaling are hyperactivated in LKB-, and AMPK deficient hosts. This is the situation in the hamartomatosis of Peutz-Jegers syndrome, one of the phakomatoses (others are Cowden's disease, tuberous sclerosis, and lymphangioleiomyomatosis) [1977a, b]. The site of the mTOR suppressor gene LKB/STK11 (*vide supra*) at chromosome short arm, region 19p is often silenced or even deleted in cases of non-small cell lung cancer [1978], and in some endometrial carcinomas rendering these tumors highly susceptible to rapamycin and rapamycin-derivative therapy [1979–1981].

Aneupolidy (Figure 65) of the transformed cell is held responsible for growth into tumors and for the resistance to immune attacks by the host, and even to chemotherapy [1982]. The digitally retained information within the DNA molecule is lost in the genomic chaos of the malignantly transformed cell [1983]. The order in the chaos is presumed to be present in malignantly transformed glioma cells and suggests as if the DNAs were not a victim of outside forces imposing the chaos, but still is in control of it [1984]. Aneuploidy (or "genomic chaos") in the oncogenic transformation of the cell was claimed to be not an absolute necessity. It is held that at least six successive somatic mutations occurring in a cell are necessary for malignant transformation. Several examples exist for the malignant transformation of a cell without aneuploidy consequential to numerous serial mutations. Human kidney epithelial cells could be transformed without the induction of extensive and continuing aneuploidy, chromosomal segmental translocations or microsatellite instability. In human fibroblasts, malignant transformation was achieved by the retroviral introduction of the SV40 large tumor antigen (LT), the *ras* proto-oncogene, and/or the hTERT telomere gene. After cellular transformation, no further secondary mutations were observed, but the transformed fibroblasts were predominantly aneuploid, but without specific chromosomal rearrangements. Kidney epithelial cells transformed without aneuploidy. However, the technology used was not sensitive enough for the recognizance of subtle deletions or inversions [1985].

Intratumor heterogeneity and chromosomal instability (CIN) characterize the cells of solid tumors. Impaired replication fork progression brings about missegregation of the chromosomes in mitosis. The loss of the long arm of the 18th chromosome (del 18q) was characteristic to many solid tumors (colon cancer cells in the study cited) [1986]. In the long arm of chromosome 18, three CIN-suppressor genes reside. The Krüppel-like zink finger family genes ZNF219/KIAA0222 (Kazusa cDNA Research Institute, Japan; AA amino acid reference characters, followed by four digits) encode 77-kDa transcriptional repressor DNA-binding proteins [1987]. One of the targets is the DNA of the HMGN1-(high mobility group nucleosome-) binding promoter [1988].

The ancestry of the MCD4 gene (multiple carboxylase deficiency) goes back to the Chlamydomonas; there it cleaves nonfunctional 3′ UTR sequences of the pre-mRNA [1989]. The insulin-like growth factor-1 regulator is the MEX3c/RKHD2 gene (malonyl co-enzyme A decarboxylase deficiency; ring finger K-homolog domain: large proteins with lysine-containing K-homolog domains). The ring finger KH-domain RNA-binding proteins were active in the ancestral sea urchins [1990]. Residing at 18q21.1 in the human genome, the gene-product proteins are involved in events of essential hypertension [1991]. These three genes, supporters of chromosomal stability, are often silenced or deleted in human malignant tumors; their absence is contributory to CIN.

Comment. Is it not an originally inscribed and inherent faculty of the primordial RNA/DNA complex to generate as many variant cellular genomes as possible, in order to fit just one of them into a drastically changed (but not accurately foreseen) environment? Once engendered in the physicochemical universe, cellular Life is to be sustained.

Selected Abbreviations

BRCA1/2 breast cancer, breast cancer susceptibility gene, tumor suppressor gene in breast and ovary; suffering loss of function mutations vertically transmitted. BCRA2 member of Fanconi anemia gene complex. Failure of the enzyme to repair double strand DNA breaks, and induction of second mutation, may lead to synthetic lethality of afflicted cells (precancerous or malignantly transformed) (Table X). Plants (Arabidopsis thaliana, and others) carry BRCA1/2 homologous gene mutations. After the division of the ancestral cells evolving separately toward plantae and animalia, the plant BRCA sequence acquired their characteristic PHD (plant homeodomain); thus, the mammalian BRCA sequences lack PHD. The acquisition of plant BRCA/PHD has had taken place in the ancestral green algae (in the ancestral Ostreococcus, and not in the chlamydomonas Chlorophyceae, or Volvocales, unless the latter genomes had, but lost, this gene). BRCA1/2 genes exist in the zebrafish and drosophila, and in the mammalian genomes. BRCA2 is part of the plants' immune system, SAR, activated by salicylic acid (Trapp O et al Front Plant Sci 2011;2:19).

WWOX double tryptophan (W) oxidoreductase, cell differentiation-inducer human tumor suppressor gene (neuroblastoma; osteogenic sarcoma, adenocarcinoma, glioma). “We named this gene WWOX because it contains two WW domains coupled to a region with high homology to the short-chain dehydrogenase/reductase family of enzymes” (Bednarek AK et al at M.D. Anderson Hospital, Figure 2 in Cancer Res 2000;60:2140-5). CpG island methylation of its promoter activates cancers (Nowakowska M et al Acta Biochim Pol 2014;61:91-7). In ovarian carcinoma, the WWOX gene promoter’s CpG islands were silenced by methylation. In breast carcinoma, BRCA1 expression positively related to axillary lymph node metastases (but not to ER, PR, HER2, Ki-67 expresssion); WWOX showed no clinicopathological correlations (Yan H & Sun J Biomed Res 2013;1:375-8; Zhang QL et al Zhonghua Bing Li Xue Za Zi 2013;42:90-4). Click, Liu CJ Int J Cancer 2014;134:811-21 for miR-134 targeting WWOX mRNA for depletion, and thus promoting metastases of head&neck cancers. Click, Xiong A et al in Biochem Biophys Res Comm 2014:443:1200-5 for suppression oh Hh oncogenes by WWOX in breast cancer genes STD3 / PBRM1 by S Peña-Liopis Cancer Res 2013;73:1-7. doi:10.1158/0006-5472.CAN-13-0360; gene PRDM in brown adipose tissue https://cn.wikipedia.org/wiki/PRDM16; gene LKB/STK serine threonine protein kinase Launonen M Human Mut 2005;26:291-7.

SET2 histone lysine methyltransferase encoded from short arm chromosome 3 specific for lysine-36 of histone H3; orthologous to S. cerevisiae; interacts with p53; involved in mismatch repair; acts as tumor suppressor (kidney carcinoma). Suppressor of variegation Enhancer of zeste Trithorax (drosophila) Dalgliesh GL et al Nature 2010;463:360–363. Li M et al Proc Natl Acad Sci USA 2006;102:17836–17841.

PRDM protein arginine demethylase. Tom Brock PhD http://www.caymanchem.com/article/2015. β-IFN genes positive regulatory domain 1-binding factor zink finger protein histone methyltransferase. Maintains embryonic stem cells; induces re-acquisition of pluripotency in somatic cells / germ cells; upregulates Sox2. Reciprocal translocation in myelodysplasia. Suppressor of large B cell and NK cell lymphoma. Thermogenesis in adipocytes (brown fat white fat). Represses c-myc/Myc. Bloomfield CD et al Blood 1986;66:1409–1413; Győry I & Fejér Gy et al J Immunology 2003;170:3125–3133; Hohenauer T & Moore AW Development 2012;139:2267–2282; Polybromo-domain PBRM1 chromatin remodeling complex tumor suppressor Varela I et al Nature 2011;469:539–542.

Demethylase gene PRDM1 encodes B lymphocyte-induced maturation protein BLIMP1 descendant of krox gene of sea urchin S. purpuratus originator of the SET domain Su(var); Enhancer of zeste; trithorax (discovered in drosophila). KDM = lysine demethylase (Hemming S et al Stem Cells 2014;32:802–815).

Part III
The Human Cerebral Cortical DNA Against Cancer DNA

Will the Human Cerebral Cortex Subdue the Oncogenome?

Combined Attack from Every Direction

Laboratory-made small targeted molecules (copies of some natural substances newly synthesized) subdue the oncoproteins; synthesized siRNAs demolish oncoprotein inducer mRNAs, other miRs induce host-friendly tumor suppressor genes. These substances chemically imitate naturally occurring molecules, but their laboratory derivatives are the products of the master bioengineers DNA/RNA working in the human brain. The two ultimate products of the master bioengineer RNA/DNA complex are the immortalized cancer cells with their faculties conserved through some two billion years or more, and the human cerebral cortex, a structure much more recently constructed. These two bioengineering masterpieces are poised antagonistically. The heterogenic cancer cell community switches from one to another survival pathways in order to resist generally cytotoxic and/or "precisely" oncogene-targeting attacks. The human cerebral cortex encoded by the same RNA/DNA↔DNA/RNA complex that encodes the incongruous oncogenes may eventually overcome its masters and generators. An achievement advancing slowly and at what price for its delayed implementation.

The oncoprotein kinases may yield to enzyme (kinase) inhibitors, inasmuch as enzyme inhibitors are widely used therapeutically for several decades in general internal medicine. The fluorophenyl-dihydroxy-methylethyl-phenyl-(phenylamino)-carbonylpyrrole-heptanoic acid, calcium trihydrate, atorvastatin, inhibitor of hydroxyl-methylglutaryl-coenzyme A is standard therapeutical for hypercholesterolemia. The carboxy-phenylpropyl-lysil-proline dihydrate, "lisinopril", the long-acting angiotensin converting enzyme inhibitor, is a standard form of treatment for hypertension. Atorvastatin, lovastatin, simvastatin promote apoptosis induction (*in vitro* in established cultures) of cancer cells. Metformin deprives cancer cells of their increased sugar intake; in combination with induced hypoglycemia, it may induce "synthetic lethality" for cancer cells.

J.G. Sinkovics, *RNA/DNA and Cancer*, DOI 10.1007/978-3-319-22279-0_25

External interventions designed for modifying (differentiation-induction with biomolecules and cytokines), or for the destruction of the oncogenome consist of **1**, therapeutic cancer vaccines; **2**, oncolytic virotherapy; **3**, viral vector-mediated gene therapy; reinsertion of tumor suppressor genes and/or pro-apoptotic genes into the transformed cells; **4**, oncoprotein-directed monoclonal antibodies and immunotoxins; **5**, small molecular inhibitors of the oncogenic tyrosine kinases; **6**, small molecular inhibitors of the "cell survival pathways"; and **7**, small inhibitory (antisense) RNAs (natural products, or their laboratory "antisense" duplicates) to degrade with precision mRNAs generated for encoding oncoproteins.

Early in inflammatory carcinogenesis [73], in the epigenome of the tumor suppressor gene, the promoter is silenced by DNA hypermethylation. At the same time, the proto-oncogenes are activated by de-acetylation of the histones in their promoters. **8**, Enzymatic demethylations and de-acetylations in the epigenome prevent the loss-of-function of the tumor suppressor gene; and methylations and reacetylations of the oncogene promoters prevent the constitutive gain-of-function or mutation and fusion of the proto-oncogenes. Observe the exemptions!

The much less selective genome-directed radiotherapy and combination chemotherapy remain the treatment for tumors of unknown genomics and oncoprotein antigenicity, that cannot be surgically removed (and the circulating tumor cell assays are positive), with the understanding that the host will suffer residual harm (including the probability of newly induced second primary tumors consequential to new genomic mutations); and new generations of treatment-resistant tumor cells are expected to arise.

Therapeutic cancer vaccines. Cancer vaccines in use prolong remissions achieved with surgery and chemoradiotherapy. By formerly held concepts, micrometastases of the original tumor remain outside the surgical field, or at distant sites. Immune reaction generated against these tumor deposits still could secure a complete remission. Therefore the vaccine is prepared against antigens expressed by the originally excised tumor (Tables III/I and XXI). According to newer concepts, it is the slowly and symmetrically dividing tumor stem cells that remain alive in the tumor-inflicted host. The genome of these tumor stem cells has undergone a process of trans-speciation and is operating immortalizing pathways quite different from that of the pre-malignant cell, or the original tumor cell. The cancer vaccines aimed at the original tumor may have improved the results of chemoradiotherapy by increasing response rates, but are doomed to failure against the tumors of late recurrence. Melanoma viral oncolysates induced remissions at Emory University, Atlanta, GA, well documented, in unique patients in non-prospectively randomized clinical trials [274]. Viral oncolysate melanoma and sarcoma vaccines increased and prolonged response rates when combined with standard therapy (occasionally maintaining very long remissions induced by surgery, and/or prolonged the life of patients with metastatic disease) [275, 276, 644, 645]. In Cassel's protocol, the NDV VO induced remissions in metastatic disease (melanoma) [274]. Tumor antigen peptide vaccines administered with immune adjuvants achieve prolongation of remissions obtained by standard therapy. These vaccines are adjuncts to standard therapy.

Table XXI Cancer Vaccines from the Era of Pre-Molecular Oncology

Viral oncolysates. *In vitro* grown autologous tumor cells lysed by oncolytic non-human pathogenic virus (Newcastle disease virus; reovirus; vesicular stomatitis virus; attenuated influenza A virus). Lysates with live virus within, are inoculated subcutaneously with or without BCG grafted, or GM-CSF injected over the site. Presumed mechanism: viral oncolysis followed by fused viral capsule/capsid/membrane antigens and tumor cell membrane antigens-induced highly immunogenic effect. Host is immunized against newly generated fusion oncoproteins. Further, active oncolytic virus to attack resting apoptosis-resistant, autophagic, and IFN-deficient and dividing tumor cells. Actually clinically tested: chemotherapy-induced remissions were sustained; occasional remission-inductions reported.[1]
Pre-immunization of bone marrow transplant donors with BRC/ABL p210 protein. In patients with BCR/ABL^+ leukemia cells, anti-leukemia cell reactive antibodies and immune T cells are generated (too weak to induce remission). Bone marrow donors pre-immunized with p210 protein (nucleic acid-free) would provide not only rescue for hematopoiesis, but antibodies co-acting with patient's NK cells, and immune T cells both cytolytic to BRC/ABL^+ leukemia cells. Proposals are not known to have had acted upon at NCI; at MD Anderson.[2]
Induction of tumor-protective genes, or Treg cells, by fetally expressed carcino-embryonic antigens and then switching off the mechanism by cyclophosphamide, methotrexate-leukovorin, or cytarabine before vaccine administration. Presume that tumor cells produce carcino-embryonic protein antigens, like fetal cells (α-fetoprotein, CEA, PSA, CA15-3, CA-119, CA-125 etc) to activate fetal cell- or tumor cell-protective protein-mediated mechanisms (cell proliferation-inducing "enhancing" antibodies; chemokine-directed Treg cells). These substances are physiologically produced by the embryo in early fetal life ("Verwandschaftreaktion zwischen Embryonal- und Krebs-Gewebe") for protection against maternal immune rejection directed against paternal antigens expressed.[3] This mechanism artificially induced by the administration of CEAs would be susceptible to switch off by pre-vaccination chemotherapy with highly selected agents (CTX;MTX; cytarabine). The cancer vaccine would be administered after such pretreatment. Theoretically considered; tested in mice,[4] but not known if applied in any investigational protocols for human patients.
1 Cassel WA & Murray DR. Med Oncol Tumor Pharmacother 1992;9: 169–71; Sinkovics JG & Horvath JC. Viral Therapy Human Cancers. Marcel Dekker, NY 2005; Sinkovics JG & Horvath JC Int J Oncology 2000;16: 81–96; Arch Immun Therapy Experiment 2008;56S: 1–59. 2 Sinkovics JG. J Natl Cancer Inst 1995;87: 1804; Acta Microbiol Immunol Hungarica 2010;57: 253–347; Sinkovics JG & Horvath JC Int J Oncol 2000;16: 81–96; Int J Oncol 2006;29: 765–777. 3 Hirszfeld L et al Zsch Immunitätsforschung 1932;75: 209–16 (cited in Buzás GM Orvosi Hetilap 2013;154: 810–9); Bahramy A et al East Mediterr Health 2012;18: 1065–9; Duffy MJ. Med Princ Pract 2013;22: 4–11; Mukhopadhyay A et al Oncotarget 2013;4: 610–21; Samineni A et al Int J Cancer 2013;133: 394–407. 4 Sinkovics JG Cabiness JR, Shullenberger CC. Cancer 1972;30: 1428–37; Anticancer Res 1989;9: 1281–90; Cancer 1996;78: 184–5; Sinkovics JG, Plager C, McMurtrey JM. Adv Med Oncol Res Educ Basis Cancer Therapy M Moore editor 12th Internat Cancer Congr Buenos Aires 1978;VI: 121–9.

Dendritic cell vaccines are expected to be of therapeutic value. The elaborate Sipuleucel (Provenge, Dendreon, Inc) was designed to be a therapeutic vaccine, even though it was aimed at a minor tumor antigen (prostatic acid phosphatase). Further, it is not a vaccine; the patient receives back his own antigen presenting cells (stimulated with GM-CSF). Sipuleucel-T fails to induce remissions; at best it prolongs the life of the patient with a few (2–4) months. Otherwise, DC vaccines are claimed to induce "robust immune reactions" against the autologous cancer, that is not necessarily translated to a clinical remission. A dendritic cell fused with a tumor cell is expected to express the tumor antigens on its surface and is presenting it to $CD4^+$ T lymphocytes in an immunogenic (not tolerogenic) manner [1992, 1993a, b]. Full proof for the efficacy of therapeutic cancer vaccines is emerging. Epigenetic interventions stimulate cancer-testis antigen expression, thus rendering transformed cells vulnerable to host immune attack directed at antigenic targets (Tables III/I, III/II and XVII).

Virally induced (exogenous) cancers (hepatocellular carcinoma hepatitis C virus-induced; squamous cell carcinoma of the uterine cervix or oral cavity (HPV-induced) are preventable with vaccines-induced immunity. The exogenously enforced virally induced cancers are reacted to with the induction of strong host immune defense, whereas the endogenously induced cancers often receive support from the subverted host, a reaction that a cancer vaccine is expected to break and reverse [1994–2000]. Some cancer vaccines caused tumor enhancement. Many cancer vaccines failed [2001a]. It is not clear at all what purpose the new review "Cancer Vaccines" served [2001b].

The NY-ESO antigens induce immune reactions, which will kill the cell expressing that antigen; however cancer cell conceal the expression the ESO-NY/MAGE antigens. By demethylating the promoter of the gene which encodes the NY-ESO antigen, the gene will induce the release of mRNAs for its translation into the antigenic proteins. The host will respond by generating $CD8^+$ T cells cytolytic to the antigen expressing cells. The dendritic cell neuroblastoma vaccine given to the patient treated with DAC induced a complete remission [2002] (*vide supra*). However, some healthy stem cells (the "hematopoietic precursors") may be naturally expressing the ESO-NY antigens (as in fetal life) may also be wiped out [2002].

Vaccines directed against the K-Ras oncoprotein, or the carcinoembryonic and mucin antigens that tumors overexpress, the excessively produced telomeres, the anti-apoptotic survivin, the heat shock proteins chaperoning oncoproteins are all responded to with anti-tumor cell immune reactions, but fail to induce durable complete remissions. VO vaccines (Figure 44; Table XXI) and their early investigators are omitted from the current reviews, or are referred to under different names, the "oncolytic vaccines" of contemporary authors. In conclusion, the numerous forms of cancer vaccines that are being tested, indicate that so far none of them excelled, but most of them are immunogenic and additive in their efficacy to standard tumor therapy, including radiochemotherapy.

Oncolytic viral therapy. A historical overview based on the author's personal experience has been provided [71]. The "Die Grundlagen der Virusforschung" in 1956 presented a chapter on "Virusvermehrung in Tumorzellen", and concluded, that in the tumor cell the oncolytic virus successfully interfered with the oncogenic virus ("So können wir bei der Verhinderung des Wachstumes der Geschwulst auch an die Interferenz des onkolytischen und des tumorbildenden Virus denken") [36c], as a testimony of the old belief that all cancers were virally induced. This author had the privilege in 1957 to visit with the tumor virologists (Charlotte Friend; Alice Moore) working at the Sloan-Kettering Cancer Center in Manhattan, New York City, NY. In the USA, Alice Moore wrote the first papers on oncolytic viruses

[2003a, b]. Kept isolated from the Western scientific world, this author in 1955 in Hungary grew NDV strains freshly isolated from cases of human conjunctivitis and infected with these viruses Ehrlich ascites carcinoma cells (the report of that work was delayed due to the Hungarian national uprising in 1956, until 1957) [71; 2004a, b, c]. The modern era of viral therapy of human cancers was introduced in a review article from Finland by Vähä-Koskela et al [2005]. Naturally oncolytic, or genetically engineered oncolytic viruses attack cancer cells. Newcastle disease virus has been the champion of naturally oncolytic viruses. It was followed soon by vesicular stomatitis virus and reovirus. In a decade or so, it has become evident that practically any virus could be rendered oncolytic, either by adaptation to tumors, or by genetical engineering (including the highly pathogenic viruses of herpes, measles, poliomyelitis, or vaccinia/myxoma), as reviewed in the Viral Therapy of Human Cancers (co-authored and edited by Sinkovics and Horvath, and published by Marcel Dekker in New York City in 2005) [2004c]. The volume includes chapters on William Cassel's work with NDV oncolysate therapy for malignant melanoma, Frank McCormicks's on the first genetically engineered oncolytic adenovirus ONYX-015, and Matthias Gromeier's Sabin' polio- and rhinovirus recombinant glioblastoma vaccine. By now, even the naturally oncolytic NDV, or reovirus strains are genetically engineered for increased and more specific oncolytic efficiency. The genetically engineered herpes simplex, or measles viruses proved to be oncolytic for melanoma cells [2006a, b]. From the simplistic beginning of intratumorally injected naturally oncolytic viruses (not supported with grant funds by the USA NIH/NCI in the 1960s–70s), an armada of genetically engineered oncolytic viruses has arisen to specifically attack defined oncogenic pathways [2007] and are becoming an attractive treatment modality in developing countries [71; 2007, 2008]. Intracellular innate defensive mechanisms, or host immunity eventually neutralize both the intra- and the the extracellular virus [2009–2011]. The host's adaptive immune system (immune T cells; virus neutralizing antibodies) may put an end to the progress of viral oncolysis. Fusogenic oncolytic viruses may spread within tumor cell syncytia, and thus escape the neutralization by host antibody of the extracellular viral particles. However, antiviral host immunity could also attack the virally infected cancer cells. If NK cells' killer receptors (NKp30/NKp46) react promptly, and NK cells kill tumor cells with immature virus particles within (before the full maturation of the new viral progeny), the process of viral replication and the spread of the oncolytic virus come to an end, and tumor growth eventually resumes. Memory cells retain the immune faculties generated for the rejection of the virally infected tumors. Mice immunized with viral oncolysates acquired immunity against challenge with non-virally infected tumor cell inocula. Yet work with viral oncolysates (NDV oncolysate, Cassel; influenza virus PR8 oncolysates, Sinkovics) received no NIH/NCI grant support in the 1970s ("approved without funding due to low priority" in NIH/NCI phraseology). However, genetically engineered influenza A virus exerts oncolytic effect in human melanoma cells. The melanoma cells expressed Ras/Raf/MEK/ERK signaling pathway and were type I IFN-deficient. The influenza virus deleted its NS1 (non-structural protein) gene and acquired an IL-15 gene; IL-15 attracted NK cells, which killed the melanoma cells [2012]. Out of a large selection, different oncolytic viruses may be used alternatingly (as suggested by Albert Sabin in 1957, cited in [71]).

Clinical trials with intravenously administered naturally oncolytic viruses (NDV, reovirus) so far did not provide the much desired (and frequently predicted) complete remissions, or unequivocal cures of metastatic disease. At the present time, effective virotherapy of human metastatic cancers is by necessity being combined with chemotherapy (allowing the use of single agents in reduced dosage). Since tumors induce host reactions protective to the tumor, such as Treg cells for the destruction of immune T cells, it is a welcome new observation that NDV reduced Treg cell generation in a tumor-bearing host [2013a, b]. Further, apoptosis-resistant tumor cells remain vulnerable to NDV infection [2013c]. In contrast, a viral epitope expressed on a cell surface, when coated with a specific antibody, could become unrecognizable to immune T cells [2013d]. However, not fully realized in the late 1960s, the antibody-coated antigen would attract FcR-bearing Mϕ or NK cells.

The field on oncolytic virotherapy from its controversial beginnings (not supported, even antagonized), has developed into a major therapeutic modality. Intravenously infused, or otherwise administered naturally oncolytic viruses (NDV, reovirus, VSV, myxoma virus) are in clinical trials, including the treatment of brain tumors (glioblastoma); reviewed in [2014a, b]. However, even the naturally oncolytic viruses (mumps and NDV, influenza- pox- and reoviruses, vesicular stomatitis virus) undergo genetic engineering to increase the extent and specificity of their oncolytic potentials. Highly pathogenic viruses (herpes-, measles-, polio- and poxviruses) became attenuated and highly oncolytic after genetic engineering. Hundreds of genetically engineered highly oncolytic adenoviruses have been constructed; reviewed in [71; 2015a, b]. A genetically engineered adenovirus has its E1A gene under the control of a 214 bp fragment of the prostate cancer oncogene DD3/AKR1C3 (tumor-derived differential RNA display/aldo-keto reductase family 1 member C3). The inserted woodchuck post-transcriptional element is the enhancer of the E1A gene expression. A CMV promoter drives the expression of the inserted PTEN gene (thus replacing the tumor suppressor gene frequently deleted from the genome of prostate cancer cells). The AdDD3.D55-PTEN virus selectively grew in prostate cancer cells and inhibited the growth of prostate cancer cell lines LNCaP and CL1, an androgen-independent cancer cell line. These cancer cells died with PTEN-induced apoptosis; the inserted PTEN also inhibited PI3K/Akt mobilization in the tumor cells [2016]. Other genetically engineered adenoviruses are driven by the PSA promoter. Anti-androgen hormone administration switches off PSA production in the cancer cells, and thus stops the replication of a PSA-driven virus. A further major problem is the host defense. After a substantial oncolytic effect, innate and adaptive immune reactions of the host neutralize the virus (view cartoon in [2015a, b]). However, the attenuated and genetically engineered herpes simplex virus armed with GM-CSF (TVEC, talimogene laherparepvec, Biovex/Amgen) kills melanoma cells and induces host immune reactions to the tumor cell [2006b] (in Appendix 1).

Immune tolerance induction to the virus could resolve the problem of antiviral immune reactions of the host. A male patient (JS) with widely metastatic melanoma entered complete remission well over a decade ago upon treatment with NDV-oncolysate preparations at Emory University, Atlanta, GA by Cassel et al [274]. He continues to receive the NDV-based melanoma cell oncolysate vaccine injections under the author's (JGS) supervision in Tampa, FL. He shows no adverse immune reactions to the much overdosed NDV (in the formula of an allogeneic melanoma viral oncolysate specially licensed in the 1990s), and maintains his clinically verified complete remission. William Cassel and this author (JGS) recognized early that viral oncolysates (VO) exert a dual action, first by viral oncolysis lysing individual tumor cell, and then by immunization of the host to combined (fused) tumor cell membrane and viral envelope antigens, thus inducing both NK cell and immune T cell reactions. Yet, citations to the "viral oncolysates" tumor vaccines have been omitted from the modern literature. Instead, VOs are renamed as "oncolytic vaccines" of the same claimed mechanisms of action, that VOs used to work with (Figure 44).

Adoptive immune T lymphocyte and NK cell therapy. Endogenously induced malignant tumors present themselves to their hosts as parts of the "self", as if deserving support, and not an attack. Nevertheless, the tumor-bearing host usually mobilizes some modalities of immune defenses, but usually inadequate for the rejection of the tumor. First, NK cells act without specific pre-immunization; second, clones of immune T cells rise, but expose themselves for elimination by Treg cells. The B lineage maturing into plasma cells produces antibodies. The antibodies may blind immune T cells by covering antigenic targets, as shown in the Author's laboratory in the case of a mouse tumor [2013d]. In favorable contrast, antibodies may activate through their Fc receptors NK cells and Mϕ/monocytes for an ADCC anti-tumor cell reaction (reviewed and photographically documented in Figures 35 and 36, and in the monograph [147]).

The HER2/neu oncoproteins usually are not mutated, just amplified. Yet anti-HER2 antibodies and/or immune lymphocytes arise. The BCR/ABL oncoprotein is recognized by its host to be a deviant from "self". These host reactions are usually totally inadequate for tumor rejections. It is presumed that patients with malignant melanoma mobilize stronger immune reactions. Circulating lymphocytes collected from the blood of these patients and activated and expanded with IL-2, upon reinfusion induce durable remissions, including complete remissions. The extracorporeally raised large populations of immune T cells with or without genetically engineered T cell receptors, and lymphokine-activated NK (LAK) cells, for adoptive immunotherapy are at the level of a practical treatment modality for malignant melanoma [2017]. The intravascular expansion of the re-infused lymphocyte population is opposed by the resident lymphocyte population. Therefore, the resident lymphocyte population is to be depleted (usually by lymphocytotoxic chemicals) before therapeutic infusions of immune T cells.

Specifically antigen-responding T cells with chimeric TCRs induce cytotoxicity *in vivo* against $CD19^+$ autologous B lymphocytic leukemia cells, including acute B lymphoblasts and B-CLL cells. A replication incompetent retroviral vector encodes the CD19-specific Cd28/CD3ζ second generation dual signaling chimeric antigen receptor (CAR) in the genetically engineered TCR in the recipient attacker T cells (19-28z CAR-modified T cells). Patients to be treated are pre-conditioned with cyclophosphamide. The treatment consists of the i.venous infusion of up to 3×10^6 autologous 19-28z$^+$ T cells/kg. The genetically altered autologous T cells rapidly expand 1000-fold. The genetically engineered T cells attack and destroy the CD19$^+$ B lymphoma cells. Each immune T cell can kill 1000 leukemia cells, so much so that acute tumor lysis syndrome may occur. The immune T cells replicate and persist to guard against relapse [2018, 2019]. Immune T cells similarly engineered attack also ovarian carcinoma cells with durable remissions so induced [1367a, b].

The heterodimer TCRαβ, the TCR:ζ (zeta), occupies a key position in immune reactivity. The Src-like adaptor protein (SLAP) and TNF downregulate TCR3ζ expression. Silencing SLAP (its mRNA inhibited by siRNA) resulted in the blockade of TNF-induced TCR3ζ chain downregulation. Targeting Jurkat T cells, TCR:ζ formed synapse interacting with CD8 and class I MHC sites independently from CD3 interaction [2020, 2021]. TCR interaction with tumor cell membranes holds the key for the effective cytotoxic reaction. Of *in vitro* expanded autologous tumor-infiltrating CD8$^+$ T lymphocytes, the subpopulations expressing the B- and T-lymphocyte attenuator negative costimulator molecule (BTLA) were the most effective attacker cells [2022]. An extensive list of antibodies targeting cell surface molecules has been provided. A cartoon provides a review of the effector phase of lymphocyte activation [2023].

Allogeneic hematopoietic stem cell transplantation *versus*-leukemia reaction [261a, b] may induce a second remission in adult patients with relapsed acute B-cell lymphoblastic leukemia. Pre-conditioning (moderate chemo-, radiotherapy; cyclophosphamide, fludarabine; bendamustine, pentostatin) is necessary to prevent the rejection of the graft. The presence of haploidentical NK cells in the graft improves the response rate. Excess T cells induce GvsH disease. *In vitro* generated tumor-specifically reactive T cells encoding chimeric Ag receptors (CAR) bypass HLA incompatibility. CAR-modified T cells attack all forms of tumor antigens (proteins, carbohydrates, glycolipids), for which a monoclonal antibody can be generated. CARs are composed of AG-binding domains fused as single chain variable fragments, most commonly to the TCR-derived CD3ζ chain. To improve reactivity, co-stimulatory signal molecules (CD28, 4-1BB, OX-40) are included in the fusion complex. These CARs deliver the double (even a third) signal required for a fully effective reaction. Gene transfers to encode CARs in the T cells are accomplished by gamma retro(lenti)viral transfer. In these procedures, no insertional leukemogenesis occurred in a decade. Infusion of autologous T cells expressing CD19-specific CD28/CD3ζ second generation dual-signaling chimeric antigen receptor (CAR) termed 19-28z, induced complete remissions in patients having residual disease. Immune T cells with the modified CAR persisted in these patients; while preventing relapse, they continued to induce B cell aplasia (due to the elimination of CD19-expressor healthy B cells). In these patients, in a frequently violent reaction resembling lymphokine storm, a complete eradication of leukemia cells can occur. The violent reaction requires corticosteroid administration for its amelioration [2024a, b, c] (more in the Appendix 1). *)

In a tumor-bearing host, immune T cell and NK/NKT cell clones naturally arise. Extracorporeally expanded populations of immunoreactive T cell clones are used for adoptive immunotherapy. Newly designed therapeutic cancer vaccines generate immunoreactive T cells. These T cells and NKT/LAK cell subpopulations generated after cancer vaccine administration offer themselves for adoptive immunotherapy. Vaccine-generated immune T cells (or NKT cells) expanded *in vitro* may be more effective against cancer cells, than the naturally generated lymphocytes. Tumors expressing the NY-ESO-1 antigen (synovial cell sarcoma; melanoma) were treated with infusions of autologous immune T cells with transduced TCR directed against NY-ESO-1 antigens. For the first time, sarcoma patients responded with partial remissions; however, melanoma patients responded with durable (over 1 year duration) complete remissions [2025]. Adoptive immune lymphocyte therapy is more effective against viremic infections (EBV, CMV in immunosuppressed patients) [2026, 2027], than against disseminated tumors.

*) *The illustrated article entitled "Chimeric antigen receptor therapy for cancer" (Barrett DM, Singh N, Porter DL, Grupp SA. June CH. Abramson Cancer Center; Perelman School of Medicine; University of Pennsylvania, Philadelphia, PA. Annu Rev Med 2014;65: 333–347) describes the procedures of human immune T cell culture and genetic engineering arriving at the clinical application of chimeric antigen receptor-carrier autologous major histocompatibility complex class I-restricted cytotoxic first, second and third generation autologous T lymphocytes. Practically any tumor antigen may be targeted. Lymphomatous tumor cells expressing the CD19 antigens were wiped out by CAR-armed immune T cells. In the autologous system, no graft-versus-host disease occurred. The use of retro (lenti)viral vectors for transduction raises the remote possibility of late leukemogenesis. In order to allow the expansion of the re-infused autologous immune T cell population, pre-infusion chemotherapy (fludara; cyclophosphamide, etc) is necessary for the depletion of the recipient's resident lymphocyte population. Acute tumor lysis syndrome, cytokine release syndrome, B cell depletion with hypogammaglobulinema require the closest attention, since these are treatment-responsive reversible conditions. Observe an overlap: virologists use CAR abbreviation to indicate cocksackie and adenovirus receptor expression on cell membranes.*

CAR T cell therapy of solid tumors begins with the destruction of the extracellular matrix. Here, the heparan sulfate proteoglycan-loaded extracellular matrix surrounds tumor cell colonies forming impenetrable stonewalls around them. Heparanase-secreting CAR T cells dissolve the barrier. The tumor cells (melanoma, mesothelioma, H&N squamous cell carcinoma, clear cell kidney carcinoma, triple-negative breast carcinoma, glioblastoma, sarcomas) express chondroitin sulfate proteoglycans. These cell surface molecules are readily targetable by the TCR CD3-ζ zeta chain fused with chondroitin proteoglycan antigen-binding mcab domains (*vide infra*), thus creating immunoreactive tumor cell-killer autologous CAR T cells (Geldres C et al Clin Cancer Res 2014;20:952–71; Caruana I et al Nat Med 2015;21:524–9).

Monoclonal antibodies (and immunotoxins). Quotation from Lancet January 1970;7638: 139–140: "*Tetraploid immunoresistant lymphoma cells in the mouse emerge by the unison of the diploid virus-producing lymphoma cell with a plasma cell producing virus-specific globulins. The resulting tetraploid cell will retain malignant growth potential and the genetically determined committedness of both parent cells – to produce leukemia virus, as coded for by the viral genome within the neoplastic cell, and to synthesize virus-specific globulins, as coded for by the plasma cell*". This first description of the "hybridoma principle" by Sinkovics in 1969 in the Lancet preceded the Köhler-Milstein discovery, as acknowledged by Milton Wainwright and others [2129a, b]. However, the independently achieved and practical Köhler-Milstein discovery provides the efficient mcabs (mAbs) for the therapy of cancer.

The production of highly oncoprotein-specific monoclonal antibodies (mcab) and immunotoxins improve their efficacy stepwise. These interventions usually neutralize specifically just one specific oncogenic pathway. The Rank ligand-inhibitor Denosumab (Amgen) reduces the incidence of lytic-blastic bone matastases. Rituximab (R, Rituxan, Genentech), the activator of Fc-receptor bearing cytolytic NK cells, may now be surpassed in efficacy by ofatumumab (Arzerra, GlaxoSmithKline). Pertuzumab (Perjeta; Roche/Genentech) inhibits dimerization of the HER oncoproteins, synergizes with lapatinib (*vide infra*) and may overcome trastuzumab in efficacy (Herceptin, Genentech) (Physicians Desk Reference, 2015). The HER2/neu is not a cellular life-sustaining molecule and its neutralization (even deletion) may not necessarily induce tumor cell death. However, trastuzumab emtansine (maytansine-derivative) (Genentech) enters the breast cancer treatment arena with the highest expectations. It retains full trastuzumab efficacy even against lapatinib-resistant breast cancers. The monoclonal antibody preparation Kadcyla (Genentech) is taken up into the cytoplasm of the targeted cell; only inside the cell is the chemotherapy molecule released [2029–2031]. However, the first mcab immunotoxin gemtuzumab ozogamicin delivering calicheamicin performed clinically weakly with severe toxicity (and was withdrawn from the market). This author remembers the first clinical trials of maytensin itself in the 1980s: due to its high toxicity and poor performance "it was shelved,".

Bevacizumab (Avastin, Genentech) inhibits tumor-induced neoangiogenesis. The anti-CD30 auristatin E-combined brentuximab immunotoxin (Seattle Genetics/Milleneum/Takeda Oncology) kills Reed-Sternberg cells of Hodgkin's disease. The German mcab obinutuzumab is expected to synergize with R in the treatment of B-CLL. These treatment modalities are not without rare, but very severe side effects (citations in the Phsicians' Desk References, PDR, years 2011–2016).

Both high dose interferon-α (Intron, Merck), and/or the CTLA4-antagonist monoclonal antibodies anti-CTLA4 ipilimumab (Yervoy), and anti-PD1 nivolumab (Opdivo, BristolMyersSquibb), and anti-PD1 tremilimumab (Pfizer) in combination with human recombinant IFNα release suppressed autoimmune reactions to attack melanoma cells, which are masquerading as "self" [2032–2034a, b, c]. Not infrequently, not-transformed (non-malignant) organs truly "self" have to sustain an autoimmune attack of varying severity during CTLA4-antagonist mcab treatment [2034a, b, c, 2035]. CTLA4 deactivates T cell immunoreactivity, especially that that would induce autoimmunity; CTLA4 blockade releases autoimmune reactivity of the T cell. CTLA4 physiologically responds to DC-expressed B7.1/2, which also binds T cell surface antigen CD28. DC in its MHC molecule presents the antigens to respond to, to TCRs. In contrast, DC in releasing PD-1 ligand (PD-L1) could induce the apoptotic death of the PD-1^{+} T cell. In oversimplification, mcab blocking CTLA4 releases that immunoreactivity; malignantly transformed cells masquerading as "self" become primary targets, but normal tissues (intestinal tract; endocrine glands) lose their physiological protection and suffer "collateral damage". In the PD-1 system (programmed death protein receptor), ligands (PD-L1/2) activate the receptor to induce apoptotic death of its own immune T cell, which would have committed autoimmune attack otherwise. Tumor cells keep their PD-1/L1/2 genes overactive in order to produce the ligands for the induction of T cell apoptosis by the activated PD-1 receptor. Mcab directed at receptor PD-1, or at its ligands block the self-apoptotic pathway and rescue the immune T cell, which will deliver its autoimmune attack. The M.D. Anderson's publication for physicians Oncolog (Feb 2014; 59/2) printed dramatic photographs on the complete regression of a large metastatic melanoma from and around the left orbit upon treatment with mcab MK-3475 (pembrolizumab, Keytruda, Merck). These mcab and the small molecular oncoprotein kinase inhibitors (*vide infra*) may be administered in combination. An outpour of similar reports is expected to appear in the literature in 2014–5 (in Immunotherapy in Oncology, Supplement to Personalized Medicine in Oncology, emanating from the AACR conferences 2014–2015). *)

*) *The most effective mcabs are not directly attacking the melanoma cell (which was the customary target of chemotherapeuticals and vaccines). The targets are the host's immune T cells that would attack any cell with self-antigens, therefore they are strongly suppressed, so much so, that these cells are physiologically coerced to surrender in apoptotic death. Inhibitor mcab (mAbs) of the apoptotic death of these cells liberates them to attack tumor cells masquerading as self in overexpressing autologous cell surface antigens. However, tumor cells also overproduce the ligands that activate the PD-1 receptor, the initiator of auto-apoptosis. Monoclonal immune globulins either covering the target receptors: anti-cytotoxic T lymphocyte-associated antigen CTLA4 (ipilimumab), or programmed death PD-1 (nivolumab), or neutralizing the activator ligands-to-PD-1 (pembrolizumab) rescue the autoimmune T lymphocyte population from apoptotic death. As instructed by their surface receptors, these autoimmune T cells attack and rapidly lyse large masses of tumors (melanomas and others). In combinations, their efficacy doubles (Lindsay CR et al Ther Adv Med Oncol 2015;7: 107–121; Romano E & Romero P J Immunother Cancer 2015;3: 15. doi:10.1186/s40425-015-0059-z; Sznol M & Longo DL N Engl J Med 2015;372: 374–5).*

Targeted small molecular inhibitors of oncogene-derived protein kinases. Of the DNA alkylators, mechlorethamine (Mustargen) was followed by cyclic phosphoramide mustard, or cyclophosphamide (Cytoxan). In the 5-FU molecule, a fluorine atom substitutes the hydrogen at carbon-5 position in the pyrimidine ring. In killing cancer cells (and some healthy cells), 5-FU synergizes with alkylators, cisplatin and MTX. Thus, the most powerful oncogenes can be destroyed by direct attack of externally administrated toxic agents. However, not selectively so: the healthy genome suffers quantitatively less, but similar damage. Some of the genomic changes are tantamount to new carcinogenic mutations. Nevertheless, combination chemotherapy has become the standard treatment of metastatic cancers. It has significantly prolonged the life of the majority of the patients, and apparently it has cured (so claimed) a smaller number of patients (sometimes with late consequences, especially in childhood). These very same drugs cause severe, often fatal, effects on unicellular life forms (kinetoplastids); developmental defects in the hydra, larvae, nymphs, insects and tadpoles; and induce severe birth defects in mammalian embryos.

A different, and much less devastating to the host, therapeutic principle emerges in the use of inhibitors of the oncoprotein tyrosine kinases (other than mcab). Antifolate analogues (aminopterin-methotrexate, MTX), 5-fluoropyrimidines (5-fluorouracil, 5-FU) showed toxicity to cancer cells. MTX is an inhibitor of the enzyme dihydrofolate reductase, which maintains the folate pool in the form of tetrahydrofolates. It damages cancer cells faster than normal cells; after a devastating attack on cancer cells, in 24 h MTX can be neutralized by leukovorin, for rescue of surviving normal cells.

Finally, small molecular inhibitors of large oncoprotein tyrosine kinases disable these oncogene product proteins. First, imatinib mesylate (Gleevec, Novartis) neutralized the $p210^{BCR/ABL}$ oncoprotein in CML. This oncoprotein is an activator of the PI3K/Akt pathway. After years of remissions, imatinib-resistant leukemia cells recur. Several different forms of imatinib-resistance are recognized. In one form, the CpG islands of one of the *hox* (homeobox) gene (HOXA4) promoter suffer hypermethylation. The fused *bcr/abl* oncogene may undergo point mutations, and return and such. A standstill of the Influx transporter Oct-1 (octamer organic cation transporter) and other solute carriers, or overactivity of the efflux pump glycoproteins (Pgp), exempt the leukemia cells from imatinib uptake. The leukemia cells may obtain resistance to apoptosis. Cytoplasmic NFκB/Rel may be released for intranuclear transfer and oncogene activation. Here, the suppressed RNA/DNA complex switched to several different pathways to sustain the transformed cell [2036–2039]. View Figure 1 in ref [2038] "multifactorial resistance phenotypes", constitutively activated NFκB, and chemotherapy-activated FoxO3a leading to cell cycle arrest and apoptosis. Nilotinib (Tasigna, Novartis), dasatinib (Sprycel, BristolMyersSquibb), bosutinib (Pfizer), and ponatinib (Iclusig, ARIAD), as BCR/ABL and SRC family kinase inhibitors are able to re-induce the remission. Ponatinib is indicated for T315I (threonine; isoleucine) point mutated Abl gene (expected citations in PDR, 2015–6).

The treatment-resistant condition, amyloidosis, yielded with rapid complete remissions to CtBorDx (cyclophosphamide, bortezomib, dexamethasone) [2040]. Major therapeutic breakthroughs are forecasted in the treatment of all B-lymphocytic malignancies, including multiple myeloma. The thalidomide-derivative, lenalidomide synergizes with rituximab; pomalidomide may surpass the efficiency of lenalidomide. The PI3Kδ inhibitor idelalisib (CAL-101/GS-111, Infinity Pharmaceuticals; Gilead), combined with rituximab (R) (with or without bendamustine), act synergistically. The production of Bruton's tyrosine kinase (BTK) oncoprotein is mightily suppressed by ibrutinib. BTK is a driver of Akt, ERK, NFκB oncoproteins and cytosine phosphate guanine-mediated mitoses. Ibrutinib (PCI-32765) synergizes with R even in patients with 17p13.1 deletion;

otherwise these patients relapse on ibrutinib. The overexpressed Bruton tyrosine kinase (BTK) is a pathognomonic driver of diffuse B cell lymphomas, and in particular mantle cell lymphoma. The small molecular inhibitor ibrutinib (PCI-32765) induces massive wide-spread apoptotic death of BTK-expressor lymphoma cells and it synergizes with proteasome inhibitors (bortezomib). The drivers of lymphoma cells (Akt, NFκB, Bcl-xL, XIAP) retreat; B cell antigen receptor signaling ceases to function and pro-apoptotic signals overcome the lymphoma cell. Normal $CD34^+$ lymphopoietic cells suffer no cytotoxicity. A very large number of new reports appear in the literature in late 2015. The FDA approved ibrutinib capsules for therapy under the name Imbruvica (not listed in Physician Desk Reference 2016).

The fusion oncogene/oncoprotein in mantle cell lymphoma is formed by the unison of the cyclin D1 and the IgH genes (another "error" that might have occurred during SHM): t(11;14)(q13;q32). The constitutively activated metalloproteinase ADAM 10 (disintegrin and metalloproteinase) cleaves into active form the TNFα. The TNFα target, NFκB, exerts anti-apoptotic effect. Upregulated mTOR and PI3K/Akt pathways drive the tumor cells. Mantle lymphoma cells suppress FOXO3 (forkhead/winged helix box class O3) within the cytoplasm. The suppression of FOXO3 is dependent on the over-expression of PI3K/Akt. For its natural tumor-suppressor effect to be exerted, FOXO3 must appear in the nucleus, in order to activate its target genes; that leads to cessation of mitoses [2041, 2042]. This heavily treated tumor (M.D. Anderson's hyper-CVAD of cyclophosphamide, vincristine, doxorubicin, dexamethasone with R; salvage high dose methotrexate with leukovorin rescue; cytarabine; stem cell transplant) had a notoriously high relapse rate [2043, 2044a, b]. It is expected that bortezomib, lenalidomide and the mTOR-targeting temsirolimus exert minimal toxicity, but durable remissions [2045]. The dual inhibitor of PI3k/Akt and mTOR, NVP-BEZ235 (patented, Novartis, Basel, Switzerland) restored bortezomib sensitivity, and suppressed the growth of mantle lymphoma cells. The NVP-BEZ325-treated mantle lymphoma cells died both extrinsically and intrinsically induced apoptotic deaths. Enzastaurine, everolimus and perifosine exerted additive effect [2046, 2047]. The 2-morpholino pyrimidine imidazoquinoline derivative pan-PI3K inhibitors, NVP-BEZ/235 and NVP-BEZ/120 with high bioavailability on *per os* administration for the latter compound, suppress all four class I PI3K isoforms, including the point-mutated oncoproteins. They decrease the cellular levels of the Akt oncoprotein. They inhibit the activation of phosphorylated Akt by mTOR on Ser473. The NVP-BEZ120 compound inhibits mTOR less effectively than NVP-BEZ235, but it is also antiangiogenic. K-Ras-mutated tumor cells gain resistance to NVP-BKM-120; however, equivalent sensitivity was observed between tumor cells with or without PTEN function. Both NVP-BKM120 without, and NVP-BKM325 with mTOR inhibitory activity are in clinical trials [2048]. The small molecular inhibitor of the anti-apoptotic BCl2/BCL_{XL} oncoproteins, ABT737 navitoclax (*vide supra*) given *per os* induces apoptotic deaths of chronic B lymphocytic leukemia cells, and also the activation of the proapoptotic *bax* gene product protein BAX (Bcl2-associated X protein). Navitoclax surpasses in efficiency the phosphorothioate antisense *bcl*2 gene antagonist Oblimersen (Genasense) (*vide supra*). Occasional complete remissions induced in patients with mantle cell lymphoma were by mcab rituximab or ofatumumab; a dendritic cell vaccine; or lenalidomide. The expectation is that ibrutinib-induced complete remissions become frequent and durable; may be curative.

An extremely useful (free of chemotherapy-related exaggerations of a past era) report on investigational targeted therapy for sarcomas is available from M.D. Anderson Hospital [2049a]. Of soft tissue sarcomas, myxoid-round cell liposarcoma (Table VII) takes use of the mTOR pathway, calling for clinical trials with mTOR inhibitors [2049b, c]. This is the tumor which has responded to ecteinascidins (Trabectedin, Yondelis) [2050a]. The use of ecteinascidin may extend to Ewing's sarcoma, and to pleomorphic sarcoma. In pediatric sarcoma cell lines with the EWS-FLI1 pathway, ecteinascidin was growth inhibitory [2050b]. In a clinical case of a large pleomorphic sarcoma, ecteinascidin given preoperatively, resulted in a fully necrotic tumor at its resection [2050c]. The common pediatric sarcomas respond better to chemotherapy, than the rare sarcomas of adult patients. This tumor entity documents best its very limited susceptibility to single agents of chemo-radio-, or immunotherapy, but combined modalities of these regimens may act additively (may be even synergistically), even against metastatic disease [275, 277] (Tables III/I and III/II). The prominent treatment for metastatic sarcomas remains the decades old combination chemotherapy of doxorubicin and ifosfamide. Recently, anti-angiogenic therapy with pazopanib emerges as an inducer of durable remissions; it significantly prolongs life. Combined with histone deacetylase inhibitors, pazopanib gaines increased efficacy. The European Organization for Research and Treatment of Cancer (EORTC) published its clinical trial in March 2014. Over 30 % of patients with metastatic soft tissue sarcomas survived progression-free, or in stable partial remissions, as "long term responders". Pazopanib is the FDA-approved oral (tablets) medication for metastatic renal cell carcinoma. The multi-tyrosine kinases inhibitor pazopanib (Votrient, GlaxoSmithKline), a complicated molecule of pyrimidinyl-amino -methylbenzene-sulfonamide, suppresses VEGF-R, PDGF-R, fibroblast GF-R, cytokine kinase receptor c-Kit tyrosine kinase (CD117, the Hardy-Zuckerman feline sarcoma retroviral c-*onc* → v-*onc* oncoprotein), and Itk, the IL-2R-inducible T-cell kinase (citations in PDR vol 65: 1610–1615, 2011; years 2012–2014). Further of sarcomas, imatinib and its second generation derivatives suppress the growth of gastrointestinal stromal tumors (GIST), and dermatofibrosarcoma protuberans. Other soft tissue and bone sarcomas so far resisted other available targeted therapy.

A combination of the monoclonal antibody trastuzumab (Herceptin, Genentech), and the small molecular oncoprotein-inhibitor lapatinib (Tykerb, GlaxoSmithKline) attack the extracellular and intracellular domains, respectively, of the HER2/*neu* oncoprotein in metastatic adenocarcinoma of the breast. The spectacular complete remission induced by gefitinib (Iressa, Astra Zeneca) in the one Oriental female patient with metastatic non-small-cell lung cancer of the L858R point mutation (leucine; arginine) was serving as a leader to erlotinib (Tarceva, Genentech). Tarceva induces remissions of lung adenocarcinomas with mutated EGF-R in codon 858, nucleotide 2819, exon 21: L858R = nucleotide T → G resulting in amino acid leucin (L) replacement by arginin (R). Regorafenib (Stivarga, Bayer) acts as an inhibitor of KIT, RET and VEGF-R1 kinases; it is licensed for treatment of metastatic colon cancer; it is also effective against GIST. Ziv-aflibercept (Zaltrap, Sanofi) is a recombinant fusion protein inhibitor of VEGF-A/B and PlGF (placental growth factor). The androgen receptor inhibitor, enzalutamide (Xtandi, Astellas) is licensed for the treatment of metastatic castration-resistant prostate cancer (citations in PDR years 2010–2013) [2051].

The Axinib (Inlyta, Pfizer) acts as an inhibitor of VEGF-R2/3 and is licensed for the treatment of renal cell carcinoma. Metastatic kidney carcinoma clear cells could be moderately suppressed by interferon-α and/or interleukin-2. Small molecular inhibitors of the VEGF-mediated oncogenesis, sunitinib and axitinib (Sutent; Inlyta, Pfizer), and mTOR-mediated (mammalian target of rapamycin) oncogenesis, everolimus (Afinitor, Novartis), and temsirolimus (Torisel, Wyeth) are capable of inducing sustained complete remissions in metastatic kidney cancers (in PDR, years 2012–2014). However, for the non-exceptional well-responding patients, the average response rates of metastatic kidney cancer to molecular inhibitors were tabulated in The Cancer Letter May 32 2013. Progression-free survivals ranged from 5 to 12 months; overall survivals from 11 to 29 months. The relapse rates document the power of the RNA/DNA complex to withstand even major refined chemical attacks.

The inexorably mortal metastatic melanoma may be stopped; rapid advancements of immunotherapy may render tabulations out-dated (Table XXII). Vemurafenib (Zelboraf, Plexxicon, Roche) induces durable remissions in V600E (valine; glutamic acid) BRAF-mutated (B homolog rat fibrosarcoma) malignant melanoma cells. In this case, the oncogenome may strike back. Point-mutated H-*ras* Q61L (glutamine; leucine) and activated MAPK pathway induce the growth of keratoacanthomas or intradermal squamous cell carcinomas in the treated and responding patients. Here, the RNA/DNA complex shows its ability to switch from one oncogenic pathway to another (PDR, years 2013, 2014). Some pilocytic astrocytomas are driven by the V600E (valine; glutamic acid) BRAF mutation; no data are available if vemurafenib could induce remissions of these astrocytomas.

The MEK kinase inhibitor trametinib (Mekinist, GlaxoSmithKline) could induce a second remission in patients with metastatic melanoma, who responded to vemurafenib, but relapsed. The second generation BRAF inhibitor dabrafenib (Tafinlar, GlaxoSmithKline) in combination with trametinib increased the remission rate and duration fourfold (as compared with vemurafenib alone); the rate of squamous carcinoma induction with vemurafenib dropped from 20 to 5 % in patients receiving the combined treatment. These drugs pass the blood-brain barrier and act upon melanoma brain metastases. (M.D. Anderson's Oncolog Feb 2014; 59/2).

The pan-histone deacetylase inhibitor panobinostat (PS) synergized with NVP-BEZ325 (*vide supra* at mantle cell lymphoma) against pancreatic carcinoma cells expressing K-ras mutation; the combined regimen suppressed Bcl-xL expression and activated the cell cycle inhibitors p21waf1 and p27kip1 [2052]. $CD133^+$ and $CD44^+$ prostate cancer stem cells have been isolated and grown in tumor spheres. The PI3K/Akt signaling pathway was the driving force of these cells; tumor stem cells with deleted (silenced, knocked-down) PTEN and/or

Table XXII Malignant Melanoma Yielding to Immunotherapy

The relapse rates of resected deep primary, or regional lymph node metastatic melanoma could be reduced by post-operative therapy with IFNα (10–15 % reduction),[1] or with viral oncolysate vaccines; especially, NDV oncolysates were effective (decades long durable remissions induced).[2,3] Preventively, the Morton allogeneic melanoma cell-BCG Canvax vaccines did not work in prospectively designed clinical trials, but it induces immune reactions and reduces circulating tumor cell numbers.[4] Oncolytic viruses directly injected into melanoma tumors induce partial remissions.[5]

Metastatic melanoma remained an essentially incurable disease with prolongation of life only in response to chemo-immunotherapy. Non-curative partial responses were induced by dacarbazine (5 %), cisplatin, IFNα, IL-2; disease terminated in relapses with deaths delayed.[6] Major clinical responses (partial remissions, durable complete remissions) are achieved with intravenous immune adoptive lymphocyte (immune T cell; NKT/LAK cells and IL-2) therapy; curative outcome is rare but possible.[7]

Targeted bio-immuno-therapy surpasses in efficacy all prior modalities.

BRAF-V600E → MAPK/MEK mutated melanoma cells replicate without apoptotic death and present themselves to their host as "self", thus switching off lymphocyte-mediated immune attacks Vemurafenib and dabrafenib inhibit BRAF-mutated cells and induce T cell-mediated anti-melanoma immunity; dabrafenib or trametinib neutralize the MAP/ERK pathway. Vemurafenib toxicity is rare but severe (dermatitis; acute renal toxicity).[8] BRAF-inhibitor-treated melanoma-bearing hosts show sharp rise of $CD8^+$ immune T cells and drop of $Fox3^+CD4^+$ Treg cells.[9] Mcab ipilimumab switches off the T cell antigen CTLA4, which is the natural blockade against autoimmune attacks. The programmed cell death PD-L1 antigen is activated by its ligands. Melanoma (and other cancer) cells express the PD-1 ligands (L) in order to induce self-apoptotic death of PD-1 autoimmune T lymphocytes; for malignantly transformed cells masquerading as self, thus to escape an autoimmune attack. Anti-PD-1 and anti-L mcab (MK-3475; nivolumab; pembrolizumab) block PD-1 and PD-L1 expression. The $PD1^+$ T lymphocytes remain active inducers of autoimmune attacks. Combined therapy with ipilimumab and nivolumab at the price of reversible gastrointestinal and hepatorenal toxicity induces durable complete remissions in metastatic disease.[10,11] Melanoma cells activating MAP/MEK pathway resist vemurafenib; BRAF inhibitor dabrafenib and MEK inhibitor trametinib induce further remissions. When targeted and mcab therapies are combined, cures become possible (MD Anderson & Moffitt protocols). Human HER2/neu^+ melanoma cells respond to trastuzumab *in vitro* and *in vivo* as xenografts.[12]

BRAF rat fibrosarcoma B oncogene, V valine, E glutamic acid, ERK extracellular regulated kinase, MAPK mitogen-activated protein kinase, CTLA4 T lymphocyte-associated antigen; PD-1 programmed death and its ligands L-1/2, pembrolizumab mcab-induced remissions, see text.

1 *Davar D et al Clin Dermatol 2013;31: 237–50.* ***2*** *Sinkovics JG Horvath JC Arch Immunol Therap Exper 2008;56S: 1–59.* ***3*** *Cassel WA et al In: Viral Therapy Human Cancers. Sinkovics JG Horvath JC authors/editors. Marcel Dekker, New York 2005; 677–89; Sinkovics JG Plager C et al In: Neoplasm Immunity RG Crispen ed Franklin Institute Press 1977; 211–19; Sinkovics JG Plager C et al In: Neoplasm Immunity RG Crispen ed Elsevier/North Holland 1980; 481–519.* ***4*** *Hashimoto S et al Ann Surg 2012;255: 357–62.*
5 *Kaufman HL & Bines SD Future Oncol 2010;6: 941–9.* ***6*** *Sinkovics JG Horvath J Melanoma Res 2003;13: 431–2.* ***7*** *Rosenberg SA et al J Immunother 2003;28: 835–93.*
8 *Regnier-Rosenecker E et al Br J Dermatol 2013 in print. doi:10.1111/bjd.12555.*
9 *Knight DA et al J Clin Invest 2013;123: 1371–81.* ***10*** *Wolchok JD et al N Engl J Med 2013;369: 122–33.* ***11*** *Sullivan RJ et al N Engl J M2013;369: 173–83.* ***12*** *Ma J et al PLoS One 2013;8(8): e7.*

FOXO3 gene product proteins produced increased tumor spheres. The dual inhibitor of PI3K and mTOR, NVP-BEZ235 (*vide supra*) inhibited tumor sphere formation and growth of prostate cancer stem cells [2053].

Interfering/inhibitory microRNAs. It is estimated that one-third of the human gene expressions is regulated by microRNAs (miRs), including the oncomirs (Table XXIII). The evolutionarily conserved polycistronic gene miR17-92 cluster resides in chromosome 13q31-q32; it releases seven miRs. One, miR-17-92, controls the off-on switch for the elongation factor E2F and proto-oncogene/oncoprotein c-*myc*/c-MYC. Overexpression of the miR-17-92 is oncogenic; accordingly, miR-17-92 is over-active in breast, colon, pancreas, prostate and stomach adenocarcinomas. When miR-17-92 is underexpressed and holds the switch off, the tumor cells (breast, ovary; melanoma) delete it [2054]. The inhibitory microRNAs render major contribution to tumor suppression; however, actively growing tumors delete them. These microRNA strands align to and actually demolish mRNAs traveling from the nucleus through the spliceosomes to the ribosomes to be translated into the gene product proteins. These gene-product proteins can be those of tumor-inducing oncogenes, or those of tumor suppressor genes. For example the *sox* 4 oncogene at 6p22 in the endometrium encodes its oncoprotein through its mRNA for the activation of the Wnt/sHh (wingless, drosophila; sonic hedgehog) cascade with Notch and TGFß signaling. The *sox*4 mRNA is suppressed by the microRNA-129-2 in the healthy endometrium. In cancerous endometrium, the CpG islands of the microRNA-129-2 gene are hypermethylated and thus silenced. The unopposed *sox*4 mRNA reaches the ribosome for the synthesis of the SOX4 oncoprotein. DNA demethylation with 2-deoxy 5-azacytidine (Vidaza, Celgene) (and histone re-acetylation with trichostatin) restores microRNA-129-2 production resulting in the inhibition of the *sox*4 mRNA, and thus the cessation of the synthesis of the Sox4 oncoprotein [2055a, b].

Of the three oncomirs, let 7a, miR-155 and miR-195, miR-195 is the most specific signature of breast cancer on board [2056]. A delicate balance of miRs regulate the resting, locomotion, metastasis production, and ETM transition of prostate cancer cells. Host-defense promotional and cancer cell suppressive are the cluster miR-23b/-27b, miR-282, miR-185, miR-200a and miR-200b-3p [2057–2059]. In contrast, mi-R182-5p and miR-143 promote cancer cell proliferation, invasiveness, and stem cell de-differentiation toward metastasizing cancer cells [2060–2062]. MicroRNAs-7, Let7, and 125b are tumor suppressors for colorectal, non-small cell lung, and hepatocellular carcinomas. These effects are measured against established cancer cell lines *in vitro* and in xenografts *in vivo* [2063–2065]. Multiple myeloma (*vide supra*) cells are driven by miR-221/222. Knock-out inhibitors of miR-221/222 stopped tumor cell growth *in vitro* and *in vivo* (in xenografts). The miR-221/222 abolished PTEN function; inhibitors of miR-221/222 restored PTEN function [2066].

Laboratory product phosphorothioate antisense oligonucleotides cognate-specific to oncogenes, or to oncogene-derived mRNAs, or to microRNAs (oncomirs) have become the physiological inhibitors of the oncogenome. Their spectacular *in vitro*, or *in vivo* (in xenografts) success rate in inhibiting tumor

Table XXIII OncomiRs

DNA emits mRNA for protein synthesis; intron emits microRNA to demolish the mRNA
Lymphoma. The seven microRNAs cluster miR-17 ~ 92 is amplified at human chromosome 13q31 in patients with certain malignant lymphomas. The mRNAs of tumor suppressor PTEN and BIM (pro-apoptotic Bcl-2 antagonist) exhibit binding sites for miR-17 ~ 92; overproduced miR-17 ~ 92 eliminate the tumor suppressive mRNAs. Mice rendered miR-17 ~ $92^{-/-}$ regress established lymphomas. The level of miR-17 ~ 92 in Burkitt's lymphoma reaches ten-fold over normal.[1]   ***Sarcoma***. Various sarcoma subtypes depend on different miR overproduction. MicroRNA biogenesis is depicted in Figure 1 of cited article.[2] Alveolar rhabdomyosarcoma cells with t(2;13)(q35;q14) of PAX3/FOXO1 exhibit upregulated myomiR-133 and suppress the myoblast differentiation inducer miR-1. The gene 13q31 is amplified. The replicatory cycle of Ewing sarcoma cells is dictated by the t(11;22)(q24;q120 EWS/Ets fusion (E 26) oncoprotein. The promoter of miR-145 is silenced, and the IGF-1R gene is overexpressed. The EWS/FLI (Friend leukemia virus insertion site gene) oncoprotein suppresses the miR-100 promoter and downregulates miR-100 production; miR-100 is a natural suppressor of IGF-1R. Synovial sarcoma (SS) cells carry the t(X;18)(p11.2;q11.2) translocation. The SYT/SS-X-1/2 oncoprotein downregulates miR-143, the repressor of oncogene SYT/SSX. Proteins SYT and SWI/SNF are united. The QPGY motifs (glutamine, proline, glycine, tyrosine) are the sites of unison. This unison induces the cell transformation. SWI and SNF are part of the SMARC protein (switched; sucrose non-fermenting; SWI/SNF-related matrix-associated actin-dependent regulator of chromatin). The overexpressed miR cluster miR-96, miR-182/3 target for repression the tumor suppressor gene EGR1 (early growth response in mammalian cells). The large miR cluster at 19q13.4 contains the overexpressed miR-let-7e, MiR-99b and miR-125a-5p targeting for deregulating HMGA2 and SMARCA5 genes (high mobility protein group A2). The FasL→FasR family member let-7/miR-98 exerts anti-apoptotic function. The high amplification of gene miRHMG1 encoding oncomiR-1 with amplified cMyc distinguished high grade neuroblastoma from rhabdomyosarcoma. Upregulated oncomiRs could be eliminated by laboratory-sythesized anti-miR oligonucleotides (the AMO antagomiRs). AMO with locked nucleic acids (AMO-LNA) first stabilizes then inactivates its complementary sequences. Adenoviral or lentiviral vectors could deliver genomic sequences for the encoding of antagomiRs. Such interventions inhibited the growth of human tumors (glioblastoma; hepatocellular carcinoma; non-small cell lung cancer) xenografted in mice.[3]   ***The miR-connections of epithelial cell human cancers***. Prostate cancer (PC) with fused oncoprotein transmembrane protease serine2, ets-related gene, E26 (TMPRSS2-Erg). The tumor suppressor miR let-7b is rendered non-functional in high risk (Gleason >8) tumors; lack of let7b liberates high mobility oncoprotein group A1 (HMGA1) to reach high levels (Tables XI/I and XI/II). The overexpressed EGR-1 protein is an anti-apoptotic driver of the cell cycle, while the p21 complex (Cip, cell cycle dependent kinase-interacting protein; Waf (wild type p53 activated fragment) brake the cell cycle and exert pro-apoptotic effect.[4] The HM group A1 oncoprotein drives breast carcinoma cells, especially the triple negatives. Silencing HMGA1, reverses transition from mesenchymal back to epithelial by repressing genes vimentin and snail. $HMGA1^{-/-}$ tumors do not grow in xenografts, and do not form the 3-dimensional stem cell mammospheres.[5] Inverse relationship in PC: ↑HMGA1, ↓miR-296. miR-296 demolishes oncogenic HMGA1 mRNA. Gleason >8 PC; ↑HMGA1.[6]   ***Melanoma.*** The adenosine deaminase acting on RNA (RNA editing, ADAR) gene is downregulated in metastatic melanoma. Consequentially, the expression of more than 100 microRNAs becomes deranged. The mRNA ADAR transcript is directly targeted by miR-17-5p and miR-432 encoded from a hypomethylated and amplified site of chromosome 14 (but the miR-17 ~ 92 cluster remains constant and not involved). $ADAR^{-/-}$ tumors were large, Ki-67 4+, and consisted of dividing cells.[7] The endothelin-3 protein transfected into melanoma cells

(continued)

Table XXIII (continued)

downregulated SPARC (secreted protein acidic cystein rich) and NFκB; the cells reduced their mitotic rate and underwent increased apoptotic events.[8]
1 *Xiao C et al Nat Immunol 2008;9: 405–14; EMBO J 2013 in print.* ***2*** *Dela Cruz F & Matushansky I Discov Med 2011;12: 307–17.* ***3*** *Hayes-Jordan A & Andrassy R Curr Opin Pediatr 2009;21: 373–6; Kato H et al J Biol Chem 2001;277: 5498–505; Renner M et al Genes Chromosomes Cancer 2012;51: 982–96; Wei JS et al Clin Cancer Res 2009;15; 5560.* ***4*** *Parra E et al Oncol Rep 2013;30: 911–6; Schubert M et al PLoS One 2013;8(6): e65064.* ***5*** *Shah SN et al PLoS One 2013;8(5): 63419.* ***6*** *Wei J-J et al Clin Cancer Res 2011;17: 1297–305.* ***7*** *Nemlich Y et al J Clin Invest 2013;123: 2703–18.* ***8*** *An XJ et al J Huazhong Univ Sch Technolog Med Sci 2013;33: 581–6.*

growth will be translated into clinical use. However, tumors removed from their natural environment and grown in tissue cultures, fed with fetal calf serum, or transplanted into a similar, but not identical new environment (a xenograft), might have switched their growth pathways. A treatment modality inhibitory for the xenografted tumor may fail to inhibit the tumor in its natural environment.

Epigenome/epigenetic control of tumor growth (silencing the promoters of the oncogene, re-activating the promoters of the tumor suppressor gene). Transfer from the laboratory to the clinics the sh/I microRNAs, which destroy the mRNAs that are to be translated into oncoproteins; rescue the suppressed microRNAs, which serve the tumor suppressor genes and the innate and adaptive anti-tumor immune reactions.

An example of many, for tumor fusion oncoprotein de-activation by epigenetic intervention, is in the cancer of prostate. The TMPRSS2-ERG fusion oncoprotein (transmembrane protease serine 2; *ets*-related gene: E-twenty-six; E26 avian erythroblastosis retrovirally acquired cellular-to-viral oncogene: c-*ets* → v-*ets*) is encoded from locus 21q22. ERG is a histone acetyl transferase (HAT) inhibitor; a de-acetylated histone cannot be reacetylated in the presence of ERG; ERG is an activator of histone deacetylases (HDAC). Histone acetylated oncogenes rest; histone deacetylated oncogenes are activated. Extrinsically administered HDAC inhibitors valproic acid or trichostatin-A allow re-acetylation of the oncogene promoter, thus silencing it. The tumor cell responds with upregulated cell cycle inhibitors $p21^{WAF1/Cip1}$ and undergoes apoptosis [2067]. (Google gives for waf "wife approval factor"; do not believe it, it stands for "wild type p53 activated fragment"; Cip, cycline-depedent kinase interacting protein).

The GTPase-activating protein, deleted in liver cancer-1 (DLC), a tumor suppressor gene product protein, extends its activity from hepatocelluler carcinoma to various adenocarcinomas and Burkitt's lymphoma. In prostate carcinoma cell lines, DLC-1 is silenced by hypermethylation of its promoter. Deacetylase inhibitors SAHA (suberoylanilide hydroxamic acid) and TSA (trichostatin) restored histone acetylation. Production of the SAHA target, RhoGTPase ceased. SAHA-treated tumor cells failed to grow as xenografts in athymic mice [2068]. The hypermethylation of tumor suppressor gene RARβ2 (retinoic acid receptor) in surgical specimens of prostate cancer is an adverse prognostic factor [2069]. The DNA methyltransferase inhibitor 5-aza-2′-deoxycytidine removes the CH_3-residues from the CpG islands.

Comment. An early assessment and listing of oncogenes and recognition of weak, but still clinically meaningful responses to the natural (not genetically engineered) lympho- cytokines (interferon-α and IL-2) predicted early that oncoprotein-driven malignancies may become responsive to biological agents [2070]. Indeed, but the response is inadequate for the control of an established malignant process. However, enzyme inhibitors control hypertension and hyperlipidemia: would oncoprotein kinases respond to enzyme inhibitors? Could de-differentiated cells be induced to re-differentiate? Author's elderly patient TS with prostatic carcinoma surgically removed, developed acute hemorrhagic promyelocytic leukemia and septic shock. He was admitted from the Emergency Room to Intensive Care. After the control of the septic condition, he entered and remained in complete remission on all-transretinoic acid (ATRA) alone in the fourth year (now transferred to under the care of medical oncologist-hematologist Christopher George). The medical history of "leukemia cells differentiating on treatment with tablets" observed in China decades ago is now re-played [2071, 2072]. While the aim is the selective elimination/destruction of cells driven by the primordial RNA/DNA complex (Re: "malignantly transformed" "carcinogenized"), the cell transformed by one special pathway, may reverse its course. This cell may differentiate and rest at the level of a mature somatic cell (or a stem cell still able to further de-differentiate). The miR-suppressed tumor cell may rest, undergo senescence (by definition: alive, metabolically active, unable to divide), or enter autophagy, and either die apoptotic/necrotic death, or eventually recover as a transformed cell, as long as a decade(s) later. This cancer cell appears mesenchymally transformed, with amoeboid features clearly exhibited (*vide supra*); or becomes neuroectodermally transspeciated (repeatedly referred to elsewhere; see "achaete scute").

As to "subduing the RNA/DNA oncogenome", two examples are presented (from the educational curriculum of medical oncologists in training). In pretreated, but not responding patients with $HER2^+$ metastatic breast cancer, the lapatinib/capecitabine (LC) regimen was compared with the Kadcyla (KC) regimen [2031]. Progression-free and overall survivals were 6.4 and 25 mo for LC, versus 9.6 and 31 mo for KC. In patients with unresectable metastatic colon carcinoma, the FOLFOXIRI/bevacizumab (5-FU; leucovorin; oxaliplatin; irinotecan, Avastin) regimen (#1) was compared with the FOLFIRI/bevacizumab regimen. (#2). The response rates (remissions) were 65 % for #1 and 53 % for #2; complete remissions were 20 % for #1 and 11 % for #2 (with late relapses occurring in both groups). Survival differences with metastatic colorectal cancers, minute for the statisticians, but significant for the patients, are being further analyzed [2073–2075].

Comment. After an early response of varying degrees and duration, a re-grouped RNA/DNA oncogenome eventually re-emerges, even after (a rest in autophagy?) of several years. The living matter expresses its primordial faculty inherent in its RNA/DNA for survival. For the first time in its milliards (billions) years of history, it is not the hostile environment, but its own product, the human cerebral cortex that forcefully disciplines the bioengineers supreme RNA/DNA complex. The response so elicited holds its secret: occasional patients remain in complete remission and are cured for their natural life time. Other patients on the same therapy relapse on a scale: rapidly and relentlessly, or extremely slowly and living with their cancers transformed into a chronic ailment. The host-tumor relationship remains highly individualized, but the new treatment modalities shifted the balance to the favor of the vast majority of the patients.

Selected abbreviations

A disintegrin and metalloproteinase. ADAM12 is a cell membrane secreted glycoprotein gelatinase. Click for breast cancer progression by ADAM12 secretion, by Roolali R Wewer UM Zurakowski D et al. Click for ADAM10-induced release of hepatocyte growth factor receptor oncogene MET from hepatocellular carcinoma cells, by Chalupský K Kanchev I Žbodáková O et al. Click for ADAM9 knock-out by specific lentiviral shRNA resulting in cessation of growth of prostate cancer cells with downregulation of REG4 (regenerating islet cell family member 4) and upregulation of $p21^{waf1/Cip1}$ (wild type p53 activated fragment/cyclin-dependent kinase interacting protein), by Liu C-M Hsieh C-L He Y-C et al.

ash/ASH achaete scute homolog. Referring to drosophila chaete, bristles; scute, horny plate, scale. Evolutionarily encoded first nervous systems in cnidaria, sponges (?), protochordates (amphioxus). Click for achaete scute homolog 1 encoding poorly differentiated human neuroendocrine carcinomas, by La Rosa S Marando A Gatti G et al. Click for achaete scute homolog 1 encoding neuroendocrine breast carcinoma, by Righi L Rapa I Votta A et al. Click for achaete scute homolog 1 encoding neuroendocrine pulmonary tumor, by Li F Zhong Z Li R et al (in Chinese; PMID 20677557). Click for genetically engineered adenovirus Ad5PTD(ASH1-SCG#-E1A (protein transduction domain; secretogranin) selectively killing ASH-expressing neuroblastoma cells, by Jin C Yu D Čančer M et al. Click for overexpression of IGF (insulin-like growth factor) in neuroblastoma cells with suppressed achaete scute homolog 1, by Li J Neumann I Volkmer I Staege MS.

BTK Bruton kinase. Bruton agammaglobulinemia tyrosine kinase. Click for BTK-like protein containing Tec domain with 38 % identity 55 % overall similarity to human Tec domain expressed in marine sponge Suberites domuncula (Demospongiae), by Helena Cetković et al, Ruđer Bošković Institute, Zagreb, Croatia (see in Introduction). Click for ibrutinib blocking B cell receptor signaling in hairy cell leukemia, by Sivina M Kreitman RJ Arona E et al. Click for ibrutinib targeting B cell signaling in CLL (chronic lymphocytic leukemia), by Cheng S Ma J Guo A et al. Click Bruton kinase in innate immune system, by López-Herrera G Vargas-Hernández A Gonzalez-Serrano ME et al. Click for targeting B-cell receptor signaling, by Fowler N & Davis E.

fli/FLI Friend erythroleukemia virus' genomic integration site proto-oncogene ets/Ets transcription factor, after E26 avian retrovirus, in human chromosome 13q14. See Ewing sarcoma rhabdomyosarcoma fusion oncogene/oncoprotein EWS/FLI1.

In Self-defense, the Cancer Cell Reactivates Its Primordial Genome Loaded with Proviruses in Order to Activate All Its Remaining Survival Pathways

From the Very Beginning on, Retrotransposons Were Transcribed into DNA, While the Introns Retained Their RNA-Identity

The world-wide virome; universal eukaryotic genome; the rhizome [14a, b]. Viruses were and remain the dominant "life forms" in the biosphere on Earth. In archaea and prokaryotes dsDNA viruses (phages) dominate, whereas animal eukaryotes harbor a most diverse virome, without a recognizable single ancestor. Animal cells harbor purely RNA viruses (replicating with RNA-dependent RNA polymerase), without a transitory DNA genome. Primarily these are the rapidly radiating picornaviruses, as the major members. The reverse-transcribed (retroid) viruses (RNA→DNA) are using viral RNA-dependent DNA polymerase, RT. The DNA→mRNA transcribed viruses are using host DNA-dependent RNA polymerase. The giant Mimiviridae family-members are the nucleocytoplasmic large DNA viruses; NCLDV also include the pox- and herpesviruses. Multicellular plants are free of dsDNA NCLDV, but the ancestral unicellular green alga harbors a NCLDV phycodnavirus. Plant cell genomes are loaded with retroelements, replicate RNA-viruses (alpha-viruses) and pararetroviruses (dsDNA→RNA→DNA), but not frank retroviruses.

Viral evolution followed one of three pathways (with possible switches): **1**, common ancestral viral progeny diverging; **2**, horizontal transfer of viral genes; **3**, parallel evolution of related viruses from related ancestral genetic elements. For monophyletic evolution of viruses, one must presume that the Last Eukaryotic Common Ancestor (LECA) harbored one (or more, different) ancestral virus(es). The picornaviruses and the NCLDV family might have each separately originated from an ancestral virus. Parallel evolution includes the viral genomes evolving simultaneously with host cell genomes. The discovery of ssDNA nanoviruses in marine picoeukaryotes opens up the field for any possible re-arrangements [13b].

J.G. Sinkovics, *RNA/DNA and Cancer*, DOI 10.1007/978-3-319-22279-0_26

Ancestral retroelements of the pre-cellular Virus World inserted into host cell genomes, as the first "cells" appeared, fused with host cell genomic segments, and might have eventually become the extant viruses. The origin of the DNA genomes from, or evolving parallel with, the pre-existing RNA sequences, without a co-existing reverse transcriptase, remains the unresolved event. Co-existing large amorphous proteins might have engulfed single RNA or DNA molecules (or an RNA/DNA complex), and thus produced inert "pseudovirions", as actually viewed by Domingo-Espin *et al* [2076]. Amorphous polypeptides, RNA and ribozymes (and possibly DNA) sequences coexisted in the precellular world. Could ribozymes catalyze synthetic reactions in such pseudovirions (and could they be subjected to laboratory experiments)?

Returning to the era, in which the Mauriceville retroplasmid was first transcribed into a new cDNA (Figure 2). Throughout the three billion/milliard years, in cellular life with eukaryotic genomes, the Mauriceville scenario was re-played on innumerable occasions. The ancestral non-LTR retrotransposons require an endonuclease to cleave a primer from the target DNA for the RNA copy to be reverse-transcribed. The LTR retrotransposons cluster into five families: the ancestral TY1/Copia and Ty3/Gypsy, and their distant radiations, which include among others, Caulimoviridae and Retroviridae [2077]. Another, more extended list of DNA transposon superfamilies of eukaryotes encompasses, among others, Merlin/IS1016, Mutator, P-element, PIF/Harbinger (P instability factor in DNA transposon system in maize), piggyBack, Tc1/Mariner, and Transib [2078]. The LTR retrotransposons are incomplete retroviruses with several ORFs for encoding the Gag, protease, RNAse, integrase and RT proteins. Complete retroviruses have ORF for encoding Env protein. The Dictyostelium's DIRS-1 (inverted repeat sequence) and the green alga Volvox's kangaroo retrotransposons have relatives in sea urchin, zebrafish/puffertfish, amphibians, and nematodes. The chlamydomonas retrotransposons structurally similar to kangaroo (both have split direct repeat termini), but the chlamydomonas' mobile elements do not encode RT (view Figure 1 of the article for the three major groups of retrotransposons). The unique integration of kangaroo and its relatives is without an endonuclease, and with a closed-circle dsDNA intermediate (Figure 5 in the cited article) [2079]. The self-spliced *Tetrahymena thermophila* 238 rRNA group I intron is an autocatalytic ribozyme. This intron assumes a tertiary structure. It is the RNA helices that fold into the specific configuration for the creation of the catalytic core. An exogenous guanosine nucleophile cleaves the 5′ splice site; the intron is released when the 5′ and 3′ exons are ligated together [2080]. Group II introns continue to be present in eukaryotic organelles of prokaryotic derivation. These introns integrate at exon junctions into intronless genomes. Group I introns encode the meganuclease to carry out the ds DNA break at the exon junction site. The dsDNA break is repaired by homologous recombination on the template provided by the invader. The ribozyme group II introns self-splice into a lariat RNA, which integrates itself as a DNA-RNA hybrid in between two exons. The template intronic RNA is reverse-transcribed, degraded and its site (the gap) is repaired by DNA polymerase. The group I intron-encoded protein is a ds endonuclease; the group II introns encode ss endonuclease and RT [2081]. Inserted retrotransposons-like elements can interrupt a protein-coding gene (in the Trypanosoma) [2082]. These are advanced, RT-mediated intracellular modes of RNA→DNA generation. These events do not lead us back to the primordial mode of the first DNA generation.

Mobile genetic elements (LTR-positive and negative retrotransposons) invaded the genome of the human blood flukes *Schistosoma mansoni, S. haematobium* and *S. japonicum*, some intact, some degenerated [2083]. The schistosoma genome passes its small non-coding RNAs through Argonaute/RISC. In contrast to one Argonaute gene in *Schizosaccharomyces pombe*, the schistosoma genome harbors four AGO genes, but only one Dicer gene [2084]. The caenorhabditis genome has 27 AGO genes; the drosophila genome operates Dicer1/2: Dicer1 for miRNAs (the Loquaceous protein for dsRNA binding), and Dicer2 for siRNAs. Most of the microRNAs derive from the intergenic regions of the schistosoma genome; endogenous small interfering endo-siRNAs originate from the retrotransposons. The Ago proteins regulate retrotransposons at the transcriptional level. Thus, these new arrivals to the genome, while they become critical regulators of gene expressions, they must pass through AGO control. Many schistosoma retrotransposons are known by name, for example, Boudicca, Curupira1/2, Perere, Nimbus, Phantom, piggy-Back. Sinbad, Pao-like, RTE clade, etc. The curupira1/2 transposons of *S. mansoni* and *S. japonicum* are in the highly mutagenic class of mutator-like DNA transposons [2085]. Extensive presence of Mutator retrotransposons reveals CHG-rich sequences (H = A, C, or T nucleotides), offering sites for methylations or recombinations, in contrast to inert CG-sparse regions. The DNA molecule bends, or undergoes B-Z (from right to left) transitions at the CG points [2086, 2087]. In the original Watson-Crick DNA model, the purine-pyrimidine base pairs (CG/GC) run in left-handed direction, but the prevalent form of DNA is the right-handed position B. Figure 1 of the cited article shows both the B- and Z-conformations of the DNA molecule. Life's most versatile master molecule could undergo zipper like actions, and could exist in a short stretch-collapse condition (S-DNA) intermediately during its B-Z transition [2088].

Outside plants, and in addition to the schistosomas, the mutator transposons were discovered in fungi (Aspergilus, Candida, Fusarium, Magnaporthe, Neurospora, Yarrowia), and in the triad of protists (Entamoeba, Giardia, Trichomonas). In the Trichomonas, the new Mutator transposons join the previously identified ones (Kolobok, Mariner, Maverick, Polinton). Of the Mutator-like elements 1–4 of *T. vaginalis* (TvMULEs), TvMULE1 is plant-related and the others are *Candida albicans*-related (TvCaMULEs) [2089], attesting that the likely horizontal spread of MULEs included an a-sexual parabasalid. The TEs (transposable elements) encode their own transposases. The cited article [2078] publishes Figures 1–6 for the phylogenetic trajectories of the Mutator, Gemini, Piglet, Hydargos, Chapka-Em1 and piggyback superfamily proteins in four species of Entamoeba. As an unexplained exception, the 22.8-Mb genome of *Plasmodium falciparum* is devoid of detectable TEs. The choanoflagellate genome (*Monosiga brevicollis*) is free of non-LTR retrotransposons, but carries the chromovirus clade Ty1/copia and Ty3/gypsy retrotransposons [2090]. The amphioxus harbors non-LTR retrotransposon [2091].

Some ancestral choanoflagellate cells were ready to form the first multicellular hosts (metazoa) in the form of sponges in the Cambrian sea some 550 million years ago. The originally intron-free mitochondrial genomes of metazoa, beginning with the sponges (Spirophorida), and by reaching the level of the cnidaria (sea anemone, Nematostella; *Hydra magnipapillata*) (Figure 37) show the presence of numerous horizontally acquired group I introns. Since mitochondrial genes travel to the nucleus, metazoan genomes from the beginning of their evolution were bombarded by alien introns and RNA retrotransposons, the latter being retrotranscribed to be DNA sequences, The phylum Nematoda possesses hundreds of genes acquired through horizontal routes from bacteria and fungi. Arthropods harbor entire bacterial genomes (the Wolbachia genome in some drosophila strains); reviewed in [27]. Totally a-sexual hosts (hydra; rotifers) collected large numbers of horizontally transferred retroviral, bacterial, fungal and other eukaryotic genes. The schistosoma specialized in populating its genome with alien retrotransposons (*vide supra*). In addition to conserved genes acquired by the ancestors, the human genome harbors a large assemblage of retroid elements, LTR-negative and LTR-positive retrotransposons and endogenous retroviruses, LINEs and SINEs (long and short interspersed elements). Figure 4 of the cited article summarizes the chimeric origin and nature of the metazoan genomes [2092].

The heavy chimeric involvement of the universal genome evolving on Earth obliterates all efforts to depict the evolution of these individual genomes in a tree-like fashion; it is rather a rhizome-like [14b]: root-stalk, a mass of roots, rhizoma, ρίζωμα. Some of the newly acquired and transcribed genes actually led to the extinction of the clade of the recipient. Other newly transcribed genes may represent those, which carry the inherent signature for the eternal continuity of cellular life. These genes render physiological services, but eventually become oncogenes.

RNA-derived mutators. RNA- and retroviruses are naturally of the mutator type. Large viral populations exist as the quasispecies of Manfred Eigen [5b, 2093a, b], due to closely related, but not identical individuals. The mutation frequencies of the individual particles could rise above the final tolerable threshold, at which point the viral particles of the entire population may die out: as an event of "error catastrophe". Malignantly transformed cells assume the mutator phenotype, when their mutational rate far exceeds that of healthy cells (a low standard mutational rate of the cellular living matter in the biosphere of the Earth). On occasion, lethal mutagenesis actually kills the cell (or a colony of such cells). However, below that threshold, there is the cell colony of rapidly mutating, replicating and expanding individual cells. Is the "genetic instability" of the cancer cell a genomic equivalent of the viral quasispecies? [2093a, b].

Does the assumption of the mutator phenotype of the individual cancer cell in a multicellular host constitute a regression to the level of the ancestral unicellular eukaryotes, which were driven by TEs (transposases, retrotransposons, retroviruses), that invaded their genome, and existed there transcribed into mutator DNA sequences? The ciliate *Oxytricha trifallax* possesses a somatic macronucleus and a diploid germline micronucleus [2094]. In the macronucleus reside innumerable nanochromosomes (some >16,000 representing >18,500 genes). These are individual single protein-coding genes. The 469 bp chromosome encodes a small, the 60 kb chromosome encodes a very large protein. Only a small fragment of the micronuclear DNA supplement is transferred to the vegetative macronucleus (Mac); the rest of the micronuclear DNA, transposons included, is destroyed. The transposons act in the germ cell micronucleus only. The Mac divides without a mitotic spindle; the germline micronucleus (Mic) forms a mitotic spindle for its own division. Mac incorporated the unusually large ~70 kb mitochondrial genome and the ~5 kb linear plasmid bearing the mitochondrial telomeres. Macronuclear genes encode the core eukaryotic proteins. The number of telomeres in the macronucleus per cell exceed tens of millions, yielding the telomere end-binding sequences for both ends of the nanochromosomes. Separate and shared telomeric proteins (telomere-binding elements) operate for the transposons harbored in the micronucleus. The micronuclear genome is polymorph due to heterozygosity (different alleles). In inbred laboratory strains of Oxytricha, the macronuclear nanochromosomes are homozygous, in contrast to wild strains of Oxytricha, which are not inbred. Inbred laboratory strains of Oxytricha may lose the micronucleus, and replicate by divisions of the macronucleus. The macronucleus harbors the cyanobacterial Anabaeana (also known as Nostoc) transposon containing ORFs. This transposon is within the common ancestral line of ciliate and metazoan transposons and it might have travelled through horizontal routes. The MULE (mutator-like) and the SWIM domains in a domesticated Oxytricha transposon are similar to the mutagenic trasnsposons of angiosperm plants. Both the Oxytricha and the dinoflagellate *Perkinsus marinus* possess this transposon. The domesticated pigyBac transposase of the Tetrahymena and Paramecium is not present in the Oxytricha. Other than the mutator type Oxytricha transposons are the Phage integrase domain (PF00589), the TBE domain (TERT-binding element; and the telomerase reverse transcriptase) Mariner family transposases. The Oxytricha telomere end-binding proteins are very diverse, appear to have evolved rapidly and include the evolutionarily different mitochondrial TBEs [380, 2094, 2095]. Guanine quadruplex terminal tetrads in parallel and anti-parallel formations by RNA/DNA hybrids are present in the G(4)T(4)G(4) telomere fragments (G, deoxyribo-guanine) of the Oxytricha [381, 382]. The omnipresent PiwiRNAs (piRNAs) (*vide supra*) are active in the diploid micronucleus in protecting DNAs to be retained for transfer into the Mac [2096]. For example, the three pMA81s (macronuclear DNA pieces) family proteins are stably inherited throughout many generations of vegetative growth after conjugation [2097]. In the Comment: The survival pathways of the Oxytricha match (or surpass) that of any cancer cell: compare with that of the now free living HeLa cells (*vide infra*).

The intruder elements undergo random, single base substitutions; these mutations are frequent (one to $10/10^6$ base pairs), heterogenous, and become clonal. Best known are these mutations in the hereditary non-polyposis colon cancer Lynch syndrome [2098]. These somatic point mutations appear as repetitive sequences, which are mismatch-repaired. While the initial mutations as oncogenic pathways were of definitive causative effect, the additional somatic point mutations occur randomly. The sites involve the locations of hairpin RNA, and the associated event can be a DNA quadruplex formation (*vide supra*). Millions of such individual cells, related but slightly different, are produced. Their genetic changes are "in the waiting" to accommodate any new environmental change, and the fittest will survive under those new conditions.

In colonic polyps, the oncogenic pathways are those of the APC deletion and BRAF and CTNNB1 (catenin-beta) mutations; whereas, the mutator phenotype is induced by numerous single nucleotide substitutions [2099].

In inflammatory carcinogenesis in the colon, miR-155 overproduction coincided with HPRT (hypoxanthine phosphoribosyltransferase) mutation and the blockade of cell cycle inhibitory tumor suppressor protein kinase WEE (small, in Scottish: active Wee kept the cell non-dividing and small). WEE kinase may be a substrate of oncogene kinase Akt (c-akt →v-akt murine thymic retroviral lymphoma) for inactivation by phosphorylation. HPRT mutation signals global genomic instability and increased frequency of several other point-mutations (referred to as increased mutator activity), Further, miR-155 overproduction inhibits DNA repair mechanisms. The cytokine cocktail of TNFα, IL1-β, IL-6 and IL-8 derived from LPS-simulated Mϕs was used to mimic the milieu of a proinflammatory environment. The proliferation rate of breast and colon cancer cell lines was increased by mRi-155, and WEE protein generation was decreased. Antisense-knockout of miR-155 production restored WEE protein production and stopped tumor cell divisions. The tumor suppressor gene-antagonist miR-155 is activated in the inflammatory environment. Due to inhibition of tumor suppressor gene product proteins FADD (Fas-associated death domain), JARID2 (Jumonji AT-rich interactive domain), SHIP1 (Src homology 2-containing inositol phosphatase-1) and WEE, miR-155 overproducing tumor cells proliferated uninhibited with an increased genomic mutation rate [2100].

Comment. To this author, the Oxytricha cell is driven by transposons-encoded mutator geno-phenotype; the chromosomal ends are short but re-capped, thus the individual cell is exempt from senescence, and its rapid cell divisions render it "immortal." The DNA guanine quadruplex propeller-type structures of Oxytricha [381] are similar to those stacked G-tetrads some oncogenes (c-myc/Myc) assume in the human genome [2101]. Otherwise, the Oxytricha double genome is different from the vertebrate mammalian genomes, which are united in one nucleus.

The "malignantly transformed cancer cells" re-cap their chromosomal ends, re-activate their ancient mutator geno-phenotype, divide frequently (without senescence and natural death); and carry heterozygous nuclei.

Endogenous retroviruses are hidden oncogenes and/or oncogene activators. In multicellular hosts (including Homo), the FV1 proteins are of erythroleukemia Friend retrovirus-derivation. The Fv-4 endogenous ecotropic retroviral genomic elements interfere with the insertion of other murine retroviral elements into the genome. Such genomic elements exert protection against other retroviral provirus insertions and mature virus-producer retroviral infections; reviewed in [261b], but unfortunately the classical article of Y Inaguma, T Yoshida U & H Ikeda was inadvertently left out *).

*) *Scheme for the generation of a truncated endogenous murine leukaemia virus, the Fv-4 resistance gene. J Gen Virol 1992;73:1925–30.*

Cats harbor related systems (*Felis cattus* endogenous retrovirus, FcEV). From the baboons up, the common ancestor of the Cercopithecinae installed such sequences, such as the recombinant (simian-endogenous and papio-endogenous retroviruses) RD114 retroviral genome, that encodes a type D envelope; this simian retrovirus is shared with cats [2102]. Retroelements-derived (gypsy; gag-related; envelope-derived syncytins), DNA transposon-derived (recombination-activating proteins RAG; mariners; harbingers) and gypsy integrases (from Hydra and Ciona) are listed and discussed [1774]. In the human genome, 85 genes encode 103 protein isoforms of retroviral *gag* gene derivation [1775]. Some of these proteins appear in the process of apoptosis regulation, or are proteins that fuse placental trophoblasts into the barrier of syncytiotrophoblasts (syncytins). Endogenous retroviral *env* genes encode the syncytins in syncytiotrophoblast of the placenta (Figure 52). The two human Syncytin proteins are encoded by the two independently acquired syncytin-1 and syncytin-2 *env* genes. The human Syncytin-1 protein also fuses osteoclasts [2103]. Its expression may cause the neuroinflammation in multiple sclerosis (*vide supra*) [2104]. Some retroviruses (mouse mammary tumor virus) and its human relative in human breast cancer tissue [2105a, b], and some herpesviruses (EBV, KSHV) may express superantigens, that contact VβTCR chains in bypassing expression by the MHC pathway [2106].

Retroviruses possess their standard *gag*, *pro*, *pol*, *env* genome for synthesis of viral structural proteins for encapsidation; for protease synthesis needed for the egress of the viral particle by budding; for encoding enzymes including RT, that insert the proviral DNA in the host cell's genome; for envelope protein synthesis for attachment to, and entry through, new host cell membranes. The HTLV-1 (founding member of the clade HTLV1,2,3,4, including now the new Central African strains) expresses its own oncogenes, *rex*, *tat*, *tax.* This virus is able to translocate c-*myc*, activate NFκB, suppress p53, and to generate FoxP3$^+$ CD4$^+$ Treg cells, but it is not a pathogen in multiple sclerosis. The retrolentivirus HIV-1, existing in subtypes A, B, C, as defined by its V3 loop in its envelope, operates a much more complicated genome with its Nef, Rev, Vpr, Tat and acidic Gag proteins overruling the human APOBEC and TRIM barriers. HIV-1 is at the borderline with oncogenic retroviruses, not only because it is immunosuppressive due to the apoptotic killing of CD4$^+$ T cells, but because it inhibits tumor suppressor gene PTEN (Figure 50; Table VIII), thus allowing a free run of the PI3K/Akt proto-oncogenic pathway [2107–2112a, b, c]. Most isolates of reticuloendothelial (aves) and retroviral (mammalian, murine, feline, bovine) viruses incorporate a host cell proto-oncogene (often by fusing with it) into their genome (and thus become acute (rapidly oncogenic) retroviruses. For example, the avian erythroblastosis-ES4 virus incorporated the *erb*B proto-oncogene; the avian MC29 virus incorporated c-*myc*: c-*myc*→v-*myc*. Others, the chronically oncogenic retroviruses (the MMTV, the Friend virus), practice insertional oncogenesis by introducing their genes into, or next to, a host cell proto-oncogene (and upon leaving the cell do not possess incorporated cellular proto-oncogenes). The replication competent avian leukosis virus (AVL) inserts its genome next to, but not into, c-*myc*.

The superfamily MOV10 RNA helicase (named after Moloney mouse leukemia virus) controls the replication of RNA viruses and both LTR^{+}/LTR^{-} endogenous mobile retroelements. It is estimated that approximately 45 % of the human genome consists of Alu/SINE/LINE short and long interspersed retroelements. The cytidine deaminase APOBEC (Figure 47) inhibits some (not all) retroviral replications in the mammalian vertebrate genomes. MOV10 and APOBECs synergize their inhibitory functions; type I IFNs are inducers of MOV10. MOV10 depletion allows a significant rise in endogenous Alu and LINE activities, whereas exogenous RNA viral replication is not accelerated in MOV10-depleted hosts; only artificially overexpressed MOV10 suppressed exogenous RNA viral replication [2113].

The retrotransposons embedded into the human genome are not inert (*vide supra*). The human genome harbors over 1 million Alu repetitive elements (SINE/LINE) (in the Appendix 1). The restriction endonclease Alu derives from *Arthrobacter luteus*; cuts DNA at tetranucleotide sequence AGCT, between G and C (Werner Arber's Nobel Prize, 1978). The originally RNA retrotransposons, and intron-embedded nascent Alu elements, are present in protein-coding genes, and consist of two ~300 nt long each, tandem monomers known as left and right arms, inserted for the stability of the mRNA (Figure 1 in cited article). The polymerase III-specific primary Alu transcripts are the small cytoplasmic Alu (scAlu) sequences, arising from the noncoding 7SL RNA gene (spliced leader). Polymerase II transcribes them into untranslated region pre-mRNAs. Alu RNAs interact with SRP proteins (signal recognition particle). These intron-encoded Alu RNAs are packaged into WDR79-associated nucleoplasmic box RNPs (tryptophan, aspartic acid, arginine). The AluACA evolutionarily most conserved small noncoding RNAs and RNP ribonucleoproteins are involved in RNA pseudouridylation (hallmark triple nucleotide; containing small nucleolar snoRNA for ribose 2′-O-methylation) (in Giardia, Figure 28) [2114a, b] (Figure 1 in the article cited in 2112b). The box H/ACA RNAs derive from pre-mRNA introns; stucturally consist of two 55–60 nt hairpins, the connecting single-stranded hinge, and the tail. Functionally, they are non-coding guide RNAs, commonly involved in telomere synthesis [2114c]. The gene encodes a leucine-rich N-terminus protein with three tryptophan/aspartic acid (WD) motifs in its C-terminus [2115]. The H/ACA snoRNAs are functional in the archaea. The biological involvement of WDR proteins extends from the chlamydomonas, plants (Arabidopsis), fission yeast, dictyostelium, caenorhabditis, drosophila, up to vertebrates and mammals, The core promoter of the LRWD (leucine, arginine, tryptophane, aspartic acid) gene expresses binding site to the p65 subunit of NFκB. These reactions commonly occur in the testis and spermatogonia [2115]. WDR79 gene is flanking TP53 ("tumor protein"), as a bidirectional promoter partner [2116]. Singe nucleotide polymorphism of the WDR79 gene was found in ER-negative breast carcinomas [2117]. WDR proteins in the human genome are involved in the epithelial-to-mesenchymal transition of malignantly transformed cells (squamous cell carcinoma) [2118]. WDR proteins are active in the reprogramming processes of stem cells [2119]. Human papillomaviral E1 helicase recruits WDR proteins in support of viral genome maintenance within keratinocytes [2120].

The African (world-wide) jaagsiekte retrovirus (JSRV, from Africaan jaag: chase, hunt; siekte: sickness) carries in its envelope its own oncoprotein encoded by its *env* gene. The MMTV-related betaretrovirus, JSRV, causes the ovine adenocarcinoma of the lung (targeting primarily the secretory epithelial Clara cells for malignant transformation; and, secondarily the fibroblast for proliferation, but not necessary for malignant transformation, even though JSRV transforms also the NIH 3T3 fibroblasts. In the mouse, the JSRV envelope oncoprotein could induce hemangiosarcomas, expressing the oncoprotein, that is detectable by histochemical staining with monoclonal antibody. The malignantly transformed epithelial cells carry and release fully mature JSRV particles, which are infectious in the sheep colony, and rapidly out-proliferate the rate of virus-neutralizing antibody production. The C-terminal transmembrane protein of the Env protein responsible for initiating the malignant transformation, apparently activating PI3K, but without binding to it; further, the cell transformation is not in any way dependent on the JSRV cell surface receptor, Hyal2. At the cell membrane, PI3K constitutively activates proto-oncogene Akt by phosphorylation. In NIH 3T3 cells, deficient for the p85 subunit of PI3K, JSRV directly activates Akt. The cascade rapidly extends to mTOR, a downstream substrate of Akt. The importance of mTOR activation is shown by the antagonistic effect of rapamycin) (product of the Easter-Island-isolate *Streptomyces hygroscopicus*). Parallel, there is activation of the Ras/Raf proteins, that induce the MAPK pathways: the extracellular signal-regulated kinase (ERK), the stress-activated protein kinase (SAPK), which includes Janus kinase activation (JNK), and MAPK p38. In addition to the mTOR inhibitor rapamycin, MEK and/or Ras inhibitors (PD98059; FTI-277) and MAPK p38 inhibition (by SB203580) antagonized JSRV-induced malignant transformation. Since Src inhibitor PP2 also exerts antagonistic effect to JSRV transformation, involvement of the Src proto-oncogene in JSRV transformation is postulated. The MMTV retroviruses induce host immune reactions through TLR4 (the LPS receptor). The TLR4 signals DCs and B lymphocytes for generating antibody-mediated immune reactions; and promotes the release of NFκB [2121–2123]. However, immunologically-mediated recovery from JSRV-induced ovine adenocarcinoma of the lung does not occur.

While the African line of Old World monkeys harbors a dominant retroviral flora (with common herpesviral, and EBV-KSHV-ancestors in the background), in the New World monkeys (marmosets, tamarins) herpesviruses have become the dominant flora (with the exemplary *Herpesvirus saimiri* in the squirrel monkey, *Saimiri sciureus*, and *Herpesvirus ateles* in the spider monkey, *Ateles paniscus*). Well recognized simian retroviruses are the Mason-Pfizer virus, the gibbon ape leukemia virus (and its relation, the Australian and Japanese koala bear retrovirus), and the woolly monkey simian sarcoma virus. The koala (*Phascolarctos cinereus*) retrovirus is referred to as an endogenous retrovirus, but it is leukemogenic in its natural host [2124, 2125a, b]. The simian sarcoma virus (SSV) incorporated into its genome the host cell gene for PDGF-R: c-*sis*→v-*sis* [2126, 2127], like practically all of the avian reticuloendothelial viruses, and feline, murine infectious retroviruses did: incorporated a host cell growth factor gene. Actually retroviral isolates helped the discovery of many proto-oncogenes. Incomplete, and enveloped budding, retroviruses readily appear in human tissues, normal and malignant [147], with preponderance in teratomas-teratocarcinomas, autoimmune diseases, lymphomas, and less frequently, in solid tumors (sarcomas, melanoma, and adenocarcinomas). These human retroviruses were not readily cultivable or propagable in vitro or in vivo, in opposition to the leukemo- and sarcomagenic retroviruses of avian, feline and murine hosts [147] (Figures 29 and 30).

The endogenous human retroviruses (view EM photographs in [147] and Figure 53) were categorized, and re-categorized (with special attention to the betaretrovirus-like HERV-K family) as "endogenous retroviruses" [2128]. The HML-1–10 group is closely related to the mouse mammary tumor virus (MMTV, Bittner). The most frequent insertions from the chimpanzee-hominid genomic divergence on, occurred with HML-2 proviruses (2500 LTRs, some recombined). These are the proviruses (K103 at 10p12.1; K113 at 19p12b) that appear in increased numbers in various human malignancies. The duplicated gene of the cancer testis antigen at Xq28 is not present in the rhesus genome, but it appeared in the genomes of orangutan, chimpanzee (the gorilla?), and humans, indicating that the provirus was inserted just before the divergence of these genomes. The proteins the gene encodes render the tumors in which the gene is activated antigenic in the host of origin. Therefore it is very useful in peptid cancer vaccine production. The sentences are phrased in the literature so, as if the malignant transformation activated these proviruses (to be a "byproduct"). The endogenous proviral genomes were considered for long as inert, unless they gained new genes. Even if the new acquisitions were truncated, by fusion with another retroviral sequence, possibly not a human genomic resident, the inserted ancient provirus could gain activity. There is rapidly increasing evidence, that the human endogenous retroviral flora is potentially pathogenic. Their pathogenicity encompasses the induction of autoimmune diseases and lymphomas. More recently, solid tumors (melanoma; adenoarcinoma) (*vide infra*) reveal re-activated endogenous retroviruses; even in KSHV-8-related Kaposi's sarcoma cells, an endogenous retrovirus may appear (Figures 29 and 30). The monograph shows budding retroviral particles in human lymphoma and sarcoma cells, and in various cells of patients with SLE [147] (Figure 53).

The budding retroviral particles may be fusogenic. Physiological cell fusions occur during ontogenesis (myoblasts; syncytiotrophoblasts). An occult underlying colony of malignant cells may gain virulence through fusion (cancer cells fusing, retrovirally mediated, with monocytes/Mϕs; lymphocytes) [26, 27]. Lymphoma cells may gain extraordinary virulence, if they form "natural hybridomas" through fusion of lymphoma cells with antibody-producing plasma cells [147, 352]. The first natural hybridoma was discovered, understood of what it was, and how it developed, some 6 years before the congenial artificial construction of hybridomas in the laboratory of Milstein and Köhler [147, 352, 2028] (*vide supra*, 2129a, b). The fusogenic agent of that particular natural hybridoma was a murine retrovirus of low leukemogenic potency. The hybridoma cells acquired immunoresistance, rapid replicatory rate, and increased invasiveness [147, 352] (Figure 72).

Figure 72 The first report, that retrospectively was "Natural Hybridoma Formation" in 1969–1970, was viewed with disbelief for decades. Reference Appendix 2, Explanations to the Figures

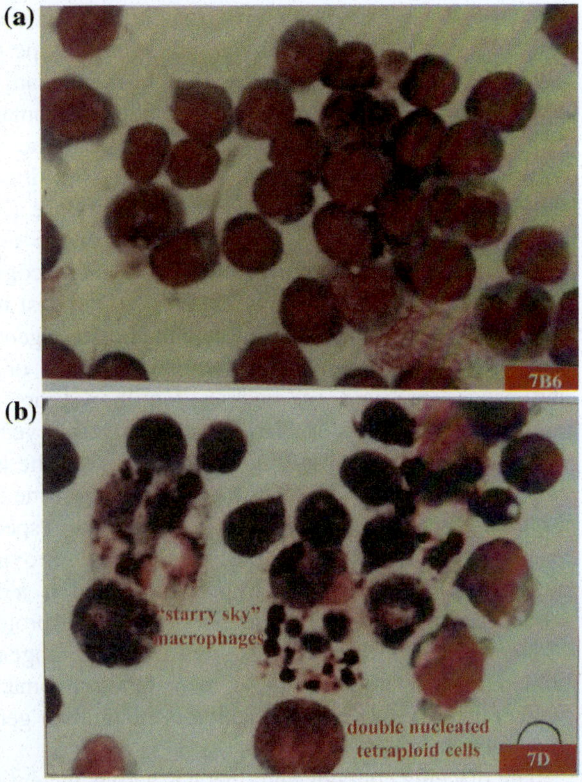

The HIV-1 retrolentivirus fused infected $CD4^+$ lymphocytes into large syncytia in which it spreads cell-to-cell without being exposed to virus-neutralizing antibodies in the extracellular spaces (Margolis *et al*) [2130a, b]. The Dittmar and Zänker's "Cell Fusion in Health and Disease. II. Cell Fusion in Disease" volume [1541] contains prominent chapters on osteoclast fusions in multiple myeloma (osteoclast-like myeloma polykaryons); cancer stem cell fusions (in the inflmmasomes); leukocyte and cancer cell fusions (contributory to Warburg-type glycolysis in melanoma cell/macrophage heterokaryons); and producing recurrent cancer stem cells in the treatment-resistant metastases [2131–2134]. What is needed to be further investigated, is the contribution of the fusogenic viruses (especially retroviruses) to the biological phenomena observed.

This author with Ferenc Györkey† surmised that during vigorous genomic activity in the process of somatic hypermutation, placentation, chronic infectious processes, or lymphomagenesis (when chromosomal fragments are cut, travel and reinsert in the process of SHM) hitherto dispersed genomic sequences of transposable elements may re-unite. If such genomic mergers happen to include an *env* gene, the endogenous retrovirus releases itself by budding and may spread again horizontally. These ideas were reviewed and discussed most recently [2], and presented in memorial lectures for Ferenc Györkey† and elsewhere [2135a–f] (Figure 53). The endogenous retrovirus may escape APOBEC and TRIM (antiviral tripartite motif) control [147, 261b]. While the avian, feline and murine hosts harbor an abundance of pathogenic (leukemogenic, sarcomagenic) retroviruses, the human host appeared to have exonerated itself of them: the pathogenic retroviral isolate for decades was confined to the HLTV-1,2 retroviruses (leading by Poiesz *et al*) [1849a–d]. This author (JGS) risked to propose [261b] that at the level of the Australopithecines, leukemo- and sarcomagenic retroviruses retreated from the human genome, leaving room for the DNA oncogenic viruses (adeno-, herpes-, papilloma-, and polyomaviruses). While the oldest retroviruses in the now human genome inserted themselves into the mammalian genomes as long as 60–70 million years ago, the "newer" insertions occurred at, or after, the divergence of the chimpanzee hominid genomes 6–7 million years ago. The "recently" inserted endogenous human retroviruses are the 800,000 years old HERV-K113 (on chromosome 19p122), and HERV-K115. However, the juvenals are the 150,000 years old almost identical HERV-K106 (on chromosome 3), and HERV-K-116. This is the time when *Homo sapiens* emerged; the new species withstood infections with these two closely related and then pathogenic retroviruses. They remain full length proviruses (the HERV-K116 provirus suffered a 2846 bp deletion in its *pol* sequence), still potentially capable of encoding proteins [2136].

Here, the cited literature lists reactivated endogenous retroviruses in numerous human tumors (teratocarcinoma, melanoma; sarcoma; adenocarcinoma) [2137]. The retroviral elements remained active in the human genome in encoding the mutator phenotypes.

The placenta contains and operates active ancient endogenous retroviral-derived LTRs as promoters for the endothelin B receptor, insulin-like factor, pleiotrophin and CYP19 (cytochrome) aromatase. Of all interspersed repetitive DNA elements in the human genome (HERV, Alu, SINE, LINE), the solitary LTRs stand out as ancient retroviral derivation, which still retain binding sites for transcription factor proteins. The first promoter exon of the human β1,3-galactosyl transferase-5 (β3Gal-T5) gene contains the LTR of an endogenous retrovirus. The β3Gal-T5 gene is active in those GI tract adenocarcinomas, which overproduce the sialyl Lewis antigen in the CA19-9 tumor marker molecule. The promoter retroviral LTR has a binding site for the hepatocyte nuclear factor-1 (HNF), and is activated by it in the LoVo colon cancer cell line (established by an MD Anderson Hospital team of Drewinko *et al* [2138]). However, the promoter is active in tissues devoid of HNF-1 expression; therefore, LTR activators other than HNF-1 exist. The LTR of a HERV is the dormancy promoter of β3Gal-T5 in the colon [2139].

Extracting guarded secrets from the Reed-Sternberg cells. The Reed-Sternberg (RS) cells of Hodgkin disease/lymphoma (Figure 48) originate from a germinal center B cell lineage, but end up trans-speciated into monocyte-like and T-lymphocyte-like cells. However, the antigens newly expressed retain some active IgG elements of B cell derivation. This author (JGS) even postulated possible natural hybridoma formation for the large (giant) RS cells, in which reactivated endogenous retroviral elements and EBV genomes co-existed [348a, b]. Indeed,

RS cells are driven by the macrophage colony stimulating growth factor (CSF-1), and express strongly its receptor [2140a, b]. The gene CSF-R is strongly expressed by RS cells (but not in B-lineage lymphocytes). RS cells in autocrine, or the heavy normal cellular infiltrates that accompany RS cells, in paracrine circuitries produce the ligand, CSF. The positive regulator of the CSF → CSF-R loop is NFκB, which usually is constitutively activated in RS cells. The MaLR family member LTR (mammalian apparent LTR-retrotransposon) [2141] in the CSF1R promoter is demethylated, and thus activated upstream. The physiological repressors of this LTR, ETO (proto-oncogene eight-twenty-one, after avian leukemogenic retrovirus 821) and transcriptional repressor CBFA2T3 (core-binding factor A, or myeloid translocation gene-related protein 2 gene) [2142], are silenced at the same time; with these repressors silenced, the LTR is liberated to act. The activated LTR is not restricted to CSF1R, it is widespread within the RS cell genome. The result is a reprogrammed cell, originally a germ cell lineage B cell, but now reprogrammed into a monocytoid RS cell. The RS cell may reprogram itself: cell fusion between a monocyte and B and T lineage lymphocytes [348] may not be necessary to form a RS cell. Thus, RS cells and KS cells are in the same category, in which a herpesvirus (EBV; KSHV) and an endogenous retrovirus form an alliance (collusion) to keep these neoplastic cells alive [2, 83, 261b]. Demethylation of the CSF1R LTR occurs also in diffuse large B cell lymphoma. The inactivation of the CBFA2T3 repressor leads to modification of the chromatin structure as well. Altering the epigenome calls to non-genotoxic gene-product protein targeting therapy, primarily directed at the silencing of the CSF1R expression is recommended [2140a, b]. The Editorial reads "The enemy within: dormant retroviruses awaken" [2143]. The possible therapeutic agents to be used against them are RT-inhibitors, monoclonal antibodies directed against the ancient retroviral Env proteins, and antiretroviral vaccinations using the ancient retroviral envelope and structural proteins as antigens [2140a, b]. Will it be possible to neutralize the mRNAs of the newly regenerated retroviral proteins by RNA interference (siRNAs)? The inbuilt anti-retroviral defense of APOBECs (Figure 47) and TRIMs apparently tolerate the reactivation of the ancient retroviral elements. Will there be a way to widen the antiviral spectrum of the APOBECs and TRIMs, which failed the human race in its recent encounter with HTLV and HIV-1/2 [261b]? For CSF-R in cancer, see Table XXV.

Reactivated ancient retroviruses, HERV-K/HML-2 (human endogenous retrovirus, human mouse mammary tumor virus-like) emerge actively replicating in patients with lymphomas, or breast cancers. Blood tests show viral RNA in high titers, viral RT, viral Gag end Env proteins and retroviral particles. Upon treatment inducing remission, the titers of these elements drop [2144a, b]. These replicating retroviruses release protein Rec, that is equivalent with protein Rev of HIV-1, and protein Rex of HTLV-1. The HERV-K/HML-2 viruses were previously found to be replicating in some human cancer cells (teratocarcinoma; melanoma, breast cancer; ovarian cancer). Recently, their presence was reported in the AIDS-independent, Mediterranean-type Kaposi sarcoma cells (in addition to KSHV-8) (Figures 29 and 30) [73, 83]. HERV-K/HMLs occur in type 1 and type 2. HML type 1 encodes protein Np9, considered to be a nuclear oncoprotein. HML type 2 encodes protein

Rec (*vide supra*). Patients with Hodgkin's disease express the HML-2 type 1, but not HML-2 type 2. HML-2 viruses co-replicate with HIV-1 in patients with AIDS [2144a, b]. In acute myelogenous leukemia of childhood, high HERV-K env gene expressions are detectable [2145]. In patients with solid tumors (breast and ovarian adenocarcinoma), expression of the HERV-K Env protein may induce antibody response (and immune T cell response). Monoclonal antibodies directed at the HERV-K Env protein block the growth of human breast cancer cell lines expressing the Env protein. In antibody-bound breast cancer cells, activated TP53 induced apoptotic death. Monoclonal antibody therapy for patients with circulating HERV-K Env protein offers possible control of the disease [2022, 2146a–c], especially, if HERV-K reactivation is a true etiological factor, and not the consequence, of a malignant transformation due to another cause.

The three malignantly transformed cell types in which coexistence of herpes (EBV) and switched on/off endogenous retroviruses occur and suggest a pathogenic collusion are the Reed-Sternberg cells, Burkitt lymphoma cells and Kaposi sarcoma cells [83, 147, 348a, b, 353, 354, 2147a]. EBV and endogenous retroviral pathogenesis is strongly suggested in SLE and multiple sclerosis [147, 2147b]. In primary effusion lymphoma cells EBV and HHV-8 (KSHV) co-exist rather in pathogenic collusion than in interference. Both viruses induce the expression of the oncogenic ubiquitin C-terminal hydrolase L1 [2147c, d]. Collaborative coexistence of HIV-1 with herpesviruses is abundantly recognized; reviewed in [261b].

Retroviral genes in the GU tract. In human prostate cancer cells, the HERV-K Gag protein (SEREX NGO-Pr-54 antigen, serological analysis of recombinant tumor cDNA expression libraries, non-government organization) is overexpressed and it is reactive with patients' sera. The same (or closely related) HERV-K Gag-protein-related antigen was expressed on the surface or in the cytoplasm in several other human tumors (bladder-, liver-, lung-, ovarian-, prostate cancers and in melanoma [2148]. Re-expressed retroviral elements in human tumors encode their own RT. Non-nucleoside RT-inhibitors of HIV-1 replication also induce growth arrest and differentiation of some neoplastic cells. A nucleoside RT inhibitor, the deoxyguanosine analogue, abacavir (Epzicom, Ziagen, Glaxo-Wellcome/SmithKline), induced cessation of growth and senescence in human prostate cancer cell lines. The senescent tumor cells eventually ended up dead. Nuclear RT inhibitors are incorporated into the newly synthesized DNA cap, thus preventing its elongation. In the reactive cells, ORF1/2 of LINE-1 are activated, whereas nuclear RT inhibitors were known to suppress LINE-1 retrotransposition. The LINE-1 segments are usually within introns, and usually non-expressed genes harbor them. Thus, stressed cells and actively proliferating neoplastic cells express more LINE-1 segments. Those LINE-1 segments that escaped 5′ truncations are able to re-express themselves. LINE re-expression can result in the induction of cellular senescence. In prostate cancer cells; LINE-1 may control endogenous retroviral reactivation and cell survival (in the Appendix 1). Abacavir by inhibition of certain reverse transcriptions attenuated the malignancy of prostate cancer cells [2149].

The enveloped and fully antigenic human endogenous retrovirus HERV-E is released from suppression in kidney clear cell carcinoma cells with inactivated VHL (von Hippel-Lindau) gene. In these cancer cells, HIF2α hypoxia-inducible factor is generated. The 5′ region of the HERV-E LTR harbors a HIF-binding site (HIF responsive element). Binding HIF-1α/β activates the retroviral LTR to encode mature enveloped retroviral particles. Host immune T cells mobilized against the retroviral antigen attack and kill clear cell type renal carcinoma cells [2150]. The corresponding author (JGS) showed autologous T cell ("small compact lymphocytes"), and large granular lymphoid cell (later: NK cells) attacks against clear cell type kidney carcinoma cells in the early 1970s, and proposed lymphokine-and lymphocyte-mediated adoptive immune therapy for kidney carcinoma [2151, 2152a–c, 2153a, b]. In the author's clinical practice, an elderly male patient MM with multiple lung metastases of clear cell carcinoma of the kidney recurrent after surgical resection, and IFNα/IFNγ therapy, received autologous dendritic cell vaccine therapy (the DC vaccine was prepared by Joseph C. Horvath). The patient experienced significant partial remission of six month duration (documented by CT scan diagnostic radiology) (presented and documented in Hungary) [2154]. Some in the ontogenesis of the mammalian (human) kidney, early embryonic anlagen naturally resolve and are replaced by the final organ, kidney cells appear to have retained a genuine special susceptibility to host chemo-lymphokines and NK cells. Indeed, clear cell kidney carcinoma showed response to various forms of immunotherapy (type I IFNs and IL-2; oncolytic viruses; vaccines; and adoptive lymphocyte therapy). Clinical responses were probably the best to lymphokine-activated killer (LAK) cells, either autologous, or allogeneic. Complete clinical responses were induced by cytokine-induced killer (CIK) cells in patients with metastatic disease. Patients mobilizing high titers of Treg cells responded the least [2155–2166].

Sal flies out of nowhere. The homologs of the *spalt/sal* gene (German for crevice, fissure, split) active in drosophila, caenorhabditis, xenopus, zebrafish, mouse and human hosts, as the *sal*-like (Sall) evolutionarily conserved metazoan genes. In the drosophila Spalt/Sal protein works in the decapentaplegic (Dpp) signaling pathway (equal to mammalian TGFβ). The drosophila Sal represses transcription in human cells [2167–2169]. After fetal and embryonic life, the expression of SAL proteins declines. However, in the human host, *sal*-like gene 4 proteins (SALL4) become re-expressed in several malignant tumors. In human cells the RNA-binding protein Lin28 retards the activity of microRNA let-7. Consequentially, some proto-oncogenes undergo a derepression process. Thus liberated are c-Myc, K-Ras, Sall4 and HMGA2 (high mobility group arginine-lysine-rich hook fitting into AT-rich minor groove of DNA). The zink-finger transcription factor Sal repress transcription, dependent on the histone deacetylase complex. The human Sall4 protein is a stemness reprogramming factor. The SALL4 gene is hypomethylated in cells of the myelodysplastic syndrome; it induces the Wnt/β-catenin pathway [2170]. The ABC transporter A3 and SALL4 confer imatinib resistance to BCR/ABL-driven CML cells. Genetic silencing of SALL4 restored susceptibility of CML cells to tyrosine kinase inhibitors [2171]. The human *spalt/sal* genes are mapped to chromosomes 16q12.1 (*sall*1), 14q11.1-q12.1 (*sall*2) and 18q23 (*sall3*). The human *sall*1 gene shows not enough sequence homology to mouse (M*sal*) or frog (X*sa*l-1) genes to conclude that the human H*sal*-1 is their orthologue; the human *sall3* gene shows sequence homology with M*sal*. The H*sall4* locus is at 20q13.13-q13.2 [2172].

The SALL4 protein is overexpressed from a hypomethylated gene in myelodysplastic syndrome [2173]. The overexpressed SALL4 protein induces the clinical picture of acute myelogenous leukemia, or the blast crisis of CML. The N-terminus of the SALL4 protein expresses a 12 aa region, that is reactive with HDAC. Downregulating SALL4 expression with shRNA (inhibiting SALL4 mRNA), or blocking HDAC with trichostatin, stopped leukemic cell growth. The natural interaction of SALL4 with HDAC results in the epigenetic inactivation of suppressor gene PTEN through the action of the histone deacetylase complex (HDAC1/2/NuRD) (histone deacetylase; nucleosome remodeling and deacetylase); thus, SALL4 suppresses PTEN. The 12 aa SALL4 peptide, or its one aa mutated replica, that inhibits the interaction of SALL4 with the histone deacetylase complex, exerts antileukemic effect in that, that it inhibits the crucial SALL4-to -HDAC interaction (histone deacetylation) and thus the inhibition of PTEN. This peptide induces apoptotic death of leukemia cells expressing SALL4. It is also effective in stopping the growth, and in inducing the death of high SALL4-expressor hepatocellular carcinoma cells. Such interventions would not work in malignantly transformed cells that irreversibly deleted PTEN. The 12aa peptide that stopped SALL4-induced PTEN inactivation in malignantly transformed cells, did not adversely affect the vitality of $CD34^+$ normal stem cells [2174a, b]. The histone chromatin-binding high mobility AT-hook protein (HMGA2) reaches high level in transformed human cells (adenocarcinoma), is engaged in DNA repair, and inhibits apoptosis by phosphorylating ataxia-telangiectasia Rad 3-kinase via checkpoint kinase 1/2 (ATR/CHK 1/2).

In prostate cancer cells miR-154 expression is suppressed, and CCND2 (cyclin D2) is overexpressed. The EMT activity of HMGA2 is inhibited by forced expression of miR-154. Epithelial cancer cells dominantly express cadherin; mesenchymally transformed adenocarcinoma cells (prostate cancer cells) overexpress vimentin [2175–2177]. So far no reports could be found showing that reactivated retrotransposon-LTRs activate the SALL4 and/or Hmga2 genes; but may be these factors were not directly looked for.

Interactions of the epigenome and chromatin with the RNA/DNA complex hold a fundamental importance in gene expression. The chromatin remodeling N-terminal domain ATPase SWI/SNF family member Lsh controls genome-wide cytosine CpG methylation in H4K4me3 (switch/sucrose non-fermentable; lymphocyte-specific helicase). Lsh is a de *novo* methylator. Cytosine methylation suppresses retroviral LTR insertions; it also silences tumor suppressor genes. Human embryonal stem cell DNAs accept methylation at sites other than CpG. In $Lsh^{-/-}$ mutants, decreased cytosine methylation leads to reactivation of endogenous retroviral LTR elements. Perinatal Lsh deletion is fatal. Lethally irradiated mice could not be reconstituted with $Lsh^{-/-}$ fetal liver or other hematopoietic cells. In $Lsh^{-/-}$ hematopoietic cells the suppression of retroviral elements in the PU.1 gene was inadequate (PU box, purine rich sequence 5′-GAGGAA-3′) The PU box elements were transcriptionally upregulated. The Ets family member (after avian erythroblastosis retrovirus E26, E-twenty six) PU.1 locus is the Friend mouse leukemia spleen focus forming virus proviral integration 1 site. The activated PU.1 transcription factor blocked terminal differentiation of erythroblasts. The activation was consequential to hypomethylation within the PU.1 locus. The PU.1 locus contains the "mammalian apparent LTR-retrotransposon" (MaLR, *vide supra*). Mice with attenuated DNMTA (DNA cytosine-5 CH_3-transferase alpha) undergo only moderate DNA methylation, and do not develop erythroleukemia. In case of the Moloney mouse leukemia retrovirus, the insertion site of the LTR is in the close vicinity of the c-*myc* gene. Indiscriminate administration of demethylating molecules (enzymes) may act negatively upon the suppressors of the ancient retroviral LTR elements; thus, potentially may initiate malignant transformation of the cells involved. Hypermethylation of specific tumor suppressor gene promoters results in leukemogenesis (murine erythroleukemia) [2178].

In summary, the *sall*4 gene encoding protein SALL4 is active in fetal life, especially in hepatoblasts; it becomes silenced in adult life. A 12 aa peptide is its natural inhibitor. When it returns in adult life, is it mutated and constitutively expressed? However, the malignant feto-oncoprotein remains susceptible to suppression by the 12 aa peptide; thus loss of the peptide allows the reactivation of the feto-oncoprotein? Is the *sall*4 gene resting in an acetylated promoter and becomes reactivated by de-acetylation? The Cancer Science institute in Singapore will have the answer.

Comments. Would the epigenome of the DNA molecule upon receiving certain 'danger signals' in a distressed host demethylate and thus, activate the potentially oncogenic (cell transforming) repetitive elements (primordial LTR inserts) in order to encode directly or indirectly, but sequentially and constitutively, all the "cell survival pathways" that the cell possesses. One in a million of these cells will be able to survive in one of the highly variable environments. The mutated and fused genes would serve the cell for its immediate needs; the numerous "passenger mutations" would be standing by "in -waiting" for any future unforeseen environmental changes, to which the cell with the right metabolism pre-arranged could immediately adapt?

This author is proposing that the inserted proviruses may be the recipient resident activators responding to stimuli for the generation of the silent mutator geno-phenotypes in their cellular hosts. The ancient faculty to encode a mutator geno-phenotype remained conserved throughout evolution in the RNA/DNA complex from the ancestor uni- and first multicellular hosts up to the fully differentiated non-dividing brain cells (see: glioblastoma). Under stress, as a final measure in preparation for survival, the descendants of the ancient retroviral proviruses activate this elementary faculty of the living cell. This is the question awaiting a confirmatory answer, if any.

The HeLa cells surrender their genome. The 60 years old HeLa cells live in adherent, and floating (anchorage-dependent, and independent suspension) cultures. The expression of the *siat*7 (sialyltransferase, Golgi II, *lama*4 laminin-α4) genes enables some of the cells to grow in suspension. The adherent culture preserved the insertion sites of the original causative papillomaviruses. Clonogenic assays (single cell forming a colony of 50 or more cells) show that HeLa cells retained some radiosensitivity. Despite the expression of E6 and E7 protein products of the HPV 18, some HeLa cells reverted from the malignant, to the non-malignant phenotype. Suspension cultures of HeLa cells could be readily infected with viruses (adeno, echo-, herpes- influenza A-, polio-, respiratory syncytial viruses), that adherent cultures could resisted. May be HeLa cells grown in suspension culture are more IFNα-deficient, than cells growing in adhesion cultures. HeLa cells produce ossicles in muscle tissues of mice due to BMP (bone morphogenetic protein) production. The 60 years old adherent HeLa cell culture is a well fed cell conglomerate, not fighting anymore rejection by an immunocompetent host; all its fighting genes had been switched off. HeLa cells in vivo (in their hosts) defended themselves against NK cell attacks by altering their FcRs (so that they will not induce an ADCC reaction), and by releasing MICA*008 soluble ligands (MHC class I-related chain A), that neutralize and downregulate NK cell receptor NKG2D expression. In vitro, NK cells do not spend energy for these unnecessary functions. HeLa cells preserved their overexpressed Twist genes, which transformed them from epithelial to mesenchymal phenotypes. HeLa cells retained susceptibility to the shhpRNAs (short hairpin), that neutralize the mRNA for the Twist oncoprotein. The HeLa cell in culture, not showing signs of senescence, is still trying to repair its genome (the dsDNA breaks), but the chromothripsis lesions remain irreparable [2179–2187a, b]. Being induced by an exogenous oncogene (the HPV-18), the HeLa cells may not represent those malignant cell colonies, which were induced by an endogenous process of transformation.

Comment. The revertant HeLa cells might have lost their immortality, and thus should have died out after a set number of cell divisions; but, no: they remained in culture for years, but expressed a non-transformed phenotype with activated tumor suppressor genes. Could these latter cells ever form a colony and a society? HeLa cells lost their immortality, when inhibitors removed PI3K expression. In one extreme, cells with devastated genomes remained immortal, while at the other end, some other cells managed to revert to close to normalcy, and live in senescence (or autophagy), or die. What HeLa cells go through, is the fate of a cell community with externally imposed (virally induced) oncogenesis. Would immortalized cell colonies with internally initiated oncogenesis expunge their mutated oncogenes and survive without them? The oncogenes function as “cell survival pathways” needed for survival in distress. Rather, the involved cells should stop constitutively expressing them. The tumor suppressor genes activated in the process of HeLa cell reversion remain unidentified. Possibly, in tumor cells fused with normal cells, a process of reversal may be induced by eliminating damaged genes and by replacing them with healthy genes.

The HeLa cells originally were transformed by the HPV-encoded oncoproteins (that neutralize p53 and Rb cellular tumor suppressor genes). Distinction between self and non-self probably did not exist when ancient chimeric cells were formed, and engulfed bacteria were accepted as self (proteobacteria for the mitochondrion; cyanobacteria for the chloroplast). Malignantly transformed cells appear to be impervious, as to the distinction between self and non-self, when it comes to perform cell fusions and/or forming hybridomas. This author proposes that at least some of the HeLa cell growing in suspension cultures are fused hybridomata.

Among its numerous piggyBack insertions, the HeLa cell can harbor piggyBats [2188]. The HeLa cells originally were transformed by the HPV-encoded oncoproteins neutralizing p53 and Rb cellular pro-apoptotic genes, in the well-known HPV pathogenesis) [1813a, b]. New technology distinguishing piggyBacs from piggyBats now unveils a hidden messenger for the activation of oncogenes. If this is so for the colon carcinoma cells and the HeLa cells, and the little brown bats will continue to harbor piggyBats, but without carcinogenesis: we shall find out how the little bats do it. Could some of the introns, preserving their RNA identity, rescue a cell entering the transformation pathway on the command of its ancient proviruses, and reverse the process either by inducing a differentiation pathway, or by issuing a command overruling all other commands for the activation of the anti-apoptotic pathways by neutralizing those mRNAs)? Could the piggyBats activate tumor suppressor genes?

More of the ciliate Oxytricha. The ciliate cells grown in suspension cultures lose their small nucleus and are driven by their macronuclei loaded with transposons/retrotransposons. The yeast can harbor both piggyBacs and piggyBats. Are hidden piggyBats driving the macronuclei for non-stop divisions of ciliate cells in suspension cultures? The little brown bat (*Myotis lucifugus*) is under close supervision for possible sudden oncogenesis, what so far is not happeneing. Its genome has recently been invaded by the DNA-only cut and paste transposons, the *piggyBat*s. The presence of the retroelements inserted through RNA intermediate transposons (converted to DNA by a RT) remains abundant: inheritance of the piggyBacs from the cabbage looper moth *Trichoplusia ni*. However firmly inserted, these transposons appeared to have been silenced. Both DNA only and RNA intermediate transposons were believed to be silent in the pre-hominoid mammalian genomes for some 37–40 million years. Now piggyBat transposons are found hidden among the numerous piggyBacs inserted into the genome of the human colon cancer cell line HCT115, and in the HeLa cells. Are they antagonistic to piggyBacks in some other settings?

The Cancer Cell Viewed as an Ancient Unicellular Parasite (an Amoeba)

Amoeboid Transformation of Tumor Cells

Locomotion, invasion, fusion or phagocytosis. Theileria bovis in itself imitates tumor cells; so do host lymphocytes infected with it. The infested bovine lymphocytes appear to have acquired a semblance of immortality as well, since they appear as if they were malignantly transformed (including activated oncogenes, among them c-*myc*). After epithelioid tumor cells undergo mesenchymal transition, their next form of transition is the assumption of amoeboid features. Not only the cytoskeleton transforms the entire cell into an amoeboid structure, the ancient genes of phagocytosis (the dictyostelium Dd*phg* phagocytotic faculty genes) are reactivated. This author observed relentless phagocytosis of smaller tumor cells by their larger partners in long term tissue cultures (in a wide range: from the neuroectodermal lung cancer cells to the mesenchymal sarcoma cells) [147, 1501]. This commonly observed phenomenon was referred to as "tumor cell cannibalism." Melanoma cells become phagocytic entities by reactivating the cell surface gene TM9SF4 (transmembrane 9 superfamily protein member 4), a human homolog of the Ddphg1A gene [2189a, b]. Tumor cells reactivate ancient genes that move the amoeba (Dictyostelium): walking, gliding, swimming [2190a, b]. In the wave-like locomotion of the Dictyostelium, two genes are named as participants: *acs* (adenylyl cyclase), and the phosphatase *myoII* (myo-[2-3H] inositol). The *myoII* gene is involved in the cell adhesion and in actin-myosin dynamics. Cells with loss of function mutation of the *aca* gene (*aca*$^{-}$) do not produce cAMP) and become round and immobile; externally introduced cAMP restores mobility [2191a, b]. However, these genes are not mentioned in tumor cells assuming amoeboid locomotion. Ezrin-rich uropods connecting plasma membrane with the cytoskeleton and blebbing occur in melanoma cells [2192]. EphA2 (ephrin) tyrosine kinase is re-expressed in melanoma cells transforming further from mesenchymal to amoeboid phenotype. Other tumor cells express the Rho/Rho-associated protein (Ras homologous protein kinase) and ROCK (Rho-associated coiled-coil containing protein kinase): the LIMs (Table XVI).

J.G. Sinkovics, *RNA/DNA and Cancer*, DOI 10.1007/978-3-319-22279-0_27

A GTPase-activating protein (GAP) is identified as the mediator of the amoeboid movement of melanoma cells; ROCK phosphorylates GAP [2193a, b, 2194]. The most ancient Rho family small GTPases are reactivated in those tumor cells which resorted to amoeboid movement, The cytoskeleton of these tumor cells underwent the mesenchymal-amoeboid transition (MAT). The actin-rich filopodia and lamellipodia appear at the leading edge of the cell. Spherical membrane herniations, the blebs, are formed. Filopodia are induced by Cdc42. Lamellipodia are driven by Rac proteins and actin polymerization within. It is possible to recognize the major driving force of tumor cell mesenchymal or amoeboid migration under different alignments of the matrix by using special inhibitors: ROCK inhibitors are H1152 and Y27632; a nonmuscle myosin ATPase inhibitor is blebbistatin; the Rho inhibitor is C3 [2195, 2196a, b]. ADP-ribosylation factors (Arfs) are highly conserved small GTPases asymmetrically distributed in the plasma membrane to regulate actin patch polarization (and thus budding) [2197a, b]. Of blebbistatins, the dictyostelium operates an UV-light inducible myosin inhibitor, the azidoblebbistatin [2198]. Tumor cells migrating in the skin or superficial subcutis might respond to a day-night cycle in their locomotion, or in the subject receiving UV light waves in a tanning bed?

MDA breast cancer cell line-231 is driven by contractility of the cells' rears [2199]. Breast cancer gene expressions are regulated by microRNAs. Triple-negative breast cancer cells are highly malignant, hormone- and chemotherapy-resistant, and thrive in a pathophysiological environment deprived of miR-708. The physiological role of miR-708 is the control of Ca^{++} storage in the endoplasmic reticulum (ER). The mRNA of the ER membrane protein neuronatin is the target of miR-708 for suppression. High neuronatin levels (with low miR-708 levels) promote Ca^{++} efflux from the ER [2200, 2201].

The Ras-like small GTPases are ancient proteins of fundamental physiological function; they act as guanine nucleotide exchange factors (GEFs) and GTPase activating proteins (GAPs). The Last Eukaryotic Common Ancestor (LECA) must have possessed RasGEFs. The entire evolutionary trajectory of Ras-like signaling pathways has been printed out [2202a, b]. If so, this author may presume that guanine nucleotides and polypeptides interacted in the precellular world (in the Precellular Virus World). However, in some fungi and in plant cells, the descendants of chloroplast-bearing green algae lost the Ras pathway. Is it domain additions and diversifications in eukaryotes, that rendered the pathway "oncogenic" at some point of the evolution, or is it simplification into an ancient pathway, when point-mutated *ras* genes begin to encode their oncoproteins?

The small GTPase members of the Ras superfamily (Jennifer Harvey's, or Werner Kirsten's rat sarcoma), consist of the Rab (Ras in brain; genes belonging to Ras family), the Rac (Rho-related C3 botulinum toxin substrate), the Ral (gene homolog Ras-related; Ras-like), and the Rap (Ras-related protein) subfamilies (*vide supra*). Predating the recognition of the Ras family, the topogenesis of intracellular protein trafficking, especially across membranes, was envisioned and experimentally documented by G. Blobel and his associates [88a, b]. (The indirect reward of these studies was the restoration of the demolished Frauenkirche in Berlin from Blobel's donation of his Nobel Prize).

Especially the membrane trafficking proteins Rabs interact as guanine nucleotide exchange factors (GEF) with domains of the evolutionarily ancient but conserved DENN protein family (differentially expressed in neoplastic versus normal cells). The first DENN protein in mammalian cells was discovered as the binding partner of the death domain of TNF-R: MAPK-activating protein containing a death domain (MADD). The human genome encodes 18 DENN domain-bearing proteins. The human DENND1A-1C protein binds clathrin, SH3 (Src homology), and Rab35. The DENN/D2B was discovered as suppressor of tumorigenicity, ST5. However, either overexpression of it, or loss of its C terminus, converts ST5 from a tumor suppressor to an activator of cell transformation. Loss-of-function mutations of the Rab and/or DENN proteins bring about severe genetic diseases [2203a, b, 2204]. Gain-of-function mutation UV light-induced of the Rac1 gene occurs in cutaneous melanoma cells. In the RAC1 protein proline is replaced by serine: RAC1(P29S). The mutated gene is continuously activated by GDP/GTP exchanges [2205].

The zeste of drosophila becomes an international agent. The EZH2 (enhancer of zeste homolog, drosophila; zeste- shaggy white locus, drosophila; enhancer of zeste homolog in human genome at 21q22.2) is a histone lysine N-methyltransferase polycomb protein; it silences the miR-708 gene [2200, 2201]. The enhancer of zeste polycomb repressive complex 2 (PRC), instead of being a silencer of miR-708, acts as a coactivator of AR in prostate cancer cells. The EZH2 acts independently from the polycomb genes [2206, 2207]. In SYT-SSX1/2/4 monophasic and biphasic synovial sarcoma cells (Figure 34), the three positive immunohistochemical stains for EZH2, H3K27m3 and Ki-67 correlated with poorly differentiated large and metastatic tumors [2208]. In mantle lymphoma cells, a corepressor complex of MYC-HDAC3 and EZH2 suppresses miR-29. EZH2 induces MYC by inhibiting the MYC suppressor miR-494; in turn, MYC upregulated EZH2 by suppressing the EZH2 inhibitor miR-26a. Combined inhibition of HDAC3 and EZH2 disrupted the MYC-EZH2-miR-29 axis, resulting in miR-29 re-expression and suppression of the mRNAs of miR-29-targeted oncogenes [2209]. These overlapping brilliant studies (Dana-Farber, Boston; Moffitt, Tampa; Weill Cornell Medical College, New York; and Semmelweis University, Budapest) forecast successful treatment of several tumors by silencing EZH2.

The zeste of Chlamydomonas reinhardtii. The enhancer of zeste homolog 2, drosophila (EZH2) is an epigenetic regulator of the Polycomb repressive complex 2 (PRC). It primarily acts as a histone methyltransferase (H3K27me3), and as a recruiter of other methyltransferases for CpG methylation in embryonic and adult stem cells, and as methylator of transcription factor GATA4 [2210]. EZH is an ancient epigenetic enzyme. The PRC2 subunits were active in the unicellular eukaryotes. The EZH of the unicellular alga *C. reinhardtii* silences transgenes and retrotransposons, methylates histone 3 lysine 4, and acetylates histone 4. It is suppressed by an iRNA [2211]. The enhancers of zeste proteins have become essential for viability in the drosophila and caenorhabditis as direct partners of the vital polycomb group proteins. The trithorax and polycomb genes are antagonists, and the enhancer of zeste is a dominant allele of the polycomb genes [2212, 2213]. The most elaborate overview of EZH2 discusses the epigenetic regulation of polycomb group (PCG) proteins, PCG 2 recruitment by EzH2; and the roles of EZH2 in embryonic, mesenchymal, hematopoietic, skeletal, and other tissue-specific stem cell regulation [2210]. The ancient protein emerged to be a histochemical marker for malignant tumors [2214].

Actin filament dynamics regulated by cofilin, and evolutionarily conserved ancient cystein rich LIM proteins are re-activated in cancer cells in locomotion; these are accounted for in Table XVI. The integrin-reactive tetraspanins mediate cell fusions (even virally mediated cell fusions) and dendritic cell lymphocyte interaction; the movements of two cells side-by-side is synchronized by tetraspanins (Table XXIV). Many tumor cells gain amoeboid locomotions by fusing with mesenchymal cells (favoring monocytes-Mϕ). The ancient growth factors are advanced to oncoproteins in vertebrate multicellular hosts (EGF & R; PDGF & R; IGF & R; EPO & R). Here, Table XXV presents the story of Mϕ CSF-R functioning as constitutive tumor cell GF.

Table XXIV Tetraspanins: Ancient Transmembrane Proteins (TSP)

Haloferax volcanii archaeasortase may be the first TSP protein.[1]
Eukaryotic TSP proteins evolved from diploblast sponges and the Anthozoan cnidaria (Nematostella). Of uro-protochordates, amphioxus harbors a cluster of 39 TSP genes (human 33 TSP genes). Orthologs* of ancient TSP genes are operational in vertebrate mammals. After 2 rounds of whole genome duplication (2R-WGD), only 25 % of ancestral pro-chordate genes retained more than two orthologs* in vertebrate lineages, but the retention rate of TSP genes (ohnologs**) is 75 %. The amphioxus, in contrast to its high transposon diversity, preserved its ancient genes in single copies without duplications and without losses (thus it remained an amphioxus). The vertebrate CNS evolved by gene duplications.[2–4] Some TSP genomic introns do not break the reading frame (phase 0 introns). TSP genomic intron/exon junctions 1 and 4 existed in the common ancestor of eukaryotes. The common ancestor of animals acquired introns 2, 3, 5 and 6. Twenty of the 33 human tetraspins originated in the common ancestor of vertebrates, but protostomes (worms and flies) did not contribute tetraspins to the human genome. Conserved (passed on) TSP genes were assigned to perform novel biological processes (neofunctionalization): mediation of cellular adhesion, cell membrane penetration, invasion, fusion, sperm-egg cell fusion. Pathogens (viral, bacterial, fungal, protozoal) expropriate the TSP proteins for cellular penetration (fungus Magnaporthe, penetrating rice leave cells). Higher fungi operate TSPs of versatile functions.[5–7]
*Homologs, genes separated by speciation. Orthologs, two genes from two different species deriving from same single gene. Paralogs, genes derived through duplication in same species, encoding similar but not identical proteins. **Ohnologs, set of gene pairs, both copies, after whole genomic duplication. Ohno S. Evolution by Gene Duplication. George Allen, London, 1970; Semi Cell Dev Biol 1999;10:517-22.
TSP and cancer. Morbilli and canine distemper virus-induced cell fusions (syncytia) are mediated by CD9 TSP. TSP proteins involved in dendritic cell and lymphocyte movements and their cell-to-cell interaction. Cancer cell interaction with stromal cells involve TSP proteins (the Villejuif studies). TSP increased colon cancer cell migration by downregulating laminin-binding

(continued)

Table XXIV (continued)

of integrin α3β1 and fibronectin-binding of integrin α5β1 and upregulating CD44 expression. TSP CD9 promotes fibrosarcoma cell migration.[8–15]
***1** Abdul-Halim MF et al Mol Microbiol 2013;88:1164-75. **2–4** Cañestro C & Albalat R Brief Funct Genomics 2012;11:131-41; Chen Y et al J Genet Genomics 2011;38:577-64; Louis A et al Brief Funct Genomics 2012;11:89-95. **5** Garcia-España A et al PLoS One 2009;4(3):e4680. **6** Huang S et al BMC Evol Biol 2010;10:306. **7** Lambou K et al BMC Genomics 2008;9:63. **8** Boucheix C et al Expert Rev Mol Med 2001;Jan:117. **9** Charrin S et al Biochm J 2009;420:133-54. **10** Gartlan KH et al Eur J Immunol 2013;43:1208-19. **11** Guo Q et al PLoS One 2012;7(6):e38464. **12** Herr MJ et al PLoS One 2013;8(6):e67766Sala. **13** Valdés M et al Expert Opin Ther Targets 2012;16:985-97. **14** Singethan K et al J Gen Virol 2006;87:1635-42. **15** Wright MD et al Tissue Antigens 2004;64:533-42.*

Table XXV The Ancient Monocyte-Macrophage Colony-Stimulating Factor Receptor (CSF-R1) Becomes an Oncoprotein

Metchnikoff's policemen. The ancient macrophage gene *fms* (deformylase) encoding CSF-R1 kept the ancestors of the lineage (monocytes, Mϕs, DCs, microglia, osteoclasts) to rise in defense of their host as shown first in the belly of *Daphnia pulex* (ИлЬя ИлЬич Мечников's macrophages mϕ).[1] Ancestral mϕ in the era of innate immunity mobilized several LPS-responsive genes. The intron2 of the c-*fms* gene harbors four DNAse-specific hypersensitive sites. CSF-R1 tyrosinases 544, 559 and 807 initiate reactive mϕ proliferation. The CSF-R1 is highly conserved through the evolutionary scale (fish, amphibians, aves, mammals, incl. murine and human). The CSF-R1 is at the nuclear membrane. Ligand CSF-to-R stimulated mϕ to induce PI3K. EBV acquired by horizontal transfer and expresses the CSF-R1 gene within viral gene BARF1 (BamHI A restriction endonuclease enzyme fragment).[2]
Specific antibody treatment acting on the mϕ FcR (initiator of antibody-induced lymphocyte-mediated cytotoxicity) instructs mϕ to target cancer cells for phagocytosis.[3,4]
Retroviral insertion and genomic incorporation of the cbl and fms macrophage genes. Retroviruses reverse transcribing and integrating their DNA provirus in the host cell genome frequently incorporated cell survival pathway genes. By expressing such genes in extraordinary multiplicities, these retroviruses were recognized as oncogenic vectors. Of mϕ genes, *cbl* at 11q23 (Cas-Br-murine ecotropic retroviral transforming sequence) and *fms* at 5q33.3 (deformylase; c-fms → v-fms McDonough feline leukemia retrovirus) have become retrovirally acquired oncogenes. Non-virally incorporated cellular *cbl* and *fms* genes may undergo loss-of-function or gain-of-function (leukemogenic; sarcomagenic) mutations.[5–9]
In the human genome, the cbl and fms genes became proto-oncogenes. The gene of the CSF-R3 undergoes gain of function mutation and the accumulating oncoproteins transform the involved stem cells into myelogenous leukemia cells (these are BCR/ABL-negative atypical chronic neutrophilic leukemia myelocytes). The membrane proximal mutations involve T615A and T618I (threonine; alanine; isoleucine). These mutations activate the JAK pathway. These leukemia cells are killed by the JAK1/2 inhibitor ruxolitinib (Jakafi). The CSF may suffer loss-of-function mutations first, resulting in neutropenia. Upon treatment with filgrastim (GCSF, Neupogen), the CSF-R may suffer truncation of its cytoplasmic tail; this results in constitutive signaling through SRC/TNK2 kinase *) manifesting clinically as an acute form of myeloid leukemia. The SRC/TNK2-mutated leukemia cells are killed by the multikinase inhibitor dasatinib (Sprycel). These are somatic and externally induced mutations; as such, they would not fall in the category of an "RNA/DNA complex initiated rescue mission". The *cbl* gene mutated at C416W (cystein, tryptophan, in the ring finger domain) and at A78V (alanine, valine, in the proline rich domain) resulting in their hypersensitivity to IL-3.[10–13]

(continued)

Table XXV (continued)

Peritoneal mesothelial cells overproduce the protein product CSF-1 (the ligand), and receptor C-FMS-expressor endometrial cells respond with overproliferation. All-trans retinoic acid (atRA) acted upon CSF-R1 and stimulated the growth and anchorage-independent invasiveness of two breast carcinoma cell lines.[14,15] (a reaction opposite to atR-induced differentiation of promyelocytic leukemia cells).
The solubilized fms-like tyrosine kinase (sFLT-1; VEGFR-1) antagonizes VEGF and thus inhibits growth of tumors (retroperitoneal fibrosarcoma; colon carcinoma; multiple myeloma, acute myelogenous leukemia; lipofectamine-vectored sFLT-1 gene delivery is not immunogenic.[16–18]
Ewing's sarcoma (ES) oncoprotein binds to mϕ-specific promoter of CSF-R1 (C-FMS encoded by c-*fms* proto-oncogene; CD115): an archetype reaction.[19] The EWS fusion oncoprotein t(11;22)(q24;q12) consists of the gene products of EWS (chromosome 22) and FLI1 (Friend leukemia virus insertion site, chromosome 11). On rare occasions FLI may be replaced by ETS (E-twentysix avian retrovirus, which snatched the host cell gene). Actinomycin D (Cosmegen) inhibits or disrupts this binding (it is a prominent chemotherapeutic agent for this tumor). The EWS/FLI1 oncoprotein suppresses intracellular lysyl oxidase (LOX); LOX re-induced exerts tumor suppressive effects.[20–22] The EWS/FLI1 oncoprotein mobilizes STEAP*) a widely-spread oncoprotein of various malignant tumors.[23] The insulin-like growth factor receptor (IGF-1R) is a major source of metabolic energy to several malignant tumors, incl EWS. The oncoprotein EWS/FLI1 is not functional without IGF-1R binding its ligand IGF-1. The mcab figitumumab (Figi) was aimed to be an antagonist of IGF-1R. In its phase III clinical trial, Figi acted rather as an agonist (not an antagonist) of the EWS IGF-1R. Figi recruited the agonist of the G protein-coupled β-arrestin to IGF-1R. The IGF-1R/β-arrestin unison induced the MAPK/ERK "cell survival pathway". Other response to the unison is IGF-1R internalization and degradation (ubiquitination). Inactivation of β-arrestin mRNA by shRNA is possible; the co-administration of MAPK inhibitors is still in the investigational phase.[24,25]
Lessons learned from rare conditions. In pigmented villonodular synovitis and giant cell tumors of the ligaments/tendons, autocrine and paracrine circuitries of overexpressed CSF receptors and ligands drive these lesions. This process is initiated and sustained in an inflammasome; it is controlled by imatinib mesylate.[26–30] The locally aggressive but not metastasizing giant tumors of the bone are driven by the RANK-ligand (receptor activator nuclear factor kappa B, B cell lymphoma, ligand) activating its receptor. The mcab denosumab blocks ligand-to-receptor binding and induces complete regression of the tumors. So respond aneurysmal bone cysts.[31–34] Some malignant tumors classified as sarcomas (alveolar soft part sarcoma; clear cell sarcoma; perivascular epithelioid cell neoplasms, pecomas) ab origine resist the doxorubicin-ifosfamide regimen, but respond to mTOR inhibitor rapamycin-derivatives (everolimus; ridaforolimus).[35,36]
***1** McTaggart SJ et al BMC Genomes 2009;10:175; Stefater JA et al Trends Mol Med 2011;17;743-52. **2** Bogunovic M et al Immunity 2009;312:513-25; Follows GA et al Nucl Acids Res 2003;31:5805-16; Garceau V et al J Leukoc Biol 2010;87:753-64; Himes SR et al J Leukoc Biol 2001;70:812-20; Kanagasundaram V et al Biochem J 1990;320:69-77; Mitrasinovic OM et al J NeuroSci 2005;25:4442-51; Moreau J et al Oncogene 1989;4:443-9; Ravasi et al J Immunol 2002;168:44-50; Strockbine LD et al J Virol 1998;72:4015-21; Wang T et al J Immunol 2008;181:3310-22; Yu W et al J Biol Chem 2012;287:13694-704; Zwaenepoel O et al FASEB 2012;26:691-706. **3** Weiskopf K et al Science 2013;341:88. **4** Kershaw MH & Smyth MJ Science 2013;341:41-2. **5** Baumbach WR et al Mol Cell Biol 1987;7:664-71. **6** Sherr CJ et al Int J Cell Cloning 1990;8S1:46-80; J Cell Biochem 1988;38:179-87. **7** Sinkovics JG CRC Crit Rev Immunol 1988;8:217-98. **8** Swaminathan G & Tsygarikov AY J Cell Physiol 2006;209:21-43. **9** Tsuruyama T et al J Leukoc Biol 2010;88:107-16. **10** Aranaz P et al Haematologica 2012;97:1234-41. **11** Beekman R et al Haematologica 2013;98:e62-3. **12** Estey EH Am J Hematol 2013;88:318-27. **13** Maxson JE et al N Engl J Med 2013;368:1781-90.*

(continued)

Table XXV (continued)

__14__ Budrys NM et al Fertil Steril 2012;97:1129-35. __15__ Sapi E et al Cancer Res 1999;59:5579-85. __16__ Amini R et al Biomed Res Int 2013:752603. __17__ Bazan-Peregrino M et al Cancer Gene Ther 2013. doi:10.1038/cgt.20123.41. __18__ Chappell JC et al ArterioScler Thromb Vasc Biol 2013;33: 1952-9. __19__ Hume DA et al J Immunol 2008;180:6733-42. __20__ Chen C et al PLoS One 2013;8(7): e69714. __21__ Agra N et al PLoS One 2013;8(6):e66281. __22__ Sinkovics JG & Plager C Canad J Surg 1977;20:542-6; Sinkovics JG et al Oncology 1980;37:114-9. __23__ Grunewald TGP et al Mol Cancer Res 2012;10;52. __24__ Zheng H et al Proc Natl Acad Sci USA 2012;109:20620-5. __25__ Wang AX & Qi XY Internat Union Biochem Mol Biol Life 2013. doi:10.1002/iub.1153. __26__ Cassier PA et al Cancer 2012;118:1649-55. __27__ Cupp JS et al Am J Surg Pathol 2007;31:970-6. __28__ Fiocco U et al Autoimmun Rev 2010;9:780-4. __29__ Galli M et al Int J Immunopathol Pharmacol 2012;25: 1131-6. __30__ Temple HT Curr Opin Oncol 2012;24:404-8. __31__ Lewin J & Thomas D Drugs Today 2013;49:693-700. __32__ Mattei TA Spine J 2014;14(6):e29-35. __33__ Pelle DW et al Transl Res 2014; pii;s1931-5244(14000089-9. __34__ van der Heijden L et al Oncologist 2014;19:550-561. __35__ Rutkowski P et al Int J Biochem Cell Biol 2014; pii:S1357-2725(14)00099-5. __36__ Ray-Coquard I & Le Cesne A. Cancer Treat Rev 2012;38:368-78.

*) *TNK = TANK-binding kinase; TANK = TRAF-family member-associated NFκB activator; TRAF- (tumor necrosis factor receptor-associated factor-) associated nuclear factor kappa B activator; STEAP, six transmembrane epithelial antigen, prostate.*

Dasatinib, the Abl-antagonist tyrosine kinase inhibitor, suppressed adhesion-mediated respiratory burst and superoxide release of neutrophil leukocytes; antagonized the leukocytes' immune complex-induced activation, and their migration toward fibrinogen-coated substrate, but it did not block LPS- or TNF signaling, stimulation by zymosan, and phagocytosis. Dasatinib blocked Syk activation (spleen tyrosine kinase, agonist of Fcγ receptors) [2215] (Table XXVI).

Comment. While dasatinib emerges as an anti-inflammatory drug, its side effects may reduce its anti-neoplastic efficacy. For example, is it inhibitory to NK cells? The mcab/NK cell anti-tumor-efficiency in the ADCC reaction must be maintained in the therapeutic protocols. Further investigations are needed in this direction (Table XXVI).

These are some of the primordial gene product proteins, which were functional in ancient unicellular life forms and remained conserved throughout evolution. If it is accepted that malignantly transformed cells of the highest order (of vertebrate mammalian hosts) descend on the evolutionary scale to the level of "simple" (not simplistic at all!) unicellular eukaryotes, it is to be recognized, that the primordial RNA/DNA complex operating in those most sophisticated biological units, was far from being simplistic. The disciplined and subservient somatic cell of the multicellular host upon its "malignant transformation" becomes a primordial amoeba loaded with reactivated cell survival pathways (called "oncogenes" and "oncoproteins" at the medical oncology clinics). The example of the colony stimulating factor and its receptor CSF-R & L), which directed the phylogenetic evolution of the large ancient phagocytic cells (monocytes, macrophages, dendritic cells, microglia, osteoclasts), can act as an oncogene-oncoprotein in human myelogenous leukemia cells, as is shown here (Table XXV). The fundamental difference between the unicellular eukaryotes primordial or extant, and the cancer cells of the

Table XXVI Interactions of Dasatinib with Immune Lymphocytes

The BCR/Abl oncoprotein is immunogenic. Ph-chromosome-bearing (Ph^+) patients with CML mobilize immunoreactive antibodies and lymphocytes against the oncoprotein without being able to reject the leukemia cells. The BCR/ABL oncoprotein was proposed to be used for vaccination of the histocompatible bone marrow donors of Ph^+ CML patients.[1] Of many choices of targeted therapy for Ph^+ CML, emerges the Bcr-Abl/Src-inhibitor dasatinib (Sprycel, BristolMyersSquibb) for the imatinib/nilotinib-resistant patient.[2] The family of *src*/Src proto-oncogenes/oncoproteins are physiological stimulators of innate and adaptive immune functions. Dasatinib in mice diminished $CD4^+/CD8^+$ T cell reactivity to ovalbumin and viral antigen, and NK cell reactivity to MHC antigen non-expressor target cells.[3] In mice, B cell lymphoma with splenomegaly induced by EBV latent membrane protein 2A and by mobilized LYN/MYC oncoproteins receded in response to treatment with dasatinib.[4] Pre-existing BCR/ABL-reactive $CD3^+CD8^+$ T, and $CD3^-CD16^+CD56^+$ NK cells in patients with Ph^+ CML significantly and mono- or oligoclonally expanded after dasatinib therapy; $TCR\gamma\delta^+$ T and NKT cells, and $CD4^+/CD8^+$ $TCR\alpha\beta$ lymphocytosis coincided with sustained remissions.[5] Dasatinib inhibited NK cell effector functions for 24 h. After dasatinib washout, physiological proto-oncogene *vav* (an *src* family member) phosphorylation occurred in the NK cells with full return of effector efficacy (cytokine response, granzyme release with anti-target cell cytotoxicity).[6] Dasatinib-treated patients with Ph^+ CML/ALL rapidly induced highly mobile NK/NKT, B and $\gamma\delta$ T cells.[7] Pneumocystis infections in frail patients on dasatinib therapy are so rare that MD Anderson Cancer Center recommends antibacterial, but not antifungal, or antiviral (CMV) prophylaxis.[8]
*1 Sinkovics JG J Natl Cancer Inst 1995;87:1804; Acta Microbiol Immunol Hung 2010;57:253-347. **2** Marin D ASH Education Program Book 2012;1:115-121. **3** Fraser CK et al Exp Hematol 2009;37:256-65. **4** Dargart JL et al Antiviral Res 2012;95:49-56. **5** Kreutzman A et al Blood 2010;116:772-82. **6** Hassold N et al Int J Cancer 2012;131:E916-927. **7** Mustjoki S et al Leukemia 2013;27:914-24. **8** Rodriguez GH et al Leuk Lymphoma 2012;53:1530-5.*

multicellular host, in activating "cell survival pathways", is that the former was left on its own in the most hostile environment, whereas the latter enabled itself throughout billion years of co-evolution to entice its host for the extension of a welcoming party, consisting of immunological and full physiological-biochemical support (Table XXV).

Astute Observers Withstand Derision

Even in the Molecular Era, It Still Looks like an Amoeboid Unicellular Parasite

Snow and Virchow call for followers! The "Krebsbacillus" of Rudolf Ludwig Carl Virchow [1821–1902] was cited in the volume "Infection and Inflammation: Impacts on Oncogenesis" edited by T. Dittmar, K.S. Zänker and A. Schmidt and published by Karger, Basel, in 2006. Virchow's "Krebsbacillus" may be viewed as the ancient conserved oncogenome, whose elements can pass horizontally and in exosomes [2216]. The pioneering work of Herbert L. Snow [1847–1930] on lymph node dissection for melanoma metastases was recalled in an article published from the Sarcoma/Melanoma Unit of The Royal Marsden Hospital, London. The article recalls Snow's views on cancer: the 46 years old Snow published his treatise on cancer. The cancer cell is our own cell reverted (underwent "devolution") into its ancestor, an amoeba (Snow H, cited in Neuhaus et al) [2217a, b]. Before any biochemistry, the astute observers looked at the cancer cell and saw an amoeba. Scanning EM pictures of the amoeba and the cancer cell are provided in the literature for comparison. The moving individual amoeba is a formidable creature in its environment; more slick, the cancer cell added great sophistication to the amoeboid locomotion by organizing it into an armada of transformed cells marching forward (in the Appendix 1).

R.N. Johnston, S.B. Pai and R.B. Pai in the era of advancing biochemistry, wrote: "…many workers have independently commented on the similarities observed between the behaviors displayed by dedifferentiated cancer cells and their embryonic precursors." "…Cancer cells behave in ways that are reminiscent of primitive eukaryotic cells." "Cancer cells might be considered to represent, with their loss of tumor suppressor inhibitory activity and elevation of oncogene stimulatory activity, a reversion to a more primitive evolutionary state capable of indeterminate growth at the expense of the host. By this analogy, the growth phenotypes displayed by cancer cells, embryonic cells, and free-living eukaryotes

J.G. Sinkovics, *RNA/DNA and Cancer*, DOI 10.1007/978-3-319-22279-0_28

are fundamentally similar." These authors turned around Ernst Haeckel's biogenetic law of "ontogeny recapitulates phylogeny" as "oncogeny reverses phylogeny" [2218].

Elena Moncevičiūtė-Eringienė conceived the idea over 20 years ago that the first living cells on Earth were endowed with an elaborate genetic apparatus for the protection against "chronic damage". These are the ATP-dependent membrane P-glycoproteins, the glutathione S-transferases, and the multiple drug resistance efflux pumps. The genes and gene product proteins of this multixenobiotic resistance are conserved from bacteria through the first eukaryotes to every cell living in multicellular host. Upon chronic damage, the reactivation of this ancient defensive apparatus changes the cell backward on the evolutionary scale "into a more primitive mode of existence." What is recognized as cancer generation is a form of "adaptation to survive in an atavistic way". The evolutionary resistance is "the evolutionary law of survival". Evolutionarily installed ancient enzymes (multixenobiotic resistance practiced in "worm, sponge, mollusc, clam, oyster, snail, fish, mammals") sustain the final capacity of the cells living in a multicellular host to revert to their ancient way of life. The inherent chemo-radiotherapy resistance of these cell populations is the proof that the impulse to revert originated from the ancient genes and gene-product proteins of the "evolutionary resistance". This "atavistic regression" converts the cancer cells to parasites (the author cites J. Huxley's work from 1938 in which he showed that chemically induced sarcomas are much more resistant to chemotherapy than spontaneously arisen tumors, and refers to cancer cells as "parasites"). B.G. Charlton's work is cited, in which the cancer cell population is described as practicing "endogenous parasitism". Moncevičiûtê-Eringienė lists oncogenes (Harvey ras; c-myc), which encode drug-resistance proteins. In case of wild-type p53, it is the mutated gene whose presence (its functional absence) is what promotes drug resistance. The anciently installed drug resistance pathways are conserved, and re-emerge as the initiators of the phenomena referred to as "oncogenesis." Primary evolutionary malignant resistance mechanisms contribute to the formation of cancer cells as a mode of adaptation for survival. The author provides a list of resistance genes and mechanisms, which throughout the evolution have been transduced from bacteria to mammalia and thus provided the basis for her theory [2219a–d]. A widely spread presence of natural IgM antibodies was discovered in different organisms and are directed to the enterobacterial common antigen "endotoxin". These antibodies were referred to as "natural, endogenous or spontaneous". The presence of these antibodies is explained by the hypothesis that "atavistic endotoxin" has a role in carcinogenesis, because antigens identical to the gram-negative bacterial endotoxin appear to be expressed in tumor cells, whereas healthy (pre-canceros) cells of the same hosts are devoid of the endotoxin. It is theorized that the regression of cells to the atavistic parasitic state induces the synthesis of the dormant endotoxin (along with evolutionarily installed resistance genes and "oncogenes)." The expression of endotoxin by proliferating cancer cells explains cancer cachexia and the clinical picture of "endotoxin septic shock" patients with advanced cancers experience [2219a–d].

Frank Arguello. http://Amazon.com. published in the summer of 2011 in paperback the essay entitled "Atavistic Metamorphosis. A New and Logical Explanation for the Origin and Biological Nature of Cancer" [2220]. In chapter "Introduction" reviews "the concept of cancer" in history from Galen and Hippocrates to the present era. Bernard Peyrille [1737–1804] was referring to the "cancerous virus" (when the word virus meant some unspecified toxin). Recently Rudolf Virchow [1821–1902] is receiving credit for recognizing that some cancers occur in inflamed tissues. Theodor Boveri [1862–1915] connected first carcinogenesis with chromosomal abnormalities. D. Hanahan and R.A. Weinberg receive credit for listing in 2000 the hallmarks of cancer. C. Croce receives credit for the discovery of oncogenes. Lynn Margulis is cited as a major contributor to the explanation how eukaryotic cells derived from endosymbiosis. The four principles elaborated on are: 1. All cells have a single common origin. 2. All relevant genetic history is conserved. 3. Cancer cells have the same phenotype as single-celled organisms. In elaborating on principle 3, the author shows that some ancestral oncogenes (p21 activated kinase, Pak; Rous sarcoma proto-oncogene Src) are operational in single-celled protozoa and early arising multicellular hosts. 4. Cancer is an atavistic reversion to single-celled life. Frank Arguello cites L. Snow [1847–1930] stating that cancer was a process of devolution, a reversion of cells to primordial amoebiform condition, in which they become parasites, or rather "autosites". Author Arguello elaborates on mechanisms, whereby atavistic reaction can be induced in a stem cell. Here he states: "Cancer is certainly a disease for the host that experiences it. However, atavistic reversion of cells to single-celled life should be seen as a last-resort mechanism of cell survival available to differentiated cells in multicellular organisms. Single-celled life has been so successful in adapting to a variety of different hostile environments on Earth that many organisms have gone through millions of years of evolution as single cells, each one sufficient unto itself in the struggle for existence. Proof of the successful existence of single-celled organisms is that they form more than half of the total biomass on earth". Thereafter Arguello elaborates on new principles of cancer therapy, combining chemo- and immunotherapy, for cancer. He travels to Mexico (Ciudad Mexico) to introduce and initiate newly designed therapeutic modalities for cancer [2220].

This author (JGS) follows suit. In the 1960s/70s at the Section of Human Tumor Virology & Immunology, M.D. Anderson Hospital, Houston, TX (*vide infra*), Sinkovics grew human cancer cells (EBV-carrier B lineage lymphoblasts; soft tissue and bone sarcoma cells, melanoma cells, adenocarcinoma, ovarian and kidney carcinoma, cells in established long-term tissue cultures [1501, 2221a–d, 2222]. These human cancer cells established in permanent cultures defied the Hayflick's rule of senescence and death, the fate setting in after a limited number of cell divisions for healthy somatic cells. In contrast, the "malignantly transformed cells" proliferated as long as nutrients were provided and waste was eliminated [70b, 724, 1501]. This author, impressed by this major biological difference, declared in the mid-1970s: "As ontogenesis appears to repeat phylogenesis, early embryos resemble mature individuals of ancestral species, but it is not known just how far back this concept may be stretched. Would gene products, i.e. antigenic proteins, expressed by the tissues of primitive species that have evolved little since archaic times, cross-react immunologically with those expressed by early embryos of the species that have advanced through evolution?" [724, p. 134]. Further: "Could the cancer cell be viewed as an individual that extricated itself from boundaries exposed on it by 500 million years of evolution (today's corrections: "imposed" and "over 3 billion years") and recaptured the immortality of unicellular organisms that existed before the social order of cell communities was established? Certainly, some cell cultures of mammals, including man, create the impression of immortality of archaic unicellular life forms, i.e. lymphoblastic cells grown in suspension culture as long as nutrients are provided and metabolic waste is removed" ([724, p 134], recited in [73]).

In a fortunate coincidence, without any knowledge either of Moncevičiūtė-Erigienė's, or of Arguello's theses (*vide supra*), in the spring of the year 2011 this author (JGS) submitted abstracts to be presented in a series of lectures at the summer and autumn conferences to be held in Budapest, Hungary [2223a, b]. These abstracts and lectures elaborated on the idea that the genuine primordial RNA/DNA complex evolved so, that it may receive and respond to danger signals of the impending demise of their multicellular hosts. Especially, severe and protracted infectious processes would emit such signals [73]. In response, inherently evolved mechanisms for the re-enactment of the life style that the unicellular ancestors practiced will be evoked. The extraordinary response thus elicited, would serve the sustenance of cellular life on Earth under any circumstances and in any form. The title of the lecture given at a conference of the Hungarian Scientific Academy was: "Carcinogenic inflammatory processes and the awakening of the primordial DNA in the form of oncogenes". The substance of lecture was published under a different title [2224].

Author's conclusion. The orderly morphology and the well-organized physiology of the extant unicellular eukaryotes appear to be different from the outside appearance and cannibalistic life style of immortalized cultures of malignantly transformed cells (view the HeLa cell line for comparison with suspension cultures of tetrahymena or trypanosoma). If malignant transformation is a descent of the cell to an ancient life style, and to the morphology associated with it, are the criteria observed in cancer cells and in extant unicellular eukaryotes valid for such a conclusion? The ancient unicellular eukaryotes are not available for a valid comparison; their descendants have to substitute for them. It is only their descendants activating their "cell survival pathways" in full gear in extreme distress that is comparable with what is called "malignant transformation". The genomics of those ancient "cell survival pathways" remained conserved and guarded, so to serve the extant unicellular eukaryotes. The larvae and the embryos of their multicellular descendants, go through their ontogenesis in a most organized manner. The ancient cell survival pathways, or physiological cell transformational events in the larvae or embryos (including events of trans-speciations), singularly and sequentially (or even in unisons), remain in action in all living cells and organisms. They are activated and switched off after they generated differentiated somatic cells in a multiplicity of organs. They preserve their germ cells and stem cells. It is the constitutive activation of those ancient cell survival pathways, amplified, truncated, mutated and fused, that leads to the irreversible transformation of the cell. Life-saving transient events of the distant ancestors are converted in the descendants into a permanent biological state with the semblance of immortality. Encountering this phenomenon in the multicellular hosts, the textbooks and the clinics diagnose and describe the mortal malady "cancer".

The tetrahymena lives in suspension cultures without senescence or death. There is no known cancer cell living *in vivo* or *in vitro* that descended to the level of the tetrahymena and displayed its silent diploid germline micronucleus separately from its transcriptionally active somatic macronucleus. When the tetrahymena cell divides asexually, the germline DNA is not transcribed. The ciliates can live asexual lives with their macronuclei, and without their micronuclei (MAC; MIC). That is comparable with the lifestyle of somatic cells in a multicellular host. The tetrahymena's regulated possession of the high mobility group protein (HMGB3) is a multifunctional methionin salvage fusion enzyme in its mitochondrium; its telomerase holoenzyme, and its conserved TEN domain (telomerase reverse transcriptase N-terminal), or the potential oncogene/oncoprotein nucleoporin Nup98 deliver regulated services in their cell community. The HMG proteins do not transform the tetrahymena into a malignant cell [2225–2230]. The centrin-containing centriolar satellites induce centrosome amplification in malignantly transformed cells [2231]. The two tetrahymena centrin homologues (human centrin homologues) do not associate with malignant transformation of the tetrahymena [2232]. The Arg/Piwi proteins accompany malignant transformation in multicellular hosts. The Tiwi12 of the tetrahymena, physiologically stabilizes the nuclear exonuclease Xm2. However, the tetrahymena metabolism reveals features similar to those a prostate cancer cells exhibits as shown in the Couvillion et al articles [125a, b, 2233] (Figures 32 and 33). Mutation of the Tetrahymena histone methyltransferase (TXR1) results in the failure to monomethylate H3K27 (lysine), but not in malignant transformation [2234]. In nature, the tetrahymena cell suffers mutations in its haploid germline genome. These are rare, recessive, but deleterious, but only in <10 % lethal [2235]. Germline mutations become evident after conjugation in the new somatic genome, generated from the germline genome. Nonlethal germ cell mutations are the driving force of evolution in the tetrahymena, and in general. The free living tetrahymena is not quite equal to a malignantly transformed cell of a multicellular host, even though its cell survival pathways may use genes which will evolve into proto-oncogenes in their multicellular descendants.

No malignantly transformed mammalian cell would look like an oxytricha. The oxytricha MAC genome is unique even among the ciliates. It consists of thousands (over ten thousand) of nanochromosomes, each representing of one single gene. Very few nanochromosomes consists of noncoding-RNA (a controversial ncRNA gene count). Its extraordinary number of telomeres (over ten millions per cell) and telomere end-binding sequences may secure its rapid, even incessant, replication, but it does not convert it into a disorganized malignantly transformed cell [2236–2238]. Its telomeric G-quadruplexes (consisting of guanine 5′-monophosphates) are self-assembled four-stranded G4-DNA G-tetramer G-tetrads. The human telomeric

G-quadruplexes (TTAGGG) are strongly conserved among the vertebrates and are structurally different from the oxytricha quadruplex (however, the differences, if any of the human c-*myc* quadruplex and the oxytricha quadruplex, are not clearly evident). If two strands are oriented in the same direction, the two guanines will show identical glycosidic conformation. If the guanine strand orientations are opposite, the glycosidic conformation will run in opposite direction (*syn-anti*, *anti-anti*, *anti-syn*, *syn-syn* conformations). For provision of free energy, the *syn-anti* steps are more favorable, than the *anti-syn* or *syn-syn* steps [2239]. A human tumor cell would use for energy gain the *syn-anti* steps. The ciliate genomes (oxytricha, paramecium, tetrahymena) practice transposons domestication by encoding both in the micro- and macronuclei PiggyBac-like domesticated transposases, which either delete thousands of alien transposase sequences from their genomes, or permanently retrotranscribe others to serve as protein coding new genes [1]. The MIC of paramecia harbor a legion of Tc1/mariner-like transposons (short intervening sequences). During the development of MAC, the domesticated PiggyBac transposase excises most of these alien transposons. DNA repair is by the non-homologous end-joining pathway [2240] (very much the same as in some cancer cells). The vertebrate genomes practice similar transposons domestication (as shown repeatedly in the text above) [1]. The oxytricha genome purges the least of its transposons (and tolerates the most of them among the ciliates) [2241]. None of these morphological or genomic faculties transform the oxytricha cell into a cancer cell. No known cancer cells of the vertebrate hosts divide their nucleus into MAC and MIC, or disperse their chromosomes into thousands of single genes (the nanochromosomes of oxytricha). No matter how many cancer-causing genes they may have contributed in the long run to their descendants, the evolutionary descent of the cancer cell is not into the direction of oxytricha or tetrahymena.

The Amoebozoans are uninucleates; by morphology (the syncytial *Physarium polycephalum*), and by physiology (locomotion, phagocytosis, genomics, cell fusions), they are closer to the cancer cell; so are Plasmodia, Naegleria (Figure 54), and Theileria. The sequenced genome of *Dictyostelium discoideum* (Figure 6) reveals a high A/T nucleotide content, retention of excessive numbers of repetitive transposons, unique SSU rRNA tree (small subunit ribosomal), and rDNA palindrome segment polyasparagine repeats for capping its chromosome ends, as a result of active tRNA/rRNA genes (within regular eukaryotic telomeres). Dd operates active actin-related genes (actin depolymerizing cofilins). Some 400 Dd proteins are directly derived from LECA. Added are a large number of polyketide synthase genes combining acyl transferases, methyltransferases, ketosynthases, sulfhydrolases, thioesterases, ketoreductases, etc. The expression of the Wnt frizzled/smoothened receptors is strongly expanded. Some Dd genes are horizontally acquired and shared with hosts *Saccharomyces cerevisiae, Schizosaccharomyces pombe, Hydra magnipapillata, Arabidopsis thaliana* and *Homo sapiens.* Dd has the ability to generate multicellular colonies. Its fruiting bodies cannot be generated under water. Dd might have been the first multicellular colony crawling out from the sea to dry land. Observe the amoeboid locomotion of several cancer cells together in unison. The reaction to UV light in the Dd cell has been established, and in its mechanism it remains related to that of all advanced eukaryotes [2242, 2243]. Dd reveals that the retinoblastoma gene-mediated cell cycle suppression was installed in its ancestors. There is a cyclin-dependent protein kinase (cdk1). There are Rb, E2F and the E2F-dimerization partner, DP proteins. However, Dd's cell cycle may overrule the blockade. The Dd cell cycle may pass through checkpoint G1→S without inhibition (a criterion of malignant transformation in advanced eukaryotic cells). Being a common ancestor of animals and plants, Dd operates a plant-like cell cycle (similar to the Arabidopsis E2FB transcription factor) [2244]. There is a great activity of SSU rRNAs (small subunit, ribosomal) [2242–2245]. Ribosomal RNA biogenesis is a target of the mTOR pathway. There, the acetylated histone H3K56ac allows RNA polymerase to carry out rRNA/rDNA processing, that results in the binding of SSU rRNA with HMG proteins (Tables XI/I and XI/II). This unison is essential for cell mitoses [2246]. Dd operates a unique SSU rRNA system. In departure from the majority of other eukaryota, Dd encoded not only ribonucleotide reductase (RNR) class I, but both classes I and II [2247]. The class I enzyme essential for the de novo synthesis of deoxyribonucleotides. Deoxyribonucleotides are needed for DNA replication and repair. B12 coenzyme is essential for the class II enzyme (operational already in Lactobacilli). The Dd RNR class I enzyme is a typical eukaryotic enzyme; however, it contributes to the malignancy of the human tumors gastric and uterine cervical cancers [2248, 2249]. As to Dd's Wnt pathway, β-catenin and GSK are present, but without Wnt and Frizzled; cyclic AMP is the activator of the GSK molecule [2250].

Colony forming fungi and metazoa share a common ancestor and form the eukaryotic clade within supergroup Opisthokonta. Amoebozoa are monophyletic with Opisthokonta and form a clade within Opisthokonta. Colony forming Choanoflagellata include *Capsaspora owczarzaki* and *Salpingoeca rosetta*, forming a sister group of Metazoa; they are considered to be ancestral to Metazoa. Opisthokonta share the protein domains Pfam (protein family containing multiple alignments in Markov model based profiles) [2251–2253]. Choanoflagellata cells (choanocytes: 'the feeding cells') in colonies communicate through intercellular bridges [2254]. Intercellular bridge formation appears in healthy embryonic cells *in vivo*, and in cancer cell colonies *in vivo* and *in vitro* [2255, 2256]. This author observed intercellular bridges between intracellular lymphocytes and their host fibroblasts [353, 717]. Single cells of the choanoflagellate *S. rosetta* differentiate into morphological subtypes and form colonies in rosettes and chains [2254]. Choanoflagellata possess the heterodimerized Myc/Max proteins within the nucleus, binding to E-box sequences of DNA [2255], and encoding arginine kinase isoforms (but so far no arginine methyltransferases were reported) [2256]. Arginine kinases are encoded by the Nematostella genome [2257], but the potentially oncogenic protein arginine methyltransferases (PAMT) remains so far undetectable below Xenopus and fish [2258]; some ascidians may be an exception for the earliest appearance of PAMT [2259] (also PRMT).

Presumably, intron loss occurred in the time interval elapsing between the eumetazoan ancestor and the appearance of the common ancestor of choanoflagellates and metazoans in the Precambrian sea over 600 years ago. Protein domains PFAM and SMART (containing 685 hidden Markov models, http//smart.embl-heidelberg) reveal protein domains exclusively shared between choanoflagellates and metazoa. The scaffold protein Shank served in the choanoflagellates, but was lost in the metazoa [2260]. An abundance of cell adhesion proteins and catenins are "in the waiting" and ready for multicellular existence. The genome expresses the precursors of the myc, *sox* and *tcf* genes (of the *tcf/lef* family genes activated by intranuclear β-catenin in eukaryota) [2261]. The Rab (*ras*-like rat brain) GTPases appeared before the split of choanoflagellates and metazoans) [2262, 2263]. The leading metazoan transcription factors are already encoded by the *C. owczarzaki* genome. These are: Rel/NFκB, p53 (p63/p73); the STAT pathway; bZIP (basic region leucine zipper); bHLH (helix-loop-helix proteins); Runx (runt-binding domain), Mef2 (monocyte/myocyte enhancer factor; eliminated from rhabdomyosarcoma cells); Hox homeobox/homeodomain; Fox (forkhead box); HMG box (high mobility group proteins); CBP/p300 (cyclic adenosine monophosphate response element-binding, CREB; LSF/GRH (late SV40 factor; grainyhead transcription factor). Some of the genes considered to be exclusively metazoan are already present in a choanoflagellate [2264, 2265]. In vertebrate mammalian hosts, the choanoflagellate LSF becomes an oncogene in hepatocellular carcinoma [2266].

To the evolution from unicellularity to multicellularity, the photosynthetic (cyanobacterial cytoplasmic residents of) sexually haploid volvocine green algae made major contributions. In the colony of *Volvox carteri*, a hierarchy of cells was formed. The small flagellated somatic cells vastly outnumbered (>2000:1) the larger non-flagellated stem cells. While the somatic cells underwent terminal differentiation, the stem cell remained undifferentiated. It was the *reg*A gene that dictated the terminal differentiation of the somatic cells. Was this an altruistic act in the interest of the race from the part of the somatic cells [2267]? Like the *Chlamydomonas reinhardtii*, volvocine green algae operate a full Rb pathway (Rb, E2F, DP) [2268, 2269]. The algal telomeres are either unique (chlamydomonas, TTTTAGGG), or also plant-like (arabidopsis, TTTAGGG), in comparison to vertebrates, including human (TTAGGG). The human Rb protein acts within the nucleus and exits from the nucleus during S phase of the cell cycle [2270]: voluntarily in normal cell divisions, forcefully expelled in malignancy. In incipient malignant tumors, the Rb (and p53; PTEN) genes are epigenetically suppressed, deleted, or their gene product proteins are ubiquitinated.

The uninuclear, hierarchial Amoebozoans, with the ability to form syncytia [2271], which crawl and phagocytose (cell cannibalism) appears to be the direction toward which the cells (de-differentiating stem or somatic cells) descend on the evolutionary scale during their so called "malignant transformation". In nature, under natural circumstances, among the extant descendents of the ancient unicellular eukaryotes, it is not possible to readily find colonies that would resemble that of the HeLa cells. Under adverse circumstances, as in the Hydractinia colony, it may happen that de-differentiated cells overgrow the colony.

Mutator features appear first in the archaea. The anaerobic crenarchaeon, *Pyrobaculum aerophilum* lives in boiling marine water holes. Its mRNA does not express 5′ untranslated regions, therefore its protein translation is initiated without a ribosome binding site ("leaderless transcripts"). It synthesizes carbohydrates through the glyoxylate cycle; and operates the citric acid cycle. It can utilize thiosulfate, but elemental sulfur blocks its metabolism. (It is lethal; did elemental sulfur killed the first Martian Pyrobaculum-like archaea?). All aminoacyl tRNA synthetases (encoded by genes separated by non-canonical introns) are operational (except for selenocysteine). The *P. aerophilum* mononucleotide repeat tracts are unstable, due to its deficient (possible lack of) mismatch repair (MMR) enzymes. The MMR enzymes of the prokaryota (*E. coli*) are not yet present in archaeota; many known archaea genomes lack genes for the proteins that carry out MMR. In consequence, the archaeota genomes suffer events of frameshifting, thus generating diversity in the surviving mutators. Enzymes other than those of MMR (the Rec/Rad enzymes) may compensate in part for the lack of MMR enzymes. The *P. aerophilum* genome encodes recombinant Rec/Rad enzymes. The cell division of the crenarchaeon *P. aerophilum* is independent from the euryarchaeal FtsZ ring (and therefore is similar to that of eukaryota cells) [42, 43, 2272]. The MMR enzymes of the prokaryota (*E. coli*) run an extraordinarily long evolutionary trajectory through the eukaryotic orders up to *Homo*. In *Homo*, mutated and vertically transferred genes of some of these enzymes encode faulty proteins (oncoproteins) known to contribute to the generation of adenocarcinomas of the colon, breast, ovary, endometrium and prostate (Table XXVII).

Of pathogens, virulent bacteria from Gram-positive cocci to Gram-negative rods (including Staphylococcus, Pseudomonas, Yersinia, etc.) and yeasts assume a mutator pheno-genotype [2273a, b]. Entamoeba carries a Mutator retrotransposon. The yeast and some parasites may undergo no, or faulty, DNA mismatch repair, and assume mutator pheno-genotypes, through which they acquire extreme virulence and drug resistance. In this stage, there may be comparability between the malignantly transformed individual cells of a multicellular host, which descended to the amoeba level on the evolutionary scale during its malignant transformation. It is possible that if it were specifically looked for in parasitic pathogens attacking the host, or even in a social amoeba under great distress, the reversion to a mutator-like pheno-genotype will match those faculties of the cancer cells both in appearance and in pathophysiology (Table XXVII). The cell survival pathways of the ancestral unicellular eukaryotes have become oncogene-generated oncoproteins in the transforming single cells of the multicellular hosts (Tables XXVIII/I and XXVIII/II).

It has become well known in the practice and science of clinical oncology that the tumor cell expropriates for its own use all the physiological cellular growth factors (Table XXIX) Inserted transposons "awaken" to seek out and activate the oncoprotein-encoding genes, which are the descendants of the genes of the ancient cell survival pathways. With their appearance, tenacious and virulent behavior, and their driver fusion protein, the *Phytophthora infestans* and the Trichomonas act in their multicellular host as if they were alien immortalized invaders (Tables XXVII and XXX; Figure 56).

The ancestors of vertebrate mammalian proto-oncogenes are functional in the ontogenesis and in the cell survival pathways of the paramecia, Giardia (Figure 28) and Trichomonas (in the Appendix 1), amphioxus (in the Appendix 1), the sea urchin (Figure 45), and the cnidaria sea anemone, and *Ciona intestinalis* (in the Appendix 1).

Table XXVII Mutator Pheno-Genotypes

History. Lack of mismatch repair (MMR) enzymes is natural in archaea: their genome is naturally mutator-like.[1] The originally discovered maize mutators were highly active transposons; so are fungal mutators in *Magnaporthe grisea, Neurospora crassa* and *Aspergillus fumigatus*. The first yeast mutator was found in *Yarrowia lipolytica*. The protists *Entamoeba dispar, E. hystolytica, E. invadens, E. moshkowskii* carry their interrelated mutator elements.[2,3] DNA replication prevents mutations by immediate editing of the elongating strand by polymerases, and correction of mismatches by MMR. The *E. coli* MMR enzymes are conserved up to Homo; failure results in "carcinogenesis" (endometrial, ovarian, etc).[1]

Yeast. Genomic replication factors recognize the primer terminus and catalize the proliferating cell nuclear antigen (PCNA), which is a potential oncoprotein in higher eukaryotes. The replication factor ELG (elongation GTPase) is a suppressor of genome instability in the sliding clamp; it responds with RAD24 to DNA damage. It prevents replication fork stalling and facilitates re-start of stalled replication forks by unloading and re-cycling PCNA by re-loading it. Deletion (or loss-of-function mutation) of ELG in budding yeast results in persistent DNA damage in the S-phase and upregulation of ribonucleotide reductase (RNR). In consequence, desoxyribonucleotide pools (DNTP) accumulate in the G1 phase. Replication fidelity is lost; events of spontaneous mutagenesis follow in a vicious circle.[4,5] Some mutants escape error-induced extinction (above a threshold). Yeast cells practice mutator suppression and escaped mutants assume "anti-mutator" phenotype. Mutator cells of multicellular eukaryotic hosts may be eliminated, or reverted to physiological functioning.[6–8]

Toxoplasma. Isolates from Uganda show evidence for gain of virulence by chromosome sorting and natural recombinations. The invasive rhoptry proteins (ROP/RON) are highly conserved toxoplasma virulence factors shared by other apicomplexans (cryptosporidia; plasmodia). The injection of these proteins into the host cell facilitates the invasion by the parasite. The *T. gondii* chromosome VII (and to a minor extent, chromosome IV) encode important ROP virulence factors. The recombinant Brazilian *T. gondii* strains display the highest degree of virulence. Several discrete genomic loci (quantitative trait loci) encode the acute virulence factors (prominently from chromosome VII).[9–17]

Plasmodia. DNA MMR and mutator gene *Pfmsh2*/MSH2 (Plasmodium falciparum muscle segment homeodomain; *Agrobacterium tumefaciens*; *Homo sapiens*): microsatellite instability established mutator phenotype in *P. berghei*; *P falciparum*. In contrast to gain-of-function mutations, the loss of the aquaglyceroporin gene is a loss-of-function mutation in recombinant plasmodia. In Thailand and Cambodia, *P. vivax* gained genetic complexity, high virulence, and multidrug resistance. In India, mutations of virulence upward, and chloroquin-resistance developed. In Tanzania, artemether-lumefantrine resistance occurred in artemisinin-resistant falciparium plasmodia, even without exposure to the drug.[18–22]

Trichomonas vaginalis. The amoeboid form, or the free-swimming flagellated trophozoite exhibit mutagenic faculties culminating in the acquisition of virulence factors, host immune-, and drug-resistance. This parasite phagocytoses bacteria and cells (including human host cells) and thus acquires genes by lateral transfer. It harbors and transmits viruses (even without the replication of viruses within it). Trichomonas is a host of the torque teno circoviruses (and may be a carrier of HIV-1). Plant-like mutator elements (MULE) are inserted into the trichomonas genome. The Kolobok superfamily mutator transposons invaded the trichomonas genome.[23–25]

Mutator cells of unicellular eukaryota and those of vertebrate mammalian hosts. In the malignantly transformed cell the very low "normal" mutation rate (<1 in 10 billion nucleotides replicating in one cell division) is replaced by an elevated rate of mutations amounting to thousands of mutations in an established cancer cell, which are not evenly dispersed among each member of the tumor's cell colony. The more complex is the genome (the term complexity encompasses faculties of the genome surpassing the number of chromosomes), the more complicated the mutator phenotype will be.[26,27] Unicellular eukaryotes assuming the mutator phenotypes may resemble the cancer cells of the vertebrate mammalian hosts, but remain much

(continued)

Table XXVII (continued)

less complicated. The uracil-DNA glycolysase null mutant *Trypanosoma brucei* exhibits a hypermutator phenotype,[28] but lacks the complexity of a mutator human cancer cell. There may be strong similarities of the unicellular host and the cancer cells of the multicellular host when both carry an overmutated genome, but the unicellular hosts will not be able to match the complexity of the HeLa cells. Much remains to be compared at the genomic level. Are the mutator unicellular hosts aneuploid and afflicted by extensive microsatellite instability comparable to that of a human cancer cell? The cancer cell may descend to the level of an amoeba, but the mutator phenotype of the real amoeba will not be able to match that of a human cancer cell.

"**Mutator phenotypes arise spontaneously** in nature". "Lesions in DNA generated by endogenous cellular metabolism".[29–31] However, microsatellite instability, mismatches (MM) during somatic hypermutation, tandem double base substitutions, error signatures of DNA polymerases, reduced MM repair efficiency, and faulty proofreading, all are considered due to "inbuilt errors" of the system.[29–31] Author's comment: For the unicellular eukaryote, gain of vitality and virulence. For select individual cells of the multicellular host: some of these events may be endogenously initiated as an inherent faculty of the RNA/DNA complex aimed at the immortalization of the cell: a "rescue attempt" inefficiently opposed, even promoted by the host; overdone in too many subclones, error-laden, but resulting in immortalized cells.

Examples of mutator human cancer cells. Endogenously arisen microsatellite instability results in hereditary or sporadic colorectal cancer (Lynch) notoriously resistant to 5-FU; methotrexate induces synthetic lethality in these cancer cells (Table X). miR-155 downregulates the MMR proteins.[32–35] "Due to endogenous DNA damage" the mutator DNA polymerase-β variant D160N (aspartic acid; asparagine) transforms gastric epithelial cells to lose anchorage dependence, form foci, proliferate and invade.[36] BRCA1/2 mutations induce mutator phenotype with chemotherapy-sensitive ovarian carcinoma.[37] Genomic instability with aneuploidy leads to mutator phenotype cancers.[38] Low grade gliomas with mutator phenotype harbor significantly more mutations than glioblastoma multiforme: the mutator phenotype generates cascades of mutations.[39] Inflammatory carcinogenesis induces mutation phenotypes of the cancer cells (see text ref 73; 2100).[40,41]

***1** Fitz-Gibbon ST et al Proc Natl Acad Sci USA 2002;99:984–9. **2** Lopes FR et al BMC Genomics 2009;10:330. **3** Pritham EJ et al Mol Biol Evol 2005;22:1751–63. **4** Bellaoui M et al EMBO J 2003;22:4304–13. **5** Davidson MB et al EMBO J 2012;31:895–907. **6** Herr et al PLoS Genet 2011;7(10):e1002282. **7** Loeb LA et al Cancer Res 2008;68:3551–7. **8** Prestin BD et al Semin Cancer Biol 2010;20:281–93. **9** Bontell IL et al Genome Biol 2009;10:R53. **10** Bradley PJ et al J Biol Chem 2005;280:32245–58. **11** Dardé ML Ann Inst Super Sanita 2004;40:57–63. **12** Dziadek B et al Exp Parasitol 2009;123:81–9. **13** Ferreira AM et al Infect Genet Evol 2006;6:22–31. **14** Lamarque MH et al PLoS One 2012;7(3):e32457. **15** Ong YC et al J Biol Chem 2010;285:28731–40. **16** Sánchez VR et al Exp Parasitol 2011;128:448–53. **17** Su C et al Proc Natl Acad Sci USA 2002;99:10753–8. **18** Bethke L et al Mol Biochem Parasitol 2007;155:18–25. **19** Kenthirapalan S et al Int J Parasitol 2012;42:1185–92. **20** Lin JT et al Am J Trop Med Hyg 2013;88:1116–23. **21** Malmberg M et al Malar J 2013;12:103. **22** Zakeri S et al Malar J 2012;11:373. **23** Hirt RP Sex Transm Infect 2013;89:439–443. **24** Lopes FR et al BMC Genomics 2009;10:330. **25** Meng Q et al J Genet Genomics 2011;38:63–70. **26** Corcos D Am J Blood Res 2012;2:160–9. **27** Schmitt MW et al Ann NY Acad Sci 2012;1267:110–6. **28** Castillo-Acosta VM et al Int J Biochem Cell Biol 2012;44:1555–68. **29** Arana ME & Kunkel TA Semin Cancer Biol 2010;20:304–11. **30** Loeb LA et al Cancer Res 2008;68:3551–7. **31** Prestin BD et al Semin Cancer Biol 2010;20:281–95. **32** Desselle F et al Rev Med Liege 2012;67:638–43. **33** Dietmaier W Pathologe 2010;S2:268–73. **34** Valeri N et al Proc Natl Acad Sci USA 2010;107:6982–7. **35** Yamamoto H et al World J Gastroenterol 2012;18:2745–55. **36** Donigan KIA et al 2012;11:381–90. **37** Yang D et al 2011;306:1557–65. **38** Kolodner RD et al Science 2011;333:942–3. **39** Misra A et al BMC Cancer 2007;7:190. **40** Tili E et al Proc Natl Acad Sci USA 2011;108:4908–13. **41** Sinkovics JG Int J Oncol 2012;40:305–49.*

Table XXVIII/I A List of Ancient "Cell Survival Pathway" Genes Which Have Become Proto-Oncogenes Encoding Oncoproteins in Vertebrate Mammalian Hosts, in Particular Homo

Proliferating cell nuclear antigen (polymerase processivity factor). PCNA is a trimeric sliding clamp that encircles the DNA strand inviting binding partners (C-terminal peptide from the cell cycle checkpoint protein p21) possessing PIP (PCNA interacting peptide) and presenting hydrophobic amino acids (isoleucine/leucine/methionine) and aromatic amino acids (phenylalanine/tryptophan)[1,2] or mono-methylations. Facilitates TLS (translesional synthesis): monoubiquitylated by Rad6-Rad18. PCNA allows access of TLS polymerase to the replication fork; de-ubiquitylated PCNA displaces TLS polymerase; the Spartan protein inhibits ubiquitin-specific protease in the deubiquitylation of PCNA, for the attachment of high fidelity polymerases to the lesion[3]. In archaea the system works under extreme environmental conditions. Crenarchaea divide eukaryota-like, but possess heterotrimeric PCNA, whereas euryarchaeal PCNAs are homotrimeric eukaryota-like (homotrimeric); original : Pyrococcus; heterotrimeric, evolved: Sulfolobus. Archaeal PCNAs are multiphyletic; eukaryotic PCNAs are monophyletic.[1] The evolution of related proteins (replication factor C, clamp loader; minichromosome maintenance complex, unwinding dsDNA) is discussed separately.[2] The E. histolytica PCNA is in the evolutionary line between yeast, protozoa and metazoa, and is stimulatory to the amoeba.[4]
PCNA in oncogenesis. A PI3K-PCNA complex stimulates DNA synthesis in tumor cells.[5] Phosphorylated RB protein allows cell divisions, and PCNA promotes DNA synthesis in glioma cells; miR-195 suppresses both.[6] PCNA (with IGF-2; MMP-7) promoted growth and metastases of gastric carcinoma cells.[7] APIM (Alk-homologue-2 PCNA-interacting motif) in a cell penetrating peptide (ATX-101) increased chemotherapy-sensitivity of cancer cells and induced apoptotic death of malignant myeloma cells.[8] Human hemangioma (HA) cells are driven by the Akt/mTOR/PCNA pathway. Adenoviral vector-delivered shpRNA induced involution of Has.[9]
High mobility group proteins (HMG). Evolutionarily highly conserved non-histone-, but chromatin-binding proteins engaged in assembly of nucleoprotein complexes, transcription regulation, embryonic development, cell differentiation (Tables XI/I and XI/II). In the amphioxus (*B. tsingtauense*) embryo, HMG proteins direct differentiation of neuroectoderm, mesoderm and mesenchyme.[10ab] The schistosoma HMGAB proteins are 35-40% identical with rodent and human HMG proteins; most active in the hatching eggs.[11] In human cancers, HMGB1 co-expressed with VEGF-C worsened the prognosis of gastric cancer,[12] In basal-like breast cancer, HMGA1 promotes metastases.[13] HMGA2 promotes autonomous growth of prostate cancer cells.[14] Interaction of HMGB1 with its receptor RAGE (receptor for advanced glycation end products) creates the inflammatory milieu for carcinogenesis (prostate cancer induction); in these lesions immune T cells and macrophages are inhibited. In prostate cancer cells, HMGB1 promotes the expression of Ets (E26) oncogene and its recombinations. Monoclonal antibodies and iRNA could target and neutralize HMGB1; HMGB1 may be applied as a vaccine.[15] In its deep evolutionary connection, HMGB1 co-acts with toll-like receptor TLR4 and NFκB.[16] HMGB1 is involved in the pathophysiology of sepsis (in a most contradictory abstracted presentation).[17]
Protein arginine methyltransferases: Evolution. Posttranscriptional methylation, acetylation, phosphorylation of basic histone tails regulate gene expressions. Protein lysine (K) methyltransferases and PRMTs apply asymmetric and symmetric di-, or monomethylations. PRMT transfers from cofactor S-adenosyl-methionine two methyl (CH_3) groups to arginine of histone H4R3 (*vide infra*). It also methylates non-histone proteins (the ER),[18] The large African Kotonkan rhabdovirus genome encodes an arginine methyltransferase.[19] Unicellular eukaryotes operate PRMTs; archaea/prokaryota seemingly lack the enzyme. The early branching Giardia does not operate PRMT.[20a] The Trypanosoma PRMT1 regulates gene expression in the mitochondria.[20bc] PRMTs act in the cytoplasm and in the nuclei posttranscriptionally regulating gene expressions, by methylating ribosomal proteins, thus generating mono- and dimethylarginines. Rmt2 operates in Saccharomyces yeasts, Candida, Aspergillus, Neurospora

(continued)

Table XXVIII/I (continued)

and Cryptococcus, but not in protozoa, and not in the human genome. PRMT1/3/5 are operational in unicellular eukaryotes (Dictyostelium, Entamoeba, Leishmania, Plasmodia, Trypanosoma). PRMT4 is shared by Cryptoccus and Toxoplasma. The Trypanosoma PRMT6 is homologous with the human PRMT6.[20a] The Plasmodium PRMTs form monomethyl-, and asymmetric dimethylarginines. It is a cell survival pathway offering itself to inactivation with anti-parasitic compounds.[21] In the Wnt-pathway, after the Wnt ligand was captured by Frizzled, Dishevelled (isoform Dvl3) is liberated to dimerize with the protein Axin, in the end result of intranuclear localization of β-catenin for the activation of proto-oncogenes Tcf/Lef (T-cell factor; lymphoid enhancer factor). In human embryonic kidney HEK 293 cells, Dvl3 is arginine-methylated by PRMTs. Methylation-deficient Dvl3 stimulates the transcription of proto-oncogenes Tcf/Lef. Thus, arginine-methylated Dvl3 antagonizes the Wnt pathway, and *vice versa* the Wnt pathway antagonizes (disrupts) Dvl3 arginine-methylation.[22] The human genome encodes PRMT6/7/8. The unicellular eukaryotes do not possess the PAD gene (peptidylarginine de-iminase), but the human genome (and fish and Xenopus genomes) encode it. PAD may be able to reverse arginine monomethylation by converting arginine to citrulline.[23]
PRMT in human oncogenesis. H4R3-dimethylation promotes the development of malignant transformation (colon and breast cancer; mixed lineage MLL-leukemia promiscuous fusion oncoprotein generation.[24] Coactivator-associated arginine methyltransferase 1 (CARM/PRMT4) activity is associated with high grade breast cancer generation.[25,26] The Rel/NFκB proto-oncogene is subject to posttranslational phosphorylation, acetylation, lysine methylation and ubiquitination. Its lysins are frequently methylated. Upon exposure to IL-1β, the p65 subunit of NFκB is dimethylated on arginine 30 by PRMT5, and thus overactivated. NFκB liberated from IkB in the cytoplasm, translocates to the nucleus to activate its target oncogenes. By itself and by its associations, PRMT is an oncogene. The suppressor genes FBXL (F box family gene; classified as FBXW, FBXL, FBXO) and GLEA (glioma-expressed antigen) fail to subdue NFκB.[27abc] The pomegranate fruit extract ellagic acid is an inhibitor of CARM1/PRMT4 in blocking methylation of H3R17, a regulator of p53-responsive pro-apoptotic tumor suppressor genes (but not H3R26; neither lysine K14/18 acetylation).[28a] Ellagic acid is ineffective in blocking the cell cycle of p53-deficient HeLa cells. In androgen-independent prostate cancer cells, ellagic acid (ellagitannins; urolithins) suppressed MMP-2 (but not MMP-9) and collagenase secretion, tumor cell proliferation and locomotion.[28bcd]
Cadherins/catenins. Cadherins and catenins are membrane-spanning complexes mediating intercellular binding adherens junctions in metazoans. Plants, fungi and the amoeba Dictyostelium do not encode cadherins/catenins. Catenin/cadherin precursors appear in choanoflagellates "in the waiting" for the evolution of metazoans. The hedgling family proteins hedgling/lefftyrins/coherins (carriers of the hedgehog signals) were present in the common ancestor of choanoflagellates and metazoans, but disappeared from Bilateria. *Nematostella vectensis* still carries the hedgling genes. The 650 million years old *Capsaspora owczarzaki* genome expresses only one cadherin gene, but *Monosiga brevicollis* and *Salpingoeca rosetta* are already enriched with 29 cadherin genes. The sponge, *Oscarella carmela* expresses the classical cytoplasmic cadherin/β-catenin complex.[29] Beta-catenin recruits PRMT2 for asymmetric dimethylation of H3R8, in addition to trimethylated lysine (H3K4me3) in the Xenopus blastula for the dorsal specification by the Wnt/β-catenin pathway in the tadpole.[30] The Wnt pathway begins with its ligand binding coreceptors Frizzled (Fzd) and LRP (low-density lipoprotein receptor-related protein). If Disheveled (Dvl) receives no signals, the cytoplasmic β-catenin is phosphorylatated in the APC/Axin/GSK complex and sent to ubiquitination. If Dvl receives the signal, the APC/Axin/GSK complex is disabled and the intact β-catenin travels to the nucleus to activate its target genes Tcf/Lef in a cell with its dickkopf DKK5 and WIF (Wnt-inhibitory factor, sequestering Wnt ligands, blocking ligand-to-Fzd binding) gene-product proteins already downregulated (Table IV). In squamous cell carcinoma of the lung, the Wnt pathway deviates

(continued)

Table XXVIII/I (continued)

from its canonical course. Dominant initiators are Wnt5a/Wnt11 in squamous, and Wnt7ab/Wnt11 in adenocarcinoma induction. Wnt11 eliminates focal adhesion protein paxillin. Overexpressed is Fzd-3 receptor (for Wnt5a) in adenocarcinoma, and Fzd-10 receptor (for Wnt7ab) in squamous cell carcinoma (in a somewhat contradictory finding).[31,32] *Fusobacterium nucleatum* is an inducer of the cadherin/β-catenin Wnt pathway in the colon, thus a promoter of colon carcinogenesis.[33]

1 *WinterJA & Bunting KA Archaea 2012;95101.* ***2*** *Chia et al PLoS One 2010;5(6):e10866.* ***3*** *Juhasz Sz Nucleic Acids Res 2012;40:10795-808.* ***4*** *Cardona-Felix CS et al Acta Crystallogr Biol 2011;67:497-505.* ***5*** *Wang G et al Mol Cancer Ther 2013;12:2100-9.* ***6*** *Hui W et al PLoS One 2013;8(1):e54932.* ***7*** *Kuang R-G et al Turkish J Gastroenterol 2012;24:99-108.* ***8*** *Müller R et al PLoS One 2013;8(7);e70430.* ***9*** *Ou JM et al Int J Immunopathol Pharmacol 2012;25:945-53.* ***10a*** *Liu Z et al Comp Biochem Physiol B Biochem Mol Biol 2004;137:131-8.* ***10b*** *Huang X et al Int J Dev Biol 2005;49:49-52.* ***11*** *Gnanasekar M et al Mol Biochem Parasitol 2006;145:137-46.* ***12*** *He W et al World J Surg Oncol 2013;11:161.* ***13*** *Peoraro S et al Oncotarget 2013;4:1293-308.* ***14*** *Müller MH et al Anticancer Res 2013;33:3069-48.* ***15*** *Gnanasekar M et al Prostate Cancer 2013;157103.* ***16*** *Weng H et al BMC Cancer 2013;13:311.* ***17*** *Diener KR Immunol Cell Biol 2013;91:443-50.* ***18*** *Zhang R et al PLoS One 2013;8(8):e72424.* ***19*** *Blasdell KR et al Virology 2012;425:143-53.* ***20a*** *Fisk JC et al J Biol Chem 2009;284:11590-600; Eukaryot Cell 2010;9:866-77; 2011;10:1013-22; Mol Cell Proteomics 2013;12:302-11.* ***20b*** *Goulah CC et al RNA 2006;12:1545-55.* ***20c*** *Pelletier M et al Mol Biochem Parasitol 2001;118:49-59; 2005;144:206-17.* ***21*** *Fan Q-i et al Biochem J 2009;421:107-18.* ***22*** *Bikkavilli RK et al Sci Rep 2012;2:805.* ***23*** *Bachand F Eukaryot Cell 2007;6:889-98.* ***24*** *Zhang R et al PLoS One 2013:8(8):e72424.* ***25*** *Cheng H et al Diagn Pathol 2013;8(1):129.* ***26*** *Davis MB et al Mol Cancer 2013;12(1):40.* ***27a*** *Wei H et al PNAS USA 2013;110:13516-21.* ***27b*** *Cui Y et al EMBO J 2011;30:2675-89.* ***27c*** *Fischer U et al Clin Exp Immunol 2001;126:206-13.* ***28a*** *Selvi BR J Biol Chem 2010;285:7143-52.* ***28b*** *Pitchakarn P et al Asian Pac Cancer Prev 2013;14:2859-63.* ***28c*** *Vicinanza R et al Evid Based Complement Alternat Med 2013;2013:3475o4.* ***28d*** *Vanella L et al Curr Pharm Des 20q13;19:2728-36.* ***29*** *Nichols SA et al PNAS USA 2012;109:13046-51.* ***30*** *Blythe SA et al Dev Cell 2010;19:220-31.* ***31*** *Bartis et al PLoS One 2013;8(3):e57393.* ***32*** *Sánchez-Hernández D et al Development 2012;139:3849-58.* ***33*** *Rubinstein MR et al Cell Host Microbe 2013;14:195-206.*

The Monosiga choanoflagellates release their *pak* gene, which renders vertebrate mammalian cells to be constitutive expressors of the PAK oncoprotein, that is transforming them [1792a, b]. Their other Scr/Abl-like tyrosine kinases activate the STAT proto-oncogenes. The virulence, vitality and pathogenicity of certain oomycetes (Phytophthora) depend on physiologically evolved genes, which have become proto-oncogenes in the vertebrate mammalian host cells (Figure 56; Table XXX). The ctenophores (Mnemiopsis; Pleurobrachia) operate a genome that would fit into a malignantly transformed cell of a mammalian vertebrate host. The ctenophore cells physiologically miss axins, and thus promote the transfer of β-catenin from cytoplasm to nucleus, where it activates the ancestral proto-oncogenes (tcfTCF/lefLEF). The suppressor of the Wnt pathway, the dickkopf, is non-existent. Instead, some four sox stem cell proto-oncogenes are operational (reference to article in Int J Oncology 2015, in the Appendix).

The ultimate bioengineers RNA/DNA are not exempt of chaotic malfunctions, such as congenital malformations in the embryo. One prominent example is reviewed in Table XXXI. Observe that oncogenesis and oncoproteins do not arise

Table XXVIII/II A List of Ancient "Cell Survival Pathway" Genes Which Have Become Proto-Oncogenes Encoding Oncoproteins in Vertebrate Mammalian Hosts, in Particular Homo. Not Oncoprotein Itself, but Oncoprotein Chaperon and Protector: HSP

Primordial Heat Shock Proteins (HSP) rescue the Archaeoglobus fulgidus and the first Eukaryotes. Cellular life on the radiation-bombarded exterior and the overheated upper inner boundaries of the ancient Earth's surface, in the overheated underground, in deep oceanic vents, in boiling water, next to pouring lava, and in hot volcanic fumes would not have been possible without chaperon and stress proteins that aligned with physiologically essential proteins to assist in their proper folding, and extended protection against physicochemical harm. Archaea and prokaryota rapidly generated such proteins and exchanged their genes through horizontal routes. The *Thermococcus kodakarensis* grows at 85°C, but possesses not only heat-inducible, but also a cold-inducible chaperonin.[1] The Group I HSP chaperonins prevent the denaturation, or the aggregation of, denatured proteins. The central cavity of the two rings of the large cylindrical chaperonins take in the misfolded proteins for ATP-dependent unfolding and productive re-folding.[2] Thaumarchaeota possessed the protective proteins DNA-K/J, DNA-J-Fer (ferredoxin domain) and the nucleotide exchange factor GroEL. The DNA-K protein is the ancestor of the eukaryotic HSP70 kD. Archaea and bacteria encoded this protein from a single gene. The nuclear genome of *Chlamydomas reinhardtii* encodes five HSP70 isotopes; some of the genes are of mitochondrial or chloroplast origin. These genes were spreading through horizontal routes, underwent frequent duplications and were inherited by the Viridiplantae and Animalia.[3abcde] ATP-driven chaperonins bring misfolded proteins to their native state.[4] The GroEL/S chaperonins are used by bacteriophages for the proper folding of their capsid proteins. The *Pseudomonas aeruginosa* giant bacteriophage EL (from Russian cyrillic capital Л for lambda; also Greek lambda λ) gene encodes the synthesis of GroEL.[2] The Paramecia (*P. caudatum*) and the Tetrahymena show that HSP70 chaperons evolved into three to four families, encoding these proteins from eleven genes through the mRNA routes. These genes are the results of duplications, functional conservation and purifying selection.[5] The pathogenic strains of *Candida albicans* use a HSP21 chaperon-stress molecule for their defense against host immunity and antifungal drugs.[6] The Leishmania operates 20 HSP genes.[7] In the sea urchin, constitutively encoded HSP70 chaperons the assembling and disassembling of the mitotic apparatus.[8] Constitutively promoting cell divisions? A HSP56 chaperonin of mitochondrial derivation is activated in sea urchin embryo; HSP70 and HSP75 are also operational.[9ab] The amphioxus is well equipped with heat inducible HSP70s.[10] Molecular evolution an co-evolutionary events in eukaryotes are extensively reviewed.[11ab] It was the accidentally overheated drosophila larvae that revealed in 1962 for the first time the existence of HSPs.[12]

HSPs: Protectors of Oncoproteins. In the cancer cell, ancient encounters of unicellular ancestors with heavy metals is re-played. Cadmium induces apoptosis in sea urchin cells. HSPs protect the cell against apoptotic death.[13ab] The native inhabitants of contaminated environments defend themselves by HSP activations: oyters, copepods, and fish; insects.[14abcd] This author envisions the intranuclear entry of specifically DNA-binding oncoproteins for the activation of the HSP genes; the amplified HSP genes overproduce their protein product proteins. Proteins regulated and protected by the HSPs are referred to as "clients." Oncoproteins become HSP clients. Elevated HSP levels in a eukaryotic cell of a multicellular host signifies underlying malignant transformation in that cell. Figure 1 of the cited article depicts the process of HSP-recruitment in a stressed or in a malignantly transforming cell. Figure 2 shows circulating HSPs in the blood of cancer-bearing patients and that the amplified expression of these proteins induces autoantibody responses. Tumor-derived exosomes and circulating tumor cells coexist in these blood samples. Here, HSP27, HSP70, HSP90 re discussed separately.[15] The level of HSP70 expression in a cancer-bearing host is directly proportional with the adversity of the prognosis.[16] HSPs are always included in the long list of oncoproteins and in the processes of stem cell reprogramming in oncogenesis including the process of EMT.[17] In the mouse, HSPs are

(continued)

Table XXVIII/II (continued)

expressed in the embryo and in embryonal carcinoma cells.[18] In neuroendorine tumors of the small intestine, in the paraphernalia of oncogenes/oncoproteins (PI3K/Akt/mTOR; SRC, SMAD, EGF-R, PDGR-R, *etc*), HSP90 is prominently present.[19] In prostate cancer, stem cell antigens and HSPs co-exist.[20] HSP are targeted for cancer therapy at many levels: small molecular inhibitors (SNX5422/2112, patented) of the protein in squamous cell carcinoma;[21] dimethyl-aminoethoamino-17-demethoxy geldanamycin for osteogenic sarcoma;[22] HSP protein vaccine including chaperone-rich cell lysates.[23abcdef] HSPs are naturally immunostimulatory for innate and adaptive immune faculties.[24]

HSP may antagonize hyperthermic therapy. Protocols for hyperthermic intraperitioneal chemotherapy for peritoneal carcinomatosis are in use. In response, intratumoral up-regulation of HSP70/72 and HSP90 occurred. At balance are tumor cell protection *versus* potential host defensive responses favorably activated by elevated temperatures. Variable temperature levels in proper sequence are essential for anti-tumor effect.[25] Antisense oligonucleotides knocking out HSP27/HSP90 in anti-hormone-refractory prostatic tumor tissue failed to induce tumor cell death.[26] In response to HSF (transcription factor) ablation, the multidrug resistance gene is activated. An ancient reaction is activated: the P-glycoprotein ATP-binding ABC cassettes expel intracellular toxins (doxorubicin). The P-gp antagonist verapamil reverses the process (retains doxorubicin loading). P-gp protects the heart against doxorubin toxicity; high HSF levels protect the tumor cells and render defenseless the cardiomyocyte. Down-regulated HSF protects the heart and exposes the tumor cell to load the chemotherapeutical.[27] HSF promotes malignant behavior of myeloma cells by increasing expression of HSPs27/40/70/90. HSF inhibition by siRNA or triptolide deprives the myeloma cell of HSP protection; such myeloma cells may succumb to apoptosis.[28] HSPs are cell survival factors that cancer cells expropriated. Working in non-malignant cells of ageing multicellular hosts, HSPs act as longevity agents. HSF is the physiological promoter of HSPs. Figure 1 of the article cited depicts the mechanisms of HSPs exerting protection from protein damage in aging cells.[29]

Ansamycin/allylamino-17-demethoxy-geldanamycin [17-AAG]: bind and inactivate HSP90 and HER2; tumor cell highly overexpressing HER2 are more susceptible to 17-AAG.[30]

1 Gao L et al Appl Environ Microbiol 2012;78:3806-15. ***2*** *Kurochkina LP et al J Virol 2012;86:10103-11.* ***3abcdf*** *Alberti S et al Cell Stress Chaperones 2003;8:225-31; Brochier-Armanet et al Nat Rev Microbiol 2008;6:245-52; Nordhues AS et al Int Rev Cell Mol Biol 2010;285:75-113; Petitjean C et al BMC Evol Biol 2012;12:226; Rohlin L et al J Bacteriol 2005;187:6046-57; Waters ER J Exp Bot 2013;64:391-403.* ***4*** *Jayasinghe M et al Biophys J 2013;103:1285-95.* ***5*** *Krenek S et al BMC Evol Biol 2013;13:49.* ***6*** *Mayer FL et al PLoS One 2013;8(3):e60417; Virulence 2013;4:119-128.* ***7*** *Fraga J et al Infect Genet Evol 2013;18:229-37.* ***8*** *Aguelli C et al Biochem J 2001;360:413-9.* ***9a*** *Roccheri NC et al J Submicrosc Cytol Pathol 1993;25:173-9; Biochem Biophys Res Commun 1997;234:646-50; 2001;287:1093-8; 2004;321:80-7.* ***9b*** *Sconzo G et al Cell Differ 1988;24:97-104; Biochem Biophys Res Commun 1997;234:24-9.* ***10*** *Li D et al Gene 2012;510:39-46.* ***11a*** *Carretero-Paulet L et al Mol Biol Evol 2013;30:2035-43.* ***11b*** *Ruiz-González- MX & Fares MA BMC Evol Biol 2013;13:156.* ***12*** *Ritossa P Experientia 1962;18:571-3.* ***13a*** *Agneilo M et al Cell Stress Chaperones 2007;12:44-50.* ***13b*** *Roccheri NC et al Biochem Biophys Res Commun 2004;321:80-7.* ***14a*** *Ivanina AV et al Aquat Toxicol 2009;91:245-54.* ***14b*** *Rhee JS et al Comp Biochem Physiol C Toxicol Pharmacol 2009;149:104-12.* ***14c*** *Padmini E Rev Environ Contam Toxicol 2010;206:1-27.* ***14d*** *Morales M et al Comp Biochem Physiol C Toxicol Pharmacol 2001;153:150-8.* ***14e*** *Singh MP Chemosphere 2010;79:577-87.* ***15*** *Seigneuric R et al Front Oncol 2011;1:37.* ***16*** *Murphy ME Carcinogenesis 2013;34:1181-8.* ***17*** *Mimeault M & Batra K Mol Aspects Med 2013 pii S0098-2997(13)00069-9.* ***18*** *Morange M et al Mol Cell Biol 1984;4:730-6.* ***19*** *Banck NS et al J Clin Invest 2013;123:2502-8.* ***20*** *Dong L et al Exp Ther Med 2013;5:1161-4.* ***21*** *Friedman JA et al Transl Oncol 2013;6:429-41.* ***22*** *Gorska M et al PLoS*

Table XXVIII/II (continued)

One 2013;8(8):e71135. ***23abcdef*** *Dong L et al Exp TherMed 2013;5:1161-4; Cancer Biother Radiopharm 2013;28:391-7; Garcia-Carbonero R et al Lancet Oncol 2013;124:e358-69; Mayer-Sonnenfeld T et al Int J Hyperthermia 2013;29:520-7; McNulty S et al Immunology 2013;139:407-15; Wang R et al Chem Biol Interact 2013;205:1-10; Weng D et al J Immunol 2013;191:755-63; J Immunol 2013;191:755-63.* ***24*** *Colaco CA et al Biomed Res Int 2013;2013:461230.* ***25*** *Pelz JO et al Ann Surg Oncol 2013;20:1105-13.* ***26*** *Ischia J et al Curr Opin Urol 2013;23:194-200.* ***27*** *Krishnamurthy K et al PNAS USA 2012;109:9023-8.* ***28*** *Heimberger T et al Br J Haematol 2013;160:465-76.* ***29*** *Calderwood SK et al Gerontology 2009;55:550-8.* ***30*** *Münster PN et al Cancer Res 2002;62:3132-7*

Table XXIX Malignantly Transformed Cells Expropriate the Systems Evolved in Their Ancestors to Serve as Organ Anlagen and Growth Promoters

Achaete scute: Evolution. Discovered in the drosophila pupae, four helix-loop-helix protein-encoding genes: chaeta, macrochaetae, sensory organ bristles; (no bristle, achaete), scute (scutum, horny plate, cuticule), lethal of scute (*l'sc),* and asense (bristle promoting) encoded the peripheral (chaetae) and central nervous (*l'sc*) systems.[1a] The achaete/scute regions are surrounded in the Calliphora genome by transposable elements (LTR and non-LTR retroposon, LINEs; cut&paste DNA transposons; helitron rolling circle transposons, and several unclassified repeats). Mariner elements and others arrived into the Calliphora genome through horizontal transfer from ancestors of other fly or ant species (*Anopheles gambiae; Drosophila mojavensis, Camponeatus floridanus*).[1b] Cnidarians and the first bilaterians harbored the first HLH-encoding nerve cells (what the sponges tried, the AmqHLH1 gene, but failed to evolve it further). The sea anemone Nematostella possesses an achaete/scute homolog gene (Ash). The evolving bilaterians left the cnidarians behind, and developed a central nervous system. In the amphioxus, the epidermal sensory neurons (ESN) develop from epidermal cells of the ventral ectoderm under the effect of Delta/Notch. Bone morphogenetic protein (BMP) converts these cells into a network. These cells connect with the notochord by long axons which transmit GABA (γ-aminobutyric acid). Ash is co-expressed with Delta in the ventral ectoderm; the basic HLH protein is of Ash-derivation. Amphi*Tlx* (tailless X, drosophila; BMP gene in amphioxus) and pan-neuronal marker gene Hu/*elav* (Hu antigen protein; embryonic lethal abnormal visual protein, drosophila) are co-expressed in ventral ectodermal cells entering early neuronal stage. The *ladybird* early/late genes direct drosophila neuro-, myo- and cardiogenesis.[1c2abcd] The amphioxus genome holds the cluster of the NK (Nirenberg and Kim) homeobox genes which were discovered in the drosophila (tinman, *tin*, NK4; bagpipe. *bap*, NK3; ladybird early, ladybird late, *lbe, lbl;* slouch, *slou*, NK1). These genes travelled through drosophila, pufferfish (Fugu) and Homo in >1,500 my (515 my from cephalochordate to vertebrates; then 420 my to Fugu, then 70 my to mouse). In the human genome, BAPX for *bap*, LBX for *ladybird*, NKx for *slou*, NKX for *tin, and Tlx/Lbx (*drosophila *tailless, ladybird)* remain inserted. Despite evolutionary dispersal of the NK cluster, NK3/4 and Tlx/Lbx remain retained together.[3]

Achaete scute: Oncogenesis. Parathyroid hormone-related peptide is a growth factor enlisted by human lung adenocarcinoma cells; the HuR protein (RNA-binding Hu protein Hu/elav class) upregulates growth factor PTHrP. Knockdown of HuR by anti-mRNA siRNA resulted in cessation of tumor growth with release of caspase-3 inducing apoptotic death of tumor cells.[4] The TLX protein is a nuclear receptor (NR, tail-less, drosophila) driving tumor stem cell proliferation in gliomagenesis, when p53 and/or PTEN suffer deletion. TLX acts as oncogene for neural stem cells and neuroblastoma cell.[5ab] Helix-loop-helix proteins feed sinonasal neoplasms; *achaete scute* gene induces primary olfactory esthesio-neuroblastomas; trans-speciates adeno-

(continued)

Table XXIX (continued)

carcinomas into relapsing neuroectodermal chemo-radiotherapy-resistant tumors.[6abc] In small cell lung cancer the neuroectodermal cells express $CD133^{hi}$, which is the achaete scute HLH protein ASCL1. Malignancy depended upon the continuous expression of $CD133^{hi}$. RNA interference against ASCL1 induced tumor cell apoptosis.[7] Bronchial epithelial cells of transgenic mice constitutively expressing ASCL1 resisted tobacco-specific nitrosamine-induced DNA damage, and cisplatin-induced apoptosis due to DNA repair induced by MMP-7, and MGMT (methylguanine-DNA methyltransferase) expression. Increasd secretion of MMP-7 and MGMT was due to direct stimulation of the gene promoters by ASCL1/HLH proteins (a controversial finding).[8]

PI3K/Akt. Phosphatidylinositol kinase-3 serves in cell survival pathways and as a virulence enhancing factor in the unicellular eukaryotes Cryptococcus, Phytophthora, Entamoeba, Giardia, Trypanosoma. Its inhibitors (LY2940002, wortmannin) stop locomotion, phagocytosis; induce encystation.[9] In advanced vertebrate mammalian hosts, PI3K is an oncogene suppressed by PTEN (Table VIII; Figure 50). PI3K/Akt/mTOR frequently escapes from under suppression. PI3K/Akt and Wnt/β-catenin (Table IV) induce the conversion of nonside population of cells in tumors into cancer stem-like population cells through phosphorylation of GSK3β, and preservation of β-catenin for its intranuclear translocation.[10] Wnt ligands activate Fzd and EGF-receptors; the non-canonical EGF-R activator is TGFα. Activated Fzl trans-activates EGF-R by MMP-mediated release of soluble EGF-R ligands. An EGF-R-to-PI3K/Akt (Jacob Fürth's AK mice retroviral thymic lymphoma c-*onc*→v-*onc*) pathway exists. Activated EGF-R phosphorylates β-catenin separating it from E-cadherin; β-catenin by-passing the GSK-complex constitutively activates its target oncogenes Tcf/Lef.[11] The Tcf/TCL (TC12-like) proto-oncogene is a more recent spinoff of the ancient ras GTPase gene superfamily; the ancestral TC10 gene renders physiological service to the protochordate sea squirt (*Ciona intestinalis*). By-passing insects and flies, the TC10 genes followed the vertebrate lineages, assuming the status of a proto-oncogene in the human genome.[11–13] Constitutive expression of TCF/LEF in human hematopoietic malignant cells (Burkitt's, histiocytic, mantle cell lymphoma, Jurkat acute T cell leukemia, K562 CML; primary human lymphoma cells) may occur independently from β-catenin, through activating transcription factor-2 effect.[14] History of the gene.[15]

Ligands-to-growth factor receptors. EGF&R; CSF&R (Table XXV); Erythropoietin EPO&R; FGF&R; Hepatocyte GF; Insulin-like GF&R; Keratinocyte GF&R; PDGF&R; VEGF&R. Malignantly transformed cells commonly expropriate and constitutively express the physiological GFs in either para- or autocrine circuitries. Examples: EPO→R and PDGF→R induce tumor cell (adenocarcinomas; sarcomas) mitoses.[16,17] Cells of tumor cell line HEPA1-6 migrate due to stimulation by EGF and cMet activation by HGF. Tetraspanin protein CD82 (Table XXIV) with active ganglioside GM3 attenuated EGFR and cMet, thus bringing HEPA1-6 cells to standstill.[18] Hormones and their receptors ER, PR, AR serve tumor cells for stimulation of growth. Examples: estrogens (breast cancer) and androgens (prostate cancer). Amplified EGF-R, HER2, FGF-R and Met proteins drive esophageal squamous cell carcinoma.[19a] FGF-R, EGF-R, HER2, and HGF-ligand drive H&N squamous cell carcinoma.[19b] Cabozantinib inhibits Met and VEGF-R in prostate cancer bone metastases.[20] Secreted proteoglycan decorin is a multi-GF inhibitor for EGF, HGF, IGF, PDGF.[21] The EGF→R unit recruits breast cancer promoting macrophages.[22]

Transforming growth factors TGFαβ. Physiological regulators of cell growth in embryogenesis and innate immunity. TGFβ-induced tolerogenesis in the cnidarian Aiptasia:* it suppressed LPS-induced nitric oxide reaction; and promoted acceptance of a dinoflagellate symbiont.[23] Caenorhabditis gene *daf7* is aTGFαβ.[24ab] Drosophila decapentaplegic is a TGFβ. Physiological antagonist of TGFβ is inhibin.[25] Antiestrogens (in breast cancer) and antiandrogens (in prostate cancer) suppress TGFβ.[26ab] The innate immune response regulator chloride intracellular channel activates TGFβ; in turn, TGFβ converts myofibroblasts into tumor-supportive cells. With Notch

(continued)

Table XXIX (continued)

and Smad 3 interacting, TGFβ generates thyroid cancer-associated fibroblasts.[27ab] TGFβ does not directly induce malignant transformation, but it supports tumor cells already induced.[28] ***TNFα and NFκB***. LPS induce a TNFα gene to react in the amphioxus. The Rel/NFκB genes/proteins are also present. The lancelots operate a highly advanced innate immune system.[29ab] In the shrimp, a defensive TNF-R superfamily exists to respond to viral (white spot syndrome virus), bacterial (*Vibrio alginolyticus*) and fungal (Candida) intrusions.[29c] The famous substance, TNF, originally causing prompt hemorrhagic necrosis in tumors, concealed not for too long its other properties, that is sustaining tumors within inflammasomes and in neoangiogenetic networks.[30] IFNγ enforced the efficacy of TNF-related apoptosis-inducing ligand (TRAIL) for apoptosis-induction in Ewing's sarcoma cells.[31] Rel/NFκB homology domain transcription factors are operational in the cnidaria Nematostella.[32] The Scleractinia corals and the Symbiodinia dinoflagellates live in symbiosis in the Caribbean sea. Their efficient innate immune system includes NFκB and MAP kinases[33], but TGFβ is tolerogenic.[34] The sea urchin coelomocytes mobilize NFκB and runt domain factor-1 in response to bacterial entry; switch off the immunosuppressive factor, GATA-2/3.[35]
Malignant cell-supporting inteferons, interleukins and chemokines. IFNγ protects CLL cells against apoptosis and acts as a CLL cell growth factor (GF). The stromal cell-derived CXC chemokine attracts CLL cells to fibroblasts, in which they take up residence and divide; CLL cells secrete TGFβ to induce tolerance for them inside the fibroblasts.[36] Melanoma cells may express IL-2R for capture of IL-2, to act as a paracrine GF.[36] The USA NCI developed a humanized mcab (Hu-Mikβ1) against CD122 fused IL-2/IL-15R to block attachment of growth stimulatory ligands to the receptor expressed by large granular T-lymphocytic leukemia cells; the mcab failed to block IL-15 and was clinically ineffective.[37] In enteropathy-associated (celiac disease) T cell lymphoma, IL-15R signals JAK3/STAT5, ERK, PI3K, and Bcl-2/BclxL pathways. Another IL-15-specific immune globulin blocked JAK3/STAT5 phosphorylation, and induced massive apoptosis of intraepithelial T-cells.[38] Mantle cell lymphoma cells express receptors IL-6^+gp80^+ for intrinsic or extrinsic circuitries of IL-6; IL-6-bound receptor signals JAK2/STAT3, and PI3k/Akt pathways, and renders mantle lymphoma cells chemotherapy-resistant. Knockdown of the receptor suppressed oncoprotein signaling and rendered lymphoma cells chemotherapy susceptible.[39] Paracrine IL-6 sustains plasma cell malignancies and hepatocellular carcinoma cells.[40] Tumor cells, including large B cell lymphoma, secrete IL-35 that recruits CD11b^+/Gr1^+ myeloid cells, which suppress anti-tumor immune reactions and induce neoangiogenesis.[41] The poly(ADP-ribos)ylation enzyme macro-PARP-14 (distinct from PARP-1, Table II) with IL-4 induces STAT6 and exerts anti-apoptotic effects in B lymphocytes undergoing transformation in lymphomagenesis.[42] Chemokine ligand CXCL13 and IL-10 reach high levels in the CS-fluid of patients with brain lymphoma.[43] Chemokine CXXC5 phosphorylating Smad (signaling mothers against decapentaplegic, TGFβ, drosophila) is a TNFα and caspase-3 activator, apoptosis-inducer, tumor suppressor.[44] In patients with diffuse large B-cell lymphomas, the NK cells are deficient in CD16 Fcγ receptor expression and thus in performing the ADCC reaction.[45] CLL cells secreting soluble BAG and BAT ligands for NK cell receptor NKp30 saturated and inactivated the receptor. When the ligand was expressed on the 50–100 nm microvesicle exosome, it activated the NK cell. NK cell receptor for BAG/BAT is NKp30; for PCNA, NKp44, for vimentin, NKp46. BAT is HLA-B-associated transcript; BAG is Bcl-2-associated gene-athano (athanogene). BAG6 levels are high in advanced CLL. High chaperone (heat shock) protein-associated BAG5 levels in prostate cancer inhibited apoptosis. BAG proteins are considered to be treatment targets for CLL, and adeno- and hepatocarcinomas.[46abcdefg]
1a *Garcia-Bellido A & de Celis JF Genetics 2009;182:631–9.* ***1b*** *Lu T-M et al Development 2012;139:2020–30.* ***1c*** *Negre B and Simpson P Mob DNA 2013;4:13.* ***2a*** *Balakirev ES et al PLoS One 2011;6(7):22613.* ***2b*** *Satoh G et al J Exp Zool 2001;291:354–64.* ***2c*** *Stepchenko AG*

(continued)

Table XXIX (continued)

et al Nucleic Acids Res 2011;39:5401–11. **2d** *Wakamatsu Y & Weston JA Development 1997;124:3449–60.* **3a** *Kim Y & Nirenberg M PNAS USA 1989;86:7716–20.* **3b** *Luke GN et al PNAS USA 2003;100:5292–5.* **4** *Lauriola L et al Histol Histopathol 2013;28:1205–16.* **5a** *Zou Y Mol Cell Biol 2012;32:4811–20.* **5b** *Zeng ZJ et al Biol Open 2012;1:527–535.* **6a** *Cordes B et al Hum Pathol 2009;40:283–92.* **6b** *Rostomily RC et al Cancer Res 1997;57:3526–31.* **6c** *Carney ME J Neurooncol 1995;26:35–43.* **7** *Jiang T et al Cancer Res 2009;69:845–54.* **8** *Wang XY et al PLoS One 2012;7(12):e52832.* **9** *Bittencourt-Silvestre J et al Arch Microbiol 2010;192:259–65; Cox SS et al BMC Microbiol 2006;18;6:45; Diaz-Gonzalez R et al PLoS Negl Trop Dis 2011; 5(8):e1297; Hernandez Y et al J Eukaryot Microbiol 2007; 54:29–32; Hu G et al J Clin Invest 2008;118:1186–97; Lu S et al Mol Plant 2013;6:1592–604; Schoijet AC et al J Biol Chem 2008;283:31541–50.* **10** *He K et al Int J Cancer 2013 in print.* **11** *HuT & Li C Mol Cancer 2010;9:236.* **12** *Sasakura et al Dev Genes Evol 2003;213:273–83* **13** *Vignal et al Mol Cancer2010;9:236.* **14** *Grumolato L et al PLoS Genet 2013;9(8):e1003603.* **15** *German S et al Hered Cancer Clin Pract 2013; 11(1):5.* **16** *Acs G & Acs P et al Cancer Res 2001;61;3561–6.* **17** *Sinkovics JG Chapter 2. In: Viral Therapy Human Cancers. Marcel Dekker, NY, 2005 Pp 149–221.* **18** *Li Y et al Int J Biochem Cell Biol 2013 in print.* **19a** *Kato H et al Int J Oncol 2013;42;1151–8.* **19b** *Singleton KR et al Mol Pharmacol 2013;83:882–93.* **20** *Lee RJ & Smith MR Cancer J 2013;19:90–8.* **21** *Baghy K et al FEBS J280:2150–64.* **22** *Nickerson NK et al Oncol Res 2013;20:303–17.* **23** *Detournay O et al Dev Comp Immunol 2012;38:525–37.* **24a** *Koga et al Development 1999;126:5387–98.* **24b** *Myers EN PLoS One 2012;7(7):e40368.* **25** *Hu J et al PLoS One2010;5(3):e9457.* **26a** *Popp SL et al Anti-cancer Res 2013;33:2473–80.* **26b** *Yamada D et al Int J Urol 2013;20:1220–7.* **27a** *Shukla A et al Oncogene 2014;33:842–50.* **27b** Zhang J et al Tumour Biol 2014;35:2795–802. **28** *Mu X et al PLoS One 2013;8(5):e63436.* **29a** *Jin P et al Fish Shellfish Immunol 2012;32:123–8.* **29b** *Yuan S t al J Immunol 2013;191:1220–30.* **29c** *Wang PH et al Dev Comp Immunol 2012;36:39–50.* **30** *Waters JP et al J Pathol 2013;230:241–8* **31** *Merchant MS et al Cancer Res 2004;64:8349–56.* **32** *Sullivan JC et al Dev Genes Evol 2007;217:63–72.* **33** *Schwarz JA BMC Genomics 2008;9:97.* **34** *Detournay O et al Dev Comp Immunol 2012;38:525–37.* **35** *Pancer Z et al Immunogenetics 1999;49:773–86.* **36** *Sinkovics JG & Horvath JC Int J Oncol 2001;19:473–88. Erratum: IJO 2005;27:33.* **37** *Waldmann TA et al Blood 2013;121:476–8.* **38** *Malamud G et al J Clin Invest 2010; 120:2131–43.* **39** *Zhang L et al Blood 2012;120:3783–92.* **40** *Gilbert LA & Hemann MT Genes Dev 2012;26:1758–68.* **41** *Wang Z et al J Immunol 2113;190:2415–23.* **42** *Cho SH et al Blood 2009;113:2416–25.* **43** *Rubenstein JL et al Blood 2013;121:4740–8.* **44** *Wang X et al Curr Mol Med 2013;13:1385–96.* **45** *Danielou-Kazareth A et al Eur J immunol 2013;43:1383–8.* **46abcdefg** *Behl C Autophagy 2011;7:795–8; Bruchmann A et al BMC Cancer 2013;p13:96; Ni W et al J Cell Biochem 2013;114:2120–30; Reiners KS Blood 2013;121:3658–65; Pogge von Strandmann E Immunity 2007;27:965–74; Yang X et al Histol Histopathol, 2013;28:1147–56; Zhu H et al Ann Hematol 2014;93:425–35;* *see copyrighted images in Google.

in the chaos; it appears as if oncogenesis required a special platform on which it advances stepwise sequentially.

Table XXXII approaches the mitochondria. Before their capture to become endosymbionts, they were independent proteobacteria dividing incessantly without senescence and natural death. In their host eukaryotic cells do they contribute to immortality (that is "malignant transformation"), or induce cell death (intrinsic mitochondrial apoptosis) (Figure 23). Table XXXIII confronts the problem of pediatric acute leukemias. Are as yet hidden human retroviruses involved and therefore it has the characteristic pathogenesis that of extrinsically induced malignancies? Table XXXIV deals with a clinical observation of experienced

Table XXX The Birth of Oncogenes. Transposon Mutagenesis → Oncogenesis

Classical proto-oncogenes. Retrovirologists isolating new leukemo- and oncogenic retroviruses in the 1960s discovered the host cell oncogenes inserted into the viral genome: *c-onc→v-onc*. These oncogenes expressed constitutively by the replicating retrovirus induced leukemias, sarcomas and other solid tumors. All oncology textbooks print long lists of these oncogenes/oncoproteins, of which *src,* myc and *ras* are the most ancestral. The choanoflagellate *Monosiga ovata* carries the ancestor of the *pak*/PAK proto-oncogene/oncoprotein as a physiological cell survival factor.[1,2] Pak (p21-activated kinase) travels upward on the evolutionary ladder through Ephydatia sponges, and fungi, reaching vertebrate mammalians. In the human genome, the prolactin receptor gene co-activates either one, or more than one kinases: the MAPK/ERK, PI3K/Akt, or JAK2/STAT5 pathways, and/or the Src family tyrosine kinases (SFK). The activity of the PAK pathway is PI3K-dependent. If the breast tissue contains a dormant tumor, these prolactin-induced kinases will enhance its proliferation and invasiveness.[3] The BAD (Bcl-2 antagonist domain) protein is pro-apoptotic (by heterodimerizing with Bcl-2/xL), except when it is phosphorylated at serines 111/112. Oncoproteins phosphorylate BAD to inactivate it: Akt phosphorylates BAD serine 136; protein kinase A phosphorylates BAD BH3 domain serine 155; and Pak directly phosphorylates BAD serine 111, and serine 112 with oncoprotein Raf.[4abc] Melanoma cells strongly overexpress PAK1 protein in the cytoplasm together with protein BRAF (B-raf rat fibrosarcoma oncogene/oncoprotein). Inhibition of PAK mRNA by miRNA, or by its special inhibitor PF-3758309 (patented) resulted in the cessation of growth of its xenografts.[5] The constitutively expressed Monosiga Pak gene encoded its protein kinase in rat fibroblasts, and in human epithelial cells and induced phenotypical and biochemical criteria of transformation. The MoPak carrier rat cells grew as tumor in nude mice.[2] Of its 128 tyrosine kinase genes, the choanoflagellate *M. brevicollis* expresses four Src homologs of the Rous sarcoma virus with the c-Src terminal kinase (*MbSrc1-4;MbCsk*). In the phosphorylation of the terminal tyrosine of MbSrc1 by MbCsk, the latter fails to shut down the former, in contrast to the mammalian gene. However, these genes were functional in cell survival pathways before the split between choanoflagellates and metazoans.[6] Fibroblasts from vertebrate hosts (rats) transfected with the *M. ovata* src-gene undergo "malignant transformation": rounded-up individually and formed colonies in soft agar.[7] For example, the Src kinase phosphorylates catenin and thus disrupts cell-to-cell adhesions and interactions. Autophosphorylated tyrosine tail inactivates the Src protein; dephosphorylation of this site activates the molecule (no switch off would mean constitutive activation). A point-mutated *src* gene would express its oncoprotein constitutively. Healthy rodent somatic cells use long telomeres and express telomerase catalytic subunits (TERT) constitutively, therefore are unusually susceptible to malignant transformation. For the Src protein to de-regulate E-cadherin, phosphorylation at Src-specific sites of the focal adhesion kinase (Table IX) is necessary.[8ab] Ras oncoprotein signaling pathway in squamous cell carcinoma of the skin includes Raf/Mek/Erk, PI3K/Akt/mTOR and Pak. Inhibiting Pak with its patented inhibitors PF3758309 or FRAX597, silenced Erk and Akt in regressing tumor.[9] Example of MYC-IgH fusion oncoprotein formation in lymphoplasmacytic lymphoma: t(8;11) (q24;q32): enzyme AID mis-directed its nick?[10]

Virulence factors and transposable elements of Phytophthora. The oomycete ("water mold") *P. infestans* and the potato originated in the Andes, spread through Mexico into the USA and from there to Europe; the ancestor of *P. infestans* was *P. andina. P. infestans*-induced blight wiped out potato crops; caused massive starvation and death in the young USA (Philadelphia) and Europe (Ireland, 1840s, Germany, !918s; East Germany, 1950s). *P. Infestans* (Pi) possesses the largest phytophthora genome (18,000 genes in 240 Mbp) loaded with transposases of all classes. The genomes of the phytophthora ancestors received cyanobacterial genes (chloroplasts; plastids); the current phytophthora cell contains mitochondria, which divide by the bacterial FtsZ-way.[11,12a] Phytophthora and plant cell mitochondria are related[12b]. Substantial genomic contributions arrived horizontally. Non-existent in eukaryotes outside the Plantae, is the lipopolysaccharide biosynthesis gene E/HGT (hypoxanthine-guanine-phospho-ribo-transferase), but the phytophthora genome has gotten it.[13,14] A virulence factor is the PI3K gene/protein[14]. There is a

(continued)

Table XXX (continued)

MAPK cell survival pathway consisting of ERK kinases. Its signal- sensing domain derives from the archaea. The Pi IRAKs (IL-1 receptor-associated kinase) predate *Plasmodium falciparum,* or *P. tetraurelia*, which have none.[15a] The CRN effectors (crinkler gene) are recombinant fusion protein virulence factors subjected to transposons insertions.[15b,16a] The Pi genome contains unusual large numbers of miRNAs, siRNAs, and piwiRNAs, and inserted transposons with or without LTRs (Gypsy/Albatross; Crypton; Copia; Dodo; Helentron, endonuclease-containing Helitron; Mariner; Mutator; Piggy-Bac; Pogo). The Dicer/AGO system is fully operational. The PiAvr3a gene encodes an RXLR (arginine, X, leucine, arginine) avirulence effector to a SINE retroelement; sRNAs silence the SINE element.[16abd] The Pi genome encodes those tyrosine kinase-like enzymes that are the properties of metazoan eukaryotic cells; with *P. infestans*, only the choanoflagellate *Monosiga brevicollis* is known to possess such tyrosine kinases. The Pi GMP-regulated protein kinases (PK) are neither yeast- or plant-related; related PK enzymes function in *Paramecium tetraurelia* and *Toxoplasma gondii.* The C terminals of some of the Pi cell cycle enzymes match Pfam domains related to those of *P. tetraurelia*, *Tetrahymena thermophila*, and *Theileria annulata.* The Pi genome is significantly diverging from all other phytophthora genomes (*P. sojae;* oak tree-pathogen *P. ramorum;* Arabidopsis-pathogen *P. parasitica;* fish-pathogen *Saprolegnia parasitica*).[17] The Pi cell is not known to succumb to mitochondria-induced apoptosis; instead, its genome encodes the SNE (suppressor of necrosis) protein for the suppression of programmed cell death in the nuclei of the plant cells.[18] Two diploid phytophthora cells of different strains fuse and become tetraploid; the fused cells created in Hawaii are superior to mononuclear cells, and display "hybrid vigor" in growth rate and infectivity.[19] Sexual recombination creates the variability of phytophthora strains. The common ancestor of the original *P. andina* (in the Andes) diverged into two lineages, which later recombined to form *P. infestans*. Similarly to other fungi (yeast; basidiomycete), *P. infestans* possesses an elementary nuclear *ras*/Ras locus.[20abc] Here, a unicellular fungal pathogen (cells from the hypha, or germinating from spores) is driven by a fusion protein virulence factor, resists apoptosis, carries proto-oncogenes, and attacks a multicellular host (the leaves and tubers of a potato plant) in a relationship that is re-played at higher levels by other unicellular eukaryotic pathogens against multicellular vertebrate hosts, including Homo (Theileria attacking the lymphocytes of buffalos; Babesia lysing red blood cells in the cattle and horses; *Trypanosoma brucei/gambiense* attacking the human brain in African sleeping sickness; toxoplasma attacking the brain in patients with AIDS; plasmodia dissolving human red blood cells for millennia in inhabitants of tropical forests; Naegleria infesting the meninges through nasal passages of children swimming in warm ponds in Florida and Louisiana). The third level is the attack of the neoplastic cells on their hosts. When exogenously induced (viral, bacterial, protozoal , schistosoma-induced urogenital), there is a vigorous host defense induced (but often in failure). When it is endogenously induced in the genome of some selected host cells, "self" is recognized, and the attacking cell population elicits the support of its doomed host.

Transposon mutagenesis. *Science* summarizes the two transposons most useful for mutagenesis: the Sleeping Beauty (SB) reconstructed from fish genome, and PiggyBac (PB) reconstructed from the genome of the cabbage looper moth (*Trichoplusia ni*).[21,22,23] These transposons carry and insert transgenes into targeted genomes, be those in plasmodia, schistosoma,[24] or in human cells. Several strains of transgenic hosts (mice) have been generated. Candidate cancer genes are discovered when the known transfected genes are oncogenic (*in vivo*; *in vitro*). Transfected oncogenes may show specific integration patterns. Tumor suppressor genes (*pten*) can be transfected for intent of cancer treatment.[25] Cellular pathogens may be attenuated by transgene insertion.[24] In the target tissue, the recombinase excises cassettes blocking expression of SB transposase; inserted SB can introduce gain-of-function (in cancer suppressor genes), or loss-of-function (in oncogenes) mutations.[26]

Transposon-activated oncogenes. Ancient inserted elements may remain alert. The ORF-1p of the LINE element (Bcl2xL interspersed nucleotide element) acts upon the AR to increase its transcriptional activity. AR so activated translocates to the nucleus to activate the PSA promoter. If

(continued)

Table XXX (continued)

this occurs in a low grade prostate cancer, the tumor cells transform into higher grade: gain anchor-independence and faster growth rate.[27] The methylated LINE-1 is silent. The hypomethylated LINE-1 changes its configuration from tetranucleosome to histone variant H2A.Z dinucleosome, which indicates gene reactivation: transcriptional activation of the L1-MET anti-sense promoter occurred by hypomethylation. Hepatocyte GF-R MET releases alternate transcripts that activate pre-existing transitional (urothelial) cell tumors of the urinary bladder.[28] Arises glioblastoma multiforme (GBM), the most malignant and least treatable human tumor. PTEN, p53, Rb, Nf, and ARF (alternating reading frame) suppressor genes are knocked out first, and amplified or gain-of-function mutated genes (EGF-R; PDGF-R; MET; PI3K/Akt; N-Myc) move in, aggravated with individual tumor genome variability distributed among many subclones of the original tumor, but without one subclone emerging as dominant.[29] The cell of origin concealed for long appears to be a neural stem cell activated and transformed via transposons-based mutagenesis. Astroglial-like cells undergo sleeping beauty (SB) transposition, get immortalized and build up to high cumulative numbers. In human cells undergoing this route, p53 is mutated ($p53^{R172H}$,arginine; histidine) and NF1 is lost. MET insertions appear (mysteriously related to a "murine stem cell virus"). Neuregulin-, IGF-, and Rac-signalings (Ras-related C3 botulinum toxin substrate guanosine triphosphatase) emerge. The role of Gli3 remains obscure; it may act as a suppressor of Hh signaling. HGF signaling to Met is active. Of kinases, the RTK (receptor tyrosine kinase) pathway dominated.[30] (This extremely complicated work carried out in Japan, Singapore, England and Houston will have to be transferred from the mouse to the human brain). Sensational recent genetically engineered or native oncolytic viral therapy induces remissions lasting until after host immunity neutralizes the oncolytic virus (adenovirus; herpesvirus; measlesvirus; myxomavirus, Newcastle virus; polio-rhinovirus; reovirus). A protocol was designed for the combination of oncolytic viral and immune lymphocyte therapy.[31ab] A four-pronged biological attack on GBM include devascularization; microglia activation; NK cell and immune T cell activation.[32]
1 Sinkovics JG Ann Clin Labor Science 1984;14:343–54; CRC Crit Rev Immunol 1988;8:217–98; Adv Exp Med Biol 2011;714:5–89. ***2*** *Watari A Oncogene 2010;29:3815–26; Yakugaku Zasshi 2012;132:1165–70.* ***3*** *Aksamitiene E et al Cell Signal 2011;23:1794–805.* ***4a*** *Tan Y et al J Biol Chem 2000;275:25865–9.* ***4b*** *Taylor SS et al Biochem Biophys Acta 2013;1834:1271–8.* ***4c*** *Ye DZ et al PLoS 2011;6(11):e27837.* ***5*** *Ong CC et al J Natl Cancer Inst 2013;105:606–615.* ***6*** *Li W et al J Biol Chem 2008;283:15491–501.* ***7*** *Segawa Y et al Proc Natl Acad Sci USA 2006;10;12021–6.* ***8a*** *Akagi T et al Proc Natl Acad Sci USA 2003;100:13567–72.* ***8b*** *Avizienyete E et al Nat Cell Biol 2002;4:632–8.* ***9*** *Chow HY et al Cancer Res 2012;72:5966–75.* ***10*** *Peker D et al Arch Pathol Lab Med 2013;137:130–3.* ***11*** *Maruyama S et al BMC Evol Biol 2009;9:197.* ***12a*** *Kiefel et al Protist 2004;155:105–15.* ***12b*** *Karlovsky P & Fartmann B J Mol Evol 1992;34:254–8.* ***13*** *Thomas A et al PNAS USA 2011;108:15258–63.* ***14*** *Lu S et al Mol Palnt 2013;6:1592–604.* ***15a*** *Judelson HS & Ah–Fong AMV BMC Genomics 2010;11:700.* ***15b*** *Haas BJ et al Nature 2009;461:393–9.* ***16a*** *Vetukuri R R et al PLoS One 2012;7(12):e51388; Fungal Biol 2011;115:1225–33.* ***16b*** *Armstrong MR et al PNAS USA 2005;102:7766–71.* ***16c*** *Qutob D et al Nat Commun 2013;4:1349.* ***16d*** *Whisson S et al Mob Genet Elements 2012;2:110–114.* ***17*** *Raffaele S et al BMC Genomics 2010;11:637.* ***18*** *Kelley BS et al Plant J 2010;62:357–66.* ***19*** *Gu YH & Ko WH Can J Microbiol 2001;47:662–6.* ***20a*** *Gómez-Alpizar L et al PNAS USA 2007;104;3306–11.* ***20b*** *Knabe N et al Eukaryot Cell 2013;12:941–52.* ***20c*** *Li Y & Wang Y J Biol Chem2013;288:11358–65.* ***21*** *Rad R et al Science 2010;330:1104–7.* ***22*** *Ivics Z, Izsvak Z Curr Gene Ther 2006;6:593–607; Ivics Z et al Nat Methods 2009;6:415; Cell 1997;91:501 (reviewed in Sinkovics JG Adv Exp Med Biol 2011;714:5–89).* ***23*** *Fraser MJ et al Insect Mol Biol 1996;5:141–51.* ***24*** *Morales ME et al FASEBJ 2007;21:3479–89.* ***25*** *Largaespada DA & Collier LS Methods Mol Biol 2008;435:95–108.* ***26*** *Janik CL & StarrTK J Vis Exp 2013;(72): e50156.* ***27*** *Lu Y et al Cell Signal 2013;25:479–89.* ***28*** *Wolff EM PLoS Genet 2010;6(4): e1000917.* ***29*** *Sottoriva A et al PNAS USA 2013;110;4009–14.* ***30*** *Koso H et al PNAS USA 2012;109;E2998–3007.* ***31ab*** *Sinkovics JG & Horvath JC Arch Immunol Ther Experiment 2008; S56:1–59; Acta Microbiol Immunol Hungar 2006;53:367–429.* ***32*** *Nieto-Sampedro M et al Clin Med Insight Oncol 2011;5:265–314.*

Table XXXI Major Faults in Complex Pathways Result in Cleft Lip and Palate

Misunifications in the top bilaterian, Homo. The incidence of the congenital malformation CL/P, with or without cleft palate, is high at 1 in 700 live birth world-wide. In the human embryo the formation of the upper lip is accomplished between days 24 to 37. The stomodeum (primitive mouth) is surrounded by derivatives of the branchial arch 1, prominences frontonasal superiorly, maxillary bilaterally and mandibular inferiorly and bilaterally. The frontonasal prominence becomes the medial nasal process and the nose. The maxillary and the mandibular prominences fuse in the midline; the upper lip is formed by this union. It is questioned if the medial nasal process contributes to the fusion in the upper lip. The processes formatted bilaterally around the stomodeum will not unite and fuse in the CL/P syndrome. ***Key genes suffer loss-of-function point mutations, and growth factors switch on an off haphazardly, or not at all.*** Type I/II TGFβ-R heterotetrameric receptors are expressed temporally and in a spatially restricted manner. The SMAD proteins (signaling mothers against decapentaplegic, drosophila) balance the act of TGFβ ligands and receptors. TGFβ-R Type I and TGFβ family member BMP4 (bone morphogenetic protein) are not expressed properly, or at all. BMP4 acts upstream of sonic Hh (hedgehog) during palatogenesis. Reduced BMP4 signaling results in upregulated FGF8 and sHh expression. The knock-out of homeobox gene Hox-7.1 in mice (named *Msx*-1,2 in drosophila and human) is a major contributor to CL/P development. TGFβ-R Type I (renamed *Alk*5, activin-like kinase) and *Msx*2 cooperate in normal upper lip development. A functional canonical Wnt-β-cateninpathway is important for the upper lip development.[1ab] Point mutated *Tbx*10 gene (T box) is added to many other suspected loss-of-function mutant genes/proteins (*Msx*1,2). The point mutations of some suspect genes have been discovered (jagged JAG A657H, alanine and histidine; Table XVI LIM homeobox LHX8, E221A glutamic acid and alanine; muscle segment homeobox MSX, P147Q proline and glutamine; the AT-rich DNA sequence-binding protein SATB2, T190A threonine and alanine; Sloan-Kettering isolate retroviruses SKI, A388V alanine and tyrosine; sprouty SPRY2, K68M lysine and methionine, or D20A aspartic acid and alanine; T box TBX10 R354Q arginine and glutamine. These mutations appear haphazardly, and are not occurring characteristically in every case of CL/P.[2] Loss of the *Dlx*5 gene (distal less, drosophila) is followed by the downregulation of FGF7 and the rise of sHh in an attempt at restoring the expansion of the stagnating edges for fusion. Sonic Hh is dependent on BMP: it acts downstream of BMP1 and upstream of BMP2. Loss of the MSX1 protein results not only in disjunction of the edges, but also in tooth agenesis.[3] The genes muscle segment homeobox (*msh* or *msx*) appeared in the ascidians (Ciona), sea anemones (Nematostella), hydra, sponges, echinodermata (Strongylocentrotus), uro- and proto-chordates (Branchiostoma), and platyhelminths. MSX proteins remained well conserved from Nematostella to Branchiostomata (a distance of 600 my). In the Branchiostoma (amphioxus), MSX proteins format neuroectodermal patterns and epithelial-mesenchymal interactions. In the lamprey's pharynx, MSX proteins format a gnathostome-type pattern, but not a jaw. Gene duplications yielded MSX1,2,3 clade proteins. In tetrapods, the MSX proteins are expressed together with sHh in limb buds. Figure 8 of the Finnerty article summarizes the evolutionary history of the MSX domains. The first and second vertebrate whole genome duplications resulted in the loss of some ancient domains of the MSX proteins. MSX1 and MSX2 diverged from one another, with domain loss in MSX2. Loss-of-function mutations of the *msx*1 human genes result in the CL/P syndrome, and even tooth agenesis. Among many point mutations, the *msx*1 P147Q (proline; glutamine) mutation was proven to be in the background of "clefting phenotype" in the Filipines.[4abcdef] ***Facial morphogenesis*** in the early embryo is under the control of canonical Wnt/FGF signaling. Loss of the Wnt9b gene results in retarded or no contact between the primordial tissue processes surrounding the stomodeum.[1b,5] A comprehensive review of all gene-suspects in close or distant involvement in CL/P (cheiloschisis; palatoschisis) syndrome, are TBX22 (T box transcription

(continued)

Table XXXI (continued)

factor), PVRL1 (poliovirus receptor-like), FGFR2 (fibroblast growth factor) and IRF6 (interferon-regulatory factor). All these genes are subject to loss-of-function point mutations: PVRL1 to W185X, tryptophan and unkown; FGFR2 to S252W, serine and tryptophane, or P253R, proline and arginine mutations.[6]

Chaotic situation. In the perioral and palatal tissues of the CL/P syndrome, several genes undergo loss-of-function point mutations, and proto-oncogenes are haphazardly switched on and off, but no increased incidence of malignant transformation is reported. The cDNA of the MSX2 homeoprotein in NIH3T3 cells suppressed v-K-*ras* and v-*raf* induced transformation; the truncated MSX2 protein interfered with the activities of intact MSX2 protein.[7]

The ultimate bioengineer RNA/DNA complex subjected to loss-of-function mutations cannot correct the deficiencies. TGFβ-R signaling in early fetal life is essential to palatal fusion. The errors leading to this deformity are the inactivation of the TGFβ1-R in the ectomesenchyme. Several other deficient gene expressions coincide. Cells of the medial nasal process and the maxillary processes fail to unite in the midline and the stagnating cells eventually succumb to apoptotic death. The epigenomics and the microRNA-mRNA conflicts are as yet unknown. No incidence of malignant transformation in the CL/P syndrome is on the record. If in damaged tissues, especially in those undergoing a chronic inflammatory process, a mechanism exists, in which reactivated retrotransposons engender malignant transformation in a few selected cells (see text), in order to immortalize, and thus rescue the living cell from an environment in which it would perish otherwise, that mechanism is not activated in the CL/P syndrome (Figure).

Comment. The CL/P syndrome refutes the credentials of the RNA/DNA complex as if it were an error-free bioengineer. In contrast, Ivan Martin and associates in Basel, Switzerland, build autologous nasal cartilage implants on scaffolds in vitro for replacement by rhinoplasty of the lost native organ. The patient's own chondrocytes format the size and shape of the implant, which is adjusted pre-implantation. The recipient host accepts the implant biologically (neovascularization) and biochemically.[8]

1a** Li WY et al Mech Dev 2008;125:874-82.* ***1b** Jin Y-R et al Development 2012;139:1821-30.* ***2** Vieira AR et al PLoS Genet 2005;1(6):e54.* ***3** Han J etal 2009;136:4225-33.* ***4abcdef** Cerny R et al PNAS USA 2010;107:17262-7; Finnerty JAR et al BMC Evol 2009;9:18; Lallemand Y et al Dev Biol 2009;331:189-98; Nagy A et al Curr Biol 1998;8:661-4; Szabo-Rogers HL et al Dev Biol 2010;341:84-94;Takahashi H et al Mol Biol Evol 2008;25:69-82.* ***5** Vieux-Rochas M et al PLoS One 2013;8(1):e51700.* ***6** Kohli SS & Kohli SK J Oral Maxilliofac Pathol 2012;15:64-72.* ***7** Takahashi C et al Leukemia 1997;3:340-3.* ***8 www.sciencedaily.com/release/2024/04/1404101 94353.htm

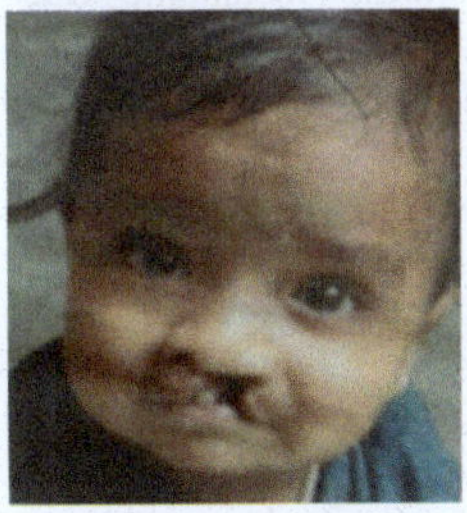

A case of cleft palate from www.SimileTrain.org a Better Business Bureau Accredited Charity

Table XXXII Mitochondia in Oncogenesis

Speculations. Either an archaea and a prokaryota created the first eukaryote by fusion (phage-mediated fusion), or the mitochondria were free-living proteobacteria before their capture by a free-living ancient unicellular eukaryote, which might have originated from the crenarchaeote lineage. In the latter scenario, phagocytosis rather than fusion might have brought them together. The original free proteobacteria divided in mature age, thus escaping senescence and natural death. They have become obedient endosymbionts retaining most of their genome (after transferring some of their primordial genes to the nucleus of their host). They continue to multiply by FtsZ prokaryotic divisional mechanism (42,43) in the cytoplasm of their host cell and pass through in the host's progeny exclusively in the ovum (absolutely not by the spermia). However, mitochondrial genes can encode for apoptotic death of their host cell. Are the apoptosis-inducing mitochondrial genes genuine, or are these genes transferred into mitochondria from their host cell nucleus? Could the original proteobacterial genes (the sirtuins) encode longevity, or even immortality, after transfer into the nucleus of their new host cell? Would this function be tantamount to so-called malignant transformation (an attempt at immortalization) in their new host cell?
Mitochondrial DNA (mtDNA). Direct repeats mediate deletions in mtDNA. Inverted repeats are subjects of mutations retained as mutated mtDNA.[1] The histone deacetylase sirtuins3-5 (silent information regulator proteins) are of mitochondrial location and function. Sirtuin3 deacetylates acetyl coenzyme A (a-CoA) and generates energy from fat and amino acid metabolism. It mobilizes the high energy group CS2 protein (Tables XI/I and XI/II). It removes ammonia (NH_3) build-up by activating ornithine transcarbamylase. Acetylated genes are silenced; deacetylated genes (including proto-oncogenes) are activated. Acetylated tumor suppressor genes are silenced; deacetylated tumor suppressor genes are activated. The enzymes silenced by hyperacetylation (knocked out Sirt3) contribute to metabolic syndrome. Sirtuin4 is an ADP-ribosylase. It is an inhibitor of insulin release from the pancreas. Sirtuin5 contributes to ammonia detoxification by deacetylating the promoter of carbamoyl P-synthase. As a consummate direction of its multi-directional effects, Sirtuin3 is inhibitory to malignant transformation, both at the level of its induction, or at its maintenance. It reverses the Warburg effect; it is pro-apoptotic; and $Sirt3^{-/-}$ mouse fibroblasts grow as sarcoma cells.[2] The mitochondrial OXPHOS (oxidative phosphorylation) genes interact with DRAM, LKB, and the Golph/Midas/GPP complex (damage-regulated autophagy modulator; liver kinase B; Golgi phospholipids shuttled by protein carriers between Golgi and mitochondria). GOLPH is overexpressed in autophagy-resistant tumor cells; GOLPH is an activator of mTOR proto-oncogene. Autophagic and mitophagic tumor-associated stromal fibroblasts overexpress DRAM and LKB, and digest caveolin-1 (via lysosomal degradation in autophagosomes). Despite mitophagy, autophagic stromal fibroblasts transfer metabolites (lactates, ketone bodies) to tumor cells to fuel their OXPHOS metabolism. If p53 protein is still present, it translocates to the nucleus to activate the DRAM gene. DRAM can activate the mitophagy gene BNIP3 (30kD Bcl-2/adenovirus E1B 19 kD protein-interacting protein). Tumor cells frequently silence the DRAM gene (or its promoter's CpG islands) by hypermethylation. The LKB1 kinase activates AMP-activated protein kinase (AMPK) by phosphorylating its threonine residue (T172), and inactivates mTOR. In conclusion, tumor cells receive paracrine promoters from fibroblasts in a HIF1- and NFκB-permeated environment. Within the tumor cell (MDA-MB-231-GFP+), GOLPH 3 activates mitochondria running OXPHOS metabolism. In the Salem et al article Figure 8 is a diagram depicting these interactions.[3abcd] The JAK tyrosine kinase STAT pathway is frequently utilized in cancer cells. A STAT chain reaction regulates electron transport in mitochondria. Mitochondrial localization MLS-STAT(S727A) (serine, alanine) decelerated tumor cell (breast cancer) growth driven by the ancient *ras* proto-oncogene; expression of MLS-STAT(S727D) (serine, aspartic acid) accelerated breast cancer cell growth. In both cases phosphorylation of serine 727 in mitoSTAT was the crucial event. Phosphorylated mitoSTAT serine transformed mouse embryo fibroblasts with reduced

(continued)

Table XXXII (continued)

ROS levels.[4] If mitochondria do not kill an incipient tumor cell by intrinsic apoptosis, they will promote immortalization of the cell in the process of advancing transformation. Mitochondria of the transforming cell and those of the stromal fibroblasts communicate in favor of malignant transformation.
1 *Yang JN et al PLoS One 2013;8(9):e73318.* ***2*** *Shih J & Donmez G Genes Cancer 2013; 4(3–4):91–6.* ***3abcde*** *Kammouni W et al Arthrit Rheum 2007;56:2854=63; Kang JS et al J Cell Biol 2008;182:497–507; Low BC et al J Biol Chem 1999;274:33123–30; Salem AF et al Cell Cycle 2012;11:2545–59; Wang EY Hypertension 2013;62:70–7.* ***4*** *Zhang Q et al J Biol Chem 2013;288:31280–8.*

oncologists-hematologists,that is not well documented by statisticians. Why and how is the brain, protected by its microglia and a blood/brain barrier, favored by metastatic tumor growth; why astrocytes and vascular endothelial cells appear to welcome the tumor cells and extend intrinsic support to them. Whereas EBV-carrier B cell lymphomas induce an immune reaction mediated by circulating NK cells and immune T cells, as well as by the microglia. Table XXXV was added to explain and contrast the similarities between Agrobacterially induced transformation in plant cells (in the Appendix 1), and the emergence of circular RNAs and noncoding ncRNAs in the generation of human neoplasms.

Small molecular targeted therapeuticals block the oncogenic kinases and induce spectacular remissions. However, some single Ph-chromosome-carrier cells resist imatinib mesylate and eventually rise to detectable numbers. The Bruton-kinase inhibitor ibrutinib induces the apoptotic death of B lymphoma and CLL cells. It probably inhibits PI3K/Akt, Bcl-2 and NFκB. Large lymphomatous tumor masses melt. Yet some B lymphoma or CLL cells survive and re-grow. These cells carry point-mutated Bruton-kinase genes: C481S (cysteine, serine); and point-mutated PCLγ2 (phosphatidylinositide-specific phospholipase gamma) gene: R665 W (arginine, tryptophane) and L45F (leucine, phenylalanine). Ibrutinib cannot bind the mutated Bruton kinases. PCLγ2 is an oncogene in another tumor (angiosarcoma). The PCLγ genes are virulence factors and cell survival pathways often constitutively expressed in unicellular eukaryotes (*Tetrahymena thermophila*, and others). Another ancient cell survival factor conserved throughout evolution becomes a proto-oncogene in the human genome (*vide supra*).

Comment. Questions on the manners of resistance acquisition persist. Viruses, bacteria and unicellular eukaryotes mutate in order to survive in the immunologically active hosts. Bacteria learned to survive in their natural habitat shared with antibiotics-producing fungi. However, bacteria co-cultured with fungi did not directly induce antibiotic production (to the disappointment of professor Selmon Waksman, discoverer of streptomycin, and many other antibiotics, including actinomycin D). He read the negative article of this author (JGS) and brought up the issue on the occasion of the introduction of this new Rockefeller Fellow (this author) appointed to his Institute in 1957. Read the story in [27]. It appears that in the large populations of microorganisms, substrains of individuals exist that possess the faculties of resistance in advance. As if their RNA/DNA genomes without a

Table XXXIII Pediatric Lymphomas/Leukemias. Network of Chaotic Mutations: Major Errors in the System? Repeated Recognizable Patterns for the Establishment of Immortalized Chemo-Radiation-Resistant Transformed Cells?

Anaplastic lymphoma of mature T cells; B cell lymphomas; lymphoplasmacytic plasmablastic lymphoma. Tumor suppressor gene-product transcriptional repressor proteins mediating terminal differentiation of lymphocytes, PRDM/BLIMP/WTp53 are deleted (positive regulatory protein; B lymphocyte-induced maturation protein). Highly aggressive dual ALK and MYC rearranged anaplastic large cell lymphoma. High mobility group protein (Tables XI/I and XI/II). Overexpression in malignant cells. MLL promiscuous recombinant oncoproteins in T cell acute leukemia lymphoblasts. ALk-mutated anaplastic T cell lymphoma responds to crizotinib. Autologous-allogeneic stem cell transplants. CD-19-specific mcab for FcγRiiiA ADCC reaction. CD-19-specific genetically engineered TCR CAR-bearing $CD8^+$ immune T lymphocytes inducing massive lymphoma cell deaths (see in text).[1–7]

Precursor B cell acute lymphoblastic leukemia. IKZF1 deletion (Ikaros family zink finger protein) worsen prognosis. Same, for protein SPRED deletion (sprouty-related protein). Erg (E26-transformng sequence-related gene) deletion compensates for IKZF1 loss. The ERG protein is notoriously resistant to small molecular inhibitors (BAY43-9006, sorafenib. (FUS-ERG fusion occurs in pediatric acute myeloid leukemia). CRLF/P2RY8 fusion oncoprotein forms (cytokine receptor-like factor; purinergic receptor G-protein complex). Janus kinase JAK2 mutation may occur (A617F, alanine; phenylalanine). IgH rearrangements. The chromatin-remodeling DNA-binding tumor suppressor factor gene/protein *arid*/ARID (AT-rich interactive domain) induces TNFα with or without apoptosis; controls cell growth and differentiation. Occasional BCR/ABL fusion oncoprotein. Overexpressed Bcl-2. In precursor B lymphoblast malignancies miR-708, miR-210, miR-18b rise; miR-345, miR-27b drop. Methylome and transcriptome in the epigenome. Bortezomib proteasome inhibitor and HDACI down-modulate PI3K/Akt/NFκB. Humanized anti-CD19 mcab Medi-551 induces NK cell-/mϕ-mediated lymphoma cell death.[8–26]

Retroviral mouse leukemias. It was possible to induce myelogenous leukemia in mice with the *bcl/abl* fused gene of the Ph chromosome. The Gross virus derived from AKR mice and caused lymphocytic leukemias. The splenomegalic Friend virus induced erythroleukemias. The Rauscher virus acted as a mixture, or as a recombinant derivative, of the Gross and Friend leukemia viruses. The spectrum of murine leukemias caused by the Graffi mouse leukemia retrovirus is extremely wide. Yet it was the genome of the Moloney murine leukemia-sarcoma (rhabdomyosarcoma) virus that underwent changes in passage in mice in acquiring the myeloproliferative leukemia virus oncogene (c-*mpl* → v-*mpl*). Inserted in the human genome (proto-oncogene c-*mpl* 1p34), c-*mpl*-acts with the JAK kinases and serves as the receptor of its ligand thrombopoietin. It may undergo point mutations. The human erythropoietin-like thrombopoietin gene is mapped at 3q26-27. Expressed in megakaryocytes, *mpl*-attached thrombopoietin induces thrombocytosis.[27–40]

Pediatric myelodysplasia, myelogenous leukemia. Major initiator of myelodysplasia is the point mutated JAK2 kinase: A617F (alanine; phenylalanine). Hematopoietic JAK2-mutated stem cells clonally expand. Current JAK2 kinase inhibitors (first ruxolitinib; others to follow) provide extraordinary symptomatic relief (qui*ck* reduction of massive splenomegaly; reduction of essential thrombocytosis, subsidence of JAK-STAT pathway), but no cure. Pediatric myelofibrosis differs from the adult variant in more extramedullary hematopoiesis (in the spleen and/or liver); more spontaneous remissions "a transient process"), less malignant transformation. On an occasion, childhood acute myelofibrosis and megakaryoblastic leukemia may "spontaneously" (after sepsis and whole blood transfusion containing T lymphocytes and NK cells) recede. Point-mutated *mpl* gene activates PI3K/Akt pathway. In mice, JAK-independent myelodysplasia with acute myelogenous leukemia was induced by retroviral insertional mutagenesis: NUP96 and HOXD13 was the fused oncoprotein at t(2;11)(q31;p15). The FUS/ERG fusion oncoprotein induced acute myelogenous leukemia in childhood.[41–48] Commentary. Chapter closed without a firm conclusion. Are pediatric (human) acute leukemias retrovirally induced (the belief of this author), or are endogenously initiated within the genome? In the latter case, the RNA/DNA complex operates

(continued)

Table XXXIII (continued)

error-laden, or is it that its inherent faculty to immortalize cells is activated due to signals released for whatever reasons (unexplained at the present time)?
1 *Boi M et al Blood 2013;122:2683-93.* **2** *Brandimarte L Blood 2013;121:5064-7.* **3** *Garcia JF et al Haematologica 2006;91:467-74.* **4** *Giulino-Roth L et al Pediatr Blood Cancer 2013;60:2018-24.* **5** *Jena et al PLoS One 2013;8(3):e57838.* **6** *Liang X et al J Pediatr Hematol Oncol 2013;35:209-13.* **7** *Lowe EJ & Gross TG Pediatr Hematol Oncol 2013;30;509-19.* **8** *Asai D et al Cancer Med 2013;2;412-10.* **9** *Bastian L et al Clin Cancer Res 2013;19:1445-57.*
10 *Bock J et al PLoS One 2013;8(1):e52872.* **11** *Busche S et al Cancer Res 2013;73:4223-36.* **12** *Chaber R et al J Pediatr Hematol Oncol 2013;35:180-7.* **13** *Clappier E et al Leukemia 2014;28;70-7.* **14** *Dallas PB et al Mol Cell Biol 2000;20:3137-46.* **15** *Gawad C et al Blood 2012;120:4407-17.* **16** *Gutiérrez-Camino A J Res Clin Oncol 2013 in print.* **17** *Joseph S et al PLoS One 2012;7(7):e42159.* **18** *King JAJ et al Biochem J 2005;388:445-54.* **19** *Krawczyk J et al Br J Haematol 2013;160:555-6.* **20** *LI X et al Pediatr Blood Cancer 2013;60:2060-7.* **21** *Olsson L et al Leukemia 2014;28:302-10.* **22** *Matlawska-Wasowska K et al Leukemia 2013;27:11263-74.*
23 *Panagopoulos I et al Oncol Rep 2014;28:70-7.* **24** *Schultz KR et al Blood 2007;109;926-35.*
25 *van der Veer A et al Blood 2013 in print.* **26** *Wisker D et al Cell Growth Differ 2002;13:95-106.*
27 *Charfi C et al Blood 2011;117:1899-910.* **28** *Daley GQ et al Science 1990;247:824-30.*
29 *Drachman JG et al J Biol Chem 2002;277:23544-53.* **30** *Foster DC et al PNAS USA 1994;912:13023-7.* **31** *Gross L Oncogenic Viruses. Pergamon Press Oxford. 1983; Pp393.*
32 *Methia N et al C R Acad Sci III 1995;318:479-82.* **33** *Moloney JB Natl Cancer Inst Monogr 1966;22:139-142.* **34** *Rumi et al Blood 2013;121:4388-95.* **35** *Sinkovics J Ann Rev Microbiol 1962;16:75-100.* **36** *Front Rad Ther Oncol 1990;24:18-31.* **37** *Skoda RC et al EMBO 1993;12:2645-53.* **38** *Souyri M Semin Hematol 1998;35:222-31.* **39** *Stacey A et al J Virol 1984;50:725-32.* **40** *Vigon I et al Oncogene 1993;8:2607-15.* **41** *Kucine N & Levine RL Ther Adv Hematol 2011;2:203-211.* **42** *DeLario MR et al Am J Hematol 2012;87:461-4.* **43** *Elhasid R et al J Pediatr Hematol Oncol 2012;34:565-8.* **44** *Panagopoulos I et al Oncol Rep 2014;32:40-4.*
45 *Pulikkan JA et al Blood 2012;20:868-79.* **46** *Slape C Cancer Res 2007;67:5148-55.* **47** *Slone JS et al J Pediatr Hematol Oncol 2013;35:559-65.* **48** *Verstovsek S et al N Engl J Med 2010;363:1117-27.*

foresight of what future circumstances will be rising, provided them with appropriate mutated genes, whose gene product proteins secure the survival of their hosts in the new environment, while other unprepared individuals of the colony perish. Means of survival other than those vertically inherited, for the defenseless individuals was the horizontal transfer and receipt of the resistance genes or plasmids. In contrast, the immunoglobulin light chain κ/λ genes are not prepared in advance for all possible antigenic challenges. They undergo the close to inexhaustible numbers of SHMs in order to specifically respond to the antigenic configurations.

Selected cells in selected individuals of a human population transform into immortalized entities with rapid growth potential. The afflicted persons receive chemotherapy with curative intent and rapidly enter remissions. However, some transformed cells are immediately resistant to the chemicals, not because of acquiring it in response to them, but because of pre-existing genomic mutations (or epigenomic adjustments). As if the genomes by an inherent faculty in them invoked innumerable minuscule changes in advance for just one of those to fit into, and thus

Table XXXIV The Brain Commonly Supports Primary or Metastatic Malignant Tumors, but Contributes to the Rejection by Biotherapy of B Cell Brain Lymphomas

The organs and viscera may or may not attempt tumor rejection; in opposition, inflammasomes generate tumor-promoter mi-RNAs. Reviewed are anti-EGF-R therapy with cetuximab, erlotinib and gefitinib;[1ab] resistance to TRAIL acquired in imatinib-treated BCR-ABL fusion oncoprotein-carrier tumor cells;[1c] tumor exosomes protect tumor cells from antibody attack;[2a] NK cell therapy;[2b] and miR-34 inducing mitotic catastrophe in tumor cells.[3] The brain frequently accepts and supports transformed cells. Patients with incipient HER2/neu-positive breast cancers, or BCR/ABL-positive myeloblasts, immunologically react to the amplified HER2/neu, or the fusion BCR/ABL oncoproteins, with generation of immune T cells and antibodies, which however fail to reject the transformed cells. Upon the expression of fetal antigens (CEA; PSA; CA125; Ca19-9), the host recognizes fetal tissue as "self" and refrains from further attack. The host mobilizes immunosuppressive Treg cells responding to the chemokine stromal-derived factor and its receptor, in order to eliminate the immunoreactive T cell clones. DCs generate tolerant T2 immunological environment. Patients with splice variant EGFRviii glioblastomas, immunologically react to the autologous tumor, but with failure of rejection. Some incipient tumors may be rejected without traceability by Burnet's immunosurveillance. NK cells attack without pre-immunization (either by direct NK cell cytotoxicity, or by an ADCC reaction). In the inflammasomes, tumor-promoter mi-RNAs (miR-155) are generated and compete with p53-promoter tumor suppressor mi-RNAs (miR-34a). Mutator miR-155 induces proto-oncogenes in the inflammasome. Resveratrol is a miR-155 antagonist. The proto-oncogenes activated in the inflammasome are recognized.[4ab] It is not clear what miRNAs rise if malignant transformation is to be supported or rejected in the brain. Are these miRs of neuronal cell or of microglial derivation?
Some endogenously originated (without external initiator) brain tumors receive full support rendering them treatment-resistant. Exogenously initiated (EBV-associated) primary brain B cell lymphomas invoke immune T cells and are highly responsive to biotherapy. Primary (malignant gliomas) and some metastatic brain tumors are received by inert microglia, and fed with growth factors provided by specialized (immature mesenchymal stem cell-like) pericytes and astrocytes.[5] The tumor's growth rate is steadily rapid, its treatment sensitivity is low, its resistance and relapse rates are high. A chronic inflammatory microenvironment with elevated innate pentraxin 3 levels is contributory to high grade gliomagenesis.[6] IFN gene polymorphisms and cyclooxygenase-2 overproduction in the brain coincide with the induction of immature myeloid cells creating the state of reduced immunosurveillance. GM-CSF promotes myeloid cell infiltrates; these cells overexpress IL-4Rα and suppress immune T cell proliferation.[7] Digoxin-induced drop of survivin and death receptor-5 levels promoting TRAIL-mediated apoptosis of glioblastoma cells.[8] Upregulation by GM-CSF of the polymorphism-inflicted IL4Rα gene in glioma-infiltrating myeloid-derived suppressor cells inhibited immune T cell proliferation.[9] Monocyte-derived DCs generated by IFNα and GM-CSF treatment failed to express cytotoxicity to TRAIL-resistant high-grade glioma cells, but treatment with ds-DNA and rIL-2 restored the DC's cytotoxicity to primary autologous or allogeneic glioma cells.[10] Treg cells physiologically prevent autoimmune reactions. Tumor cells are masquerading as "self" for inviting Treg cell protection. CpG-depleted Treg cells secreting IL-10 and TGFβ are further recognized by their expression of CTLA-4, FoxP3 and OX40 molecules. Direct injections of CpG and low dose of either anti-OX40 or anti-CTLA-4 antibodies triggered both local and distant systemic immunomodulation resulting in the depletion of Treg cell infiltration.[11]
Diffuse large B cell lymphomas of the brain are highly vulnerable to biotherapy of low toxicity. In contrast to primary malignant gliomas, or metastatic brain tumors, diffuse large B lineage primary brain lymphomas often express EBV antigens; retain high sensitivity to infiltrating immune T cell clones and rituximab inducing an ADCC reaction mediated by NK cells. High dose methotrexate with leukovorin rescue; dexamethasone; rituximab; low dose full brain

(continued)

Table XXXIV (continued)

radiotherapy with much reduced toxicity, and as needed consolidation with cytarabine secure long complete remissions (cures).[12] The rare proven diagnosis of the aggressive primary B cell brain lymphomas arising in lesions of tumefactive demyelinaton, or in amyloid angiopathy, so far prevented the clear understanding of their etiology.[13]
Blood/brain barrier exists in primary or metastatic brain tumors? Or tumor-induced neoangiogenesis by-passed it? Ready transfer of ipilimumab and vemurafenib into melanoma brain metastases with clinical response and long survival.[14] Tumor cells communicate via the release of exosomes, which deliver proteins and microRNAs. The up-take of exosomes deriving from metastatic brain tumors endows recipient non-metastatic tumor cells with altered adhesive (more or less) and invasive (more) capabilities. Exosome uptake influences the recipient cells' propensity for, or against, apoptosis. T lymphocytes taking up tumor-derived exosomes undergo apoptosis.[15] Debated if anti-neoangiogenesis therapy (bevacizumab) reduces access of chemo-immunotherapy to the metastatic tumor?
1a *Mendelson J Baselga J Semin Oncol 2006;33:369-65.* ***1b*** *Halatsch ME et al Cancer Treat Rev 2006;32:74-89.* ***1c*** *Kuroda I et al Oncogene 2013;32:1670-81.* ***2a*** *Battke C et al Cancer immunol Immunother 2022;60:639-48.* ***2b*** *Poli et al J Immunol 2013;190:5355-62.* ***3*** *Kofman AV et al Cell Cycle 2013;12:3500-11.* ***4a*** *Sinkovics JG J Natl Cancer Inst 1995;87:1894; Cancer 1996;78:184-5; Int J Oncol 2012;40:305-349.* ***4b*** *Tili E et al J Immunol 2007;179:5082-9; Int Rev Immunol 2009;28:264-84; Carcinogenes 2010;31:1561-6; Proc Natl Acad Sci USA 2011;108:4908-13; Immun Rev 2012;253:167-84.* ***5*** *Diéras V & Pierga JY Bull Cancer 2011;98;385-9; Ochs K et al J Neuroimmunol 2013;285:106-16.* ***6*** *Locatelli M et al J Neuroimmunol 2013;260:99-106.* ***7*** *Okada H et al Ann N Y Acad Sci 2013;1284:17-23.* ***8*** *Lee DH et al Anticancer Drugs 2013 in print.* ***9*** *Kohanbasch G et al Cancer Res 2013 in print.* ***10*** *Tyrinova TV et al Cell Immunol 2013;284:146-53.* ***11*** *Marabelle A et al J Clin Invest 2013;123:2447-63.* ***12*** *Rubenstein JL et al Blood 2013;122:2318-30; J Clin Oncol 2013;31:3061-8. Morris PG et al J Clin Oncol 2013 in print.* ***13*** *Ohe Y et al Intern Med 2012;52:483-8.* ***14*** *Balakan O et al Intern Med 2012;51:2819-23.* ***15*** *Camacho L et al PLoS 2013;8(9):e73790.*
The glucocorticoid-induced OX40/CD134/ACT35 molecule is a tumor necrosis factor receptor-related factor (TRAF) expressed on CD 4/8 T lymphocytes including $CD4^{+}Fox3^{+}$ Treg cells. Its ligand is of mesenchymal cell-origin, and TLR-dependent. T cells with activated OX40 may either enhance or suppress tumor cell growth (Moran AE et al Curr Opin Immunol 2013;25 (2):10.1016/j.coi.2013.01.004; Shibahara I et al Mol Cancer 2015;14:41.

survive, a new environmental conflict. The chemotherapy resistant cells will eventually emerge and take over.

The so called malignantly transformed cells (cancer cells) of a multicellular eukaryotic host (including Homo) have descended on a reversed phylogenetic ladder to the level of those low-branching unicellular eukaryotic ancestors (amoeba, giardia, trichomonads), which were in the stage of transition from their independent life style to parasitism. The extant descendants of these ancestral parasites and the cancer cells are on the same phylogenetic level: devoid of interferon production (but responsive to it) [770, 1242, 2070]; and in the full possession of a large supply of ancient chemokines, cytokines and toll-like receptors and their ligands. Their versatile genomes rather induce their hosts' support, while averting occasional rejections. Their antigenic patterns readily change upon an attack; they switch off NK cells, and neutralize dendritic cells and immune T cells of their hosts at evolutionarily installed checkpoints (CTLA4; PD1). They convert attacker M1 macrophages to M2 supportive partners providing growth factors. Such an example is

Table XXXV Non-Coding ncRNA, Complementary cDNA: Relation to Proto-oncogenes/proteins. Circular RNAs

ncRNA-derived cDNA libraries generated. The human genome transcribes >450,000 ncRNAs. While mRNAs serve as templates for protein-coding, the function of ncRNAs, if any, is different (not translation into proteins). RNA-binding proteins form ribonucleoprotein (RNP) particles. RNPs presumably existed in the pre-cellular world. Now unannotated ncRNAs are present in intronic or in intergenic regions of the genome. In cellular life, the RNA-attachments (miRNA, snoRNA) guide the proteins to nucleic acid (either RNA or DNA) targets. RNPs in glycerol gradients reveal their size. Co-immunoprecipitation with antibodies directed at RNA-binding proteins identify the associated ncRNAs. ncRNAs were converted into cDNAs by reverse transcriptase using anchor primers and amplified by PCR using complementary primers to 5′ linkers and poly(C) tails, forward and reverse. cDNAs are cloned into plasmid vectors. Pyrosequencing is carried out with alternative pair of primers, forward and reverse. cDNA clones with or without ORF are identified.[1] Most ncRNAs lack ORFs, are transcribed by polymerase II, spliced, and polyadenylated at 3′ ends, and contain incompletely processed transcripts (3′ untranslated regions, UTRs). The most versatile RNA processing events reveal mRNA intron → snoRNA → miRNA pathways. The bioinformatic computational pipeline APART (Automated Pipeline for Analysis of RNA Transcripts) creates ncRNA cDNA libraries in archaea, yeast and mammalian subjects.[2]

Agrobacterial oncogenesis (from the Introduction). The agrobacterial Ti T-DNA oncogenic plasmid integrates itself into the genome of plants cells and induces malignant transformation resulting in crown gall tumor formation. Inter-kingdom oncogene transfers occur between agrobacteria and plants; bartonella (*B. bacilliformis)* causing hemolytic anemia (oroya fever) and verruga peruana in the Andes, and helicobacter (*H. pylori*) and vertebrate mammalian hosts (including Homo).[3–5] Oncogene-silencing transgenes protect grapewine embryos (*Vitis* species) against the Ti oncogene.[4–6] Crown gall tumor cells in *Arabidopsis thaliana* overexpress abscisic acid, ethylene and salicylic acid, major protectors against drought stress and cell death. Abscisic acid induces promoter methylation. The activated crown gall cell oncogenes (encoded by the T-DNA) Ipt, IaaH, and IaaM, are un- or hypomethylated. The less methylated are the promoters, the larger are the tumors. siRNA-mediated methylation switched off these oncogenes resulting in cessation of tumor cell growth. Hypermethylation (cytosine, rather than CG) by methyltransferases silenced the oncogenes. Crown gall tumor cells with active oncogenes overproduce growth hormones auxin and cytokinin and undergo endoreduplication, cytoskeletal re-organization (mediated by kinesin motor protein) and rapid replication.[5] The agrobacterial virulence F-box protein VirF and VirE2 unite with the plant host cells's VIP1 protein; agrobacterial genes can induce plant host cells' F box protein. The host cell's MAPK pathway helps the intranuclear entry of the agrobacterial T-DNA.[6] Agrobacterial ncRNAs include ribosomal rRNAs, transfer tRNAs, cis-antisense asRNAs and trans-small sRNAs; some ncRNAs with internal transcriptional start site (TSS) derive from mRNAs. Agrobacterial VirA sensor kinase recognizes the plant product phenolic acetosyringone. VirA phosphorylates VirG in a cascade of gene activations. Virulence gene E encodes the periplasmic sugar-binding protein ChvE. Plant cell ncRNAs initiate agrobacterial Ti plasmid replication. The ncRNAs involved are transcribed from the opposite strand of protein-coding genes and from intergenic regions of the *A. tumefaciens* genome. Riboswitches are in the 5′ UTRs of the agrobacterial mRNAs. The result is accelerated and enlarged thiamine synthesis in the bacterium.[7]

ncRNAs in oncogenesis. The pluripotency regulator human Sox2 proto-oncogene is embedded within an intron of a long spliced ncRNA mapping to locus 3q26.3-q27 (see in text).[8] ncRNAs in extremely high numbers emerge as background regulators of chromatin remodeling, gene transcription, mRNA translation and protein function. "Aberrant" ("dysregulated") ncRNA expression is highly promotional to the malignant transformation of selected cells, while large cell communities remain intact by-standers. These processes may be ancient inherent RNA/DNA

(continued)

Table XXXV (continued)

functions transferred from the precellular to the intracellular arena. The number of their nucleotides categorizes ncRNAs. The short ncRNAs are 17-30 nt in length and include micro miRNAs, piwi-interacting piRNAs and transcription initiation tiRNAs. Middle size ncRNAs are between 20 to 200 nt long and include the small nucleolar snoRNAs.The long lncRNAs (metastasis-associated lung adenocarcinoma transcript1 MALAT1, and HOX antisense intergenic RNA, HOTAIR) are over 200 nt long. miRNAs work physiologically with Argonaute proteins in RISC (RNA slicing-silencing complex). The miR-5/16 cluster at 13q14 is deleted in some cases of CLL. The miRNA-17-92 cluster at 13ORF25 regulates cell survival and as such it is amplified in certain malignant lymphomas as a promoter of the c-Myc oncoprotein. Some members of the cluster act as suppressors of the AIB1 oncogene (amplified in breast cancer). A p53 protector miR17-3p represses MDM. Adenocarcinoma cells prevail when they suppress miR-101; it restores E-cadherin's control over β-catenin. The anti-RAS let-7 miR family members act as tumor suppressors. The pro-apoptotic p53 protein exerts its activity through the activation of the miR-34 family members. The p53 antagonist c-Myc is a repressor of the miR-34 family. The lncRNA Xist transcript (X-inactive-specific transcript) silences the X chromosome by H3K27 methylation. MALAT1 represses anti-metastasis genes in favor of activating pro-metastasis genes. The prometastasis HOTAIR methylates H3K27 and demethylates H3K4. The PI3K/Akt suppressor PTEN gene (Table VIII; Figure 50) is under the control of its pseudogene (P1). The telomeric repeat-containing RNA (TERRA) is an inhibitor of telomerase, thus acting as a tumor suppressor.
Circular RNAs are relics of precellular viroids and resist degradation by ribonucleases; thus they may function post-transcriptionally. Presumably, the intron-rich Last Common Eukaryotic Ancestor (LECA) operated spliceosomes able to perform circular splicing of RNA: a downstream splice donor exon spliced to an upstream splice acceptor site. From unicellular eukaryotes metazoans inherited this faculty. CircRNAs resist digestion by exonucleases. CircRNAs express numerous microRNA binding sites, thus functioning as miR sponges.[9,10] An enormous field moves from a hidden background to the foreground.
1 *Mammalian Gene Collection Program Team Proc Natl Acad Sci USA 2002;99:16899–903; Numata K et al Genome Res 2003;12:1301–6; Rederstoff M et al Nucleic Acids Res 2010;38(10):e113; Methods Mol Biol 2012;925:211–8. Strausberg RL et al Proc Natl Acad Sci USA 2002;99:16899–903.* ***2*** *Zywicki M et al Nucleic Acids Res 2012;40:4013–24.* ***3*** *Gyémánt N et al Acta Microbiol Immunol Hung 2004;51:321–32.* ***4*** *Galambos A Plant Cell Rep 2013;32:1751–7.* ***5*** *Gohlke J et al PLoS Genet 2013;9(2):e1003287.* ***6*** *Zaltsman A et al Cell Host Microbe 2010;7:197–209.* ***7*** *Lee K et al PLoS One 2013;8(8):e707220.* ***8*** *Amaral PP et al RNA 2009;15:2013–27.* ***9*** *Deng G & Sui G Int J Mol Sci 2013;14:18319–49.* ***10*** *Wang PL et al PLoS One 2014;9(3):e90859.*

the CD47 ligand of the leiomyosarcoma cell converting signal regulatoy proteins of M1 to M2 in the host Mϕs [2274]. Research to be initiated may prove that the ancient parasites used first the PD1 ligands in order to induce the apoptotic death of the attacking immune T cells of their hosts. The cancer cells just revert to that inherited level [1961].

Comment. This author repeatedly expressed his admiration for the performance of the ultimate bioengineer RNA/DNA complex. Even the endogenously initiated so-called malignant transformation of selected cells in a distressed multicellular host is described as if it were an inherent faculty of the primordial RNA/DNA complex, fundamental to cellular life self-engendered in a physicochemical universe. The ultimate bioengineers were not spared from exposing some of their faults and failures (misrepaired DNA; AID/APOBEC-induced lymphoma-genesis; progeria).

Without distorting the balance, the RNA/DNA complex remains a superior system still evolving. Table XXXI is added to show that the supreme bioengineers may fall victim to chaos induced by cascades of externally, or internally superimposed mutagenesis manifested in loss-of-function point mutations of genes crucial in encoding basic anatomy. However, the chaotic background did not favor oncogenesis, as if the process of oncogenesis were regulated by its own rules (deletion of tumor suppressor genes; gain-of-function point mutations; gene fusions, etc). *)

*) *Extraordinary genomics assay for congenital maxillofacial deformities (Sólya K et al Orvosi Hetilap Budapest 2015;156:1483–1490. doi:10.1556/650.2015. 30240). No increased incidence of malignant tumors in 8093 patients with oral clefts (Bille C et al Am J Epidemiol 2005;161:1047–1055).*

Comment. The RNA/DNA complex proved itself to be the consummate bioengineer of the living matter. Were it based on exact intron counts (it is not), an intron-rich RNA/DNA complex would be held dominant in the unicellular and early multicellular eukaryotes (extinct and extant), and in selected single cells of advanced multicellular eukaryotes (including Homo), which undergo the so-called 'malignant transformation'. This process could be viewed as a descent to the lower phylogenetic levels of the ancestors, in the form of a reversed ontogenesis. In it, cell proliferation would outweigh apoptosis. Whereas, in advanced multicellular hosts (including Homo), a highly evolved DNA/RNA complex would hold under control the large mass of subservient somatic cells. Disciplinary apoptotic events would outweigh proliferation. Resting and activated stem cells would operate the switches between DNA/RNA and RNA/DNA complecti. In the fertilized zygote, the initial RNA/DNA complex would mature into the DNA/RNA complex of the adult individual.

Senescence and natural death terminate the existence of the individual. In the most complex and versatile rescue operation of individual cells, an RNA/DNA complex would regain the dominance. Under the present biological circumstances, that ends in the death both of the host, and within it the driver, the RNA/DNA complex, in the "oncogenome-driven" vehicles (the cancer cells). The immortalized single cells are not able to replicate in the soil or in the oceans, and could not so far create in the nature new societies for themselves. In cultures maintained in the laboratory, these cells live as long as they are cared for. In permanent cultures, these cells maintain a violent community. Well fed, and exempted of an immunological attack of a host, each individual cell is for itself, and the stronger and larger cells engulf their smaller and weaker neighbors (as this author watched his long term human cancer cell cultures in his laboratory, where HeLa cells were not carried in order to avoid the cross-contaminations overrunning all other cultures) [1501]. Genes are snatched from each other and recombine; could they eventually create new free-living individuals? Deep-frozen in liquid nitrogen (at the temperature of outer space), the RNA/DNA complex may preserve its vitality "forever". It may re-start evolution in another favorable environment. Cryopreservation and retrieval of livable cells is possible with appropriate technology (in tissue culture media with dimethylsulfoxide and 20 % fetal calf serum, in the laboratory).

Some factual information may become available on the long term fate of the living matter in the Universe, whether in organized or disorganized and doomed communities, when other planets may be explored. Inhabitants of the Earth have the advantage (or the disadvantage, if *Homo sapiens* totally degenerates or even annihilates itself), that the RNA/DNA complex also encoded the human cerebral cortex (*vide supra*): another consummate achievement of bioengineering (but not without errors). The primordial and the genetically engineered RNA/DNA ↔ DNA/RNA complex may save life on Earth, or even in the Universe.

Selected Abbreviations

HERV human endogenous retrovirus. HML human mouse mammary virus like. Click for coding open reading frames of HERV-K including the expression of full-length post-translationally processed env proteins that could be incorporated into HIV-1 particles, by Brinzevich D Young GR Sebra R et al J Virol 2014;88:6213-23. Click for syncytin-1 env protein-expressing endogenous retrovirus in human pituitary gland cells, by Buslei R Strissel PL Henke C et al Neuropathol Appl Neurobiol 2014; Mar 17. doi:10.111/nan.12136. Click for chromosome X-inserted HERV-Fc1 releasing viral mRNA in blood in patients with multiple sclerosis, by de la Hera B Varadé J García-Montojo M et al PLoS One 2014 3;9(3):e90182. Click for 16th International Conference on Human Retrovirology highlights from Montreal, Canada, by Barbeau B Hiscott J Bazarbachi A et al Retrovirology 2114;11:19. doi:10.1186/1742-4890-11-19.

PI-PCL phsphoinositide-specific phospholipase Ca^{++} -mobilizing. For PI-PCL in Tetrahymena, click Leondantis G Sarri T Dafnis I et al Eukaryot Cell 2011;10:412-22. For PLCγ1 mutation in angiosarcoma, click Behjati S Tarpey PS Sheldon H et al Nature Genetics 2014;46:376-9. For ibrutinib-resistant Bruton kinase, click Woyach JA Furman RR Liu T-M et al N Engl J Med 2014;370:2286-94.

NFκB/STAT Nuclear factor kappa B; signal transduction and transcription. A parallel drawn in manuscript between this ancient cell survival pathway in the Nematostella, and in the oncogene-oncoprotein of human diffuse B lymphoma cells including Reed-Sternberg (RS) cells, pointing out that both subjects harbor latent viral flora: endosymbiotic algal viruses in the Nematostella, EBV and endogenous retrovirus in the lymphoma (RS) cells (in the Appendix; in Int J Oncology Oct 2015;47:1211–1229).

"Making Cancer History" at M.D. Anderson Hospital

In the 1950s: Most Cancers Defeat Surgeons, Radiotherapists, Hematologists; Medical Oncologists Do not Exist

Cancer chemotherapy and BCG. In the late 1950s and the early 1960s, the Department of Medicine (DM) at The University of Texas MD Anderson Hospital, Houston, TX (later MD Anderson Hospital Comprehensive Cancer Center, MDACCC) struggled to establish that tumor-specific immunotherapy may become a (or: the) major treatment modality for cancer. Arrived to MDACCC in the mid-1960s, a very prominent NIH/NCI (National Institutes of Health; National Cancer Institute; Bethesda, Maryland) detachment, set up as the Department of Developmental Therapeutics (DDT). The most prominent and highly supported DDT succeeded in proving that combination chemotherapy, even in its early phase, and despite its major, sometimes even fatal, toxicity, practically without remedy for that at that time, was more efficient in inducing remissions, than any of the earliest futile attempts at immunotherapy (consisting of various "cancer vaccines", or "immune sera", arrived at on immunizations with irradiated cancer cells). Non-tumor-specific immunostimulation with live attenuated bacteria (BGC, Corynebacteria) was accepted in combination with the chremotherapy. The leader of this trend, Georges Mathé of Villejuif, France, was receiving credit worldwide, a citation in [261b], for immune-stimulating the chemotherapy-recipient leukemia patients with Bacille Calmette-Guerin (BCG). For the same purpose, TICE-BCG of R.G. Crispen was used in the USA. It has become a vogue to show that BCG improved chemotherapy results 10 % plus. The newly appointed chief of the medical breast cancer service at the DM (George Blumenschein) and his resident physician associate of the DDT (Gabriel Hortobagyi) united their services, and as a result reported the highly desired extra 10 % benefits in remission induction for those patients with metastatic breast cancer, who received BCG scarifications in addition to their combination chemotherapy regimens [2275a, b]. Cancer history

J.G. Sinkovics, *RNA/DNA and Cancer*, DOI 10.1007/978-3-319-22279-0_29

was made! Members of the DDT followed suit showing very similar results in practically every tumor category (from leukemias to solid tumors, including especially malignant melanoma), in which BCG was used in addition to standard chemotherapy. However, BCG-enforced regimens were then abruptly abandoned, except for intravesicular instillations for superficial transitional cell carcinoma of the urinary bladder, discovered and practiced elsewhere, not at the DDT (A. Morales *et al*, University of Texas Health Science Center, San Antonio, TX).

In many cancer institutions worldwide, numerous exemplary DDTs were established. The outstanding merit to these worldwide DDTs, among them MDACCC's DDT first, was deserved, for their steadfast efforts to diminish the adverse side effects of combination chemotherapy. There was a gradual introduction of new antibiotics; red cell, platelet and leukocyte transfusions; natural bone marrow stimulatory biomolecules for erythro-, leuko-, and thrombocytopoiesis, CNS-acting antinausea medications; and anti-GvHD medications for patients receiving allogeneic bone marrow transplants. While none of these interventions was actually discovered at the MDACC DDT, they were immediately applied, evaluated and reported on from there. NIH/NCI grants highly favored the DDT projects.

The late mutagenic effects of chemo-radiotherapy remained unpreventable. Saving a child's life with chemo-radiotherapy carries most severe late consequences (St. Jude Hospital data in JAMA June 12, 2013). Pediatric patients becoming survivors of cancer chemo-radiotherapy received in their childhood, in reaching middle age, at age 45 years, in 98 % of the cases suffer a severe chronic ailment. In percentiles, 65 % pulmonary, 62 % hypothalamic-pituitary axis (endocrine), 56 % cardiac (arrhythmia, valvular defects, ischemia), and 48 % neurocognitive disorders (CNS dysfunctions) are dominating. Hepatic, renal, hematopoietic, osteoporotic pathological findings amount up to 13 %. Of these disorders, 80 % is considered to be disabling to life threatening. Some of the disabling conditions are unique, and are not known to have occurred naturally. Of 1713 adult subjects, hundreds developed one or more subsequent new neoplasia (335 solid tumors, 13 hematologic neoplasms) [2276, 2277]. In the pediatric population mesenchymal tumors (hemato-lymphopoietic and sarcomatous) prevail. At the present time, it has not been resolved how diagnostic CT scans could possibly be replaced by PET/MRI scans, and if combination chemotherapy could possibly be replaced by targeted therapy and immunotherapy (vaccines; monoclonal antibodies; immune lymphocytes; genetically engineered oncolytic virotherapy). At The Ohio State University College of Medicine, Timothy Cripe wrote an editorial in the *Molecular Therapy* about the possible use of oncolytic viral therapy for some pediatric malignancies [2278]. Indeed, for neuroblastoma and sarcomas this treatment modality applied postoperatively appears to be effective; for acute leukemias, in which an immediate complete remission is the most efficient prognostic event, only new targeted modalities of treatment could possibly replace combination chemotherapy.

The non-dividing nerve cells of the adult brain suffer long-lasting physico-chemical damage during chemotherapy. The "chemobrain syndrome" is not a psychosis rooted in fear consequentially to adverse propaganda. It is against the dogma, that rapidly dividing cells succumb to chemotherapy, but resting non-dividing cells survive intact. The metabolically active, but not dividing brain cells undergo as yet not clearly defined pathophysiology. *)

*) *Is it possible that androgen hormone suppressive therapy of prostate cancer increases the incidence of Alzheimer's disease? (Nead KT et al J Clin Oncol 2015 Dec 7 pii:JCO636266).*

The damage is detectable by highly sensitive functional fMRIs, and near-infra-red spectroscopy. The major early conclusion is that chemotherapy (combination chemotherapy for breast cancer) induced in young adult patients the cognitive decline that characterizes the aging brain (the Jeffrey Wefel and Charles Cleeland data, as presented in the MDACCC publication to physicians Oncolog, vol 58/9 Sept 2013). In the September 2013 issue of the J Clinical Oncology, European oncologist-hematologists report 20+ years late brain damage consisting of decreased white matter integrity, neuropsychological dysfunction, and other characteristics of the prematurely aging brain in those cancer survivors, who in their younger years received brain-directed radiotherapy and/or intrathecal chemotherapy. F. Daniel Armstrong of the University of Miami, FL, Jackson Memorial Medical Center, in his editorial, cannot immediately abandon life-saving chemotherapeutic modalitites (and so informs parents and child patients). He recognizes that the "wake up call" was sounded, and that the medical community must respond to the challenge (meaning that effective and less harmful replacement treatment modalities will have to be found).

Here it is on the screen on April 27, 2015! Announcement from Targeted Oncology http://targetedonc.com/articles at AACR Annual Meeting, Philadelphia. The chimeric receptor CAR T cell therapy JCAR017 elicited 91 % flow cytometry-proven complete remission rate in 20 of 22 pediatric patients with relapsed/refractory $CD19^+$ acute T cell lymphoblastic leukemia. Neurotoxicity and/or cytokine release syndrome occurred in 8 patients. Breakthrough therapy designed Protocol JCAR015 treats patients with relapsed/refractory B cell ALL with similar efficacy. Hans Bishop & Michael Jensen, Juno Therapeutics, Ben Towne Center for Childhood Cancer Research at Seattle Cancer Research Institute (more in the Appendix 1). Observe: virologists use the CAR abbreviation for cocksackie and adenoviral receptors expressed on cell membranes.

The high incidence of fatal neutropenic septic shock after combination chemotherapy given in the 1950s–60s has substantially receded. Newly established cancer institutes in Florida (North Ridge Hospital, Ft. Lauderdale; St. Joseph's Hospital, Tampa) invited this author (JGS) for discussions on the incidence and treatment of chemotherapy-associated septic infections (Howe CD & Sinkovics JG. Life-threatening infections. Project M26/gm9. Research Report, University of Texas M.D. Anderson Hospital 1970; 1972. Sinkovics JG & Smith JP. Bacteroides and Salmonella septicemia. Cancer 1969;24:631–6; 1970;25:663-71). Sinkovics JG. Infectious Complications of Cancer. J Inter-American Medicine 1979;4:10–14; In: The Immunobiology and Immunopharmacology of Endotoxins. H. Friedman & A. Szentivanyi, editors, Plenum, New York, 1986;269–279. Sinkovics JG. Medical Oncology An Advanced Course, Marcel Dekker, New York, 2nd ed. pp 1557–1738, 1986). Shullenberger CC, Leavens ME, Sinkovics JG.

Viruses in malignancy-associated encephalopathies. Project M23/gm20. Research Report The University of Texas MD Anderson Hospital, pp 455-6 1974-5.

Patients with advanced stage Hodgkin lymphoma are treated with the BEACOPP (escalating dosage bleomycin, etoposide, doxorubicin, cyclophosphamide, vincristine, procarbazine, prednisone) regimen. Treatment-related mortality is in the 2 % range, except for aging (>40–50 y old) patients with declining performance status; there it reaches 7 %. In contrast, 8 courses of this regimen results in 82 % relapse-free 10 y survival [2279].

Another triumph for chemoradiotherapy (rather biotherapy) is the steadily improving survival with primary B cell brain lymphomas. The brain particularly welcomes a primary or a metastatic tumor. While the microglia remains inert, pericytes and astrocytes have been recognized to support tumor cells by providing them with growth factors. Exceptions are the primary cerebral B-lineage lymphomas, which attract T cell infiltrates (either immune T cells, or Treg cells, or both), especially, when the lymphoma cells express EBV antigens. Primary B lineage (large cell undifferentiated) brain lymphomas respond to high dose MTX with LV rescue, dexamethasone, rituximab, and low dose whole brain radiotherapy with (or without) consolidation with cytarabine. Long tumor-free survival with minimal neurotoxicity is the ever improving results (Table XXXIV). The contribution of MDACCC's Neurosurgery Department to brain tumor immunology/immunotherapy (viral oncolysis) have been cited in the text [71, 2014a, b]. The oncolytic Delta-24 glioblastoma viral therapy 'moonshot project' of the Neurosurgery Department of MDACCC is not aware of the fact that the first viral oncolysate therapy project for metastatic human sarcoma and melanoma was initiated at the MD Anderson Hospital some 50 years ago [71, 2014b], and that patients' immune lymphocytes induced autophagy of autologous sarcoma cells (Figure 68).

In the 1960s and 70s it has been firmly established that the standard treatment for metastatic cancers (including subclinical disease) both in children and adults should be combination chemotherapy (and radiotherapy, as indicated). Even patients suspected of having subclinical undetectable metastases received preventive (adjuvant) chemotherapy. For example, women with breast cancer producing lymphovascular invasion of tumor cells, but with negative axillary lymph nodes, are subjected to late metastases, due subclinical spread of the primary tumor. After postoperative adjuvant chemotherapy to seemingly tumor-free patients, but some of them presumably harboring undetectable subclinical disease, the incidence of late metastases dropped significantly: as shown in the survival graphs in the Nottingham, UK, report [2280].

Natural IFN-alpha for cancer biotherapy. Dick (Frank) Rauscher for the American Cancer Society and Mary Lasker released funds in the 1970s for the purchase and investigational use in cancer therapy of a human natural (leukocyte) IFN-alpha preparation. The natural (leukocyte) interferon was prepared by Kari Cantell at the State Serum Institute and Central Public Health Laboratory, Helsinki, Finland, and was used first by Hans Strander at the Karolinska Institutet, Stockholm, Sweden, most prominently, for the elimination of subclinical metastases from osteosarcoma of the limbs, after the surgical removal of the primary tumor [2281a, b]. A connection between tumor suppressor gene activation and IFNα production was proposed [2282]. At MDACCC, the American Cancer Society funds supported in full the DDT for the use of the Finnish natural leukocyte IFNα preparation for the treatment of human cancers. The DM, working already with an in-house-made mouse IFN preparation, which was tested against a retrovirally induced mouse leukemia [2283a–d], was completely left out from the investigational clinical trials of both the natural (leukocyte), and later the first industrially genetically engineered IFNα preparations (IFNα2B, Intron, Schering; IFNα1B, Roferon, Hoffmann La Roche). In the DDT trials, the human IFN preparations worked in inducing stabilization of disease, or partial remissions (falling short of being curative) for hairy cell leukemia, some lymphomas and myeloma, and some "solid tumors" (breast and kidney carcinoma; and especially, for melanoma). Donor Lasker, and recipients Clark and Rauscher appointed Jordan Gutterman of the DDT to conduct the clinical trials with IFNs [2284a–c]. Cancer history was made! The DM could only express future expectations, that biologicals may replace chemotherapy, at least in the adjuvant setting [2070]. The DM left out of the pioneering IFN-trials, made an attempt to extract natural human IFN from the blood of patients who received viral oncolysate therapy (registered Project M27/gm23, in the MDACC Research Report 1978, p 332). Due to lack of financial support, the project could not be initiated. However, the IFN project at the DDT was successfully concluded [2284c].

Refined combination chemotherapy continues its rapid advance with moderate temporary efficiency for metastatic solid tumors and leukemias. However, sizeable numbers of patients achieving remission of originally metastatic solid or liquid tumors (or leukemias) eventually relapsed, sometimes years after entering remissions. If relapses occurred several months after remission induction, the relapsed tumors retained some of their original chemotherapy sensitivity and could be re-induced into remission (of short duration). If relapses occurred years later, the tumors proved to be chemo-radiotherapy resistant. Some prominent oncologists claimed that relapses occurred due to under-dosed chemotherapy (raising the immediate attention of malpractice attorneys). It could be recognized only recently, that the late relapses were due to the emergence of chemo-radiotherapy-resistant trans-speciated stem cell tumors, consistent with the natural biology of the cancer cell. A basic biological faculty of the cancer cell was recognized much belatedly. The debate concerning chemotherapy dosage continues: should massively overweight patients receive chemotherapy dosages calculated to their actual weight and height in full (and bear the toxic side effects of that dosage), or reduced to what their "ideal weight" would call for. However, even against the full dosage, the inherent cancer cell biology could find its escape route in the form of a trans-speciated tumor recurrence. The amphisbaenic cancer cell upon encountering an obstacle can readily change the direction of its locomotion.

The NCI does not support cancer immunotherapy, or clinical trials with oncolytic viruses. Work on cancer vaccines to sustain remissions induced by chemotherapy (prevent relapse from residual micrometastases), or on therapeutic cancer vaccines to actually induce durable remissions, remained way behind in the experimental realms.

Attempts at oncolytic viral therapy received no NIH/NCI funding in the 1960s and 1970s: at the best it was "approved without funding due to low priority". Oncolytic viral therapy neither at the DM/MDACCC (a DM project), nor at Emory University, Atlanta, Ga (William Cassel's project) received any NCI funding. However, just in the last decade (late 2009s, early 2010s), genetically engineered poxviruses expressing human tumor antigens were proven to induce powerful and specific immune reactions against tumors (prostatic carcinoma) with significant retardation of disease progression [2285] (*vide infra*). These clinical trials were not initiated or tested at the DDT of MDACCC. However, full durable complete remission inductions have not as yet been achieved with these viral antigen-carrier tumor cell vaccines.

Of the protocols initiated at the DM, that the NIH/NCI refused to support in the mid-1970s, was especially prominent the viral oncolysate vaccine in combination with chemotherapy, as proposed by this author (JGS). The proposal was a regimen for the treatment of metastatic melanoma and sarcoma. It was not funded by the NIH/NCI. The NCI consented to non-specific immunostimulation (BCG) to be combined with cancer chemotherapy, but was opposed to, and refused to fund, attempts at tumor-specific immunotherapy in combination with chemotherapy. However, new protocols for the combination of immunotherapy with viral tumor vaccines and chemotherapy, are now (with decades of delay) receiving full NCI support. This is now in sharp contrast to the theoretical expectations prevailing at the NCI in the 1970s, that these two modalities of treatment would extinguish each other. The carefully selected active tumor-specific immunizations in addition to certain modalities of chemotherapy are now proven to be at least additive (may be even synergistic) in clinical results. Even some chemotherapeutic agents (cyclophosphamide; paclitaxel), or even radiotherapy-afflicted tumor cells, may exert immunotherapeutic effects. Tumor cells damaged or killed by radio-chemotherapy induce immune reactions in their hosts. Merit for this change of attitude goes to the NIH/NCI department of Jeffrey Schlom, for the design and supervision of these clinical trials, especially for patients with inoperable prostate cancers [2285]. The combination of a p53 vaccine with chemotherapy was at least additive in clinical results for metastatic small cell carcinoma of the lung conducted at the H.L. Moffitt Comprehensive Cancer Center [2286].

After the "cancer history" with BCG and interferon, all other tumor immunotherapy measures emerged outside the MDACCC. It was combination chemotherapy that continued to receive a disproportionately overwhelming support at the DDT of MDACC and elsewhere. Discoveries made outside the MDACCC (doxorubicin; cytosine arabinoside, or cytarabine) received prompt supportive funds to be tested in clinical trials at the DDT of MDACCC, while support for tumor immunotherapy was left way behind. Early efforts at it failed to receive any support from the NCI/NIH at the DM through the entire 1970s. Cancer chemotherapy proved itself to be a remission-inducer modality of treatment, and thus it was supported in full. The pharmaceutical industry started to pay consultation fees to academic physician principal investigators of chemotherapy protocols at academic institutions. Emil Frei III, the head of the DDT, published in Cancer Research in 1985 his famous paper "Curative cancer chemotherapy", appearing after he left M.D. Anderson for Dana-Farber [2287].

Finally the Basic Elements of Tumor-Specific Immunotherapy Appear on the Scene

Immune T cells and NK cells are recognized; *monoclonal antibodies appear*. Much belatedly, tumor-specific immunotherapy achieved its first firm adjunctive status to chemotherapy with the discovery of monoclonal antibodies (rituximab and trastuzumab, etc), but none of these original monoclonal antibody discoveries were made at the DDT of MDACCC; however they were immediately applied there in clinical trials. The CHOP chemotherapy regimen for malignant lymphomas was designed at the DDT: it has made worldwide medical history. Now, with rituximab added, it can induce durable (>5 y) complete remissions (CHOP alone 67 %; CHOP-R 100 %, for example in British Columbia, Vancouver, Canada) and elsewhere. Rituxan fails only in the case of defective NK cells, which do not express Fcγ globulin receptors, and cannot perform the ADCC reaction.

T cell and NK cell transfusions with lymphokine support (first with IL-2) provided unequivocal evidence for the mobilization of natural host defenses against autologous cancers, as these interventions could induce durable complete remissions (especially in cases of malignant melanoma and kidney carcinoma in endogenously lymphocyte-depleted patients). The depletion of the resident lymphocyte population (with fludara or cyclophosphamide chemotherapy) allowed the expansion of the activated and by infusion-donated immune T or NKT cells [2017, 2288]. Lately, genetically engineered immune TCRs (CAR: chimeric antigen receptor) increased the efficacy of the procedure (*vide infra*). The flagship was the NIH/NCI Bethesda, MD, tumor immunology/Immunotherapy department under the directorship of Steve A. Rosenberg. Standing in line, is the Philadelphia and Seattle, and Sloan-Kettering protocols inducing rapid complete remissions in B cell malignancies. Genetically engineered TCRs (the CAR receptors) specifically and most powerfully activate *in vitro* autologous immune T cells for the destruction of malignant B cell lymphomas upon their re-infusions to the patients (the ASCO POST interview with R.J. Brentjens of the Sloan-Kettering Cancer Center on May 15 2013). The CD19-specific CD28/CD3ζ (zeta) second-generation dual-signaling chimeric antigen receptors (CAR 19-28z) mobilize the replicating immune T cells to kill the very last chemotherapy-resistant ALL tumor cells [2289] (more in the Appendix 1). Antibodies and immune lymphocytes finally receive the attention they always should have deserved, may be decades earlier [147].

At the Department of Medicine private donors support combined chemo- and tumor-specific immunotherapy. There was a tendency at the DM in the 1960s to individualize treatment outside of the rigidity of the earliest NCI/DDT-designed and NIH/NCI-approved chemotherapy protocols. If in the original ongoing protocols of the 1960s, patients in a treatment arm did better than patients in the control arms, the statisticians (and principal investigators) of that early era, forbad transfer patients from the control arm to the treatment arm. Thus, at the time of the final calculations, the treatment arm stood out splendidly shining over the control arm. Such a protocol was the one, which proved that continuous infusion of cytarabine was more effective, than bolus cytarabine. This protocol was not accepted as written at the DM: there eventually all patients received cytarabine in continuous infusion. Convinced by the superior efficacy and better control of toxicity of subsequent protocols designed at the DDT, also allowing the transfer of patient from the failing to the better performing arms of the protocol, patients with metastatic cancers received full dose combination chemotherapy regimens by registered protocols from the late-1960s on at the Department of Medicine. Modern protocols allow transfer of patients from treatment arms performing inferiorly to the treatment arms performing best throughout the course of an investigational treatment protocol.

The Head of the DM, Clifton Dexter Howe, believed in cancer immunotherapy. He continued his efforts to promote cancer immunotherapy at the DM. His belief was that cancer will eventually be vaccinated against, and will be treatable with antibiotics and antibodies, such as "immune sera", and white blood cells. However, he observed an analogy between tropical (protozoal) infectious diseases and cancer, and agreed that chemotherapy was useful and necessary, when the cellular pathogens were not the body's own. His hopes were that chemotherapy will eventually be replaced by biological means of treatment. He observed the early years of chemotherapy with great concerns. The early years of combination chemotherapy were truly very dismal. A young patient with acute leukemia could die in leukopenic septic shock (most frequently due to pseudomonas septicemia) in 48 h after receiving a VAMP-like regimen (vincristine, amethopterin/methotrexate, mercaptopurin-6, prednisone) in full dosage [73]. In tantalizing sharp contrast, a unique patient managed to recover from severe life-threatening myelosuppression, and survive in complete remission, most often temporarily, but exceptionally in a durable form of it. The possibility was raised, that actual cure of acute leukemia, on those early protocols, on rare occasions, was truly achievable. Yet Clifton Howe declared that "combination chemotherapy was totally irrational" and that "the administration of cytotoxins destroys or damages both cancer cells and the normal cells indiscriminately" (cited in 2290, *vide infra*). However, the promoters of combination chemotherapy administered in top dosage at the DDT claimed that the effects were not "indiscriminate", in that, the rapidly dividing cancer cells were much more sensitive to it, than the resting healthy cells; thus, recovery from toxicities (cell deaths) from the remaining intact healthy cell compartments was possible. The DDT received the full support of hospital president Randolph Lee Clark, who has originally recruited the financially highly supported DDT from the NCI/NIH. That was the turning point in medical history, when the granting agency NIH/NCI, and the developing and rapidly expanding pharmaceutical cancer chemotherapy industry devoted their full support to combination chemotherapy for the cure of metastatic cancers. Combination chemotherapy in full dosage was established to be the standard treatment for metastatic cancers in any age groups of patients. Some promoters of it went as far as to accuse medical oncologists not to have used chemotherapy in full dosage, when the treated patients failed to respond, or relapsed. That is, the basic nature of "malignant transformation" (the natural history of relapses and recurrencies of trans-speciated tumor stem cells) was not allowed to be recognized earlier. Cancer cell dormancy was thought to occur due to underdosed chemotherapy. The failure to cure cancer with chemotherapy was attributed to reduced dosages for attenuating the side effects. The naturally chemotherapy-resistant *ur*-stem cell escaped recognition and its special targeting has become applicable much belatedly. That is, now, when imatinib mesylate was administered first to patients with Ph-chromosome-mutated chronic myelogenous leukemia. Predating that event were the epoch-making discoveries made in China: trans-retinoic acid and arsenicals inducing remissions especially in acute promyelocytic leukemia by re-differentiating the leukemia cell [2291a–c].

Cancer chemotherapy from its empirical application has become a scientifically designed modality. It was the result of cooperative professional work between pharmacists and clinicians. It has become an established scientific exercise how to reason for the efficiency of combination chemotherapy protocols. The efficient, but toxic, chemicals were combined without overlapping toxicities (if that was at all possible). Also, the chemicals were to act upon different biochemical pathways of the tumor cell (but most of them indiscriminately destroyed the DNA). The aim was the destruction of the malignantly transformed and rapidly replicating DNA in the tumor cells, without affecting the resting healthy cells. Unfortunately the physiologically rapidly replicating normal hematopoietic cells, and the daily regenerating mucous membranes in the GI tract, scalp hair roots and nail beds were unavoidably co-targeted (suffered "collateral damage"). Alkylators attach the terminal carbon of their ionized ethylene immonium ($-N\text{-}CH_2CH_2-$) in nuclear DNA as their targeted sites. Even if repair enzymes correct the breaks, the error-prone repairs prohibit the function of the DNA strand as a valid template. Either loss of function of mutated cells, or eventually cell death (by various mechanisms) follow. Broken off genes travel to fuse with another gene creating entirely new genomic modalities. The prodrug, cyclophosphamide is converted first by liver cell enzymes (microsomal cytochrome P450), and then in the tumor (and other healthy) cells into phosphoramide mustard. However, the reduction of chemotherapy toxicity usually resulted in decreased anti-cancer efficacy (truly). This recognition led to the design and use of "dose-dense" chemotherapy regimens, in which the individual dosages of the drugs were not increased, but they were administered in shorter intervals.

This author, as a medical oncologist in the adult clinical service at the DM, delivered combination chemotherapy in full dosage by the protocol (and mastered all measures of the "supportive care" to counteract life-threatening toxicities). He shared the vision of his department head, Clifton Howe, but had to realize that effective cancer immunotherapy was not available (non-existent), whereas the chemotherapy protocols promised the induction of an immediate remission. However, hope for eventual cancer immunotherapy was not ever abandoned, inasmuch as it was clearly shown that the inflicted hosts attempted the rejection of their tumors by making antibodies and generating immunoreactive lymphocytes with weak, but detectable anti-tumor efficiency (*vide infra*, The Section).

A major setback at the DM should now be revealed. Cis-platinum was discovered when platinum electrodes released it into an electrical field of ammonium chloride solution, where it inhibited bacterial growth. Rapidly replicating bacteria were inhibited to divide by the cis-diamminedichloroplatinum in the solution.

The Department of Medicine received the original cis-platinum manuscript from the journal Science for review. Clifton Howe opposed the publication of the original cis-platinum manuscript. He felt that there was a world of difference between rapidly growing genetically identical bacteria and a highly diverse population of malignant cancer cells; he wanted new experiments actually showing that the growth of cancer cells was inhibited in the same system, and normal healthy cells were spared. Howe worded a negative reply to be sent to Science; the negative reply was sent to Science. The manuscript was not published in Science (Science must have received other opposing evaluations, as well, because the astute editors did not base their final decision on just one reviewer's opinion). Soon, a revised cis-platinum manuscript appeared in print in Nature. With the support of Alexander Haddow, Nature published that manuscript [2292]. Disregarding its initial quite severe side effects: intractable nausea and vomiting; neurotoxicity; rare second mutations, inducing acute leukemia (this author observed this latter event in one patient with ovarian carcinoma, who responded to cisplatin, but died with acute leukemia), cisplatin has become one of the major cancer chemotherapeutic drugs. It, and its chemical derivatives, were developed into widely used cancer chemotherapeutic agents (but not without toxicity). In aqueous solution, cis-platinum sheds its chloride atoms, and thus becomes a DNA-binding reactive electrophile. It was Melvin Samuels at the DM/MDACCC, who induced the first complete remissions of metastatic testicular germ cell carcinomas with cis-platinum therapy. He exclaimed: "cis-platinum is the penicillin for cancer!" This author regrets, that he was involved in the negative evaluation of the first cisplatin manuscript sent from the Department of Medicine to Science for rejection. However, ultimately, cancer history was made again (but not at MDACCC). Decades later, cisplatin remains an important component of many combination chemotherapeutic regimens; especially, it keeps its firm hold on germ cell cancers (with bleomycin and etoposide, the BEC regimen). Indeed, cisplatin contributed to the cure of metastatic germ cells tumors, while with etoposide, it induces seemingly complete remissions in cases of small (oat) cell lung cancer, but with a very high relapse rate.

Up to the 1980s, the NIH/NCI did not support clinical projects in which active tumor-specific immunotherapy (tumor vaccines) were combined with chemotherapy. A tumor vaccine might have been acceptable only if it were directed against a specifically identified "cancer antigen". The oncolytic viral vaccine therapy for the treatment of metastatic cancers (melanoma and sarcoma) at the DM was not supported either. The viral oncolysate melanoma vaccine of W.A. Cassel at Emory University, Atlanta, Ga, was also flatly and repeatedly rejected for support by the NIH/NCI. There, NDV oncolysate immunotherapy was administered to patients with institutional support in itself, without chemotherapy [274]. In combination with chemotherapy, a PR8 influenza viral oncolysate was given at the DM/MDACCC [71]. The NIH/NCI's reasoning against support was the strong presumption prevailing then there, that these treatment modalities would antagonize/extinguish each other. Even though, this argument could not be held against the Cassel protocol, which contained no chemotherapy, only NDV oncolysate [274], that was also rejected for NCI support, nevertheless.

This NCI principal investigator of retroviral mouse leukemia research (JGS) experienced an unexplained series of denials for the continuation of his research grants. Initiated much before the arrival of the DDT, his grant renewal applications were now declared to be "duplication of research" and at the best "renewed without funding due to low priority". In a much belittled and suppressed state of existence, the DM still exerted its efforts toward developing immunotherapeutic protocols for the treatment of cancer. This author (JGS) as a Rockefeller Fellow at the Waksman Institute, Rutgers, New Brunswick, NJ (where streptomycin, actinomycin D, and several externally applicable antibacterial antibiotics were discovered), visited from there M.D. Anderson in 1957. After completing his internship and medical residency at Cook County Hospital in Chicago and at M.D. Anderson as an American Cancer Society fellow in 1959, he was appointed assistant professor of the University of Texas M.D. Anderson Cancer Center in 1962. He received NIH/NCI grants for the study of the anti-leukemia effect of the graft-*vs*-host disease [261a, b], and became recipient of a NIH/NCI cancer research career award (5-K-CA-1647). His several specialty board examinations were passed in good order, and impeccable performance of his clinical duties advanced this author to full professorship at MDACCC. In the mid-1960s, he was appointed chief of the research laboratory Section of Clinical Tumor Virology and Immunology (the "Section"). His clinical appointments were to lead the then united Sarcoma/Melanoma Clinical Service. Ideally, the Section could prepare a treatment modality (an autologous "viral oncolysate" cancer vaccine) promptly for a patient treated at the Clinical Service. Not prospectively randomized patients with metastatic sarcoma or melanoma received active tumor-specific immunotherapy with a preferably autologous preparation of the viral oncolysate vaccine in combination with standard chemotherapy; in addition, one group of patients received BCG scarified in the skin over the viral oncolysate inoculation site. The efficacy of the investigational vaccine was uncertain; the efficacy of the chemotherapeutic regimen (as to remission induction, but without a cure) was proven. It would have been unethical to withhold chemotherapy. Chemotherapy was given in full dose. When the blood counts started to recover, the vaccinations were administered.

This author is grateful to thoracic surgeon (and outstanding drosophila investigator) Walter J. Burdette, who within his authority released intra-institutional funds to the Section. As one of the institutional members excelling in basic sciences outside the operating room (out of his many prominently published papers, one was submitted to Cancer Research on Feb 8, 1972 and was accepted for publication on Feb 8 1972) [2293a–c], Walter Burdette openly declared his support for the DM. However, on one day, Walter was summarily dismissed from the institution.

DM head Clifton Howe mobilized private donors in support of the viral oncolysate project. The Hospital's Surveillance Committee approved the project with the support of Jan van Eys, head of the MDACC Department of Pediatrics, chairman of the Surveillance Committee. Staff physician medical oncologists-hematologists Nicholas Papadopoulos and Carl Plager joined this author for the clinical conduct of the project; resident fellows rotating in their oncology training program, co-attended the patients. Staff surgeon-oncologists Ed White, and especially Richard Martin, Marion McMurtrey, and Marvin Magnus Romsdahl referred their patients to the project. However, Vice President Robert Hickey took the position as to withhold his support (and institutional funds) from the DM. The DDT received the full support of the NIH/NCI and the MDACCC. Yet DDT staff member medical oncologist Jeffrey Gottlieb in a warm complimentary professional gesture released his sarcoma patients for treatment to the Sarcoma Clinics of the DM within the project (serving as chemotherapy only control patients). DDT staff member medical oncologist Jules E. Harris was readily available for consultation. Jules was a firm believer in cancer immunotherapy [70a, b, 724]. There was a palpable personal conflict between the Head of the DM, Clifton Howe, and the Vice President, Robert Hickey. Emil Frei, head of the DDT, openly declared that he will be an immediate replacement successor of MDACCC president Randolph Lee Clark. It was with the support and supervision of DM head Clifton Howe (his frequent visits on the premises of the Section and the Clinics with requests for verbal reports, and bed-side patients' chart reviews), that the project was conducted openly, and that it could be brought to a preliminary (6 months) conclusion.

The initial sarcoma/melanoma immunotherapy protocol with viral oncolysates was registered as an institutional project (Project m27 gm13/gm15/gm17); its continuation as the chemo-immunotherapy protocol remained under the same registration code. Preliminary reports were given at several national and international conferences (one is cited here, emphasizing the laboratory work on tumor immunology) [2294]. The final results were prepared by Carl Plager and this author based on the review of all individual patients one by one, and published [71, 276, 277, 2294–2296]. The patients treated in this protocol were not prospectively randomized. This could have been a disadvantage of the report. In the meantime prospective randomization of patients on investigational protocols has become obligatory. The manuscript was not rejected by any prominent journals, because by choice it was not submitted to that high level of publication. Clifton Howe would have favored the New England Journal of Medicine; instead, this author published it as a transcript of an invited lecture at the Chicago oncology symposium [2296]. It was recognized that the work was not supported by the NIH/NCI, and that private donations supported the expenses. However, by this act, the material lost its rank of an original work not yet published elsewhere, a prime stipulation for its acceptance by the NEJM. Clifton Howe declined his co-authorship in the book chapter; in friendly terms, for the only reason that the first report of the work was not published in his choice, the New England Journal of Medicine. Thus, the results of this remarkable clinical trial have remained long unrecognized. An NIH/NCI-supported second trial of it with the continued support of Clifton Howe could have received the attention it deserved. Nevertheless, by time elapsed, it is fitting in "Cancer History", inasmuch as similar reports appeared subsequently in proof that immunotherapy and chemotherapy could be additive, or even synergistic, in their clinical results. As such, this pioneering clinical trial was carried out at the DM/MDACCC, with the Hospital Surveillance Committee approval and support.

It occurred in the pre-CT scan era, when patients with incurable disease were enlisted in a not prospectively randomized clinical trial, and were evaluated by standard examinations of that time period. All candidate patients were treated (considered to be a disadvantage by today's statistical standards). The vaccine was an autologous viral oncolysate preparation given in combination with full dosage of chemotherapy: a first such a trial against a human cancer. In the 1960s and 70s, this was a combined chemo- and tumor-specific immunotherapy trial, overriding the opposition of the NIH/NCI to such trials. Since then, chemo-immunotherapy in prospectively randomized clinical trials has been proven effective [277, 2285, 2286, 2294]. By now, well over a dozen clinical trials prove that tumor-specific immunotherapy (a "cancer vaccine") and chemotherapy could be clinically efficient. Figure 73 is a reproduction of the original tabulation of the preliminary results as prepared in 1977 by Carl Plager and this author [277, 2295–2297a, b].

However, with time passing, the results slowly and steadily shifted towards relapses, slightly less and occurring later in the chemo-immunotherapy cadre of patients, than what the only chemotherapy-recipient patients experienced. Rare patients with metastatic sarcomas treated with the CyVADIC chemotherapy protocol without immunotherapy (at MDACCC and within the Southwest Oncology Group) could remain in remission for several years [2297c]. Achievement of complete remission with chemotherapy only stayed some 15 % below the figure what was reached in the chemo-immunotherapy receiving patients. The proof was clear that chemoimmunotherapy did not result in worsening of the final outcome of the disease. Statisticians debated the value of this difference; it was referred to as a constant "favorable trend" for immunotherapy added to chemotherapy, statistically

INTERNATIONAL JOURNAL OF ONCOLOGY 29: 765-777, 2006 767

Table I. Chemoimmunotherapy of patients with sarcomas at the Department of Medicine, M.D. Anderson Hospital, 1974-6.

Treatment	Date of evaluation	Patients[a]	Complete remission (no evidence of disease)	Partial remission and stable	Progression (death)	% Progressors
ChemoRx	Oct '76	48	10	6	32	66
	Feb '77	49	10[b]	3	36 (17)	72
ChemoRx	Oct '76	10	1	5	4	40
BCG	Feb '77	19	3[c]	6	10 (4)	53
ChemoRx	Oct '76	14	6	6	2	14
BCG, VO	Feb'77	19	4[d]	9	6 (3)	32

[a]In treatment more than 6 months. Rendered NED surgically [b]5, [c]3, [d]3. ChemoRx: vincristine, cyclophosphamide, adriamycin/actinomycin D, dimethyl triazeno imidazole carboxamide, days 1-5 Q28 days. ImmunoRx: BCG (Chicago) and sarcoma viral oncolysates, days 17 and 24. Reproduction of Table II from the Chicago Symposium, February 24-25, 1977 sponsored by the University of Illinois at the Medical Center and the Illinois Cancer Council, Chicago, IL; from The Franklin Institute Press, Philadelphia, PA, with permission. Library of Congress Catalog Card #77-777-7. Patients with metastatic sarcomas were treated and evaluated at the Melanoma-Sarcoma Service, Department of Medicine, The University of Texas M.D. Anderson Hospital, Houston, TX, by fellows and resident physicians supervised by staff physicians Drs Carl Plager and Joseph Sinkovics (16).

Figure 73 The Intimidated Authors Did Not Submit the First Manuscript Showing Additive Results of Tumor-specific Immunization and Chemotherapy to the New England Journal of Medicine, because not the NIH/NCI, but Private Donations Supported the Work in the 1970s. Reference Appendix 2, Explanations to the Figures

not certain to be significant. It occurred in both groups of patients with metastatic sarcoma, or melanoma [276, 2296, 1635–1637a, b]. These follow-up results were reported by B.S. Yap, J.G. Sinkovics, M.A. Burgess *et al* at ASCO (American Society Clinical Oncology) annual conferences, in New Orleans (year 1979) and San Diego (years 1980, 1983). The year 1983 ASCO presentation was entitled "The curability of advanced soft tissue sarcoma in adults with chemotherapy?" It should have been "chemoimmunotherapy".

For a reasonable conclusion, viral oncolysate immunotherapy was recognized to be not a curative cancer vaccine; it was able to retard relapses from those cancer (sarcoma and melanoma) cells, which remained residually present in the form of minuscule metastases after surgery and radio-chemotherapy [71]. Current knowledge allows the explanation that malignant cells resting for a prolonged period in the state of autophagy (Figure 68) relapse with a newly activated oncogenic survival pathway different from what initiated their primary transformation. These relapses were delayed, but not necessarily eliminated, by viral oncolysate immunotherapy. Different scheduling and dosages were expected to further improve these results. Even in the era of molecular biology of cancer, there is still room for the conduct of a properly designed viral oncolysate chemo-immunotherapy project.

Some two decades of productive work were lost in this area due to a misconception prevailing at the NCI/NIH even in the "Conquest of Cancer" era of President Nixon's initiative and financial support. The grant application for the renewal, continuation and refinement of these studies in prospectively randomized patients was submitted in the mid-1970s to the NIH/NCI. It was returned with this remark: "approved without funding due to low priority". The "low priority" qualification was given due to the failure to specifically identify the tumor antigens against which the immunotherapy was aimed.

This author acknowledges the endorsement of the value of the viral oncolysate vaccines in the tumor-specific immunotherapy of gynecological cancers, as it is has been referred to in the publications of C.G. Ioannides and Ralph Freedman [2298a–c]. To the invitation of C.G. Ioannides, the Editor-in-Chief of the International Reviews of Immunology, an article entitled "Viral oncolysates…" was published [2299a]. A similar invitations followed by Demetrios Spandidos for lectures at his international conferences in Athens and on the Island of Crete [645, 2297a, b, 2299b]. At the Krebsforschungszentrum Heidelberg, Germany, Volker Schirrmacher, himself a major contributor to the immunotherapy of cancer with viral oncolysates (he used the Ulster strain NDV), generously gave credit to this author in his reviews [2300]. The New York City publisher Marcel Dekker requested a historical overview on the escalation of viral oncolytic therapy from the original intratumorally injected viruses to the active tumor-specific vaccination with viral oncolysates [71, 275, 1242, 2301]. The state of viral oncolysate cancer vaccine therapy received a detailed and supportive review, without any involvement of this author (JGS) in the preparation of that review, in a volume published from the NCI, yet with this acknowledgment: "Sinkovics was the first investigator to provide use of *in vitro* prepared viral oncolysate vaccines for patients with cancer" [2302]. However, William Cassel and associates for their work [274] received no recognition in this review; it was not mentioned that Sinkovics' grant applications at the NCI received an "approved without funding due to low priority" evaluation.

As a much belated reward, the NIH/NCI invited the presentation of the original project for its September 22–23, 2011 conference entitled "Cancer Immunology and Immunotherapy: Building on Success". There was an opportunity to list the major collaborators of the original work in this abstract [2303]. The conference was chaired by Steve Rosenberg, Director of the Center of Excellence in Immunology, Center for Cancer Research, NIH/NCI.

Basic science research at The Section/DM. The NIH/NCI grant-supported retroviral mouse leukemia work at the Section of Clinical Tumor Virology & Immunology included the early study of graft-versus-leukemia reactions in Rauscher virus-infected mice [261a, b].

Specific antibody-producer healthy plasma cells (secreting mouse leukemia virus-neutralizing antibody) and leukemia/lymphoma cells were observed to undergo fusion *in vivo*. The fused cells continued the production of the specific antibody, as the fused cells were grown either in suspension cultures, or as ascites tumors in mice. The antibody-producer fused cells were so maintained for several (>10) years (with Jose Trujillo). This phenomenon, the very first in its category, was presented in 1969 at the MDACCC "Leukemia-Lymphoma" conference [278] and reviewed again recently [147, 352]. Chief pathologist José Trujillo was engaged to confirm these completely original observation [2304]. H. David Kay was invited to repeat the experiments, in which diploid, retrovirally infected lymphoma cells (cell line #640) fused with specific antibody-producer diploid plasma cells in the peritoneal cavity of mice (Figure 72). Dave did succeed to confirm the basic experiment; he wrote an Introduction to the monograph [147]. This discovery made at the Section in the late 1960s preceded with some 6 years [147, 278, 352, 1904a, b, 2305–2307] the congenial in-laboratory technology of monoclonal antibody-production in plasma cells fused with malignant lymphoma cells. Nobel Prize-winners G. Köhler and C. Milstein conducted and reported those experiments in the mid- and late-1970s in the Nature (but without ever referring to the preceding Lancet and other publications of Sinkovics *et al*). Professor Milton Wainwright, Sheffield, England, gave the credit to this author (JGS) for discovering the "hybridoma principle" and for recognizing first natural hybridoma formation *in vivo* [2129a, b]. For the verbatim quotation of how this phenomenon, never seen before, was worded in December 1969–January 1970 in the Lancet [2305], *vide supra* [2028a, 2129ab].

The retroviral particles released from the antibody-producer fused cells were often incomplete and deformed [348a]. Ferenc Györkey showed by immuno-electron microscopy immune globulin molecules attached to the budding retroviral particles [348a, b]. Benjamin Drewinko was engaged to do cell volume measurements using the earliest not-yet-marketed prototype of the flow cytometry Coulter counter. He confirmed the doubling to quadrupling of the cell volumes after the fusion of the originally diploid cells [2305]. The publisher (Year Book Medical Publishers, Chicago) of the volume on the M.D. Anderson Hospital Leukemia-Lymphoma Conference of 1969 put the simplified depiction in drawing of this phenomenon on its outside title cover page [147, 278, 1904b] (Figure 74).

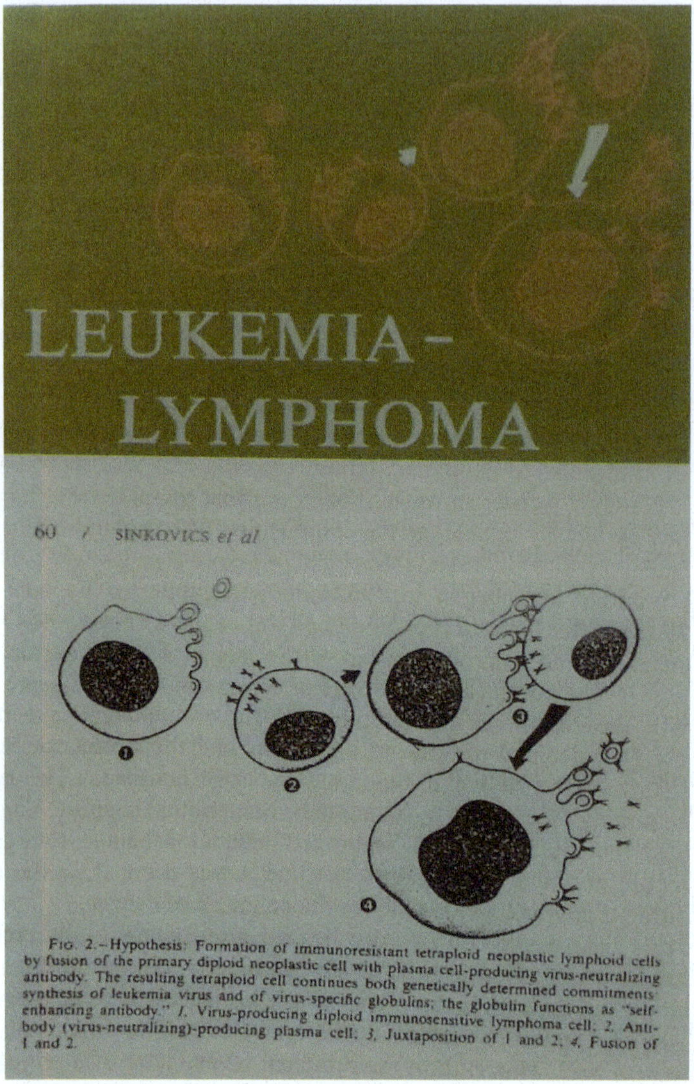

Figure 74 Book Editors Realized the Importance of Fusion Between Antibody-Producing Plasma Cells and Lymphoma Cells Way Before (1969–70) the Granting Agency NIH/NCI Has Come to It. Reference Appendix 2, Explanations to the Figures

This is the first depiction of a hybridoma formation. The editors of the Cancer Research, reviewed genuine laboratory documents and accepted that the existence of an antibody-secreting natural hybridoma in the late 1960s was documented to their satisfaction [2306]. This author was invited to present the principles of "natural hybridoma" formation at the Monoclonal Antibodies Conference in San Francisco. The editors, J.M. Vaet and J.L. Meyer of the Karger volume issued on the contents of the conference, published the author's presentation in full. Karger Publisher again used the original drawing of the cell fusion on the outside front table of the volume [2307]. Professors Gheorghe Benga and Victor Cristea of the Haţieganu University Medical School in Cluj Napoca, Romania (formerly Kolozsvár, Hungary) extended a great honor to this author (JGS) for the discovery of natural hybridomas [1904b]. Professor Sándor Eckhardt requested a Hungarian publication [2308]. The presentations given at the annual conferences of the Hungarian Medical Association of America were published by Professor Julian Ambrus [2309a, b]. The Hungarian Acta published the last report on this antibody-secreting natural hybridoma, summarizing all preceding events [352]. Referrals to the search for natural hybridoma formation in human lymphomas were reported with high likelihood of their occurrence [147, 352, 1904a, b]. José Trujillo and this author maintained the fused cells in suspension cultures for some 10 years [26], however, nobody requested a sample of it. Yet, the author's poster presentation entitled "The earliest work with fused lymphoid cells continuously secreting a specific antibody" at the Second International Congress on Monoclonal Antibodies in Cancer in Banff, Alberta, Canada, in 2002, was widely attended.

What is biotechnology! Making of monoclonal antibodies' http://www.whatisbiotechnology.org/exhibitions/milstein/monoclonals (Google).

One of the earliest was Joseph Sinkovics, a Hungarian clinical pathologist and laboratory clinical virologist based at M.D. Anderson Hospital in Texas, USA. In the mid-1960s he reported the successful development of a cell line of antibodies with known specificity by the fusion of antibody producing plasma cells with lymphoma cells. These he had managed to grow in continuous cultures in spinner bottles or in fluid found in the abdominal cavity of mice known as ascites tumors. The method was published in J.G. Sinkovics et al, "A System of Tissue Cultures for the Study of a Mouse Leukemia Virus", Journal of Infectious Diseases 119 (1969), 19–38. Lack of funding, however, prevented Sinkovics and his team from developing the technique any further.

Work at the Section on Kaposi's sarcoma [83] received a highly complimentary editorial in the New England Journal of Medicine from Dorothea Zucker-Franklin [533]. This author (JGS) received the Kaposi Award in Budapest, Hungary, from professor Attila Horváth [895c, 2310].

In another project at the Section, cytotoxicity to tumor cells of two distinct types of human lymphocytes was first observed and preserved in original photographs taken in the late 1960s. Some of the very first pictures of the large granular lymphocytes attracted to cancer cells are shown in the article [278] cited above in the text of this volume. Unfortunately, the chief lymphoma pathologist James Butler referred to these cells at first sight as "monocyte-like" ("M", personal communication), even though he was told that these cells did not phagocytose bacteria. This discovery is detailed in the author's monograph (page 77 [147]). The cytotoxic "small compact lymphocytes" were immune T cells. The very first patient (MDAH#73587) showing circulating small compact T lymphocytes able to dissolve his tumor cells *in vitro* within 24 h had a large localized chondrosarcoma [30a, 147, 279]. In the Lichtman article [280], the references are to the chondrosarcoma cell line that was attacked by the large granular lymphocytes of the healthy control, JGS, shown also in the monograph, Ref. [147, p. 77], and reproduced here (Figures 35 and 36); and to patient (MDAH#62557), whose case history was presented as a new entity "cytotoxic lymphoma" at the First Meeting of the International Society of Haematology European Division in Milano, Italy in 1971. The abstract of this presentation was re-published in the Acta Microbiologica Immunologica Hungarica in 1997 (cited here in Ref. [30a]). The cytotoxic "large granular lymphocytes" were the natural killer cells (not referred to them as such at the time of the first encounter with them; as they were called "Burnet's surveillance cells" in the Sinkovics Lab). The attack of these cells on human tumor cells (both autologous and allogeneic) was shown in the late 1960s in chamber-slide preparations, in which the encounter of tumor cells and lymphocyte was actually visualized. The radioactivity-releasing cytolytic assays, or tumor cell colony inhibitory assays in other laboratories (Herberman and Oldham; the Hellströms) entirely missed this phenomenal sight. The small compact lymphocytes were the privilege of patients with cancer; the large granular lymphocytes were shared between patients and healthy donors [147]. The cytotoxicity of the small compact lymphocytes could be blocked by pretreatment of the tumor cells with patients' sera, shown in graphs, reproduced in this volume from Ref. [147]. In contrast, patients' sera (and healthy donors' sera) intensified the cytotoxicity of the large granular lymphoid cells [30a, b, 1631a]. That was an ADCC reaction, since NK cells express Fc receptors, but T cells are devoid of them. Cameron Tebbi produced "pure" lymphocyte preparations extracted from buffy coats; the separation of granulocytes from the lymphocytes was quite successful (iron granule-loaded granulocytes rapidly sedimented); very few monocytes remained in the lymphocyte preparations; and attempts at separating the small compact lymphocytes from the large granular lymphocytes (in lack of specific antibody-coated beads in the late 60s and early 70s) required sedimentations and centrifugation technology, that proved to be inadequate [30b].

The project site visitors of the NIH/NCI, led by the one from Seattle, who was working with the colony inhibition assay, which failed to distinguish subclasses of lymphocytes, declared that the phenomena involving the "large granular lymphocytes" of healthy donors (first this author's) must be regarded as "*in vitro* artifacts", because there should not exist any immune reactivity in nature without specific pre-immunization. These NCI project site visitors missed observing the distinction between the small compact T cells and the large granular NK cells, because of their use of a crude colony inhibition assay (instead of the sophisticated and refined chamber-slide assay used in the Sinkovics Lab). Other laboratories (Herberman and Oldham) used the radioactivity release assay for cytolysis, that was also unable to distinguish between cancer cell lysis by the immune T cells, or by NK cells. Those assays did not actually visualize the encounter of tumor cells with different types of lymphocytes, what the "old-fashioned and primitive" chamber slide assay, as reproduced in Figure 8 in the article published in the International Journal of Oncology 2005;27:5-47, showed in the early 1970s: the clear morphological differences between small compact immune T cells and large granular NK cells. The more lenient project site visitors suggested that the large cytotoxic lymphocytes might have "learned" to react *in vitro*, as if they represented a "blastic reaction" (and that would be acceptable for continuing the funding of the project). However, when this author (JGS) adhered to his view, that some lymphoid cells, those which are able to attack without pre-immunization, do exist naturally, the two NCI project site visitors (Gary R. Pearson & Paul H. Levine) [30a], who originally issued the NIH/NCI contract for the "Study of tumor-specific human cytotoxic lymphocytes" to the Section headed by this author (JGS) [2294], arbitrarily and abruptly, transferred the contract to a NIH/NCI intramural laboratory (the R. Herberman, R. Oldham Lab) "for the study of the in vitro artifacts" generated in the chamber-slide cultures (no other reasons for this unjust act were given; application for another related but different grant was encouraged). There, "natural killer cells" killing mouse and human tumor cells without pre-immunization (!) were years thereafter, but eventually "re-discovered" by Julie Djeu and H. David Kay (a former associate of Sinkovics, and a salaried employee of the contract) [30a, 147]. It is medical history, that when trained immunologists first met NK cells, they did not believe in their existence, and rather sacrificed the laboratory, where NK cells were first observed, photographed, and recognized, than to divest themselves from their petrified indoctrination. MDACCC Vice President Robert Hickey's remarks to this author (JGS) upon the transfer of that NIH/NCI contract from the MDACCC to the NIH/NCI intramural laboratory: "Dr. Clark and I will not forget that you have lost that contract," Nevertheless, Robert Hickey credentialed this author (in his handwriting) as "Ethical" in a later document.

For personal gratification to this author, active immunizations with sarcoma or melanoma cell "viral oncolysates" significantly increased the numbers of both types of attacker lymphocytes per units in the patients' blood samples [71, 147] (Figure 36). Professor Marshall Lichtman in his chapter restored the actual events, and graciously attributed the true discovery of NK cells to this author (JGS) [280]: "*In 1969, Joseph (József) Géza Sinkovics (b. 1924), a physician-scientist and Hungarian émigré, working in the Section of Clinical Tumor Virology and Immunology at the M.D. Anderson Cancer Center, observed that his unprimed lymphocytes killed allogeneic tumor cells in vitro. He reported his findings shortly thereafter [117]. He also showed that the cells involved in attaching to and lysing tumor cells were large granular lymphocytes. Although his report was met with skepticism and rejection (a project site-visitor proclaimed "There is no immune reaction without preimmunization"), this work was confirmed and expanded to indicate that the unprimed lymphocytes also could kill virus-infected cells. In his monograph [118], Sinkovics reviewed the reaction of the scientific community to his finding that tumor reactive lymphoid cells might be present without preimmunization of their host. He provided the first photomicrographs and tumor cell growth-inhibitory graphs describing such cells. At about the same time (1970), Sinkovics described a lymphoproliferative disease, which in retrospect may have had a natural killer cell phenotype [119]*". Lichtman cites these three references "117. Sinkovics JG, Shirato E, Martin RG, Cabiness JR, White BC. Chondrosarcoma. Immune reactions of a patient to autologous tumor. Cancer 1971;27:782–93. 118. Sinkovics JG. Cytolytic Immune Lymphocytes. Passau:

Schenk Verlag; 2008. pp. 1–383. 119. Sinkovics JG. Malignant lymphoma arising from natural killer cells. Report of the first case in 1970 and newer developments in the FASL-FASR system. Acta Microbiol Immunol Hungar 1997;44:295–307." This last citation is Ref. [716] in this volume; it has an Erratum: "in the original text, the Fas receptor was correctly printed as CD95".

In the case of a pair of female twins, one with rhabdomyosarcoma, one healthy, the healthy twin's buffy coat lymphocytes were used to treat the patient. The donor lymphocytes attacked the rhabdomyosarcoma cells *in vitro*; figures are shown in [340a, 341a, 1631a, b], but could not be expanded in culture *in vitro* in the pre-IL-2 era. Used repeatedly as "buffy coat" lymphocytes for adoptive lymphocyte therapy *in vivo*, with cyclophosphamide and vincristine chemotherapy to the patient (in the pre-adriamycin era), the treatment induced a remission. However, the patient's original tumor burden was enormous (with circulating tumor cells in the blood: "a rhabdomyosarcoma cell acute leukemia"), thus the donor lymphoid cells (including large granular lymphoid cells, later recognized to be NK cells) must have been numerically inadequate. This patient relapsed with multiple brain metastases and died [340a, 341a, 1631b].

Before the discovery of exosomes (*vide supra*), malignant ascites and pleural effusions provided the arena for direct confrontation between large numbers of tumor cells and all classes of defensive cells, in particular lymphocytes. Subfractions of lymphoid cells termed LAK cells deriving from malignant effusions gained high levels of cytotoxicity towards autologous tumor cells in response to treatment with with IL-2 [2311]. Since then, immune T cells with or without LAK cells (lymphokine-acivated killer NKT cells) induce long stabilization of disease, or partial and complete remissions in lymphocyte-depleted patients with metastatic malignant melanoma at the NIH/NCI, Bethesda, MD [2017, 2288]. In contrast, exosomes of cancer cell origin may transmit miRs, that destroy mRNAs released for generation of tumor suppressor proteins (a topic deserving a detailed discussion elsewhere).

It has been under consideration, if low dose radiotherapy to the abdomen would improve the results of lymphocyte therapy [2312]. It appears that gut flora induces strong innate immunity in regional lymph nodes in the mesentery, if the intestinal tract receives low dose gamma rays irradiation (explained in pp 264–266 [147], citing work from the Section). The recipient tumor-bearing patient is expected to experience a much favorable response to adoptive lymphocyte therapy against the background of a pre-activated innate immune system. At the Section it was found in the late 1960s (and incredulously observed) that the peritoneal cavity of low-dose pre-irradiated mice was better protected against intraperitoneally inoculated lymphoma cells, than that of intact non-pre-irradiated mice [147, 2313, 2314]. This result remained "unexplainable" (received with disbelief by project site visitors of the NIH/NCI) until after innate immune reactions induced by gut flora in the mesenteric lymph nodes were discovered in irradiated mice; explained in pp 264–266 [147].

The Section entered into a close working relationship with the Department of Virology, headed by Joseph Melnick, at the University Medical School Baylor, Houston, TX (where this author served as visiting professor). The top electron microscopist Baylor professor, Ferenc Györkey, chief physician at the Pathology

Department at the Veterans' Administration Hospital, Houston, TX, and this author (JGS) repeatedly observed budding virions of endogenous retroviruses in human pathological materials (cellular specimens from patients with autoimmune diseases, especially SLE, AIDS, lymphoma and sarcoma). In lectures and abstracts, Ferenc Györkey, Jenő (Gino) Szakacs, and this author (JGS) concluded that enveloped endogenous human retroviruses are frequently of the fusogenic type (inducing fusions of lymphoid cells and tumor cells), become reactivated at, during and after, malignant transformation of their host cells, and are potentially pathogenic in these clinical entities; reviewed in [147, 2135a–f].

Irwin H. Krakoff, the former Sloan-Kettering professor, the new Chairman of the Division of Medicine at MDACCC (who followed both Emil Frei and Clifton Howe), with the full endorsement of the new MDACCC president Charles A. LeMaistre (follower of Randolph Lee Clark), retained this author in the status of adjunct consulting professor at the external staff of MDACCC for the next two decades. During this time, The Southern Medical Association in San Antonio, TX, honored this author with its gold medal, for "achievements in cancer research and tumor virology".

After leaving MDACCC in 1980 for the private practice of medical oncology/hematology in Tampa, FL (with teaching university affiliation), this author published (for his own education) the volumes "Medical Oncology an Advanced Course" 2nd edition vol I/II pp. 2050 with well over 10,000 references condensed in it in abbreviated format (Marcel Dekker, New York, 1986) [2315]. The dean of the University of South Florida medical school, Andor Szentivanyi, and the chairman of the Department of Medical Microbiology and Immunology, Herman Friedman, requested and published a chapter from this author in their volume "Immunological Revolution…" [2316].

Farewell. This author is highly rewarded, as he has survived to see the appearance of the molecular targeted therapy with its highly increased precision and reduced toxicity, as compared with combination chemotherapy. It was in the late 1960s to the early 1970s, when this author working at the Department of Medicine at M.D. Anderson Hospital discovered a spontaneously formed antibody-producing "natural hybridoma" in the mouse. For the first time he photographed human lymphocytes (both immune T and NK cells) in the act of killing autologous (and allogeneic) human tumor cells. With his clinical associates, he immunized patients with a viral oncolysate vaccine and observed the rise of these lymphocytes both in numbers and in biological activity. Immunotherapy of cancer is now moving forward on all four of its frontiers: newer, possibly "therapeutic" cancer vaccines epigenetically designed; genetically engineered viruses oncolytic to tumor cells and immunogenic to the hosts (what viral oncolysates were designed for); highly effective monoclonal antibodies of hybridomas; and above all, immune lymphocytes (both immune T cells and NK/NKT cells) effectively targeting tumor cells. With modified (molecularly engineered) CAR TCRs, these lymphocytes replicate and wipe out large tumor masses with such a vigor, that the patients need to be premedicated in order to avert a tumor lysis syndrome. The immune lymphocyte population becomes permanently installed and ready to attack again residual or recurrent tumors (Matthew Stenger's interview with Renier J. Brentjens in The ASCO POST. May 15 2013) (more in the Appendix 1).

The events antecedent to, and surrounding the work, concerning the genetical engineering of TCRs by retrolentiviral vectors of selected genetic sequences are reviewed in a narrative fashion in Science, the issues being devoted to cancer immunotherapy (June 28, 2013 pp. 1514–1518).

The discoveries of the checkpoints, whereby the expression of "self" repels all autoimmune attacks, and that an inbuilt faculty of the transformed cell (the cancer cell) sets it in to favor the exemption from immune rejection, could have been made over 50 years ago. Physiologically, these checkpoints are double-reinforced (*vide supra*). Even if the CTLA4-specific ipilimumab fails to expose the self-masquerading cancer (melanoma) cell to the immune T cells/NK cells, blockade of PD-1 by nivolumab, or PD-1 ligands (PD-L1/L2, CD274/273) by different monoclonal antibodies (pembrolizumab), rescues the T lymphocyte from its own apoptotic death and activates it for an attack on the malignant cell (melanoma), which is disguised of being a non-attackable "self" entity due to its excessively expressed PD-1 ligands [2317]. The Abramson Cancer Center at the University of Pennsylvania, Philadelphia, PA, and the H.L. Moffitt Comprehensive Cancer Center at the University of Morsani College of Medicine in Tampa, FL, extended the efficacy of pembrolizumab to lung small cell, adeno- and squamous cell carcinomas. The trajectory of lymphocyte-mediated cancer immunotherapy from the first photographs of such cytolytic lymphocytes taken in the late 1960s at the DM of MDACC in Houston TX [147] encompassed over 30–40 years to reach its target (more in the Appendix 1).

In the Memory of Clifton Dexter Howe

If we were still together. Clifton is not alive to observe the spectacular achievements of the era of molecular medicine, and cancer immunotherapy. He died of lung cancer, no doubt, brought upon him by cigarette-smoking. If we still could have our weekly DM conferences, and the report on the adjuvant chemotherapy of 100,000 women with early adverse prognosis breast cancer were discussed [2318], what would Clifton Howe say? Relapses were reduced in one third of the patients: adjuvant chemotherapy worked (if the recipients truly had postoperatively concealed undetectable metastatic breast cancer). Chemotherapy has failed in 70 % of the treated but relapsed subjects. Deaths due to septic shock or acute leukemia were not accounted for. All those deaths were counted as "non-breast cancer deaths" thus, probably favoring the statistics. Clifton Howe would say: "An effective breast cancer vaccine is needed to reduce relapse rates in those women, who harbor postoperatively hidden residual disease". He would request Joe to review recent publications on cancer vaccines and to put together a critical review of all breast cancer vaccine trials that are in the investigational stage [2319–2321]. He would add: "Make sure you include those potential viral antigens" (meaning the MMT-like retroviral agent). Confronted with targeted therapy in combination with, or for replacement of, chemotherapy, what would Clifton Howe say? The graph on p 2392 of the N England J Med June 20, 2013 issue [2322] shows crizotinib out-performing combination chemoradiotherapy in metastatic lung cancer (caused by the mutated ROS1-fused ALK, anaplastic lymphoma kinase gene), and with substantially less immediate and late toxic side effects. At 2 years, the tumor-free survivors are 10 % for combination chemotherapy, and 20 % for crizotinib. The molecularly targeted treatment failed in 80 % of the patients with lung cancer. In the same issue of the N England J Med, there is another article describing acquired resistance of the cancer cell to crizotinib [2323]. It is a new point-mutation of the ROS1 gene (reactive oxygen) with glycine-to-arginine substitution in gene codon 2032, that confers the resistance to crizotinib. The target still eludes the bullet.

Clifton Howe would ask the DM's chief of hematology, Charles C. Shullenberger to explain the price of curing Hodgkin's disease with chemoradiotherapy. Dose-dependently, patients who received subdiaphragmati radiotherapy and procarbazine, face the increased risk of developing stomach cancer [2324].

In his straightforward manner (as a USA liaison officer was reporting to the British general Bernard Montgomery, the Viscount of El-Alamein, in North Africa during WWII), Clifton Howe would say: "Cancer vaccines and cancer immunotherapy, what not the laboratory, but the host generates, the natural product, will prevent and cure cancers. It should have been available at the very beginning, but now without further delay, must be employed ('launched') in full immediately!"

Respectfully addressing Clifton Howe: Will any cancer vaccine ever match what the barrage of the combined attack directed at the amplified (not mutated) oncogene/oncoprotein HER2/neu (glutamyl-tRNA amidotransferase subunit) could achieve as to durable remission induction: lapatinib, neratinib; trastuzumab, trastuzumab emtansine; pertuzumab, with mobilization of the NK cells for the ADCC reaction [2325, 2326]? Yes, the host's own weak immune reactions indicate that a vaccine could enhance them. Nevertheless, against something so fundamentally immortal as a cancer cell is, at the expense of all attenuated other cells, certain chemotherapeuticals should retain their role in carefully designed combinatorial regimens. To stay alive on Earth has never been a risk-free adventure.

Would have the aging Clifton Howe immediately stopped smoking, and started taking aspirin, with or without resveratrol, cyclooxygenase inhibitors, and/or metformin, as cancer preventive medications? After all, all these substances attack, disarm and potentially kill (through apoptosis induction) cancer cells *in vitro* (or *in vivo* in xenografts) [2327–2330a, b]? Simple substances against a formidable machinery! The 1962 DNA Nobel Prize winner James Watson has just announced in his lecture at the H.L. Moffitt Comprehensive Cancer Center in Tampa, FL, that he is on daily aspirin and metformin.

He stands among the best men of his era. Apart from cancer, Clifton Howe willingly expressed his political views to this refugee from Communist Hungary: "The liberation of Central Europe should have been advanced from North Africa through Greece and the Balkan peninsula through Hungary and Poland up to the Baltic States" (thus, keeping the communist Soviet Union out of Central Europe). He knew: he was in North Africa during WWII facing Erwin Rommel's "Afrika Korps" and the Italian generals Pietro Badoglio and Rodolfo Graziani committing one of the most atrocious act in history in occupying the many thousand years old koptic Christian state of Abyssinia. Hearing Clifton Howe say this in 1957, this 33 years old job-applicant visiting with him in Houston, TX, from Rutgers, NJ, for the first interview, uncontrollably extended both of his arms toward him for a mighty hug. The Hungarians occupied either from the West, or from the East (from as far as from Mongolia, or from both sides), and from the South (the Ottoman Empire) identified themselves with the Baltic people tormented and occupied from the East and from the West (Prit Buttar "Between Giants", Osprey, 2013). The Hungarians underwrite every word the USA Ambassador to Hungary (1933–1941), John Flournoy Montgomery wrote in his book ("Hungary: The Unwilling Satellite", Devin-Adair Company, 1947), of which Clifton Howe was quite aware.

For this author, it was an honor and a great privilege and pleasure (personal and professional) to serve with Clifton Dexter Howe for two decades at the Department of Medicine of the young MDACCC! (Howe CD & Sinkovics JG A retrospective analysis of preoperative medical clearances and life-threatening infections (endotoxin-shock included) in patients with cancer. Project M26/gm9. Research Report The University of Texas MD Anderson Hospital pp 486-94 1972).

Over 40 years after the above cited work was carried out on a 14–16 h/d non-stop schedule at the Section of Clinical Tumor Virology & Immunology and at the Sarcoma/Melanoma Clinics of the DM in the fledgling MDACCC, the history of the high-flying MDACCC is narrated now in a volume under the title "Making Cancer History" by Olson [2290]. The highest praises so abundantly extended to the DDT in that volume are fully deserved for their achievements in practicing combination chemotherapy, as that era demanded and dictated it. However, Clifton Dexter Howe, the head of the DM, received some depreciatory comments, as he is falsely depicted as an ignorant opponent of chemotherapy without any ideas of his own. In that "history book", his associate (the author of this chapter) was dismissed to the depths of bottomless oblivion. His works received not one single word of

Figure 75 Dr. Clifton Dexter Howe Opponent of the First Highly Toxic Combination Chemotherapy Protocols Was Transferred from His Rank of Head of the Department of Medicine to an Administrative Position. Reference Appendix 2, Explanations to the Figures

citation in that volume. However, their publications remain preserved in the printed volumes of M.D. Anderson Hospital's annual symposia and bi-annual Research Reports from the 1960s to the 1980s on the bottom shelves of closed-down storage rooms in the cellar. The flooding of the Texas Medical Center by hurricane Allison in June 2001 could have damaged this material, as it has devastated some cell lines left behind by this author under the care of friends Jose Trujillo and Marvin Romsdahl (both deceased). The subdued and misinformed writer of that "historical account", the dying MDACCC cancer-patient Olson, cannot name his advisors or editors, or remedy his omissions. Even though Olson might have tried to schedule an interview with Sinkovics, that encounter was canceled by him and it has not taken place. This is the last opportunity to recollect the events and memories of those two decades from the late 1950s to the early 1980s.

This account of debit and credit is given here, as an honest attempt at the attenuation of that grossly inattentive treatment with what the early history of a noble institution was dealt with [2290]. To those who still remember Clifton Dexter Howe, or read about the work conducted, and the original observations made at The Section at the DM, which pre-dated their time in medical history, but eventually turned out to be all correct ("OK"): Their advocacy is highly appreciated!

Figure 75. Photograph (in the mid-1970s). From the left to the right: Sebron Culpepper Dale, Associate Professor of Medicine, internist; employees' health care. Clifton Dexter Howe, Head, Department of Medicine, Professor of Medicine, medical oncologist. Joseph G. Sinkovics, Chief, Section of Clinical Tumor Virology and Immunology; Chief, Clinical Melanoma-Sarcoma Service (clinics and in-patient services); newly appointed Professor of Medicine, infectious diseases specialist, medical oncologist/hematologist. Charles C. Shullenberger, Chief, Clinical Hematology Service (clinics and in-patient services), Professor of Medicine, hematologist. M.D. Anderson Hospital Comprehensive Cancer Center, The University of Texas Medical School and College of Medicine, Austin, Galveston & Houston, TX.

Summary

Considerations elaborated on in this volume support the concept that the malignantly transformed cell is driven by an ancient, but well conserved substance of the RNA/DNA complex. In the pre-cellular world, RNA was the progenitor of the first DNA molecules. The one-exon Sox2 sequence remains preserved embedded in a long noncoding ncRNA strand in the human genome at 3q26.3-q27, as a relic of their ancient biophysical precellular relationship. The primordial RNA/DNA complex remains preserved in the archaea, and in the cells of unicellular, early multicellular, or advanced multicellular eukaryotic hosts as chloroplasts and mitochondria. Their primordial genes were donated to the newly formed nucleus of their hosts. When the mitochondria sense the rise of the incipient so-called malignant transformation of their host cell, they either contribute to it, or kill their host cell by inducing apoptosis. Now acting together in the cell, the RNA retains the rights of its ancestry and priority over DNA. It persists in the forms of mRNAs, rRNAs, tRNAs, microRNAs, noncoding but not inert ncRNAs, and circular RNAs. The entity of the RNA/DNA complex regains its primordial position for the direction of the overall metabolism, fully resistant and ever-lasting, of the transformed cell. The transformed cells returned to the metabolism of their distant ancient unicellular ancestors. The encoding of the ancient "cell survival pathways" includes the silencing of the so-called "cell suicidal genes" (now referred to as "tumor suppressor genes"), followed by sequential gain-of-function point mutational events, constitutively expressed, and primarily involving the cell cycle. The dsDNA breaks are repaired rapidly (may be not error-free). If there is a mismatch repair-deficiency, the survivor cell will live with a mutator pheno-genotype. Genes are amplified, point-mutated and newly generated by duplications and fusions. As in the ancient unicellular life forms, rapid divisions at young mature age with recapped telomeres prevent senescence and natural death. The transformed cells now recover even from mitotic catastrophe. LUCA/LECA very likely were able to do that. In processes of aneuploidy and trans-speciation, the transformed extant cells invade their multicellular hosts. They are able to revert from adenocarcinoma to mesenchymal sarcoma-like entities; assume amoeboid characteristics of locomotion and phagocytosis, or vascular blood vessel endothelial cell-like forms, or convert into neuroectodermal geno-phenotypes, thus resisting physico-chemical attacks. The ancient helix-loop-helix achaete-scute proteins generated the first nerve

J.G. Sinkovics, *RNA/DNA and Cancer*, DOI 10.1007/978-3-319-22279-0_30

cells in the ancestors of sea anemones (Nematostella) and the protochordate amphioxus (Branchiostoma). The derivatives of these primordial genes encode now neuroblastomas (the esthesioneuroblastoma) and transform adenocarcinoma cells into highly resistant neuroectodermal/neuroendocrine malignant cellular entities. Proteins essential for life in the archaea, prokaryota and unicellular eukaryota (proliferating cell nuclear antigen; high mobility group proteins; arginine methyl-transferases; cadherins/catenins) rise in selected single cells of the multicellular host to induce, or promote, their "malignant transformation". These proteins (the heat shock proteins that supported unicellular life on the overheated ancient Earth) now chaperone, in providing a supportive scaffold to, the oncoproteins encoded by the oncogenes. The ancestors of the oncogenes were the "cell survival pathway" genes of the first unicellular life forms (eukaryota); these genes remain conserved in their extant descendants. The beta-catenin-dependent Wnt pathway encodes the body axes in unicellular and early multicellular eukaryotes (choanoflagellates; Ciona; Nematostella, Spongiae). In the human genome the Wnt pathway is a proto-oncogene. The NFkappaB/STAT pathway of the Nematostella becomes a functional oncoprotein in human B lymphocytes (including Reed-Sternberg cells).

Ancient and hitherto dispersed endogenous retroviral genes unite in the nuclei of lymphoid cells during somatic hypermutation (or sarcoma/carcinoma cells during rapid replications, or cell fusions). When acquiring an *env* gene, these elements reappear as budding retroviruses. They mediate cell fusions, thus producing "natural hybridomas", fully armed for self-defense and prolonged survival. Many malignantly transformed (lymphoma, melanoma, carcinoma) cells of the human host appear to have been induced by the reactivated genes of ancient retroviral inserts. Prominent among them is the LINE pathway, which is upregulated in the Nematostella, and in human so-called cancer cells, alike. Are these ancient proviral DNAs the inherent elements responding to the "danger signals" of a stressed host, and thus alarm the RNA/DNA complex to transform?

Taken from their multicellular hosts, the transformed extant cells live independent lives without forming organized multicellular communities. The larger cells move like amoebas, engulf the smaller ones (practicing "cell-cannibalism"); or they expropriate their useful genes by horizontal transfer (as their ancient ancestors did). Essential pathways of cellular life, the ligand-to-receptor pathways, will be expropriated in the form of constitutively expressed oncoprotein receptors and ligands. The malignantly transformed cells clearly exhibit a semblance of immortality. The HeLa cell line in six decades in culture preserved its original papillomavirus insertion sites, but otherwise has become a disorganized cell conglomerate, that is well fed, and thus switched off all its genes fighting for survival in its original host. It is still trying to repair its genome, but the events of chromothripsis are irreparable. However some HeLa cells obey apoptotic signals and die (credit Rocky Mountains Laboratory; Google/Wikipedia). In fresh tumor cells emerging in the inflammasome, the "mutator phenotype" is activated for survival. There, the retrotransposon-derived proviral DNAs encode the "mutator phenotype" (in Lynch syndrome; in glioblastoma multiforme). Retrotransposon-induced malignant transformations probably are the cell survival pathways constitutively installed. The piggyBacs may be the messengers, which activate the oncogene promoters in most newly arising transformed cells. The domesticated Sleeping Beauty transposons cannot expunge all the proviral inserts.

The process called "oncogenesis" may be an inherent defense reaction of the primordial RNA/DNA complex in response to a threatening demise of their multicellular host. The attempted rescue of the transformed cells is accomplished through those ancient faculties that secured the survival of the first unicellular eukaryotic life forms in independence and in a semblance of immortality. The transformed cells eventually kill the multicellular host, and die within it. If these cells are transferred to an academic tissue culture laboratory and are cared for there, they will prove their immortality; in xenografts, they will grow as malignant tumors in experimental animal hosts. This endogenous form of oncogenesis (invited by its host, misinformed, or truly in demise) is cooperated with, and supported by, the original multicellular hosts. In contrast, exogenous forms of oncogenesis imposed upon the host by mutagenesis consequential to chemical or radiation exposure, or especially by biologically acting oncogenic viruses, bacteria, or parasites, are resisted by their multicellular hosts.

The master bioengineer RNA/DNA complex may commit mistakes resulting in cleft palates in the newborn, or in accidental forms of oncogenesis. Examples are the error-prone enzymatic repair of dsDNA breaks, or faulty removal of overaccumulated Okazaki-fragments. Misguided nicks by the AID/RAG/APOBEC enzymes during somatic hypermutation, or elsewhere, occur. In the process of somatic hypermutation, instead of uniting newly generated light chain genes with the heavy chain gene (IgH), the oncogenes c-*myc*, or *bcl*-2 fuse with the heavy chain gene. Lymphomas (or other tumors) arise as the consequence of "faulty installments" of the faculties of the adaptive immune system, not yet corrected by the evolution still in progress. Whatever the initiating event is, 1, a rescue mission to save immortalized cells; or 2, a failure to correct an error of installment, or a dsDNA break; or 3, an exogenous agent(s), the result is the same. The reactivated primordial RNA/DNA complex becomes a re-ignited "rocket engine" of its space travel vehicle, the "tumor cell". These transformed cells are endowed with the natural aptitudes for independence and possible immortality (survival in liquid nitrogen at the temperature of the outer space), that characterizes life self-engendered in a physicochemical universe.

The transformed cells' immortality is annulled by knocking out its anti-apoptotic machinery. Demolishing the mRNA released to encode the anti-apoptotic proteins, most prominent of these being the Bcl-2/Bcl-xL, is effective. Either the pro-apoptotic guardian WTp53, the anti-PI3K PTEN, the extrinsic cell membrane FasL → FasR, or the intrinsic mitochondrial Bak/Bax/caspase3 pathway switch into action to abolish the transformed cells' invulnerability to the stimuli of apoptotic death. Most of the oncogenomes disabled and neutralized for months (or years) lingering in the state of cellular autophagy are still able to recover by initiating new mutations and gene fusions. Examples are the adenocarcinoma cells of breast and prostate in the state of autophagy, which mobilize the most ancient achaete-scute helix-loop-helix protein, or the medkine protein, for the trans-speciation of these cells, into neuroectodermally transformed malignant cells, highly resistant to physico-chemical insults. The living matter on Earth is *ab origine* chemo-radio-resistant and highly mutable for survival. So remain the "malignantly transformed" stem, or even the de-differentiated somatic, cells of a multicellular host.

Is it presumable that the human cerebral cortex has been encoded by a consummate DNA/RNA complex replacing the ancient RNA/DNA complex, the encoder of cell survival pathways and oncogenomes for the rescue of unicellular life forms (in whatever form or shape) on Earth? The human cerebral cortex, despite some faulty installments, is at the verge of overcoming the consummate bioengineers of the primordial cell survival pathways, which result in the process termed 'oncogenesis'. The acts of "fits-all" chemo-radiotherapy with their inbuilt harmful consequences to the tumor-bearing host, are complemented, or even attempted at being replaced, by therapeutic measures precisely targeting the oncogenome, at its epigenome, and at its product oncoproteins. The oncogenes may be silenced, or abolished by an armada of special methylator-acetylator enzymes and their antagonists, acting on the oncogene promoters, or in the epigenome. When the oncogene issues a mRNA to be translated into an oncoprotein in the ribosome, the genomic introns are able to release microRNAs to align specifically with that mRNA sequence and demolish it. Thus, no oncoprotein will be synthesized. However, the same type of silencing rules may apply to the protein products of the so-called tumor suppressor genes, which are silenced by methylation and acetylation, and activated by demethylation and deacetylation. Deleted tumor suppressor genes may be replaced in the cell by vectors of gene therapy, the vectors being genetically engineered oncolytic viruses. The large oncoprotein kinases may be disabled by the small molecular enzyme inhibitors fitting into their "side pockets". Monoclonal antibodies and immunotoxins neutralize the oncoproteins. These disabled large protein kinases are marked for ubiquitination in the proteasomes. However, the same rules apply to the protein products of the so-called tumor suppressor genes. Thus, frequently the oncogenome still prevails.

The most prominent cancer driver genes have been identified by the Cancer Genome Atlas/Cancer Gene Census research networks in the genomes of 3205 human malignant tumors, representing 12 different cancer types. Of 291 most prominent cancer driver genes, 16 genes are confined to one specific tumor type; the other oncogenes are shared by many histologically different, or even identical, cancers. A characteristically lymphoma gene (ALK), or melanoma gene (BRAF) may drive either lung cancer or lymphoma cells. The healthy RNA/DNA genomes can generate mutations in response to demands (in the cocoons of insects; in mammals returning to the sea; in reptiles or mammals beginning to fly; create new immunoglobulin light chains by somatic hypermutations in the plasma cell in response to an antigenic challenge). The oncogenome carries out its measured-out sequential tasks (silence or eliminate pro-apoptotic and other tumor suppressor genes; amplify or constitutively activate anti-apoptotic and other oncogenes in a manner whereby a suppressed oncogene/oncoprotein can be replaced by another activated pathway immediately, or with a delay in an autophagic transformed cell;

and create new fusion oncogenes). In addition to the prominent cancer driver genes, cancer cell genomes seemingly haphazardly install thousands of different new somatic point-mutations of metabolic types. One in the millions of transformed cells endowed with different innumerable additional metabolically vital point-mutations will be able to fit into an unforeseen new environment, survive in there, and may even re-organize into colonies.

The task to overcome an inherent faculty that evolved originally in the first living cells in order to sustain cellular life on Earth, borders the extinction of stem cell-based cellular life itself. The inherent inbuilt faculty for the sustenance of cellular life in utmost distress, so far prevails (disorganized, but alive when transferred from the dying host into tissue culture vessels). "Oncogenesis", and its sequential molecular mechanisms, reveals the thaumaturgic abilities of the ancient RNA/DNA complex, admired in the academia (as reflected to in this essay). However, the superb achievements of natural and laboratory bioengineering rooted in the cerebral cortex hold the promise for the control of life's basic rules. The process of "oncogenesis", even if it is a basic faculty for the rescue of cellular life, cannot be tolerated to run its course at our clinics. The human cerebral cortex is to triumph over the oncogenome. Avoid mutagenic toxic exposures and utilize the consumption of natural substances (phytochemicals) that prevent random and even inherent genomic mutations. Manipulate the epigenome in favor of the tumor suppressor genes. Laboratory-made miRs will neutralize the mRNAs of the oncoproteins; circulating oncoprotein kinases will be marked for ubiquitination by small molecular inhibitors fitted into their specific domains (their "side pockets"); monoclonal antibody-complexed oncoproteins will be naturally eliminated (by ubiquitination); tumor suppressor genes will be vectored by lentiviruses; and tumor masses will be dissolved by targeted small molecular inhibitors or oncolytic viruses. Avoid indiscriminate attacks on the genomes; target the oncogene by highly specialized gene therapy; target the oncoprotein by specific immunotherapy; and combine these modalities of treatment. Use inhibitory monoclonal antibodies to release NK and immune T cells from the evolutionarily installed checkpoints (CTLA4; PD1) for the attack on malignantly transformed cells masquerading as 'self'. Intensify CAR therapy for solid tumors. The multicellular host is to be rescued intact, not just as a disorganized cell community preserved in a tissue culture flask.

Credo. A one sentence quotation from the Vogelstein et al article (Science March 28 2013) deserves citation: "The vast majority of our knowledge of the function of driver genes has been derived from the study of pathways through which their homologs work in nonhuman organisms". The author's present volume hereby submitted to the reader was inspired by identical motivations.

Medical oncologists of the author's era stood at the bed side of patients with inoperable chemo- and radiotherapy-resistant malignant tumors, unable to act in terms of a remission induction. Acting forcefully, they faced the devastating side effects of massive dosages of chemo-radiotherapy. Today, non-genotoxic targeted therapy induces durable remissions in those formerly untreatable malignancies: acute phase blast crisis CML; GIST, known then as "leiomyosarcomas of the stomach and small intestine"; devastating and mutilating basal cell carcinomas; chemo-immunotherapy-resistant metastatic melanoma; IL-2-resistant metastatic clear cell carcinoma of the kidney; young women suffocating with lymphangioleiomyomatosis; devastating amyloidososis; acute B lymphoblastic, or chronic lymphocytic leukemias; and the list is expanding toward some adenocarcinomas. Most rapidly fatal cancers, if not cured, are being converted into chronic illnesses. Patrick Hwu and his team, the head of M.D. Anderson's Melanoma/Sarcoma Medical Oncology Service (where this author JGS served in the 1960s–70s), print the pictures of the spectacular regression of a large malignant melanoma metastatic to the left orbit in an elderly gentleman (OncoLog MD Anderson's Report to Physicians, Feb 2014; volume 59 No 2). The antibody MK-3475 blocked the immune T cell's PD-1 receptor (programmed cell death protein-1), thus its natural ligand could not activate it. Another antibody could block the ligand for the same purpose. The tumor cells were masquerading as "self" in order to escape an autoimmune attack. The real identity of the tumor cell now revealed, it succumbs to an immune attack. The activated T cells attack and kill the melanoma cells. Beneath the immense practical health benefits harvested now, the ancient RNA/DNA ↔ DNA/RNA genomes still expand to fulfill their inherent faculty of sustaining *retrogradely immortalized cellular life* on Earth and in the Universe (Sinkovics JG Invertebrate Survival Journal, March 2016;13:68-75).

The medical oncologists of the past era experienced both the glorifying uplifts upon successfully inducing complete remissions, and the devastating downfalls when confronted by the relapses. Consolation for keeping the faith was gained from the ever increasing numbers of the patients sustaining their remissions. There remained the late onset adverse side effects to live with. Working naturally, targeted and immunobiological therapeutics are expected to secure cures with less adverse side effects.

Appendix 1*) For Essential Information Very Well Illustrated in Google and Wikipedia

*) *in Support of the Text with Literature Citations. Referrals to illustrations in Appendix 2.*

Cancer in the Plant. The insertion of the Agrobacterium tumefaciens circular plasmid T (transferred) DNA into the genome of its new host, the plant (Gelvin BS. Microbiol Molecular Biol Rev 2003;67:16–37). The plant cancer "crown gall" (agrocallus; Agrobacterial crown gall) consists of malignantly transformed cells replicating the agrobacterial T DNA plasmid (reviewed in postscript Table XXXV). For reference: Koncz C Mayerhofer R Koncz-Kálmán Zs et al EMBO J 1990;9:1337–1346. Transfer of potentially oncogenic bacterial genes and proteins to patients: Septicemic Bacteroides enterotoxigenic (Sinkovics J G & Smith JP Cancer 1970;25:663–671; Viljoen KS et al PLoS One 2015;10(3):e0119462); Bartonella bacilliformis etc (Guy L et al PLoS Genet 2013;9(3):e1003393; Harms A & Dehio C Clin Microbiol Rev 2012;25:42–78; Llosa M et al Trends Microbiol 2012;20:355–9; Minnick MF et al PLoS Negl Trop Dis 2014;6(7):e2919); Helicobacter pylori (Bonsor DA et al J Biol Chem 2015;pii:jbc.M115.641829; Su YL et al J Immunol 2015;194:3997–4007; Vaziri F et al Pathog Dis 2015;73(3). pii.ftu021); Porphyromonas gingivalis (Katz J et al Int J Oral Sci 2011;3:209–215); Tuberculous infections with *A. tumefaciens* in patients (Ramirez FC et al Clin Infect Dis 1992;15:938–940).

DNA-binding Antibodies. DNA- (or RNA-) binding proteins use zink finger motifs, leucine zippers and winged (beta-sheet loops) helix-turn helix motifs (HTH, two helices separated by the loop, RNA/DNA-binding domains) in recognition of RNA/DNA receptors for attachment. Pro-apoptotic protein p53 is commonly attached to DNA (vide infra). Advanced multicellular organism established a fundamental relationship between DNA genes and circulating antibodies, which overlapped the pre-existing miR activity for posttranslational gene expression. One of the best examples is played out in the pathogenesis of the T cell lymphoma mycosis fungoides and Sézary syndrome (that may also hide the underlying retroviral pathogen/passenger presumed to be there by Dorothea Zucker Franklin) (Proc Natl Adad Sci 1991;88:7630–7634; 1997;94:6403–6407). One of the inducer oncogenes of mycosis fungoides is TOX (thymocyte selection-associated high mobility group protein box gene), which normally regulates T cell differentiation.

J.G. Sinkovics, *RNA/DNA and Cancer*, DOI 10.1007/978-3-319-22279-0

Overexpressed TOX induces mycosis fungoides tumor cell proliferation and migration. The TOX protein mRNA is under the control of miR-223 through specific domain alignment to its 3′ UTR (untranslated region) in the AGO complex. High blood levels of miR-223 bring about the cessation of mycosis fungoides tumor cell growth. The production of tumor initiator high mobility group (HMG) DNA-binding protein is targeting its own DNA genes of T cell dedifferentiation and homing; it is switched off by a specific miR (Huang Y et al Oncotarget 2014;5:4418–4425; McGirt L et al J Invest Dermatol 2014; 134:1101–1107; O'Flaherty E & Kaye J BMC Genomics 2003;4:13).

Continuous Replication of the Trypanosoma in liquid media. Not in the insect's gut, but in the blood stream of its mammalian host, the slender Trypanosoma cell is fed primarily on sugar. This protozoan is also capable to synthesize glucose-6-phosphate. Growing in vitro in suspension, its speed of replication is dependent on the concentration of sugar in the liquid media. The stress responsive heat shock proteins (Hsp90) work in cell survival pathways in the protozoa, but became oncoproteins in multicellular, incl. human, hosts. Questions still incompletely answered: In the rapidly growing T. brucei, sugar metabolism is warburgian? Insulin-like growth factor receptor (IGFR) is overexpressed? Hsp90 promotes growth; is there growth dependence on Hsp90? Intranuclear Ki-67 is overexpressed? Telomers are capped after each cell division? Hsp90 inhibitors (geldanamycin analogs) induced trypanosomal death due to heat shock. Positive answers would mean that the unicellular protozoan's rapid growth is comparable to that of a cancer cell. Cooperative studies of Ludwig Maximilian University (LMU); Institut National des Sciences Appliqueés (INSA), Université Toulouse; Centre National de la Recherche Scientifique (CNRS), Université Bordeaux. Allmann S et al Biol Chem 2013;288:18494–18505. Van Reet N et al Exp Parasitol 2011;128:285–290. Meyer KJ & Shapiro TA J Infect Dis 2013;208:489–499 http://www.pasteur.fr/ip/easysite/pasteur/en/research/scientific-department/parasitology; http:/phys.org/news/2013-07-parasite-sugar.html; http://www.ncbi.nlm.nih.gov/pubmed/?term=trypanosoma+liquid++culture Faculty Biology, Ludwig-Maximilians-Universität München; Biozentrum, Grosshadernerstrasse 2–4, D82152 Martinsried, Germany. The Johns Hopkins University School of Medicine, Baltimore, USA.

Stem Cells Ascend from the bottom of the Lieberkühn crypt to renew themselves and release somatic cells to function as epithelial cells of the mucous membrane (reviewed in Table I).

S. Yamanaka Lentivirally Transferred four stem cell genes (oct4; sox2; klf4; c-myc) for reprogramming human differentiated cells into pluripotential stem cells; re-differentiation of pluripotential stem cells into differentiated cell could be induced. http://syntheticdaisies.blogspot.com/2012/10/the-spawn-of-gurdons-frogs.html.

Dictyostelia amoebae. The Amoebozoan Dictyostelium discoideum exists in many life forms: in sexual cycle, in vegetative cycle, and in social cycle. Its discoidin is a fibronectin; its tiger genes encode histocompatibility complexes; it uses

ERK/MAPKK signaling "cell survival pathways" (extracellular signal regulated kinase; mitogen-activated protein kinase kinase). MAPKK becomes a proto-oncogene in vertebrate mammalian (human) cells: in ras- (rat sarcoma) transformed cancer cells. Members of the D. discoideum clade (D. purpureum, D. fasciculatum) are comparable in their genomics and proteomics. Evolutionarily they separated from their common ancestor 600–800 mya and in comparison to the yeast, doubled their number of protein-coding genes up to 12,000 (Loomis WE Methods Mol Biol 2013;983:39–58). The other social amoeba, Acytostelium subglobosum emerges as a different species, which increased its chromosome number up to 18, but not because it has become diploid. The starving D. discoideum forms a conditionally multicellular fruiting body with spore ball on a stalk. It is the migratory slug cells that either form the fruiting body or the stalk. The fruiting body represents the germ cells and the stalk consists of the terminally differentiated somatic cells. The A. subglobosum does not differentiate its stem cells to a stalk; its stalk is a-cellular. The cellulose synthase gene builds the cell-free stalk. If "The differentiation of mortal or sacrificed somatic cells from reproductive germ cells was the key event for the establishment and diversification of multicellular systems" (Urushihara H et al BMC Genomics 2015;16(1):80), then D. subglobosum should be regarded evolutionarily more ancient than D. discoideum. A. subglobosum and D. discoideum are casting light to the beginning of a decisive process in the development of multicellularity. The Dd-MRP4 mitochondrial ribosomal protein is able to induce in human cancer cells the process of programmed cell death (apoptosis). Mitochondrial ribosomes are of ancient bacterial derivation. The Dd-mrp4 gene product protein in D. discoideum is the major initiator of cell differentiation (induced by starvation). Differentiation induction is achieved by the Dd-MRP4 protein through its entry into the nucleus for activation of its target genes (miR genes?). Another unrelated and different D. discoideum ribosomal protein is S4 (Dd-RPS4). It is homologous to the human chromosome X-linked cytoplasmic ribosomal protein S4 (Hs-RPS4X) in 66% identity and 92% similarity in their aa sequence (Maeda Y Biomolecules 2015;5:113–120).

Amphioxus I, prominent among them is Branchiostoma floridae. Possesses notochord and primordial nervous system consisting of intraabdominal sensory cells and notochord nerve cells connected with axons and synapses. An advanced neurochemical network releases and circulates cholinergic and GABAergic/glycinergic (gamma aminobutyric acid) molecular mediators. The amphioxus' achaete scute homologues (Ash) exist conserved up to Homo (Lu TM et al Development 2012;139:2020–2030). The Ash genes may act as proto-oncogenes subject to malignant transformation. The achaete scute homolog basic helix loop helix gene (ash) and the delta ligand Notch pathway evolved at this site. The adult amphioxus preserves its notochord, whereas tunicate larvae of Ciona lose their cordate characteristics through metamorphosis into adulthood. The amphioxus frontal eye spot reveals photoreceptor Joseph cells with many homologies to camera-type lens eyes of jawless (lamprey) to jawed (sharks) vertebrates (Candiani S et al BMC Neurosci 2012;13:59; Lu TM et al Development 2012;139:2020–2030; Vopalensky P et al

Proc Natl Acad Sci 2012;109:15383–15388). The amphioxus and the sponge Amphimedon qeenslandica share an ancestral TGFβ gene (Song X et al Genomics 2014;103:147–153). An advanced miR system controls amphioxus gene expression (Zhou X et al Gene 2012;508:110–116). The amphioxus genome preserved inserted six clades of non-LTR retrotransposons named among a long list of non-LTR elements of other sources including those from Ciona intestinalis (Permanyer J et al Int J Biol Sci 2006;2:48–53). The Lu TM team worked with B. floridae specimens collected in the Tampa Bay, Florida (W.P. Hall http://forums.about.com/ab-christianity/messages?msg=10917.2462). As oncogenes, Ash induce esthesioneuroblastoma*), medullary carcinoma of the thyroid, small cell undifferentiated carcinoma of the lung, or transforms pre-existing adenocarcinoma cells into chemo-radiotherapy-resistant neuroectodermal malignant cells (Wang XY et al Lab Invest 2007;87:527–539) Single nucleotide polymorphism in the 3′ untranslated region of miRNAs (hsa-miR-1827) occur in small cell lung cancers, in which the L-myc1 (1p34) and achaete scute human homolog genes (11p15.1; 12q22-q23) are involved (Xiong F et al Cancer Res 2011;71:5175–5181).

*) ***Esthesioneuroblastoma***. *The orbits and sinuses contain a large, enhancing, and expansile mass occupying the ethmoid air cells. The cribriform plate is invaded. The left anterior cranial fossa is broken into. Preoperative platinum-based chemotherapy, surgical resection, and postoperative radiotherapy is the standard treatment of Hyams' high grade and Kadish advancing stage tumors. From Mayo Clinics: USA, McElroy EA Jr et al Neurosurgery 1998;42:1023–1027; from Institut Gustave Roussy, Villejuif, France, Modesto A et al Oral Oncol 2013;49:830–834; from Children Hospital, Rabat, Morocco, El Kababri M et al Pediatr Hematol Oncol 2014;36:91–95. Comment: Treated high grade tumors lie dormant and relapse.*

Choanoflagellates (Monosiga brevicollis; ovata). Choanoflagellate genomic segments containing cell survival pathway Pak (p21 activated kinase); when transfected into somatic cells of metazoan multicellular vertebrate hosts (mice; rats; human), pak/Pak induces "malignant transformation" (Watari A et al Oncogene 2010;29:3815–3826). The merlin-encoding tumor suppressor gene neurofibromatosis NF2, the suppressor of cyclin D1 (or Bcl-1) promoter, is regulated by the pak/Pak pathway. PAK1 stimulates the cyclin D1 promoter; cyclin D1 mRNAs rise. Merlin inhibits the cell cycle in G1 by suppressing cyclin D1 and CDK4 (cyclin-dependent kinase) and dephosphorylating the retinoblastoma RB protein. The Rac/Pak axis is an oncogene (Xiao G-H et al Mol Cell Biol 2005;25:2384–2394).

Blooming Dinoflagellate Symbiodinium (zooxanthella) cells survive self-generated toxins that are highly lethal to multicellular life forms up to fishes and Florida manatees. ATP-cassette ABC transporter protein pumps expel the toxins by mechanisms similar to those of tumor cells gaining resistance to chemotherapy (Table I). Symbiodinium cells parasitize the cytoplasm of cnidaria corals, where they are replicating very rapidly. There is a mass exodus of flagellated mastigote Symbiodinium cells from the parasitized host cells. "Bleached" (whitened) coral colonies are left behind. The motile mastigotes discard they flagellae and assume non-motile metabolically quiescent coccoid morphology ranging in size of 6–13 μm. Both mastigotes and coccoid cells undergo mitosis generating two mastigotes. Dinoflagellates carry reticulated chloroplasts. The light-harvesting center is in the thylakoid membrane; it contains a chlorophyll-protein complex. Symbiodinium cells carry also mitochondria retaining some of their original genes. The intranuclear DNA is kept supercoiled within the chromosomes. In the S-phase of the cell cycle, the uncoiled DNA duplicates. Symbiodinium cells rapidly replicate in cultures in the laboratory. The rapidly replicating genomes are of interest: what genes are constitutively activated? Are intranuclear Ki67 proteins overproduced? The Symbiodinium genome operates over 42,000 protein-coding genes; duplicated genes; basic eukaryotic and horizontally acquired bacterial genes. It is spliceosomal intron-rich: 18.6 introns / gene; its spliceosomal snRNA genes (U1-U6) are clustered (Jeong HJ et al J Eukaryot Microbiol 2014;61:75–94; Shinzato C et al PLoS One 2014;9(1):e85182; Shoguchi E Curr Biol 2013;23:1399–1408). In Summary, some cnidarians carry virally infected endosymbiotic algae and respond with an active NFκB/STAT pathway (in Sinkovics JG Int J Oncology 2015;47:1211–1229).

Tibor Gánti's Chemoton. The chemoton is a self-reproducing automatic machinery; it operates autocatalytic cycles of chemicals in a liquid base surrounded by a bilipid membrane. Chemotons grow by metabolism and reproduce by biological fission, but as non-genetic entities. Osmotic pressure within the microsphere (the chemoton) induces invagination of the membrane leading to fission. The chemical reactions within the chemoton are not dependent on enzymes. Divided chemoton units are identical without hereditary variation. Cycle stochiometry characterizes mathematically and expresses quanitatively the chemical reactions occurring within the chemoton. Incorporation of a new external molecule (molecule T) into the system occurs through small gaps of the membrane. The amphipathic dynamic molecule advances the internal metabolism of the chemoton toward the acquisition of heritable enzymes (ribozymes being the first). Chemotrons are the primordial first cells of the living matter (Gánti T: “The Principles of Life”. Oxford University Press, 2003; Chemoton Theory. I. Theoretical foundation of fluid machineries. II. Theory of living systems. Kluwer Academic/Plenum Publishers 2003. Csendes T. A simulation study of the chemotron. Kybernetes 1984;13:79–85).

An Advanced Oparin's Coacervate (коацерват, Опарин) would be an imaginary protobiont, which would engulf either proteobacteria for mitochondrial symbionts (in animal cells), or cyanobacteria for plastids with eye spot for photosynthesis (in plant cells). Visiting lecturer Nasif Nahle, at Aguascalientes, Monterrey, Mexico in 2004, at the Conference of Astrobiology http://biocab.org/Astrobiology.html.

Reverse Transcriptase (D. Baltimore/H.Temin) transforms RNA into complementary cDNA. mRNA is used as template. The ssDNA strand is made into a dsDNA strand on a primer by DNA polymerase. Taq (Thermus aquaticus) heat stable polymerase (K. Mullins) synthesizes complementary DNA strands from primer in the polymerase chain reaction.

The Extant DNA/RNA Complex Reverts back to the Primordial RNA/DNA Complex. Predecessor cytogeneticist Barbara McClintock laid down the morphological foundations of chromosome science: deletions, inversions, translocations of mobile elements, ring formation, broken ends and fusions, that restructure the genome. Resurrection from 'genome shock'. Insertion of related elements at distinct loci and remaining under coordinated regulation (McClintock B. Significance of response of the genome to challenge. Science 1964;226:792–800). Re-emerges the primordial dependence of DNA on RNA, and on protein enzymes. SINE (short interspersed nucleotide elements) releasing mRNAs. Retrotransductions of LINEs (long interspersed nucleotide element) consisting of RNA and cDNA molecules. A combination of exons and introns in inverted terminal repeats. MITE DNA transposons (miniature inverted-repeat transposable elements). Inter-prokaryotic, intereukaryotic, including inter-kingdom, horizontal gene transfers. Symbiotic cellular unisons (cyanobacterial plastids in plant cells; alpha-proteobacterial mitochondria in animal cells). Tertiary symbioses in dinoflagellates, thus earning energy from photosynthesis. Evolution of genomes by whole genome doublings (WGD, repeatedly so) (Shapiro JA. Mobile DNA and evolution in the 21st century. Mob DNA 2019;1.4). The human genome is interspersed by transposable elements (TE), which are either DNA transposons, or RNA retrotransposons. The retrotransposons are SINEs, LINEs, LTRs (long terminal repeats), and Alus. TEs dictate the long term evolution of the genome. Satellite S (micro-, or macroS) DNAs are high copy number tandem repeats. The snoRNAs (small nucleolar) are the ancestors of the α-Satellate RNAs. Some ncRNAs are expressed from repetitive repRNAs. Noncoding ncRNAs of highly repetitive regions form a large reservoir capable to evolve into novel functional RNAs. The 300 bp Alu repeats (Arthrobacter luteus Alu enzyme that cuts DNA) are SINE family member ncRNAs. SINEs derive from tRNAs. SINE's RNA polymerase III transcribes Alus. Alu sequences remain embedded in protein-coding genes. Small nuclear snRNAs act like TEs and end up as pseudogenes (Matylla-Kulinska K et al Functional repeat-derived RNAs often originate from retrotransposon-propagated ncRNAs. Wiley Interdiscip Rev RNA 2014;5:591–600). MicroRNAs (miR-661; miR-885-3p) balance p53 and MDM protein expressions by controlling their mRNAs. In another level, Alus embedded in the 3'UTR (untranslated region) of mRNAs, as co-recipients of the aligning microRNAs, may accept or reject alignment (Hoffman Y et al microRNAs and Alu elements J Mol Cell Biol 2014;6:192–197). The primate-related SINE-Alu sequences appeared first in the Branchiostoma amphioxus (Holland LZ A SINE in the genome of the cephalochordate Int J Biol Sci 2006;2:61–65). In the Argonaute (AGO) protein family, PIWI proteins bind PIWI-interacting piRNAs (P element-induced wimpy testis, drosophila); it is one of the three small RNAs (micro miRNA; small interfering siRNA). AGO proteins also bind miRNAs and siRNAs. PIWI proteins are primarily expressed in the germline, but somatic tissues are not exempt of their presence either; that includes stem cells. The PIWI/piRNA pathway suppresses transposons, thus saving the integrity, and the regulation with chromatin, of the protein-coding genes (Mani SR & Juliano CE Untangling the web Mol Reprod Dev 2013;80:632–654).

Acholeplasma (Mycoplasma) laidlawii was grown in culture and its fusogenic phage particles were purified and concentrated (www.lookfordiagnosis.com). Author's first encounter with A. laidlawii occurred in the early 1970s at M.D. Anderson Hospital, where A. laidlawii was found to have contaminated a human sarcoma established cell line. The contaminated fluid was transferred to Ferenc Györkey of the Pathology Department at the Veterans Administration Hospital, Houston, TX for further studies in cell fusion experiments (unpublished). Similar contaminated human cancer cell cultures were reported from Russia (Polinskaia GG et al Tsitologiia 2000;42:190–195; 794–801). A swine pathogenic Acholeplasma exists in Hungary (Stipkovits L et al Acta Vet Acad Sci Hung 1973;23:307–313; 361–368; J Hyg (London) 1974;72:289–296). Cells of the pathogenic PG8 strain of Acholeplasma laidlawii (the causative agent of phytomycoplasmosis in Oryza sativa L plants infected through their roots) release extracellular membranous vesicles. These vesicles contained virulence proteins, RNA nucleotide sequences, including rRNAs (Chernov VM, Chernova O A et al Scientific World J 2011;11:1120–1130; 2012;3154–3174; J Proteomics 2014;110:117–128). The presence of rRNAs indicates to this author that the extracellular membranous vesicles are ribosome-like exosomes. The phenomenon of exosome release is universal in the cellular world. Observations without actual precise counts indicate that virulent pathogenic or malignantly transformed cells release excessive numbers of exosomes. This author suggested that DNA-free but RNA-loaded exosomes (including rRNA) represent a retrograde evolutionary descent of the cell to the precellular era of the living matter. In the RNA World exosome-like units might have populated the Earth. Precellular ribozyme-armed and readily mutable rRNA-containing ribosome-like units might have existed with simple ribonucleoproteins (RNP) within (Sinkovics JG Europ J Microbiol Immunol 2015;5:25–43). Metabolomic analysis shows some unique differences between A. laidlawii (strain PG-8A) and Mycoplasma pneumoniae in that, three amino acids (alanine, asparagine, valine) present in M. pneumoniae are missing from A. laidlawii. It is possible that alanin is very rapidly incorporated into proteins through aminoacyl-tRNA synthetase, thus apparently disappearing from the metabolome. A. laidlawii operates a unique lysine decarboxylase not possessed by any other Mycoplasmataceae. In riboflavin biosynthesis, in addition to the customary bifunctional riboflavin kinase, only A. laidlawii possessed enzymes for the conversion of diamino-hydroxyphosphoribosyl-aminopyrimidines (Vanyushkina AA et al PLoS One 2014;9(3):e89312). Virally infected (phage-carrier) JA1 strains of A. laidlawii reveal acute infections with cell lysis, and chronic non-lytic non-cytocidal infection with rapid growth of infected cells. A. laidlawii has no bacterial cell wall; it is surrounded by a single lipoprotein cell membrane, similarly to spheroplasts, or to eukaryotic cells of uni- or multicellular hosts (Liss A & Ritter BE J Gen Microbiol 1985;131:1713–1718). After lysis of the membrane with antibody and complement, the remaining cytoplasms were referred to as ghosts

(Brunner H et al Infect Immun 1976; 13:1671–1677), but without stating if the membrane-free cytoplasms could metabolize and to what extent. The membrane-free E. coli cytoplasms observed by this author were presented as examples of "cell-free living matter", "plasma" (even though they were lost for further studies in 1956) (Sinkovics J Biol Közlem (Hungary) 1954;2:69–81; 1957;5:19–27; Acta Microbiol Hung 1957;4;59–75; Die Grundlagen der Virusforschung, Hung Acad Sci, 1956). However, extracellular E. coli ribosomes remained functional in vitro in the M W Nirenberg and J H Matthaei experiments (Proc Natl Acad Sci USA 1961;47:1580–1586; 1588–1602). Now, A. laidlawii exosomes emerge; however, they are membrane-bound cytoplasmic units (vide supra).The extant Mollicute A. laidlawii strains exhibit a high degree of genomic and proteomic deviation among themselves; the deepest branching member of the clade remains A. laidlawii (Stepens EB et al Yale J Biol Med 1983;56:729–735; Kube M et al J Mol Microbiol Biotechnol 2014;24:19–35). The ancestors of the unique Mollicute mycoplasma A. laidlawii and its fusogenic virus coexisted with archaea (crenarchaea) spheroplasts, a scenario that readily offers itself for theories of primordial cell fusions that might have created the first eukaryotic cells; a situation possibly amenable to reproduction in the laboratory (Sinkovics JG Acta Microbiol Immunol Hung 2001;48:115–127).

Fusogenic Viruses. Not all phages are lysogenic. An Acholeplasma phage (see Text) might have been the ancestor of eukaryotic fusogenic viruses, which induce coalescence of their host cells' membranes. Of numerous eukaryotic fusogenic viruses, stand out the Mus cervicolor leukemia retrovirus M813 fusing T lymphoma cells; the EnvV syncytin-producer retroviruses fusing syncytiotrophoblasts of the placenta; HIV-1 fusion protein fusing infected lymphocytes; influenza viral fusion peptides fusing lipid bilayer vesicles; parainfluenza and paramyxoviral cell fusion proteins; corona virus-induced syncytia; measles virus producing multinucleated giant cells; Sendai virus-fused cells; herpes simplex viral cell-fusogenic glycoprotein M; env gene/protein-acquiring endogenous retroviruses budding from human tumor cells (lymphoma, Reed-Sternberg cells, Kaposi sarcoma, adenocarcinoma, melanoma) and inducing cell fusions. Virus-producer fused tumor cells may gain growth advantages; may express new antigenic targets to immune defenses or to immunotherapy (see Text). Upon cellular encounters, the identity or chemical relatedness of cell membrane or viral capsid/envelope structures leads to cell fusions (Ogle BM et al Nature Reviews Molecular Cell Biology 2005;6:567–575). A candidate fusion partner is the ancestral mycoplasma Acholeplasma, and the primordial mediator is its fusogenic virus (phage) (see in Text).

The much Debated Tree of Life was set into three domains by Ernst Haeckel: "Monophyletischer Stammbaum der Organismen" consisting of Protista, Plantae and Animalia. Extraordinary acquisition of genes through horizontal (lateral) routes invalidates the evolutionary pathways that were constructed by true vertical inheritance. Among Bacteria, Mycoplasmataceae vie with Aquifex aeolicus for the basic position; regression of Mycoplasmataceae from Bacteroides is unacceptable

(Klenk H-P et al J Mol Evol 1999;48;528–541; Woese CR Maniloff J Zablen LB Proc Natl Acad Sci USA 1980;77:494–498).

After long debates, it has been fully accepted that cyanobacteria entered as endoplasmic symbionts of the eukaryotic algae-to-plants lineage; and that ancestral purple proteobacteria have become the eukaryotic cytoplasmic endosymbiont mitochondria of animalia.

Archaea are more accurately divided as phyla Crenarchaeota, Euryarchaeota, Korarchaeota, and Nanoarchaeota. In evolutionary order, for the position of the first domain, vie the prokaryotes Mycoplasmataceae or Aquifex, with Nanoarchaea. Recent isolates include the Lokiarchaeota loaded with genes, whose signatures later become dominant in the eukaryotic proteomic repertoire (Spang A et al Nature 2015;521;173–179). The TACK archaeal superphylum consists of Thaumarchaeota/Aigarchaeota/Crenarchaeota/Korarchaeota existing in sister relationship with Euryarchaeota, both being ancient carriers of genomes/proteomes that reappear in eukaryota (Raymann K et al Proc Natl Acad Sci USA 2015;12:6670–6675). A hyperthermophilic extremophile archaea isolated from the hot springs of the Yellowstone National Park carries a virus with a circular dsDNA genome, which encodes elaborate turret-like structures in its capsid. These structures of the Sulfolobus turreted icosahedral virus re-appear throughout evolution in bacteriophages, algal viruses and adenoviruses. Its gene might have originated in a common ancestor that existed before the division of the three domains of life (Rice G et al Proc Natl Acad Sci USA 2004;101:7716–20).

Joseph, Rhawn: Genetics and evolution of the life from other planets. J Cosmology 2009;1:150–200). In contrast, the present author prefers to envision the origin of life on the ancient planet Earth. Between the first protocells, virally-mediated fusion of two equal partners: a spheroplast of an archaeon (a crenarchaeota) and a spheroplast of a prokaryota (a mycoplasma), followed by the later engulfment of the α-proteobacterium to become the mitochondrium, or a cyanobacterium to become a photosynthetic unit (plastid). Phage-mediated fusions of cell wall-free mycoplasma-like cells (the Acholeplasma) do occur. The equal fusion partner of the prokaryota would be a crenarchaeota (both in the form of spheroplasts), but would they yield to the same fusogenic phage? These basic fundamental experiments of nature could possibly be reproduced in a laboratory. The candidate fusion partner archaea might have been thermophilic. Thermophilic archaea populated (still do) hot submarine vents. Light energy from the Sun does not penetrate the depths of the seas. There, geothermal radiation provides the energy to anaerobic photosynthetic pigments (Beatty JT et al Proc Natl Acad Sci USA 2005;102:9306–10). Some nanoarchaea (N. equitans and Ignicoccus hospitalis) form unisons without fusion. Such unisons are expected to have been formed among the deepest branch-offs (Stetter KO: A brief history of the discovery of hyperthermophilic life. Biochem Soc Trans 2013;41:416–429. Jahn U et al Nanoarchaeum equitans and Ignicoccus hospitalis: New insights into a unique, intimate association of two archaea. J Bacteriol 208;190:1743–1750. Podar M et al A genomic analysis of the archaeal system Ignicoccus hospitalis-Nanoarchaeum equitans. Genome Biol 2008;9(11):R158). Fused cells of equal ranks created the first eukaryote; the united large cell was able to engulf single-celled bacteria, which surrendered to endosymbiosis in a subservient position as mitochondria and plastids. The mitochondria transferred most of their genes to the host cell nucleus, but retained the faculty of the endogenous apoptosis induction in their host cells. The widely practiced lateral (horizontal) gene transfers between archaea and bacteria (unidirectional; bidirectional?) resulted in species diversification, but transfer-resistant macromolecular genes preserved the dominant vertical routes of evolution (Abby SS et al Proc Natl Acad Sci USA 2012;109:4962–4967).

The Unicellular Ciliate Protozoon, Paramecium tetraurelia, possesses nuclear dimorphism: a germline genome in its micronucleus (MIC) and a somatic genome in its macronucleus (MAC). Its cytochrome b5 genes carry out the ubiquitous electron transports. Upon its three or four rounds of whole genome duplications, the largest numbers of reduplicated and retained, and the smallest numbers of lost, genes are the cytochrome b5 genes (of the whole macronuclear and mitochondrial genomes consisting of some 40,000 genes). Among others, sequenced were the tRNA trp and tRNA tyr (tryptophan; tyrosine) genes. The paramecium germ line nucleus is transcriptionally silent, and the somatic macronucleus is transcriptionally active. Germ line-specific sequences (internal eliminated sequences, IESs) are excised by the thousands, but with great precision (as in other ciliates). One IES is located in an intron. Gene expressions are carried out with extraordinary replication fidelity, especially in their DNA polymerases, resulting in nucleotides conserving amino acid sites (Sung W et al Proc Natl Acad Sci USA 2012;109;19339–19344; Vayssié l et al Nucleic Acids Res 1997;25:1036–1041; Wu KJ et al Genet Mol Res 2013;12:1882–1896).

Conjugating Paramecia. After the sexual event, the new somatic MAC submits the germline MIC to a series of rearrangements. The rearrangements consist of the precise excision from the diploid germline genome of inserted parasitic DNA elements. Unique-copy Internal Eliminated Sequences (IESs) are excised by certain long ago inserted, but by now domesticated transposases, such as from the piggyBack (from cabbage looper moth Trichoplusia ni) to the PiggyMac element. Targeted are the Tc1/mariner, sardine and anchois transposons. All IESs are remnants of these, or other pre-inserted transposons. Some IES inserts shorten (<159 bp) and decay. Epigenetic ncRNAs locate the targets; ncRNAs remain on board to purify eukaryotic genomes of RNA-to-DNA inserts. Two dinucleotide TA sequences flank the IES to be excised. Mutated flanking TA dinucleotides with cis-acting Mendelian mutations prevent their excision; otherwise mutation in one of the two flanking TA dinucleotides pose an absolute requirement for IES excision. The Oxytricha genome is heavily infested by inserted transposons. The endonuclease required for the cleavage of the Oxytricha genome is a transposase derived from its germline TBE (telomere-bearing elements) transposons. The piggyback domain5 (PGBD) element was domesticated in the amphioxus 500 mya. It remains conserved in all vertebrates, but in the human genome it lost its catalytic motif and acquired restrictive-silencer factors, thus it is non-functional (Arnaiz O et al PLoS Genet20112;8(10):e1002984; Pavelitz T et al Mob DNA 2013;4(1):23).

The Vegetative Giardia Trophozoites are binucleated. The mitosomes of the Giardia and the hydrogenosomes of the trichomonads are remnants of pre-existing mitochondria (Dolezal P et al Proc Natl Acad Sci USA 2005; 102;10924–9). The giardial genome operates 6470 ORFs. They encode ubiquitin-activating enzymes. During the process of encystation, many proteins are eliminated by ubiquitination. Polo-like kinases, initiators of mitosis, are also degraded (Nino CA et al Microbiology 2013;2:525–539). The low branching unicellular eukaryotic diplomonads, Giardia, operate noncoding ncRNAs, small nucleolar snoRNAs, spliceosomal small nuclear snRNAs, and inhibitory micro miRNAs. Giardias preserved the key ribozymes/RNPs for pre-rRNA processing (mitochondrial RNA processing in an amitochondriate). The H/ACA box snoRNAs (adenine cytosine adenine) are for pseudouridylation (posttranscriptional isomerization of uridine to pseudouridine ψ) of mRNA/rRNA/snRNA. The ψ synthases are guided to their targets by the small snRNAs already in the archaea. The location of pseudouridine nucleotides has been conserved throughout evolution. H/ACA-like RNAs, the first stem-loop, are shown (Chen XS et al BMC Genomics 2011;12:550; Kiss-László Z et al Cell 1996;85:1077–1088). The nucleolar localization elements, the single-stranded domain H box stands for ANANNA, N for any nucleotide; a 3' hairpin is followed by the adenine cytosine adenine ACA box (Fourmann J-B et al PLoS One 2013;8(7):e70313). The three PI3K genes encode the kinases for the cell survival pathways. Giardia PI3Ks disable host dendritic cells rendering them from immunogenic to tolerogenic entities (Cox SS et al BMC Microbiol 2006;6:45; Hernandez Y et al J Eukaryot Microbiol 2007;54:29–32; Kamda JD & Singer SM Infect Immunol 2009;77:685–693). Giardia lamblia is expressing its cell surface antigens upon inhibition of iRNA (rendering inhibitory RNA machinery defective). Hugo D. Luján, Catholic University, Córdoba, Argentina. http://medgadger.com/2008/12/ is studying_the_chameleonlike_properties_of_giardia_parasites

Chlamydomonas reinhardtii single-celled-green-alga. The bi-flagellated photosynthetic green alga C. reinhardtii operates some 15.000 genes, cyclic AMP and GMP prominent among them, and generates oxygen in its chloroplasts. The ancestors of some genes of flowering plants and vertebrate mammalians (humans) are operational in its genome, including those of the trichomonads, and mammalian vertebrate cells' myb proto-oncogene. AlgaePath home page (Zheng H-Q et al BMC Genomics 2014;15:196 pp 1–11. doi:10.1186/1471-2164-15-196,http://AlgaePath.itps.ncku.edu.tw

The Colony of Volvox Cells swimming in accord toward light consists of the large (1 millimeter) gelatinous body surrounded by hundreds of non-reproductive cells. Inside the gelatinous body are the reproductive cells (gonidia) practicing isogamy.

The related and more advanced Pleodorina gametes distinguish sperm-producing male individuals from ovum-holding female individuals, and thus practice anisogamy. The male Pleodorina gametes possess the active intranuclear otokogi (chivalry) gene. The chlamydomonas mid/MID (minus dominance) gene, the forerunner of the otokogi gene, is operational in the otherwise indistinguishable plus female individuals; the males are recognized as the the minus individuals; plus-females rendered MID-negative minus males (but not vice versa) (Hisayoshi Nozaki, Biological Sciences Graduate School of Science, University of Tokyo, Japan. Nikon Corporation, 2011).

Of Retroviruses, the reticuloendotheliosis viruses (REV) excel as the most prominent collectors of cellular proto-oncogenes (c-onc → v-onc). None of this was known when Vilhelm Ellermann and Oluf Bang isolated the first chicken leukosis retrovirus in the first years of the last century, and when J.W. Beard, Dorothy Beard and Ursula Heine collected and maintained by passages all new isolates of leukemia-and sarcomagenic reticuloendotheliosis viruses at Dukes' in Durham N.C. Hidesaburo Hanafusa in New York (beginning with the Rous sarcoma virus), and Peter K. Vogt and associates in Los Angeles CA, identified the oncoproteins encoded by the avian reticuloendotheliosis viral isolates. J.M. Bishop and H.E. Varmus and associates (D. Stehelin, P.K. Vogt), and H. Hanafusa established the cellular origin of these oncoproteins. Phylogenetic studies of the REVs reveal that they are of mammalian retroviral origin; their ancestral genomic sequences are conserved in the mammalian genomes. A REV- infected plasmodium (P. lophurae) was distributed in various laboratories in the 1930s for the study of anti-malaria vaccination. This plasmodium infected first laboratory birds with REV. From REV-infected birds a wide array of wild bird population picked up REV, which rapidly spread world-wide. The most mutagenic and promiscuous REV genomes inserted themselves within coinfected cells first into the Marek's herpesvirus of turkeys (gallid herpesvirus type 2), and more recently into the avian fowlpox virus (Niewiadomska AM & Gifford RJ PLoS Biol 2013;11(8):e1001642). The Bittner mouse mammary tumor (MMTV) retrovirus may spread horizontally with the milk, or may insert its gene into its host's genome for vertical spread. A MMTV-relative is detectable in human breast cancer tissue. MMTV encodes transmembrane glycoprotein superantigens (Sag). Sag-expressor infected B lymphocytes spread the virus within their host. TCR Vβ CD 4^+ lymphocytes respond to the horizontally spread MMTV infection. In lack of its own oncogene, the MMTV genome inserts itself into cellular proto-oncogenes, targeting the int, wnt and fgf/FGF, or Notch genes. An avian REV erythroblastosis virus-incorporated the EGFR gene which is expressed in some human breast cancer cells (Her2/neu). The amplified HER protein induces an antibody and T-lymphocyte-mediated immune reaction (Punkosdy GA et al Proc Natl Acad Sci USA 2011;108:3677–3682 george.punkosdy@emory.edu; Holt MP et al Front Oncol 2013;3:287; Szakacs JG & Livingston SK Ann Clin Lab Sci 1994;24:324–338). Human endogenous retroviral gene-encoded proteins and fully mature enveloped retroviral particles are frequently expressed in human leukemia and solid tumor cancer cells. The endogenous retroviral particles often induce the fusion of the malignant cell with a non-transformed host cell (monocyte-Mϕ) (Januszkieweicz-Lewandowska D et al Acta Haematol 2013;1229:232–237; Mullins CS & Linnebacher M World Gastroenterol 2012;18:6027- 6035; Wang-Johanning F et al J Natl Cancer Inst 2012;104:189–210).

Exogenous viruses force neoplastic transformation upon their host, usually overcoming host resistance. Exogenous virally induced neoplastic transformation for a cell is an attack, not a "rescue mission", and as such, it is responded to immunologically. In contrast, domesticated ancient retroviral inserts (endogenous retroviruses or retrotransposons) may internally (endogenously) induce cell transformations, that may be considered as an attempt at regressing the cell into a state of primordial immortality, and as such, it may be supported by the host. Not all endogenous retroviruses are reticuloendotheliosis viruses. Soft-shell clams (Mya arenaria) harbor the leukemogenic retrotransposon Steamer of the gypsy family. One clone of leukemic cells has been spreading among the clams. The leukemic cells express high levels of Steamer retrotransposon mRNAs. Steamer DNA is inserted in the cells' genomic DNA, and reaches high copy numbers in leukemic haemocytes. The Steamer ORF encodes a large Gag-Pol protein, but not an env protein. Thus, Steamer does not spread as an infectious retrovirus. The leukemic cells show increased RT expression. Disabled p53 levels are high, and no apoptotic haemocytes are reported. Leukemic haemocytes spread the disease as they are able to travel in sea water. Similar leukemias exist in mussels (Mytilus) (see in text), and oysters (Crassostrea) (Arriagada G et al Proc Natl Acad Sci USA 2014;111:14175–14180; Metzger MJ et al Cell 2015;161:255–263).

Ancestral Unicellular and Early Multicellular Eukaryotes Utilized an Array of Transposable Elements (TE), LINE-1 (L1) Being Prominent, in Their Cell Survival Pathways, as Shown Conserved in the Extant Descendants. Malignantly Transformed (Cancer) Cells Also Utilize High LINE-1 Activity and HSP. The amoeba Dictyostelium discoideum operates a long list of repeated transposable elements (Glöckner G et al Genome Res 2001;11:685–694). The yeast Saccharomyces cerevisiae harbors the non- LTR retrotransposons LINE-1 (Dong C et al Genetics 2009;181:301–311). All ancient sexually developed phyla harbor LINE-1 and gypsy-like reverse transcriptases (RT) and mariner transposases. Listed in Table 1 of the cited article, these are the Cnidarian Hydra, ctenophores, and Spongiae. Only the asexually evolved bdelloid rotifers miss LINE and gypsy retrotransposons and their encoded RTases (but harbor mariners), indicating that in all other taxa LINE-and gypsy transposons are sexually transmitted. In absence of env/Env, these sequences are are not readily transmitted horizontally. Higher up on the evolutionary scale (Branchiostoma floridae; Danio rerio; Xenopus; Mus musculus), the genomic copy numbers of the LINE and gypsy sequences rise (Arkhipova I & Meselson M Proc Natl Acad Sci USA 2000;97:14473–14477).

The gypsy family LINE-1 retrotransposons remained active in, and form cca 17% of the human genome. These elements express ORF 1 and 2 and are either non-coding, or are able to encode endonucleases, RTases, and RNA-binding proteins, most of which are truncated and immobile. However, rapidly proliferating cells overexpress RT induction of LINE-1-derivation. Transposable elements often appear as non-coding ncRNAs. and miRNAs. Within them embedded are the TE sequences at their transcription starting sites, whereas protein-coding exons are free of the TE sequences. RNAi (inhibitory RNA) by downregulating LINE-1, exerted anti-proliferative effect to human melanoma xenografts by inducing their differentiation. Upon cessation of RNAi administration, the tumor cells regained their aggressive undifferentiated phenotype and re-expressed L1 and RNA:DNA hybrids. The role of L1-encoded RT in the maintenance of the malignantly transformed state of the cell is essential (Spadafora C Ann N Y Acad Sci 2015;1341:164–171; Sciamanna I et al Oncotarget 2014;5:8039–8051). An example of extraordinary LINE-1 retrotransposon activity in cancer cells is the human colorectal carcinoma. L1 retrotransposition is carried by the process of target-primed reverse transcription (TPRT) with the help of ORF1/2-encoded endonuclease and RTase. The L1 elements practice insertional mutagenesis in genes protein tyrosine phosphatase receptor type M (PTPRM), odd Oz/ten-m homolog 3 (ODZ3), roundabout axon guidance receptor homolog 2 (ROBO2), pericentriolar material 1 (PCM1) and cadherin -11 (CDH11), or ICDH12. The genes mutagenized by the human L1H (human) elements fulfill driver function for the transformed cell. Due to attempted DNA repair mechanisms, the inserted L1 elements suffer truncation, nevertheless sustain rapid cell cycle turnovers (elevated cell division) rates (Solyom S et al Genome Res 2012;22:2328–2338). L1 retrotransposition are prelude to pseudogene generation. In that, the Trichomomas cells and the malignantly transformed (cancer) cells imitate each other (vide infra). In comparison to differentiated resting cells, de-differentiated active cells overexpress these retroelements. Both low branching unicellular, or early multicellular eukaryotes, ancestral and extant, and selected single cells of advanced multicellular hosts, in the process of malignant transformation, utilize alike L1 sequences for survival. Termed 'malignant' only, because transformed single cells of advanced multicellular hosts inadvertently consume and kill their host. Questions arise: where is the origin of endogenous LINE-encoded RT overproduction? It was in the first unicellular and early multicellular eukaryotes, where it was in use in the cell survival pathways in support of survival in distress LINE-1 together with the ancient heat shock proteins (HSP) (Nicosia A et al PLoS One 2014;9(9):e 105908), were utilized in response to exogenous physicochemical environmental inflicts. In a form of retrograde phylogenesis, selected cells of multicellular extant hosts may return to that stage of evolution by reactivating genomic faculties conserved from the ancestors, and thus resume their phenotype of rapid replication and resistance to physicochemical inflictions. Is the cancer-free longevity of the blind or naked mole rats (Spalax; Heterocephalus) consequential to the absence of viable transposable elements (no LINE-1) in their genomes? (Manov I et al BMC Biology 2013;11:91. doi:10.1186/1741-7007-11-91).

Gene Methylators-Demethylators; Acetylators-Deacetylators. Early discovered histone lysine (K) methyltransferase protein domains are those from Paramecium bursaria chlorella virus, and trithorax from drosophila's suppressor of variegation Su (var), and enhancer of zeste (SET); further, the myeloid translocation protein, nervy, and deformed epidermal autoregulatory factor (DEAF; MYND; SMYD*).

*) *The two transcript variant human SMYD (SET and MYND domains) gene maps on 1q44 (Luo XG et al J Biosci Bioeng 2007;103:444–450). Highly over-expressed in cancer cells (Hamamoto R et al Cancer Sci 2006;97:113–118). SET and SMYD transactivate the promoter of telomerase reverse transcriptase (hTERT) by di- or trimethylating its H3K4. This process is highly and constitutively upregulated in malignantly transformed human cells. It is targeted for inhibition by small iRNA. The hTERT promoter is subject to demethylation and repression; and acetylation and phosphorylation (Liu C et al Cancer Res 2007;67:2626–31). Hypermethylation of the CpG island of the cycline-dependent kinase inhibitor p16/CDKN3A tumor suppressor gene at 9p21 (or deletion of this gene) is a pre-disposing event for the development of sporadic malignant melanoma (Gonzalgo ML et al Cancer Res 1997;57:5336–47). These are early recognized examples of epigenetic proto-oncogene transactivation and tumor suppressor gene de-activation by gene silencer/promoter methylation/demethylation (Dillon SC et al Genome Biol 2005;6(8):227).*

Histone deacetylases are upregulated in malignantly transformed cells. The histone deacetylase inhibitor suberoylanilide hydroxamic acid (SAHA) induced differentiation of murine erythroleukemia cells (Richon VM et al Proc Natl Acad Sci USA 1998;95:3003–7). Histone deacetylase 1 is an activator of the translocated fusion oncogene/oncoprotein TMPRSS2/ERG in prostate cancer (androgene-dependent transmembrane serine protease; E26 ETS-related gene ERG) (Iljin K et al Cancer Res 2006;66:10242–6). The deacetylated constitutively expressed testosterone-driven androgen receptor genome drives prostate cancers. Histone deacetylase inhibitors re-acetylate and silence the androgen gene. Prostate cancer cells so treated undergo apoptotic death. Of numerous histone deacetylase inhibitors (depsipeptide, Na-butyrate, trichostatin, valproic acid), SAHA appears to be the most efficient (Marroco DL et al Mol Cancer Ther 2007;651–60). In black C57 mice, SAHA subdued allogeneic graft-vs-host disease with preservation of the graft-vs-leukemia effect (Reddy P et al Proc Natl Acad Sci USA 2004;101:3921–6).

Missense Mutations. Mismatch Repair. Rare missense point mutations (non-synonymous substitutions in a single nucleotide of a codon) may not be causally cancer-associated; recurrent missense mutations are frequently cancer-related. In a nonsense mutation, the stop codon encodes a truncated protein. The encoded serine/threonine protein kinase of B-raf (rat fibrosarcoma) gene residing at 7q34 is BRAF, activator of the ancient MAP/ERK pathway. In the V600E mutation in codon 600 valine is replaced by glutamic acid (E). Somatic mutations prevalently associated with the cause of cancers, while germ line mutations rather spell predisposition. For distinction, computational algorithms have been designed (Kaminker et al Cancer Res 2007;67:465–73). Frameshift mutations consist of deletions and insertions of several nucleotides resulting in the sequential position of codons, thus altering the triplets, and thus translation of the genetic code. Frameshift indel mutations in the tumor necrosis factor superfamily gene 10 in prostate cancer cells disrupt its open reading frame disabling it to induce apoptosis (Xu X et al Gene 2013;519:343–7). The common single nucleotide polymorphisms (snips) consist of single nucleotide replacements (thymine for cytosine). In the two parental copies of the genome, single nucleotide polymorphic regions with different bases coexist in heterozygosity. Loss of one parental region leaves one copy remaining: hemizygosity for that region, due to loss of heterozygosity. Tumor suppressor genes losing one parental copy remain functional; loss of the second allele (Knudson's second hit for the retinoblastoma gene) opens up unopposed oncogenic pathways. Microsatellites are repetitious up to six DNA base-pair sequences (simple sequence repeats, SSR). Instability of microsatellites occurs during faulty mismatch repair, leading to hypermutability of the genes involved (as in xeroderma pigmentosum; prominently in colorectal adenocarcinomata, etc). The evolutionarily conserved mismatch repair mechanism excises the faulty DNA segment and replaces it with the correct sequence (that can be over 1000 base pair long). The mismatch repair enzymes MutS, MutH, MutL recruit other enzymes (methyl transferase, helicase, exo- and endonuclease) to jointly carry out the nicking, separating and replacing the involved segments. The process of mismatch repair may fail due to point mutations of the enzymes: mismatch repair deficiency. The familial non-polyposis colon cancer Lynch syndrome is characterized by germ line mutations of the mismatch repair enzymes (Boland CR Gastroenterology 2013;144:868–70).

The 'driver mutations' conferring selective growth advantages of the tumor cell usually appear in a set sequential order of tumor suppressor gene eliminations and proto-oncogene point mutations (the Vogelstein sequence in colon cancer). Thereafter, very large numbers of haphazard somatic gene mutations occur. Mismatch repair deficient tumors excel in the pursuit of this latter pattern. In this phase of host-tumor relationship, the tumor loses the support of its host, and is subjected to excessive lymphocyte-mediated immune reactions. Innumerable new protein mutations induce the attacker lymphocytes. In response, the tumor cell overproduces PD-1 ligands for the induction of apoptotic death in the attacker lymphocytes, and thus it prevails. External help with monoclonal antibodies nivolumab and pembrolizumab protect PD-1 from contact with its ligands; the intact lymphocytes attack and destroy the tumor. Thus is the extraordinary susceptibility of mismatch repair deficient tumors to antibodies that protect PD-1 from contact with its apoptosis-inducer ligands (Le DT et al New England J Med 2015;372:2509–20).

The First Encounter with NK Cells. The NCI project site visitors arbitrarily transferred the healthy control investigator's (JGS) research funds from M.D. Anderson Hospital to the NCI intramural Herberman laboratory for the "clarification of the in vitro artifact reactions". In the clarification efforts, the Herberman laboratory used the modern and accurate "radioactivity-release assay", instead of the "out-moded and clumsy" chamber-slide assay. However the radioactivity-release assay did not distinguish between different types of cytolytic lymphocytes. Neither the Hellström, nor the Herberman laboratories visualized the morphological differences that attacking lymphocytes clearly exhibited in the most meticulous and reliable chamber-slide assay (in use in the Sinkovics Lab). Several years later the Herberman Lab declared their discovery of "natural killer cells", a subclass of lymphocytes, that killed autologous virally-infected or malignantly transformed cells without pre-immunization. The NK cells appeared as large granular lymphocytes (as shown earlier in the chamber-slide cultures in the Sinkovics Lab). Both healthy donors and patients with cancer possessed NK cells, but only the patients generated the small compact immune T lineage lymphocytes lytic to autologous (and related allogeneic) tumor cells (Lichtman MA: Historical landmarks in the understanding of the lymphocytes. In P.H. Wiernik et al (eds) "Neoplastic Diseases of the Blood". Springer Science, New York, 2013; Pp 789–83. Sinkovics JG & Horvath JC. Internat J Oncol 2005;27:5–47; Sinkovics JG: "Cytolytic Immune Lymphocytes…" Schenk Buchverlag, Passau & Budapest, 2008). From the late 1960s on, at the Section of Human Tumor Virology & Immunology of M.D. Anderson Hospital, this author (JGS) observed lymphocyte-mediated cytotoxicity expressed to various target cells, prominent among them human sarcoma, melanoma and kidney carcinoma cells. Lymphocytes killed human tumor cells by cytoplasmic lysis or by nuclear clumping (later recognized as apoptosis) illustrated in photographs and graphs. Patients with metastatic melanoma or sarcoma circulated small compact lymphocytes (later: immune T cells) and large granular lymphocytes (later: NK cells), that killed autologous (and related allogeneic) tumor cells in vitro (Sinkovics JG: "Cytolytic Immune Lymphocytes…" Schenk Buchverlag Passau/Budapest 2008). While the patients' lymphocytes killed tumor cells in vitro, in vivo the tumors could advance. Treatment of disseminated rhabdomyosarcoma-afflicted twin with buffy coat lymphocytes of the healthy twin sister was attempted in 1968. The healthy twin yielded large granular lymphocytes attacking the afflicted twin's rhabdomyosarcoma cells. Remission was induced; that was lost later to relapse (Sinkovics JG Expert Rev Anticancer Ther 2007;7:31–56; 183–210). The failure of immune T cells to eradicate a tumor in vivo was attributed (in other laboratories) to the induction of Treg cells that were directed to eliminate immune T cells. The chemokine stromal cell-derived factor-1 as ligand-to-receptor CXCL→R, acted in a network to direct Treg cells. Depleting the host's resident lymphocyte population allowed autologous immune T cells (and NK cells) to attack tumor cells in vivo and to induce remissions. S.A. Rosenberg & associates (Patrick Hwu now M.D. Anderson professor) at the NIH/NCI established the efficacy of immune T (and NK) cell infusions into lymphocyte-depleted hosts by inducing durable remissions of metastatic tumors (especially melanomas) (Rosenberg SA Cancer Immunology: Building on Success, NIH/NCI Sept 22–23 2011). Consideration was given to low dose intraabdominal radiotherapy for the induction of innate immunity in lymph nodes in response to gut flora invasion of the lymph nodes (explained in Sinkovics JG "Cytolytic Immune Lymphocytes…" Schenk Buchverlag, Passau/Budapest pp 264–5, 2008). Recently discovered autologous immune T cells with genetically engineered TCRs (the CAR chimeric antigen receptors), and replicating in their tumor-bearing hosts in vivo, are able to completely eradicate metastatic tumors (lymphocytic leukemia-lymphoma cells; solid tumor cells). Massive tumor cell lysis occurs so rapidly that patients suffer, and are to be protected from, acute tumor lysis syndromes (Morgan A et al Cancer regression in patients after transfer of genetically engineered lymphocytes. Science 2006;314:126–129; Robbins PF et al Tumor regression in patients with metastatic synovial cell sarcoma and melanoma using genetically engineered lymphocytes reactive with NY-ESO-1. J Clin Oncol 2011;29:917–924; Frigault MJ et al including Carl H June. Identification of chimeric antigen receptors that mediate constitutive or inducible proliferation of T cells. Cancer Immunol Res 2015;3:356–367), that is tumor-specific immune T cells. Resting continuously (as long as residual tumors exist; or even after the eradication of all tumors), the immune T cells survive. Their renewed attacks on relapsed tumor cells sustain complete remissions for years: a result tantamount to cure (earliest results reviewed in Science Dec 20 2013).

Some Misconceptions. In 1968, the landmark paper of Hellström & Hellström appeared to have set the basic rules of cancer immunology. In the case of pediatric neuroblastoma, granulocyte-cleared peripheral blood "lymphocyte suspensions" of patients (and their mothers) specifically inhibited the growth of autologous ("autochthonous") tumor cells (but not fibroblasts). The sera of the donors, who yielded 'colony inhibitory lymphocytes', contained antibodies, which were specifically inhibitory to tumor cell colonies in a strictly complement-dependent manner. "In no case did lymphocyte from healthy controls, from patients with non-neoplastic disease or from patients with other kinds of tumor selectively inhibit the growth of neuroblastoma cells." Patients' sera and complement acted similarly to the lymphocytes in tumor cell colony inhibition assays. Both the text and the summary contains the fateful statement: "In no case were tumor cells selectively inhibited by lymphocytes from the controls, which included children with other tumors, children with non-neoplastic diseases, or mothers of healthy children…" (Hellström IE, Hellström KE, Pierce GE, Bill AE: Demonstration of cell-bound and humoral immunity against neuroblastoma cells. Proc Natl Acad Sci USA 1968;60:1231–1238). The work was supported by USA NIH and American Cancer Society grants. The manuscript was sponsored for publication by USA National Academy of Sciences member Frank L. Horsfall Jr. The paper failed to explain that if the control lymphocytes never killed tumors selectively, could they kill tumors 'non-selectively'? In the late 1960s, at the Section of Clinical Tumor Virology and Immunology in the M.D. Anderson Hospital, this author (JGS) was engaged in the establishment of human tumor cell lines (lymphoblasts-lymphocytes in suspension cultures; solid tumors, especially sarcoma, melanoma and kidney carcinoma) (Sinkovics JG et al Methods Cancer Res 1978;14: 243–323). Against these cultured tumor cells, the patients' and healthy controls' ficoll-hypaque purified lymphocytes (prepared by Cameron Tebbi and Jerry Cabiness) and sera were tested for non-specific and specific immune reactions. The chamber-slide assay was used; stained slides were viewed to actually visualize the lymphocytes' interactions with tumor cells (Sinkovics JG & Horvath JC. Internat J Oncol 2005;27;5–47). It has become immediately evident that healthy individuals, first this author (and patients with tumors) circulate a class of lymphocytes (large granular lymphocytes) that attached to allogeneic tumor cells and killed them. Under the effect of the Hellström paper (and on the recommendation of the first author of that paper, who actually rendered a NCI-appointed project site visit in this author's (JGS) laboratory), it was declared that this author's chamber-slide assay was outmoded, unreliable, and thus yielding "laboratory artifacts". The author's NCI grant was transferred to Ron Herberman's in-house NCI laboratory in order to clarify what the "in vitro artifacts" meant. To the NCI project site visitors, it was unacceptable that lymphocytes would render an immune reaction without their specific pre-immunization. Even though, in reply, this author showed that his lymphocytes killed female breast and ovarian carcinoma cells, and that the lymphocytes responsible for this reaction were not the compact small T cells, but the large granular cells (referred to as 'Burnet's immune surveillance cells' in the author's laboratory). View convincing original photographs in Sinkovics JG Cytolytic Immune Lymphocytes. Schenk Buchverlag, Passau/Budapest, 2008. However, the incorrect "Hellström doctrine" prevailed until after Herberman's and other laboratories proved the existence of "natural killer cells" (same as the allogeneic large granular lymphocytes of this author). Marshall Lichtman credited this author (JGS) for the first observation of tumor cell-killer human large granular lymphocytes (later: NK cells) (Lichtman M A in P.H. Wiernik et al Neoplastic Diseases of the Blood. Springer Science New York, 2013. doi:10.1007/978-1-4614-3764-2_39.

HMG Proteins Drive the Amoeba Nucleus. High mobility group helix-turn-helix DNA-binding proteins are shared by the amoeba and its parasite mimiviruses. Prokaryotes, archaea and unicellular eukaryotes share HMG proteins. LUCA, the last universal common (cellular) ancestor, presumably operated HTH HMG proteins for carbamoyl phosphate synthetase. In the precellular RNA World the ancestors of these proteins functioned in ribonucleoproteins, ribozymes and riboswitches. The amoeba expresses intranuclear HMG14/17 proteins (Anantharaman V Koonin EV Aravind. Nucl Acids Res 2002;30:1427–1464). The HMGB1/2/3 transcription proteins of higher multicellular eukaryotes derive from the amphioxus' single HMGB gene. Tumor cells of the highest eukaryotes utilize HTH HMGB proteins as auxiliary oncoproteins (Tables XI/I and XI/II). For amoeboid movement of invading tumor cells, see text.

Ribozymes Appeared. Tetrahymena thermophila revealed its group I intron ribozymes in Thomas Czeh's laboratory. Ribozymes catalyze their substrates ("like protein enzymes") intra- and intermolecularly. The tetrahymena ribozyme's intramolecular catalysis consists of self-splicing (Golden BL et al Howard Hughes Medical Institute, Department of Chemistry and Biochemistry, University of Colorado, Boulder, CO. Science 1998;282:259–264). Alu ribonucleoproteins consists of polymerase III-transcribed Alu sequences and signal recognition proteins (SRP9/14) united. In the ribosome, these units inhibit IRES-mediated (internal ribosome entry site) translation initiation (Ivanova E et al Nucleic Acids Res 2015;43:2874–2887).

Spliceosomes. Spliceosome, the intranuclear organ of eukaryotes, consists of five small nuclear snRNAs and their associated proteins, the small nuclear ribonucleic proteins (snRNP). Each pre-mRNA/hnRNA (heterogeneous RNA) can generate the assemblage of its own spliceosome. Each spliceosome will contain five snRNAs marked as U1, U2, U4, U5, U6. In the spliceosome, ribozymal catalytic reactions remove (excise) the introns from nuclear pre-mRNAs and reconnect (ligate) the flanking exons: the splicing reaction of Philip Sharp. The assembly of snRNP particles occurs stepwise on hnRNA substrate. The first transesterification takes place in the bulging duplex of the U2 snRNP and the branch region of the hnRNA. The U2 snRNA auxiliary factor recruits U2 snRNP to the branch region. An altered conformation of RNA-RNA duplex is shaped by the peudouridine residue of the U2 snRNA. Splicing begins in the 2' OH of the bulging adenosine. U4/U5/U6 snRNPs form a triplet. U2 snRNA binds to the branch point sequence and RNA-RNA interactions displace protein-RNA interactions. The stem loops U4/U6 dissociate with the elimination of U4. Stem loop U6 base pairs with U2 snNA, and appears as the U2/U6 helix. Alternative splicing is performed when two different exons are recombined. BS = branch site ISL = internal stem loop http://www.cityofhope.org/the-catalytic-core-of-the-spliceosome Excision of intron in the spliceosome (SM). Copyright McGraw-Hill Companies. McGraw-Hill Higher EducationMcGraw-Hill Online Learning Center Test mhh.co, Spliceosome abbreviations: LSm protein: like SM spliceosomal; Prp genes: precursor RNA processing (not "platelet-rich plasma"); Brr: balanced repeated replication (not "bad response to refrigeration"; not "bean root rot fusarium"); Snu: snRNP "snurp" proteins, snurps). The supraspliceosomes remove the introns and ligate the exons. They are composed of four united active native spliceosomes equipped with components for processing intronic microRNAs and changing the 5'-end and the 3'-end pri- into pre-miRNAs. All five-spliceosomal U1, U2, U4, U5, U6 snRNPs are present. The RNA-editing enzymes ADAR1,2 are actively involved (dsRNA adenosine deaminase). The enzymes Drosha/DGCR8 are the major microprocessors (DiGeorge chromosomal region). Stop codon-mediated suppression of splicing (SOS) is installed. Latent splicing initiated by a mutated translation initiation codon (AUG) is suppressed by an initiator tRNA construct (Shefer K, Sperling J, Sperling K Comput Struct Biotechnol J 2014;11:113–122). Some degenerative diseases due to RNA/DNA processing errors occur (retinitis pigmentosa; spinal muscular atrophy) due to misregulation of splicing. Mutation of U2 snRNP results in defective snRNP assembly in CCL and myelodysplasia (Matera AG & Wang Z Nat Rev Mol Cell Biol 2914;15:108–121). Indeed, these malignant transformations are due to functional error of the RNA/DNA complex.

Drosophila Brain Tumors and Leukemia. The drosophila genome encodes the ontogenesis of the larva to adulthood with precision, but not without occasional lapses. Carcinogenesis appears to be included into the physiology of the genome: anlage lay down, differentiation, organ formation and polarity, stem cell maintenance with asymmetric release of somatic cells, provisions for progeny, cell senescence and death. Not in the case of every individual, but in selected (gain-of-function mutated) hosts, regression of a part of the soma to germline level, or evolutionarily deeper down below, to the independent life style of immortalized single cells, takes its place. "Carcinogenesis" is recognized. Most commonly involved in this latter event are the hedgehog, notch, runt, polycomb, trithorax and wingless genes. Gene expressions are regulated by mono-dimethylases at H3K4me3 (trimethylated by the cara mitad enzyme); a demethylase acts also at H3K27. The polycomb repressive complex 2 (PRC) is the chromatin modifying enzyme of H3K27. The SET domain methyl-transferases (Su-var9/ enhancer of zeste/ trithorax) are the most prominent. Most drosophila genes have their orthologs downward and upward on the evolutionary scale. The drosophila trithorax (histone lysine methyl-transferase) and trithorax-related genes are orthologous to the mll1-mll2/MLL1-MLL2 (mixed lineage leukemia, lymphocytic, myelocytic) genes and gene product proteins. In human lymphomas and lymphocytic blastic leukemias the overproduced MLL proteins derive from a promiscuously fused mll gene (Chauhan C et al Development 2012;139:1997–2008; Geissler K & Zach O Ann Haematol 2012;91:645–669; Kanda H et al Mol Cell Biol 2013;33:1702–1710; Patel A et al J Biol Chem 2014;289:868–884; Rain AN et al Mol Cell Biol 2013;33:4844–4856; Xie J et al OncoTargets Ther 2013;6:1425–1435). The increased ectopic expression of meiosis-related gene-product proteins (cancer-testis, CT; synaptonemal complexes, SYCP; PIWI) in cancers, both in drosophila and human, indicates a soma-to-germline transitional shift in cancer (Feichtinger J et al Int J Cancer 2014;134:2359–2365). Receptor tyrosine kinase RTK (one of them EGFR), and the PI3K/Akt pathways, are the initiators of glioma induction in the drosophila brain. Right open ORF serine/threonine kinases 1/2 (RIOK) respond and form complexes with mTOR. Forced reduced expression of RIOK1/2 (immune iRNA) allowed p53-induced cessation of cell cycle and apoptosis to take place. The inhibitor of neuroblast differentiation (symmetrical cell divisions) is the early growth response EGR transcriptional regulator klu/Klumpfuss protein. Transplantable neuroblast (NB) tumor cells overexpress Klu. Drosophila larval type I NB cells divide into a large stem cell and a smaller ganglion mother cell, which symmetrically divides into two postmitotic ganglion cells. Type II NB cells divide asymmetrically into a stem cell and a smaller intermediate neural progenitor (INP) cell. Deadpan (Dpn) and Asense (Ase) proteins differentiate INP cells; without Ase and Dpn expression, INP cells remain stem cells. Gain-of-function mutated drosophila brain tumor gene (brat) and drosophila lethal(3) malignant brain tumor (l3mbt) gene induce malignantly transformed cell clones. Differentiation inducers are the Notch inhibitor Numb, and the cell cycle inhibitor Pro (prospero). Loss-of-function mutated numb and pro genes fail to induce neuroblast cell differentiation (Berger C et al Cell Rep 2010;2:407–418; Read RD et al PLoS Genet 2013;9(2):e1003253; springer com.article/10.1186% 2F1471-21645-24). The neurotrophin receptor p75NTR reacts to early growth response (EGR) transcriptional regulators. Klu is an EGR agent. EGR proteins contain three Zn finger motifs allowing them to positively or negatively regulate cell proliferation (apoptosis versus cell proliferation). In mammalian cells, EGR-1 is an inhibitor of TNF-related apoptosis-inducing ligand (TRAIL) expression in NK cells. The extracellular signal-regulated oncogenic ERK pathway in lung adenocarcinoma cells induces the egr-1/EGR-1 gene/protein. EGR-1 acting as a DNA-binding protein, activates the VEGF-A promoter. Both the mutually exclusive K-RAS and EGFR mutated proteins also activate EGR-1. Activated EGR-1 overrules its inhibitor NAB (nerve growth factor binding) protein, when it is epigenetically silenced (unless NABs are constitutively expressed) (Balzarolo M et al Immunol Lett 2013;151:61–67; Gao X et al Mol Cell Neurosci 2007;36:501–514; Shimoyamada H et al Am J Pathol 2010;177:70–83; Thiel G & Cibelli G J Cell Physiol 202;193:287–292). Terminal differentiation of blast stage myeloid leukemia cells is blocked by c-Myb or by C-Myc. EGR-1 could abrogate the c-Myc, but not the c-Myb blockade (Gibbs JD Cell Cycle 2007;6:1205–1209; Leukemia 2008;22:1909–1916). EGRs are evolutionarily conserved. A recent new discussion from another angle (Sinkovics JG Europ J Microbiol Immunol 2015;5:25–43). The ancestral oncogene / oncoprotein of human carcinoma of the thyroid, RET (rearranged in transformation) suffers gain-of-function mutation in the drosophila. Abnormal neural epithelium restored to normal by the administration of small molecular Ret inhibitor ZD6474, which is in human clinical trials as Zactima (Marcos Vidal & Ross Cagan, Mount Sinai School of Medicine) http://www.cancer.gov/NCIcancerbulletin/archive/2009/012709/page6

Reversed Ontogenesis. Hydractinia echinata and H. symbiolongicarpus colonies consist epithelial cells, stem cells or interstitial I cells (originally called Urkeimzellen, Stammzellen by August Weismann in 1883), and defensive stinging cells (microbasic mastigophores). The epithelial cells form the mass of the colony; these cells may de-differentiate into stem cells, or tumor stem cells. The genuine stem cells are in the interstitium in between epithelial cells in the epidermal compartment. The stem cells provide germ cells. Silenced and activated stem cell genes include ancestral Wnt elements (from frizzled to β-catenin); Pin, encoding POU proteins and possible predecessors of Oct4; Myc; Nanog; Piwi. The stinging cells protect the integrity of the colony (practice allorecognition) and kill invader cells upon attempt at fusion by another, but incompatible, colony. POU proteins are pituitary/Oct-octamer/undifferentiated or uncontrolled (see text). Cnidaria consisting of anthozoa (the “flower animals”, sea anemones, the Nematostella; Hydra and Hydractinia); and Medusozoa (medusa, the jelly fish) practice bidirectional regeneration, transdifferentiation, and reverse development, or reversal of ontogenesis. In transdifferentiation, a differentiated cell assumes the appearance and functioning of another differentiated cell. In the reversal of ontogenesis, a mature cell assumes the appearance and functioning of one of its embryonic ancestors. Reverse development may be viewed as a physiological process of rejuvenation: senescence postponed, life span extended (Carla EC et al Tissue & Cell 2003;35:213–222; Khalturin K et al Dev Biol 2007;309:32–44; Plickert G et al Int J Dev Biol Sci 2012;56:519–534; Schmich J et al Int J Dev Biol 2007;51:45–56). Induced de-differentiation (reprogramming) of somatic cells in a vertebrate multicellular host (a mammal) to pluripotent stem cells is a form of reverse evolution encoded by the classical stem cell product proteins (Oct4, Kfl4, Myc, Sox2: the OKMS). Dox (doxycycline) is an inducer of stem cells (breast cancer stem cells, BCSC). In the background, BMP/SMAD (bone morphogenetic protein; signaling mothers against decapentaplegic, drosophila), and miR-200 induce mesenchymal-to-epithelial transition (MET), including the activation of stage-specific embryonic antigen 1 (SSEA) and pluripotency regulators Nanog and Sall4 (spalt-like transcription factor). The induction and the maintenance of the state of pluripotency is encoded by two different arrangements of the stem cell genes (Golipour A et al Cell Stem Cell 2012;11:769–783). The cell rejuvenation process is an inherent ancient faculty of the RNA/DNA complex; mature somatic cells of a multicellular host are able to resort to it. If the promoters are constitutively amplified without a gain-of-function mutation, is the process excessive but still physiological? If the inhibitors (suppressors) suffer a loss-of-function mutation, is the process pathological? Certain "rejuvenating mutations" in the RNA/DNA complex may be viewed inherent and physiological, inasmuch as distant ancestors of cellular life (the medusa) practice it. The overgrowth of rejuvenated cell colonies may be lethal to their host. Further discussed (Sinkovics JG Europ J Microbiol Immunol 2015;5:25–43).

Nematostella vectensis, is the early branching metazoan cnidarian without mesoderm. The sea anemone, anthozoan cnidaria, N. vectensis, is among the first to operate the venus kinase receptor (vkr) gene, which characterized the first bilaterian phyla (Annelida, Mollusca, Echinodermata, Platyhelminthes). The receptor captured extracellular amino acids and eventually formed the venus fly trap module of the larvae in invertebrate hosts up to the the metazoan cnidaria. Its possible role in cell-to-cell communication in higher ranking metazoan is unknown, however, the vkr gene was lost at the caenorhabditis and drosophila level (Vanderstraete M et al BMC Genomics 2013;14:361). The Wnt signaling pathway is very active in Nematostella larvae; its activity extends to TCF (T cell factor) signaling by intranuclear β-catenin. The hedgehog Hh pathway is active as well. The faultless physiological pathways establish the correct anterior-posterior axis and the dorsal-ventral polarity of the embryo (Marlow H et al Dev Biol 2013;380:324–334; Matus DQ et al Dev Biol 2008;313:501–518; Röttinger E et al PLoS Genet 2012;8 (12):e10031164). The Wnt/Hh/Tcf signaling pathways may malignantly transform human cells into adenocarcinomas. LIM homeobox Lhx transcription factors (lineage, caenorhabditis; isl, insulin-like rat gene enhancer binding protein; med-3, mechanosensory, cnidaria) of the cnidaria with proteins Rho and ROCK (rhombotin; Rho-associated kinase) form fusion oncoproteins in human T cells. The hydra (H. magnipapillata) genome operates six Lhx genes (Boehm T et al Proc Natl Acad Sci USAc 1991;88:4367–4371; Srivastava M et al BMC Biol 2010;8:4). The achaete scute (ash) gene product proteins regulate the early neurogenesis in the cnidaria Nematostella (and in the early chordate Branchiostoma amphioxus); four such genes in Nematostella (Layden MJ et al Development 2012;139:1013–1022). Human ash genes encode transformation of adenocarcinoma cells into neuroectodermal cancer cells, or induce primary esthesioneuroblastomas (Linnoila RI et al Exp Lung Res 2005;31:37–55; Wang XY et al Lab Invest 2007;87:527–539). Cnidarian "small silencing" microRNAs (87 identified, including PIWI-interacting piRNAs) are extremely active in cleavage (post-transcriptional regulation / translational inhibition) of mRNAs, thus regulating gene expression. Cnidaria miRNAs/siRNAs are ancestral to those of both animalia (Bilateria) and plantae. Plant miRNAs are methylated (CH_3) at their 3′ ends; animal miRNAs are not. The HOXD (homeobox) gene/protein cluster harbors its regulator, the intron-embedded miR-2026. The cnidarian miR-156 shows sequence homology with moss and Arabidopsis miRs reflecting back to a common ancestor of animalia and plantae (rule out horizontal transfers). The cnidarian miRs slice their mRNA targets directly, that is different from the mRNA slicing of Bilaterian miRs. The cnidarian ur-bilaterian miRNA miR-100 remains conserved in all vertebrates up to Homo (Moran Y et al Genome Res 2014;24:651–663). The human miR-100 often expresses tumor suppressive activity (hepatocellular carcinoma; osteogenic and chondrosarcoma; malignant lymphoma (Chen P et al Mol Cell Biochem 2013;383:49–58; Huang J et al Tumour Biol 2014;35:1095–1100; Li XJ et al Br J Cancer 2012;109:189–198). In non-small cell lung cancer, tumor suppressor miR-100 targets polo-like kinase (Liu J et al BMC Cancer 2012;12:519). Dihydroxyvitamin D and miR-100 suppress prostate cancer (Giangreco AA et al Cancer Prev Res 2013;6:483–494). miR-100 suppresses growth of transitional cancer cells of urinary bladder (Oliveira JC et al Asian Pac J Cancer Prev 2011;12:3001–3004). In small cell lung cancer, HOXA gene expressions favored the host by inducing tumor cell growth arrest and increased chemotherapy sensitivity and apoptosis. High expression levels of miR-100 reduced HOXA expression. miR-100 targeted the 3’UTR region of the HOXA protein mRNA (Xiao F et al Eur J Cancer 2014;50:1541–1554). Devious expressions of “involved” “regulated” without specifying up or down characterize the contradictory results reported on miR-100 and polo-like kinase interactions in human cancers. This ancestral interaction was established in the cnidaria. In cnidarian cells, the existence and physiological functioning of human disease-associated genes, including those of cancer, is remarkable. Prominent are 11 of the 12 human Wnt genes, and close to human-identical numbers and repeats of the BRCA2 and RAD51 genes (Sullivan JC & Finnerty JR Genome 2007;50:689–692). Add bone morphogenetic protein (BMP2/4 and TGFβ drosophila decapentaplegic orthologs with the mammalian counterparts) and the BMP antagonist chordin (Technau T & Steele RE Development 2011;138:1447–1456). Fully mature cnidarian cells practice bidirectional regeneration and reverse development, a form of transdifferentiation (TD). In TD, a fully differentiated cell functioning with a limited number of active genes, silences these genes, and activates a number of previously silenced genes, thus assuming the appearance and functioning of another mature cell (Graf T & Enver T Nature 2009;462:587–594). A reversal of ontogenesis consists of the de-differentiation of a fully matured cell. Cnidaria cells practice reverse development. De-differentiation of a mature somatic cell of set limited functioning in a vertebrate mammalian host is tantamount to a descent on the evolutionary scale to far distant ancestors? Cnidarians accept each-others plasmid DNA and laboratory-made transgenic cnidarians (Hydra; Hydractinia) exist (reviewed in Technau T & Steele RE Development 2011;138:1447–1458). Recently, a comparison is being made between the cnidarian (N. vectensis) NFκB/STAT combined pathway and its more advanced derivative operational in human cells. The cnidarian cell survival NFκB/STAT pathway's human derivative is highly oncogenic especially in certain malignant lymphoma cells. Extensively so in Reed-Sternberg cells of Hodgkin lymphoma, and mediastinal diffuse large B cell lymphoma with constitutive expression of these oncoproteins, and to a lesser degree in other B cell lymphomas and in the T cell lymphoma mycosis fungoides. A correlation with viral initiation in both case was suggested, inasmuch that cnidaria are often infected with herpes-like viruses that may be predecessors of the Epstein-Barr HHV4. The cnidarian endosymbiont algae carry these viruses. The cnidarian genomes carry predecessors of the human proto-oncogenes/oncoproteins, the runt-related Runx, Hedgehog and Wnt (wingless, drosophila; integrated, mouse); Sox (sex determination region Y, SRY box); and the achaete scute predecessors, that will encode the first nervous system in the amphioxus, and become a proto-oncogene in the human genome, as inducers of esthesioneuroblastoma, small cell undifferentiated lung carcinoma and others (Sinkovics JG Europ J Mirobiol Immunol 2015;5:25–43; Abstracts, Second NorHGT & LUCA Conference, Sohan Jheeta, University Leeds, England; M D Anderson Hospital Symposium on Cancer Research, Oct 9–10 2014. Referenced manuscript in print Internat J Oncology 2015;47:1211–1229). In advanced eukaryocytes of multicellular (human) hosts, the Janus kinase-related STAT is activated by phosphorylation of its tyrosine residue Tyr705, dimerizes and translocates intranuclearly. There, it activates genes involved in cell survival, self-renewal, neo-angiogenesis, locomotion, and immune evasion. Constitutively activated STATs induce the malignant transformation of the cell. In human T lymphoma cells, Jak3 activates STAT3/5 DNA-binding proteins, that suppress the promoter gene of tumor suppressor miR-22, thus inducing constitutive replication of the involved T lymphocytes in absence of miR-22. Anti-Jak3 histone deacetylase inhibitors, or anti-Jak3 curcumin, reverse the process (Sibbesen NA et al Oncotarget 2015;6:20555–69).

The ancestral mena proteins appeared first in the cnidarians (Hydra; Nematostella), and in the yeast (Saccharomyces); and are fully operational in the Spemann-Mangold gastrula organizer blastomeres of the Xenopus. This is the site where the LIM homeobox gene lhx1/lim1, noggin and chordin (nog, chd) interact (Figure 38; Table XVI) (Krogan NJ et al Nature 2006;440:637–43; Insenser MR et al J Proteomics 2010;73:1183–95; Pham TK & Wright PC Expert Rev Proteomics 2007;4:793–813; Rentzsch F et al Proc Natl Acad Sci USA 2007;104: 3249–54; Yasuoka Y et al Development 2009;136:2005–14). View Internet/Google images of 'mena proteins in Spemann blastomeres'. The vertebrate mammalian (including Homo) menaINV proteins (mammalian enabled; mammalian invasive; actin microfilament regulatory) remodel the cytoskeleton, regulate cellular apical orientation, determine cell polarity, and direct lamelli- and filopodial cell migration. The human ortholog is encoded from locus 1q42.12. When the gene is amplified, or is gain-of-function mutated in malignantly transformed cells (adenocarcinoma cells), the gene-product proteins appear in multiple isoforms, are excessively produced, and dictate accelerated multicellular streaming locomotion, thus, invasiveness and metastatic spread. Mena protein overproducing cancer cells exhibit highly malignant behavior, express vasodilator phosphoproteins, create the microanatomic Tumor Microenvironment of Metastasis environment, and significantly shorten patients' survival (Forse CL et al BMC Cancer 2015;15:483. doi:10.1186/s12885-015-1468-6; Rohan TE J Natl Cancer Inst 2014;106:135. doi:10.1093/jnci/dju136). Macrophages support mena protein-expressor tumor cells, which withstand humoral and immune CD8 T cell-mediated host immune reactions (Di Modungo F et al Int J Cancer 2004;109:909–18; Roussos ET et al J Cell Sci 2011;124:2120–31; Wychoff JB et al Cancer Res 2007;67:2649–56). Comment. Here, a fundamentally basic physiological embryonic event is turned into a constitutively expressed cell survival pathway, for the rescue of some individual cells, while the host of these cells succumbs in the process.

Ciona intestinalis. The ascidian urochordate sea squirt, vase tunicate, C. intestinalis (Chordata/Tunicata) demonstrates extraordinary genomic versatility, as it transforms from swimming tadpoles into tubular organisms fixed to the floor in the bottom of the sea (Satou Y et al BMC Genomics 2012;13:208). The embryonic ciona swims with appropriate musculature. The mature ciona colony lives attached to the sea floor. In the swimming ciona larva or tadpole, the notochord and its pigment cell sensory organs, the photoreceptive ocellus and the melanin granule-containing otolith receive their activation by fibroblast growth factor FGF3 and its receptor, and the MAPK pathway. The MAPK proteins activate the ets1/2 transcription factor (named after the avian erythroblastosis retrovirus E26 E-twentysix, that originally incorporated the host cell proto-oncogene ets). The ETS protein activates the T cell factor gene (tcf/TCF) and the lymphocyte enhancer factor gene (lef/LEF). The tcf/lef genes are proto-oncogenes in the human genome, serving as targets to β-catenin. TCF/LEF are downstream members of the Wnt pathway through the intranuclear transfer of β-catenin. Another activator of the Wnt pathway is the downregulation of GSK3β (glycogen synthase kinase) gene product protein. Ciona operates TGFβ and its antagonist Smad (signaling mothers against decapentaplegic, drosophila); and hedgehog and JAK/STAT genes, active in a switched on and off mechanism in the metamorphosis of the tadpole into adult organisms (Hino K et al Dev Genes Evol 2003;213:264–272; Shi W et al Development 2009;136:23–28; Squarzoni P et al Development 2011;138:1421–1432). The Ciona cell survival Wnt pathway is correlated with the human oncogenic Wnt pathway (Sinkovics JG Europ J Microbiol Immunol 2015;5:25–43). All the ancient pathways named are conserved in vertebrate mammalian cells in switched on and off sequences in embryonic ontogenesis. All the ancient pathways named become oncogenes/oncoproteins, when either amplified or constitutively expressed with or without gain-of-function mutations in vertebrate mammalian hosts (Sinkovics JG Europ J Microbiol Immunol 2015;5:25–43). FGF-R overexpression with or without mutations or fusion acts as an oncoprotein in diverse (adenocarcinomas, including thyroid tumors) human cancers. FGF-induced lymphangiogenesis promotes tumor cell invasiveness and metastasis (Larrieu-Lahargue F et al PLoS One 2012;7(6):e39540; Redler A et al PLoS One 2013;8(8):e72224). Otherwise, the LIM (Table XVI) and Hox homeodomain genes were lost from the

Ciona genome. Gene product proteins in Ciona are ancestral to thyroid hormones and their receptors (ciona nuclear receptor-1 as thyroid hormone receptor-associated protein) in vertebrate mammalian (human) hosts. The ciona nkx/Nkx gene/protein is referred to as TTF-1 (thyroid transcription factor, or natural killer homeobox protein, Nkx). The ortholog of mammalian vertebrate thyroid stimulating hormone activator of the NIS system (natrium-iodine symporter) and its mRNA in the ciona is Nkx/TTF-1. The amphioxus harbors ancestors of the vertebrate mammalian thyroid transcription factor-1. In ascidian larvae (Ciona), thyroid hormone precursors regulate larval metamorphosis in selecting cells to be retained, or to be eliminated by apoptosis (Carosa E et al Poc Natl Acad Sci USA 1998;95:1152–1157. Di Fiore et al Life Sci 1997;61:623–629. Krishnan A et al Gene 2013;526:122–133. Ogasawara M et al J Exp Zool 199;285:158–169. Patricolo E et al J Exp Zool 2001;290:426–430. Tóth G & Noszál B Acta Pharm Hung 2013;83:35–45). In distress (heat; endoplasmic reticulum stress) ciona mobilizes cell survival pathways. It releases heat shock proteins (Hsp) as chaperones, and the BAG (Bcl-2-associated athanogene 3) co-chaperone proteins. Observe ciona cell responses to brefeldin, thapsigargin and tunicamycin (FujikawaT et al Cell Stress Chaperons 2010; 15: 193–204). The stressed ciona cell metabolizes like a tumor (colon cancer) cell (Yang X et al Histol Histopathol 2013;28:1147–1156). For the differentiating notochord-ephrins, Ciona cells establish the proper FGF/MAPK balance (Shi W & Levine M Development 2008;135:931–940). This overexpressed pathway is frequently operational in human cancer cells. Germ layer segregation in the ciona embryo is directed by β-catenin (Hudson et al Curr Biol 2013;23:491–495), a tcf/lef proto-oncogene-activator in human cells. Reviewed with historical details (Sinkovics JG European J Microbiol Immunol 2015;5:25–43). Are some thyroxin-related proteins physiological in the Ciona, become oncoproteins in human thyroid carcinoma? The other tunicate, Oikopleura dioica, compacted its genome even more, and reduced the numbers and sizes of its introns (down to 70 Mb with 18,000 genes with short introns of <50 nt, in comparison to the amphioxus 520 Mb genome operating 22,000 genes). Most lost O. dioica introns (spliced out in spliceosomes) are being recently replaced in different new positions. Of many retrotransposons, the O. dioica genome retained two prominent ones, Odin and Tor, and eliminated many others. O. dioica cells exposed to natural UV light at the surface of the oceans evolve fast due to an accelerated rate of point-mutations and to limited faculties of DNA repair. Is the O. dioica genome a representative of the mutator geno-phenotypes? So is the genome of C. savignyi, that split from C. intestinalis some 180 mya (Berná L & Alvarez-Valin F Genome Biol Evol 2014;6: 1724–1738).

Botryllus. The tunicate urochordates Botryllus live in colonies, and operate a 580 Mbp genome consisting of some 14,000 intron-containing, and some 13,500 intron-less genes. The Botryllus genomics has not as yet been correlated with those of the highly evolved multicellular eukaryotes (Homo included). Calling for "Botryllus Wnt, Hedgehog, NFkappaB/STAT " etc etc etc, the answer is "no item found". Cursory mention is available to correlation of Botryllus genes with mutated counterparts in human diseases (cataracts; deafness; myopathy; pregnancy-specific glycoproteins), but not cancer. Native immunity and hematopoiesis (an NK-like cell in the common haemolymph of each zooid cluster) are accurately addressed (Khalturin K et al Proc Natl Acad Sci USA 2003;100:622–627. Voskoboynik A et al Elife 2:e00569). The origin of NK cells in the Botryllus; the correlation of the human maternal NK cells with the placenta and the fetus; and the imitation of the fetus by cancer cells were the subject matters of one of the discussions this author initiated in the early 2000s (Sinkovics JG & Horvath JC Int J Oncol 2005;27:5–47; Sinkovics JG, Horvath JC Kay HD Book of Abstracts, The 56th Annual Symposium Fundamental Cancer Research, The University of Texas M.D. Anderson Cancer Center, Houston TX 2003 p 117 #1–33). A more sophisticated presentation followed (Lightner A et al Clin Dev Immunol 2008:631920. doi:10.1155/2008/631920).

Amphioxus II. Integrated in the amphioxus genome resides an ancient herpesviral genomic sequence related to oyster, and more distantly, to the gastropod abalone herpesvirus (Savin KW et al Virol J 2010;7:308. doi:10.1186/1743-422X-7-308). One of the genes in the amphioxus waiting to be recruited by the forthcoming adaptive immune system is the rag/RAG sequence. The protein expresses a retroviral type II nuclease motif, localizes in the nucleus, and performs V(D)J-like recombination if brought together with a mouse RAG2 protein (Yu C et al J Immunol 2005;174:3493–3500; Zhang Y et al Proc Natl Acad Sci USA 2014;111:397–402). A newly isolated bat betaherpesvirus encodes proteins homologous with human herpesvirus' 6 U94, the incorporated parvoviral sequence; MHC class II homologous proteins; NK cell receptor-like lectins; and immunoglobulin beta-sandwich domains (Zhang H et al J Virol 2012;86:8014–8030). Herpesviruses ancestral to EBV were claimed to have incorporated V(D)J-like genomic elements (Dreyfus DH PLoS One 2009;4:e5778; Nillen et al Acta Microbiol Immunol Hung 2004:51:469–484), and thus initiated adaptive immunity in their hosts, which were living at the innate immunity level. Could the amphioxus herpesviral genome be the first possessor of primordial genes of adaptive immunity? It is also possible that herpesviruses incorporated the genes of adaptive immunity from hosts, which already possessed these genes firmly in use; this could be the case for some bat herpesviruses and EBV (see text).

Viral Oncolysates. In an interview, Sinkovics stated (in replying the question: "VOs cure sarcomas or melanomas?") that metastatic tumors were not cured with VO treatment alone (however lives were prolonged). Unfortunately, the interviewer published the interview under the title: "Viral oncolysates are ineffective". Most recent references avoid mentioning "viral oncolysates" by name; instead re-name them as "oncolytic vaccines", and thus entirely omit referrals to their predecessors (Elsedawy NB & Russell SJ Expert Rev Vaccines 2013;12:1155–1172); however, claim and illustrate their rationality for high immunogenicity. The original theory in support of VOs as cancer vaccines rested on the expectation that the immunized tumor-bearing host will recognize antigens of cytoplasm and cell membranes and viral capsids or other structural proteins fused at viral egress, as non-self. Therefore vaccinated subjects should respond to their altered self-antigens immunologically. Such immune responses by antibodies and immune T cells to autologous tumor cells (sarcoma, melanoma, adenocarcinoma) were documented in the chamber slide cultures in vitro in patients immunized with autologous VOs (Sinkovics JG Cytolytic Immune Lymphocytes. Schenk Buchverlag Budapest/Passau 2008), but the formation of such fused antigens was not experimentally documented at that time. The cancer cell surface antigens against which the host directed its immune reactions were not identified. For that reason, the NIH/NCI denied grant support for VO cancer immunotherapy. However, later elsewhere it has been well documented that such fused antigens are formed. Influenza viral structural proteins fuse not only with cell surface sialic acid proteins, but also with the host cells' endosomal CD81 proteins (He J et al PLoS Pathog 2013; 9(10):e1003701). Thus, the influenza virus particles emerging from cancer cells expresses immunogenic cancer cell antigens fused with viral structural proteins. Properly prepared VOs (re-named "oncolytic vaccines") may remain in use for cancer immunotherapy.

Lamprey lymphocytes. Leucine-rich repeat cassettes and variable lymphocyte receptors provide adaptive immune reactions without immunoglobulins. Release chemokines and interleukins and their receptors; express TCR (Mayer WE Proc Natl Acad Sci USA 2002;99:14350–14355; Pancer Z et al Proc Natl Acad Sci USA 2004;101:13273–13278; Uinuk-Ool T et al Proc Natl Acad Sci USA 2002;99:14356–14361;

Gnathostomata Collected and United the Genomics of the Adaptive Immune System. The by now extinct armored fish Placoderms predated the cartilagenous sharks and their followers, the bony fishes. While they preserved the innate immune system (Toll-like receptors with a chemokine, microRNA and some lympho- and cytokine network; phagocytic leukocytes, monocytes and macrophages, dendritic cells; invariable NK cells; but no immunoglobulins), the followers of the placoderms, the chondrichtyes gnathostomata sharks, possess and practice a full adaptive immune system consisting of MHC cell-presented antigens, B and T lymphocytes with variable receptors and an extensive immunoglobulin network. Some of the genes of the complex adaptive immune system (V and RAG) appear singly "in the waiting" in the amphioxus and sea urchin, but assigned to tasks other than those of the forthcoming adaptive immune system. The adaptive immune system worked in unison for the first time at the placoderm/shark level. Vertically inherited and horizontally transferred genes generated the adaptive immune system. Either one or both, retrotransposons (harbingers, helitrons, politrons, transibs), and/or elements inserted into ancient herpesviral genomes (vdj, rag) and thus passed from host-to-host, encountered for the first time in united activity in the lymphoid tissues encircling the intestinal tracts of ancient sharks. Phylogenetic vertical inheritance interrupted by episodes of horizontal gene transfers secured the existence and reinforced the performance of the system up to the highest level of vertebrate mammalian hosts (Dreyfus DH et al Virol J 2011;8:422. doi:10.1186/1743-422X-8-422; Fugmann SD Semin Immunol 2010;22:10–16; Kapitonov VV & Jurka J PLoS Biol 2005;3(6):e181; Litman GW et al Nat Rev Immunol 2010;10:543–553; Niller HH et al Acta Microbiol Immunol Hung 2004;51:469–484). The adaptive immune system had to reconcile with the placenta of the vertebrate mammalian hosts. The placenta acts as a "pseudomalignant organ" inducing tolerance in its host (the mother) (Sinkovics JG Horvath JC International J Oncol 1999;14:615–646).

Ferroplasma acidarmanus. Tetraether-linked membranes enable the hyper-acidophil Ferroplasma acidarmanus to live and metabolize in hyperacidic (pH 0–1) environments. Zinc reaches extremely high concentrations in heavy metal-polluted environments inhabited by the arsenic-hypertolerant, iron-oxidizing extremophil archaeon and its relatives. These microorganisms suffer oxidative stress and produce methanethiol and volatile sulfur. In response, heavy metal exporter and zinc homeostasis genes and gene product proteins are constitutively expressed and upregulated in order to keep intracellular zinc levels low constant. The DNA strand separations in F. acidarmanus are performed by DNA helicases, which are recognized to be ancestral Rad3 xeroderma pigmentosum (F. Hebra & M. Kaposi, 1874) group D (FacXPD) proteins (related to the human enzymes in the pathological entity XP). These helicases possess unique iron-sulfur cluster (FeS) for their insertion to perform NER (DNA unwinding in nucleotide excision repair) and to initiate transcription. The FeS cluster recognizes the enzyme's substrate, the ssDNA-dsDNA junction; thus, ssDNA-dependent ATP hydrolysis and ssDNA translocation are performed. The FacRd3 enzyme is a ssDNA translocase (Mangold S et al Extremophiles 2013;17:75–85. Pugh RA et al J Biol Chem 2008;283:1732–1743; J Mol Biol 2008;383:982–998). The FeS clusters of the human XPD enzymes are subject to point mutations. A common XPD point mutations consists of the substitution of serine to cysteine. In the recessive human disorder XP: a common XPD point mutation is R683W (arginine; tryptophan) at 19q13, inhibiting NER. Other consequences of this mutation are trichothiodystropy and Cockayne syndrome. The risk for melanoma is increased. Megakaryoblastic leukemia may occur. Risk for breast cancer in Egyptian women is increased (Hussein YM et al Mol Biol Rep 2012;39:1895–1901; Janjetovic S et al Acta Haematol 2013;129:121–125; Paszkowszka-Szczur K et al Int J Cancer 2013;133:1094–1100; Ueda T et al J Exp Med 2009;206:3031–3046). Triplex DNA formations (H-DNA) develop dsDNA breaks with the formation of γH2AX (histone variant; ataxia telangiectasia) foci, which recruit XPD. The phosphorylated serine 139 and tyrosine 142 residues of H2AX induce p53-mediated apoptosis (Tiwari MK & Rogers FA Nucleic Acids Res 2013;41:8979–8994), thus preventing transformation. Mutations of helicases other than XPD (RecQ4; BLM; WRN) lead to Fanconi anemia; Bloom syndrome; Werner syndrome (see text). The descendant of the archaean enzyme in the human genome suffers point mutations; the mis-repaired DNA either translocates (c-myc in Burkitt's lymphoma in the H-DNA triplex, cited by Tiwari & Rogers), or encodes an oncoprotein.

p53/MDM. The pro-apoptotic p53 gene product protein is neutralized by the anti-apoptotic MDM oncoprotein. The gene and gene-product protein mdm/MDM (murine double minute; human counterpart) exerts ancient opposition to the p53 family members: the MDM proteins fused with the p53/p73 proteins direct the complex to ubiquitination. The MDM protein is a ubiquitin ligase. Thus the MDM protein can transform cells (rat embryo fibroblasts) into focus-forming entities, which grow into tumors in nude mice. The human mdm gene from its locus 12q13-15 encodes its MDM protein. Human sarcoma cells amplify the mdm gene and overproduce the MDM protein. In human malignant tumors the MDM4 protein is the most active suppressor of p53/p73. In the MDM/p53 protein complex, the p53 transactivator pro-apoptotic protein is inactivated (Barak Y & Oren M EMBO J 1992;11:2115–2121; Momand J et al Cell 1992;69:1237–1245; Oliner JD et al Nature 1992;358:80–83. Bert Vogelstein, Clayton Professor of Oncology; Johns Hopkins University Medical Institutions; Ludwig Center at Johns Hopkins; Sidney Kimmel Comprehensive Cancer Center; Baltimore, MD). The sea squirt (Ciona intestinalis) and the amphioxus operate the first mdm/MDM1 genes/proteins (only one mdm gene, which also encodes a p53 family member). Choanoflagellates did not as yet reveal a mdm/MDM system; they may be devoid of it. However a genome-protector p53 family ancestor exists in choanoflagellates. A substitute different from mdm/MDM suppresses p53 in the caenorhabditis. The drosophila is devoid of the classical MDM proteins, yet both drosophila and caenorhabditis experience constitutively replicating cells tantamount to malignant transformations. The mdm genes underwent duplications in cartilaginous fishes (chondrichtyes); since then, MDM2/4 remain conserved. The p53 proteins were DNA-binding, whereas the MDM proteins do not target the p53 gene; the MDM proteins form complexes with the p53 proteins (Momand J et al Gene 2011;486:23–30). The mussel (mollusk) Mytilus trossulus expresses the ancestor of the MDM2/4 molecule (before its divergence) and the p53 ancestor p63. The mussel MDM molecule forms complexes with its p63/63 proteins. The N-terminal deleted p63 (ΔNp63) isoform was up-, and the transactivation domain p63 (TAp63) isoform was down-regulated. Not quantitatively, but by its codon 13 mutation, the mussel haemocytes' ancestral ras gene dictates uncontrolled cell divisions (Muttray AF et al Comp Biochem Physiol B Biochem Mol Biol 2010;156:298–308). In plants (maize; rice), the MDM1/2 genomic elements reveal long terminal inverted repeats (LTIR), possible functional transposase-binding sites, as if they were originally Mutator Elements; their 8–9 bp termini are those of mutator-like elements (Yang G & Hall TC J Mol Evol 2003;56:255–264).

G 4 Quadruplex DNA. All eukaryotic G-quadruplexes assemble from G-rich single-stranded (ss) DNA sequences when dsDNA is unwound. The human telomeres contain TTAGGG repeats in dsDNA 2–15 kb strands ending in a G-rich single-stranded 3′ overhang consisting of 50–400 nucleotides. Intramolecular G-qadruplexes in the template inhibit DNA polymerases; DNA synthesis is stalled. Translesion DNA polymerases are recruited to the sites of replication blockade. Thus, misincorporation of nucleotides occurs: mutagenesis ensues. Single strand gaps lead to ds breaks, the hallmark of pre-cancerous state. Some helicases may succeed or fail to resolve G-quadruplex structures (Fanconi, Bloom, Werner syndrome helicases) (Edwards DN et al PLoS One 2014;9(1):e80664). Proto-oncogenes/oncogenes often assume constitutive G-quadruplex formations and encode oncoproteins (KIT, BCL2, VEGF, KRAS, MYC, SRC). Recruitment of the DNA damage marker γH2AX (ataxia) reveals the site involved. G-quadruplex-stabilizing ligands exist in human cancer cells (breast adenocarcinoma) (Lam EY et al Nat Commun 2013;4:1796). In unicellular eukaryotes (oxytricha), G-4 DNA structures function as a cell survival pathway. In the differentiated somatic cells of multicellular hosts, G-quadruplexes maintain telomeres and induce dsDNA mutations: they act as oncogenes. G-4 quadruplexes are targeted for cancer therapy. Natural and synthesized G-quadruplex-intercalating agent telomestatin (S2T1-6OTD) disables c-Myc in medulloblastoma and rhabdoid tumor cells (Shalaby T et al Mol Cancer Ther 2010 9;167. doi:10.1158/1535-7163.MTC-09-0586). An acridinium methosulfate derivative (RHPS4) brings primitive neuroectodermal tumor, glioblastoma and medulloblastoma cells to standstill at G1 by anti-proliferative effect (Lagah S et al PLoS One 2014;9(1):e86187). Enantiomer NiP (nickel; a non-superposable stereoisomer) induced telomere uncapping and G-overhang degradation in tumor (but not in normal human fibroblast or umbilical vein endothelial) cells. The lesions attracted γH2AX. Some of the 3′ overhang-protective shelterin proteins were removed. The tyr707-phosphorylated hTERT protein translocated from nucleus to cytoplasm and was reduced in amount. Cell cycle inhibitors (p16; p21) rose. G-quadruplex remained stable; it could not protect the tumor cells against apoptosis: apoptosis may overrule the G-quadruplex (Wang J et al Nucleic Acids Res 2014;42: 3792–3802). After their discovery (Gellert M et al Proc Natl Acad Sci USA 1962;48: 2013–2016), DNA G-quadruplex formations were recognized to have occurred within human proto-oncogenes (bcl-2; K-ras; myb; c-myc; hTERT) (Ambrus A et al Biochemistry 2005;44:2048–2058; Dai J et al Nucleic Acids Res 2006;34:5133–5144; J Am Chem Soc 2006;128:1096–1098; Cogoi S & Xodo LE Nucleic Acids Res 2006:34:2536–2549; Palumbo SL et al J Am Chem Soc 2009:131:10878–10891). These structures first downregulate transcription. However, mutations occur within the quadruplex DNAs, which abrogate the negative effects, and induce increased gene expressions: loss of quadruplex stability abolishes the gene silencing effect of the quadruplex formation. The promoter gene of the human telomerase is a subject of DNA G-quadruplex mutations in glioblastoma and melanoma cells. The central quadruplex of the structure is affected. For example, guanine-to-adenine mutations produce new binding sites for the ETS oncoprotein (product of the E26 retroviral c-onc→v-onc proto-oncogene). Such mutations in a G4 DNA-quadruplex within the hTERT (human telomeric reverse transcriptase) promoters in glioblastoma and melanoma cells increases those gene expression (Chaires JB et al PLOS 2014;9(12): e115580). Telomeric overexpressions in malignantly transformed cells are the major contributors to the cells' immortality. However, G-quadruplex-forming G-rich oligodeoxynucleotides (GQ-ODNs), and the telomeric G-tail oligodeoxynucleotides (TG-ODNs) may exert antiproliferative activity in tumor cells, leading to apoptosis, mediated by ataxia telangiectasia mutated ATM-dependent c-Jun NH_2-terminal kinase (JNK) (Qi H et al Cancer Res 2006;66:11808-16).

Theileria-Infested Bovine Lymphocytes support the parasite by assuming "malignant transformation". The tick-transmitted protozoan T. parva undergoes a major transitional life cycle in the lymphoid cells of the infected ruminants. The gametes transform in the guts of the tick into into zygotes, the fused male and female gamonts. Kinetes derive from gamonts and occupy cellular elements of the tick's salivary glands. There, the infectious sporoblasts and sporozoites are formed. In the mammalian (bovine, ovine) lymphocytes sporozite-derived schizonts undergo merogony and emerge as merozoites. The schizonts-containing lymphocytes de-differentiate into lymphoblasts and multiply rapidly by clonal expansion expressing constitutively certain proto-oncogenes . The lymphoblasts and the schizonts divide synchronously. The lymphoblasts exert cytotoxicity to parenchymal host cells. Immune T cells of the host recognize the infected lymphoblasts and try to kill them. The killed lymphoblasts release the pathogen in its piroplasma stage. The piroplasmas enter red blood cells and upon tick bites of the host enter the ticks' gut, where the piroplasmas convert into sporozoites; the sporozoites will infect the mammalian hosts. The African East Coast fever endemic in cattle (Faculty of Veterinary Science, University of Zimbabwe, Harare. Assistance to Veterinary Service of Zambia. Control of East Coast Fever in Zambia. Control Strategy for theileriosis in Rwanda. Centre for Ticks and Tick-borne Diseases in Malawi. ITM Antwerp Annual Report 2002 D. Geysen http://www.itg.be/internet/jaarverslag02/en/animalhealth.html.

Life Cycle of Malaria Plasmodium. The analysis of plasmodia mitochondrial genomes indicates that human ancestors were infected with plasmodia of gorilla origin in median estimate 365,000 years ago; human-adapted plasmodia radiated thereafter. P. malariae might have been the first human malaria pathogen, and P. falciparum is the last most recent acquisition, while human infestation with P. vivax might have originated even more recently from Southeast Asian macaques (Baron JM et al J Mol Evol 2011;75:297–304; Escalante AA et al Proc Natl Acad Sci USA 2005;102:1980–5; Sabbatini et al Infez Med 2010;18:56–74). The P. falciparum genome harbors a phage integrase-like domain referred to as Pf-Int atv locus Mal13P1.42. Pf-Int is a tyrosine recombinase; the mRNA encoded enzyme is a 490 aa polypeptide. The enzyme is not essential for the life cycle of the plasmodium, but it contributes to its genetic diversity and virulence (Ghorbal M et al PLoS One 2012;7(10):e46507). Targeting the ubiquitin/proteasome system kills the plasmodia (Chung DW & Le Roch KG Infect Disord Drug Targets 2010;10:158–164). The proteasome inhibitors bortezomib and carfilzomib kill the human myeloma cell (Jurczyszyn A et al Contemp Oncol (Poznan) 2014;18:17–21).

Pinealoma. The circadian rhythms of the pineal gland and their ancient beginnings, and their oncologic connotations, are present in the pinealoma. This neuroendocrine organ receives its adrenergic innervations from the hypothalamic suprachiasmatic nucleus. Peripheral photosensitive cells (some of them in the retina) provide input to the suprachiasmatic nucleus (Hafner et al PLoS Comput Biol 2012;8(3): e1002419. Parenchymal cells of the gland produce from serotonin the nocturnal hormone melatonin. Other constituents are germ cells and neuroectodermal cells. All cellular elements are synchronized by the circadian rhythm. Pineal gland tumors are of parenchymal, germ cell, and neuroectodermal histopathology. Tumors may de-differentiate into pineoblastomas, anaplastic xanthoastrocytomas or papillary cancers. All pinealomas are interconnected with the clock genes of the circadian rhythm. Long noncoding RNAs are metabolically and transcriptionally extremely active. Leading genes are for proteins BMAL (brain and muscle Arnt-like), Clock, Cry and PER (circadian molecular clocks; cryptochromes; period), CREB/ETS (cAMP-response element-bindig protein; E-twentysix retroviral proto-oncogene), c-Myc at 17q (myelocytosis), CaMP/CaKM (calcium/calmodulin-dependent protein kinase), Wee (small), cyclin D cell circle genes (Coon SL et al Proc Natl Acad Sci USA 2012;109:13319–13324; Kees UR et al Cancer Cytogenet 1998;100:159–164; Kelleher FC et al Cancer Lett 2014;342:9–18; Tovin A et al PLoS Genet 2012;8(12):e1003116.

Neurospora crassa, operates active CREB proteins (vide supra). Superoxide anions through protein phosphatase 2 activate the positive elements of the circadian clock (Gyöngyösi N et al Free Radic Biol Med 2013;58:134–143).The signalosome integrity is essential for hyphal growth. The circadian clock activates the MAPK/ERK (MAK) pathway; it induces robust activation of genes ccg (clock-controlled genes) and mitochondrial gene U07465 (Bennett LD et al Eukaryot Cell 2013;12:59–69; Zhou Z et al PLoS 2012;8(5):e1002712). The protein arginine methyl transferase PRMT in a cell survival pathway protects against oxidative stress (Feldman D et al PLoS One 2013;8(11):e80756). The CaMK proteins phosphorylate frq (frequency), thus accelerating cell growth (Gooch VD et al J Biol Rhythms 2014;29:38–48; Yang Y et al J Biol Chem 2001;276:41064–41072).The Δcamk-2 mutant CaKM protein remained active in oxidative stress and exhibited thermotolerance. The Δcamk-1,3,4 mutant genes reduced growth rate and hyphal growth and failed to protect against oxidative stress (Kumar R & Tamuli R Arch Mikrobiol 2014;196:295–305; Yang Y et al J Biol Chem 2001;276:41064–41072). A genome-wide characterization of blue-light-regulated genes of N. crassa is available (Wu C et al G3 Bethesda 2914;4:1731–1745). The N. crassa genes encode kinases for the phosphorylation of amino acids serine, threonine, tyrosin in order to activate proteins; and phosphatases to deactivate proteins by removal of the phosphate groups (dephosphorylation). The class 2 phosphatases (PP2A) dephosphorylate, thus inactivate MAPK. Loss-of function-mutated PP2A phosphatases allow MAPK overactivity to its constitutive expression. Hyphal overgrowth and hyphal cell fusions follow. The proliferating hyphae fail to respond with cessation of growth to the inhibitory chemical butyl-hydroxy-peroxide (t-BuOOH, Sigma) (Bennett LD et al Eukaryot Cell 2013;12:59–69; Ghosh A et al G3 Bethesda 2014;4:349–365). Response to the light-induced circadian rhythm activates MAPK. Recently reviewed (Sinkovics JG Europ J Microbiol Immunol 2015; 5:25–43).

Comment. The human CaKMs promote cancer cell (breast, endometrium) growth and are targeted for therapy (Bitschgi A et al Proc Natl Acad Sci USA 2013;110: E1026–1034; Takai N et al Cancer Lett 2009;277:235–243).The Neurospora genome contains the elements antecedent to protooncogenes (Creb; MAPK) and tumor cell promoter (frq; PRMT; CaKM; Δcamk-2) and tumor suppressor genes (Δcamk-1,3,4). In some human cancers cells (colon cancer; neuroblastoma; myeloid leukemia) some of the CaKM pathways were apoptosis/necroptosis- or differentiation-inducers (Jutong S et al J Clin Invest 2007;117:1412–1421; Kim do Y et al J Agric Food Chem 2009;57:10573–10578; Nomura M et al Cancer Res 2014;74:1056–1066). Pineal region masses: differential diagnosis (Smirniotopoulos JG et al RadioGraphics 1992; 12: 577). Correlate with circadian rhythm of Neurospora crassa (in Appendix 2 Explanations to the Figures, Figure 63).

Trichomonas. The parabasalian trichomonads enriched their genomes through receptions by horizontal transfers and insertion of prokaryotic genes, prominent among them the GTPase signaling system (Anantharaman V et al Gene 2011;475:63–78). The eukaryotic GTP-binding signal transductions are conducted by ancient monomeric proteins of the Ras superfamily. The enzymes catalyzing GTP to GDP emerged in prokaryota and archaea, and are conserved in eukaryota. The ribbon plots and crystal structures of the GTPases are very well known. The ancestral GTPase MnmE (methylaminomethyl E) was acquired in eukaryotes from a protomitochondrial endosymbiont. The ancestral E. coli Ras-like protein (Era) is present in the Aquifex aeolicus. An ancestral Era might have existed in precellular ribosomes? Chlamydia, Mycobacteria and fungi lost Mnm and Era. In plants and fungi, the Era-related GTPase (Erg) is operational. The human Era homolog ERAL (Era-like) remains located in the mitochondrial matrix. Overexpressed ERAL induces mitochondrial apoptotic death of HeLa cells. Related and conserved GTPases conduct ribosome assembly, cell cycle regulation and stress response. The vast majority of bacteria (over 75% of the species) share 13 GTPases. Many of these GTPases remain conserved in eukaryotes functioning as translation factors, or signal recognition associates. Of translation factors, IF (initiation factors) expand from archaea to human cells. Some of the signal recognition associates remained in chloroplasts, where they mutually activate each other (Verstraeten N et al Microbiol Mol Biol Rev 2011;75:507–542). The prokaryotic MgLA (CH3-galactoside transport, L-arabinose) proteins are considered to be the ancestors of their eukaryotic descendants (Notley-McRobb L & Ferenci T Genetics 2000;156:1493–1501; Dong JH et al Gene 2007;396:116–124). N-ras appears to be a derivative of prokaryotic cold-shock domain proteins (Doniger J et al New Biol 1992;4:389–395). Early multicellular eukaryotes with two ecto/endodermal cell layers without mesoderm (N. vectensis, C. intestinalis) operate ancestral GTPase ras genes. Ras family genes are absent in plants and alveolates (P. falciparum). Rho family genes replace them. Rab family genes operate in amebozoans and ciliates. The ras gene-coded monomeric GTP-binding proteins of 181 aa appear in the T. vaginalis (Xu MY et al Biochem Cell Biol 2007;85:239–245). In the human genome 167 ras superfamily sequences are operational. Some basic science reviewers omitted all oncogenic connotations. For the phylogenetic tree of human Ras superfamily, see Figure 3 of the article (Rojas AM et al J Cell Biol 2012;196;189–205). The Harvey (H) and Kirsten (K) ras sarcoma genes were found in the Moloney murine leukemia/sarcoma retrovirus strains passaged in rats: ras rat sarcoma. The low branching unicellular eukaryote Trichomonas in its very large 160 mb genome operates ancestral Ras subfamily protein-encoding intronless genes (Rsp) (Xu MY et al Biochem Cell Biol 2007;85:239–245). Transfer from its ancient free-living entity to being a parasitic microorganism has come with major genomic changes in the trichomonads: replacement of lost genes by horizontally acquired novel genes. Indeed, trichomonads share numerous genes with bacteria, yeast fungi, Entamoeba, Giardia, and the human host. The ras genes are not named in the list of laterally acquired genes (Singh S et al, Bioinformation 2012;8:189–195). The Trichomonas genome harbors three myb (myeloblastosis, avian) transcription factor genes, whose gene-product proteins attach to Myb protein recognition sites (MRE) in its genome. The iron-responsive promoter region of the ap65-1 gene (adhesion protein, adhesin) is the target of the Myb proteins. Cytoadherence is essential for the pathogenicity and virulence of trichomonas cells. The adhesion DBD (DNA-binding domain) protein is a pyruvate:ferredoxin oxidoreductase homologous to a similar gene of Entamoeba histolytica (Moreno-Brito V Cell Microbiol 2005;7:245–58). The trichomonas Myb3 protein displays 83% aa identity with the human c-Myb. Its binding to DNA is similar to that of the FOXO (forkhead box) protein (Wei S-Y et al Nucleic Acids Res 2012;40:449–460). The most widely spread ras proto-oncogenes/oncoproteins served in cell survival pathways of bacteria and archaea before moving into eukaryotes. In multicellular hosts, the cell survival pathways have become proto-oncogenic transformators. In unicellular eukaryotes (amoeba; ciliates; trichomonas) appear the ancestral myb/Myb pathways not known to be constitutively expressed, but essential for virulence. Trichomonas may coexist with pathogenic viruses (HPV, BK polyoma virus, XMRV, the xenotropic murine leukemia virus-related virus, HIV-1), in the human female and male genitourinary tract (including the prostate), and remains a suspect for the initiation of malignant transformation.

The Trichomonas genome is large, constantly active due to multiple gene acquisitions; readily trans-speciates the cell (amoeboid to flagellar, to and from); carries the ancestors of several human proto-oncogenes, primarily myb/Myb; and is loaded with pseudogenes (1354 annotated pseudogenes). The total gene numbers are between 46,000 to 60,000 in six chromosomes. Transposable elements make up 45%, pseudogenes 5% of the genome. There are 668 ribosomal rRNAs and 468 transfer tRNAs; snRNAs, snoRNAs, miRNAs, long non-coding lncRNAs are all operational in the nuclear spliceosome and the cytoplasmic AGO/Dicer complex. Many lncRNAs reveal significant sequence similarities to genomic loci annotated as protein-coding genes. Thus, these lncRNAs are descendant pseudogenes. The fact that prokaryotes and archaea do not possess spliceosomes indicates that these organs appeared first in eukaryotes. Yet the Trichomonas introns appear to support the "introns first" hypothesis. In opposition, eukaryotic introns derive from the presumably intron-rich LECA (last eukaryotic common ancestor) (Rogozin IB et al Biol Direct 2012;7:11; Vaňáčková S et al Proc Natl Acad Sci USA 2005;102:4430–4435; Woehle C et al BMC Genomics 2014;15:906). Pseudogenes persist as copies of ancestral protein-coding genes deriving from gene duplications or retrotranspositions of mRNA sequences. Thus, pseudogenes may yield endogenous miRNAs to inhibit the mRNA of a related fully operational gene; they may be transcribed into dominant-negative polypeptides. All in these, the trichomonas genome is similar to an oncogenome. In this last aspect, human cancer cells accumulate and utilize pseudogenes for their own promotion. Prominent example is the stem cell proto-oncogene oct/Oct4 encoded from gene POU5F1 (Pit-1, pituitary; Oct1,2; Caenorhabditis unc, uncoordinated) at chromosome 6p21.3, and its numerous pseudogenes in glioblastoma and hepatocellular carcinoma, where they upregulate the PI3K/Akt pathway (Zhao QW et al Oncol Rep 2015;33:1521–1529). In prostate cancer, both ETS-positive and ETS-negative cells express the gene CXADR, but only ETS-negative cancer cells express its pseudogene CXADR-Ψ (retrovirus Ets/E26; coxsackie virus adenovirus receptor, CAR). In breast cancer, the ATP8A2 (adenosinetriphosphate) gene and its pseudogene ATP8A2Ψ compete. The pseudogene-expressor tumor cells gain replication and migration advantage (Kalyanja-Subaram S et al Cell 2012;149:1622–1634). Thus, the Trichomonas pseudogenes may also serve (not yet documented) as factors of increased virulence. llustrations can be found at Science Photo Library Trichomonas vaginalis_parasites_computer_artwork_z100202.jp. The avian retrovirus myeloblastosis E26 (Ets E twentysix) incorporated its host's c-myb gene. The human c-myb oncogene/Myb oncoprotein is expressed in HeLa cells, as shown by staining with fluorescent polyclonal IgG antibody ab167537 of ABCAM 888-7722226.

Epithelial-to-Mesenchymal Transition. Tumor cells may recruit fibroblast-, or monocyte-Mϕ-like cells to fuse with them and thus assume a highly mobile slick elongated shape. Accordingly, the cytoskeleton is transformed. The cytoplasm is vimentin-positive. These tumor cells locally invade tissues, or enter into lymph vessel or vascular endothelial cells, into the lumen of lymphatic and blood vessels, in order to be able to travel by the lymph or blood stream and settle at, and colonize, distant sites (Dittmar T & Zänker KS "Cell Fusion in Health and Disease. II. Cell Fusion in Disease", 2011. Advances in Experimental Medicine and Biology, 2010; 714. Springer Dordrecht Heidelberg London New York. Sinkovics JG Int J Oncol 2009;35:441–465). The biological meaning (metastasis formation), early and reliable recognition and the value of immediate treatment for circulating tumor cells (CTC) are under extensive investigation Epithelioid breast cancer cells leave the tumor mass, convert into elongated mesenchymal-like cells to enter blood vessels (Karabacak NM et al Nat Protoc 2014;9:694–710; Ma X et al Tumour Biol 2014;35:5551–5560; Shao C et al PLoS One 2014;9(2):e88967).

Oxytricha fallax and trifallax. These large (1 mm) unicellular ciliates ingest nutrients in food vacuoles (a "digestive tract"; osmoregulate their cytoplasm with contractile vacuoles (a kidney-like "urinary tract"), and evacuate waste through their "cytoproct". In their macronucleus they carry over 25 million minichromosomes (fragmented chromosomes) and over 30,000 genes. Massive apoptosis occurs when post-conjugation the germline micronucleus (Mic) transforms ("differentiates") into the somatic macronucleus (Mac): transposons and non-coding DNA sequences are eliminated. Transposases excise and do not reinsert the transposons. The fragmented genome is precisely reconstructed based on ssRNA copies of each minichromosome gene. The oxytricha genomes, cap their chromosomes' ends after each binary fission of vegetative clones (karyonides), or conjugation. Prior fission, the Mac chromosomes end-reduplicate. Telomeres are re-caped. Thus, they experience no double break repair, or breakage-fusion-bridge catastrophes (first described by Barbara Mc McClintock in 1938). The tandem repeat G-rich 20 bp sequence telomeres are de novo polymerized, templated by an RNA component, and primed by a 3'OH end. Telomeres maintained in full length prevent senescence of the cell (Nowacki M et al Nature 2007;451:153–158; Science 2009;324:935–938; Williams KR et al BMC Genet 2002;3:16). The oxytricha mitochondrial genes are arranged in two transcriptional directions (in contrast to paramecia mitochondrial genes arranged in just one transcriptional direction). The oxytricha mt genome contains a large linear plasmid. These mitochondria have recently acquired this plasmid, because its ORFs have not as yet engage in the tryptophan codon usage (the tryptophan bias). The mt genome encodes 11 tRNA genes, more than other ciliates'. It operates both RNA and DNA polymerases; one RNA polymerase is of bacteriophage-derivation. Thus, the ancient proteobacteria before became mitochondria harbored phages. Others are distantly related to DNA polymerases of the amoebozoan D. discoideum, or of the oomycete Phytophthora (Table XXX). The mitochondrial genes are capped by 35 bp telomeres (Swart EC et al Genome Biol Evol 2012;4:136–154).

Pluricellular Advancement of a Cancer Cell Armada. By morphological appearance, the crawling locomotion of cancer cells imitates that of the amoeba. A pseudopod protrudes first and attaches to the ground; the body elongates; the rear of the cell is retracted. The malleable cytoplasm can negotiate narrow spaces between fibers, but the nucleus retains its firm substance, except for some limited elongation. The cancer cell nuclei are usually larger (7 μm^2 cross section) than that of normal cells (T lymphocyte, 4 μm^2; neutrophil leukocyte, 2 μm^2). Extracellular collagen matrix can be dissolved by metalloproteinases. The sedentary cuboid cancer cell restructures its cytoskeleton, becomes sarcomatoid and mobile and activates the descendants of the native amoebal genes, that were in use for cell locomotion since ancient times (see text). What is not available for the solitary amoeba, the colonies of cancer cell have acquired: multicellular mechanocoupling moving cell sheets forward in a coordination arranged between leader and follower cells, in which the leader cell applies a traction force to the follower cells. The Rho GTPase-driven leader cell is able to drag follower cells by peripheral pluricellular actin-myosin cables. Multicellular migration follows and results in collective cancer cell invasion (Alexander S et al Curr

Opin Cell Biol 2013;25:659–671; Friedl P et al Nat Cell Biol 2014;16:208–210; Reffay M et al Nat Cell Biol 2014;16:217–223; Wolf K et al J Cell Biol 2013;201:1069–1084). The pseudopods of Amoeba proteus: one pseudopod forms two arches and three points of contact with the substratum (Cameron I, Rinaldi RA, Kirby G, Davidson D. University Texas Health Science Center, Department Cellular Structural Biology; Southwest Research Institute, San Antonio, TX; University Colima, Colima, Mexico; Tulane University School Medicine, Department Structural Cellular Biology, New Orleans Louisiana. Cell Biology International 2007;31:759–776). Invasive human breast cancer cell driven by protein tTG (tissue transglutaminase, vide infra) chaperoned by Hsp70 protrudes its leading edge. The tTG protein protect cells from serum-starvation-induced apoptosis, and activates c-Scr and PI3K/mTOR complex proto-oncogenes (Boroughs LK, Antonyak MA, Cerione RA. Johnson JL Cornell University, Department Molecular Medicine, Ithaca, NY. J Biol Chemistry 2014;289:10115–10125).

Tissue Transglutaminase (tTG). tTGs form cross-linking isopeptide bonds between two proteins in embryogenesis, apoptosis, angiogenesis, and cancer. tTGs evolved through unicellular eukaryotes, crustaceans to vertebrate mammalians, incl Homo. Increased levels occur in malignantly transformed mammalian vertebrate cells: adenocarcinomas of esophagus, pancreas, kidney, prostate; melanoma (Ashour AA et al J Cell Mol Med 2014;18:2235–2251; Erdem S et al World J Urology 2014 Dec 17 SpringerLink PMID 25515319; Han AL et al Eur J Cancer 2014;50:1685–1696; Kim HJ et al Biomol Ther 2014;22:207–212; Leicht DT et al J Thorac Oncol 2014;9:872–881). In clear cell carcinoma of the kidney, strong TG2 expression correlated with metastatic disease and increased mortality (Park MJ et al J Pathol Transl Med 2015;49:37–43). tTG is a cell survival promoter, as it acts as an anti-apoptotic agent and PI3K/mTOR complex activator (Boroughs KL et al J Biol Chem 2014;289:10115–10125). tTG inhibitors suppress cancer cell growth (Ku BM et al J Cancer Res Clin Oncol 2014;140:757–767; Yakubov B PLoS One 2014;90 (2):e89285). The slime mold Physarum polycephalum pertains the life style of unicellular eukaryotes, such as amoeba and plasmodia. It possesses mammalian type tTGs, also related to the tTGs of Ciona, crustaceans, Drosophila, Caenorhabditis, fishes, and mammals, including Homo (Wada F et al Eur J Biochem 2002;269:3451–3460). Comment: Another example of a life survival pathway operational in a unicellular eukaryote, that has become a human proto-oncogene.

Neuroblastoma (NB). Tumor suppressor gene and immunotherapy for NB. There is an extensive discussion in the text on biological measures that are able to induce differentiation in neuroblastoma cells. Toward tumor antigen ganglioside GD2-redirected, chimeric antigen receptor-armed (CAR) (vide supra), immune T cells, synergized with oncolytic adenovirus Ad5Δ24 expressing chemokine RANTES (regulated on activation normal T cells expressed and secreted; CCL5 from chromosome 17q1) in solid tumor immunotherapy (Nishio N et al Cancer Res 2014;74:5195–5205). Vα24 anti-GD2-expressor CAR-armed invariant natural killer cells (NKT) attacked NB tumors, thus extending patients' survival without inducing GvH disease (Heczey A et al Discov Med 2013;16:287–294; Blood 2014;124:2824–2833). Advancing NB in patients was due to the elimination of the FOXP1 (forkhead box) tumor suppressor gene in chromosome 3p14.1. The loss was not due to potentially reversible CpG gene hypermethylation. In tumor cell lines, doxycycline could induce re-expression of FOXP1 gene. These, and FOXP1-transgenic NB cells, ceased their growth and underwent either apoptosis or a neuronal differentiation pathways (Ackerman S et al BMC Cancer 2014;14(1):840. doi:10.1186/1471-2407-14-840). New monoclonal antibodies, new vaccines and adoptive transfer of immune T cells or NK cells (even allogeneic from parents or siblings) are in the investigational use (Croce M et al Immunotherapy 2015;7:285–300; Seidel D et al Department Pediatric Hematology Oncology, University Medicine Greifswald, Germany. Cancer Immunol Immunother 2015; Epub PMID 25711293 SpringerLink). The NK cell line NK-92-scFV(ch14.18)-zeta carries the genetically engineered chimeric antigen receptor (CAR) specific to GD2 (disialoganglioside). It attacks and lyses GD2-expressor drug-resistant NB-cells. MYCN-amplified tumor-bearing patients should receive immunotherapy in addition to standard surgery, radio-chemotherapy (Csanády M et al Int Pediatr Otorhinolaryngol 2014;78:2103–2106). Neuroblastoma cells may be induced to re-differentiate. Comment. Gene therapy aimed at the oncogene, and immunotherapy aimed at the oncoprotein are to be combined. Some malignantly transformed tumor cells (acute promyelocytic leukemia, PML cells) carrying fusion oncoprotein with retinoic acid receptor (RAR) at t(15;17, or losing tumor suppressor genes (in NB, FOXP1), may recover from the stages of "malignant transformation" and resume their service as members of organized cell communities. Unexplained re-differentiation of chondrosarcoma cells in vitro in co-existence with lymphocytes of a healthy donor was observed (Sinkovics JG Pathol Oncol Res 2004;10:174–187). Recently reviewed (Sinkovics JG Europ J Microbiol Immunol 2015;5:25–43).

Accelerated Life Style of the Ctenophores. Prior to the bilaterians, there existed four clades of metazoans: 1 Cnidaria corals, sea anemones (the Nematostella; Hydra/Hydractinia), Medusozoa jellyfish). 2 Porifera sponges, (Amphimedon; Suberites). 3 Placozoans (Trichoplax). 4 Ctenophores represented by the predatory omnivore Mnemiopsis leidyi (warty sea walnut) and Pleurobrachia pileus, the comb jelly. The ctenophores' outer epidermal layer is separated by mesoglea from the inner gastrodermal layer. Mesenchymal smooth muscles are operational in the body wall, tentacles (the specialized colloblasts for capturing prey), and the pharynx. For swimming, there are eight longitudinal rows of cilia alongside the body. The ctenophore cydippid larva is a free-swimming entity. The mature adult is a hermaphrodite harboring both ova and sperms for self-fertilization. Freshly shed eggs are immediately fertilized. The ctenophores possess photocells with opsins mediating light detection, operate TGFβ, homeodomain Hox and Frizzled Wnt pathways, and a neurosensory system (Feuda R et al Genome Biol Evol 2014;6:1954–1971; Moroz LL et al Nature 2014;510:109–114; Pang K et al EvoDevo 2010;1:10; Ryan JF et al Science 2013;342(6164):1342592; EvoDevo 2010;1:9; Biol Direct 2007;2:37), but operate a much premature microRNA system (Maxwell EK et al

BMC Genomics 2012;13:714; Moran Y et al Mol Biol Evol 2013;30:2541–2552). The ctenophores' nervous system functions without serotonin (in contrast to that of the cnidarians). It consists of a subepithelial net, neurons in the mesoglea and innervation of the tentacles (Moroz LL et al Nature 2014;519:109–114). Ctenophores are usually captured in the Eel Pond of Woods Hole MA for genomic, metabolic, and innate immunological studies to be conducted at Whitney Lab for Marine Bioscience, University of Florida, St. Augustine, FL; Rice University, Houston, TX, and at many other sites worldwide. From the western Atlantic Ocean, this extremely vigorous predator species is rapidly invading the Black, Caspian and North Seas (Ryan JF et al Science 2013;342/6164. doi:10.1126/1242592).

The ctenophore genome acquired several of its mitochondrial genes; especially tRNAs had been depleted; thus the M. leidyi mitochondria are small (10 kb, 10,326 bp, 33 ORF). Some (34) mt ribosomal proteins of M.leidyi are homologous to human MRPs, what is half of what A. queenslandica (61) and N. vectensis (62) possess. Just like the alga, Chlamydomonas reinhardtii, M.leidyi also transferred its mitochondrial gene atp (adenosine triphosphate) to its nucleus. This continued to be like this in Cnidaria, Placozoa, Bilateria, and all Eumetazoa (including Homo). The mitochondrial atp genes were retained in the mitochondria in fungi, choanoflagellates and poriferan spongi (Pett W et al Mitochondrial DNA 2011;22:130–142).

M. leidyi operates some 16,545 protein-coding genes. Among the encoded proteins, there are glycoprotein hormones of the cystine knot growth factor protein superfamily .These are the TGFβ; nerve growth factor, platelet-derived growth factor (without platelets), bone morphogenetic protein antagonists, which are the predecessors of the bursicons of insects, gremlin, neuroblastoma suppressor of tumorigenicity, and the Wnt-activator and BMP/TGFβ-inhibitor norrin (without bones). Norrin is one of the ligands to the LGR (leucine-rich repeat-containing G protein-coupled receptor) (Deng C et al J Cell Sci 2013;126:20602–20608. Roch GJ & Sherwood NM Genome Biol Evol 2014;6:1466–1479. Xu S et al PLoS Biol 2012;10(30;E10011286). The ctenophore M. leidyi expresses the LIM homeobox proteins (Table XVI) (Simmons DK et al EvoDevo 2012;3:2. doi:10.1186/2014-9139-3-2). Cnidarians (N. vectensis; H. magnipapillata), ctenophores, and the placozoan Trichoplax adhaerens possess the ribonuclease family III Dicers operating in the RISC (RNA-induced silencing complex), but choanoflagellates (Monosiga brevicollis) are devoid of the system (Gao Z et al PLoS One 2014;9(4): e95350). The presence of mesenchymal genes (fibroblast growth factor, notch, hedgehog, nodal) is expected on account of the smooth muscles, but these are not detectable; the gli gene is present, but not the mesenchymal cells, rather the nerve cells express it. The hedgehog Hh and JAK/STAT pathways are not operational (Ryan JF Science 2013;342:1242502). Instead, the Wnt pathway with Frizzled proteins is highly upregulated, as it is acting without its customary inhibitors. Some introns of the Wnt genes are conserved in number, size and location up to the mammalian Wnt genes. Wnt ligands are accepted by the Frizzled (Fz) and LRP5/6 (lipoprotein receptor-related protein), and Dishevelled (Dvl/Dsh) receptors. Thus GSK-3 (glycogen synthase kinase) could be inhibited through phosphorylation by Dvl/Dsh. Failure of GSK-3 to phosphorylate β-catenin allows for the accumulation of the latter in the cytoplasm and for its eventual transfer into the nucleus, there, for the activation of the targeted tcfTCF/lefLEF gene complex, whose gene product proteins activate promoter DNAs of numerous other genes. The descendants of the tcfTCF/lefLEF genes and their protein products will often form oncogenic pathways in vertebrate mammalian (human) cells. This very same cascade is that of a constitutive oncogenic pathway in many human cancers (especially adenocarcinomas). The ctenophores' "β-catenin-destructive complex" is deficient, in that it lacks axin, and its suppressor gene APC (adenomatous polyposis coli) is defective in expressing all its domains. The "destructive complex" in malignantly transformed human cells is notoriously rendered inactive due to loss-of-function mutations or deletions of its components (axin, APC, GSK). This is naturally so in ctenophores (M. leidyi). Further, the physiological inhibitors of the canonical Wnt pathway working in mammalian cells, are absent in ctenophore cells. These are primarily the dickkopf (drosophila) gene product proteins, Wnt inhibitory factor WIF (and Cerberus/CER, present only in vertebrates). The secreted frizzle-related protein MlSfrp lacks the netrin domain (Jager M et al PLoS One 2013;8(12): e84363, Pang K et al EvoDevo 2010;1:10. doi:10.1186/2014-9130-1-10; Schenkelaars Q et al Evol Dev 2015;17:160–169). In addition to its naturally overactive Wnt pathway, the ctenophore genome expresses several sox/Sox genes and high mobility group gene-product proteins, from the larval cydippid stage on. The Sox proteins are encoded by the regular mRNA pathway and are subject to miRNA/siRBA control. The mammalian sox genes form five groups (BCDEF). The ancestors of the BCEF sox genes are present in the ctenophore genome (M. leidyi; P. pileus). Some of these sox genes are shared with the Hydra and the Choanoflagellata (Schnitzler CE et al EvoDevo 2014;5:15). Comment: Genomic and protein pathways (wnt/Wnt; sox/Sox HMGP), that will become mammalian vertebrate proto-oncogenes/oncoproteins, are physiologically hyperactive ("upregulated") in the ctenophore genomes. Their suppressors are non-functional, as they are eliminated alike in the "malignantly transformed" cells of mammalian vertebrate (human) hosts. The ctenophore cells are physiologically driven by the predecessors of the proto-oncogenes Wnt/β-catenin (Sinkovics JG Europ J Microbiol Immunol 2015;5:25–43; Sinkovics JG Int J Oncology 2015;47:1211–1229).

The Evolution of Ribosomes. This author would like to further promote the idea that precellular ribozyme-armed, functional rRNA-containing precellular ribosomes existed on, and populated the ancient Earth. Thus, ribonucleoproteins (RNP) were synthesized with substantial genetic errors. If so, these structures might have synthesized the successors of the viroids, and provided the platform on which precellular viruses could have existed and grow by fusion. JH Matthaei and MW Nirenberg worked well with extracellular E. coli ribosomes in the late 1950s, early 1960s (Proc Natl Acad Sci USA 1961;47:1580–1586; 1588–1602). However, the ribosome assembly in bacteria (E. coli) requires RNA helicases (Peil L et al FEBS J

2008;275:3772–3782). Nevertheless, ribosomal assembly in vitro is slow, but possible (Shajani Z et al Annu Rev Biochem 2011;80:501–526). The in vitro 30S ribosome assembly cofactors are recognized (20 cofactors in bacteria), including rRNA-binding proteins, such as essential GTPases. Figure 1 of the cited article shows the Nomura assembly map. In the 30S biogenesis, the role of Rims (RimM, RimP, ribosome maturation factor) has been clarified. The kinetic interactions of the Era (E. coli ras-like protein, see Trichomonas) with 16S rRNA and nucleotide cofactor GTP could be determined. The final ribosome biogenesis is depicted in Figure 4 of the article (Brunner AE et al J Mol Biol 2010;398:1–7). When singled out cells of multicellular hosts undergoing the process of the so-called malignant transformation in the form of de-differentiation to their ancestral life forms, could the excessive release of RNA-loaded exosomes be a manifestation of that process? Could the exosomes be viewed as the extant replicas of their predecessors, the precellular ribosomes? (Sinkovics JG Europ J Microbiol Immunol 2015;5:25–43). In ribosomal 60S subunits stalled translation recruits a ribosome quality control complex (RQC). In turn, RQC enlists alanine- and threonine- charged tRNAs. Consequentially, tRNAs elongate the nascent chain independently from mRNA. The incompletely synthesized nascent chains are degraded by ubiquitylation (Shen PS et al Science 2015;347:75–78). Comment: Are the natural ribosomes punctilious?

Problems with unexplained abbreviations. Lenghty complicated nomenclature is commonly cited in abbreviated form; the abbreviations often remain unexplained and not spelled out. In this Text, Legends, Tables and Appendix the abbreviations received extraordinary attention: they are often spelled out repeatedly. An exemplary article deserves to be cited here for its clarity of style and for its elaborate listing of spelled out abbreviations: Futosi K, Fodor Sz, Mócsai A: Neutrophil cell surface receptors and their intracellular signal transduction pathways (Int Immunopharmacol 2013;17:638–650). An extensive alphabetical list of abbreviation for oncology terminology is printed in Weinberg RA: The Biology of Cancer. 2nd Ed Garland Science, New York & London, 2014. The volume publishes a very large number of superb illustrations in color.

A Contrast. Cancer chemotherapy. In the years 1960–70s combination chemotherapy has become the officially accepted, established, and firmly implemented standard treatment of leukemias and metastatic cancers. Non-specific immunization primarily with BCG was accepted (by now discontinued), but tumor-specific immunotherapy received practically no support. The concepts and aim of high dose combination chemotherapy were that the higher the dosage was raised, the better the chances will be to kill the last cancer cell, thus resulting in cure of the disease. Without recognized specific cancer antigens, attempts at immunizing against cancer were considered to be futile (but the general immune state of the patients stimulated by coryne-, or mycobacterial vaccines was allowed). This concept was developed by Georges Mathé at Gustave Roussy's in Villejuif, France (le nécrologie; décédé 2010; in Acta Microbiol Immunol Hung 2010;57:322) and promoted at the National Institutes of Health / National Cancer Institute in Bethesda MD.

It was most prominently practiced at the Department of Developmental Therapeutics (DDT) at M.D. Anderson Hospital. The DDT was a detachment of the NIH/NCI, and as such received the highest support from the NCI and from the Pharmaceutical Industry. Truly, partial and complete remissions for the first time were induced by combination chemotherapy in patients with leukemias and solid tumors. However, not all patients receiving the same treatment responded. The actual cure rate remained low, due to frequent relapses. Toxic side effects (severe protracted nausea; leukopenic septic infections, often fatal septic shock; hemorrhagic events; neurological and endocrine side effects; retarded mentation) were slowly and steadily diminished by new antibiotic combinations, platelet transfusions and anti-nausea medications. Intensified general medical and nursing care, and specialized attendance allowed the administration of full dosage chemotherapy as calculated in the protocols. Exceptional patients in durable complete remission with minimal residual side effects were prominently publicized. Private practice medical oncology clinics flourished on the high profits of combination chemotherapy. The efforts to cure leukemia and metastatic cancers culminated in the STAMP program initiated by the former head of the DDT, Emil Frei III. The STAMP regimen consisted of very large (triple) dosages of combination chemotherapy followed by "stem cell rescue" (autotransplants of stem cells collected from the circulating blood, or the bone marrow). Various aspects of the natural biology of the cancer cell were either unknown or ignored. The unusual natural biological features of the cancer cell consisted of its extraordinary radiation- and chemotherapy-resistance, manifesting in the form of recurrences after periods of "tumor dormancy" (now autophagy); the ability to circulate in the blood or lymph; to emit exosomes; to circulate its DNA; to recruit-supporting mesenchymal cells from the bone marrow or in its microenvironment; and to overcome immune reactions of the host, if any. The fact that the cancer-bearing patient generated some immune reactions against the autologous tumor was recognized. Some immunizations with irradiated tumor cells could be conducted, but were abandoned as ineffective. Vaccinations with viral oncolysates, in which a non-self viral antigen fused with a self cell surface oncoprotein, could have immunized the patient, but these projects received no grant support (because the targeted tumor antigen was not specifically identified). Instead, the highly upheld curative aim of combination chemotherapy was persuasively promoted (because it was known how it damaged the DNA in general, but without knowing exactly which oncogenes were targeted). The possibility that treatment-acquired endocrine or central and peripheral neurological deficiencies (especially in children and adolescents) may remain permanent, was listed among collateral damages acceptable for the cure. However, the end results of the STAMP program revealed that it was much more toxic, but not more effective than the previous standard dosage chemotherapy. The STAMP program was discontinued (Berry DA et al J Clin Oncol 2011; 29:3214–3223). In addition to its failure to improve treatment results (overall survival), a faked STAMP-like program was subjected to major falsifications of its results for the better: a most dismal event in the history of medical oncology (Bezwoda WR Eur J Cancer Care 1997;6:10–15; Bezwoda W et al Oncology 1990;47:4–8. Vickers A & Christos P. Bezwoda: "evidence of fabrication" in original article (J Clin Oncol 2000;15:2933).

Cancer Immunotherapy. Clifton Dexter Howe, the Head of the Department of Medicine (DM) at M.D. Anderson Hospital, was an open opponent of high dose combination chemotherapy (in the era when supportive measures were non-existent); eventually, he was demoted into an administrative position. This author (JGS) serving as a medical oncologist at the DM of M.D. Anderson Hospital (1959; 1962–1980) received no NCI support for human cancer immunotherapy, or virotherapy; he treated his patients at the Melanoma-Sarcoma Clinics and In-patient Service with standard full dose combination chemotherapy by approved institutional protocols. His proposal of immunotherapy with viral oncolysates was based on the idea, that viral structural, capsid or membrane proteins fuse with tumor cell membrane proteins at the egress of the viral particle, and that these fused proteins are "non-self" to the host, thus are immunogenic. In the treated subjects, viral oncolysis would be followed by anti-cancer immunization. Viral oncolysates induced strong anti-cancer immunity in mice. Influenza virus particles replicating in, and thus lysing cancer cells, would construct such fused, and thus immunogenic antigens. However, such fused and thus immunogenic proteins were not actually produced in the author's laboratory. Their existence was documented elsewhere (in the literature). It remains a solid fact; so is the fusion of the influenza viral hemagglutinin with cellular sialic acid receptors expressed in the cell membrane (Hamilton BS et al Viruses 2012;4:1144–1168). The viral oncolysate immunotherapy protocol was approved by the M.D. Anderson Hospital Institutional Surveillance Committee in 1967; it was funded from private donations (not by the NCI); and its course was reported in the biannual M.D. Anderson Hospital's Research Reports (as summarized in Sinkovics JG & Horvath JC Archives Immunologiae Therapiae Experimentalis Birkhäuser Verlag, Poland/Switzerland 2008;56:3–59). This author recognized that large granular non-specific (later NK cells), and small compact specific (later immune T cells) lymphocytes are naturally generated against autologous tumor cells in tumor-bearing patients, and that these reactions quantitatively increased in patients receiving viral oncolysate vaccine. In the pre-IL-2 era, large live populations of these lymphocytes for treatment could not be produced and provided. Later in the 1990s in academic medicine-affiliated private practice of medical oncology at St. Joseph's Hospital in Tampa, FL, this author with Joseph C. Horvath could apply viral oncolysate immunotherapy with standard combination chemotherapy given in full dosage, and use autologous immune lymphocyte infusions prepared in IL-2-containing liquid media for the treatment of metastatic melanoma. This author declined participation in any STAMP-like high dose chemotherapy programs, which were installed also in Tampa by private practitioners of medical oncology, and was carried out at St. Joseph's Hospital under the egis of a nationally approved private company (discontinued in total failure after a few years).

After many decades of trials and errors, the faculties of virotherapy and immunotherapy of cancer have by now succeeded in proving their extraordinary value in the treatment of lymphomas,-leukemias and certain metastatic cancers (especially melanoma). The basic elementary knowledge that the host recognizes and attempts the rejection of incipient cancers by antibody- and lymphocyte-mediated reactions has been well recognized and elaborated on, but without the practical means of effective therapeutic translation of this knowledge into practice (Harris JE & Sinkovics JG. The Immunology of Malignant Disease, 2nd ed Mosby, St. Louis, 1976; Sinkovics JG Cytolytic Immune Lymphocytes. Schenk Buchverlag, Passau-Budapest, 2008; Sinkovics JG Acta Microbiol Immunol Hung 2010;57:253–347; Sinkovics JG & Horvath JC Acta Microbiol Immunol Hung 2006;53:367–419; Arch Immunol Ther Exper 2008;56:S1–59). Monoclonal antibodies were the first agents which could suppress oncoproteins. Rituximab and trastuzumab were the flagships followed by even more effective new agents. They combine well with standard dosage chemotherapy. The CHOP-R regimen is highly effective (probably curative) in diffuse B cell and other malignant lymphomas. Pertuzumab and trastuzumab attack the breast cancer oncoprotein Her2/neu synergistically. Natural oncoprotein-reactive immune T lymphocytes without or with genuinely tumor cell-reactive NK cells extracted from the patient and grown to billion-size populations in IL-2-containing media, upon re-infusion attack and lyse large tumors (primarily melanomas). The National Cancer Institute program of Steven A Rosenberg proved that effectively oncoprotein-reactive immune T cells are mobilized in limited numbers in patients with cancer. Retrieved from the patient, expanded by lymphokines in vitro, and re-infused, these autologous immune T cells lyse large tumors (but not without failures or relapses). Further, oncoprotein-reactive autologous immune T cells with the genetically engineered T cell receptors CAR induce complete remissions of curative effect in patients with lymphomatous malignant tumors (Frigault M et al Cancer Immunol Res 2015;3:356–367). Tumor cell destruction is so extensive and rapid that clinically it manifests in a tumor lysis syndrome with a lymphokine storm ensuing. Tumor lysis syndrome is treatable with rasburicase. IL-6 is the leader of the lymphokine storm. Major participants in the lymphokine storm are IL-2, IL-6, IL-10, TNFα and IFNγ. Monoclonal antibodies tocilizumab or siltuximab promptly neutralize IL-6 with life-saving efficacy. Ipilimumab protects from apoptotic death the CTLA-4$^+$ T lymphocytes: they are ready to attack malignantly transformed melanoma cells. Potentially autoimmune T cells are further withheld by their natural inhibitor PD1 (programmed death). When the PD1 ligand activates receptor PD1, the PD1$^+$ T lymphocyte kills itself in apoptosis. Malignant tumor cells masquerading as self, produce the PD1 ligands excessively, and thus eliminate PD1-expressor, autoimmune-active, T cells from their environment. The mcab pembrolizumab neutralizes the ligand and the PD1$^+$ cell is free to attack the transformed cell; another mcab neutralizing the receptor PD1 (nivolumab), also rescues the T cell from its apoptotic death. At the potential price of some autoimmune reactions (thyroiditis; hypophysitis), a massive destruction of tumor (melanoma; bronchogenic carcinoma) cells ensues (Luke JJ & Ott PA Oncotarget 2015;6:3479–3492; Mahoney KM et al Clin Ther 2015. doi:10.1016/jclinthera.2015.02.018). Intratumoral Newcastle disease virus inoculations (oncolytic viral therapy that was denied research funding at the NIH/NCI in the 1960–70s) combined with CTLA-4 checkpoint blockade mutually intensify the efficacy of each other reactions (Zamarin D et al Sci Transl Med 2014;6(226):2264ra32). Many times what worked in the rodent host, like this experiment, failed in the human host. Nevertheless, clinical trials on human patients with ipilimumab and oncolytic herpesvirus talimogene laherparepvec are ready to go (Clinical Trials.gov; NCT01740297). In metastatic prostate cancer, docetaxel chemotherapy added to androgen-depriving hormonal therapy may be additively effective. The Bavarian Nordic avian poxvirus-based prostate cancer vaccine rilimogene galvacirepvec glafolivec increased the survival of chemotherapy-naïve patients with metastatic prostate cancer from 18 months to 31 months; and combined with ipilimumab, to 37 months (Report from the 2015 Genitourinary Cancers Symposium, The ASCO POST 2015;6:1, 3–4). A new generation of oncologists rises to catch up with decades of delays that the practice of cancer immunotherapy suffered, a replica of a natural faculty, undeservedly; while high dose combination chemotherapy (utterly alien to the treated host) was favored. Accept, that combination chemotherapy saved and prolonged many lives. It will have to be kept in use as long as it contributes to survival with acceptable and reversible side effects. Targeted chemotherapy, with especially oncoprotein-targeted therapies (small kinase enzyme inhibitor molecules aiming at a special domain of oncoproteins); and monoclonal antibodies prominently remain in use. Therapeutic cancer vaccines based on live oncolytic viruses will emerge. Genetically engineered oncolytic viruses may surpass in efficacy naturally oncolytic viruses; some of them may replace lost tumor suppressor genes.

The initial natural defense of an organized cell community against malignantly transformed cells within, led by the NK cells and tumor suppressor gene-product proteins of the host, seldom eliminates a tumor without outside help. Tumor biology reveals that the multicellular host may promote the co-existence of internally induced proliferation of transformed cells. In contrast, the host is opposed to the proliferation of externally induced transformed cells. These host immune reactions can be augmented to curative strength. The number of spontaneously regressed tumors, if any, is unknown. The malignantly transformed cells in multicellular, especially vertebrate mammalian, hosts acquired placenta-like defensive reactions, that induce Th2-type tolerance in their host. Further, when the malignant transformation occurs in cells living in a multicellular community, those cells, in addition to their ancient inherited cell survival pathways, have evolved new faculties, that neutralize or eliminate the tumor suppressor proteins of the host. These cells mobilize mesenchymal and vasculo-endothelial cells of their host to render full support to them, while they are undergoing a reversal into the life forms of their primordial ancestors. The genes which frequently undergo fusion in extant malignantly transformed cells are well conserved from their primordial ancestors and frequently function in stem cells (Narsing S et al Department Molecular Medicine, The University South Florida Morsani College of Medicine, Tampa, Fl. Cancer Genetics Cytogenetics 2009;191:78–84). Just practicing an inherent and conserved faculty of the primordial RNA/DNA complex, serving blindly and determinedly for the sustenance of living matter in the Universe.

Appendix 2 Explanations to the Figures*)

*) *In elaborated legends, the author is trying to render service to the readers.*

Figure 1. The A and B X-ray diffraction spatial (3-dimensional) configurations of the double helix deoxyribose DNA molecule are from Rosalind Franklin's lab (1951). The A-DNA B-DNA photographs are from Baianu IC et al. Maurice Wilkins' lab yielded the C configuration in 1958. DNA A and B are natural (derive from living cells); DNA C configuration is an artifact formed in the drying process of a lithium-DNA salt. RNA the predecessor of DNA, is equipped with the 2'OH group on each of its ribose molecules; it frequently functions single-stranded (ss), but in certain circumstances (ds RNA viruses) it can assume a double-stranded (ds) helical structure (A-RNA to Z-RNA). The interferon-induced enzyme dsRNA adenosine deaminase acting on RNA (ADAR) promotes the A-to-Z left-handed helical dsRNA transition; ADAR specifically with high affinity binds to ds Z-RNA/Z-DNA.

RNA molecular constituents adenine, guanine, cytosine, uracil form from formamide under the impacts of high energy laser imitating extraterrestrial bombardments of the ancient Earth over four billion years ago (Ferus M et al J Phys Chem 2014;118:719–736; Proc Natl Acad Sci USA 2015;112:657–662).

Figure 2. Reverse transcriptases encode in Neurospora crassa the ancient mitochondrial Mauriceville plasmid (Chen B & Lambowitz AM J Mol Biol 1997;271:311–332; Baidyaroy D et al Fungal Biol 2012;116:919–931) and the telomeres (Wu C et al Genetics 2009;181:1129–45). The chromosome harbors the Pogo transposable elements. Distinct subtelomeric elements and the TLH telomere-linked helicase are uniquely absent. The biological clock is circadian rhythm-dependent. Longevity versus senescence is determined in the T-loop structure of the TTAGGG duplex region. The ancestral proto-oncogene ras is known in N. crassa as the band (bd) gene; it is re-enforced by the longevity assurance gene lag/LAG. Macroconidia with gain-of-function mutated ras and lag genes live 120 days, versus 24 days median life span physiologically (Case ME et al Ecol Evol 2014;4:3494–3507). More in Figure 63.

J.G. Sinkovics, *RNA/DNA and Cancer*, DOI 10.1007/978-3-319-22279-0

Figure 3. Sex determining region Y-box 2 (SRY) pluripotent transcription factors are active in embryonic stem cells. SOX high mobility group (HMG) proteins express evolutionarily conserved DNA-binding domain (Tables XI/I and XI/II). The human gene sox2 maps to chromosome 3q26.33. The so-called LIF (leukemia inhibitory factor) protein activates sox2 (and Krüppel-like factor 4, Klf4) genes. Proteins Oct4, Nanog and Sox2 maintain pluripotency for their stem cells. These gene product proteins form complexes with protein NPM1 (nucleophosmin; protein B23), a regulator of NFκB. The Sox2 protein inhibits mesenchymal, and promotes neuroectodermal, germ layer differentiation. Stem cells with active Sox2 protein renew themselves and release differentiated neural cell by asymmetric divisions. Hypermethylation of the sox2 gene (or its promoter) deprives the gene of its pluripotent faculty. Amplified (or constitutively expressed) Sox2 protein drives cancers (squamous cell carcinoma lung; high Gleason grade prostate adenocarcinoma; undifferentiated colorectal adenocarcinoma). More in Ctenophores: The cell survival pathways of the comb jelly ctenophores are driven by ancient sox/SOX genes/proteins (see more in the Appendix 1).

Figure 4. Deinococcus radiodurans. Withstanding high dose radiation of canned food, there it was discovered alive and replicating. Its radioresistance is effective against 1.5 million rads. Survives 5000 Gray (500,000 rads) ionizing radiation; upon an exposure of 15000 Gy, it retains 37% of its viability. For comparison, 2.5–5 Gy (250–500 rads) whole body exposure to the human subject is considered to be lethal. D. radiodurans metabolizes nuclear waste. Its ring-shaped DNA does not fall apart upon radiation exposure and retains high sensitivity to its extremely effective DNA repair enzymes. Carries two circular chromosomes 2.65 million base pairs long consisting of 3195 genes; and two plasmids 117000 and 46000 base pairs in size. Presented here in its tetrad form. Its DNA can migrate within the tetrads. The Mycoplasma laboratorium artificially constructed utilizes some D. radiodurans DNA repair enzymes.

Figure 5. The Coelacanth's fins evolving toward hind legs, its gills transforming into lungs and its brain enlarging were postulates read out of its fossils. When the extant coelacanth re-appeared, those postulates had to be seriously doubted and eventually discarded. The transitional pathway from fish to amphibians is characterized by missing links. The maximum parsimony, neighbor-joining and maximum likelihood of the nuclear ribosomal genes 28S of coelacanths and lungfishes indicate that they are monophyletic and are related to the first land vertebrates. The mitochondrial cytochrome oxidase subunit genes' nucleotide to amino acid sequences indicate coelacanth/lungfish and lungfish/tetrapod clades (overruling an immediate coelacanth/tetrapod clade) (Zardoya R & Meter A Proc Natl Acad Sci USA 1996;93:5449–5454; Yokobori S et al J Mol Evol 1994;38:602–609). Intron-containing Interferon gene type I in fish and amphibians, and intron-less IFN gene type III in amphibians (Xenopus) diverged during a transitional period of their hosts migrating from their aquatic environment to land. Intron-less type I IFN genes of reptiles, birds and mammals lost the intron-containing type I IFN genes due to an retroposition event. The receptors of the Xenopus type III IFN are distinct from those of type I IFNs (Qi Z et al J Immunol 2010;184:5038–5046). The ancestors of the amphibians were the sarcopterygii lobe-finned bony fishes (osteichtyes: Latimeria coelacanths; Dipnoi lungfishes), whose pectoral and pelvic paired fins articulate with their pectoral and pelvic girdles using a single bone, that was the ancestor of the humerus in descendent tetrapod amphibians. Their shared 28S rRNA genes make coelacanths and lungfishes sister groups. Of the recombination activation genes rag1/2, one amino acid is deleted from the aminoterminal region of the Rag2 protein in lungfishes, chondrichtyes and amphibians, but not in the coelacanths. This deletion makes lungfishes and amphibians sister groups. The entire rag gene sequences suggest lungfishes to be the ancestors the amphibians, and all tetrapods, in the monophyly of sarcopterygii (Brinkmann H et al Proc Natl Acad Sci USA 2004;101:4900–4905). Illustrations: West Indian coelancant (Latimeria chalumnae) caught in 1974, Comoro Island.

Figure 6. The Amoebozoan, Dictyostelium discoideum exists in many life forms: in sexual cycle, in vegetative cycle, and in social cycle. Its discoidin is a fibronectin; its tiger genes encode histocompatibilty complexes; it uses an ERK/MAPKK signaling "cell survival pathway" (extracellular signal regulated kinase; mitogen-activated protein kinase kinase). MAPKK becomes a proto-oncogene in vertebrate mammalian cells: in ras- (rat sarcoma-) transformed cancer cells. For vital functions of D. discoideum cells, the expression of the ancestral PTEN phosphatase gene is essential; it suppresses phosphatidyl inositol by dephosphorylation, like its ortholog human gene. D. discoideum's cell cycle passes through G1 as if it were non-existent (the Rb protein fails to stabilize it); the protein-product of the retinoblastoma gene ortholog inhibits the S-phase of the cell cycle by interacting with transcription factor E2F (cyclin kinase E2 phosphorylator of the retinoblastoma protein); it also dictates the differentiation of spores (Lusche DF et al PLoS One 2014;9(9):e108498; Strasser K et al PLoS One 2012;7(6):e39914). See more in Appendix 1. Illustration: The Amoebozoan life cycle of Dictyostelium discoideum. Schaap, Pauline, College Life Sciences University Dundee. Evolutionary crossroads in developmental biology: Dictyostelium discoideum. Development 2011; 138: 387–396. doi:10.1242/dev.048934, p.schaap@dundee.ac.uk

Figure 7. The Wnt (wingless; integration site, drosophila; mouse) pathway. DSH = disheveled; FRZ = frizzled; LPR = low density lipoprotein receptor-related protein; APC = adenomatosis polyposis coli; CK = cell cycle; GSK = glycogen synthase kinase; LEF = lymphoid enhancer-binding factor (associated with TCF = T cell factor, proto-oncogenes) The Wnt signaling pathway is ancient. It was and it is operational in the ancestral and in the extant Ciona and Nematostella. It is vital for the developing embryo in regulating cell proliferation and differentiation. In healthy cells, the Axin, APC, GSF proteins regulate β-catenin in the cytoplasm by marking it to ubiquitynation. In malignantly transformed cells (colon cancer cells with deleted APC gene, or otherwise defective "destruction complex"), β-catenin escapes demolition in the proteasome, and transfers from cytoplasm to nucleus.

Being a DNA-binding protein, β-catenin activates proto-oncogenes TCF/LEF. In the genome of comb jelly ctenophores, the "destruction complex" is physiologically defective (lacks Axin), thus β-catenin readily enters the nucleus to activate its target genes (which are predecessors of vertebrate mammalians' proto-oncogenes, tcf/lef); dickkopf, a major inhibitor of the Wnt pathway, is missing. The ctenophore cells physiologically function like Wnt-transformed entities (in the Appendix 1). Jörg Huesken, Walter Birchmeier, Max Delbrück Centrum, Berlin-Buch, Germany.

Figure 8. Staging of colon cancer. Educational videos. Tony Talebi & Caio Rocha Lima at The University of Miami Sylvester Comprehensive Cancer Center, Miami FL http://www.hemonc101.com/colon-cancer. With permission of copyright owner Terese Winslow. All rights reserved.

Figure 9. Warburg glycolysis (Vander Heiden MG, Cantley LC, Thompson CB. Understanding the Warburg Effect: The Metabolic Requirements of Cell Proliferation. Science 2009;324(5930):1029–1033). In osseous metastases of hormone-resistant prostate cancer cells enhanced aerobic Warburg glycolysis is the main source of energy. The process is inhibited by simvastatin (mevalonate pathway inhibitor) and metformin (adenosine monophosphate-activated protein kinase activator) combined, resulting in cell cycle arrest in G1, and leading to the autophagy-apoptosis-necrosis pathway of the tumor cells (Babcock MA et al Cell Death Dis 2014;5(11):e1536).

Figure 10. Neuron with axons and synapses in their surroundings. (Boundless Textbook chapter 62.3 open contents http://en.wikipedia.org/wiki/Neuron).

Figure 11. Biomolecules. Lower left column from bottom up shows the synthesis of amino acids from simple organic molecules. These natural processes were reproduced in the laboratory by Stanley Miller and their occurrence was documented further very recently (Parker ET et al PNAS 2011;108;5526–5531; Orig Life Evol Biosph 2012;41:569–574). The self-replicating and hypercycling polypeptides and the chiroselective peptide replicator are cited from work carried out at the Ghadiri Laboratory, Scripps Research Institute (Lee DH et al Nature 1996;382:525–528; 1997;390:591–594; Saghatelian A et al Nature 2001;409:797–801). The right lower column from bottom up shows the formation of nucleotides (example: cytosine) from simple cyanide molecules. RNA, ribozymes, ribonucleoproteins (RNP), ribosomes and riboswitches naturally existed in the pre-cellular/protocellular world. RNA and polypeptide molecules hypercycled; carbohydrates and lipids formed a-new, but without hypercycling. Vesicle-forming lipids arrived via meteorites (Murchison). RNA strands self-replicated in hypercycles in hydrothermal vents (Eigen M & Schuster P The Hypercycle. Springer Verlag). Arguments are presented for the coexistence of RNA and DNA and the complex formations thereof already in the precellular era. The ingredients of life started to interact in unison in double lipid-walled vesicles (Figure 12). There tRNAs captured amino acids and lined them up in chains. In spheroplast-like protocells, DNA has taken over from RNPs the initiation of protein synthesis. The genomes enriched themselves by capturing new laterally entering genes and then propagated them vertically. Genomes increasing in size were condensed into individual chromosomes, which were separated from, but biochemically interacted with, compartments (especially spliceosomes and ribosomes) formed in the nuclei and cytoplasm. The fusion of crenarchaeal and prokaryotic (mycoplama-like) spheroplasts as equal partners was very likely mediated by a fusogenic virus (phage, marked with red arrow). The ancestor of the Acholeplasma laidlawii phage is the fusogenic suspect (Figures 15–18). The complex large cells phagocytosed proteo- and/or cyanobacteria for ever-lasting intracellular symbiosis in the ancestral animal and plant cells. Viral particles synthesized in the precellular ribosomes invaded the newly formed cells. A mutual exchange of host cell and viral genes had taken place. In phase one, viral particles lost the majority of their genes in favor of the host cell genome, and thus viruses became totally dependent for replication on their host cells. Viruses retained their "virus hallmark genes" not shared with their host cells (Koonin VE, Senkevich O, Dolja VV Biol Direct 2006;1:29). In phase two, viruses are retrieving host cell genes in order to inhibit host cell death (apoptosis) prior to the full maturation of the viral progeny; and present themselves to the host as "self" in order to alleviate host immune reactions. This illustration was presented first in Sinkovics J. Studia Physiologica Scientia, Budapest, 2001. Text ref 2.

Figure 12. Protocells. It is not known how protocells were formed. Gánti's biochemical machine is the chemoton without a genome. Biological units were envisioned as Oparin's coacervats; advanced coacervats became protobionts. Self-replicating and hypercycling RNA/RNPs and polypeptides interacted. Advanced biological units might have formed a conglomerate of adherent spheroplasts. Within the double lipid membrane of the spheroplasts, RNPs catalysed the generation and fusion of nucleotides and polypeptides. Short sequences existing in long non-coding RNA strands were to be transformed enzymatically into cDNAs (creating a natural cDNA library). Ribonucleotide reductase-like proteins converted RNA into deoxyribonucleotides (DNA), either in precellular, or in protocellular ribosomes. If DNA existed in the precellular environment, it managed to enter the protocells together with RNA ribozymes. RNA/DNA complexes with their supportive proteins condensed into chromosomes. Cytoplasmic compartments (precellular ribosomes might have existed) were formed internally, or acquired horizontally from the outside. Horizontal acquisitions and exchanges of genes, even operons, were extensive. The first eukaryotes were fused cells with nuclei. Eukaryotes acquired mitochondria and/or chloroplasts. These organelles engaged in a workable unison with their host cells' nuclei and cytoplasm. Nucleocentrists and cytoplasmists debate the priority of a genome, or a metabolizing cytoplasm with ribosomes in the origin of life pre-cellularly, or in the first protocells. Could any one of these units exist without the other? Figure was first introduced in Sinkovics J, Studia Physiologica Scientia, Budapest, 2001. Text ref 2.

Further developments: Szostak et al generating multi-lamellar vesicles; dividing long threadlike vesicles; protocells with native and synthesized ribozymes. Mann S

Acc Chem Res 2012;45:21231–31241. Jia TZ et al Orig Life Evol Biosph 2014;44:1–12. Schrum JP et al Cold Spring Harb Perspect Biol 2010;(9):a002212. Teresawa H et al Proc Natl Acad Sci USA 2012;109:5942–5947. Sohan Jheeta: Emergence of cellular life from LUCA (Last Universal Cellular Ancestor). Sohan Jheeta: Life (Basel) 2013;2:518-523; 2015;5:1446-1453.

Figure 13. Alpha-Helical Peptides in autocatalytic reactions chiroselectively templated and propagated their own synthesis in hyper- or supercycles. Template-directed condensation of peptide fragments occurs. In the presence of carboyl sulfide (as in volcanic fumes) in aqueous solution amino acids are condensed into peptides. Thioester peptid nucleic acids (tPNA) cross-pair with RNA and DNA strands, and as such interact with side chain functionalities of the molecules. tRNA do not assume self-replicatory reactions (unless a catalyser for such a reaction is found) (Lee DH et al Nature 1997;390:591–594; Saghatelian A et al Nature 2001;409;797–801)

Figure 14. Plasma Alive. Sealed aged bouillon-cultures of E. coli yielded pleuropneumonia-like units ("Pleuropneumonie-ähnliche Eigenheiten") that could regenerate vegetative cells. Through ruptured cell walls, aged E. coli cells released plasma with electron-dense units (bacterial chromosome fragments vs phage particles). Some larger units (>350 mμ) measured by filtration (Ferenc Fornosi's membranes) regenerated only after fusion, as visualized in light, dark field and electron microscopy technology (Iván Hollós). Large (>350 mμ) cytoplasmic units passed through collodion membrane filters of 600 mμ pore size (Ferenc Fornosi's membranes) and regenerated into vegetative E. coli cultures. In dark field- and electron-microscopy, units passing the filter were seen to fuse. Other (fused) units physically attached to the surface of co-cultured Sarcina lutea cells. Regenerating units contained electron-dense formations (bacterial chromosomes? fusogenic phage?) (Sinkovics J Die Grundlagen der Virusforschung Ungarischen Akademie der Wissenschaften, 1956; Acta Microbiol Hung 1957;4:59–75; Nature 1958;181:566–567; Acta Microbiol Immunol Hung 2001;48:115–127; Studia Physiologica Scientia Budapest 2001;9:1–151; Hungarian (Magyar) Epidemiology 2008;5/3–4:237–255; Idem 2009;6:60; Sinkovics JG & Horvath JC Viruses ancient biomolecules. Cancer Research at the Millenium. The University of Texas M.D. Anderson Cancer Center, Houston, TX, 2000; pp 122–123). Cited in I.C. Gunsalus & R.Y. Stanier: The Bacteria, Academic Press N.Y., 1960; vol I: 345–6; A.J. Salle: Fundamental Principles of Bacteriology, 5th ed., McGraw Hill, N.Y. 1961; p 77. Author presented the cell wall-free plasma as a representative of primordial life forms (Sinkovics J: Pre-cellular forms of life in microbiology (in Hungarian). Biológiai Közlemények (Biological Publications) 1954;2:69–81. Gratitude to Elek Farkas† for allowing the conduct of these experimentations in 1954–5 at his Department of Virology in the State Institute of Public Health, (OKI) Budapest.

Figure 15. Individual Cells of Acholeplasma laidlawii Gamaleya PG8 strain release exosomes (Chernov VM Chernova OA et al, Kazan Institute Biochemistry

Biophysics, Russian Academy of Sciences, Kazan, Russian Federation. The Scientific World J 2012; article ID 315474). The 400–600 nm diameter Acholeplasma cells (like other prokaryota) release extracellular vesicles of bilayered membranes in sizes of 70–200 nm diameter enriched in virulence proteins (Chernov VM et al The Scientific World J 2011;11:1120–1130; J Proteomics 3014;110:117–128). Acholeplasma cells carry a single circular chromosome of 1497 kb, 1.5 Mb; its rRNA operons and flanking tRNAs are depicted. A map of its genetic elements is constructed. Glycolysis from D-glucose is its ATP generating pathway (Kube M et al J Microbiol Biotechnol 2014;24:19–36). In addition to its lytic phages, the group 1 acholeplasmavirus is non-cytocidal, non-lytic; the cells remain chronically infected and undergo fusion (Brunner H et al Zbl Bakteriologie A 1980;248:120–128; Liss A & Ritter BE J Gen Microbiol 1885;131:1713–1718).

History of pleuropneumonia Mollicutes and L-forms of bacteria. Sabin AB. The filtrable microörganisms of the pleuropneumonia group. Bacteriol Rev 1941;5:1–67. Sinkovics J. A new chapter in bacteriology: Pleuropneumonia-like growth. Orvosi Hetilap 1955;96:1401–1408 (illustrated). Sinkovics JG. Die Grundlagen der Virusforchung. Ungarischen Akademie der Wissenschaften, Budapest, 1956.

Figure 16. Acholeplasma laidlawii Культура клеток микоплазмы. Ponomareva Anastasiya A. Kazan Scientific Centre Russian Academy of Sciences, Kazan, Russia. The cell membrane of the mycoplasma Acholeplasma laidlawii is cholesterol-free; its genome consists of 1,469,992 base pairs containing 34 tRNA genes and 1,469,992 ORFs; it carries two mRNA operons. Source: gratitude to Anastasiya Ponomareva A.

Figure 17. Nonlytic MVL-3 Viral (Phage) Particles form a chain of single particles aligned alongside the inner aspect of the cell membrane of an Acholeplasma laidlawii cell (Gourlay RN Wyld SG Bland PA. J Gen Virol 1979;42:315–322). Such cells fuse (Haberer K, Maniloff J, Gerling D. J Virol 1980;36:264–270; Brunner H, Schaar H, Krauss H. Zbl Bakteriologie Hygiene I Abt Orig, Gustav Fischer Verlag, Berlin, Jena, Stuttgart, Germany; New York USA.1980;248:120–128).

Demonstration by electron microscopy of intracellular virus in Acholeplasma laidlawii infected with either MV-L3 or a similar but serologically distinct virus (BN1 virus).

Figure 18. Illustration 1: Thin section of two Acholeplasma laidlawii cells attached to each other by one non-lytic MVL3 phage particle before fusion. Fused cells appeared as giant cells. Mature MVL3 particles formed single (or double) lines just beneath the cell membranes, with the viral tails pointed toward the inner cell aspect of the membrane (Haberer K Maniloff J Gerling D J Virol 1980;36:264–270). From the Department of Microbiology, University Rochester Medical Center, Rochester, NY. Publisher: Journal of Virology.

Illustration 2/3: The JA1 Acholeplasma strain harbors a nonlytic phage that fused the cells into large conglomerates. The conglomerates were 20-fold larger than the individual cells. Fusion of mycoplasma cells, not endomitoses, formed the large conglomerates. Longitudinal formations of fused cells extrude long filamentous structures, assume "tumor-like growth" patterns, and retain the enveloped virus particles (Brunner H, Schaar H, Krauss H. Zentralblatt für Bakteriologie Hygiene 1980;248:120–128). Shown here with gratitude to Professor H Brunner.

Further morphological studies on highly polymorph Acholeplasma cells: Polak-Vogelzang AA, Samson RA, De Leeuw GT Canad J Microbiol 1979;25:1373–1380.

Figure 19. T2 bacteriophages attach to the surface of a bacterial cell before penetration of the cytoplasm; the infected cell will be lysed. Phage http://en.wikipedia.org/wiki/File:Phage.jpg by Dr Graham Beards http://en.wikipedia.org/wiki/User:Graham_Beards is licensed under CC BY-SA 3.0 http://creativecommons.org/licenses/by-sa/3.0/.

Bacteriophages of archaea and prokaryota transferred horizontally host cell genes; established temperate non-lytic relationships with some of their hosts; and laid down the basic biochemistry of virus-host relationship (Félix D'Herelle: "Le Bacteriophage", Monographs Pasteur Institut, Masson, Paris 1921; Hugó von Preisz: "Die Bakteriophagie" Fischer, Germany, 1925. Luria SE & Delbrück M Genetics 1943;28:491–511; Lwoff A, by Morange M J Biosci 2005;30:591–594).

Figure 20. CRISPR (clustered regularly interspaced short palindromic repeats). The First Antiviral (Antiphage) Defense of Bacteria CRISPR [2342]. Schematic representation of the crRNA maturation/processing and Cas9-crRNA-tracrRNA complex assembly pathways (see text). (A). The wild-type CRISPR locus is transcribed as a long pre-crRNA molecule. Cas9 promotes pre-crRNA: tracrRNA duplex formation. The tracrRNA /pre-crRNA duplex regions are recognized and cleaved by the host RNase III to generate effector complexes that undergo further trimming at the 5' end by unknown nuclease(s) to produce mature Cas9-crRNA-tracrRNA complexes. (B) Short pre-crRNA transcripts produce functional effector complexes in the absence of RNase III (Karvelis T et al RNA Biol 2013;10:841-851).

Figure 21. Prokaryotes (and archaea) evolved upon a network of laterally transferred genes (Kunin V & C.A, Ouzounis Genome Research 2003;13:1589–1594). In the spheroplast community of the first unicellular microorganisms, the genomes were wide open to gene losses and gains. Discarded or lost genes were replaced by sought after, readily accepted and put-into-service newly acquired genes. Apparently, at that stage, the gene product proteins caused no major immunological conflicts. Genes that existed within the clade were transferred by routes of conjugation (J. Lederberg). Bacteriophages and viruses were the vectors of large transferred genes. Viral cell surface receptors might have posed a restriction, but once in the cytoplasm, the new genes readily entered the chromosomes (since nuclei and nuclear membranes did not as yet exist). Cell-fusogenic viruses must have mightily contributed to genomic transfers (Sinkovics JG. Acta Microbiol Immunol Hung 2001;48:115–121). In case of endosymbiosis in unicellular eukaryotes, the symbionts (mitochondria; plastids) exchanged their genes with those of the host cell nuclei, usually from the symbiont to the host. In many unicellular eukaryotes, there is a genetic micro-, and a somatic macronucleus (MIC/MAC) . The micronuclei carry numerous retrotransposons; these are meticulously excised and expunged when, after cell divisions or conjugations, the genomic micronucleus forms the genome of the somatic macronucleus. This is the line above which the widely spread promiscuity of horizontal gene transfers were curtailed, but never eliminated (Sinkovics JG Int J Oncol 2009;35:441–465; Adv Exp Med Biol 2011;714:5–89). Network of Horizontally Transferred Genes in Prokaryota and Archaea. With kind permission ©Barth F. Smeths PhD All rights reserved 2015.

Figure 22. Phylogenetic and symbiogenetic tree of living organisms. Recommended for reading: Koonin EV. Carl Woese's vision of cellular evolution and the domains of life. RNA Biology 2014;11:197–204. Kutschera U. From the scala naturae to the symbiogenetic and dynamic tree of life. Biol Direct 2011;6:33. doi:10.1186/1745-6150-6-33. Phylogenetic and Symbiogenetic Tree of Living Organisms. Konstantin Mereschkowsky's Symbiogenesis by Cell Fusions in 1910 http://commons.wikimedia.org/wiki/File:Tree_of_Living_Organisms_2.png by Maulucioni http://es.wikipedia.org/wiki/Usuario:Maulucioni and Doridi licensed under CC BY-SA 3.0.

Figure 23. Mitochondria, © AbD Serotec. All rights reserved. The mitochondrial inner and outer membranes undergo permeability transition. In this reaction, ANT (adenine nucleoside translocator) of the inner membranes activates voltage-dependent anion channels in the outer membrane. Water-soluble cytosolic substances enter the mitochondrial matrix. The membranes rupture with the release of cytochrome C and Bax (Bcl-2-associated X protein). Bax blocks Bcl-XL (B-cell lymphoma-leukemia gene-2, extra-large). CARD (caspase recruitment domain) in Apaf (apoptotic protease-activating factor) in the apoptosome releases and cleaves the procaspases. Caspase cascade is initiated with intranuclear fragmentation of the DNA. The mitochondrial internal apoptosis differs from the FasL→FasR-initiated (apoptosis stimulating fragment) cell surface-localized external apoptosis. The bio-philosophical connotations of the apoptotic phenomena ("cell suicide") are widely disputed. In ontogenesis (in the cocoons of insects; in precursor organs of the mammalian kidney, etc.) tissue anlagen are readily dissolved by apoptosis, to be replaced by the permanently functional organ. In malignantly transformed cells, the RNA/DNA complex forcefully inhibits the apoptotic processes. Bcl-2 gene anti-sense inhibitors, microRNA Bcl-2 mRNA inhibitors, or BCL-2 enzyme-protein small molecular inhibitors occasionally overcome the blockade and force the initiation of cancer cell apoptosis. Anti-cancer cell lymphocyte-mediated immune reactions naturally (or by genetic engineering of the lymphocytes' TCR) aiming at apoptosis induction in the cancer cell, referred to in the late 1960s early 1970s as "nuclear clumping" (photographed and discussed in J.G. Sinkovics: "Cytolytic Immune Lymphocytes…" Schenk Buchverlag, Passau/Budapest, 2008).

Figure 24. Chloroplast Stroma. The thylakoid lumen. Phospholipid thylakoid membranes. Of 335 thylakoid proteins, 89 are operational in the lumen. Antenna proteins (photosystems I/II chlorophylls, carotenoids, phycobiliproteins) harvest light. From the chloroplasts, chlorophyll genes were transferred into the cell nucleus. Light-induced water oxidation (splitting) by photosystem II is generating oxygen, protons and electrons in the oxygen evolving complex; protons pumped across the thylakoid membrane from stroma to lumen; electron transport in the cytochrome b67 complex; plastoquinone reduced by ferredoxin; reduced plastoquinol oxidized by cytochrome b67 protein; ATP synthase activated by the proton gradient generates ATP. Proton concentration gradient across thylakoid membrane is 10,000-fold up, as compared to a 10-fold gradient across mitochondrial inner membrane.

Figure 25. Ribosomes. Did precellular sites of protein synthesis exist? Ribozymes and RNA-binding proteins appear to have been in existence and in action. Indeed, a globe-wide network of a Ribonucleoprotein World was fully operational. Ribozyme- and microRNA-packed exosomes released from extant cells (prokaryota; uni- and multicellular eukaryota) represent a form of retrograde evolution to the precellular ribosome level? (Sinkovics JG European J Microbiol Immunol 2015;5:25–43). Viroids are relics of that world. Fusing viroids might have formed the first precellular large viruses. Did DNA join in, or appeared first only in the protocells? What primordial enzymes carried out the deoxyribosylation of the RNA molecule, or could the DNA molecule be synthesized from nucleotides same as the RNA had come to its existence eons earlier? Matthaei and Nirenberg utilized E. coli ribosomes working extracellularly for the discovery of the gene code (Proc Natl Acad Sci USA 1961;47:1580–1586; 1961;47:1588–1602). The RNA-binding proteins (domains, RBD) appear to have existed in the precellular world, predating LUCA, the last universal common ancestor. The RNA methylases of the Rossmann-fold class are assumed to be of pre-LUCA origin. Figure 1 of the cited article shows the RNA-binding THUMP domain (thiouridine synthase, methylase, pseudouridine synthase) being present in LUCA, as it was transferred from there into each one of the three derivative kingdoms. These base N-methylases share the [N/D]PP[Y/F] motif (asparagine/aspartic acid; proline, proline; tyrosine/phenylalanine) at the end of their strand 4 (Anantharaman V, Koonin EV, Aravind L. Nucleic Acids Res 2002;30:1427–1467). For ribosome evolution, see Appendix 1. mRA interactions http://commons.wikimedia.org/wiki/File:mRNA-interaction.png by Sverdrup licensed under CC BY-SA 3.0.

Figure 26. The Argonautes represent highly conserved primordial gene silencing systems. The argonaute (AGO) proteins mediate gene silencing by RNA interference (iRNA) in the RNA-induced silencing complex (RISC). Of four human argonautes (hAGO), only hAGO2 slices naturally RNA, but both hAGO1 and hAGO4 could be transformed into RNA-slicers. The slicer enzyme is guided to its mRNA target by a small interfering siRNA. When inactive, the guide strand rests in the Mid-domain (middle) by its 5' end and in the PAZ domain (PIWI/Argonaute/Zwille, ZLL) by its 3' end. The active guide strand is phosphorylated at its 5' end. The PIWI domain (wimpy testicle, drosophila) is an RNAse. The triad DDH (aspartate- aspartate-histidine) signifies catalysis. The catalytic tetrad is DEDH (aspartic acid, glutamic acid, histidine). In the hAgo1-let-7 complex, nucleotides 21 and 22 of miR let-7 are bound to the PAZ domain. hAgo2 forms complexes with the N domain (duplex unwinding in RISC assembly) and with miR-20a; and the N domain (the "duplex wedge") makes contact with let-7 (letter; lethal). The N domain of hAgo2 activates slicing in hAgo3. The PIWI domain is subject to mutation L674F (leucine, phenylalanine) in hAGO1. The double mutation F676L and Q677P (glutamine, leucine) disabled (no slicing) hAGO2 (Faehnie CR et al Cell Rep 2013;3:1901–1909. W.M. Keck Structural Biology Laboratory; Greg Hannon's Laboratory; Howard Hughes Medical Institute Cold Spring Harbor Laboratory; Leemor Joshua-Tor Laboratory). RNAse endonucleases, the intranuclear Drosha class II (invertebrate Pasha), and the cytoplasmic Dicer class III process (shorten pri-miRs into 60–70 hairpin sequences) and load (processed sequences) into RISC (Laboratory RNA Biology Max Planck Institute Biochemistry, Am Klopferspitz 18, Martinsried 82152 Germany). The ancestors of the green alga, Chlamydomonas reinhardtii, already possessed and forwarded the zwille (zll) genes into plants (Arabidopsis thaliana) to encode AGO proteins (ZLL) that maintain shoot meristem stem cells for new organ (leaves, branches, angiogenesis, flowers) formation in plant embryos. Zll-mutant plants overexpress the erecta receptor (ER) kinase, and its ligands, regulators of vascular cell organization and replication (Czyzewicz N et al J Exp Bot 2013;64:5281–5296; Tucker et al BMC Genomics 2013;14:809; Uchida N & Tasaka M J Exp Bot 2013;64:5335–5343). Downstream of ER, a MAPK cascade drives proliferation of some selected cell clones. The activated ER sequence expresses a multiplicity of introns. The Wuschel homeobox ligands promote vascular cell proliferation (Etchells JP et al Development 2013;140:2224–2234; Karve R et al RNA 2011;17:1907–1921; Meng X et al Plant Cell 2012;24:4948–4960). The AGO proteins engage in small smRNA silencing. Should Ago proteins silence a gene(s) encoding cell proliferation inhibitory factors, they may indirectly promote cell proliferations; should the proliferating cells carry a gain-of-function mutation, those cells may end up as a malignantly transformed stem cell clone. Illustration: Faehnle CR, Elkayam E, Haase AD, Hannon GJ, Joshua-Tor L. The Making of a Slicer: Activation of Human Argonaute-1. *Cell Reports* 2013;3 (6):1901–1909. doi:10.1016/j.celrep.2013.05.033. RISC RNA-induced silencing complex [2343].

Figure 27. MicroRNAs. The intranuclear Drosha releases pri-miRNA →pre-miRNA to the cytoplasm. The cytoplasmic RNAse Dicer generates the duplex dsRNAs (miRNA/miRNA). The corresponding strand within RISC (RNA-induced silencing complex) aligns to the untranslated region (UTR) of the targeted mRNA and cleaves it. miRNAs may act as oncogenes, or as tumor suppressor genes. Federica Calore & Muller Fabbri (Department Molecular Virology, Immunology, Medical Genetics. Ohio State University Columbus, OH 43210).

MicroRNAs and Cancer. Atlas of Genetics and Cytogenetics in Oncology Haematology 2011 pp 1–27. The microRNAs Drosha & Dicer miRNA http://commons.wiikimedia.org/wiki/File:miRNA.svg by Kelvinsong licensed under CC BY 3.0.

Figure 28. Pseudouridylation [2344] H/ACA-box signature hairpin adenine cytosine adenine. Functional in archaea. Appear in first stem loop in Ciona intestinalis. Described in Giardia and Trichomonas. Human box H/ACA analyzed (Chen XS Penny D Collins LJ BMC Genomics 2011;12:550. Kiss AM Jády B Bertrand E Kiss T Mol Cell Biol 2004;21:5797–5807).

Figure 29. Kaposi Sarcoma. The most prominent herpesviral particles in KS cells are those of the KSHV/HHV-8 agents (illustrated in: Sinkovics JG "Cytolytic Immune Lymphocytes…" Schenk Buchverlag Passau/Budapest 2008) (147). The chaperon heat shock proteins mightily contributed to the survival of primordial archaeons, prokaryotes and unicellular eukaryotes in the overheated environment of the ancient Earth. Cellular and virally encoded oncoproteins continue to enlist a special form of HSP for their own protection: the enriched-in-tumor-cells etHsp90 (Moulick K et al Nat Chem Biol 2011;7:818–826). These transformed cells successfully practice anti-apoptotic defense and encode proto-oncogenic pathways (NFκB; PI3K/Akt; Bcl-2). Antagonized are FLIP, FLICE-inhibitory proteins, FADD-like IL-1β converting enzyme; Fas-associated death domain; Fas, fragment apoptosis signaling. Tumor cells induced and/or promoted by gammaherpesviruses Epstein-Bar virus (HHV4) and/or Kaposi sarcoma-associated herpesvirus (HHV8) encode oncoproteins that engage etHsp90 for their own protection. In primary effusion lymphoma (PEL), KSHV/HHV-8 with or without EBV, with or without HIV-1, is the leading pathogen (Matolcsy A Nádor RG Cesarman E Knowles DM Am J Pathol 1998;153:1609–1614). In PEL cell lines and in human PEL cells grown in mice, investigational small molecular inhibitor PU-H71 destabilized the oncoprotein protective effect of etHsp90 on KSHV/HHV8-induced oncoproteins. The BCL2 oncoprotein antagonist obatoclax (patented GX15-070) synergized in PEL cell-killing through autophagy and apoptosis with the patented PU-H71 etHsp90 antagonist (Najar U et al Blood 2013;122:2837–2847; Ojala PM Blood 2013;122:2767–2768). Herpesvirally infected cells often induce the release of latent retroviruses. This is very evident in KSHV/HHV8-infected KS cells of endothelial cell origin, even in the classical Mediterranean KS cell lines. The released retrovirus particles are not those of HIV-1, or related viruses, neither morphologically, nor immunohistochemically. In immunohistochemistry, the KS-related retroviral structural proteins were not identical with those of the human T cell lymphotrophic retrovirus (HTLV-1), but were related to those retroviral particles that occasionally appeared in Reed-Sternberg cells of Hodgkin's disease (F Györkey† unpublished). In indirect immunofluorescence assays, patients with KS yielding these cell lines appeared to have circulating antibodies reacting with budding retroviral envelope antigens (Sinkovics JG. Cytolytic Immune Lymphocytes… Schenk Buchverlag, Passau/Budapest, 2008; Sinkovics JG & Horvath JC in "Kaposi sarcoma a breeding ground of herpesviruses" Int J Oncol 1999;14:615–646, Errata: Int J Oncol 2005;27:5–47) (Figure 30).

Figure 30. Endogenous Retrovirus Activated in KS. Herpesvirally infected cells often induce the release of latent (endogenous) retroviruses (reviewed in Sinkovics JG Acta Microbiol Immunol Hung 2010;57:253–347). This is very evident in KSHV/HHV-8-infected KS cells of endothelial origin. The released retrovirus particles are not those of HIV-1, or related viruses, neither morphologically, nor immunohistochemically. In immunohistochemistry, the KS-related retroviral structural proteins were not identical with those of the human T cell lymphotrophic retrovirus (HTLV-1). Immunohistochemical relationship with retroviral particles budding from Reed-Sternberg cells of Hodgkin's disease was found (unpublished, F Györkey †). In indirect immunofluorescence assays, patients with KS yielding these cell lines circulated antibodies reacting with budding retroviral envelope antigens (Sinkovics JG "Cytolytic Immune Lymphocytes…" Schenk Buchverlag, Passau/Budapest 2008; Sinkovics JG & Horvath JC Kaposi sarcoma… herpesviruses. Int J Oncol 199;14:615–646; Errata: Int J Oncol 2005;27:5–47) (Figure 29). Illustration. A male patient of the Veterans' Administration Hospital, Houston, TX, originally diagnosed with classical Mediterranean KS, that later turned out to be AIDS-associated, yielded cells of sarcomatoid morphology in primary tissue culture. The cells concealed herpesvirus-like particles, but yielded budding retroviral particles. The retroviral particles were neither those of HIV-1 nor HTLV-1. The patient circulated immunoglobulins reacting in indirect immunofluorescence assays with the envelopes of the budding retroviral particles. First published in Sinkovics JG et al Orvosi Hetilap 2006;147:617–621; re-published as Fig 4.18a, b in Sinkovics JG: "Cytolytic Immune Lymphocytes…" Schenk Buchverlag, Passau/Budapest, 2008 (147).

Figure 31. Entamoeba histolytica [2345] trophozoite by confocal microscopy. Endoplasmic reticulum stained green (Parasitology Research Springer Link). Genome clues (Matt Berriman, Wellcome Trust Sanger Institute). Complete genome sequencing for E. histolytica is available. The intracellular parasite Entamoeba invadens operates extrachromosomal ribosomal RNA genes. Adaptations to different hosts (E. invadens to reptiles) results in non-circular transcripts of ncRNAs deriving from the rDNA locus (Northern hybridization). In contrast, human-adapted intracellular E. histolytica works with circular transcripts (Ojha S et al Mol Biochem Parasitol 2013;pii:S0166-6851(13)00150-3). The large number of protein phosphatases propagating intracellular signals makes up to 3.1% of the proteome of E. histolytica (Anwar T & Gourinath S PLoS One 2013;8(11):e78714). E. histolytica evolutionarily conserves polyadenylation sequences in its mRNAs 3'-UTRs. Some of its mRNAs are subjected to decay by iRNAs (López-Camatrillo C et al Wiley Interdiscip Rev RNA 2013 Nov 8). Under oxygen-restricted conditions, the aerobic ATP-generating mitochondria of E. histolytica undergo transformations into mitochondrion-related organelles (mitosomes, hydrogenosomes) (Makiuchi T

& Nozaki T Biochimie 2014;10:3–17). The amoeboid locomotion is comparable to that of tissue-invading cancer cells (see text).

Figure 32. *Tetrahymena* Molecular Biology I. Twi12 association with tRNA fragments. (A) The schematics show the strategy of ZZTwi12 transgene knock-in and endogenous TWI12 locus knockout; the latter is evidenced with Southern blot data using NheI-digested genomic DNA at right. The position of the Southern blot probe is indicated in the TWI12 locus schematic with a thin gray line. (B) RNAs copurified with Twi12 in vegetative growth (veg) or after 12 h of starvation (st12) were stained directly with SYBR Gold in comparison with size-selected total RNA from the same cultures. (C) The composition of sRNA library sequencing reads is depicted for reads with either a perfect match to the genome or a single internal mismatch as indicated. (D) Read numbers for tRNA fragments copurified with Twi12 were plotted relative to tRNA gene copy number as an approximation of full-length tRNA abundance. The labels of some data points at left were removed for clarity. mtGluTCC, SecTCA, ThrCGT, mtHisGTG. Pseud indicates combined reads from the 11 annotated tRNA pseudogenes, and mt indicates mitochondrial. This verbatim quotation from the article is very complex. The reader is advised to print out the entire text of the article (Couvillion MT, Sachidanandam R, Collins K. Department of Molecular and Cell Biology, University of California, Berkeley, CA. Genes Dev 2010;24:2742–2747) (125b).

Figure 33. *Tetrahymena* Molecular Biology II. Retention of the tRNA 3' fragment as the tightly bound strand. (A) Blot hybridization of Twi12-bound tRNAs was performed using probes complementary to the 5' or 3' end of several abundant tRNAs. Direct SYBR Gold staining of Twi12-bond sRNAs is shown at left for reference. (B) Arrowheads and lines indicate the potential processing positions and detected sRNAs, respectively, for two of the tRNAs from A. The 3' CCA nucleotide of a mature tRNA are not illustrated. (C) Twi12-associated sRNAs were directly stained after RNP isolation using two different gentle, nondenaturing buffer conditions. (D) Twi12-associated sRNAs were 5' end-labeled after isolation using stringent purification conditions, with or without prior cross-linking; different concentrations of urea were used for washes prior to RNA extraction. (E-G) Twi12-cross-linked sRNAs washed with 2 M urea were used for blot hybridization to detect specific tRNA fragmentation (Shown in E) and for library construction. The length distribution of sequencing reads (F) is shown for 3' fragments that match the genome perfectly or have untemplated 3' C, CC, or CCA, with the combined nucleotide frequency at each position (G) revealing a 5' sequence signature. This verbatim quotation from the article is very complex. The reader is advised to print out the entire text of the article (Couvillion MT, Sachidanandam R, Collins K Department of Molecular and Cell Biology, University of California, Berkeley, CA. Genes Dev 2010;24:2742–2747) (125b).

Figure 34. The Synovial Sarcoma fusion oncoprotein was recognized as New York NY-ESO cancer testis class (CTA) antigen (Ayyoub M et al Cancer Immun 2004

4:7). The fusion oncogene t(X;18)(p11.2;q11.2) was originally described in A.A. Sandberg's laboratory in the 1980s. The cadherin-binding protein, β-catenin (β-C) connects the cytoplasmic tail of E-cadherin to the actin cytoskeleton. When β-catenin translocates into the nucleus, the sarcoma cell assumes spindle cell morphology. With β-C remaining in the cell membrane and cytoplasm, the sarcoma cell exhibits the epithelioid phenotype. The SSX1→Snail zink finger fusion protein is a strong antagonist of the repressor of the E-cadherin promoter. The SSX2→Slug fusion protein is a weak antagonist of the repressor of the E-cadherin promoter. When the repressor of the E-cadherin promoter prevails (suppressing E-cadherin), synovial sarcoma cells assume the monophasic spindle cell state (reviewed in Sinkovics JG Expert Rev Anticancer Ther 2007;7:31–56). Cadherin genes and proteins (lefftyrin, coherin, hedgeling) appeared first in the last common ancestor of choanoflagellates and metazoans, and in the single celled choanoflagellates (C. owczarzaki; Salpingoeca rosetta), before the divergence of the metazoan lineage. The sponge, Oscarella carmela possesses cytoplasmic β-catenin-binding cadherins with cytoplasmic tail (Nichols SA et al Proc Natl Acad Sci USA 2012;109:13046–13051). The basic evolution of cadherins predated that of the metazoans. The pre-metazoan pre-choanoflagellates' cadherins can not be recognized beyond the first bilaterians. The cadherins in the cytoplasmic bridges of S. rosetta are different from the metazoan cadherins. Tyrosine kinases appeared first in C. owczarzaki as ligands in search of receptors; their receptors appeared later in single cells of multicellular metazoans (Suga H et al Sci Signal 2012;5(222)ra36). Illustration : Nichols SA Department of Biological Sciences, University of Denver, Denver, CO 80208; Roberts BW Department of Molecular and Cell Biology, University of California, Berkeley, CA 94720. Within the Wnt pathway (Figure 7), the position of Dishevelled (Dsh) at the animal pole of the female pronucleus in the unfertilized egg of the sea anemone (Nematostella vectensis) is opposed to the vegetal pole, where Dsh is degraded. The position of Dsh determines the site of gastrulation, while β-catenin stabilizes the endoderm (Lee PN et al Dev Biol 2007;310:169–86). The ancient physiologically installed β-catenin becomes a DNA-binding proto-oncogene activator protein in the vertebrate mammalian hosts within the Wnt oncogenic pathway [2346].

Figure 35. Medical History. In the late 1960s, a middle aged male patient MDAH#73587 yielded established cell lines from his large locally invasive but not known to be metastatic chondrosarcoma of his right hip. His blood contained circulating small compact T lineage lymphocytes (later immune T cells), which promptly surrounded his tumor cells and lysed them in vitro. A healthy control individual (this author, JGS) yielded small compact circulating lymphocytes (later: immune T cells), which practiced emperipolesis over the patient's allogeneic chondrosarcoma cells, but refrained from attacking them. Instead, the healthy control released large granular lymphocytes (later: natural killer NK cells), which firmly attached to the allogeneic chondrosarcoma cells and eventually lysed them (Sinkovics JG Pathol Oncol Res 2004;10:174–187). NCI project site visitors considered that the healthy donor rejected a latent incipient chondrosarcoma (without knowing it). The healthy male donor (JGS) then documented that his blood contained large granular lymphocytes, which attached to and lysed female breast cancer and ovarian carcinoma cells (Sinkovics JG, Reeves WJ, Cabiness JR. J Natl Cancer Inst 1972:48:1145–1149; Proc 63rd Ann Meet Am Assoc Cancer Res 1972; #57). Then the NCI project site visitors declared that the chamber-slide assay in use produced "in vitro artifacts", since no specific immune reactions should occur without preimmunization and that such reactions were not observed in the Hellströms' laboratory. However, that laboratory used a colony inhibitory assay for gross inspection, and without the actual visualization of individual attacker lymphocytes. In contrast, the chamber-slide assay in the author's Lab visualized and clearly individually distinguished the attacker lymphocytes (in Appendix 1). Illustration: Figures 1–2. The chondrosarcoma cell line, advanced passages, in the late 1960s. Figures 3–4. The patient's small compact T lymphocytes collected and purified from peripheral blood killed in vitro autologous chondrosarcoma cells in 24–48 h. Figures 5 and 6. Healthy donor's (JGS) small compact lymphocytes practice emperipolesis, but refrain from attacking allogeneic chondrosarcoma cells. Large granular lymphoid cells (arrows) firmly attach to, and eventually lyse, allogeneic chondrosarcoma cells. These pictures are among the first photographs taken in the Sinkovics Lab of "large granular lymphocytes" attacking allogeneic human tumor cells without any pre-immunization (J.G. Sinkovics et al in: "Leukemia-Lymphoma" M.D. Anderson Hospital 14th Clinical Symposium in 1969. Year Book Medical Publishers, Chicago, 1970; Cancer 1971;27:782–793; The 7th San Francisco Cancer Symposium, Front Radiat Ther Oncol 1972;7:141–154). Further medical historical event: upon first sight of the large granular lymphoid cells (later: NK cells), chief pathologist James Butler suggested that they looked "monocyte-like" (even though the large granular cells failed to phagocytose bacteria) (Figure 27 citation) (278).

Illustration: Reproduction of microphotographs from the early 1970s from J.G. Sinkovics In: "Cytolytic Immune Lymphocytes" Schenk Buchverlag, Passau/Budapest 2008, page 120. 10.10. (147) Patient's autologous lymphocytes attack chondrosarcoma cells and induce "nuclear clumping" (later: apoptosis). Several attacker lymphocytes lose their morphological integrity. 10.11. Autologous small compact lymphocytes (later: immune T cells) kill by cytoplasmic lysis squamous carcinoma cells of the nasopharynx. 10.12. Two autologous immune T cells kill by total disintegration a melanoma cell. 10.13 and 10.14. Immune T cells of patients with rhabdomyosarcoma attack and lyse allogeneic rhabdomyosarcoma cells. Using immune T cells (and not NK cells) in an allogeneic encounter indicates tumor-specific immune reaction to a shared tumor antigen in rhabdomyosarcomas.

Figure 36. Medical History concerning NK cells. Small compact lymphocytes (recognized later as immune T cells) of the patient with chondrosarcoma surrounded and killed the autologous tumor cells (in color). Of the control lymphocytes deriving from this author (black and white), the small compact lymphocytes refrained from attack, instead large granular lymphocytes (later recognized as natural killer NK cells) attached to the tumor cells to eventually lyse them (see previous Figure 35). Reviewed with dates of the photographs cited (Sinkovics JG Pathology Oncology Research 2004;10:174–187). The tumor antigens, that autologous immune T cells or autologous and allogeneic NK cells were attacking, were not identified (presumed were amplified fetal antigens; mutated oncoproteins; NY-ESO antigens; HiWi antigens).

Illustrations: 1, Large granular lymphocytes (NK cells) and small compact lymphocytes (immune T cells) of a patient (LN, MDAH#90641) with metastatic liposarcoma in remission after receiving chemoimmunotherapy including viral oncolysate (VO) added, attack in vitro a sarcoma cell deriving from the allogeneic sarcoma cell line #3743 from which the VO was prepared (Presented at the 22nd Annual Clinical Conference at M.D. Anderson Hospital in 1978; cited in ref 147). In Illustration 2, a graph prepared by H David Kay and Harikishan Thota in 1971 in the Sinkovics Lab clearly shows that the healthy author's (JGS) large granular lymphocytes killed allogeneic tumor cells in vitro: an immune reaction without pre-immunisation (30c). Reprinted in the "Atlas. Effectors of Anti-Tumor Immunity" by M Kiselevsky. Springer Verlag, 2008.

Figure 37. The Hydra. The entire genome of the hydra (H. magnipapillata; H. vulgaris) has been sequenced. These diploblasts lack mesoderm, but the genes ancestral to the generation of mesoderm (snail, twist and their high mobility group protein forkhead, FoxA), are present. Other genes "in the waiting" are those of the photoreceptors, may be ancestral to the eye-encoding pax6/Pax6 cluster to come. Homologs of the vertebrate stem cells (Klf4, Oct4, Sox2, Nanog) are not detectable, but cells interstitially located in the parenchyma are multipotent stem cells. The cnidaria hydra possess and express the LIM protein complex (LIM homeobox Table XVI). LIM and ROCK (Rho-associated coiled coil kinase) proteins drive the amoeboid locomotion in human cancer (breast and prostate) cells. Of the six LIM homeobox clusters, the Hydra utilizes four (Srivastava M et al BMC Biol 2010 18;8:4). The hox/Hox gene/protein clusters of bilaterians to follow later are predated by a very active Wnt signaling pathway (both Disheveled and β-catenin). The Wnt system encodes the regeneration of mutilated body parts, both in Hydra, Hydractinia and Nematostella vectensis. The bone morphogenetic pathway (BMP2/4 and TGFβ/Dpp transforming growth factor decapentaplegic, drosophila) and its antagonists chordin and gremlin, are operational in the Hydra/Hydractinia, even though morphological symmetry in these hosts is not yet present. This system will encode the Spemann-Mangold organizer in the newt (*vide infra*). Plasmid DNA injections created transgenic Hydra and Hydractinia species (Trevino M et al Dev Dyn 2011;240:2673–2679; Technau U & Steele RE Development 2011;138:1447–1466). In the evolution of its genome, the hydra experienced invasions of transposable elements, lateral gene transfers (some large bacterial genes inserted) and the retention of stem cells for its physiological ontogenesis, may be for its reversal from mature to immature cells, and for the rapid regeneration of lost body parts (like its head) (Chapman et al Nature 2010;464:592–596.)

Figure 38. The Spemann-Mangold Organizer [2347]. Cells of the gray crescent in the gastrulating amphibian (frog; newt) embryo form the notochord. Cells of the gray crescent of the newt blastula transplanted ventrally into another newt blastocoel induced the development of another notochord located ventrally. It was not the transplanted gray crescent cells that grew into the second notochord; the transplanted gray crescent cells induced the recipient host's ectodermal cells to develop into notochords. Otherwise host bone morphogenetic proteins (BMP) would have converted the ectodermal cells into skin cells (epidermis). The transplanted gray crescent cells produced the host BMP antagonist proteins chordin and noggin, that prevented the trans-speciation of ectoderm into epidermis; thus, in absence of BMP, the ectoderm continued its predetermined physiological course to form notochords and central nervous systems. Other factors interacting, especially in the drosophila (insects, crustaceans, arachnids, annelid worms), are ADMP (anti-dorsaling morphogenetic protein), ACVR2A (activin A receptor type 2A), protein Nodal, and its antagonists, proteins Cerberus and Lefty. Another Nodal and Wnt antagonist is DKK-1 (dickkopf); SOG (short gastrulation) interacts with SMAD (signaling mothers against decapentaplegic, dpp, TGFβ). In the drosophila, the dorsally acting dpp (TGFβ) is blocked by the ventrally acting SOG. Insects develop ventral nerve cords (Inui M et al Proc Natl Acad Sci USA 2012;109:15354-15359; Spemann H & Mangold H (neé H. Pröscholdt: Transplantation Versuche… Über Induktion von Embryonalanlagen durch Implantation…Universität Freiburg-Breisgau). Induction of embryonic primordia by implantation of organizers from a species, 1923-4. Reprinted Int J Dev Biol 2001;45:13–38)

Figure 39. Caenorhabditis. A nematode living in the soil and feeding on bacteria with a difficult name to apply: caenos for recent, rhabdos for a wand (slender stick). Hermaphrodite females (XX) in vast majority of the population consist of 959 individually recognized cells; extremely rare males (X0) operate 1031 cells. The first totipotent cells of the embryo hatching from the eggs are asymmetrically dividing stem cells. In ontogenesis, some cells are eliminated by apoptosis. The 100-million base pairs C. elegans genome is condensed into six chromosomes; mitochondrial genomes further contribute. Individual genes frequently function arranged in operons. Distressed (starving) larvae enter the dauer stage: resist stress without ageing. Orthologs of the anaplastic lymphoma kinase (alk/ALK) proto-oncogene and TGFβ (DAF7) signaling establish the state of dauer in the larva. The scd gene (suppressor of constitutive dauer) fails to oppose. Further pathways activated in the dauer state are the IGF, TGFβ, SMAD, and v-Ski (Sloan-Kettering avian sarcoma retroviral isolates). Is the state of dauer in a larva analogous to a human pediatric tumor? Some mRNAs are destroyed by RNA interference (iRNA). There is H3K4 methylation and H4 acetylation. The miRNA let-7 (lethal, letter 7) induces cell differentiation and in vertebrate mammalian hosts is a suppressor of ras/Ras oncogenes/oncoproteins (deletion of let-7 may be lethal). In the healthy caenorhabditis, let-7 and argonaute (argonaute-like gene and protein, alg/ALG) synergize in the elimination of potentially harmful mRNAs. Let-7 is essential for the differentiation of the uterus and vulva. Let-7 deficient nematodes experience the replication of undifferentiated vulvar cells and the rupture of the hypercellular organ (a lethal outcome). Are these "ulcers that do not heal" cancers? Insulin-like growth factor-(IGF-) blocked individuals extend their life-span three-fold. Wild Caenorhabditis strains are infected with positive-sense RNA nodaviruses (Le Blanc, Orsay, Santeuil strains). A state of viral carriership is established (Franz CJ et al J Virol 2012;86:11940. González-Aguilera C et al Brief Funct Genomics 2013 Dec 10. Hunter SE et al PLoS 2013;9(3):e1003353. Reiner DJ et al Curr Biol 2008;18:1101–1109. Gal, TZ; Glazer, I; Koltai, H (2004). "An LEA (protein late embryogenesis abundant) group 3 family member is involved in survival of C. elegans during exposure to stress". FEBS Letters 2004;577: 21–26. doi:10.1016/j.febslet.2004.09.049. PMID 15527756. http://en.wikipedia.org/wiki/File:Caenorhabditis_elegans_hermaphrodite_adult-en.svg by KDS444 licensed under CC BY-SA 3.0.

Figure 40. SOX2 Gene amplification and protein overexpression foretell more favorable clinical course of squamous cell lung cancer. Strong nuclear immunohistochemistry in squamous lung carcinoma cells in a sample (a). Lung squamous carcinoma cells with amplified and non-amplified Sox2 (b). No SOX2 amplification in tumor cells (c). Normal bronchial epithelial cells show limited confined SOX2 expression (d) (Wilbertz T et al, Mod Pathol 2011;24:944–953).

Figure 41. The Human Stem Cell Transcription factor SOX2 protein is encoded from chromosome 3q26.3-q27. The intranuclear SOX2 protein binds a receptive high mobility group (HMG) protein DNA domain (Tables XI/I and XI/II). Co-expressed with embryonic stem cell factors NANOG and Oct3/4, the SOX2 protein reprograms terminally differentiated cells into pluripotent cells. Certain human parenchymal cells expressing high levels of amplified SOX2 protein assume constitutive and incessant mitotic activity and grow into tumors with propensity to be invasive and metastatic. Sinonasal undifferentiated and inverted papillomatous carcinomas were driven by highly amplified SOX2 oncoprotein. The relapse rates of these tumors were increased. The early expression of SOX2 protein indicates inductive rather than promotional role for SOX2. SOX2-positive tumors developed more lymph node metastases than SOX2-negative tumors (Schröck A et al, PLoS One 2013;8(3):e59)

Figure 42. Notch [2350]. Gene mastermind is a notch signal effector in the drosophila. The human homolog of the drosophila notch gene is TAN-1 (T cell leukemia translocation-associated Notch homologue). The translocation is at t(7;9)(q34;q34.3). The locus 9q34.3 is homologous with the drosophila notch gene. The constitutively activated gene encodes the overproduction of gamma-secretase. The extracellular domain of TAN-1 is EGF-like; the intracellular domain is ankyrin-rich and as such is inhibitory to DNA-binding of the p50 subunit of NFκB. Exceptional IkB protein Bcl-3 acts like TAN-1. TAN-1 and Bcl-3 reverse transcriptional inhibition imposed on them by high levels of p50 dimers. In this case, TAN-1 promotes NFκB-dependent gene expression resulting in T cell activation; so does a truncated

TAN-1 protein (Aster J et al Cold Spring Harb Symp Quant Biol 1994;59:125–136. Ellisen LW et al Cell 1991;66:649–661. Guan E et al J Exp Med 1996;183:2025–2032. Pear WS & Aster JC Curr Opin Hematol 2004;11:426–433). Porphyromonas gingivalis induces proliferation of vascular smooth muscle cells by activating TGFβ and Notch signaling (Zhang B et al BMC Genomics 2013;14:770). See Notch activation in pediatric brain tumors (ependymoma) in the text.

Figure 43. An Annelid. The ancestral bilaterian protostome marine annelids are represented by Platynereis dumerilii. The triploblastic polychaete, P. dumerilii displays a far advanced neuropeptid network. Its Wnt and FGF ligand-to-receptor systems are highly active in its embryo. Its germ cells provide the progeny; the stem cell compartments lay down the organs consisting of somatic cells. No uncontrolled constitutively replicating stem cell clones (tantamount to tumors) have so far been reported (Conzelmann M et al BMC Genomics 2013;14(1):906; Denes AS et al Cell 2007:129:277–288). The achaete scute neurogenic genome (see Nematostella; Amphioxus) in its vertical evolutionary ascent uses P. dumerilii as one of its stations (Simionato E et al BMC Evol Biol 2006;8:170). Illustration: Unpublished illustration of Dr Nicole Rebscher, University of Marburg, Germany; shown with gratitude.

Figure 44. Viral Oncolysates (VO) with, or without chemotherapy, were effective cancer vaccines before the era of molecular virology. H Koprowski and J Lindenmann independently prepared viral lysates of cultured tumor cells and used the cell-free lysates for immunization of tumor-bearing mice. Viral oncolysates contained the live oncolytic virus and a mixture of unidentified tumor antigens. Some of the antigens were recombined proteins of the viral and cell membrane envelopes/capsids. As such, these antigens were immunogenic in the hosts that would not have responded to their own self cell surface antigens. Tumor-bearing mice, rendered tumor-free upon immunization with viral oncolysates, specifically rejected challenge with large doses of live cells of the same tumor (oncolytic virus-free inocula) that they were bearing, but rejected upon pre-immunization with their viral lysates (Koprowski H et al Texas Rep Biol Med 1957;15:11–128; Lindenmann J & Klein PA. "Immunological Aspects of Viral Oncolysis". Springer Verlag, Berlin-Heidelberg-New York, 1967). WA Cassel (Emory University, Atlanta, GA), and JG Sinkovics (University of Texas MD Anderson Hospital, Houston TX) prepared viral oncolysates in the 1970s for clinical use in patients with metastatic melanoma (Cassel WA, VO in itself without chemotherapy), and metastatic melanoma and sarcoma (Sinkovics JG, in combination of VO with chemotherapy). In not prospectively randomized patient populations, both investigators reported partial remissions with prolongation of life. Cassel and Murray DR excelled in being able to record durable complete remissions. JG Sinkovics and Plager C were first to report synergism of VO and chemotherapy, manifesting in prolongation of remission duration and life over chemotherapy alone (vide infra). These investigators concluded that viral oncolysates first performed viral oncolysis, and then tumor cells undergoing lysis induced anti-tumor immunity. Sinkovics,

Cabiness J, Kay HD, Thota H, and clinical associates reported antibody and lymphocyte-mediated immune reactions quantitatively increasing in responding (and in non-responding) patients (Cassel WA et al In: "Viral Therapy of Human Cancers" edited/authored by JG Sinkovics & JC Horvath, Marcel Dekker, New York, 2005; Pp 677–689; Sinkovics JG "Cytolytic Immune Lymphocytes…" Schenk Buchverlag, Passau/Budapest, 2008; Sinkovics J G & Horvath JC Arch Immunol Ther Experiment, Wroclaw, Poland; Birkhäuser Verlag, Basel 2008;56S:1–59). Citations from other investigators using viral oncolysates for human cancer therapy (P Hersey; MK Wallack; V Schirrmacher) are provided (in "Viral Therapy of Human Cancers", vide supra). Illustration: Tumor cells are shown undergoing viral lysis by PR8 influenza A virus, or NDV, before removal of all cellular debris. The final cell-free products were reached by carrying out a rigid triple-controlled protocol, providing a cell debris-free supernatant. Dieter Gröschel ascertained bacteriological sterility (including mycoplasma) and freedom from endotoxins (by limulus assay) of the final preparations. Not UV-irradiated preparates were used for human inoculations. However, even UV-irradiated VO preparations contained residual live oncolytic viruses (Horvath JC, Horak A, Sinkovics JG. Proc Am Assoc Cancer Res 86th Ann Meet Toronto Canada 1995;36:493#2619). Autologous viral oncolysates were preferred over allogeneic preparations. Non-prospectively randomized patients in need of treatment were accepted and were tested clinically and by laboratory technology for response (or failure thereof) (Sinkovics JG & Horvath JC Int J Oncol 2006;29:765-777).

Figure 45. Coelomocytes [2351]. Innate immune coelomocytes of sea urchin Strongylocentrotus purpuratus. The polygonal P cells possess Sp-185-333-positive perinuclear vesicles. The small S phagocytic cells are Sp-185-333 positive. The discoid D phagocytic cells are Sp-185-333 negative (Brockton V et al George Washington University, Washington DC Cell Sci 2008;121:339–348). Progenitors of the variable leucine-rich repeats receptors (LRR/VLR) of the lamprey are present in the sea urchin genome in addition to 222 Toll-like receptors (TLR). The Toll-interleukin 1 receptor (TIR) and the MYd88 (myelogeous) domains are functional. IL-1 and IL-17 and their receptors are present. The recombination activator genes and proteins rag/RAG1/2 and the Immunoglobulin variable domain V genes are present for an unidentified (not adaptive immunity) function. The Transib family transposons' DNA inserts encoded first the V(D)J/RAG1-2/RSS complex (variable/diversity/joining; recombination activating genes; recombination signal sequences). In the genomes of the sea urchin, amphioxus, hydra and starlet sea anemone (Nematostella) elements of the system occur singly and unlinked to one another. The entire system worked in unison first in the sharks (chondrichtyes) (Kapitonov V & Jurka J PLoS Biol 2005;3(6):e181; Fugmann S et al Proc Natl Acad Sci USA 2006;103:3728–3733; Rast JP et al Science 2006;314:952–956). The herbicide glyphosate induces DNA-damage in sea urchin embryos, however no excessive abnormal cell proliferations resulted (Bellé R et al J Soc Biol 2007;201:317–327). The sea urchin telomerases, contrary to the rules that invariant identity of crucial catalytic enzymes maintains cell survival lengths, are highly hypervariable (Wells TG etal Mol Biol Cell 2009;20:464–480). Yet the life span of individual sea urchins extends 100 years.

Figure 46. The urochordate Botryllus (B. schlosseri) lives in colonies, in which individual members of the colony share a tunic and a circulatory system within it. In the haemolymph, one of the circulating cells expresses recognition toward self versus non-self. A colony approaching to fuse with a sedentary colony, may be recognized as non-self. The intruder cells will be killed by a process of dissolution. The process of rejection leaves behind scars in each colonies. The cells responsible to kill the allo-incompatible colony are the ancestors of the vertebrate mammalian natural killer (NK) cells (Khalturin K et al Proc Natl Acad Sci USA 2003;100:622–627). Poster presentation at M.D. Anderson Hospital's Symposium "Cancer Immunity: Challenges for the Next Decade" by Sinkovics JG, Horvath JC, Kay HD on "From the origin of NK cells in the Botryllus to their interaction with syncytiotrophoblasts in the placenta" Abstracts 2003; #1–33:117. The fusion-histocompatibility gene locus consists of two cFuHC genes, one secreted, one membrane-bound (both referred to as fosmid). A third related gene encodes the Botryllus histocompatibility factor (BHF), representing the faculty of the general tunicate allorecognition. Colonies able to fuse must carry histocompatible BHFs. Cells of blood vessels express BHF the most. The reaction of fusion versus rejection is determined by BHF in an as yet unknown mechanism. In the fused chimeric colonies, the stem cells of each colony strive to take over by eliminating the other colonies stem cells. Reconciliation in a mixed genotype is possible (Voskoboynik A et al Science 2013;341:384–387). Observe tunicate genes organized into operons and transcribed in single mRNAs in contrast to amphioxus and vertebrates operating operon-free genomes (Holland LZ J Exp Zool B Mol Dev Evol 2014. doi:10.1002/jez.b.22569). Addition in Appendix 1.

Figure 47. Of the AID/APOBEC enzymes [2352], the activation-induced cytidine deaminase (AID) converts cytosine to uracil. This enzyme is active in somatic hypermutation and class-switch recombination in adaptive immunity (see text). The eight human APOBECs (apolipoprotein B mRNA editing enzyme, catalytic polypeptide-like) primarily restrict retrotransposons and retroviral genomic expressions. Human HIV-1 and papillomaviral genomes induce APOBEC over-expression (without or with mutagenesis induction). APOBECs specifically target as their exclusive substrate ssDNA, inducing in them cytosine to guanine (C→G) and cytosine to thymine (C→T) substitutions. The C→G and C→T substitutions are recognized as APOBEC signature mutations. Clustered mutations are termed kataegis, Greek for escalating thunderstorms (originally written in Russian Cyrillic letters). APOBEC signature mutations in exomes are common in adenocarcinomas (non-existent in acute myeloid leukemias). The figure shows APOBEC-signature mutations in four breast cancer subtypes (luminal A, luminal B, basal-like, and HER2-enriched). The HER2 protein was considered to be amplified, not mutated; however, it contains high copy number variations due to frequent APOBEC signature mutations. APOBEC cytidine deaminases are a significant source of mutagenesis in the genomes of human cancers, to the extent that they may act as human cancer-driver mutations (Roberts SA et al Nat Gen 2013 Sept 45(9):970-976; Sinkovics JG Acta Microbiol Immunol Hung 2010;57:253–347). Author's comment. The archaic APOBECs induce point mutations "in the waiting" to blindly prepare single cells for survival in any future environmental changes? The AID/APOBEC enzymes are installed to provide antiviral immunity to the cell, but by mis-mutating ssDNAs bring about an eventually fatal outcome to the host by the overreproduction of immortalized cells? Or, is the initiation of immortalized cells an inherent faculty of the RNA/DNA complex carried out in this case by the AID/APOBEC protein?

Figure 48. Reed-Sternberg (RS) Cells of Hodgkin's lymphoma (HL) are originally mononucleated, but eventually become bi- and multinucleated giant cells. Sinkovics and Shullengerger CC observed these cells in primary bone marrow or lymph node cultures at standstill for several weeks: no RS cells in mitosis could be seen and photographed. Sinkovics repeatedly proposed in lectures (Honolulu, Hawaii; Padua, Italy; M.D. Anderson Hospital, Houston, TX; ASCO, Los Angeles, CA) that RS cells were the products of cell fusions mediated by EBV and an endogenous retrovirus, and thus they were "natural hybridomas" (Sinkovics JG et al Blood 1964;24:389–401; Lancet 1970; 1 (7638):139–140; 1975;2(7933):506–7; Crit Rev Immunol 1991;11:33–63; Leukemia 1992; 6S3:495–535). The cell fusion theory of RS cells was cited by P. Mauch & J. O. Armitage in Holland & Frei "Cancer Medicine" 5th ed. B.C. Decker Inc. Hamilton London 2000 p 2014. However, the German literature forcefully opposed cell fusion in RS cells: "Cell fusion is not involved in the generation of … Reed-Sternberg cell line" (Re D et al Am J Hematol 2001;67:6–9). "Evidence that Reed-Sternberg cells do not represent cell fusions" (Küppers R et al Blood 2001;97:818–821). Endomitosis without cell divisions was suggested for the generation of giant multinucleated RS cells. Michels and Sinkovics independently argued that the cell fusion theory had experimental support (Michels KB Eur J Cancer Prev 1995;4:379–388; Sinkovics JG Acta Microbiol Immunol Hung 2005;52:1–40). Long term time-lapse microscopy recently revealed that fusion and re-fusion of the HL mononuclear cells are the main route of RS cell formation. Mononuclear RS cells separated by division re-unite by fusions (Rengstl B et al Proc Natl Acad Sci USA 2013; 110:20729–20734; Commun Integr Biol 2014;7:e38602). However, the viruses (EBV; endogenous retrovirus) suspected to induce these cell fusions (Sinkovics JG Crit Rev Immunol 1991;11:33–63) received no attention at all. The RS cell-rich syncytial variant HL cells were repeatedly described (Lee SG et al Int J Hematol 2015;101:107–108), but without adequate virological studies performed. The RS cells possess a malignantly transformed genome and the V(D)J immunoglobulin pathway (however defective, due to loss of normal B cell characteristics) of the original B cells. The expression of B cell receptor in RS cells is lost due to loss-of-function mutations in its promoter. These are the SPI/POU protein mutations (serine protease inhibitor SPI proto-oncogene; Pit-1, pituitary-specific; Oct-1 octamer-binding; Unc-86 uncoordinated protein, caenorhabditis) (Vockerodt M et al Chin J Cancer 2014;33:591–597). The classical hybridoma cells possess the transformed genome of a malignantly transformed cell and the V(D)J hypersomatic mutation of a normal immunoglobulin-producer B cell. The RS cell possesses both criteria: a transformed genome and the basic B cell V(D)J pathway; thus, it qualifies to the pathognomonic position of a natural hybridoma. It frequently openly expresses EBV, and silently an endogenous retrovirus. Its endogenous retroviral expression is suppressed by acetylation of its inserted proviral genome, which can be activated by deacetylation to express the endogenous retrovirus (Kewitz S & Staege MS Front Oncol 2013;3:179).

Illustration: Thirty years old male patient (MR) at St. Joseph's Hospital Tampa FL in mid-1990s with Reed-Sternberg cells of lymphocyte-rich mixed cellularity Hodgkin's disease in lymph nodes and in the bone marrow. His RS cells expressed EBV antigens (no specifics available). History revealed clinically diagnosed infectious mononucleosis in young age with full recovery. Patient achieved uneventful complete remission on ABVD (doxorubicin, bleomycin, vinblastine, dacarbazine) chemotherapy. Later, upon a fall at his home resulting in multiple leg fractures and moving about on crutches, he was found in relapse with bone marrow aspirate positive for many mononuclear and multinuclear RS cells. His circulating T cell population presumably contained EBV-reactive $CD4/8^+$ memory cells. He was started on M.D. Anderson Hospital's low dose paclitaxel chemotherapy (Younes A et al Ann Oncol 2001;12:923–927). He was infused with autologous $CD4/8^+$ T cells ($5x10^8$ cells/mL x8). Bone marrow aspirates showed apoptotic RS cells and mononuclear RS cells dissolving upon contact with immune T cells (shown). Patient sustained good granulocyte count in afebrile state. He was considered to be in second complete remission. Later in 2001 he was admitted in the night through the Emergency Room to the Intensive Care Unit in Gram-negative septic and hemorrhagic shock; was attended immediately and in full. Expired a few hours later. No necropsy was performed (Sinkovics JG & Horvath JC Internat J Oncol 2001;19:473–488; Erratum 2005;27:5–47).

Illustration: Elongated double-nucleated large RS cell surrounded by small, compact lymphocytes (immune T cells); one lymphocyte overlying lower end of the RS cell (Sinkovics JG et al Br J Med 1970;1:172–173). Copyright Printer William Lewis Ltd, Cardiff, Great Britain). This is an example of emperipolesis; no cytoplasmic lysis or nuclear clumping (apoptosis) in the RS cell is evident. RS cells acquire resistance to immune T cells. Mononuclear RS cells are lysed by autologous cytotoxic lymphocytes; multinucleated RS cells are resistant to, or are protected by Treg cells against, cytotoxic lymphocytes (shown) (Sinkovics JG et al Brit Med J Jan 17 1970;1:172–173; Int J Oncol 2012;40:305–349). RS cells harbor herpesviral (EBV) and endogenous retroviral particles (Sinkovics JG & Györkey F J Med 1973;4:283–317; Sinkovics JG & Horvath JC Int J Oncol 1999;14:615–646. Erratum idem 2005;27:5–47; Sinkovics JG "Cytolytic Immune Lymphocytes…" Schenk Buchverlag, Passau/Budapest, 2008). RS cells offer a wide array of antigenic targets for monoclonal antibodies, NK- and immune T cells. NK cells attack RS cells in the antibody-directed cell-mediated cytotoxicity reaction with the help of lymphokines IL-2/IL-12 (Jahn T et al PLoS One 2012;7(9):e4448). The anti-CD30 moab conjugated to auristatin E was generated for therapy (Stein H, Diehl V Hematol Oncol Clin North Am 2014;28(1);1–11. doi:10.1016/j.hoc.2013.10.007).

Figure 49. G4 quadruplex DNA. The DNA-protein complex 6-base pair repeat telomeres are synthesized by the telomerase reverse transcriptase (TERT). Re-capped chromosomal ends are protected from end-to-end fusions. The ciliate, Stylonychia lemnae renews its telomeres of 36 nucleotides in its dividing macronucleus. The G-rich 3'-overhang is formed by the last 16 nucleotides. The overhang adopts a subtelomeric guanine-quadruplex DNA structure. The G-rich quadruplex DNA structures are both parallel- and antiparallel-stranded. Two overhangs mediate end-to-end association in four-stranded conformations. The formations are stabilized by hydrogen-bonded guanine quartets. Specific single-chain fragments (scFv) recognize the parallel G-quadruplexes in millions of minichromosomes in the macronuclei. The telomeres receive their caps in the replication bands of the quadruplexes in the macronucleus at each cell division. Telomere capping is accomplished by t-loops, TEBP (telomere-binding protein), and by the G-quadruplex; TEBP and G-4 are interacting (Schaffitzel C et al Proc Natl Acad Sci USA 2001;98:8572–8577). Illustration: G-quadruplex DNA in the macronuclear telomeres of Stylonychia lemnae. Christiane Schaffitzel et al, Proc Natl Acad Sci USA. 2001 July 17;98(15):8572–8577.

Figure 50. PTEN. The tumor suppressor gene and gene product protein PTEN (phosphatase tensin homolog deleted on chromosome ten, human 10q23.3) (Table VIII), is the natural inhibitor of oncogenes PI3K/Akt (phosphatidyl inositol kinase 3; protein kinase 3; phosphatidylinositol 3 phosphate; Jacob Fürth's AK mouse strain thymic lymphoma retroviral oncogene; phosphoinosite-dependent kinase). The PTEN protein inactivates PIP3 by dephosphorylation; PDK1 (phosphoinositol-dependent kinase) is not recruited to the plasma membrane to activate Akt by phosphorylation. Sarah M. Planchon et al, The nuclear affairs of PTEN. *J Cell Sci 2008 121:249–253*. doi:10.1242/jcs.022459

Highly malignant adenocarcinoma cells (prostate cancer; triple-negative breast cancer; colorectal cancer) with deleted PTEN (Carver BS et al Nat Genet 2009;41:619–624; Karseladze AI et al Arkh Pathol (Russian) 2010;72:20–23; Kato S et al Int J Cancer 2007;121:1771–1778; Perrone et al Ann Oncol 2009;20:84–90; Reid AH et al Mod Pathol 2012; 25:902–910; Sircar K et al J Pathol 2009;218:505–513; Yoshimoto M et al Genes Chromosomes Cancer 2012;51:149–160). MicroRNA-122 is a promoter of the PI3K/Akt oncogenic pathway (Lian JH et al Asian Pac J Cancer Prevent 2013;14:5017–5021). The mRNA of the tumor suppressor gene GAS1 (growth arrest specific) is targeted for destruction by overexpressed miR-34a. The GAS1-blockade results in the overproduction of PI3K/Akt/Bad (Bcl-2 antagonist of cell death) with cancers (papillary carcinoma of thyroid) developing (Ma Y et al Biochem Biophys Res Commun 2013;441:958–963). The activated (phosphorylated) Akt oncoprotein targets intranuclear DNA oncogenes (mammalian target of rapamycin mTOR; mitogen-activated protein kinase MAPK; vascular endothelial growth factor VEGF) (Shafee N et al Cancer Lett 2009;282:109–115; Teng HF et al Mol Cell Endocrinol 2014;383:147–158). PI3K/Akt inhibitors in clinical trials are bortezomib, resveratrol, and small molecular inhibitors of oncogenes/oncoproteins (Ge J et al Biomed Environ Sci 2013;26:902–911; Lin L Int J Oncol 2014;44:557–561; Polivka J & Janku F Pharmacol Ther 2013; pii:50163–7258(13)992404; Qazi AK et al Anticancer Agents Mol Chem 2013;13:1552–1564). PI3K/AKT/mTOR patented small molecular inhibitors (LY2949; GDC-0980; NVR-BEZ-235 etc) are awaiting their clinical trials.

Figure 51. Ancient Chinese Yin (black) Yang (white). Forces (energies) in the Universe act and counteract, expand and contract, change hot to cold, light to dark and back in continuous movement. Immune T cells and regulatory T cells balance each other (Bluestone JA: The yin and yang of interleukin-2-mediated immunotherapy. N Engl J Med 2011;365:2129–2131).

Figure 52. Placental syncytins. Endogenous retroviral syncytin genes format the two-layered multinucleated murine placental syncytiotrophoblasts at the feto-maternal interface. Two distinct retroviral genes were retroinscribed into the ancestral rodent genome. These proviral genes are expressed in unison, but work independently in the fetus to format the fetal organ placenta. The mouse, rabbit and human syncytin genes are related. Anne Dupressoir A et al, Proc Natl Acad Sci USA 2011;108:E1164–1173 PNAS. doi:10.1073/pnas.1112304108

Figure 53. The Team of Ferenc Györkey†, Joseph (József) Sinkovics and Jeno (Jenő) Szakacs has engaged for several decades in the search of retroviral particles in human tissue specimens submitted to their laboratories. Illustration 1. Györkey and Sinkovics discovered endoplasmic tubuloreticular structures in lymphocytes and parenchymal cells of patients with systemic lupus erythematosus and pre-AIDS era classical Mediterranean Kaposi's sarcoma cells (shown) (Gyorkey F & Sinkovics JG Lancet 1971;1(7690):131–132; Gyorkey F et al Am J Med 1972;53:148–158; Lancet 1982;2(8305):984–5). Cited in ref 147, fig 4.1: in kidney glomerular cell of patient with SLE. Mistakenly, they regarded the tubuloreticular structures to be incomplete retroviral particles developing in the endoplasmic reticulum. In association with these structures, budding retroviral particles were also observed. Dorothea Zucker-Franklin has also seen an association between the tubuloreticular structures and the retroviral particles in lupus and in KS cells (Zucker-Franklin D N Engl J Med 1983;300;837–838). However, the tubuloreticular structures proved to be nucleic acid-free; they were formed in the presence of excessive type I IFN production. Once the tubuloreticular structures proved to be not of retroviral origin, literature citations ceased for both the tubuloreticular structures and the co-existing, genuine retroviral particles (shown in SLE cells). A seemingly valid co-existence of the tubuloreticular structures and budding retroviral particles in human sarcoma cells remains present (shown). Illustrations 2/3/4: endogenous retroviral particles budding from kidney glomerular cells of patients with SLE. Cited in ref 147 page 48 Fig 4.3/4/5. Illustrations 5/6: human soft tissue sarcoma cells with budding retroviral particles. Cited in ref 147 pages 251, 252, Fig 1212 and Fig 1213. The co-existence of HHV8/KSHV with endogenous retroviral particles in KS cells has been repeatedly documented (Sinkovics J G et al Orvosi Hetilap (Hungary) 2006;147:617–621; Sinkovics JG "Cytolytic Immune Lymphocytes…" Schenk Buchverlag Passau/Budapest) (Figures 29 and 30). Clarification: Szepesházi K Lapis K et al Orvosi Hetilap Budapest 1977;118:315–318.

Figure 54. The Naegleria. The life cycle of N. fowleri consists of three stages: flagellate (converting a trophozoite), amoeba, and cyst. Moving on highly mobile by lobodium pseudopodia, the amoeboid trophozoites infect the meninges. Mitochondria secure the aerobic metabolism of the trophozoites. The most versatile genome of the N. gruberi is able to re-format its entire microtubular cytoskeleton when it transforms from an amoeba into a flagellate. To form the flagella, it synchronizes some 82 genes. Not the flagellar, but its centriole gene cluster consists of 55 genes. Some of its centriol genes are conserved up to the human genome. Its centrosome-localized conserved centriole component contains genes orthologous with some of those expressed in HeLa cells (Fritz-Laylin KL & Cande WZ J Cell Sci 2010;123:4024–4031) http://jcs.biologists.org/cgi/content/full/123/23/4024/DC1.

Figure 55. Two Different Tumors with Fusion Oncoproteins: African Burkitt's Lymhoma (BL) with, and Congenital Mesoblastic Nephroma (Fibrosarcoma) (CMN/IFS) without, a Host Immune Response. Epstein-Barr virus attacks a host immunologically overburdened by plasmodia and assorted pathogens. The endemically spreading external viral pathogen EBV induces the malignant transformation in African teenagers. Chromosomal breaks form the c-myc/IgH fusion oncogene/oncoprotein by the translocation t(8:12)(q24;q32). In the attacked B lymphocytes, class switching and V(D)J somatic hypermutations are not accomplished. EBV EBNA1-6 and LMP1-2 immortalize the B lymphocytes in the viral latency III program. Additional oncogenes (NFκB; T cell leukemia 1 from chromosome 14q32.1) are recruited. Chemical inhibitor of c-Myc 10058-F inhibited the growth of BL cells. The "starry sky" (shown) histological pattern is that of macrophages phagocytosing antibody-coated lymphoma cells as reproduced as such in the mouse (Sinkovics JG et al J Inf Dis 1969;120:250–254). There are host immune reactions to BL cells. Combination chemotherapy induces remissions (with relapses). Natural measles virus infection may induce remission due to viral oncolysis. In non-endemic sporadic Western Hemisphere BL cells, the same translocation can occur without EBV. BL occurs in HIV-1 infected patients. Diffuse large B-cell lymphomas occur with or without EBV, but with t(8:12)(q24;q32. In this entity, c-Myc inhibits NFκB and IFN-response (Burmeister T et al Mol Oncol 2013;7:850–858; Faumont N et al J Virol 2009;83:5014–5027; Kiss C et al Proc Natl Acad Sci USA 2003;100:4813–4818; J Virol Methods 2003;107:195–203). EBV is known to physiologically persist in memory B lymphocytes; the pathogenesis of BL involves the memory B lymphocytes. The EBV-encoded nuclear antigen (EBNA1) escapes recognition by immune T cells of the host. However, EBNA1 is associated with latent membrane protein antigens' (LMP1/2) expression. BL cells constitutively express EBV oncoprotein LMP1 (Hochberg D et al Proc Natl Acad Sci USA

2004;101:139–44). EBNA2 is responsible for the latent immortalization of EBV-infected BL cell by activating Notch signaling. The EBNA2 transcriptional transactivator TAT protein downregulates LMP1/2 production and induces cessation of BL cell proliferation (Farrell et al Proc Natl Acad Sci USA 2004;101:4625–30). In proliferating BL cells, the promoters of cell cycle suppressor genes $p16^{INK4A}$ and $p14^{ARF}$ are hypermethylated and thus inactivated (Nagy E et al Eur J Cancer 2003;39:2298–2300; Anticancer Res 2005;25:2153–2160). In contrast, the constitutively activated CDK4/6 is de-activated by its molecular inhibitor palbociclib.

The ETV6/NTRK3 fusion oncoprotein (ETS translocation variant of reticuloendotheliosis bird retrovirus E twenty-six; neurotrophin tyrosine receptor kinase) induces secretory basal cell breast cancers in premenarcheal teenager girls, or teenager boys (Yorozya K et al Jpn J Clin Oncol 2012;42:208–211; Szanto J et al Breast 2004;13:439–442). The congenital CMN/IFS kidney tumors express the ETV6/NTRK3 translocation-encoded t(12;15)(p13;q25) oncoprotein with or without trisomy of chromosome 11. Whether these two tumors are identical or different is still debated. The fusion of bands 12p13 of gene ETV6 and band 15p15 of gene NTRK3 is the hallmark of these tumors (they do not occur in adulthood fibrosarcomas). The tumor's energy is carbohydrate-intake dependent with insulin-like growth factor and insulin receptor substrate (IGF2; IRS1) overexpression (Adem C et al Mod Pathol 2001;14:1246–1251; Argani P et al Mod Pathol 2000;13;29–36; Gisselsson D et al Mod Pathol 2001;14:1246–1251; Morrison et al Oncogene 2002;21:5684–5695; Watanabe N et al Cancer Genet Cytogenet 2002;136:10–16). Transmission electron microscopy so far revealed no viral particles (PCR, Immunohistochemistry for viral-encoded proteins; Y. Chang's technology for the first isolation of KSHV and Merkel cell polyomavirus have not as yet been applied) (Knezevich SR et al Cancer Res 1998;58:5046–5048; Rubin BP et al Am J Pathol 1998;153:1451–1458; Ramachadran C et al Pediatr Dev Pathol 2001;4:402–411). The fetal innate and maternal passive immune faculties of the fetus and newborn apparently mount no immune reactions to these tumor cells: they appear to be accepted as self. Illustration: Sinkovics JG Pienta JR Trujillo J Ahearn MJ University of Texas MD Anderson Hospital, Houston, Texas. An immunological explanation for the starry sky histological pattern of a lymphoma. J Inf Dis 1969;120:250–254. Illustration: David Gisselsson, University Hospital, Lund, Sweden. FISH: G-banded chromosome 12 with ETV6 and KRAS2 probes.

Figure 56. The Life Cycle of Phytophthora infestans [2357]. P. infestans gains virulence and drug resistance by gene fusion. Fusion occurs between the disease resistant recombinase gene (DRR206) and β-glucoronidase (GUS). Cis-regulatory elements involved are the wound/pathogen inducible box genes (W/P box). Fusion with the Fusarium solani phaseolae DNAse gene (FsphDNase) provided resistance against Pseudomonas syringae (Choi JJ et al Phytopathology 2004;94:651–660). The class WRKY proteins, sometimes with GQK (trp, arg, lys, tyr; glyc, gln, lys), are segmentally duplicated DNA-binding transcription factors in plant cells (Arabidopsis thaliana; Solanum lycopersicum, the tomato; Petroselinum crispum, the parsley; Vitis vinifera, the grape; Nicotiana benthamiana, the tobacco; Erysiphe necator, the mildew); appearing first in Chlamydomonas reinhardtii (but absent in animals). Occur in the amoeba: Dictyostelium discoideum. Act as first defensive signalers (Eulgem T et al EMBO J 1999;18:4689–4699; Guo C et al J Exp Bot 2014. doi:10.1093/jxb/eru007). WRKY proteins protect Phytophthora infestans against its pathogens. It is the mitogen-activated protein kinase (MAPK) pathway that activates WRKY proteins by phosphorylation (Ishihama N et al Plant Cell 2011;23:1153–1170). The MAPK is a frequently used proto-oncogenic pathway in transformed eukaryotic cells. Large peripheral vesicle proteins transiently fused by dynamine interaction in some oomycetes in the Phytophthora genus. These proteins are secreted as an introductory act to host tissue invasion. Similarly, fibrosarcoma cells are known to practice cargo release (exocytosis) from superficial vesicles (Jaiswal Rivera Simon et al Cell 2009;137:1308–1319; Zhang W et al PeerJ 2013;1:e221).

Figure 57. Showing some of the first human NK cell pictures. Ewing sarcoma cells carry the EWS-FLI-1 translocation t(11;22)(q24;q12) at Friend leukemia insertion site (Fli), express cancer-testis antigens, and are attacked by host lymphocytes, both immune T cells and NK cells. The first human NK cells viewed in the late 1960s are shown; cited from Sinkovics JG et al Leukemia Lymphoma. Yearbook Medical Publisher Pp 53–92, 1970) (278). The first Ewing sarcoma cell line was established by author JGS and associates from 10 year old female patient (MDAH#79610) at the Section of Clinical Tumor Virology & Immunology of M.D. Anderson Hospital in 1970 (shown in its 50th in vitro passage). At that time it has just become known that patients' small compact lymphocytes (later: immune T cells) reacted to autologous (and related allogeneic) tumors, and large granular lymphocytes (later: NK cells, red arrows) reacted to autologous and allogeneic tumors (shown in Figures 35 and 36) and that healthy tumor-free individuals also possessed these latter cells (Sinkovics JG Györkey F et al Methods Cancer Res Academic Press 1978;XIV:243–323; Sinkovics JG et al Leukemia-Lymphoma Year Book Medical Publishers 1970; 53–92; Management Bone & Soft Tissue Tumors. Year Book Medical Publishers 1977;361–410; Sinkovics JG, Plager C et al Canad J Surg 1977;20:542–546; Oncology 1980;37:114–119). Since then, the chemo-radiotherapy-resistant Ewing sarcomas have become targets of immunotherapy with anti-IGF-R or anti-CD99 monoclonal antibodies, and biological interventions (EWS-FLI gene silencing; αTNF/TRAIL induction; TGFβ induction with rapamycin and mTOR inhibition; PARP-cleavage by the pan-cycline kinase inhibitor flavopiridol). Autologous immune T cells and/or allogeneic NK cells attack Ewing sarcoma cells. Type I IFNs induce apoptosis of EWS cells through the activation of TNFα-inducing ligand (TRAIL). Expression of ligand NKG2D (stimulated by histone deacetylase inhibitors, or by lentiviral transduction) renders EWS cells susceptible to native or activated NK cell therapy. Chondromodulin-, or enhancer of zeste-stimulated HLA-A0201+ dendritic cells rendered allo-restricted HLA-A201− CD8+ T lymphocytes cytotoxic to EWS cells. Anchor-residue reinforced EWS-FLI-1 peptides induced T lymphocytes to attack EWS xenografts (Berhuis D et al Clin Sarcoma Res 2012;2(1):8; Cho D et al Clin

Cancer Res 2010;16:3901–3909; Evans CH et al Clin Cancer Res 2012;18:5341–5351; Kelleher FC & Thomas DM Clin Sarcoma Res 2012;2(1):6; Lehner M et al PLoS One 2012;7(2):e31210; Mahlendorf DE & Staege MS Cancer Biol Ther 2013;14:254–261; Thiel U et al Br J Cancer 2011;104:948–956). FLI = Friend leukemia insertion site. Illustrations: 1, 2. Sarcoma cell attacked by autologous small compact lymphocytes (later: immune T cells) and large granular lymphocytes (later: NK cells) shown in 1969 at M.D. Anderson Hospital Symposium "Leukemia-Lymphoma" (Sinkovics JG et al Year Book Medical Publishers, Chicago, 1970 Pp 53–92) (278). Picture of medical historic value showing for the first time NK cells attacking a human cancer cell. 3. Ewing sarcoma cell line established from patient MDAH#79610) identified as that of a Ewing's sarcoma cell with t(11;22) (24;q12) by pathologist Bruce McKay who prepared the transmission electron microscopic picture (Sinkovics JG, Plager C et al Immunotherapy of human sarcomas. In: "Management of Primary Bone & Soft Tissue Tumors". Year Book Medical Publishers, Chicago, 1977 Pp 361–410.

Figure 58. BRAF mutations. The rat fibrosarcoma retrovirus incorporated in its genome the host cell proto-oncogenes rafABC: c-raf → v-raf (Rapp UR et al Proc Natl Acad Sci USA 1983;80:4218–4222). The raf (rat fibrosarcoma) proto-oncogenes are widely spread in vertebrate mammalian hosts. The Braf gene is subject to gain-of-function point mutations in its nucleotide BRAF V600E (valine; glutamic acid in exon 11, codon 600). The mutated genome is constitutively active in many different human tumors: melanoma (60%), hairy cell leukemia (100%), splenic B cell lymphoma, histiocytic sarcoma, platelet-derived growth factor receptor PDGFR-negative gastrointestinal stromal tumors, some (40%), papillary thyroid carcinoma, some colon carcinoma (10%). Constitutively activated MEK and ERK pathways (MAPK/ERK kinase; mitogen-activated protein kinases; extracellular signal-regulated kinase) result in non-stop cell divisions. This mutation does not trans-speciate the cell in which it is active (morphologically different tumor cells carry the same mutations and retain their native morphology). Melanoma and hairy cell leukemia showed clinical response to type I IFNα. All BRAF-mutated tumor cells respond with cessation of growth and apoptotic death to the small molecular inhibitor vemurafenib. Patients with melanoma treated with vemurafenib may develop squamous cell carcinoma of the skin. These tumors commonly remain localized and are curable with surgical excision. On rare occasion, these tumors display eruptive features and malignant behavior. It is not yet clear what genomic mutations induce these tumors (Mays R et al J Cutan Med Surg 2013;17:419–422 Illustrations: Human melanoma cells with melanosome. EBV particles released from hairy leukemia cells (Sinkovics JG et al Methods Cancer Res 1978;XIV:243–323; Lancet 1975;1:749–750; Ann Int Med 1976;85:532–533; N Engl J Med 1976;295:1016).

Figure 59. FasL → FasR-Induced Mitosis. Short term (several months) in vitro established human melanoma cells deriving from patients with metastatic disease expressed the Fas receptor (R, fragment apoptosis signaling). These melanoma cells were exposed to highly purified Fas ligand (FasL). Figures shown were prepared by

Joseph C. Horvath for publication (vide infra). Instead of apoptotic cell death, as expected, FasL induced mitoses of FasR$^+$ melanoma cells. Melanoma cells are known to undergo chromosomal breaks involving chromosome 1, with the locus of granulocyte colony stimulating factor receptor at 1p32-34 (Tweardy DJ et al Blood 1991;79:1148–54), and that of chromosome 10, with the locus of FasR (receptor, CD95) at 10q23-26 (Lichter P et al Genomics 1992;14:179–180). In order to explain how FasL-to-FasR could induce a mitotic response, it was postulated that the broken chromosomes 1 and 10 fused and generated a new fusion product gene of FasR and GCSF-R as t(1;10)(p32;q23-26). The signal conduction pathway of the gene product protein placed the FasR on the cell surface. Upon capturing FasL by FasR at the cell surface, signaling traveled through the cytoplasm into the G-CSF-R fragment of the fusion protein, that induced mitosis within the nucleus (Sinkovics JG & Horvath JC Int J Oncol 2001;19:473–488; Horvath JC et al Proc 89th Ann Meet Am Assoc Cancer Res 1998;38:2371 #584; Sinkovics JG Acta Microbiol Immunol Hung 1997;44:295–307. Erratum. In the MS FasR was correctly spelled CD95). The generation of this pathway in vivo as proposed receives support from experiments in which effective signals depended on the cytoplasmic region of chimeric FasR and G-CSF-R receptors: FasR cytoplasmic region induced apoptosis and G-CSF-R cytoplasmic region induced mitosis of leukocytes carrying the artificially prepared chimeric receptors (Takahashi T et al J Biol Chem 1996;271:17555–17560). Karyotyping of the melanoma cells expressing pathologically formed chimeric receptors with FasR on the cell surface and G-CSF-R in the cytoplasm would provide the final proof. Human B cell lymphoblasts release exosomes constitutively that induce apoptotic death in CD4+ T lymphocytes (Klinker MW et al Front Immunol 2014.doi:10.3389/fimmu.2014.00144). Illustrations: Published figure showing soluble FasL stimulating mitoses of FasR$^+$ melanoma cells and anti-FasL antibody abrogating the effect (Sinkovics JG & Horvath JC Internat J Oncology 2001;19:473–488. Errata: Typographic error in the cartoon. Idem 2005;27:33). FasL-to-FasR. Ralph Budd, Vermont Center for immunobiology and Infectious Diseases. University of Vermont. Abbreviations: FLICE: Fas-associated death domain-like IL-1β-converting enzyme (caspase 8). TRAF: tumor necrosis factor receptor TNFR-associated factor. GCSF-R human granulocyte colony stimulating factor receptor (M Kimmel & S Corey in Frontiers in Oncology, 2013. doi:10.3389/fonc.2013.00089)

Figure 60. HHV6.*) History.The replication of human herpesvirus 6 is within the host cell nucleus. Immature viral particles bud out of the nucleus and become tegumented in the cytoplasm. Tegumented virions are taken up by the Golgi apparatus to receive their glycoprotein envelope. The egress of mature virus particles is through the cell membrane. Infected children experience the skin rash of roseola (exanthema subitum) consequential of viremia, also involving the cerebrum and meninges. Thereafter the viral genome rests in infected cells (commonly in lymphocytes; some parenchymal cells) in a stage of life-long latency. In latency, the viral genome integrates itself into chromosomes and is passed to cellular progenies, germ cells included. Cells taken up HHV-6A genomes will not allow the integration of HHV-6B genomes. During its integration via homologous recombination, the HHV-6 genome loses 79 of its nucleotides. The chromosomal subtelomeric region is the favored viral integration site. The viral gene U94 encodes protein RepH6; enzyme RepH6 functions as an endonuclease, helicase, or ATPase. HHV-6 RepH6 might have derived from an adenoassociated viral gene incorporated into an ancestral HHV-6 genome now recognized as HHV-6 gene U94. HHV-6-infected lymphocytes may form syncytia. The gene product protein of HHV-6 gene U24 share amino acid identity with the myelin basic protein. Immune T cells attacking HHV-6 may attack also cells expressing myelin basic protein and thus induce the clinical picture of multiple sclerosis. During HIV-1-induced AIDS, HHV-6 becomes reactivated and mature virions appear in the blood. The inherited familial HHV-6 syndrome is a serious ailment (Arbuckle JH et al Proc Natl Acad Sci USA 2020;107:5563–5568; Virology 2013;442:3–11). Peter Medveczky considers familial HHV-6 syndrome to be a clinical form of the chronic fatigue syndrome. Illustration: FISH documentation of subtelomeric HHV-6 genomic integrat. *) *Isao Miyoshi's group finds HHV6 in Burkitt's lymphoma (Daibata M et al Br J Haematol 1998;102:1307–1313). Isao Miyoshi and this author served together in fellowship in 1959 at MD Anderson Hospital (cited in [147]).*

Figure 61. H. H. Niller's Presention at his inaugural honorary membership lecture at the 14th International Congress of the Hungarian Society of Microbiology in 2003 at Balatonfüred, Hungary. The topic was a particular genomic segment of the Epstein-Barr virus. This viral genomic segment located within the EBER1 promoter appeared to be co-linear with a part of a sequence in the human immunoglobulin locus V(D)J. The co-linear human genomic sequence is within the 5'-end of the human lambda light chain immunoglobulin locus. Niller proposed that these two genomic segments derive from a common ancestor and reached the human genome through herpesviral vectoring. The original acquisition of the sequence in question might have occurred horizontally at the era of the first gnathostomata, from where the genomic segment was passed vertically up to Homo. Niller's idea fits into a scenario whereby protocells on one side of the ping-pong table evolved and acquired DNA replicons from viruses evolving on the opposite side of the ping-pong table. There, tRNAs evolved into transposable element, retroviruses and DNA viruses (without explaining how RNA has become DNA). In the possession of DNA replicons of viral derivation, the protocells separated their soma (cytoplasm) from their germ line (nucleus) (Niller HH et al Acta Microbiol Immunol Hung 2004;51:469–484). The long ago formed separation of the RNA and retroviral camp of virologists may argue with the DNA virus camp in proposing that retrotransposons provided the basic genomic elements to the gnathostomata sharks (Kapitonov W & Jurka J PLoS Biol 3(6):e181), and that herpesviral ancestors acquired their adaptive viral genomic elements more recently from hosts higher up on the evolutionary scale already being in the possession of these elements well established. The ancient herpesviral genomic segments inserted into the amphioxus genome (Savin KW et al Virol J 2010;7:308) may hold the trump card.

Figure 62. Class I Retrotransposons use RNA-mediated, Class II DNA transposons use DNA-mediated mechanisms for their transposition. Retrotransposons far outnumber DNA transposons in the human genome. Non-LTR (long terminal repeats) short and long interspersed nuclear elements (SINE; LINE) are the most numerous transposons. The two ORFs of the 6 kbp LINEs encode an endonuclease and a reverse transcriptase. The endonuclease performs the nick on the ssDNA and the reverse transcriptase converts the RNA into DNA for insertion. The excised autonomous Class II DNA transposons encode the transposase that fits to their own inverted terminal repeats (ITR). The ITRs of a non-autonomous Class II DNA transposon depend on another transposon's transposase for insertion (trans-supplementing the transposase protein). The first dormant transposase reconstructed from the fish genome is the "Sleeping Beauty" (Z. Ivics, Zs Izsvák, R.H. Plasterk et al Cell 1997;901:501–510; Nat Methods 2009;6:415–422). Reconstructed natural Class II DNA transposons from insects, fish and amphibians are the Sleeping Beauty, Frog Prince, Hsmart, Mariner, Minos, Tol2, piggyBack (from the insect cabbage looper Trichoplusia ni) and piggyBat (from the little brown bat Myotis lucifugus), and a synthesized mammalian Class I non-LTR-retrotransposon (ORFeus for mice). Disregarding for practical purposes, in the evolutionary drive that transposons exerted since the generation of life (Dupeyron M et al Mob DNA 2014;29:5:4; Feschotte C & Pritham EJ Annu Rev Genet 2007;41:331–368; Pace JK II & Feschotte C Genome Res 2007;17:422–432; Skipper KA et al J Biomed Sci 2013;20:92), they may serve as vehicles of inserted genes and thus for gene therapy. In contrast to virally-vectored gene therapy, the gene-carrier transposons induce no neutralizing antibodies or immune reactions in the treated hosts, and their cargo capacity extend to large transgenes (Hong IS et al PLoS One 2014;9(1):e86324; Swierczek M, Izsvák Z, Ivics Z Expert Opin Biol Ther 2012;12(2):139–153; Human Gene Ther 2011;22(9):1043–1051). They may contribute to gene recombinations and mutations (known in the Daphnia; in crustaceans) (Schaack S et al Genome Biol 2010;1104):R46). They may spread horizontally (extensively, known in the drosophila) (Bartolomé C et al Genome Biol 2009;20(2):R22; Dupeyron M et al MobDNA 2014;5(1):4; Schaack S et al Trends Ecol Evol 2010;25:537–546). Zoltán Ivics, Zsuzsanna Izsvák et al Max Delbrück Center for Molecular Medicine, Berlin, Germany. Nat Methods June 2009;6:415–422. SBTS3 http://commons.wikimediua.org/wiki/File:SBTS3.png by Perryhackett licensed under CC BY.

Figure 63. Neurospora crassa. Barbara McClintock identified first that the chromosomes in the fungal genome (Neurospora) were eukaryotic. E.T. Tatum and G.W. Beadle won the Nobel prize for showing in 1941 that the Neurospora haploid genome works by the "one gene one enzyme" basic rule. Neurospora cells harbor mitochondria. The mitochondrial genomic introns revealed the process of self-splicing, and that interfering RNA (iRNA) could eliminate mRNAs issued by DNA genes. Neurospora cells' haploid genome contains approx 10,000 genes. Neurospora cells change their haploid genome to diploid at conjugation. The genome returns to haploid state by methylating, thus silencing, the cytosines in both copies of duplicated sequences. Neurospora cells operate PARP (poly(ADP-ribose) polymerase) (Table II) orthologs of the mammalian cells' PARPs for ss/dsDNA break repairs and polyADP-ribozylation (Kothe GO et al Fungal Genet Biol 2010;47:297–309). In hyperosmotic stress (or due to mutations), the evolutionarily highly conserved PP2A class of phosphatases interact with the MAPK pathway; PP2A dephosphorylates and thus inactivates MAPK. Loss-of-function mutated phosphatases fail to stop MAPK. Hyperactive MAPKs induce excessive hyphal growth. The unregulated cell proliferation does not respond to inhibitors (the tert-butylhydroperoxide, t-BuOOH (Sigma), becomes ineffective). Inactivation by dephosphorylation of gene cdc-14 (cell division cycle) leads to unopposed activity of cdk-1 (cycle dependent kinase) resulting in incessant mitoses. In these cells the microtubule-attaching mitosis inhibitor benomyl becomes ineffective (Bennett LD et al Eukaryot Cell 2013;12:59–69; Ghosh A et al G3: Genes Genomes Genetics 2014;4:349–365). The Neurospora cell photoreceptors VIVID and White Collar (WC) activate the cells' circadian rhythm, which is regulated by the clock genes/proteins ATF/CREB (activating transcription factor; cyclicAMP-responsive element-binding protein; adenosine monophosphate). These genes/proteins synchronize Neurospora cell divisions (Hong CI et al Proc Natl Acad Sci USA 2014;11: 1397–1402; Lamb TM et al Fungal Genet Biol 2012;49:180–188). The Neurospora clock gene frq (frequency) is shared with other fungi (Magnaporthe, Podospora). The frq and catp (clock ATPase) genes regulate histones in the nucleosome. The ccg2 (clock-controlled gene) encodes the spore surface rodlets. The bd (band) and fluffy genes induce high GTP (guanine triphosphate) levels and conidia (asexual spores) formation; they are activated by ROS (reactive oxygen species) and by receptor WC. Comment. The two Neurospora bd genes are ancestral to eukaryotic proto-oncogenes ras (rat sarcoma). Mutated Neurospora hyphal cells grow like chemotherapy-resistant human cancer cells. The descendants of Neurospora clock gene mRNAs and clock proteins respond to circadian rhythm in human cancer cells (naevus; cutaneous and uveal melanoma) (Lengyel Z et al Tumour Biol 2013;34: 811–819; Gamaleia NF et al Exp Oncol 2006;28;54–60; Otálora BB et al J Pineal Res 2008;44:307–315; Riabykh TP et al Vest Ross Akad Med Nauk 200;8:30–33). In the mammalian genome, target sites for the CREB proteins are proto-oncogenes AKT, Fos and Jun (AK mouse retroviral thymic lymphoma oncogene; Finkel murine osterosarcoma retroviral oncogene; bird reticuloendothelial retroviral oncogene 17; seventeen in Japanese is jun). Comment. Vertebrate mammalian (human) cells in the process of their malignant transformation regress to the level of the cell survival pathways of ancient eukaryotes. Ancient cry (circadian rhythm) genes may be distant orthologs of vertebrate mammalian (human) genes. The human tumor pinealoma possibly shares circadian rhythm genomic features with the ancestral genes operating in Neurospora crassa (in Appendix 1).

Figure 64. The Guardian of the Genome. The p53p63 genes underwent over one billion years of evolution from the sea anemone to the human genome. The caenorhabditis' hybrid gene p53p73 (C. elegans p53, CEP-1) duplicated to produce the vertebrates' p53 gene (Belyi VA et al Cold Spring Harb Perspect Biol 2010;2: a001 198). The p53 gene is a target of DNA-binding proteins that regulate its activation or suppression. Anti-apoptotic microRNA-125b (oncomir miR-125b) (Table XXIII) by suppressing tumor suppressors CDKN2A/ink4a/14ARF (cyclin-dependent kinase; alternative reading frame), upregulates murine double minute (MDM) and its homologs (Amir S et al PLoS One 2013;8:e61064). The human gene product protein p53 is encoded from 17p13.1 and is synthesized in the ribosome. The p53 protein either succumbs to ubiquitination by MDM (or is guided into mitochondria by the MDM antagonist small molecule nutlin) (Lee S-Y et al Int J Oncol 2013; 10:1027–1035). Within mitochondria the p53 protein activates the intrinsic mitochondrial apoptotic death of the cell (The p53 protein may enter the nucleus in order to activate its numerous target genes (Vogelstein B et al Nature 2000;408:307–310), among them PUMA (p53-upregulated mediator of apoptosis), and pro-apoptotic BAX (Bcl2-associated X protein). The NADH-quinone oxidoreductase translocates protons through membranes. It started its activity in prokaryotes and continued it in the mitochondria of eukaryota. In dysfunctional mitochondria, p53 activity is reversibly (by chloramphenicol) suppressed; apoptosis induction is stalled; and cells are protected against gamma-irradiation- (γIR-) induced death. Cancer cells gain survival advantage (Compton S et al J Biol Chem 2011;286:20297–20312). The ancient NDUFA gene encodes the MWFE protein (abbreviations explained in text) (Au HC et al Proc Natl Acad Sci USA 1999;96: 4354–4359) . Constitutively overproduced (due to such NDUFA promoters, resulting in MWFE mRNA overexpression), or mutated MWFE accelerate cell growth (Yadava N et al J Biol Chem 2002;277:21221–21230). Illustration: p53 protein reacts with its target DNA. Cho et al Science 1994;265:346. Springer ML et al Genes Dev 1992;6:346–355.

Figure 65. Aberrant Mitoses lead to Aneuploidy. The consequences may be apoptotic cell death, autophagy, or malignant transformation. Illustration: Kops GJPL, Weaver BAA, Cleveland DW. Nature Reviews Cancer 2005;5:773–786.

Figure 66. Pettenkoferien. Hugo von Preisz, professor & chairman of the Institute of Bacteriology at the Péter Pázmány University in Budapest, Hungary devoted two decades for the study to the phenomenon of "Pettenkoferien", that was first observed by Ph. Kuhn and K. Sternberg. He then published a lavishly illustrated monumental article on his studies in 1937 in the Zentralblatt für Bakteriologie (Erste Abteilung / Originale Juli 12 1937; Band 139, Heft 5/6 Seiten 225–268 mit XIII Tafeln; Verlag von Gustav Fischer in Jena). What he believed he was observing was not a symbiotic relationship between bacteria and a protozoal parasite, the imaginary Pettenkoferia. It was the response of the bacterial genome to an attack by low dose LiCl put in the media. The living matter responded with the creation of an inexhaustible variety of morpho-biochemical life forms from "punktförmige Körperchen mit Wachstum", "größere Keimlinge", and the most spectacular "Riesenformen". For this author, what Professor Preisz viewed with astonishment was the ancient scenario of primordial spheroplasts surviving "at the edge" in the hostile environment of the ancient Earth, as replayed in his culture vessels (Table XIV).

Figure 67. Vascular Mimicry. Tumor cells recruit host vascular endothelial cells in the form of primitive blood vessels by secreting the VEGFs to be captured by VEGF-Rs on the neighboring endothelial cells. Tumor-friendly M2 Mϕs may be producing the VEGFs. The bone marrow upon receiving appropriate signals from tumor cells releases immature vascular endothelial cells, which travel to the tumor bed and line up into vascular channels. Certain tumor cells are able to genetically transspeciate themselves into vascular endothelial-like cells, which form primitive channels, that can circulate blood into the tumor bed. Gene activations and silencing include EphA2 receptors, cadherin VE, focal adhesion kinase (FAK), extracellular regulated kinase ERK1/2, and PI3K (Döme et al Am J Pathol 2007;170:1–15; Hess AR et al Dev Dyn 2007;236:3283–3296; Sharma N et al Prostate 2002;50:189–201). Uveal melanoma cells in the process of transformation into vasculogenic morphology, expressed the stem cell marker CD271 (NGFR, p45NTR nerve growth factor neurotrophin receptor) (Valyi-Nagy K et al Mol Vis 2012;18:588–592). In the placenta of the fetus, syncytiotrophoblasts are known to transspeciate themselves into artery-like blood vessels. In choriocarcinoma, an abundance of endotheial cell-like syncytiotrophoblasts provide the tumor's major blood supply (Shih I-M Mod Pathol 2011;24:646–652). Illustration: Uveal melanoma cells in culture display vasculogenic mimicry pattern and express CD271 (shown in Fig 2 of the cited article) (Valyi-Nagy K et al Mol Vis 2012;18:588–592)

Figure 68. Autophagy. Under lymphocyte attack, human sarcoma cells extensively vacuolized their cytoplasm, but remained metabolically alive. These unusual cells were viewed and photographed in the late 1960s. Then the phenomenon of autophagy was not recognized as such. Some sort of tumor cell dormancy was considered (Sinkovics JG J Med edited by Julian Ambrus Sr 1970;1:15–25; 313–326; reproduced in Sinkovics JG "Cytolytic Immune Lymphocytes…" Schenk Buchverlag, Passau/Budapest, 2008) (147).

Figure 69. Neuroblastoma (NB). Undifferentiated high grade tumor cells (showing ganglioneuromatous features; or rosette-formation elsewhere) are shown. The tumor cells express chromogranin, synaptophysin, vimentin, neurofilament and neuron-specific enolase. In the genome there is N-myc amplification; deletion of subtelomeric region at 1p36.33; low expression of maturation factor TrkA (tropomyosin receptor kinase). In sporadic (not the familial) NB tumors the NBPF gene (neuroblastoma breakpoint-family member) undergoes copy number variations in 1q21.1 deletion, or in 1q21.1 duplication. The entire NBPF gene underwent duplication during primate evolution. A human endogenous retroviral LTR has been inserted in one of the introns of the NBPF3 gene. On occasion, gene NBPF1 fuses with gene ACCN: t(1;17)(p36.2;q11.2) (ACCN = ASIC, acid sensing proton-gated ion channel) (Diskin SJ et al Nature 2009;459:987–991; Vandepoele K et al Mol Biol Evol 2005;22:2265–2274; PLoS One 3(5):e2207; Virology 2007;362:1–5). The LMO1 gene is duplicated. The homeobox LMO proteins (LIM Only) (Table XVI) are evolutionarily conserved (schistosoma; caenorhabditis; drosophila). LMO2 is a human leukemia oncogene/oncoprotein (El

Omari K Blood 2001;117:2145–2154; Matthews JM et al Nat Rev Cancer 2013;13: 111–122; Smith S et al PLoS One 9(1):e85883). The amoeba and the amoeboid cancer cells use LIM domain proteins for their locomotion (Guetta D et al PLoS One 2010;5(12):e15249). Additional information in the Appendix 1.

Figure 70. Pediatric Rhabdomyosarcoma of the orbit in the 1960s at M.D. Anderson Hospital's Sarcoma Service (Author's material). Treated with surgery (enucleation; orbital exenteration), radiotherapy and chemotherapy (actinomycin D, cyclophosphamide, vincristine); later, doxorubicin (Adriamycin), ifosfamide (Ifex), with prolongation of life with without cure. One eye lost, life saved in the 2010s. Intergroup Rhabdomyosarcoma Study Group / Children's Oncology Group report: sarcoma-free survival 92% at median 7 years; survival 82% at 5 years after relapse from remission(s) (Ranet B Pediatric Blood Cancer 2013;60:371–376). For relapses intensive chemoradiotherapy is recommended still with "curative intent". Late general health adverse side effects of chemoradiotherapy are significant (Hudson MM et al JAMA 2013;309:2371–81; Hofman MC et al J Clin Oncol 2013;31:2799–2805; St. Jude's Pediatric Oncology Hospital annual reports). Possible targeted epigenetic therapy emerges. The BMP2 (bone morphogenetic protein) promoter is silenced, especially in a-rhabdomyosarcoma. Demethylation by 5-Aza-2'-deoxycytidine (Decitabine) released BMP production with reduced viability of rhabdomyosarcoma cells resulting (Wolf S et al Int J Oncol 2014 Feb 27. doi:10.3892/ijo.2014/2312). Osteo-, rhabdomyo-, and Ewing-sarcoma cells treated with 5-Aza-2'dc overexpressed their MHC and NY-ESO-1 antigens (MAGE-A1/3) and become susceptible to the hosts' cytotoxic immune T lymphocytes (Krishnadas DK et al Tumour Biol 2014;35:5753–5762). Sirtuin (silent information regulators) histone deacetylases and ADP-ribosyl transferases prolong cellular life and are highly upregulated in synovial and rhabdomyosarcoma cells (and in several other malignantly transformed, such as leukemia and glioblastoma, cells). Tenovin-6- and anti-Sirt siRNA-treated sarcoma cells ceased proliferating due to increased levels of cell cycle inhibitors, inhibited autophagic flux, and restored p53-mediated apoptosis, but did not interact with the fusion oncoprotein FOXO/PAX (Ma L et al Cell Death Dis 2014;5(10):e1483). For cytotoxicity of NK and immune T lymphocytes to autologous sarcoma cells discovered in the late 1960s, see Sinkovics JG "Cytolytic Immune Lymphocytes…" 2008 Schenk Buchverlag, Passau/Budapest. Illustration: Sinkovics JG, Pp 417–488 in Monograph on Recent Trends in Cancer Chemotherapy. JR Chowdhury S Mitra. Chittaranjan National Cancer Research Centre, Calcutta, India. Terese Winslow. All rights reserved.

Figure 71. The Cancer Stem Cell is a mutated normal progenitor cell, or stem cell. The mutation constitutively turns on the stem cell genes (FAK, Nanog, c-Myc, Oct, Sox, Kfl, etc) (Table IX). The mutated stem cell will not differentiate. It may stop its asymmetric divisions (a self renewed stem cell, and a differentiating somatic daughter cell), and instead produces two undifferentiated stem cells. It has reactivated the genetically encoded ancient "cell survival pathways", which have been encoded by proto-oncogenes. The process is dominated by the constitutive expression of faculties for resistance to apoptosis, and to chemical and radiation exposure. The transforming cell assumes the locomotion of an amoeba by remodeling its cytoskeleton in the process of epithelial-to-mesenchymal transition. It hides its MHC-expressed antigens like a parasitic unicellular pathogen (Entamoeba, Giardia; Trichomonas; Naegleria; Plasmodia; Theileria; Trypanosoma; Neurospora; Phytophthora). It demands blood supply like a syncytiotrophoblast by inducing host fibroblasts and macrophages for its exocrine supply; or trans-speciates itself in imitating vascular endothelial cells in lining up into a primitive tubular structure with red blood cells inside (vasculogenic mimicry). Mutated cancer stem cells were generated in therapeutically γ-irradiated human breast cancer tissue. No indication for abandoning cancer radiotherapy (Pajonk F Dept Radiology Oncology David Geffen School of Medicine University of California Los Angeles CA). Terese Winslow.

Figure 72. A Murine Hybridoma. In the 1960s at the Section of Clinical Tumor Virology & Immunology of The University of Texas M.D. Anderson Hospital, the leukemogenic retrovirus-producer murine lymphoma cell line #620 was maintained by cell passages in vivo and in tissue cultures in vitro. In the mid-1960s in the in vitro cultures large round lymphoid cells appeared spontaneously and grew in suspension cultures (cell line #818). The large round cells were tetra-, or polyploid. In mice these cells grew in large haemorrhagic ascitic and invasive lymphomas. It was determined that these large round cells were the fusion products of the diploid #620 cells and normal plasma cells, which secreted immunoglobulins specific to the retroviral particles budding from the surface of #620 cells. In the fusion product #818 cell, its immune plasma cell component secreted the immunoglobulins, which neutralized the retrovirus released from the #620 cells (Sinkovics JG et al In: Leukemia-Lymphoma. YearBook Medical Publisher 1970; pp 53–92; J Inf Dis 1969;119:19–38; Lancet 1970;1:139–140; Cancer Res 1981;41:1246–1247; Rev Inf Dis 1983;5:9–34; Bull Mol Med 2005;26:61–80; Acta Microbiol Immunol Hung 2005;52:1–40). The virus-neutralizing antibodies were precipitated by Roman Pienta, who also confirmed the validity of the virus neutralization experiments (M.D. Anderson Hospital Research Report, vide infra). Eiichi Shirato found these antibodies weakly effective in inhibiting the growth of #620 cells (Shirato E et al Oncology 1972;26:80–86). The tetra- or polyploid #818 cells formed a permanent suspension culture cell line in vitro, which was maintained in the Author's Lab and also by M.D. Anderson's chief pathologist Jose Trujillo (Cancer Res 1970;30:540–545). Macrophages attacked and lysed the antibody-coated tetra- or polyploid #818 cells, thus portraying the "starry sky" phenomenon (Butler J et al Am J Pathol 1967;51:629–637; Sinkovics JG et al J Inf Dis 1969;120:250–254). The actual fusion of #620 diploid lymphoma cells with antibody-secreting immune plasma cells could be reproduced in the peritoneal cavity of mice inoculated separately but simultaneously with #620 lymphoma cells and immune plasma cells (see Illustration). This was independently confirmed by H. David Kay (Research Report of M.D. Anderson Hospital 1970; pp 331–336; 1972; pp 473–476; 1974 pp 447–455). Milton Wainwright (University Sheffield, Sheffield, UK) after interviewing H.

D. Kay, Jose Trujillo and this author (JGS), studying laboratory note books, and contemporary laboratory photographs, declared that the first natural hybridoma and the hybridoma principle (continuous antibody secretion by immune plasma cells fused with lymphoma cells) was discovered by J.G. Sinkovics and first presented in lectures in 1969 (Wainright M Perspect Biol Med 1992;35:372–379). Similar credit Sinkovics received from plenary session speaker M. D. Scharff at the 18th Annual Conference of the Infectious Diseases Society of America in 1980 in New Orleans (Yelton DE & Scharff MD Annu Rev Biochem 1981;50:657–680), and through the invitation of Jerome Vaeth for a comprehensive presentation on natural hybridomas at the 24th San Francisco Cancer Symposium (Sinkovics JG Front Rad Ther Oncol "The earliest concept of the hybridoma principle…" 1990;24:18–31.*)

*) *The editors of the Cancer Research accepted documentation of the discovery and existence of a murine natural hybridoma in 1960–70s (Sinkovics JG Cancer Research 1981;41:1296–1. The Iuliu Haţieganu University of Cluj Napoca (Kolozsvár) Romania gave Sinkovics the "Out-Nobel Award, 2005" for the discovery of the first "natural hybridoma" in the mid-1960s (Bull Mol Med 2005;26:61–80). In ten years of its availability in culture, neither Sinkovics nor Trujillo have ever received a request for the #620/#818 cell lines for independent studies, and the work with these cell lines was not cited in the literature.*

Figure 73. Combined Chemoimmunotherapy. The first report on this initially discouraged modality of treatment for metastatic sarcomas was generated at the Department of Medicine, The University of Texas M.D. Anderson Hospital, as first reported by Sinkovics JG, Plager C, Romero J, in Neoplasm Immunity: Solid Tumor Therapy. A University of Chicago Symposium, February 24/25 1977. Crispen RG editor. Franklin Institute Press, Philadelphia Pp 211–219, 1977 (Library of Congress Catalog Card Number 77-777-79. International Standard Book Number 0-89168-000-4). This article is cited in Sinkovics JG & Horvath JC Internat J Oncol 2006;29:765–777. Patients (odd numbers reporting for treatment) in the third group receiving cyclophosphamide/doxorubicin-based chemotherapy, BCG and viral oncolysate (VO) immunotherapy were directly compared with patients in the second group (even numbers reporting for treatment) receiving identical chemotherapy with BCG (without VO). Patients in the first group received standard cyclophosphamide/doxorubicin-based chemotherapy only (no VO; no BCG) at the same Institution preceding and during the administration of the chemoimmunotherapy protocol to the alternatingly registered patients of the second and third groups (cohorts). Patients in the first group could be considered to be contemporaneous historical controls of the second and third groups of patients. This clinical trial was supported by private donations raised by the Head of the Department of Medicine, Clifton Dexter Howe, who supervised/supported this clinical trial. C D Howe originally intended the publication of this clinical trial in the N Engl J Med with his co-authorship. The first author (JGS) was discouraged: for a clinical trial not supported and funded by the NCI, he expected adverse criticism and humiliating rejection. A preliminary presentation in the volume of a distinguished Chicago conference was a guarantee for a publication. That publication could have served as a platform for a NCI-approved second randomized clinical trial. C D Howe agreed; he did not oppose the presentation of the material at the Chicago symposium, reserved his co-authorship and offered his support in a NCI-approved future randomized clinical trial, but declined his co-authorship for the current publication as a book chapter. The patients enlisted in this clinical trial were co-attended by assistant internist-medical oncologist Nicholas Papadopoulos.

Figure 74. The Front Cover of the Volume containing the presentations given at the 14th Annual Clinical Conference on Cancer in 1969 at The University of Texas M. D. Anderson Hospital and Tumor Institute at Houston, Texas. The illustration was taken from the printed presentation entitled Relationship between Lymphoid Neoplasms and Immunologic Functions, by J.G. Sinkovics, E. Shirato, F. Gyorkey, J.R. Cabiness and C.D. Howe, Pp 53–92 (278). This article contains the first view of a murine natural hybridoma and the first photographs of human large granular lymphoid cells (later: NK cells) attacking human tumor cells. Published by Year Book Medical Publishers, Inc, Chicago, 1970 (Library of Congress Catalog Card Number 74–119578. International Standard Book Number 0-8151-0207-0). Cited in Sinkovics JG Bulletin Molecular Medicine 2005;26:61–80; Sinkovics JG Cytolytic Immune Lymphocytes.Schenk BuchVerlag, Passau/Budapest, 2008. View similar illustration of hybridoma formation on the cover of the volume "The 24th San Francisco Cancer Symposium" (cited in Figure 71). *)

*) *The Author's two invited presentations at The San Francisco Cancer Symposia were 1, on cytotoxic large granular human lymphocytes; and 2, on natural hybridomas (Front Rad Therap Oncol 1972;7:141–154; 1990;24:18–31).*

Figure 75. Of the Staff, Department of Medicine, The University of Texas M.D. Anderson Hospital and Tumor Institute, mid-1970s. From left to right, Sebron Culpepper Dale, associate internist, internal medicine, staff health care. Clifton Dexter Howe, board-certified internist, professor of medicine, medical oncologist, Head, Department of Medicine. Joseph Géza Sinkovics, professor of medicine; board-certified specialist in clinical pathology, laboratory medical virology, public health, internal medicine, infectious diseases, and medical oncology; medical oncologist, board eligible hematologist; Chief, Section (Laboratories) Clinical Tumor Virology & Immunology; Chief, Melanoma/Sarcoma Clinics and in-Patient Services. C. Charles Shullenberger, professor of medicine, Mayo Clinics-trained board-certified internist, hematologist; Chief, Section of Hematology (Laboratories; Clinics; in-Patient Services).

References

1. Bushman F. Lateral DNA Transfer. Mechanisms and Consequences. Cold Spring Harbor Laboratory Press, Cold Spring Harbor, New York, i-xiv pp. 1–448, 2002.
2. Sinkovics J. Vírusok az élő anyag körforgásában. Viruses in the cyclorama of the living matter. In: Hungarian with English summary and 600 references. Budapest: Studia Physiologica Scientia Publisher 2001;9:5–151.
3. Leach S, Smith IW, Cockell CS. Introduction: conditions for the emergence of life on the early Earth. Philos Trans R Soc London B Biol Sci. 2006;361:1675–9.

4a. Parker ET, Cleaves HJ, Dworkin JP, Glavin DP, Callahan MP, Aubrey A, Lazcano A, Bada R. Primordial synthesis of amines and amino acids in a 1958 Miller H_2S-rich spark discharge experiment. Proc Natl Acad Sci USA. 2011;108:5526–31.

4b. Parker ET, Cleaves HJ, Callahan MP, Dworkin JP, Glavin DP, Lazcano A, Bada JL. Enhanced synthesis of alkyl amino acids in Miller's 1958 H_2S experiment. Orig Life Evol Biosph. 2011;41:569–74.

5a. Eigen M, Schuster P. The hypercycle. Berlin: Springer; 1979.

5b. Eigen M. Error catastrophe and antiviral strategy. Proc Natl Acad Sci USA. 2002;99:13374–6.

5c. Eigen M, Schuster P. The hypercycle. A principle of natural self-organization. Emerg Hypercycle Naturwissenschaften. 1977;64:541–65.

6. Lee DH, Granja JAR, Martinez JA, Severin K, Ghadiri MR. A self-replicating peptide. Nature. 1996;382:525–8.
7. Lee DH, Severin K, Yokobayashi Y, Ghadiri MR. Emergence of symbiosis in peptide self-replication through a hypercyclic network. Nature. 1997;390:591–4.
8. Saghatelian A, Yokobayashi Y, Soltani K, Ghadiri MR. A chiroselective peptide replicator. Nature. 2001;409:797–801.
9. Cech TR. Crawling out of the RNA world. Cell. 2009;136:599–603.
10. Darnell J. RNA and the beginning of life. In: RNA life's indispensable molecule. Cold Spring Harbor, NY: Cold Spring Harbor Laboratory Press; 2011. p. 351–396.
11. Diener TO. Circular RNAs: relics of precellular evolution. Proc Natl Acad Sci USA. 1989;86:9370–4.
12. Flügel RM. The precellular scenario of genovirions. Virus Genes. 2010;40:151–4.

13a. Koonin EV. On the origin of cells and viruses: primordial virus world scenario. Ann NY Acad Sci. 2009;1178:47–64.

13b. Dolja VV, Koonin EV. Common origin and host-dependent diversity of plant and animal viruses. Curr Opin Virol. 2011;1:322–31.

14a. Raoult D, Forterre P. Redefining viruses: lessons from Mimivirus. Nat Rev Microbiol. 2008;6:3125–9.

14b. Raoult D. The post-Darwinist rhizome of life. Lancet. 2010;375(9709):104–5.

15. Oparin AI. Evolution of the concepts of the origin of life. 1924–1974. Orig Life. 1976;7:3–8.

J.G. Sinkovics, *RNA/DNA and Cancer*, DOI 10.1007/978-3-319-22279-0

16. Gánti T. The priciples of life. New York: Oxford University Press; 2003.
17. Szathmáry E. The origin of replicators and reproducers. Philos Trans R Soc London B Biol Sci. 2006;361:1761–76.
18. Koonin VE, Senkevich TG, Dolja VV. The ancient virus world and evolution of cells. Biol Direct. 2006;1:29. doi:10.1186/1745-6150-1-29.
19. Wahba L, Amon JD, Koshland D, Vuica-Ross M. RNase and multiple RNA biogenesis factors cooperate to prevent RNA:DNA hybrids from generating genome instability. Mol Cell. 2011;44:978–88.
20. Nakamura H, Oda Y, Iwai S, Inoue H, Ohtsuka E, Kanaya S, Kimura S, Katsuda C. How does RNase H recognize a DNA.RNA hybrid? Proc Natl Acad Sci USA. 1991;88:11535–9.
21. Woese CR. The universal ancestor. Proc Natl Acad Sci USA. 1998;95:6854–9.
22. Woese CR. On the evolution of cells. Proc Natl Acad Sci USA. 2002;99:8742–7.
23. Forterre P, Philippe H. Where is the root of the universal tree of life? Bioessays. 1999;21:871–9.
24. Glansdorff N, Xu Y, Labedan B. The last universal common ancestor: emergence, constitution and genetic legacy of an elusive forerunner. Biol Direct. 2008;3:29. doi:10.1186/1745-6150-3-29.
25. Glansdorff N, Xu Y, Labedan B. The origin of life and the last universal common ancestor: do we need a change of perspective? Res Microbiol. 2009;160:522–8.
26. Sinkovics JG. Horizontal gene transfers and cell fusions in microbiology, immunology and oncology. Int J Oncol. 2009;35:441–65.
27. Sinkovics JG. Horizontal gene transfers with or without cell fusions in all categories of the living matter. In: Cell fusion in health and disease. II. Cell fusion in disease. Springer. Adv Exper Med Biol. 2011;714:5–89. doi:10.1007/978-94-007-0782-5_2.
28. Takemura M. Poxviruses and the origin of the eukaryotic nucleus. J Mol Evol. 2001;52:419–25.
29a. Villarreal LP. Viruses and the evolution of life. Herndon, VA, USA: American Society Microbiology Press; 2005.
29b. Villarreal LP, Witzany G. Viruses are essential agents within the roots and stem of the tree of life. J Theor Biol. 2010;262:698–710.
30a. Sinkovics JG, Horvath JC. Human natural killer cells: a comprehensive review. Int J Oncol. 2005;27:5–47.
30b. Sinkovics JG, Tebbi K, Cabiness JR. Cytotoxicity of lymphocytes to established cultures of human tumors: evidences for specificity. Natl Cancer Inst Monograph "Conference and Workshop on Cellular Immune Reactions to Human Tumor-Associated Antigens" 1973;37:9–18.
30c. Sinkovics JG. Adoptive immunotherapy of human cancers. Flagmen signal first "open road" then "road blocks". A narrative synopsis. In: Kiselevsky MV, editor. Atlas. Effectors of antitumor immunity. Springer; 2008. p. 1–23
31. Ferguson AL, Falkowska E, Walker LM, Seaman MS, Burton DR, Chakraborty A. Computational prediction of broadly neutralizing HIV-1 antibody epitopes from neutralization activity data. PLoS One. 2013;8(12):e80562.
32. Temin HM, Baltimore D. RNA-directed DNA synthesis and RNA tumor viruses. Adv Virus Res. 1972;17:129–86.
33. El Kharroubi A, Piras G, Zensen R, Martin MA. Transcriptional activation of the integrated chromatin-associated human immunodeficiency virus type 1 promoter. Mol Cell Biol. 1998;18:2535–44.
34. Lemasson I, Polakowski NJ, Laybourn PJ, Nyborg JK. Transcription factor binding and histone modifications on the integrated proviral promoter in human T-cell leukemia virus-I-infected T-cell. J Biol Chem. 2002;277:49459–65.

35. Easley R, Carpio L, Guendel I, Klase Z, Choi S, Kahn-Hall K, Brady JN, Kashanchi F. Human T-lymphotrophic virus type 1 transcription and chromatin-remodeling complexes. J Virol. 2010;84:4755–68.
36a. Sinkovics J. Pleuropneumonie-ähnliche Eigenheiten in alten E. coli Kulturen. Acta Microbiologica Hungarica. 1957;4:59–75.
36b. Sinkovics J. A bakteriológia új fejezete. A new chapter in bacteriology: Pleuropneumonia-like growth (In Hungarian). Orvosi Hetilap (Budapest) Wkly Med J. 1955;96:1401–8.
36c. Sinkovics J. Die Grundlagen der Virusforschung. Budapest: Verlag der Ungarischen Akademie der Wissenschaften; 1956.
37. Sinkovics J. Occurrence and filterability of protoplast-like elements in aged bacterial cultures. Nature. 1958;181:566–7.
38. Sinkovics JG. Up-dating the monograph "Cytolytic Immune Lymphocytes". Part I. Basic science. Magyar Epidemiológia (Hungarian Epidemiology, in English). 2008;5:237–255. Addendum. *Idem*, 2009;6:60.
39. Sinkovics JG. The place of viruses in the "Tree of Life". Acta Microbiol Immunol Hung. 2001;48:115–27.
40. Klenk H-P, Meier T-D, Durovic P, Schwass V, Lottspeich F, Dennis PP, Zillig W. RNA polymerase of Aquifex pyrophilus: implications for the evolution of the bacterial rpoBC operons and extremely thermophilic bacteria. J Mol Evol. 1999;48:528–41.
41. Av Kukekova, Malinin AY, Ayala JA, Borkhsenius SN. Characterization of Acholeplasma laidlawii ftsZ gene and its gene product. Biochem Biophys Res Commun. 1999;282:44–9.
42. Lindås A-C, Karlsson EA, Lindren MT, Ettema TJG, Bernander R. A unique cell division machinery in Archaea. Proc Natl Acad Sci USA. 2008;105:18942–6.
43. Cox CJ, Foster PG, Hirt RP, Harris SR, Embley TM. The archaebacterial origin of eukaryotes. Proc Natl Acad Sci USA. 2008;105:20356–61.
44. Woese CR, Maniloff J, Zablen LB. Phylogenetic analysis of the mycoplasmas. Proc Natl Acad Sci USA. 1980;77:494–8.
45. Miller S, Lazcano A. The origin of life—Did it occur at high temperatures. J Mol Evol. 1995;41:689–92.
46. Forterre P. A hot topic: the origin of hyperthermophiles. Cell. 1996;65:789–92.
47. Korolev S, Nayal M, Barnes WM, Di Cerca E, Waksman G. Crystal structure of a large fragment of Thermus aquaticus DNA polymerase at 2.5Å resolution: structural basis for thermostability. Proc Natl Acad Sci USA. 1995;92:9264–8.
48. Wei W, Davis RE, Jomantiene R, Zhao Y. Ancient, recurrent phage attacks and recombinations shaped dynamic sequence-variable mosaics at the root of phytoplasma genome evolution. Proc Natl Acad Sci USA. 2008;105:11827–32.
49. Koui T, Natsuaki T, Okuda S. Antiserum raised against gyrase A of Acholeplasma laidlawii reacts with phytoplasma proteins. FEMS Microbiol Lett. 2002;206:169–74.
50. Maniloff J. Evolution of wall-less prokaryotes. Ann Rev Microbiol. 1983;37:477–99.
51. Lazarev VN, Levitskii SA, Basovskii YI, Chukin MM, Akopian TA, Vereshchagin VV, Kostrjukova ES, Kovaleva GY, Kazanov MD, Malko DB, Vitreschak AG, Sernova NV, Gelfand MS, Demina IA, Serebryakova MV, Galyamina MA, Vyturin NN, Rogov SI, Alexeev DG, Ladygina VG, Govorun VM. Complete genome and proteome of Acholeplasma laidlawii. J Bacteriol. 2011;193:4943–53.
52. Tarshis M, Salman M, Rottem S. Fusion of mycoplasmas: the formation of cell hybrids. FEMS Microbiol Lett. 1991;66:67–71.
53. Brunner H, Schaar H, Krauss H. Giant cell formation in Acholeplasma laidlawii strain JA1. Zentralbl Bakteriol A. 1980;248:120–8.
54. Haberer K, Frösch D. Lateral mobility of membrane-bound antibodies on the surface of Acholeplasma laidlawii: evidence for virus-induced cell fusion in a prokaryote. J Bacteriol. 1982;152:471–8.

55. Sladek TL, Nowak JA, Maniloff J. Mycoplasma restriction: identification of a new type of restriction specify for DNA containing 5-methylcytosine. J Bacteriol. 1986;165:219–25.
56. Gourlay RN. Isolation of a new virus infecting strain of Mycoplasma laidlawii. Nature. 1970;225:1165.
57. Gourlay RN. Mycoplasmatales virus laidlawii 2, a new virus isolated from Acholeplasma laidlawii. J Gen Virol. 1971;12:65–7.
58. Gourlay RN, Wyld SG. Isolation of mycoplasmatales virus-laidlawii 3, a new virus infecting Acholeplasma laidlawii. J Gen Virol. 1973;19:279–83.
59. Dybwig K, Maniloff J. Integration and lysogeny by an enveloped mycoplasma virus. J Gen Virol. 1983;64:1781–5.
60. Dybwig K, Sladek TL, Maniloff J. Isolation of mycoplama virus L2 insertion variants and miniviruses. J Virol. 1986;59:584–90.
61. Maniloff J, Kampo GJ, Dascher CC. Sequence analysis of a unique temperature phage: mycoplasma virus L2. Gene. 1994;141:1–8.
62. Just W, de Silva Cardoso M, Lorenz A, Klotz G. Release of mycoplasmavirus L1 upon transfection of Acholeplasma laidlawii with homologous and heterologous viral DNA. Arch Virol. 1989;107:1–13.
63. Sladek TL, Maniloff J. Polyethylene glycol-dependent transfection of Acholeplasma laidlawii with mycoplasma virus L2 DNA. J Bacteriol. 1983;155:734–41.
64. Just W, Klotz G. Terminal redundancy and circular permeation of mycoplasma virus L3 DNA. J Gen Virol. 1990;71:2157–62.
65. Lorenz A, Just W, de Silva Cardoso M, Klotz G. Electroporation-mediated transfection of Acholeplasma laidlawii with mycoplasma virus L1 and L3 DNA. J Virol. 1988;62:30503052.
66. Haberer K, Klotz G, Maniloff J, Kleinschmidt AK. Structural and biological properties of Mycoplasmavirus MLV3: an unusual virus-procaryote interaction. J Virol. 1979;32:268–75.
67. Haberer K, Haberer AI, Cadden SP, Maniloff J. Isolation and characterization of mycoplasma virus L3 temperature-sensitive mutants. Microbios. 1990;64:111–25.
68. Congdon AL, Kenny GE. Alteration of colonial morphology of Acholeplasma laidlawii and Acholeplasma modicum by infection with mycoplasmatales viruses. J Bacteriol. 1979;138:962–8.
69. Franzoso G, Dimitrov DS, Blumenthal R, Barile MF, Rottem S. Fusion of Mycoplasma fermentans strain incognitus with T lymphocytes. FEBS Lett. 1992;303:251–4.
70a. Sinkovics JG. Immunology of tumors in experimental animals. In: The Immunology of Malignant Disease. 2nd ed. In: Harris JE, Sinkovics JG, editors. St. Louis: C.V. Mosby; 1978. p. 93–282.
70b. Sinkovics JG, Harris JE. Immunology and immunotherapy of human tumors. In: The Immunology of Malignant Disease. 2nd ed. In: Harris JE, Sinkovics JG, editors. St. Louis: C.V. Mosby; 1976. p. 410–579.
71. Sinkovics JG, Horvath JC. Natural and genetically engineered viral agents for oncolysis and gene therapy of human cancers. Arch Immunol Ther Exp. 2006;56:3s–50s.
72. Tsai S, Wear DJ, Shih JW, O SC. Mycoplasmas and oncogenesis: persistent infection and multistage malignant transformation. Proc Natl Acad Sci USA. 1995;92:10197–201.
73. Sinkovics JG. Molecular biology of oncogenic inflammatory processes. I. Non-oncogenic and oncogenic pathogens, intrinsic inflammatory reactions without pathogens, and microRNA/DNA interactions. Int J Oncol. 2012;40:305–49.
74. Wiedenheft B, Stedman K, Roberto F, Willits D, Gleske A-K, Zoeller L, Snyder J, Douglas T, Young M. Comparative genomic analysis of hyperthermophilic archaeal fuselloviridae viruses. J Virol. 2004;78:1954–61.
75. Prangishvili D, Garrett RA, Koonin EV. Evolutionary genomics of archaeal viruses: unique viral genomes in the third domain of life. Virus Res. 2006;117:52–7.

76. Pina M, Bize A, Forterre P, Prangishvili D. The archaeoviruses. FEMS Microbiol Rev. 2011;35:1035–54.
77. Brodt A. Lune-Weinberger MN, Gophna U. CRISPR loci reveal networks of gene exchange in archaea. Biol Direct. 2011;21.6(1):65.
78. Lintner NG, Kerou M, Brumfield SK, Graham S, Liu H, Naismith JH, Sdano M, Peng N, She Q, Copié V, Young MJ, White M, Lawrence CM. Structural and functional characterization of an archaeal clustered regularly interspaced short palindromic repeat (CRISPR)-associated complex for antiviral defence (CASCADE). J Biol Chem. 2011;286:21643–56.
79. Makarova KS, Wolf YI, Snir S, Koonin EV. Defense islands in bacterial and archaeal genomes and prediction of novel defense systems. J Bacteriol. 2011;193:6039–56.
80. Wiedenheft B, Sternberg SH, Doudna JA. RNA-guided genetic silencing systems in bacteria and archaea. Nature. 2012;482:331–8.
81. Takeuchi N, Wolf YI, Makarova KS, Koonin E. Nature and intensity of selection pressure on CRISPR-associated genes. J Bacteriol. 2012;194:1216–25.
82. Forterre P. A new fusion hypothesis for the origin of Eukarya: better than previous ones, but probably also wrong. Res Microbiol. 2011;162:77–91.
83. Sinkovics JG, Horvath JC. Kaposi's sarcoma: breeding ground of herpesviridae—A tour de force over viral evolution. Int J Oncol. 1999;14:615–646. Errata: *Idem* 2005;27:5–47.
84. Al-Attar S, Westra ER, van der Oost J, Brouns SJ. Clustered regularly interspersed short palindromic repeats (CISPRs): the hallmark of an ingenious antiviral defense mechanism in prokaryotes. Biol Chem. 2011;392:277–89.
85. Horvath P, Barrangou R. CRISPR/Cas, the immune system of bacteria and archaea. Science. 2010;327:167–70.
86. Makarova KS, Wolf YI, van der Oost J, Koonin EV. Prokaryotic homologs of argonaute proteins are predicted to function as key components of a novel system of defense against mobile genetic elements. Biol Direct. 2009;4:29. doi:10.1186/1745-6150-4-29.
87. Portillo MC, Gonzalez JM. CRISPR elements in the Thermococcales: evidence for associated horizontal gene transfer in Pyrococcus furiosus. J Appl Genet. 2009;50:421–30.
88a. Blobel G. Intracellular protein topogenesis. Proc Natl Acad Sci USA. 1980;77:1496–500.
88b. Blobel G. Protein targeting. Chembiochem. 2000;1:86–102.
89a. Margulis L. Recombination of non-chromosomal genes in Chlamydomonas: assortment of mitochondria and chloroplasts? J Theor Biol. 1970;26:337–42.
89b. Schaechter M. Lynn Margulis (1938–2011). Science. 2012;335:302.
90a. Palmer CS, Osellame LD, Stojanovski D, Ryan MT. The regulation of mitochondrial morphology: intricate mechanisms and dynamic machinery. Cell Signal. 2011;23:1534–45.
90b. Osellame LD, Blacker TS, Duchen MR. Cellular and molecular mechanisms of mitochondrial function. Best Pract Res Clin Endocrinol Metab. 2012;26:711–23.
91. Matthaei JH, Nirenberg MW. Characteristics and stabilization of DNAase-sensitive protein synthesis in E. coli cell extracts. Proc Natl Acad Sci USA. 1961;47:1580–6.
92. Nirenberg MW, Matthaei JH. The dependence of cell-free protein synthesis in E. coli upon naturally occurring or synthetic polyribonucleotides. Proc Natl Acad Sci USA. 1961;47:1588–602.
93. Taylor WR. Transcription and translation in an RNA world. Philos Trans R Soc London B Biol Soc. 2006;361:1751–60.
94. Taylor WR. A molecular model for the origin of protein translation in an RNA world. J Theor Biol. 2006;243:393–406.
95. Bernhardt HS, Tate WP. The transition from noncoded to coded protein synthesis and coding mRNAs arise from stability-enhancing binding partners to tRNAs. Biol Direct 2010;5:16. doi:10.1186/1745-6150-5-16.

96. Catania F, Lynch M. Evolutionary dynamics of a conserved sequence motif in the ribosomal genes of the ciliate Paramecium. BMC Evol Biol. 2010;10:229. doi:10.1186/1471-2148-10-129.
97. Chen XS, Collins LJ, Biggs PJ, Penny D. High thoroughput genome-wide survey of small RNA from the parasitic protists Giardia intestinalis and Trichomonas vaginalis. Genome Biol Evol. 2009;1:165–75. doi:10.1093/gbe/evp017.
98. Garlapati S, Saraiya AA, Wang CC. A La autoantigen homologue is required for the internal ribosome entry site mediated translation of giardiavirus. PLoS One. 2001;6(3): e18263. doi:10.1971/journal.pone.0018263.
99. Wei L, Saraiya AA, Wang CC. Gene regulation in Giardia lamblia involves a putative microRNA derived from a small nucleolar RNA. PLoS Negl Trop Dis. 2011;5(10): e1338. doi:10.1371/journal.pntg.0001338.
100. Prucca CG, Lujan MD. Antigen variation in Giardia lamblia. Cell Microbiol. 2009;11:1706–15.
101. Molnár A, Schwach F, Studholme DJ, Thuenermann EC, Baulcombe DC. mRNAs control gene expression in the single-cell alga Chlamydomonas reinhardtii. Nature. 2007;477:1126–9.
102. Godman JE, Molnár A, Baulcombe DC, Balk J. RNA silencing of hydrogenase(-like) genes and investigation of their physiological roles in the green alga Chlamydomonas reinhardtii. Biochem J. 2010;431:345–51.
103. Boyd JB, Lamb MR, Dieckmann CL. Miniature- and multiple-eyespot loci in Chlamydomonas reinhardtii define new modulators of eyespot photoreception and assembly. Genes Genomes Genetics. 2011;1:489–98.
104. Förstemann K, Horwich MD, Wee LM, Tomari Y, Zamore PD. Drosophila microRNAs are sorted into functionally distinct Argonaute protein complexes after their production by Dicer-1. Cell. 2007;130:287–97.
105. Miura S, Nozawa M, Nei M. Evolutionary changes of the target sites of two microRNAs encoded in the Hox gene cluster of Drosophila and other insect species. Genome Biol Evol. 2011;3:129–39.
106. Siu RWC, Fragkoudis R, Simmonds P, Donald CL, Chase-Topping ME, Barry G, Attarzadeh-Yazdi G, Rodrigues-Andres J, Nash AA, Merits A, Fazakerley JK, Kohl A. Antiviral RNA interference responses induced by Semliki forest virus infection of mosquito cells: characterization, origin, and frequency-dependent functions of virus-derived interfering RNAs. J Virol. 2011;85:2907–17.
107. Félix M-A, Ashe A, Piffaretti J, Wu G, Nuez I, Bélicard T, Jiang Y, Zhao G, Franz CJ, Goldstein LD, Sanroman M, Miska EA, Wang D. Natural and experimental infection of Caenorhabditis nematodes by novel viruses related to nodaviruses. PLoS Biol. 2011;9(1): e1000586. doi:10.1371/journal.pbio.1000586.
108. Rechavi O, Minevich G, Hobert O. Transgenerational inheritance of an acquired small RNA-based antiviral response in C. elegans. Cell. 2011;147:1248–56.
109. Iskra S, Kalla M, Delecluse H-J, Hammerschmidt W, Moosmann A. Toll-like receptor agonists synergically increase proliferation and activation of B cells by Epstein-Barr virus. J Virol. 2010;8:3612–23.
110. Gottwein E, Cullen BR. Viral and cellular microRNAs as determinants of viral pathogenesis and immunity. Cell Host Microbe. 2008;3:375–87.
111. Hasnoot J, Berkhout B. RNAi and cellular miRNAs in infections by mammalian viruses. Methods Mol Biol. 2011;721:23–41.
112. Sachdeva M, Liu Q, Cao J, Lu Z, Mo Y-Y. Negative regulation of miR-145 by C/EBP-beta through the Akt pathway in cancer cells. Nucleic Acid Res. 2012. doi:10.1093/nar/gks324.

113. Skalsky RL, Corcoran DL, Gottwein E, Frank CL, Kang S, Hafner M, Nusbaum JD, Feederle R, Delecluse H-J, Luftig MA, Tuschl T, Ohier U, Cullen BR. The viral and cellular microRNA targetome in lymphoblastoid cell lines. PLoS Pathog 2012;8(1): e1002484. doi:10.1371/journal.ppat1002484.
114. Samols MA, Skalsky RL, Maldonado AM, Riva A, Lopez MC, Baker HV, Renne R. Identification of cellular genes targeted by KSHV-encoded microRNAs. PLoS Pathog. 2007;3(5):e65. doi:10.1371/journal.ppat.0030065.
115. Hinas A, Reimegård J, Wagner EGH, Nellen W, Ambros VR, Söderbom F. The small RNA repertoire of Dictyostelium discoideum and its regulation by components of the RNAi pathway. Nucleic Acids Res. 2007;35:6714–26.
116. Kapitonov VV, Jurka J. The esterase and PHD domains in CR1-like non-LTR retrotransposons. Mol Biol Evol. 2003;20:38–46.
117. Goodwin TJD, Poulter RTM. A new group of tyrosine recombinase-encoding retrotransposons. Mol Biol Evol. 2004;21:746–59.
118. Gimenez G, Bertelli, Moliner C, Robert C, Raoult D, Fournier P-E, Greub G. Insight into cross-talk between intra-amoebal pathogens. BMC Genomics. 2011;12:542. doi:10.1186/1471-2164-12-542.
119. Boyer M, Azza S, Barrassi L, Klose T, Campocasso A, Pagnier I, Fournous C, Borg A, Robert C, Zhang X, Desnues C, Henrissat B, Rossmann MG, La Scola B, Raoult D. Mimivirus shows dramatic genome reduction after intraamoebal culture. Proc Natl Acad Sci USA. 2011;108:10296–301.
120. Boyer M, Yutin N, Pagnier I, Barrassi L, Fournous G, Espinosa L, Robert C, Azza S, Sun S, Rossmann M, Suzan-Monti M, La Scola B, Koonin EV, Raoult D. Giant Marseillevirus highlights the role of amoebae as a melting pot in emergence of chimeric microorganisms. Proc Natl Acad Sci USA. 2009;106:21848–53.
121. Arsian D, Legendre M, Seltzer V, Abergel C, Claverie JM. Distant Mimivirus relative with larger genome highlights the fundamental features of Megaviridae. Proc Natl Acad Sci USA. 2011;108:17486–91.
122. Moliner C, Raoult D, Fournier P-E. Evidence that the intra-amoebal Legionella drancourtii acquired a strerol reductase from eukaryotes. BMC Res Notes. 2009;2:51. doi:10.1186/1756-0500-2-51.
123. MacFarlane RC, Shah PH, Singh U. Transcriptional profiling of Entamoeba histolytica trophozoites. Int J Parasitol. 2005;35:533–42.
124. Fritsch T, Horn M, Wagner M, Herwig RP, Schleifer K-H, Gautom RK. Phylogenetic diversity among geographically dispersed Chlamydiales endosymbionts recovered from clinical and environmental isolates. Appl Environ Microbiol. 2000;66:2613–9.
125a. Couvillion MT, Lee SR, Hogstad B, Malone CD, Tonkin LA, Sachidanandam R, Hannon GJ, Collins K. Sequence, biogenesis, and function of diverse small RNA classes bound to the Piwi family proteins of Tetrahymena thermophila. Genes Dev. 2009;23:2016–32.
125b. Couvillion MT, Sachidanandam R, Collins K. A growth-essential Tetrahymena Piwi protein carries tRNA fragment cargo. Genes Dev. 2010;24:2742–7.
126. Yu L, Gorovsky MA. Constitutive expression, not a particular primary sequence, is the important feature of the H3 replacement variant hv2 in Tetrahymena thermophila. Mol Cell Biol. 1997;17:6303–10.
127. Wilusz JE, Whipple JM, Phizicky EM, Sharp PA. tRNAs marked with CCACCA are targeted for degradation. Science. 2011;334:817–21.
128. Kawaji H, Nakamura M, Takahashi Y, Sandelin A, Katayama S, Fukuda S, Daub CO, Kai C, Kawai J, Carninci P, Hayashikazi Y. Hidden layers of human small RNAs. BMC Genomics. 2008;9:157. doi:10.1186/1471-2164-9-157.
129. Jenkins VK, Trentin JJ, Speirs RS, McGarry MP. Hemopoietic colony studies. J Cell Physiol. 1972;79:413–22.

129a. Trentin JJ, Curry JL, Wolf N, Cheng V. Factors controlling stem cell differentiation and proliferation: the hematopoietic inductive microenvironment (HIM). In: The proliferation and spread of neoplastic cells. Baltimore: The Williams and Wilkins Company; 1968, p. 713–727. Discussion 727–731.

130. Thomson T, Lin H. The biogenesis and function of PIWI proteins and piRNAs: progress and prospect. Annu Rev Cell Dev Biol. 2009;25:395–376.

131. Szakmary A, Reedy M, Qi H, Lin H. The Yb protein defines a novel organelle and regulates male germline stem cell self-renewal in Drosophila melanogaster. J Cell Biol. 2009;185:613–27.

132. Smulders-Srinivasan TK, Szakmary A, Lin H. A drosophila chromatin factor interacts with the piwi-interacting RNA mechanism in niche cells to regulate germline stem cell self-renewal. Genetics. 2019;186:573–83.

133. Juliano C, Wang J, Lin H. Uniting germline and stem cells: the function of Piwi proteins and the piRNA pathway in diverse organisms. Annu Rev Genet. 2011;45:447–69.

134. Kowalczykiewicz D, Wrzesinski J. The role of piRNA and Piwi proteins in regulation of germline development. Postepy Biochem (Poland). 2011;57:249–356.

135. Woolley AG, Algie M, Samuel W, Harfoot R, Wiles A, Hung NA, Tan P-H, Hains P, Valova VA, Huschtscha L, Royds JA, Perez D, Yoon H-S, Cohen SB, Robinson PJ, Bay B-H, Lasham A, Braithwaite AW. Prognostic association of YB-1 expression in breast cancers: a matter of antibody. PLoS One. 2011;6(6):e20603. doi:10.1371/journal.pone.0020603.

136. Ye Y, Yin D-T, Chen L, Zhou Q, Shen R, He G, Yan G, Tong Z, Issekutz A, Shapiro CL, Barsky SH, Lin H, Li J-J, Gao J-X. Identification of Piwil2-like (PL2L) proteins that promote tumorigenesis. PLoS. 2010;5(10):e13406. doi:10.1371/journal.pone,0013406.

137. Lee JH, Schütte D, Wulf G, Füzesi L, Radzun HJ, Schweyer S, Engel, Nayernia K. Stem-cell protein Piwil2 is widely expressed in tumors and inhibits apoptosis through activation of Stat3/Bcl-XL pathway. Hum Mol Genet. 2006;15:201–11.

138. Lu Y, Zhang K, Li C, Yao Y, Tao D, Liu Y, Zhang S, Ma Y. Piwi2 suppresses p53 by inducing phosphorylation of signal transducer and activator of transcription 3 in tumor cells. PLoS One. 2012;7(1):e30999. doi:10.1371/journal.ponr.0030999.

139. He W, Wang Z, Wang Q, Fan Q, Shiu C, Wang J, Giercksky KE, Nesland JM, Suo Z. Expression of HIWI in human esophageal squamous cell carcinomas is significantly associated with poorer prognosis. BMC Cancer. 2009;89:426. doi:10.1186/1471-2407-9-426.

140. Taubert H, Würl P, Greither T, Kappler M, Bache M, Bartel F, Kehlen A, Lautenschläger C, Harris LC, Kaushal D, Füssel S, Meye A, Böhnke A, Schmidt H, Holzhausen HJ. Stem cell-associated genes are extremely poor prognostic factors for soft-tissue sarcoma patients. Oncogene. 2007;26:7170–4.

141. Zeng Y, L-k Qu, Meng L, Liu C-Y, Dong B, Xing X-F, Wu J, Shou C-C. HIWI \expression profile in cancer cells and its prognostic value for patients with colorectal cancer. Chinese Med J. 2011;124:2144–9.

142. Lacerda I, Pusztai L, Woodward WA. The role of tumor-initiating cells in drug resistance of breast cancer: implications for future therapeutic approaches. Drug Resist Updat. 2010;13:99–108.

143. Smith HA, Cronk RJ, Lang JM, McNeel DG. Expression and immunotherapeutic targeting of the SSX family of cancer-testis antigens in prostate cancer. Cancer Res. 2011;71:6785–95.

144. Liu W, Zhai M, Wu Z, Qi Y, Wu Y, Dai C, Sun M, Li L, Gao Y. Identification of a novel HLA-A2-restricted cytotoxic T lymphocyte epitope from cancer-testis antigen PLAC1 in breast cancer. Amino Acids. 2011;PMID:21710262.

145. Yuan J, Adamov M, Ginsberg BA, Rasaian TS, Ritter E, Gallardo HF, Xu Y, Pogonier E, Terzulli SL, Kuk D, Panageas KS, Riter G, Sznol M, Halanban R, Jungbluth AA, Allison JP, Old LJ, Wolchok JD, Gnjatic S. Integrated NY-ESO-1 antibody and CD8+ T-cell responses correlate with clinical benefit in advanced melanoma patients treated with ipilimumab. Proc Natl Acad Sci USA. 2011;108:16723–8.
146. Robbins OPF, Morgan RA, Feldman SA, Yang JC, Sherry RM, Dudley ME, Wunderlich JAR, Nahvi AV, Helman LJ, Mackall CL, Kammula US, Highes MS, Restifo NP, Raffeld M, Lee C-CR, Levy CL, Li YF, El-Gamil M, Schwarz SL, Laurencot C, Rosenberg SA. Tumor regression in patients with metastatic synovial sarcoma and melanoma using genetically engineered lymphocytes reactive with NY-ESO-1. J Clin Oncol. 2011;29:917–24.
147. Sinkovics JG. Cytolytic immune lymphocytes in the armamentarium of the human host. Passau and Budapest: Schenk Buchverlag; 2008.
148. Noon KR, Guymon R, Crain PF, McCloskey JA, Thomm M, Lim J, Cavicchioli R. Influence of temperature on tRNA modification in archaea: Metanococcides burtonii optimum growth temperature (Topt 23°C) and Stetteria hydrogenophila (Topt 95°C). J Bacteriol. 2003;5483–5490.
149. Guymon R, Pomerantz SC, Crain PF, McCloskey JA. Influence of phylogeny on posttranscriptional modification of rRNA in thermophilic prokaryotes: the complete modification map of 16S rRNA of Thermus thermophilus. Biochemistry. 2006;45:4888–99.
150. Du S, Pal D, Ghosh SK. Entamoeba histolytica: computational identification of putative microRNA candidates. Exp Parasitol. 2006;113:239–43.
151. Huang PJ, Lin WC, Chen SC, Lin YH, Sun CH, Lyu PC, Tang P. Identification of putative miRNAs from the deep-branching unicellular flagellates. Genomics. 2012;99:101–7.
152. Wei FY, Tomizawa K. Functional loss of Cdkal1, a novel tRNA modification enzyme, causes the development of type 2 diabetes. Endocr J. 2011;58:819–25.
153. Wei KF, Wu LJ, Chen J, Chen YF, Xie DX. Structural evolution and functional diversification analyses of argonaute protein. J Cell Biochem. 2012;113:2576–85.
154. Zhang H, Ehrenkaufer GM, Pompey JM, Hackney JA, Singh U. Small RNA with 5'-polyphosphate termini associate a Piwi-related protein and regulate gene expression in the single-celled eukaryote Entamoeba histolytica. PLoS Pathog. 2008;4(11):e1000219.
155. Bouhouche K, Gout J-F, Kapusta A, Bétermier M, Meyer E. Functional specialization of Piwi proteins in Paramecium tetraaurelia from posttranscriptional gene silencing to genome remodeling. Nucleic Acid Res. 2011;39:4249–84.
156. Mallick B, Ghosh Z. Chakrabarti J. MicroRNA switches in Trypanosoma brucei. Biochem Biophys Res Commun. 2008;372:459–63.
157. Shi H, Chamond N, Djikeng A, Tschudi C, Ullu E. RNA interference in Trypanosoma brucei. Role of the terminal RGG domain and the polyribosome association of Argonaute. J Biol Chem. 2009;284:36511–20.
158. Garcia-Silva MR, Frugier M, Tosar JP, Correa-Dominguez A, Ronalte-Alves L, Parodi-Talice A, Rovira C, Robello C, Goldenberg S, Cayota A. A population of tRNA-derived small RNAs is actively produced in Trypanosoma cruzi and recruited to specific granules. Mol Biochem Parasitol. 2010;171:64–73.
159. Kolev NG, Tschudi C, Ullu E. RNA interference in protozoan parasites: achievements and challenges. Eukaryot Cell. 2011;10:1156–63.
160. Heimberg AM, Sempere LF, Moy VN, Donoghue PCJ, Peterson KHJ. MicroRNAs and their advent of vertebrate morphological complexity. Proc Natl Acad Sci USA. 2008;105:2946–50.
161. Li Y, Liu X, Huang L, Guo H, Wang XJ. Potential coexistence of both bacterial and eukaryotic small DNA biogenesis and functional related protein homologs in Archaea. J Genet Genomics. 2010;37:493–503.

162. Medina M. Genomes, phylogeny, and evolutionary systems. Proc Natl Acad Sci USA. 2005;102:6630–5.
163. Cerutti H, Casas-Moliano JA. On the origin and functions of RNA-mediated silencing: from protists to man. Curr Genet. 2006;50:81–99158.
164. Tucker RP, Beckman J, Leachman NT, Schöler J, Chiquet-Ehrisman R. Phylogenetic analysis of the teneurins: conserved features and premetazoan ancestry. Mol Biol Evol. 2012;29:1019–29.
165. Kenzelmann D, Chiquet-Ehrismann R, Leachman NT, Tucker RP. Teneurin-1 is expressed in interconnected regions of the developing brain and is processed in vivo. BMC Dev Biol. 2008;8:30. doi:10.1186/1471-213X-8-30.
166. Abedin M, King N. The premetazoan ancestry of cadherins. Science. 2008;319:946–8.
167. van 'T Veer LJ, van Kessel AG, van Heerikhuizen H, van Ooyen A, Nusse R. Molecular cloning and chromosomal assignment of the human homolog of int-1, a mouse gene implicated in mammary tumorigenesis. Mol Cell Biol 1984;4:2532–2534.
168. Chien AJ, Conrad WH, Moon RT. A Wnt survival guide: from flies to human disease. J Invest Dermatol. 2009;129:1614–27.
169. Philipp I, Aufschnaiter R, Özbek S, Pontasch S, Jenewein M, Watanaba H, Rentzsch F, Holstein TW, Hobmayer B. Wnt/ß-catenin and noncanonical Wnt signaling interact in tissue evagination in the simple eumetazoan Hydra. Proc Natl Acad Sci USA. 2009;106:4290–5.
170. Cselenyi CS, Jernigan KK, Tahinci E, Thorne CA, Lee LA, Lee E. LRP6 transduces a canonical Wnt signal independently of Axin degradation by inhibiting GSK3's phosphorylation of ß-catenin. Proc Natl Acad Sci USA. 2008;105:8032–7.
171. Glinka A, Wu W, Delius H, Monaghan AP, Blumenstock C, Niehrs C. Dickkopf-1 is a member of a new family of secreted proteins and functions in head induction. Nature. 1998;391:357–62.
172. Liao G, Tao Q, Kofron M, Chen J-S, Schloemer A, Davis RJ, Hsieh J-C, Wylie C, Heasman J, Kuan C-Y. Jun NH2-terminal kinase (JNK) prevents nuclear ß-catenin accumulation and regulates axis formation in Xenopus embryos. Proc Natl Acad Sci USA. 2006;103:16313–8.
173. Takadera T, Nakajima Y, Kanai Y. Colchicine-induced apoptosis was prevented by glycogen synthase kinase-3 inhibitors in PC12 cells. Cell Mol Neurobiol. 2010;30:863–8.
174. Arce L, Pate KT, Waterman ML. Groucho binds two conserved regions of LEF-1 for HDAC-dependent repression. BMC Cancer. 2009;9:159. doi:10.1186/1471-2407-9-159.
175. Guder C, Pinho S, Nacak TG, Schmidt HA, Hobmayer B, Niehrs C, Holstein TW. An ancient Wnt-Dickkopf antagonism in Hydra. Development. 2006;133:901–11.
176. Yang X-D, Huang S, Lo M-CV, Mizumoto K, Sawa H, Xu W, Robertson S, Lin R. Distinct and mutually inhibitory binding by two divergent ß-catenins coordinates TCF levels and activity in C. elegans. Development. 2011;138:4255–65.
177. Broun M, Gee L, Reinhardt B, Bode HR. Formation of the head organizer in hydra involves the canonical Wnt pathway. Development. 2005;132:2907–16.
178. Stanczak A, Stec R, Bodnar I, Olszewski W, Cichowicz M, Kozlowski W, Szczylik C, Pietrucha T, Wieczorek M, Lamparska-Przybysz M. Prognostic significance of Wnt-1. ß-catenin and E-cadherin expression in advanced colorectal carcinoma. Pathol Oncol Res. 2011;17:955–63.
179. Lepourcelet M, Chen YN, France DS, Wang H, Crews P, Petersen F, Bruseo C, Wood AW, Shivdasani RA. Small molecule antagonists of the oncogenic Tcf/ß-catenin protein complex. Cancer Cell. 2004;5:91–102.
180. Hallett RM, Kondratyev MK, Giaconelli AO, Nixon AML, Girgis-Gabardo A, Ilieva D, Hassell JA. Small molecule antagonists of the Wnt/beta-catenin signaling pathway target breast tumor-initiating cells in a Her2/Neu mouse model of breast cancer. PLoS One. 2012;7(3):e33976. doi:10.1371/journal.pone.0033976.

181. Choi HY, Lim JE, Hong JH. Curcumin interrupts the interaction between the androgen receptor and Wnt/ß-catenin signaling pathway in LNCaP prostate cancer cells. Prostate Cancer Prostatic Dis. 2010;13:3493–349.
182. Lu W, Tinsley HN, Keeton A, Qu Z, Piazza GA, Li Y. Suppression of Wnt/ß-catenin signaling inhibits prostate cancer cell proliferation. Eur J Pharmacol. 2009;602:8–14.
183. Jia X, Li X, Xu Y, Zhang S, Mou W, Liu Y, Liu Y, Ly D, Liu CH, Tan X, Xiang R, Li N. SOX2 promotes tumorigenesis and increases the apoptotic property of human prostate cancer cell. J Mol Cell Biol. 2011;3:230–8.
184. Hutchison DA, Hanson MI, Moore KB, Le TT, Brown NL, Vetter ML. HLH-dependent and independent modes of Ath5 gene regulation during retinal development. Development. 2005;132:829–39.
185. Burns CJ, Vetter ML. Xath5 regulates neurogenesis in the Xenopus olfactory placode. Dev Dyn. 2002;225:536–43.
186. Agathocleous M, Iordanova I, Willardson MI, Xue XY, Vetter ML, Harris WA, Mootre KB. A directional Wnt/ß-catenin Sox2-proneural pathway regulates the transition from proliferation to differentiation in the Xenopus retina. Development. 2009;136:3289–99.
187. Sun Y, Kanekar SL, Vetter ML, Gorski S, Jan Y-N, Glaser T, Brown NL. Conserved and divergent functions of Drosophila, amphibian and mammalian Ath5 genes. Evol Dev. 2003;5:532–41.
188. Fire A, Xu Montgomery MK, Kostas SA, Driver SE, Mello CC. Potent and specific genetic interference by double-stranded RNA in Caenorhabditis elegans. Nature. 1998;391:806–11.
189. Lee RC, Ambros V. An extensive class of small RNAs in Caenorhabditis elegans. Science. 2001;294:862–4.
190. Massirer KS, Perez SG, Mondol V, Pasquinelli AE. The miR-35-41 family of microRNAs regulates RNAi sensitivity in Caenorhabditis elegans. PLoS. 2012;8(3): e1002536. doi:10.1371/journal.pgen.1002536.
191. Jérôme T, Laurie P, Louis B, Pierre C. Enjoy the silence: the story of let-7 microRNA and cancer. Curr Genomics. 2007;8:220–33.
192. Resnick TD, McCulloch KA, Rougvie AE. miRNAs give worms the time of their lives: small RNAs and temporal control in Caenorhabditis elegans. Dev Dyn. 2010;239:1477–89.
193. Conine CC, Batista PJ, Gu W, Claycomb JM, Chaves DA, Shirayama M, Mello CC. Argonautes ALG-3 and ALG-4 are required for spermatogenesis-specific 26G-RNAs and thermotolerant sperm in Caenorhabditis elegans. Proc Natl Acad Sci USA. 2010;107:3588–93.
194a. Amaral PP, Neyt C, Wilkins SJ, Akarian-Amin ME, Sunkin SM, Perkins AC, Mattick JS. Complex architecture and regulated expression of the Sox2ot locus during vertebrate development. RNA. 2009;15:2013–27.
194b. Amaral PP, Dinger ME, Mercer TR, Mattick JS. The eukaryocytic genome as an RNA machine. Science. 2008;319:1787–9.
195a. Zhong X, Li N, Liang S, Huang G, Coukos G, Zhang L. Identification of microRNAs regulating reprogramming factor LIN28 in embryonic stem cells and cancer cells. J Biol Chem. 2010;285:41961–71.
195b. Godshalk SE, Melnik-Martinez KV, Pasquinelli AE, Slack NFJ. MicroRNAS and cancer. Epigenetics. 2010;5:164–8.
196. Félix MA, Asho A, Piffaretti J, Wu G, Nuez I, Bélicard T, Jiang Y, Zhao G, Franz CJ, Goldstein LD, Sanroman M, Miska EA, Wang D. Natural and experimental infection of Caenorhabditis nematodes by novel viruses related to nodaviruses. PLos Biol. 2011;9(1): e1000586.
197. Lu R, Yigit E, Li WX, Dong AW. An RIG-I-like RNA helicase mediates antiviral RNAi downstream of viral siRNA biogenesis in Caenorhabditis elegans. PLoS Pathog. 2009;5 (2):e1000286.

198. Rizki G, Iwata TN, Li J, Riedel CG, Picard CL, Jan M, Murphy CT, Lee SS. The evolutionarily conserved longevity determinants HCF-1 and SIR-2.1/SIRT1 collaborate to regulate DAF16/FOXO. PLoS Genet. 2011;7(9):e12002235. doi:10.1371/journal.pgen.1002235.
199. Knapp P, Chabowski A, Blachnio-Zabielska A, Jarzabek K, Wolczynski S. Altered peroxisome-proliferator activated receptors expression in human endometrial cancer. PPAR Res. 2012; doi:10.1155/2012/471524.
200. Peters JM, Foreman JF, Gonzalez FJ. Dissecting the role of peroxisome proliferator-activated receptor-ßδ (PPARßδ) in colon, breast, and lung carcinogenesis. Cancer Metastasis Rev. 2011;30:619–40.
201. Spanier B, Rubio-Aliaga I, Hu H, Daniel H. Altered signaling from germline to intestine pushes daf-2;pept-1 Caenorhabditis elegans into extreme longevity. Aging Cell. 2010;9:636–46.
202. Honnen SJ, Büchter C, Schröder V, Hoffmann M, Kohara Y, Kampkötter A, Bossinger OC. C. elegans VANG-1 modulates life span via insulin/IGF-1-like signaling. PLoS One. 2012;7(2):e32183. doi:10.137/journal.pone.0032183.
203. Bowman SK, Rolland V, Betschinger J, Kinsey KA, Emery G, Knoblich JA. The tumor suppressor Brat and Numb regulate trans-amplifying neuroblast lineages in Drosophila. Dev Cell. 2008;14:535–46.
204. Kurth P, Preiss A, Kovall RA, Maier D. Molecular analysis of the Notch repressor-complex in Drosophila: characterization of potential hairless binding sites on suppressor of hairless. PLoS One. 2011;6(11):e27986. doi:10.1371/journal.pone.0027986.
205. Dang TP. Notch, apoptosis and cancer. Adv Exp Med Biol. 2012;727:199–209.
206. Hu YY, Zheng MH, Zhang R, Liang YM, Han HJ. Notch signaling pathway and cancer metastasis. Adv Exp Med Biol. 2012;727:186–98.
207a. Huang Y. A mirror of two faces: Lin28 as a master regulator of both miRNA and mRNA. Wiley Interdiscip Rev RNA. 2012;3:483–94. doi:10.1002/wrna.1112.
207b. Huang B, Zhaon J, Unkeless JC, Feng ZH, Xiong H. TLR signaling by tumor and immune cells: a double-edged sword. Oncogene. 2008;27:218–24.
208. Shi W, Hendrix D, Levine M, Haley B. A distinct class of small RNAs arises from pre-miRNA-proximal region in a simple cordate. Nat Struct Mol Biol. 2009;16:183–9.
209. Zeller RW. Computational analysis of Ciona intestinalis. Integr Comp Biol. 2010;50:75–85.
210. Chen JS, Pedro MS, Zeller RW. miR-124 function during Ciona intestinalis neuronal development includes extensive interaction with the Notch signaling pathway. Development. 2011;138:49343–4953.
211. Sureban SM, May R, Lightfoot SA, Hoskins AB, Lerner M, Brackett DJ, Postier RG, Ramanujam R, Mohammed A, Rao CV, Wyche JH, Anant S, Houchen CW. DCAMKL-1 regulates epithelial-mesenchymal transition in human pancreatic cells through a miR-200a-dependent mechanism. Cancer Res. 2011;71:2328–38.
212a. Wang Z, Banerjee S, Ahmad A, Li Y, Azmi AS, Gunn JAR, Kong D, Bao B, Ali S, Gao J, Mohammed RM, Miele L, Korc M, Sarkar FH. Activated K-ras and INK4a/Arf deficiency cooperate during the development of pancreatic cancer by activation of Notch and NFκB signaling pathways. PLoS One. 2011;6(6):e20537.
212b. Wang Z, Ahmad A, Li Y, Azmi AS, Miele L, Sarkar FH. Targeting notch to eradicate pancreatic cancer stem cells for cancer therapy. Anticancer Res. 2011;31:1105–13.
213. Santra M, Santra S, Buller B, Santra KL, Nallani A, Chopp M. Effect of doublecortin on self-renewal and differentiation in brain tumor stem cells. Cancer Sci. 2011;102:1350–7.
214. Liu N, Zhang S, Liu Z, Gaowa S, Wang Y. Characterization and expression of gamma-interferon-inducible lysosomal thiol reductase (GILT) gene in amphioxus Branchiostoma belcheri with implications for GILT in innate response. Mol Immunol. 2007;2631–2637.

215. Srinivasan P, Maric M. Signal transducer and activator of transcription negatively regulates constitutive gamma interferon-inducible lysosomal thiol reductase expression. Immunology. 2011;132:209–16.
216. O'Donnell PW, Hague A, Klemsz MJ, Kaplan MH, Blum JS. Cutting edge: induction of the antigen-processing enzyme IFN-gamma-inducible lysosomal thiol reductase in melanoma cells is STAT1-dependent but CIITA-independent. J Immunol. 2004;173:731–5.
217. Cui XW, Xiao W, Ke Z, Liu X, Xu XZ, Zhang SQ. Cloning and expression analysis of interferon-γ-inducible-lysosomal thiol reductase gene in South African clawed frog (Xenopus laevis). Int Immunopharmacol. 2011;11:2091–7.
218. Yang M, Yuan S, Huang S, Li J, Xu L, Huang H, Tao X, Peng J, Xu A. Characterization of btTICAM from amphioxus suggests the emergence of MyD88-independent pathway in basal chordates. Cell Res. 2011;21:1410–23.
219. Chen X, Li Q, Wang J, Guo X, Jiang X, Ren Z, Weng C, Sun G, Wang X, Liu Y, Ma L, Chen JY, Wang J, Zen K, Zhang J, Zhang CY. Identification and characterization of novel amphioxus microRNAs by Solexa sequencing. Genome Biol. 2009;10:R78.
220. Dai Z, Chen Z, Ye H, Zhou L, Cao L, Wang Y, Peng S, Chen L. Characterization of microRNAs in cephalochordates reveals a correlation between microRNA repertoire homology and morphological similarity in chordate evolution. Evol Dev. 2009;11:41–9.
221. Li G, Zhang J, Sun Y, Wang H, Wang Y. The evolutionary dynamic IFN-inducible GTPase proteins play conserved immune functions in vertebrates and cephalochordates. Mol Biol Evol. 2009;26:1619–30.
222. Robertsen B. The interferon system of teleost fish. Fish Shelfish Immunol. 2006;20:172–91.
223. Hemming OJ, Lutfalla G, Levraud JP, Hartmann R. Crystal structure of zebrafish interferons I and II reveals conservation of type I interferon structure in vertebrates. J Virol. 2011;85:8181–7.
224. Zou X, Xiang X, Chen Y, Peng T, Luo X, Pan Z. Understanding inhibition of viral proteins on types I IFN signaling pathways with modeling and optimization. J Theor Biol. 2010;265:691–703.
225. Bell DM, Roberts NJ Jr, Hall CB. Different antiviral spectra of human macrophage interferon activities. Nature. 1983;305:319–21.
226. Gallwitz M, Reimer JM, Hellman L. Expansion of the mast cell chymase locus over the past 200 million years of mammalian evolution. Immunogenetics. 2006;58:655–69.
227. Wang S-Z, Roberts RM. The evolution of the gene product, a little known protein implicated in stress responses and type I interferon signaling in vertebrates. BMC Evol Biol. 2005;v.5. doi:10.1186/1471-2148-5-13.
228. Icardi L, Lievens S, Mori R, Piessevaux J, De Cauwer L, De Bosscher K, Tavernier J. Opposed regulation of type I IFN-induced STAT3 and ISGF3 transcriptional activities by histone deacetylases (HDACs) 1 and 2. FASEB. 2012;26:240–9.
229. Xiong H, Du W, Zhang YJ, Hong J, Su WY, Tang JT, Wang YC, Lu R, Fang JY. Trichostatin A, a histone deacetylase inhibitor, suppresses JAK2/STAT3 signaling via inducing the promoter-associated histone acetylation of SOCS1 and SOCS3 in human colorectal cancer. Mol Carcinog. 2012;51:174–84.
230. Banu SK, Lee J, Stephen SD, Nithy TK, Arosh JA. Interferon tau regulates PGF2α release from the ovine endometrial epithelial cells via activation of novel JAK/EGFR/ERK/EGR-1 pathways. Mol Endocrinol. 2010;24:23125–2330.
231. Kim J, Villadsen R, Sørlie T, Fogh L, Grenlund SZ, Fridriksdottir AJ, Kuhn I, Rank F, Wielenga VT, Solvang H, Edwards PA, Børresen-Dale AL, Rønnov-Jessen L, Bissel MJ, Petersen OW. Tumor initiating but differenciated luminal-like breast cancer cells are highly invasive in the absence of basal-like activity. Proc Soc Natl Acad Sci USA. 2012;109:6124–9.

232. Wang X, Kruithof-de Julio M, Economides KD, Walker D, Yu H, Halili MV, Hu YP, Price SM, Abate-Shen C, Shen MM. A luminal epithelial stem cell that is the origin of prostate cancer. Nature. 2009;461:495–500.
233. Wagner K-U, Schmidt JW. The two faces of Janus kinases and their respective STATs in mammary gland development and cancer. J Carcinog. 2011;10:32. doi:10.4103/1477-3163-90677.
234. Xu H, Yang YJ, Qian HY, Wang YD, Wang H, Zhang Q. Rosuvastatin treatment activates JAK-STAT pathway and increases efficacy of allogeneic mesenchymal stem cell transplantation in infarcted heart. Circ J. 2011;75:1476–85.
235. Duan W, Yang Y, Yan J, Yu S, Liu J, Zhou J, Zhang J, Jin Z, Yin D. The effects of curcumin post-treatment against myocardial ischemia and reperfusion by activation of the JAK2/STAT3 signaling pathway. Basic Res Cardiol. 2011;107:1–12.
236. Houssain S, Singh N, Salam Bandil K, Yuvaraj M, Akbar Bhat M, Muzaffar Mir M, Siddiqi MA, Sobti RC, Bharadwaj M, Das BC. Methylation-mediated gene-silencing of suppressor of cytokine signaling-1 (SOCS-1) gene in esophageal squamous cell carcinoma patients of Kasmir valley. J Recept Signal Transduct Res. 2011;31:147–56.
237. Souma Y, Nishida T, Serada S, Iwahon K, Takahashi T, Fujimoto M, Ripley B, Nakajima K, Miyazaki Y, Mori M, Dokin Y, Sawa Y, Naka T. Antiproliferative effect of SOCS-1 through the suppression of STAT3 and p38 MAPK activation in gastric cancer cells. Int J Cancer. 2012;131:1287–1296. doi:10.1002/ijc.27350.
238. Qin J, Xin H, Nickoloff BJ. Specifically targeting ERK1 or ERK2 kills melanoma cells. J Transl Med. 2012;10:15. doi:10.1186/1479-5876-10-15.
239. Xiong H, Zhang Z-G, Tian X-Q, Sun D-F, Liang Q-C, Zhang Y-J, Lu R, Chen Y-X, Fang J-Y. Inhibition of JAK1, 2/STAT3 signaling induces apoptosis, cell cycle arrest, and reduces tumor cell invasion in colorectal cancer cells. Neoplasia. 2006;10:287–97.
240. Huang C, Yang G, Jiang T, Huang K, Gao J, Qiu Z. Effects of IL-6 and AG490 on regulation of Stat3 signaling pathway and invasion of human pancreatic cancer cells in vitro. J Exp Clin Cancer Res. 2010;29(1):51. doi:10.1186/1756-9966-29-51.
241. Mohanty S, Jermyn KA, Early A, Kawata T, Aubry L, Ceccarelli A, Scaap P, Williams JG, Firtel RA. Evidence that the Dictyostelium Dd-STATa protein is a repressor that regulates commitment to stalk differentiation and is required for efficient chemotaxis. Development. 1999;126:3391–405.
242. Copf T, Goguel V, Lampin-Saint-Amaux A, Scaplehorn N, Preat T. Cytokine signaling through the JAK/STAT pathway is required for long-term memory in Drosophila. Proc Natl Acad Sci USA. 2011;108:8059–64.
243. Bina S, Wright VM, Fisher KH, Milo M, Zeidler MP. Transcriptional targets of Drosophila JAK/STAT pathway signaling as effectors of haematopoietic tumour formation. EMBO Rep. 2010;11:201–7.
244. González I, Simón R, Busturia A. The polyhomeotic protein induces hyperplastic tissue overgrowth through the activation of the JAK/STAT pathway. Cell Cycle. 2009;8:4103–11.
245. Karsten P, Häder S, Zeidler MP. Cloning and expression of Drosophila SOCS36E and its potential regulation by the JAK/STAT pathway. Mech Dev. 2002;117:343–6.
246. Callus BA, Mathey-Prevot B. SOCS36E, a novel Drosophila SOCS protein, suppresses JAK/STAT and EGF-R signaling in the imaginal wing disc. Oncogene. 2002;21:4812–21.
247. Silver DL, Montell DJ. Paracrine signaling through the JAK/STAT pathway activates invasive behavior of ovarian epithelial cells in Drosophila. Cell. 2001;107:831–41.
248. Van de Bor V, Zimniak G, Cérézo D, Schaub S, Noselli S. Asymmetric localization of cytokine mRNA is essential for JAK/STAT activation during cell invasiveness. Development. 2011;138:1383–93.

249. Migita K, Komori A, Torigoshi T, Maeda Y, Izumi Y, Jiuchi Y, Miyashita T, Nakamura M, Motokawa S, Ishibashi H. CP690,550 inhibits oncostatin M-induced JAK/STAT signaling pathway in rheumatoid synoviocytes. Arthritis Res Ther. 2011;13 (3):R72. doi:10.1186/ar3333.
250. Gavasso S, Gjertsen B, Anderssen E, Myhr K, Vedeler C. Immunogenic effects of recombinant interferon-beta therapy disrupt the JAK/STAT pathway in primary immune cells from patients with multiple sclerosis. Mult Scler. 2012;18:1116–24.
251. Kim MO, Si Q, Zhou JN, Pestell RG, Brosnan CF, Locker J, Lee SC. Interferon-beta activates multiple signaling cascades in primary human microglia. J Neurochem. 2002;81:1361–71.
252. Qin H, Yeh WI, De Sarno P, Holdbrooks AT, Liu Y, Muldowney MT, Reynolds SKL, Yanagissawa LL, Fox TH 3rd, Park K, Harrington LE, Raman C, Benveniste EN. Signal transducer and activator of transcription-3/suppressor of cytokine signaling-3 (STAT/SOCS3) axis in myeloid cells regulates neuroinflammation. Proc Natl Acad Sci USA. 2012;109:5004–9.
253. Ramgolam VS, Markovic-Plese S. Regulation of suppressors of cytokine signaling as a therapeutic approach in autoimmune diseases, with an emphasis on multiple sclerosis. J Signal Transduct. 2011: 635721. doi:10.1155/2011/635721.
254. Eghtedar A, Verstovsek S, Estrov Z, Burger J, Cotes J, Bivins C, Faderl S, Ferrajoli A, Borthakur G, George S, Scherle PA, Newton RC, Kantarjian HM, Ravandi F. Phase II study of the JAK kinase inhibitor ruxolitinib in patients with refractory leukemias, including post myeloproliferative neoplasms (MPN) acute myeloid leukemia (AML). Blood. 2012;119:4614–8.
255a. Harrison C, Verstovsek S, McMullin MF, Mesa R. Janus kinase inhibition and its effect upon the therapeutic landscape for myelofibrosis: from palliation to cure? Brit J Haematol. 2012;157:426–37. doi:10.1111/j.1365-2141.2012.09108.x.
255b. Harrison C, Kiladjian J-J, Al-Ali HK, Gisslinger H, Waltzman R, Stablovskaya V, Mc Quitty M, Hunter DS, Levy R, Knoops L, Cervantes F, Vannucchi AM, Barbui T, Barosi G. Jak inhibition with ruxolitinib versus best available therapy for myelofibrosis. N Engl J Med. 2012;366:787–98.
256. Iványi JL, Marton E, Plander M. Significance of the JAK2V617F mutation in patients with chronic myeloproliferative neoplasia. Orvosi Hetilap (Budapest). 2011;152:1795–1780.
257. Tefferi A. Challenges facing JAK inhibitor therapy for myeloproliferative neoplasms. N Engl J Med. 2012;366:844–6.
258. Oh SJ, Ahn JY, Chung DH. Comparison of invariant NKT cells with conventional T cells by using gene set enrichment analysis (GSEA). Immune New. 2011;11:406–11.
259. Inakufu M, Li C, Kanda Y, Kaewamura T, Takeda K, Oku H, Watanabe H. Beta-glucosylceramide administration (i.p.) activates natural killer T cells in vivo and prevents tumor metastasis in mice. Lipids. 2012 Epub March 20.
260. Alvarez-Breckenridge CA, Yu J, Price R, Wei M, Wang Y, Nowicki ND, Ha YP, Bergin S, Hwang C, Fernandez SA, Kaur B, Caliguri MA, Chiocca EA. The HDAC inhibitor valproic acid lessens NK cell action against virus-infected glioblastoma cells with inhibition of STAT5/T-BET signaling and IFNγ generation. J Virol. 2012: 702839. doi:10.1155/2012/702839.
261a. Sinkovics JG. The graft-versus-leukemia (GvL) reaction: Its early history, value in human bone marrow transplantation and recent developments concerning its mechanism and induction. In: Dicke KA, Keating A, authors/editors. Autologous Marrow and Blood Transplantation. Proceedings Seventh International Symposium. Arlington, TX: The Cancer Treatment Research and Educational Institute; 1994. p. 305–318.
261b. Sinkovics JG. Antileukemia and antitumor effects of the graft-versus host disease: a new immunovirological approach. Acta Microbiol Immunol Hung. 2010;57:253–347.

262a. Chaidos A, Patterson S, Szydlo R, Chaudhry MS, Dazi F, Kanfer E, MacDonald D, Marin D, Milojkovic D, Pavlu J, Davis J, Rahemlulla A, Rezvani K, Goldman J, Roberts L, Apperley J, Karadimitris A. Graft invariant natural killer T cell dose predicts risk of acute graft-versus-host disease in allogeneic hematopoietic stem cell transplantation. Blood. 2012;119:5030–6.

262b. Detre C, Keszei M, Barrido-Mesa N, Kis-Toth K, Castro W, Agyermang AF, Veerapen N, Besra G, Garrott MC, Tsokos GC, Wang N, Leadbetter EA, Terhorst C. SAP expression in invariant NKT cells is required for cognate help to support B-cell responses. Blood. 2012;120:122–9.

263. Kawasaki M, Fujishiro M, Yamaguchi A, Nozawa K, Kaneko H, Takasaki Y, Takamori K, Ogawa H, Sekigawa I. Possible role of the JAK/STAT pathways in the regulation of T cell-interferon related genes in systemic lupus erythematosus. Lupus. 2011;20:1231–9.

264. Tanaka Y, Maeshima Y, Yamaoka K. In vitro and in vivo analysis of a JAK inhibitor in rheumatoid arthritis. Ann Rheum Dis. 2012;71S2:i70–4.

265. Sørensen PS. Effects of neutralizing antibodies to interferon beta in multiple sclerosis: a logical paradox. Mult Scler. 2012;18:131–2.

266. Baker BJ, Park KW, Qin H, Ma XC, Benveniste EN. IL-27 inhibits OSM-mediated TNF-alpha and iNOS gene expression in microglia. Glia. 2010;58:1082–93.

267. Oliver-Martos B, Órpez T, Pinto-Medel MJ, Mayorga C, Garcia-León JA, Maldonado-Sanchez R, Suardiaz M, Guerrero M, Luque G, Leyva L, Fernández O. Gene expression in IFN-beta signaling pathway differs between monocytes, CD4 and CD8 T cells from MS patients. J Neuroimmunol. 2011;230:153–9.

268. Ping L, Ogawa N, Zhang Y, Sugai S, Masuki Y, Weiguo X. p38 Mitogen-activated protein kinase and nuclear factor-kB facilitate CD40-mediated salivary epithelial cell death. Rheumatol 2012;39:1256–64.

269. Koch L, Hofer S, Weigand MA, Frommhold D, Poeschl J, Ruef P. Inhibition of LPS-induced activation of coagulation by p38 MAPK inhibitor. ISRN Hematol. 2012:762614. doi:10.5402/2012/76614.

270. Pauls E, Shpiro N, Peggie M, Young ER, Sorcek RJ, Tan L, Choi HG, Cohen P. An essential role for IKKbeta in the production of type 1 inteferons by plasmacytoid dendritic cells. J Biol Chem. 2012;287:19216–28. doi:10.1074/jbc.M112.345405.

271. Koprowski H, Love R, Koprowska I. Enhancement of susceptibility to viruses in neoplastic tissues. Texas Rep Biol Med. 1957;15:111–28.

272. Isaacs A, Lindenmann J. Virus interference. I. The interferon. By A. Isaacs and J. Lindenmann, 1957. J Interferon Res. 1987; 7:429–438.

273. Lindenmann J, Klein PA. Immunological Aspects of Viral Oncolysis. Berlin-Heidelberg, New York: Springer Verlag; 1967. p. 1–84.

274. Cassel WA, Murray DR, Olkowski ZL. Newcastle disease virus oncolysate in the management of stage III malignant melanoma. In: Sinkovics JG, Horvath JC, editors. Viral therapy of human cancers. Marcel Dekker, New York. 2005; 677–689.

275. Sinkovics JG. New biological therapeutics: competitors or collaborators of viral therapy for human cancers. In: Viral Therapy of Human Cancers. In: Sinkovics JG, Horvath JC, editors. New York: Marcel Dekker; 2005. p. 149–453.

276. Sinkovics JG, Plager C, McMurtrey M, Papadopoulos NE, Waldinger R, Combs S, Romero J, Romsdahl MM. Adjuvant chemoimmunotherapy for malignant melanoma. In: Crispen RG, editor. Neoplasm immunity: experimental and clinical. New York, Amsterdam, Oxford: Elsevier/North-Holland; 1980. p. 481–519.

277. Sinkovics JG, Horvath JC. Evidence accumulating in support of cancer vaccines combined with chemotherapy: a pragmatic review of past and present efforts. Int J Oncol. 2006;29:765–77.

278. Sinkovics JG, Shirato E, Gyorkey F, Cabiness J, Howe CD. Relationship between lymphoid neoplasms and immunologic functions. In: Leukemia-Lymphoma. Fourteenth Annual Clinical Conference on Cancer, 1969 at the University of Texas M.D. Anderson Hospital and Tumor Institute, Houston, Texas. Year Book Medical Publishers, Chicago. 1970; 53–92.
279. Sinkovics JG. Chondrosarcoma cell differentiation. Pathol Oncol Res. 2004;10:174–87.
280. Lichtman MA. Historical landmarks in the understanding of the lymphomas. In: Neoplastic Diseases of the Blood. In: Wiernik PH, Goldman JH, Dutcher JM, Kyle RA, editors. New York: Springer; 2013. p. 789–833.
281. Savage HE, Rossen RD, Hersh EM, Freedman RS, Bowen JM, Plager C. Antibody development to viral and allogeneic tumor cell-associated antigens in patients with malignant melanoma and ovarian carcinoma treated with lysates of virus-infected tumor cells. Cancer Res. 1986;46:2127–33.
282a. Chen W, Masterman K-A, Basta S, Haeryfar SM, Dimopoulos N, Knowles B, Bennink JAR, Yewdell JW. Cross-priming of CD8+ T cells by viral and tumor antigens is a robust phenomenon. Eur J Immunol. 2004;34:194–9.
282b. Chen W, Pang K, Masterman K-A, Kennedy G, Basta S, Dimopoulos N, Hornung F, Smyth M, Bennink JR, Yewdell JW. Reversal in the immunodominance hierarchy in secondary CD8+ T cell responses to influenza A virus: roles for cross-presentation and lysis-independent immunodomination. J Immunol. 2004;173:5021–7.
283. Hiss DC, Fiedling BC. Optimization and preclinical design of genetically engineered viruses for human oncolytic therapy. Exp Opin Biol Ther. 2012;12:1427–47.
284. Johnson TJ, Höti N, Liu C, Chowdhury WH, Li Y, Zhang Y, Lupold SE, Deweese SE, Rodriguez R. Bicalutamide-activated oncolytic adenovirus for the adjuvant therapy of high-risk prostate cancer. Cancer Gene Ther. 2013; doi:10.1038/cgt.2013.34.
285. Basta S, Bennink JAR. A survival game of hide and seek: cytomegaloviruses and MHC class I antigen presentation pathways. Viral Immunol. 2003;16:231–42.
286. Endo Y, Sakai R, Ouchi M, Onimatsu H, Hioki M, Kagawa S, Uno F, Watanabe Y, Uratya Yanaja N, Fujuwara T. Virus-mediated oncolysis induces danger signal and stimulates cytotoxic T-lymphocyte activity via proteasome activator upregulation. Oncogene. 2008;27:2375–81.
287. Kaur B, Chiocca EA, Cripe TP. Oncolytic HSV-1 virotherapy: clinical experience and opportunities for progress. Curr Pharm Biotechnol. 2012;13:1842–51.
288. Li Q-X, Liu G, Zhang X. Fusogenic oncolytic herpes simplex viruses as a potent and personalized cancer vaccine. Curr Pharmaceut Biotech. 2010;123:1773–85.
289. Terwilliger DP, Buckley KM, Mehta D, Moorjani PG, Smith LC. Unexpected diversity displayed in cDNAs expressed by the immune cells of the purple sea urchin, Strongylocentrotus purpuratus. Physiol Genomics. 2006;26:134–44.
290a. Dhelly NM, Birch D, Nair SV, Raftos DA. Ultrastructural localization of highly variable 185/333 immune response proteins in the coelomocytes of the sea urchin, Heliocidaris erythrogramma. Immunol Cell Biol. 2011;89:861–9.
290b. Dheilly NM, Haynes PA, Raftos DA, Nair SV. Time course proteomic profiling of cellular responses to immunological challenge in the sea urchin, Heliocidaris erythrogramma. Dev Comp Immunol. 2012;37:243–56.
291. Haire RN, Miracle AL, Rast JP, Litman GW. Members of the Ikaros gene family are present in early representative vertebrate. J Immunol. 2000;165:306–12.
292. Mayer WR, O'Huigin C, Tichy H, Terzic J, Saraga-Babic M. Identification of two Ikaros-like transcription factors in lamprey. Scand J Immunol. 2002;55:162–70.
293. Herrin BR, Cooper MD. Alternative adaptive immunity in jawless vertebrates. J Immunol. 2010;185:1367–74.
294. Khalturin K, Becker M, Rinkevich B, Bosch TC. Urochordates and the origin of natural killer cells: identification of a CD94/NKR-P1-related receptor in blood cells of Botryllus. Proc Natl Acad Sci USA. 2003;100:622–7.

295. Morales HDF, Robert J. Characterization of primary and memory CD8 T-cell responses against ranavirus (FV3) in Xenopus laevis. J Virol. 2007;81:2240–8.
296. Morales H, Robert JK. In vivo and in vitro techniques for comparative study of antiviral T cell responses in the amphibian Xenopus. Biol Proced Online. 2008;10:1–8.
297. Goyos A, Robert J. Tumorigenesis and anti-tumor immune responses in Xenopus. Front Biosci. 2009;14:167–76.
298. McKinnell RG, Carlson DL. Lucké renal adenocarcinoma, anuran neoplasm, studies at the interface of pathology, virology, and differentiation. J Cell Physiol. 1997;173:115–8.
299. Gantress J, Maniero GD, Cohen N, Robert J. Development and characterization of a model system to study amphibian immune responses to iridoviruses. Virology. 2003;311:254–62.
300. Campo-Paysaa F, Sémon M, Cameron RA, Peterson KJ, Schubert M. microRNA complements in deuterostomes: origin and evolution in microRNAs. Evol Dev. 2011;13:15–27.
301. Janvier P. microRNAs revive old views about jawless vertebrate divergence and evolution. Proc Natl Acad Sci USA. 2010;107:19137–8.
302. Heimberg AM, Cowper-Sal-lari R, Sémon M, Donoghue PC, Peterson KJ. microRNAs reveal the interrelationship of hagfish, lampreys, and gnathostomes and the nature of the ancestral vertebrate. Proc Natl Acad Sci USA. 2010;107:19379–83.
303a. Hertel J, Stadler PF. Hairpins in a haystack: recognizing microRNA precursors in comparative genomics data. Bioinformatics. 2006;22:197–202.
303b. Hertel J, Lindemeyer M, Missal K, Fried C, Tanzer A, Flamm C, Hofacker IL, Stadler P. the Students. The expansion of the metazoan microRNA repertoire. BMC Genomics v.7; 2006;7:25.
304. Moran Y, Fredman D, Prasher D, Li XZ, Wee LM, Rentzsch FG, Zamore PD, Technau U, Seitz H. Cnidarian microRNAs frequently regulate targets by cleavage. Genome Res. 2014;24:651–63.
305. Saha NR, Smith J, Amemiya CT. Evolution of adaptive immune recognition in jawless vertebrates. Semin Immunol. 2010;22:25–33.
306. Lada AG, Iyer LM, Rogozin IB, Aravind L, Pavlov YI. Vertebrate immunity, mutator proteins and their evolution. Genetika. 2007;43:1311–27.
307. Rogozin IB, Iyer LM, Liang L, Glazko GV, Liston VG, Pavlov YI. Evolution and diversification of lamprey antigen receptors: evidence for involvement of an AID APOBEC family cytosine deaminase. Nat Immunol. 2007;8:647–56.
308. Falkner S, Emdin SO, Ostberg Y, Mattisson A, Sjöbeck ML, Fänge R. Tumor pathology of hagfish, Myxine glutinosa, and the river lamprey, Lampetra fluviatilis. A light microscopal study with particular reference to the occurrence of primary liver carcinoma, islet cell tumors, and epidermal cysts of the skin. Prog Exp Tumor Res. 1976;20:217–2550.
309. Falkner S, Marklund S, Mattisson PE, Rappe C. Hepatomas and other neoplasms in the atlantic hagfish (Myxine glutinosa): a histopathologic and chemical study. Ann N Y Acad Sci. 1978;298:342–55.
310a. Mechtcheriakova D, Svoboda M, Meshcheryakova A, Jensen-Jarolim E. Activation-induced cytidine deaminase (AID) linking immunity, chronic inflammation, and cancer. Cancer Immunol Immunother. 2012;61:1591–8. doi:10.1007/s00262-012-1255-z.
310b. Mechtcheriakova D, Sobanov Y, Holtappels G, Bajna E, Svoboda M, Jaritz M, Bachert C, Jensen-Jarolim E. Activation-induced cytidine deaminase (AID)-associated multigene signature to assess impact of AID in etiology of diseases with inflammatory component. PLoS One. 2011;6(10):e25611.
311. Kapitonov VV, Jurka J. Rag1 core and V(D)J recombination signal sequences were derived from Transib transposons. PLoS Biol. 2005;3(6):e181.

312. Dreyfus DH. Paleo-immunology: evidence consistent with insertion of a primordial herpes virus-like element in the origins of acquired immunity. PLoS One. 2009;4(6): e5778.
313a. Fugmann SD, Messier C, Novack LA, Cameron RA, Rast JP. An ancient evolutionary origin of the Rag1/2 gene locus. Proc Natl Acad Sci. 2006;103:3728–33.
313b. Fugmann SD. The origin of the RAG genes—from transposition to V(D)J recombination. Semin Immunol. 2010;22:10–6.
314. Baughn LB, Kalis SL, MacCarthyT Wei L, Fan M, Bergman A, Scharff MD. Recombinase-mediated cassette exchange as a novel method to study somatic hypermutation in Ramos cells. MBio. 2011;2(5):E00186–11. doi:10.1128/mBio.00186-11.
315. Chowdhury M, Forouhi O, Dayal S, McCloskey N, Gould HJ, Felsenfeld G, Fear DJ. Analysis of intergenic transcription and histone modification across the human immunoglobulin heavy chain locus. Proc Natl Acad Sci USA. 2008;105:15872–7.
316. Pham P, Bransteitter R, Goodman MF. Reward versus risk: DNA cytidine deaminases triggering immunity and disease. Biochemistry. 2005;44:2703–15.
317. Fugmann SD. Rag-2 unleashed: lymphocytes beware. Immunity. 2011;34:137–9.
318. Raghavan SC, Swanson PC, Ma Y, Lieber MR. Double-strand break formation by the RAG complex at the BCL-2 major breakpoint region is cleaved by the RAG complex. Nature. 2004;428:88–93.
319. Robbiani DF, Bothmer A, Callen E, Reina-San-Martin B, Dorsett Y, Difilippantonio S, Bolland DJ, Chen DJ, Corcoran AE, Nussenzweig A, Nussenzweig MC. Activation induced deaminase is required for the chromosomal breaks in c-myc that lead to c-myc/IgH translocation. Cell. 2008;135:1028–38.
320. Zhang L, Reynolds TL, Shan X. Coupling of V(D)J recombination to the cell cycle suppresses genome instability and lymphoid tumorigenesis. Immunity. 2011;34:163–74.
321. Jing YZ, Wang Y, Jia YP, Luo B. Polymorphisms of Epstein-Barr virus BHRF1 gene, a homologue of bcl-2. Chin J Cancer. 2010;29:1000–5.
322. Seitz V, Butzhammer P, Hirsch B, Hecht J, Gütgemann I, Ehlers A, Lenze D, Oker E, Sommerfeld A, von der Wall E, König C, Zinser C, Spang R, Hummel M. Deep sequencing of Myc DNA-binding sites in Burkitt lymphoma. PLoS One. 2011;6(11): e26837. doi:10.1371/journal.pone.0026837.
323. Knezevich S, Ludkovski O, Salski C, Lestou V, Chhanabhai M, Lam W, Klasa R, Connors JM, Dyer MJ, Gascoyne RD, Horsman DE. Concurrent translocation of BCL2 and MYC with a single immunoglobulin locus in high-grade B-cell lymphomas. Leukemia. 2005;19:659–63.
324. Martin-Subero JI, Odero MD, Hernandez R, Ciqudosa JC, Agirre X, Saez B, Sanz-García E, Ardanaz MT, Novo FJ, Gascoyne RD, Calasanz MJ, Siebert R. Amplificatiion of IGH/MYC fusion in clinically aggressive IGH/BCL2-positive germinal center B-cell lymphomas. Genes Chromosomes Cancer. 2005;43:414–23.
325. Koduru S, Wong E, Strowig T, Sunaram R, Zhang L, Strout MP, Flavell RA, Schatz DG, Dhodapkar KM, Dhodapkar MV. Dendritic cell-mediated activation-induced cytidine deaminase (AID)-dependent induction of genomic instability in human myeloma. Blood. 2012;119:2302–9.
326. Roux S, Meignin V, Quillard J, Meduri G, Guiochon-Mantel A, Fermand JP, Milgrom E, Mariette X. RANK (receptor activator of nuclear factor-kappa B) and RANKL expression in multiple myeloma. Br J Haematol. 2002;117:86–92.
327. Santini D, Schiavon G, Vincenzi B, Gaeta L, Pantano F, Russo A, Ortega C, Porta C, Galluzzo S, Armento G, La Verde N, Caroti C, Treilleux I, Ruggiero A, Perrone G, Addeo R, Clezardin P, Netti Muda A, Tonini G. Receptor activator of NF-kB (RANK) expression in primary tumors associates with bone metastasis occurrence in breast cancer patients. PLoS One. 2011;6(4):e19234. doi:10.1371/journal.pone.0019234.
328. Kothapalli NR, Fugmann SD. Targeting of AID-mediated sequence diversification to immunoglobulin genes. Curr Opin Immunol. 2011;23:184–9.

329. Deriano L, Caumeil J, Coussens M, Mullanin A, Chou Y, Alekseyenko AV, Chang S, Skok JA, Roth DB. The RAG2 C terminus suppresses genomic instability and lymphomagenesis. Nature. 2011;471:119–23.

330. Roa S, Li Z, Peled JU, Zhao C, Edelmann W, Scharff MD. MSH2/MSH6 complex promotes error-free repair of AID-induced dU: G mispairs as well as error-prone hypermutations of A: T sites. PLoS One. 2010;5(6):e11182. doi:10.1371/journal.pone.0011182.

331. Hasham MG, Donghia NM, Cofey E, Maynard J, Snow KJ, Ames J, Wilpan RY, He Y, King BL, Mills KD. Widespread genomic breaks generated by activation-induced cytidine deaminase are prevented by homologous recombinations. Nat Immunol. 2010;11:820–6.

332. Jankovic M, Robbiani DF, Dorsett Y, Eisenreich T, Xun Y, Tarakhovsky A, Nussenzweig A, Nussenzweig MC. Role of the translocation partner in protection against AID-dependent chromosomal translocations. Proc Natl Acad Sci USA. 2010;107:187–12.

333. McBride KM, Gazumyan A, Woo EM, Schwickert TA, Chait BT, Nussenzweig NC. Regulation of class switch recombination and somatic mutation by AID phosphorylation. J Exp Med. 2008;205:2585–94.

334. de Yébenes VG, Belver L, Pisano DG, González S, Villasante A, Croce C, He L, Ramiro AR. miR-181b negatively regulates activation-induced cytidine deaminase in B cells. J Exp Med. 2008;205:2199–206.

335. Dorsett Y, McBride KM, Jankovic M, Gazumyan A, Thai TH, Robbiani DF, Di Virgilio M, Reina San-Martin B, Heidkamp G, Schwickert TA, Eisenreich TA, Rajewsky K, Nussenzweig MC. MicroRNA-155 suppresses activation-induced cytidine deaminase-mediated Myc-IgH translocation. Immunity. 2008;28:630–8.

336. Teng G, Hakimpour P, Landgraf P, Rice A, Tischl T, Casellas R, Papavasilou FN. microRNA-155 is a negative regulator of activation-induced cytidine deaminase. Immunity. 2008;28:621–9.

337. Wong KY, So CC, Loong F, Chung LP, Lam WWL, Liang R, L GKH, Jin D-Y, Chim CS. Epigenetic inactivation of the miR-124-1 in haematologic malignancies. PLoS One. 2011;6(4):e19027. doi:10.1371/journal.ponr.0019027.

338. Borchert GM, Holton NW, Larson ED. Repression of human activation-induced cytidine deaminase by miR-93 and miR-155. BMC Cancer. 2011;11:347. doi:10.1186/1471-2407-11-347.

339. Kawamata T, Lu J, Sato T, Tanaka M, Nagaoka H, Agata Y, Yoyoshima T, Yokoyama K, Oyaizu N, Nakamura N, Ando K, Tojo A, Kotani A. Imatinib mesylate directly impairs class switch recombination through down-regulation of AID: its potential efficacy as an AID suppressor. Blood. 2012;119:3123–7.

340a. Sinkovics JG. Adult human sarcomas. I. Basic science. Expert Rev Anticancer Ther. 2007;7:31–56.

340b. Gonzalez F, Tzobari S, Sinkovics JG, Campos LT, Howe CD. Immunological evidence for the viral etiology of human mesenchymal neoplasms. 65th Annual Meeting Am Assoc Cancer Res 10th Annual Meeting Am Soc Clinical Oncol March 27-30 1974 Houston TX. Proceedings Am Assoc Cancer Res 1974;15:62 #247.

340c. Sinkovics JG, Szakacs JE. Kaposi's sarcoma. In: Lapis K, Eckhardt S, editors. Lectures & Symposia 14th Internatinal Cancer Congress Budapest. Cancer Research and Treatment Today: Results, Trends, Frontiers. Akadémiai Kiadó, Budapest, 1986;1:223–244.

341a. Sinkovics JG. Adult human sarcomas. II. Medical oncology. Expert Rev Anticancer Ther. 2007;7:183–210.

341b. Sinkovics JG, Györkey F, Melnick JL, Gyorkey P. Acquired immune deficiency syndrome: speculations about its etiology and comparative immunology. Rev Infect Dis. 1984;6:745–60.

342. Busch K, Keller T, Fuchs U, Yeh RF, Harbott J, Klose I, Wiemels J, Novosel A, Reiter A, Burkhardt A. Identification of two distinct MYC breakpoint clusters and their association with various IGH breakpoint regions in the t(8;14) translocation in sporadic Burkitt lymphoma. Leukemia. 2007;21:1789–1751.
343. Szymanowska N, Klapper W, Gesk S, Küppers R, Martin-Subero JI, Siebert R. BCL2 and BCL3 are recurrent translocation partners of the IGH locus. Cancer Genet Cytogenet. 2008;186:110–4.
344. Slovak ML, Bedell V, Hsu YH, Estrine DB, Nowak NJ, Delioukina ML, Weiss LM, Smith DD, Forman SJ. Molecular karyotypes of Hodgkin and Reed-Sternberg cells at disease onset reveal distinct copy number alterations in chemosensitive versus refractory Hodgkin lymphoma. Clin Cancer Res. 2011;17:3443–54.
345. Re D, Benenson E, Beyer M, Gresch O, Draube A, Diehl V, Wolf J. Cell fusion is not involved in the generation of giant cells in the Hodgkin-Reed Sternberg cell line L1236. Am J Hematol. 2001;67:6–9.
346. Küppers R, Bräuninger A, Müschen M, Distler V, Hansmann M-L, Rajewsky K. Evidence that Hodgkin and Reed-Sternberg cells in Hodgkin disease do not represent cell fusions. Blood. 2001;97:818–21.
347. Drexler HG, Gignac SM, Hoffbrand AV, Leber BF, Norton J, Lok MS, Minowada J. Characterization of Hodgkin's disease derived cell line HDLM-2. Recent Res Cancer Res. 1989;117:75–82.
348a. Sinkovics JG. Hodgkin's disease revisited: Reed-Sternberg cells as natural hybridomas. Crit Rev Immunol. 1991;11:33–63.
348b. Sinkovics JG, Gonzalez F, Gyorkey F. Viral expressions in Reed-Sternberg cells. Leukemia. 1992;6S:49–53.
349. Michels KB. The origin of Hodgkin's disease. Eur J Cancer Prev. 1995;4:379–88.
350. Gupta RK, Norton AJ, Thompson IW, Lister TA, Bodmer JG. P53 expression in Reed-Sternberg cells in Hodgkin's disease. Br J Cancer. 1992;66:649–52.
351. Gupta RK, Patel K, Bodmer WF, Bodmer JG. Mutation of p53 in primary biopsy material and cell lines from Hodgkin disease. Proc Natl Acad Sci USA. 1993;90:2817–21.
352. Sinkovics JG. A notable phenomenon recapitulated. A fusion product of a murine lymphoma cell and a leukemia virus-neutralizing antibody-producer host plasma cell formed spontaneously and secreting the specific antibody continuously. Acta Microbiol Immunol Hung. 2005;52:1–40.
353. Sinkovics JG, Györkey F. Hodgkin's disease: involvement of viral agents in the etiology. J Med. 1973;4:282–317.
354. Sinkovics JG, Shullenberger CC. Hodgkin's disease. Lancet. 1975;2:506–7.
355. Ushmorov A, Ritz O, Hummel N, Leithäuser F, Möller P, Stein H, Wirth T. Epigenetic silencing of the immunoglobulin heavy-chain gene in classical Hodgkin lymphoma-derived cell lines contributes to the loss of immunoglobulin expression. Blood. 2004;104:3326–34.
356. Pedrazzoli F, Chrysantzas I, Dezzani L, Rosti V, Vinctorio M, Sitar G. Cell fusion in tumor progression: the isolation of cell fusion products by physical methods. Cancer Cell Int. 2011;11:32. doi:10.1186/1475-2867-11-32.
357. Martinez-Climent JA, Fontan L, Gascoyne RD, Siebert R, Prosper F. Lymphoma stem cells: enough evidence to support their existence? Haematologica. 2010;95:293–302.
358. Schmitt C, Balogh B, Grundt A, Buchholtz C, Leo A, Benner A, Hensel M, Ho AD, Leo E. The bcl-2/IgH rearrangement in a population of 204 healthy individuals: occurrence, age and gender distribution, breakpoints, and detection method validity. Leuk Res. 2006;30745–750.
359. Marusawa H, Takal A, Chiba T. Role of activation-induced cytidine deaminase in inflammation-associated cancer development. Adv Immunol. 2011;111:109–41.
360. Endo Y, Marusawa H, Chiba T. Involvement of activation-induced cytidine deaminase in the development of colitis-associated colorectal cancers. J Gastroenterol. 2011;46S1:6–10.

361. Shimizu T, Marusawa H, Endo Y, Chiba T: Inflammation-mediated genomic instability: role of activation-induced cytidine deaminase in carcinogenesis. Cancer Sci. 2012; doi:10.1111/j.1349-7006.2012.02293.x.
362. Humby F, Bombardieri M, Manzo A, Kelly S, Blades MC, Kirkham B, Spencer J, Pitzolis C. Ectopic lymphoid structures support ongoing production of class-switched autoantibodies in rheumatoid synovium. PLoS Med. 2009;6:56–75. doi:10.1371/journal.pmed.0060001.
363a. Baker-Austin C, Dopson M, Wexler M, Sawers RG, Stemmler A, Rosen BP, Bond PL. Extreme arsenic resistance by the acidophilic archaeon Ferroplasma acidarmanus. Extremophils. 2007;11:425–34.
363b. Baker-Austin C, Potrykus J, Wexler M, Bond PL, Dopson M. Biofilm development in the extremely acidophilic Ferroplasma acidarmanus Fer1. Extremophils. 2010;14:485–91.
364a. Baumler DJ, Hung KF, Jeong KC, Kaspar CW. Production of metanethiol and volatile sulfur compounds by the archaeon "Ferroplasma acidarmanus". Extremophils. 2007;11:841–51.
364b. Sato D, Nozaki T. Methionine gamma-lyase: The unique reaction mechanism, physiological roles, and therapeutic applications against infectious diseases and cancer. IUBMB Life 2009;61:1018–28.
365. DeLong EF. Life on the thermodynamic edge. Science. 2007;317:327–8.
366. Mouttaki H, Nanny MA, McInerney MJ. Metabolism of hydroxylated and fluorinated benzoates by Syntrophus aciditrophicus and detection of fluorodiene metabolite. Appl Environ Microbiol. 2009;75:998–1004.
367. Schmier B, Seetharaman J, Deutscher MP, Hunt JF, Malhorta A. The structure and enzymatic properties of a novel RNAase II family enzyme from Deinococcus radiodurans. J Mol Biol. 2012;415:547–59.
368. Abreyava XC, Paulino-Lima IG, Galante D, Rodriguez F, Mauas PJ, Lage A. Comparative survival of Deinococcus radiodurans and the Haloarchaea natrialba magadii and Halofex volcanii exposed to vacuum ultraviolet irradiation. Astrobiology. 2001;11:1034–49.
369. Kubista M, Andrade JM, Bengtsson M, Forootan A, Jonák J, Lind K, Sindelka R, Sjöback R, Sjögreen B, Strömbom L, Ståhlberg A, Zoric N. The real-time polymerase chain reaction. Mol Aspects Med. 2006;27:95–125.
370. Williams KR, Doak TG, Herrick G. Telomere formation on macronuclear chromosomes of Oxytricha trifallax and O. fallax: alternatively processed regions have multiple telomere addition sites. BMC Genet > v.3; 2002. doi:10.1186/1471-2156-3-16.
371. McClintock B. The fusion of broken ends of sister half chromatids following chromatid breakage at meiotic anaphases. Missouri Agric Exp Station Res Bulletin. 1938;290:1–48.
372. Zhuang XY, Tang J, Hao YH, Tan Z. Fast detection of quadruplex structure in DNA by the intrinsic fluorescence of a single-stranded DNA binding protein. J Mol Recognit. 2007;20:386–91.
373. Du Z, Zhao Y, Li N. Genome-wide colonization of gene regulatory elements by G4 DNA motifs. Nucleic Acids Res. 2009;37:6784–98.
374. Zheng K-w, Chen Z, Y-h Hao, Tan Z. Molecular crowding creates an essential environment for the formation of stable G-quadruplexes in long double-stranded DNA. Nucleic Acids Res. 2010;38:327–38.
375. Yadav VK, Abraham JK, Mani P, Kulshrestha R, Chowdhury S. QuadBase: Genome-wide database of G4 DNA—occurrence in human, chimpanzee, mouse and rat promoters and 146 microbes. Nucleic Acids Res. 2008;36:D381–5.
376. Giraldo R, Suzuki M, Chapman L, Rhodes D. Promotion and parallel DNA quadruplexes by a yeast telomere binding protein: a circular dichronism study. Proc Natl Acad Sci USA. 1994;91:7658–62.
377. Frantz JD, Gilbert W. A novel yeast gene product, G4p1, with a specific affinity for quadruplex nucleic acids. J Biol Chem. 1995;270:20692–7.

378. Capra JA, Paeschke K, Singh M, Zakian VA. G-quadruplex DNA sequences are evolutionarily conserved and associated with distinct genomic features in Saccharomyces cerevisiae. PLoS Comput Biol. 1000;6(7):e1000891. doi:10.1371/journal.pcb.861.
379. Smith JA, Chen Q, Yatsunyk LA, Nicoludis KM, Garcia MS, Kranaster R, Balasubramanian S, Monchaud D, Teulade-Fichou M-P, Abramowitz L, Schultz DC, Johnson FB. Rudimentary G-quadruplex-based telomere capping in Saccharomyces cerevisiae. Nat Struct Mol Biol. 2011;18:478–85.
380. Cavalcanti AR, Dunn DM, Weiss R, Herrick G, Landweber LF, Doak TG. Sequence features of Oxytricha trifallax (class Spirotrichea) macromolecular telometric and subtelometric sequences. Protist. 2004;155:311–22.
381. Vondrusková J, Kypr J, Kejnovská I, Fialová M, Vorlickiová M. Guanine quadruplex formation by RNA/DNA hybrid analogs of Oxytricha telomere G(4)T(4)G(4) fragment. Biopolymers. 2008;89:797–806.
382. Swart EC, Nowacki M, Shum J, Stiles H, Higgins BP, Doak TC, Schotanus K, Magrini VJ, Minx P, Mardis ER, Landweber LF. The Oxytricha trifallax mitochondrial genome. Genome Biol Evol. 2012;4:136–54.
383. Prescott JD, DuBois ML, Prescott DM. Evolution of the scrambled germline gene encoding alpha-telomere binding protein in three hypertrichous ciliates. Chromosoma. 1998;107:293–303.
384. Witherspoon DJ, Doak TG, Williams KR, Seegmiller A, Seger J, Herrick G. Selection on the protein-coding genes of the TBE1 family of transposable elements in the ciliates Oxytricha fallax and O. trifallax. Mol Biol Evol. 14:696–706.
385. Akhmanova A, Voncken F, van Alen T, van Hoek A, Boxma B, Vogels G, Veenhuis M, Hackstein JH. A hydrogenosome with a genome. Nature. 1998;396:527–8.
386. Sedwick C. Anna Akhmanova: great tips on microtubules. J Cell Biol. 2011;195:168–9.
387. Bryan TM, Sperger JM, Chapman KB, Cech TR. Telomerase reverse transcriptase genes identified in Tetrahymena thermophila and Oxytricha trifallax. Proc Natl Acad Sci USA. 1996;95:8479–84.
388. Hu L, Lim K, Bouaziz S, Phan AT. Giardia telomeric sequence d(TAGGG)4 forms two intramolecular G-quadruplexes in K+ solution: effect of loop length and sequence on the folding topology. J Am Chem Soc. 2009;131:16824–31.
389. Burke WD, Malik HS, Rich SM, Eickbush TH. Ancient lineages of non-LTR retrotransposons in the primitive eukaryote, Giardia lamblia. Mol Biol Evol. 2002;19:619–30.
390. Pardue M-L, DeBaryshe PG, Lowenhaupt K. Another protozoan contributes to the understanding telomeres and transposable elements. Proc Natl Acad Sci USA. 2001;98:14195–7.
391. Vanacova S, Liston D, Tachezy J, Johnson PJ. Molecular biology of the amitochondriate parasites, Giardia intestinalis, Entamoeba histolytica, and Trichomomas vaginalis. Int J Parasitol. 2003;33:235–55.
392. Müller M. The hydrogenosome. J Gen Microbiol. 1993;139:2879–89.
393. Vellai T, Takács K, Vida G. A new aspect to the origin and evolution of eukaryotes. J Mol Evol. 1998;46:499–507.
394. Embley TM. Multiple secondary origins of the anaerobic lifestyle in eukaryotes. Philos Trans R Soc London B Biol Sci. 2006;36:1055–67.
395. Wu G, Müller M. Glycogen phosphorylase sequences from the amitochondriate protists, Trichomonas vaginalis, Mastigamoeba balamuthi, Entamoeba histolytica and Giardia intestinalis. J Eukaryot Microbiol. 2003;50:366–72.
396. Henze K, Horner DS, Suguri S, Moore DV, Sánchez LB, Müller M, Embley TM. Unique phylogenetic relationship of glucokinases and glucosephosphate isomerase of the amitochondriate eukaryotes Giardia intestinalis, Spironucleus barkhanus and Trichomonas vaginalis. Gene. 2001;281:123–31.

397. Mitra S, Cui J, Robbins PW, Samuelson J. A deeply divergent phosphoglucomutase (PGM) of Giardia lamblia has both PGM and phosphomannomutase activities. Glycobiology. 2010;20:1233–124061.
398. Gladyshev EA, Meselson M, Arkhipova IR. A deep-branching clade of retrovirus-like retrotransposons in bdelloid rotifers. Gene. 2007;390:136–45.
399. Gorinšek B, Gubenšek F, Kordiš D. Evolutionary genomics of chromoviruses in eukaryotes. Mol Biol Evol. 2004;21:781–98.
400. Bittencourt-Silvestre J, Lemgruber L, de Souza W. Encystation process of Giardia lamblia; morphological and regulatory aspects. Arch Microbiol. 2010;192:259–65.
401. Cox SS, van der Giezen M, Tarr SJ, Crompton MFR, Tovar J. Evidence from bioinformatics, expression and inhibition studies of phosphoinositide-3 kinase signaling Giardia intestinalis. BMC Microbiol. 2006;18(6):45.
402. Kamda JD, Singer SM. Phosphoinositide-3 kinase-dependent inhibition of dendritic cell interleukin-12 production by giardia lamblia. Infect Immun. 2009;77:685–93.
403. Stiles BL. Phosphatase and tensin homologue deleted on chromosome 10 extending its PTENtacles. Int J Biochem Cell Biol. 2009;41:757–61.
404. Zhao H, Cui Y, Dupont J, Sun H, Henninghausen L, Yakar S. Overexpression of the gene tumor suppressor gene phosphatase and tensin homologue partially inhibits Wnt-1-induced mammary tumorigenesis. Cancer Res. 2005;65:6864–73.
405. Whelan JT, Kellogg A, Shewchuk BM, Hewan-Lowe K, Bertrand FE. Notch-1 signaling is lost in prostate adenocarcinoma and promotes PTEN gene expression. J Cell Biochem. 2009;107:992–1001.
406. Hart JR, Liao L, Yates JR 3rd, Vogt PK. Essential role of Stat3 in PI3K-induced oncogene transformation. Proc Natl Acad Sci USA. 2011;108:13247–53.
407. Vogt PK, Hart JR. PI3K and STAT3: a new alliance. Cancer Discov. 2011;1:481–6.
408. Ortutay C, Nore BF, Vihinen M, Smith CI. Phylogeny of Tec family kinases identification of a premetazoan origin of Btk, Bmx, Itk, Tec, Txk, and the Btk regulator SH3BP5. Adv Genet. 2008;64:51–80.
409. Alegado RA, Ferriera S, Nussbaum C, Young SK, Zeng Q, Imamovic A, Fairclough SR, King N. Complete genome sequence of Algoriphagus sp. PR1, bacterial prey of a colony-forming choanoflagellate. J Bacteriol. 2911;193:1485–1486.
410. Kutach AK, Villaseñor AG, Lam D, Belunis C, Janson C, Lok S, Hong LN, Liu CM, Deval J, Novak TJ, Barnett JW, Chu W, Shaw D, Kuglstatter A. Crystal structures of IL-2-inducible T cell kinase complexed with inhibitors: insights into rational drug design and activity regulation. Chem Biol Drug Des. 2010;76:154–83.
411. Iyer AS, Morales JL, Huang W, Ojo F, Ning G, Wills E, Baines JD, August A. Absence of Tec family kinases interleukin-2 inducible T cell kinase (ltk) and Bruton's tyrosine kinase (Btk) severely impairs Fc epsilonRI-dependent mast cell responses. J Biol Chem. 2011;286:9503–13.
412. Sahu N, August A. ITK inhibitors in inflammation and immune-mediated disorders. Curr Top Med Chem. 2009;9:690–703.
413. Young SL, Diolaiti D, Conacci-Sorrell M, Ruiz-Trillo I, Eisenman RN, King N. Premetazon ancestry of the Myc-Max network. Mol Biol Evol. 2011;28:2961–1971.
414. van Leeuwen F, Taylor MC, Mondragon A, Moreau H, Gibson W, Kieft R, Borst P. Beta-D-glucosyl-hydroxymethyluracil is a conserved DNA modification in kinetoplastid protozoans and is abundant in telomeres. Proc Natl Acad Sci USA. 1996;95:2366–71.
415. Sandhu R, L B. Examination of the telomere G-overhang structure in Trypanosoma brucei. J Vis Exp 2011;(47).PII.1959. doi:10.3791/1959.
416a. Dreesen O, Cross GAM. Telomere length in Trypanosoma brucei. Exp Parasitol. 2008;118:103–10.
416b. Dreesen O, Cross GAM. Telomerase-independent stabilization of short telomeres in Trypanosoma brucei. Mol Cell Biol. 2006;26:4911–9.

417a. Van Reet N, Pyana PP, Debourggraeve S, Büscher P, Claes F. Trypanosoma brucei gambiense: HMI-9 medium containing methylcellulose and human serum supports the continuous axenic in vitro propagation of the bloodstream form. Exp Parasitol. 2011;128:285–90.

417b. Van Reet N, Pyuana P, Rogé S, Claes F, Büscher P. Luminescent multiplex viability assay for Trypanosoma brucei gambiense. Parasit Vectors. 2013;6:207. doi:10.1186/1756-3306-6-207.

418. RawaL P, Kummarasetti VBR, Ravindran J, Kumar Halder K, Sharma R, Mukerji M, Das SK, Chowdhury S. Genome-wide prediction of G4 DNA as regulatory motifs: role in Escherichia coli global regulation. Genome Res. 2006;16:644–55.

419. Shum KT, Lui EL, Wong SC, Yeung P, Sam L, Wang Y, Watt RM, Tanner JA. Aptamer-mediated inhibition of Mycobacterium tuberculosis polyphosphate kinase 2. Biochemistry. 2011;50:3261–71.

420. Gaubler P, Wu HJ, Laudié M, Delseny M, Grellet E. A chlorophyll synthetase gene from Arabidopsis thaliana. Mol Gen Genet. 1995;249:58–64.

421. Naydov IA, Mubarakshina MM, Ivanov BN. Formation kinetics and H2O2 distribution in chloroplasts and protoplasts of photosynthetic leaf cells of higher plants under illumination. Biochemistry (Moscow). 2012;77:143–51.

422. Niessen M, Krause K, Horst I, Staebler N, Klaus S, Gaertner S, Kebeish R, Araujo WL, Fernie AR, Peterhansel C. Two alanine aminotransferases link mitochondrial glycolate oxidation to the major photorespiratory pathway in Arabidopsis and rice. J Exp Biol. 2012;63:2705–16.

423. Zhou W, Zhou T, Li MX, Zhao CL, Jia N, Wang XX, Sun YZ, Li GL, Xu M, Zhou RG, Li B. The Arabidopsis J-protein AtDjB1 facilitates thermoresistance by protecting cells against heat-induced oxidative damage. New Phytol. 2012 Feb 22. doi:10.1111/j.1469-8137.2012.04070.x.

424. Flanary BE, Kletetschka G. Analysis of telomere length and telomerase activity in tree species of various lifespans, and with age in the bristlecone pine Pinus longaeva. Rejuvenation Res. 2005;9:61–63; Biogerontology. 2005;6:101–111.

425. Bingham MA, Simard SW. Mycorrhizal networks affect ectomycorrhizal fungal community similarity between conespecfic trees and seedlings. Mycorrhiza. 2012;22:317–26.

426. Gellert M, Lipsett MN, Davies DR. Helix formation by guanylic acid. Proc Natl Acad Sci USA. 1962;48:2013–8.

427. Ambrus A, Chen D, Dai J, Bialis T, Jones RA, Yang D. Human telomeric sequence forms a hybrid-type intramolecular G-quadruplex structure with mixed parallel/antiparallel strands in potassium solution. Nucleic Acids Res. 2006;34:2723–35.

428. Phan AT. Human telomeric G-quadruplex: structures of DNA and RNA sequences. FEBS J. 2010;277:1107–17.

429. Qin Y, Hurley LH. Structures, folding patterns, and functions of intramolecular DNA G-quadruplexes found in eukaryotic promoter regions. Biochimie. 2008;90:1149–71.

430. Burge S, Parkinson GN, Hazel P, Todd AK, Neidle S. Quadruplex DNA: sequence, topology and structure. Nucleic Acids Res. 2006;34:5402–15.

431. De S, Michor F. DNA secondary structures and epigenetic determinants of cancer genome evolution. Nt Struct Mol Biol. 2011;18:950–5.

432. Patel DJ, Phan AT, Kuryavy V. Human telomere, oncogenic promoter and 5'-UTR G-quadruplexes: diverse higher order DNA and RNA targets for cancer therapeutics. Nucleic Acids Res. 2007;35:7429–55.

433. Tsai YC, Qi H, Lin COP, Lin RK, Kerrigan JE, Rzuczek SG, laVoie EJ, Rice JE, Pilch DS, Lyu YL, Liu LF. A G-quadruplex stabilizer induces M-phase cell cycle rest. J Biol Chem. 2009;284:22536–43.

434. Nielsen MC, Ulven T. Macrocyclic G-quadruplex ligands. Curr Med Chem. 2010;17:3438–4348.

435. Linder J, Garner TP, Williams HE, Searle MS, Moody CJ. Telomestatin: formal total synthesis and cation-mediated interaction of its seco-derivatives with G-quadruplexes. J Am Chem Soc. 2011;133:1044–51.
436. Huang W, Smaldino PJ, Zhang Q, Miller LD, Cao P, Stadelman K, Wan M, Giri B, Lei M, Nagamina Y. Yin Yang 1 contains G-quadruplex structures in its promoter and 5-UTR and its expression is modulated by G4 resolvase 1. Nucleic Acids Res. 2012;40:1033–49.
437. Moriuchi M, Moriuchi H, Margolis DM, Fauci AS. USF/c-Myc enhances, while Yin-Yang suppresses, the promoter activity of CXCR4, a coreceptor for HIV-1 entry. J Immunol. 1999;162:5986–92.
438. Moriuchi M, Moriuchi H. YY1 transcription factor down-regulates expression of CCR5, a major coreceptor for HIV-1. J Biol Chem. 2003;278:13003–7.
439. He G, Margolis DM. Counterregulation of chromatin deacetylation and histone deacetylase occupancy at the integrated promoter of human immunodeficiency virus type 1 (HIV-1) by the HIV-1 repressor YY1 and HIV-1 activator Tat. Mol Cell Biol. 2002;22:2965–73.
440. Yliastigui L, Kaur R, Johnson H, Volker J, He G, Hansen U, Margolis D. Mitogen-activated protein-kinases regulate LSF occupancy at the human immunodeficiency virus type 1 promoter. J Virol. 2005;79:5952–62.
441. Inayoshi Y, Okino Y, Miyake K, Mizutani A, Yamamoto-Kishikawa J, Kinoshita Y, Morimoto Y, Imamura K, Morshed M, Kono K, Itoh T, Nishijima K-I, Iijima S. Transition factor YY1 interacts with retroviral integrases and facilitates integration of Moloney murine leukemia virus cDNA into the host chromosome. J Virol. 2010;84:8250–61.
442. Zalani S, Coppage A, Holley-Guthrie E, Kenney S. The cellular YY1 transcription factor binds a cis-acting, negatively regulating element in the Epstein-Barr virus BRLF1 promoter. J Virol. 1997;71:3268–74.
443. Chang PJ, Chen LW, Shih YC, Tsai PH, Liu AC, Hung CH, Liou JY, Wang SS. Role of the cellular transcription factor YY1 in the latent-lytic switch of Kaposi's sarcoma-associated herpesvirus. Virology. 2011;41:194–204.
444. Daigle D, Gradoville L, Tuck D, Schultz V, Wand'ondu R, Ye J, Gorres K, Miller G. Valproic acid antagonizes the capacity of other histone deacetylase inhibitors to activate Epstein-Barr virus lytic cycle. J Virol. 2011;85:5626–5643.
445. Paul AG, Sharma-Walia N, Kerur N, White C, Chandran B. Piracy of PGE2EP receptor mediated signaling by Kaposi's sarcoma associated herpes virus (KSHV/HHv-8) for latency expression: Strategy of a successful pathogen. Cancer Res. 2010;70:3697–708.
446. Du C, Wen B, Li D, Lin Y, Zheng Y, Peng X, Hong C, Chen J, Lin W, Hong X, Xie L, Wu M. Downregulation of Epstein-Barr virus-encoded latent membrane protein-1 by arsenic trioxide in nasopharyngeal carcinoma. Tumori. 2006;92:140–8.
447. He G, Wang Q, Zhou Y, Wu X, Wang L, Duru L, Kong X, Zhang P, Wan B, Sui L, Guo Q, Li JJ, Yu L. YY1 is a novel potential therapeutic target for the treatment of HPV infection-induced cervical cancer by arsenic trioxide. Int J Gynecol Cancer. 2011;21:1097–104.
448. Bheda A, Creek KE, Pirisi L. Loss of p53 induces epidermal growth factor receptor promoter activity in normal human keratinocytes. Oncogene. 2008;27:4315–23.
449. Darnell GA, Antalis TM, Rose BR, Suhrbier A. Silencing of integrated human papillomavirus type 18 oncogene transcription in cell expressing serpin B2. J Virol. 2005;79:4246–56.
450. Kadeppagari R-K, Sankar N, Thimmapaya B. Adenovirus transforming protein E1A induces c-Myc in quiescent cell by a novel mechanism. J Virol. 2009;83:4810–22.
451. Lackner DF, Muzyczka N. Studies of the mechanism of transactivation of the adeno-associated virus p129 promoter by Rep protein. J Virol. 2002;76:8225–35.

452. Grönroos E, Terentiev AA, Punga T, Ericsson J. YY1 inhibits the activation of the p53 tumor suppressor in response to genotoxic stress. Proc Natl Acad Sci USA. 2004;101:12165–70.
453. Erkeland SJ, Valkhof M, Hejams-Antoissen C, Delwel R, Valk PJ, Hermans MH, Touw IP. The gene encoding the transcriptional regulator Yin Yang 1 (YY1) is a myeloid transforming gene interfering with neutrophilic differentiation. Blood. 2003;101:1111–7.
454. Voronina EN, Kolokol'tsova TD, Slyn'ko NM, Nechaev EA, Filipenko ML. Transcription factor YY1 participates in activation transcription of the human ribosomal protein L11 gene. Mol Biol (Moskva). 2006;42:110–6.
455. Ferreira AO, Myers CR, Gordon JS, Vencato M, Collmer A, Wehling MD, Alfano JAR, Moreno-Hagelsieb G, Lamboy WF, DeClerck G, Schneider DJ, Cartinhour SW. Whole-genome expression profiling defines the HrpL regulon of Pseudomonas syringae pv. Tomato DC3000, allows de novo reconstruction of the Hrp cis element, and identifies novel coregulated genes. Mol Plant Microbe Interact. 2006;19:1167–79.
456. Chen L, Shioda T, Coser KR, Lynch MC, Yang C, Schmidt EV. Genome-wide analysis of YY2 versus YY1 target genes. Nucleic Acids Res. 2010;38:4011–26.
457. Chen QR, Yu LR, Tsang P, Wei JS, Sing YK, Cheuk A, Chung JY, Hewitt SM, Veenstra TD, Khan J. Systematic protein analysis identifies transcription factor YY1 as a direct target of miR-34a. J Proteome Res. 2011;10:479–87.
458. Jiang L, Huang Q, Chang J, Wang E, Qiu X. MicroRNA HSA-miR-125a-5p induces apoptosis by activating p53 in lung cancers. Exp Lung Res. 2011;37:387–98.
459. Fabbri M, Bottoni A, Shimizu M, Spizzo R, Nicoloso MS, Rossi S, Barbarotto E, Cimmino A, Adair B, Wojcik SE, Valeri N, Calore F, Sampath D, Fanini F, Vannini I, Musuraca G, Dell'Aquila M, Alder H, Davuluri RV, Rassenti LZ, Negrini M, Nakamura T, Amadori D, Kay NE, Rai KR, Keating MJ, Kipps TJ, Calin GA, Croce CM. Association of microRNA/TP53 feedback circuitry with pathogenesis and outcome of B-cell chronic lymphocytic leukemia. JAMA. 2011;305:59–67.
460. Joshi B, Rastogi S, Morris M, Carastro LM, Decook C, Seto E, Chellappan SP. Differential regulation of human YY1 and caspase 7 promoters by prohibitin through E2F1 and p53 binding sites. Biochem J. 2007;401:155–66.
461. Rizkallah R, Alexander KE, Kassardjian A, Luscher B, Hurt MM. The transcription factor YY1 is a substrate for Polo-like kinase 1 at the G2/M transition of the cell cycle. PLoS One. 2011;6(1):e15928. doi:10.1371/journal.pone.0015928.
462. O'Brien SJ. The platypus genome unraveled. Cell. 2008;133:953–5.
463. Casey A, Daish T, Grützner F. Platypus chain reaction: directional and ordered meiotic pairing of the multiple sex chromosome chain in Ornithorhynchus anatinus. Reprod Fertil Dev. 2009;21:976–84.
464. Veyrunes F, Waters PD, Miethke OP, Rens W, McMillan D, Alsop AE, Grützner F, Deakin JE, Whittington CM, Schatzkamer K, Kreinitzky CL, Graves T, Ferguson-Smith MA, Warren W. Marshall Graves JA. Bird-like sex chromosomes of platypus imply recent origin of mammalian sex chromosomes. Genome Res. 2008;18:965–73.
465. Edwards CA, Rens W, Carke O, Mungall AJ, Hore T, Graves JA, Dunham I, Ferguson-Smith AC, Ferguson-Smith MA. The evolution of imprinting: chromosomal mapping of orthologues of mammalian domains in monotreme and marsupial mammals. BMC Evol Biol. 2007;6(7):157.
466. Killian JK, Nolan CM, Stewart N, Munday BK, Andersen NA, Nicol S, Jirtle RL. Monotreme IGF2 expression and ancestral origin of genomic imprinting. J Exp Zool. 2001;29:205–12.
467. Killian JK, Nolan CM, Wylie AA, Li KT, Vu TH, Hoffman AR, Jirtle RL. Divergent evolution in M6P/IGF2R imprinting from the Jurassic to the Quaternary. Hum Mol Genet. 2001;10:1721–8.

468. Pask AJ, Papenfuss AT, Ager EI, McColl KA, Speed TP, Renfree MB. Analysis of the platypus genome suggests a transposons origin for mammalian imprints. Genome Biol 2009; 10(1):R1. doi:10.1186/gb-2009-10-1-r1.
469. Lux H, Flammann H, Lux A. Genetic and molecular analyses of PEG10 reveal new aspects of genomic organization, transcription and translation. PLoS One. 2010;5(1): e8686. doi:10.1371/journal.pone.000868.
470. Cornelis G, Heidmann O, Bernard-Stoecklin S, Reynaud K, Véron G, Mulot B, Dupressoir A, Heidmann T. Ancestral capture of syncytin-Car1, a fusogenic endogenous retroviral gene involved in placentation and conserved in Carnivora. Proc Natl Acad Sci USA. 2012;109:432–41.
471. Vernochet C, Heidmann O, Dupressoir A, Corneliad O, Dessen P, Catzefkis F, Heidmann T. A syncytin-like endogenous retrovirus envelope gene of the guinea pig specifically expressed in the placenta junctional zone and conserved in Cavimorpha. Placenta. 2011;32:885–92.
472. Suzuki S, Shaw G, Kaneko-Ishino T, Ishino F, Renfree MB. The evolution of mammalian genomic imprinting was accompanied by the acquisition of novel CpG islands. Genome Biol Evol. 2011;3:1276–8.
473. Zeh JA, Zeh DW. Viviparity-driven conflict: more to speciation than meets the fly. Ann N Y Acad Sci. 2008;1133:126–48.
474. Zhang XY, Loflin PT, Gehrke CW, Andrewes P, Ehrlich M. Hypermethylation of human DNA sequences in embryonal carcinoma cells, and somatic tissues, but not in sperm. Nucleic Acids Res. 1987;15:9429–49.
475. Harris LK, Crocker IP, Baker PN, Aplin JD, Westwood M. IGF2 actions on trophoblast in human placenta are regulated by the insulin-like growth factor 2 receptor, which can function as both a signaling and clearance receptor. Biol Reprod. 2011;84:440–6.
476. Koukoura O, Sifakis S, Soufla G, Zaravinos A, Apostolidou S, Jones A, Widschwendter M, Spandidos DA. Loss of imprinting and aberrant methylation of IGF2 in placentas from pregnancies complicated with fetal growth restriction. Int J Mol Med. 2011;28:481–7.
477. Sun XM, Shao Y, Wang XL, Wu WZ. Expression and clinical significance of constitutive androstane receptor in placental syntrophoblast of intrahepatic cholestasis of pregnancy. Zsionghua Fu Chan Ke Za Zhi. 2011;46:338–41.
478. Ala-Honkola O, Friman E, Lindström K. Costs and benefits of polyandry in a placental poeciliid fish Heterandria formosa are in accordance with the parent-offspring conflict theory of placentation. J Evol Biol. 2011;24:2600–10.
479. Meredith RW, Mn Pires, Reznik DN, Springer MS. Molecular phylogenetic relationships and the coevolution of placentotrophy and superfetation in Poecilla (Poecilliidae: Cyprinodontiformes). Mol Phylogenet Evol. 2011;59:148–57.
480. Murphy BF, Thompson MB. A review of the evolution of viviparity in squamous reptiles: the past, present and future role of molecular biology and genomics. J Comp Physiol. 2011;181:575–94.
481. Avise JC, Liu JX. Multiple mating and its relationship to brood size in pregnant fishes versus pregnant mammals and other viviparous vertebrates. Proc Natl Acad Sci USA. 2011;108:7091–5.
482. Rengaraj D, Lee BR, Lee SI, Seo HW, Han JY. Expression patterns and miRNA regulation of DNA methyltransferases in chicken primordial germ cells. PLoS One. 2011;6(5):e19524.
483. Green AE, O'Neil S, Swan KE, Bohm RP, Ratterree MS, Henson MC. Leptin receptor transcripts are constitutively expressed in placenta and adipose tissue with advancing baboon pregnancy. Proc Soc Exp Biol Med. 2000;223:362–6.
484. Carranza PG, Luján HD. New insights regarding the biology of Giardia lamblia. Microbes Infect. 2010;12:71–80.

485. Csaba G, Gaál A, Kovács P, Simon G, Kőhidai L. Prolonged elevation of insulin content in the unicellular tetrahymena after insulin treatment: induction of insulin production or storage? Cell Biochem Funct. 1999;17:165–73.
486. Christopher GK, Sundermann CA. Intracellular insulin binding in Tetrahymena pyriformis. Tissue Cell. 1996;28:427–37.
487. Olinski RP, Dahlberg C, Thorndyke M, Hallbook F. Three insulin-relaxin-like genes in Ciona intestinalis. Peptides. 2006;27:2535–46.
488. Somorjai IM, Somorjai RL, Garcia-Fernández J, Escrivá H. Vertebrate-like regeneration in the invertebrate chordate amphioxus. Proc Natl Acad Sci USA. 2012;109:517–22.
489. Pashmforoush M, Chan SJ, Steiner DF. Structure and expression of the insulin-like peptide receptor from amphioxus. Mol Endocrinol. 1996;10:857–66.
490. Guo ZY, Shen L, Gu W, Wu AZ, Ma JG, Feng YM. In vitro evolution of amphioxus insulin-like peptide-to-mammalian insulin. Biochemistry. 2002;41:10603–10507.
491. Zhang W, Thompson BJ, Hietakangas V, Cohen SM. MAPK/ERK signaling regulates insulin sensitivity to control glucose metabolism in Drosophila. PLoS Genet 2011. 2011;7(12):e1002429.
492. Sandmann T, Girardot C, Brehme M, Tongprasil W, Stolc V, Furlong EE. A core transcriptional network for early mesoderm development in Drososphila melanogaster. Genes Dev. 2007;21:436–49.
493. Liu J, Uygur B, Zhang Z, Shao L, Romero D, Vary C, Ding Q, Wu WS. Slug inhibits proliferation of human prostate cancer cells via downregulation of cyclin d1 expression. Prostate. 2010;70:1768–77.
494. Yee DS, Tang Y, Li X, Liu Z, Guo Y, Ghaffar S, McQueen P, Atreya D, Xie J, Simoneau AR, Hoang BH, Zi X. The Wnt inhibitory factor 1 restoration in prostate cancer cells was associated with reduced tumor growth, decreased capacity of cell migration and invasion and a reversal of epithelial to mesenchymal transition. Mol Cancer. 2010;23(9):162.
495. Uygur B, Wu WS. SLUG promotes prostate cancer cell migration and invasion via CXCR4/CXCL12 axis. Mol Cancer. 2011;10(10):139.
496a. Liu YN, Yin JJ, Abou-Kheir W, Hynes PG, Casey OM, Fang L, Yi M, Stephens RM, Seng V, Sheppard-Tillman H, Martin P, Kelly K. Mir-1 and miR-209 inhibit EMT via Slug-dependent and tumorigenesis via Slug-independent mechanisms. Oncogene. 2012 Feb 17. doi:10.1038/onc.2012.58.
496b. Liu YN, Abou-Kheir W, Yin JJ, Fang L, Hynes P, Casey O, Hu D, Wan Y, Seng V, Sheppard-Tillman H, Martin P, Kelly K. Critical and reciprocal regulation of KLF4 and SLUG in transforming growth factor β-initiated prostate cancer epithelial-mesenchymal transition. Mol Cell Biol. 2012;32:941–53.
497a. Park SK, Link CD, Johnson TE. Life span extension by dietary restriction is mediated by NLP-Z signaling and coelomocyte endocytosis in C. elegans. FASEB J. 2010;24:383–92.
497b. Park D, Jones KL, Lee H, Snutch TP, Tabrt S, Riddle DL. Repression of a potassium channel by nuclear hormone receptor and TGF-β signaling modulates insulin signaling in Caenorhabditis elegans. PLoS Genet. 2012;8(2):e1002519.
498a. Narasimhan SD, Am Mukhopadhyay, Issenbaum HA. Inactivation of insulin/IGF-1 signaling by dephosphorylation. Cell Cycle. 2009;8:3878–84.
498b. Narasimham SD, Yen K, Bansal A, Kwon ES, Padmanabban S, Tissenbaum HAS. PDP-links the TGF-β and IIS pathways to regulate longevity, development, and metabolism. PLoS Genet. 2011;7(4):e1001377.
499. Crozatier M, Vincent A. Drosophila: a model for studying genetic and molecular aspects of haematopoiesis and associated leukaemias. Dis Model Mech. 2011;4:439–45.
500. Pinkston-Gosse J, Kenyon C. DAF-16/FOXO targets genes that regulate tumor growth in Caenorhabditis elegans. Nat Gen. 2007;39:1403–9.

501. Witte K, Schuh AL, Hegermann J, Sarkeshik A, Mayers JR, Schwarze K, Yates JR 3rd, Eimer S, Audhya A. TGF-1 function in protein secretion and oncogenesis. Nat Cell Biol. 2011;13:550–8.
502. Breckenridge DG, Kang BH, Xue D. Bcl-2 proteins EGL-1 and CED-9 do not regulate mitochondrial fission or fusion in Caenorhabditis elegans. Curr Biol. 2009;19:768–73.
503. Shirinian M, Varshney G, Lorén CE, Grabbe C, Palmer RH. Drosophila anaplastic lymphoma kinase regulates Dpp signaling in the developing embryonic gut. Differentiation. 2007;75:418–26.
504. Yang HL, Eriksson T, Vernersson E, Vigny M, Hallberg B, Palmer RH. The ligand jelly belly (jeb) activates the Drosophila Alk RTK to drive PC12 cell differentiation, but is unable to activate the mouse ALK RTK. J Exp Zool B Mol Dev Evol. 2007;308:269–82.
505. Weiss JB, Xue C, Benice T, Xue L, Morris SW, Raber J. Anaplastic lymphoma kinase and leukocyte tyrosine kinase: functions and genetic interactions in learning, memory and adult neurogenesis. Pharmacol Biochem Behav. 2012;100:566–74.
506. Gouzi JY, Moresis A, Walker JA, Apostolopoulou AA, Palmer RH, Bernards A, Skoulakis EM. The receptor tyrosine kinase Alk controls neurofibromin functions in Drosophila growth and learning. PLoS Genet. 2011;7(9):e1002281.
507. Galietta A, Gunby RH, Redaelli S, Stanom P, Carniti C, Bachi A, Tucker PW, Tartari CJ, Huang CJ, Colombo E, Pulford K, Puttini M, Piazza RG, Ruchatz H, Villa A, Donella-Deana A, Marin O, Perotti D, Gambacorti-Passerini C. NPM/ALK binds and phosphorylates the RNA/DNA-binding protein PSF in anaplastic large cell lymphoma. Blood. 2007;119:2600–9.
508. Paznekas WA, Okajima K, Schertzer M, Wood S, Jabs EW. Genomic organization, expression, and chromosome location of the human snail gene (SNAI1) and a related processed pseudogene (SNAI1P). Genomics. 1999;15:62-42-49.
509. Emadi Baygi M, Soheili ZS, Essmann F, Deezagi A, Engers R, Goering W, Schulz WA. Slug/SNAI2 regulates cell proliferation and invasiveness of metastatic prostate cancer cell lines. Tumor Biol. 2010;31:297–307.
510. Wainszelbaum MJ, Charon AJ, Kong C, Kirkpatrick DS, Srikanth P, Barbieri MA, Gygi SP, Stahl PD. The hominoid-specific oncogene TBC1D3 activates Ras and modulates epidermal growth factor receptor signaling and trafficking. J Biol Chem. 2008;283:13233–42.
511. Wainszelbaum MJ, Liu J, Kong C, Srikanth P, Samovski S, Su X, Stahl PD. TBC1D3, a hominoid-specific gene delays IRS-1 degradation and promotes insulin-signaling by modulating p70 S6 kinase activity. PLoS One. 2012;7(2):e31225.
512. Pei L, Peng Y, Yang Y, Ling XB, Van Eyndhoven WG, Nguyen KC, Rubin M, Hoey T, Powers S, Li J. PRC17, a novel oncogene encoding a Rab GTPase-activating protein, is amplified in prostate cancer. Cancer Res. 2002;62:5420–4.
513. Paulding CA, Ruvolo M, Haber DA. The Tre (USP6) oncogene is a hominoid-specific gene. Proc Natl Acad Sci USA. 2003;100:2507–11.
514. Lumb JH, Field MC. Rab23 is a flagellar protein in Trypanosoma brucei. BMC Res Notes. 2011;4:190.
515. Lumb JH, Leung KF, Dubois KN, Field MC. Rab28 function in trypanosomes: interactions with retromer and ESCRT pathways. J Cell Sci. 2011;124:3771–83.
516. Leung KF, Dacks JB, Field MC. Evolution of the multivesicular body ESCRT machinery: retention across the eukaryotic lineage. Traffic. 2008;9:1698–716.
517. Chen MC, Hong MC, Huang YS, Liu MC, Cheng YM, Fang LS. ApRab11, a cnidarian homologue of the recycling regulatory protein Rab11, is involved in the establishment and maintenance of the Aiptasia-Symbodium endosymbiosis. Biochem Biophys Res Commun. 1995;338:1607–16.
518. Kim J, Jang SS, Song K. Different levels of Bfa1/Bub2/ GAP activity are required to prevent mitotic exit of budding yeast depending on the type of perturbation. Mol Biol Cell. 2008;19:4328–40.

519. Fraschini R, D'Ambrosio C, Venturetti M, Lucchini G, Piatt S. Disappearance of the budding yeast Bub2-Bfa1 complex from the mother-bound spindle pole contributes to mitotic exit. J Cell Biol. 2006;172:335–46.
520. Kim J, Luo G, Bahk YY, Song K. Cdc5-dependent asymmetric localization of Bfa1 fine tunes timely mitotic exit. PLoS Genet January 2012;8(1):e1002450. doi:10.1371/journal.pgen.1002450.
521. Wang Y, Ng T-Y. Phosphatase 2A negatively regulates mitotic exit in Saccharomyces cerevisiae. Mol Biol Cell. 2006;17:89–89.
522. Conner SD, Wessel GM. Syntaxin, VAMP, and Rab3 are selectively expressed during sea urchin embryogenesis. Mol Reprod Dev. 2001;58:22–9.
523. Shi A, Chen CC, Banerjee R, Glodowski D, Audhya A, Rongo C, Grant BD. EHBP-1 functions with RAB-10 during endocytic recycling in Caenorhabditis elegans. Mol Biol Cell. 2010;21:2930–43.
524. Pei L, Peng Y, Yang Y, Ling XB, van Eyndhoven W, Nguyen KCQ, Rubin M, Hoey T, Powers S, Li J. PRC17, a novel oncogene encoding a Rab GTPase-activating protein, is amplified in prostate cancer. Cancer Res. 2002;62:5420–33.
525. Masuda-Robens JM, Kutney SN, Qi H, Chou MM. The TRE17 oncogene encodes a component of a novel effector pathway for Rho GTPase Cdc42 and Rac 1 and stimulates actin remodeling. Mol Cell Biol. 2003;23:2151–61.
526. Chen S, Murphy J, Toth R, Campbell DG, Morrice NA, MacKintosh C. Complementary regulation of TBC1D1 and AS160 by growth factors, insulin and AMPK activators. Biochem J. 2008;409:449–59.
527. Albalat R, Marti-Solans J, Cañestro C. DNA methylation in amphioxus: from ancestral functions to new roles in vertebrates. Brief Funct Genomics. 2012;11:142–55.
528. Banki K, Maceda J, Hurley E, Ablonczy E, Mattson DH, Szegedy L, Hung C, Perl A. Human T cell lymphotropic virus (HTLV)-related endogenous sequence, HRES-1, encodes a 28-kD protein: a possible autoantigen for HTLV-1 gag-reactive antibodies. Proc Natl Acad Sci USA. 1992;89:1939–43.
529. Sinkovics JG. Viral induction of autoimmune disease and neoplasia. Lancet. 1966;2:229–30.
530. Sinkovics JG. Working hypothesis. Viral etiology of autoimmune diseases. N Engl J Med. 1969;280:903–4.
531. Gyorkey F, Sinkovics JG, Min KW, Gyorkey PA. A morphological study on the occurrence and distribution of structures resembling viral nucleocapsids in collagen diseases. Am J Med. 1972;53:148–58.
532. Gyorkey F, Sinkovics JG, Gyorkey P. Tubuloreticular structures in Kaposi's sarcoma. Lancet. 1982;2:984–5.
533. Zucker-Franklin D. "Looking" for the cause of AIDS. N Engl J Med. 1983;308:837–8.
534a. Perl A, Fernandez D, Telarico T, Phillips PE. Endogenous retroviral pathogenesis in lupus. Curr Opin Rheumatol. 2010;22:483–92.
534b. Caza TN, Fernandez DR, Talaber G, Oaks Z, Haas M, Madaio MP, Lai ZW, Miklossy G, Singh RR, Chudakov DM, Malorni W, Middleton F, Banki K, Perl A. HRES-1/Rab4-mediated depletion of Drp1 impairs mitochondrial homeostasis and represents a target for treatment of SLE. Ann Rheum Dis 2013; doi:10.1136/annrheumdis-2013-203794.
535. Alvino CL, Ong SC, McNail KA, Delaine C, Brooker GW, Wallace JC, Forbes BE. Understanding the mechanism of insulin and insulin growth factor (IGF) receptor activation by IGF-II. PLoS One. 2011;6(11):e27488.
536. Bhagrava R, Beriwal S, McManus K, Dabbs DJ. Insulin-like growth factor receptor-1 (IGF-1R) expression in normal breast, proliferative breast lesions, and breast carcinoma. Appl Immunohistochem Mol Morphol. 2011;19:218–25.

537. Davison Z, de Blacquière GE, Westley BR, May FE. Insulin-like growth factor-dependent proliferation and survival of triple-negative breast cancer cells: implications for therapy. Neoplasia. 2011;13:504–15.
538. Fox EM, Miller TW, Balko JM, Kuba MG, Sánchez V, Smith RA, Liu S, González-Angulo AM, Mills GB, Ye F, Shyr Y, Manning HC, Buck E, Arteaga CL. A kinome-wide screen identifies the insulin/IGF-1 receptor pathway as a mechanism of escape from hormone dependence in breast cancer. Cancer Res. 2011;71:6773–84.
539. Litzenburger BC, Creighton CJ, Tsimelzon A, Chan BT, Hilsenbeck SG, Wang T, Carboni JM, Gottards MM, Huang F, Chang JC, Lewis MT, Rimawi MF, Lee AV. High IGF-IR activity in triple-negative breast cancer cell lines and tumor grafts correlates with sensitivity to anti-IGF-IR therapy. Clin Cancer Res. 2011;17:2314–27.
540. Peiró G, Adrover E, Sánchez-Tejada L, Lerma E, Planeles M, Sánchez-Payá J, Aranda FI, Giner D, Gutiérrez-Aviño FJ. Increased insulin-like growth factor-1 receptor mRNA expression predicts poor survival in immunophenotypes of early breast carcinoma. Mol Pathol. 2011;24:201–8.
541. Richardson AE, Hamilton N, Davis W, Brito C, De León D. Insulin-like growth factor-2 (IGF-2) activates estrogen receptor-α and -β via the IGF-1 and the insulin receptors in breast cancer cells. Growth Factors. 2011;29:82–93.
542. Sarfstein R, Pasmanik-Chor M, Yeheskei A, Edry L, Shomrion N, Warman N, Wertheimer E, Maor S, Shochat L, Werner H. Insulin-like growth factor-I receptor (IGF-IR) translocates to nucleus and autoregulates IGF-IR gene expression in breast cancer cells. J Biol Chem. 2012;287:2766–76.
543. Tamimi RM, Colditz GA, Wang Y, Collins LC, Hu R, Rosner B, Irie HY, Connolly JL, Schnitt SJ. Expression of IGF-IR in normal breast tissue and subsequent risk of breast cancer. Breast Cancer Res Treat. 2011;128:2430–50.
544. Pengchong H, Tao H. Expression of IGF-IR, VEGF-C and D2-40 and their correlation with lymph node metastasis in endometrial adenocarcinoma. Eur J Gynecol Oncol. 2011;32:660–4.
545. Shu S, Yang Y, Li X, Li T, Zhang Y, Xu C, Liang C, Wang X. Down-regulation of IGF-1R expression inhibits growth and enhances chemosensitivity of endometrial carcinoma in vitro. Mol Cell Biochem. 2011;353:225–33.
546. Tennant MK, Thrasher JB, Twomey OPA, Drivdahl PH, Birnbaum RS, Plymate SR. Protein and messenger ribonucleic acid (mRNA) for the type 1 insulin-like growth factor (IGF) receptor is decreased and IGF-II mRNA is increased in human prostate carcinoma compared to benign prostate epithelium. J Clin Endocrinol Metab. 1996;81:3774–82.
547. Ma Y, Cheng Q, Ren Z, Xu L, Zhao Y, Sun J, Hu S, Xiao W. Induction of IGF-1R expression by EGF-1 facilitates the growth of prostate cancer. Cancer Lett. 2012;317:150–6.
548. Rojas A, Liu G, Coleman I, Nelson PS, Zhang M, Dash R, Fisher PB, Plymate SR, Wu JD. IL-6 promotes prostate tumorigenesis and progression through autocrine cross-activation of IGF-IR. Oncogene. 2011;30:2345–55.
549. Donohoe CL, Doyle SL, McGarrigle S, Cathcart MC, Daly E, O'Grady A, Lysaght J, Pidgeon GP, Reynolds JV. Role of the insulin-like growth factor 1 axis and visceral adiposity in esophageal adenocarcinoma. Br J Surg. 2012;99:387–96.
550. Ohashi H, Adachi Y, Yamamoto Taniguchi H, Noso K, Suzuki H, Arimura Y, Imai K, Carbone DP, Shinomura Y. Insulin-like growth factor receptor expression is associated with aggressive phenotypes and has therapeutic activity in biliary tract cancers. Cancer Sci. 2012;103:252–61.
551. Yang JM, Chen WS, Liu ZP, Luo YH, Liu WW. Effects of insulin-like growth factors IR and -IIR antisense gene transfections on the biological behavior of SMMC-7721 human hepatoma cells. J Gastroenterol Hepatol. 2003;18:296–301.

552. Inno A, Di Salvatore M, Cenco T, Martini M, Orlandi A, Strippoli A, Ferrara AM, Bagalá C, Cassano A, Larocca LM, Barone C. Is there a role for IGF-1R and c-MET pathways in resistance to cetuximab in metastatic colorectal cancer? Clin Colorectal Cancer. 2011;10:325–32.
553. Shan HB, Zhang R, Li Y, Xu GL, Luo GY, Gao XY, Yang HL. Expression of IGF-1R in colorectal polyps and its role in colorectal carcinogenesis. Technol Cancer Res Treat. 2011;10:381–9.
554. Zhao D, Bakirtzi K, Zhan Y, Zeng H, Koon HW, Pothoulakis C. Insulin-like growth factor-1 receptor transactivation modulates the inflammatory and proliferative responses of neurotensin in human colonic epithelial cells. J Biol Chem. 2011;286:6092–9.
555. Badzio A, Wynes MW, Dziadziuszko R, Merrick DT, Pardo M, Rzyman W, Kowalczyk A, Singh S, Ranger-Moore J, Manniquez G, Gaire F, Jassem J, Hirsch FR. Increased insulin-like growth factor 1 receptor protein expression and gene copy number in small cell lung cancer. J Thorac Oncol. 2010;5:1905–11.
556. Zhang X, Feng J, Sha H, Huang J, Han B. Correlations between IGF-IR expression and clinicopathological factors and prognosis in patients with lung adenocarcinoma. Zhongguo Fei Ai Za Zhi. 2010;13:113–1117.
557. Zuo Q, Luo RC. Relationship between the insulin-like growth factor 1 receptor signaling pathway and the resistance of nasopharyngeal carcinoma to cetuximab. Zhonhua Zhing Liu Za Zhi. 2010;32:575–9.
558. Molhoek KR, Shada AL, Smolin M, Chowbina S, Papin J, Brautigan DL, Slingluff CL Jr. Comprehensive analysis of receptor tyrosine kinase activation in human melanomas reveals autocrine signaling through IGF-IR. Melanoma Res. 2011;19:890–900.
559. Abraham J, Prajapati SI, Nishijo K, Schaffer BS, Taniguchi E, Kilcoyne A, McCleish AT, Nelon LD, Giles FG, Efstratiadis A, LeGallo RD, Nowak BM, Rubin BP, Malempati S, Keller C. Evasion mechanisms to Igf1r inhibition in rhabdomyosarcoma. Mol Cancer Ther. 2011;10:697–707.
560. Burrow S, Andrulis IL, Pollak M, Bell RS. Expression of insulin-like growth factor receptor, IGF-1, and IGF-2, in primary and metastatic osteosarcoma. J Surg Oncol. 1998;89:21–7.
561. Demicco EG, Tiorres KE, Ghadimi MP, Colombo C, Bolshakov S, Hoffman A, Peng T, Bovée JV, Wang WL, Lev D, Lazar AJ. Involvement of the PI3K/Akt pathway in myxoid/round cell liposarcoma. Mol Pathol. 2012;25:212–21.
562. Martin AS, Ordónez JL, Amaral AT, Prins F, Floris G, Debiec-Rychter M, Hogendoom PC, de Alava E. IGF-IR signaling in Ewing sarcoma is shaped by clathrin-/caveolin-dependent endocytosis. PLoS One. 2011;6(5):e19846.
563. Ohlsson C, Kley N, Werner H, Leroith D. p53 regulates insulin-like growth factorIR (IGF-I) receptor expression and IGF-I-induced tyrosine phosphorylation in an osteosarcoma cell line: interaction between p53a and Sp1. Endocrinology. 1998;139:1101–7.
564. Wang YH, Han XD, Qiu Y, Xiong J, Wang B, Zhu ZZ, Qian BP, Chen YX, Wang SF, Shi HF, Sun X. Increased expression of insulin-like growth factor-1-receptor is correlated with tumor metastasis and prognosis in patients with osteosarcoma. J Surg Oncol. 2012;105:235–43.
565. He Q, Li X, Zhang Z, Zhang QX, Yang LP. The type I insulin-like growth factor receptor highly expressed in clonal cells of myelodysplastic syndromes. Zhonghua Xue Ye Xue Za Zhi. 2011;32:744–7.
566. Medyouf H, Gusscott S, Wang H, Tseng JC, Wai C, Nemirovsky O, Trumpp A, Pflumio F, Carboni J, Gottardis M, Pollak M, Kung AL, Aster JC, Holzenberger M, Weng AP. High level IGF1R expression is required for leukemia-initiating cell activity in T-ALL and is supported by Notch signaling. J Exp Med. 2011;208:1809–22.
567. Pollak M. The insulin and insulin-like growth factor receptor family in neoplasia: an update. Nat Rev Cancer. 2012;12:159–69.

568. Shi Y, Yan H, Frost P, Gera J, Lichtenstein A. Mammalian target of rapamycin inhibitors activate the AKT kinase in multiple myeloma cells by up-regulating the insulin-like growth factor receptor/insulin receptor substrate: 1/phosphatidyl-inositol 3-kinase cascade. Mol Cancer Ther 2005;4:1533-1540
569. Vishwamitra D, Shi P, Wilson D, Manshouri R, Vega F, Schlette EJ, Amin HM. Expression and effects of inhibition of type I insulin-like growth factor receptor tyrosine kinase in mantle cell lymphoma. Haematologica. 2011;96:871–80.
570. Attias-Geva Z, Bentov I, Ludwig DL, Fishman A, Bruchim I, Werner H. Insulin-like growth factor-I receptor (IGF-IR) targeting with monoclonal antibody cixutumumab (IMC-A12) inhibits IGF-I action in endometrial cancer cells. Eur J Cancer. 2011;47:1717–26.
571. Dong J, Sereno A, Snyder WB, Miller BR, Tamraz S, Doern A, Favis M, Wu X, Tran H, Langley E, Joseph I, Boccia A, Kelly R, Wortham K, Wang Q, Berquist L, Huang Gao SM, Hariharan K, Demarest SJ. Stable IgG-like bispecific antibodies directed toward the type I insulin-like growth factor receptor demonstrate enhanced ligand blockade and anti-tumor activity. J Biol Chem. 2011;286:4703–17.
572. Durfort T, Tkach M, Meschaninova MI, Rivas MA, Elizalde PV, Venyaminova AG, Schillaci R, François JC. Small interfering RNA targeted to IGF-IR delays tumor growth and induces proinflammatory cytokines in a mouse breast cancer model. PLoS One. 2012;7(1):e29213.
573. Ganguly S, Basu B, Shome S, Jadhav T, Roy S, Majumdar J, Dasguopta PS, Basu S. Dopamine, by acting through its D2 receptor, inhibits insulin-like growth factor-I (IGF-I) -induced gastric cancer cell proliferation via upregulation of Krüppel-like factor 4 through down-regulation of IGF-IR and AKT phosphorylation. Am J Pathol. 2010;177:2701–7.
574. Hartog H, Van Der Graaf WT, Boezen HM, Wesseling J. Treatment of breast cancer cells by IGF-R tyrosine kinase inhibitor combined with conventional systemic drugs. Anticancer Res. 2012;32:1309–18.
575. Höpfner M, Huerther A, Sutter AP, Baradan V, Schuppan D, Scherübl H. Blockade of IGF-1 receptor tyrosine kinase has antineoplastic effects in hepatocellular carcinoma cells. Biochem Pharmacol. 2006;71:1435–48.
576. Iozzo RV, Buraschi S, Genua M, Xu SQ, Solomides CC, Peioer SC, Gomelia LG, Owens RC, Morrison A. Decorin antagonizes IGF receptor I (IGF-IR) function by interfering with IGF-IR activity and attenuating downstream signaling. J Biol Chem. 2011;286:34712–21.
577. Liu B, Fan Z, Edgerton SM, Yang X, Lind SE. Thor Ad. Potent anti-proliferative effects of metformin on trastuzumab-resistant breast cancer cells via inhibition of erb2/IGF-1 receptor interactions. Cell Cycle. 2011;10:2959–66.
578. Liu X, Allen JD, Arnold JT, Blackman MR. Lycopene inhibits IGF-I signal transduction and growth in normal prostate epithelial cells by decreasing DHT-modulated IGF-I production in co-cultured reactive stromal cells. Carcinogenesis. 2008;29:816–23.
579. Mao Y, Shang Y, Pham VC, Ernst JA, Lill JAR, Scales SJ, Zha J. Polyubiqitination of insulin-like growth factor I receptor (IGF-IR) activation loop promotes antibody-induced receptor internalization and down-regulation. J Biol Chem. 2011;286:41852–61.
580. Birmann BM, Neuhouser ML, Rosner B, Albanes D, Buring JE, Giles GG, Lan Q, Lee IM, Purdue MP, Rothman N, Severi G, Yuan JM, Anderson KC, Pollak M, Rifai N, Hartge P, Landgren O, Lessin L, Virtamo J, Wallace RB, Manson JE, Colditz GA. Prediagnosis biomarkers of insulin-like growth factor-1, insulin, and interleukin-6 dysregulation and multiple myeloma risk in the Multiple Myeloma Cohort Consortium. Blood. 2012;120:4929–37.
581. Tagnon CE, Sorensen PH. Targeting the insulin-like growth factor 1 receptor (IGF1R) signaling pathway for cancer therapy. Expert Opin Ther Targets. 2012;16:33–48.

582. Weickhardt A, Doebele R, Oton A, Lettieri J, Maxson D, Reynolds M, Brown A, Jackson MK, Dy G, Adjei A, Fetterly G, Lun X, Franklin W, Varella-Garcia M, Hirsch FR, Wynes MW, Youssoufian H, Adjei A, Camidge DR. A phase I/II study of erlotinib in combination with anti-insulin-like growth factor-1 receptor monoclonal antibody IMC-A12 (cixutumumab) in patients with advanced non-small cell lung cancer. J Thorac Oncol. 2012;7:419–26.
583. D'Amario D, Cabral-Da-Silva MC, Zheng H, Fiorini C, Goichberg P, Steadman E, Ferreira-Martins J, Sanada F, Piccoli M, Cappetta D, D'Alessandro DA, Michler RE, Hosoda T, Anastasia L, Rota M, Leri A, Anversa P, Kajstura J. Insulin-like growth factor-1 receptor identifies a pool of human cardiac stem cells with superior therapeutic potential for myocardial regeneration. Circ Res. 2011;108:1467–81.
584. Vlashi E, Lagadec C, Vergnes L, Matsutani T, Masui K, Poulou M, Popescu R, Della Donna L, Evers P, Dekmezian C, Reue K, Christofk H, Mischel PS, Pajonk F. Metabolic state of glioma stem cells and nontumorigenic cells. Proc Natl Acad Sci USA. 2011;108:16062–7.
585. Le A, Lane AN, Hamaker M, Bos S, Gouw A, Barbi J, Tsukamoto T, Rojas CJ, Slusher BS, Zhang H, Zimmermann LJ, Liebler FDC, Slebos RJ, Lorkiewicz PK, Higashi RM, Fan TW, Fang CV. Glucose-independent glutamine metabolism via TCA cycling for proliferation and survival in B cells. Cell Metab. 2012;15:110–21.
586. Cheng T, Sudderth J, Yang C, Mullen AR, Jin ES, Matés JM, DeBerardinis R. Pyruvate carboxylase is required for glutamine-independent growth of tumor cells. Proc Natl Acad Sci USA. 2011;108:8674–9.
587. Chen JK, Xio J, Jiang Z, Wang B, Wang Y, Hu X. Shikonin and its analogs inhibit cancer cell glycolysis by targeting tumor pyruvate kinase-M2. Oncogene. 2011;30:4297–2306.
588. Kretschmer N, Rinner B, Deutsch AJ, Lohberger B, Nausz H, Kunert O, Blunder M, Boechzelt H, Schaider H, Bauer R. Naphthokinones from Onosma paniculata induce cell-cycle arrest and apoptosis in melanoma cells. J Nat Prod. 2012;75:865–9.
589. Gong K, Li W. Shikonin, a Chinese plant-derived naphthoquinone, induces apoptosis in hepatocellular carcinoma cells through reactive oxygen species: a potential new treatment for hepatocellular carcinoma. Free Radic Biol Med. 2011;51:2259–71.
590. Selig M, Xavier KB, Santos H, Schönheit P. Comparative analysis of Embden-Meyerhof and Entner-Doudoroff glycolytic pathways in hyperthermophilic archaea and the bacterium Thermotoga. Arch Microbiol. 1997;167:217–232.
591. Malki L, Yanku M, Borovok I, Cohen G, Mevarech M, Aharonowitz Y. Identification and characterization of gshA, a gene encoding the glutamate-cysteine ligase in the halophilic archaeon Haloferax volcanii. J Bacteriol. 2009;191:5196–5204.
592. Maaty WS, Wiedenheft B, Tarlykov P, Schaff N, Heinemann J, Robison-Cox J, Valenzuela J, Blum P, Lawrence CM, Douglas T, Young MJ, Bothner B. Something old, something new, something borrowed; how the thermoacidophilic archaeon Sulfolobus solfataricus responds to oxidative stress. PLoS One. 2009;4(9):e6964. doi:10.1371/journal.pone.0006964.
593. Nunn CE, Johnsen U, Schönheit P, Fuhrer T, Sauer U, Hough DW, Danson MJ. Metabolism of pentose sugars in the hyperthermophilic archaea Sulfolobus solfataricus and Sulfolobus acidocaldarius. J Biol Chem. 2010;285:33701–3370.
594. Pickl A, Johnsern U, Schönheit P. Fructose degradation in the haloarchaeon Haloferax volcanii involves a bacterial type phosphoenolpyruvate-dependent phosphotransferase system, fructose-1-phosphatase kinase, and class II fructose-1,6-bisphosphate aldolase. J Bacteriol. 2012;194:3088–97.
595. Rogers PV, Luo S, Sucic JF, Rutherford CL. Characterization and cloning of glycogen phosphorylase 1 from Dictyostelium discoideum. Biochem Biophys Acta. 1991;1129:262–71.

596. Szijgyártó Z, Garedew A, Azevedo C, Saiardi A. Influence of inositol pyrophosphates on cellular energy dynamics. Science. 2011;334:802–5.
597. Slamovits CH, Keeling PJ. Pyruvate-phosphate dikinase of oxymonads and parabasalia and the evolution of pyrophosphate-dependent glycolysis in anaerobic eukaryotes. Eukaryot Cell. 2006;5:148–54.
598. Cervantes-Sandoval I, Serrano-Luna J, Pecheco-Yépez J, Silva-Olivares A, Tsutsumi V, Shibayama M. Differences between Naegleria fowleri and Naegleria gruberi in expression of mannose and fucose glycoconjugates. Parasitol Res. 2010;106:695–701.
599. Lizundia R, Chaussepied M, Huerre M, Werling D, Di Santo JP, Langsley G. c-Jun NH2-terminal kinase/c-Jun signaling promotes survival and metastasis of B lymphocytes transformed by Theileria. Cancer Res. 2006;66:6105–10.
600. Dessauge F, Hialy S, Baumgartner M, Blumen B, Werling D, Langsley G. c-Myc activation by Theileria parasites promotes survival of infected B-lymphocytes. Oncogene. 2005;24:1075–83.
601. Guergnon J, Chaussepied M, Sopp P, Lizundia R, Moreau MF, Blumen B, Weling D, Howard CJ, Langsley G. A tumour necrosis factor alpha autocrine loop contributes to proliferation and nuclear factor-kappaB activation of Theileria parva-transformed B cells. Cell Microbiol. 2003;5:709–16.
602. Jensen K, Makins GD, Kalisewska A, Hulme MJ, Paxton E, Glass EJ. The protozoan parasite Theileria annulata alters the differentiation state of the infected macrophage and suppresses musculoaponeurotic fibrosarcoma oncogene (MAF) transcription factors. Int J Parasitol. 2009;39:1099–108.
603. Ahmed JS, Glass EJ, Salih DA, Seitzer YU. Innate immunity to tropical theileriosis. Innate Immun. 2008;14:5–12.
604. Connelley TK, MacHugh ND, Pelle R, Weir W, Morrison WI. Escape from CD8+ T cell response by natural variants of an immunodominant epitope from Theileria parva is predominantly due to loss of TCR recognition. J Immunol. 2011;187:5910–20.
605. Latif AA, Hove T. History and critical review of Theileria parva (Boleni), the vaccine stock against Zimbabwean cattle theileriosis. Ticks Tick Borne Dis. 2011;2:163–7.
606. Lizundia R, Werling D, Langsley G, Ralph SA. Theileria apicoplast as a target for chemotherapy. Antimicrob Agents Chemother. 2009;53:1213–7.
607. Haller D, Mackiewicz M, Gerber S, Beyer D, Kullmann B, Schneider I, Ahmed JS, Seitzer U. Cytoplasmic sequestration of p53 promotes survival in leukocytes transformed by Theileria. Oncogene. 2010;29:3079–86.
608. Ralser M, Wamelink MM, Kowald A, Gerisch B, Heeren G, Struys E, Klipp E, Jakobs C, Breitenbach M, Lehrach H, Krobitsch S. Dynamic rerouting of the carbohydrate flux is key to counteracting oxidative stress. J Biol. 2007;6(4):10. doi:10.1186/jbiol61.
609. Emadi A, Ross AE, Cowan KM, Fortenberry YM, Vuica-Ross M. A chemical genetic screen for modulators of asymmetrical 2,2'-dimeric naphthoquinones cytotoxicity in yeast. PLoS One. 2010;5(5):e10846. doi:10.1371/journal.pone.0010846.
610. Yang H, Zhou P, Huang H, Chen D, Ma N, Cui CQ, Shen S, Dong W, Zhang X, Lian W, Wang X, Dou QP, Liu J. Shikonin exerts antitumor activity via proteasome inhibition and cell death induction in vitro and in vivo. Int J Cancer. 2009;124:2450–9.
611. Yanai I, Wolf YI, Koonin EV. Evolution of gene fusions: horizontal transfer versus independent events. Genome Biol 2002; 3(5): research 0024.1-research0024.13.
612. Furuta Y, Yahara K, Hatakeyama M, Kobayashi I. Evolution of cagA oncogene of Helicobacter pylori through recombination. PLoS One. 2011;6(8):e23499. doi:10.1371/journal.pone.0023499.
613. Richards TA, Cavalier-Smith T. Myosin domain evolution and the primary divergence of eukaryotes. Nature. 2005;436:113–1118.
614. Richards TA, Dacks JB, Campbell SA, Blanchard JL, Foster PG, McLeod R, Roberts CW. Evolutionary origin of the eukaryotic shikimate pathway: gene fusions, horizontal gene transfer, and endosymbiotic replacements. Eukaryot Cell. 2006;5:1517–31.

615. Rogozin IB, Basu MK, Csűrös M, Koonin EV. Analysis of rare genomic changes does not support the unikont-bikont phylogeny and supports cyanobacterial symbiosis as the point of primary radiation of eukaryotes. Genome Biol Evol. 2009;1:99–113.
616. Kondrashov FA, Koonin EV, Morgunov IG, Finogenova TV, Kondrashova MN. Evolution of glyoxylate cycle enzymes in metazoa: evidence of multiple horizontal transfer events and pseudogene formation. Biol Direct. 2006;1:36. doi:10.1186/1745-6150-131.
617. Stover NA, Dixon TA, Cavalcanti ARO. Multiple independent fusions of glucose-6-phosphate dehydrogenase and enzymes in the pentose phosphate pathway. PLoS One. 2011;6(8):e22269. doi:10.1371/journal.pone.0022269.
618. Slavic K, Straschl U, Reininger L, Doering C, Morin C, Towari R, Krishna S. Life cycle studies of the hexose transporter of Plasmodium species and genetic validation of their essentiality. Mol Microbiol. 2010;75:1402–13.
619. Müller I, Diaz de Escauriaza M, Lajoie P, Theis M, Jung M, Müller A, Burgard C, Greiner M, Snapp EL, Dudek J, Zimmermann R. Evolutionary gain of function for the ER membrane protein Sec62 from yeast to humans. Mol Biol Cell. 2010;21:691–703.
620. Pick C, Ebersberger I, Spielmann T, Bruchhaus I, Burmester T. Phylogenomic analyses of malaria parasites and evolution of their exported proteins. BMC Evol Biol. 2001;11:167. doi:10.1186/1471-2148-11-167.
621. Guruprasad L, Tanneeru K, Guruprasad K. Structural rationale for the recognition of arginine at P3 in PEXEL motif containing proteins of Plasmodium falciparum by plasmepsin V. Protein Pept Lett. 2011;18:634–41.
622. Falk I, de Carvalho EG, Kun JF. Parasite-host interaction in malaria: genetic clues and copy number variations. Genome Med 2009; 1(9):82. doi:10.1186/gm52.
623. Sabbatini S, Manfredi R, Fiorino S. Malaria infection and human evolution. Infect Med. 2010;18:56–74.
624. Papazahariadou M, Athanasiadis GI, Papadopoulos E, Simeonidou I, Hatzistiljanou M, Castellani ML, Bhattacharya K, Shanmugham LN, Conti P, Frydas S. Involvement of NK cells against tumors and parasites. Int J Biol Markers. 2007;22:144–53.
625. Vetukuri RR, Tian Z, Avrova AO, Savenkov EI, Dixelius C, Whisson SC. Silencing of the PiAvr3a effector-encoding gene from Phytophthora infestans by transcriptional fusion to a short interspersed element. Fungal Biol. 2011;115:1225–33.
626. Morris PF, Schlosser LR, Onasch KD, Wittenschlaeger T, Austin R, Provart N. Multiple horizontal gene transfer events and domain fusions have created novel regulatory and metabolic networks in the oomycete genome. PLoS One. 2009;4(9):e6133. doi:10.1371/journal.pone.0006133.
627. Dong S, Yu D, Cui L, Qutob D, Tedman-Jones J, Kale SD, Tyler BM, Wang Y, Gijzen M. Sequence variants of the Phytophthora sojae RXLR effector Avr3a/5 are differentially recognized by Rps3a and Rps5 in soybean. PLoS One. 2011;6(7):e20172. doi:10.1371/journal.pone.0020172.
628. Kay J, Meiler HJG, ten Have A, van Kan J. The aspartic protease family of three Phytophthora species. BMC Genomics. 2011;12:254. doi:10.1186/1471-2164-12-254.
629. Feng BZ, Li PQ, Fu L, Sun BB, Zhang XG. Identification of 18 genes encoding necrosis-inducing proteins from the plant pathogen Phytophthora capsici (Pythiaceae: Oomycetes). Genet Mol Res. 2011;10:910922.
630. Gyetvai G, Sønderkaer M, Göbel U, Basekow R, Ballvora A, Imhoff M, Kersten B, Nielsen KL, Gebhardt C. The transcriptome of compatible and incompatible interactions of potato (Solanum tuberosum) with Phytophthora infestans revealed by DeepSAGE analysis. PLoS One. 2012;7(2):e31526. doi:10.1371/journal.pone.0031526.
631. Bozkurt TO, Schornack S, Win J, Shindo T, Ilyas M, Oliva R, Cano LM, Jones AM, Huitema E, van der Hoorn RA, Kamoun. Phytophthora infestans effector AVRblb2 prevents secretion of plant immune protease at the haustorial interface. Proc Natl Acad Sci USA. 2011;108:20832–7.

632. Bouwmeester K, de Sain M, Weide R, Gouget A, Klamer S, Canut H, Govers F. The lectin receptor kinase LecRK-I.9 is a novel Phytophthora resistance component and a potential host target for a RXLR effector. PLoS Pathog. 2011;7(3):e10011327. doi:10.1371/journal.ppat.1001327.
633. Chen Y, Liu Z, Halterman DA. Molecular determinants of resistance activation and suppression by Phytophthora infestans effector IPI-O. PLoS Pathog. 2012;8(3):e1002595. doi:10.1371/journal.ppat.1002595.
634. Dimitriadis D, Koumandou VL, Trimpalis P, Kossida S. Protein functional links in Trypanosoma brucei, identified by gene fusion analysis. BMC Evol Biol. 2011;11:193. doi:10.1186/1471-2148-11-193.
635. Nara T, Hashimoto M, Hirawake H, Liao CW, Fukai Y, Suzuki S, Tsubouchi A, Morales J, Takamiya S, Fujimura T, Taka H, Mineki R, Fan CK, Inaoka DK, Inoue M, Tanaka A, Harada S, Kita K, Aoki T. Molecular interaction of the first 3 enzymes of the de novo pyrimidine biosynthetic pathway of Trypanosoma cruzi. Biochem Biophys Res Commun. 2012;418:140–3.
636. Hoek M, Engstler M, Cross GAM. Expression-site-associated gene 8 (ESAG8) of trypanosome brucei is apparently essential and accumulates in the nucleolus. J Cell Sci. 2000;113:3959–68.
637. Sirri V, Urcuqui-Inchima S, Roussel P, Hernandez-Verdun D. Nucleolus: the fascinating nuclear body. Histochem Cell Biol. 2008;129:13–31.
638. Makiuchi T, Annoura T, Hashimoto T, Murata E,. Aoki, Nara T. Evolutionary analysis of synteny and gene fusion for pyrimidine biosynthetic enzymes in Euglenozoa: an extraordinary gap between kinetoplastids and diplonemids. Protists 200;159:459-470.
639. Tsagrasoulis D, Danos V, Kissa M, Trimpalis P, Koumandoun VL, Karagouni AD, Tsakalidis A, Kossida S. SAFE software and FED database to uncover protein-protein interactions using gene fusion analysis. Evol Bioinform Online. 2012;8:47–60. doi:10.4137/EBO.S8018.
640. Salim HMW, Negritto MC, Cavalcanti ARO. 1+1=3: A fusion of 2 enzymes in the methionine salvage pathway of Tetrahymena thermophila creates a trifunctional enzyme that catalyzes 3 steps in the pathway. PLoS One. 2009;5(10):e1000701. doi:10.1371/journal.pgen.1000701.
641. Salim HMW, Koire AM, Stover NA, Cavalcanti AR. Detection of fused genes in eukaryotic genomes using gene deFuser: analysis of the Tetrahymena thermophila genome. BMC Bioinform. 2011;12:279. doi:10.1186/1471-2105-12-279.
642a. Schuldt NJ, Aldhamen YA, Appledom DM, Seregin SS, Kousa Y, Godbehere S, Amalfitano A. Malaria vaccines. Vaccine platforms combining circumsporozoite proteins and potent immune modulators. PLoS One. 2011;6(8):e24147.
643. Ojan F, Reiter K, Zhang Y, Shimp RL Jr, Nguyen V, Aebig JA, Rausch KM, Zhu D, Lambert L, Mullen GE, Martin LB, Long CA, Miller LH, Narum DL. Immunogenicity of self-assembled aggregates and chemically cross-linked conjugates of the 42 kDa Plasmodium falciparum merozoite surface protein-1. PLoS One. 2012;7(6):e36999.
644. Horvath JC, Horak A, Sinkovics JG, Pritchard M, Pendleton S, Horvath E. Cancer vaccines with emphasis on a viral oncolysate melanoma vaccine. Acta Microbiol Immunol Hung. 1999;46:1–20.
645. Sinkovics JG, Horvath JC. Vaccination against human cancers. Int J Oncol. 2000;16:81–96.
646. Andersen BH, Ohlfest JAR. Increasing the efficacy of tumor cell vaccines by enhancing cross priming. Cancer Lett. 2012;325:155–64.
647. Bolli E, Quaglino E, Arigoni M, Lollini PL, Calogero R, Forni G, Cavallo F. Oncoantigens for an immune prevention of cancer. Am J Cancer Res. 2011;1:255–64.
648. Turcotte S, Rosenberg SA. Immunotherapy of metastatic solid cancers. Adv Surg. 2011;45:341–60.

649. Daley GQ, Van Etten RA, Baltimore D. Induction of chronic myelogenous leukemia in mice by the P210bcr/abl gene of the Philadelphia chromosome. Science. 1990;247:824–30.
650. Daley GQ, Van Etten RA, Baltimore D. Blast crisis in a murine model of chronic myelogenous leukemia. Proc Natl Acad Sci USA. 1991;88:11336–8.
651. Sinkovics JG. Active tumor-specific immunization of the bone marrow (progenitor stem cell) donor. J Natl Cancer Inst. 1995;87:1804.
652. Kaufmann I, Martin G, Friedlein A, Keller W. Human Fip1 is a subunit of CPSF that binds to U-rich elements and stimulates poly(A) polymerase. EMBO. 2004;23:616–26.
653. Li B, Zhang G, Li C, He D, Li X, Zhang C, Tang F, Deng X, Lu J, Tang Y, Li R, Chen Z, Duan C. Identification of JAK2 as a mediator of FIP1L1-PDGFRA-induced eosinophil growth and function in CEL. PLoS One. 2012;7(4):e349122. doi:10.1371/journal.pone.0034912.
654. Messie K, Vovor A, Kueviakoe IM, Sallah LK, Agbetiafa K, Segbena AY. Clonal hypereosinophilic syndrome: two cases report in black men from sub-Saharan Africa and literature reviews. ISRN Hematol. 2011;2011:974609. doi:10.5402/2011/974609.
655. Tang TC, Chang H, Chuang WY. Complete response of myeloid sarcoma with FIP1L1-PDGFRA-associated myeloproliferative neoplasms to imatinib mesylate monotherapy. Acta Haematol. 2012;128:83–7.
656. Montano-Almendras CP, Essaghir A, Schoemans H, Varis I, Noel LA, Velghe AI, Latinne D, Knoops L, Demoulin JB. ETV6-PDGFRB and FIP1L12-PDGFRA stimulate human hematopoietic progenitor proliferation and differentiation into eosinophils: role of NFκB. Haematologica 2012;97:1064–72.
657. Domer PH, Fakharzadeh SS, Chen CS, Jockel J, Johansen L, Silverman GA, Kersey JH, Korsmeyer SJ. Acute mixed lineage leukemia t(4;11)(q21;q23) generates an MLL0-AF4 fusion product. Proc Natl Acad Sci USA. 1999;90:7884–8.
658. Harper DP, Aplan PD. Chromosome rearrangement leading to MLL gene fusions: clinical and biological aspects. Cancer Res. 2008;68:10024–7.
659. Valverde R, Pozdnyakova I, Kajander T, Venkatraman J, Regan L. Fragile X mental retardation syndrome: structure of the KH1-KH2 domains of fragile X mental retardation protein structure. 2007;15:1090–1098.
660. Schrier M, Severijnen LA, Reis S, Rife M, van Padje S, van Cappellen G, Oostra BA, Willemsen R. Transport kinetics of FMRP containing the I304N mutation of severe fragile X syndrome in neurites of living rat PC12 cells. Exp Neurol. 2004;189:343–53.
661. Darnell JC, Richter JD. Cytoplasmic RNA-binding proteins and the control of complex brain function. Cold Spring Harbor Perspect Biol 2012 Jun 20pii:cshperspect.a012344v1. doi:10.1101/sshperspect.a012344.
662. Sandberg G, Schalling M. Effect of in vitro promoter methylation and CGG repeat expansion on FMR1 expression. Nucleic Acids Res. 1997;25:2883–7.
663. Xu S, Poidevin M, Han E, Bi J, Jin P. Circadian rhythm-dependent alterations of gene expression in drosophila brain lacking X mental retardation protein. PLoS One. 2012;7(5):e37937. doi:10.1371/journal.pone.0037937.
664. Darnell JC, Fraser CE, Mostovetsky O, Darnell RB. Discrimination of common and unique RNA-binding activities among fragile X mental retardation protein paralogs. Hum Mol Genet. 2009;18:3164–77.
665. Hollingworth D, Candel AM, Nicastro G, Martin SR, Briata P, Gherzi R, Ramos A. KH domains with impaired nucleic acid binding as a tool for functional analysis. Nucleic Acids Res 2012 29. doi:10.1093/nar/gks368.
666. Siomi H, Matunis MJ, Michael WM, Dreyfuss G. The pre-mRNA binding K protein contains a novel evolutionarily conserved motif. Nucleic Acids Res. 1993;21:1193–8.
667. Siomi H, Siomi MC, Nussbaum RL, Dreyfuss G. The protein product of the fragile X gene, FMR1, has characteristics of an RNA-binding protein. Cell. 1993;74:291–8.

668. Anantharaman V, Koonin EV, Aravind L. Comparative genomics and evolution of proteins involved in RNA metabolism. Nucleic Acids Res. 2002;30:1427–64.
669. Palenchar JB, Bellofatto V. Gene transcription in trypanosomes. Mol Biochem Parasitol. 2006;145:135–41.
670. Wurth L. Versatility of RNA-binding proteins in cancer. Comp Funct Genomics. 2012;2012:178525. doi:10.1155/2012/178525.
671. Brykailo M, McLane LM, Fridovich-Keil J, Corbett AH. Analysis of a predicted nuclear localization signal: implications for the intracellular localization and function of the Saccharomyces cerevisiae RNAp-binding protein Scp160. Nucleic Acids Res. 2007;35:6862–9.
672. Deshpande G, Calhoun G, Schedl P. The drosophila fragile X protein dFMR1 is required during early embryogenesis for pole cell formation and rapid nuclear division. Genetics. 2006;174:1287–96.
673. Bueno C, Montes R, Catalina P, Rodriguez R, Menendez P. Insights into the cellular origin and etiology of the infant pro-B acute lymphoblastic leukemia with MLL-AF4 rearrangement. Leukemia. 2011;25:400–10.
674. Sun Y, Varambally S, Maher CA, CanQ Chockley P, Toubai T, Malter C, Nieves S, Tawara I, Wang Y, Ward PA, Chinnayan A, Reddy P. Targeting of microRNA-142-3p in dendritic cells regulates endotoxin-induced mortality. Blood. 2011;117:6177–83.
675. Schreiber V, Mooq-Lutz C, Régnier CH, Chenard MP, Boeuf H, Vonesch JL, Tomasetto C, Rio MC. Lasp-1, a novel type of actin-binding protein accumulating in cell membrane extensions. Mol Med. 1998;4:675–87.
676. Strehl S, Borkhardt A, Slany R, Fuchs UE, Konig M, Haas OA. The human Lasp1 gene is fused to MLL in an acute myeloid leukemia with t(11;17)(q23;q21). Oncogene. 2003;22:157–60.
677. Frietsch JJ, Grunewald TG, Jasper S, Kammerer U, Herterich S, Kapp M, Honig A, Butt E. Nuclear localization of LASP-1 correlates with poor long-term survival in female breast cancer. Br J Cancer. 2010;102:1645–53.
678. Payne VA, Au WS, Gray Nora Ed, Rahman SM, Sanders R, Hadaschik D, Friedman JE, O'Rahilly S, Rochford JJ. Sequential regulation of diacylglycerol acetyltransferase 2 expression by CAAT/enhancer-binding protein beta (C/EBPbeta) and C/EBPalpha during adipogenesis. J Biol Chem. 2007;282:21005–14.
679. Esteves CL, Kellly V, Bay V, Man TY, Morton NM, Leutz A, Seckl JAR, Chapman KE. Regulation of adipocyte 11β-hydroxysteroid dehyrogenase type1 (11β-HSD1) by CCAAT/enhancer-binding protein (C/EBP) β isoforms, LIP and Lap. PLoS One. 2012;7(5):e37953.
680. Göranson M, Andersson MK, Forni C, Stahlberg A, Andersson C, Olofsson A, Mantovani R, Aman P. The myxoid liposarcoma FUS-DDIT3 fusion oncoprotein deregulates NF-kappaB target genes by interaction with NFKBIZ. Oncogene. 2009;28:270–8.
681. Forni C, Minuzzo M, Virdis E, Tamborini E, Simone M, Tavecchio M, Erba E, Grosso F, Gronchi A, Amari P, Casali P, D'Icalci M, Pilotti S, Mantovani R. Trabectedin (ET-743) promotes differentiation in myxoid liposarcoma tumors. Mol Cancer Ther. 2009;6:449–57.
682. Peng C, Guo W, Yang Y, Zhao H. Downregulation of SS18-SSX1 expression by small interfering RNA inhibits growth and induces apoptosis in human synovial sarcoma cell line HS-SY-II in vitro. Eur J Cancer Prev. 2008;17:392–8.
683. Brassesco MS, Cortez MA, Valera ET, Engel EE, Nogueira-Barbosa MH, Becker AP, Scridell CA, Tone LG. Cryptic SYT/SXX1 fusion gene in high-grade biphasic synovial sarcoma with unique complex rearrangement and extensive BCL2 overexpression. Cancer Genet Cytogenet. 2010;196:189–93.

684. Picchione F, Pritchard C, Lagutina I, Janke L, Grosveld GC. IRIZIO: a novel gene cooperating with PAX3-FOXO1 in alveolar rhabdomyosarcoma. Carcinogenesis. 2011;32:452–61.
685. Tonelli R, McIntyre A, Camerin C, Walters ZS, Di Leo K, Selfe J, Purgato S, Missiaglia E, Portori A, Renshaw J, Astolfi A, Taylor KR, Serravalle S, Bishop R, Nasnni C, Valentin LJ, Faccini A, Euschner I, Formica S, Reis-Filho JS, Ambrosini V, Thway K, Franzoni M, Summersgill B, Marchelli R, Hralia P, Cantelli-Forti G, Fanti S, Corradini R, Pession A, Shipley J. Antitumor activity of sustained N-myc reduction in rhabdomyosarcomas and transcription block by antigen therapy. Clin Cancer Res. 2012;18:796–807.
686. Peron M, Bonvini P, Rosolen A. Effect of inhibition of the ubiquitin-proteasome system and Hsp90 on growth and survival of rhabdomyosarcoma cells in vitro. BMC Cancer. 2012;12(1):233.
687. von Levetzow C, Jiang X, Gwye Y, von Levetzow G, Hung L, Cooper A. Modeling initiation of Ewing sarcoma in human neural crest cells. PLoS One. 2011;6(4):e19305. doi:10.1371/journal.pone.0019305.
688. Sankar S, Lessnick SL. Promiscuous partnership in Ewing's sarcoma. Cancer Genet. 2011;204:351–65.
689. Brodie SG, Stocketr SJ, Wardlaw JC, Duncan MH, Mc Connell TS, Feddersen RM, Williams TM. EWS and WT-1 gene fusion in desmoplastic round small cell tumor in the abdomen. Hum Pathol. 1995;26:1370–4.
690. Rekhi B, Basak R, Desai SB, Jambhekar NA. A t(11;22)(p13;q120 EWS-WT1 positive desmoplastic small round cell tumor of the maxilla: an unusual case indicating the role of molecular diagnosis in round cell sarcomas. J Postgrad Med. 2010;56:201–5.
691. Spahn L, Ptermann R, Siligan C, Schmid JA. Interaction of the EWS NH2 terminus with BARD1 links the Ewing's sarcoma gene to a common tumor suppressor pathway. Cancer Res. 2002;62:4583–7.
692. Kim S, Lim B, Kim J. EWS-Oct-4b, an alternative EWS-Oct-4 fusion gene, is a potent oncogene linked to human epithelial tumors. Br J Cancer. 2010;102:436–46.
693. French CA, Miyoshi I, Kubonishi I, Grier HE, Perez-Atayde AR, Fletcher JA. BRD4-NUT fusion oncogene a novel mechanism in aggressive carcinoma. Cancer Res. 2003;63:304–7.
694. Hsieh MS, French CA, Liang CW, Hsiao CH. NUT midline carcinoma: case report and review of the literature. Int J Surg Pathol. 2011;19:808–12.
695. Braese JC, Johannes M, Mannsperger H, Fälth M, Metzger J, Kacprzyk LA, Andrasiuk T, Gade S, Meister M, Sirma H, Sauter G, Simon R, Sckomm T, Beissbarth T, Korf U, Kunor R, Sultmann H. TMPRSS2-ERG-specific transcriptional modulation is associated with prostate cancer biomarker and TGF-β signaling. BMC Cancer. 2011;11:507. doi:10.1189/1471-2407-11-507.
696. Luedeke M, Linnert CM, Hofer MD, Surowy HM, Rinckleb AE, Hoegel J, Kuefer R, Rubin MA, Vogel W, Maier C. Predisposition for TMPRSS2-ERG fusion in prostate cancer by variants in DNA repair genes. Cancer Epidemiol Biomarkers Prev 2999;18:3030–3035.
697. Ribeiro FR, Paulo P, Costa VL, Barros-Silva JD, Ramalho-Carvalho J, Jerónimo C, Henrique R, Lind GE, Skotheim RI, Lothe RA, Teixeira MR. Cysteine-rich secretory protein-3 (CRISP3) is strongly up-regulated in prostate carcinomas with the TMPRSS2-ERG fusion gene. PLoS One. 2011;6(7):e22311. doi:10.1371/journal.pone.0022317.
698. Li Y, Kong D, Wang Z, Ahmad A, Bao B, Padhye S, Sarkar FH. Inactivation of AR/TMPR2-ERG/Wnt signaling networks attenuates the aggressive behavior of prostate cancer cells. Cancer Prev Res. 2011;4:1495–506.

699. Lynch JA, Özüak O, Abderrahman K, Abouheif E, Despan C, Roth S. The phylogenetic origin of oskar coincided with the origin of maternally provisioned germ plasma and pole cells at the base of the Holometabola. PLoS Genet. 2011;7(4):e1002029. doi:10.1371/journal.pgen.1002029.

700. Chuma S, Hosokawa M, Kitamura K, Kasai S, Fujioka M, Hiyoshi M, Takamune K, Noce T, Nakatsuji N. Tdrd1/Mtr-1, a tudor-related gene is essential for male germ-cell differentiation and nuage/germinal granule formation in mice. Proc Natl Acad Sci USA. 2006;103:15894–9.

701. Tanaka Tosokawa M, Vagin VV, Reuter M, Hayashi E, Mochizuki AL, Kitamura K, Yamanaka H, Komdoh G, Okawa K, Kuramochi-Miyagawa S, Nakano T, Schidanandam R, Hannon GJ, Pillai RS, Nakatsuji N, Chuma S. Tudor domain containing 7 (Tdrd7) is essential for dynamic ribonucleoprotein (RNP) remodeling of chromatoid bodies during spermatogenesis. Proc Natl Acad Sci USA. 2011;108:10579–84.

702. Yabuta Y, Ohta H, Abe T, Kurimoto K, Chuma S, Saitou M. TDRD5 is required for retrotransposons silencing chromatoid body assembly and spermiogenesis in mice. J Cell Biol. 2011;192:781–95.

703. Kobayashi A, Okuda H, Xing F, Pandey PR, Watabe M, Hirota S, Pai SK, Liu W, Fukuda K, Chambers C, Wilber A, Watabe K. Bone morphogenetic protein 7 in dormancy and metastasis of prostate cancer stem cells in bone. J Exp Med. 2011;208:2641–55.

704. Secondini C, Wetterwald A, Schwanninger R, Thalmann GN, Cecchini MG. The role of the BMP signaling antagonist noggin in the development of prostate cancer osteolytic bone metastasis. PLoS One. 2011;6(1):e160778. doi:10.1371/journal.pone.0016073.

705. Célia-Terrassa T, Meca-Cortés O, Mateo F, de Paz AM, Rubio N, Amal-Estapé A, Ell BJ, Bermudo R, Diaz A, Guerra-Rebollo M, Lozano JJ, Estarás C, Ulloa C, Álvarez-Simón D, Milá J, Vilela R, Paciucci R, Marínez-Balbás M, de Hereros AG, Gomis RR, Kang Y, Blanco J, Fernández PL, Thomson TM. Epithelial-mesenchymal transition can suppress major attributes of human epithelial tumor-initiation cells. J Clin Invest. 2012;122:1849–68.

706. Pernicová Z, Slabáková E, Kharaishvili G, Bouchal J, Kraal M, Kunickvá Z, Machjala M, Kiozubik A, Souček K. Androgen depletion induces senescence in prostate cancer cells through down-regulation of Skp2. Neoplasia. 2011;13:526–36.

707. Rickman DS, Pflueger D, Moss B, VanDoren VE, Chen CX, de la Taille A, Kuefer R, Tewari AK, Setlur SR, Demichelis F, Rubin MA. SLC45A3-ELK4 is a novel and frequent erythroblast transformation-specific fusion transcript in prostate cancer. Cancer Res. 2009;69:2734–8.

708. Zhang Y, Gong M, Yuan H, Park HG, Frierson HF, Li H. Chimeric transcript generated by cis-splicing of adjacent genes regulates prostate cancer cell proliferation. Cancer Discov. 2012;2:598–607.

709. Schaefer K-L, Brachwitz K, Wai DH, Braun Y, Diallo R, Korsching E, Eisenacher M, Voss R, van Valen F, Baer C, Selle B, Spahn L, Liaio S-K, Lee KAW, Hogendoorn PCW, Reifenberger G, Gabbert HE, Poremba C. Expression profiling of t(12;22) positive clear cell sarcoma of soft tissue cell lines reveals characteristic up-regulation of potential new marker genes including ERBB3. Cancer Res. 2004;64:3395–405.

710. Hisaoka M, Ishida T, Kuo TT, Matsumaya A, Imamura T, Nishida K, Kuroda H, Imayama Y, Oshiro H, Kobayashi H, Nakajima T, Fukuda T, Ae K, Hashimoto H. Clear cell sarcoma of soft tissue: a clinicopathologic, immunohistochemical, and molecular analysis of 33 cases. Am J Surg Pathol. 2008;32:452–60.

711. Wang WL, Mayordomo E, Zhang W, Hernandez VS, Tuvin D, Garcia L, Lev DC, Lazar AJ, López-Terrada D. Detection and characterization of EWSR1/ATF1 and EWSR1/CREB chimeric transcripts in clear cell sarcoma (melanoma of soft parts). Mol Pathol 2009;22:1291–9. doi:10.1038/modpathol.2009.85.

712. Lichter OP, Walczak H, Weitz S, Behrmann I, Krammer PH. The human APO-1 (APT) antigen maps to 10q21, a region that is syntenic with mouse chromosome 19. Genomics 992;14:179–180.

713. Sokova OI, Kirichenko OP, Mukeria AF, Demidiov LV, Chebotarev AN, Koonin BP. Enhanced expression of 1p32 and 1p22 fragile sites in lymphocytes in cutaneous malignant melanomas. Cancer Genet Cytogenet. 1992;58:24–8.

714. Okamoto I, Pirker C, Bilban M, Berger W, Losert D, Marosi C, Haas OA, Wolff K, Pehamberger H. Seven novel and stable translocations associated with oncogenic gene expression in malignant melanoma. Neoplasia. 2005;7:303–11.

715. Tweardy DJ, Anderson K, Cannizzaro LA, Steinman RA, Croce CM, Huebner K. Molecular cloning of cDNA for the human granulocyte colony-stimulating factor receptor from HL-60 and mapping of the gene to chromosome region 1p32-34. Blood. 1992;79:1148–54.

716. Sinkovics J. Malignant lymphoma arising from natural killer cells: report of the first case in 1970 and newer developments in the FasL-FasR system. Acta Microbiol Immunol Hungar. 1997;44:295–307. Erratum. The FasR was written in the MS as CD95.

717. Sinkovics JG, Horvath JC. Virological and immunological connotations of apoptotic and anti-apoptotic forces in neoplasia. Int J Oncol 2001;19:473-488. Erratum. Artist's typographic error in cartoon: as if c-Kit receptor captured hepatocyte growth factor/scatter factor as its ligand. Correctly, these ligands react with receptor c-Met; c-Kit's (CD117) ligand is the stem cell factor. Idem. 2006;27:33.

718. Horvath JC, Horvath E, Sinkovics JG, Horak A, Pendleton S, Mallah J: Human melanoma cells (HMC) eliminate autologous host lymphocytes (Ly $^{FasR+}$) and escape apoptotic death by utilizing the FasL system as an autocrine growth loop (HMC $^{FasL\rightarrow FasR}$). Proc American Assoc Cancer Res. 1998;39:584 #3971.

719. Takahashi T, Tanaka M, Ogasawara J, Suda T, Murakami H, Nagata S. Swapping between Fas and granulocyte colony-stimulating factor receptor. J Biol Chem. 1996;271:17555–60.

720. Shinohara H, Yagita H, Ikawa Y, Oyaizu N. Fas drives cell cycle progression in glioma cells via extracellular signal-regulated kinase activation. Cancer Res. 2000;60:1766–72.

721. Koonin EV. Darwinian evolution in the light of genomics. Nucleic Acids Res. 2009;37:1011–34.

722. Uddin M, Sturge M, Peddle L, O'Reilly DD, Rahman P. Genome-wide signatures of 'rearrangement hotspots' within segmental duplications in humans. PLoS One. 2011;6 (12):e28853. doi:10.1371/journal.pone.0028853.

723. Dameria RR, Knickelbein KE, Kepchia D, Jackson A, Armitage BA, Eckert KA, Opresko PL. Telomeric repeat mutagenicity in human somatic cells is modulated by repeat orientation and G-quadruplex stability. DNA Repair. 2010;9:1119–29.

724. Harris JE, Sinkovics JG. The Immunology of Malignant Disease. JE Harris, JG Sinkovics. In: Mosby CV, editor. 2nd ed. Saint Louis 1976.

725. Gleason JE, Eisenmann DM. Wnt signaling controls the stem cell-like asymmetric division of the epithelial stem cells during C. elegans larval development. Dev Biol. 2010;348:58–66.

726. Geissler K, Zach O. Pathways involved in Drosophila and human cancer development: the Notch, Hedgehog, wingless, Runt, and Trithorax pathway. Ann Hematol. 2012;91:645–69.

727. Saini N, Reichert H. Neural stem cells in Drosophila: molecular genetic mechanisms underlying normal neural proliferation and abnormal brain tumor formation. Stem Cells 2012;2012:486169. doi:10.1155/2012/486169.

728. Gafuik C, Steller H. A gain-of-function germline mutation in Drosophila ras 1 affects apoptosis and cell fate during development. PLoS One. 2011;6(8):e23535. doi:10.1371/journal.pone.0023535.

729. Smadja C, Shi P, Butlin RK, Robertson HM. Large gene family expansion and adaptive evolution for odorant and gustatory receptors in the pea aphid, Acyrthosiphon pisum. Mol Biol Evol. 2009;26:2073–86.
730. Sloan DB, Alverson AJ, Chuckalovcak JP, Wu M, McCauley DE, Palmer JD, Taylor DR. Rapid evolution of enormous, multichromosomal genomes in flowering plant mitochondria with exceptionally high mutation rates. PLoS Biol. 2012;10(1):e1001241. doi:10.1371/journal.pbio.1001241.
731. Frary A, Nesbitt TC, Grandillo S, van der Knaap E, Cong B, Liu J, Meller J, Elber R, Alpert KB, Tanksley SD. *fw2.2*: A quantitative trait locus key to the evolution of tomato fruit size. Science. 2000;289:85–8.
732. D'Aniello S, Irimia M, Maeso I, Pascual-Anaya J, Jiménez-Delgado S, Bertrand S, Garcia-Fernández J. Gene expansion and retention leads to a diverse tyrosine kinase superfamily in amphioxus. Mol Biol Evol. 2008;25:1841–54.
733. Yu Li , Y XZ, L TS, Song LX, Chen PL, Suo TL, Li YH, Wang SD, Chen Y, Ren YM, Zhang SP, Chang ZJ, Fu XY. A novel protein tyrosine kinase NOK that shares homology with platelet-derived growth factor/fibroblast growth factor receptors induces tumorigenesis and metastasis in nude mice. Cancer Res. 2004;64:3491–3499.
734. Shiu SH, Li WH. Origins, lineage-specific expansion, and multiple losses of tyrosine kinases in eukaryotes. Mol Biol Evol. 2004;21:828–40.
735. Fasano O, Birnbaum D, Edlund L, Fogh J, Wigler M. New human transforming genes detected by a tumorigenicity assay. Mol Cell Biol. 1984;4:1695–705.
736. Fukui M, Yamamoto T, Kawai S, Maruom K, Toyoshima K. Detection of raf-related and two other transforming DNA sequences in human tumors. Proc Natl Acad Sci USA. 1985;82:5954–8.
737. Liu E, Hjelle B, Bishop JM. Transforming genes in chronic myelogenous leukemia. Proc Natl Acad Sci USA. 1988;85:1952–6.
738. O'Bryan JP, Frye RA, Cogswell PC, Neubauer A, Kitch B, Prokopp C, Espinosa R III, Le Beau MM, Earp HS, Liu ET. *axl,* A transforming gene isolated from primary human myeloid leukemia cells, encodes a novel receptor tyrosine kinase. Mol Cell Biol. 1991;11:5016–31.
739. Goruppi S, Charuttini C, Ruaro ME, Varnum B, Schneider C. Gas6 induces growth, β-catenin stabilization, and T-cell factor transcriptional activation in contact-inhibited C57 mammary cells. Mol Cell Biol. 2011;21:902–15.
740. Vajkoczy P, Knyazev P, Kunkel A, Capelle HH, Behrndt S, von Tengg-Kobligk H, Kiessling F, Eichelsbachter U, Essig M, Read TA, Erber R, UllricH A. Dominant-negative inhibition of the Axl receptor tyrosine kinase suppresses brain tumor cell growth and invasion and prolongs survival. Proc Natl Acad Sci USA. 2006;103:5799–804.
741. Shiozawa Y, Pedersen EA, Patel LR, Ziegler AM, Havens AM, Jung Y, Wang J, Zalucha S, Loberg RD, Pienta KJ, Taichman RS. GAS6/AXK axis regulates prostate cancer invasion, proliferation, and survival in the bone marrow. Neoplasia. 2010;12:116–27.
742. Turnay J, Lecona E, Fernández-Lizarbe S, Gusman-Aránquez M, Olmo N, Lizarbe MA. Structure-function relationship in annexin A13, the founder member of the vertebrate family of annexins. Biochem J. 2005;389:899–911.
743. Jung Y, Shiozawa Y, Wang J, McGregor N, Dai J, Park SI, Barry JE, Havens AM, Joseph J, Kim JK, Patel L, Carmeliet P, Daignault S, Keller ET, McCauley LK, Pienta KJ, Taichman RS. Prevalence of prostate cancer metastases after intravenous inoculation provide clues into the molecular basis of dormancy in the bone marrow microenvironment. Neoplasia. 2012;14:429–39.
744. Rentala S, Mangamoor LN. Isolation, characterization and mobilization of prostate cancer tissue derived CD133+ MDR1+ cells. J Stem Cells. 2010;5:75–81.

745. Ibrahim T, Sacanna E, Gaudion M, Mercatali KL, Scarpi E, Zoli W, Serra P, Kang Y, Amadori D. Role of RANK, RANKL, OPG, and CXCR4 tissue markers in predicting bone metastases in breast cancer patients. Clin Breast Cancer. 2011;11:369–75.
746. Balogh I, Hafizi S, Stenhoff J, Hanssen K, Dahlback B. Analysis of Gas6 in human platelets and plasma. Arterioscler Thromb Vasc Biol. 2005;25:1280–6.
747. Schmidt T, Ben-Batalla I, Schultze A, Loges S. Macrophage-tumor crosstalk: role of TAMR tyrosine kinase receptors and their ligands. Cell Mol Life Sci. 2012;69:1391–414.
748. Weinger JG, Brosnan CE, Loudig O, Goldberg MF, Macian F, Arnett HAS, Prieto AL, Tsiperson V, Shafit-Zagardo B. Loss of the receptor tyrosine kinase Axl leads to enhanced inflammation in the CNS and delayed removal of myelin debris during experimental autoimmune encephalomyelitis. J Neuroinflammation. 2011;8:49. doi:10.1186/1742-2094-8-49.
749. Clauser S, Meilac O, Biéche I, Raynal P, Bruneval P, Michel JB, Borgel D. Increased secretion of Gas6 by smooth muscle cells in human atherosclerotic carotid plaques. Thromb Haemost. 2012;107:140149.
750. Camenish TD, Koller BH. Earp HS, Matsushima GK. A novel receptor tyrosine kinase, Mer, inhibits TNFα production and lipopolysaccharide-induced endotoxic shock. J Immunol. 1999;162:3498–503.
751. Weinger JG, Omafri KM, Marsden K, Raine CS, Shafit-Zagardo B. Up-regulation of soluble Axl and Mer receptor tyrosine kinases negatively correlates with Gas6 in established multiple sclerosis lesions. Am J Pathol. 2009;175:283–93.
752. Wu J, Ekman C, Jönsen A, Sturfelt G, Bengtsson AA, Gottsäfer A, Lindblat B, Lindquist E, Saxne T, Dahlbäck B. Increased plasma levels of the soluble Mer tyrosine kinase receptor in systemic lupus erythematosus relate to disease activity and nephritis. Arthritis Res Ther. 2011;13:R62. doi:10.1186/ar3316.
753. Linger RM, DeRyckere D, Brandão L, Sawczyn KK, Jacobsen KM, Liang X, Keating AK, Graham DK. Mer receptor tyrosine kinase is a novel therapeutic target in pediatric B-cell acute lymphoblastic leukemia. Blood. 2009;114:2678–87.
754. Migdall-Wilson J, Bates C, Schlegel J, Brandão L, Linger RM, DeRyckere D, Graham DK. Prolonged exposure to a Mer ligand in leukemia: Gas6 favors expression of a partial Mer glycoform and reveals a novel role for Mer in the nucleus. PLoS One. 2012;7(2)e31635. doi:10.1371/journal.pone.0031635.
755. Huang X, Finerty P, Walker JAR, Butler-Cole C, Vedadi M, Schapira M, Parker SA, Turk BE, Thompson DA. Dhe-Paganon. Structural insights into the inhibited states of the Mer receptor tyrosine kinase. J Struct Biol. 2009;165:88–96.
756. Dutta A, Wang L-H, Hanafusa T, Hanafusa H. Partial nucleotide sequence of Rous sarcoma virus-29 provides evidence that the original Rous sarcoma virus was replication defective. J Virol. 1985;55:728–35.
757. Jia R, Mayer BJ, Hanafusa T, Hanafusa H. A novel oncogene, v-ryk, encoding a truncated receptor tyrosine kinase is transduced into RPL30 virus without loss of viral sequences. J Virol. 1992;66:5975–87.
758. Jia R, Hanafusa H. The proto-oncogene of v-eyk (v-ryk) is a novel receptor-type protein tyrosine kinase with extracellular Ig/FN-III domains. J Biol Chem. 1994;269:1839–44.
759. Stehelin D, Varmus HE, Bishop JM, Vogt P. DNA related to the transforming gene(s) of avian sarcoma viruses is present in normal avian DNA. Nature. 1976;260(5547):170–3.
760. Sinkovics JG, Bertin BA, Howe CD. Occurrence of low leukemogenic but immunizing mouse leukemia virus in tissue culture. National Cancer Institute Monograph. 1965;22:349–67.
761. Sinkovics JG. The causative viruses of murine leukemia and their identification through immune responses of the host. Addition. In: Carcinogenesis: A Broad Critique. The University of Texas M.D. Anderson Hospital and Tumor Institute, Houston, Texas 1967; pp 157–175.

762. Sinkovics JG, Shullenberger CC, Howe CD, Bertin BA. Leukemogenic effect of human leukemic materials in combination with inactivated mouse leukemia virus. Cancer. 1967;20:846–50.
763. Saharinen P, Bry M, Alitalo K. How do angiopoietins Tie in with vascular endothelial growth factors? Curr Opin Hematol. 2010;17:198–205.
764. Seegar TC, Eller B, Tzvetkova-Robey D, Kolev MV, Henderson SC, Nikolov DB, Barton WA. Tie1Tie2 interactions mediate functional differences between angiopoietin ligands. Mol Cell. 2010;37:643–55.
765. Brown LF, Dezube BJ, Tognazzi K, Dvorak HF, Yancopoulos GD. Expression of Tie1, Tie2, and angiopoietins 1, 2, and 4 in Kaposi's sarcoma and cutaneous angiosarcoma. Am J Pathol. 2000;156:2179–83.
766. Györkey F, Sinkovics JG, Melnick JL, Gyorkey P. Retroviruses in Kaposi sarcoma cells. N Engl J Med. 1984;311:1183–4.
767. Sinkovics J, Szakács J, Horváth J, Györkey F, Györkey P, Gergely AS. Kaposi sarcoma. Orvosi Hetilap (Budapest). 2006;147:617–21.
768. Sidky YA, Borden EC. Inhibition of angiogenesis by interferons: effects on tumor- and lymphocyte-induced vascular responses. Cancer Res. 1987;47:51555161.
769. Folkman J. Successful treatment of an angiogenetic disease. N Engl J Med. 1989;320:1211–2.
770. Sinkovics JG. Interferons: antiangiogenesis agents. Canadian J Inf Dis. 1992;3 (B):128B–32B.
771. Haselstein JAR, Kasmerchak K, Buehler D, Hafez GR, Cleary K, Moody JS, Kozak KR. Efficacy of Tie2 receptor antagonism in angiosarcoma. Neoplasia. 2012;14:131–40.
772. Tartari CJ, Donadoni C, Manieri E, Mina PD, Villa A, Gambacorti-Passerini C. Dissection of the RET/β-catenin interaction in the TPC1 thyroid cancer cell line. Am J Cancer Res. 2011;1:716–25.
773. Hoffmeyer K, Raggioli A, Rudloff S, Anton R, Hierholzer A, Del Valle I, Hein K, Vogt R, Kemler R. Wnt/β-catenin signaling regulates telomerase in stem cells and cancer cells. Science. 2012;336:1549–54.
774. Matus DG, Magie C, Pang K, Martindale MQ, Thomsen GH. The Hedgehog gene family of the cnidarians, Nematostella vectensis, and implications for understanding metazoan Hedgehog pathway evolution. Dev Biol. 2008;313:501–18.
775. Greider CW. Wnt regulates TERT—putting the horse before the cart. Science. 2012;336:1519–20.
776. Fonar Y, Gutkovich YE, Root H, Malyarova A, Aamar E, Golubovskaya VM, Elias S, Elkouby YM, Frank D. Focal adhesion kinase protein regulates the gene expression to control cell fate specification in the developing neural plate. Mol Biol Cell. 2011;22:2409–21.
777. Morton JP, Mylant KB, Sansom OJ. A FAK-PI-3K-mTOR axis is required for driven intestinal regeneration and tumorigenesis. Cell Cycle. 2011;10:173–5.
778. Wu Y, Sommers JA, Loiland JA, Kitaon H, Kuper J, Kisker C, Brosh RM. The Q motif of FANCJ DNA helicase regulates its dimerization, DNA binding and DNA repair function. J Biol Chem. 2012;287:21699–716. doi:10.1074/jbc.M112.351338.
779. Sarlós K, Gyimesi M, Kovács M. RecQ helicase translocates along single-stranded DNA with a moderate processivity and tight mechanochemical coupling. Proc Natl Acad Sci USA. 2012 Jun 4 doi:10.1073/pnas.1114468.109.
780. Raymond J, Monnat JAR. Human RECQ helicase: roles in DNA metabolism, mutagenesis and cancer biology. Semin Cancer Biol. 2010;20:329–39.
781. Suhasini AN, Brosh RM Jr. Fanconi anemia and Bloom's syndrome crosstalk through FANCJ-BLM helicase interaction. Trends Genet. 2012;28:7–13.
782. Phillips LG, Sale JE. The Werner's syndrome protein collaborates with Rev1 to promote replication fork progression on damaged DNA. DNA Repair (Amst). 2010;9:1064–72.

783. Sarkies P, Murat P, Phillips LG, Patel KJ, Balasubramanian S, Sale JE. FANCJ coordinates two pathway that maintain epigenetic stability at G-quadruplex DNA. Nucleic Acids Res. 2012;40:1485–98.
784. Yu X, Wu LC, Bowcock AM, Aronheim A, Baer R. The C-terminal (BRCT) domain of BRCA1 interact in vivo with CtIP, a protein implicated in the CtBP pathway of transcriptional repression. J Biol Chem. 1998;273:25388–92.
785. Jansen JG, Tsaalbi-Shtylik A, Langerak P, Calléja F, Meijers CM, Jacobs H, de Wind N. The BRCT domain of mammalian Re1 is involved in regulating DNA translesion synthesis. Nucleic Acids Res. 2005;33:356–65.
786. Pozo FM, Oda T, Sekimoto T, Murakumo Y, Masutani C, Hanaoka F, Yamashita T. Molecule chaperone Hsp90 regulates REV1-mediated mutagenesis. Mol Cell Biol. 2011;31:3396–409.
787. Dumstorf CA, Mukhopadhyay S, Krishnan E, Haribabu B, McGregor WG. REV1 is implicated in the development of carcinogen-induced lung cancer. Mol Cancer Res. 2009;7:247–54.
788. Xie K, Doles J, Hemann MT, Walker GC. Error-prone translesion synthesis mediates acquired chemoresistance. Proc Natl Acad Sci USA. 2010;107:20792–7.
789. Debska S, Kubicka J, Czyzykowski R, Habib M, Potemski P. PARP inhibitors – theoretical basis and clinical application. Postepy Hig Med Dosw. 2012;30:311–21.
790. Passetto ZY, Yan Y, Bessho T, Natarajan A. Inhibition of BRCT(BRCA1)-phosphoroprotein interaction enhances the cytotoxic effect of olaparib in breast cancer cells: a proof of concept study for synthetic lethal therapeutic option. Breast Cancer Res Treat. 2012;134:511–7. doi:10.1007/s10549-912-2079-4.
791. Sharma S, Hicks JK, Chute CL, Brennan JAR, Ann JY, Glover TW, Canman CE. REV1 and polymerase ζ facilitate homologous recombination repair. Nucleic Acids Res. 2012;4:682–91.
792. Yap TA, Sandhu SK, Carden CP, de Bono JS. Poly(ADP-Ribose) polymerase(PARP) inhibitors: exploiting a synthetic lethal strategy in the clinic. CA; A Cancer J Clinicians. 2011;61:31–49.
793. Rudner J, Ruiner D-E, Handrick R, Eibl H-J, Jendrossek V. The Akt-inhibitor Erufosine induces apoptotic cell death in prostate cancer cells and increases the short term effect of ionizing radiation. Radiat Oncol. 2010;5:108664. doi:10.1186/1748-717X-5-108.
794. Kim MY, Zhang T, Kraus WL. Poly(ADP-ribosyl)ation by PARP-1: 'Par-laying' NAD into a nuclear signal. Genes Dev. 2005;19:1951–7. doi:10.1101/gad.1331805.
795. Virag L. PARP-1 and the shape of cell death. In: Burkle A, editor. Poly(ADP-ribosyl) ation. Landes Bioscience, Georgetown, TX, 2005;141–154.
796. Ivanov A, Cragg MS, Erenpreisa J, Emzinsh D, Lukman H, Illidge TM. Endopolyploid cells produced after severe genotoxic damage have the potential to repair DNA double strand breaks. J Cell Sci. 2003;116:4095–106.
797. Lyakhovich A, Ramirez MJ, Castellanos A, Castella M, Simons AM, Parvin JD, Surralles J. Fanconi anemia protein FANCD2 inhibits TRF1 polyADP-ribosylation through tankyrase1-dependent manner. Genome Integr. 2011;122(1):4. doi:10.1186/2041.2.4.
798. Meng L, Hsu JK, Zhu Q, Lin T, Tsai RY. Nucleostemin inhibits TRF1 dimerization and shortens its dynamic association with the telomere. J Cell Sci. 2011;124:3706–14.
799. Zhang Yang J, Liu Y, Chen X, Yu T, Jia J, Liu C. PR55alpha, a regulatory subunit of PP2A, specifically regulates PP2A-mediated beta-catenin dephosphorylation. J Biol Chem. 2009;284:22649–226556.
800. Yang L, Chen Y, Cui T, Knösel T, Zhang Q, Kai A, Huber O, Petersen I. Desmoplakin acts as a tumour suppressor by inhibition of the Wnt/β-catenin signaling pathway in human lung cancer. Carcinogenesis. 2012;33:1863–70.

801. Luo WR, Chen XY, Li SY, Wu AB, Yao KT. Neoplastic spindle cells in nasopharyngeal carcinoma show features of epithelial-mesenchymal transition. Histopathology. 2012;61:113–22.

802. Hay E, Biczkowski T, Marty, Da Nascimento S, Sonnet P, Marie PJ. Peptide-based mediated disruption of N-cadherin-LRP5/6 interaction promotes Wnt signaling and bone formation. J Bone Miner Res. 2012;27:1852–63. doi:10.1002/jbmr.1956.

803. Jiang Y, He X, Howe PH. Disabled-2 (Dab2) inhibits Wnt/β-catenin signaling by binding LRP6 and promoting its interaction through clathrin. EMBO. 2012;10:31(10):22336–49. doi:10.1038/emboj.2012.83.

804. Gwak J, Hwang SG, Park HS, Choi SR, Park SH, Kim H, Ha NC, Bae SJ, Han JK, Kim DE, Chom JW, Oh S. Small molecule-based disruption of Axin/β-catenin protein complex regulates mesenchymal stem cell differentiation. Cell Res. 2012;22(1):237–47. doi:10.1038/cr.2011 Aug 9.

805. Putnam NH, Srivastava M, Hellsten U, Dirks S, Chapman J, Salamov A, Terry A, Shapiro M, Lindquist E, Kapitonov VV, Jurka J, Genikhovich, Grigoriev IV, Lucas SM, Steele RE, Finnerty JR, Technau U, Martindale MQ, Rokhsar DS. Sea anemone genome reveals ancestral eumetazoan gene repertoire and genomic organization. Science. 2007;317:86–94

806. Erwin DE, Davidson EH. The last common bilateral ancestor. Development. 2002;129:3021–32.

807. Acosta H, López SL, Revinsky DR, Carrasco AE. Notch destabilizes maternal beta-catenin and restricts dorsal-anterior development in Xenopus. Development. 2011;138:2567–79.

808. Fujimi TJ, Hatayama M, Aruga J. Xenopus Zic3 controls notochord and organizer development through suppression of the Wnt/β-catenin signaling pathway. Dev Biol. 2012;36:220–31.

809. Lyons JP, Miller RK, Zhou X, Weidinger G, Deroo T, Denayer T, Park JI, Ji H, Hong JY, Li A, Moon RT, Jones EA, Vleminckx K, Vize PD, McCrea PD. Requirement of Wnt/beta-catenin signaling in pronephric kidney development. Mech Dev. 2009;126:142–59.

810. Wang Y, Fu Y, Gao L, Zhu G, Liang J, Gao C, Huang B, Fenger U, Niehrs C, Chen YG, Wu W. Xenopus skip modulates Wnt/beta-catenin signaling and functions in neural crest induction. J Biol Chem. 2010;285:10890–901.

811. Bläker H, Helmchen B, Bönisch A, Aulmann S, Penzel R, Otto HF, Rieker RJ. Mutational activation of the RAS-RAF-MAPK and Wnt pathway in small intestinal adenocarcinomas. Scand J Gastroenterol. 2004;39:748–53.

812. Clark PE, Polosukhina S, Love H, Correa H, Coffin C, Perlman EJ, de Caestecker M, Moses HL, Zent R. β-Catenin and K-RAS synergize to form primitive renal epithelial tumors with features of epithelial Wilms tumors. Am J Pathol. 2011;179:3045–55.

813. Torre C, Benhamouche S, Mitchell C, Godard C, Veber P, Latourneur F, Cagbard N, Jacques S, Finzi L, Perret C, Colnot S. The transforming growth factor-α and cyclin D1 genes are direct targets of β-catenin signaling in hepatocyte proliferation. J Hepatol. 2011;55:86–95.

814. Yamazaki K, Hanami K, Nagao T, Asoh A, Sugano I, Ishida Y. Increased cyclinD1 expression in cancer of the ampula of Vater: relevance to nuclear beta-catenin accumulation and K-ras gene mutation. Mol Pathol. 2003;56:336–41.

815. Zhang B, Ougolkov A, Yamashita K, Takahashi Y, Mai M, Minamoto T. beta-Catenin and ras oncogenes detect most human colorectal cancers. Clin Cancer Res. 2003;9:3079–3079.

816. Holland LZ. Heads or tails? Amphioxus and the evolution of anterior-posterior patterning in deuterostomes. Dev Biol. 2002;241:209–28.

817. Onai T, Takai Akira, Setimarga DH, Holland LZ. Essential role of Dkk3 for head formation by inhibiting Wnt/β-catenin and Nodal/Vg1 signaling pathways in the basal chordate amphioxus. Evol Dev. 2012;14:338–50.

818. Bao J, Zheng JJ, Wu D. The structural basis of DKK-mediated inhibition of Wnt/LRP signaling. Sci Signal. 2012;5(224):22.

819. Dellinger TH, Planutis K, Jandial DD, Eskander RN, Martinez ME, Zi X, Monk BJ, Holcombe RE. Expression of the Wnt antagonist Dickkopf-3 is associated with prognostic clinicopathologic characteristics and impairs proliferation and invasion in endometrial cancer. Gynecol Oncol. 2012;126:259–67.

820. Li L, Mao J, Sun L, Liu W, Wu D. Second cysteine-rich domain of Dickkopf-2 activates canonical Wnt signaling pathway via LRP-6 independently of dishevelled. J Biol Chem. 2002;277:5977–81.

821. Xu WH, Liu ZB, Yang C, Qin W, Shao ZM. Expression of dickkopf-1 beta-catenin related to the prognosis of breast cancer patients with triple negative phenotype. PLoS One. 2012;7(5):e37624.

822. Dihlmann S, Kloor M, Fallsehr C, von Knebel Doeberitz M. Regulation of AKT1 expression by beta-catenin/Tcf/Lef signaling in colorectal cancer cells. Carcinogenesis. 2005;26:1503–12.

823. Yang ZR, Dong WG, Lei XF, Liu M, Liu QS. Overexpression of Dickkopf-3 induces apoptosis through mitochondrial pathway in human colon cancer. World J Gastroenterol. 2012;18:1590–601.

824. Hernandez Y, Zamora G, Ray S, Chapoy J, Chavez E, Valvarde R, Williams E, Aley SB, Das S. Transcriptional analysis of three major putative phosphatidylinositol kinase genes in a parasitic protozoan, Giardia lamblia. J Eukaryot Microbiol. 2007;54:29–32.

825. Szentpetery Z, Szakacs J, Bojjireddy N, Tai AW, Balla T. Genetic and functional studies of phosphatidyl-inositol 4-kinase type IIIa. Biochim Biophys Acta. 2011;1811:476–83.

826. Li S, Lin P, Fayad LE, Lennon PA, Miranda RN, Yin CC, Lin E, Medeiros LJ. B-cell lymphomas with MYC/8q24 rearrangements and IGH/BCL2/t(14;18)(q32;q21): an aggressive disease with heterogenous histology, germinal center B-cell immunophenotype and poor outcome. Mol Pathol. 2012;25:145–56.

827. Yan Y, Park SS, Janz S, Eckhardt LA. In a model immunoglobulin heavy-chain (IGH)/MYC translocation, the IgH 3' regulatory region induces MYC expression at the immature stage of B cell development. Genes Chromosomes Cancer. 2007;46:950–9.

828. Zaheen A, Boulianne B, Parsa JY, Ramachandran S, Gommerman JL, Martin A. AID constrains germinal center size by rendering B cells susceptible to apoptosis. Blood. 2009;114:547–54.

829. Phan RT, Saito M, Kitagawa Y, Means AR, Dalla-Favera R. Genotoxic stress regulates expression of the proto-oncogene Bcl6 in germinal center B cells. Nat Immunol. 2007;8:1132–9.

830. Ranuncolo SM, Polo JM, Dierov J, Singer M, Kuo T, Greally J, Green R, Carroll M, Melnick A. Bcl-6 mediates the germinal center B cell phenotype and lymphomagenesis through transcriptional repression of the DNA-damage sensor ATR. Nat Immunol. 2007;8:705–14.

831. Hangaishi A, Kurokawa M. Blimp-1 is a tumor suppressor gene in lymphoid malignancies. Int J Hematol. 2010;91:46–53.

832. Küçük C, Iqbal J, Hu X, Gaulard P, De Leval L, Srivastava G, Au WY, McKeithan TW, Chan WC. PRDM1 is a tumor suppressor gene in natural killer cell malignancies. Proc Natl Acad Sci USA. 2011;108:20119–24.

833. Lin J, Lwin T, Zhao JJ, Tam W, Choi YS, Moscinski LC, Dalton WS, Sotomayor EM, Wright KL, Tao J. Follicular dendritic cell-induced microRNA-mediated upregulation of PRDM1 and downregulation of BCL-6 in non-Hodgkin's B-cell lymphomas. Leukemia. 2011;25:145–52.

834. Sherman MH, Kuraishy AI, Deshpande C, Hong JS, Cacalano NA, Gatti RA, Manis JP, Demore MA, Pellegrini M, Teitell MA. AID-induced genotoxic stress promotes B cell differentiation in the germinal center via ATM and LKB1 signaling. Mol Cell. 2010;39:873–85.
835. Shioh Y. ATM (ataxia telangiectasia mutated): expanding roles in the DNA damage response and cellular homeostasis. Biochem Soc Trans. 2001;29(Pt 6):661–6.
836. Martelli AM, Charini F, Evangelisti C, Ognibene A, Bressanin D, Bill AM, Manzolin L, Cappellini, McCubrey JA. Targeting the liver kinase B1/AMP-activated protein kinase pathway as a therapeutic strategy for hematological malignancies. Expert Opin Ther Targets. 2012;16:720–42.
837. Liu L, Siu F-M, Che C-M, Xu A, Wang Y. Akt blocks the tumor suppressor activity of LKB1 by promoting phosphorylation-dependent nuclear retention through 14-3-3 proteins. M J Transl Res. 2012;4:175–86.
838. Fumasoni I, Meani N, Rambaldi D, Scafetta G, Alcalay M, Ciccarelli FD. Family expansion and gene rearrangements contributed to the functional specialization of PRDM genes in vertebrates. BMC Evol Biol. 2007;7:187. doi:10.1186/1471-2148-7-187.
839. X-s Shu, Geng H, Li L, Ying J, Ma C, Wang Y, Poon FF, Wang X, Ying Y, Yeo W, Srivastava G, Tsao SW, Yu J, Sung JJY, Huang S, Chan ATC, Tao Q. The epigenetic modifier PRDM5 functions as a tumor suppressor through modulating WNT/β-catenin signaling and is frequently silenced in multiple tumors. PLoS One. 2011;6(11):e27346. doi:10.1371/journal.pone.0027346.
840. Sun W, Qiao L, Liu Q, Chen L, Ling B, Sammynaiken S, Yang J. Anticancer activity of the PR domain of tumor suppressor RIZ1. Int J Med Sci. 2011;8:161–7.
841. Münk C, Willemsen A, Bravo IG. An ancient history of gene duplications, fusions and losses in the evolution of APOBEC3 mutators in mammals. BMC Evol Biol. 2012;12(1):71.
842. Muranyi W, Flügel RM. Analysis of splicing patterns of human spumaretrovirus by polymerase chain reaction reveals complex RNA structure. J Virol. 1991;65:727–35.
843. Flügel RM, Rethwilm A, Maurer B, Darai G. Nucleotide sequence analysis of the env gene and its flanking regions of the human spumaretrovirus reveals two novel genes. EMBO J. 1987;6:2077–87.
844. Flügel RM. Spumaviruses: a group of complex retroviruses. J Acquired Immune Defic Syndr. 1992;4:739–59.
845. Russelln RA, Wiegand HK, Moore MD, Schäfer A, Mc Clure MC, Cullen BP. Foamy virus Bet proteins function as novel inhibitors of the APOBEC3 family of innate antiretroviral defense. J Virol. 2005;79:8724–31.
846. Maurer B, Flügel RM. The 3'-orf protein of human immunodeficiency virus 2 shows sequence homology with the bel3 gene of the human spumaretrovirus. FEBS Lett. 1987;222:286–8.
847. Liniai ML. Foamy viruses are unconventional retroviruses. J Virol. 1999;73:1747–55.
848. D'Errico I, Gadaleta G, Saccone C. Pseudogenes in metazoan: origin and features. Brief Funct Genomic Proteomic. 2004;3:157–67.
849. Permanyer J, Albalat R, Gonzáles-Duarte R. Getting closer to a pre-vertebrate genome: the non-LTR retrotransposon. Int J Biol Sci. 2006;2:48–53.
850. Moreau-Aubry A, Le Guiner S, Labarrière N, Gesnel MC, Jotereau F, Breathnach R. A processed pseudogene codes for a new antigen recognized by CD8(+) T cell clone on melanoma. J Exp Med. 2000;191:1617–24.
851. Csűrös M, István M. Statistical alignment of retropseudogenes and their functional paralogs. Mol Biol Evol. 2005;22:2457–71.
852. Salmena L, Poliseno L, Tay Y, Kats L, Pandolfi PP. A ceRNA hypothesis: the Rosetta Stone of a hidden RNA language? Cell. 2011;146:353–6.

853. Kalyanma-Sudaram S, Kumar-Sinha C, Shankar S, Robinson DR, Wu YM, Cao X, Asangani IA, Kothari V, Prensner KR, Lonigro RJ, Mk Iyer, Barrette T, Shanmugam A, Dhanasekaran SM, Palanisamy N, Chinnaiyan AM. Expressed pseudogenes in the transcriptional landscape of human cancers. Cell. 2012;149:1622–34.
854. Verschoor EJ, Langenhuijzen S, Bontjer I, Fagrouch Z, Niphuis H, Warren KS, Eulenberger K, Heney JL. The phylogeography of orangutan foamy viruses supports the theory of ancient repopulation of Sumatra. J Virol. 2004;78:12712–6.
855. Winkler IG, Flügel RM, Asikainen K, Flower RL. Antibody to human foamy virus not detected in individuals treated with blood products or in blood donors. Vox Sang 2000;79:118–9.
856. Dang Y, Wang X, Esselman WJ, Zheng YH. Identification of APOBEC3DE as another antitretroviral factor from the human APOBEC family. J Virol. 2006;80:10522–33.
857. Nascimento MC, Sumita LM, Souza VU, Weiss HA, Oliveira J, Mascheretti M, Quiroga M, Vela RA, Sabino E, Pannuti CS, Mayaud P. Seroprevalence of Kaposi sarcoma-associated herpesvirus and other serologic markers in the Brazilian Amazon. Emerg Infect Dis. 2009;15:663–7.
858. Rogozin IB, Iyer LM, Liang L, Glazko GV, Liston VG, Palov YI, Aravind L, Pancer Z. Evolution and diversification of lamprey antigen receptors: evidence for involvement of an AID-APOBEC family cytosine deaminase. Nat Immunol. 2007;8:647–56.
859. Severi F, Chiocca A, Conticello SG. Analysis of reptilian APOBEC1 suggests that RNA editing may not be its ancestral function. Mol Biol Evol. 2011;28:1125–9.
860. Kohli RM, Maul RW, Guminski AF, McClure RL, Gajula KS, Saribasak H, McMahon MA, Siciliano RF, Gearhart PJ, Stivers JT. Local sequence targeting in the AID/APOBEC family differentially impacts retroviral restriction and antibody diversification. J Biol Chem. 2010;285:40956–64.
861. Bernacchi S, Mercenne G, Tournaire C, Marquet R, Paillart JC. Importance of the proline-rich multimerization domain on the oligomerization and nucleic acid binding properties of HIV-1 Vif. Nucleic Acids Res. 2011;39:2404–15.
862. Oroszlan S, Luftig RB. Retroviral proteases. Curr Top Microbiol Immunol. 1990;157:153–85.
863. Fehér A, Boross P, Sperka T, Miklóssy G, Kádas J, Bagossi P, Oroszlan S, Weber IT, Tőzsér J. Characterization of the murine leukemia virus protease and its comparison with the human immunodeficiency virus type 1 protease. J Gen Virol. 2006;87:1321–30.
864. Ali A, Wang J, Nathans RS, Cao H, Sharova N, Stevenson M, Rana TM. Synthesis and structure-activity relationship studies of HIV-1 virion infectivity factor (Vif) inhibitors that block viral replication. Chem Med Chem. 2012;7:1217–29.
865a. Paprotka T, Delviks-Frankenberry KA, Cingöz O, Martinez A, Kung HJ, Pepper CG, Hu WS, Fivash MJ Jr, Coffin JM, Pathak VK. Recombinant origin of the retrovirus XMRV. Science. 2011;333:97–101.
865b. Paprotka T, Venkatachan NJ, Chaipan C, Burdick R, Delviks-Frankenberry KA, Hu WS, Pathak VK. Inhibition of xenotropic murine leukemia virus-related virus by APOBEC3 proteins and antiviral drugs. J Virol. 2010;84:5719–29.
866. Hohn O, Bannert N. Origin of XMRV and its demise as a human pathogen associated with chronic fatigue syndrome. Viruses. 2011;3:13121319.
867. Del Prete GQ, Kearney MF, Spindler J, Wiegand A, Chertova E, Roser JD, Estes JD, Hao XP, Trubey CM, Lara A, Lee K, Chaipan C, Bess JW Jr, Nagashima K, Keele BE, Macallister R, Smedley J, Pathak VK, Kewalramani VN, Coffin JM, Lifson JD. Restricted replication of xenotropic murine leukemia virus-related virus in pigtail macaques. J Virol. 2012;86:3152–66.
868. Yamanaka S, Balestra ME, Ferrell LD, Fan J, Arnold KS, Taylor S, Taylor JM, Innerarity TL. Apolipoprotein B mRNA-editing protein induces hepatocellular carcinoma and dysplasia in transgenic animals. Proc Natl Acad Sci USA. 1995;82:8483–7.

869. Greeve J, Lellek H, Apostel F, Hundoegger K, Barialai A, Kirsten R, Welker S, Greten H. Absence of APOBEC-1 mediated mRNA editing in human carcinomas. Oncogene. 1999;18:6357–65.

870. Cesana D, Sgualdino J, Rudilosso L, Merella S, Naldini L, Montini. Whole transcriptome characterization of aberrant splicing events induced by lentiviral vector integration. J Clin Invest. 2012;122:1667–76.

871a. Bandyopadhyay S, Cookson MR. Evolutionary and functional relationships within the DJ-1 superfamily. BMC Evol Biol 2004;4:6. doi:10.1186/1471-2148-4-6.

871b. Maita C, Maita H, Iguchi-Ariga SMM, Ariga H. Monomer DJ-1 and its N-terminal sequence are necessary for mitochondrial localization of DJ-1 mutants. PLoS One. 2013;8(1):e54087. doi:10.1371/journal.pone.0054087.

871c. Martin SE, Sausen M, Joseph A, Kingham BF, Martin ES. Identification of a HMGA2-EFCAB6 gene rearrangement following next-generation sequencing in a patient with a t(12;22)(q14.3;q13.2) and JAK2V617F-positive myeloproliferative neoplasm. Cancer Genet. 2012;205:295–303.

872a. Wu Y, Deford J, Benjamin R, Lee MG, Ruben L. The gene family of EF-hand calcium-binding proteins from the flagellum of Trypanosoma brucei. Biochem J. 1994;304:833–41.

872b. Wu Y, Haghighat NG, Ruben L. The predominant calcimedin from Trypanosoma brucei comprises a family of flagellar EF-hand calcium-binding proteins. Biochem J. 1992;287 (187–193):873.

873. Abe N, Watanabe T, Matsumoto M, Masaki T, Mori T, Sugiyama M, Ghiasappetta G, Fusco A, Atomi Y. An increased high mobility group A2 expression level is associated with malignant phenotype in pancreatic exocrine tissue. Br J Cancer. 2003;89:2104–9.

874a. Mansouri A, Ridgway LD, Korapati A, Zhang G, Tian L, Wang Y, Siddik ZH, Mills GS, Claret FX. Sustained activation of JNK/p38 MAPK pathways in response to cisplatin leads to Fas ligand induction and cell death in ovarian carcinoma cells. J Biol Chem. 2003;278:19245–56.

874b. Takahashi K, Taira T, Niki T, Seino C, Iguchi-Ariga SMM, Ariga H. DJ-1 positively regulates the androgen receptor by impairing the binding of PIASxα to the receptor. J Biol Chem. 2001;276:37556–63.

874c. Pei X-j Wu, T-t Li B, X-y Tian, Li Z, Q-x Yang. Increased expression of macrophage migration inhibitory factor and DJ-1 contribute to cell invasion and metastasis of nasopharyngeal carcinoma. Int J Med Sci. 2014;11:106–15.

875. Lee H, Choi SK, Ro JY. Overexpression of DJ-1 and HSP90α, and loss of PTEN associated with invasive urothelial carcinoma of urinary bladder: possible prognostic markers. Oncol Lett. 2012;3:507–12.

876. Hornbruch-Freitag C, Griemert B, Buttgereit D, Renkawitz-Pohl R. Drosophila Swiprosin-1 / EFHD2 accumulates at the prefusion complex stage during Drosophila myoblast fusion. J Cell Sci. 2011;124:3266–78.

877. Avramidou A, Kroczek C, Lang C, Schuh W, Jäck HM, Mielenz D. The novel adaptor protein Swiprosin-1 enhances BCR signals and contributes to BCR-induced apoptosis. Cell Death Diff. 2007;14:1936–47.

878. Kaneko-Ishino T, Ishino F. The role of genes domesticated from LTR retrotransposons and retroviruses in mammals. Front Microbiol. 2012;3:262.

879. Villesen P, Aagaard L, Wiuf C, Pederson FS. Identification of endogenous retroviral reading frames in the human genome. Retrovirology. 2004;1:32. doi:10.1186/1742-4690-1-32.

880. Kämmerer U, Germeyer A, Stengel S, Kapp Denner J. Human endogenous retrovirus K (HERV-K) is expressed in villous and extravillous cytotrophoblast cells of the human placenta. J Reprod Immunol. 2011;91:1–8.

881. Buzdin A. Human-specific endogenous retroviruses. Sci World J. 2007;7:1848–68.

882. Solyom Sz, Kazazian HH. Mobile elements in the human genome: implications for disease. Genome Med. 2012;4(2):12. doi:10.1185/gm311.
883. Dupressoir A, Lavialle C, Heidmann T. From ancestral infectious retroviruses to bona fide cellular genes. Role of the captured syncytins in placentation. Placenta. 2012;33:663–71.
884. Dunk CE, Gerllhaus A, Drewlo S, Baczyk D, Pötgens AJ, Winterhagen E, Kingdom JC, Lye SJ. The molecular role of connexin 43 in human trophoblast cell fusion. Biol Reprod. 2012;86:115. doi:10.1095/bioreprod.111.096925.
885. Bjerregaard B, Holck S, Christensen IJ, Larsson LI. Syncytin is involved in breast cancer endothelial cell fusion. Cell Mol Life Sci. 2006;63:1906–11.
886. Strick R, Ackermann S, Langbein M, Swiatek J, Schubert SW, Hashemolhossieini S, Koscheck T, Fasching PA, Schild RL, Beckmann MW, Strissel PL. Proliferation and cell-cell fusion of endometrial carcinoma are induced by the human endogenous retroviral syncytin-1 and regulated by TGF-beta. J Mol Med. 2007;85:23–38.
887. Gimenez J, Montgiraud C, Pichon J-P, Arsac M, Ruel M, Bouton O, Mailet F. Custom human endogenous retrovirus dedicated microarray identifies self-induced HERV-W family elements reactivated in testicular cancer upon methylation control. Nucleic Acids Res. 2010;38:2229–46.
888. Bonnaud B, Beliaeff J, Bouton O, Oriol G, Duret L, Mallet F. Natural history of the ERvWE1 endogenous retroviral locus. Retrovirology. 2005;2:57. doi:10.1186/1742-4690-2-57.
889. Cáceres M, Comparative NISC. Sequencing Program, Thomas JW. The gene of retroviral origin syncytin 1 is specific to hominids and is inactive in Old World monkeys. J Hered. 2006;97:100–6.
890. Roebke C, Wahl S, Laufer G, Stadelmann C, Sauter M, Mueller N. An N-terminally truncated envelope protein encoded by a human endogenous retrovirus HERV-W from Xq22.3. Retrovirology. 2010;7:69. doi:10.1186/1742-4690-7-69.
891. Blaise S, de Parseval N, Bénit L, Heidmann T. Genomwide screening for fusogenic human endogenous retrovirus envelopes identifies syncytin 2, a gene conserved on primate evolution. Proc Natl Acad Sci USA. 2003;100:1301–13018.
892. de Parseval N, Diop G, Blaise S, Helle F, Vasilescu A, Matsuda F, Heldmann T. Comprehensive search for intra- and inter-specific sequence polymorphisms among coding envelope genes of retroviral origin found in the human genome: genes and pseudogenes. BMC Genomics. 2005;5:117. doi:10.1186/1471-2164-6-117.
893. Malassiné A, Frendo J-L, Blaise S, Handschuh P, Tsatsaris V, Heidmann T. Human endogenous retrovirus-FRD envelope protein (syncytin 2) expression in normal and trisomy 21-affected placenta. Retrovirology. 2008;5:6. doi:10.1186/1742-4690-5-6.
894. Esnault C, Priet S, Ribet D, Vernochet C, Bruis T, Lavialle C, Weissenbach J, Heidmann T. A placenta-specific receptor for the fusogenic, endogenous retrovirus-derived, human syncytin-2. Proc Natl Acad Sci USA. 2006;105:17532–7.
895a. Sinkovics JG, Horvath JC, Gyorkey F. From evolutionary coexistence to criminal collusion of herpes- and retro(lenti)viruses in human pathological entities. Program & Abstracts, 15th Annual Clinical Virology Symposium, Pan American Society for Virology & Department of Medical Microbiology & Immunology, The University of South Florida College of Medicine, Clearwater, Florida, May 1999 #T32.
895b. Sinkovics JG, Horvath JC. Ancient viruses coexisted with the Universal Ancestor(s) and coevolved with each other and with their host cells. Program & Abstracts, 16th Clinical Virology Symposium, Pan American Society for Clinical Virology & Department of Medical Microbiology & Immunology, The University of South Florida College of Medicine, Clearwater, FL, May 2000 #M35.

895c. Sinkovics JG, Horvath JC. Criminal collusion of herpes- and retro(lenti)viruses. 2nd International Conference of M Kaposi Research Foundation, Budapest, Aug 31–Sept 1 2000. Magyar Venerologiai Archivum (Hungarian Archives Venerology) 2000;4/2-3:105–106.

896. Laufer G, Mayer J, Mueller BF, Mueller-Lantzsch N, Ruprecht K. Analysis of transcribed human endogenous retrovirus W *env* locus clarifies the origin of multiple sclerosis -associated retrovirus *env* sequences. Retrovirology. 2009;5:37. doi:10.1186/1742-4690-6-37.

897. Perron H, Germi R, Bernard C, Garcia-Montojo M, Deluen C, Farinelli L, Faucard R, Veas F, Stefas I, Fabriek BO, Van-Horssen J, Van-der-Valk P, Gerdill C, Mancuso R, Saresella M, Clerici M, Marcel S, Creange A, Cavaretta R, Caputo D, Arru G, Morand P, Lang AB, Sotgiu S, Ruprecht K, Rieckmann P, Villoslada P, Chofflon M, Boucraut J, Pelletier J, Hartung HP. Human endogenous retrovirus type W envelope expression in blood and brain cells provides new insights into multiple sclerosis disease. Mult Scler. 2012;18:1721–36.

898. Iba H, Takeya T, Cross FR, Hanafusa T, Hanafusa H. Rous sarcoma virus variants that carry the cellular src gene instead of the viral src gene cannot transform chicken embryo fibroblasts. Proc Natl Acad Sci USA. 1984;81:4424–8.

899. Sinkovics JG, Szakacs A, Butler JJ, Hirano M, Shullenberger CC, Howe CD. Pathological conditions (leukemias and sarcomas, leukemoid reactions, runt disease, and immunization without disease) associated with the inoculation of a long-term tissue culture line of a murine leukemia virus. In: Clifford P, Linsell CA, Timms GL, editors. "Cancer in Africa" East African Med J (East African Publishing House, Nairobi, Kenya); 1968;143–153.

900. Sinkovics JG, Howe DC, Shullenberger CC. Studies on virus attenuation in murine leukemia. In: International Symposium on Comparative Leukemia Research. Paris, July 12–31, 1967. Bibliotheca Haematol (Karger, Basel, New York) 1968;31:73–75.

901. Huebner RJ, Todaro GJ. Oncogenes of RNA tumor viruses as determinants of cancer. Proc Natl Acad Sci USA. 1969;64:1087–94.

902. Pani AM, Mullarkey EE, Aronowicz J, Assimacopoulos S, Grove EA, Lowe CJ. Ancient deuterostome origins of vertebrate brain signaling centres. Nature. 2012;483:289–94.

903. Perrimon N, Pitsouli C, Shilo BZ. Signaling mechanisms controlling cell fate and embryonic patterning. Cold Spring Harb Perspect Biol. 2012;4(8):005975. doi:10.1101/cshperspect.a005975.

904. Vincent S, Perrimon N, Axelrod JD. Hedgehog and wingless stabilize but do not induce cell fate during Drosophila dorsal embryonic epidermal patterning. Development. 2998;135:27r67-2775.

905. Lin G, Chen Y, Slack JM. Transgenic analysis of signaling pathways required for Xenopus tadpole spinal cord and muscle regeneration. Anat Rec (Hoboken) 2012. doi:10.1002/ar.22437.

906. Nusse R, Theunissen H. Waginjer vE, Rijsewik F, Gennissen A, Otto A, Schuuring E, van Ooyen A. The Wnt-1 (int-1) oncogene promoter and its mechanism of activation by insertion of proviral DNA of the mouse mammary tumor virus. Mol Cell Biol. 1990;10:4170–9.

907. Pond AC, Herschkowitz I, Schwertfeger KL, Weim B, Zhang Y, York B, Cardiff RD, Hilsenbeck S, Perou CM, Creighton CJ, Lloyd RE, Rosen JM. Fibroblast growth factor receptor signaling dramatically accelerates tumorigenesis and enhances oncoprotein translation in the mouse mammary tumor virus-Wnt-1 mouse model of breast cancer. Cancer Res. 2010;70:4868–79.

908. Cui J-W, Y-j Li, Sarkar A, Brown J, Tan Y-H, Premyslova M, Michaud C, Iscove C, Wang G-J, Ben-David Y. Retroviral insertion activation of the *Fli*-3 locus in erythroleukemias encoding a cluster of microRNAs that convert Epo-induced differentiation to proliferation. Blood. 2007;110:2631–40.

909. Wang CL, Wang BB, Bartha G, Li L, Channa N, Klinger M, Killeen N, Wabl M. Activation of an oncogenic microRNA cistron by provirus integration. Proc Natl Acad Sci. 2006;103:18680–4.
910. Kashif M, Hellwig A, Hashemolhosseini S, Kumar V, Bock F, Wang F, Shahzad K, Ranjan S, Wolter J, Madhusudhan T, Bierhaus A, Nawroth P, Isermann B. Nuclear factor erythroid-derived 2 (Nfe2)regulates JunD DNA-binding activity via acetylation: a novel mechanism regulating trophoblast differentiation. J Biol Chem. 2012;287:5400–11.
911. Arany I, Jozsa Z, Ember I. Further studies on migration and colonization of leukemic lymphoblasts in AKR and HSS inbred mice. Cancer Biochem Biophys. 1987;9:149–54.
912. Hays EF, Bristol GC, McDougall S, Klotz JL, Kronenberg M. Development of lymphoma in the thymus of AKR mice treated with the lymphomagenic virus SL3-3. Cancer Res. 1989;49:4225–30.
913. Glud SZ, Sørensen AB, Andrulis M, Wang B, Kondo E, Serfling RS, Palmetshofer A, Pedersen FS. A tumor suppressor function for NFATc3 in T-cell lymphomagenesis by murine leukemia virus. Blood. 2005;106:3546–52.
914. Beá S, Tort F, Pinyol M, Puig X, Hernández L, Hernández S, Fernandez PL, van Lohuizen M, Colomer D, Campo E. BMI-1 gene amplification and overexpression in hematological malignancies occur mainly in mantle cell lymphoma. Cancer Res. 2001;61:2409–12.
915a. Pinyol M, Bea S, Plá L, Ribrag V, Bosq J, Rosenwald A, Campo E, Jares P. Inactivation of RB1 in mantle cell lymphoma detected by nonsense-mediated mRNA decay pathway inhibition and microarray analysis. Blood. 2007;109:5422–9. doi:10.1182/blood-2006-11-057208.
915b. Enjuanes A, Fernández V, Hernández L, Navarro A, Beá S, Pinyol M, López-Guillermo A, Rosenwald A, Ott G, Campo E, Jares P. Identification of methylated genes associated with aggressive clinicopathological features in mantle cell lymphoma. PLoS One. 2011;6 (5):e19736.
916. Guo BH, Feng Y, Zhang R, Xu LH, Li MZ, Kung HE, Song LB, Zeng MS. Bmi-1 promotes invasion and metastasis and its elevated expression is correlated with advanced stage of breast cancer. Mol Cancer. 2011;10(1):10.
917. Song W, Tao K, Li H, Jin C, Song Z, Li J, Shi H, Li X, Dang Z, Dou K. Bmi-1 is related to proliferation, survival and poor prognosis in pancreatic cancer. Cancer Sci. 2010;101:1754–60.
918. Yang GF, He WP, Cai MY, He LR, Luo JH, Deng HX, Guan XY, Zeng MS, Zeng YX, Xie S. Intensive expression of Bmi-1 is a new independent predictor of poor outcome in patients with ovarian carcinoma. BMC Cancer. 2010;10:133.
919a. Yu Q, Meng XX, Liu B, Liu DD, Zhang W, Liu WH, Yang G. Effects of antisense Bmi-1 RNA on the proliferation of lung cancer cell line A5491. Zhongua Bing Li Xue Za Zhi. 2009;38:829–32.
919b. Yu CC, Lo WL, Chen YW, Huang PI, Hsu HS, Tseng LK, Hung SC, Kao SY, Chang CJ, Chiou SH. Bmi-1 regulates snail expression and promotes metastasis ability in head and neck squamous cancer-derived ALDH1 positive cells. J Oncol 2011; 2011.pii;609259.
920a. Jiang Y, Su B, Meng X, Liu C, Liu B, Liu D, Fan Y, Yang Y. Effect of siRNA-mediated silencing of Bmi-1 gene expression on HeLa cells. Cancer Sci. 2010;101:379–86.
920b. Jiang L, Wu J, Yang Y, Liu L, Song L, Li J, Li M. Bmi-1 promotes the aggressiveness of glioma via activating the NF-κB/MMP-9 signaling pathway. BMC Cancer. 2012;12:406.
921. Smith MR, LI H, Gordon L, Gascoyne RD, Paietta E, Forero-Torres A, Kahl BS, Advani R, Hong F, Horning SJ. Phase II study of rituximab plus cyclophosphamide, doxorubicin, vincristine, and prednisone immunochemotherapy followed by yttrium-90-ibritumomab tiuxetan in untreated mantle-cell lymphoma: Eastern Cooperative Oncology Group study E1499. J Clin Oncol. 2012;30:3119–26.

922a. Fiskus W, Rao R, Balusu R, Ganguly S, Tao J, Sotomayor EM, Mudunuru U, Smith JE, Hembruff SL, Atadja PW, Marquez VE, Bhalla KN. Superior efficacy of a combined epigenetic therapy against human mantle cell lymphoma (MCL) cells. Clin Cancer Res. 2012;18:6227–38.

922b. Zhang X, Zhao X, Fiskus W, Lin J, Lwin T, Rao R, Zhang Y, Chan JC, Fu K, Marquez VE, Chen-Kiang S, Moscinski LC, Stao E, Dalton WS, Weight KL, Sotomayor E, Bhalla K, Tao J. Coordinated silencing of MYC-mediated miR-29 by HDAC3 and EZH2 as a therapeutic target of histone modification in aggressive B-cell lymphomas. Cancer Cell. 2012;22:506–23.

923. Tátrai P, Szepesi Á, Matula Z, Szigeti A, Buchan G, Mádi A, Uher E, Német K. Combined introduction of Bmi-1 and hTERT immortalizes human adipose tissue-derived stromal cells with low risk of transformation. Biochem Biophys Res Commun. 2012;422:28–35.

924. Pandhare-Dash J, Mantri CK, Gong Y, Chen Z, Dash C. XMRV accelerates cellular proliferation, transformational activity, and invasiveness of prostate cancer cells by downregulating p27(Kip1). Prostate. 2012;72:886–97.

925. Niller H-H, Wolf H, Ay E, Minarovits J. Epigenetic dysregulation of Epstein-Barr virus latency and development of autoimmune disease. Adv Exp Med Biol. 2011;711:82–102.

926. Niller H-H, Wolf H, Minarovits J. Viral hit and run oncogenesis: genetic and epigenetic scenarios. Cancer Lett. 2011;305:200–17.

927. Morrison JA, Klingelhutz AJ, Raab-Traub N. Epstein-Barr virus latent membrane protein 2A activates beta-catenin signaling in epithelial cells. J Virol. 2003;77:12276–84.

928. Takacs M, Segesdi J, Banati F, Koroknai A, Wolf H, Niller HH, Minarovits J. The importance of epigenetic alterations in the development of Epstein-Barr virus-related lymphomas. Mediterr J Hematol Infect Dis. 2009;1(2):e2009012. doi:10.4084/MJHID.2009.012.

929. Portal D, Zhao B, Calderwood MA, Sondermann T, Johannsen E, Kieff E. EBV nuclear antigen EBNALP dismisses transcription repressors NCo and RBPJ from enhancers and EBNA2 increases NCoR-deficient RBPJ DNA binding. Proc Natl Acad Sci USA. 2011;108:7808–13.

930. Zuo J, Thomas WA, Haigh TA, Fitzsimmons L, Long HM, Hislop AD, Taylor GS, Rowe M. Epstein-Barr virus evades $CD4^{+}T$ cell responses in lytic cycle through BZLF1-mediated downregulation of CD74 and the cooperation of vBcl-2. PLoS Pathog. 2011;7(12):e1002455. doi:10.1371/journal.ppat.1002455.

931. Palmeri D, Carroll KD, Gonzalez-Lopez O, Lukac DM. Kaposi's sarcoma-associated herpesvirus Rta tetramers make high-affinity interactions with repetitive DNA elements in the Mta promoter to stimulate DNA binding of RBP-Jk/CSL. J Virol. 2011;85:11901–15.

932a. Lu M, Suen J, Frias C, Pfeiffer R, Tsai MH, Chuang E, Zeichner SL. Dissection of Kaposi's sarcoma-associated herpesvirus gene expression program by using the viral DNA replication inhibitor cidofovir. J Virol. 2004;78:13637–52.

932b. Lu J, Verma SC, Cain Q, Saha A, Dzeng RK, Robertson ES. The RBP-Jk binding sites within the RTA promoter regulate KSHV latent infection and cell proliferation. PLoS Pathog. 2012;8(1):e10002479. doi:10.1371/journal.ppat.1002479.

933. Martinez-FP Tang O. Leucine zipper domain is required for Kaposi sarcoma-associated herpesvirus (KSHV) K-bZIP protein to interact with histone deacetylase and is important for KSHV replication. J Biol Chem. 2012;287:15622–34.

934. Naranatt PP, Krishnan HH, Svojanovsky SR, Bloomer C, Mathur S, Chandran B. Host gene induction and transcriptional reprogramming in Kaposi's sarcoma-associated herpesvirus (FSHV/HHV-8)-infected endothelial, fibroblast, and B cells: insight into modulation events early during infection. Cancer Res. 2004;64:72–84.

935. Ritthipichai K, Nan Y, Bossis I, Zhang Y. Viral FLICE inhibitory protein of rhesus monkey rhadinovirus inhibits apoptosis by enhancing autophagosome formation. PLoS One. 2012;7(5):e39438. doi:10.1371/journal.pone.0039438.

936. Klemsz MJ, McKercher SR, Celada A, Van Beveren C, Maki RA. Pillars article: the macrophage and B cell-specific transcription factor PU.1 is related to the ets oncogene. Cell. 1990;61:113–24.
937. Moreau-Gachelin F. Multi-stage Friend murine erythroleukemia: molecular insights into oncogene cooperation. Retrovirology. 2008;4(5):99.
938. Ballon G, Chen K, Perez R, Tam W, Cesarman E. Kaposi sarcoma herpesvirus (KSHV) vFLIP oncoprotein induces B cell transdifferentiation and tumorigenesis in mice. J Clin Invest. 2011;121:1141–53.
939. Cavallaro U, Gasparini G, Soria MR, Maier JA. Spindle cells isolated from Kaposi's sarcoma-like lesions of BKV/tat-transgenic mice co-express markers of different cell types. AIDS. 1996;10:1211–129.
940. Majerciak V, Zheng Z-M. Kaposi's sarcoma-associated herpesvirus ORF57 in viral RNA processing. Front Biosci. 2009;14:1516–28.
941. Yoo J, Kang J, Lee HN, Aguilar B, Kafka D, Lee S, Choi I, Lee J, Ramu S, Haas J, Koh CJ, Hong Y-K. Kaposin-B enhances the PROX1 mRNA stability during lymphatic reprogramming of vascular endothelial cells by Kaposi's sarcoma herpes virus. PLoS Pathog 2010; 6(8):e1001046. doi:10.1371/journal.ppat.1001046.
942. Gottwein E, Corcoran DL, Mukherjee N, Skalsky RL, Hafner M, Nusbaum JD, Shamulailatpam P, Love CL, Dave SS, Tuschl T, Ohler U, Cullen BR. Viral microRNA targetome of KSHV-infected primary effusion lymphoma cell lines. Cell Host Microbe. 2011;10:515–26.
943. Morissette G, Flamand L. Herpesviruses and chromosomal integration. J Virol. 2010;84:12100–9.
944. Kniazhanskaia ES, Kondrashina OV, Gottikh MB. Approaches towards directed DNA integration by the use of retroviral integrases and transposases. Mol Biol (Moskva). 2011;45:931–48.
945. Aubert M, Ryu BY, Banks L, Rawkings DJ, Scharenberg AM, Jerome KR. Successful targeting and disruption of an integrated reporter lentivirus using the engineered homing endonuclease Y2 I-Anil. PLoS One. 2011;6(2):e16825.
946. Flamand L, Komaroff AL, Arbuckle JH, Medveczky PG, Ablashi DV. Review, Part I. Human herpesvirus-6-basic biology, diagnostic testing, and antiviral efficacy. J Med Virol. 2010;82:1560–8.
947. Kaufer BB, Arndt S, Trapp S, Osterrieder N, Jarosinski KW. Herpesvirus telomerase RNA (vTR) with a mutated template sequence abrogates herpesvirus-induced lymphomagenesis. PLoS Pathog. 2011;7(10):e1002333. doi:10.1371/journal.ppat.1002333.
948. Arbuckle JH, Medveczky MM, Luka J, Hadley SH, Luegmayr A, Ablashi D, Lund TC, Tolar J, De Meirleir K, Montoya JG, Komaroff AL, Ambros PF, Medveczky PG. The latent human herpesvirus-6A genome specifically integrates in telomeres of human chromosomes in vivo and in vitro. Proc Natl Acad Sci USA. 2010;107:5563–8.
949. Kaufer BB, Jarosinski KW, Osterrieder N. Herpesvirus telomeric repeats facilitate genome integration into host telomeres and mobilization of viral DNA during reactivation. J Exp Med. 2011;208:605–15.
950a. Daibata M, Taguchi T, Taguchi H, Miyoshi I. Integration of human herpesvirus 6 in Burkitt's lymphoma cell line. Br J Haematol. 1998;102:1307–13.
950b. Daibata M, Taguchi T, Nemoto Y, Taguchi H, Miyoshi I. Inheritance of chromosomally integrated human herpesvirus 6 DNA. Blood. 1999;94:1545–9.
951. Murakami Y, Tanimoto K, Fujiwara H, An J, Suemori K, Ochi T, Hasegawa H, Yasukawa M. Human herpesvirus 6 infection impairs Toll-like receptor signaling. Virol J. 2010;7:91. doi:10.1186/1743-422X-7-91.
952. Cermelli C, Cenacchi V, Beretti F, Pezzini F, Luca DD, Blasi E. Human herpesvirus-6 dysregulates monocyte-mediated anticryptococcal defences. J Med Microbiol. 2006;55:695–702.

953a. Arbuckle JH, Pantry S, Medveczky PG. The mechanism and significance of integration and vertical transmission of human herpesvirus 6 genome. In: Berencsi III Gy, editor. Maternal fetal transmission of human viruses and their influence on tumorigenesis. Springer Verlag 2912. p. 171–194.

953b. Arbuckle KH, Medveczky PG. The molecular biology of human herpesvirus 6 latency and telomere integration. Microbes Infect. 2011;13:7831–41.

954. Tuddenham L, Jung JS, Chane-Woon-Ming B, Dölken L, Pfeffer S. Small RNA deep sequencing identifies microRNAs and other small noncoding RNAs from human herpesvirus 6B. J Virol. 2012;86:1638–49.

955. Li Z, Yamauchi Y, Kamakura M, Murayama T, Goshima F, Kimura H, Nishiyama Y. Herpes simplex virus requires poly(ADP-ribose) polymerase activity for efficient replication and induces extracellular signal-related kinase-dependent phosphorylation and ICP0-dependent nuclear localization of tankyrase. J Virol. 2012;86:492–503.

956. Deng ZH, Atanasiu C, Zhao K, Marmorstein R, Sbodio JI, Chi N-W, Lieberman PM. Inhibition of Epstein-Barr virus OriP function by tankyrase, a telomere-associated poly-ADP ribose polymerase that binds and modifies EBNA. J Virol. 2005;79:4640–50.

957. Dregalla RC, Zhou J, Idate RP, Battaglia CLR, Liber HL, Bailey SM. Regulatory roles of tankyrase 1 at telomeres and DNA repair: suppression of T-SCE and stabilization of DNA-PKcs. Aging. 2010;2:691–708.

958. Bao R, Christova T, Song S, Angers S, Yan X, Attisano L. Inhibition of tankyrase induces axin stabilization and blocks Wnt signaling in breast cancer cells. PLoS One 2012;7(11):e48670. doi:10.1371/journal.pone.0048670.

959. James RG, Davidson KC, Bosch KA, Biechele TL, Robin NC, Taylor RJ, Major MB, Camp ND, Fowler K, Martins TJ, Moon RT. WIKI4, a novel inhibitor of tankyrase and Wnt/β-catenin signaling. PLoS. 2012;7(12):e50457. doi:10.1371/journal.pone.0050457.

960. Gunaydin H, Gu Y, Huang X. Novel binding mode of a potent and selective tankyrase inhibitor. PLoS One. 2012;7(3):e33740. doi:10.1371/journal.pone.0033740.

961. Azumi K, De Santis R, De Tomaso A, Rigoutsos I, Yoshizaki F, Pinto MR, Marino R, Shida K, Ikeda M, Ikeda M, Arai M, Inoue Y, Shimizu T, Satoh N, Rokssaar DS, Du Pasquier L, Kasahara M, Satake M, Noriala M. Genomic analysis of immunity in a Urochordate and the emergence of the vertebrate immune system: "waiting for Godot". Immunogenetics. 2003;55:570–81.

962. Dishaw LJ, Flores-Torres JA, Mueller MG, Karrer CR, Skapura DP, Melillo D, Zucchetti I, De Santis R, Pinto MP, Litman FW. A basal chordate model for studies of gut microbial immune interactions. Front Immunol. 2012;3:96. doi:10.3389/fimmu.2012.00096.

963. Wilson DR, Norton DD, Fugmann SD. The PHD domain of the sea urchin RAG2 homolog, SpRAG2L, recognizes dimethylated lysine 4 in histone H3 tails. Dev Comp Immunol. 2008;32:1221–30.

964. De Tomaso AW. Sea squirts and immune tolerance. Dis Model Med v. 2009;2:440–5.

965. Rinkevich B, Douek J, Rabinowitz C, Paz G. The candidate Fu/HC gene in Botryllus schlosseri (Urochordata) and ascidians' historecognition – an oxymoron? Dev Comp Immunol. 2112;36:718–727.

966. Chernakhovskaya Y, Shaghijan HS, Slavina RG, Svet-Moldavsky GJ. Helminths and allotransplantation. Rev Eur Etud Clin Biol. 1972;17:395–9.

967. Dales RP. Second-set graft rejections: do they occur in invertebrates? Symp Soc Exp Biol. 1978;32:203–19.

968. Orta AJ, Sullivan JT. Short-term immunoisolation of incompatible xenografts in a snail, Biomphalaria glabrata. Dev Comp Immunol. 2000;24:543–51.

969a. Smith LC, Clow LA, Terwilliger DP. The ancestral complement system in sea urchins. Immunol Rev. 2001;180:16–34.

969b. Smith LC, Davidson EH. The echinoderm immune system. Characters shared with vertebrate immune systems and characters arising later in deuterostome phylogeny. Ann N Y Acad Sci. 1994;15:213–26.
970. Detournay O, Schnitzler CE, Poole A, Weis VM. Regulation of cnidarian-dinoflagellate mutualisms: Evidence that activation of a host TGFβ innate immune pathway promotes tolerance of the symbiont. Dev Comp Immunol. 2012; 38:525-37. doi:10.1016/j.dci.2012.08008.
971. Rosengarten RD, Moreno MA, Lakkis FG, Buss LW, Dellaporta SL. Genetic diversity of the allodeterminant alr2 in Hydractinia symbiolongicarpus. Mol Biol Evol. 2011;28:933–47.
972. Gloria-Soria A, Moreno MA, Yund PO, Lakkis HG, Dellaporta SL, Buss LW. Evolutionary genetics of the Hydroid allodeterminant alr2. Mol Biol Evol. 2012;29:3921–32.
973. Millane RC, Kanska J, Duffy DJ, Seoighe C, Cunningham S, Plickert G, Frank U. Induced stem cell neoplasia in a cnidaria by ectopic expression of a POU domain transcription factor. Development. 2011;138:2429–39.
974. Duffy DJ, Millane RC, Frank U. A heat shock protein and wnt signaling crosstalk during axial patterning and stem cell proliferation. Dev Biol. 2012;362:271–81.
975. Seipp S, Schmich J, Leitz T. Apoptosis—a death-inducing mechanism tightly linked with morphogenesis in Hydractinia echinata (Cnidaria, Hydrozoa). Development. 2001;128:4891–8.
976. Wang WS, Hung SW, Lin TH, Tu CY, Wong ML, Chiou SH, Shieh MT. The effect of five different glycans on innate immune responses by phagocytes of hybrid tilapia and Japanese eels, Anguilla japonica. J Aquat Anim Health. 2007;19:49–59.
977. Hossein MM, Kawai K, Osima S. Immunogenicity of pressure inactivated Edwardsiella tarda bacteria in Anguilla japonica (Japanese eel). Pak J Biol Sci. 2011;14:755–67.
978. Hozumi N, Tonegawa S. Evidence for somatic rearrangement of immunoglobulin genes coding for variable and constant regions. J Immunol. 2004;173:4260–4.
979. Gellert M. V(D)J recombination: RAG proteins, repair factors, and regulation. Annu Rev Biochem. 2002;71:101–32.
980. Grundy GJ, Ramón-Maiques S, Dimitriadis EK, Kotova S, Biertümpfel C, Heymann JB, Steven AC, Gellert M, Yang W. Initial stages of V(D)J recombination: the organization of RAG1/2 and RSS DNA in the postcleavage complex. Mol Cell. 2009;35:217–27.
981. Hiom K, Malek M, Gellert M. DNA transposition by the RAG1 and RAG2 proteins: possible source of oncogenic translocations. Cell. 1998;94:463–770.
982. Onozawa M, Aplan PD. Illegitimate V(D)J recombination involving nonantigen receptor loci in lymphoid malignancy. Genes Chromosomes Cancer. 2012;51:525–35.
983. Kulski JK, Shigenari A, Shiina T, Ota M, Hosimichi K, James I, Inoko H. Human endogenous retrovirus (HERVK9) structural polymorphism with haplotypic HLA-A allelic associations. Genetics. 2008;180:445–57.
984. Sugamata R, Suetake H, Kikuchi K, Suzuki Y. Teleost B7 expressed on monocytes regulates T cell responses. J Immunol. 2009;182:6799–806.
985. Criscitiello MF, Ohta Y, Graham MD, Eubanks JO, Chen PL, Flajnik MF. Shark class II invariant chain reveals ancient conserved relationships with cathepsins and MHC class II. Dev Comp Immunol. 2012;36:521–33.
986. Flajnik MF, Tlapakova T, Criscitiello MF, Krylov V, Ohta Y. Evolution of the B7 family: co-evolution of B7H6 and NKp30, identification of a new B7 family member, B7H7, and B7's historical relationship with the MHC. Immunogenetics. 2012;64:571–90.
987. Holen E, Lie KK, Araujo P, Olsvik PA. Pathogen recognition and mechanisms in Atlantic cod (Gadus morhua) head kidney cells: bacteria (LPS) and virus (poly I:C) signals through different pathways and affect distinct genes. Fish Shellfish Immunol. 2012;33:267–76.

988. Magnadottir B, Audunsdottir SS, Bragason BT, Gisladottir B, Jonsson ZO, Gudmundsdottir S. The acute phase response of Atlantic cod (Gadus morhua): humoral and cellular response. Fish Shellfish Immunol. 2011;3:1124–30.
989. Star B, Nederbragt AJ, Hentoft S, Grimholt U, Malstrøm M, Gregers TF, Rouinge TB, Paulsen J, Solbakken MH, Sharma A, Wetten OF, Lanzén A, Winer R, Kight J, Vogel JH, Aken B, Andersen O, Lagesen K, Tooming-Klunderud A, Edvardsen RB, Tiria KG, Espelund M, Nepal C, Previti C, Karlsen BO, Moum T, Skage M, Berg PR, Gjøen T, Kuhl H, Thorsen J, Malde K, Reinhardt R, Du L, Johansenm SD, Searle S, Lkien S, Nilsen F, Jonassen T, Omholt SW, Sternseth NC, Jakobsen KS. The genome sequence of Atlantic cod reveals a unique immune system. Nature. 2011;477:207–10.
990a. Dreyfus DH. Evidence suggesting an evolutionary relationship between transposable elements and immune system recombination sequences. Mol Immunol. 1992;29:807–10.
990b. Dreyfus DH, Jones JF, Gelfand EW. Asymmetric DDE (D35E)-like sequences in the RAG proteins: implications for V(D)J recombination and retroviral pathogenesis. Med Hypotheses. 1999;52:545–9.
990c. Dreyfus DH, Nagasawa M, Kelleher CA, Gelfand EW. Stable expression of Epstein-Barr virus BZLF-1 encoded ZEBRA protein activates p53-dependent transcription in human Jurkat T-lymphoblastoid cells. Blood. 2000;96:625–34.
991. McGeoch DJ, Cook SD, Colan A, Jamieson FE, Elford EA. Molecular phylogeny and evolutionary timescale for the family of mammalian herpesviruses. J Mol Biol. 1995;247:443–58.
992. Niller HH, Salamon D, Rahmann S, Ilg K, Koroknai A, Bánáti F, Schwarzmann F, Wolf H, Minárovits J. A 30kb region of the Epstein-Barr virus genome is colinear with the rearranged human immunoglobulin gene loci: implications for a "ping-pong evolution" model for persisting viruses and their hosts. A review. Acta Microbiol Immunol Hung. 2004;51:469–84.
993. Forterre P. Defining life: the virus viewpoint. Orig Life Evol Biosci. 2010;40:151–60.
994a. Niller HH, Salamon D, Ilg K, Koroknai A, Banati F, Baum G, Rucker O, Schwarzmann F, Wolf GH, Minarovits J. The in vivo binding site for oncoprotein c-Myc in the promoter for Epstein-Barr virus (EBV) encoding RNA (EBER) 1 suggests a specific site for EBV in lymphomagenesis. Med Sci Monit. 2003;9:1–9.
994b. Szenthe K, Koroknai A, Banati F, Bathori Z, Lozsa R, Burgyan J, Wolf H, Salamon D, Nagy K, Niller HH, Minarovits J. The 5' regulatory sequences of active miR-146a promoters are hypomethylated and associated with euchromatic histone modification markers in B lymphoid cells. Biochem Biophys Res Commun. 2013;433:489–95.
995. Dreyfus DH, Liu Y, Ghoda LY, Chang JT. Analysis of an ankyrin-like region in Epstein Barr virus encoded (EBV) BZLF-1 (ZEBRA) protein: implications for interactions with NFκB and p53. Virol J. 2011;8:42.
996a. Dishaw LJ, Mueller MG, Gwatney N, Cannon JP, Haire RN, Litman RT, Amemiya CT, Ota T, Rowen L, Glusman G, Litman GW. Genomic complexity of the variable region-containing chitin-binding proteins in amphioxus. BMC Genet. 2008;9:78.
996b. Dishaw LJ, Haire RN, Litman GW. The amphioxus genome provides unique insight into the evolution of immunity. Brief Funct Genomics. 2012;11:167–76.
997. Goodwin TJ, Poulter RT. A group of deuterostome Ty3/gypsy-like retrotransposons with Ty1/copia-like pol-domain orders. Mol Genet Genomics. 2002;267:481–91.
998. Savin KW, Cocks BG, Wong F, Sawbridge T, Cogan N, Savage D, Warner S. A neurotropic herpesvirus infecting the gastropod, abalone, shares ancestry with oyster herpesvirus and a herpesvirus associated with the amphioxus genome. Virol J. 2010;10 (7):308.
999. Arzul I, Renault T, Lipart C. Experimental herpes-like viral infections in marine bivalves: demonstration of interspecies transmission. Dis Aquat Organ. 2001;46:1–6.

1000. Batista FM, Arzul I, Pepin JF, Ruano F, Friedman CS, Boudry P, Renault T. Detection of ostreid herpesvirus 1 DNA by PCR in bivalve mollusks: a critical review. J Virol Methods. 2007;139:1–11.
1001. Gs Richards, Simionato E, Perron M, Adamska M, Vervoort M, Degnan SM. Sponge genes provide new insight into the evolutionary origin of the neurogenic circuit. Curr Biol. 2008;18:1156–61.
1002a. Davison AJ. Herpesvirus systematics. Vet Microbiol. 2010;143:52–69.
1002b. Davison AJ, Davison MD. Identification of structural proteins of channel catfish virus by mass spectrometry. Virology. 1995;206:1035–43.
1002c. Davison AJ, Sauerbier W, Dolan A, Addison C, McKinnell RG. Genomic studies of the Lucké herpesvirus (RaHV-1). J Cancer Res Clin Oncol. 1999;125:232–8.
1002d. Davison AJ, Trus BL, Cheng N, Steven AC, Watson MS, Cunningham C, Le Deuff RM, Renault T. A novel class of herpesvirus with bivalve hosts. J Gen Virol. 2005;86:41–53.
1003a. van Beurden SJ, Bossers A, Voorbergen-Laarman MH, Haenen OL, Peters S, Abma-Henkens MH, Peeters BP, Rottier PJ, Engelsma MY. Complete genome sequence and taxonomic position of anguillid herpesvirus 1. J Gen Virol. 2010;91:880–7.
1003b. van Beurden SJ, Leroy B, Wattiez R, Haenen OL, Boeren S, Vervoort JJ, Peeters BP, Rottier PJ, Engelsma MY, Vanderplasschen AF. Identification and localization of the structural proteins of anguillid herpesvirus 1. Vet Res. 2011;42(1):105. doi:10.1186/1297-9716-42-105.
1003c. van Beurden SJ, Gatherer D, Kerr K, Galbraith J, Herzyk P, Peeters BP, RottierOJ Engelsma MY, Davison AJ. Anguillid herpesvirus 1 transcriptome. J Virol. 2012;86:10150–61.
1004. Waltzek TB, Kelley GO, Alfaro ME, Kurobe T, Davison AJ, Hedrick RP. Phylogenetic relationships in the family Alloherpesviridae. Dis Aquat Organ. 2009;84:179–94.
1005. Papermaster BW, Condie RM, Finstad J, Good RA. Evolution of the immune response. I. The phylogenetic development of adaptive immunologic responsiveness in vertebrates. J Exp Med. 1964;119:105–30.
1006. Granoff A. Herpesvirus and the Lucké tumor. Cancer Res. 1973;33:1431–3.
1007. Govett PD, Harms CA, Johnson AJ, Latimer KS, Wellahan JF, Fatzinger MH, Christian LS, Kelly TR, Lewbart GA. Lymphoid follicular cloacal inflammation associated with a novel herpesvirus in juvenile alligators (Alligator mississippiensis). J Vet Diagn Invest. 2005;17:474–9.
1008. McCowan C, Shepherdley C, Slocombe RF. Herpesvirus-like particles in the skin of a saltwater crocodile (Crocodylus porosus). Aust Vet J. 2004;82:357–77.
1009. Stacy BA, Wellehan JF, Foley AM, Coberley SS, Herbst LH, Manire CA, Garner MM, Brookins MD, Childress AL, Jacobson ER. Two herpesviruses associated with disease in wild Atlantic loggerhead sea turtles (Caretta caretta). Vet Microbiol. 2008;126:63–73.
1010. Arbello M, Sierra E, Esperón F, Watanabec TT, Bellière EN, Espinosa de los Monteros A, Fernández. Herpesvirus infection with severe lymphoid necrosis affecting a beaked whale stranded in Canary Islands. Dis Aquat Organ. 2010;89:261–4.
1011. Fossum E, Friedel CC, Rajagopala SV, Titz B, Baiker A, Schmidt T, Kraus T, Stellberger T, Frutenberg Sutham S, Bandyopadhyay S, Rose D, von Brunn A, Uhlmann M, Zeretzke C, Dong YA, Boulet H, Koegl M, Bailer SM, Koszinowski U, Ideker T, Uetz P, Zimmer R, Haas J. Evolutionarily conserved herpesviral protein interaction networks. PLoS Pathog. 2009;5(9):1000570.
1012. Hamblin C, Hedger RS. Prevalence of neutralizing antibodies to bovid herpesvirus 2 in African wildlife. J Wildl Dis. 1982;18:429–36.
1013. Roychoudhury S, Mukherjee D. A detailed comparative analysis on the overall codon usage pattern in herpesviruses. Virus Res. 2010;148:31–43.
1014. Maness H, Nollens HH, Jensen ED, Goldstein T, LaMere S, Childress A, Sykes J, St Leger J, Lacave G, Latson FE, Wellehan JF Jr. Phylogenetic analysis of marine mammal herpesviruses. Vet Microbiol. 2011;149:23–9.

1015. Miyoshi K, Nishida S, Sone E, Tajim Y, Makara M, Yoshioka M, Nakamura M, Yamada TK, Koike H. Molecular identification of novel alpha-and gammaherpesviruses from cetaceans stranded on Japanese coasts. Zoolog Sci. 2011;28:126–33.
1016. Dong M, Fu Y, Yu C, Su J, Huang S, Wu X, Wei J, Yuan S, Shen Y, Xu A. Identification and characterization of a homolog of an activation gene for the recombination activating gene 1 (RAG1) in amphioxus. Fish Shellfish Imunol. 2005;19:165–74.
1017. Athineos D, Marshall L, White RJ. Regulation of TFIIIB during F9 differentiation. BMC Mol Biol. 2010;12(11):212.
1018. Codina A, Love JD, Li Y, Lazar MA, Neuhaus D, Schwabe JW. Structural insights into the interaction and activation of histone deacetylase 3 by nuclear receptor coreceptors. Proc Natl Acad Sci USA. 2005;102:6009–14.
1019. Euskirchen G, Auerbach RK, Snyder M. SWI/SNF chromatin-remodeling factors; multiscale analyses and diverse functions. J Biol Chem. 2012;287:30897–905.
1020. Ghisletti S, Huang W, Jepsen K, Benner C, Hardiman G, Rosenfeld MG, Glass CK. Cooperative NCoR/SMRT interactions establish a corepressor-based strategy for integration of inflammatory and anti-inflammatory signaling pathways. Genes Dev. 2009;23:681–93.
1021. Kapitonov W, Jurka J. Harbinger transposons and an ancient HARBI1 gene derived from a tansposase. DNA Cell Biol. 2004;23:311–24.
1022. Noguchi C, Rapp JB, Skorobogatko YV, Bailey LD, Noguchi E. Swi1 associates with chromatin through CCT domain and recruits Swi3 to preserve genomic integrity. PLos One. 2012;7(8):e3988.
1023a. Sinzelle L, Izsvák Z, Ivics Z. Molecular domestication of transposable elements: from detrimental parasites to useful host genes. Cell Mol Life Sci. 2009;66:1073–93.
1023b. Sinzelle L, Kapitonov VV, Grzela DP, Jursch T, Jurka J, Izsvák Z, Ivics Z. Transposition of a reconstructed Harbinger element in human cells and functional homology with two transposons-derived cellular genes. Proc Natl Acad Sci. 2008;105:4715–20.
1024. Yu J, Li Y, Ishizuka T, Guenther MG, Lazar MA. A SANT motif in the SMRT corepressor interprets the histone code and promotes histone deacetylation. EMBO J. 2003;22:3403–10.
1025. Zhang H, Kruk JA, Reese JC. Dissection of coactivator reqirement at RNR3 reveals unexpected contributions from TFIID and SAGA. J Biol Chem. 2008;283:27360–8.
1026. Baker AM, Fu Q, Hayward W, Lindsay SM, Fletcher TM. The Myb/SANT domain of the telomere-binding protein TRF2 alters chromatin structure. Nucleic Acids Res. 2009;37:5019–31.
1027. Ryan DP, Sundaramoorthy R, Martin D, Singh V, Owen-Hughes T. The DNA-binding domain of the Chd1 chromatin-remodeling enzyme contains SANT and SLIDE domains. EMBO J. 2011;30:2596–609.
1028a. Ivics Z. Recombinase technology for gene therapy. Curr Gene Ther. 2011;11(5):331.
1028b. Ivics Z, Izsvák Z. Nonviral gene delivery with the sleeping beauty transposons system. Hum Gene Ther. 2011;22:1043–51.
1029. Lv B, Shi T, Wang X, Song Q, Zhang Y, Shen Y, Ma D, Lou Y. Overexpression of the novel human gene, nuclear apoptosis-inducing factor1, induces apoptosis. Int J Biochem Cell Biol. 2006;38:671–83.
1030. Smith JJ, Sumiyama K, Ameyima CT. A living fossil in the genome of a living fossil: Harbinger transposons in the Coelacanth genome. Mol Biol Evol. 2012;29:985–93.
1031. Chiu L-W, Zhou X, Burke S, Wu X, Prior RL, Li L. The purple cauliflower arises from activation of a MYB transcription factor. Plant Physiol. 2010;154:1470–80.
1032. Dmochowski L, Grey CE, Burmester BR, Walter WG. Submicroscopic morphology of avian neoplasms. II. Studies on granuloblastosis (myeloblastosis). Proc Soc Exp Biol Med. 1958;98:666–9.

1033a. Beard D, Beaudreau GS, Bonar RA, Sharp DG, Beard JW. Virus of avian erythroblastosis. III. Antigenic constitution and relation to the agent of avian myeloblastosis. J Natl Cancer Inst. 1957;18:231–59.

1033b. Beard JW, Chabot JF, Beard D, Heine U, Houts GE. Renal neoplastic response to leukosis virus strain BAI A avian myeloblastosis virus and MC29. Cancer Res. 1976;35:339–53.

1034a. Heine U, Bonar RA, Beard JW. Aberrant cytoplasmic structures in avian virus tumor cells. Exp Mol Pathol. 1965;76:81–97.

1034b. Heine U, Beard D, Beard JW. Ultracytochemical studies on aberrant structures in avian virus tumor cells. Cancer Res. 1968;28:585–94.

1035. Castellano M, Golay J, Mantovani A, Introna M. Detection of a transcriptional block in the first intron of the human c-myb gene. Int J Clin Lab Res. 1992;22:159–64.

1036. Quelen C, Lippert E, Struski S, Demur C, Soler G, Prade N, Delabesse E, Broccardo C, Dastugue N, Mahon FX, Brousset P. Identification of a transforming MYG-GATA1 fusion gene in acute basophilic leukemia: a new entity in male infants. Blood. 2011;117:5719–22.

1037. Miyazaki T, Pan Y, Joshi K, Purohit D, Hu B, Demir H, Mazumder S, Okabe S, Yamori T, Viapiano M, Shin-ya K, Seimiya H, Nakano I. Telomestatin impairs glioma stem cell survival and growth through the disruption of telomeric G-quadruplex and inhibition of the proto-oncogene, c-Myb. Clin Cancer Res. 2012;18:1268–80.

1038. Chu CH, Chang LC, Hsu HM, Wei SY, Liu HW, Lee Y, Kuo CC, Indra D, Chen C, Ong SJ, Tai JH. A highly organized structure mediating nuclear localization of a Myb2 transcription factor in the protozoan parasite Trichomonas vaginalis. Eukaryot Cell. 2011;10:1607–17.

1039. de Souza RF, Iyer LM, Aravind L. Diversity and evolution of chromatin proteins encoded by DNA viruses. Biochem Biophys Acta. 2010;1799:302–18.

1040. Colson P, de Lamballerie X, Fournous G, Raoult D. Reclassification of giant viruses composing a fourth domain of life in the new order of Megavirales. Intervirology. 2012;55:321–32.

1041. Thomas V, Bertelli C, Collyn F, Casson N, Telenti A, Goesmann A, Coxatto A, Greub G. Lausannevirus, a giant amoebal virus encoding histone doublets. Environ Microbiol. 2011;13:1454–66.

1042. Williams TA, Embley TM, Heinz E. Informational gene phylogenies do not support a fourth domain of life for nucleocytoplasmic large DNA viruses. PLoS One. 2011;6(6): e21080.

1043. Abhiman S, Iyer LM, Aravind L. BEN: a novel domain in chromatin factors and DNA viral proteins. Bioinformatics. 2008;24:458–61.

1044. Gram H, Rüger W. The alpha-glucosyltransferases of bacteriophages T2, T4, and T6. A comparison of their primary structures. Mol Gen Genet. 1986;202:467–70.

1045. Yoshida T, Claverie JM, Ogata H. Mimivirus reveals MRE11/Rad50 fusion proteins with a sporadic distribution in eukaryotes, bacteria, viruses and plasmids. Virol J. 2011;8:427.

1046. de Jager M, Trujillo KM, Sung P, Hopfner KP, Carney JP, Tainer JA, Connelly JC, Leach DR, Kanaar R, Wyman C. Differential arrangements of conserved building blocks among homologs of the Rad50/Mre11 DNA repair protein complex. J Mol Biol. 2004;339:937–49.

1047. Connelly JC, de Leau ES, Leach DR. Nucleolytic processing of a protein-bound DNA end by the E. coli SbcCD (MR) complex. DNA Repair (Amst). 2003;2:795–897.

1048. Darmon E, Dorenbos R, Meens J, Freudi R, Antelmann H, Hecker M, Kuiper OP, Bron S, Quax WJ, Dubois JY, van Dijl JM. A disulfide bond-containing alkaline phosphatase triggers a BDbC-dependent secretion stress response in Bacillus subtilis. Appl Environ Microbiol. 2006;72:6876–85.

1049. Fujisawa M, Wada Y, Tsuchiya T, Ito M. Characterization of Bacillus subtilis YfkE (ChaA): a calcium-specific Ca2+/H+ antiporter of the CaCA family. Arch Microbiol. 2009;191:649–57.
1050. Curtis FA, Reed P, Sharples GJ. Evolution of a phage RuvC endonuclease for resolution of both Holiday and branched DNA junctions. Mol Microbiol. 2005;55:1332–45.
1051. Storvik KA, Foster PL. The SMC-like protein complex SbcCD enhances DNA polymerase IV-dependent spontaneous mutation in Escherichia coli. J Bacteriol. 2011;193:660–9.
1052. Yamada K, Fukuoh A, Iwasaki H, Shinagawa H. Novel properties of the Thermus thermophilus RuvB protein, which promotes branch migration of Holiday junctions. Mol Gen Genet. 1999;261:1001–11.
1053. Zahradka K, Zahradka D, Petranović M. Loss of lambda prophage recombinogenicity in UV-irradiated Escherichia coli: the role of host genes ruvA, ruvB, ruvC, and ruvG. Res Microbiol. 2001;152:873–81.
1054. Leipe DD, Koonin EV, Aravind L. STAND, a class of P-lop NTPases including animal and plant regulators of programmed cell death: multiple, complex domain architectures, unusual phyletic patterns, and evolution by horizontal gene transfer. J Mol Biol. 2004;343:1–28.
1055. Qian C, Zhou MM. SET domain protein lysine methyltransferase: Structure, specificity and catalysis. Cell Mol Life Sci. 2006;63:2755–63.
1056. Zhu X, Ma H, Chen Z. Phylogenetics and evolution of Su(var)3-9 SET genes in land plants: rapid diversification in structure and function. BMC Evol Biol. 2011;11:63.
1057. Aravind L, Abhiman S, Iyer LM. Natural history of the eukaryotic chromatin protein methylation system. Prog Mol Biol Transl Sci. 2011;101:105–16.
1058. Mujtaba S, Manzur KL, Gurnon JR, Kang M, Van Etten JL, Zhou MM. Epigenetic transcriptional repression of cellular genes by a viral SET protein. Nat Cell Sci. 2008;10:1114–22.
1059. O'Connell PA, Madureira PA, Berman JN, Liwski RS, Waisman DM. Regulation of S100A10 by the PML-RAR-alpha oncoprotein. Blood. 2011;117:4095–105.
1060. Chen SJ, Zhou GB, Zhang XW, Mao JH, de Thé H, Chen Z. From an old remedy to a magic bullet: molecular mechanism underlying the therapeutic effects of arsenic in fighting leukemia. Blood. 2011;117:6425–37.
1061. He P, Liu Y, Zhang M, Wang X, Xi J, Wu D, Li J, Cao Y. Interferon-gamma enhances promyelocytic leukemia protein expression in acute promyelocytic cells and cooperates with allotrans-retinoic acid to induce maturation of NB4 and NB4-R1 cells. Exp Ther Med. 2012;3:776–80.
1062. Iriyama N, Yuan B, Hatta Y, Horikoshi A, Yoshino Y, Toyoda H, Aizawa S, Takeuchi J. Granulocyte colony-stimulating factor potentiates differentiation induction by all-trans retinoic acid and arsenic trioxide and enhances arsenic uptake in the acute promyelocytic leukemia cell line HT93A. Oncol Rep. 2012;28:1875–82.
1063. Ghaffari SH, Momeny M, Bashash D, Mirzaei R, Ghavamzadeh A, Alimoghaddam K. Cytotoxic effect of arsenic trioxide on acute promyelocytic leukemia cells through suppression of NFkB-dependent induction of hTERT due to down-regulation of Pin1 transcription. Hematology. 2012;17:198–206.
1064. Suliman BA, Xu D, Williams BR. The promyelocytic leukemia zink finger protein: two decades of molecular oncology. Front Oncol. 2012;2:74. doi:10.3389/fonc.2012.00074.
1065. Lynch EA, Langille MG, Darling A, Wilbanks EG, Haltiner C, Shao KS, Starr MO, Teiling C, Harkins TT, Edwards RA, Eisen JA, Facciotti MT. Sequencing of seven haloarchaeal genomes reveals patterns of genomic flux. PLoS One. 2012;7(7):e41389.
1066. Andam CP, Harlow TJ, Papke RT, Gogarten JP. Ancient origin of the divergent forms of leucyl-tRNA synthesases in the Halobacteriales. BMC Evol Biol. 2012;12:85.
1067. Aparicio T, Megías D, Méndez J. Visualization of the MCM DNA helicase at replication factories before the onset of DNA synthesis. Chromosoma. 2012;121:499–507.

1068. de Castro IF, Volonté L, Risco C. Virus factories: biogenesis and structural design. Cell Microbiol. 2013;15:24–34. doi:10.1111/cmi.12029.
1069. Erickson KD, Bouchet-Marquis C, Heiser K, Szomolanyi-Tsuda E, Mishra R, Lamothe B, Hoenger A, Garcea RL. Virion assembly factories in the nucleus of polyoma-infected cells. PLoS Pathog. 2012;8(4):e1002630.
1070. Netherton CL, Wileman T. Virus factories, double membrane vesicles and viroplasm generated in animal cells. Curr Opin Virol. 2011;1:381–7.
1071. Song W, Pascal JM, Ellenberger T, Tomlinson AE. The DNA binding domain of human DNA ligase I interacts with both nicked DNA and the DNA sliding clamps, PCNA and hRad9-hRad1-hHus1. DNA Repair (Amst). 2009;8:912–9.
1072. Gupta A, Mehra P, Dhar SK. Plasmodium falciparum origin recognition complex subunit 5: functional characterization and role in DNA replication foci formation. Mol Microbiol. 2008;69:646–65.
1073. Kaufmann D, Gassen A, Maiser A, Leonhardt H, Janzen CJ. Regulation and spatial organization of PCNA in Trypanosoma brucei. Biochem Biophys Res Commun. 2012;419:698–702.
1074. Kilbey BJ, Fraser I, McAleese S, Goman M, Ridley RG. Molecular characterization and stage-specific expression of proliferating cell nuclear antigen (PCNA) from the malarial parasite, Plasmodium falciparum. Nucleic Acids Res. 1993;21:239–43.
1075. Kumar P, Wang CC. Dissociation of cytokinesis initiation from mitotic control in a eukaryote. Eukaryot Cell. 2006;5:92–102.
1076. Minocha N, Kumar D, Rajanala K, Saha S. Characterization of Leishmania donovani MCM4: expression patterns and interaction with PCNA. PLoS One. 2011;6(7):e23107.
1077. Acherya N, Klassen R, Johnson RE, Prakash L, Prakash S. PCNA binding domain in all three subunits of yeast DNA polymerase delta modulate its function in DNA replication. Proc Natl Acad Sci USA. 2011;108:17927–32.
1078. Li Z, Umeyama T, Wang CC. The Aurora kinase in Trypanosoma brucei plays distinctive roles in metaphase-anaphase transition and cytokine initiation. PLoS Pathog. 2009;5(9):e1000575.
1079. Hammarton TC, Kramer S, Tetley L, Boshart M, Mottram JC. Trypanosoma brucei Polo-like kinase is essential for basal body duplication, kDNA segregation and cytokinesis. Mol Microbiol. 2007;65:1229–48.
1080. Saeki T, Ouchi M, Ouchi T. Physiological and oncogenic Aurora-A pathway. Int J Biol Sci. 2009;5:758–62.
1081. Eckerdt F, Yuan J, Strebhardt K. Polo-like kinases and oncogenesis. Oncogene. 2005;24:267–76.
1082. Jang YJ, Kim YS, Kim WH. Oncogenic effect of Polo-like kinase 1 expression in human gastric carcinomas. Int J Oncol. 2006;29:589–94.
1083. Takai N, Hamanaka R, Yoshimatsu J, Miyakawa I. Polo-like kinases (Plks) and cancer. Oncogene. 2005;24:287–91.
1084. Weichert W, Schmidt M, Gekeler V, Denkert C, Stephan C, Jung K, Loening S, Dietel M, Kristiansen G. Polo-like kinase1 is overexpressed in prostate cancer and linked to higher tumor grades. Prostate. 2004;60:240–5.
1085. Khan J, Khan S, Attaullah S, Ali I, Khan SN. Aurora kinase-CT191D is constitutively active mutant. BMC Cell Biol. 2012;13:8. doi:10.1186/1471-2121-13-8.
1086. den Hollander J, Rimpi S, Doherty JAR, Rudelius M, Buck A, Hoellein A, Kremer M, Graf N, Scheerer M, Hall MA, Goga A, von Bubnoff N, Duyster J, Peschel C, Cleveland JL, Nilsson JA, Keller U. Aurora kinases A and B are up-regulated by Myc and are essential for maintenance of the malignant state. Blood. 2010;116:1496–505.
1087a. Ricke RM, Jeganathan KB, van Deursen JM. Bub1 overexpression induces aneuploidy and tumor formation through Aurora B kinase hyperactivation. J Cell Biol. 2011;193:1049–64.

1087b. Ricke RM, van Deursen JM. Aurora B hyperactivation by Bub1 overexpression promotes chromosome missegregation. Cell Cycle. 2011;10:3645–51.

1088. Lazzaro F, Novarina D, Amara F, Watt DL, Stone JE, Cistanzo V, Burgers PM, Kunkel TA, Plevani P, Muzi-Falconi M. RNAase H and postreplication repair protect cells from ribonucleotides incorporated in DNA. Mol Cell. 2012;45:99–110.

1089. Muramoto T, Chubb JR. Live imaging of the Dictyostelium cell cycle reveals widespread S phase during development, a G2 bias in spore differentiation and a premitotic checkpoint. Development. 2008;135:1647–57.

1090. Zamir L, Zaretsky M, Fridman Y, Ner-Gaon H, Rubin E, Aharoni A. Tight coevolution of proliferating cell nuclear antigen (PCNA)-partner interaction network in fungi leads to interspecies network incompatibility. Proc Natl Acad Sci USA. 2012;109:E406–14.

1091. Lin S, Carpenter EJ. Identification and preliminary characterization of PCNA gene in the phytoplankton Dunaliella tertiolecta and Isochrysis galbana. Mol Mar Biol Biotechnol. 1998;7:62–71.

1092. Zhang H, Hou Y, Lin S. Isolation and characterization of proliferating cell nuclear antigen from the dinoflagellate Pfiesteria piscida. J Eukaryot Microbiol. 2006;53:142–50.

1093. Philipova R, Whitaker M. Active ERK1 is dimerized in vivo: bisphosphodimers generate peak kinase activity and monophosphodimers maintain basal ERK1 activity. J Cell Sci. 2005;118:5767–76.

1094. Jeffery WR. Role of PCNA and ependymal cells in ascidian neural development. Gene. 2002;287:97–105.

1095. Gerenday A, Fallon AM. Increased levels of the cell cycle inhibitor protein, decapo, accompany 20-hydroxyecdysone-induced GH1 arrest in a mosquito cell line. Arch Insect Biochem Physiol. 2011;8:61–73.

1096. Ruike T, Takeuchi R, Takata K, Oshige M, Kasai N, Shimanouchi K, Kanai Y, Nakamura R, Sugawara F, Sakaguchi K. Characterization of a special proliferating cell nuclear antigen (PCNA2) from Drosophila melanogaster. FEBS J. 2006;273:5062–73.

1097. Boerckel J, Walker D, Ahmed S. The Caenorhabditis elegans Rad 17 homolog HRP-17 is required for telomere replication. Genetics. 2007;176:703–9.

1098. Mahler M, Miyachi K, Peebles C, Fritzler MJ. The clinical significance of autoantibodies to the proliferating cell nuclear antigen (PCNA). Autoimmun Rev. 2012;11:771–5.

1099. Ku DH, Travali S, Calabretta B, Huebner K, Baserga R. Human gene for proliferating cell nuclear antigen has pseudogenes and localizes to chromosome 20. Somat Cell Mol Genet. 1989;15:297–307.

1100. Stoimenov I, Lagerqvist A. The PCNA pseudogene in the human genome. BMC Res Notes. 2012;5:87.

1101. Webb G, Parsons P, Chenevix-Trench G. Localization of the gene for human proliferating nuclear antigen/cyclin by in situ hybridization. Hum Genet. 1990;86:84–6.

1102. Zhao H, Ho PC, Lo YH, Espejo A, Bedford MT, Hung MC, Wang SC. Interaction of proliferation cell nuclear antigen (PCNA) with c-Abl in cell proliferation and response to DNA damages in breast cancer. PLoS One. 2012;7(1):e29416.

1103. Gioeli D, Wuderlich W, Sebolt-Leopold J, Bekiranov S, Wulfkufle JD, Petricoin EF 3rd, Conaway M, Weber MJ. Compensatory pathways induced by MEK inhibition are effective drug targets for combination therapy against castration-resistant prostate cancer. Mol Cancer Ther. 2011;10:1581–90.

1104. Thomas RS, Sarwar N, Phoenix F, Coombes RC, Ali S. Phosphorylation at serine 104 and 106 by Erk1/2 MAPK is important for estrogen receptor-alpha activity. J Mol Endocrin. 2008;40:173–84.

1105. Lin S-L, Yan L-Y, Zhang X-T, Yuan J, Li M, Qiao J, Wang Z-Y, Sun Q-Y. ER-alpha36, a variant of ER-α, promotes tamoxifen agonist action in endometrial cancer cells via the MAPK/ERK and PI3K/Akt pathways. PLoS One. 2010;5(2):e9013.

1106. Chen Z, Liu P, Li C, Luo Y, Chen I, Liang W, Chen X, Feng Y, Xia H, Wang F. Deregulated expression of the clock genes in gliomas. Technol Cancer Res Treat. 2012 Aug 10 Epub.
1107. Glading A, Chang P, Lauffenburger DA, Wells A. Epidermal growth factor receptor activation of calpain is required for fibroblast motility and occurs via an ERK/MAP kinase signaling pathway. J Biol Chem. 2000;275:2390–6.
1108. Philipova R, Kisielewska J, Lu P, Larman M, Huang JY, Whitaker M. ERK1 activation is required for S-phase onset and cell cycle progression after fertilization in sea urchin embryos. Development 132:579–589.
1109. Gali H, Juhasz S, Morocz M, Hajdu I, Fatyol K, Szukacsov V, Burkovics P, Haracska L. SUMO modification of human PCNA at stalled replication fork. Nucleic Acids Res. 2012;40:6049–59.
1110. Xu ZW, Wang FM, Gao MJ, Chen XY, Shan NN, Cheng SX, Mai X, Zala GH, Hu WL, Xu RC. Cardiotonic steroids attenuate ERK phosphorylation and generate cell cycle arrest to block human hepatoma cell growth. J Steroid Biochem Mol Biol. 2011;126:181–91.
1111. Rosental B, Brusilovsky M, Hadad U, Oz D, Appel MY, Afergan F, Yossef R, Rosenberg LA, Aharoni A, Cerwenka A, Campbell KS, Braiman A, Porgador A. Proliferating cell nuclear antigen is a novel inhibitory ligand for the natural cytotoxicity receptor NKp44. J Immunol. 2011;187:5693–792.
1112. Chalamcharia VR, Curcio MJ, Belfort M. Nuclear expression of a group II intron is consistent with spliceosomal intron ancestry. Genes Dev. 2010;15:827–36.
1113. Bachvaroff TR, Place AR. From stop to start: tandem gene arrangement, copy number and trans-splicing sites in the dinoflagellate Amphidium carterae. PLoS One. 2008;3: e2929.
1114. Adl SM, Simpson AGB, Lane CE, Lukes J, Bass D, Bowser SS, Brown MW, Burki F, Dunthorn M, Hampl V, Heiss A, Hoppenrath M, Lara E, le Gall L, Lynn DH, McManus H, Mitchell EAD, Mozley-Stanridge SE, Parfrey LW, Pawlowski J, Rueckert S, Shadwick L, Schoch CL, Smirnov A, Spiegel FW. The revised classification of Eukaryotes. J Eukaryiot Microbiol. 2012;59:4290514.
1115. Roy SW, Hudson AJ, Joseph J, Yee J, Russell AG. Numerous fragmented spliceosomal introns.: AT-AC splicing, and an unusual dynein gene expression pathway in Giardia lamblia. Mol Biol Evol. 2012;29:43–9.
1116a. Kamikawa R, Inagaki Y, Tokoro M, Roger AJ, Hashimoto T. Split introns in the genome of Giardia intestinalis are excised by spliceosome-mediated trans-splicing. Curr Biol. 2011;21:311–5.
1116b. Kamikawa R, Inagaki Y, Roger AJ, Hashimoto T. Splintrons in Giardia intestinalis; spliceosomal introns in a split form. Commun Integr Biol. 2011;4:464–6.
1117. Nixon JE, Wang A, Morrison HG, McArthur AG, Sogin ML, Loftus BJ, Samuelson J. A spliceosomal intron in Giardia lamblia. Proc Natl Acad Sci USA. 2002;99:3701–5.
1118. Vanácova S, Yan W, Carlton JM, Johnson PJ. Spliceosomal introns in the deep-branching eukaryote Trichomonas vaginalis. Proc Natl Acad Sci USA. 2005;102:4430–2235.
1119. Russell AG, Shutt TE, Watkins RF, Gray MW. An ancient spliceosomal intron in the ribosomal protein L7a gene (Rpl7a) of Giardia lamblia. BMC Evol Biol. 2005;5:45.
1120. Hudson AJ, Moore A, Elniski D, Joseph J, Yee J, Russell AG. Evolutionarily divergent spliceosomal snRNA and a conserved non-coding RNA processing motif in Giardia lamblia. Nucleic Acids Res. 2012; 40:10996–1006. doi:10.1093/nar/gks887.
1121. Deng XL, Xu MY, XuXY Ba-Thein W, Zhang RL, Fu YC. A 25-bc ancient spliceosomal intron in the TvRab1a gene of Trichomonas vaginalis. Int J Biochem Cell Biol. 2009;41:417–23.

1122. Simoes-Barbosa A, Meloni D, Wohlschlegel JA, Konarska MM, Johnson PJ. Spliceosomal snRNAs in the unicellular eukaryote Trichomonas vaginalis are structurally conserved but lack a 5'-cap structure. RNA. 2008;14:1617–31.
1123. Mukherjee M, Brown MT, McArthur AG, Johnson PJ. Proteins of the glycerine decarboxylase complex in the hydrogenosome of Trichomonas vaginalis. Eukaryot Cell. 2006;5:2062–71.
1124. Singh S, Singh G, Sagar N, Yadav PK, Jain PA, Gautam B, Wadhwa G. Insight into Trichomonas vaginalis genome evolution through metabolic pathway comparison. Bioinformation. 2012;8:189–95.
1125. Smith AJ, Lauwaet T, Davida BJ, Gilin FD. Giardia lamblia Nek1 and Nek2 kinases affect mitosis and excystation. Int J Parasitol. 2012;42:411–9.
1126. Siol O, Spaller T, Schiefner J, Winckler T. Genetically tagged TRES-A retrotransposons reveal high amplification rates and authentic target site preference in the Dictyostelium discoideum genome. Nucleic Acids Res. 2011;39:6608–19.
1127. Wikmark OG, Haugen P, Haugli K, Johansen SD. Obligatory group I introns with unusual features at positions 1949 and 2449 in nuclear LSU rDNA of Didymiaceae myxomycetes. Mol Phylogenet Evol. 2007;43:596–604.
1128. Cech TR. Ribozymes, the first 20 years. Biochem Soc Trans. 2002;30:1162–6.
1129. Wan T, Suh H, Russell R, Herschlag D. Multiple unfolding events during native folding of the Tetrahymena group I ribozyme. J Mol Biol. 2010;400:1067–77.
1130. Oganesian L, Bryan TM. Physiological relevance of telomeric G-quadruplex formation: a potential drug target. Bioassays. 2007;29:155–65.
1131. Scierer T, Henderson E. A protein from Tetrahymena thermophila that specifically binds parallel-stranded G4-DNA. Biochemistry. 1994;33:2240–6.
1132. Phan AT, Modi YS, Patel DJ. Two-repeat tetrahymena telomeric d(TGGGGTTGGGGT) sequence interconverts between asymmetric dimeric G-quadruplex in solution. J Mol Biol. 2004;338:93–102.
1133. Ban G, Song MS, Lee SW. Cancer cell targeting with mouse TERT-specific group I introns of Tetrahymena thermophila. J Microbiol Biotechnol. 2009;19:1070–6.
1134. Madinger CL, Collins K, Fields LG, Taron CH, Benner JS. Constitutive secretion in Tetrahymena thermophila. Eukaryot Cell. 2010;9:674–81.
1135. Pap E. The role of microvesicles in malignancies. In: Dittmar T, Zänker KS, editors. Cell fusion in health and disease. II. Cell Fusion in disease. Springer Verlag 2011. p. 183–199.
1136. Doležal D, Jirků M, Maslov DA, Lukeš J. Phylogeny of the bodonid flagellates (Kinetoplastida) based on small-subunit rRNA gene sequences. Int J System Evol Microbiol. 2000;50:1943–51.
1137. von der Heyden S, Cavalier-Smith T. Culturing and environmental DNA sequencing uncover hidden kinetoplastid biodiversity and a major marine clade within ancestrally freshwater Neobodo designis. Int J System Evol Microbiol. 2005;55:2605–21.
1138a. Palfi Z, Bindereif A. Immunological characterization and intracellular localization of trans-spliceosomal small nuclear ribonucleoproteins in Trypanosoma brucei. J Biol Chem. 1992;267:20159–63.
1138b. Palfi Z, Jaé N, Preusser C, Kaminska KH, Bujnicki JM, Lee JH, Günzl A, Kambach C, Urlaub H, Bindereif A. SMN-assisted assembly of snRNP-specific Sm cores in trans-spliceosomal. Genes Dev. 2009;23:1650–64.
1139. Siegel TN, Gunasekera K, Cross GA, Ochsenreiter T. Gene expression in Trypanosoma brucei: lessons from high throughput RNA sequencing. Trends Parasitol. 2011;27:434–41.
1140. Procházka E, Franko F, Polákova S, Sulo P. A complete sequence of Saccharomyces paradoxus mitochondrial genome that restores the respiration in S. cerevisiae. FEMS Yeast Res 2012; doi:10.1111/jk.1567-1364.2012.00833.
1141. Hughes SS, Buckley CO, Neafsey DE. Complex selection on introns size in Cryptococcus neoformans. Mol Biol Evol. 2008;25:247–53.

1142. Nielson CB, Friedman B, Birren B, Burge CB, Galagan JE. Patterns of introns gain and loss in fungi. PLoS Biol. 2004;2(12):e422.
1143. van der Burgt A, Severing E, de Wit PJ, Collemare J. Birth of new spliceosomal introns by multiplication of introner-like elements. Curr Biol. 2012;22:1260–5.
1144. Denoeud F, Henriet S, Mungpakdee S, Aury JM, da Silva C, Brinkmann H, Mikhaleva J, Olsen LC, Jubin C, Cañestro C, Bouquet JM, Danks G, Poulain J, Campsteijn C, Adamski M, Cross I, Yadetie F, Muffao M, et al. Plasticity of animal genome architecture unmasked by rapid evolution of a pelagic tunicate. Science. 2010;330:1381–8.
1145. Roy SW, Irimia M. Genome evolution: where do new introns come from? Curr Biol. 2012;22:R529–31.
1146. Moabbi AM, Agarwal N, El Kaderi B, Ansari A. Role for gene looping in intron-mediated enhancement of transcription. Proc Natl Acad Sci USA. 2012;109:8505–10.
1147. Slamovits CH, Keeling PJ. A high density of ancient spliceosomal introns in oxymonad excavates. BMC Evol Biol. 2006;6:34.
1148. Roy SW, Penny D. A very high fraction of unique intron positions in the intron-rich diatom Thalassiosira pseudonana indicates widespread intron gain. Mol Biol Evol. 2007;24:1447–57.
1149. Martin W, Koonin EV. Introns and the origin of nucleus-cytosol compartmentalization. Nature. 2006;440:41–5.
1150. Jékely G. Origin of the nucleus and Ran-dependent transport to safeguard ribosome biogenesis in a chimeric cell. Biol Direct. 2008;3:31.
1151a. Csurös M, Rogozin IB, Koonin EV. A detailed history of intron-rich eukaryotic ancestors inferred from a global survey of 100 complete genomes. PLoS Comput Biol. 2911;7(9): e1002150.
1151b. Csuros M, Rogozin IB, Koonin EV. Extremely intron-rich genes in the alveolate ancestors inferred with a flexible maximum-likelihood approach. Mol Biol Evol. 2008;25:903–11.
1152a. Jeffares DC, Penkett CJ, Bähler J. Rapidly regulated genes are intron poor. Trends Genet. 2008;24:375–8.
1152b. Penny D, Hoeppner MP, Poole AM, Jeffares DC. An overview of the introns-first theory. J Mol Evol. 2009;69:527–40.
1153. Attoui H, Jaafar FM, Belhouchet M, de Micco P, de Lamballerie X, Brussard CP. Micromonas pusilla reovirus: a new member of the family Reoviridae assigned to a novel proposed genus (minoreovirus). J Gen Virol. 2006;87:1375–83.
1154. Brussard CP, Noordeloos AA, Sandaa RA, Heldal M, Bratbak G. Discovery of a dsRNA virus infecting the marine photosynthetic protist Micromonas pusilla. Virology. 2004;31:280–91.
1155. Chen F, Suttle CA. Evolutionary relationship among large double-stranded DNA viruses that infect microalgae and other organisms as inferred from DNA polymerase genes. Virology. 1996;21:170–8.
1156. Clerissi C, Desdevises Y, Grimsley N. Prasinoviruses of the marine green algae Ostreococcus tauri are mainly species specific. J Virol. 2012;86:4611–9.
1157. Rogozin IB, Carmel L, Csuros M, Koonin EV. Origin and evolution of spliceosomal introns. Biol Direct. 2012;7:11. doi:10.1186/1745-6150-7-11.
1158. Lin K, Zhang DY. The excess of 5' introns in eukaryotic genomes. Nucleic Acids Res. 2005;33:6522–7.
1159. McManus HA, Lewis LA, Fučikova K, Haugen P. Invasion of protein coding genes by green algal ribosomal group I introns. Mol Phylogenet Evol. 2012;62:109–16.
1160. Kong S, Liu X, Fu L, Yu X, An C. I-PfoP31: a novel nicking HNH homing endonuclease encoded in the group I intron of the DNA polymerase gene in Phormidium foveolarum phage PF-WMP3. PLoS One. 2012;7(8):e43738. doi:10.1371/journal.pone.0043738.

1161. Taylor GK, Stoddard BL. Structural, functional and evolutionary relationship between homing endonucleases and proteins from their host organisms. Nucleic Acids Res. 2912;40:5189-5200.

1162a. Nozaki H, Ohta N, Yamada T, Takano H. Characterization of rbcL group IA introns from two colonial volvocalean species (Chlorophyceae). Plant Mol Biol. 1998;37:77–85.

1162b. Nozaki H, Takahara M, Nakazawa A, Kita Y, Yamada T, Takano H, Kawano S, Kato M. Evolution of rbcL group IA introns and intron open reading frames within the colonial Volvocales (Chlorophyceae). Mol Phylogenet Evol. 2002;23:326–38.

1163. Turmel M, Boulanger J, Lemieux C. Two group I intron with long internal open reading frames in the chloroplast psbA gene of Chlamydomonas reinhardtii. Nucleic Acids Res. 1989;17:3875–87.

1164. Wikmark OG, Einvik C, De Jonckheere JF, Johansen SD. Short term sequence evolution and vertical inheritance of the Naegleria twin ribozyme group I intron. BMC Evol Biol. 2006;6:39. doi:10.1186/1471-2148-6-39.

1165. Gogarten JP, Hilario E. Inteins, introns, and homing endonucleases: recent revelations about the life cycle of parasitic genetic elements. BMC Evol Biol 2006;6:94. doi:10.1186/1471-2148-6-94.

1166. Zimmerly S, Hausner G, Wu XC. Phylogenetic relationships among group II intron ORFs. Nucleic Acids Res. 2001;29:1238–50.

1167. Anraku Y, Hirata R, Wada Y, Ohya Y. Molecular genetics of the yeast vacuolar HATPase. J Exp Biol. 1992;172:67–81.

1168. Butler MI, Gray J, Goodwin TJ, Pouler RT. The distribution and evolutionary history of the PRP8 intein. BMC Evol Biol. 2006;5:42. doi:10.1186/147-2148-6-42.

1169. Rivero F, Dislich H, Glöckner G, Noegel AA. The Dictyostelium discoideum family of Rho-related proteins. Nucleic Acids Res. 2001;29:1068–79.

1170. Ogawa S, Yoshino R, Angata K, Iwamoto M, Pi M, Kuroe K, Matsuo K, Morio T, Urushihara H, Yanagisawa K, Tanaka Y. The mitochondrial DNA of Dictyostelium discoideum: complete sequence, gene content and genome organization. Mol Gen Genet. 2000;263:514–9.

1171. Angata K, Ogawa S, Yanmagashiwa K, Tanaka Y. A group-I intron in the mitochondrial large-subunit ribosomal RNA-encoding gene of Dictyostelium discoideum: same site localization in alga and in vitro self-splicing. Gene 1995;15349–55.

1172. Neuvéglise C, Marck C, Gaillardin C. The introns of budding yeasts. C R Biol. 2011;334:662–70.

1173. Yun C-S, Nishida H. Distribution of introns in fungal histone genes. PLoS One. 2011;6(1):e416548.

1174. Lee SK, Yu SL, Garcia MX, Alexander H, Alexander S. Differential developmental expression of the rep B and rep D xeroderma pigmentosum related DNA helicase genes from Dictyostelium discoideum. Nucleic Acids Res. 1997;25:2365–74.

1175. Szcześniak MW, Ciomborowska J, Nowak W, Rogozin IB, Makalowska I. Primate and rodent specific intron gains and the origin of retrogenes with splice variants. Mol Biol Evol. 2011;28:33–7.

1176. Au HC, Seao BB, Matsuno-Yagi A, Yasgi T, Scheffler IE. The NDUFA1 gene product (MWFE protein) is essential for activity of complex I in mammalian mitochondria. Proc Natl Acad Sci USA. 1999;96:4354–9.

1177. Xu G, Guo C, Shan H, Kong H. Divergence of duplicate genes in exon-intron structure. Proc Natl Acad Sci USA. 2012;109:1187–92.

1178. Elliott NE, Cleveland SM, Grann V, Janik J, Waldmann TA, Davé UP. FERM domain mutations induce gain of function in JAK3 in adult T-cell leukemia/lymphoma. Blood. 2011;118:3911–21.

1179. Schrago CG. On the time scale of New World primate diversification. Am J Phys Anthropol. 2007;132:344–54.

1180a. Brosius J. Many G-protein-coupled receptors are encoded retrogenes. Trends Genet. 1999;15:304–5.
1180b. Schmitz J, Brosius J. Exonization of transposed elements: a challenge and opportunity for evolution. Biochemie. 2011;93:1928–34.
1181. Ayyoub M, Pignon NY, Dojcinovic D, Raimbaud D, Old LJ, Luescher I, Valmori D. Assessment of vaccine-induced CD4 T cell responses to the 119-143 immunodominant region of the tumor specific antigen NY-ESO-1 using DRB1*0101 tetramers. Clin Cancer Res. 2010;16:4607–15.
1182. Sharma P, Shen Y, Wen S, Bajorin DF, Reuter VE, Old LJ, Jungbluth AA. Cancer testis antigens: expression and correlation with survival in human urothelial carcinoma. Clin Cancer Res. 2006;12:5442–7.
1183a. Boukerche H, Su ZZ, Emdad L, Sarkar L, Sarkar D, Fisher PB. MDA-9/syntenin regulates the metastatic phenotype in human melanoma cells by activating nuclear factor-kappaB. Cancer Res. 2007;67:1812–22.
1183b. Boukerche H, Aissaoul H, Prévost C, Hirbec H, Das SK, Su ZZ, Sarkar D, Fisher PB. Src kinase activation is mandatory for MDA-9/syntenin-mediated activation of nuclear factor-kappa B. Oncogene. 2010;29:3054–66.
1183c. Boukerche H, Su ZZ, Prévost C, Sarkar D, Fisher PB. MDA-9/syntenin promotes metastasis in human melanoma cells by activating c-Src. Proc Natl Acad Sci USA. 2008;105:15914–14919.
1184. Flynn M, Saha O, Young P. Molecular evolution of the LNX gene family. BMC Evol Biol. 2011;11:235.
1185. Han M, Wang H, Zhang HT, Han Z. The PDZ protein TIP-1 facilitates cell migration and pulmonary metastasis of human invasive breast cancer cells in athymic mice. Biochem Biophys Res Commun. 2012;422:139–45.
1186. Wawrzyniak AM, Vermeiren E, Zimmerman P, Ivarsson Y. Extension of PSD-95/discs large/ZO-1 (PDZ) domains influence lipid binding and membrane targeting of syntenin-1. FEBS Lett. 2012;586:1445–51.
1187. Lazić D, Hofbauer M, Zigrino P, Buchholz S, Kazem S, Feltkamp MC, Mauch C, Steger G, Pfister H, Akgül B. Human papillomavirus type 8 E6 oncoprotein inhibits transcription of the PDZ protein syntenin-2. J Virol. 2012;86:7043–952.
1188. Oulion S, Bertrand S, Belgacem MR, Petillon YL, Escriva H. Sequencing and analysis of the Mediterranean amphioxus (Branchiostoma lanceolatum) transcriptome. PLoS One. 2012;7(5):e36554.
1189. Irimia M, Royo JL, Burguera D, Maeso I, Gómez-Skarmeta JL, Garcia-Fernandez J. Comparative genomics of the Hedgehog loci in chordates and its origins of Shh regulatory novelties. Sci Rep. 2012;2:433.
1190. Sing X, Jin P, Hu J, Qin S, Chen L, Li-Ling J, Ma F. Involvement of AmphiREL, a Rel-like gene identified in Branchiostoma belcheri, in LPS-induced response: Implications for evolution of Rel subfamily genes. Genomics. 2012;99:361–9.
1191. Short S, Kozmik Z, Holland LZ. The function and developmental expression of alternatively spliced isoforms of amphioxus and Xenopus laevis Pax2/5/8 genes: traveling divergence at the invertebrate to vertebrate transition. J Exp Zool B Mol Dev Evol. 2012;318:555–71.
1192. Jiménez-Delgado S, Crespo M, Permanyer J, Garcia-Fernández J, Manzanares M. Evolutionary genomics of the recently duplicated amphioxus Hairy genes. Int J Biol Sci. 2006;2:66–72.
1193. Ou Q, Morris SC, Han J, Zhang Z, Liu J, Chen A, Zhang X, Shu D. Evidence for gill slits and a pharynx in Cambrian vetulicolians: implications for the early evolution of deuterostomes. BMC Biology. 2012;10:81. doi:10.1186/1741-7007-10-81.
1194. Barros P, Jordan P, Matos P. Rac1 signaling modulates BCL-6-mediated expression of gene transcription. Mol Cell Biol. 2009;29:4156–66.

1195a. Murthy S, Ryan A, He C, Mallampali RK, Carter AB. Rac1-mediated mitochondrial H2O2 generation regulates MMP-9 gene expression in macrophages via inhibition of SP-1 and AP-1. J Biol Chem. 2010;285:25062–73.

1195b. Murthy S, Ryan AJ, Carter AB. SP-1 regulation of MMP-9 expression requires Ser586 in the PEST domain. Biochem J. 2012;445:229–36.

1196. Stone DM, Murone M, Luoh S, Ye W, Armanini MP, Gurney A, Phillips H, Brush J, Goddard A, de Sauvage FJ, Rosenthal A. Characterization of the human suppressor of fused, a negative regulator of the zinc-finger transcription factor Gli. J Cell Sci. 1999;112:4437–48.

1197. Polizio AH, Chinchilla P, Chen X, Manning DR, Riobo NA. Sonic hedgehog activates the GTPases Rac1 and RhoA in a Gli-independent manner through coupling of smoothened to Gli-proteins. Sci Signal. 2011;4(200):7.

1198a. Hudson SG, Garrett MJ, Carlson JW, Micklem G, Celniker SE, Goldstein ES, Newfeld SJ. Phylogenetic and genomewide analyses suggest a functional relationship between kayak, the Drososphila fos homolog, and fig, a predicted protein phosphatase 2c nested within a kayak intron. Genetics. 2007;177:1349–61.

1198b. Hudson SG, Goldstein ES. The gene structure of the Drosophila melanogaster proto-oncogene kayak, and its nested gene, fos-intronic gene. Gene. 2008;420:76–81.

1199. Lin K, Zhang DY. The excess of 5' introns in eukaryotic genomes. Nucleic Acids Res. 2005;33:6522–7.

1200. Reijnen MJ, Hamer KM, den Blaauwen JL, Lambrechts C, Schoneveld I, van Driel R, Otte AP. Polycomb and bmi-1 homologs are expressed in overlapping patterns in Xenopus embryos and are able to interact with each other. Mech Dev. 1995;53:35–46.

1201. Xia R, Jia H, Fan J, Liu Y, Jia J. USP8 promotes smoothened signaling by preventing its ubiquitination and changing its subcellular localization. PLoS Biol. 2012;10(1):e1001238.

1202. Hanawa-Suetsugu K, Kukimoto-Niino M, Mishima-Tsamagari C, Akasaka R, Ohsawa N, Sekine S, Ito T, Tochio N, Koshiba S, Kigawa T, Terada T, Shirouzu M, Nishikimi A, Uruno T, Katakai T, Kinashi T, Kohda D, Fukui Y, Yokoyama S. Structural basis for mutual relief of the Rac guanine nucleotide exchange factor DOCK2 and its partner ELMO1 from their autoinhibited forms. Proc Natl Acad Sci USA. 2012;109:3305–10.

1203. Sadia H, Siddiqui RT, Nasim A. A unique BCR-ABL1 transcript with the insertion of intronic sequence from BCR and ABL1 genes in a patient with Philadelphia-positive chronic myeloid leukemia: a case study. Cancer Genet Cytogenet. 2010;201:57–61.

1204. Romero P, Colombatti A. The first intron of the BCR gene contains two minor alternative exons. Leuk Res. 1995;19:963–70.

1205. Fioretos T, Panagopoulos I, Lassen C, Swedin A, Billström R, Isaksson M, Strömbeck B, Olofsson T, Mitelman F, Johansson B. Fusion of the BCR and the fibroblast growth factor receptor-1 (FGFR1) genes as a result of t(8;22)(p11;q11) in a myeloproliferative disorder: the first fusion gene involving BCR but not ABL. Genes Chromosomes Cancer. 2001;32:302–10.

1206. Göppner D, Leverkus M. Basal cell carcinoma: from the molecular understanding of the pathogenesis to targeted therapy of progressive disease. J Skin Cancer 2011:650256.

1207a. LoRusso PM, Rudin CM, Reddy JC, Tibes R, Weiss GJ, Borad MJ, Hann CL, Brahmer JAR, Chang J, Darbonne WC, Graham RA, Zerivitz KL, Low JA, Von Hoff DD. Phase I trial of hedgehog pathway inhibitor vismodegib (GDC-0449) in patients with refractory, locally advanced or metastatic solid tumors. Clin Cancer Res. 2011;17:2502–11.

1207b. LoRusso PM, Piha-Paul SA, Mita M, Colevas AD, Malhi V, Colburn D, Yin M, Low JA, Graham RA. Co-administration of vismodegib with rosiglitazone or combined oral contraceptive in patients with locally advanced or metastatic solid tumors: a pharmacokinetic assessment of drug-drug interaction potential. Cancer Chemother Pharmacol. 2013;71:193–202.

1208a. Uhmann A, Niemann H, Lammering B, Henkel C, Hess I, Nitzky F, Fritsch A, Prüfer A, Rosenberger A, Cullin C, Schraepler A, Reifenberger J, Schweyer S, Pietsch T, Strutz F, Schulz-Schaeffer W, Hahn H. Antitumoral effects of calcitriol in basal cell carcinomas involve inhibition of hedgehog signaling and induction of vitamin D receptor signaling and differentiation. Mol Cancer Ther. 2011;10:2179–88.

1208b. Uhmann A, Nieman H, Lammering B, Henkel C, Heß I, Rosenberger A, Dullin C, Schraepler A, Schultz-Schaeffer W, Hahn H. Calcitriol inhibits hedgehog signaling and induces vitamin D signaling and differentiation in the patched mouse model of embryonal rhabdomyosarcoma. Sarcoma. 2012:357040.

1208c. Ecke I, Petry F, Rosenberger A, Tauber S, Mönkemeyer S, Hess I, Dullin C, Kimmina S, Pirngruber J, Hohnen SA, Uhmann A, Nitzky F, Wojnowski L, Schulz- Scaeffer W, Witt O, Hahn H. Antitumor effects of a combined 5-aza-2'deoxycytidine and valproic acid treatment on rhabdomyosarcoma and medulloblastoma in Ptch mutant mice. Cancer Res. 2009;69:887–95.

1209. Lozina-Lozinsy LK, Bychenkova VN, Zaar EI, Lavin VL, Rumyantseva VM. Some potentialities of living organisms under simulated Martian conditions. Life Sci Space Res. 1971;9:159–65.

1210. Michael LE, Westerman BA, Ermilov AN, Wang A, Ferris J, Liu J, Blom M, Ellison DW, van Lohuizen M, Dlugiosz AA. Bmi1 is required for hedgehog pathway-driven medulloblastoma expansion. Neoplasia. 2006;10:1343–9.

1211. Wang X, Venugopal C, Manranian B, McFarlane N, O'Farrell E, Nolte S, Gunnarsson T, Hollenberg R, Kwiecien J, Northcott F, Taylor MD, Hawkins C, Singh SK. Sonic hedgehog regulates Bmi1 in human medulloblastoma brain tumor-initiating cells. Oncogene. 2012;31:187–99.

1212. Siegelin MD, Siegelin Y, Habel A, Ami A, Gaiser T. KAAD-cyclopamine augmented TRAIL-mediated apoptosis in malignant glioma cells by modulating the intrinsic and extrinsic apoptotic pathway. Neurobiol Dis. 2009;34:259–66.

1213. Fareh M, Turchi L, Virolle V, Debruyne D, Almairac F, de-la-Forest Divonne S, Paquis P, Preynat-Seauve O, Krause KH, Chneiweiss H, Virolle T. The miR 302-367 cluster drastically affects self-renewal and infiltration properties of glioma-initiating cells through CXCR4 repression and consequent disruption of the SHH-GLI-NANOG network. Cell Death Differ. 2912;19:232–244.

1214. Barco R, Garcia CB, Eid JE. The synovial sarcoma-associated SYT-SSX2 oncogene antagonizes the polycomb complex protein Bmi1. PLoS One. 2009;4(4):e5060.

1215. Bálint E, Reisman D. Increased rate of transcription contributes to elevated expression of the mutant p53 gene in Burkitt's lymphoma. Cancer Res. 1996;56:1648–53.

1216. Reisman D, Takahashi P, Polson A, Boggs K. Transcriptional regulation of the p53 tumor suppressor gene in S-phase of the cell cycle and the cellular response to DNA damage. Biochem Res Int. 2012:808934.

1217. Moore SW, Zaahl MG. Intronic RET gene variants in Down syndrome-associated Hirschsprung disease in an African population. J Pediatr Surg. 2012;47:299–302.

1218. Moore SW, Zaahl M. The Hirschsprung's multiple endocrine neoplasia connection. Clinics (Sao Pauo). 2012;67S1:63–7.

1219. Cenik C, Chua HN, Zhang H, Tarnawsky SP, Akef A, Derti A, Tasan M, Moore MJ, Palazzo AF, Roth FP. Genome analysis reveals interplay between 5'UTR introns and nuclear mRNA export for secretory and mitochondrial genes. PLoS Genet. 2011;7(4): e1001366.

1220. Smits M, Wurdinger T, van het Hof B, Drexhage JA, Geerts D, Wesseling P, Noske DP, Vandertop WP, de Vries HE, Reijerkerk A. Myc-associated zinc-finger protein (MAZ) is regulated by miR-125b and mediates VEGF-induced angiogenesis in glioblastoma. FASEB. 2012;26:2639–47.

1221. Fiesel FC, Weber SS, Supper J, Zell A, Kahle OJ. TDP-43 regulated global translational yield by splicing of exon junction complex component SKAR. Nucleic Acids Res. 2012;40:2668–82.

1222a. Johnson SA, Ubberley G, Bentley DL. Cotranscriptional recruitment of the mRNA export factor Yra1 by direct interaction within the 3' end processing factor Pcf11. Mol Cell. 2009;33:215–26.

1222b. Johnson SA, Kim H, Erickson B, Bentley DL. The export factor YRA1 modulates mRNA 3' end processing. Nat Struct Mol Biol. 2011;18:1164–71.

1223. Ma XM, Yoon SO, Richardson CJ, Jülich K, Blenis J. SKAR links pre-mRNA splicing to mTOR/S6K1-mediated enhanced translation efficiency of spliced mRNAs. Cell. 2008;133:303–13.

1224. Smyk A, Szuminska M, Uniewicz KA, Graves LM, Kozlowski P. Human enhancer of rudimentary is a molecular partner of PDIP46/SKAR, a protein interacting with DNA polymerase delta and S6K1 and regulating cell growth. FEBS J. 2006;273:4728–41.

1225. Han E, Phan D, Lo P, Poy MN, Behringer R, Najjar SM, Lin SH. Differences in tissue-specific and embryonic expression of mouse CEACAM1 and CEACAM2 genes. Biochem J. 2001;355:417–23.

1226. Piekielko-Witkowska A, Wiszomirska H, Wojcicka A, Poplawski P, Boguslawska J, Tanski Z, Nauman A. Disturbed expression of splicing factors renal cancer affects alternative splicing of apoptosis regulators, oncogenes, and tumor suppressors. PLoS One. 2010;5(10):e23690.

1227. Han WL, Li WD, Hu J, Rusidanmu A, Chen LF, Shen L, Zheng SS. Expression of the recepteur d'originenantais receptor tyrosine kinase in non-small cell lung cancer and its clinical significance. Clin Med J (Engl). 2012;125:1110–4.

1228. Poliseno L, Salmena L, Riccardi L, Fornari A, Song MS, Hobbs RM, Spotoletti P, Varmeh S, Egia A, Fedele G, Rameh L, Loda M, Pandolfi PP. Identification of the miR-106b≈25 microRNA cluster as a proto-oncogenic PTEN-targeting intron that cooperates with its host gene MCM7 in transformation. Sci Signal. 2010;3(117):29.

1229. Mimeault M, Johansson SL, Batra SK. Pathobiological implications of the expression of EGFR, pAkt, NF-κB and MIC-1 in prostate cancer stem cells and their progenies. PLoS One. 2012;7(2):e31919.

1230. Deng JH, Deng Q, Kuo CH, Delaney SW, Ying SY. MiRNA targets of prostate cancer. Methods Mol Biol. 2013;936:357–69.

1231. Kaufman DS, Zietman AL, McDougal WS, Dahl DM, Harisinghani MG, Wu CL. Case records of the Massachusetts General Hospital. Case 9-2010. A 67-year old man with a persistently elevated PSA level. N Engl J Med. 2012;366:1143–50.

1232. Krauss DJ, Hayek S, Amin M, Ye H, Kestin LL, Zadora S, Vicini FA, Cotant M, Brabbins DS, Ghilezan MI, Gustafson GS, Martinez AA. Prognostic significance of neuroendocrine differentiation in patients with Gleason score 8-10 prostate cancer treated with primary radiotherapy. Int J Radiat Oncol Biol Phys. 2011;81(3):E119–25.

1233. Majid S, Dar AA, Saini S, Arora S, Shahryani V, Zaman MS, Chang I, Yamamura S, Tanaka Y, Deng G, Dahiya R. MicroRNA-23b represses proto-oncogene Src kinase, functions as a methylation-silencer tumor suppressor with diagnostic and prognostic significance in prostate cancer. Cancer Res. 2012;72:6435–46.

1234. Singh D, Chan JM, Zoppoli P, Niola F, Sullivan R, Castano A, Liu EM, Reichel J, Porrati P, Pellegatta S, Qiu K, Gao Z, Ceccarelli M, Riccardi R, Brat DJ, Guha A, Aldape K, Golfinos JG, Zagzag D, Mikkelsen T, Finocchiaro G, Lasorella A, Rabadan R, Iavarone A. Transforming fusions of FGFR and TACC genes in human glioblastoma. Science. 2012;337:1231–5.

1235a. Agrawal N, Frederick MJ, Pikering CR, Bettegowda C, Chang K, Fakhry C, Xie TX, Zhang J, Wang J, Zhang N, El-Naggar AK, Jasser SA, Weinstein JN, TreviñoL Drummind JA, Muzny DM, Wu Y, Wood LD, Hruban RH, Westra WH, Koch WM, Califano JA, Gibbs RA, Sidransky D, Vogelstein B, Velculescu VE, Papadopoulos N, Wheeler DA, Kinzler KW, Myers JN. Exone sequencing of head and neck squamous cell carcinoma reveals inactivating mutations in NOTCH1. Science. 2011;333:1154–7.

1235b. Agrawal N, Jiao Y, Bettegowda C, Hutfless SM, Wang Y, David S, Cheng Y, Twaddell WS, Latt NL, Shin EJ, Wang LD, Wang L, Yang W, Velculescu VE, Vogelstein B, Papadopoulos N, Kinzler KW, Meltzer SJ. Comparative genomic analysis of esophageal adenocarcinoma and squamous cell carcinoma. Cancer Discov. 2012;2:899–905.

1236. Amakawa R, Jing W, Ozawa K, Matsunami N, Hamaquchi Y, Matsuda F, Kawaichi M, Honjo T, Human JK. Recombination signal binding protein (IGKJRB): comparison with its mouse homologue. Genomics. 1993;17:306–15.

1237. Bargo S, Raafat A, McCurdy D, Amirjazil I, Shu Y, Traicoff J, Plant J, Vonderhaar BK, Callahan R. Transforming acidic-coiled-coil protein-3 (Tacc3) acts as a negative regulator of Notch signaling through binding to CDC10/Ankyrin repeats. Biochem Biophys Res Commun. 2010;400:606–12.

1238. Hood FE, Royle SJ. Pulling it together: the mitotic function of TACC3. Bioarchitecture. 2011;1:105–9.

1239. Minoguchi S, Ikeda T, Itohara S, Kaneko T, Okaichi H, Honio T. Studies on the cell-type specific expression of RBP-L a RBL-L family member, by replacement insertion of beta-galactosidase. J Biochem. 1999;126:738–47.

1240. Wang SS, Chang PJ, Chen LW, Chen LY, Hung CH, Liou JY, Yen JB. Positive and negative regulation in the promoter of the ORF46 gene of Kaposi's sarcoma-associated herpesvirus. Virus Res. 2012;165:157–69.

1241. Shamay M, Liu J, Li R, Liao G, Greenway M, Hu S, Zhu J, Xie Z, Ambinder RF, Qian J, Zhu H, Hayward SD. A protein array screen for Kaposi's sarcoma-associated herpesvirus LANA interactions links LANA to TIP60, PP2A activity and telomere shortening. J Virol. 2012;86:5179–91.

1242. Sinkovics JG. Progressive development of viral therapy of human cancers: a personal narrative account. In: Sinkovics JG, Horvath JC, authors/editors. Viral Therapy of Human Cancers. Marcel Dekker, New York; 2005. pp 1–147.

1243. Hayward SD. Viral interactions with the Notch pathway. Semin Cancer Biol. 2004;14:387–96.

1244. Ma W, Du J, Chu Q, Wang Y, Liu L, Song M, Wang W. hCLP46 regulates U937 cell proliferation via Notch signaling pathway. Biochem Biophys Res Commun. 2011;408:84–8.

1245. Kohn A, Dong Y, Mirando AJ, Jesse AM, Honjo T, Zuscik MJ, O'Keefe RJ, Hilton MJ. Cartilage-specific RBPjk-dependent and independent Notch signals regulate cartilage and bone development. Development. 2012;139:1198–212.

1246. Tzeng CW, Frolov A, Frolova N, Jhala NC, Howard JH, Vickers SM, Buchsbaum DJ, Heslin MJ, Arnoletti JP. Pancreatic cancer epidermal growth factor receptor (EGFR) intron 1 polymorphism influences postoperative patient survival and in vitro erlotinib response. Ann Surg Oncol. 2007;14:2150–8.

1247. Frolov A, Liles JS, Kossenkov AV, Tzeng CW, Jhala N, Kulesza P, Varadarajulu S, Sloubeidi M, Heslin MJ, Arnoletti JP. Epidermal growth factor receptor (EGFR) intron 1 polymorphism and clinical outcome in pancreatic adenocarcinoma. Am J Surg. 2010;200:398–405.

1248. Beres TM, Masui T, Swift GH, Shi L, Henke RM, MacDonald RJ. PTF1 is an organ-specific and Notch-independent basic helix-loop-helix complex containing the mammalian Suppressor of Hairless (RBP-J) or its paralogue, RBP-L. Mol Cell Biol. 2006;26:117–30.

1249. Ehm O, Göritz C, Covic M, Schäffner I, Schwarz TJ, Karaca E, Kempkes B, Kremmer E, Pfrieger FW, Espinosa L, Bigas A, Giachino C, Taylor V, Frisén J, Lie DC. RBPJkappa-dependent signaling is essential for long-term maintenance of neural stem cells in the adult hippocampus. J Neurosci. 2010;30:13794–807. doi:10.1523/JNEUROSCI.1567-10.2010.

1250. Feng F, Wang YC, Hu XB, Liu G, Chen YR, Wang L, He F, Dou GR, Liang L, Zhang HW, Han H. The transcription factor RBP-J-mediated signaling is essential for dendritic cells to evoke efficient anti-tumor immune responses in mice. Mol Cancer. 2010;9:90.

1251. Oswald F, Liptay S, Adler G, Schmidt RM. NF-kappaB2 is a putative target gene of activated Notch-1 via RBP-Jkappa. Mol Cell Biol. 1998;18:2077–88.

1252. Fu T, Zhang P, Feng L, Wang XH, Zheng MH, Qin HY, Chen DL, Wang WZ, Han H. Accelerated acute allograft rejection accompanied by enhanced T-cell proliferation and attenuated Treg function in RBP-J deficient mice. Mol Immunol. 2011;48:751–9.

1253. Wang H, Zou J, Zhao B, Johannsen E, Ashworth T, Wong H, Pear WS, Schug J, Blacklow SC, Arnett KL, Bernstein BE, Kieff E, Aster JC. Genome-wide analysis reveals conserved and divergent features of Notch1/RBPJ binding in human and murine T-lymphoblastic leukemia cells. Proc Natl Acad Sci USA. 2011;108:14908–13.

1254. Yenerall P, Zhou L. Identifying the mechanisms of intron gain: progress and trends. Biol Direct. 2012;7:29. doi:10.1186/1745-6150-7-29.

1255. Hellsten U, Aspden JL, Rio DC, Roksar DS. A segmental genomic duplication generates a functional intron. Nat Commun. 2011;2:454. doi:10.1038/ncomms1461.

1256. Szczepankowska AK, Prestel E, Mariadassou M, Bardowski JK, Bidnenko E. Phylogenetic and complementation analysis of a single-stranded DNA binding protein family from lactococcal phages indicates a non-bacterial origin. PLoS One. 2011;6(11):e26942.

1257. Dlakić M, Mushegian A. Prp8, the pivotal protein of the spliceosomal catalytic center, evolved from a retroelement-encoded reverse transcriptase. RNA. 2011;17:799–808.

1258. Boonsong A, Marsh S, Rooney PH, Stevenson DA, Cassidy J, McLeod HL. Characterization of the topoisomerase Iocus in human colorectal cancer. Cancer Genet Cytogenet. 2000;12:56–60.

1259. Chang JY, Liu JE, Juang SH, Liun TW, Chen LT. Novel mutations of topoisomerase I in rendering cells resistant to camptothecin. Cancer Res. 2002;62:3716–21.

1260. Wesierska-Gądek J, Składanowski A. Therapeutic intervention by the simultaneous inhibition of DNA repair and type I or type II DNA topoisomerase: one strategy, many outcomes. Future Med Chem. 2012;4:51–72.

1261. Gongora C, Vezzio-Vie N, Tuduri S, Denis V, Causse A, Auzanneau C, Collod-Beroud G, Coquelle A, Pasero P, Pourquier P, Martineau P, Del Rio M. New topoisomerase I mutations are associated with resistance to camptothecin. Mol Cancer. 2011;10:64. doi:10.1186/1476-4598-10-64.

1262. de Souza RF, Iyer LM, Aravind L. Diversity and evolution of chromatin proteins encoded by DNA viruses. Biochem Biophys Acta. 2010;1799:302–18.

1263. Aravind L, Anantharaman V, Zhang D, de Souza LM, Iyer LM. Gene flow and biological conflict systems in the origin and evolution of eukaryotes. Front Cell Infect Microbiol. 2012;2:89. doi:10.3389/fcimb.2012.00089.

1264. Du Z, Park KW, Yu H, Fan Q, Li L. Newly identified prion linked chromatin-remodeling factor Swi1 in Saccharomyces cerevisiae. Nat Genet. 2008;40:460–5.

1265. Kaul-Ghanekar R, Singh S, Mamgain H, Jalota-Badhwar A, Paknikar KM, Chattopadhyay S. Tumor suppressor protein SMAR1 modulates the roughness of cell surface: combined AFM (atomic force microscopy) and SEM (scanning electron microscopy) study. BMC Cancer. 2009;9:350.

1266. Koonin EV. The origin of introns and their role in eukaryogenesis: a compromised solution to the introns-early versus introns-late debate? Biol Direct. 2006;1:22. doi:10.1186/1745-6150-1-22.

1267. Wurster AL, Precht P, Becker KG, Wood WH III, Zhang Y, Wang Z, Pazin MJ. IL-10 transcription is negatively regulated by BAF180, a component of the SWI/SNF chromatin remodeling enzyme. BMC Immunol. 2012;13:9. doi:10.1186/1471-2172-13-9.
1268. Zhu X. M H, Chen Z. Phylogenetics and evolution of Su(var)3-9 SET genes in land plants: rapid diversification in structure and function. BMC Evol Biol. 2011;11:63. doi:10.1186/1471-2148-11-63.
1269. Charney S. Behavior genetics and postgenomics. Behav Brain Sci. 2012;35:331–58.
1270. Schanker BD. Epigenetic regulation of brain-derived neurotrophic factor: implications in neurodevelopment and behavior. Behav Brain Sci. 2012;35:377–8.
1271. Nagy GR, Győrffy B, Nagy B, Rigó J Jr. Lower risk for Down syndrome associated with longer oral contraceptive use: a case-control study of women of advanced maternal age presenting for prenatal diagnosis. Contraception. 2012;87:455–458; pii:S0010-7824(12) 00806-2. doi:10.016/j.contraception.2012.08.0440.
1272. Zhong Q, Layman LC. Genetic considerations in the patient with Turner-syndrome-45, X with or without mosaicism. Fertil Steril. 2012;98:775–9.
1273. Gorley A, Wilkie AO. Paternal age effect mutations and selfish spermatogonil selection: causes and consequences for human disease. Am J Hum Genet. 2012;90:175–200.
1274. Lin TY, Cheng YC, Yang HC, Lin WC, Wang CC, Lain PL, Shieh SY. Loss of the candidate tumor suppressor BTG3 triggers acute cellular senescence via the ERK-JMJD3-p16(INKa) signal axis. Oncogene. 2012;31:3287–97.
1275. Majid S, Dar AA, Shahryani V, Hirata H, Ahmad A, Saini S, Tanaka Y, Dahiya AV, Dahiya R. Genistein reverses hypermethylation and induces active histone modifications in tumor suppressor gene B-cell translocation gene 3 in prostate cancer. Cancer. 2010;116:66–76.
1276. Koonin EV, Wolf Y. Evolution of microbes and viruses: a paradigm shift in evolutionary biology? Front Cell Infect Microbiol. 2012;2:119. doi:10.3389/fcimb.2012.00119.
1277. Duesberg P, Mandrioli D, McCormack A, Nicholson JM. Is carcinogenesis a form of speciation? Cell Cycle. 2011;10:2100–14.
1278a. Ricke RM, Jaganathan KB, van Deursen JM. Bub1 overexpression induces aneuploidy and tumor formation through Auora B kinase hyperactivation. J Cell Biol. 2011;193:1049–64.
1278b. Ricke RM, van Deursen JM. Aurora B hyperactivation by Bub1 overexpression promotes chromosome missegregation. Cell Cycle. 2011;10:3645–51.
1279. Saeki T, Ouchi M, Ouchi T. Physiological and oncogenic Aurora-A pathway. Int J Biol Sci. 2009;5:758–62.
1280. Khan J, Khan S, Attaullah S, Ali I, Khan SN. Aurora kinase-C-T191D is constitutively active mutant. BMC Cell Biol. 2012;13:8. doi:10.1186/1471-2121-13-8.
1281. Macurek L, Lindquist A, Medema RH. Aurora-A and B for a joint the game of polo. Cancer Res. 2009;69:4555–8.
1282. Alsford S, Horn D. Cell-cycle regulated control of VSG expression site silencing by histones and histone chaperones ASF1A and CAF-1b in Trypanosoma brucei. Nucleic Acid Res. 2012;40:10150–60.
1283. Barry JD, Hall JP, Plenderleith L. Genome hyperevolution and the success of a parasite. Ann NY Acad Sci. 2012;1267:11–7.
1284. Kanayake DK, Cipriano MJ, Sabatini R. Telomeric co-localization of the modified base J and contingency genes in the protozoan parasite Trypanosoma cruzi. Nucleic Acid Res. 2007;35:6367–77.
1285. Li B. A newly discovered role of telomeres in an ancient organism. Nucleus. 2010;1:260–3.

1286. Moraes Barros RR, Marini MM, Antônio CR, Cortez DR, Miyake AM, Lima FM, Ruiz JC, Bartholomeu DC, Chiurillo JB, Ramirez JL, da Siveira JF. Anatomy and evolution of telomeric and subtelomeric regions in the protozoan parasite Trypanosoma cruzi. BMC Genomics. 2012;13:299. doi:10.1186/1471-2164-13-229.
1287. Schuster FL, Sullivan JJ. Cultivation of clinically significant Hemoflagellates. Clin Microbiol Rev. 2002;15:374–89.
1288. DuBois KN, Alsford S, Holden JM, Buisson J, Swiderski M, Bart J-M, Ratushny A, Wan Y, Bastin P, Barry JD, Navarro M, Horn D, Aitchison JD, Rout MP, Field Mc. NUP-1 is a large coiled-coil nucleoskeletal protein in Trypanosomes with lamin-like functions. PLoS Biol. 2012;10(3):e1001287n. doi:10.1371/journal.pbio.1001287.
1289. Krüger A, Batsios P, Baumann O, Luckert E, Schwartz, Stick R, Meyer I. Characterization of Ne8q1, the first lamin-like nucleoskeletal protein in a unicellular organism. Mol Biol Cell. 2012;23:360–70.
1290. Lam BD, Anthony EC, Hordijk PL. Analysis of nucleo-cytoplasmic shuttling of the proto-oncogene SET/I2PP2A. Cytometry. 2012;81:81–9.
1291. Yurin YB, Iourov IY, Vorsanova SG, Liehr T, Kolotii AD, Kutsev SI, Pellestor F, Beresheva AK, Demidova IA, Kravets VS, Monakhov VV, Soloviev IV. Aneuploidy and confined chromosomal mosaicism in the developing human brain. PLoS One. 2007;2(6): e558. doi:10.1371/journal.pone.000558.
1292. Senovilla L, Vitale I, Martins I, Tailler M, Pailleret C, Michaud M, Galluzzi L, Adjemian, Kepp O, Niso-Santano M, Shen S, Mariño G, Criollo A, Boilève, Job B, Ladoire S, Ghiringhelli F, Sistigu A, Yamazaki T, Rello-Varona S, Locher C, Poirier-Colame V, Talbot, Valent A, Berardinelli F, Antoccia A, Ciccosanti F, Fimia GM, Piacentini M, Fueyo A, Messina NL, Li M, Chan CJ, Sigl V, Pourcher G, Ruckenstuhl C, Carmona-Gutierrez D, Lazar V, Penninger G, Madeo F, López-Otín C, Smyth MJ, Zitvogel L, Castedo M, Kroemer G. An immunosurveillance mechanism controls cancer cell ploidy. Science. 2012;337:1678–1684.
1293. Zanetti M, Mahadevan NR. Immune surveillance from chromosomal chaos? Science. 2012;337:1616–7.
1294. von Preisz H. Studien über Pettenkoferien. Zbl f Bakteriologie, Parasitenkunde u Infektionskrankheiten, I. Abt/ Originals 1937;139:225-270 Tafeln I-XIII.
1295. Kimura A. Regulation of ribonucleic acid synthesis in spheroplasts, cold-shocked cells, and toluene-treated cells of Escherichia coli. J Bacteriol. 1976;128:123–9.
1296. Althoff S, Selinger D, Wise JA. Molecular evolution of SRB cycle components: functional implications. Nucleic Acids Res. 1994;22:1933–47.
1297. Rojas AM, Fuentes G, Rausell A, Valencia A. The Ras protein superfamily: evolutionary tree and role of conserved amino acids. J Cell Biol. 2012;196:189–201.
1298. Sowa MP, Coulter LJ, Tait A, Hide G. A novel gene encoding a ras-like GTP-binding protein from Trypanosoma brucei: an evolutionary ancestor of the ras and rap genes of higher eukaryotes? Gene. 1999;230:155–61.
1299. Xu MY, Liu JL, Zhang RL, Fu YC. Isolation of a novel ras gene from Trichomonas vaginalis: a possible evolutionary ancestor of the Ras and Rap genes of higher eukaryotes. Biochem Cell Biol. 2007;85:239–45.
1300. Pacheco MA, Ryan EN, Poe AC, Basco L, Udhayakumar V, Collins WE, Escalante AA. Evidence for negative selection on the encoding rhoptry-associated protein 1 (Rap-1) in Plasmodium spp. Infect Gen Evol. 2010;10:655–61.
1301. Bolourani P, Spiegelman G, Weeks G. Ras proteins have multiple functions in vegetative cells of Dictyostelium. Eukaryot Cell. 2010;9:1728–33.
1302. Xie XQ, Guan Y, Ying SH, Feng MG. Differential functions of Ras 1 and Ras 2 proteins in regulating the germination, growth, conidiation, multi-stress tolerance and virulence of Baeauveria bassiana. Environ Microbiol 2012; doi:10.1111/j.1462-2920.2012.02871.x.
1303. Mavromatakis YE, Tomlinson A. The role of the small GTPase Rap in Drosophila R7 photoreceptor specification. Proc Natl Acad Sci USA. 2012;109:3844–9.

1304. Frische EW, Pellis van Berkel W, van Haaften G, Cuppen E, Plasterk RH, Tijsterman M, Bos JL, Zwartkruis FJ. RAP-1 and the RAL-1/exocyst pathway coordinate hypodermal cell organization in Caenorhabditis elegans. EMBO. 2007;26:5083–92.
1305. Freeman SA, Lein V, Dang-Lawson M, Mizuno K, Roskelley CD, Gold MR. Cofilin-mediated F-actin severing is regulated by the Rap GTPase and controls the cytoskeletal dynamics that drive lymphocyte spreading and BCR microcluster formation. J Immunol. 2011;187:5887–900.
1306. Beaudoin GM 3rd, Schofield CM, Nuwal T, Zang K, Ullian EM, Huang B, Richardt LF. Afadin, a Ras/Rap effector that controls adherin function, promotes spine and excitatory synapse density in the hippocampus. J Neurosci. 2012;32:99–110.
1307. Wu J, Zhang Y, Frilot N, Kim JI, Kim WJ, Daaka Y. Prostaglandin E2 regulates renal cell carcinoma invasion through the EP4 receptor-Rap GTP-ase signal transduction pathway. J Biol Chem. 2011;286:33954–62.
1308. Gorbushin AM, Panchin YV, Iakovleva NV. In search of the origin of FREPs: characterization of Aplysia californica fibrinogen-related proteins. Dev Comp Immunol. 2010;34:465–73.
1309. Weber F, Aldred MA, Morrison C, Plass C, Frilling A, Broelsch CE, Waite K, Eng C. Silencing of the maternally imprinted tumor suppressor ARHI contributes to follicular thyroid carcinogenesis. J Clin Endocrin Metabol. 2005;90:1149–55.
1310. Baljuls A, Beck M, Oenel A, Robubi A, Kroschewski R, Hekman M, Rudel T, Rapp UR. The tumor suppressor DiRas3 forms a complex with H-Ras and C-RAF proteins and regulates localization, dimerization, and kinase activity of C-RAF. J Biol Chem. 2012;287:23128–40.
1311a. Huang S, Chang IS, Lin W, Ye W, Luo RZ, Lu Z, Lu Z, Lu Y, Liao WS-L, Tao T, Bast RC Jr, Chen X, Yu Y. ARHI (DIRAS3), an imprinted tumor suppressor gene, binds to importins and blocks nuclear import of cargo proteins. Biosci Rep. 2009;30:159–68.
1311b. Badgwell DB, Lu Z, Le K, Gao F, Suh GK, Bao J-J, Das P, Andreef M, Chen W, Yu Y, Ahmed AA, Liao WS-L, Bast RC Jr. The tumor suppressor gene ARHI (DIRAS3) suppresses ovarian cancer cell migration through inhibition of the STAT3 and FAK/Rho signaling pathways. Oncogene. 2012;31:68–79.
1311c. Lu Z, Bast RC Jr. The tumor suppressor gene ARHI (DIRAS3) inhibits ovarian cancer cell migration through multiple mechanisms. Cell Adh Migr. 2013;7:232–6.
1312. Lin D, Cui F, Bu Q, Yan C. The expression and clinical significance of GTP-binding RAS-like 3 (ARHI) and microRNA 221 and 222 in prostate cancer. J Int Med Res. 2011;39:1870–5.
1313. Chen Y, Zaman MS, Deng G, Majid S, Saini S, Liu J, Tanaka Y, Dahiya R. MicroRNAs 221/222 and genistein-mediated regulation of ARHI tumor suppressor gene in prostate cancer. Cancer Prev Res (Phila). 2011;4:76–86.
1314. Liu X, Li X, Yin L, Ding J, Jin H, Feng Y. Genistein inhibits placental choriocarcinoma cell line JAR invasion through ERβ/MTA3/Snail/E-cadherin pathway. Oncol Lett. 2011;2:891–7.
1315. Murugan AK, Munirajan AK, Tsuchida N. Ras oncogene in oral cancer: the past 20 years. Oral Oncol. 2012;48:383–92.
1316. Turvey SE, Broide DH. Chapter 2: Innate immunity. J Allergy Clin Immunol. 2010;125: S24–32.
1317. Fineran PC, Charpentier E. Memory of viral infections by CRISP-Cas adaptive immune systems: acquisition of new information. Virology 2012; pii: S0042-6822(12)00501-6. doi:10.1016/j.virol.2012.10.003.
1318. Fischer S, Maier LK, Stoll B, Brendel J, Fischer E, Pfeiffer F, Dyall-Smith M, Marchfelder A. An archaeal immune system can detect multiple protospacer adjacent motifs (PAMs) to target invader DNA. J Biol Chem. 2012;287:33351–63.

1319. Westra ER, Swarts DC, Staals RH, Jore MM, Brouns SJ, van der Oost J. The CRISPRs, they are a-changin': how prokaryotes generate adaptive immunity. Annu Rev Genet. 2012;46:311–39.
1320. Jorth P, Whiteley M. An evolutionary link between natural transformation and CRISPR adaptive immunity. MBio. 2012;3(5):e00309–12. doi:10.11228/mBio.00309-12.
1321. McCurley N, Hirano M, Das S, Cooper MD. Immune related genes underpin the evolution of adaptive immunity in jawless vertebrates. Curr Genomics. 2012;13:86–94.
1322. Boehm T. Evolution of vertebrate immunity. Curr Biol. 2012;22:R722–32.
1323. Gu X, Shivaov V, Strout MP. The role of activation-induced cytidine deaminase in lymphomagenesis. Curr Opin Hematol. 2012;19:292–8.
1324. Magrath I. Epidemiology: clues to the pathogenesis of Burkitt lymphoma. Br J Haematol. 2012;156:744–56.
1325. Gonin J, Larousserie F, Bastard C, Picquenot J-M, Couturier J, Vacher-Lavenu M-C, Devergne O. Epstein-Barr virus-induced gene 3 (EBI3): a novel diagnosis marker in Burkitt lymphoma and diffuse large B-cell lymphoma. PLoS One. 2011;6(9):e24617. doi:10.1371/journal.pone.0024617.
1326. Visekruna A, Volkov A, Steinhoff U. A key role for NF-κB transcription factor c-Rel in T-lymphocyte-differentiation and effector functions. Clin Dev Immunol. 2012;2912:239368. doi:10.1155.2012/239368.
1327. El-Mallawany NK, Frazer JK, Van Vlierberghe P, Ferrando AA, Perkins S, Lim M, Chu Y, Cairo MS. Pediatric T- and NK-cell lymphomas: new biological insights and treatment strategies. Blood Cancer J. 2012;2(4):e65. doi:10.1038/BCJ.2012.8.
1328. Freguja R, Gianesin K, Zanchetta M, De Rossi A. Cross-talk between virus and host innate immunity in pediatric HIV-1 infection and disease progression. New Microbiol. 2012;35:249–57.
1329. Han D, Cai X, Wen J, Matheson D, Skyler JS, Kenyon NS, Chen Z. Innate and adaptive immune gene expression profiles as biomarkers in human type 1 diabetes. Clin Exp Immunol. 2012;170:131–8.
1330. Allcock RJ, Barrow AD, Forbes S, Beck S, Trowsdale J. The human TREM gene cluster at 6.21.1 encodes both activating and inhibitory single IgV domain receptors and includes NKp44. Eur J Immunol. 2003;33:567–77.
1331. Ubagai T, Tansho S, Ieki R, Ono Y. Evaluation of TREM1 gene expression in circulating polymorphonuclear leukocytes and its inverse correlation with the severity of pathophysiological conditions in patients with acute bacterial infections. Jpn J Infect Dis. 2012;65:376–82.
1332. Wang YS, Li XJ, Zhao WO. TREM-1 is a positive regulator of TNF-α and IL-8 production in U937 foam cells. Bosn J Basic Med Sci. 2012;12:94–101.
1333. Wu J, Li J, Salcedo R, Mivechi NF, Trinchieri G, Horuzsko A. The proinflammatory myeloid cell receptor TREM-1 controls Kupfer cell activation and development of hepatocellular carcinoma. Cancer Res. 2012;72:3977–86.
1334. Ko SY, Barengo N, Ladanyi A, Lee JS, Marini F, Lengyel E, Naora H. HOXA9 promotes ovarian cancer growth by stimulating cancer-associated fibroblasts. J Clin Invest. 2012;122:3603–17.
1335. Nieman KM, Kenny HA, Penicka CV, Ladanyi A, Buell-Gutbrod R, Zillhardt MR, Romero IL, Carey MS, Mills GB, Hotamisligil GS, Yamada SD, Peter ME, Gwin K, Lengyel E. Adipocytes promote ovarian cancer metastasis and provide energy for rapid tumor growth. Nat Med. 2011;17:1498–503.
1336. Boyerinas B, Park SM, Murmann AE, Gwin K, Montag AG, Zillhardt M, Hua YJ, Lengyel E, Peter ME. Let-7 modulates acquired resistance of ovarian cancer to taxanes via IMP-1-mediated stabilization of multidrug resistance 1. Int J Cancer. 2012;130:1787–97.

1337. Park SM, Shell S, Radjabi AR, Schickel R, Feig C, Boyerinas B, Dinulescu DM, Lengyel E, Peter ME. Let-7 prevents early cancer progression by suppressing expression of the embryonic gene HMG-A2. Cell Cycle. 2007;6:2585–90.
1338. Park SM, Gaur AB, Lengyel E, Peter ME. The miR-200 family determines the epithelial phenotype of cancer cells by targeting the E-cadherin repressors ZEB1 and ZEB2. Genes Dev. 2008;22:894–907.
1339. Colas AR, McKeithan WL, Cunningham TJ, Bushway PJ, Garmine LX, Duester G, Subramariam S. Whole-genome microRNA screening identifies let-7 and mir-18 as regulators of germ layer formation during early embryogenesis. Genes Dev. 2012;26:2567–79.
1340. Qin B, Xiao B, Liang D, Li Y, Jiang T, Yang H. MicroRNA let-7 inhibits Bcl-xl expression and regulates ox-LDL-induced endothelial apoptosis. BMB Rep. 2012;45:464–9.
1341. Thornton JE, Gregory RI. How does Lin28 let-7 control development and disease. Trends Cell Biol. 2012;22:474–82.
1342. Hertel J, Bartschat S, Wintsche A, Otto C. The Students of the Bioinformatics Lab, Stadler PF. Evolution of the let-7 microRNA family. RNA Biol. 1012;9:231–41.
1343. Philippe H, Brinkmann H, Copley RR, Moroz LL, Nakano H, Poustka AJ, Wallberg A, Peterson KJ, Telford MJ. Acoelomorph flatworms are deuterostomes related to Xenoturbella. Nature. 2011;470:255–8.
1344. Schmidt C, Bladt F, Goedecke S, Brinkmann V, Zschiesche W, Sharpe M, Gherardi E, Birchmeier C. Scatter factor/hepatocyte growth factor is essential for liver development. Nature. 1995;373:699–702.
1345. Uehara Y, Minowa O, Mori C, Shiota K, Kuno J, Noda T, Kitamura N. Placental defect and embryonal lethality in mice lacking hepatocyte growth factor /scatter factor. Nature. 1995;373:702–5.
1346. Yang Y, Wang Y, Zeng X, Ma XJ, Zhao Y, Qiao J, Cao B, Li YX, Wang YL. Self-control of HGF regulation on human trophoblast cell invasion via enhancing c-Met receptor shedding by ADAM10 and ADAM17. J Clin Endocrinol Metab. 2012;97: E1390–401.
1347. Baek JH, Birchmeier C, Zenka M, Hieronymus T. The HGF receptor/Met tyrosine kinase is a key regulator of dendritic cell migration in skin immunity. J Immunol. 2012;189:1699–709.
1348. Nam HJ, Chae S, Jang SH, Cho H, Lee JH. The PI3K-Akt mediates oncogenic Met-induced centrosome amplification and chromosome instability. Carcinogenesis. 2010;31:1531–40.
1349. Szabo R, Rasmussen AL, Moyer AB, Kosa P, Schafer JM, Molinolo AA, Gutkind JS, Bugge TH. C-Met-induced epithelial carcinogenesis is initiated by the serin protease matripase. Oncogene. 2011;30:2003–16.
1350. Buurman R, Gürlevik E, Schäffer V, Eilera M, Sandbothe M, Kreipe H, Wilkens L, Schlegelberger B, Kühnel F, Skawran B. Histone deacetylases activate growth factor signaling by repressing microRNA-449 in hepatocellular carcinoma cells. Gastroenterology. 2012;143:811–20.
1351. Bishop EA, Lengyel ER, Yamada SD, Montag A, Temkin SM. The expression of hepatocyte growth factor (HGF) and c-Met in uterine serous carcinoma. Gynecol Oncol. 2011;121:218–23.
1352. Lengyel E. Ovarian cancer development and metastasis. Am J Pathol. 2010;177:1053–64.
1353. Mitra AK, Sawada K, Tiwari P, Mui K, Gwin K, Lengyel E. Ligand-dependent activation of c-Met by fibronectin and α(5)β(1)-integrin regulates ovarian cancer invasion and metastasis. Oncogene. 2011;30:1566–76.

1354. Zillhardt M, Park SM, Romero IL, Sawada K, Montag A, Krausz T, Yamada SD, Peter ME, Lengyel E. Foretinib (GSK1363089), an orally available multikinase inhibitor of c-Met and VEGFR-2, blocks proliferation, induces anoikis, and impairs ovarian cancer metastasis. Clin Cancer Res. 2011;17:4042–51.
1355. Chen L, Park SM, Tumanov AV, Hau A, Sawada K, Feig C, Turmer JR, Fu YX, Romero IL, Lengyel E. CD95 promotes tumour growth. Nature. 2010;465:492–6.
1356. Lin HC, Lai PY, Lin YP, Huang JY, Yang BC. Fas ligand enhances malignant behavior of tumor cells through interaction with Met, hepatocyte growth factor receptor, in lipid rafts. J Biol Chem. 2012;287:20664–73.
1357. O'Reilly LA, Tai L, Lee L, Kruse EA, Grabow S, Fairlie WD, Haynes NM, Tarlinton DM, Zhang JG, Belz GT, Smyth MJ, Bouillet P, Robb L, Strasser A. Membrane-bound Fas ligand only is essential for Fas-induced apoptosis. Nature. 2009;461:659–63.
1358. Ohta Y, Hayakawa S, Karasaki-Suzuki M, Sugita K, Komine S, Chishima F, Hatta Y, Horie T, Seo N, SDheijh A, Nemoto N, Yamamoto T. Granulocyte colony-stimulating factor suppresses autologous tumor killing activity of the peripheral blood lymphocytes in the patients with ovarian carcinoma. Am J Reprod Immunol 204;52:81–87.
1359. Wang Xie X, Lu WG, Ye DF, Cheng HZ, Li X, Cheng Q. Ovarian carcinoma cells inhibit T cell proliferation of IL-2 receptor beta and gamma expression and their JAK-STAT signaling pathway. Life Sci. 2004;74:1739–49.
1360. Baral E, Nagy E, Krepart GV, Lotocki RJ, Unruh HW, Berczi I. Antiestrogens sensitize human ovarian and lung carcinoma for lysis by autologous killer cells. Anticancer Res. 2000;20:2027–31.
1361. Leveque L, Deknuydt F, Bioley G, Old LJ, Matsuzaki J, Odunsi K, Ayyoub M, Valmori D. Interleukin2-mediated conversion of ovarian cancer-associated CD+ regulatory T cells into proinflammatory interleukin 17-producing helper T cells. J Immunother. 2009;32:101–8.
1362. Chang XH, Cheng HY, Cheng YXC, Ye X, Guo HF, Fun TY, Zhang L, Zhang G, Cui H. Specific immune cell therapy against ovarian cancer in vivo and in vitro. Ai Zheng. 2008;27:1244–50.
1363. Wright SE, Rewers-Felkins KA, Quinlin IS, Phillips CA, Townsend M, Philip R, Dobrzanski MJ, Lockwood-Cooke PR, Robinson W. Cytotoxic T-lymphocyte immunotherapy for ovarian cancer: a pilot study. J Immunother. 2012;35:196–204.
1364. Michaud NR, Jani JP, Hilleman S, Tsaparikos KE, Barbacci-Tobin EG, Knauth E, Putz H, Campbell M, Karam GA, Chrunyk B, Gebhard DF, Green LL, Xu JJ, Dunn MC, Coskran TM, Lapointe JM, Cohen BD, Coleman KG, Bedian V, Vincent P, Kajiji S, Steyn SJ, Borzillo GV, Los G. Biochemical and pharmacological characterization of human c-Met neutralizing monoclonal antibody CE-355621. MAbs. 2012;4:710–23.
1365. Feng Y, Ma PC. Anti-MET targeted therapy has come of age: the first durable complete response with MetMAB in metastatic gastric cancer. Cancer Discov. 2011;1:550–4.
1366a. Sinkovics JG, Horvath JC, Crumpton C, Lay S, Pritchard M. Immunotherapy for ovarian carcinoma. Poster presentation at the MD Anderson/Sloan-Kettering Hospitals 5th International Conference on Ovarian Cancer. Houston TX Dec 1–4 2004, Abstract #17.
1366b. Manjunath SR, Ramanan G, Dedeepiya VD, Terunuma H, Deng X, Baskar S, Senthilkumar R, Thamaraikannen P, Seinivasan T, Preethy S, Abraham SJ. Autologous immune enhancement therapy in recurrent ovarian cancer with metastases: a case report. Case Rep Oncol. 2012;5:114–8.
1367a. Kandalaft LE, Powell DJ, Singh N, Coukos G. Immunotherapy of ovarian cancer: what is next? J Clin Oncol. 2011;29:925–33.
1367b. Kandalaft LE, Powell DJ, Coukos G. A phase I clinical trial of adoptive transfer of folate receptor-alpha redirected autologous T cell for recurrent ovarian cancer. J Transl Med. 2012;10:157. doi:10.1186/1479-5876-10-157.

1368a. Fritz-Laylin LK, Cande WZ. Ancestral centriole and flagella proteins identified by analysis of Naegleria differentiation. J Cell Sci. 2010;123:4024–31.

1368b. Fritz-Laylin LK, Ginger ML, Walsh C, Dawson SC, Fulton C. The Naegleria genome: a free-living microbial eukaryote lends unique insights into core eukaryotic cell biology. Res Microbiol. 2011;162:607–18.

1369. Leidel S, Gönczy P. Centrosome duplication and nematodes: recent insights from an old relationship. Dev Cell. 2005;9:317–25.

1370. Strnad P, Leidel S, Vinogradova T, Euteneuer U, Khodjakov A, Gönczy P. Regulated HsSAS-6 levels ensure formation of a single procentriole per centriole during the centrosome duplication cycle. Dev Cell. 2007;13:203–13.

1371. Maruyama S, Misawa K, Iseki M, Watanabe M, Nozaki H. Origins of a cyanobacterial 5-phosphogluconate dehydrogenase in plastid-lacking eukaryotes. BMC Evol Biol. 2008;8:151.

1372. Rüdinger M, Volkmar U, Lenz H, Groth-Malonek M, Knoop V. Nuclear DYW-type PPR gene families diversify with increasing RNA editing frequencies in liverworth and moss mitochondria. J Mol Evol. 2012;74:37–51.

1373. Knoop V, Rüdinger M. DYW-type PPR proteins in a heterolobosean protist: Plant RNA editing factors involved in an ancient horizontal gene transfer? FEBS Lett. 2010;584:4287–91.

1374. Robbins JC, Heller WP, Hanson MR. A comparative genomics approach identifies a PPR-DYW protein that is essential for C-to-U editing of the Arabidopsis chloroplast accD transcript. RNA. 2009;15:1142–53.

1375. Rüdinger M, Fritz-Laylin L, Polsakiewicz M, Knoop V. Plant-type mitochondrial RNA editing in the protist Naegleria gruberi. RNA. 2011;17:2058–62.

1376. Zehrmann A, Verbitsky D, van der Merwe JA, Brennicke A, Takenaka M. S DYW domain-containing pentatricopeptide repeat protein is required for RNA editing at multiple sites in mitochondria of Arabidopsis thaliana. Plant Cell. 2009;21:558–67.

1377. Solgi R, Niyyati M, Haghighi, Mojarad EN. Occurrence of thermotolerant Hartmannella vermiformis and Naegleria spp in hot springs of Ardebil province, Northwest Iran. Iran. J Parasitol. 2012;7:47–52.

1378. Zbikowska E, Walczak M, Krawiec A. Distribution of Legionella pneumophila bacteria and Naegleria and Hartmanella amoebae in thermal saline baths used in balneotherapy. Parasitol Res. 2012;112:77–83.

1379. Brugerolle G, Simpson AG. The flagellar apparatus of heteroloboseans. J Eukaryot Microbiol. 2004;51:96–1071384.

1380. Desmond E, Brochier-Armanet C, Gribaldo S. Phylogenomics of the archaeal flagellum: rare horizontal transfer in a unique motility structure. BMC Evol Biol. 2007;7:106.

1381. Diniz MC, Pacheco AC, Farias KM, de Oliveira DM. The eukaryotic flagellum makes the day: novel and unforeseen roles uncovered after post-genomics and proteomics data. Curr Protein Pept Sci. 2012.

1382. Elias M, Archibald JM. The RJL family of small GTPases is an ancient eukaryotic intervention probably functionally associated with flagellar apparatus. Gene. 2009;442:63–72.

1383. Heiss AA, Walker G, Simpson AG. The ultrastructure of Ancyromonas, a eukaryote without supergroup affinities. Protist. 2011;162:373–93.

1384. Ng SY, Chaban B, Jarrell KF. Archaeal flagella, bacterial flagella and type IV pili: a comparison of genes and posttranslational modifications. J Mol Microbiol Biotechnol. 2006;11:167–91.

1385. Vonderviszt F, Sajó R, Dobó J, Závodszky P. The use of a flagellar export signal for the secretion of recombinant proteins in Salmonella. Methods Mol Biol. 2012;824:131–43.

1386. Graham SP, Honda Y, Pallé R, Mwangi DM, Glew EJ, de Villiers EP, Shah T, Bishop R, van der Bruggen P, Nene V, Taracha EL. A novel strategy for the identification of antigens that are recognized by bovine MHC class I restricted cytotoxic T cells in a protozoan infection using reverse vaccinology. Immunome Res. 2007;3:2.

1387. Hayashida K, Kajino K, Hattori M, Wallace M, Morrison I, Greene MI, Sugimoto C. MDM2 regulates a novel form of incomplete neoplastic transformation of Theileria parva infected lymphocytes. Exp Mol Pathol. 2012; pii:S0014-4800(12)00122-0. doi:10.1016/j.yexmp.2012.08.008.

1388. Schmuckli-Maurer J, Casanova C, Schmied S, Affentranger S, Parvanova I, Kang'a S, Nene V, Katzer F, Keever D, Müller J, Bishop R, Pain A, Dobbelaere DA. Expression analysis of the Theilera parva subtelomeric-encoded variable secreted protein gene family. PLoS One. 2009;4(3):4839. doi:10.1371/journal.pone.0004839.

1389. Steinaa L, Saya R, Awino E, Toye P. Cytotoxic T lymphocytes from cattle immunized against Theileria parva exhibit pronounced cross-reactivity among different strain-specific epitopes of the Tp1 antigen. Vet Immunol Immunopathol. 2012;145:571–81.

1390. von Schubert C, Xue G, Schmuckli-Maurer J, Woods KL, Nigg EA, Dobbelaere DA. The transforming parasite Theileria co-opts host cell mitotic and central spindles to persist in continuously dividing cells. PLoS Biol. 2010;8(9):pii: e10000499.

1391. Farley RD. Ultrastructure of book gill development in embryos and first instars of the horseshoe crab Limulus polyphemus L (Chelicerate, Xiphosura). Front Zool. 2012;9:4. doi:10.1186/1742-9994-9-4.

1392. Way M, Sanders M, Chafel M, Tu YH, Knight A, Matsudaira P. beta-Scruin, a homologue of the actin crosslinking protein scruin, is localized to the acrosomal vesicle of Limulus sperm. J Cell Sci. 1995;108:3155–62.

1393. Kamenz C, Staude A, Dunlop JA. Sperm carriers in Silurian sea scorpions. Naturwissenschaften. 2011;98:889–96.

1394. Rehm P, Pick C, Bomer J, Markl J, Burmester T. The diversity and evolution of chelicerate hemocyanins. BMC Evol Biol. 2012;12:19.

1395. Wirkner CS, Prendini L. Comparative morphology of the haemolymph vascular system in scorpions – a survey using corrosion casting, MicroTC, and 3D-reconstruction. J Morphol. 2007;268:401–13.

1396. Klußmann-Fricke BJ, Prendini L, Wirkner CS. Evolutionary morphology of the hemolymph vascular system in scorpions: a character analysis. Arthropod Struct Dev. 2012; pii: S1467-8039(12)00056-4. doi:10.1016/j.asd.2012.06.002.

1397. Coates CJ, Whalley T, Naim J. Phagocytic activity of Limulus polyphemus amoebocytes in vitro. J Invertebr Pathol. 2012;111:205–10.

1398. Kawabata S. Immunocompetent molecules and their response network in horseshoe crabs. Adv Exp Med Biol. 2010;708:122–36.

1399. Kawabata S, Muta T. Sadaaki Iwanaga: Discovery of the lipopolysaccharide- and beta-1,3-D-glucan-mediated proteolytic cascade and unique proteins in invertebrate immunity. J Biochem. 2010;147:611–8.

1400. Braverman H, Leibovitz L, Lewbart GA. Green algal infection of American horseshoe crab (Limulus polyphemus) exoskeletal structures. J Invertebr Pathol. 2012;111:90–3.

1401. Milsom WK. New insights into gill chemoreception: receptor distribution and roles in water and air breathing fish. Respir Physiol Neurobiol. 2012;184:326–39.

1402. Turko AJ, Cooper CA, Wright PAS. Gill remodeling during terrestrial acclimation reduces aquatic respiratory function of the amphibious fish Kryptolebias marmoratus. J Exp Biol. 2012;215:3973–80.

1403. Val AL. Oxygen transfer in fish: morphological and molecular adjustments. Brazil J Med Biol Res. 1995;28:1119–27.

1404. Pan TC, Burggren WW. Onset and early development of hypoxic ventilator response and branchioal neuroepithelial cells in Xenopus laevis. Comp Biochem Physiol A Mol Integr Physiol. 2010;157:382–91.
1405. Eo SH, Doyle JM, Hale MC, Marra NJ, Ruhl JD, DeWoody JA. Comparative transcriptomics and gene expression in larval tiger salamander (Ambystoma tigrinum) gill and lung tissues as revealed by pyrosequencing. Gene. 2012;492:329–38.
1406. Chen AK, Hedrick MS. Role of glutamate and substance P in the amphibian respiratory network during development. Respir Physiol Neurobiol. 2008;162:24–31.
1407. Torday JS, Ihida-Stansbury K, Rehan VK. Leptin stimulates Xenopus lung development: evolution in a dish. Evol Dev. 2009;11:219–24.
1408. Brundage CM, Cartagena CM, Potter EA, Taylor BE. Nicotine elicits a developmentally dependent depression in bullfrog neuroventillatory response to CO2. Respir Physiol Neurobiol. 2010;170:226–35.
1409. Kertesz N, Krasnoperov V, Reddy R, Leshanski L, Kumar SR, Zozulya S, Gill PS. The soluble extracellular domain of EphB4 (sEphB4) antagonizes EphB4-EphrinB2 interaction, modulates angiogenesis, and inhibits tumor growth. Blood. 2006;107:2330–8.
1410. Pennisi A, Ling W, Li X, Khan S, Shaughnessy JD Jr, Barlogie B, Yaccoby S. The ephrinB2/EphB 4 axis is dysregulated in osteoprogenitors from myeloma patients and its activation affects myeloma bone disease and tumor growth. Blood. 2009;114:1803–12.
1411. Jürgens KD, Gros G. Phylogeny of gas exchange systems. Anesthesiol Intensivmed Notfallmed Schmerzther. 2002;37:185–98.
1412. Tickle PG, Norell MA, Codd JAR. Ventilatory mechanics from maniraptoran theropods to extant birds. J Evol Biol. 2012;25:740–7.
1413a. Codd JAR, Klein W. Breathing, locomotion and everything in between. Comp Biochem Physiol A Mol Integr Physiol. 2010;156:301–2.
1413b. Klein W, Codd JAR. Breathing and locomotion: comparative anatomy, morphology and function. Respir Physiol Neurobiol. 2010;173S:S26–532.
1414. Perry SF, Sander M. Reconstructing the evolution of the respiratory apparatus in tetrapods. Respir Physiol Neurobiol. 2004;144:125–39.
1415. Truman JW, Riddiford LM. The origin of insect metamorphosis. Nature. 1999;401:447–52.
1416. Pandiarajan J, Cathrin BP, Pratheep T, Krishnan M. Defense role of the cocoon in the silk worm Bombyx mori. Rapid Commun Mass Spectrom. 2011;25:3203–6.
1417a. Tian L, Guo E, Diao Y, Zhou S, Peng Q, Cao Y, Ling E, Li S. Genome-wide regulation of innate immunity by juvenile hormone and 20-hydroxyecdysone in the Bombyx fat body. BMC Genomics. 2010;11:549.
1417b. Tian L, Liu S, Liu H, Li S. 20-hydroxyecdysone upregulates apoptotic genes and induces apoptosis in the Bombyx fat body. Arch Insect Biochem Physiol. 2012;79:207–19.
1418. Baum JS, Arama E, Steller H, McCall K. The drosophila caspase strica and dronc function redundantly in programmed cell death during oogenesis. Cell Death Differ. 2007;14:1508–17.
1419. Klapper R, Stute C, Schomaker O, Strasser T, Janning W, Renkawitz-Pohl R, Holz A. The formation of syncytia within the visceral musculature of the Drosophila midgut is dependent on duf, ans and mbc. Mech Dev. 2002;110:85–96.
1420. Susic-Jung L, Hornbruch-Freitag C, Kuckwa J, Rexer KH, Lammel U, Renkawitz-Pohl R. Multinucleated smooth muscles and mononucleated as well as multinucleated striated muscles develop during establishment of the male reproductive organs of Drosophila melanogaster. Dev Biol. 2012;370:86–97.
1421. Gaspar-Perira S, Fullard N, Townsend PA, Banks PS, Ellis EL, Fox C, Maxwell AG, Murphy LB, Kirk A, Bauer R, Caamaño JH, Figg N, Foo RS, Mann J, Mann DA, Oakley F. The NF-κB subunit c-Rel stimulates cardiac hypertrophy and fibrosis. Am J Pathol. 2012;180:929–39.

1422. Eblaghie MC, Song SJ, Kim JY, Akita K, Tickle C, Jung HS. Interactions between FGF and Wnt signals and Tbx3 gene expression in mammary gland initiation in mouse embryos. J Anat 204;205:1-13.
1423. Takeuchi JK, Koshiba-Takeuchi K, Suzuki T, Kamimura M, Ogura K, Ogura T. Tbx5 and Tbx4 trigger limb initiation through activation of the Wnt/Fgf signaling cascade. Development. 2003;130:2729–39.
1424. Liss AS, Tiwari R, Kralova J, Bose HR. Jar. Cell transformation by v-Rel reveals distinct roles of AP-1 family members in Rel/κNF-B oncogenesis. Oncogene. 2010;29:4925–37.
1425. Lee HH, Norris A, Weiss JB, Frasch M. Jelly belly protein activates the receptor tyrosine kinase Alk to specify visceral muscle pioneers. Nature. 2003;425:507–12.
1426. Inoue H, Imamura T, Ishidou Y, Takase M, Udagawa Y, Oka Y, Tsuneizumi K, Tabata T, Miyazono K, Kawabata M. Interplay of signal mediators of decapentaplegic (Dpp): molecular characterization of mothers against dpp, Medea, and daughters against vdpp. Mol Biol Cell. 1998;9:2145–56.
1427. Akong K, McCartney BM, Peifer M. Drosophila APC2 and APC1 have overlapping roles in the larval brain despite their distinct intracellular localizations. Dev Biol. 2002;250:71–90.
1428. Pyatkov KI, Shostak NG, Zelentsova ES, Lyozin GT, Melekhin MI, Finnegan DJ, Kidwell MG, Evgen'ev MB. Penelope retroelements from Drosophila virilis are active after transformation of Drosophila melanogaster. Proc Natl Acad Sci USA. 2002;99:16150–5.
1429. Volff JN, Hornung U, Schartl M. Fish retrotransposons related to the Penelope element of Drosophila virilis define a new group of retrotransposable element. Mol Genet Genomics. 2001;265:711–20.
1430. Gordon SD, Stand MR. The polyembryonic wasp Copidosoma floridanum produces two castes by differentially parceling the germ line to daughter embryos during embryo proliferation. Dev Genes Evol. 2009;219:445–54.
1431. Smith MS, Milton J, Strand MR. Phenotypically plastic traits regulate caste formation and soldier function in polyembryonic wasps. J Evol Biol. 2010;23:2677–84.
1432. Mukherjee K, Fischer R, Vilcinskas A. Histone acetylation mediates epigenetic regulation of transcriptional reprogramming in insects during metamorphosis, wounding and infection. Front Zool. 2012;9(1):25. doi:10.1186/1742-9994-9-25.
1433. Jiang H, Vilcinkas A, Kanost MR. Immunity in lepidopteran insects. Adv Exp Med Biol. 2010;708:181–204.
1434. Sabri A, Hance T, Leroy PD, Frère I, Haubruge E, Destain J, Compère P, Thonart P. Placenta-like structure of the aphid endoparasitic Aphidius ervi: a strategy of optimal resources acquisition. PLoS One. 2011;6(4):e18847. doi:10.1371/journal.pone.0018847.
1435. Ambros V. The functions of animal microRNAS. Nature. 2004;431:350–5.
1436. Zhang X, Zheng Y, Jagadeeswaran G, Ren R, Sunkar R, Jiang H. Identification and developmental profiling of conserved and novel microRNAs in Manduca sexta. Insect Biochem Mol Biol. 2012;42:381–95.
1437. Kemp NE. Ameloblastic secretion and calcification of the enamel layer in shark teeth. J Morphol. 1985;184:215–30.
1438. Nanci A, Bringas P Jr, Samuel N, Slavkin HC. Selachian tooth development: III. Ultrastructural features of secretory amelogenesis in Squalus acanthias. J Craniofac Genet Dev Biol. 1983;3:53–73.
1439. Sawada T, Inoue S. Mineralization of basement membrane mediates dentogingival adhesion in mammalian and nonmammalian vertebrates. Calcif Tissue Inf. 2003;73:186–95.

1440. Ibarretxe G, Aurrekoetxea M, Crende O, Badiola I, Jimenez-Rojo L, Nakamura T, Yamada Y, Unda F. Epiprofin/Sp6 regulates Wnt-BMP signaling and the establishment of cellular junctions during the bell stage of tooth development. Cell Tissue Res. 2012;350:95–107.
1441. DeVilliers P, Suggs C, Simmons D, Murrah V, Wright JT. Microgenomics of ameloblastoma. J Dent Res. 2011;90:463–9.
1442. Spaulding M, O'Learyb MA, Gatesy J. Relationship of Cetacea (Artiodactyla) among mammals: increased taxon sampling alters interpretations of key fossils and character evolution. PLoS One. 2009;4(9)e7062. doi:10.1371/journal.pone.0007062.
1443. Shen T, Xu S, Wang X, Yu W, Zhou K, Yang G. Adaptive evolution and functional constraint at TLR4 during the secondary aquatic adaptation and diversification of cetaceans. BMC Evol Biol. 2012;12:39.
1444. Milinkovitch MC, Meyetr A, Powell JAR. Phylogeny of all major groups of cetaceans based on DNA sequences from the mitochondrial genes. Mol Biol Evol. 1994;11:939–48.
1445. Boisserie JR, Fisher RE, Lihoreau F, Weston EM. Evolving between land and water: key questions on the emergence and history of the Hippopotamidae (Hippopotamidea, Cetancodonta, Cetartiodactyla). Biol Rev Camb Philos Soc. 2011;86:601–25.
1446. Orliac M, Boisserie JR, Maclatchy L, Lihoreau F. Early Miocene hippopotamids (Cetartiodactyla) constrain the phylogenetic and spatiotemporal settings of hippopotamid origin. Proc Natl Acad Sci USA. 2010;107:11871–6.
1447. Carney RM, Vinther J. Shawkey MD. D'Alba L, Ackermann J. New evidence on the colour and nature of the isolated Archaeopteryx feather. Nat Commun 2012;3:637. doi:10.1038/ncomms1642.
1448. Erickson GM, Rogers KC, Yerby SA. Dinosaurian growth patterns and rapid avian growth rates. Nature. 2001;412:429–33.
1449. Bhullar BA, Marugán-Lobón J, Racimo F, Bever GS, Rowe TB, Norell MA, Abzhanov A. Birds have paedomorphic dinosaur skulls. Nature. 2012;487:223–6.
1450. Matke A, Churakov G, Berkes P, Arms EM, Kelsey D, Brosius J, Kriegs JO, Schmitz J. Retroposon insertion patterns of neoavian birds: strong evidence for an extensive incomplete lineage sorting era. Mol Biol Evol. 2012;29:1497–501.
1451. Hickford D, Frankenberg S, Shaw G, Renfree MB. Evolution of vertebrate interferon inducible transmembrane proteins. BMC Genomics. 2012;13:155. doi:10.1186/1471-2164-13-155.
1452. Li Q, Gao KQ, Meng Q, Clarke JA, Shawkey MD, D'Alba L, Pei R, Ellison M, Norell MA, Vinther J. Reconstruction of microraptor and the evolution of iridescent plumage. Science. 2012;335:1215–9.
1453. Longrich NR, Takaryk T, Field DJ. Mass extinction of birds at the Cretaceous-Paleogene (K-Pg) boundary. Proc Natl Acad Sci USA. 2011;108:15253–7.
1454. Werner J, Griebeler EM. Reproductive biology and its impact on body size: comparative analysis of mammalian, avian and dinosaurian reproduction. PLoS One. 2011;86(12): e28442.
1455. Wang Z, Dong D, Ru B, Young RL, Han N, Guo T, Zhang S. Digital gene expression tag profiling of bat digits provides robust candidates contributing to wing formation. BMC Genomics. 2010;11:619.
1456. Zeng J, Xiang, Jiasng L, Jones G, Zheng Y, Liu B, Zhang S. Moth wing scales slightly increase the absorbance of bat echolocation calls. PLoS One. 2011;6(11):e27190.
1457. Cooper KL, Tabin CJ. Understanding of bat wing evolution takes flight. Genes Dev 208;22:121–124.
1458a. Muijres FT, Henningsson P, Stuiver M, Hedenström A. Aerodynamic flight performance in flap-gliding birds and bats. J Thor Biol 306:120–128.

1458b. Muijres FT, Johansson LC, Bowlin MS, Winter Y, Hedenström A. Comparing aerodynamic efficiency in birds and bats suggests better flight performance in birds. PLoS One. 2012;7(5):e37335.
1459. Thomas DB, Ksepka DT, Fordyce RE. Penguin heat-retention structures evolved in a greenhouse Earth. Biol Lett. 2011;7:461–4.
1460. Williams CL, Sato K, Shiomi K, Ponganis PJ. Muscle energy stores and stroke rates of emperor penguins: implications for muscle metabolism and dive performance. Physiol Biochem Zool. 2012;85:120–33.
1461. Ksepka DT, Thomas DB. Multiple Cenozoic invasions of Africa by penguins (Aves, Sphenisciformes). Proc Biol Sci. 2012;279:1027–32.
1462. Reichert H. Drosophila neural stem cells: cell cycle control of self-renewal, differentiation, and termination in brain development. Results Probl Cell Differ 201;53:529-546.
1463. Wang H, Somers GW, Bashirullah A, Heberlein U, Yu F, Chia W. Aurora-A acts as a tumor suppressor and regulates self-renewal of Drosophila neuroblasts. Genes Dev. 2006;20:3453–63.
1464. Gleason J, Eisenmann DM. Wnt signaling controls the stem cell-like asymmetric division of the epithelial seam cells during C. elegans larval development. Dev Biol. 2010;384:58–66.
1465. Perez-Reyes R. Life cycle of Trypanosoma montezumae, a frog flagellate parasite with asymmetric division. Rev Latinoan Microbiol Parasitol (Mexico). 1969;11:57–60.
1466a. Fisk JC, Sayegh J, Zurita-Lopez C, Menon S, Presnyak V, Clarke SG, Read LK. A type III protein arginine methyltransferase from the protozon parasite Trypanosoma brucei. J Biol Chem. 2009;284:11590–600.
1466b. Fisk JC, Zurita-Lopez C, Sayegh J, Tomasello DL, Clarke SG, Read LK. TbPRMT6 is a type I protein arginine methyltransferase that contributes to cytokinesis in Trypanosoma brucei. Eukaryot Cell. 2010;9:866–77.
1467. Pasternack DA, Sayegh J, Clarke S, Read LK. Evolutionarily divergent type II protein arginine methyltransferase in Trypanosoma brucei. Eukaryot Cell. 2007;6:1665–81.
1468. Fan Q, Miao J, Cui L, Cui L. Characterization of PRMT1 from Plasmodium falciparum. Biochem J. 2009;421:107–18.
1469. Takahashi Y, Daitoku H, Hirota K, Tamiya HJ, Yokoyama A, Kako K, Nagashima Y, Nakamura A, Shimada T, Watanabe S, Yamagata K, Yasuda K, Ishii N, Fukamizu A. Asymmetric arginine dimethylation determines life span in C. elegans by regulating forkhead transcription factor DAF-16. Cell Metab. 2011;13:505–16.
1470. Mitchell TRH, Glenfield K, Jevanthan K, Zhun XD. Arginine methylation regulates telomere length and stability. Mol Cell Biol. 2009;29:4918–34.
1471. Baldwin RM, Morettin A, Paris G, Goulet I, Côté J. Alternatively spliced protein arginine methyltransferase 1 isoform PRMT1v2 promotes the survival and invasiveness of breast cancer cells. Cell Cycle. 2012;11(24).
1472. Kleinschmidt MA, de Graaf P, van Teeffelen HA, Timmers HT. Cell cycle regulation by the PRMT 6 arginine methyltransferase through repression of cyclin-dependent kinase inhibitors. PLoS One. 2012;55:7978–87.
1473a. Gu Z, Li Y, Lee P, Liu T, Wan C, Wang Z. Protein arginine methyltransferase 5 functions in opposite ways in the cytoplasm and nucleus of prostate cancer cells. PLoS One. 2012;7(8):e44033. doi:10.1371/journal.pone.0044033.
1473b. Gu Z, Gao S, Zhang E, Wang Z, Ma W, Davis RE, Wang Z. Protein arginine methyltransferase 5 is essential for growth of lung cancer cells. Biochem J. 2012;446: 235–41.
1474. Kim JD, Kako K, Kakiuchi M, Park GG, Fukamizu A. EWS is a substrate of type I protein arginine methyltransferase, PRMT8. Int J Mol Med. 2008;22:309–15.
1475. Kirk DL. Germ-soma differentiation in volvox. Dev Biol. 2001;238:213–23.

1476. Nozak H, Takahara M, Nakazawa A, Kita Y, Yamada T, Takano H, Kawano S, Kato M. Evolution of rbcL group IA introns and intron open reading frames within the colonial volvocales (Chlorophyceae). Mol Phylogenet Evol. 2002;23:326–38.
1477. Blifernez O, Wobbe L, Niehaus K, Kruse O. Protein arginine methylation modulates light-harvesting antenna translation in Chlamydomonas reinhardtii. Plant J. 2011;65:119–30.
1478. Cheng Q, Fowler R, Tam LW, Edwards L, Miller SM. The role of GIsA in the evolution of asymmetric cell divisions in the green alga Volvox carteri. Dev Genes Evol 003;213:328–335.
1479a. Herron MD, Michod RE. Evolution of complexity in the volvocine algae: transitions in individuality through Darwin's eye. Evolution. 2008;62:436–51.
1479b. Dick JM, Shock EL. A metastable equilibrium model for the relative abundances of microbial phyla in a hot spring. PLoS One. 2013;8(9):72395.
1480. Debec ZA, Sullivan W, Bettencourt-Dias M. Centrioles: active players or passengers during mitosis? Cell Mol Life Sci. 2010;67:2173–94.
1481. Fábián Á, Veréb G, Szöllősi J. The hitchhikers guide to cancer stem cell theory: pathways and therapy. Cytometry A. 2013;83:62-71. doi:10.1002/cyto.a.22206.
1482. Molina-Peña R, Álvarez MM. A simple mathematical model based on the cancer stem cell hypothesis suggests kinetic commonalities in solid tumor growth. PLoS One. 2012;7(2):e26233.
1483. Jeter CR, Liu B, Liu X, Chen X, Liu C, Calhoun-Davis T, Repass J, Zaehres H, Shen JJ, Tang DG. NANOG promotes cancer stem cell characteristics and prostate cancer resistance to androgen deprivation. Oncogene. 2011;30:3833–45.
1484. Jia X, Li X, Xu Y, Zhang S, Mou W, Liu Y, Lv CH, Tan, Xiang R, Li N. SOX2 promotes tumorigenesis and increases the anti-apoptotic property of human prostate cancer. J Mol Cell Biol. 2011;3:230–8.
1485. Qin J, Liu X, Laffin B, Chen X, Choy G, Jeter CR, Calhoun-Davis T, Li H, Palapattu GS, Pang S, Lin K, Huang J, Ivanov I, Li W, Suraneni MV, Tang DG. The PSA(-/lo) prostate cancer cell population harbors self-renewing long-term tumor propagating cells that resist castration. Cell Stem Cell. 2012;10:556–69.
1486. Takahashi K, Narita M, Yokura M, Ichisaka T, Yamanaka S. Human induced pluripotent stem cells on autologous feeders. PLoS One. 2009;4(12):e8067.
1487. Petty AP, Wright SE, Rewers-Felkins KA, Yederrozos MA, Vorderstrasse BA, Lindsey JS. Targeting migration inducing gene-7 inhibits carcinoma cell invasion, early primary tumor growth, and stimulates monocyte oncolytic activity. Mol Cancer Ther. 2009;8:2412–23.
1488. Lin Ci DuJ, Shen WT, Whang EE, Donner DB, Griff N, He F, Jar Moor FD, Clark OH, Ruan DT. Mitogen-inducible gene-6 is a multifunctional adaptor protein with tumor suppressor-like activity in papillary thyroid cancer. J Clin Endocrinol Metab. 2011;96:E554–65.
1489. Shih IM. Trophoblastic vasculogenic mimicry in gestational choriocarcinoma. Mod Pathol. 2011;24:646–52.
1490. Döme B, Hendrix MJ, Paku S, Tóvári J, Tímár J. Alternative vascularization mechanism in cancer: pathology and therapeutic implications. Am J Pathol. 2007;170:1–15.
1491. Hess AR, Seftor EA, Seftor RE, Hendrix MJ. Phosphoinosite 3-kinase regulates membrane type 1-matrix metalloproteinase (MMP) and MMP-2 activity during melanoma cell vasculogenic mimicry. Cancer Res. 2003;63:4757–62.
1492. Karroum A, Mirshahi P, Faussat AM, Therwath A, Mishahi M, Hatmi M. Tubular network formation by adriamycin-resistant MCF-7 breast cancer cells is closely linked to MMP-9 and VEGFR-2/VEGF-3 over-expression. Eur J Pharmacol. 2012;685:1-7.
1493. Vartanian A, Stapanova E, Grigorieva I, Solomko E, Baryshnikov A, Lichinitser M. VEGFR1 and PKCα signaling control melanoma vasculogenic mimicry in a VEGFR2 kinase-independent manner. Melanoma Res. 2011;21:91–8.

1494. Frank NY, Schatton T, Kim S, Zhan Q, Wilson BJ, Ma J, Saab KR, Osherov V, Widlund V, Gasser M, Waage-Gasser AM, Kupper TS, Murphy GF, Frank MH. VEGFR-1 expressed by malignant melanoma-initiating cells is required for tumor growth. Cancer Res. 2011;71:1474–85.

1495a. Valyi-Nagy K, Kormos B, Ali M, Shukla D, Valyi-Nagy T. Stem cell marker CD271 is expressed by vasculogenic mimicry-forming uveal melanoma cells in three-dimensional cultures. Mol Vis. 2012;18:588–92.

1495b. Valyi-Nagy K, Dosa S, Kovacs SK, Bacsa S, Voros A, Shukla D, Folberg R, Valyi-Nagy T. Identification of virus resistant tumor cell subpopulations in three-dimensional uveal melanoma cultures. Cancer Gene Ther. 2010;127:223–34.

1496. Wu X, Zhou J, Rogers AM, Jänne PA, Benedettini E, Loda M, Hodi FS. C-Met, epidermal growth factor receptor, and insulin-like growth factor-1 receptor are important for growth in uveal melanoma and independently contribute to migration and metastatic potential. Melanoma Res. 2012;22:123–32.

1497. Desbois-Mouthon C, Baron A, Blivet-Van Eggelpoël MJ, Fartoux L, Venot C, Bladt F, Housset C, Rosmorduc O. Insulin-like growth factor-1 receptor inhibition induces a resistance mechanism via the epidermal growth factor receptor/HER3/AKT signaling pathway: rational basis for cotargeting insulin-like growth factor-1 receptor and epidermal growth factor receptor in hepatocellular carcinoma. Clin Cancer Res. 2008;15:5445–56.

1498. Paznekas WA, Okajima K, Schertzer M, Wood S, Jabs EW. Genomic organization, expression, and chromosome location of the human SNAIL gene (SNAI1) and a related processed pseudogene (SNAI1P). Genomics. 1999;62:42–9.

1499. Rangel MC, Karasawa H, Castro NP, Nagaoka T, Salomon DS, Bianco C. Role of Cripto-1 during epithelial-to-mesenchymal transition in development and cancer. Am J Pathol. 2012;189:2188–200.

1500. Yoon CH, Kim MJ, Lee H, Kim RK, Lim DEJ, Too KC, Cui YH, Oh YS, Gye MC, Lee YY, Park IC, An S, Hwang SG, Park MJ, Suh Y, Lee SJ. PTTG1 oncogene promotes tumor malignancy via epithelial to mesenchymal transition and expansion of cancer stem cell population. J Biol Chem. 2012;287:19516–27.

1501. Sinkovics JG, Györkey F, Kusyk C, Siciliano MJ. Growth of human tumor cells in established cultures. In: Busch H, editor. Methods in cancer research. Vol XV. New York: Academic Press; 1978. pp 243–323.

1501a. DiSaia PJ, Sinkovics JG, Rutledge FN, Smith JP. Cell-mediated immunity to human malignant cells. A brief review and further studies with two gynecologic tumors. Am J Obstet Gynecol. 1972;114:979–89.

1502. Fong MY, Kakar SS. The role of cancer stem cells and the side population in epithelian ovarian cancer. Histol Histopathol. 2010;25:113120.

1503. Shah PP, Kakatr SS. Pituitary tumor transforming gene induces epithelial to mesenchymal transition of Twist, Snail, Slug, and E-cadherin. Cancer Lett. 2011;311:66–76.

1504. Sun T, Sun BC, Zhao XK, Zhao N, Dong XY, Che N, Yao Z, Ma YM, Gu Q, Zong WK, Liu ZY. Promotion of tumor cell metastasis and vasculogenic mimicry by way of transcription coactivation by Bcl-2 and Twist1: a study of hepatocellular carcinoma. Hepatology. 2011;54:1690–709.

1505. Arzumanyan A, Friedman T, Kotei E, Ng IO, Lian Z, Feitelson MA. Epigenetic repression of E-cadherin expression by hepatitis B virus X antigen in liver cancer. Oncogene. 2012;31:563–72.

1506a. Ogunwobin OO, Liu C. Hepatocyte growth factor upregulation promotes carcinogenesis and epithelial-mesenchymal transition in hepatocellular carcinoma via Akt and COX-2 pathways. Clin Exp Metastasis. 2011;28:721–31.

1506b. Ogunwobin OO, Wang T, Zhang L, Liu C. Cyclooxygenase-2 and Akt mediate multiple growth-factor-induced epithelial-mesenchymal transition in human hepatocellular carcinoma. J Gastroenterol Hepatol. 2012;27:566–78.

1506c. Ogunwobin OO, Liu C. Therapeutic and prognostic importance of epithelial-mesenchymal transition in liver cancers: insights from experimental models. Crit Rev Oncol Hematol. 2012;83:319–28.

1507. Tong ZT, Cai MY, Wang XG, Kong LL, Mai SJ, Liu YH, Zhang HB, Liao YJ, Zheng F, Zhu W, Bian XW, Guan XY, Lin MC, Zeng MS, Zeng YX, Kung HF, Xie D. EZH2 supports nasopharyngeal carcinoma cell aggressiveness by forming a co-repressor complex with HDAC1/HDAC2 and Snail to inhibit E-cadherin. Oncogene. 2012;31:583–94.

1508. Luo GQ, Li JH, Cao L, Zho YH, Wen JF. Activator protein-1 involvement in proliferation inhibition by gene silencing of Twist in gastric cancer cells. Pathology. 2011;43:697–701.

1509. Ishida M, Miyamoto M, Naitoh S, Tatsuda D, Hasegawa T, Nemoto T, Yokozeki H, Nishioka K, Matsukage A, Ohki M, Ohta T. The SYT-SSX fusion protein down-regulates the cell proliferation regulator COM1 in t(x;18) synovial sarcoma. Mol Cell Biol. 2007;27:134–1355.

1510. Lubieniecka JM, de Bruijn DR, Su L, van Dijk AH, Subramanian S, van de Rijn M, Poulin N, van Kessel AG, Nielsen TO. Histone deacetylase inhibitors reverse SS18-SSX-mediated polycomb silencing of the tumor suppressor growth response 1 in synovial sarcoma. Cancer Res. 2008;68:4303–10.

1511. Suehara Y, Tochigi N, Kubota D, Kikuta K, Nakavama R, Seki K, Yoshida A, Ichikawa H, Hasegawa T, Kaneko K, Chuman H, Beppu Y, Kawai A, Kondo T. Sacernin-1 as a novel prognostic biomarker candidate of synovial sarcoma revealed by proteomics. J Proteomics. 2011;74:829–42.

1512. Shirley SH, Greene VR, Duncan LM, Cabala CA, Grimm EA, Kusewitt DF. Slug expression during melanoma progression. Am J Pathol. 2012;180:2479–89.

1513. Pierrat MJ, Marsaud V, Mauviel A, Javelaud D. Expression of microphthalmia-associated transcription factor (MITF), which is critical for melanoma progression, is inhibited by both transcription factor GLI2 and transforming growth factor-β. J Biol Chem. 2012;287:17996–8004.

1514. Hao L, Ha JR, Kuzel P, Garcia E, Persad S. Cadherin switch from E- to N-cadherin in melanoma progression is regulated by the PI3K/PTEN pathway through Twist and Snail. Br J Dermatol. 2012;166:1184–97.

1515. Poell JB, Van Haastert RJ, de Gunst Schultz IJ, Gommans WM, Verheul M, Cerisoli F, van Noort PI, Prevost GP, Cuppen E. A functional screen identifies specific microRNAs capable of inhibiting human melanoma cell viability. PLoS One. 2012;7(8):e43569. doi:10.1371/journal.pone.0043569.

1516. Blumenschein GR, Mills GB, Gonzalez-Angulo AM. Targeting the hepatocyte growth factor-cMET axis in cancer therapy. J Clin Oncol. 2012;30:3287–96.

1517. Stites EC. The response of cancers to BRAF inhibition underscores the importance of cancer systems biology. Sci Signal. 2012;5(246):46. doi:10.1126/scisignal.2003354.

1518. Straussman R, Morikawa T, Shee K, Barzily-Rokni M, Qian ZR, Du J, Davis A, Mongare MM, Gould J, Frederick DT, Cooper ZA, Chapman PB, Solit DB, Ribas A, Lo RS, Flaherty KT, Ogino S, Wargo JA. Tumour micro-environment elicits innate resistance to RAF inhibitors through HGF secretion. Nature. 2012;487:500–4.

1519. Cui JJ, McTigue, Nambu M, Tran-Dubé M, Pairish M, Shen H, Jia L, Cheng H, Hoffman J, Le P, Jalale M, Goetz GH, Ryan K, Grodsky N, Deng YL, Parker M, Timofeevski S, Murray BW, Yamazaki S, Aguirre S, Li Q, Zou H, Christensen J. Discovery of a novel class of exquisitely selective mesenchymal-epithelial transition factor (cMET) protein kinase inhibitors and identification of the clinical candidate 2-(4-(1-(quinolin-6-methyl)-1H-[1,2,3]triazolol[4,5-b]pyrazin-6yl)-1H-pyrazol-1-ethanol (PF-04217903) for the treatment of cancer. J Med Chem. 2012;55(18):8091–8109. Erratum 2012;55(22):1031. doi:10.1021/JM3014647.

1520. Yang SP, Zhang XW, Ai J, Gan LS, Xu JB, Wang Y, Su ZS, Wang L, Ding J, Geng MY, Yue JM. Potent HGF/c-MET axis inhibitors from Eucalyptus globules: the coupling of phloroglucinol and sesquiterpenoid is essential for the activity. J Med Hem. 2012;55:8183–7.

1521. Lue HW, Yang X, Wang R, Qian W, Xu RZ, Lyes R, Osunkova AQ, Zhou BP, Vessella RL, Zayzafoon M, Liu ZR, Zhau HE, Chung LW. LIV-1 promotes prostate cancer epithelial-to-mesencymal transition and metastasis through HB-EGF shedding and EGFR-mediated ERK signaling. PLoS One. 2011;6(11):e27720.

1522. Unno J, Satoh K, Hirota M, Kanno A, Hamada S, Masamune A, Tsukamoto N, Motoi F, Egawa S, Unno M, Horiii A, Shimosegawa T. Liv-1 enhances the aggressive phenotypes through the induction of epithelial to mesenchymal transition in human pancreatic carcinoma cells. Int J Oncol. 2009;35:813–21.

1523. Albino D, Longoni N, Curti L, Mello-Grand M, Pinton S, Civenni G, Thalmann G, D'Ambrosio G, Sarti M, Sessa F, Chiorino G, Catapano CV, Carbone GM. ESE3/EHF controls epithelial cell differentiation and its loss leads to prostate tumors with mesenchymal and stem-like features. Cancer Res. 2012;72:2889–900.

1524. Powell MJ, Casimiro MC, Cordon-Cardo C, He X, Yeow WS, Wang C, McCue PA, McBurney MW, Pestell RG. Disruption of a Sirt1-dependent autophagy checkpoint in the prostate results in prostate intraepithelial neoplasia lesion formation. Cancer Res. 2011;71:964–75.

1525. Byles V, Zhu L, Lovaas JD, Chmilewski LK, Wang J, Faller DV, Dai Y. SIRT-1 induces EMT by cooperating with EMT transcription factors and enhances prostate cancer cell migration metastasis. Oncogene. 2012;31:4619–29.

1526. Kurahara H, Takao S, Maemura K, Mataki Y, Kuwahata T, Maeda K, Ding Q, Sakoda M, Lino S, Ishigami S, Ueno S, Shinchi H. Natsugoe S. Epithelial-mesenchymal transition and mesenchymal-epithelial transition via regulation of ZEB-1 and ZEB-2 expression in pancreatic cancer. J Surg Oncol. 2012;105:655–61.

1527. Mulholland DJ, Kobayashi N, Ruscetti M, Zhi A, Tran LM, Huang J, Gleave M, Wu H. PTEN loss and RAS/MAPK activation cooperate to promote EMT and metastasis initiated from prostate cancer stem/progenitor cells. Cancer Res. 2012;72:1878–89.

1528. Sun Y, Wang BE, Leong KG, Yue P, Li L, Jhunjhunwala S, Chen D, Seo K, Modrusan Z, Gao WQ, Settleman J, Johnson L. Androgen deprivation causes epithelial-mesenchymal transition in the prostate: implications for androgen-deprivation therapy. Cancer Res. 2012;72:527–36. doi:10.1158/0008-5472.CAN-11-3004.

1529. Slabáková E, Pernicová Z, Slavičková E, Staršichová A, Kozubik A, Souček K. TYGF-β1-induced EMT of non-transformed prostate hyperplasia cells is characterized by early induction of SNAI2/Slug. Prostate. 2011;71:1332–43.

1530a. Liu DF, Wub JT, Wang JM, Liu QZ, Zl Gao, Liu YX. MicroRNA expression profile analysis reveals diagnostic biomarker for human prostate cancer. Asian Pac J Cancer Prev. 2012;13:3313–7.

1530b. Liu J, Huang J, He Y, Liu J, Liao B, Liao G. Genetic variants in the integrin gene predicted microRNA-binding sites were associated with the risk of prostate cancer. Mol Carcinog. 2012; doi:10.1002/mc.21973.

1531a. Shankar S, Chen Q, Siddiqui I, Sarva K, Srivastava RK. Sensitization of TRAIL-resistant LNCaP cells by resveratrol (3,4',5 trihydroxystilbene): molecular mechanisms and therapeutic potential. J Mol Signal. 2007;2:7.

1531b. Shankar S, Siddiquin I, Srivastava RK. Molecular mechanisms of resveratrol (3,4,5-trihydroxy-trans-stilbene) and its interaction with TNF-related apoptosis inducing ligand (TRAIL) in androgen-insensitive prostate cancer cells. Mol Cell Biochem. 2007;304:273–85.

1532. Phillip CJ, Giardina CK, Bilir B, Cutler DJ, Lai YH, Kucuk O, Moreno CS. Genistein cooperates with histone deacetylase inhibitor vorinostat to induce cell death in prostate cancer cells. BMC Cancer. 2012;12:145. doi:10.1186/1471-2407-12-145.

1533. Chua HL, Bhat-Nakashatri P, Clare SE, Morimiya A, Badve S, Nakshatri H. NF-kappaB represses E-cadherin expression and enhances epithelial to mesenchymal transition of mammary epithelial cells: potential involvement of ZEB-1 and ZEB-2. Oncogene. 2007;26:711–24.

1534. Dong C, Wu Y, Yao J, Wang Y, Yu Y, Rychahou PG, Evers BM, Zhou BP. G9a interacts with Snail and is critical for Snail-mediated E-cadherin repression in human breast cancer. J Clin Invest. 2012;122:1469–96.

1535. Moes M, Le Béchec A, Crespo I, Laurini C, Halavatyl A, Vetter G, Del Sol A, Friederich E. A novel network integrating a miRNA-203/SNAI1 fedback loop which regulates epithelial to mesenchymal transition. PLoS One. 2012;7(4):e35440.

1536a. Han M, Liu M, Wang Y, Mo Z, Bi X, Liu Z, Fan Y, Chen X, Wu C. Re-expression of miR-21 contributes to migration and invasion by inducing epithelial-mesenchymal transition consistent with cancer stem cell characteristics in MCF-7 cells. Mol Cell Biochem. 2012;363:427–36.

1536b. Han M, Wang Y, Liu M, Bi X, Bao J, Zeng N, Zhu Z, Mo Z, Wu C, Chen X. miR-21 regulates epithelial-mesenchymal transition phenotype and hypoxia-inducible factor-1α expression in third-sphere forming breast cancer stem cells. Cancer Sci. 2012;103:1058–64.

1536c. Han M, Liu M, Wang Y, Chen X, Xu J, Sun Y, Zhao L, Qu H, Fan Y, Wu C. Antagonism of miR-21 reverses epithelial-mesenchymal transition and cancer stem cell phenotype through AKT/ERK1/2 inactivation by targeting PTEN. PLoS One. 2012;7(6): e39520.

1537a. Conley SJ, Wicha MS. Antiangiogenic agents: fueling cancer's hypoxic roots. Cell Cycle. 2012;11:1265–6.

1537b. Conley SJ, Gheordunescu E, Kakarala P, Newman B, Korkaya H, Heath AN, Clouthier SG, Wicha MS. Antiangiogenic agents increase breast cancer stem cells via the generation of tumor hypoxia. Proc Natl Acad Sci USA. 2012;109:2784–9.

1538. Polyak K, Vogt PK. Progress in breast cancer research. Proc Natl Acad Sci USA. 2012;109:2715–7.

1539a. Santisteban M, Reiman JM, Asiedu MK, Behrens MD, Nassar A, Kalli KR, Haluska P, Ingle JN, Hartmann LC, Manjili MH, Radisky DC, Ferrone S, Knutson KL. Immune-induced epithelial to mesenchymal transition in vivo generates breast cancer stem cells. Cancer Res. 2009;69:2887–95.

1539b. Santisteban M, Reynolds C, Barr Fritcher EG, Frost MH, Vierkant RA, Anderson SS, Degnim AC, Visscher DW, Pankratz VS, Hartmann LC. Ki67: a time-varying biomarker of risk of breast cancer in atypical hyperplasia. Breast Cancer Res Treat. 2010;121:431–7.

1540a. Morel A-P, Liévre M, Thomas C, Hinkal G, Ansieau S, Puisieux A. Generation of breast cancer stem cells through epithelial-mesenchymal transition. PLoS One. 2008;3(8): e2888.

1540b. Morel AP, Lievre M, Thomas C, Hinkal G, Ansieau S, Puisieux A. Generation of breast cancer stem cells through epithelial-mesenchymal transition. PLoS One. 2008;3:e28888.

1541. DittmarT, Zänker KS, authors/editors. Cell fusion in health and disease, II. Cell fusion in disease. Advances in experimental medicine and biology. Vol 714. New York: Springer;2011. p. 1–208.

1542. Fan Y, Bergmann A. The cleaved-caspase-3 antibody is a marker of caspase-9-like DRONC activity in drosophila. Cell Death Differ. 2010;17:534–9.

1543. Lee TV, Fan Y, Wang S, Srivastava M, Broemer M, Meier P, Bergmann A. Drosophila IAP1[-mediated ubiquitylation controls activation of the initiator caspase DRONC independent of protein degradation. PLoS Genet. 2011;7(9):e1002261. doi:10.1371/journal.pgen.1002261.

1544. Bender CE, Fitzgerald P, Tait SW, Llamni F, McStay GP, Tupper DO, Pellettier J, Sánchez-Alvarado A, Salversden GS, Green DR. Mitochondrial pathway of apoptosis is ancestral in metazoans. Proc Natl Acad Sci USD. 2012;109:4904–9.

1545. Dichtel-Danjoy ML, Ma D, Dourien P, Chatelain G, Napoletano F, Robin M, Corbet M, Levet C, Hafsi H, Hainaut P, Ryoo HD, Bourdon JC, Mollereau B. Drosophila p53 isoforms differentially regulate apoptosis and apoptosis-induced proliferation. Cell Death Differ. 2013;20:108–16.
1546. Dorstyn L, Kumar S. A biochemical analysis of the activation of the drosophila caspase DRONC. Cell Death Differ. 2008;15:461–70.
1547. Rutkowski R, Hofman K, Gartner A. Phylogeny and function of the invertebrate p53 superfamily. Cold Spring Harb Perspect Biol. 2010;2(7):a001131.
1548. Snipas SJ, Drag M, Stennice HR, Salversen GS. Activation mechanism and substrate specificity of the drosophila initiator caspase DRONC. Cell Death Differ. 2008;15:938–45.
1549. Zmasek CM, Zhang Q, Ye Y, Godzik A. Surprising complexity of the ancestral apoptosis network. Genome Biol. 2007;8:R226. doi:10.1186/gb-2007-8-10-r226.
1550. Chen J, Xie F, Zhang L, Jiang WG. iASPP is over-expressed in human non-small cell lung cancer and regulates the proliferation of lung cancer cells. BMC Cancer. 2010;10:694. doi:10.1186/1471-2407-10-694.
1551. González JM, Esteban M. A poxvirus Bcl-2-like gene family involved in regulation of host immune response: sequence similarity and evolutionary history. Virol J. 2010;7:59. doi:10.1186/1743-422X-7-59.
1552. Douglas AE, Corbett KD, Berger JM, McFadden G, Handel TM. Structure of M121L: a myxoma virus structural homolog of the apoptosis inhibitor, Bcl-2. Protein Sci. 2007;16:695–703.
1553. Deng X, Li X, Shen Y, Qiu Y, Shi Z, Shao DF, Jin Y, Chen H, Ding C, Li L, Chen P, Ma Z. The Meq oncoprotein of Marek's disease virus interacts with p53 and inhibits its transcriptional and apoptotic activities. Virol J. 2010;7:348. doi:10.1186/1743-422X-7-348.
1554. Inoue Y, Iemura S, Natsume T, Miyazawa K, Imamura T. Suppression of p53 activity through the cooperative action of Ski and histone deacetylase SIRT1. J Biol Chem. 2011;286:6311–20.
1555. Gyöngyösi E, Szalmás A, Ferenczi A, Kónya J, Gergely L, Veress G. Effects of human papillomavirus (HPV) type 16 oncoproteins on the expression of involucrin in human keratinocytes. Virol J. 2012;9:36. doi:10.1186/1743-422X-9-36.
1556. Karunakaran K, Mehlitz A, Rudel T. Evolutionary conservation of infection-induced cell death inhibition among Chlamydia. PLoS One. 2011;6(7):e22528. doi:10.1371/journal.pone.0022528.1557.
1557. Sinkovics JG, Horvath JC, Kay HD. NKT cells emerge to restrain the trophoblast: Tumor cells imitating the trophoblast evoke NKT cell reactions. In: Cancer immunity: challenges for the next decade. Program & abstracts. The 56th annual symposium on Fundamental Cancer Research, The University of Texas M.D. Anderson Cancer Center, Houston, TX, 14–17 Oct 2003. p. 117, I-33.
1558. Sinkovics JG, Horvath JC. The evolution of natural killer cells for the practice of antiviral and anti-cellular defense. In: Abstracts, The 20th annual clinical virology symposium & annual meeting PanAmerican Society for Clinical Virology, Department of Clinical Microbiology & Immunology, The University of South Florida College of Medicine, Clearwater, FL 25–28 April 2004. #M20.
1559. Berencsi III GY, editor. Maternal fetal transmission of human viruses and their influence on tumorigenesis. Dordrecht: Springer Science;2012. p. v–vi, 1–463.
1560. Momand J, Villegas A, Belyi VA. The evolution of the MDM2 family genes. Gene. 2011;486:23–38.
1561. Feng XL, Liu QT, Cao RB, Zhou B, Ma ZY, Deng WL, Wei JC, Qiu YE, Wang FQ, Gub JY, Wang FJ, Zheng QS, Ishag H, Chen PY. Identification and characterization of novel immunomodulatory bursal-derived pentapepti de-II (BPP-II). J Biol Chem. 2012;287:3798–807.

1562. Belyi VA, Ak P, Markert E, Wang H, Hu W, Puzio-Kuter A, Levine J. The origins and evolution of the p53 family of genes. Cold Spring Harb Perspect Biol. 2010;2(6): a001198. doi:10.1101/cshperspect.a001198.

1563. Feng Z, Zhang C, Kang HJ, Sun Y, Wang H, Nagvi A, Frank AK, Rosenwaks Z, Murphy ME, Levine AJ, Hu W. Regulation of female reproduction by p53 and its family members. FASEB J. 2011;25:2245–55.

1564. Hu W, Feng Z, Atwal GS, Levine AJ. P53: a new player in reproduction. Cell Cycle. 2008;7:848–52.

1565. Gu L, Zhu N, Findley HW, Zhou M. MDM2 antagonist nutin-3 is a potent inducer of apoptosis in pediatric acute lymphoblastic leukemia cells with wild-type p53 and overexpression of MDM2. Leukemia. 2008;22:730–9.

1566. Renouf B, Hollville E, Pujals A, Tétaud C, Garibal J, Wiels J. Activation of p53 by MDM2 antagonist has differential apoptotic effects on Epstein-Barr virus (EBV)-positive and EBV-negative Burkitt's lymphoma cells. Leukemia. 2009;23:1557–63.

1567. Miyachi M, Kakazu N, Yagyu S, Katsumi Y, Tsubai-Shimizu S, Kikuchi K, Tsuchiya K, Iehara T, Hosoi H. Restoration of p53 pathway by nutlin-3 induces cell cycle arrest and apoptosis in human rhabdomyosarcoma cells. Clin Cancer Res. 2009;15:4077–84.

1568. Muttray AF, O'Toole TF, Morrill W, Van Beneden RJ, Baldwin SA. An invertebrate mdm homolog interacts with p53 and is differentially expressed together with p53 and ras in neoplastic Mytilus trossulus haemocytes. Comp Biochem Physiol Biochem Mol Biol. 2010;156:298–306.

1569. Wojcik IO, Szybka M, Golanska E, Rieske P, Blonski JZ, Robak T, Bartkowiak J. Abnormalities of the P53, MDM2, BCL2 and BAX genes in acute leukemias. Neoplasma. 2005;52:318–24.

1570. Walker CW, Böttger SA. A naturally occurring cancer with molecular connectivity to human disease. Cell Cycle. 2008;7:2286–9.

1571. Jessani N, Niessen S, Mueller BM, Cravatt BE. Breast cancer cells grown in vivo. Cell Cycle. 2005;5:253–5.

1572. Brinkley BR, Beall PT, Wible LJ, Mace ML, Turner DS, Cailleau RM. Variations in cell form and cytoskeleton in human breast carcinoma cells in vitro. Cancer Res. 1980;40:3118–29.

1573a. Sinkovics JG, Reeves WJ, Cabiness JR. Cell- and antibody-mediated immune reactions of patients to cultured cells of breast carcinoma. J Natl Cancer Inst. 1972;48:1145–9.

1573b. Sinkovics JG, Ahmed N, Cabiness JR, Reeves WJ. Serum factors specifically antagonistic or synergistic with cytotoxic lymphocytes in patients with neoplastic disease. In: Proceedings of 63rd annual meeting of American Association for Cancer Research. 4–6 May, 1072 Boston, MA, 1972;13:109 #436.

1573c. Sinkovics JG, Reeves WJ, Cabiness JAR. Reactions of leukocytes and sera of patients with cultured tumor cells. In: Ceglowski W, Friedman H, editors. Conference on virus tumorigenesis and immunogenesis. New York: Academic Press;1973. p. 373–386.

1574. Walker C, Böttger SA, Mulkern J, Jerszyk E, Litvaitis M, Lesser M. Mass culture and characterization of tumor cells from a naturally occurring invertebrate cancer model: applications for human and animal disease and environmental health. Biol Bull. 2009;216:23–39.

1575a. Editorial. Special issue in honour of Stanley J. Korsmeyer. Cell Death Differ. 2006;13:1245–142.

1575b. Oltvai ZN, Milliman CL, Korsmeyer SJ. Bcl-2 heterodimerizes in vivo with a conserved homolog, Bax, that accelerates programmed cell death. Cell. 1993;74:609–19.

1576. Wolenski FS, Garbati MR, Lubinski J, Traylor-Knowles N, Dresselhaus E, Stefanik DJ, Goucher H, Finnerty JAR, Gilmore TD. Characterization of the core elements of the NF-κB signaling pathway of the sea anemone Nematostella vectensis. Mol Cell Biol. 2011;31:1076–87.

1577. Srivastava N, Manvati Srivastava A, Pal R, Kalaiarasan P, Chattopadhyay S, Gochhait S, Dua R, Bamezai ENK. miR-24-2 controls H2AFX expression regardless of gene copy number alteration and induces apoptosis by targeting antiapoptosis gene BCL-2: a potential for therapeutic intervention. Breast Cancer Res. 2011;13(2):R39. doi:10.1186/bcr2861.

1578. Roberts AW, Seymour JE, Brown JR, Wierda WG, Kipps TJ, Khaw SL, Carney DA, He SZ, Huang DC, Xiong H, Cui Y, Busman TA, McKeegan EM, Krivoshik AP, Enschede SH, Humerickhouse R. Substantial susceptibility of chronic lymphocytic leukemia to BCL2 inhibition; results of a phase I study of navitoclax in patients with relapsed or refractory disease. J Clin Oncol. 2012;30:488–96.

1579. Hampl V, Silberman JD, Stechmann A, Diaz-Triviño S, Johnson PJ, Roger Aj. Genetic evidence for a mitochondriate ancestry in the 'amitochondriate' flagellate Trimastix pyriformis. PLoS One. 2008;3(1):e1383. doi:10.1371/journal.pone.0001383.

1580. Walensky LD. From mitochondrial biology to magic bullet: navitoclax disarms BCL-2 in chronic lymphocytic leukemia. J Clin Oncol. 2012;30:554–7.

1581. Erdélyi P, Borsos E, Takács-Vellai K, Kovács T, Kovács AL, Sigmond T, Hargitai B, Pásztor L, Sengupta T, Dengg M, Pécsi I, Tóth J, Nilsen H, Vértessy BG, Vellai T. Shared developmental roles and transcriptional control of autophagy and apoptosis in Caenorhabditis elegans. J Cell Sci. 2011;124:1510–8.

1582. Goehe RW, Bristol ML, Wilson EN, Gewirtz DA. Autophagy, senescence, and apoptosis. Methods Mol Biol. 2013;962:31–48.

1583. Walsh CM, Edinger AL. The complex interplay between autophagy, apoptosis, and necrotic signals promotes T-cell homeostasis. Immunol Rev. 2010;236:95–109.

1584. Yuan H, Perry CN, Huang C, Iwai-Kanai E, Carreira RS, Glembotski CC, Gottlieb RA. LPS-induced autophagy is mediated by oxidative signaling in cardiomyocytes and is associated with cytoprotection. Am J Physiol Heart Circ Physiol. 2009;296:H470–9.

1585. Heath RJ, Xavier RJ. Autophagy, immunity and human disease. Curr Opin Gastroenterol. 2009;25:512–20.

1586. Blagosklonny MV. Hypoxia. MTOR and autophagy. Autophagy. 2013;9:1–3.

1587. Coppé JF, Rodier F, Patil CK, Freund A, Desprez PY, Campisi J. Tumor suppressor and aging biomarker p16(INK4a) induces cellular senescence without the associated inflammatory secretory phenotype. J Biol Chem. 2011;286:36396–403.

1588. Capparelli C, Chiavarina B, Whitaker-Menenzes D, Pestell TG, Pestell FG, Hulit J, Andó, Howell A, Martinez-Outschoorn UE, Sotgia F, Lisanti MP. CDK inhibitors (p16/p19/p21) induce senescence and autophagy in cancer-asociated fibroblasts, "fueling" tumor growth via paracrine interactions, without an increase in neoangiogenesis. Cell Cycle. 2012;11:3599–610.

1589. Kang HT, Lee KB, Kim HR, Park SC. Autophagy impairment induces premature senescence in primary human fibroblasts. PLoS One. 2011;6(8):e23367n. doi:10.1371/journal.pone.0023387.

1590. Kongara S, Karantza V. The interplay between autophagy and ROS in tumorigenesis. Front Oncol. 2012;2:171. doi:10.3389/foncv.2012.00171.

1591. Yue ZY, Jin SK, Yang CW, Levine AJ, Heintz N. Beclin 1, an autophagy gene essential for early embryonic development, is a haploinsufficient tumor suppressor. Proc Natl Acad Sci USA. 2003;100:15077–82.

1592. Mohapatra P, Preet R, Das D, Satapathy SR, Choudhuri T, Wyatt MD, Kundu CN. Quinacrine-mediated autophagy and apoptosis in colon cancer cells is through a p53- and p21-dependent mechanism. Oncol Res. 2012;20:81–91.

1593. Chaabane W, User SD, El-Gazzah M, Jaksik R, Saijadi E, Rzeszowska-Wolny J, Los MJ. Autophagy, apoptosis, mitoptosis and necrosis: interdependence between those pathways and effects on cancer. Arch Immunol Ther Exp (Wroclav/Warszow). 2013;61:43–58.

1594. Nehs MA, Lin CI, Kozono DE, Whang EE, Cho NL, Zhu K, Moalem J, Moore FD, Ruan DT. Necroptosis is a novel mechanism of radiation-induced thyroid and adrenocortical cancers. Surgery. 2011;150:1032–9.
1595. Han W, Xie J, Fang Y, Wang Z, Pan H. Nec-1 enhances shikonin-induced apoptosis in leukemia cells by inhibition of RIP-1 and ERK1/2. Int J Med Sci. 2012;13:7212–25.
1596. Qased AB, Yi H, Liang N, Ma S, Qiao S, Liu X. MicroRNA-18a upregulates autophagy and ataxia telangiectasia mutated gene expression in HCT116 colon cancer cells. Mol Med Report 2012; doi:10.3892/mmr.2014.
1597. Golden EB, Pelliciotta I, Demaria S, Barcellos-Hoff MH, Formenti SC. The convergence of radiation and immunogenic cell death signaling pathways. Front Oncol. 2012;2:88. doi:10.3389/fonc.2012.0008.
1598. Hegyi H, Buday L, Tompa P. Intrinsic structural disorder confers cellular viability on oncogenic fusion proteins. PLoS Comput Biol. 2009;5910:e1000552. doi:10.1371/journal.pcbi.100055.
1599. Bateman A, Birney E, Durin R, Eddy SR, Howe KL, Sonnhammer EL. The Pfam protein families database. Nucl Acids Res 2000;28:263–6.
1600. Punta M, Coggill PC, Eberhardt RY, Mistry J, Tate J, Boursnell C, Pang N, Forslund K, Ceric G, Clements J, Heger A, Holm L, Sonnhammer ELL, Eddy SR, Bateman A, Finn RD. The Pfam protein families database. Nucl Acids Res. 2012;50:D290–301.
1601. Billington WD. The immunological problem of pregnancy: 50 years with the hope of progress. A tribute to Peter Medawar. J Reprod Immunol. 2003;60:1–11.
1602. Chakraborty S, Valiya Veetti M, Sadagopan S, Paudel N, Chandran B. c-Cbl-mediated selective virus-receptor translocations into lipid rafts regulate productive Kaposi's sarcoma-associated herpesvirus infection in endothelial cells. J Virol. 2011;85:12410–30.
1603. Jain SK, Langdon WY, Varticovski L. Tyrosine phosphorylation of p120cbl in BCR/abl transformed hematopoietic cells mediates enhanced association with phosphatidyl inositol 3-kinase. Oncogene. 1997;14:1127–2228.
1604. Goga A, McLaughlin J, Afar DE, Safran DC, Witte ON. Alternative signals to RAS for hematopoietic transformation by the BCR-ABL oncogene. Cell. 1995;82:981–8.
1605. Kao HW, Sanada M, Liang DC, Lai CL, Lee EH, Kuo MC, Lin TL, Shib YS, Wu JH, Huang CF, Ogawa S, Shih LY. A high occurrence of acquisition and/or expansion of C-CBL mutant clones in the progression of high-risk myelodysplastic syndrome to acute myeloid leukemia. Neoplasia. 2011;13:1035–42.
1606a. Weinkauf C, Pereiraperrin M. Trypanosoma cruzi promotes neuronal and glial cell survival through the neurotrophic receptor TrkC. Infect Immun. 2009;77:1368–75.
1606b. Weinkauf C, Salvador R, PreiraPerrin M. Neurotrophin receptor TrkC is an entry receptor for Trypanosoma cruzi in neural, glial, and epithelial cells. Infect Immun. 2011;79:4081–7.
1607. Sasahira T, Ueda N, Yamamoto K, Bhawal UK, Kurihara M, Kirita T, Kuniyasu H. Trks are novel oncogenes involved in the induction of neovascularization, and nodal metastasis in oral squamous cell carcinoma. Clin Exp Metastasis 2013;30:165–76. doi:10.1007/s10585=012-0525-x.
1608. Dutton-Regester K, Aoude LG, Nancarrow DJ, Stark MS, O'Connor L, Lanagan C, Pupo GM, Tembe V, Carter CD, O'Rourke M, Soolyer RA, Mann GJ, Schmidt CW, Herington A, Hayward NK. Identification of TGF (TRK-fused gene) as a putative metastatic melanoma tumor suppressor gene. Genes Chromosomes Cancer. 2012;51:452–61.
1609. Poell JB, van Haastert RJ, de Gunst T, Schultz IJ, Gommans WM, Verheul M, Cerisoli F, Prevost GP, Schaapveld RQJ, Cuppen E. A functional screen identifies specific microRNAs capable of inhibiting human melanoma cell viability. PLoS One. 2012;7(8): e43569. doi:10.1371/journal.pone.0043569.
1610. Kelly BL, Singh G, Alyar A. Molecular and cellular characterization of an AT-hook protein from Leishmania. PLoS One. 2011;6(6):e21412. doi:10.1371/journal.pone.00214112.

1611. Yun J, Kim YS, Jung JH, Seo PJ, Park CM. The AT-hook motif-containing protein AHL22 regulates flowering initiation by modifying flowering locus T chromatin in Arabidopsis. J Biol Chem. 2012;287:15307–16.
1612. Moliterno AR, Resar LMS. AKNA: another AT-hook transcription factor "hooking-up" with inflammation. Cell Res. 2011;21:1528–30.
1613. Perales G, Burguete-García AI, Dimas J, Bahena-Román M, Bermúdez-Morales VH, Moreno J, Madrid-Marina V. A polymorphism in the AT-hook motif of the transcriptional regulator AKNA is a risk factor for cervical cancer. Biomarkers. 2010;15:470–4.
1614. Takeuchi I, Takaha N, Nakamura T, Hongo F, Mikami K, Kamoi K, Okihara K, Kawauchi A, Miki T. High mobility group protein AT-hook 1 (HMGA1) is associated with the.development of androgen independence in prostate cancer cells. Prostate. 2012;72:1124–32.
1615. Murga-Penas EM, Callet-Bauchu E, Ye H, Gazzo S, Berger F, Schilling G, Albert-Konetzny N, Vettorazzi E, Salles G, Wlodarska I, Du MQ, Bokemeyer C, Dierlamm J. The t(134;18)(q32;q21)/IGH-MALT 1 translocation in MALT lymphoma contains template nucleotide insertions and a major breakpoint region similar to follicular and mantle cell lymphoma. Blood. 2010;115:2214–9.
1616. Baró C, Espinet B, Salidom M, Garcia M, Sánchez B, Florensa L, Bellosillo S, Solé F. Cryptic IGH/BCL2 rearrangements with variant FISH patterns in follicular lymphoma. Leuk Res. 2011;35:256–9.
1617. Nambiar M, Raghavan SC. Mechanism of fragility at BCL2 gene minor breakpoint cluster region during t(14:;18) chromosomal translocation. J Biol Chem. 2012;287:8688–701.
1618. Khanim F, Dawson C, Meseda CA, Dawson J, Mackett M, Young LS. BHRF1, a viral homologue of the Bcl-2 oncogene, is conserved at both the sequence and functional level in different Epstein-Barr virus isolates. J Gen Virol. 1997;78:2987–99.
1619. Jing YZ, Wang Y, Jia YP, Luo B. Polymorphisms of Epstein-Barr virus BHRF-1 gene, a homologue of bcl-2. Chin J Cancer. 2010;12:1000–5.
1620. Bisogno G, Soloni P, Conte M, Podda M, Ferrari A, Garaventa A, Luksch R, Cecchetto G. Esthesioneuroblastoma in pediatric and adolescent age. A report from the TREP project in cooperation with the Italian Neuroblastoma and Soft Tissue Sarcoma Committee. BMC Cancer. 2012;12:117. doi:10.1186/1471-2407-12-117.
1621. Corvi R, Amier LC, Savelyeva L, Gehring M, Schwab M. MYCN is retained in single copy at chromosome 2 band p23-24 during amplification in human neuroblastoma cells. Proc Natl Acad Sci USA. 1994;91:5523–7.
1622. Gostissa M, Ranganath S, Bianco JM, Alt FW. Chromosomal location targets different MYC family gene members for oncogenic translocations. Proc Natl Acad Sci USA. 2009;106:2265–70.
1623. Szabó PM, Pintér M, Szabó DR, Zsippai A, Patócs A, Falus A, Rácz K, Igaz P. Integrative analysis of neuroblastoma and pheochromocytoma genomics data. BMC Med Genomics. 2012;5:48. doi:10.1186/1755-8794-5-48.
1624. Choudhury SR, Karmakar S, Banik NL, Ray SK. Targeting angiogenesis for controling neuroblastoma. J Oncol. 2012;2012:782020. doi:10.1155/2012/782020.
1625. Stallings RL, Foley NH, Bray IM, Das S, Buckley PG. MicroRNA and DNA methylation alterations mediating retinoic acid induced neuroblastoma cell differentiation. Semin Cancer Biol. 2011;21:283–90.
1626. Zheng B, Han M, Y-n Shu, Y-j Li, S-b Miao, X-h Zhang, H-j Shi, Zhang T, Wen J-K. HDAC2 phosphorylation-dependent Llf5 deacetylation and RARα acetylation induced by RAR agonist switch the transcription regulatory programs of p21 in VSMCs. Cell Res. 2011;21:12487–1508.

1627. Maresca G, Natoli M, Nardella M, Arisi I, Trisciuglio D, Desideri M, Brandi R, D'Aquanno S, Nicotra MR, D'Onofriom M, Urbani A, Del Bufalo D, Felsani A, D'Agnano I. LMNA knock-down affects differentiation and progression of human neuroblastoma cells. PLoS One. 2012;7(9):e45513. doi:10.1371/journal.pone.0045513.

1628. Pollack S, Heppner G, Brawn RJ, Nelson K. Specific killing of tumor cells in vitro in the presence of normal lymphoid cells and sera from host immune to the tumor antigens. Int J Cancer. 1972;9:316–23.

1629. Pollack S, Nelson K. Specific, "arming" of normal lymph-node cells by sera from tumor-bearing mice. Early appearance of a lymphoid arming factor and cytotoxic lymph-node cells after tumor induction. Int J Cancer. 1974;14:522–9.

1630a. Sinkovics JG, Dryer DA, Shirato S, Cabiness JR, Shullenberger CC. Cytotoxic lymphocytes. I. Destruction of neoplastic cells by lymphocytes of human origin. Texas Rep Biol Med. 1971;29:227–42.

1630b. Sinkovics JG, Ahmed N, Cabiness JR, Hrgovcic MJ, Wilbur JR. Cytotoxic lymphocytes. II. Antagonism and synergism between serum factors and lymphocytes of patients with sarcoma, as tested against cultured tumor cells. Texas Rep Biol Med. 1972;30:347–60.

1631a. Sinkovics JG, Cabiness JR, Shullenberger CC. Monitoring in vitro of immune reactions to solid tumors. In: 7th annual San Francisco cancer symposium "the interrelationship of the immune response and cancer." Front Radiat Ther Oncol Karger Basel New York1972;7:141–154.

1631b. Sinkovics JG, Cabiness JR, Shullenberger CC. In vitro cytotoxicity of lymphocytes to human sarcoma cells. Bibliotheca Haematologica. 1973;39:846–51.

1632. Sinkovics JG, Cabiness JR, Shullenberger CC. Disappearance after chemotherapy of blocking serum factors as measured in vitro with lymphocytes cytotoxic to tumor cells. Cancer. 1972;30:1428–37.

1633. Sinkovics JG. Monitoring in vitro of cell-mediated immune reactions to tumors. Methods Cancer Res. 1973;8:107–175 (Harris Bush editor New York: Academic Press).

1634. Sinkovics JG, Campos LT, Kay HD, Loh KK, Cabiness JR, Ervin F. The concept of immunotherapy of human sarcomas: the need for an in vitro monitoring assay. Ann Clin Lab Sci. 1973;3:414–23.

1635. Sinkovics JG, Williams DE, Campos LT, Kay HD, Romero JJ. Intensification of immune reactions of patients to cultured sarcoma cells: attempts at monitored immunotherapy. Sem Oncol. 1974;1:351–65.

1636. Sinkovics JG, Thota H, Loh KK, Gonzalez F, Campos LT, Romero JJ, Kay HD, King D. Prospectives for immunotherapy of human sarcomas. In: The 19th annual clinical conference "cancer chemotherapy: fundamental concepts and recent advances". The University of Texas M.D. Anderson Hospital and Tumor Institute, Houston TX. Year Book Medical Publisher, Chicago. 1975;417–443.

1637a. Sinkovics JG, Plager C, McMurtrey M, Romero JJ, Romsdahl MM. Immunotherapy of human sarcomas. In: The 21st annual clinical conference on cancer, 1976, The University of Texas System Cancer Center M.D. Anderson Hospital and Tumor Institute, Houston, Texas. Year Book Medical Publishers, Chicago London. 1977;361–410.

1637b. Sinkovics JG, Plager C, Papadopoulos N, McMurtrey MJ, Romero JJ, Waldinger RR, Romsdahl MM. Immunology and immunotherapy of human sarcomas. In: Immunotherapy of Human Cancer. 22nd Annual Clinical Conference, 1977. The University of Texas M.D. Anderson Hospital and Tumor Institute, Houston, TX. Raven Press, New York 1978; 267–288.

1638. Tarek N, Le Luduec JB, Gallagher MM, Zheng J, Venstrom JM, Chamberlain E, Modak S, Heller G, Dupont B, Cheung NK, Hsu KC. Unlicensed NK cells target neuroblastoma following anti-GD2 antibody treatment. J Clin Invest. 2012;122:3260–70.

1639. Koehn TA, Trimble LL, Alderson KL, Erbe AK, McDowell KA, Grzywacz B, Hank JA, Sondel PM. Increasing the clinical efficacy of NK and antibody-mediated cancer immunotherapy : potential predictors of successful clinical outcome based on observations in high-risk neuroblastoma. Front Pharmacol. 2012;3:91. doi:10.3389/fphar.2012.00091.

1640. Cheung IY, Hsu K, Cheung NK. Activation of peripheral-blood granulocytes is strongly correlated with patient outcome after immunotherapy with anti-GD2 monoclonal antibody and granulocyte-macrophage colony stimulating factor. J Clin Oncol. 2012;30:426–32.

1641. Sheard MA, Asgharzadeh S, Liu Y, Lin TY, Wu HW, Groshen S, Lee DA, Seeger RC. Membrane-bound TRAIL supplements natural killer cell cytotoxicity against neuroblastoma cells. J immunother. 2013;36:319–29.

1642. Costa FF, Bischof JM, Vanin EF, Lulla RR, Wang M, Sredni ST, Rajaram V, de Fátima Bonaldo M, Wang D, Goldman S, Tomita T, Soares MB. Identification of microRNAs as potential prognostic markers in ependymoma. PLoS One. 2011;6(10):e25114. doi:10.1371/journal.pone.0025114.

1643. Johnson RA, Wright KD, Poppleton H, Mohankumar KM, Finkelstein D, Pounds SB, Rand V, Leary SES, White E, Eden C, Hogg T, Northcott P, Mack S, Neale G, Wang Y-D, Coyle B, Atkinson J, DeWire M, Kranenburg TA, Gillespie Y, Allen JC, Merchant T, Boop FA, Sanford RA, Gajjar A, Ellison DW, Taylor ND, Grundy RG, Gilbertson RJ. Cross-species genomics matches driver mutations and cell compartments to model ependymoma. Nature. 2010;466:632–6.

1644. Yeung JT, Hamilton RL, Okada H, Jakacki RI, Pollack IF. Increased expression of tumor-associated antigens in pediatric and adult ependymomas: implication for vaccine therapy. J Neurooncol. 2013;111:103–11.

1645. Lu Z, Zhang Y, Li Z, Yu S, Zhao G, Li M, Wang Z, Wang Q, Yang Y. Overexpression of the B-types Eph and ephrin genes correlates with progression and pain in human pancreatic cancer. Oncol Lett. 2012;3:1207–12.

1646. Nakada M, Anderson EM, Demuth T, Nakada S, Reavie LB, Drake KL, Hoelzinger DB, Berens ME. The phosphorylation of ephrin-B2 ligand promotes glioma cell migration and invasion. Int J Cancer. 2010;126:1155–65.

1647. Mateo Dopeso H, Lozano S, Mazzolini R, Rodrigues P, Lagares-Tena L, Ceron J, Romero J, Esteves M, Landolfi S, Hernández-Losa J, Castaño J, Wilson AJ, Ramon Y, Cajal S, Mariadason JM, Schwartz S Jr, Arango D. The receptor tyrosine kinase EPHB4 has tumor suppressor activities in intestinal tumorigenesis. Cancer Res. 2009;69:7430–8.

1648. MacQuarrie KL, Yao Z, Fong AP, Diede SJ, Rudzinski ER, Hawkins DS, Tapscott SJ. Comparison of genome-wide binding of MyoD in normal human myogenic cells and rhabdomyosarcomas identifies regional and local suppression of pro-myogenic transcription factors. Mol Cell Biol. 2013;33:773–84.

1649. Murata T, Noda C, Saito S, Kawashima D, Sugimoto A, Isomura H, Kanda T, Yokoyama KK, Tsurumi T. Involvement of Jun dimerization protein 2 (JDP2) in the maintenance of Epstein-Barr virus latency. J Biol Chem. 2011;286:22007–16.

1650. Magnusson S, Gisselsson D, Wiebe T, Kristofferson U, Borg Å, Olsson H. Prevalence of germline TP53 mutations and history of Li-Fraumeni syndrome in families with childhood adrenocortical tumors, choroid plexus tumors, and rhabdomyosarcoma: a population-based survey. Pediatr Blood Cancer. 2012;50:846–53.

1651. Ladanyi M, Cha C, Lewis R, Jhanwar SC, Huvos AG, Healey JH. MDM2 gene amplification in metastatic osteosarcoma. Cancer Res. 1993;53:16–8.

1652. Keleti J, Quezado MM, Abaza MN, Raffeld M, Tsokos M. The MDM2 oncoprotein is overexpressed in rhabdomyosarcoma cell lines and stabilizes ld-type p53 protein. Am J Pathol 1996;149:143–51.

1653. Schwienbacher C, Sabbioni S, Campi M, Veronese A, Bernardi G, Menegatto A, Hatada I, Mukai T, Ohashi H, Barbanti-Brodano G, Croce CM, Negrini M. Transcriptional map of 170-kb region at chromosome 11p15.5: identification and mutational analysis of the BWR1A gene reveals the presence of mutations in tumor samples. Proc Natl Acad Sci USA 1998;95:3873–3878.
1654. Walter D, Satheesa S, Albrecht P, Bornhauser BC, D'Alessandro V, Oesh SM, Rehrauer H, Leuschner J, Koscielniak E, Gengler C, Moch H, Bernasconi M, Niggli FK, Schäfer BW. CWS, Study Group. CD133 positive embryonal rhabdomyosarcoma stem-like cell population is enriched in rhabdosphere. PLoS One. 2011;65:e19506. doi:10.1371/journal.pone.0019506.
1655. Langenau DM, Keefe MD, Storer NY, Guyon JAR, Kutok JL, Le X, Goessling W, Neuberg DS, Kunkel LM, Zon Li. Effects of Ras on the genesis of embryonal rhabdomyosarcoma. Genes Dev. 2007;21:1382–95.
1656. Rao PK, Missiaglia E, Shields L, Hyde G, Yuan B, Shepherd CJ, Shipley J, Kodish HF. Distinct roles for miR-1and miR-133a in the proliferation and differentiation of rhabdomyosarcoma cells. FASEB J. 2010;24:3427–37.
1657. Chang L, Kiriazis H, Gao XM, Du XJ, El-Osta A. Cardiac genes show contextual SWI/SNF interactions with distinguishable gene activities. Epigenetics. 2011;6:760–8.
1658. Chew CS, Chen X, Parente JA Jr, Tarrer S, Okamoto C, Qin HY. Lasp-1 binds to non-muscle F-actin in vitro and is localized within multiple sites of dynamic actin assembly in vivo. J Cell Sci. 2002;115:4787–99.
1659a. Grunewald TG, Kammerer U, Kapp M, Eck M, Dietl J, Butt E, Honig A. Nuclear localization and cytosolic overexpression of LASP-1 correlates with tumor size and nodal-positivity of human breast carcinoma. BMC Cancer. 2007;7:198.
1659b. Grunewald TG, Butt E. The LIM and SH3 domain protein family: structural proteins or signal transducers or both? Mol Cancer. 2008;7:31. doi:10.1186/1476-4598-7-31.
1660. Hamelin M, Scott Ian M, Wayb JC, Culottin JG. The mec-7 β-tubulin gene of Caenorhabditis elegans is expressed primarily in the touch receptor neurons. EMBO J. 1992;11:2885–2893.
1661. Jayo A, Parsons M, Adams JC. A novel Rho-dependent pathway that drives interactions of fascin-1 p-Lin-11/Isl-1?Mec-3 kinases (LIMK) 1/2 to promote fascin-1/actin binding and filopodi stability. BMC Biol. 2012;10:72. doi:10.1186/1741-7007-10-72.
1662. Liu J, Walp ER, May CL. Elevation of transcription factor Islet-1 levels in vivo increases β-cell function but not β-cell mass. Islets. 2012;4:199–206.
1663. Marri S, Gupta BP. Dissection of lin-11 enhancer region in Caenorhabditis elegans and other nematodes. Dev Biol. 2009;325:402–11.
1664. Savage N, Linn D, McDonough C, Donohoe JM, Franco A, Reuter A, Biddinger PW, Eaton KW, Biegel JA, Sharma S. Molecularly confirmed primary malignant rhabdoid tumor of the urinary bladder: implications of accurate diagnosis. Ann Diagn Pathol. 2012;16:504–7.
1665. Wahlström G, Vartianen M, Yamamoto L, Mattila PK, Lappalainen P, Heino TI. Twinfilin is required for actin-dependent developmental processes in drosophila. J Cell Biol. 2001;155:787–96.
1666. San Juan BP, Andrade-Zapata Baonza A. The bHLH factors Dpn and members of the E (spl) complex mediate the function of Notch signaling regulating cell proliferation during wing S-disc development. Biol Open. 2012;1:667–76.
1667. Heisig J, Weber D, Englberger E, Winkler A, Kneitz S, Sung WK, Wolf E, Eilers M, Wei CL, Gessler M. Target gene analysis by microarrays and chromatin immunoprecipitation identifies HEY proteins as highly redundant bHLH repressors. PLoS Genet. 2012;8(5):e1002728. doi:10.1371/journal.pgen.1002728.

1668. Johnsen SA, Güngör C, Prenzel T, Riethdorf S, Riethdorf L, Taniguchi-Ishigaki N, Rau T, Tursun B, Furlow JD, Sauter G, Scheffner M, Pantel K, Gannon E, Bach I. Regulation of estrogen-dependent transcription by the LIM cofactors CLIM and RLIM in breast cancer. Cancer Res. 2009;69:128–36.
1669. Numata M, Morinaga S, Watanabe T, Tamagawa H, Yamamoto N, Nakamura Y, Kamea Y, Okawa S, Rino Y, Akaike M, Masuda M, Miyagi Y. The clinical significance of SWI/SNF complex in pancreatic cancer. Int J Oncol. 2013;42:403–10. doi:10.3892/ijo.2012.1723.
1670. Raman D, Sai J, Neelm NF, Chew CS, Richmond A. LIM and SH3 protein-1 modulates CXCR2-mediated cell migration. PLoS One. 2010;5(4):e10050.
1671. Hanashima A, Kubokawa K, Kimura S. Characterization of amphioxus nebulin and the similarity to human nebulin. J Exp Biol. 2009;212:668–72.
1672. Sumegi J, Nishio J, Nelson M, Frayer RW, Perry D, Bridge JA. A novel t(4;22)(q31;q12) produces an EWSR1-SMARCA5 fusion in extraskeletal Ewing sarcoma/primitive neuroectodermal tumor. Mol Pathol. 2011;24:333–42.
1673. Bourdeaut F, Biéche I. Chromatin remodeling defects and cancer: the SWI/SNF example. Bull Cancer. 2012;99:1133–40.
1674. Wang X, Sansam CS, Cs Thom, Metzger D, Evans JA, Nguyen PTL, Roberts CWM. Oncogenesis caused by loss of the SNF 5 tumor suppressor is dependent upon activity of BRG1, the ATPase of the SWI/SNF chromatin remodeling complex. Cancer Res. 2009;69:8094–101.
1675. Naidu SR, Love IM, Imbalzano AN, Grossman SR, Androphy EJ. The SWI/SNF chromatin remodeling subunit BRG1 is a critical regulator of p53 necessary for proliferation of malignant cells. Oncogene. 2009;28:2492–501.
1676. Bryant KE, Colgrove RC, Knipe DM. Cellular SNF2H chromatin-remodeling factor promotes herpes simplex virus 1 immediate-early gene expression and replication. MBio. 2011;2(1):e00330–10. doi:10.1128/mBio.00330-10.
1677. Wang L, Pal S, Sif S. Protein arginine methyltransferase 5 suppresses the transcription of the RB family of tumor suppressors in leukemia and lymphoma cells. Mol Cell Biol. 2008;28:6262–77.
1678. Caldas H, Holloway MP, Hall BM, Qualman SJ, Altura RA. Survivin-directed RNA interference cocktail is a potent suppressor of tumour growth in vivo. J Med Genet. 2006;43:119–28.
1679. Kang Z, Sun SY, Cao L. Activating death receptor DR5 as a therapeutic strategy for rhabdomyosarcoma. ISRN Oncol. 2012;2012:395952. doi:10.5402/2012/395952.
1680. Ramírez-Peinado S, Alcázar-Limones F, Lagares-Tena L, Mjiyad NE, Caro-Maldonado A, Tirado OM, Muñoz-Pinedo C. 2-Deoxyglucose induces noxa-dependent apoptosis in alveolar rhabdomyosarcoma. Cancer Res. 2011;71:6796-806 doi:10.1158/0008-5472. CAN-11-0759.
1681. MacFarlane M, Robinson GL, Cain K. Glucose—a sweet way to die. Cell Cycle. 2012;11:3919–25.
1682. Robinson GL, Dinsdale D, MacFarlane M, Cain K. Switching from aerobic glycolysis to oxidative phosphorylation modulates the sensitivity of mantle lymphoma cells to TRAIL. Oncogene. 2012;31:4996–5006.
1683. Rooswinkel RW, van de Kooij B, Verheij M, Borst J. Bcl-2is a better ABT737 target than Bcl-xL or Bcl-w and only Noxa overcomes resistance mediated by Mcl-1, Bfl-1, or Bcl-B. Cell Death Dis. 2012;3(8):e366.
1684. Fulda S. Cell death pathways as therapeutic targets in rhabdomyosarcoma. Sarcoma. 2012;2012:326210.
1685. Fulda S. Targeting c-FLICE-like inhibitory protein (CFLAR) in cancer. Expert Opin Ther Targets. 2012;17:195–201.
1686. Regenbrecht CR, Jung M, Lehrbach H, Adjave J. The molecular basis of genistein-induced mitotic arrest and exit of self-renewal in embryonal carcinoma and primary cancer cell lines. BMC Med Genomics. 2008;1:49. doi:10.1186/1755-8794-1-49.

1687. Kuci S, Rettinger E, Voß, Weber G, Stais M, Kreyenberg H, Willasch A, Kuci Z, Koscielniak E, Klöss S, von Laer D, Klinbebiel T, Bader T. Efficient lysis of rhabdomyosarcoma cells by cytokine-induced killer cells: implications for adoptive immunotherapy after allogeneic stem cell transplantation. Haematologica. 2010;95:1579–1586.

1688. Rettinger E, Meyer V, Kreyenberg H, Volk A, Kuci S, Willasch S, Koscielniak E, Fulda S, Wels WS, Boenig H, Klingebiel T, Bader P. Cytotoxix capacity of IL-15-stimulated cytokine-induced killer cells against human acute myeloid leukemia and rhabdomyosarcoma in humanized preclinical mouse model. Front Oncol. 2012;2:32. doi:10.3389/fonc.2012.00032.

1689. Kong Y, Cao W, Xi X, Ma C, Cui L, He W. The NKG2D ligand ULBVP4 binds to TCRγ9/б2 and induces cytotoxicity to tumor cells through both TCRγб and NKG2D. Blood. 2009;114:310–7.

1690. Guo H, Wang L, Peng L, Zhou ZH, Deng H. Open reading frame 33 of gammaherpesvirus encodes a tegument protein essential for virion morphogenesis and egress. J Virol. 2009;83:10582–95.

1691. Toth Z, Maglinte DT, Lee SH, Lee H-R, Wong L-Y, Brulois KF, Lee s, Buckley J, Laird PW, Marquez VE, Jung JU. Epigenetic analysis of KSHV latent and lytic genomes. PLoS Pathog 201;6(7):e1001013. doi:10.137/journal.ppat.1001013.

1692. Yoo AS, Sun AX, Li L, Shcheglovitov A, Portmann T, Li Y, Lee-Messer C, Dolmetsch RE, Tsien RW, Crabtree GR. MicroRNA-mediated conversion of human fibroblasts to neurons. Nature. 2011;476:228–31.

1693. Müller S, Zocher G, Steinle A, Stehle T. Structure of the HCMV UL16-MICB complex elucidates select binding of a viral immunoevasion to diverse NKGD2D ligands. PLoS Pathog. 2010;6(1):e1000723.

1694. Orentas RJ, Lee DW, Mackall C. Immunotherapy targets in pediatric cancer. Front Oncol. 2012;2:3. doi:10.3389/fonc.2012.00003.

1695. Lai JP, Robbins PF, Raffeld M, Aung PP, Tsokos M, Rosenberg SA, Miettinen MM, Lee CC. NY-ESO-1 expression in synovial sarcoma and other mesenchymal tumors: significance for NY-ESO-1-based targeted therapy and differential diagnosis. Mod Pathol. 2012;25:854–8.

1696. Lubieniecka JM, de Bruin DR, Su L, van Dijk AH, Subramanian S, van de Rijn M, Poulin N, van Kessel AG, Nielsen TO. Histone deacetylase inhibitors reverse SS128-SSX-mediated polycomb silencing of the tumor suppressor early growth response in synovial sarcoma. Cancer Res. 2008;68:4303–10.

1697a. Ohta H, Hashii Y, Yoneda A, Takizawa S, Kusuki S, Tokimasa S, Fukuzawa M, Tsuboi A, Murao A, Oka Y, Oji Y, Aozasa K, Nakatsuka S, Sugiyama H, Ozono K. WT1 (Wilms tumor1) peptide immunotherapy for childhood rhabdomyosarcoma: a case report. Pediatr Hematol Oncol. 2009;26:74–83.

1697b. Krishnadas KD, Shapiro T, Lucas K. Complete remission following decitabin/dendritic cell vaccine for relapsed neuroblastoma. Pediatrics. 2013;131:331–41.

1698. Chou J, Voong LN, Mortales CL, Towlerton AM, Pollack SM, Chen X, Yee C, Robbins PF, Warren EH. Epigenetic modulation to enable antigen-specific T-cell therapy of colorectal cancer. J Immunother. 2012;35:131–41.

1699. Cho D, Shook DR, Shimasaki N, Chang Y-H, Fujisaki H, Campana D. Cytotoxicity of activated natural killer cells against pediatric solid tumors. Clin Cancer Res. 2010;16:3901–9.

1700. Pérez-Martinez A, de Prada Vicente I, Fernández L, González-Vicent M, Valentín J, Martin R, Maxwell H, Sevilla J, Vicario JL, Díaz MÁ. Natural killer cells can exert a graft-vs-tumor effect in haploidentical stem cell transplantation for pediatric solid tumors. Exp Hematol. 2012;40:882–91.

1701. Moretta A, Pende D, Locatelli F, Moretta L. Activating and inhibitory killer immunoglobulin-like receptors (KIR) in haploidentical haematopoietic stem cell transplantation to cure high-risk leukaemias. Clin Exp Immunol. 2009;157:325–31.

1702. Simon-Keller K, Paschen A, Eichmüller S, Gattenlöhner S, Barth S, Koscieliak E, Leuschner I, Stöbel P, Hombach A, Abken H, Marx A. Adoptive T-cell therapy of rhabdomyosarcoma. Pathologe. 2010;S2:215–20.

1703. Herrmann D, Seitz G, Fuchs J, Armeanu-Ebinger S. Susceptibility of rhabdomyosarcoma cells to macrophage-mediated cytotoxicity. Oncoimmunology. 2012;1:279–86.

1704. Marshall AD, Grosveld GC. Alveolar rhabdomyosarcoma—the molecular drivers PAX3/7-FOXO1-induced tumorigenesis. Skeletal Muscle. 2012;2:25. doi:10.1186/2044-5040-2-25.

1705. de Souza RR, Oliviera ID, Caran EM, Alves MT, Abib SD, Toledo SR. Investigation of PAX3/7-FKHR fusion genes and IGF2 gene expression in rhabdomyosarcoma tumors. Growth Horm IGF Res. 2012;22:245–9.

1706. Armeanu-Ebinger S, Herrmann D, Bonin M, Leuschner I, Warmann SW, Fuchs J, Seitz G. Differential expression of miRNAS in rhabdomyosarcoma and malignant rhabdoid tumor. Exp Cell Res. 2012;318:2567–77.

1707. Liu L, Wang YD, Wu J, Cui J, Chen T. Carnitine palmytoyltransferase 1A (CPT1A): a transcriptional target of PAX3-FKHR and mediates PAX3-FKHR-dependent motility in alveolar rhabdomyosarcoma cells. BMC Cancer. 2012;12:154. doi:10.1186/1471-2407-12-154.

1708. Burnet FM. The concept of immunological surveillance. Proc Exp Tumor Res. 1970;13:1–27.

1709. Barkan D, Green JE, Chamber AF. Extracellular matrix: a gatekeeper in the transition from dormancy to metastatic growth. Eur J Cancer. 2010;46:1181–8.

1710. Bragado P, Sosa MS, Keely P, Condeelis J, Aquirre-Ghiso JA. Microenvironment dictating tumor cell dormancy. Recent Res Cancer Res. 2012;195:2539.

1711. Layden MJ, Boekhout M, Martindale MQ. Nematostella vectensis achaete-scute homolog NvAshA regulates embryonic ectodermal neurogenesis and represents an ancient component of the metazoan neural specification pathway. Development. 2012;139:1013–22.

1712. Lu TM, Luo YJ, Yu JK. BMP and Delta/Notch signaling control the development of amphioxus epidermal sensory neurons: insights into the evolution of the peripheral sensory system. Development. 2012;139:2020–30.

1713a. Borello U, Coletta M, Tajbakhsh S, Leyns L, De Romerts EM, Buckingham M, Cossu G. Transplacental delivery of the Wnt antagonist Frzb1 inhibits development of caudal paraxial mesoderm and skeletal myogenesis in mouse embryo. Development. 1999;126:4247–4255.

1713b. Borello U, Berarducci B, Murphy P, Bajard L, Buffa V, Piccolo S, Buckingham M, Cossu G. The Wnt/beta-catenin pathway regulates Gli-mediated Myf5 expression during somitogenesis. Development. 2006;133:3723–32.

1714. Liu LF, Shen WJ, Zhang ZH, Wang LJ, Kraemer FB. Adipocytes decrease Runx2 expression in osteoblastic cells: roles of PPARγ and adiponectin. J Cell Physiol. 2010;225:837–45.

1715. Kamimura K, Ueno K, Nakagawa J, Hamada R, Saitone M, Maedea N. Perlecan regulates bidirectional Wnt signaling at the drosophila neuromuscular junction. J Cell Biol. 2013;200:219–33. doi:10.1083/jcb.201207036.

1716. Eichhorn PJ, Rodón L, Gonzáles-Juncá A, Dirac A, Gili M, Martinez-Sáez E, Aura C, Barba I, Peg V, Prat A, Cuartas I, Jimenez J, Garcia-Dorado D, Sahuguillo J, Bernards R, Baselga J, Seoane J. USP15 stabilizes TGF-β receptor I and promotes oncogenesis through the activation of TGF-β signaling in glioblastoma. Nat Med. 2012;18:429–35.

1717. Kawamata S, Matsuzaki K, Murate M, Seki T, Matsuoka K, Iwao Y, Hibi T, Okazaki K. Monogenic Smad3 signaling induced by chronic inflammation is an early event in ulcerative colitis-associated carcinogenesis. Inflamm Bowel Dis. 2011;17:683–95.
1718. Lampropoulos P, Zizi-Sermpetzoglou A, Rizos S, Kostakis A, Nikiteas N, Papavassiliou AG. TGF-beta signaling in colon carcinogenesis. Cancer Lett. 2012;314:1–7.
1719. Tarasewicz E, Jeruss JS. Phospho-specific Smad 3 signaling impact on breast oncogenesis. Cell Cycle. 2012;11:2443–51.
1720. Horrocks P, Kilbey BJ. Physical and functional mapping of the transcriptional start sites of Plasmodium falciparum proliferating cell nuclear antigen. Mol Biochem Parasitol. 1996;82:207–15.
1721. Juhasz S, Balogh D, Hajdu I, Burkovics P, Vilamil MA, Zhuang Z, Haracska L. Characterization of human Spartan/C1orf124, an ubiquitin-PCNA interacting regulator of DNA damage tolerance. Nucl Acids Res. 2012;40:10795–808.
1722. Takawa M, Chon HS, Hayami S, Tohyokawa G, Kogure M, Yamane Y, Iwai Y, Maejuima K, Ueda K, Masuda A, Dohmae N, Field HI, Tsunoda T, Kobayashi T, Akasu T, Sugiyama M, Ohnuma S, Atomi Y, Ponder BA, Nakamura Y, Hamamoto R. Histone lysine methyltransferase SETD8 promotes carcinogenesis by deregulating PCNA expression. Cancer Res. 2012;72:3217–27.
1723. Gloor JW, Balakrishnan L, Campbell JL, Bambara RA. Biochemical analyses indicate that binding and cleavage specificities define the ordered processing of human Okazaki fragments by Dna2 and FEN1. Nucl Acids Res. 2912;40:6774–6786.
1724. Masuda-Sasa T, Imamura O, Campbell JL. Biochemical analysis of human Dna2. Nucl Acids Res. 2006;34:1865–75.
1725. Zheng L, Shen B. Okazaki fragment maturation: nucleases take center stage. J Mol Cell Biol. 2011;3:23–30.
1726. Zhang Z, Zhang S, Lin SH, Wangb X, Wu L, Lee EY, Lee MY. Structure of monoubiquitinated PCNA: implications for DNA polymerase switching and Okazaki fragment maturation. Cell Cycle. 2012;11:2128–36.
1727. Liu L, Zhou C, Zhou L, Peng L, Li D, Zhang X, Zhou M, Kuang P, Yuan Q, Song X, Yang M. Functional FEN1 genetic variants contribute to risk of hepatocellular carcinoma, esophageal cancer, gastric cancer and colorectal cancer. Carcinogenesis. 2012;33:119–23.
1728. Falini B, Flenghi L, Fagioli M, Stein H, Schwarting R, Riccardi C, Manocchio I, Pileri S, Pelicci PG, Lanfrancone L. Evolutionary conservation in various mammalian species of the human proliferation-associated epitope recognized by the K-67 monoclonal antibody. J Histochem Cytochem. 1989;37:1471–8.
1729. Gerdes J. Ki-67 and other proliferation markers useful for immunohistological diagnostic and prognostic evaluations in human malignancies. Semin Cancer Biol. 1990;1:199–206.
1730. Scholzen T, Gerdes J. The Ki-67 protein: from the known and the unknown. J Cell Physiol. 2000;182:311–22.
1731. Silva CTM, dos Santos Souza S, de Almeida TF, Andrade ZA. Ki-67 is expressed in multiplying forms of Schistosoma mansoni, but not in snail host tissues. Memór Instituto Oswaldo Cruz. 2007;102:651–3.
1732. Sundara YT, Jordanova ES, Hernowo BS, Gandamihardja S, Fleuren GJ. Decidual infiltration of FoxP3+ regulatory T cells, CD58+ decidual natural killer cells and Ki-67 trophoblast cells to hydatidiform mole compared to normal and ectopic pregnancies. Mol Med Report. 2012;5:275–81.
1733. Winking H, Gerdes J, Traut W. Expression of the proliferation marker Ki-67 during early mouse development. Cytogenet Genome Res. 2004;105:251–6.

1734. Santisteban M, Reynolds C, Barr Fritscher EG, Frost MH, Vierkant RA, Anderson SS, Degnim AC, Visscher DW, Pankratz VS, Hartmann LC. Ki-67: a time-varying biomarker of risk of breast cancer in atypical hyperplasia. Breast Cancer Res Treat. 2010;212:431–7.
1735. Ali HR, Dawson SJ, Blows FM, Provenzano E, Leung S, Nielsen T, Pharoah PD, Caldas C. A Ki67/BCL2 index based on immunohistochemistry is highly prognostic in ER-positive breast cancer. J Pathol. 2012;226:97–107.
1736. Mineta H, Miura K, Ogino T, Takebayashi S, Misawa K, Ueda Y. Vascular endothelial growth factor (VEGF) expression correlates with p53 and Ki-67 expression in tongue squamous cell carcinoma. Anticancer Res. 2002;22:1039–44.
1737. Liu J, Fang L, Cheng Q, Li L, Su C, Zhang B, Pei D, Yang J, Li W, Zheng J. Effects of G250 promoter controlled conditionally replicative adenovirus expressing K67-siRNA on renal cancer cell. Cancer Sci. 2012;103:1880–8.
1738. Fisk JC, Li J, Wang H, Aletta JM, Qu J, Read LK. Proteomic analysis reveals diverse classes of arginine methylproteins in mitochondria of trypanosomes. Mol Cell Proteomics. 2012;12:302–211.
1739. Guo W, Yang D, Xun H, Zhang Y, Huang J, Yang Z, Chen X, Huang Z. Mutations in the D-loop region and increased copy number of mitochondrial DNA in human laryngeal squamous cell carcinoma. Mol Biol Rep. 2013;40:13–20.
1740. Lin CS, Lee HT, Lee SY, Shen YA, Wang LS, Chen Yj, Wei YH. High mitochondrial DNA copy number and bioenergetic function are associated with tumor invasion of esophageal squamous cell carcinoma cell lines. Int J Mol Sci. 2012;13:11228–46.
1741. Ding C, Li R, Wang P, Jin P, Li S, Guo Z. Identification of sequence polymorphism in the D-loop region of mitochondrial DNA as a risk factor of lung cancer. Mitochondrial DNA. 2012;23:251–4.
1742. Lu TM, Luo YJ, Yu JK. BMP and delta/Notch signaling control the development of amphioxus epidermal sensory neurons; insights into the evolution of the peripheral sensory system. Development. 2012;139:2020–30.
1743. Grens A, Mason E, Marsh JL, Bode HR. Evolutionary conservation of a cell fate specification gene: the Hydra achaete-scute homolog has proneural activity in Drosophila. Development. 1995;121:4027–35.
1744. Bergmann A, Steller H. Apoptosis, stem cells, and tissue regeneration. Sci Signal. 2010;3 (145):8. doi:10.1126/scisignal.3145re8.
1745. Fraune J, Alsheimer M, Volff JN, Busch K, Fraune S, Bosch TC, Benavente R. Hydra meiosis reveals unexpected conservation of structural synaptonemal complex proteins across metazoans. Proc Natl Acad Sci USA. 2012;109:16588–93.
1746. Plickert G, Frank U, Müller WA. Hydractinia, a pioneering model for stem cell biology and reprograming somatic cells to pluripotency. Int J Dev Biol. 2012;56:519–34.
1747. Pankow S, Bamberger C. The p53 tumor suppressor-like protein nvp63 mediates selective germ cell death in the sea anemone Nematostella vectensis. PLoS One. 2007;2 (9):e782.
1748. Simionato D, Kerner P, Dray NJ, Le Gouar M, Ledent V, Arendt D, Vervoort M. Achaete-scute-related genes in the Platynereis dumeilii: insights into the evolution of neural basic helix-loop-helix genes. BMC Evol Biol. 2008;8:170. doi:10.1186/1471-2148-8-170/.
1749. Zhou Q, Zhang T, Xu W, Yu L, Yi Y, Zhang Z. Analysis of four achaete-scute homologs in Bombyx mori reveals new viewpoints of the evolution and functions of this gene family. BMC Genet. 2008;9:24. doi:10.1186/1471-2156-9-24.
1750. Fromental-Ramain C, Taquet N, Ramain P. Transcriptional interactions between two pannier isoforms and the cofactor U-shaped during neural development in Drosophila. Mech Dev. 2010;127:442–57.

1751. Cordes B, Williams MD, Tirado Y, Bell D, Rosenthal DI, Al-Dharhri SF, Hanna EY, El-Naggar AK. Molecular and phenotypic analysis of poorly differentiated sinonasal neoplasms: an integrated approach for early diagnosis and classification. Hum Pathol. 2009;40:283–92.
1752. Rostomily RC, Bermingham-McDonogh O, Berger MS, Tapscott SJ, Reh TA, Olson JM. Expression of neurogenic basic helix-loop-helix genes in primitive neuroectodermal tumors. Cancer Res. 1997;57:3526–31.
1753. Carney ME, O'Reilly RC, Sholevar B, Bujakova OI, Lowry LD, Keane WM, Margolis FL, Rothstein JL. Expression of the human achaete-scute 1 gene in olfactory neuroblastoma (esthesioneuroblastoma). J Neurooncol. 1995;26:35–43.
1754. Mhawech P, Berczy M, Asaly M, Hermann F, Bouzourene H, Allal AS, Dulguerov P, Schwaller J. Human achaete-scute homologue (hASH1) mRNA level as a diagnostic marker to distinguish esthesioneuroblastoma from poorly differentiated tumor arising in the sinonasal tract. Am J Clin Pathol. 2004;122:100–5.
1755. Jiang T, Collins BJ, Jin N, Watkins DN, Brock MV, Matsu W, Nelkin BD, Ball DW. Achaete-scute complex homologue 1 regulates tumor-initiating capacity in human small cell lung cancer. Cancer Res. 2009;69:845–54.
1756. Zhu R, Yang Y, Tian Y, Bai J, Zhang X, Li X, Peng Z, Chen Y, Qian C, Bian X, Wang R. Ascl2 knockdown results in tumor growth arrest by miRNA-302b-related inhibition of colon cancer progenitor cells. PLoS One. 2012;7(2):e32170.
1757. Osada H, Tomida S, Yatabe Y, Tatematsu Y, Takeuchi T, Murakami H, Kondo Y, Sekido Y, Takahashi T. Roles of achaete-scute homologue 1 in DKK1 and E-cadherin repression and neuroendocrine differentiation in lung cancer. Cancer Res. 2008;68:1647–55.
1758a. Wang H, Lambowitz AM. The Mauriceville plasmid reverse transcriptase can initiate cDNA synthesis de novo and may be related to the reverse transcriptase and DNA polymerase progenitor. Cell. 1993;75:1071–81.
1758b. Kennell JC, Wang H, Lambowitz AM. The Mauriceville plasmid of Neurospora spp. uses novel mechanisms for initiating reverse transcription in vivo. Mol Cell Biol. 1994;14:3094–107.
1758c. Chiang CC, Lambowitz AM. The Mauriceville retroplasmid reverse transcriptase initiates cDNA synthesis de novo at the 3' end of tRNA. Mol Cell Biol. 1997;17:4526–35.
1758d. Chen B, Lambowitz AM. De novo and DNA primer-mediated initiation of cDNA synthesis by the Mauriceville retroplasmid reverse transcriptase involve recognition of a 3' CCA sequence. J Mol Biol. 1997;271:311–32.
1759. Antal Z, Manczinger L, Kredics L, Kevel F, Nagy E. Complete DNA sequence and analysis of a mitochondrial plasmid in the mycoparasitic Trichoderma harzarium strain T95. Plasmid. 2002;47:1348–152.
1760. Simpson EB, Ross SL, Marchetti SE, Kenell JC. Relaxed primer specificity associated with reverse transcriptases by the pFOXC retroplasmids of Fusarium oxysporum. Eukaryot Cell. 2004;3:1589–600.
1761. Belfort M, Curcio MJ, Lue NF. Telomerase and retrotransposons: reverse transcriptase that shaped genomes. Proc Natl Acad Sci USA. 2011;108:20304–10.
1762. Beauregard A, Curcio MJ, Belfort M. The take and give between retrotransposable elements and their hosts. Annu Rev Genet. 2008;42:587–617.
1763. Gladyshev EA, Arkhipova IR. A widespread class of reverse transcriptase-related cellular genes. Proc Natl Acad Sci USA. 2011;108:20311–6.
1764. Woodrow P, Ciarmiello LF, Fantaccione S, Annunziata MG, Pontecorvo G, Carillo P. Ty1-copia group retrotransposons and the evolution of retroelements in several angiosperm plants: evidence of horizontal transmission. Bioinformation. 2012;8:267–71.
1765. Barry JD, Hall JPJ, Plenderleith L. Genome hyperevolution and the success of a parasite. Ann N Y Acad Sci. 2012;1267:11–7.

1766. Pardue ML, DeBaryshe PG. Drosophila telomeres: two transposable elements with important roles in chromosomes. Genetica. 1999;107:189–96.
1767. Beilstein MA, Bruinegar AE, Shippen DE. Evolution of the Arabidopsis telomerase RNA. Front Genet. 2012;3:188.
1768. Wojtyla A, Gladych M, Rubis B. Human telomerase activity regulation. Mol Biol Rep. 2011;38:3339–49.
1769. Das R, Shimamoto T, Zahid Hosen SM, ArIfuzzaman M. Comparative study of different msDNA (multicopy single-stranded DNA) structures and phylogenetic comparison of reverse transcriptases (RTS): evidence for vertical inheritance. Bioinformation. 2011;7:176–9.
1770. Zhaxybayeva O, Swithers KS, Foght J, Green AG, Bruce D, Detter C, Han S, Teshima H, Han J, Woyke T, Pitluck S, Nolan M, Ivanova N, Pati A, Land ML, Dlutek M, Doolittle WF, Noll KM, Nesbe CL. Genome sequence of the mesophilic Thermotogales bacterium Mesotoga prima MesG1.Ag.4.2 reveals the largest Thermotogales genome to date. Genome Biol Evol. 2012;4:700–708.
1771. Wiedenheft B, Stedman K, Roberto F, Willits D, Gleskev AK, Zoeller L, Snyder J, Douglas T, Young M. Comparative genomic analysis of hyperthermophilic archaeal Fuselloviridae viruses. J Virol. 2004;78:1954–61.
1772. Geslin C, Gaillard M, Flament D, Roault K, Le Romancer M, Prieur D, Erauso G. Analysis of the first genome of a hyperthermophilic marine virus-like particle, RAV1, isolated from Pyrococcus abyssi. J Bacteriol. 2007;189:4510–9.
1773. Gorlas A, Koonin EV, Benven N, Prieur D, Geslin C. TPV1, the first virus isolated from the hyperthermophilic genus Thermococcus. Environ Microbiol. 2012;14:503–16.
1774. Chalopin D, Galiana D, Volff JN. Genetic innovation in vertebrates: gypsy integrase genes and other genes derived from transposable elements. Int J Evol Biol 2012:724519. doi:10.1155/2012/724519.
1775. Campillos M, Doerks T, Shah PK, Bork P. Computational characterization of multiple Gag-like human proteins. Trends Genet. 2006;22:585–9.
1776. Williams AJ, Blacklow SC, Collins. The zinc finger-associated SCAN box is a conserved oligomerization domain. Mol Cell Biol. 1999;19:8526–35.
1777. Zwolinska K. Retroviruses-derived sequences in the human genome. Human endogenous retroviruses (HERVs). Postepy Hig Med Dosw. 2006;60:637–6521774.
1778a. Sinkovics JG, Horvath JC. Transposons → Retrotransposons → Retro(lenti)viruses: their role in genomics and proteomics during evolution, ontogenesis, and morbidity. In: 18th clinical virology symposium annual meeting, PanAmerican Society Clinical Virology, University South FL Department Medical Microbiology & Immunology, Clearwater, FL, May 2002;#M17.
1778b. Sinkovics JG, Horvath JC. Acqusition of new genomic sequences by retro(lenti)viruses and by advanced mutated transposons-retrotransposons. In: 92nd annual meeting American Association Cancer Research, New Orleans, LA, March 2001, Proceedings 2001;42:115 #622.
1779. Sinkovics J, Molnár E. Specific paralysis of antibody production in mice infected with lymphocytic choriomeningitis virus (in Hungarian). Kisérletes Orvostudomány Exp Med Sci. 1955;6:647–53.
1780. Punkosdy GA, Blain M, Glass DD, Lozano MM, O'Mara L, Du JP. Regulatory T-cell expansion during chronic viral infection is dependent on endogenous retroviral superantigens. Proc Natl Acad Sci USA. 2011;108:3677–382.
1781. Sutkowski N, Conrad B, Thorley-Lawson DA, Hubert BT. Eptein-Barr virus transactivates the human endogenous retrovirus HERV-K18 that encodes a superantigen. Immunity. 2001;15:579–89.
1782. Rodriguez AJ, Seipel SA, Hamill DR, Romancino DP, Di Carlo D, Suprenant KA. Seawi — a sea urchin piwi/argonaute family member is a component of MT-RNP complexes. RNA. 2005;11:646–56.

1783. Klingel S, Morath I, Strietz J, Menzel K, Holstein TW, Gradl D. Subfunctionalization and neofunctionalization of vertebrate Lef/Tcf transcription factors. Dev Biol. 2012;368:44–53.

1784. Mao CD, Byers SW. Cell-context dependent TCF/LEF expression and function: alternative tales of repression, de-repression and activation potentials. Crit Rev Eukaryot Gene Exp. 2011;21:207–36.

1785. Lavra L, Rinaldo C, Ulivieri A, Luciani E, Fidanza P, Giacomelli L, Bellotti C, Ricci A, Trovato M, Soddu S, Bartolazzi A, Sciacchitano S. The loss of the p53 activator HIPK2 is responsible for galectin-3 overexpression in well differentiated thyroid carcinomas. PLoS One. 2011;6(6):e20665. doi:10.1371/journalmpone.0020665.

1786. Resch AM, Palakodeti D. Small RNA pathways in Schmidtea mediterranea. Int J Dev Biol. 2012;56:67–74.

1787. Holder T, Basquin C, Ebert J, Randel N, Jollivet D, Conti E, Jékely G, Bono F. Deep-transcriptome-sequencing and proteome analysis of the hydrothermal vent annelid Alvinella pompejana identifies the CvP-bias as a robust measure of eukaryotic thermostability. Biok ir. 2013;8:2. doi:10.1186/1745-6150-6-2.

1788. Darricarrère N, Liun N, Watanabde T, Lin H. Function of Piwi, a nuclear Piwi/Argonaute protein, is independent of its slicer activity. Proc Natl Acad Sci USA. 2013.

1789. Olovnikov I, Aravin AA, Fejes-Toth K. Small RNA in the nucleus: the RNA-chromatin ping-pong. Curr Opin Genet Dev. 2012;22:164–71.

1790. Vagin VV, Hannonn GJ, Aravin AA. Arginine methylation as a molecular signature of the Piwi small RNA pathway. Cell Cycle. 2889;15:4003–4004.

1791. Siomi MC, Sato K, Pezic D, Aravin AA. PIWI-interacting small RNAs: the vanguard of genome defence. Nat Rev Mol Cell Biol. 2011;12:246–58.

1792a. Watari A, Iwabe N, Masuda H, Okada M. Functional transition of Pak proto-oncogene during early evolution of metazoans. Oncogene. 2010;29:3815–26.

1792b. Watari A. Nature of cancer explored from the perspective of the functional evolution of proto-oncogenes. Yakugaku Zasshi. 2012;132:1165–70.

1793. Kumar A, Molli PR, Pakala SB, Nguyen TB, Rayala SK, Kumar R. PAK thread from amoeba to mammals. J Cell Biochem. 2009;107:579–85.

1794. Koga H, Matsui S, Hirota T, Takebayashi S, Okumura K, Saya H. A human homolog of drosophila lethal(3)malignant brain tumor (I(3)mbt) protein associates with condensed mitotic chromosomes. Oncogene. 1999;18:3799–809.

1795. Manning G, Young SL, Miller WT, Zhai Y. The protist, Monosiga brevicollis, has a tyrosine kinase signaling network more elaborate and diverse than found in any known metazoan. Proc Natl Acad Sci USA. 2008;105:9674–9.

1796. Segawa Y, Suga H, Iwabe N, Oneyama C, Akagi T, Miyata T, Okada M. Functional development of Src tyrosine kinase during evolution from a unicellular ancestor to multicellular animals. Proc Natl Acad Sci USA. 2006;103:12921–121026.

1797. de Mendoza A, Suga H, Ruiz-Trillo I. Evolution of the MAGUK protein gene family in premetazoan lineages. BMC Evol Biol. 2010;10:93. doi:10.1186/1471-2148-10-93.

1798. Roberts S, Delury C, Marsh E. The PDZ protein discs-large (DLG): the 'Jekyll and Hyde' of the epithelial polarity proteins. FEBS J. 2012;279:3549–58.

1799. Bonilla-Delgado J, Bulut G, Liu X, Cortés-Malagón EM, Schlegel R, Flores-Maldonado C, Contreras RG, Chung SH, Lambert PF, Uren A, Gariglio P. The E6 oncoprotein from HPV16 enhances the canonical Wnt/β-catenin pathway in skin epidermis in vivo. Mol Cancer Res. 2012;10:250–8.

1800. Pim D, Bergant M, Boon SS, Ganti K, Kranjec C, Massimi P, Subbaiah VK, Thomas VK, Tomaić V, Banks L. Human papillomavirus and the specificity of PDZ targeting. FEBS J. 2012;279:3530–7.

1801. Baumgartner M, Radziwill G, Lorger M, Weiss A, Moelling K. c-Src-mediated epithelial cell migration and invasion regulated by PDZ binding site. Mol Cell Biol. 2008;28:642–55.

1802. Hou SW, Zhi HY, Pohl N, Loesch M, Qi XM, Li RS, Basir Z, Chen G. PTPH1 dephosphorylates and cooperates with p38gamma MAPK to increase ras oncogenesis through PDZ-mediated interaction. Cancer Res. 2010;70:2901–10.

1803. Subbaiah VK, Kranjec C, Thomas M, Banks L. PDZ domains: the building blocks regulating tumorigenesis. Biochem J. 2011;439:195–205.

1804. Sensui N, Yoshida M, Tachibana K. Role of Mos/MEK/ERK cascade and Cdk12 in Ca2 + oscillations in fertilized ascidian eggs. Dev Biol. 2012;367:208–15.

1805a. Dumollard R, Levasseur M, Hebras C, Huttorei P, Carroll M, Chambon JP, McDougall A. Mos limits the number of meiotic divisions in urochordate eggs. Development. 2011;138:885–95.

1805b. McDougall A, Chenevert J, Lee KW, Hebras C, Dumollard R. Cell cycle in ascidian eggs and embryos. Results Probl Cell Differ. 2011;53:153–69.

1806a. Moloney JB. Biological studies on a lymphoid-leukemia virus extracted from Sarcoma 31. I. Original and introductory investigations. J Natl Cancer Inst. 1960;24:933–51.

1806b. Moloney JB. A virus-induced rhabdomyosarcoma in mice. Natl Cancer Inst Monogr. 1966;22:139–42.

1807a. Schlom J, Moloney JB, Groupé V. Immunological and pathological manifestations of Moloney sarcoma virus (Moloney) infections. Cancer Res. 1970;30:2955–61.

1807b. Schlom J, Moloney JB, Groupé V. Evidence for the rapid decrease in leukemogenic potential of Rauscher leukemia virus in cell culture. Cancer Res. 1971;31:260–4.

1808a. Sinkovics JG, Howe CD, Shullenberger CC. Studies on virus attenuation in mouse leukemia.In: International symposium on comparative Leukemia Research, Paris, 1967. Bibliotheca Haematol (Karger, Basel) 1968;31:73–76.

1808b. Sinkovics JG, Gyorkey F, Groves GF, Gyorkey P. Malignant transformation in a tissue culture infected with a mouse leukemia virus. Arch ges Virusforsch. 1968;23:169–74.

1809. Caubet JF, Mathieu-Mahul D, Bernheim A, Larsen CJ, Berger R. Human proto-oncogene c-mos maps to 8q11. EMBO J. 1985;4:2245–8.

1810a. Sinkovics JG, Horvath JC. Cell-free human sarcoma fluids induced temporary focus formation (FF) and antigenic conversion (AC) in cultured human embryonic cells (CHEC). Does siRNA inhibit retroviral or cellular oncogenes encoding fusion oncoproteins in this system? In: Twenty-first annual clinical virology symposium; annual meeting PanAmerican Society Clinical Virology; University of South FL College of Medicine. Clearwater FL, 8–11 May 2005;Abstracts TA49 & TA50.

1810b. Sinkovics JG, Horvath JC. A retroviral paradox: sarcomagenic retroviruses related to those of animals have not been isolated (as yet) from human sarcomas. In: International Symposium. Association Comparative Research Leukemia & Related Diseases. Heidelberg, Germany, 2–6 July 2005;Abstract #49 p 58.

1811. Polyak K. Heterogeneity in breast cancer. J Clin Invest. 2011;121:3786–37.

1812. Eswaran J, Horvath A, Godbole S, Reddy SD, Mudvan P, Ohshiro K, Cyanam D, Nair S, Fugua SAW, Polyak K, Florea LD, Kumar R. RNA sequencing of cancer reveals novel splicing alterations. Sci Rep. 2013;3:1689.

1813a. Szalmás A, Gyöngyösi E, Ferenczi A, László B, Karosi T, Csomor P, Gergely E, Veress G, Kónya J. Activation of Scr, Fyn and Yes non-receptor tyrosine kinases in keratinocytes expressing human papillomavirus (HPV) type 16 E7 oncoprotein. Virol J. 2013;10:79. doi:10.1186/1743-422X-10-79.

1813b. Ferenczi A, Gyöngyösi E, Szalmás A, Hernádi Z, Tóth Z, Kónya J, Veress G. Sequence variation of human papillomavirus type 31 long control region: phylogenetic and functional implications. J Med Virol. 2013;85:852–9.

1814. Li YC, Deng YH, Guo ZH, Zhang MM, Zhu J, Pu CL, Xiang CP, Guo CB. Prognostic value of hedgehog signal component expressions in hepatoblastoma patients. Eur J Med Res. 2010;15:468–74.

1815. Gupta K, Rane S, Das A, Marwaha RK, Menon P, Rao KL. Relationship of β-catenin and postchemotherapy histopathologic changes with overall survival in patients with hepatoblastoma. J Pediatr Hematol Oncol. 2012;34(8):e320–8. doi:10.1097/MPH.0b013e318280471.
1816. Purcell R, Childs M, Maibach R, Miles C, Turner C, Zimmermann A, Sullivan M. HGF/c-Met related activation of β-catenin in hepatoblastoma. J Exp Clin Cancer Res. 2011;30(1):96. doi:10.1186/1756-9966-30-96.
1817. Chattopadhyay S, Mukherjee S, Boler A, Sharma A, Biswas SK. Hepatoblastoma in the neonatal period. An unusual presentation. J Cytol. 2012;29:252–4.
1818. Lovvorn HN III, Ayers D, Zhao Z, Hilmes M, Prasad P, Shinall MC Jr, Berch B, Neblett WW III, O'Neill JA Jr. Defining hepatoblastoma responsiveness to induction therapy as measured by tumor volume and serum α-fetoprotein kinetics. J Pediatr Surg 201;45:121–129.
1819. Black CT, Cangir A, Choroszy M, Andrassy RJ. Marked response to preoperative high-dose cis-platinum in children with unresectable hepatoblastoma. J Pediatr Surg. 1991;26:1070–3.
1820. Malogolowkin MH, Katzenstein H, Krailo MD, Chen Z, Bowman L, Reynolds M. Intensified platinum therapy is an ineffective strategy for improving outcome in pediatric patients with advanced hepatoblastoma. J Clin Oncol. 2006;24:2879–84.
1821. Collett GP, Goh XF, Linton EA, Redman CW, Sargent IL. RhoE is regulated by cyclic AMP and promotes fusion of human BeWo choriocarcinoma cells. PLoS One. 2012;7(1):e30453. doi:10.1371/journal.pone.0030453.
1822. Xu F, Wang H, Zhang X, Liun T, Liu Z. Cell proliferation and invasion ability of human choriocarcinoma cells lessened due to inhibition of Sox2 expression by microRNA-145. Exp Ther Med. 2013;5:77–84.
1823. Pang RTK, Leung CON, Lee C-L, Lam KKW, Ye T-M, Chiu PCN, Yeung WSB. Micro-RNA-34a is a tumor suppressor in choriocarcinoma via regulation of Delta-like 1. BMC Cancer. 2013;13:25. doi:10.1186/1471-2407-13-25.
1824. Shih I-M, Kuo K-T. Power of the eternal youth. Am J Pathol. 2008;173:911–4.
1825. Siu MKY, Wong ESY, Chan HY, Ngan HYS, Chan KYK, Cheung ANY. Overexpression of NANOG in gestational trophoblastic disease. Am J Pathol. 2008;173:1165–72.
1826. Ferretti C, Bruni L, Dangles-Marie V, Pecking AP, Bellet D. Molecular circuits shared by placental and cancer cells, and their implications in the proliferative, invasive and migratory capacities of trophoblasts. Hum Reprod Update. 2007;13:121–14.
1827. Siu MKY, Yeung MCW, Zhang HJ, Kong DSH, Ho JW, Ngan HY, Chan DC, Cheung AN. p21-Activated kinase-1 promotes aggressive phenotype, cell proliferation, and invasion in gestational trophoblastic disease. Am J Pathol. 2010;176:3015–22.
1828. Zhang HJ, Siu MK, Yeung MC, Jiang LL, Mak VC, Ngan HY, Wong OG, Zhang HQ, Cheung AN. Overexpressed PAK4 promotes proliferation, migration and invasion of choriocarcinoma. Carcinogenesis. 2011;32:765–71.
1829. Mak VCY, Lee L, Siu MKY, Wong OGW, Lu X, Ngan HYS, Wong ESY, Cheung ANY. Downregulation of ASPP1 in gestational trophoblastic disease: correlation with hypermethylation, apoptotic activity and clinical outcome. Mod Pathol. 2011;24:522–32.
1830a. Mok SC, Wong KK, Chan RK, Lau CC, Tsao SW, Knapp RC, Berkowitz RS. Molecular cloning of differentially expressed genes in human epithelial ovarian cancer. Gynecol Oncol. 1994;52:247–52.
1830b. Mok SC, Chan WY, Wong KK, Cheung KK, Lau CC, Ng SW, Baldini A, Colitti CV, Rock CO, Berkowitz RS. DOC-2, a candidate tumor suppressor gene in human epithelial ovarian cancer. Oncogene. 1998;16:2381–7.

1831. Fulop V, Colitti CV, Genest D, Berkowitz RS, Yiu GK, Ng SW, Szepesi J. Mok SC. DOC-2/hDAB2, a candidate tumor suppressor gene involved in the development of gestational trophoblastic disease. Oncogene. 1998;17:419–24.
1832. Asanoma K, Matsuda T, Kondo H, Kato K, Kishino T, Niikawa N, Wake N, Kato H. NECC1, a candidate choriocarcinoma suppressor gene that encodes a homeodomain consensus motif. Genomics. 2003;82:15–25.
1833. Sinkovics JG. Human chorionic gonadotropin-beta subunit gene expression in cultured human fetal and cancer cells of different types and origins. Cancer. 1996;78:184–5.
1834. Männik J, Vaas P, Rull K, Teesalu P, Laan M. Differential placental expression profile of human growth hormone/chorionic somatomammotropin genes in pregnancies with pre-eclampsia and gestational diabetes mellitus. Mol Cell Endocrinol. 2012;355:180–7.
1835. Vereczkey I, Csermák E, Olasz J, Konya Z, Szentirmay Z, Tóth E. Renal choriocarcinoma: gestational or germ cell origin? Int J Surg Pathol. 2012;20:623–8.
1836a. Jékely G. Origin of the nucleus and Ran-dependent transport to safeguard ribosome biogenesis in a chimeric cell. Biol Direct. 2008;3:31. doi:10.1186/1745-6150-3-31.
1836b. Jékely G. Origin of eukaryotic endomembranes: a critical evaluation of different model scenarios. Adv Exp Med Biol. 2007;607:38–51.
1837. Hurtado CA, Rachubinski RA. YIBMH1 encodes a 14-3-3 protein that promotes filamentous growth in the dimorphic yeast Yarrowia lipolytica. Microbiology. 2002;148:3725–35.
1838. Lightner A, Schust DJ, Chen YB, Barrier BF. The fetal allograft revisited: does the study of an ancient invertebrate species shed light on the role of natural killer cells at the maternal-fetal interface? Clin Dev Immunol. 2008;631920. doi:10.1155/2006/631920.
1839. Cichoki F, Felices M, McCullar V, Presnell SR, Al-Attar A, Lutz CT, Miller Js. Cutting edge: microRNA-181 promotes human NK cell development by regulating Notch signaling. J Immunol. 2011;187:6171–5.
1840. Yamanaka Y, Tagasawa H, Takahashi N, Watanabe A, Guo YM, Iwamoto K, Yamasita J, Saitoh H, Kameoka Y, Shimizu N, Ichinohasama R, Sawada K. Aberrant overexpression of microRNAs activate Akt signaling via downregulation of tumor suppressors in natural killer-cell lymphoma/leukemia. Blood. 2009;114:3265–75.
1841. Paik JH, Jang JY, Jeon YK, Kim WY, Kim TM, Heo DS, Kim CW. MicroRNA-146a downregulates NFκB activity via targeting TRAF6 and functions as a tumor suppressor having strong prognostic implications in NK/T cell lymphoma. Clin Cancer Res. 2011;17:4761–71.
1842. Várady G, Sarkadi B, Fátyol K. TTRAP is a novel component of the non-canonical TRAF6-TAK1 TGF-β signaling pathway. PLoS One. 2011;6(9):e25548. doi:10.1371/journal.pone.0025548.
1843a. Nador RG, Cesarman E, Chadburn A, Dawson DB, Ansari MQ, Said J, Knowles DM. Primary effusion lymphoma: a distinct clinicopathologic entity associated with the Kaposi's sarcoma-associated herpes virus. Blood. 1996;88:645–56.
1843b. Nador RG, Cesarman E, Knowles DM, Said JW. Herpes-like DNA sequences in a body cavity-based lymphoma in an HIV-negative patient. N Engl J Med. 1996;333:943.
1844. Matolcsy A, Nádor RG, Cesarman E, Knowles DM. Immunoglobulin VH gene mutational analysis suggests that primary effusion lymphomas derive from different stages of B cell maturation. Am J Pathol. 1998;153:1609–14.
1845. Bhatt AP, Damania B. AKTivation of PI3K/AKT/mTOR signaling pathway by KSHV. Front Immunol. 20122;3.401. doi:10.3389/fimmu.2012.00401.
1846. Graham C, Matta H, Yang Y, Yi H, Suo Y, Tolani B, Chaudhary PM. Kaposi's sarcoma-associated herpesvirus oncoprotein K13 protects against B cell receptor-induced growth arrest and apoptosis through NF-κB activation. J Virol. 23013;87:2242–2252.

1847a. Ay E, Buzas K, Banati F, Minarovits J. Recent results on the development of fetal immune system: self, epigenetic regulation, fetal immune responses. In: Berencsi III G, author/editor. Maternal fetal transmission of human viruses and their influence on tumorigenesis. Dordrecht: Springer; 2012. p. 51–82.

1847b. Berencsi III G, Takács M. Barriers of the human organism and their Achilles' heels. In: Berencsi III G, author/editor. Maternal fetal transmission of human viruses and their influence on tumorigenesis. Dordrecht: Springer; 2012. p. 1–50.

1848a. Lund TC, Prator PC, Medveczky MM, Medveczky PG. The Lck binding domain of herpesvirus saimiri tip-484 constitutively activates Lck and STAT3 in T cells. J Virol. 1996;73:1689–94.

1848b. Kjellen P, Amdjadi K, Lund TC, Medveczky PG, Sefton BM. The herpesvirus saimiri tip484 and tip488 proteins both stimulate lck tyrosine protein kinase activity in vivo and in vitro. Virology. 2002;297:281–8.

1848c. Hartley DA, Amdjadi K, Hurley TR, Lund TC, Medveczky PG, Sefton BM. Activation of the Lck tyrosine protein kinase by the Herpesvirus saimiri tip protein involves two binding interactions. Virology. 2000;276:339–48.

1849a. Poiesz BJ, Ruscetti W, Reitz MS, Kalyanaraman VS, Gallo RC. Isolation of a new type C retrovirus (HTLV) in primary uncultured cells of a patient with Sézary T-cell leukemia. Nature. 1981;294:268–71.

1849b. Gallo RC, Poiesz BJ, Ruscetti FW. Regulation of Human T-cell proliferation: T-cell growth factor and isolation of a new class of type-C retroviruses from human T cells. Haematol Blood Transfus. 1981;16:502–14.

1849c. Poiesz BJ, Dube S, Jones B, Bryz-Gornia V, Dean MT, Spicer T, Hengst J, Sayre K, Esteban W, Ferrer JF. Comparative performances of enzyme-linked immunoabsorbent, western blot, and polymerase chain reaction assays for human T-lymphotropic virus type II infection that is endemic among Indians of the Gran Chaco region of South America. Transfusion. 1997;37:52–9.

1849d. Poiesz BJ, Poiesz MJ, Choi D. The human T-cell lymphoma-leukemia viruses. Cancer Invest. 2003;21:253–77.

1850a. Sinkovics JG, Gyorkey F, Gyorkey P, Wang CH. Long term cultures of hairy cells. Ann Internal Med. 1976;85:532–3.

1850b. Sinkovics JG, Gyorkey F, Gyorkey P. Lekemic reticuloendotheliosis: cultured hairy cells. N Engl J Med. 1976;295:1016.

1850c. Sinkovics JG, Wang CH, Gyorkey F. Epstein-Barr virus in cultured hairy leukemia cells. Lancet. 1975;1:749–50.

1850d. Lemon SM, Pagano JS, Utsinger PD, Sinkovics JG. Infection of cultured "hairy cells" with Epstein-Barr virus: evidence of B lymphocyte origin. Ann Internal Med. 1979;90:5455.

1851a. Bhargava R, Barbashina V, Filippa DA, Teruya-Feldstein J. Epstein-Barr virus positive large B-cell lymphoma arising in a patient previously treated with cladribine for hairy cell leukemia. Leuk Lymphoma. 2004;45:1043–8.

1851b. Lenz G, Golf A, Rüdiger T, Hiermann W, Haferlach T. Epstein-Barr virus-associated B-cell non-Hodgkin lymphoma following treatment of hairy cell leukemia with cladribine. Blood. 2003;102:3457–8.

1851c. Huang AT, Silverstein L, Gonias SL, Rundles RW, Raab-Traub N. Transformation of hairy cell leukemia to EBV genome-containing aggressive B cell lymphoma. Leukemia. 1987;1:369–72.

1852. Miyoshi I, Hiraki S, Tsubota T, Hikita T, Masuji H, Nishi Y, Kimura I. Hairy cell leukemia: establishment of a cell line and its characteristics. Cancer. 1981;47:60–5.

1853. Schiller JH, Bittner G, Meisner LF, Oberley TD, Norback D, Schwabe M, Faltynek CR, Raab-Taub N. Establishment and characterization of an Epstein-Barr virus spontaneously transformed lymphocytic cell line derived from a hairy cell leukemia patient. Leukemia. 1991;5:399–407.

1854. Quesada JR, Reuben J, Hopfer RL, Mundon FK, Hersh EM. Serologic studies in hairy cell leukemia: high prevalence of Epstein-Barr and cytomegalovirus antibodies and absence of human T-cell lymphotrophic viruses antibodies. Leuk Res. 1986;10:1169–73.

1855a. Tiacci E, Liso A, Piris M, Falini B. Evolving concepts in the pathogenesis of hairy cell leukemia. Nat Rev Cancer. 2006;6:437–48.

1855b. Tiacci E, Schiavoni G, Martelli MP. Boveri E, Pacini R, Tabarrini A, Zibellini S, Santi A, Pettirossi V, Fortini E, Ascani S, Arcaini L, Inghirami G, Paulli M, Falini B. Constant activation of the RAF-MEK-ERK pathway as a diagnostic and therapeutic target in hairy cell leukemia. Haematologica. 2013; 98:635–639.

1856. Sári E, Nagy Zs , Demeter J. Recurrent somatic mutation in hairy cell leukemia. Orvosi Hetilap Budapest. 2013;154:123–7.

1857. Manns A, Blattner WA. The epidemiology of the human T-cell lymphotrophic virus type I and type II: etiologic rule in human disease. Transfusion. 1991;31:67–75.

1858a. Moreira L, Balaguer F, Lindor N, de la Chapelle A, Hampel H, Aaltonen LA, Hopper JL, Le Marchand L, Gallinger S, Newcomb PA, Haile R, Thibodeau SN, Gunawardene S, Jenkins MA, Buchanan DD, Potter JD, Baron JA, Ahnen DJ, Moreno V, Andreu M, Ponz de Leon M, Rustgi AK, Castellis A, EPICOLON Consortium. Identification of Lynch syndrome among patients with colorectal cancer. JAMA. 2012;308:1555–1565.

1858b. Bonadona V, Bonaïti B, Olschwang S, Grandjouan S, Huiart L, Longy M, Guimbaud R, Buecher B, Bignon YJ, Caron O, Colas C, Nigués C, Lejeune-Dumoulin S, Olivier-Faivre L, Polycarpe-Osaer F, Nguyen TD, Desseigne F, Saurin JC, Berthet P, Leroux D, Duffour J, Manouvrier S, Frébourg T, Sobol H, Lasset C, Banaïti-Pellié, French Cancer Genetics Network. Cancer risk associated with germline mutations in MLH1, MSH2, and MSH6 genes in Lynch syndrome. JAMA. 2011;305:2304–2310.

1859a. D'Antonio M, Pendino V, Sinha S, Ciccarelli FD. Network of cancer genes (NCG 3.0): integration and analysis of genetic and network properties of cancer genes. Nucl Acid Res. 2012; 40(database issue) D978-83. doi:10.1093/nar/gkr952.

1859b. D'Antonio M, Ciccarelli D. Modification of gene duplicability during the evolution of protein interaction network. PLoS Comput Biol. 2011;7(4):e1002029. doi:10.1371/journal.pcbi.1002029.

1860. Tan W, Zhou W, Yu HG, Luo HS, Shen L. The DNA methyltransferase inhibitor zebularine induces mitochondria-mediated apoptosis in gastric cancer cells in vitro and in vivo. Biochem Biophys Res Commun. 2013;4430:250–6.

1861. Wu CY, Tang ZH, Jiang L, Li XE, Jiang ZS, Liu LS. PCSK9 siRHNA inhibits HUVEC apoptosis induced by ox-LDL via Bcl/Bax-caspase9-caspase3 pathway. Mol Cell Biochem. 2012;359:347–58.

1862. Tang Y, Luo Y, Jiang Z, Ma Y, Lin CJ, Kim C, Carter MG, Amano T, Park J, Kish S, Tian XC. Jak/Stat3 signaling promotes somatic cell reprogramming by epigenetic regulation. Stem Cells. 2012;30:2645–56.

1863. Strand DW, Degraff DJ, Jiang M, Sameni M, Franco OE, Love HD, Hayward WJ, Lin-Tsai O, Wang AY, Cates JM, Sloane BF, Matusik RJ, Hayward SW. Deficiency in metabolic regulators PPARγ and PTEN cooperates to drive keratinizing squamous metaplasia in novel models of human tissue regeneration. Am J Pathol. 2013;182:449–59.

1864. Yoshimoto M, Ludkovski O, deGrace D, Williams JL, Evans A, Sircar K, Bismar TA, Nuin P, Squire JA. PTEN genomic deletions that characterize aggressive prostate cancer originate close to segmental duplications. Genes Chromosomes Cancer. 2012;51:149–60.

1865. Sayed D, He M, Hong C, Gao S, Rane S, Yang Z, Abdellatif M. MicroRNA-21 is a downstream effector of AKT that mediates its antiapoptotic effects via suppression of Fas ligand. J Biol Chem. 2010;285:20281–90.

1866. Sun J, Zhang D, Zheng Y, Zhao Q, Zheng M, Kovacevic Z, Richardson DR. Targeting the metastasis suppressor, NDRG1, using novel iron chelators: regulation of stress fiber-mediated tumor cell migration via modulation of the ROCK1/pML:C2 signaling pathway. Mol Pharmacol. 2013;83:454–69.

1867. Dixon KM, Lui GY, Kovacevic Z, Zhang, Yao M, Chen Z, Dong, Assinder SJ, Richardson DR. Dp44mT targets the AKT,TGF-β and ERK pathways via the metastasis suppressor NDRG1 in normal prostate cells and prostate cancer cells. Br J Cancer. 2013. doi:10.1038/bjc.2012.582.

1868a. Yu Y, Suryo Rahmanton Y, Richardson DR. Bp44mT: an orally active iron chelator of the thiosemicarbazone class with potent anti-tumor efficacy. Br J Pharmacol. 2012;165:146166.

1868b. Yu Y, Gutierrez E, Kovacevic Z, Saletta F, Obeidy P, Suryo Rahmanto Y, Richardson DR. Iron chelators for the treatment of cancer. Curr Med Chem. 2012;19:2689–702.

1869. Bakrania P, Robinson DO, Bunyan DJ, Salt A, Martin A, Crolla JA, Wyatt A, Fielder A, Ainsworth J, Moore A, Read S, Uddin J, Laws D, Pascuel-Salcedo D, Avuso C, Allen L, Collin JR, Ragge NK. Sox2 anophthalmia syndrome: 12 cases demonstrating broader phenotype and high frequency of large gene deletions. Br J Ophthalm. 2007;91:1471–6.

1870. Fantes J, Ragge NK, Lynch SA, McGill NI, Collin JAR, Howard-Peebles PN, Hayward C, Vivian AJ, Williamson K, van Heynigen V, Fitzpatrick DR. Mutations in Sox2 cause anophthalmia. Nat Genet. 2003;33:461–3.

1871. Hussenet T, du Manoir S. SOX2 in squamous cell carcinoma: amplifying a pleiotropic oncogene along carcinogenesis. Cell Cycle. 2010;9:1480–6.

1872. Williamson KA, Hever AM, Rainger J, Rogers RC, Magee A, Fiedler Z, Keng WT, Sharkey FH, McGill N, Hill CJ, Schneider A, Messina M, Turnpenny PD, Fantes JA, Van Heyningen V, Dr Fitz Patrick. Mutations in Sox2 cause anophthalmia-esophageal-genital (AEG) syndrome. Human Mol Genet. 2006;15:1413–22.

1873. Tay Y, Zhang J, Thomson AM, Lim B, Rigoutsos I. MicroRNAs to Nanog, Oct4, and Sox2 coding regions modulate embryonic stem cell differentiation. Nature. 2008;455:1124–8.

1874. Huang YW, Liu JC, Deatherage DE, Luo J, Mutch DG, Goodfellow PJ, Miller DS, Huang TH. Epigenetic repression of microRNA-129-2 leads to overexpression of SOX4 oncogene in endometrial cancer. Cancer Res. 2009;69:9038–46.

1875. Mahajan K, Mahajan NO. PI3K-independent AKT activation in cancers: a treasure trove for novel therapeutics. J Cell Physiol. 2012;227:3178–84.

1876. Clark JM, Alaksiyadis K, Martin A, McNamee K, Tharmalingam T, Williams RO, Mémel S, Cope AP. Inhibitor of kappa B epsilon (IκBε) is a non-redundant regulator of c-Rel-dependent gene expression in murine T and B cells. PLoS One. 2011;6(9): e324504.

1877. Guan H, Zhang H, Cai J, Wun J, Yuan J, Li J, Huang Z, Li M. IKBKE is over-expressed in glioma and contributes to resistance of glioma cells to apoptosis via activation of NFκB. J Pathol. 2011;223:436–45.

1878a. Guo J-P, Coppola D, Cheng JQ. PKBKE protein activates Akt independent of phosphatidylinositol 3-kinase /PDK1/mTORC2 and pleckstrin homology domain to sustain malignant transformation. J Biol Chem. 2011;186:37389–98.

1878b. Guo J, Kim D, Gao J, Kurtyka C, Chen H, Yu C, Wu D, Mittal A, Beg AA, Chellappan SP, Haura EB, Cheng JQ. IKBKE is induced by STAT3 and tobacco carcinogen and determines chemosensitivity in non-small lung cancer. Oncogene. 2013;32:151–9.

1879. Robson JE, Eaton SA, Underhill P, Williams D, Peters J. MicroRNA 296 and 298 are imprinted and part of the GNAS/Gnas cluster and miR-296 targets IKBKE and Tmed9. RNA. 2012;18:135–44.

1880. Berencsi K, Gyulai Z, Gönczöl E, Pincus S, Cox WI, Michelson S, Kari L, Meric L, Cadoz M, Zahradnik J, Starr S, Plotkin S. A canarypox vector-expressing cytomegalovirus (CMV) phosphoprotein 65 induces long-lasting cytotoxic T cell responses in human CMV-seronegative subjects. J Inf Dis. 2001;183:1171–9.

1881. Bernstein DI, Schleiss MR, Berencsi K, Gonczol E, Dickey M, Khoury P, Cadoz M, Meric C, Zahradnik J, Duliege AM, Plotkin S. Effect of previous or simultaneous immunization with canarypox expressing cytomegalovirus (CMV) glycoproteinB (gB) on response to subunit gB vaccine plus MF59 in healthy CMV-seronegative adults. J Infect Dis. 2002;185:686–90.

1882. Gonczol E, Berencsi K, Pincus S, Endresz V, Méric C, Paoletti E, Plotkin SA. Preclinical evaluation of an ALVAC (canarypox)-human cytomegalovirus glycoprotein B vaccine candidate. Vaccine. 1995;13:1080–5.

1883. Gonczol E, Plotkin S. Development of a cytomegalovirus vaccine: lessons from recent clinical trials. Expert Opin Biol Ther. 2001;1:401–12.

1884. Arav-Boger R, Wojcik GL, Duggal P, Ingersoll RG, Beaty T, Pass RF, Yolken RH. Polymorphisms in Toll-like receptor genes influence antibody responses to cytomegalovirus glycoprotein B vaccine. BMC Res Notes. 2012;5:140.

1885. Stein MF, Lang S, Winkler TH, Deinzer A, Erber S, Nettelbeck DM, Naschberger E, Jochmann R, Stürzl M, Slany RK, Werner T, Steinkasserer A, Knippertz I. Multiple IRF- and NFκB-sites cooperate in mediating cell type and maturation-specific activation of the human CD83 promoter in dendritic cells. Mol Cell Biol. 2013;33:1331–44.

1886. Berencsi K, Endresz V, Klurfeld D, Kari L, Kritchevsky D, Gönczöl E. Early atherosclerotic plaques in the aorta following cytomegalovirus infection in mice. Cell Adhes Commun. 1998;5:3947.

1887. Heltai K, Kis Z, Burian K, Endresz V, Veres A, Ludwig E, Gönczöl E, Valyi-Nagy I. Elevated antibody levels against Chlamydia pneumoniae, human HSP60 and mycobacterial HSP65 are independent risk factors in myocardial infarction and ischemic heart disease. Atherosclerosis. 2004;173:339–46.

1888. Gravel SP, Servant MJ. Roles of an IkappB kinase-related pathway in human cytomegalovirus-infected vascular smooth muscle cells: a molecular link in pathogen-induced proatherosclerotic conditions. J Biol Chem. 2005;280:7477–88.

1889. Seifert M, Schltysik R, Küppers R. Origin and pathogenesis of B cell lymphomas. Methods Mol Biol. 2013;971:1–25.

1890. Muto T, Muramatsu M, Taniwaki M, Kinoshita K, Honjo T. Isolation, tissue distribution, and chromosomal localization of the human activation-induced cytidine deaminase (AID) gene. Genomics. 2000;68:85–8.

1891a. Jardin F, Ruminy P, Bastard C, Tilly H. The BCL6 proto-oncogene: a leading role during germinal center development and lymphomagenesis. Pathol Biol (Paris). 2007;55:73–63.

1891b. Jardin F, Sahota SS, Ruminy P, Parmentier F, Piquenot JM, Rainville V, Buchonnet G, Leprêtre S, Tily H, Bastard C. Novel Ig V gene features of t(14;18) and t(3;14) de novo diffuse large B-cell lymphoma displaying germinal center B cell-like and non-germinal center B cell-like markers. Leukemia. 2006;20:2070–4.

1892. Chang PP, Barral P, Fitch J, Pratama A, Ma CS, Kallies A, Hogan JJ, Cerundolo V, Tangye SG, Bittman R, Nutt SL, Brink R, Godfrey DI, Batista FD, Vinuesa CG. Identification of BCL-6-dependent follicular helper NKT cells that provide cognate help for B cell responses. Nat Immunol. 2011;13:35–43.

1893. Kim J, Kim DW, Chang W, Choe J, Kim J, Park CS, Song K, Lee I. Wnt5a is secreted by follicular dendritic cells to protect germinal center B cells via Wnt/Ca2+/NFAT/NF-kB-B cell lymphoma 6 signaling. J Immunol. 2012;188:182–169.

1894. Duan S, Cermak L, Pagan JK, Rossi M, Martinengo C, di Celle PF, Chapuy B, Shipp M, Chiarle R, Pagano M. FBXO11 targets BCL5 for degradation and is inactivated in diffuse large B-cell lymphoma. Nature. 2012;481:90–3.

1895. Conticello SG. Creative deaminases, self-inflicted damage, and genome evolution. Ann NY Acad Sci. 2012;1267:79–85.

1896. Shukla N, Kobos R, Renaud T, Treya-Feldstein J, Price A, McAllister-Lucas L, Steinherz P. Successful treatment of refractory metastatic histiocytic sarcoma with alemtuzumab. Cancer. 2012;118:3719–24.

1897. Kondratiev S, Duraisamy, Unitt CL, Green MR, Pinkus GS, Shipp MA, Kutok JK, Drapkin RI, Rodig SJ. Aberrant expression of the dendritic cell marker TNFAIP2 by the malignant cells of Hodgkin lymphoma and primary mediastinal large B-cell lymphoma distinguishes these tumor types from morphologically and phenotypically similar lymphomas. Am J Surg Pathol. 2011;35:1531–1439.

1898. Hure MC, Elco CP, Ward D, Hutchinson L, Meng X, Dorfman DM, Yu H. Histiocytic sarcoma arising from clonally related mantle cell lymphoma. J Clin Oncol. 2912;30(5): e46-53. doi:10.1200/JCO.2011.38.8553.

1899. Shao H, Xi L, Raffeld M, Feldman AL, Ketterling RP, Knudson R, Rodriguez-Canales J, Hanson J, Pittaluga S, Jaffe ES. Clonally related histiocytic/dendritic cell sarcoma and chronic lymphocytic leukemia/small lymphocytic lymphoma: a study of seven cases. Mod Pathol. 2011;24:1421–32.

1900a. Zhang D, McGuirk J, Ganguly S, Persons DL. Histiocytic/dendritic cell sarcoma arising from follicular lymphoma involving the bone: a case report. Int J Hematol. 2009;89:529–32.

1900b. Zhang D. Histiocytic sarcoma arising from lymphomas via transdifferentiation pathway during clonal evolution. Leuk Lymphoma. 2010;51:739–40.

1901. Pruneri G, Mssullo M, Renne G, Taccagni G, Manzotti G, Luin A, Viale G. Follicular dendritic cell sarcoma of the breast. Wirchows Arch. 2002;44:194–9.

1902. Shimizaki K, Ohsima K, Haraoka S, Suzumiya J, Nakamura N, Kikuchin M. Accessory cell tumor: a clinicopathological study of 16 aggressive tumours containing EBV-positive Hodgkin and Reed-Sternberg-like giant cells. Histopathology. 2002;40:12–21.

1903. Wang S, Rowe M, Lundgren E. Expression of the Epstein-Barr virus transforming protein LMP1 causes rapid and transient stimulation of the Bcl-2 homologue Mcl-1 levels in B-cell lines. Cancer Res. 1996;56:4610–3.

1904a. Sinkovics JG, Dreesman G. Monoclonal antibodies of hybridomas. Rev Inf Dis. 1983;5:9–34.

1904b. Sinkovics JG. The first observation (in the late 1960s) of fused lymphoid cells continuously secreting a specific antibody. Bull Mol Med. 2005;26:61–80.

1905. Gross L. Oncogenic viruses. 3rd ed. Oxford/Pergamon Press, Oxford. 1983; ppXi, 393 pp.

1906a. Hanafusa H, Hanafusa T. Determining factor in the capacity of Rous sarcoma virus to induce tumors in mammals. Proc Natl Acad Sci USA. 1966;55:532–8.

1906b. Hanafusa H, Baltimore D, Smoler D, Watson KF, Yaniv A, Spiegelman S. Absence of polymerase protein in virions of alpha-type Rous sarcoma virus. Science. 1972;177:1188–91.

1906c. Hanafusa H, Aoki T, Kawai S, Miyamoto T, Wilsnack RE. Presence of antigen common to avian tumor viral envelope antigen in normal chicken embryo cells. Virology. 1973;56:22–32.

1906d. Duesberg PH, Kawai S, Wang LH, Vogt PK, Murphy HM, Hanafusa H. RNA of replication-defective strains of Rous sarcoma virus. Proc Natl Acad Sci USA. 1975;72:1569–73.

1906e. Hanafusa H. Cellular origin of transforming genes of RNA tumor viruses. Harvey Lect. 1979;75:255–75.

1906f. Lerner TL, Hanafusa H. DNA sequence of the Bryan high-titer strain of Rous sarcoma virus: extent of env deletion and possible genealogical relationship with other virus strains. J Virol. 1984;49:549–56.

1907a. Varmus HE, Bishop JM, Nowinski RC, Sarker NH. Mammary tumor virus specific nucleotide sequences in mouse DNA. Nature New Biol. 1972;238:189–91.

1907b. Varmus HE, Weiss RA, Friis RR, Levinson W, Bishop JM. Detection of avian tumor virus specific nucleotide sequence in avian cell DNAs. Proc Natl Acad Sci USA. 1972;69:20–4.

1907c. Stehelin D, Varmus HE, Bishop JM, Vogt PK. DNA related to the transforming gen(e) of avian sarcoma viruses is present in normal avian DNA. Nature. 1976;260:170–3.
1908. Sinkovics JG, Horvath J. Viral contaminants of poliomyelitis vaccines. Acta Microbiol Immunol Hung. 2000;47:471–6.
1909. Rizzo P, Di Resta I, Powers A, Ratner H, Carbone M. Unique strains of SV40 in commercial poliovaccines from 1955 not readily identifiable with current testing for SV40 infection. Cancer Res. 1999;59:6103–8.
1910. Butel JS. Patterns of polyomavirus SV40 infections and associated cancers in humans: a model. Curr Opin Virol. 2012;2:508–14.
1911. Gross L. A filterable agent recovered from Ak leukemia extracts causing salivary gland carcinoma in C3H mice. Proc Soc Exp Biol Med. 1953;83:414–21.
1912. Stewart SE, Eddy BE, Gochenour AM, Borgese NG, Grubbs GE. The induction of neoplasms with a substance released from mouse tumors by tissue culture. Virology. 1957;3:380–400.
1913. Piña-Oviedo S, De León-Bojorge B, Cuesta-Mejias T, White MK, Ortiz-Hidalgo C, Khalili K, Del Valle L. Glioblastoma multiforme with small cell neuronal-like component: association with human neurotropic JC virus. Acta Neuropathol. 2006;111:388–96.
1914. Shullenberger CC, Leavens ME, Sinkovics JG. Project M23/gm 20: Viruses in malignancy-associated encephalopathies. Research Report 1974. The University of Texas Cancer Center M.D. Anderson Hospital, Houston TX. The University of Texas Press, Austin TX 1974; 455–456.
1915. Feng H, Shuda M, Chang Y, Moore PS. Clonal integration of a polyomavirus in human Merkel cell carcinoma. Science. 2008;319:1096–100.
1916. Kassem A, Schöpflin A, Diaz C, Weyers W, Stickeler E, Werner M, zur Hausen A. Frequent detection of Merkel cell polyomavirus in human Merkel cell carcinomas and identification of a unique deletion in the VP1 gene. Cancer Res. 2008;68:5009–13.
1917. Varga E, Kiss M, Szabó K, Kemény L. Detection of Merkel cell polyomavirus DNA in Merkel cell carcinomas. Br J Dermatol. 2009;161:930–2.
1918. Horváth KB, Pankovics P, Battyáni Z, Kálmán E, Reuter G. A probable etiological role for Merkel cell polyomavirus in the development of Merkel cell carcinoma. Orvosi Hetilap (Budapest). 2013;154:102–12.
1919. Bertrand M, Mirabel X, Desmedt E, Vercambre-Darras S, Martin de Lassalle E, Bouchindhomme B, Giuerreschi P, Martinot V, Mortier L. Merkel cell carcinoma: a new radiation-induced cancer? Ann Dermatol Venereol. 2013;140:41–45.
1920. Gomez BP, Wang C, Viscidi RP, Peng S, He L, Wu T-C, Hung C-F. Strategy for eliciting antigen-specific CD8+ T cell-mediated immune response against a cryptic CTL epitope of Merkel cell polyomavirus large T antigen. Cell Biosci. 2012;2:36. doi:10.1186/2045-3701-2-36.
1921. Iyer JG, Afanasiev OK, McClurkan C, Paulson K, Nagase K, Jing L, Marshak JO, Dong L, Carter J, Lai I, Farrar E, Byrd D, Galloway D, Yee C, Koelle DM, Nghiem P. Merkel cell polyomavirus-specific CD8+ and CD4+ T-cell responses identified in Merkel cell carcinomas and blood. Clin Cancer Res. 2011;17:6671–80.
1922. Sihto H, Joensjuu H. Tumor-infiltrating lymphocytes and outcome in Merkel cell carcinoma, a virus-associated cancer. Oncoimmunology. 2012;1:1420–1.
1923. Nokes BT, Cunliffe HE, Lafleur B, Mount DW, Livingston RB, Futscher BW, Lang JE. In vitro assessment of the inflammatory breast cancer cell line SUM149: discovery of 2 single nucleotide polymorphisms in the RNase L gene. J Cancer. 2013;4:104–16.
1924. Van Laere SJ, Ueno N, Finetti P, Vermeulen PB, Lucci A, Robertson FM, Marsen M, Iwamoto T, Krishnamurthy S, Masuda H, Van Dam PA, Woodward WA, Viens P, Cristofanilli M, Birnbaum D, Dirix LY, Reuben JM, Bertucci F. Uncovering the molecular secrets of inflammatory breast cancer biology: an integrated analysis of three distinct Affymetrix gene expression data sets. Clin Cancer Res. 2013;19:4685–96.

1925. Okano H, Nakamura M, Yoshida K, Okada Y, Tsuji D, Nori S, Ikeda E, Yamanaka S, Miura K. Steps towards safe cell therapy using induced pluripotential stem cells. Circ Res. 2013;112:523–33.
1926. Duesberg P, McCormack A . Immortality of cancers: a consequence of inherent karyotypic variations and selections for autonomy. Cell Cycle. 2013;12(5).
1927. Contreras CM, Akbay EA, Gallardo TD, Haynie JM, Sharma S, Tagao O, Bardeesy N, Takahashi M, Settleman J, Wong K-K, Castillon DH. Lkb1 inactivation is sufficient to drive endometrial cancers that are aggressive yet highly responsive to mTOR inhibitor monotherapy. Dis Model Mech. 2010;3:181–93.
1928. Shivakumar L, Minna J, Sakamaki T, Pestell R, White MA. The RASSF1A tumor suppressor blocks cell cycle progression and inhibits cyclin D1 accumulation. Mol Cell Biol. 2002;22:4309–18.
1929. Carvalho S, Raposo AC, Martins FB, Grosso AR, Sridhara SC, Rino J, Carmo-Fonseca M, de Akmeida SF. Histone methyltransferase SETD2 coordinates FACT recruitment with nucleosome dynamics during transcription. Nucleic Acids Res. 2013;41:2881–93.
1930. Chen Z, Yan CT, Dou Y, Viboolsittisen SS, Wang JH. The role of a newly identified SET domain-containing protein, SETD3, in oncogenesis. Haematologica. 2012; doi:10.3324/haematol.2012.066977.
1931. Fog CK, Galli GG, Lund AH. PRDM proteins: important players in differentiation and disease. Bioassays. 2012;34:50–60.
1932. Pawlowski R, Mühl SM, Sulser T, Krek W, Moch H, Schraml P. Loss of PBRM1 expression is associated with renal cell carcinoma. Int J Cancer. 2013;132:E11–7. doi:10.1002/ijz.27822.
1933. Strahl BD, Grant PA, Briggs SD, Sun Z-W, Bone JAR, Caldwell JA, Mollah S, Cook RG, Shabanowitz j, Hunt DF, Allis CD. Set2 is a nucleosomal histone H3-selective methyltransferase that mediates transcriptional repression. Mol Cell Biol. 2002;22:1298–306.
1934. Rasmussen MH, Jensen NF, Tarpgaard LS, Qvortrup C, Remer MJ, Stanwang J, Hansen TOP, Christensen LL, Lindebjerg J, Hansen F, Jensen BV, Hansen TF, Pfeiffer P, Brünner N, Orntoft TF, Andersen CL. High expression of microRNA-625-3p is associated with poor response to first-line oxaliplatin-based treatment of metastatic colorectal cancer. Mol Oncol. 2013; pil. S1574-7891(13)j00040-9. doi:10.1016/j.molonc.2013.02.016.
1935a. Acs G, Acs P, Beckwith SM, Pitts RL, Clements E, Wong K, Verma A. Erythropoietin and erythropoietin receptor expression in human cancer. Cancer Res. 2001;61:3561–5.
1935b. Acs G, Zhang PJ, McGrath PJ, Acs P, McBroom J, Mohyeldin A, Liu S, Lu H, Verma A. Hypoxia-induced erythropoietin signaling in squamous dysplasia and squamous cell carcinoma of the uterine cervix and its potential role in cervical carcinogenesis and tumor progression. Am J Pathol. 2003;162:1789–1.
1936. Malenda A, Skrobanska A, Issat T, Winiarska M, Bil J, Oleszczak B, Sinski M, Firczuk M, Bujnicki JM, Chlebowska J, Staruch AD, Glodkowska-Mrowka E, Kunikowska J, Krolicki L, Szablewski L, Gaciong Z, Koziak K, Jakobisiak M, Golab J, Nowis DA. Statins impair glucose uptake in tumor cells. Neoplasia. 2012;14:311–23.
1937. Safdie F, Brandhorst S, Wei M, Wang W, Lee C, Hwang S, Conti PS, Chen TC, Longo VD. Fasting enhances the response of glioma to chemo- and radiotherapy. PLoS One. 2012;7(9):e44603. doi:10.1371/journal.pone.0044603.
1938. Ito E, Honma R, Imai J, Azuma S, Kanno T, Mori S, Yoshie O, Nishio J, Iwasaki H, Yoshida K, Gohda J, Inoue J, Watanabe S, Semba K. A tetraspanin-family protein, T-cell acut lymphoblastic leukemia-associated antigen 1, is induced by the Ewing's sarcoma-Wilms' tumor 1 fusion protein of desmoplastic small round cell tumor. Am J Pathol. 2003;163:2165–72.

1939. De J, Zanjani R, Hibbard M, Davis BH. Immunophenotypic profile predictive of KIT activating mutations in AML1/ETO leukemia. Am J Clin Pathol. 2007;128:550–7.
1940. Erickson P, Gao J, Chang KS, Look T, Whisenant E, Raimondi S, Lasher R, Trujillo J, Rowley J, Drabkin H. Identification of breakpoints in t(8:21) acute myelogenous leukemia and isolation of a fusion transcript, AML1/ETO, with similarity to Drosophila segmentation gene, runt. Blood. 1992;80:1825–31.
1941. Nordin SL, Anderson C, Bjermer L, Bjartell A, Mörgelin M, Egesten A. Midkine is part of the antibacterial activity released at the surface of differentiated bronchial epithelial cells. J Innate Immun. 2013;5:519–30.
1942. Obama H, Matsubara S, Guénet JL, Muramatsu T. The midkine (MK) family growth/differentiation factors: structure of an MK-related sequence in a pseudogene and evolutionary relationships among members of the MK family. J Biochem. 1994;1215:516–22.
1943. Svensson SL, Pasupuleti M, Walse B, Malmten M, Mörgelin M, Sjrgen C, Olin AI, Collin M, Schmidtchen A, Palmer R, Egesten A. Midkine and pleiotrophin have bactericidal properties: preserved antibacterial activity in a family of heparin-binding growth factors during evolution. J Biol Chem. 2010;285:16105–15.
1944. Toledano-Katchalski H, Nir R, Volohonsky G, Volk T. Post-transcriptional repression of the Drosophila midkine and pleiotrophin homolog miple by HOW is essential for correct mesoderm spreading. Development. 2007;134:3473–81.
1945. Cohen S, Shishana OY, Zelman-Toister E, Maharshak N, Binsky-Ehrenreich I, Gordin M, Hazan-Halevy I, Herishanu Y, Shvidel L, Haram M, Leng L, Bucala R, Harroch S. Shacher I The cytokine midkine and its receptor RPTPζ regulate B cell survival in a pathway induced by CD74. J Immunol. 2012;188:259–69.
1946. DaSilva JO, Amorino GP, Casarez EV, Pemberton B, Parsons SJ. Neuroendocrine-derived peptides promote prostate cancer cell survival through activation of IGF-1R signaling. Prostate. 2013;73:801–12. doi:10.1002/pros.22624.
1947. Kuo AH, Stoica GE, Riegel AT, Wellstein A. Recruitment of insulin receptor substrate-1 and activation of NF-kappaB essential for midkine growth signaling through anaplastic lymphoma kinase. Oncogene. 2007;26:859–69.
1948. Nordin A, Wang W, Welén K, Damber JE. Midkine is associated with neuroendocrine differentiation in castration-resistant prostate cancer. Prostate. 2012; doi:10.1002/pros.22607.
1949. You Z, Dong Y, Kong X, Beckett LA, Gandour-Edwards R, Melamed J. Midkine is a NF-kappaB-inducible gene that supports prostate cancer cell survival. BMC Med Genomics. 2008;1:6. doi:10.1189/1755-87694-1-6.
1950. Dai LC, Wang X, Yao X, Lu YL, Ping JL, He JF. Antisense oligonucleotides targeting midkine induced apoptosis and increased chemosensitivity in hepatocellular carcinoma cells. Acta Pharmacol Sin. 2006;27:1630–6.
1951. Wang Q, Huang Y, Ni Y, Wanmg H, Hou Y. siRNA targeting midkine inhibits gastric cancer cells growth and induces apoptosis involved caspase 3,8,9 activation and mitochondrial depolarization. J Biomed Sci. 2007;14:783–95.
1952. Xu Y, Qu X, Zhang X, Luo Y, Zhang Y, Luo Y, Hou K, Liu Y. Midkine positively regulates the proliferation of human gastric cancer cells. Cancer Lett. 2009;279:137–44.
1953. Khan MO, Ather MH. Chromogranin A—serum marker for prostate cancer. J Pak Med Assoc. 2011;61:108–11.
1954. Voss MJ, Möller MF, Powe DG, Niggemann B, Zänker KS, Entschladen F. Luminal and basal-like breast cancer cells show increased migration induced by hypoxia, mediated by an autocrine mechanism. BMC Cancer. 2011;11:158. doi:10.1186/1471-2407-11-158.
1955. Izerasdjene K, Hingorani SR. Targets, trials, and travails in pancreas cancer. J Natl Compr Canc NETW. 2007;5:1042–53.

1956a. Pernick NL, Sarkar FH, Philip PA, Arlauskas P, Shields AF, Vaitkevicius VK, Dugan MC, Adsay NV. Clinicopathologic analysis of pancreatic adenocarcinoma in African Americans and Caucasians. Pancreas. 2003;26:28–32.

1956b. Pernick NL, Sarkar FH, Tabaczka P, Kotcher G, Frank J, Adsay NV. Fas and Fas ligand expression in pancreatic adenocarcinoma. Pancreas. 2002;25(3):e36–41.

1957. Koster R, Timmer-Bosscha H, Bischoff R, Gietema JA, de Jong S. Disruption of the MDM2-p53 interaction strongly potentiates p53-dependent apoptosis in cisplatin-resistant human testicular carcinoma cells via the Fas/FasL pathway. Cell Death Dis. 2011;2:e148. doi:10.1038/cddis.2011.33.

1958. Lin Y, Liu L, Zhang T, Liu J. Functional investigation of Fas ligand expression in human non-small cell lung cancer cells and its clinical implications. Ann Thorac Surg. 2013;95:412–8.

1959. Sinkovics JG, Horvath JC, Horvath E, Horak A. Fas system-generated immune-privileged state, BCL2-mediated apoptosis resistance and autocrine growth factor-guided proliferative signal transduction in human melanoma cells. 50th Annual Symposium on Fundamental Cancer Research: "Molecular Determinants of Cancer Metastasis." The University of Texas MD Anderson Cancer Center, Houston TX, Oct 28-31, 1997 VI-111 pp 212-213. Erratum: "…FasR+ autologous host lymphocytes by apoptosis…." (in the abstract).

1960. Cai Z, Yang F, Yu Z, Jiang L, Wang Q, Yang Y, Wang L, Cao X, Wang L. Activated T cell exosomes promote tumor pathways or invasion via Fas signaling pathway. J Immunol. 2012;188:5954–61.

1961. Sinkovics JG. The cell survival pathways of ancient unicellular (Amoebae; Choanoflagellates) and multicellular (Amphioxus; Spongiae) eukaryotes were conserved in their extant descendants and serve as proto-oncogenes in vertebrate hosts. Posters, in: Illuminating Genomic Dark Matter ncRNA in Disease and Cancer. Symposia on Cancer Research. The University of Texas MD Anderson Cancer Center, Houston, TX Oct 9–10, 2014. Eur J Microbiol Immunol. 2015;5:25–43.

1962. Peng P, Yan Y, Keng S. Exosomes in the ascites of ovarian cancer patients: origin and effects on anti-tumor immunity. Oncol Rep. 2011;25:749–62.

1963. Wahlgren J, de Karlson TL, Glader P, Telemo E, Valadi H. Activated human T cells secrete exosomes that participate in IL-2 mediated immune response signaling. PLoS One. 2012;7(11):e49723. doi:10.1371/journal.pone.0049723.

1964. Whiteside TL. Immune modulation of T-cell and NK (natural killer) cell activities by TEXs (tumour-derived exosomes). Biochem Soc Trans. 2013;41:245–51.

1965. Robinson DR, Wu YM, Kalyanap-Sundaram S, Cao X, Lonigro RJ, Sung TS, Chen CL, Zhang L, Wang R, Su F, Iyer MK, Roychowdhury S, Siddiqui J, Pienta KJ, Kunju LP, Talpaz M, Mosquera JM, Singer S, Schuetze SM, Antonescu CR, Chinnalyan A. Identification of recurrent NAB2-STAT6 gene fusions in solitary fibrous tumor by integrative sequencing. Nat Genet. 2013;45:180–5.

1966. Wolkers MC, Gerlach C, Arens R, Janssen EM, Fitzgerald P, Schumacher TN, Medema JP, Green DR, Schoenberger SP. Nab2 regulates secondary CD8+ T-cell responses through control of TRAIL expression. Blood. 2012;119:798–804.

1967. Penel N, Amela EY, Decanter G, Robin Y-M, Marec-Berard P. Solitary fibrous tumors and so-called hamangiopericytoma. Sarcoma. 2012;690251. doi:10.1155/2012/690251.

1968. Cuello J, Brugés R. Malignant solitary fibrous tumor of the kidney: report of the first case managed with interferon. Case Rep Oncol Med. 2013:564980. doi:10.1155/2013/564980.

1969. Quek R, Wang Q, Morgan JA, Shapiro GI, Butrunski JE, Ramaiya N, Huftalen T, Jederlinic N, Manola J, Wagner AJ, Demetri GD, George S. Combination mTOR and IGF-1R inhibition: phase 1 trial of everolimus and figitumumab in patients with advanced sarcomas and other solid tumors. Clin Cancer Res. 2011;17:871–9.

1970. O'Reilly MS, Boehm T, Shing Y, Fukai N, Vasios G, Lane WS, Flynn E, Birkhead JAR, Olsen BR, Folkman J. Endostatin: an endogenous inhibitor of angiogenesis and tumor growth. Cell. 1997;88:277–85.

1971a. Folkman J. Role of angiogenesis in tumor growth and metastasis. Semin Oncol. 2002;29 (S16):15–8.

1971b. Folkman J. Antiangiogenesis in cancer therapy – endostatin and its mechanisms of action. Exp Cell Res. 2006;312:594–607.

1971c. Folkman J. Strategies for the discovery of endogenous angiogenesis inhibitors. Ernst W. Bertner Award Presentation and Lecture. 50th Annual Symposium on Fundamental Cancer Research. The University of Texas MD Anderson Cancer Center: Molecular Determinants of Cancer Metastasis, Houston TX Oct 28–31, 1997 pp 5–7.

1971d. Folkman J. Antiangiogenic therapy: experimental and clinical studies. 42nd Annual Clinical Conference & 52nd Annual Symposium on Fundamental Cancer Research. The University of Texas MD Anderson Comprehensive Cancer Center: Cancer Research at the Millenium. Houston TX Jan 9–12, 2000 pp 56–57.

1972. Horak A, Horvath JC, Sinkovics JG. Cancer therapy by anti-angiogenesis. 30th Annual Scientific Symposium, Hungarian Medical Association of America, Sarasota FL Oct 23–30, 1998. Arch Hung Med Assoc Am. 1998;6/2:9.

1973. Dell'Eva R, Pfeffer U, Indraccolo S, Albini A, Noonan D. Inhibition of tumor angiogenesis by angiostatin L: from recombinant protein to gene therapy. Endothelium. 2002;9(1):3–10.

1974. Szentirmai O, Baker CH, Bullain SS, Lin N, Takahashi M, Folkman J, Mulligan RC, Carter BS. Successful inhibition of intracranial human glioblastoma multiforme xenograft via systemic adenoviral delivery of soluble endostatin and soluble vascular endothelial growth factor receptor-2: laboratory investigation. J Neurosurg. 2008;108:979–88.

1975. Gerlinger M, Rowan AJ, Horswell S, Larkin J, Endesfelder D, Gronroos E, Martinez P, Matthews N, Stewart A, Tarpey P, Vareira I, Phillimore B, Begum S, McDonald NQ, Buler A, Jones D, Raine K, Latietr C, Santos CR, Nohadfani M, Eklund AC, Spencer-Dene B, Clark G, Pickering L, Stamp G, Gore M, Szallasi Z, Downwars J, Futreal PA, Swanton C. Intratumor heterogeneity and branched evolution revealed by multiregion sequencing. N Engl J Med. 2012;366:883–92.

1976. Hutz JE, Manning WA, McLeod HL. Genomwide analysis of inherited variations associated with phosphorylation of PI3K/Akt/mTOR signaling proteins. PLoS One. 2011;6(9):e24873. doi:10.1371/journal.pone.0024873.

1977a. Shackelford DB, Shaw RJ. The LKB1-AMPK pathway: metabolism and growth control in tumour suppression. Nat Rev Cancer. 2009;9:563–75.

1977b. Shackelford DB, Vasquez DS, Cebeil J, Wu S, Leblanc M, Wu C-L, Vera DR, Shaw RJ. mTOR and HIF-1α-mediated tumor metabolism in an LKB1 mouse model of Peutz-Jeghers syndrome. Proc Natl Acad Sci USA. 2009;106:11137–42.

1978. Gill RK, Yang SH, Meerzaman D, Mechanic LE, Bowman ED, Jeon HS, Roy Chowdhuri S, Shakoori A, Dracheva T, Hong KM, Fukuoka J, Zhang JH, Harris CC, Jen J. Frequent homologous deletion of the LKB1/STK11 gene in non-small cell lung cancer. Oncogene. 2011;30:3784–91.

1979. Diaz-Padilla I, Duran I, Clarke BA, Oza AM. Biologic rationale and clinical activity of mTOR inhibitors in gynecologic cancer. Cancer Treat Rev. 2012;38:767–75.

1980. Hill EK, Dizon DS. Medical therapy of endometrial cancer: current status and promising novel treatments. Drugs. 2012;72:705–13.

1981. Weigelt B, Banerjee S. Molecular targets and targeted therapeutics in endometrial cancer. Curr Opin Oncol. 2012;24:554–5563.

1982. Rasnick D. Aneuploidy theory explains tumor formation, the absence of immune surveillance, and the failure of chemotherapy. Cancer Genet Cytogenet. 2002;136:66-72.

1983. Heffner DK. Chaotic tumors and 2 mistakes of molecular oncologists. Ann Diagn Pathol. 2005;9:61–7.

1984. Huse JT, Phillips HS, Brennan CW. Molecular subclassification of diffuse gliomas: seeing order in the chaos. Glia. 2011;59:1190–9.

1985. Zimonic D, Brooks MW, Popescu N, Winberg RA, Hahn WC. Derivation of human tumor cells without widespread genomic instability. Cancer Res. 2001;61(8838):8844.

1986. Burrell RA, McClelland SE, Endesfelder D, Groth P, Wellar MC, Shaikh N, Domingo E, Kanu N, Dewhurst SM, Gronroos E, Chew SK, Rowan AJ, Schenk A, Sheffer M, Howell M, Kschischo M, Behrens A, Helleday T, Bartek J, Tomlinson IP, Swanton C. Replication stress links structural and numerical cancer chromosomal instability. Nature. 2013;494:492–6.

1987. Okazaki N, F-Kikuno R, Ohara R, Inamoto S, Koseki H, Hiraoka S, Saga Y, Seino S, Nishimura M, Kaisho T, Hoahino K, Kitamura H, Nagase T, Ohara O, Koga H. Prediction of the coding sequences of mouse homologues of KIAA gene: IV. The complete nucleotide sequences of 500 moist KIAA-homologous cDNAs identifies by screening of terminal sequences of CDNA clones randomly sampled from size-fractionated libraries. DNA Res. 2004;11:205–18.

1988. Sakai T, Inom K, Wada S, Maeda H. Identification of the DNA binding specificity of the human ZNF219 protein and its function as a transcription repressor. DNA Res. 2003;10:155–65.

1989. Rymarkis LA, Webster BR, Stern DB. The nucleus-encoded factor MCD4 participates in degradation of nonfunctional 3' UTR sequences by cleavage of pre-mRNA in Chlamydomonas chloroplasts. Mol Genet Genomics. 2007;277:329–40.

1990. Röttinger E, Besnardeau L, Lepage T. Expression pattern of three putative RNA-binding proteins during early development of the sea urchin Paracentrotus lividus. Gene Expr Patterns. 2006;6:864–72.

1991. Guzmán B, Cormand B, Ribasés M, González-Núñez D, Botey A, Poch E. Implications of chromosome 18 in the hypertension by sibling pair and association analyses: putative involvement of the RKHD2 gene. Hypertension. 2006;48:883–91.

1992. Morse MA, Chapman R, Powderly J, Blackwell K, Keler T, Green J, Riggs R, He LZ, Ramakrishna V, Vitale L, Zhao B, Butler SA, Hobelka A, Osada T, Davis T, Clay T, Lyerly HK. Phase I study utilizing a novel antigen-presenting cell-targeted vaccine with Toll-like receptor stimulation to induce immunity to self-antigens in cancer patients. Clin Cancer Res. 2011;17:4844–53.

1993a. Rosenblatt J, Glotzbecker B, Mills H, Vasir B, Tzachanis D, Levine JD, Joyce RM, Wellenstein K, Keefe W, Schickler M, Rotem-Yehudar R, Kufe D, Avigan D. PDF-1 blockade by CT-011, anti-PD-1 antibody, enhances ex vivo T-cell responses to autologous dendritic cell/myeloma fusion vaccine. J Immunother. 2011;34:409–18.

1993b. Rosenblatt J, Vasir B, Uhl L, Blotta S, Macnamara C, Somaiya P, Wu Z, Joyce R, Levine JD, Dombagoda D, Yuan YE, Francoeur K, Fitzgerald K, Richardson P, Weller E, Anderson K, Kufe D, Munshi N, Avigan D. Vaccination with dendritic cell/tumor fusion cells results in cellular and humoral antitumor immune responses in patients with multiple myeloma. Blood. 2011;117:393–402.

1994. Mansilla C, Berraondo P, Durantez M, Martinez M, Casares N, Arrbillaga L, Rudilla F, Fioravanti J, Lozano T, Villanueva L, Sarobe P, Borrás P, Leclerc, Prieto J, Lasarte JJ. Eradication of large tumors expressing human papilomavirus E7 protein by therapeutic vaccination with E7 fused to the domain a from fibronectin. Int J Cancer. 2012;131:641–51.

1995. Paolini F, Massa S, Manni I, Franconi R, Venuti A. Immunotherapy in new pre-clinical models of HPV-associated oral cancers. Hum Vaccin Immunother. 2013;9(3).

1996. Kawana K, Adachi K, Kojima S, Kozuma S, Fuji T. Therapeutic human papillomavirus (HPV) vaccines: a novel approach. Open Virol J. 2012;6:264–9.

1997. Bellier B, Klatzmann D. Virus-like particle-based vaccines against hepatitis C virus infection. Expert Rev Vaccines. 2013;12:143–54.

1998. Baumont E, Poingeard P. Prospects for prophylactic hepatitis c vaccines based on virus-like particles. Hum Vaccin Immunother. 2013;9(5).

1999. Lauer GM. Immune responses to hepatitis C virus (HCV) infection and the prospects for an effective HCV vaccine or immunotherapies. J Infect Dis. 2013;207:S1–12.

2000. Hui E, Taylor GS, Jia H, Ma BB, Chan SL, Ho R, Wong W, Wilson S, Johnson BF, Edwards C, Stocken DD, Rickinson AB, Steven NM, Chan AT. Phase 1 trial of recombinant modified vaccinia Ankara (MVAS) encoding Epstein-Bar viral tumor antigens in nasopharyngeal carcinoma patients. Cancer Res. 2013;73:1676–88.

2001a. Yaddanapudi K, Mitchell RA, Eaton JW. Cancer vaccines. Oncoimmunology. 2013;2(3): e23403.

2001b. Man S. Back to the future: learning from cancer vaccine trials in Cardiff. Z Appl immunohistochem Mol Morphol. 2013;21:113–6.

2002. Krishnadas DK, Shapiro T, Lucas K. Complete remission following decitabine/dendritic cell vaccine for relapsed neuroblastoma. Pediatrics. 2013;1312:e336–41.

2003a. Toolan HW, Moore A. Oncolytic effect of Egypt virus on a human epidermoid carcinoma grown in X-irradiated rats. Proc Soc Exp Biol Med. 1952;79:697–701.

2003b. Moore AE. Present status of oncolytic viruses and their effect on tumors in mouse, man and tissue culture. Acta Unio Int Contra Cancrum. 1956;12:491–4.

2004a. Sinkovics JG. Enhancement of the carcinostatic activity of Newcastle disease virus (NDV) associated with adaptation of NDV to suckling mouse brain. Bact Proc. 1957;57, 96.

2004b. Sinkovics J. Studies on the biological characteristics of the Newcastle disease virus (NDV) adapted to the brain of newborn mice. Arch ges Virusforsch. 1957;7:242–57.

2004c. Sinkovics JG, Horvath JC (authors/editors). Viral therapy of human cancers. New York: Marcel Dekker.

2005. Vähä-Koskela MJ, Heikkila JE, Hinkkanen AE. Oncolytic viruses in cancer therapy. Cancer Lett. 2007;254:178–216.

2006a. Kaufman JK, Bossow S, Grossardt C, Sawall S, Kupsch J, Erbs P, Hassel JC, van Kalle C. Enk AH, Nettelbeck DM, Ungerechts G. Chemovirotherapy of malignant melanoma with a targeted and armed oncolytic measles virus. J Invest Dermatol. 2012; doi:10.1038/jid.2012.459.

2006b. Kaufman JK, Bines SD. OPTIM trial: a phase III trial of an oncolytic herpes virus encoding GM-CSF for unresectable stage III or IV melanoma. Future Oncol. 2010;6 (6):941-949. doi:10.22117/fon.10.66.

2007. Patel MR, Kratzke RA. Oncolytic virus therapy for cancer: the first wave of translational clinical trials. Transl Res. 2013;pii:S1931-5244(12)OO446-X. doi:10.1016/k.trsl.2012. 12.o10.

2008. Zeyaullah M, Patro M, Ahmad I, Ibraheem K, Sultan P, Nehal M, Ali A. Oncolytic viruses in the treatment of cancer: a review of current strategies. Pathol Oncol Res. 2012;18:771–87.

2009. Ferguson MS, Lemoine NR, Wang Y. Systemic delivery of oncolytic viruses: hopes and hurdles. Adv Virol. 2012;2012:805629. doi:10.1155/2012/805629.

2010. Wojton J, Kaur B. Impact of tumor microenvironment on oncolytic viral therapy. Cytokine Growth Factor Rev. 2010;21:127–34.

2011. Wong HH, Lemoine NR, Wang Y. Oncolytic viruses for cancer therapy: overcoming the obstacles. Viruses. 2010;2:78–106.

2012. von Rikxoort M, Michaelis M, Wolschek M, Muster T, Egorov A, Seipelt J, Doerr HW, Cinatl Jr. Oncolytic effects of a novel influenza A virus expressing interleukin-15 from the NS reading frame. PLoS. 2012;7(5):e36506. doi:10.1371/journal.pone.0036506.

2013a. Fournier P, Bian H, Szeberényi J, Schirrmacher V. Analysis of three properties of Newcastle disease virus for fighting cancer: tumor-selective replication, antitumor cytotoxicity, and immunostimulation. Methods Mol Biol. 2012;797:177-204.

2013b. Fournier P, Arnold A, Wilden H, Schirrmacher V. Newcastle disease virus induces pro-inflammatory conditions and type I interferon for counter-acting Treg activity. Int J Oncol. 2012;40:840-850.

2013c. Mansour M, Palese P, Zamarin D. Oncolytic specificity of Newcastle disease virus is mediated for selectivity for apoptosis-resistant cells. J Virol. 2011;85:6015-6023.

2013d. Sinkovics JG, Howe CD. Superinfection of tumors with viruses. Experientia. 1969;25:733–434.

2014a. Sinkovics JG. Glioblastoma multiforme. In: Sinkovics JG, Horvath JC, authors, editors. Viral therapy of human cancers. New York: Marcel Dekker; 2005. p. 410–451.

2014b. Sinkovics JG, Horvath JC. The molecular biology and immunology of glioblastoma multiforme (GBM) with the presentation of an immunotherapy protocol for a clinical trial. Acta Microbiol Immunol Hung. 2006;53:367–429. Erratum. Due to error, the mislabeled Figure 1 should be invalidated.

2015a. Cerullo V, Koski A, Vähä-Koskela M, Hemminki A. Chapter 8: Oncolytic adenoviruses for cancer immunotherapy: data from mice, hamsters and humans. Adv Cancer Res. 2012;115:265–31.

2015b. Cerullo V, Vähä-Koskela M, Hemminki A. Oncolytic adenoviruses. A potent form of immunovirotherapy. Oncoimmunology. 2012;1(6):979–98.

2016. Ding M, Cao X, Xu HN, Fan JK, Huang HL, Yang DQ, Li YH, Wang J, Li R, Liu XY. Prostate cancer-specific and potent antitumor effect of a DD3-controlled oncolytic virus harboring the PTEN gene. PLoS One. 2012;7(4):e35153. doi:10.1371/journal.pone.0035153.

2017. Somerville RP, Devillier L, Parkhurst MR, Rosenberg SA, Dudley ME. Clinical scale rapid expansion of lymphocytes for adoptive cell transfer therapy in the Wave® bioreactor. J Transl Med. 2012; doi:10.118/1479-5876-10-69.

2018. Kalos M, Levine BL, Porter DL, Katz S, Grupp SA, Bagg A, June CH. T cells with chimeric antigen receptors have potent antitumor effects and can establish memory in patients with advanced leukemia. Sci Transl Med. 2011;3(95):95ra73. doi:10.1126/scitranslmed.3002842.

2019. Porter DL, Levine BL, Kalos M, Bagg A, June CH. Chimeric antigen receptor-modified T cells in chronic lymphocytic leukemia. N Engl J Med. 2011;365:725–33.

2020. Érsek B, Molnár V, Balogh A, Matkó J, Cope AP, Buzás EI, Falus A, Nagy G. CD3ζ-chain expression of human T lymphocytes is regulated by TNF via Src-like adaptor protein-dependent proteasomal degradation. J Immunol. 2012;189:1602–10.

2021. Roszik J, Sebestyén Z, Govers C, Guri Y, Szöor A, Pályi-Krekk Z, Vereb G, Nagy P, Szöllősi J, Debets R. T-cell synapse formation depends on antigen recognition but not CD3 interaction: studies with TCRζ, a candidate transgene forTCR gene therapy. Eur J Immunol. 2011;41:1288–97.

2022. Radvanyi LG, Bernatchez C, Zhang M, Fox PS, Miller P, Chacon J, Wu R, Lizee G, Mahoney S, Alvarado G, Glass M, Johnson VE, McMannis JD, Shpall E, Prieto V, Papadopoulos N, Kim K, Homsi J, Bedikian A, Hwu WJ, Patel S, Ross MI, Lee JE, Gershenwald JE, Lucci A, Royal R, Cormier JN, Davies MA, Mansaray R, Fulbright OJ, Toth C, Ramachandran R, Wardell S, Gonzalez A, Hwu P. Specific lymphocyte subsets predict response to adoptive cell therapy using expanded autologous tumor-infiltrating lymphocytes in metastatic melanoma patients. Clin Cancer Res. 2012;18:6758–70.

2023. Yao S, Zhu Y, Chen L. Advances in targeting cell surface signaling molecules for immune modulation. Nature Rev. 2013;12:130–46.

2024a. Brentjens RJ, Santos E, Nikhamin Y, Yeh R, Matsushita M, La Perle K, Quintás-Cardama A, Larson SM, Sadelain M. Genetically targeted T cells eradicate systemic acute lymphoblastic leukemia xenografts. Clin Cancer Res. 2007;13:5426–5435.

2024b. Brentjens RJ, Curran HJ. Novel cellular therapies for leukemia. CAR-modified T cells targeted to the CD19 antigen. Hematology Am Soc Hematol Educ Program. 2012;2012:143–151. doi:10.1182/asheducatiion-2012.1,143.

2024c. Brentjens RJ, DavilaML, Riviere I, Park J, Wang X, Cowell LG, Bartido S, Stefanski J, Taylor C, Olszewska M, Borquez-Ojeda O, Qu J, Wasielewska T, He O, Bernal Y, Rijo IV, Hedvat C, Kobos R, Curran K, Steinherz P, Jurcic J, Rosenblat T, Maslak P, Frattini M, Sadelain M. CD-19-targeted T cells rapidly induce molecular remissions in adults with chemotherapy-refractory acute lymphoblastic leukemia. Sci Transpl Med. 2013;5(177):177ra38. doi:10.1126/scitranslmed.3005930.

2025. Robbins PF, Morgan RA, Feldman SA, Yang JC, Sherry RM, Dudley ME, Wunderlich JAR, Nahvi AV. HelmanLJ, Mackall CL, Kammula US, Hughes MS, Restifo NP, Raffeld M, Lee CC, Levy CL, Li YF, El-Gamil M, Schwarz SL, Laurencot C, Rosenberg SA. Tumor regression in patients with metastatic synovial cell sarcoma and melanoma using genetically engineered lymphocytes reactive with NY-ESO-1. J Clin Oncol. 2011;29:917–24.

2026. Clancy LE, Blyth E, Simms R, Micklethwaite KP, Kris Ma CK, Burgess J, Antonenas V, Shaw P, Gottlieb DJ. CMV-specific cytotoxic T lymphocytes can be efficiently expanded from G-CSF mobilized haemopoietic progenitor cell products ex vivo and safely transferred to stem cell transplant recipient to facilitate immune reconstruction. Biol Blood Marrow Transpl. 2013; pii:S1083-8791(13)00057–8.

2027. Papadopoulos EB, Ladanyi M, Emanuel D, Mackinnon S, Boulad F, Carabasi MH, Castro-Malaspina H, Childs BH, Gillio AP, Small TN, et al. Infusion of donor leukocytes to treat Epstein-Barr virus-associated lymphoproliferative disorders after allogeneic bone marrow transplantation. N Engl J Med. 1994;330:1185–91.

2028a. Cotton RGH. The road to monoclonal antibodies. In: Gordon S, editor. The legacy of cell fusion. Oxford: Oxford University Press; 1994. p 153-166.

2028b. Sinkovics JG, Groves G, Bertin B, Shullenberger CC. A system of tissue cultures for the study of a mouse leukemia virus. J Infect Dis. 1969;119:19-38.

2029. Bender BC, Schaedeli-Stark F, Koch R, Joshi A, Chu YW, Rugo H, Krop IE, Girish S, Friberg LE, Gupta M. A population pharmacokinetic/pharmacodynamic model of thrombocytopenia characterizing the effect of trastuzumab emtansine (T-DM1) on platelet counts in patients with Her2-positive metastatic breast cancer. Cancer Chemother Pharmacol. 20122;70:591–601.

2030. Junttila TT, Li G, Parsons K, Phillips GL, Slikowksi MX. Trastuzumab –M1 (T-DM1) retains all the mechanisms of action of trastuzumab and efficiently inhibits growth of lapatinib in sensitive breast cancer. Breast Cancer Res Treat. 2011;128:347–56.

2031. Hurvitz SA, Kakkar R. The potential for trastusumab emtansine in human epidermal growth factor receptor 2 positive metastatic breast cancer: latest evidence and ongoing studies. Ther Adv Med Oncol. 2012;4:235–45.

2032. Tarhini AA, Cherian J, Moschos SJ, Tawbi HA, Shuai Y, Gooding WE, Sander C, Kirkwood JM. Safety and efficacy of combination immunotherapy with interferon alfa-2b and tremelimumab in patients with stage IV melanoma. J Clin Oncol. 2012;30:322–8 Erratum. J Clin Oncol 2012;30:3903.

2033. Hersh EM, O'Day SJ, Powderly J, Khan KD, Pavlick AC, Cranmer LD, Samlowski WE, Nichol GM, Yellin MJ, Weber JS. A phase II multicenter study of ipilimumab with or without dacarbazine in chemotherapy-naïve patients with advanced melanoma. Invest New Drugs. 2011;29:489–96.

2034a. Weber J. Ipilimumab: controversies in its development, utility and autoimmune adverse effects. Cancer Immunol Immunother. 2009;58:823–830.

2034b. Weber J. Immune checkpoint proteins: a new therapeutic paradigm for cancer-preclinical background: CTLA-4 and PD0-1 blockade. Semin Oncol. 2010:37:430–439.

2034c. Weber JS, Kähler KC, Hauschild A. Management of immune-related adverse events and kinetics of response with ipilimumab. J Clin Oncol. 2012;30:2691–2697.

2035. Bakacs T, Mehrishi JN, Moss RW. Ipilimumab (Yervoy) and the TGN1412 catastrophe. Immunobiology. 2012;217:583–9.

2036. Elias MH, Baba AA, Husin A, Sulong S, Hassan R, Sim GA, Wahid SFA, Ankathil R. HOXA4 gene promoter hypermethylation as an epigenetic mechanism mediating resistance to imatinib mesylate in chronic myeloid leukemia patients. Biomed Res Int. 2012;2013:129715. doi:10.1155/2013/129715.

2037. Hegedűs C, Özvegy-Laczka C, Apáti AÁ, Magócsi M, Német K, Orfi L, Kéri G, Katona M, Takáts Z, Váradi A, Szakács G, Sarkadi B. Interaction of nilotinib, dasatinib and bosutinib with ABCB1 and ABCG2: implications for altered anti-cancer effects and pharmacological properties. Brit J Pharmacology. 2009;158:1153–64.

2038. Mullard A. FDA drug approvals. Nature Rev. 2012;2913(12):87–91.

2039. Nestal de Moraes G, Souza PS, Costas F, Vasconcelos FC, Reis FR, Maia RC. The interface between BCR-ABL-dependent and -independent resistance signaling pathways in chronic myeloid leukemia. Leuk Res Treatment. 2012;2012:671702 article ID 671702. doi:10.1155/2012/671702.

2040. Mikhael JAR, Schuster SR, Jimenez-Zepeda VH, Bello N, Spong J, Reeder CB, Stewart AK, Bergsagel PL, Fonseca R. Cyclophosphamide-bortezomib-dexamethasone (CtBorD) produces rapid and complete hematological response in patients with AL amyloidosis. Blood. 2012;110:4391–43942034.

2041. Armanous H, Gekiebart P, Anand M, Belch A, Lai R. Constitutive activation of metalloproteinase ADAM19 in mantle cell lymphoma promotes cell growth and activates the TNFα/NFκB pathway. Blood. 2011;117:6237–46.

2042. Obrador-Hevia AQ, Serra-Sitjar M, Rodriguez J, Villalonga P. Fernández de Matos S. The tumour suppressor FOXO3 is a key regulator of mantle cell lymphoma proliferation and survival. Br J Haematol. 2012;156:334–45.

2043. Hill BT, Rybicki L, Smith S, Dean R, Kalaycio M, Pohlman B, Sweetenham J. Tench, Sobecks R, Andresen S, Copelan E, Bolwell BJ. Treatment with hyperfractionated cyclophosphamide, vincristine, doxorubicin, and dexamethasone combined with cytarabine and methotrexate results in poor mobilization of peripheral blood stem cells in patients with mantle cell lymphoma. Leuk Lymphoma. 2011;52:986–93.

2044a. Romaguera JE, Fayad LE, F eng L, Hartig K, Weaver P, Rodriguez MA, Hagemeister FB, Pro B, McLaughlin P, Younes A, Samaniego F, Goy A, Cabanillas F, Kantarjian H, Kwak L, Wang M. Ten year follow up after intense chemoimmunotherapy with rituximab, hyperCVAD alternating with rituximab-high dose methotrexate /cytarabine (R-MA) and without stem cell transplantation in patients with untreated aggressive mantle cell lymphoma. Br J Haematol. 2010;150:200-208. Erratum. Br J Haematol. 2010;151;111.

2044b. Romaguera JE, Fayad LE, McLaughlin P, Pro B, Rodriguez A, Wang M, Weaver P, Hartig K, Kwak L-W, Feldman T, Smith J, Ford P, Goldberg S, Pecora A, Goy A. Phase I trial of bortezomib in combination with rituximab-hyperCVAD alternating with rituximab, methotrexate and cytarabine for untreated aggressive mantle cell lymphoma. Br J Haematol. 2010;151:47–53.

2045. Brett LK, Williams ME. Current and emerging therapies in mantle cell lymphoma. Curr Treat Options Oncol. 2013;14:198–211.

2046. Kim A, Park S, Lee JE, Jang WS, Lee WS, Kang HJ, Lee SS. The dual PI3K and mTOR inhibitor NVP-BEZ235 exhibits anti-proliferative activity and overcomes bortezomib resistance in mantle cell lymphoma cells. Leuk Res. 2012;36:912–20.

2047. Civallero M, Cosenza M, Marcheselli L, Pozzi S, Sacchi S. NVP-BEZ235 alone and in combination in mantle cell lymphoma: an effective therapeutic strategy. Expert Opin Investig Drugs. 2012;21:1597–606.

2048. Saveur-Michel M, Pecchin S, Huang A, Burger M, Knapp M, Sterker D, Schnell C, Guthy D, Nagel T, Wiesman M, Brachman S, Fritsch C. Dorsch, Chéne P, Shoemaker K, De Over A, Menezes D, Martiny-Baronm G, FabbroD, Wilson CJ, Schlegel R, Hofmann F, Garcia-Echevarria C, Sellers WR, Voliva CF. Identification and characterization of NVP-BKM120, an orally available pan-class I PI3-kinase inhibitor. Mol Cancer Ther. 2012;11:317. doi:10.1158/1535-7163.MCT-11-0474.

2049a. Somaiah N, von Mehren M. New drugs and combinations for the treatment of soft-tissue sarcoma: a review. Cancer Manag Res. 2012;4:397–411.

2049b. Demetri GD, Chawla SP, Ray-Coquard I, Le Cesne A, Staddon AP, Milhem MM, Penel N, Riedel RF, Bui-Nguyen B, Cranmer LD, Reichardt P, Bompas E, Alcindor T, Rushing D, Song Y, Lee R-m, Ebbinghause S, Eid JE, Loewy JW, Haluska FG, Dodion PF, Bay J-Y. Results of an international randomized phase III trial of the mammalian target of rapamycin inhibitor ridaforolimus versus placebo to control metastatic sarcomas in patients after benefit from prior chemotherapy. J Clin Oncol. 2013;31:2485–2492.

2050a. Germano G, Frapolli R, Frapolli R. Antitumor and anti-inflammatory effects of trabectedin on human myxoid liposarcoma cells. Cancer Res. 2010;70:2235–2244.

2050b. Grohar PJ, Griffin LB, Yeung C, Chen QR, Pommier Y, Khanna C, Khan J, Helman LJ. Ecteinascidin 743 interferes with the activity of EWS-FLI1 in Ewing sarcoma cells. Neoplasia. 2011;13:145–153.

2050c. Schuler MK, Rickter S, Platzek I, Beuthien-Baumann B, Wieczorek K, Hamann C, Mohm J, Ehninger G. Trabectidin in the neoadjuvant treatment of high-grade pleomorphic sarcoma: report of a rare case and literature review. Case Reports Oncol Med. 2013;ID320797.

2051. Pal SK, Stain CA, Sartor O. Enzalutamide for the treatment of prostate cancer. Expert Opin Pharmacother. 2013;14:679–85.

2052. Venkannagari S, Fiskus W, Peth K, Atadia P, Hidalgo M, Maitra A, Bhalla KN. Superior efficacy of co-treatment with dual PI3K/mTOR inhibitor NVP-BEZ235 and pan-histone deacetylase inhibitor against human pancreatic cancer. Oncotarget. 2012;3:1416–27.

2053. Dabrovska A, Kim S, Salamone RJ, Walker JAR, Maira SM, García-Echevarria C, Schultz PG, Reddy VA. The role of PTEN/AKT/PI3K signaling in the maintenance and viability of prostate cancer stem-like cell population. Proc Natl Acad Sci USA. 2009;106:268–73.

2054. Aguda BD, Kim Y, Piper-Hunter C, Friedman A, Marsh CB. MicroRNA regulation of cancer network: consequences of the feedback loops involving miR-17-92, E2F, and Myc. Proc Natl Acad Sci USA. 2008;105:19678–83.

2055a. Huang YW, Liu JC, Deatherage DE, Luo J, Mutch DG, Goodfellow PJ, Miller DS, Huang TH. Epigenetic repression of microRNA-129-2 leads to overexpression of SOX4 oncogene in endometrial cancer. Cancer Res. 2009;69:9038-9046.

2055b. Huang YW, Ruiz CR, Eyler EC, Lin K, Meffert MK. Dual regulation of mRNA biogenesis generates target specificity in neurotrophin-induced protein synthesis. Cell. 2012;148:933-946.

2056. Heneghan HM, Miller N, Kelly R, Newell J, Kerin MJ. Systemic miRNA-195 differentiates breast cancer from other malignancies and is a potential biomarker for detecting noninvasive and early stage disease. Oncologist. 2010;15:673–82.

2057. He M, Liu Y, Deng X, Qi S, Sun X, Liu G, Liu Y, Liu Y, Zhao M. Down-regulation of miR-200b-3p by low p73 contributes to the androgen-independence of prostate cancer cells. Prostate. 2913; doi:10.1002/pros.22652.

2058. Ishteiwy RA, Ward TM, Dykxhoom DM, Burnstein KL. The microRNA-23b/-27b cluster suppresses the metastatic phenotype of castration-resistant prostate cancer cell. PLoS One. 2012;7(12)e52106. doi:10.1371/journal.pone.0052106.

2059. Qu F, Cui X, Hong Y, Wang J, Li Y, Chen L, Liu Y, Gao Y, Xu D, Wang Q. MicroRNA-185 suppresses proliferation, invasion, migration, and tumorigenicity of human prostate cancer cells through targeting androgen receptor. Mol Cell Biochem. 2013;377:121–30.

2060. Fan X, Chen X, Deng W, Zhong G, Cai Q, Lin T. Up-regulated microRNA-143 in cancer stem cells differentiation promotes prostatic cancer cells metastasis by modulating FNDC3B expression. BMC Cancer. 2013;13(1):61.

2061. Hirata H, Ueno K, Shahryan V, Deng G, Tanaka Y, Tabatacai ZL, Hinoda Y, Dahiya R. MicroRNA-182-5p promotes cell invasion and proliferation by down-regulating FOXF, RECK and MTSS1 genes in human prostate cancer. PLoS One. 2013;8(1)e455592. doi:10.1371/journal.pone.0055502.

2062. Qu Y, Li WC, Hellem MR, Rostad K, Popa M, McCormack E, Oyan AM, Kalland KH, Ke XS. MiR-182 and miR-203 induce mesenchymal to epithelial transition and self sufficiency of growth signals via repressing SNAI2 in prostate cells. Int J Cancer. 2012; doi:10.1002/IJC.28056/.

2063. Han HB, Gu J, Zuo HJ, Chen ZG, Zhao W, Li M, Ji DB, Lu YY, Zhang ZQ. Let-7c functions as a metastasis suppressor by targeting MMP11 and PBX3 in colorectal cancer. J Pathol. 2012;226:544–55.

2064. Liang L, Wong CM, Ying Q, Fan DN, Huang S, Ding J, Yao J, Yan M, Li J, Yao M, Ng IO, He X. MicroRNA-125b suppressed human liver cancer cell proliferation and metastasis by directly targeting oncogene LIN28B2. Hepatology. 2010;52:1731–40.

2065. Xiong S, Zheng Y, Jiang P, Liu R, Liub X, Chu Y. MicroRNA-7 inhibits the growth of human non-small cell lung cancer A549 cells through targeting BCL-2. Int J Biol Sci. 2011;7:805–8142021.

2066. Di Martino MT, Gullá A, Cantafio ME, Lionetti M, Leone E, Amodio N, Guzzi PH, Foresta U, Conforti F, Cannataro M, Neri A, Giordano A, Tagliaferro P, Tassone P. In vitro and in vivo anti-tumor activity of miR-221-222 inhibitors in multiple myeloma. Oncotarget. 2013;4:242–55.

2067. Fortson WS, Kayarthodi S, Fujimura Y, Xu H, Matthews R, Grizzle WE, Rao VN, Bhat GK, Reddy ES. Histone deacetylase inhibitors, valproic acid and trichostatin-A induce apoptosis and affect acetylation status of p53 in ERG-positive prostate cancer cells. Int J Oncol. 2011;39:111–9.

2068. Zhou X, Yang XY, Popescu NC. Synergistic antineoplastic effect of DLC1 tumor suppressor protein and histone deacetylase inhibitor, suberoylanilide hydroxamic acid (SAHA), on prostate and liver cancer cells: perspectives for therapeutics. Int J Oncol. 2019;36:999–1005.

2069. Rosenbaum E, Hoque MO, Cohen Y, Zahurak M, Eisenberger MA, Epstein JI, Partin AW, Sidransky D. Promoter hypermethylation as an independent prognostic factor for relapse in patients with prostate cancer following radical prostatectomy. Clin Cancer Res. 2005;11:8321. doi:10.1158/1078-0432.CCR-05-1183.

2070. Sinkovics JG. Oncogenes and growth factors. CRC Crit Rev Immunol. 1988;8:217–98.

2071. Liu YH, Xue SP. Potency of induced differentiation in vitro of a human promyelocytic leukemia cell mutant (HL-60-AR) with characteristics of resistance to 8-AG and deficiency in HGPRT. Zhongguo Yi Xue Ke Xue Yuan Xue Bao. 1986;8:461–5.

2072. Wu X, Wang X, Qien X, Liu H, Ying J, Yang Z, Yaom H. Four years' experience with the treatment of all-trans retinoic acid in acute promyelocytic leukemia. Am J Hematol. 1993;43:183–9.

2073. de Melo JV, de Melo V, Abad MH. Clinical and radiological response with FOLFOXIRI and bevacizumab as third-line therapy after FOLFOXx6 and FOLFIRI failure. Anticancer Drugs. 2011;S2:S19-20.

2074. Fornaro L, Vasile E, Masi G, Loupakis F, Baldi GG, Allegrini G, Salvatore L, Cupini S, Cortesi E, Tuzi A, Granetto C, Brunetti IM, Ricci S, Falcone A. Outcome of second-line treatment after first-line chemotherapy with the GONO FOLFOXIRI regimen. Clin Colorectal Cancer. 2012;11:71–6.
2075. Stein A, Clockzin G, Wienke A, Arnold D, Edelmann T, Hildbrandt B, Hollerbach S, Illerhaus G, Königsrainer A, Richter M, Schlitt HJ, Schmoll HJ. Treatment with bevacizumab and FOLFOXIRI in patients with advanced colorectal cancer: presentation of two novel trials (CHARTA and PERIMAX) and review of the literature. BMC Cancer. 2012;12:356. doi:10.1186/1471-2407-12-356.
2076. Domingo-Espin J, Vazquez E, Ganz J, Conchillo O, García-Fruitós E, Cedano J, Unzueta U, Petegnief V, Gonzalez-Montalbán N, Planas AM, Daura X, Peluffo H, Ferrer-Miralles N, Villaverde A. Nanoparticle architecture of protein-based artificial viruses is supported by protein-DNA interactions. Nanomedicine. 2011;6:1047–61.
2077. Llorens C, Munoz-Pomer A, Bernad L, Botella H, Moya A. Network dynamics of eukaryotic LTR retroelements beyond phylogenetic trees. Biol Direct. 2009;4:41. doi:10.1186/17645-6150-4-41.
2078. Pritham EJ, Feschotte C, Wessler SR. Unexpected diversity and differential success of GNA transposons in four species of Entamoeba protozoans. Mol Biol Evol. 2005;22:1751–63.
2079. Duncan L, Bouckaert K, Yeh F, Kirk DL. Kangaroo, a mobile element from Volvox carteri, is a member of a newly recognized third class of retrotransposons. Genes. 2002;162:1617–30.
2080. Esnault C, Casella JF, Heidmann T. A Tetrahymena thermophila ribozyme-based indicator gene to detect transposition of marked retroelements in mammalian cells. Nucleic Acids Res. 2002;30(11):e49.
2081. Zerbato M, Holic N, Moniot-Frin S, Ingrao D, Galy A, Perea J. The brown alga PI.LSU/2 group II intron-encoded protein has functional reverse transcriptase and maturase activities. PLoS One. 2013;8(3):e58263.
2082. Affolter M, Rindisbacher L, Braun R. The tubulin gene cluster of Trypanosoma brucei starts with an intact beta-gene and ends with a truncated beta-gene interrupted by a retrotransposons-like sequence. Gene. 1989;80:177–83.
2083. Laha T, Kewgrai N, Loukas A, Brindley P.J. Characterization of SR3 reveals abundance of non-LTR retrotransposons of the RTE clade in genome of the human blood fluke, Schistosoma mansoni. BMC Genomics. 2005;g:154. doi:10.1186/1471-2164-6-154.
2084. Cai P, Piao X, Hou N, Liu S, Wang H, Chen Q. Identification and characterization of argonaute protein, AGO2 and its associated small RNAs in Schistosoma japonicum. PLoS Negl Trop Dis. 2012;6(7):e1745. doi:10.1371/journal.pntd.0001745.
2085. Jacinto DS. Muniz Hdos S, Venancio TM, Wilson RA, Verjovski-Almeida S, Demarco R. Curupira-1 and curupira-2, two novel mutator-like DNA transposons from the genomes of human parasites Schistosoma mansoni and Schistosoma japonicum. Parasitology. 2011;138:1124–33.
2086. Ferguson AA, Jiang N. Mutator-like elements with multiple long terminal inverted repeats in plants. Comp Fund Genomics. 2012;2912:695827. doi:10.1155/2012/695827.
2087. Jiang N, Ferguson AA, Slotkin RK, Lisch D. Pack-Mutator-like transposable elements (Pack-MULES) induce directional modification of genes. Proc Natl Acad Sci USA. 2011;108:1537–42.
2088. Moradi M, Babin V, Roland C, Sagui C. Reaction path ensemble of the B-Z-DNA transition: a comprehensive atomistic study. Nucleic Acids Res. 2013;41:33–43.
2089. Lopes FR, Silva JC, Benchimol M, Costa GG, Preira GA, Carareto M. The protist Trichomonas vaginalis harbors multiple lineages of transcriptionally active Mutator-like elements. BMC Genomics. 2009;10:330. doi:10.1156/1471-2164-10-330.

2090. Carr M, Nelson M, Leadbeater BS, Baldauf SL. Three families of LTR retrotransposons are present in the genome of the choanoflagellate Monosiga brevicollis. Protist. 2008;159:579–90.
2091. Albalat R, Permanyer J, Cañestro C, Martinez-Mir A, González-Angulo O, González-Duarte R. The first non-LTR retrotransposons characterized in the cephalochordate amphioxus, BfCR1, shows similarities to CR1-like elements. Cell Mol Life Sci. 2003;60:803–9.
2092. Ramulu HG, Raoult D, Pontarotti P. The rhizome of life: what about metazoan? Front Cell Infect Microbiol. 2012;2.50. doi:10.3389/fcimb.2012.00050.
2093a. Fox EJ, Loeb LA. Lethal mutagenesis: targeting the mutator phenotype in cancer. Semin Cancer Biol. 2010;20:353–359.
2093b. Prindle MJ, Fox EJ, Loeb LA. The mutator phenotype in cancer: molecular mechanisms and targeting strategies. Curr Drug Targets. 2010;11:1296–1303.
2094. Swart EC, Bracht JAR, Magrini V, Minx P, Chen X, Zhou Y, Khurana JS, Goldman AD, Nowacki M, Schotanus K, Jung S, Fulton RS, Ly A, McGrath S, Haub K, Wiggins JL, Storton D, Matese JC, Parsons L, Chang WJ, Bowen MS, Stover NA, Jones TA, Eddy SR, Herrick GA, Doak TG, Wilson RK. Mardis Er, Landweber LF. The Oxytricha trifallax macronuclear genome: a complex eukaryotic genome with 16,000 tiny chromosomes. PLoS Biol. 2013;11(1):e1001473. doi:10.1371/journal.pbio.1001473.
2095. Horvath MP. Evolution of telomere-binding proteins. In: Nosek J, Tomáska L, editors. Origin and evolution of telomeres. Austin, TX: Landes Bioscience; 2008. p. 83–99.
2096. Fang W, Wang X, Bracht JAR, Nowacki M, Landweber LF. Piwi-interacting RNAs protect DNA against loss during Oxytricha genome rearrangement. Cell. 2012;151:1243–55.
2097. Cartinhour SW, Herrick GA. Three different macronuclear DNAs in Oxytricha fallax share a common sequence block. Mol Cell Biol. 1984;4:931–8.
2098. Jang E, Chung DC. Hereditary colon cancer: Lynch syndrome. Gut Liver. 2010;4:151–69.
2099. Nicolaev SI, Sotiriou SK, Pateras IS, Santoni F, Sougioultzis S, Edgren H, Almusa H, Robyr D, Guipponi M, Saarela J, Gorgoulis VG, Antonarakis SE, Halazonetis TD. A single-nucleotide substitution mutator phenotype revealed by exome sequences of human colon adenomas. Cancer Res. 2012;72:6279–89.
2100. Tili E, Michaile JJ, Wernicke D, Alde H, Costinean S, Volinia S, Croce CM. Mutator activity induced by microRNA-155 (miR-155) links inflammation and cancer. Proc Natl Acad Sci USA. 2011;108:4908–13.
2101. Yang D, Hurley LH. Structure of the biologically relevant bG-quadruplex in the c-Myc promoter. Nucleosides Nucleotides Nucleic Acids. 2006;25:951–68.
2102. van der Kuyl AC, Dekker JT, Goudsmit J. Discovery of a new endogenous type C retrovirus (FcEV) in cats. Evidence for RD-114 being an $FcEV^{gagpol}$/baboon endogenous virus $BaEV^{Env}$ recombinant. J Virol. 1999;73:7994–8002.
2103. Søe K, Andersen TL, Hobolt-Pedersen AS, Bierregaard B, Larsson LI, Delaissé JM. Involvement of human endogenous retroviral syncytin-1 in human osteoclast fusion. Bone. 2011;48:837–46.
2104. Antony JM, Deslauriers AM, Bhat RK, Ellestad KK, Power C. Human endogenous retroviruses and multiple sclerosis: innocent bystanders or disease determinants? Biochim Biophys Acta. 2011;1812:162–76.
2105a. Pogo BG, Holland JF, Levine OH. Human mammary tumor virus in inflammatory breast cancer. Cancer. 2010;116S:295–297.
2105b. Pogo BG, Melana SM, Moran F, Holland JF. Presence of MMTV-like env gene sequences in human breast cancer. Breast Cancer Res Treat. 2011;125:295–297.
2106. Nelson PN, Hooley P, Ejtehadi HD, Rylance P, Warren P, Martin J, Murray PG. Molecular Immunology Research Group. Human endogenous retroviruses: transposable elements with potential? Clin Exp Immunol. 2004;138:1–9.

2107. Dabrowska A, Kim N, Aldovini A. Tat-induced FOXO3a is a key mediator of apoptosis in HIV-1-infected human CD4+ lymphocytes. J Immunol. 2008;181:8460–77.

2108. Eizert H, Bander P, Bagossi P, Sperka T, Miklóssy G, Boross P, Eber IT, Tözsér J. Amino acid preference of retroviral proteases for amino-terminal positions in a type 1 cleavage site. J Virol. 2998;82:10111–10117.

2109a. Huliák I, Sike A, Zencir S, Boros IM. The objectivity of reporters: interference between physically unlinked promoters affects reporter gene expression in transient transfection experiments. DNA Cell Biol. 2012;31:1580–1584.

2109b. Imaizumi Y, Sugita S, Yamamoto K, Imanishi D, Kohno T, Tomonaga M, Matsuyama T. Human T cell leukemia virus type-I Tax activates human macrophage inflammatory protein-3a/CCL20 gene transcription via the NF-κB pathway. Int Immunol. 2002;14:147–155.

2110. Kim Y, Hollenbaugh JA, Kim D-H, Kim B. Novel PI3K/Akt inhibitors screened by the cytoprotective function of human immunodeficiency virus type 1 Tat. PLoS One. 2011;6 (7):e21781. doi:10.1371/journal.pone.0021781.

2111a. Nagy K, Barabás É, Várkonyi V, Horváth A. Determination of HIV-1 subtypes in Hungary by synthetic peptides representing the V3 loop of env. Pathol Oncol Res. 1996;2:268–271.

2111b. Nagy K, Young M, Baboonian C, Merson J, Whittle P, Oroszlán S. Antiviral activity of human immunodeficiency virus type 1 protease inhibitors in a single cycle of infection: evidence for a role in the early phase. J Virol. 1994;68:757–765.

2112a. Perl A, Banki K. Genetic and metabolic control of the mitochondrial transmembrane potential and reactive oxygen intermediate production in HIV disease. Antioxid Redox Signal. 2000;2:551–573.

2112b. Perl A. Role of endogenous retroviruses in autoimmune diseases. Rheum Dis Clin North Am. 2003;29:123–143.

2112c. Perl A. Fernandez D, Telarico T, Phillips PE. Endogenous retroviral pathogenesis in lupus. Curr Opin Rheumatol. 2010;22:483–492.

2112d. Caza TN, Fernandez DR, Talaber G, Oaks Z, Haas M, Madaio MP, Lai Z-w, Miklossy G, Singh RR, Chudakow DM, Malorni W, Middleton F, Banki K, Perl A. HRES-1/Rab4-mediated depletion of Drp1 impairs mitochondrial homeostasis and represents a target for treatment in SLE. Ann Rheum Dis. 2014;73:1888–97.

2113. Arjan-Odedra S, Swanson CM, Sherer NM, Wolinsky SM, Malim MH. Endogenous MOV19 inhibits the retrotransposition of endogenous retroelements but not the replication of exogenous retroviruses. Retrovirology. 2012;9:53 doi:10.1186/1742-4690-9-53.

2114a. Kiss-László Z, Henry Y, Bachellerie J-P, Caizergues-Ferrer M, Kiss T. Site-specific ribose methylation of ribosomal RNA: a novel function for small nucleolar RNAs. Cell. 1996;85:1077-1088.

2114b. Kiss-László Z, Henry Y, Kiss T. Sequence and structural elements of methylation guide snoRNAs essential for site-specific ribose methylation of pre-rRNA. EMBO J. 1998;17:797–807.

2114c. Jády B, Ketele A, Kiss T. Human intron-encoded AluRNAs are processed and packaged into Wdr9-associated nucleoplasmic box H/ACA RNPs. Genes Dev. 2012;26:1897–1910.

2115. Teng YN, Chuang PJ, Liu YW. Nuclear factor-κB (NF-κB) regulates the expression of human testis-enriched leucine-rich repeats and WD repeat domain containing 1 (LRWD1) gene. Int J Med Sci. 2012;14:625–39.

2116. Polson A, Durrett E, Reisman D. A bidirectional promoter reporter vector for the analysis of the p53/WDR79 dual regulatory element. Plasmid. 2011;66:169–79.

2117. Garcia-Closas M, Kristensen V, Langerød A, Qi Y, Yeager M, Burgett L, Welch R, Lissowska J, Peplonska B, Brinton L, Gerhard DS, Gram IT, Perou CM. Børrsen -Dale AL, Chanock S. Common genetic variation in TP53 and its flanking genes, WDR79 and ATP1B2, and susceptibility to breast cancer. Int J Cancer. 2007;121:2532–8.

2118. Wang Q, Ma C, Kemmner W. Wdr66 is a novel marker for risk stratification and involved in epithelial-mesenchymal transition of esophageal squamous cell carcinoma. BMC Cancer. 2013;13:137. doi:10.1186/1471-2407-13-137.

2119. Ang Y-S, Tsai S-Y, Lee D-F, Monk J. SuJ, Ratnakumar K, Ding J, Ge Y, Darr H, Chang B, Wang J, Rendl M, Bernstain E, Schaniel C, Lemischka IR. Wdr5 mediates self-renewal and reprogramming via the embryonic stem cell core transcriptional network. Cell. 2011;145:183–97.

2120. Côté-Martin A, Moody C, Fradet-Turcotte A, D'Abramo CM, Lerhoux M, Joubert S, Poirier GC, Coulombe B, Laimins LA, Archambault J. Human papillomavirus E1 helicase interacts with the WD repeat protein p80 to promote maintenance of the viral genome in keratinocytes. J Virol. 2008;82:1271–83.

2121. DeMartini JC, Bishop JV, Allen TE, Jassim FA, Sharp JM, de las Heras M, Voelker DR, Carlson JO. Jaagsiekte sheep retrovirus proviral clone $JSRV_{JS7}$, derived from the JS7 lung tumor cell line induces ovine pulmonary carcinoma and is integrated into the surfactant protein gene. J Virol. 2001;75:4239–4246.

2122. Hofacre A, Fan H. Jaagsiekte sheep retrovirus biology and oncogenesis. Viruses. 2010;2:2618–48.

2123. Wootton SK, Metzger MJ, Hudkins KL, Alpers CS, York D, DeMartini JC, Miller AD. Lung cancer induced in mice by the envelope protein of jaagsiekte sheep retrovirus (JSRV) closely resembles lung cancer in sheep infected with JSRV. Retrovirology. 2006;3:94. doi:10.1186/1742-4690-3-94.

2124. Hanger JJ, Bromham LD, McKee JJ, O'Brien TM, Robinson WF. The nucleotide sequence of koala (Phascolarctos cinereus) retrovirus: a novel type C endogenous virus related to gibbon ape leukemia virus. J Virol. 2000;74:4264–72.

2125a. Tarlinton R, Meers J, Young P. Biology and evolution of the endogenous koala retrovirus. Cell Mol Life Sci. 2008;65:3413–3421.

2125b. Tarlinton R, Meers J, Hanger J, Young P. Real time reverse transcriptase PCR for endogenous koala retrovirus reveals an association between plasma viral load and neoplastic disease in koalas. J Gen Virol. 2005;86:783–787.

2126. Boensch C, Huang SS, Connolly DFT, Huang JS. Cell surface retention sequence binding protein-1 interacts with the v-sis gene product protein-derived growth factor beta-type receptor in simian sarcoma virus –transformed cells. J Biol Chem. 1999;274:10582–9.

2127. Pierce GF, Shawver LK, Milner PG, Yeh HJ, Thomason A, Deuel TF. Identification and purification of PDGF/sis-like proteins from nuclei of simian sarcoma virus-transformed fibroblasts. Oncogene Res. 1988;2:235–44.

2128. Subramanian RP, Wildschutte JH, Russo C, Coffin JM. Identification, characterization, and comparative genomic distribution of the HERV-K (HML-2) group of human endogenous retroviruses. Retrovirology. 2011;8:90. doi:10.1186/1742-4690-8-90.

2129a. Wainwright M. The Sinkovics hybridoma. The discovery of the first "natural hybridoma". Prospect Biol Med. 1992;35:372–379.

2129b. Wainwright M, Lederberg J. History of microbiology. In: Encycopedia of Microbiology, Vol 2. London New York: Academic Press; 1992. p. 419–437.

2130a. Margolis LB, Glushakova S, Baibakov B, Zimmerberg J. Syncytium formation in cultured human lymphoid tissue; fusion of implanted HIV glycoprotein 120/41-expresssing cells with CD4+ cells. AIDS Res Hum. Retroviruses. 1995;11:697–704.

2130b. Glushakova S, Baibakov B, Zimmerberg J, Margolis LB. Experimental HIV infection of human lymphoid tissue: correlation of CD4+ T cell depletion and virus syncytium-inducing/non-syncytium-inducing phenotype in histocultures inoculated with laboratory strains and patient isolates of HIV type 1. AIDS Res Human Retroviruses. 1997;13:461–471.

2131. Lazova R, Chakraborty A, Pawelek JM. Leukocyte-cancer cell fusion initiator of the Warburg effect in malignancy. In: Dittmar T, Zänker KS, editors. Cell fusion in health and disease. II. Cell fusion in disease. Dordrecht Heidelberg London New York: Springer Verlag; 2011. p. 151–172.

2132. Lu X, Kang Y. Cell fusion hypothesis of the cancer stem cell. In: Dittmar T, Zänker KS, editors. Cell fusion in health and disease. II. Cell fusion in disease. Dodrecht Heidelberg London New York: Springer Verlag; 2011. p. 129–140.

2133. Nagler C, Zänker KS, Dittmar T. Cell fusion. Drug resistance and recurrent cancer. In: Dittmar T, Zänker KS, editors. Cell fusion in health and disease. II. Cell fusion in disease. Dordrecht Heidelberg London New York: Springer Verlag; 2011. p. 173–182.

2134. Silvestris F, Ciavarella S, Strippoli S, Dammaco F. Cell fusion and hyperactive osteoclastogenesis in multiple myelomas. In: Dittmar T, Zänker KS, editors. Cell fusion in health and disease. II. Cell fusion in disease. Dordrecht Heidelberg London New York: Springer Verlag; 2011. p. 113–128.

2135a. Sinkovics JG, Szakacs J,E, Györkey F. Thirty-year search for human retroviruses. American College of Physicians Florida Scientific Meeting. Abstracts Sept 19-20 1992.

2135b. Sinkovics JG, Györkey F, Gyorkey P. Immunovirology of Reed-Sternberg cells. Eighth Internationational Congress Immunology. Budapest, Hungary, Aug 23–28, 1992 Abstracts 1992; p 416 #66-19.

2135c. Györkey F, Gyorkey P, Sinkovics JG. Immunovirology of Kaposi sarcoma cells. Eighth International Congress Immunology, Budapest, Hungary Aug 23-28 1992 Abstracts 1992; p 415 #66-7.

2135d. Gyorkey F†, Gyorkey P, Sinkovics JG, Szakacs J, Gergely S. Thirty-year search for human retroviruses. In Memoriam Ferenc Gyorkey MD. First International Conference on Human Retroviruses, Washington DC, Dec 12–16 1993 Abstracts p. 95 #233.

2135e. Sinkovics JG, Szakacs JE, Gergely AS, Gyorkey P, †Gyorkey F. Unidentified human retroviruses. Presentation dedicated to the Memory of Professor Ferenc Györkey. Infectious Disease Society of America Annual Meeting. Orlando, FL 1994. Book of Abstracts.

2135f. Sinkovics JG. Contradictory concepts in the etiology and regression of Kaposi's sarcoma. The Ferenc Györkey Memorial Lecture. Pathol Oncol Res. 1996;2:249–267.

2136. Jha AR, Nixon DF, Rosenberg MG, Martin JN, Deeks SG, Hudson RR, Garrison KE, Pillai SK. Human endogenous retrovirus K106 (HERV-K-106) was infectious after the emergence of anatomically modern humans. PLoS One. 2011;6(5):e20234. doi:10.1371/journal.pone.0020234.

2137. Mullins CS, Linnebacher M. Human endogenous retroviruses and cancer: causality and therapeutic possibilities. World J Gastroenterol. 2012;18:6027–35.

2138. Drewinko B, Romsdahl MM, Yang LY, Ahearn MJ, Trujillo JM. Establishment of a human carcinoembryonic antigen-producing colon adenocarcinoma cell line. Cancer Res. 1976;36:467–75.

2139. Dunn CA, Medstrand P, Mager DL. An endogenous retroviral long terminal repeat is the dominancy promoter for human beta1,3-galactosyltransferase 5 in the colon. Proc Natl Acad Sci USA. 2003;100:12841–6.

2140a. Lamprecht B, Bonifer C, Mathas S. Repeat-element driven activation of proto-oncogenes in human malignancies. Cell Cycle. 2010;9:4276–4281.

2140b. Lamprecht B, Walter K, Kreher S, Kumar R, Hummel M, Lenze D, Köchert K, Bouhlel K, Richter J, Soler E, Stadhouders R, Jöhrens K, Wurster KD, Callen DF, Harte MF, Giefing M, Barlow R, Stein H, Anagnostopoulos I, Janz M, Cockerill PN, Siebert R, Dörken B, Bonifer C, Mathas S. Derepression of an endogenous long terminal repeat activates CSF1R proto-oncogene in human lymphoma. Nat Med. 2010;16:571–579.

2141. Bonnaud B, Beliaeff J, Bouton O, Oriol G, Duret L, Mallet F. Natural history of the ERVWE1 endogenous retroviral locus. Retrovirology. 2005;22(2):57. doi:10.1186/1742-4690-2-57.

2142. Ajore R, Kumar P. Singh Dhanda R, Gullberg U, Olsson I. The leukemia associated nuclear corepressor ETO homologue genes MTG16 and MYGR1 are regulated differently in hematopoietic cells. BMC Mol Biol. 2012;13:11. doi:10.1186/1471-2199-13-11.

2143. Engel ME, Hiebert SW. The enemy within: dominant retroviruses awaken. Nat Med. 2010;16:517–8.

2144a. Contreras-Galindo R, Kaplan MH, Leissner P, Verjat T, Ferlenghi I, Bagnoli F, Giusti F, Dosik MH, Hayes DF, Gitlin SD, Markovitz DM. Human endogenous retrovirus K (HML2) elements in the plasma of people with lymphoma and breast cancer. J Virol. 2008;82:9329–9336.

2144b. Contreras-Galindo R, Kaplan MH, Contreras-Galindo AC, Gonzalez-Hernandez MJ, Ferlenghi I, Giusti F, Lorenzo E, Gitlin SD, Dosik MH, Yamamura Y, Markovitz DM. Characterization of human endogenous retroviral elements in the blood of HIV-1-infected individuals. J Virol. 2112;86:262–276.

2145. Januszkiewitz-Lewandowska D, Nowicka K, Rembowska J, Fichna M. Zurawek, Derwich K, Nowak J. Env gene expression of human endogenous retrovirus-K and human endogenous retrovirus-W in childhood acute leukemia cells. Acta Haematol. 2013;129:232–7.

2146a. Wang-Johanning F, Liu J, Rycay K, Huang M, Tsai K, Rosen DG, Chen DT, Lu DW, Barnhart KF, Johanning GL. Expression of multiple human endogenous retrovirus surface envelope proteins in ovarian cancer. Int J Cancer. 2007;120:81–90.

2146b. Wang-Johanning F, Rycaj K, Plummer JB, Li M, Yin B, Frerich K, Garza JG, Shen J, Lin K, Yan P, Glynn SA, Dorsey TH, Hunt KK, Ambs S, Johanning GL. Immunotherapeutic potential of anti-human endogenous retrovirus-K envelope protein antibodies in targeting breast tumors. J Natl Cancer Inst. 2012;104:189–210.

2146c. Zhao J, Rycaj k, Geng S,, Li G, Plummer JB, Yin B, Liu H, Xu X, Zhang Y, Yan Y, Glynn SA, Dorsey TH, Ambs S, Johanning GL, Lin G, Wang-Johanning F. Expression of human endogenous retrovirus type K . Genes Cancer. 2011;2:914–922.

2147a. Yaniv A, Gotlieb-Stematsky T, Vonsover A, Perk K. Evidence for type-C retrovirus production by Burkitt's lymphoma-derived cell line. Int J Cancer. 1980;25:205–211.

2147b. Dreyfus DH. Autoimmune disease: A role for new anti-viral therapies? Autoimmun Rev. 2011;11:88–97.

2147c. Crane GM, Ambinder RF, Shirley CM, Fishman EK, Kasamon YL, Taube JM, Borowitz MJ, Duffield AS. HHV-8-positive and EBV-positive intravascular lymphoma: an unusual presentation of extracavitary primary effusion lymphoma. Am J Surg Pathol. 2014;38:426–432.

2147d. Bentz GL, Bheda-Malge A, Wang L, Shackelford J, Damania B, Pagano JS. KSNV LANA and EBV LMP1 induce the expression of UCH-L1 following viral transformation. Virology. 2014;448:293–302.

2148. Ishida T, Obata Y, Ohara N, Matsushita H, Sato S, Uenaka A, Saika T, Miyamura T, Chayama K, Nakamura Y, Wada H, Yamashita T, Morishi A t, Old LJ, Nakayama E. Identification of the HERV-R gag antigen in prostate cancer by SEREX using autologous patient serum and its immunogenicity. Cancer Immun. 2008;8:15.

2149. Carlini F, Ridolfi B, Molinari A, Parisi C, Bozzulo G, Toccacieli L, Formisano G, De Orsi D, Paradisi S, Grober OM, Rayo M, Weisz A, Arcien R, Vella S, Gaudi S. The reverse transcription inhibitor abacavir shows anticancer activity in prostate cancer cell lines. PLoS One. 2010;5(12):e14221.

2150. Cherkasova E, Malinzak E, Rao S, Takahashi Y, Senchenko VN, Kudryavtseva AV, Nickerson ML, Merino M, Hong JA, Schrump DS, Srinivasan R, Linehan WM, Tian X, Lerman MI, Childs RW. Inactivation of the von Hippel-Lindau tumor suppressor leads to selective expression of a human endogenous retrovirus in kidney cancer. Oncogene. 2011;30:4697–706.

2151. Sinkovics JG, Samuels ML, Thota H, Johnson DE. Kidney carcinoma: a chemotherapy-resistant tumor probably amenable to chemoimmunotherapy. Proc Am Assoc Cancer Res & Am Soc Clin Oncol Joint Ann Meeting San Diego CA May 7–11 1975 16:123 #491.

2152a. Horvath JC, Szabó-Szabari M, Sinkovics JG. Autologous activated lymphocyte production for cancer immunotherapy. Sixteenth Symposium International Association Comparative Research Leukemia and Related Diseases. July 11–16 1993, Montreal, Canada, Abstracts 1993 #79.

2152b. Horvath J, Szabó-Szabari M, Sinkovics JG. Autologous activated lymphocyte therapy in a community hospital. Acta Microbiol Hung. 1994;41:205–214.

2152c. Horvath JC, Sinkovics JG. Adoptive immunotherapy with activated peripheral blood lymphocytes. Leukemia. 1994;8S1:121–126.

2153a. Sinkovics JG, Horvath JC. Combination of active tumor-specific immunization (ATSI) and adoptive immunotherapy (AIT) with LAK cells and /or TIL expanded with IL-2. XVIth International Cancer Congress, New Delhi, India, Collection of Abstracts, 1994.

2153b. Sinkovics JG, Horvath JC, Horak A. Lymphocytes cytotoxic to human tumor cells. Proceedings Plenary Lectures XVIth International Cancer Congress, New Delhi, India. Monduzzi Editore, Bologna, Italy, 1994;93–102.

2154. Horvath JC, Sinkovics JG. Dendritoma. Presentations, 50th Annual Conference Hungarian Microbiological Society, Budapest, Hungary. Acta Microbiol Immunol Hung 50th Jubilee Issue 2003;50:250–251.

2155. Amato RJ, Shetty A, Lu Y, Ellis R, Low PS. A phase I study of folate immune therapy (EC90 vaccine administered with GPI-0100 adjuvant followed by EC17) in patients with renal cell carcinoma. J Immunother. 2013;36:268–75.

2156. Bedke J, Gouttefangeas C, Singh-Jasuja H, Stevanović S, Behnes CL, Stenzl A. Targeted therapy in renal cell carcinoma: moving from molecular agents to specific immunotherapy. World J Urol. 2014;32:31–8.

2157. Childs R, Chernoff A, Contentin N, Bahceci E, Schrump D, Leitman S, Read EJ, Tisdale J, Dunbar C, Linehan WM, Young NS, Barrett AJ. Regression of metastatic renal cell carcinoma after nonmyeloablative allogeneic peripheral blood stem cell transplantation. N Engl J Med. 2005;353:2477–90.

2158. Galluzzi L, Vacchelli E, Eggermont A, Fridman WH, Galon J, Sautés-Fridman C, Tartour E, Zitvogel L, Kroemer G. Trial watch: adoptive cell transfer immunotherapy. Oncoimmunology. 2012;1:306–15.

2159. Inman BA, Harrison MR, George DJ. Novel immunotherapeutic strategies in development for renal cell carcinoma. Eur Urol. 2013;63:881–9.

2160. Jäkel CE, Hauser S, Rogenhofer S, Müller SC, Brossart P, Schmidt-Wolf IG. Clinical studies applying cytokine-induced killer cells for the treatment of renal cell carcinoma. Clin Dev Immunol. 2012;2012:473245. doi:10.1155/2012/473245.

2161. Lawson KA, Morris DG. Oncolytic virotherapy for renal cell carcinoma: a novel treatment paradigm? Expert Opin Biol Ther. 2012;12:891–903.

2162. Quesada JAR, Swanson DA, Trindade A, Gutterman JU. Renal cell carcinoma: antitumor effects of leukocyte interferon. Cancer Res. 1983;43:940–7.

2163. Rini BI, Zimmerman T, Stadler WM, Gajewski TF, Vogelzang NJ. Allogeneic stem-cell transplantation of renal cell cancer after nonmyeloablative chemotherapy: feasibility, engraftment, and clinical results. J Clin Oncol. 2002;20:2017–24.
2164. Schwarzer A, Wolf B, Fisher JL, Schwaab T, Olek S, Baron U, Tomlinson CR, Seigne JD, Crosby NA, Gui J, Hampton TH, Fadul CE, Heaney JA, Ernstoff MS. Regulatory T cells and associated pathways in metastatic renal cell carcinoma (mRCC) patients undergoing DC-vaccination and cytokine-therapy. PLoS One. 2012;7(10): e46600. doi:10.1371/journal.pone.0046600.
2165. Topolian SL, Rosenberg SA. Therapy of cancer using the adoptive transfer of activated killer cells and interleukin-2. Acta Haematol. 1987;78S:75–6.
2166. Wolf B, Schwarzer A, Coté AL, Hampton TH, Schwaab T, Huarte E, Tomlinson CR, Gui J, Fisher JL, Fadul CE, Hamilton JW, Ernstoff MS. Gene expression profile of peripheral blood lymphocytes from renal cell carcinoma patients treated with IL-2, interferon-α and dendritic cell vaccine. PLoS One. 2012;7(12):e50221.
2167. Camp E, Hope R, Kortschak RD, Cox TC, Lardelli M. Expression of three spalt (sal) genes in zebrafish embryos. Dev Genes Evol. 2003;213:35–43.
2168. de Celis JE, Barrio R. Regulation and function of Spalt proteins during animal development. Int J Dev Biol. 2009;53:1385–95.
2169. Sánchez J, Talamillo A, González M, Sánchez-Pulido L, Jimenéz S, Pirone L, Sutherland JD, Barrio R. Drosophila Sal and Salr are transcriptional repressors. Biochem J. 2011;438:437–45.
2170. Shuai X, Zhou DF, Shen T, Wu Y, Zhang X, Wang X, Li Q. Overexpression of the novel oncogene SALL4 and activation of the Wnt/beta-catenin pathway in myeloproliferative syndrome. Cancer Genet Cytogenet. 2009;194:119–24.
2171. Hupfeld T, Chapuy B, Schrader V, Beutler M, Veltkamp C, Koch R, Cameron S, Aung T, Haase D, Larosee P, Truemper L, Wulf GG. Tyrosine kinase inhibition facilitates cooperation of transcription factor SALL4 and ABC transporter A3 towards intrinsic CML cell drug resistance. Br J Haematol. 2013;161:204–13.
2172. Kohlhase J, Hausmann S, Stojmenovic G, Dixkens C, Bink K, Schulz-Schäeffer W, Altmann M, Engel W. SALL3, a new member of the human spalt-like gene family, maps to 18q23. Genomics. 1999;62:216–22.
2173. Lin J, Qian J, Yao DM, Qian W, Yang J, Wang CZ, Chai HY, Ma JC, Deng ZQ, Li Y, Chen Q. Aberrant hypomethylation of SALL4 gene in patients with myelodysplastic syndrome. Leuk Res. 2013;37:71–5.
2174a. Gao C, Kong NR, Li A, Tatetsu H, Ueno S, Yang Y, He J, Yang J, Ma Y, Kao GS, Tenen DG, Chai L. SALL4 is a key transcription regulator in normal human hematopoiesis. Transfusion. 2013;53:1037-49.
2174b. Gao C, Dimitrov T, Yong KJ, Tatetsu H, Jeong HW, Luo HR, Bradner JE, Tenen DG, Chai L. Targeting transcription factor SALL4 in acute myeloid leukemia by interrupting its interaction with an epigenetic complex. Blood. 2013;121:1413–1421.
2175. Natarajan S, Hombach-Klonisch S, Dröge P, Klonisch T. HMGA2 inhibits apoptosis through interaction with ATR-CHK1 signaling complex in human cancer cells. Neoplasia. 2013;15:263–80.
2176. Luo, Li W, Liao H. HMGA2 induces epithelial-to-mesenchymal transition in human hepatocellular carcinoma cells. Oncol Lett. 2013;5L1353–1356.
2177. Zhu C, Li J, Cheng G, Zhou H, Tao L, Cai H, Li P, Cao Q, Ju X, Meng X, Wang M, Zhang Z, Qin C, Hua L, Yin C. Shao. miR-154 inhibits EMT by targeting HMGA2 in prostate cancer cells. Mol Cell Biochem. 2013;379:69–75.
2178. Fan T, Schmidtmann A, Xi S, Briones V, Zhu H, Suh HC, Gooya J, Keller JAR, Xu H, Roayaei J, Anver M, Ruscetti S, Muegge K. DNA hypermethylation caused by Lsh deletion promotes erythroleukemia development. Epigenetics. 2008;#:134–142.

2179. Ashiru O, Boutet P, Fernández-Messina L, Agüera-González S, Skepper JN, Valés-Gómez M, Reyburn HT. Natural killer cell cytotoxicity is suppressed by exposure to the human NKG2D ligand MICA*008 that is shed by tumor cells in exosomes. Cancer Res. 2010;70:481–9.

2180. Boylan MO, Athanassiou M, Houle B, Wang Y, Zarbl H. Activation of tumor suppressor genes in nontumorigenic revertants of the HeLa cervical carcinoma cell line. Cell Growth Differ. 1996;7:725–35.

2181. Chu C, Lugovtsev V, Golding H, Betenbaugh M, Shiloach J. Conversion of MDCK cell line to suspension culture by transfecting with human siat7e gene and its application for influenza virus production. Proc Natl Acad Sci USA. 2009;106:14802–7.

2182. Cinatl J Jr, Cinatl J, Wichelhaus T, Weber B, Rabenau H, Gümbel HO, Komhuber B, Doerr HW. Suspension culture of HeLa cells in protein-free medium: sensitivity to human pathogenic viruses. Intervirology. 1994;37:361–6.

2183. Hou H, Zhang Y, Huang Y, Yo Q, Lv L, Zhang T, Chen D, Hao Q, Shin Q. Inhibitors of phosphatidylinositol 3'-kinase promote mitotic death in HeLa cells. PLoS One. 20122;7 (4):e35665.

2184. Kochanowska IE, Wlodarski K, Wojtowicz A, Kinsner A, Ostrowski K. BMP-4 and BMP-6 involvement in the osteogenic properties of the HeLa cell line. Exp Biol Med. 2002;227:57–62.

2185. Landry JJ, Pyl PT, Rausch T, Zichner T, Tekkedil MM, Stütz AM, Jauch A, Aiyar RS, Pau G, Delhomme N, Gagneur J, Korbel JO, Huber W, Steinmetz LM. The genomic and transcriptomic landscape of a HeLa cell line. G3 (Genes/Genomes/Genetics) Bethesda. 2013;3:1213-1224. pii.g3.113.005777v6. doi:10.1534/g3.113.005777.

2186. Li M, Zheng H, Duan Z, Liu H, Hu D, Bode A, Dong Z, Cao Y. Promotion of cell proliferation and inhibition of ADCC by cancerous immunoglobulin expressed in cancer cell lines. Cell Mol Immunol. 2012;9:54–61.

2187a. Rafehi H, Orlowski C, Georgiadis GT, Ververis K, El-Osta A, Karagiannis TC. Clonogenic assay: adherent cells. J Vis Exp. 2011;13;(49). Pii:2573. doi:10.3791/2573.

2187b. Zhu K, Chen L, Han X, Wang J, Wang J. Short hairpin RNA targeting Twist1 suppresses cell proliferation and improves chemosensitivity to cisplatin in HeLa human cervical cancer cells. Oncol Rep. 2012;27:1027–1034.

2188. Mitra R, Li X, Kapusta A, Mayhew D, Mitra RD, Feschotte C, Craig NL. Functional characterization of piggyBat from the bat Myotis lucifugis unveils an active mammalian DNA transposons. Proc Natl Acad Sci USA. 2013;110:234–9.

2189a. Lozupone C, Perdicchio M, Brambilla D, Borghi M, Meschini S, Barca S, Marino ML, Logozzi M, Federici C, Iessi E, de Milito A,, Fais S. The human homologue of Dictyostelium discoideum phag1A is expressed by human metastatic melanoma cells. EMBO Rep. 2009;10:1348–1354.

2189b. Lozupone C, Faust K, Raes J, Faith JJ, Frank DN, Zaneveld J, Gordon JI, Knight R. identifying genomic and metabolic features that can underlie early successional and opportunistic lifestyles of human gut symbionts. Genome Res. 2012;22:1974–1984.

2190a. Van Haastert PJ. How cells use pseudopods for persistent movement and navigation. Sci Signal. 2011;4(159):pe6.doi:10.1126/scisignal.2001708.

2190b. Van Haastert PJ. Amoeboid cells use protrusions for walking, gliding and swimming. PLoS One. 2011;6(11):e27532. doi:10.1371/journal.pone.0027532.

2191a. Driscoll M, Kopace R, Li L, McCann C, Watts J, Fourkas JT, Losert W. The adventures of Dicty, the Dictyostelium cell. Chaos 2009;19(4):041110. doi:10.1063/1.3212926.

2191b. Driscoll MK, McCann C, Kopace R, Homan T, Fourkas JT, Parent C, Losert W. Cell shape dynamics: from waves to migration. PLoS Comput. iol 2012;8(3):e1002392. doi:10.1371/journal.pcbi.1002392.

2192. Lorentzen A, Bamber J, Sadok A, Elson-Schwab I, Marshall CJ. An ezrin-rich rigid uropod-like structure directs movement of amoeboid blebbing cells. J Cell Sci. 2011;124:1256–67.

2193a. Parri M, Taddei ML, Bianchini F, Calorini L, Chiarugi P. EphA2 induces metastatic growth regulating amoeboid motility and clonogenic potential in prostate carcinoma cells. Mol Cancer Res. 2011;9:149–160.

2193b. Parri M, Chiarugi P. Rac and Rho GTPases in cancer cell motility control. Cell Commun Signal. 2010;8:23. doi:10.1186/1478-811X-8-23.

2194. Saito K, Ozawa Y, Hibino K, Ohta Y. FilGAP, a Rho/Rho-associated protein kinase-regulated GTPase-activating protein for RAC, controls tumor cell migration. Mol Biol Cell. 2012;23:4739–50.

2195. Panková K, Rösl D, Novotný K, Brábek J. The molecular mechanisms of transition between mesenchymal and amoeboid invasiveness in tumor cells. Cell Mol Life Sci. 2010;67:63–71.

2196a. Provenzano PP, Inman DR, Eliceiri KW, Trier SM, Keely PJ. Contact guidance mediated three-dimensional cell migration is regulated by Rho.ROCK-dependent matrix reorganization. Biophys. 2008;95:5374–5384.

2196b. Provenzano PP, Eliceiri KW, Keely PJ. Shining new light on 3D cell motility and the metastatic process. Trends Cell Biol. 2009;19:638–648.

2197a. Tsai SC, Adamik R, Haun RS, Moss J, Vaughan M. Effects of brefeldin A and accessory proteins on association of ADP-ribozylation factors 1,3, and 5 with Golgi. J Biol Chem. 1993;268:10820–10825.

2197b. Tsai PC, Hsu JW, Liu YW, Chen KY, Lee FJ. Arl1p regulates special membrane organization at the trans-Golgi network through interaction with Arf-GEF Gea2p and flippase Drs2p. Proc Natl Acad Sci USA. 2013;110:E666–677.

2198. Képíró M, Várkuti BH, Bodor A, Hegyi G, Drahos L, Kovács M, Málnási-Csizmadia A. Azidoblebbistatin, a photoreactive myosin inhibitor. Proc Natl Acad Sci USA. 2112;109:9402–9407.

2199. Poincloux R, Collin O, Lizárraga F, Romao M, Debray M, Piel M, Chavrier P. Contractility of the cell rear drives invasion of breast tumor cells in 30 Matrigel. Proc Natl Acad Sci USA. 2011;108:1943–8.

2200. Ryu S, McDonnell K, Choi H, Gao D, Hahn M, Joshi N, Park SM, Catena R, Do Y, Brazin J, Vahdat LT, Silver RB, 2026. Ryu S, McDonnell K, Choi H, Gao D, Hahn M, Joshi N, Park SM, Catena R, Do Y, Brazin J, Vahdat LT, Silver RB, Mittal V. Suppression of miRNA-708 by polycomb group promotes metastases by calcium-induced cell migration. Cancer Cell. 2013;23:63–76.

2201. Mittal V. Suppression of mRNA-708 by polycomb group promoters metastases by calcium-induced cell migration. Cancer Cell. 2013;23:63–76.

2202a. van Dam TJ, Rehmann H, Bos JL, Snel B. Phylogeny of the CDC25 homology domain reveals rapid differentiation of Ras pathways between early animals and fungi. Cell Signal. 2009;21:1579–1585.

2202b. van Dam TJ, Bos JL, Snel B. Evolution of the Ras-like small GTPases and their regulators. Small GTPases. 2011;2:4–16.

2203a. Marat AL, Ioannou MS, McPherson PS. Connecdenn 3/DENND1C binds actin linking Rab35 activation to the actin cytoskeleton. Mol Biol Cell. 2012;23:163–175.

2203b. Marat AL, Dokainish H, McPherson PS. DENN domain proteins: regulators of Rab GTPases. J Biol Chem. 2011;286:13791–13800.

2204. Zhang D, Iyer LM, He F, Aravind L. Discovery of novel DENN proteins: implications for the evolution of eukaryotic intracellular membrane structures and human disease. Front Genet. 2012;3:283. doi:10.3389/fgene.2012.00283.

2205. Davis MJ, Ha BH, Holman EC, Halaban R, Schlessinger J, Boggon TJ. RAC1 P29S is a spontaneously activating cancer-associated GTPase. Proc Natl Acad Sci USA. 2013;110:912–7.

2206. Cavalli G. Molecular biology: EZH2 goes solo. Science. 2012;338:1430–1.

2207. Xu K, Wu ZJ, Groner AC, He HH, Cai C, Lis RT, Wu X, Stack EC, Loda M, Liu T, Xu H, Cato L, Thornton JE, Gregory RI, Morrissey C, Vesella RL, Montinori RL, Magi-Galuzzi C, Kantoff PW, Balk SP, Liu XS, Brown M. EZH2 oncogenic activity in castration-resistant prostate cancer cells is polycomb-independent. Science. 2012;338:1465–1469.

2208. Changchien YC, Tátrai P, Papp G, Sápi J, Fónyad L, Szendrői M, Pápai Z, Sápi Z. Poorly differentiated synovial sarcoma is associated with high expression of enhancer of zeste homologue 2 (EZH2). J Transl Med. 2012;10:216. doi:10.1186/1479-5876-10-216.

2209. Zhang X, Zhao X, Fiskus W, Lin J, Lwin T, Rao R, Zhang Y, Chan JC, Fu K, Marquez VE, Chen-Kiang S, Moscinski LC, Seto E, Dalton WS, Wright K, Sotomayor E, Bhalla K, Tao J. Coordinated silencing of MYC-mediated miR-29 by HDAC3 and EZH2 as a therapeutic target of histone modification in aggressive B-cell lymphomas. Cancer Cell. 2012;22:506–23.

2210. Chen YH, Hung MC, Li LY. EZH2: a pivotal regulator in controlling cell differentiation. Am J Transl Res. 2012;4:3640375.

2211. Shaver S, Casas-Mollano JA, Cerny RL, Cerutti H. Origin of the polycomb repressive complex 2 and gene silencing by an E(z) homologue in the unicellular alga Chlamydomonas. Epigenetics. 2010;5:301–12.

2212. Bajusz I, Sipos L, Györgypál Z, Carrington EA, Jones RS, Gausz J, Gyurkovics H. The trithorax-mimic allele of enhancer of zeste renders active domains of target genes accessible to polycomb-group-dependent silencing in Drosophila melanogaster. Genetics. 2001;159:1135–50.

2213. Jones CA, Ng J, Peterson AJ, Morgan K, Simon J, Jones RS. The Drosophila esc and E (z) proteins are direct partners in polycomb group-mediated repression. Mol Cell Biol. 1996;18:2825–34.

2214. Hajósi-Kalcakosz S, Dezső K, Bugyik E, Bödör C, Paku S, Pávai Z, Halász J, Schlachter K, Schaff Z. Enhancer of zeste homologue 2 (EZH2) is a reliable immunohistochemical marker to differentiate malignant and benign hepatic tumors. Diagn Pathol. 2012;786. doi:10.1186/1746-1596-7-86.

2215a. Futosi K, Németh T, Pick R, Vántus T, Walzog B, Mócsai A. Dasatinib inhibits proinflammatory functions of mature human neutrophils. Blood. 2012;110:4981–4991.

2215b. Futosi K, Fodor Sz, Mócsai A. Neutrophil cell surface receptors and their intracellular signal transduction pathways. Int Immunopharmacol. 2013;17:638–650.

2216. Schmidt W, Weber OF. In memoriam of Rudolf Virchow: a historical retrospective including aspects of inflammation, infection, and neoplasia. Contrib Microbiol. 2006;13:1–15.

2217a. Snow H. A treatise, practical and theoretic on cancers and the cancer process. London: JA Churchill; 1893., cited in 2217b.

2217b. Neuhaus SJ, Clark MA, Thomas JM. Dr. Herbert Lumley Snow MD, MRCS (1847-1930): the original champion of elective lymph node dissection in melanoma. Ann Surg Oncol. 2004:11:875–878.

2218. Johnston RN, Pai SB, Pai RB. The origin of the cancer cell: oncogeny reverses phylogeny. Biochem Cell Biol. 1992;70:831–4.

2219a. Moncevičiūtė-Eringiene E. Evolutionary malignant resistance of cells to damaging factors as common biological defence mechanism in neoplastic development. Review of conception. J Exp Clin Cancer Res. 2000;19:335–348.

2219b. Moncevičiūte-Eringiene E. Neoplastic growth: the consequence of evolutionary malignant resistance to chronic damage for survival of cells (review of a new theory of the origin of cancer). Med Hypotheses. 2005;65:595–604.

2219c. Moncevičiūtė-Eringiene E, Kazbariene B, Milasiene V, Characiejus D, Kemekliene R. Natural antibodies to endotoxin in experimental and clinical oncology. Exp Oncol. 2006;28:89–91.

2219d. Moncevičiūtė-Eringienė E. The evolutionary atavistic endotoxin and neoplastic growth. Med Hypotheses. 2011;76:126–131.

2220. Arguello F. Atavistic metamorphosis. A new and logical explanation for the origin and biological nature of cancer. Amazon. 2011;1–126.

2221a. Sinkovics JG, Howe CD, Shullenberger CC. Cellular activities in tissue cultures of leukemia human bone marrow. Blood. 1964;24:389–401.

2221b. Sinkovics JG, Györkey F, Gyorkey P, Jones S, Shullenberger CC, Dreyer DA. Adaptation of lymphoid cells in human and murine leukemia tissue cultures to in vitro environment through a phase of intracellular residence within fibroblasts. 18th Annual Meeting Tissue Culture Association Proceedings June 4–7 1967 Philadelphia PA 1967; 26-27 #68.

2221c. Sinkovics JG, Sykes JA, Shullenberger CC, Howe CD. Patterns of growth in cultures deriving from human leukemic sources. Texas Rep Biol Med. 1967;25:446–467.

2221d. Sinkovics JG. Lymphoid cells in long term cultures. Med Records Annals (Houston). 1968;61;50–56.

2222. Wang C-H, Sinkovics JG, Kay HD, Gyorkey F, Shullenberger CC. Growth of permanent lymphoid cell; cultures from human source: 10th anniversary. Texas Rep Biol Med. 1975;33:213–250. Erratum. Spelling errors.

2223a. Sinkovics JG. Inflammatory carcinogenesis. I. The wild-type original and the tamed contemporary DNA. II. Inflammatory molecular mediators promote carcinogenesis. 2nd International Conference Cancer Immunotherapy & Immunomonitoring (CITIM) Redefining Cancer Therapy. May 2–5 2011, Budapest Hungary, 96–98.

2223b. Sinkovics JG. Inflammatory carcinogenesis. I. Generating tumor cells in the inflammasome. II. The cellular DNA is induced to resume its ancient formation existing in the era of the "Primordial Gene Pool" and in the genomes of the first unicellular eukaryotes. In the 16th International Congress Hungarian Society for Microbiology. July 20-22 Budapest, Hungary. Acta Microbiol Immunol Hung. 2011;58:S211–213. Errata *idem* "in the multicellular host. Of DNA mutations," "…close to indestructible. Of many…".

2224. Sinkovics JG. The molecules of life and death. Hungarian Epidemiology 2012; IX/1:35–65.

2225. Cassidy-Hanley DM. Tetrahymena in the laboratory: strain resources, methods for culture, maintenance, and storage. Methods Cell Biol. 2012;109:237–76.

2226. Eckert B, Collins K. Roles of telomere reverse transcriptase N-terminal domain in assembly and activity of tetrahymena telomerase holoenzyme. J Biol Chem. 2012;287:12805–14.

2227. Iwamoto M, Asakawa H, Ohtsuki C, Osakada H, Koujin T, Hiraoka Y, Haraguchi T. Monoclonal antibodies recognize gly-leu-phe-gly repeat of nucleoporin nup98 of tetrahymena, yeasts and humans. Monoclon Antib Immunodiagn Immunother. 2013;32:81–90.

2228. Jiang J, Miracco EJ, Hong K, Eckert B, Chan H, Cash DD, Min B, Zhou ZH, Collins K, Feigon J. The architecture of Tetrahymena holoenzyme. Nature. 2013;496:187–92.

2229. Nakano T, Ohki KI, Yokota A, Ashida H. Min BD is a multifunctional fusion enzyme in the methionine salvage pathway of Tetrahymena thermophila. PLoS One. 2013;8(7): e7385.

2230. Xu J, Tian H, Liu X, Wang W, Liang A. Localization and functional analysis of HmgB3p, a novel protein containing high-mobility-group-box domain from Tetrahymena thermophila. Gene. 2013;526:87–95.

2231. Löffler H, Fechter A, Liu FY, Poppelreuther S, Krämer A. DNA damage-+induced centrosome amplification occurs via excessive formation of centriolar satellites. Oncogene. 2013;32:2963–71.

2232. Vonderfecht T, Cookson MW, Giddings TH Jr, Clarissa C, Winey M. The two human centrin homologues have similar but distinct function at Tetrahymena basal bodies. Mol Biol Cell. 2012;234766–4777.

2233. Couvillion MT, Bounova G, Purdom E, Speed TPO, Collins K. A Tetrahymena Piwi bound to mature tRNA 3' fragments activates the exonuclease Xrn2 for RNA processing in the nucleus. Mol Cell. 2012;48:509–20.

2234. Gao S, Xiong J, Zhang C, Beruist BR, Yang R, Zhao M, Molascon AJ, Kwiatkowski SY, Yuan D, Qin Z, Wen J, Kapler GM, Andrews PC, Miao W, Liu Y. Impaired replication elongation in Tetrahymena mutants deficient in histone H3 Lys 27 monomethylation. Genes Dev. 2013;27:1662–79.

2235. Long HA, Paixão T, Azevedo RB, Zufall RA. Accumulation of spontaneous mutations in the ciliate Tetrahymena thermophila. Genetics. 2013;195:527–40.

2236. Jung S, Swart EC, Minx PJ, Magrini V, Mardis ER, Landweber LF, Eddy SR. Exploiting Oxytricha trifallax nanochromosomes to screen for non-coding RNA genes. Nucleic Acids Res. 2011;39:7529–47.

2237. Swart EC, Bracht JAR, Magrini V, Minx P, Chen X, Zhou Y, Khurana JS, Goldman AD, Nowacki M, Schotanus K, Jung S, Fulton RS, Ly A, McGrath S, Haub K, Wiggins JL, Storton D, Mateze JC, Parsons L, Chang WJ, Ms Bowen, Stover NA, Jonea TA, Eddy SR, Harrick GA, Doak TG, Wilson RK, Mardis ER, Landweber LF. The Oxytricha trifallax macronuclear genome: a complex eukaryotic genome with 16,000 tiny chromosomes. PLoS Biol. 2013;11(1):e1001473.

2238. Zahler AM, Neeb ZT, Lin A, Katzman S. Mating of the stichotrichous ciliate Oxytricha trifallax induces production of a class of 27 nt small RNAs derived from the parental micronucleus. PLoS One. 2012;7(8):e42371.

2239. Cang X, Šponer J, Cheatham T III. Explaining the varied glycoside conformations, G-tract length and sequence preference for anti-parallel G-quadruplexes. Nucleic Acids Res. 2011;39:4499–512.

2240. Dubois E, Bischerour J, Marmington A, Mathy N, Régnier V, Bétermier M. Transposon invasion of the Paramecium germline genome countered by a domesticated PiggyBac transposase and the NHEJ pathway. Int: J Evol Biol; 2012. p. 436196.

2241. Vogt A, Goldman AD, Mochizuki K, Landweber LE. Transposon domestication versus mutualism in ciliate genome rearrangement. PLoS One. 2013;9(8):e1003659.

2242. Heidel AJ, Lawai HM, Felder M, Schilde C, Helps NR, Tunggal B, Rivero F, John U, Schleicher M, Eichunger L, Platzer M, Noegel AA, Schaap P, Glöckner G. Phylogeny-wide analysis of social amoeba genomes highlights ancient origins for complex intercellular communication. Genome Res. 2011;21:1882–91.

2243. Moore HM, Bai B, Boisvert FM, Latonen L, Rantanen V, Simpson JC, Pepperkok R, Lamond AI, Laiho M. Quantitative proteomics and dynamic imaging of the nucleolus reveal distinct responses to UV and ionizing radiation. Mol Cell Proteomics. 2011;(10): M111.009241.

2244. Magyar Z, De Veylder L, Atanassova A, Bakó L, Inzé D, Bögre L. The role of the Arabidopsis E2FB transcription factor in regulating auxin-dependent cell division. Plant Cell. 2005;17:2527–41.

2245. Strasser K, Bloomfield G, MacWilliams A, Ceccarelli A, MacWillims H, Tsang A. A retinoblastoma orthologue is a major regulator of S-phase, mitotic, and developmental gene expression in Dictyostelium. PLoS. 2012;7(6):e39914.

2246. Chen H, Fan M, Pfeffer LM, Laribee RN. The histone H3 lysine 56 acetylation pathway is regulated by target of rapamycin (TOR) signaling and functions directly in ribosomal RNA biogenesis. Nucleic Acids Res. 2012;40:6534–46.

2247. Crona M, Avesson L, Sahlin M, Lundin D, Hinas A, Klose R, Söderbom F, Sjöberg BM. A rare combination of ribonucleotide reductases in the social amoeba Dictyostelium discoideum. J Biochem. 2013;288:8198–208.

2248. Wang Q, Liu X, Zhou J, Huang Y, Zhang S, Shen J, Loera S, Yuan X, Chen W, Jin M, Shibada S, Liu Y, Chu P, Wang L, Yen Y. Ribonucleotide reductase large subunit m1 predicts poor survival due to modulation of proliferative and invasive ability of gastric cancer. PLoS One. 2013;8(7):e70191.

2249. Kunos CA, Winter K, Dicker AP, Small W Jr, Abdul-Karim FW, Dawson D, Jhingran A. Valicenti, Widha JB, Gaffney DK. Ribonucleotide reductase expression in cervical cancer: a radiation therapy oncology group translational science analysis. Int J Gynecol Cancer. 2013;23:615–21.
2250. Croce JC, McClay DR. Evolution of Wnt pathways. Methods Mol Biol. 2008;469:3–18.
2251. Fairclough SR, Dayel MJ, King N. Multicellular development in a choanoflagellate. Curr Biol. 2010;20:R875–8.
2252. Sonnhammer ELL, Eddy SR, Birney E, Bateman A, Durbin R. Pfam: multiple sequence alignments and HMM-profiles of protein domains. Nucleic Acids Res. 1998;26:320–2.
2253. Torruella G, Derelle R, Paps J, Lang J, Roger BF, Shalchian-Tabrizi K, Ruiz-Trillo I. Phylogenetic relationship within the opisthokonta based on phylogenomic analyses of conserved single-copy protein domains. Mol Biol Evol. 2012;29:531–44.
2254. Dayel MJ, Alegado RA, Fairclough SR, Levin TC, Nichols SA, McDonald K, King N. Cell differentiation and morphogenesis in the colony-forming choanoflagellate Salpingoeca rosetta. Dev Biol. 2011;357:73–82.
2255. Young SL, Diolaiti D, Conacci-Sorrell M, Ruiz-Trillo I, Eisenman RN, King N. Premetazoan ancestry of the Myc-Max network. Mol Biol Evol. 2011;28:2961–71.
2256. Hoffman GG, Ellington WR. Arginine kinase isoforms in the closest protozoan relative of metazoan. Comp Biochem Physiol Part D Genomics Proteomics. 2011;6:171–7.
2257. Uda K, Ellington WR, Suzuki T. Adverse array of creatine kinase and arginine kinase isoform genes is present in the starlet sea anemone Nematostella vectensis, a cnidarians model system for studying developmental evolution. Gene. 2012;497:214–27.
2258. Wang YC, Li C. Evolutionarily conserved protein arginine methyltransferase in non-mammalian animal systems. FEBS J. 2012;279:932–45.
2259. Hung CM, Li C. Identification and phylogenetic analyses of the protein arginine methyltransferase gene family in fish and ascidians. Gene. 2004;340:179–87.
2260. Alié A, Manuel M. The backbone of the post-synaptic density originated in a unicellular ancestor of choanoflagellates and metazoans. BMC Evol Biol. 2010;10:34.
2261. King N & 35 co-authors. The genome of the choanoflagellate Monosiga brevicollis and the origin of metazoans. Nature. 2008;451:783–788.
2262. Jékely G. Small GTPases and the evolution of the eukaryotic cell. Bioessays. 2003;25:1129–38.
2263. Mackiewicz P, Wyroba E. Phylogeny and evolution of Rab7 and Rab9 proteins. BMC Evol Biol. 2009;9:101.
2264. Sebé-Pedrós A, de Mendoza A, Lang BF, Degnan BM, Ruiz-Trillo I. Unexpected repertoire of metazoan transcription factors in the unicellular holozian Capsaspora owczarzaki. Mol Biol Evol. 2011;28:1241–54.
2265. Traylor-Knowles N, Hansen U, Dubuc TQ, Martindale MQ, Kaufman L, Finnerty JAR. The evolutionary diversification of LSF and Grainyhead transcription factors preceded the radiation of basal animal lineages. BMC Evol Biol. 2010;10:101.
2266. Santhekadur PK, Rajasekaran D, Siddiq A, Gredler R, Chen D, Schaus SE, Hansen U, Fisher PB, Sarkar D. The transcription factor LSF: a novel oncogene for hepatocellular carcinoma. Am J Cancer Res. 2012;2L:269–265.
2267. Michoud RE. Evolution of individuality during the transition from unicellular to multicellular life. Proc Natl Acad Sci USA. 2007;104:8613–8.
2268. Bradley JS, Olson C, Oberholzer M, Lin Y, Zones JM, Kohll HS, Bisova K, Fang S-C, Meisenheider J, Hunter T, Umen JG. Regulation of the Chlamydomonas cell cycle by a stable, chromatin-associated retinoblastoma tumor suppressor complex. Plant Cell. 2010;22:3331–47.
2269. Hellmann A. Key elements of the retinoblastoma tumor suppressor pathway in Volvox carteri. Commun Integr Biol. 2009;2:396–9.
2270. Szekely L, Úzvölgyi E, Jiang WQ, Durko M, Wiman KG, klein G, Sümegi J. Subcellular localization of the retinoblastoma protein. Cell Growth Differ. 1991;2:287–95.

2271. Solnica-Krezel L, Burland TG, Dove WF. Variable pathways for developmental changes of mitosis and cytokines in Physarium polycephalum. J Cell Biol. 1991;113:591–604.
2272. Fitz-Gibbon ST, Ladner H, Kim U-J, Stetter KO, Simon MI, Miller JH. Genome sequence of the hyperthermophilic crenarchaeon Pyrobaculum aerophilum. Proc Natl Acad Sci USA. 2002;99:984–9.
2273a. Fehér T, Bogos B, Méhi O, Fekete G, Csörgő B, Kovács K, Pósfai G, Papp B, Hurst LD, Pál C. Competition between transposable elements and mutator genes in bacteria. Mol Biol Evol. 2012;29:3153–9.
2273b. Davidson MB, Katou Y, Keszthelyi A, Sing TL, Xia T, Ou j, Vaisica JA, Thevakumuran N, Marjavaara L, Myers CL, Chabes A, Shirahige K, Brown GW. Endogenous DNA replication stress results in expansion of dNTP pools and mutator phenotype. EMBO J. 2012;31:895–907.
2274. Edris B, Weiskopf K, Volkmer AK, Volkmer J-P, Willingham SB, Contreras-Trujillo H, Liu J, Majeti R, West RB, Fletcher JA, Beck AH, Weissman IL, van de Rijn. Antibody therapy targeting the CD47 protein is effective in a model of aggressive metastatic leiomyosarcoma. Proc Natl Acad Sci USA 2012;109:6656–61.
2275a. Hortobagyi GN, Gutterman JU, Blumenschein GR, Buzdar A, Burgess MA, Richman SP, Tashima CK, Schwarz M, Hersh E. Chemoimmunotherapy of advanced breast cancer with BCG. Recent Results Cancer Res. 1977;62:143–50.
2275b. Hortobagyi GN, Gutterman JU, Blumenschein GR, Tashima CK, Burgess GN, Einhorn L, Buzdar AU, Richman SP, Hersh EM. Combination chemotherapy of metastatic breast cancer with 5-fluorouracil, adriamycin, cyclophosphamide, and BCG. Cancer. 1979;44:1955–62.
2276. Hudson MM, Ness KK, Gurney JG, Mulrooney DA, Chemaitilly W, Krull KR, Green DM, Armstrong GT, Nottage KA, Jones KE, Sklar CA, Srivastava DK, Robinson LL. Clinical ascertainment of health outcomes among adults treated for childhood cancer. JAMA. 2013;309:2371–81.
2277. Hoffman MC, Mulrooney DA, Steinberger J, Lee J, Baker S, Ness KK. Deficits in physical function among young childhood cancer survivors. J Clin Oncol. 2013;31:2799–805.
2278. Cripe T. Differing approaches to experimental therapeutics: are we a world apart? Mol Ther. 2012;20:1649–50.
2279. Wongso D, Fuchs M, Plütschow A, Klimm B, Sasse S, Hertenstein B, Maschmeyer G, Vieler T, Dührsen U, Lindemann W, Aulitzky W, Diehl V, Borchmann P, Engert A. Treatment-related mortality in patients with advanced-stage Hodgkin lymphoma: an analysis of the German Hodgkin Study Group. J Clin Oncol. 2013;31:2819–24.
2280. Lee AH, Pinder SE, Macmillan RD, Mitchell M, Ellis IO, Elston CW, Blamey RW. Prognostic value of lymphovascular invasion in women with lymph node negative invasive breast carcinoma. Eur J Cancer. 2006;42:357–62.
2281a. Strander H. Antitumor effects of interferon and its possible use as an antineoplastic agent. Tex Rep Biol Med. 1981;41:621–5.
2281b. Strander H. Interferon treatment of human neoplasia. Adv Cancer Res. 1986;46:1–265.
2282. Lengyel P. Tumor-suppressor genes: news about the interferon connection. Proc Natl Acad Sci USA. 1993;90:5893–5.
2283a. Sinkovics JG, Howe CD. Growth characteristics, virus yield and interferon assay of leukemic mouse spleen tissue cultures. J Infect Dis. 1964;114:359–72.
2283b. Sinkovics JG, Howe CD, Shullenberger CC. Interferon production, antibody response and homograft rejection type defense mechanisms in viral mouse leukemia. Proc Am Assoc Cancer Res 35th Annual Meeting April 9–11 1964, Chicago IL 1964;5/1:59 #239.
2283c. Sinkovics JG, Bertin BA, Howe CD. Latency of mouse leukemia virus in tissue cultures as related to cell growth and interferon. 16th Annual Meeting Tissue Culture Association, May 31–June 3 1965. Miami Beach FL. Section A 1965 Abstract #1 p21.

2283d. Sinkovics JG, Groves GF, Howe CD. Actions of interferon in tissue cultures harboring mouse leukemia virus. Experientia 1968;24:927.

2284a. Gutterman JU, Cardenas JO, Blumenschein GR, Hortobagyi G, Burgess MA, Livingston RB, Mavligit GM, Freireich EJ, Gottlieb JA, Hersh EM. Chemoimmunotherapy of advanced breast cancer: prolongation of remission and survival with BCG. Br Med J. 1976;2:1222–5.

2284b. Gutterman JU, Blumenschein GR, Alexanian R, Yap HY, Buzdar AU, Cabanillas P, Hortobagyi GN, Hersh EM, Rasmussen GR, Harmon M, Kramer M, Pestka S. Leukocyte interferon-induced tumor regression in human metastatic breast cancer, multiple myeloma and malignant lymphoma. Ann Int Med. 1980;93:399–408.

2284c. Gutterman JU. Cytokine therapeutics: lessons from interferon alpha. Proc Natl Acad Sci USA. 1994;91:1198–205.

2285. Schlom J. Therapeutic cancer vaccines: current status and moving forward. J Natl Cancer Inst. 2012;104:599–613.

2286. Antonia SJ, Mirza N, Fricke I, Chiappori A, Thompson P, Williams N, Bepler G, Simon G, Janssen SW, Lee JH, Menander K, Chada DS, Gabrilovich DI. Combination of p53 cancer vaccine with chemotherapy in patients with extensive small cell lung cancer. Clin Cancer Res. 2006;12:878–89.

2287. Frei E III. Curative cancer chemotherapy. Cancer Res. 1985;45:6523–37.

2288. Rosenberg SA, Yang JC, Robbins PF, Wunderlich JH, Hwu P, Sherry RM, Schwartzentruber DJ, Topalian SL, Restifo NP, Filia A, Chang R, Dudley ME. Cell transfer for cancer: lessons from sequential treatments of patients with metastatic melanoma. J Immunother. 2003;28:385–93.

2289. Brentjens RJ, Davila ML, Riviera I, Park J, Wang X, Cowell LG, BartiDo S, Stafanski J, Taylor C, Olszewska M, Borquez-Ojeda O, Qu J, Wasilewska T, He Q, Bernal Y, Rijo IV, Hedvat C, Kobos R, Curran K, Steinherz P, Jurcic J, Rosenblat T, Maslak P, Frattini M, Sadelain M. CD19-targeted T cells rapidly induce molecular remissions in adults with chemotherapy-refractory acute lymphoblastic leukemia. Sci Transl Med. 2013;5(177):177RA38.

2290. Olson JS. Making Cancer History. Baltimore: Johns Hopkins University Press; 2009.

2291a. Huang ME, Ye YC, Wang ZY. Treatment of 4 APL patients with all-transretinoic acid. Chin J Intern Med. 1987;26:220–322.

2291b. Huang ME, Ye YC, Chen SR, Chai JR, Lu JX, Zhoa L, Gu LJ, Wang ZY. Use of all-transretinoic acid in the treatment of acute promyelocytic leukemia. Blood. 1988;72:567–72.

2291c. Wang ZY. Ham-Wasserman lecture: treatment of acute leukemia by inducing differentiation and apoptosis. Hematology Am Soc Hematol Educ Program 2003;1-13.

2292. Rosenberg B, Van Camp L, Krigas T. Inhibition of cell division in Escherichia coli by electrolysis products from a platinol electrode. Nature. 1965;205:698.

2293a. Burdette WJ, Evans C. Management of coin lesions and carcinoma of the lung. Ann Surg. 1965;161:649–73.

2293b. Burdette WJ. Tumors, hormones, and viruses in Drosophila. Natl Cancer Inst Monogr. 1969;31:303–21.

2293c. Burdette WJ. Hormonal heterophyllly, invertebrate endocrinology, and phytohormones. Cancer Res. 1972;32:1088–90.

2294. Sinkovics JG, Thota H, Kay HD, Loh KK, Williams DE, Howe CD, Shullenberger CC. Intensification of immune reactions by immunotherapy: attempts at measuring sarcoma-specific reactions in vitro. In: University of Illinois at the Medical Center, R. Crispen editor. ITR 904 W Adams, Chicago. Library Congress Catalog LC75-18796 ISBN 0-915311-02-9.

2295. Sinkovics JG. Immunotherapy of sarcomas with viral oncolysates. J Am Med Assoc. 1977;237:869. *)

*) *Patients were co-attended by Carl Plager and Clifton D Howe, whose names were withheld for freedom to submit intended final paper to NE J Med as 'unpublished elsewhere'.*

2296. Sinkovics JG, Plager C, Romero J. Immunology and immunotherapy of patients with sarcomas. In: Crispen R, editor. Chicago symposium: immunotherapy of solid tumors. Philadelphia PA: Franklin Institute Press; 1977. p 211–219. Library Congress Catalog 77-777-79 ISBN 0-89168-000-4.

2297ab. Sinkovics JG, Horvath JC, Plager C. Coaxing human sarcomas to surrender. I. Basic science. II. Clinical science. Eleventh World Congress Advances in Oncology & 9th International Symposium Molecular Medicine. D. Spandidos, organizer. Crete, Greece, Oct 12–14 2006. Int J Mol Med 2006;18:S1/S37 #231 #232.

2297c. Sinkovics JG. Complete remissions lasting over three years in adult patients treated for metastatic sarcomas. In: Crispen R, editor. Tumor progression. Elsevier North Holland; 1981. p. 315–331.

2298a. Ioannides CG, Platsoucas CD, O'Brian C, Patenia R, Bowen JM, Wharton JT, Freedman RS. Viral oncolysates in cancer treatment: immunological mechanisms of action. Anticancer Rev. 1989;9:535–44.

2298b. Ioannides CG, Platsoucas CD, Patenia R, Kim PL, Bowen JM, Morris M, Edwards C, Wharton JT, Freedman RS. T-cell functions in ovarian cancer patients treated with viral oncolysates. I. Increased helper activity to immunoglobulin production. Anticancer Res. 1990;10:645–54.

2298c. Ioannides CG, Platsoucas CD, Freedman RS. Immunological effects of tumor vaccines. II. T cell responses directed against cellular antigens in the viral oncolysates. In Vitro. 1990;4:14–24.

2299a. Sinkovics JG. Viral oncolysates as human tumor vaccines. Intern Rev Immunol. 1991;7:259–87.

2299b. Sinkovics JG. Oncogenes-antioncogenes and virus therapy of cancer. Anticancer Res. 1989;9:1281–90.

2299c. Sinkovics JG, Horvath JC. Can virus therapy of human cancer be improved by apoptosis induction? Med Hypotheses 1995;44:359–68.

2300. Schirrmacher V, Fournier P. Chapter 30. Newcastle disease virus: a promising vector for viral therapy, immune therapy, and gene therapy of cancer. In: Walther W, SS Stain US, editors. Methods in molecular biology, gene therapy of cancer. Humana Press, Part of Springer Science. Book ID 1449271-02/01/229 lines 1–1444.

2301. Sinkovics JG, Horvath JC. Epilogue: "Damn the torpedoes! Full speed ahead!" In: Sinkovics JG, Horvath JC, authors/editors. Viral therapy of human cancers. New York: Marcel Dekker; 2005; 1–147; 149–457; 797–811.

2302. Sivanandham M, Wallack MK. Viral oncolysates. In: DeVita VT, Hellman S, Rosenberg SA editors. Biological therapy of cancer. 2nd ed. Philadelphia: J.B. Lippincott Company; 1995. p. 659–667.

2303. Sinkovics JG. The beginnings of cancer chemo-immunotherapy in the 1960-70s at the Department of Medicine, MD Anderson Hospital, Houston, TX. In: Cancer Immunology and Immunotherapy: Building on Success. Center of Cancer Research, National Cancer Institute, Sept 22–23 2011 Poster #88.

2304. Trujillo J, Ahearn MJ, Pienta RJ, Gott C, Sinkovics JG. Immunocompetence of leukemic murine lymphoblasts: ultrastructure, virus and globulin production. Cancer Res. 1970;30:540–5.

2305. Sinkovics JG, Drewinko B, Thornell EW. Immunoresistant tetrapoid lymphoma cells. Lancet. 1970;1:139–40.

2306. Sinkovics JG. Early history of specific antibody-producing lymphocyte hybridomas. Cancer Res. 1981;41:511–8.

2307. Sinkovics JG. The earliest concept of the 'hybridoma principle' recognized in 1967-1968. Front Rad Ther Oncol. (JM Vaeth, JL Meyer editors) 1990;24:18–31.

2308. Sinkovics J. Lymphocyta hybridomák specifikus ellenanyagtermelése (Specific antibody production of lymphocyte hybridomas). Magyar Onkológia (Sándor Eckhardt editor). 1982;26:1–6.

2309a. Sinkovics JG. Lymphocyte hybridomas in cancer research. J Med (Julian Ambrus editor). 1981;12:109–126.

2309b. Sinkovics JG. Discovery of the hybridoma principle in 1968-9. Immortalization of specific antibody-producing cell by fusion with a lymphoma cell. J Med (Julian Ambrus editor). 1985;16:509-524.

2310. Sinkovics JG. Retroviruses R Us. Kaposi Award Presentation at the Kaposi Sarcoma Foundation, Hungarian Scientific Academy, Apr 12, 2002. Magyar Venerologiai Archivum/Arch Venerol Hung (Attila Horvath editor). 2002;5/4:231–233.

2311. Blanchard DK, Kavanagh JJ, Sinkovics JG, Cavanagh D, Hewitt SM, Djeu JY. Infiltration of interleukin-2-inducible killer cells in ascitic fluid and pleural effusions of advanced cancer patients. Cancer Res. 1988;48:6321–7.

2312. Dudley ME, Yang JC, Sherry R, Hughes MS, Royal R, Kammula UI, Robbins PE, Huang J, Citrin DE, Leitman SE, Wunderlich J, Restifo NP, Thomasian A, Downey SG, Smith FO, Klapper J, Morton K, Laurence C, White DE, Rosenberg SA. Adoptive cell therapy for patients with metastatic melanoma: evaluation of intensive myeloablative chemoradiation preparative regimens. J Clin Oncol. 2008;26:5233–9.

2313. Sinkovics JG, Hirano M, Szakacs A, Shullenberger CC, Howe CD. Growth of lymphoreticular malignant cells in the preconditioned murine host. A Collection of Papers Presented at the 21st Annual Symposium on Fundamental Cancer Research, The University of Texas M.D. Anderson Hospital, Houston, TX. Williams & Wilkins, Baltimore 1967, pp 615–639.

2314. Sinkovics JG. Up-dating the Monograph "Cytolytic Immune Lymphocytes…" Acta Microbiol Immunol Hung. 2008;55:371–382. Errata AMIH 2009;56:197.

2315. Sinkovics JG. Medical oncology an advanced course. 2nd ed vol I/II. New York: Marcel Dekker; 1986. p. 2062.

2316. Sinkovics G. On the threshold of the door of "No Admittance". In: Szentivanyi A, Friedman H, editors. The immunologic revolution: facts and witnesses. Boca Raton FL: CRC Press; 1994. p. 241–86.

2317. Wolchok JD, Kluger H, Callahan MK, Postow MA, Rizvi NA, Lesokhin AM, Segal NH, Ariyan CE, Gordon R-A, Reed K, Burke MM, Caldwell A, Kronenberg SA, Agunwamba BU, Zhang X, Lowy I, Inzunza HD, Feely W, Horak CE, Hong Q, Korman AJ, Wigginton JM, Gupta A, Sznol M. Nivolumab plus ipilumab in advanced melanoma. N Engl J Med. 2013;369:122–33.

2318. Peto R, Davies C, Godwin J, Gray R, Pan HC, Clarke M, Cutter D, Darby S, McGale P, Taylor C, Wang YC, Bergh J, Di Leo A, Albain K, Swain S, Piccart M, Pritchard K. Comparisons between different polychemotherapy regimens for early breast cancer: meta-analyses of long-term outcome among 100,999 women in 123 randomized trials. Lancet. 2012;379:432–44.

2319. Cegolon L, Salata C, Weiderpass ER, Vineis P, Palu G, Mastrangelo G. Human endogenous retroviruses and cancer prevention: evidence and prospects. BMC Cancer. 2013;13:4.

2320. Emens LA. Breast cancer immunobiology driving immunotherapy: vaccines and immune checkpoint blockade. Expert Rev Anticancer Ther. 2012;12:1597–611.

2321. Lazzeroni M, Serrano D. Potential use of vaccines in the primary prevention of breast cancer in high-risk patients. Breast Care. 2012;7:281–7.

2322. Shaw AT, Kim D-W, Nakagawa K, Seto T, Crinó L, Ahn M-j, De Pas T, Besse B, Solomon BJ, Blackhall F, Wu Y-L, Thomas M, O'Byrne KJ, Moro-Sibilot D, Camidge DR, Mol T, Hirsh V, Riely GJ, Iyer S, Tasssell V, Polli A, Wilner KD, Jänne PA. Crizotinib versus chemotherapy in advanced ALK-positive lung cancer. N Engl J Med. 2013;368:2385–94.

2323. Awad MM, Katayama R, McTigue M, Liu W, Deng Y-L, Brooun A, Friboulet L, Huang D, Falk MD, Timofeevski S, Wilner KD, Lockerman EL, Khan TM, Mahmood S, Gainor JF, Digumarthy SR, Stone JAR, Mino-Kenudson M, Christensen JG, Lafrate AJ, Engelman JA, Shaw AT. Acquired resistance to crizotinib from a mutation in CD74−ROS1. N Engl J Med. 2013;368:2395–401.
2324. Morton LM, Dores GM, Curtis RE, Lynch CF, Stovall Hall P, Gilbert ES, Hodgson DC, Storm HH, Johanneden TB, Smith SA, Weathen RE, Andersson M, Possa SD, Hauptman M, Holowaty EJ, et al. Stomach cancer risk after treatment for Hodgkin lymphoma. J Clin Oncol. 2013;31:3369–77.
2325. Cao Y, Marks JD, Huang Q, Rudnick SI, Xiong C, Hittelman WN, Wen X, Marks JW, Cheung LH, Boland K, Lim C, Adams GP, Rosenblum MG. Single-chain antibody-based immunotoxins targeting Her2/neu: design optimization and impact of affinity on antitumor efficacy and off-target toxicity. Mol Cancer Ther. 2012;11:143.
2326. Diessner J, Bruttel V, Becker K, Pawliok M, Stein R, Häusler S, Dietl J, Wischhusen J, Hönig A. Targeting breast cancer stem cells with HER2-specific antibodies and natural killer cells. Am J Cancer Res. 2013;3:211–20.
2327. Jurczyszyn A, Legieć W, Helbig G, Hus G, Kyrcz-Krzemień S, Skotnicki AB. New drugs in multiple myeloma—role of carfilzomib and pomalidomide. Contemp Oncol. 2014;18:17–21.
2328. Gutterman JU, Blumenschein GR, Alexanian R, Yap HT, Buzdar AU, Cabanillas F, Hortobagyi GN, Hersh EM, Rasmussen SL, Harmon M, Kramer M, Pestka S. Leukocyte interferon-induced tumor regression in human metastatic breast cancer, multiple myeloma, and malignant lymphoma. Ann Int Med. 1980;03:200–406.
2329. Lissa D, Senovilla L, Reno-Varona S, Vitale I, Michaud M, Pietrocola F, Boilève A, Obrist F, Bordenave C, Garcia P, Michels J, Jermaà M, Kepp O, Castedo M, Kroerner G. Resveratrol and aspirin eliminate tetraploid cells for anticancer chemoprevention. Proc Natl Acad Sci USA. 2014;111:3020–5. doi:10.1073/pnas.1318440111.
2330. Choi BH, Chakraborty G, Baek K, Yoon HS. Aspirin-induced Bcl-2 translocation and its phosphorylation in the nucleus trigger apoptosis in breast cancer cells. Exp Mol Med. 2013;45:e47. doi:10.1038/emm.2013.91.
2331. Vosooghi M, Amini M. The discovery and development of cyclooxygenase-2 inhibitors as potential anticancer therapies. Expert Opin Drug Discov. 2014;9:255–67. doi:10.1517/17460441.2014.883377.
2332. Nakamura M, Ogo A, Yamura M, Yamaguchi Y, Nakashima H. Metformin suppresses sonic hedgehog expression in pancreatic cancer cells. Anticancer Res. 2014;34:1765–9.
2333. Anisimov VN. Metformin for aging and cancer prevention. Aging. 2010;2:760–74.
2334. Tan HK, Moad AIH, Tan ML. The mTOR signaling pathway in cancer and the potential mTOR inhibitory activities of natural phytochemicals. Asian Pacific J Cancer Prevent. 2014;15:6463–75.
2335. Kuiper MTR, Sabourin JR, Lambowitz AM. Identification of the reverse transcriptase encoded by the Mauriceville and Varkud mitochondrial plasmids of Neurospora. J Biol Chem. 1990;265:6936–43.
2336. Schaap P. Evolutionary crossroads in developmental biology: Dictyostelium discoideum. Development. 2011;138:387–96. doi:10.1242/dev.048934.
2337. Jörg Huesken, Walter Birchmeier, Max-Delbrück Centrum, Berlin-Buch, Helmholtz Gemeinschaft, mdc-berlin.de https://www.mdc-berlin.de/8048310/de/news/archive/2007/20070912-die_balance_halten_neue_einblicke_.
2338. Vander Heiden MG, Cantley LC, Thompson CB. Understanding the Warburg effect: the metabolic requirements of cell proliferation. Science. 2009;324(5930):1029–33. doi:10.1126/science.1160809.

2339. Chernov VM, Chernova OA, Alexey A. Mouzykantov AA, Baranova NB, Gorshkov OV, Trushin MV, Nesterova TN, Ponomareva AA. Extracellular membrane vesicles and phytopathogenicity of Acholeplasma laidlawii PG8. Sci World J. 2012; 2012:315474. doi:10.1100/2012/315474.

2340. Gourlay RN, Wyld SG, Bland AP. Demonstration by electron microscopy of intracellular virus in Acholeplasma laidlawii infected with either MV-L3 or a similar but serologically distinct virus (BN1 virus). J Gen Virol. 1979;42(2):315–22.

2341a. Haberer K, Maniloff J, Gerling D. Adsorption, capping, and release of a complex bacteriophage by mycoplasma cells. J Virol. 1980;36:264–70.

2341b. Brunner H, Schaar H, Krauss H. Giant cell formation in Acholeplasma laidlawii, strain JA1. Zentralbl Bakteriol A. 1980;248:120–8.

2342. KarvelisT Gasiunas G, Miksys A, Barrangou R, Horvath P, Siksnys V. crRNA and tracrRNA guide Cas9-mediated DNA interference in Streptococcus thermophilus. RNA Biology. 2013;10(5):841–51. doi:10.4161/rna.24203.

2343. Faehnle CR, Elkayam E, Haase AD, Hannon GJ, Joshua-Tor L. The making of a slicer: activation of human Argonaute-1. Cell Rep. 2013;3(6):1901–9. doi:10.1016/j.celrep.2013.05.033.

2344. Karijolich J, Yu YT. Spliceosomal snRNA modifications and their function. RNA Biology. 2010;7(2):192–204. doi:10.4161/rna.7.2.11207.

2345. Teixeira JE, Huston CD. Evidence of a continuous endoplasmic reticulum in the protozoan parasite Entamoeba histolytica. Eukaryotic Cell. 2008;7:1222–6. doi:10.1128/EC.00007-08.

2346. Nichols SA, Roberts BW, Richter DJ, Fairclough SR, King N. Origin of metazoan cadherin diversity and the antiquity of the classical cadherin/β-catenin complex. Proc Natl Acad Sci USA. 2012;109:13046–51. doi:10.1073/pnas.1120685109.

2347. http://users.rcn.com/jkimball.ma.ultranet/BiologyPages/S/Spemann.html.

2348. Wilbertz T, Wagner P, Petersen K, Stiedl AC, Scheble VJ, Maier S, Reischl M, Mikut R, Altorki NK, Moch H, Fend F, Staebler A, Bass AJ, Meyerson M, Rubin MA, Soltermann A, Lengerke C, Perner S. SOX2 gene amplification and protein overexpression are associated with better outcome in squamous cell lung cancer. Modern Pathology. 2011;24(7):944–53. doi:10.1038/modpathol.2011.49.

2349. Schröck A, Göke F, Wagner P, Bode M, Franzen A, Braun M, Huss S, Agaimy A, Ihler S, Menon R, Kirsten R, Kristiansen G, Bootz F, Lengerke C, Perner S. Sex determining region Y-Box 2 (SOX2) amplification is an independent indicator of disease recurrence in sinonasal cancer. PLoS One. 2013;8(3):e59201. doi:10.1371/journal.pone.0059201.

2350. Kankel MW, Hurlbut GD, Upadhyay G, Yajnik V, Yedvobnick B, Artavanis-Tsakonas S. Investigating the genetic circuitry of mastermind in Drosophila, a Notch signal effector. Genetics. 2007;177:2493–505. doi:10.1534/genetics.107.080994.

2351. Smith LC. Diversification of innate immune genes: lessons from the purple sea urchin. Dis Model Mech. 2010;3:274–9. doi:10.1242/dmm.004697.

2352. Roberts SA, Lawrence MS, Klimczak LJ, Grimm SA, Fargo D, Stojanov P, Kiezun A, Kryukov GV, Carter SL, Saksena G, Harris S, Shah RR, Resnick MA, Getz G, Gordenin DA. An APOBEC cytidine deaminase mutagenesis pattern is widespread in human cancers. Nat Gen. 2013;45:970–6. doi:10.1038/ng.2702.

2353. Schaffitzel C, Berger I, Postberg J, Hanes J, Lipps HJ, Plückthun A. In vitro generated antibodies specific for telomeric guanine-quadruplex DNA react with Stylonychia lemnae macronuclei. Proc Natl Acad Sci USA. 2001;98:8572–7. doi:10.1073/pnas.141229498.

2354. Planchon SM, Waite KA, Eng C. The nuclear affairs of PTEN. J Cell Sci. 2008;121:249–53. doi:10.1242/jcs.022459.

2355. Dupressoir A, Vernochet C, Harper F, Guégan J, Dessen P, Pierron G, Heidmann T. A pair of co-opted retroviral envelope syncytin genes is required for formation of the two-layered murine placental syncytiotrophoblast. Proc Natl Acad Sci USA. 2011;108: E1164–73. doi:10.1073/pnas.1112304108.
2356. Adem C, Gisselsson D. Dal Cin P, Nascimento AG. ETV6 rearrangements in patients with infantile fibrosarcomas and congenital mesoblastic nephromas by fluorescence in situ hybridization. Mod Pathol. 2001;14:1246–51.
2357. Judelson HS, Blanco FA. The spores of Phytophthora: weapons of the plant destroyer. Nat Rev Microbiol. 2005;3:47–58. doi:10.1038/nrmicro1064.
2358. Virgili A, Brazma D, Reid AG, Howard-Reeves J, Valgañón M, Chanalaris A, De Melo VA, Marin D, Apperley JF, Grace C, Nacheva EP. FISH mapping of Philadelphia negative BCR/ABL1 positive CML. Mol Cytogenet. 2008;18:1–14. doi:10.1186/1755-8166-1-14.
2359. Springer ML, Yanofsky C. Expression of con genes along the three sporulation pathways of Neurospora crassa. Genes Dev. 1992;6:1052–7. doi:10.1101/gad.6.6.1052.
2360. Cho Y, Gorina S, Jeffrey PD, Pavletich NP. Crystal structure of a p53 tumor suppressor-DNA complex: understanding tumorigenic mutations. Science. 1994;265:346–55.
2361. Kops GJPL, Weaver BAA, Cleveland DW. On the road to cancer: aneuploidy and the mitotic checkpoint. Nat Rev Cancer. 2005;5:773–86.

Index

Note: Page numbers followed by f and t indicate figures and tables, respectively

J.G. Sinkovics, *RNA/DNA and Cancer*, DOI 10.1007/978-3-319-22279-0

I

V

W

X

Y

Z

The manufacturer's authorised representative in the EU is Springer Nature Customer Service Centre GmbH, Europaplatz 3, 69115 Heidelberg, Germany. If you have any concerns regarding our products, please contact ProductSafety@springernature.com

Printed and bound by CPI Group (UK) Ltd, Croydon, CR0 4YY
01/10/2026
02225872-0020